AIRCRAFT POWERPLANTS

AVIATION TECHNOLOGY SERIES

Ralph D. Bent and James L. McKinley

Aircraft Powerplants
Aircraft Maintenance and Repair
Aircraft Basic Science
Aircraft Electricity and Electronics

AIRCRAFT POWERPLANTS

● **FIFTH EDITION** ●

RALPH D. BENT
JAMES L. McKINLEY

GREGG DIVISION
McGRAW-HILL BOOK COMPANY

New York   Atlanta   Dallas   St. Louis   San Francisco
Auckland   Bogotá   Guatemala   Hamburg   Johannesburg   Lisbon
London   Madrid   Mexico   Montreal   New Delhi   Panama   Paris
San Juan   São Paulo   Singapore   Sydney   Tokyo   Toronto

Sponsoring Editor: **Peggy Adams**
Editing Supervisor: **Richard Caleskie**
Design and Art Supervisor: **Patricia Lowy**
Production Supervisor: **Priscilla Taguer**

Text Designer: **Phyllis Lerner**
Cover Photograph: **Arthur Meyerson/The Image Bank**

Library of Congress Cataloging in Publication Data

Bent, Ralph D.
   Aircraft powerplants.

   (Aviation technology series)
   Includes index.
   1. Airplanes—Motors.   2. Propellers, Aerial.
   3. Aircraft gas-turbines.   I. McKinley, James L.
   II. Title.
TL701.N6   1985   629.134′35   85-113
ISBN 0-07-004797-9

Aircraft Powerplants, Fifth Edition

1 2 3 4 5 6 7 8 9 0   SEMSEM   8 9 2 1 0 9 8 7 6 5

ISBN   0-07-004797-9

CONTENTS

ACKNOWLEDGMENTS

The authors wish to express appreciation to the following organizations for their generous assistance in providing illustrations and technical information for this text:

AiResearch Manufacturing Company, Division of the Garrett Corporation, Torrance, California; Allison Gas Turbine Operations, Division of General Motors Corporation, Indianapolis, Indiana; American Hall of Aviation History, Northrop University, Inglewood, California; American Society of Mechanical Engineers, New York, New York; Avco Lycoming, Avco Corporation, Williamsport, Pennsylvania; Aviation Maintenance Foundation, Basin, Wyoming; Beech Aircraft Corporation, Wichita, Kansas; Bell Helicopter Textron, Fort Worth, Texas; Bendix Corporation, Energy Controls Division, South Bend, Indiana; Bendix Corporation, Engine Products Division, Jacksonville, Florida; Bendix Corporation, Fluid Power Division, Utica, New York; Boeing Commercial Airplane Company, Seattle, Washington; Bray Oil Company, Los Angeles, California; Cessna Aircraft Company, Wichita, Kansas; Champion Spark Plug Company, Toledo, Ohio; Dee Howard Company, San Antonio, Texas; Douglas Aircraft Company, Division of the McDonald Douglas Corporation, Long Beach, California; Dowty Rotol Ltd., Gloucester, England; Elcon Division, Icore International, Inc., Sunnyvale, California; Facet Aerospace Products Company, Jackson, Tennessee; Federal Aviation Administration, Washington, D.C.; Fenwall, Inc., Ashland, Massachusetts; Garrett Turbine Engine Company, Phoenix, Arizona; General Electric Company, Aircraft Engine Group, Cincinnati, Ohio; Hamilton Standard Division, United Technologies, Windsor Locks, Connecticut; Hartzell Propeller Division, TRW, Piqua, Ohio; Howell Instruments, Inc., Fort Worth, Texas; Howmet Turbine Components Corporation, Greenwich, Connecticut; Hughes Helicopters, Inc., Culver City, California; Walter Kidde and Company, Belleville, New Jersey; McCauley Accessory Division, Cessna Aircraft Company, Dayton, Ohio; Northrop University, Inglewood, California; Piper Aircraft Company, Lock Haven, Pennsylvania; Prestolite Division, Eltra Corporation, Toledo, Ohio; Pratt & Whitney Aircraft Group, United Technologies, East Hartford, Connecticut; Pratt & Whitney Aircraft of Canada, Ltd., Longueuil, PQ, Canada; Rajay Industries, Long Beach, California; Rolls-Royce, Ltd., Derby, England; Santa Monica Propeller Services, Santa Monica, California; SGL Auburn Spark Plug Company, Auburn, New York; Slick Electro, Inc., Rockford, Illinois; Systron Donner Corporation, Berkeley, California; Teledyne Continental Motors, Mobile, Alabama; Woodward Governor Company, Rockford, Illinois.

In addition to the above, the authors wish to thank the many aviation technical schools and instructors for providing valuable suggestions, recommendations, and technical information for the revision.

PREFACE

This textbook is the fifth edition of *Aircraft Powerplants,* two editions of which had the title, *Powerplants for Aerospace Vehicles.* This edition contains all the essential information contained in earlier editions plus additional material relating to new powerplants. *Aircraft Powerplants,* Fifth Edition, is designed to assist the student in attaining the proficiency levels defined in current Federal Aviation Regulations when used as a study text in connection with classroom discussions, demonstrations, and practical application.

In preparing this edition, the authors have reviewed all documents published by the Federal Aviation Administration relating to the training of aviation maintenance students and the operation, maintenance, and overhaul of aircraft powerplants. Manufacturers' manuals, bulletins, and other documents have been examined to ensure that accurate information is presented. FAR Part 147 and FAA Advisory Circulars AC 65-2 and AC 43-13-1 have been employed for reference to ensure that FAA requirements are met. The FAA Written Test Guide, Advisory Circular AC 65-22, has been reviewed carefully so the inclusion of all technical data needed by the student to pass FAA written and oral examinations is covered. In addition, suggestions and recommendations have been solicited and received from aviation instructors, aircraft engine manufacturers, aviation operators, and maintenance specialists. The new information added to the text includes descriptions of some of the latest powerplants, accessories, and systems.

Aircraft Powerplants is one of four textbooks in the McGraw-Hill Aviation Technology Series. The other books in the series are *Aircraft Basic Science, Aircraft Maintenance and Repair,* and *Aircraft Electricity and Electronics.* When used together in an Airframe and Powerplant maintenance training program, these texts encompass information on all phases of airframe and powerplant technology.

Each topic in *Aircraft Powerplants* has been explained in as much detail as possible to ensure that the student may gain the knowledge of theory and maintenance principles at a level which meets the requirements of the Federal Aviation Regulations. To ensure that the needs and desires of the majority of students and instructors are met, it has been the policy for this revision to provide more information than may actually be required for proficiency levels 1 and 2.

In addition to the use of this book as a classroom text, it is valuable for home study and as an on-the-job reference for the technician. The subjects are so organized that instructors in public and private technical schools, training departments of aircraft manufacturers and airlines, vocational schools, high schools, and industrial education divisions of colleges and universities are provided with a wealth of classroom material.

Review questions have been added at the end of each chapter so students may check their knowledge of the material presented.

Ralph D. Bent
James L. McKinley

AIRCRAFT POWERPLANTS

1 POWERPLANT PROGRESS

The development of aviation powerplants is the result of utilizing principles that were employed in the design of earlier internal-combustion engines. During the latter part of the nineteenth century a number of successful engines were designed and built, and these were used to operate machinery and to supply power for "horseless carriages." The challenge to aviation was to design engines that had high power-to-weight ratios. This was accomplished first with light-weight piston engines and then, more effectively, with gas-turbine engines.

The industrial revolution, which took place in the late eighteenth century and continued into part of the twentieth century, was largely the result of the ability of human beings to find ways whereby energy sources could be used to develop power. Before these times, work was accomplished solely by animal and human power. In continental Europe and England it was apparent that power sources other than human were needed to drive the new machines which were being invented, largely in the textile manufacturing field. Among the energy sources developed were wind, water, steam, and the fuel for internal-combustion engines.

Wind energy has been used for hundreds of years to supply the power to meet certain needs. In its most direct application, wind energy has been employed to propel sailing vessels over the waters of the earth from the beginning of recorded history. Among its uses it has served to power grist mills, operate pumps, and generate electricity. In recent years engineers have been designing and building giant wind machines to drive generators capable of supplying the electrical needs of small cities.

Waterpower, used extensively in the past to operate many kinds of machines, is still being used to drive turbines, which, in turn, drive generators of electric power. Like wind power, waterpower has the advantage of being an inexhaustible supply, and it produces no by-products which pollute the air or water.

Steam engines were developed in the eighteenth century and became particularly important for operating machinery, propelling steamships, and serving as locomotives for pulling trains. Steam engines and turbines are still used for power production, and steam plants supply a large amount of the electric power used throughout the world. Today many of the steam powerplants are operated by means of heat produced through nuclear fission. The same principle which led to the atomic bomb is now used to produce heat for the operation of steam turbines. The nuclear powerplant has the advantage of being a very clean power source, since there is no emission of the products of combustion, as is the case with a powerplant burning coal, oil, or gas. Furthermore, a nuclear powerplant can produce very large amounts of heat energy from very little fuel. A properly designed nuclear powerplant also has the capability of producing additional fuel while it generates heat for the production of steam. This type of nuclear powerplant utilizes a **breeder reactor** and is described as being "self-proliferating."

It is expected that a time will come when engineers will develop aircraft engines which can be operated from nuclear energy. At the present time, practical methods for adapting nuclear energy to aircraft have not been worked out, but research is continuing in the field. This is important because a time will come when fossil fuels (petroleum and gas) will be exhausted and new sources of energy will be needed.

Another possibility for a source of power for which the fuel supply is inexhaustible is the hydrogen-oxygen engine. Since water is composed of hydrogen and oxygen, and the world's supply of water is abundant, human beings will undoubtedly find a practical and economical means for using hydrogen and oxygen to operate internal-combustion engines. Such engines would be suitable for both aircraft and automobiles and would produce little or no air pollution.

Early Engines

Development of the internal-combustion engine took place largely during the nineteenth century. One of the first such engines was described in 1820 by the Reverend W. Cecil in a discourse before the Cambridge Philosophical Society in England. This engine operated on a mixture of hydrogen and air. In 1838 the English inventor William Barnett built a single-cylinder gas engine which had a combustion chamber at both the top and bottom of the piston. This engine burned gaseous fuel rather than the liquid fuel used in the modern gasoline engine.

The first practical gas engine was built in 1860 by a French inventor named Jean Joseph Étienne Lenoir. This engine utilized illuminating gas as a fuel, and ignition of the fuel was provided by a battery system. Within a few years approximately 400 of these engines had been built to operate a variety of machinery, such as lathes and printing presses.

The first four-stroke-cycle engine was built by August Otto and Eugen Langen of Germany in 1876. As a result, the four-stroke-cycle engine is often called an Otto-cycle engine. Otto and Langen also built a two-stroke-cycle engine.

In America, George B. Brayton, an engineer, built an engine using gasoline as fuel and exhibited it at the 1876 Centennial Exposition in Philadelphia. The first truly successful gasoline engine operating according to the four-stroke-cycle principle was built in Germany in 1885 by Gottlieb Daimler, who had previously been associated with Otto and Langen. A similar gasoline engine was built by Karl Benz of Germany in the same year. The Daimler and Benz engines were used in early automobiles, and the engines used today are similar in many respects to the Daimler and Benz engines.

The First Successful Airplane Engine

Inasmuch as the first powered flight in an airplane was made by the Wright brothers on December 17, 1903, it is safe to say that the first successful gasoline engine for an airplane was the engine used in the Wright airplane. This engine was designed and built by the Wright brothers and their mechanic, Charles Taylor. The engine had the following characteristics: (1) water-cooled; (2) four cylinders; (3) bore, $4\frac{3}{8}$ inches (in) [11.11 centimeters (cm)], and stroke, 4 in [10.16 cm]; displacement, 240 cubic inches (in³) [3932.9 cm³]; (4) 12 horsepower (hp) [8.94 kilowatts (kW)]; (5) weight, 180 pounds (lb) [82 kilograms (kg)]; (6) cast-iron cylinders with sheet-aluminum water jackets; (7) valve-in-head with the exhaust valve mechanically operated and the intake valve automatically operated; (8) aluminum-alloy crankcase; (9) carburetion by means of fuel flow into a heated manifold; and (10) ignition by means of a high-tension magneto. A picture of an early Wright engine is shown in Fig. 1-1.

World War I Engines

The extensive development and use of airplanes during World War I contributed greatly to the improvement of engines. One type of engine that found most extensive use was the air-cooled, rotary-type radial engine. In this engine the crankshaft is held stationary, and the cylinders rotate around the crankshaft. Among the best-known rotary engines were the LeRhone shown

FIG. 1-2   Le Rhone rotary engine.

in Fig. 1-2, the Gnôme-Monosoupape shown in Fig. 1-3, and the Bentley, which has a similar appearance. With these engines, the crankshaft is secured to the aircraft engine mount, and the propeller is attached to the engine case.

Even though the rotary engines powered many World War I airplanes, they had two serious disadvantages: (1) The torque and gyro effects of the large rotating mass of the engines made the airplanes difficult to control. (2) The engines used castor oil as a lubricant, and since the castor oil was mixed with the fuel of the engine in the crankcase, the exhaust of the engines contained castor-oil fumes which were often nauseating to the pilots.

A number of in-line and V-type engines were also developed during World War I. Among these were the Hispano-Suiza shown in Fig. 1-4, a 90° V-8 engine; the Rolls-Royce V-12 engine; the American-made Liberty

FIG. 1-1   Early Wright engine.

FIG. 1-3   Gnôme-Monosoupape rotary engine.

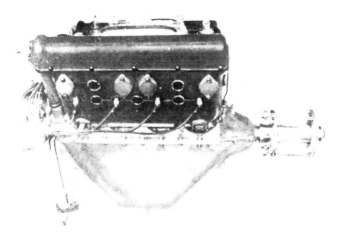

FIG. 1-4 Early Hispano-Suiza engine.

V-12 engine shown in Fig. 1-5; and several German engines, including the Mercedes, Benz, and BMW.

After World War I many different engine designs were developed. Some of those with rather unusual configurations are shown in Fig. 1-6. They are identified as follows: (a) Szekeley, three-cylinder radial, (b) Italian MAB, four-cylinder fan-type engine, (c) British Napier "Rapier," 16-cylinder H-type engine, (d) British Napier "Lion," 12-cylinder W-type engine, (e) U.S. Viking, 16-cylinder X-type engine.

A popular American engine was the Curtiss OX-5 engine manufactured during and after World War I. This engine powered the Curtiss "Jennie" (JN-4) trainer plane used for training U.S. military aviators. After the war, many were sold to the public, and the majority of these were used in the early barnstorming days for air shows and passenger flights. An OX-5 engine is shown in Fig. 1-7.

FIG. 1-5 Liberty engine.

Other engines developed in the United States between World War I and World War II were the Wright Hisso (an American-built Hispano-Suiza), the Packard V-12, the Curtiss D-12 (a V-12 engine), the Wright Whirlwind radial engines, and the Pratt & Whitney Wasp and Hornet engines, which are air-cooled radial types. Numerous smaller engines were also designed and built, including radial, opposed-cylinder, and in-line types.

Progress in Design

Engineers who specialize in the design of aircraft powerplants have used light alloy metals for construction of the engines and have adopted weight-saving cylinder arrangements, such as placing the cylinders radially in one or more rows around the crankshaft, with the result that today the weight per horsepower on several engines is below 1.2 lb [0.54 kg] and on some less than 1 lb [0.45 kg].

Airplanes have increased in size, carrying capacity, and speed. With each increase has come a demand for more power, and this has been met by improvements in engine and propeller design and by the use of turbojet and turboprop engines. As piston engines increased in power, they became more complicated. The early powerplant engineers and mechanics had only a few comparatively simple problems to meet, but the modern powerplant specialist must be familiar with the principles of the internal-combustion engine; the classification, construction, and nomenclature of engines; their fuel and carburetion systems; supercharging and induction systems; the lubrication of powerplants; engine starting systems; ignition systems; valve and ignition timing; engine-control systems; and propellers, both wood and metal. The specialist who works with turbine engines must be familiar with the construction and operation of these engines and the complex fuel-control units and systems required for such engines. In addition to understanding the design, construction, and operation of powerplants, the specialist must know the proper procedures for inspection, maintenance, overhaul, and repair.

Fundamentally, the reciprocating internal-combustion engine that we know today is a direct descendant of the first Wright engine. It has become larger, heavier, and much more powerful, but the basic principles are essentially the same. However, the modern reciprocating aircraft engine has reached a stage in its development where it is faced with what is commonly called the **theory of diminishing returns.** More cylinders are added to obtain more power, but the resulting increases in size and weight complicate matters in many directions. For example, the modern conventional engine may lose more than 30 percent of its power in dragging itself and its nacelle through the air and in providing necessary cooling.

The improvement in reciprocating engines has become quite noticeable in the smaller engines used for light aircraft. This has been accomplished chiefly with the opposed-type four- and six-cylinder engines. Among the improvements developed for light engines during and since World War II are geared propellers, superchargers, and direct fuel-injection systems. Whereas

FIG. 1-6 Different engine configurations developed after World War I.

FIG. 1-7 Curtiss OX-5 engine.

light airplanes were once limited to flight at comparatively low altitudes, today many of them are capable of cruising at well over 20,000 feet (ft) [6096 meters (m)] altitude.

Jet Propulsion

During World War II, the demand for increased speed and power expedited the progress which was already taking place in the development of jet-propulsion powerplants. As a result of the impetus given by the requirements of the War and Navy departments in the United States and by similar demands on the part of our British allies, engineers in England and the United States designed, manufactured, and tested in flight an amazing variety of jet-propulsion powerplants. It must be noted also that the German government was not trailing behind in the jet-propulsion race, because the first flight by an airplane powered with a true jet engine was made in Germany on August 27, 1939. The airplane

FIG. 1-8 Experimental Whittle turbojet engine.

FIG. 1-9 Whittle W1 engine.

was a Heinkel He 178 and was powered by a Heinkel HeS 3B turbojet engine.

The first practical turbojet engines in England and the United States evolved from the work of Sir Frank Whittle in England. The experimental Whittle engine is shown in Fig. 1-8. The success of experiments with this engine led to the development and manufacture of the Whittle W1 engine shown on a test stand in Fig. 1-9. Under agreement with the British government, the United States was authorized to manufacture an engine of a design similar to the W1. The General Electric Company was given a contract to build the engine because of their extensive experience with turbine manufacture and with the development of turbosuperchargers used for military aircraft in World War II. Accordingly, the General Electric GE I-A engine was built and successfully flown in a Bell XP-59A airplane.

The successful development of jet propulsion is beyond question the greatest single advance in the history of aviation since the Wright brothers made their first flight. The speed of aircraft has increased from below Mach 1, the speed of sound, to speeds of more than Mach 3. Commercial airliners now operate at more than 600 miles per hour (mph) [965.6 kilometers per hour (km/h)] rather than at high speeds of around 350 mph [563.26 km/h] which is usually about the maximum

for conventional propeller-driven airliners. Commercial jet airliners designed to operate at supersonic speeds have been developed by the U.S.S.R. and through a combined effort by France and Great Britain. These aircraft, the Russian Tu-144 and the French-British Concorde, commenced regularly scheduled service in 1976.

The reciprocating engine will be retained for many years for use in airplanes where low power and relatively low speed are expected. The gas-turbine engine is used both as a jet engine and for driving propellers and helicopter rotors. Small gas-turbine engines have been developed to power luxury-type personal and business aircraft. Some operate as pure jets, and others are of the turboprop configuration.

The design and construction of jet engines have advanced most rapidly, and it is difficult for any book to keep up with the latest developments. Nevertheless, the basic principles of theory, construction, operation, and maintenance of gas-turbine engines are covered in this text.

2 INTERNAL-COMBUSTION ENGINE PRINCIPLES, PERFORMANCE, AND OPERATION

Two general types of aircraft engines are in common use for the propulsion of almost all types of powered aircraft. These are the **reciprocating engine,** so called because the pistons reciprocate (move back and forth or up and down) as the engine operates and the **gas-turbine engine,** in which a turbine is driven or rotated by the hot, high-velocity gases. Both types of engines are termed **heat engines** because they utilize heat for the development of power.

Basically, an engine is a device for converting a source of energy to useful work. In heat engines, the source of energy is the fuel that is burned to develop heat. The heat, in turn, is converted to power (the rate of doing work) by means of the engine. The reciprocating engine uses the heat to expand a combination of gases (air and the products of fuel combustion) and thus create a pressure against a piston in a cylinder. The piston, being connected to a crankshaft, causes the crankshaft to rotate, thus producing power and doing work. In the gas-turbine engine, the heat is used to expand the gas (air) as it moves through the engine, with the result that the velocity of the gases is greatly increased. The high-velocity flow of gases is directed through a turbine which rotates to produce shaft power. With a **turbojet** engine, the jet of gases from the engine exhaust results in thrust that is used to propel the aircraft. In a **turboshaft** or **turboprop** engine, a large portion of the energy is converted to shaft horsepower; however, the exhaust of the engine often produces from 15 to 20 percent of the total thrust.

It is the purpose of this chapter to explain the basic principles and operation of the reciprocating engine. The principles and operation of gas-turbine engines will be discussed in Chap. 19.

● FUNDAMENTALS

Cycle

A **cycle** is a complete sequence of events returning to the original state. That is, a cycle is an interval of time occupied by one round, or course, of events repeated in the same order in a series—such as the cycle of the seasons, with spring, summer, autumn, and winter following each other and then recurring.

An **engine cycle** is the series of events that an internal-combustion engine goes through while it is operating and delivering power. In a four-stroke five-event cycle, these events are intake, compression, ignition, combustion, and exhaust. An **internal-combustion** engine, whether it be a piston-type or gas-turbine engine,

is so called because the fuel is burned inside the engine rather than externally, as with a steam engine. Since the events in a piston engine occur in a certain sequence and at precise intervals of time, they are said to be **timed.**

Most piston-type engines operate on the four-stroke five-event-cycle principle originally developed by August Otto in Germany. There are four strokes of the piston in each cylinder, two in each direction, for each engine operating cycle. The five events of the cycle consist of these strokes plus the ignition event. The four-stroke five-event cycle is called the **Otto cycle.** Other cycles for heat engines are the **Carnot cycle,** named after Nicolas-Leonard-Sadi Carnot, a young French engineer; the **Diesel cycle,** named after Dr. Rudolf Diesel, a German scientist; and the **Brayton cycle,** named for George B. Brayton, an American engineer mentioned in Chap. 1. All the cycles mentioned pertain to the particular engine theories developed by the men whose names are given to the various cycles. All the cycles include the compression of air, the burning of fuel in the compressed air, and the conversion of the pressure and heat to power.

Stroke

The basic power-developing parts of a typical gasoline engine are the **cylinder, piston, connecting rod,** and **crankshaft.** These are shown in Fig. 2-1. The cylinder has a smooth surface such that the piston can, with the aid of piston rings and a lubricant, create a seal so that no gases can escape between the piston and the cylinder walls. The piston is connected to the crankshaft by means of the connecting rod so that the rotation of the crankshaft causes the piston to move with a reciprocating motion up and down in the cylinder. The distance through which the piston travels is called the **stroke.** During each stroke, the crankshaft rotates 180°. The limit of travel to which the piston moves *into* the cylinder is called **top dead center,** and the limit to which it moves in the opposite direction is called **bottom dead center.** For each revolution of the crankshaft there are two strokes of the piston, one up and one down, assuming that the cylinder is in a vertical position. Figure 2-2 shows that the stroke of the cylinder illustrated is 5.5 in [13.97 cm] and that its bore (internal diameter) is also 5.5 in. An engine having the bore equal to the stroke is often called a **square** engine.

It is important to understand top dead center and bottom dead center because these positions of the piston are used in setting the timing and determining the

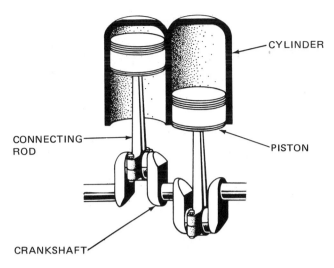

FIG. 2-1　Basic parts of a gasoline engine.

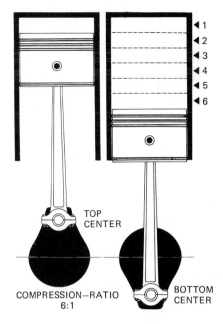

FIG. 2-3　Top and bottom dead center.

valve overlap. **Top dead center** (TDC or TC) may be defined as the point which a piston has reached when it is at its maximum distance from the center line of the crankshaft. In like manner, **bottom dead center** (BDC or BC) may be defined as the position which the piston has reached when it is at a minimum distance from the center line of the crankshaft. Figure 2-3 illustrates the piston position at top dead center and at bottom dead center. This figure also illustrates compression ratio, which is explained later in this chapter.

The Four-Stroke Five-Event Cycle

The four strokes of a four-stroke-cycle engine are the intake stroke, the compression stroke, the power stroke, and the exhaust stroke. In a four-stroke-cycle engine the crankshaft makes two revolutions for each complete cycle. The names of the strokes are descriptive of the nature of each stroke.

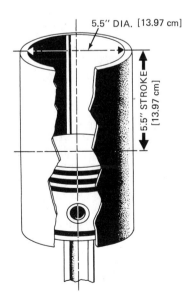

FIG. 2-2　Stroke and bore.

During the **intake** stroke the piston starts at top dead center with the intake valve open and the exhaust valve closed. As the piston moves downward, a mixture of fuel and air, sometimes called the **working fluid,** from the carburetor is drawn into the cylinder. The intake stroke is illustrated in Fig. 2-4.

When the piston has reached bottom dead center at the end of the intake stroke, the piston moves back toward the cylinder head. The intake valve closes as much as 60° of crankshaft rotation after bottom center in order to take advantage of the inertia of the incoming fuel-air mixture, thus increasing volumetric efficiency. Since both valves are now closed, the fuel-air mixture is compressed in the cylinder. For this reason the event illustrated in Fig. 2-5 is called the **compression** stroke. A few degrees of crankshaft travel *before* the piston reaches top dead center on the compression stroke, the **ignition** event takes place. Ignition is caused by a spark plug which produces an electric spark in the fuel-air mixture. This spark ignites the fuel-air mixture, thus creating heat and pressure to force the piston downward toward bottom dead center. The reason that the ignition is timed to occur a few degrees before top dead center is to allow time for complete combustion of the fuel. When the fuel-air mixture and the ignition timing are correct, the combustion process will be complete just after top center at the beginning of the power stroke, producing maximum pressure. If the ignition occurred at top dead center, the piston would be moving downward as the fuel burned, and a maximum pressure would not be developed. Also, the burning gases moving down the walls of the cylinder would heat the cylinder walls, and the engine would develop excessive temperature.

The stroke during which the piston is forced down, as the result of combustion pressure, is called the **power** stroke because this is the time when power is developed in the engine. The movement of the piston downward causes the crankshaft to rotate, thus turning the

7

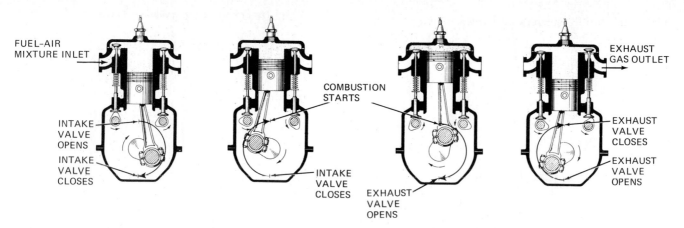

FUEL-AIR
MIXTURE INLET

INTAKE
VALVE
OPENS

INTAKE
VALVE
CLOSES

COMBUSTION
STARTS

INTAKE
VALVE
CLOSES

EXHAUST
VALVE
OPENS

EXHAUST
GAS OUTLET

EXHAUST
VALVE
CLOSES

EXHAUST
VALVE
OPENS

FIG. 2-4 Intake stroke.   **FIG. 2-5 Compression stroke.** **FIG. 2-6 Power stroke.**   **FIG. 2-7 Exhaust stroke.**

flywheel or propeller being driven by the engine. The power stroke illustrated in Fig. 2-6 is also called the **expansion** stroke because of the gas expansion which takes place at this time.

Well before the piston reaches bottom dead center on the power stroke, the exhaust valve opens, and the hot gases begin to escape from the cylinder. The pressure differential across the piston drops to zero, and the gases that remain in the cylinder are forced out the open exhaust valve as the piston moves back toward top dead center. This is the **exhaust** stroke and is also called the **scavenging** stroke because the burned gases are scavenged (removed from the cylinder) during the stroke. The exhaust stroke is illustrated in Fig. 2-7.

We may summarize the complete cycle of the four-stroke-cycle engine as follows: **Intake stroke**—the intake valve is open and the exhaust valve closed, the piston moves downward, drawing the fuel-air mixture into the cylinder, and the intake valve closes; **compression stroke**—both valves are closed, the piston moves toward top dead center, compressing the fuel-air mixture, and ignition takes place near the top of the stroke; **power stroke**—both valves are closed, the pressure of the expanding gases forces the piston toward bottom dead center, and the exhaust valve opens well before the bottom of the stroke; **exhaust stroke**—the exhaust valve is open and the intake valve closed, the piston moves toward top dead center, forcing the burned gases out through the open exhaust valve, and the intake valve opens near the top of the stroke.

The five-event sequence of intake, compression, ignition, power, and exhaust is a cycle which must take place in the order given if the engine is to operate at all, and it must be repeated over and over for the engine to continue operation. None of the events can be omitted, and each event must take place in the proper sequence. For example, if the gasoline supply is shut off, there can be no power event. The mixture of gasoline and air must be admitted to the cylinder during the intake stroke. Likewise, if the ignition switch is turned off, there can be no power event because the ignition must occur before the power event can take place.

It should be emphasized at this point that each event of crankshaft rotation does not occupy exactly 180° of crankshaft travel. The intake valve begins to open sub-

stantially before top center, and the exhaust valve closes after top center. This is called **valve overlap** and is designed to take advantage of the inertia of the out-flowing exhaust gases to provide more complete scavenging and to allow the entering mixture to flow into the combustion chamber at the earliest possible moment, thus greatly improving **volumetric efficiency.** Volumetric efficiency is discussed later in this chapter.

Near bottom center, valve opening and closing is also designed to improve volumetric efficiency. This is accomplished by keeping the intake valve open substantially past bottom center to permit a maximum charge of fuel-air mixture to enter the combustion chamber. The exhaust valve opens as much as 60° before bottom center on the power stroke to provide for optimum scavenging and cooling.

Details of valve mechanism design and operation are explained in Chap. 3

An engine cannot normally start until it is rotated to begin the sequence of operating events. For this reason a variety of starting systems have been employed. Among such systems are hand cranking of the engine, hand-cranked inertia starters, electric inertia starters, combustion starters, and direct-cranking electric starters. These are described in Chap. 11.

Conversion of Heat Energy to Mechanical Energy

Energy is the capacity for doing work. There are two kinds of energy: kinetic and potential. **Kinetic** energy is the energy of motion, such as that possessed by a moving cannon ball, falling water, or a strong wind. **Potential** energy, or stored energy, is the energy of position. A coiled spring has potential energy. Likewise, the water behind the dam of a reservoir has potential energy, and gasoline has potential energy.

Energy can be neither created nor destroyed. A perpetual-motion machine cannot exist because even if friction and the weight of the parts were eliminated, a machine can never have any more energy than that which has been put into it.

Energy cannot be created, but it can be transformed from one kind into another. When a coiled spring is wound, work is performed. When the spring unwinds, its stored (potential) energy becomes kinetic energy.

When a mixture of gasoline and air is ignited, the combustion process increases the kinetic energy of the molecules in the gases. When the gas is confined, as in a reciprocating-engine cylinder, this results in increased pressure (potential energy), which produces work when the piston is forced downward.

Heat energy can be transformed into mechanical energy, mechanical energy can be transformed into electrical energy, and electrical energy can be transformed into heat, light, chemical, or mechanical energy.

Boyle's Law and Charles's Law

Boyle's law states that *the volume of any dry gas varies inversely with the absolute pressure sustained by it, the temperature remaining constant*. In other words, increasing the pressure upon a volume of confined gas reduces its volume correspondingly. Thus, doubling the pressure reduces the volume of the gas to one-half, trebling the pressure reduces the volume to one-third, etc. The formula for Boyle's law is

$$\frac{V_1}{V_2} = \frac{P_2}{P_1}$$

Charles's law states that *the pressure of a confined gas is directly proportional to its absolute temperature*. Therefore, as the temperature of the gas is increased, the pressure is also increased as long as the volume remains constant. The formula for Charles's law is

$$\frac{P_1 V_1}{P_2 V_2} = \frac{T_1}{T_2}$$

These laws may be used to explain the operation of an engine. The mixture of fuel and air burns when it is ignited and produces heat. The heat is absorbed by the gases in the cylinder, and they tend to expand. The increase in pressure, acting on the head of the piston, forces it to move, and the motion is transmitted to the crankshaft through the connecting rod.

A further understanding of engine operation may be gained by examining the theory of the **Carnot cycle**. The Carnot cycle explains the operation of an "ideal" heat engine. The engine employs a gas as a working medium, and the changes in pressure, volume, and temperature are in accordance with Boyle's and Charles's laws. A detailed study of the Carnot cycle is not essential to the present discussion; however, if students desire to pursue the matter further, they can find a complete explanation in any good college text on physics.

As explained previously, the conventional piston engine utilizes a connecting rod to transmit the movement of the piston to the crankshaft in order to produce the rotary motion of the crankshaft. An ordinary sewing-machine band wheel and treadle, illustrated in Fig. 2-8, afford a good example of the conversion of straight-line (reciprocating) motion to rotary motion. One end of a connecting rod is fastened at *A* to the treadle. The other end of the connecting rod is fastened at *B*, a point on the band wheel which is off center. When the treadle is pushed down with the toe of the

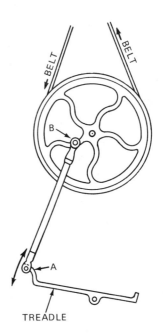

FIG. 2-8 Converting reciprocating motion to rotary motion.

foot, the end of the connecting rod marked *A* moves downward and pulls the other end of the connecting rod *B* around for a part of a circle. If the operator continues to push down with a toe, the band wheel will stop at dead center, but if the operator pauses a moment, the inertia of the band wheel will carry it past the dead-center position. The operator then pushes with the heel on the rear edge of the treadle, and the end of the connecting rod is pushed upward, causing the point *B* on the band wheel to move upward. As the point *B* nears the top-dead-center position, the operator pauses again and allows the band wheel to be carried past the dead-center position by inertia. Thus, to keep the band wheel turning, the operator pushes down with the toe, pauses, pushes down with the heel, pauses, and repeats the process indefinitely. The energy of the operator is exerted up and down; that is, it is a reciprocating motion, but a rotary motion is imparted to the band wheel.

The Two-Stroke Cycle

Although present-day aircraft engines of the reciprocating type usually operate on the four-stroke-cycle principle, a few small engines (such as those used on ultralight aircraft) operate on the two-stroke-cycle principle. The differences are the number of strokes per operating cycle and the method of admitting the fuel-air mixture into the cylinder. The two-stroke-cycle engine is mechanically simpler than the four-stroke-cycle engine, but it is less efficient and is more difficult to lubricate; hence, its use is restricted. The operating principle of the two-stroke-cycle engine is illustrated in Fig. 2-9.

Like the four-stroke-cycle engine, the two-stroke-cycle engine is constructed with a cylinder, piston, crankshaft, connecting rod, and crankcase; however, the valve arrangement and fuel intake system are con-

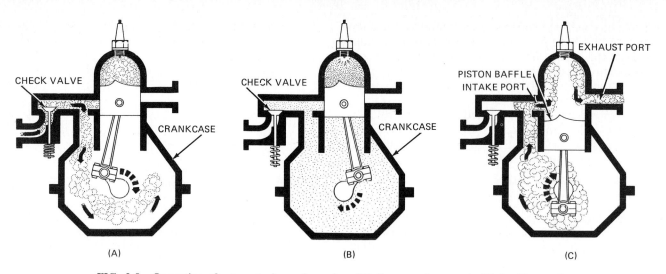

FIG. 2-9 Operation of a two-stroke-cycle engine. (A) Compression event. (B) Ignition and power events. (C) Exhaust and intake events.

siderably different. The upward movement of the piston in the cylinder of the engine creates low pressure in the crankcase. This reduced pressure causes a suction which draws the fuel-air mixture from the carburetor into the crankcase through a check valve. When the piston has reached top dead center, the crankcase is filled with the fuel-air mixture and the inlet check valve is closed. The piston then moves downward in the cylinder and compresses the mixture in the crankcase. As the piston reaches the lowest point in its stroke, the intake port is opened to permit the fuel-air mixture which is compressed in the crankcase to flow into the cylinder. This is the **intake event.**

The piston then moves up in the cylinder, the intake port is closed, and the fuel-air mixture in the cylinder is compressed. While this is happening, a new charge of fuel and air is drawn into the crankcase through the check valve. This is the **compression event** and is shown in Fig. 2-9(A).

The piston continues to move up in the cylinder, and when it is almost at the top of the stroke, a spark is produced at the gap of the spark plug, thus igniting the fuel-air mixture. This is the **ignition event.** As the fuel-air mixture burns, the gases of combustion expand and drive the piston down. This is the **power event** and is shown in Fig. 2-9(B). During the power event, the fuel-air mixture in the crankcase is pressurized.

When the piston approaches the bottom point of its travel, the exhaust port is opened to allow the hot gases to escape. This occurs a fraction of a second before the intake port opens to allow the pressurized fuel-air mixture in the crankcase to flow through the intake port into the cylinder. As the exhaust gases rush out the exhaust port on one side of the cylinder, the fuel-air mixture flows into the other side. A baffle on the top of the piston reflects the incoming mixture toward the top of the cylinder, thus helping to scavenge the exhaust gases and reduce the mixing of the fuel-air mixture with the exhaust gases. It can easily be seen that the exhaust and intake events take place almost simultaneously, with the exhaust event leading by a small fraction of the piston stroke. This is illustrated in Fig. 2-9(C).

It will be noted that there are five events in the two-stroke engine cycle, but at one point, two of the events happen at approximately the same time. During the time that the combined exhaust and intake events are occurring, some of the fuel-air mixture is diluted with burned gases retained from the previous cycle, and some of the fresh mixture is discharged with the exhaust gases. The baffle on the top of the piston is designed to reduce the loss of the fresh mixture as much as possible.

It is important to understand that two strokes of the piston (one complete crankshaft revolution) are required to complete the cycle of operation. For this reason, all cylinders of a multicylinder two-stroke-cycle engine will fire at each revolution of the crankshaft. Remember that the four-stroke-cycle engine fires only once in two complete revolutions of the crankshaft.

The operation of the two-stroke-cycle engine may be summarized as follows: The piston moves upward and draws a fuel-air mixture into the crankcase through a check valve, the crankcase being airtight except for this valve; the piston moves downward and compresses the mixture in the crankcase; the intake port is opened and the compressed fuel-air mixture enters the cylinder; the piston moves upward and compresses the mixture in the combustion chamber; near the top of the piston stroke, the spark plug ignites the mixture, thus causing the piston to move down; near the bottom of the stroke, the exhaust port is opened to allow the burned gases to escape, and the intake port opens to allow a new charge to enter the cylinder. Note that as the piston moves down during the power event, the fuel-air mixture is being compressed in the crankcase. As the piston moves upward during the compression event, the fuel-air mixture is being drawn into the crankcase.

The two-stroke cycle has three principal disadvantages: (1) There is a loss of efficiency as a result of the fuel-air charge mixing with the exhaust gases and the loss of some of the charge through the exhaust port; (2) the engine is more difficult to cool than the four-stroke-cycle engine, chiefly because the cylinder fires at every revolution of the crankshaft; and (3) the engine is somewhat difficult to lubricate properly because the lubricant must be introduced with the fuel-air mixture through the carburetor. This is usually accomplished by mixing the lubricant with the fuel in the fuel tank.

● THE DIESEL ENGINE

The **operating principle** of the four-stroke-cycle diesel engine superficially resembles that of the four-stroke-cycle gasoline engine except that the pure diesel engine requires no electrical ignition. Also, the diesel engine operates on fuel oils that are heavier and cheaper than gasoline.

On the **intake stroke** of the diesel engine, only pure air is drawn into the cylinder. On the **compression stroke,** the piston compresses the air to such an extent that the air temperature is high enough to ignite the fuel without the use of an electric spark. As the piston approaches the top of its stroke, the fuel is injected into the cylinder under a high pressure in a finely atomized state. The highly compressed hot air already in the cylinder ignites the fuel. The fuel burns during the **power stroke,** and the waste gases escape during the **exhaust stroke** in the same manner as they do in a gasoline engine. On many diesel engines, particularly those employed in automobiles, glow-plug igniters are installed to aid in starting the combustion of the fuel. These igniters are not in operation after the engine is running.

The **compression ratio,** to be discussed more fully later, is the ratio of the volume of space in a cylinder when the piston is at the bottom of its stroke to the volume when the piston is at the top of its stroke. The compression ratio of a diesel engine may be as high as 14:1 as compared with a maximum of 10 or 11:1 for conventional gasoline engines. It is common for a gasoline engine to have a compression ratio of about 7:1; however, certain high-performance engines have a higher ratio. The compression ratio of a conventional gasoline engine must be limited because the temperature of the compressed gases in the cylinder must not be high enough to ignite the fuel. The high-octane and high-performance-number fuels have made it possible to utilize higher compression ratios for conventional engines. If the compression ratio is too high for the fuel being used, **detonation** (explosion) of the fuel will occur, thus causing overheating, loss of power, and probable damage to the pistons and cylinders. Detonation will be described more fully later in this text.

Like the gasoline internal-combustion engine, the diesel engine may be either a two-stroke-cycle or a four-stroke-cycle engine.

● POWER CALCULATIONS

Power

Power is the rate of doing work. A certain amount of work is accomplished when a particular weight is raised a given distance. For example, if a weight of 1 ton [907.2 kg] is raised vertically 100 ft [30.48 m] we may say that 100 ton feet (ton·ft) [27 651 kilogram-meters (kg·m)] of work has been done. Since a ton is equal to 2000 lb [907.2 kg], we can also say that 200,000

foot pounds (ft·lb) [27 651 kg·m] of work has been done. When we speak of power, we must also consider the *time* required to do a given amount of work. Power depends on three factors: (1) the force extended, (2) the distance the force moves, and (3) the time required to do the work.

James Watt, the inventor of the steam engine, found that an English workhorse could work at the rate of 550 foot pounds per second (ft·lb/s) or 33,000 foot pounds per minute (ft·lb/min) [4563 kilogram-meters per minute (kg·m/min)] for a reasonable length of time. From his observations came the **horsepower,** which is the unit of power in the English system of measurements.

When a 1-lb [0.45 kg] weight is raised 1 ft [0.3048 m], 1 ft·lb [0.14 kg·m] of work has been performed. When a 1000-lb weight is lifted 33 ft, 33,000 ft·lb of work has been performed. If the 1000-lb weight is lifted 33 ft in 1 min, 1 hp [0.745 kW] has been expended. If it takes 2 min to lift the weight through the same distance, $\frac{1}{2}$ hp [372.85 W] has been used. If it requires 4 min, $\frac{1}{4}$ hp [186.43 W] has been used. **One horsepower equals 33,000 ft·lb/min of work, or 550 ft·lb/s of work.** The capacity of automobile, aircraft, and other engines to do work is measured in horsepower. In the metric system, the unit of power is the *watt* (W). One kilowatt (kW) is equal to 1.34 hp.

Piston Displacement

In order to compute the power of an engine, it is necessary to determine how many foot pounds of work can be done by the engine in a given time. To do this, we must know various measurements, such as cylinder bore, piston stroke, and piston displacement.

The **piston displacement** of one piston is obtained by multiplying the area of a cross section of the cylinder bore by the total distance that the piston moves during one stroke in the cylinder. Since the volume of any true cylinder is its cross-sectional area multiplied by its height, the piston displacement can be stated in terms of cubic inches of volume. The piston displacement of one cylinder can be determined if the bore and stroke are known. For example, if the bore of a cylinder is 6 in [15.24 cm] and the stroke is 6 in, we can find the displacement as follows:

$$\text{Cross-sectional area} = \pi r^2$$
$$= 28.274 \text{ in}^2$$
$$[182.41 \text{ cm}^2]$$

$$\text{Displacement} = 6 \times 28.274 = 169.644 \text{ in}^3$$
$$[2.779 \text{ L}]$$

The total piston displacement of an engine is the total volume displaced by all the pistons during one revolution of the crankshaft. It equals the number of cylinders in the engine multiplied by the piston displacement of one piston. Other factors remaining the same, the greater the total piston displacement, the greater will be the maximum horsepower that an engine can develop.

Displacement is one of the many factors in power-plant design which are subject to compromise. If the cylinder bore is too large, fuel will be wasted and the intensity of the heat and the restricted flow of the heat may be so great that the cylinder may not be cooled properly. If the stroke (piston travel) is too great, excessive dynamic stresses and too much angularity of the connecting rods will be the undesirable consequences.

It has been found that a "square" engine provides the proper balance between the dimensions of bore and stroke. Remember that a **square engine** has the bore and stroke equal. Increased engine displacement can be obtained by adding cylinders, thus producing an increase of power output. The addition of cylinders produces what is known as a **closer spacing of power impulses,** which increases the smoothness of engine operation.

In addition to the method shown previously for determining piston displacement using bore and stroke, we can use the formula $\frac{1}{4}\pi D^2 = A$ for determining the cross-sectional area of the cylinder. This formula can also be written $A = \pi D^2/4$, where A is the area in square inches and D is the diameter of the bore.

If a piston has a diameter of 5 in [12.70 cm], its area is $\frac{1}{4}\pi \times 25$, or 19.635 in^2 [126.68 cm^2]. In place of $\frac{1}{4}\pi$ we can use 0.7854, which is the same value.

If the piston mentioned above is used where the stroke is 6 in [15.24 cm], the displacement of the piston is 6×19.635, or 117.81 in^3 [1.92 L]. If the engine has 14 cylinders, the total displacement of the engine is $117.81 \times 14 = 1649.34$ in^3 [27.03 L]. This engine would be called an **R-1650** engine, the R standing for *radial*.

One typical radial engine has both a bore of 5.5 in [13.97 cm] and a stroke of 5.5 in. The cross-sectional area of the cylinder is then $5.5^2 \times 0.7854 = 23.758$ in^2 [153.28 cm^2]. The displacement of one piston is $5.5 \times 23.758 = 130.669$ in^3 [2.142 L]. The engine has 14 cylinders; so the total displacement is $14 \times 130.669 = 1829.366$ in^3 [29.98 L]. This is called an **R-1830** engine.

Compression Ratio

The **compression ratio** of a cylinder is the ratio of the volume of space in the cylinder when the piston is at the bottom of its stroke to the volume when the piston is at the top of its stroke. For example, if the volume of the space in a cylinder is 120 in^3 [1.97 L] when the piston is at the bottom of its stroke and the volume is 20 in^3 [0.33 L] when the piston is at the top of its stroke, the compression ratio is 120:20. Stated in the form of a fraction, it is $\frac{120}{20}$, and when the larger number is divided by the smaller number, the compression ratio is therefore shown as 6:1. This is the usual manner for expressing a compression ratio. In Fig. 2-3, the piston and cylinder provide a compression ratio of 6:1.

As stated previously, the compression ratio of any internal-combustion engine is a factor which controls the maximum horsepower which can be developed by the engine, other factors remaining the same. Within reasonable limits, the maximum horsepower increases as the compression ratio increases. However, it has been found by experience that, if the compression ratio of the internal-combustion engine is much greater than 10:1, preignition (premature ignition of the fuel-air charge) or detonation may occur and cause overheat-

ing, loss of power, and damage to the engine. If the engine has a compression ratio as high as 10:1, it is necessary that the fuel used have a high antiknock characteristic (high octane rating or high performance number).

The **maximum compression ratio** of an engine, as indicated above, is limited by the detonation characteristics of the fuel used. Detonation has been mentioned briefly before and will be discussed more thoroughly in later sections of the text, but it is sufficient to state here that detonation occurs when 75 to 80 percent of the fuel-air mixture burns normally and then the remainder burns with explosive rapidity, or, in effect, explodes.

In addition to the detonation characteristics of the fuel used, the maximum compression ratio of an aircraft engine is also limited by the design limitations of the engine, the availability of high-octane fuel, and the degree of supercharging.

An increase in the compression ratio of an engine may be accomplished by installing "higher" pistons, by using longer connecting rods, or by installing a crankshaft with a greater throw. In an engine which has a removable cylinder head, the combustion space in the head may be reduced by "shaving" the surface of the cylinder head where it mates with the top of the cylinder. This, of course, increases the compression ratio.

Increasing the compression ratio of an engine causes a lower specific fuel consumption (pounds of fuel burned per hour per horsepower) and a greater thermal efficiency. **Thermal efficiency** is the ratio of the heat converted to useful work to the heating value of the fuel consumed.

The Indicator Diagram

Indicated horsepower (ihp) is based on the theoretical amount of work done according to calculations made from the actual pressure recorded in the form of a diagram on an indicator card, as illustrated in Fig 2-10. This particular indicator diagram shows the pressure rise during the compression stroke and after ignition. It also shows the pressure drop as the gases expand during the power stroke. It clearly emphasizes the fact that the force acting on the piston during the combustion (power) stroke of the engine is not constant, because the fuel-air mixture burns almost instantaneously, with a resulting high pressure at the top of the stroke and a decreasing pressure as the piston descends.

Power computations for indicated horsepower are somewhat simplified by using the average pressure acting on the piston throughout the working stroke. This average pressure, often called the **mean effective pressure,** is obtained from the indicator diagram. The indicator diagram is drawn on the indicator card by a mechanical device atttached to the engine cylinder. Modern engine manufacturers utilize much more sophisticated instrumentation with computers to determine the various engine measurements, or parameters. Figure 2-10 does, however, provide a graphic illustration of the process.

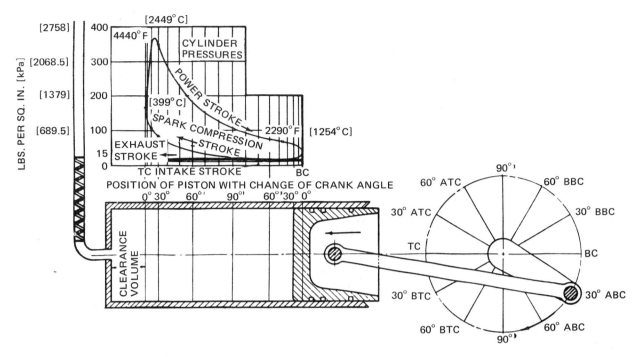

FIG. 2-10   Cylinder-pressure indicating diagram.

Indicated Horsepower

Indicated horsepower (ihp) is the horsepower developed by the engine, i.e., the total horsepower converted from heat energy to mechanical energy.

If the characteristics of an engine are known, the indicated-horsepower rating can be calculated. The total force acting on the piston in one cylinder is the product of the indicated mean effective pressure (imep) P and the area A of the piston head in square inches (found by the formula which states that the area of a circle is πr^2 or $\frac{1}{4}\pi D^2$).

The distance through which this total force acts in 1 min multiplied by the total force gives the number of foot pounds of work done in 1 min. The work done in 1 min by one piston multiplied by the number of cylinders in operation gives the amount of work done in 1 min by the entire engine. This product is divided by 33,000 (the number of foot pounds per minute in 1 hp) to obtain the indicated horsepower rating of the engine.

The length of the stroke in feet is represented by L, the area of the piston in square inches by A, the indicated mean effective pressure in pounds per square inch (psi) by P, the number of working strokes per minute per cylinder by N, and the number of cylinders by K. Indicated horsepower can then be computed by the formula

$$\text{ihp} = \frac{PLANK}{33,000}$$

The foregoing formula can be made clear by remembering that **work** is equal to *force times distance* and that **power** is equal to *force times distance divided by time*. *PLA* is the *product of pressure, distance, and area*, but *pressure times area equals force;* hence, *PLA* = *FD*. In the formula, *PLANK* is the number of foot pounds per minute produced by an engine because N represents the number of working strokes per minute for each cylinder, and K is the number of cylinders. To find horsepower, it is merely necessary to divide the number of foot pounds per minute by 33,000 since 1 hp = 33,000 ft·lb/min [1 W = 6.12 kg·m/min].

Brake Horsepower

Brake horsepower (bhp) is the actual horsepower delivered by an engine to a propeller or other driven device. It is the ihp minus the friction horsepower. **Friction horsepower** (fhp) is that part of the total horsepower necessary to overcome the friction of the moving parts in the engine and its accessories. The relationship may be expressed: bhp = ihp − fhp.

It may also be stated that the bhp is that part of the total horsepower developed by the engine which can be used to perform work. On many aircraft engines it is between 85 and 90 percent of the indicated horsepower.

The bhp of an engine can be determined by coupling the engine to any power-absorbing device, such as an electric generator, in such a manner that the power output can be accurately measured. A **dynamometer** is an apparatus for measuring force; it can be used to determine the power output of an engine. If an electric generator is connected to a known electric load and the efficiency of the generator is known, the bhp of the engine driving the generator can be determined. For example, assume that an engine is driving a generator producing 110 volts (V) and that the load on the generator is 50 amperes (A). Electric power is measured in watts and is equal to the voltage multiplied by the amperage. Therefore, the electrical power developed by the generator is 50×110 or 5500 watts (W). Since 1 hp = 746 W, 5500 W = 7.36 hp. If the generator is 60 percent efficient, the power required to drive it is equal to 7.36/0.60 or 12.27 hp [9.17 kW]. Therefore, we have determined that the engine is developing 12.27 bhp to drive the generator.

The Prony Brake

The **prony brake,** or dynamometer, illustrated in Fig. 2-11, is a device used to measure the **torque,** or turning moment, produced by an engine. The value indicated by the scale is read before the force is applied, and the reading is recorded as the **tare.** The force F, produced by the lever arm, equals the weight recorded on the scale minus the tare. The known values are then F, the distance L, and the rpm of the engine driving the prony brake. To obtain the bhp, these values are used in the following formula:

$$ bhp = \frac{F \times L \times 2\pi \times rpm}{33,000} $$

In the foregoing formula, the distance through which the force acts in one revolution is the circumference of the circle of which the distance L is the radius. This circumference is determined by multiplying the radius L by 2π. In the formula we see, then, that the force acts through a given distance a certain number of times per minute, and this gives us the foot pounds per minute. When this is divided by 33,000, the result is bhp.

If a given engine turning at 1800 revolutions per minute (rpm) produces a force of 200 lb [889.6 N] on the scales at the end of a 4-ft [1.22-m] lever, we can compute the bhp as follows:

$$ bhp = \frac{200 \times 4 \times 2\pi \times 1800}{33,000} $$

$$ = 274 \ [204.3 \ kW] $$

The Torque Nose

It has often been the practice with large engines used in transport aircraft to equip one or more of the engines with a torque indicating system. This system consists of a mechanism in the nose section of the engine which applies pressure to oil in a closed chamber in proportion to engine torque. Since the planetary gears of a propeller reduction-gear system must work against a large stationary ring gear, the ring gear can be used to develop an indication of engine torque. The outside of the ring gear is constructed with helical teeth which fit into similar helical teeth in the nose case. When torque is developed, the ring gear tends to move forward. This movement is transmitted to hydraulic pistons which are connected to a pressure gage in the cockpit or on the flight engineer's panel. The gage may be calibrated to read directly the **brake mean effective pressure** (bmep), which, in turn, is used to compute the bhp.

Brake mean effective pressure can be derived mathematically from the bhp and vice versa. When the bhp has been determined by means of a dynamometer, the bmep can be computed by means of the following formula:

$$ bmep = bhp \times \frac{33,000}{LAN} $$

where L = stroke, ft
A = area of bore, in^2
N = number of working strokes per min

In a four-stroke-cycle engine, $N = \frac{1}{2}$ rpm of the engine multiplied by the number of cylinders.

Power Ratings

The **takeoff power** rating of an engine is determined by the maximum rpm and manifold pressure at which the airplane engine may be operated during the process of taking off. The takeoff power may be given a time limitation, such as a period of 1 to 5 min. **Manifold pressure** is the pressure of the fuel-air mixture in the intake manifold between the carburetor or internal supercharger and the intake valve. The pressure is given in inches of mercury (inHg) above absolute zero pressure. Standard sea-level pressure is 29.92 inHg [101.34 kilopascals (kPa)], so the reading on the manifold pressure gage may be either above or below this figure. As manifold pressure increases, the power output of an engine increases, provided that the rpm remains constant. Likewise, the power increases as rpm increases, provided that the manifold pressure remains constant.

The takeoff power of an engine may be about 10 percent above the maximum continuous power-output allowance. This is the usual increase of power output permitted in the United States, but in British aviation the increase above maximum cruising power may be as much as 15 percent. It is sometimes referred to as the **overspeed** condition. The maximum continuous power is also called the maximum except takeoff (METO) power.

During takeoff conditions with the engine operating at maximum takeoff power, the volume of air flowing around the cylinders is restricted because of the low speed of the airplane during takeoff, and the initial carburetor air temperature may be very high in hot weather. For these reasons the operator of the airplane must exercise great care, especially in hot weather, to avoid overheating the engine and damaging the valves, pistons, and piston rings. The overheating may cause

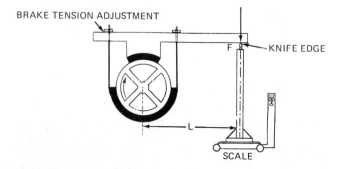

FIG. 2-11 Prony brake.

detonation or preignition, with a resultant loss of power in addition to engine damage.

The **rated power,** also called the **standard engine rating,** is the maximum horsepower output which can be obtained from an engine when it is operated at specified rpm and manifold pressure conditions established as safe for continuous operation. This is the power guaranteed by the manufacturer of the engine under the specified conditions and is the same as the METO power.

Maximum power is the greatest power output that the engine can develop at any time under any condition.

Critical Altitude

The **critical altitude** is the highest level at which an engine will maintain a given horsepower output. For example, an aircraft engine may be rated at a certain altitude which is the highest level at which rated power output can be obtained from the engine at a given rpm. Turbochargers and superchargers are employed to increase the critical altitude of engines. These applications are discussed in later chapters.

● ENGINE EFFICIENCY

Mechanical Efficiency

The **mechanical efficiency** of an engine is measured by the ratio of the brake horsepower, or shaft output, to the indicated horsepower, or power developed in the cylinders. For example, if the ratio of the bhp to the ihp is 9:10, then the mechanical efficiency of the engine is 90 percent. In determination of mechanical efficiency, only the losses suffered by the energy that has been delivered to the pistons is considered. The word "efficiency" may be defined as the ratio of output to input.

Thermal Efficiency

Thermal efficiency is a measure of the heat losses suffered in converting the heat energy of the fuel into mechanical work. In Fig. 2-12 the heat dissipated by the cooling system represents 25 percent, the heat carried away by the exhaust gases represents 40 percent, the mechanical work on the piston to overcome friction and pumping losses represents 5 percent, and the use-

ful work at the propeller shaft represents 30 percent of the heat energy of the fuel.

The thermal efficiency of an engine is the ratio of the heat developed into useful work to the heat energy of the fuel. It may be based on either bhp or ihp and is represented by a formula in this manner:

$$\text{Indicated thermal efficiency} = \frac{\text{ihp} \times 33{,}000}{\text{wt of fuel burned per min} \times \text{heat value (Btu)} \times 778}$$

The formula for brake thermal efficiency (bte) is the same as that given above with the word "brake" inserted in place of "indicated" in both sides of the equation.

If we wish to find the brake thermal efficiency of a particular engine, we must first know the following quantities: the bhp, the fuel consumption in pounds per minute, and the heat value of the fuel in British thermal units (Btu). In this case, let us suppose that the engine develops 104 bhp at 2600 rpm and burns 6.5 gallons per hour (gal/h) [24.61 liters per hour (L/h)] of gasoline. The heat value of the fuel is 19,000 to 20,000 Btu [20 045 000 to 21 110 000 joules (J)].

First we must convert gallons [liters] per hour to pounds [kilograms] per minute. Since there are 60 min/h, we divide 6.5 by 60 to obtain 0.108 gal/min [0.41 L/min]. Since each gallon of fuel weighs approximately 6 lb [2.72 kg], we multiply 0.108 by 6 to obtain 0.648 lb/min [0.29 kg/min]. The formula then becomes

$$\text{bte} = \frac{104 \times 33{,}000}{0.648 \times 20{,}000 \times 778}$$
$$= \frac{3{,}432{,}000}{10{,}080{,}000} = 0.34$$

Hence, the brake thermal efficiency is 34 percent.

To explain the formula, we must know only that the energy of 1 Btu is 778 ft·lb [107.6 kg·m]. The product of 104 × 33,000 provides us with the total foot·pound output. The figures in the denominator give us the total input energy of the fuel. The fraction then represents the ratio of input to output.

In the foregoing problem, if the engine burns 100 gal [378.54 L] of gasoline, only 34 gal [128.7 L] is converted to useful work. The remaining 66 percent of the heat produced by the burning fuel in the engine cylinders is lost by being exhausted through the exhaust manifold or through the cooling of the engine. This is an excellent value for many modern aircraft engines running at full power. At slightly reduced power, the thermal efficiency may be a little greater, and by the use of high compression with high-octane fuels, an engine may be made to produce as high as 40 percent brake thermal efficiency. This is not normal, however, and for mechanical reasons is not necessarily desirable.

Although a thermal efficiency of 34 percent may not appear high, it is excellent when compared with other types of engines. For example, the old steam locomotive had a thermal efficiency of not much more than

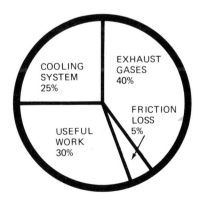

FIG. 2-12 Thermal efficiency chart.

5 percent. The thermal efficiency of many diesel engines is 35 percent if run at an output of one-half to three-fourths full power, but when the output is increased to full power, the thermal efficiency of the diesel drops to less than one-half that of the usual carburetor-type engine. This is because of an incomplete combustion of fuel when large amounts of excess air are no longer present. Thermal efficiencies as high as 45 percent have been obtained under favorable conditions in low-speed stationary or marine engines. The diesel engine has been used successfully in airplanes, but in its present state of development it lacks many of the advantages of the carburetor-type aircraft engine.

Volumetric Efficiency

Volumetric efficiency is the ratio of the volume of the fuel-air charge burned by the engine at atmospheric pressure and temperature to the piston displacement. If the cylinder of an engine draws in a charge of fuel and air having a volume at standard atmospheric pressure and temperature which is exactly equal to the piston displacement of the cylinder, the cylinder has a volumetric efficiency of 100 percent.

In a similar manner, if a volume of 95 in^3 [1.56 L] of the fuel-air mixture is admitted into a cylinder of 100-in^3 [1.64-L] displacement, the volumetric efficiency is 95 percent. Volumetric efficiency may be expressed as a formula thus:

$$\text{Vol eff} = \frac{\text{vol of charge at atm pressure}}{\text{piston displacement}}$$

Factors that tend to decrease the mass of air entering an engine have an adverse effect on volumetric efficiency. Typical factors that have this effect are (1) improper valve timing, (2) engine rpm, (3) carburetor air temperature, (4) design of the induction system, and (5) high combustion chamber temperature. A combination of these factors can exist at any one time.

Improper timing of the valves affects volumetric efficiency because the degree of opening of the intake valve influences the amount of airflow into the cylinder and because the timing of the opening and closure of the exhaust valve affects the outflow of exhaust gases. The intake valve must be open as wide as possible during the intake stroke, and the exhaust valve must close precisely at the instant that exhaust gases stop flowing from the combustion chamber. Valve timing is explained in Chap. 3.

High engine rpm can limit volumetric efficiency because of the air friction developed in the intake manifold, valve ports, and carburetor. As intake air velocity increases, friction increases and reduces the volume of airflow. At very high engine rpm the valves may "float" (not close completely), thereby affecting airflow.

Carburetor air temperature affects volumetric efficiency because as air temperature increases, the density decreases. This results in a decreased mass (weight) of air entering the combustion chambers. High combustion chamber temperature is a factor because it affects the density of the air.

Maximum volumetric efficiency is obtained when the throttle is wide open and the engine is operating under a full load.

A **naturally aspirated** (unsupercharged) engine always has a volumetric efficiency of less than 100 percent. On the other hand, the supercharged engine often is operated at a volumetric efficiency of more than 100 percent because the supercharger compresses the air before it enters the cylinder. The volumetric efficiency of a naturally aspirated engine is less then 100 percent for two principal reasons: (1) The bends, obstructions, and surface roughness inside the intake system cause substantial resistance to the airflow, thus reducing air pressure below atmospheric in the intake manifold; and (2) the throttle and the carburetor venturi provide restrictions across which a pressure drop occurs.

● ENGINE REQUIREMENTS

To be satisfactory for use in an airplane, the engine must possess certain characteristics. The most important of these are light weight, reliability, economy of operation, flexibility, and balance. Each of these characteristics is important and will be discussed individually.

Weight

The weight of an engine must be related to its power output in order to determine its desirability for use in an airplane. The weight is therefore included in the factor called the **weight-power ratio,** which may be defined as the weight in pounds per horsepower output. If an engine weighs 2400 lb [1088.61 kg] and the METO horsepower is 2200 [1640 kW], then the weight-power ratio is 2400/2200 or 1.09 lb/hp [1.51 kg/kW]. The overall usefulness of an engine must be considered with the necessary accessories and systems required for operation. When all the items of the powerplant are included, we have the all-up (flying) weight of the powerplant.

Powerplant weight must be kept as low as possible in order to leave a reasonable amount of the gross weight of the airplane available for a useful load and also to provide a margin of safety. One of the principal advantages of a gas-turbine engine is the excellent weight-power ratio compared with that of a piston engine. Turboshaft engines have been built with weight-power ratios of more than 8:1, while piston engines rarely attain a ratio of more than 1:1. The ratio of the weight to power may be expressed either as a weight-power ratio or a power-weight ratio. The latter is usually employed with turbine engines.

The weight per horsepower depends in part on the size of the engine, but this, in turn, depends on the choice of metals and alloys, the engine accessories that are added, the stress analysis factors, the use of high-performance fuels, and the absence or presence of a supercharger to give a better performance with a lower fuel consumption. Use of the power recovery turbine by which a portion of the exhaust energy is returned to the crankshaft on the Wright R-3350 engine has proved to be a most effective method of improving the weight-power ratio and also the specific fuel consumption.

Small engines are usually built for small airplanes which are operated by private pilots who have a limited budget. To keep down costs, the manufacturer may select less expensive materials and follow cheaper manufacturing processes, with the result that the typical small engine has a greater weight per horsepower than larger engines have.

Large engines are normally built for airplanes operated by the airlines or by the government; hence, the original cost of the engine is not as important to the buyer as the efficiency of its performance over a long period of time. Horsepower can be increased by the use of superchargers, higher crankshaft speeds, and other devices and methods. The selection of lighter weight but more expensive materials lowers the weight; consequently, the ratio of horsepower to weight is greater than in the typical small engine. In recent years, use of direct fuel injection and supercharging on small engines has greatly improved their efficiency, weight-power ratio, and specific fuel consumption.

Reliability

An aircraft engine is **reliable** when it can be depended on to do what it is rated to do by the manufacturer. It should not fail if properly inspected, maintained, and operated by skilled personnel. It should have a long life, with the maximum of **time between overhaul** (TBO) periods. It should also have a low fuel consumption at cruising speeds. Its units, parts, and accessories should serve well for a long period of time, and they should be easily, quickly, and inexpensively replaced after a reasonable time of service.

Economy of Operation

It is apparent that if an airplane engine is mechanically reliable, it costs less to operate than one that constantly needs overhauling or other maintenance. In addition, low fuel and oil consumption must be considered, not only from the standpoint of their direct cost but also from the standpoint of reducing weight in flight.

As previously mentioned, **specific fuel consumption** (sfc) is a measure of the power developed to fuel consumed; hence, it is also a measure of fuel economy. Specific fuel consumption is obtained by dividing the weight of the fuel burned per hour by the horsepower developed by the engine. **Brake specific fuel consumption** (bsfc) is the term used when the horsepower value is brake horsepower.

The specific fuel consumption for a turbojet or turbofan engine is found by dividing the weight of the fuel burned per hour by the thrust of the engine. In this case, the term **thrust specific fuel consumption** (tsfc) may be used. For a turboprop engine, the divisor in the formula is the **equivalent shaft horsepower** (eshp). This is the actual shaft horsepower developed by the engine plus a value determined to be equivalent to the thrust produced by the jet exhaust of the engine. In this case the term used could be **equivalent specific fuel consumption** (esfc) to indicate the fuel economy of the engine.

Flexibility

Flexibility is the ability of the engine to run smoothly and perform in the desired manner at all speeds from idling to full power output and through all variations of atmospheric conditions. Flexibility also includes the ability to operate efficiently at all altitudes at which the aircraft is designed to fly.

Balance

Balance has several possible meanings in a discussion of the fundamental requirements for aircraft engines, but the principal factor is *freedom from vibration*. Since the airplane is light and rather flexible, engine vibrations may reduce the life of certain parts or units and may cut down the life of the airplane as a whole. In addition, vibration is fatiguing to the pilot and passengers, thus reducing the desirability of flight. Tubing, cables, conduits, instruments, and accessories may be severely damaged and, at least, their functions may be impaired by excessive engine vibration.

One of the methods of reducing vibration is to design the engine with a large number of cylinders, thus reducing the total effect of the pulsating torque delivered by the separate cylinders. Rotating masses, such as crankshafts, may be balanced by the use of counterweights, dynamic dampeners, or a Vibratory Torque Control (VTC), a patented feature employed on Continental Tiara engines.

In addition to proper balancing of the engine and its parts, the engine mounting structure attached to the airplane is designed to reduce the effects of vibration by the use of rubber or synthetic rubber mounting attachments. These prevent the engine vibration from being transmitted to the engine mount and thence to the aircraft structure. These mountings are designed so that the entire engine weight is supported on rubber. **Dynafocal** mounts provide dynamic suspension of the engine and are so designed that the stress axes extended pass through the center of gravity (c.g.) of the engine and provide the most effective dampening of engine vibration by *dynamically* supporting the engine at the *focal* point, or c.g., of its vibrating mass. Thus, dynamic suspension greatly aids in isolating the normal vibration of the engine from the aircraft.

● FACTORS AFFECTING PERFORMANCE

Earlier in this chapter we discussed engine power, mean effective pressure, rpm, displacement, and other factors involved in the measurement of engine performance. We will explore these areas in greater depth and apply them to actual engine operation.

Manifold Pressure

As we have explained previously, **manifold pressure,** or manifold absolute pressure (MAP), is the absolute pressure of the fuel-air mixture immediately before it enters the intake port of the cylinder. **Absolute pressure** is the pressure above a complete vacuum and is often indicated in *pounds per square inch absolute* (psia) or in *inches of mercury* (inHg). In the metric system, MAP may be indicated in kilopascals (kPa). The pressure we read on an ordinary pressure gage is the pressure above ambient atmospheric pressure and is often called **gage pressure** or **pounds per square inch gage** (psig). Manifold pressure is normally indicated on a pressure gage in inches of mercury instead of pounds per square inch gage; hence, the reading on a manifold gage at sea level when an engine is not running will be about 29.92 inHg [101.34 kPa] when conditions are standard. When the engine is idling, the gage may read from 10 to 15 inHg [33.87 to 50.81 kPa] because the manifold pressure will be considerably below atmospheric pressure owing to the restriction of the throttle valve.

Manifold pressure is of primary concern to the operator of a high-performance engine because such an engine will often be operating at a point near the maximum allowable pressure. It is essential, therefore, that any engine which can be operated at an excessive manifold pressure be equipped with a manifold pressure gage so that the operator can keep the engine operation within safe limits.

The operator of an aircraft engine must take every precaution to avoid operating at excessive manifold pressure or incorrect MAP-rpm ratios because such operation will result in excessive cylinder pressures and temperatures. Excessive cylinder pressures are likely to overstress the cylinders, pistons, piston pins, valves, connecting rods, bearings, and crankshaft journals. Excessive pressure usually is accompanied by excessive temperature, and this leads to detonation, preignition, and loss of power. Detonation usually results in engine damage if continued for more than a few moments. Damage may include piston failure by cracking or burning, failure of cylinder base studs, cracking of the cylinder head, and burning of valves.

Naturally aspirated engines using variable-pitch propellers *must be equipped with MAP gages in order to assure safe operation*. Such engines, when equipped with fixed-pitch propellers, do not require the use of a MAP gage because on these engines the MAP is a function of the throttle opening.

Detonation and Preignition

We already have mentioned detonation and preignition; however, it is well to review these conditions in connection with our discussions of engine performance.

Detonation is caused when the temperature and pressure of the compressed mixture in the combustion chamber reaches levels sufficient to cause instantaneous burning (explosion) of the mixture. Excessive temperatures are caused by high manifold pressure, high intake air temperature, or an overheated engine. A principal cause of detonation is operating an engine either on a fuel whose octane rating is not sufficiently high for the engine or on a high-combustion-rate fuel. A high-octane fuel can withstand greater temperature and pressure before igniting than can a low-octane fuel.

When detonation occurs, the fuel-air mixture may burn properly for a portion of its combustion and then explode as the pressure and temperature in the cylinder increase beyond their normal limits. Detonation will further increase the temperature of the cylinders and pistons and may cause the head of a piston to melt. Extended detonation will often cause valves to burn and this, of course, leads to serious power loss. Detonation cannot be detected in an aircraft engine as easily as preignition.

Preignition is caused when there is a hot spot in the engine that ignites the fuel-air mixture before the spark plug fires. The hot spot may be red-hot spark-plug electrodes or carbon particles which have reached burning temperature and are glowing. Preignition is indicated by roughness of engine operation, power loss, and high cylinder-head temperature. To prevent preignition and detonation, operate the engine with the proper fuel and within the correct limits of manifold pressure and cylinder head temperature.

Mean Effective Pressure

Mean effective pressure, mentioned previously, is a computed pressure derived from power formulas in order to provide a measuring device for determining engine performance. For any particular engine operating at a given rpm and power output, there will be a specific **indicated mean effective pressure** (imep) and a corresponding **brake mean effective pressure** (bmep).

Mean effective pressure may be defined as an average pressure inside the cylinders of an internal-combustion engine based on some calculated or measured horsepower. It increases as manifold pressure increases. Imep is the mean effective pressure derived from indicated horsepower, and bmep is the mean effective pressure derived from brake horsepower output.

The pressure in the cylinder of an engine throughout one complete cycle is indicated in the curve of Fig. 2-13. This curve is not derived from any particular engine; it is given to show the approximate pressures that exist during the various events of the cycle. It will be observed that ignition takes place shortly before TDC, and then there is a rapid pressure rise which reaches maximum shortly after TDC. Thus, the greatest pressure on the cylinder occurs during the first 5 to 12° after TDC. By the end of the power stroke very little pressure is left, and this is being rapidly dissipated through the exhaust port.

The indicated horsepower (ihp) of an engine is the result of the imep, the rpm, the distance through which

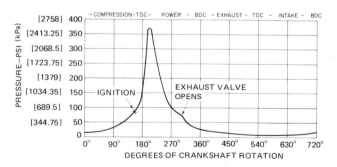

FIG. 2-13 Curve to show cylinder pressure.

the piston travels, and the number of cylinders in the engine. The formula for this computation was previously given as

$$\text{ihp} = \frac{PLANK}{33,000} \qquad (1)$$

where P = imep
L = length of the stroke, ft
A = area of the piston
N = rpm divided by 2
K = number of cylinders

The foregoing formula can also be given as

$$\text{ihp} = \frac{PLAN}{33,000} \qquad (2)$$

where N is the number of power strokes per minute. The number of power strokes per minute is equal to the rpm/2 times the number of cylinders. In this latter formula the N includes both the N and K of the previous formula.

If we can obtain the brake horsepower output of an engine by means of a dynamometer or prony brake, we can determine the bmep by means of a formula derived from the power formula given above. By simple transposition, the formula set up for brake horsepower becomes

$$P(\text{bmep}) = \frac{33,000 \times \text{bhp}}{LAN} \qquad (3)$$

In order to simplify the use of the formula, we can convert the length of the stroke and the area of the piston to the displacement of one cylinder and then multiply by the number of cylinders to find the total displacement of the engine. In the formula, L is equal to the length of the stroke in feet, A is the area of the piston. (The area of the piston is calculated with the formula $A = \pi r^2$ or $\pi d^2/4$), and N is the number of cylinders times the rpm divided by 2. Since we must multiply the area of the piston by the length of the stroke in inches in order to obtain piston displacement in cubic inches, we may use S for length of stroke in place of L and express S in inches. Then $S/12$ is equal to L because L is expressed in feet. For example, if the stroke (S) is 6 in, then it is equal to $\frac{6}{12}$ or $\frac{1}{2}$ ft. With these adjustments in mind, we find that LAN becomes

$$\frac{SA}{12} \times \text{no. of cylinders} \times \frac{\text{rpm}}{2}$$

or

$$\frac{SA \times \text{no. of cylinders} \times \text{rpm}}{12 \times 2}$$

Since $SA \times$ no. of cylinders is the total displacement (disp) of the engine, we can express the above value as

$$\frac{\text{Displacement} \times \text{rpm}}{12 \times 2}$$

Substituting this value in formula (3),

$$\begin{aligned}
\text{bmep} &= \frac{33,000 \times \text{bhp}}{(\text{disp} \times \text{rpm})/(12 \times 2)} \\
&= \frac{24 \times 33,000 \times \text{bhp}}{\text{disp} \times \text{rpm}} \\
&= \frac{792,000}{\text{disp}} \times \frac{\text{bhp}}{\text{rpm}} \qquad (4)
\end{aligned}$$

If an R-1830 engine is turning at 2750 rpm and developing 1100 hp [820.27 kW], we can find the bmep as follows:

$$\begin{aligned}
\text{bmep} &= \frac{792,000}{1830} \times \frac{1100}{2750} \\
&= 173 \text{ psi } [1192 \text{ kPa}]
\end{aligned}$$

For any particular engine computation, the value of 792,000/disp may be considered as a constant for that engine and given the designation K. Formula (4) then becomes

$$\text{bmep} = K \times \frac{\text{bhp}}{\text{rpm}}$$

To find the bhp of an engine, the foregoing formula is rearranged as follows:

$$\text{bhp} = \frac{\text{bmep} \times \text{rpm}}{K}$$

The constant K is often called the **K factor** of the engine.

It must be noted that the foregoing formula may be used for imep as well as for bmep if ihp is used in the formula instead of bhp. It is easy to determine the bhp of an engine by means of a dynamometer or prony brake; hence, the bmep computation is more commonly employed to determine engine performance.

Power and Efficiency

The efficiency of an engine is the ratio of output to input. For example, if the amount of fuel consumed should produce 300 hp [223.71 kW] according to its Btu rating and the output is 100 hp [74.57 kW], then the thermal efficiency is $\frac{100}{300}$ or $33\frac{1}{3}$ percent.

An engine producing 70 hp [52.20 kW] burns about 30 lb/h [13.61 kg/h] of gasoline or $\frac{1}{2}$ lb/min [0.23 kg/min]. One-half pound of gasoline has a heat value of about 10,000 Btu, and since 1 Btu can do 778 ft·lb [107.60 kg·m] of work, the fuel being consumed should produce 778 × 10,000 ft·lb [1383 kg·m] of work per min. Then

$$\text{Power} = \frac{778 \times 10,000}{33,000} = 235 \text{ hp } [175.24 \text{ kW}]$$

The fuel being consumed has a total power value of 235 hp, but the engine is producing only 70 bhp. The thermal efficiency is then $\frac{70}{235}$, or approximately 30 percent.

We may ask the question: What happens to the other 70 percent of the fuel energy? The answer is that the largest portion of the fuel energy is dissipated as heat and friction. The distribution of the fuel energy is approximately as indicated in the following:

Brake horsepower	30 percent
Friction and heat loss from engine	20 percent
Heat and chemical energy in exhaust	50 percent

Brake Specific Fuel Consumption

As stated previously, one of the measures of engine performance is **brake specific fuel consumption** (bsfc). Bsfc is the number of pounds of fuel burned per hour for each bhp produced. The bsfc for modern reciprocating engines is *usually* between 0.40 and 0.50 lb/hp/h [0.18 and 0.226 kg/kW/hr]. The bsfc depends upon many elements of engine design and operation, volumetric efficiency, rpm, bmep, friction losses, etc. In general, we may say that bsfc is a direct indicator of overall engine efficiency. The best values of bsfc for an engine are obtained at a particular cruising setting, usually at a little over 70 percent of a maximum power. During takeoff the bsfc may increase to a value almost double what it is at the best economy setting, because a richer mixture must be used for takeoff and because engine efficiency decreases with the higher rpm needed for maximum power.

Weight-Power Ratio

Another important indicator of engine performance is the **weight-power ratio.** This is the ratio of the weight of the engine to the bhp at best power settings. For example, if the basic weight of the engine is 150 lb [68.04 kg] and the power output is 100 hp [74.57 kW], the weight-power ratio is 150:100 or 1.5 lb/hp [0.91 kg/kW]. Since weight is a prime consideration in the design of any aircraft, the weight-power ratio of the engine is always an important factor in the selection of the airplane powerplant. Weight-power ratios for reciprocating engines vary between 1.0 and 2.0 lb/hp [0.61 and 1.22 kg/kW], with the majority of high-performance engines having ratios between 1.0 and 1.5 lb/hp.

● ENGINE PERFORMANCE CURVES

A large number of different curves may be developed for an engine to indicate a wide variety of operating conditions. For the purpose of this text we shall employ curves that deal with specific fuel consumption, power output, manifold pressure, rpm, air density, and propeller load.

Manifold Pressure and RPM

The effects of manifold pressure and rpm on the power output of an engine are shown in the curves of Fig. 2-14. These values were developed from the performance of Continental 0-470-K and -L engines with *atmospheric conditions at the sea-level standard* of pressure (P_s) and temperature (T_s). From this chart the

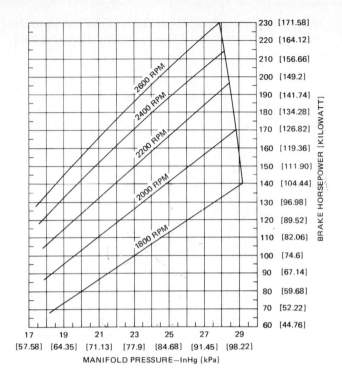

FIG. 2-14   Bhp vs. intake manifold pressure vs. revolutions per minute.

horsepower of the engine can be determined for any normal values of rpm and manifold pressure. For example, if the engine is operating at 23 inHg [77.9 kPa] manifold pressure and 2200 rpm, the power delivered is about 149 hp [111.11 kW]. For full power output at 2600 rpm and 27.9 inHg [94.5 kPa] manifold pressure, the power is 230 hp [171.58 kW].

It is of interest to note from the chart that volumetric efficiency decreases as rpm increases above 1800. At 1800 rpm the manifold pressure can be as high as 29.2 inHg [98.9 kPa]; but as rpm increases, the maximum manifold pressure decreases to 27.9. Volumetric efficiency (VE) decreases as rpm increases because the velocity of the air or fuel-air mixture increases, with a resultant increase in air friction and gas-inertia effects through the manifold passages.

Maximum VE is always less than 100 percent for a naturally aspirated engine, and at full power is likely to be about 75 percent. VE is greatly increased by valve overlap, up to the point where valve overlap would allow exhaust pressure to act against intake pressure or where there would be a reversal of exhaust flow back into the cylinder.

The power values shown in the chart of Fig. 2-14 will be as indicated only at sea-level standard conditions. If temperature or pressure changes, there will be a corresponding change in power output at a given rpm and manifold pressure.

Propeller Load

It can be readily understood that it requires more power to drive a propeller at high speeds than at low speeds, and that a certain propeller being driven at a given speed will absorb a specific amount of power.

Actually, the power required to drive a propeller varies as the cube of the rpm. This is expressed in the formula

$$hp = K \times rpm^3$$

where K is a constant whose value depends on the propeller type, size, pitch, and number of blades. Another formula that can be used to express the same principle is

$$hp_2 \, [W_2] = hp_1 \, [W_1] \left(\frac{rpm_2}{rpm_1}\right)^3$$

This means that it will require eight times as much power to drive a propeller at a given speed than it will require to drive it at half that speed. Also, if the speed of a propeller is tripled, it will require 27 times as much power to drive it than it did at the original speed.

Propeller load curves are shown in the chart of Fig. 2-15. This chart shows the manifold pressure, the power output, and the bsfc at different rpms when the engine is operated at full throttle with a particular fixed-pitch propeller.

At the top of the chart, it will be noted that manifold pressure decreases at full throttle as rpm increases. This is in keeping with the observation made from the chart of Fig. 2-14. From the prop load curve at the top of the chart, we can see that the propeller can be turned at 1950 rpm with a manifold pressure of 20 inHg [67.74 kPa], at 2200 rpm at a manifold pressure of 22 inHg [74.51 kPa], and at 2600 rpm with a manifold pressure of 27.8 inHg [94.16 kPa]. This is the maximum output available with this propeller because the load curve meets the manifold pressure curve at this rpm.

From the curves at the middle portion of the graph, we can see that the engine power output increases as rpm increases. The increase is not proportional because of the decrease in manifold pressure which takes place as rpm increases. We also note from the prop load curve that the propeller can be driven at 2100 rpm with 142 hp [105.89 kW], at 2400 rpm with 202 hp [150.63 kW], and at 2600 rpm with 248 hp [184.93 kW]. Another way of saying the same thing is that the propeller absorbs 248 hp at 2600 rpm.

The curves at the bottom of the graph in Fig. 2-15 show the specific fuel consumption under various conditions of rpm and prop load. It will be observed that the best fuel consumption takes place at approximately 2200 rpm when the propeller is absorbing 160 hp [119.31 kW]. The bsfc at this point is about 0.52 lb/hp/h [0.316 kg/kW/h]. If the engine were operated at full throttle with the rpm at 2200, the bsfc would be about 0.61 lb/hp/h [0.371 kg/kW/h].

Effects of Altitude on Performance

It has been mentioned previously that air density will affect the power output of an engine at a particular rpm and manifold pressure. Since air density depends on pressure, temperature, and humidity, these factors must be taken into consideration in determining the exact performance of an engine. In order to obtain the power of an engine from a power chart, it is necessary

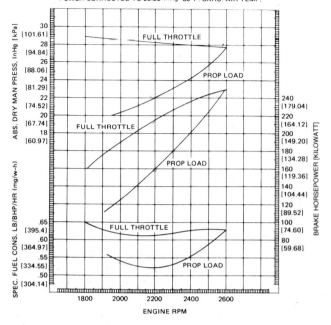

FIG. 2-15 Bhp, intake manifold pressure, and bsfc vs. revolutions per minute.

to find what corrections must be made. In the first place, the **density altitude** must be determined by applying approximate corrections to the pressure altitude as shown on a standard barometer. A chart for converting pressure altitude to density altitude is shown in Fig. 2-16. If the temperature at a particular altitude is the same as standard (T_s) at that altitude, no cor-

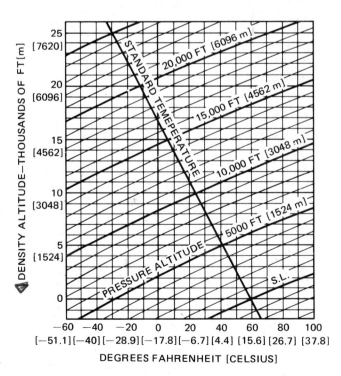

FIG. 2-16 Chart to convert pressure altitude to density altitude.

rection for density will be required unless humidity is a factor.

The charts of Fig. 2-17 are used to determine the power output of a Continental 0-470-M engine at altitude. The chart at the left shows engine output at sea-level standard conditions with no ram air pressure applied to the carburetor intake. The chart at the right shows the effect of altitude and is used in conjunction with the first chart. The points corresponding to engine rpm and manifold pressure are located on both charts. The horsepower indicated on the sea-level chart is transferred to an equivalent point C on the altitude chart. Then a straight line is drawn from the point A to the point C to establish the altitude correction. The intersection of this line with the density altitude line (D in the example) establishes the power output of the engine. The horsepower should be corrected by adding 1 percent for each 6°C (42.8°F) temperature decrease below T_s (standard temperature) and subtracting 1 percent for each 6°C temperature increase above T_s.

Effects of Fuel-Air Ratio

Thus far we have considered engine performance under fixed conditions of fuel-air ratio and without reference to other variables which exist under actual operating conditions. There are two fuel-air ratio values which are of particular interest to the operator of an engine. These are the best power mixture and the best economy mixture. The actual fuel-air ratio in each case will also depend upon engine rpm and manifold pressure.

The **best power mixture** for an aircraft engine is that fuel-air mixture which permits the engine to develop maximum power at a particular rpm. The **best economy mixture** is that fuel-air mixture which provides the lowest brake specific fuel consumption (bfsc). This is the setting which would normally be employed by a pilot in attempting to obtain maximum range for a certain quantity of fuel.

The specific effects of various fuel-air mixtures under different conditions of operation are discussed at length in Chap. 4.

Other Variables Affecting Performance

The performance of an aircraft engine is affected by a number of conditions or design features not yet mentioned. However, these must be taken into account if an accurate evaluation of engine operation is to be made. Among these conditions are carburetor air-intake ram pressure, carburetor air temperature, water-vapor pressure, and exhaust back pressure.

Ram air pressure at the carburetor air scoop is determined by the design of the scoop and the velocity of the air. Ram air pressure has the effect of supercharging the air entering the engine; hence, the actual power output will be greater than it would be under standard conditions of rpm, pressure, and temperature. An empirical formula for ram is

$$\text{Ram} = \frac{V^2}{2045} - 2$$

where ram is in inches of water and V is air velocity in miles per hour.

Carburetor air temperature (CAT) is of considerable importance because it affects the density, and hence the quantity, of the air taken into the engine and, if it is too high, will result in detonation.

If an engine is equipped with a supercharger, the manifold mixture temperature should be observed rather than carburetor air temperature because the temperature of the mixture actually entering the engine is the factor governing engine performance. A standard rule to correct for the effects of temperature is to add 1 percent to the chart horsepower for each 6°C below T_s and to subtract 1 percent for each 6°C above T_s.

Water vapor pressure effects must be determined when an engine is required to operate at near maximum

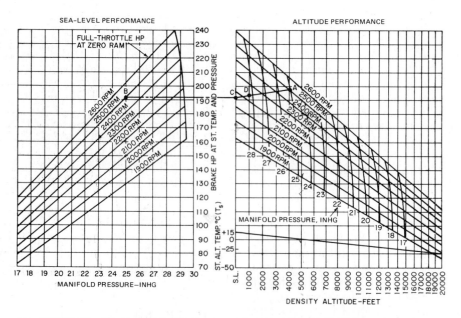

FIG. 2-17 Finding actual horsepower from sea-level and altitude charts.

power output under conditions of high humidity. In extreme cases an engine may lose as much as 5 percent of maximum rated power; hence, an allowance must be made for takeoff distance and other critical factors. At altitudes above 5000 ft [1524 m], water-vapor pressure is considered inconsequential.

Exhaust back pressure has a decided effect on the performance of an engine because any pressure above atmospheric at the exhaust port of a cylinder will reduce volumetric efficiency. The design of the exhaust system is therefore one of the principal items to be considered by both the engine manufacturer and the manufacturer of the exhaust system. Exhaust-back-pressure effect begins at the cylinder with the exhaust port. Both the size and shape of the opening and passages will affect the pressure. From the exhaust port onward, the exhaust stacks and sound-reduction devices will produce varying amounts of back pressure, depending upon design.

Engineers have developed exhaust-augmenting systems to assist in reducing exhaust back pressure and to utilize the ejected exhaust gases for the production of additional thrust. These devices have proved effective for increasing engine performance on the airplane. Such systems usually consist of one or more tubes into which the exhaust stacks from the engine are directed. The engine exhaust passing through the tubes through which ram air is also flowing results in a reduced pressure against the exhaust and an increased thrust because the jet of exhaust gases is directed toward the rear of the aircraft. **Exhaust augmentors** with inlets inside the engine nacelle also increase airflow through the nacelle, thus improving cooling.

● ENGINE OPERATION

Operating Requirements

The operation of any reciprocating engine requires that certain precautions be observed and that all operations be kept within the limitations established by the manufacturer. Among the conditions which must be checked during the operation of an engine are the following:

1. Engine oil pressure
2. Oil temperature
3. Cylinder-head temperature
4. Engine rpm
5. Manifold pressure
6. Drop in rpm when switching to single magneto operation
7. Engine response to propeller controls, if a constant-speed or controller-pitch propeller is used with the engine
8. Exhaust gas temperature

No engine should be operated at high-power settings unless its oil pressure and temperature are within satisfactory limits; otherwise, oil starvation of bearings and other critical parts will occur, thus potentially doing permanent damage to the engine. For this reason, a reciprocating engine must be properly warmed up before full-power operation is undertaken. When the engine is started, the oil pressure gage should be observed to see that the oil pressure system is functioning satisfactorily. *If no oil pressure is indicated within 30 seconds after starting, the engine must be shut down and the malfunction located.* If the engine is operated without oil pressure for much more than 30 seconds (s), damage is likely to result.

Generally speaking, if an engine has been warmed up for a few minutes and it can be accelerated to full power and rpm without showing excessive oil pressure, and if it operates smoothly, then it is likely that a safe takeoff can be made. In any event, however, the directions given in the airplane owner's handbook should be followed.

Prior to takeoff, the reciprocating engine should be given an ignition check and a full-power test. This is usually done while the airplane is parked just off the end of the takeoff runway in a warm-up area. For the magneto check, the throttle is moved slowly forward until the engine rpm is at the point recommended by the manufacturer. This is usually from 1500 to 1600 rpm, although it may vary from this range. To make the check, the ignition switch is turned from the BOTH position to the LEFT MAGNETO position and the tachometer is observed for rpm drop. The amount of drop is noted, and then the switch is turned back to the BOTH position for a few seconds until the engine is again running smoothly at the full-test rpm. The switch is then turned to the RIGHT MAGNETO position for a few seconds so the rpm drop can be noted. The engine should not be allowed to operate for more than a few seconds on a single magneto because of possible plug fouling.

The permissible rpm drop during the magneto test varies, but it is usually in the range between 50 and 125 rpm. In all cases the instructions in the operator's manual should be followed. Usually the rpm drop will be somewhat less than the maximum specified in the instructions.

When the magnetos are checked on an airplane having a constant-speed or controllable-pitch propeller, it is essential that the propeller be in the full HIGH-RPM (low-pitch) position; otherwise, a true indication of rpm drop may not be obtained.

After the magneto check is completed, the engine should be given a brief full-power check. This is done by slowly advancing the throttle to the full forward position and observing the maximum rpm obtained. If the rpm level and the manifold pressure are satisfactory and the engine runs smoothly, the throttle is slowly retarded until the rpm has returned to the desired idling speed.

When making the full-power check, the pilot must make sure that the airplane is in a position which will not direct the propeller blast into another airplane or into any other area where damage or inconvenience to another person may be caused. The pilot should also make sure that the brakes are on and the elevator control pulled back if the airplane has conventional landing gear.

Power Settings and Adjustments

During the operation of an airplane the engine power settings must be changed from time to time for various types of operation. The principal power settings are those for *takeoff, climb, cruises* (from maximum to minimum), *letdown,* and *landing.*

The methods for changing power settings differ according to the type of engine, the type of propeller, whether the engine is equipped with a supercharger, the type of carburetion, and other factors. The operator's manual will give the proper procedures for a particular airplane-engine combination.

The following rules generally apply to most airplanes and engines:

1. Always move the throttle slowly for either a power increase or power decrease. "Slowly" in this case means that the throttle movement from full open to closed or the reverse should require 2 to 3 s rather than the fraction of a second required to "jam" the throttle forward or "jerk" it closed.

2. Reduce the power setting to the climb value as soon as practical after takeoff if specified climb power is less than maximum power. Continued climb at maximum power can produce excessive cylinder-head temperatures and detonation. This is particularly true if the airplane is not equipped with a cylinder-head temperature gage.

3. Do not reduce power suddenly when the cylinder head temperature is high (at or near the red line on the cylinder-head temperature gage). The sudden cooling which occurs when power is reduced sharply will often cause the cylinder head to crack. When preparing to let down, it is well to reduce the power slowly by increments to allow for a gradual reduction of temperatures.

4. When operating an airplane with a constant-speed propeller, *always reduce manifold pressure with the throttle before reducing rpm* with the propeller control. Conversely, always increase rpm with the propeller control before increasing manifold pressure. If the engine rpm setting is too low and the throttle is advanced, it is possible to develop excessive cylinder pressures with the consequences explained previously. The operator of an engine should become thoroughly familiar with the maximums allowable for the engine and then make sure that the engine is operated within these limits. It must be remembered that a constant-speed propeller holds engine rpm to a particular value in accordance with the position of the propeller control. When the throttle is moved forward, the propeller blade angle increases and the manifold pressure increases, but the rpm remains the same.

5. During a prolonged glide with power low (throttle near closed position) "clear the engine" occasionally to prevent spark-plug fouling. This is done by advancing the throttle to a medium-power position for a few seconds. If the engine runs smoothly, the power may be reduced again.

6. Always place the manual mixture control in the FULL RICH position when the engine is to be operated at or near full power. This is to aid in preventing overheating. The engine should be operated with the mixture control in a LEAN position only during cruise, in accordance with the instructions in the operator's manual. When power is reduced for letdown and preparatory to landing, the mixture control should be placed in the FULL RICH position. Some mixture controls and carburetors do not include a full-rich setting. In this case the mixture control is placed in the RICH position for high power and takeoff.

7. If there is any possibility of ice forming in the carburetor at the time that power is reduced for a letdown preparatory to landing, it is necessary to place the carburetor heat control in the HEAT ON position. This is a precautionary measure and is common practice for all engines in which carburetor icing may occur.

8. At high altitude, adjust the mixture control to a position less rich than that used at low altitudes. The density of the air at high altitudes is less than it is at lower altitudes; hence, the same volume of air will contain less oxygen. If the engine is supercharged, the increase in altitude will not be of particular consequence until the capacity of the supercharger is exceeded. Usually the manifold pressure gage will provide information helpful for proper adjustment of mixture control; however, an accurate exhaust gas temperature (EGT) gage is considered essential for leaning the mixture for cruise power at altitudes normally flown.

● CRUISE CONTROL

Cruise control is the adjustment of engine controls to obtain the results desired in range, economy, or flight time. Since an engine consumes more fuel at high power settings than it does at the lower settings, it is obvious that maximum speed and maximum range or economy cannot be attained with the same power settings. If a maximum distance flight is to be made, it is desirable to conserve fuel by operating at a low-power setting. On the other hand, if maximum speed is desired, it is necessary to use maximum-power settings with a decrease in range capability.

Range and Speed Charts

The charts of Fig. 2-18 were developed for the operation of the Piper PA-23-160 Apache aircraft. The chart on the left shows the effects of power settings on range, and the chart on the right shows how power settings and true airspeed (TAS) are related. From these charts we can easily determine the proper power settings for any flight within the range of the airplane, taking into consideration the flight altitude, flight distance, and desired flight time.

If we wish to make a flight of 900 miles [1448.40 kilometers (km)] at an altitude of 6500 ft [1981.20 m], we can determine the flight values for maximum speed or maximum range or we can choose to select a compromise setting. If we wish to make the flight in the shortest possible time, we must use 75 percent of the engine power. With this setting (2400 rpm and full throttle) the TAS will be about 175 mph [281.58 km/h] and the flight will take 5.14 h, assuming no tail wind or head wind. At this setting the fuel consumption will be 18.8 gal/h [71.17 L/h]; hence, the flight will require

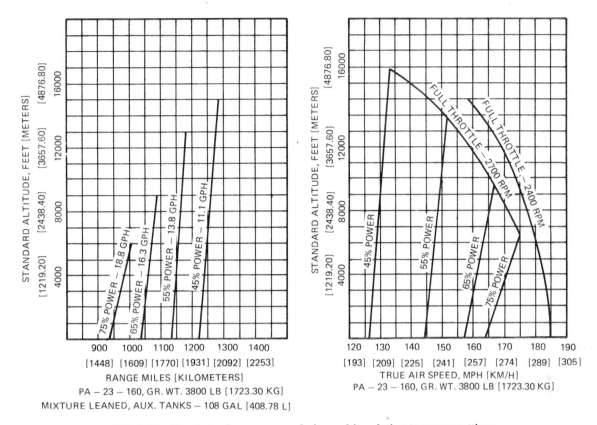

FIG 2-18 Charts to show range and airspeed in relation to power settings.

96.7 gal [366.05 L] of fuel. If we wish to make this same flight with maximum economy, we may operate the engines at 45 percent of power with the mixture control leaned as far as good engine operation will permit. At this power setting, the TAS will be about 128 mph [205.95 km/h] and the fuel required for the trip will be about 77.7 gal [294.13 L]. The time required for the flight is about 7 h.

We would seldom actually operate the engine at the extremes mentioned above because the recommended power setting for cruise conditions is 65 percent of power. Flying at an altitude of 9500 ft [3306 m], this would provide a TAS of about 166 mph [267.09 km/h]. If we wished to operate more economically or with greater range, we would probably use a power setting of about 55 percent of maximum.

Power Settings

In order to set the controls of an engine for a particular power output, we adjust manifold pressure and rpm according to density altitude when the airplane is equipped with constant-speed propellers. Table 2-1 shows the settings for the Lycoming O-320-B opposed engine. This table is adjusted for the use of pressure altitude at standard temperature (T_s) instead of density altitude. Observe the following facts regarding the settings for manifold pressure and rpm:

1. At a given rpm and a given power setting, manifold pressure must be decreased as altitude increases. This is because the T_s of the air decreases and the density therefore increases. Thus, a given volume of air at a certain pressure will have a greater weight as

altitude increases, and manifold pressure must be reduced in order to maintain constant power.

2. When the engine is operated at higher rpm, a lower manifold pressure is used in order to maintain the same power.

3. At a certain level of altitude the manifold pressure can no longer be maintained because of the reduction in atmospheric pressure. This is the point in the chart shown as FT, meaning full throttle.

4. At 55 percent rated power, the power can be maintained up to 15,000 ft [4572 m] pressure altitude. An output of 75 percent power can be maintained only up to about 7000 ft [2133.6 m] pressure altitude.

5. Manifold pressure settings must be adjusted to maintain a particular power output if outside air temperature is above or below the standard given in the chart.

For large engines equipped with torquemeters, a power table of the type shown in Table 2-1 is not required. Here, a conversion chart is used to change the torquemeter readings to horsepower. The formula used with the Pratt & Whitney R-2800 engine is bhp = torque pressure × rpm × torque constant.

● **ENGINE STARTING AND STOPPING**

Starting Procedures

The starting of an aircraft engine is a relatively simple procedure; however, certain precautions must be taken in order to obtain the best results and to avoid damage to the engine.

TABLE 2-1 Power setting table—Lycoming Model O-320-B, 160-hp [119.31-kW] engine

Press. alt. 1000 ft [304.80 m]	Std. alt. temp., °F [°C]	88 hp [65.62 kW]—55% rated Approx. fuel 7 gal/h [26.50 L/h] rpm & man. press.				104 hp [77.55 kW]—65% rated Approx. fuel 8 gal/h [30.28 L/h] rpm & man. press.				120 hp [89.48 kW]— 75% rated Approx. fuel 9 gal/h [34.07 L/h] rpm & man. press.		
		2100	2200	2300	2400	2100	2200	2300	2400	2200	2300	2400
SL	59 [15.0]	22.0	21.3	20.6	19.8	24.4	23.6	22.8	22.1	25.9	25.2	24.3
1	55 [12.8]	21.7	20.0	20.3	19.6	24.1	23.3	22.5	21.8	25.6	24.9	24.0
2	52 [11.1]	21.4	20.7	20.1	19.3	23.8	23.0	22.3	21.5	25.0	24.3	23.5
3	48 [8.9]	21.1	20.5	19.8	19.1	23.5	22.7	22.0	21.2	25.3	24.6	23.8
4	45 [7.2]	20.8	20.2	19.6	18.9	23.1	22.4	21.7	21.0	24.7	24.0	23.2
5	41 [5.0]	20.5	19.9	19.3	18.6	22.8	22.1	21.4	20.7	FT	23.7	23.0
6	38 [3.3]	20.2	19.6	19.0	18.4	22.5	21.8	21.2	20.5		FT	22.7
7	34 [1.1]	19.9	19.3	18.8	18.2	22.2	21.5	20.9	20.2			FT
8	31 [−0.56]	19.5	19.0	18.5	18.0	FT	21.2	20.6	19.9			
9	27 [−2.8]	19.2	18.8	18.3	17.7		FT	20.3	19.7			
10	23 [−5.0]	18.9	18.5	18.0	17.5			FT	19.4			
11	19 [−7.2]	18.6	18.2	17.8	17.3				FT			
12	16 [−8.9]	18.3	17.9	17.5	17.0							
13	12 [−11.1]	FT	17.6	17.3	16.8							
14	9 [−12.8]		FT	17.0	16.6							
15	5 [−15.0]			FT	16.3							

To maintain constant power, correct manifold pressure approximately 0.15 in Hg for each 10°F variation in carburetor air temperature from standard altitude temperature. Add manifold pressure for air temperatures above standard; subtract for temperatures below standard.

Before an engine is started prior to flight, the airplane should be given a standard preflight inspection in accordance with the operator's handbook. During this inspection a small amount of fuel should be drained from each drain valve to remove sediment and water. The fuel and oil quantity should be checked to make sure that all tanks are properly filled. The ground or pavement near the propellers should be checked for loose items which might be drawn into the propellers.

After all preliminary inspections and cockpit checks have been made, the engine may be started according to the procedure set forth in the operator's manual. The following starting procedure is recommended for the engines on a Cessna 310F airplane:

1. Turn ignition switches on.
2. Open the throttle approximately ½ in [0.127 cm].
3. Set the propeller pitch lever full forward for HIGH RPM.
4. Set the mixture lever full forward for FULL RICH.
5. Clear the propeller.
6. Turn the auxiliary fuel pump switch to PRIME position. Avoid leaving the auxiliary fuel pump switch in either the PRIME or ON position for more than a few seconds unless the engine is running.
7. Turn the ignition switch to START when the fuel flow reaches 2 to 4 gal/h [7.57 to 15.14 L/h]. (Read the fuel pressure gage.) If the engines are warm, turn the ignition switch to START first, then turn the auxiliary pump switch to PRIME.
8. Release the ignition switch as soon as the engine fires.
9. Turn off the auxiliary fuel pump switch when the engine runs smoothly. During very hot weather, if there is an indication of vapor in the fuel system (indicated by fluctuating fuel flow) with the engine running, turn the auxiliary fuel pump switch to ON until the system is purged.

10. Check for an oil-pressure indication within 30 s in normal weather and 60 s in cold weather. If no indication appears, shut off the engine and investigate.
11. Disconnect the external power source if used.
12. Warm up the engine at 800 to 1000 rpm.

Starting procedures will vary to some extent for different types of engine installations. There are certain general rules which almost always apply, however, and these should be noted carefully.

1. The propeller control should be placed in the HIGH-RPM (low-pitch) position.
2. The mixture control should be placed in the FULL RICH position unless otherwise specified.
3. If the engine cowling is equipped with cowl flaps, the flaps should be open.
4. A radial engine should always be turned through several revolutions before attempting to start in order to clear cylinders of possible hydraulic lock due to oil drainage in the lower cylinders. It is good practice to turn any engine through a few times if the engine is cold in order to provide a small amount of prelubrication to the moving parts. Ignition switch must be checked OFF.
5. If the engine is equipped with oil cooler flaps which have a manual control, the control should be placed in the CLOSED position until the engine is warm.

Stopping Procedure

Usually an aircraft engine has cooled sufficiently for an immediate stop because of the taxi time required to move the airplane into the parking area. It is good practice, however, to observe the cylinder-head temperature (CHT) gage to see that the CHT is somewhat under 400°F [204.24°C] before stopping. If the engine is equipped with an idle cutoff on the mixture control, the engine should be stopped by placing the control in the IDLE CUTOFF position. Immediately after the engine stops, the ignition switch must be turned off. If the airplane is equipped with cowl flaps, the flaps should be left in the open position until after the engine has cooled.

If an airplane is equipped with a Hamilton Standard counterweight-type propeller, the propeller should be placed in the LOW-RPM (high-pitch) position shortly before stopping the engine in order to move the propeller cylinder rearward where it will cover the piston. This prevents corrosion of the piston and keeps sand and other dirt from collecting on the piston surface. In the LOW-RPM position, the cylinder is rearward and the oil in the cylinder has been returned to the engine. This helps prevent congealing of oil in the cylinder during very cold weather.

After stopping the engine, check all switches in the cockpit to OFF. This is especially important with respect to the ignition switches and the master battery switch. Check to ensure that all wheel chocks are installed and release the parking brake to prevent undue stress to the brake system.

● REVIEW QUESTIONS

1. Why are reciprocating engines for aircraft called heat engines?
2. Describe and explain the four-stroke five-event cycle of a piston engine.
3. Name and describe the function of each of the basic parts of a typical piston engine.
4. Explain *bore* and *stroke*.
5. Why are the TDC and BDC positions of the piston important?
6. What are the positions of the intake and exhaust valves at the end of the power stroke?
7. At what point in the operating cycle of an engine does the ignition event take place?
8. Why is ignition timed to take place at this point?
9. Why are reciprocating engines designed with valve overlap?
10. Explain the energy conversion in a piston engine.
11. Briefly discuss Boyle's law and Charles's law with respect to the operation of a piston engine.
12. What is the function of a connecting rod?
13. Describe the operation of a two-stroke-cycle engine.
14. How does the valve action in a two-stroke-cycle engine differ from the valve action in a four-stroke-cycle engine?
15. Why is a two-stroke-cycle engine less efficient than a four-stroke-cycle engine?
16. Describe the operation of a diesel engine.
17. Compare the compression ratio of a diesel engine with that of a conventional gasoline engine.
18. Define *power*.
19. Define *compression ratio*.
20. How would you determine the piston displacement of an engine? The compression ratio?
21. What is *indicated horsepower?*
22. Compute the horsepower output of the following described engine operating at 2000 rpm: bore, 3.5

in [8.89 cm]; stroke, 4 in [10.16 cm]; number of cylinders, 6; bmep, 140 psi [965.3 kPa].
23. Compute the piston displacement of a radial engine having nine cylinders, a bore of 5 in [12.7 cm], and a stroke of 5 in.
24. Compute the compression ratio of an engine which has a bore of 5 in [12.7 cm] and a stroke of 5 in [12.7 cm] when the volume of the combustion chamber is 16.36 in³ [0.268 L] with the piston at top dead center (TDC).
25. What factors limit the compression ratio of an engine?
26. Discuss the relation between compression ratio and fuel octane rating.
27. In what way are indicated horsepower, friction horsepower, and brake horsepower related?
28. Explain the operation of a dynamometer or prony brake.
29. What is the function of a torque nose on an engine?
30. Compute the bmep of an engine when the output is 450 hp [335.57 kW], the rpm is 2300, the bore and stroke are each 5.5 in [13.97 cm], and the engine has nine cylinders.
31. How is the power output of an engine affected by manifold pressure?
32. Explain the *rated power* of an engine.
33. Explain *mechanical* and *thermal* efficiencies.
34. What is the *critical altitude* of an engine?
35. Define *volumetric efficiency*.
36. Give the causes of reduced volumetric efficiency.
37. Compare the weight-power ratio of a reciprocating engine with that of a gas-turbine engine.
38. Explain *brake specific fuel consumption* and *thrust specific fuel consumption*.
39. List the most important requirements of an aircraft engine. Explain the meaning of each.
40. List some of the methods by which crankshaft balance is obtained to reduce engine vibration.
41. What is the purpose of dynamic suspension in an engine installation?
42. Explain the importance of manifold pressure in the operation of a piston engine.
43. Why does an increase in manifold pressure result in an increase in power output if the rpm of an engine remains constant?
44. Explain the difference between detonation and preignition.
45. List likely causes of detonation.
46. If an R-2800 engine is delivering 2000 hp [1491.4 kW] at 2700 rpm, what is the bmep of the engine?
47. Approximately what power output would you expect to obtain from a good aircraft engine consuming 60 lb [27.22 kg] (10 gal [37.85 L]) per hour of gasoline? Use a heat value of 20,000 Btu per lb of fuel.
48. At approximately what percent of full power output is a piston aircraft engine likely to be most efficient?
49. If you operate a Continental 0-470-K engine at 2200 rpm and 25 inHg [84.68 kPa] manifold pressure, what power would the engine develop at sea level (standard conditions)?
50. Why is the volumetric efficiency of a naturally-aspirated engine always less than 100 percent?

51. If it requires 50 hp [37.29 kW] to drive a certain fixed-pitch propeller at 600 rpm, what power is required to drive it at 1800 rpm?
52. Compare density altitude with pressure altitude.
53. If a Continental 0-470-M engine is operated at 5000 ft [1524 m], density altitude, with a power setting of 2400 rpm and 23 inHg [77.91 kPa] manifold pressure, what is the actual power output?
54. Explain the difference between best power mixture and best economy mixture.
55. Discuss the effects of carburetor air temperature (CAT) on engine operation.
56. What effect does carburetor air-intake ram pressure have on engine operation?
57. How does exhaust back pressure affect engine power? Why?
58. What features of an exhaust system affect exhaust back pressure?
59. What is the purpose of exhaust augmentors?
60. Discuss the conditions which should be observed when an engine is in operation.
61. What is likely to occur if a piston (reciprocating) engine is operated at high power before it is properly warmed up?
62. In the operation of an engine with a variable-pitch or constant-speed propeller, what sequence must be followed in changing power settings? Why?
63. What precaution is taken to prevent carburetor icing when power is reduced for a letdown?
64. What is meant by *cruise control*?
65. How is maximum range obtained in the operation of an aircraft engine?
66. Give the general rules which apply to the starting of a reciprocating engine.

3 ENGINE CLASSIFICATION, CONSTRUCTION, AND NOMENCLATURE

Since the first internal-combustion engine was successfully operated, many different types of engines have been designed. Many have been suitable for the operation of automobiles and/or aircraft and others have been failures. The failures have been a result of poor efficiency, lack of dependability (owing to poor design and to materials which could not withstand the operation conditions), high cost of operation, excessive weight for the power produced, and similar deficiencies.

The engine which emerged as the most practical and dependable for aircraft use from before World War I until after World War II was the conventional piston, or reciprocating, engine; today most light and medium-light aircraft use this type of engine. The gas-turbine engine, developed since the 1950s, has advanced to the point where it has proved a superior powerplant for both military and commercial aircraft for which high power and high speed are important. The gas-turbine engine can produce much more power for its weight than a piston engine. For example, a piston engine usually produces less than 1 hp/lb of engine weight, whereas some turboprop engines can produce more than 4 hp/lb of engine weight.

In this chapter we will examine the design and construction of various types of piston engines, particularly those that are still in use.

Conventional piston engines are classified according to a variety of characteristics, including cylinder arrangement, cooling method, and number of strokes per cycle. The most satisfactory classification, however, is by cylinder arrangement. This is the method usually employed because it is more completely descriptive than the other classifications. Gas-turbine engines are classified according to construction and function; these classifications will be discussed in Chap. 21.

Cylinder Arrangement

Although some engine designs have become obsolete, we shall mention the types most commonly constructed throughout the history of powerplants. Aircraft engines may be classified according to cylinder arrangement with respect to the crankshaft as follows: (1) in-line, upright; (2) in-line, inverted; (3) V-type, upright; (4) V-type, inverted; (5) double-V- or fan-type; (6) opposed- or flat-type; (7) X-type; (8) radial-type, single row; (9) radial-type, double row; (10) radial-type, multiple row or "corncob." The simple drawings of Fig. 3-1 illustrate some of these arrangements. Photographs of different types are shown in Chap. 1.

The double-V- and fan-type engines have not been in use for many years, and the only piston engines in extensive use for aircraft in the United States at the present time are the opposed and radial types. A few V-type and in-line engines may still be found in operation, but these engines are no longer manufactured in the United States for general aircraft use.

Early Designations

Most of the early aircraft engines, with the exception of the rotary types, were water-cooled and were of either in-line or V-type design. These engines were often classified as liquid-cooled in-line engines, water-cooled in-line engines, liquid-cooled V-type engines, or water-cooled V-type engines. As air-cooled engines were developed, they were referred to as air-cooled plus the cylinder arrangement on the crankcase.

Classification or Designation by Cylinder Arrangement and Displacement

Current designations for reciprocating engines generally employ letters to indicate the type and characteristics of the engine, followed by a numerical indication of displacement. The following letters usually indicate the type or characteristic shown:

L   **Left-hand rotation** for counter-rotating propeller
T   **Turbocharged** with turbine operated device
V   **Vertical,** for helicopter installation with the crankshaft in a vertical position
H   **Horizontal,** for helicopter installation with the crankshaft horizontal
A   **Aerobatic;** fuel and oil systems designed for sustained inverted flight
I   **Fuel injected;** continuous fuel-injection system installed
G   **Geared** nose section for propeller rpm reduction
S   **Supercharged;** engine structurally capable of operating with MAP of more than 30 inHg [101.61 kPa] and equipped with either a turbine-driven supercharger or an engine-driven supercharger
O   **Opposed cylinders**
R   **Radial engine;** cylinders arranged radially around the crankshaft

It should be noted that many engines are not designated by the foregoing standardized system. For example, the Continental W-670 engine is a radial-type, whereas the A-65, C-90, and E-225 are all opposed-type engines. V-type engines and inverted in-line engines have such designations as V and I. In every case, the technician working on an engine must interpret the designation correctly and utilize the proper information for service and maintenance.

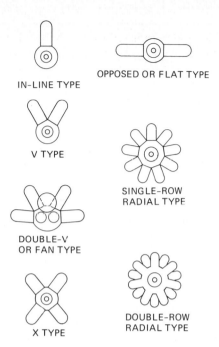

FIG. 3-1 Engines classified according to cylinder arrangement.

The three-digit numbers in the second part of the engine designation indicate displacement to the nearest 10 in³. An engine with a displacement of 471 in³ [7.72 L] is shown as 470, as is the case with the Teledyne Continental O-470 opposed engine.

A system of suffix designations has also been established to provide additional information about engines. The first suffix letter indicates the type of power section and the rating of the engine. This letter is followed by a number from 1 to 9, which gives the design type of the nose section. Following the nose-section number is a letter indicating the type of accessory section, and after this letter is a number which tells what type of counterweight application is used with the crankshaft. This number indicates the mode of vibration, such as 4, 5, or 6. The mode number will be found on the counterweights or dynamic balances on the crankshaft.

The final character in the designation suffix may be a letter indicating the type of magneto utilized with the engine. The letter D indicates a dual magneto.

An example of the standard designation for an engine is as follows.

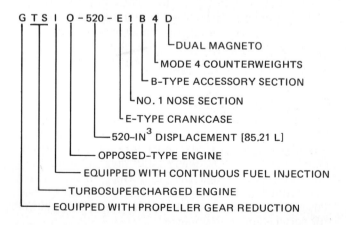

In some cases, the displacement number will end with a figure other than zero. In such a case, this is a special indication to reveal a characteristic such as an integral accessory drive.

Radial engines generally employ only the letter R followed by the displacement. For example, the R-985 is a single-row radial engine having a displacement of approximately 985 in³ [16.14 L].

In-Line Engines

The cylinders of **in-line** engines are arranged in a single row parallel with the crankshaft. The cylinders are either upright above the crankshaft or inverted, that is, below the crankshaft. The inverted configuration is generally employed. A typical inverted in-line engine is shown in Fig. 3-2. The engine shown is a Menasco "Pirate," Model C-4. The number of cylinders in an in-line engine is usually limited to six in order to facilitate cooling and to avoid excessive weight per horsepower. There is generally an even number of cylinders in order to provide a proper balance of firing impulses. The in-line engine utilizes one crankshaft. The crankshaft is located above the cylinders in an inverted engine. The engine may be either the air-cooled or the liquid-cooled type; however, liquid-cooled types are seldom utilized at present.

Use of the in-line-type engine is largely confined to low- and medium-horsepower applications for small aircraft. The engine presents a small frontal area and is therefore adapted to streamlining and a resultant low-drag nacelle configuration. When the cylinders are mounted in the inverted position, greater pilot visibility and a shorter landing gear are possible. However, the in-line engine has a greater weight-to-horsepower ratio than most other types. When the size of an aircraft engine is increased, it becomes increasingly difficult to cool it if it is the air-cooled in-line type; hence, this engine is not suitable for a high horsepower output.

V-Type Engines

The **V-type** engine has the cylinders arranged on the crankcase in two rows (or banks) forming the letter V, with an angle between the banks of 90, 60, or 45°. There is always an even number of cylinders in each row.

FIG. 3-2 Inverted in-line engine.

Since the two banks of cylinders are opposite each other, two sets of connecting rods can operate on the same crankpin, thus reducing the weight per horsepower as compared with the in-line engine. The frontal area is only slightly greater than that of the in-line type; hence, the engine cowling can be streamlined to reduce drag. If the cylinders are above the crankshaft, the engine is known as the **upright-V-type,** but if the cylinders are below the crankshaft, it is known as an **inverted-V-type.** Better pilot visibility and a short landing gear are possible if the engine is inverted.

Opposed, Flat, or O-Type Engine

The opposed-type engine is most popular for light conventional aircraft and helicopters and is manufactured in sizes delivering from less than 100 hp [74.57 kW] to more than 400 hp [298.28 kW]. These engines are the most efficient, dependable, and economical types available for light aircraft. Gas-turbine engines are being installed in some light aircraft, but their cost is still prohibitive for the average, private airplane owner.

The **opposed-type** engine is usually mounted with the cylinders horizontal and the crankshaft horizontal; however, in some helicopter installations the crankshaft is vertical. The engine has a low weight-to-horse-power ratio, and because of its flat shape it is very well adapted to streamlining and to horizontal installation in the nacelle. Another advantage is that it is reasonably free from vibration. Figure 3-3 illustrates a modern opposed engine for general aircraft use.

Radial Engines

The radial engine has been the workhorse of military and commercial aircraft ever since the 1920s, and during World War II they were used in all United States bombers and transport aircraft and in most of the other categories of aircraft. They were developed to a peak of efficiency and dependability; and even today, in the jet age, many of them are still in operation throughout the world in all types of duty.

A **single-row radial** engine has an odd number of cylinders extending radially from the centerline of the crankshaft. The number of cylinders is usually from five to nine. The cylinders are arranged evenly in the same circular plane, and all the pistons are connected to a single-throw 360° crankshaft, thus reducing the number of working parts and also reducing the weight.

A **double-row radial** engine resembles two single-row radial engines combined on a single crankshaft as shown in Fig. 3-4. The cylinders are arranged radially in two

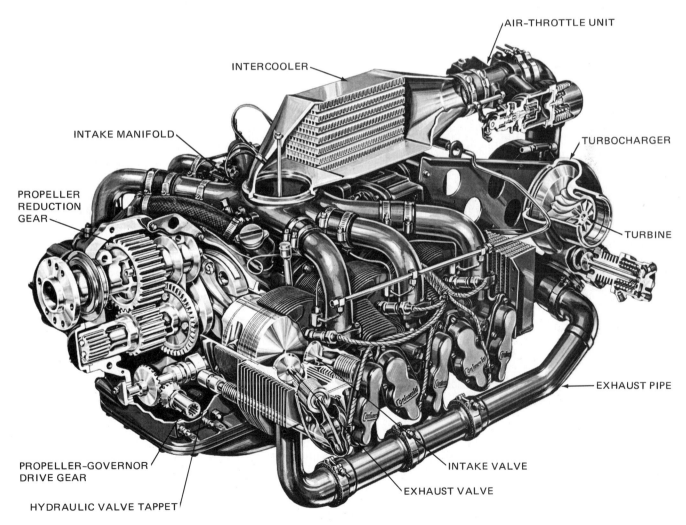

FIG. 3-3 Teledyne Continental six-cylinder opposed engine. *(Teledyne Continental)*

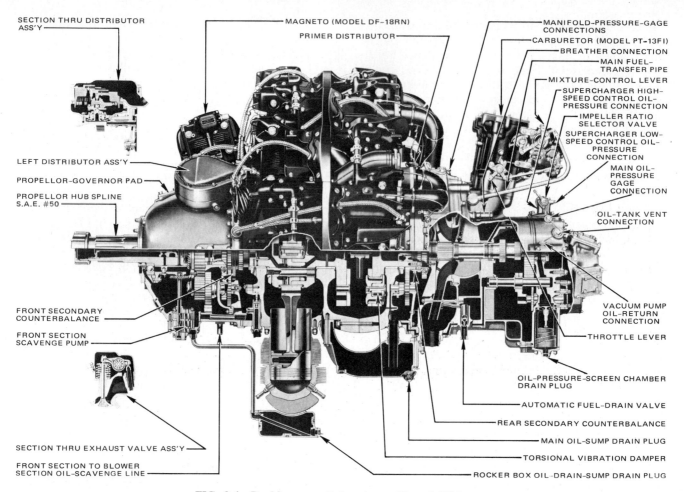

FIG. 3-4 Double-row radial engine. *(Pratt & Whitney)*

rows, and each row has an odd number of cylinders. The usual number of cylinders used is either 14 or 18, which means that the same effect is produced as having either two seven-cylinder engines or two nine-cylinder engines joined on one crankshaft. A two-throw 180° crankshaft is used to permit the cylinders in each row to be alternately staggered on the common crankcase. That is, the cylinders of the rear row are located directly behind the spaces between the cylinders in the front row. This allows the cylinders in both rows to receive ram air for the necessary cooling.

The radial engine has the lowest weight-to-horsepower ratio of all the different types of piston engines. It has the disadvantage of greater drag because of the area presented to the air, and it also has some problems in cooling. Nevertheless, the dependability and efficiency of the engine have made it the most widely used type for large aircraft equipped with reciprocating engines.

Multiple-Row Radial Engine

The 28-cylinder Pratt & Whitney R-4360 engine was used extensively at the end of World War II and afterward for both bombers and transport aircraft. This was the largest and most powerful piston-type engine built and used successfully in the United States. A photograph of this engine is shown in Fig. 3-5. Because of the development of the gas-turbine engine, the very large piston engine has been replaced by the more powerful, lighter-weight turboprop and turbojet engines. Since it has few moving parts compared with the piston engine, the gas-turbine engine is more trouble-free and the maintenance cost is reduced. Furthermore, the time between overhauls (TBO) is greatly increased.

Engine Cooling

Aircraft engines may be cooled either by air or by liquid; however, there are few liquid-cooled engines still in operation in the United States. We shall therefore devote most of our discussion to the air-cooled types.

Excessive heat is undesirable in any internal-combustion engine for three principal reasons: (1) It adversely affects the behavior of the combustion of the fuel-air charge, (2) it weakens and shortens the life of the engine parts, and (3) it impairs lubrication.

If the temperature inside the engine cylinder is too great, the fuel mixture will be preheated and combustion will occur before the proper time. Premature combustion causes detonation, "knocking," and other undesirable conditions. It will also aggravate the overheated condition and is likely to result in failure of pistons and valves.

The strength of many of the engine parts depends on their heat treatment. Excessive heat weakens such parts and shortens their life. Also, the parts may be-

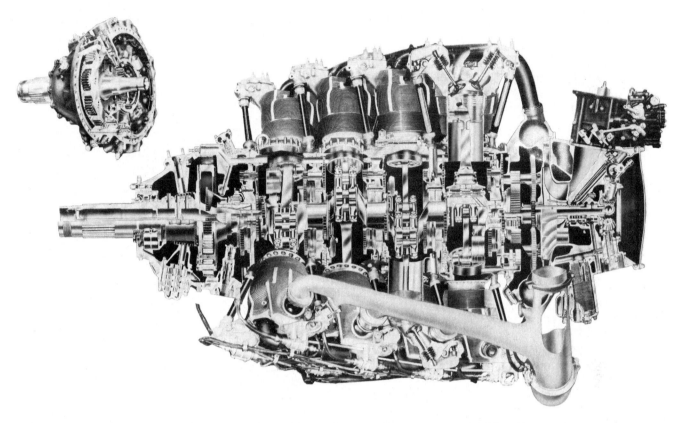

FIG. 3-5 Pratt & Whitney R = 4360 engine. *(Pratt & Whitney)*

come elongated, warped, or expanded to the extent that they freeze or lock together and stop the operation of the engine.

Excessive heat "cracks" the lubricating oil, lowers its viscosity, and destroys its lubricating properties.

Air Cooling. In an air-cooled engine, thin metal fins project from the outer surfaces of the walls and heads of the engine cylinders. When air flows over the fins, it absorbs the excess heat from the cylinders and carries it into the atmosphere. Deflector **baffles** fastened around the cylinders direct the flow of air to obtain the maximum cooling effect. The baffles are usually made of aluminum sheet. They are called **pressure baffles** because they direct airflow caused by ram air pressure. A cylinder with baffles is shown in Fig. 3-6. The operating temperature of the engine can be controlled by movable **cowl flaps** located on the engine cowling. On some airplanes, these cowl flaps are manually operated by means of a switch which controls an electric actuating motor. On other airplanes they can be operated either manually or by means of a thermostatically controlled actuator. Cowl flaps are illustrated in Fig. 3-7.

In the assembly of the engine baffling system, great care must be taken to see that the pressure baffles around the cylinders are properly located and secured. An improperly installed or loose baffle can cause a hot spot to develop, with the result that the engine may fail. The proper installation of baffles around the cylinders of a twin-row radial engine is illustrated in Fig. 3-8. Baffling for an opposed-type engine is shown in Fig. 3-9. It will be observed that the baffling maintains

a high-velocity airstream close to the cylinder and through the cooling fins. The baffles are attached by means of screws, bolts, spring hooks, or special fasteners.

Cylinder cooling is accomplished by carrying the heat from the inside of all the cylinders to the air outside the cylinders. Heat passes by conduction through the metal walls and fins of the cylinder assembly to the cooling airstream which is forced into contact with the fins by the baffles and cowling. The fins on the cylinder head are made of the same material as the head and are forged or cast as part of the head. Fins on the steel cylinder barrel are of the same metal as the barrel in most instances and are machined from the same forging as the barrels. In some cases the inner part of the cylinder is a steel sleeve and the cooling fins are made as a part of a muff or sleeve shrunk on the outside of the inner sleeve. A large amount of the

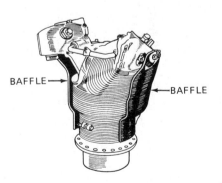

FIG. 3-6 Cylinder with pressure baffles for cooling.

33

FIG. 3-7 Cowl flaps.

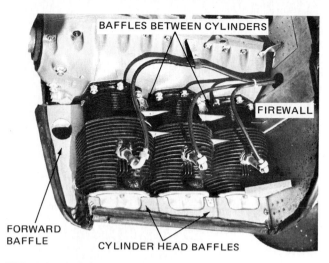

FIG. 3-9 Baffling for an opposed-type engine.

heat developed in an engine cylinder is carried to the atmosphere with the exhaust. This amount varies from 40 to 50 percent, depending upon the design of the engine. The proper adjustment of valve timing is the *most critical* factor in heat rejection through the exhaust.

In the operation of a helicopter, the ram air pressure is usually not sufficient to cool the engine, particularly when the craft is hovering. For this reason, a large engine-driven fan is installed in a position to maintain a strong flow of air across and around the cylinders and other parts of the engine. Helicopters powered by turbine engines do not require the external cooling fan.

The principal advantages of air cooling are that (1) the weight of the air-cooled engine is usually less than that of a liquid-cooled engine of the same horsepower because the air-cooled engine does not need a radiator, connecting hoses and lines, and the coolant liquid; (2) the air-cooled engine is less affected by cold-weather operations; and (3) the air-cooled engine in military airplanes is less vulnerable to gunfire. If an enemy bullet or bomb fragment strikes the radiator, hose, or lines of a liquid-cooled engine, it is obvious that its cooling system will leak and soon cause a badly overheated engine.

Liquid Cooling. Liquid-cooled engines are rarely found in U.S. aircraft today; however, the powerplant technician should have some understanding of the principal elements of such systems.

A liquid cooling system consists of the liquid passages around the cylinders and other hot spots of the engine (see Fig. 3-10), a radiator by which the liquid is cooled, a thermostatic element to govern the amount of cooling applied to the liquid, a coolant pump for circulating the liquid, and the necessary connecting pipes and hoses. If the system is sealed, a relief valve is required to prevent excessive pressure and a sniffler valve is necessary to allow the entrance of air to prevent negative pressure when the engine is stopped and cooled off.

Water was the original coolant for liquid-cooled engines. Its comparatively high freezing point (32°F) [0°C] and its relatively low boiling point (212°F) [100°C], made it unsatisfactory for the more powerful engines used in military applications. The liquid most commonly used for liquid-cooled engines during World War II was either **ethylene glycol** or a mixture of ethylene glycol and water. Pure ethylene glycol has a boiling point of about 350°F [176°C] and a slush-forming freezing point of about 0°F [−17.78°C] at sea level. This combination

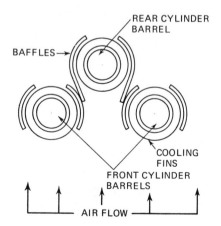

FIG. 3-8 Baffles around the cylinders of a twin-row radial engine.

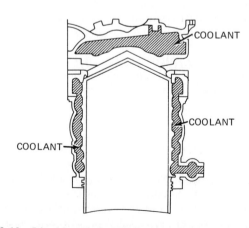

FIG. 3-10 Liquid-cooled cylinder and jacket.

of high boiling point and low freezing point made it a satisfactory coolant for aircraft engines.

● THE CRANKCASE

The **crankcase** of an engine is the housing that encloses the various mechanisms surrounding the crankshaft; hence, it is the foundation of the engine. The functions of the crankcase are as follows: (1) The crankcase must support itself, (2) it contains the bearings in which the crankshaft revolves, (3) it provides a tight enclosure for the lubricating oil, (4) it supports various internal and external mechanisms of the powerplant, (5) it provides mountings for attachment to the airplane, (6) it provides support for the attachment of the cylinders, and (7) by reason of its strength and rigidity, it prevents the misalignment of the crankshaft and its bearings.

Crankcases are of many sizes and shapes and may be of one-piece or multipiece construction. Most aircraft engine crankcases are made of aluminum alloy because it is both light and strong, but some engines which develop a great power output have crankcases made of forged steel. Although the variety of crankcase designs makes any attempts at classification difficult, they may be divided into three broad groups for discussion: (1) in-line and V-type crankcases, (2) opposed-engine crankcases, and (3) radial-engine crankcases.

In-Line and V-Type Engine Crankcases

Large in-line and V-type engine crankcases usually have four major sections: (1) the front, or nose, section; (2) the main, or power, section; (3) the fuel-induction and -distribution section; and (4) the accessory section.

The front, or nose, section is directly behind the propeller in most tractor-type airplanes. A **tractor-type** airplane is one in which the propeller ''pulls'' the airplane forward. The **nose section** may be cast as part of the main, or power, section, or it may be a separate construction with a dome or conical shape to reduce drag. Its function is to house the propeller shaft, the propeller thrust bearing, the propeller reduction-gear train, and sometimes a mounting pad for the propeller governor. In a very few arrangements where the nose section is not located close to the engine, the propeller is connected to the engine through an extension shaft and the reduction-gear drive has its own lubricating system. This same arrangement is found in some turboprop engines.

The **main**, or **power**, **section** varies greatly in design for different engines. When it is made up of two parts, one part supports one-half of each crankshaft bearing and the other supports the opposite half of each bearing. The cylinders are normally mounted on and bolted to the heavier of the two parts of this section on an in-line engine, and the crankshaft bearings are usually supported by reinforcing weblike partitions. External mounting lugs and bosses are provided for attaching the engine to the engine mount.

The **fuel-induction and -distribution section** is normally located next to the main, or power, section. This section houses the diffuser vanes and supports the in-

ternal blower impeller when the engine is equipped with an internal blower system. The induction manifold is located between the fuel-induction and -distribution section and the cylinders. The housing of this section has an opening for the attachment of a manifold-pressure-gage line, and it also has internal passages for the fuel drain valve of the blower case. The **fuel drain valve** is designed to permit the automatic drainage of excess fuel from the blower case.

The **accessory section** may be a separate unit mounted directly on the fuel-induction and -distribution section, or it may form a part of the fuel-induction and -distribution section. It contains the accessory drive-gear train and has mounting pads for the fuel pump, coolant pump, vacuum pump, lubricating-oil pumps, magnetos, tachometer generator, and similar devices operated by engine power. The material used in constructing this section is generally either an aluminum-alloy casting or a magnesium-alloy casting.

Opposed-Engine Crankcase

The crankcase for a six-cylinder opposed engine is shown in Fig. 3-11. This assembly consists of two matching, reinforced aluminum-alloy castings divided vertically at the center line of the engine and fastened together by means of a series of studs and nuts. The mating surfaces of the crankcase are joined without the use of a gasket, and the main bearing bores are machined for the use of precision-type main bearing inserts. Machined mounting pads are incorporated into the crankcase for attaching the accessory housing, cylinders, and oil sump. Opposed engines with propeller reduction gearing usually incorporate a separate nose section to house the gears.

The crankcase of the opposed engine contains bosses and machined bores to serve as bearings for the camshaft. During overhaul it is important to inspect these areas for excessive wear. On the camshaft side of each crankcase half are the tappet bores which carry the hydraulic valve tappet bodies.

Essential portions of the lubricating system are contained in the crankcase. Oil passages and galleries are drilled in the sections of the case to supply the crankshaft bearings, camshaft bearings, and various other moving parts which require lubrication. During overhaul, the technician must make sure that all oil

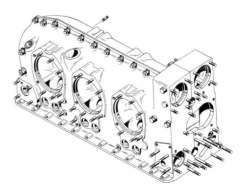

FIG. 3-11 Crankcase for a six-cylinder opposed engine. *(Teledyne Continental)*

passages are free of foreign matter and that passages are not blocked by gaskets during assembly.

Radial-Engine Crankcase

Radial-engine crankcases (see Fig. 3-12) may have as few as three or as many as seven principal sections, the number depending upon the size of the engine and its type, although the large engines usually have more sections than the small ones. For the purpose of describing radial-engine crankcases, it is customary to assume that the typical radial-engine crankcase has four major sections, although this is not necessarily true.

The **front,** or **nose, section** is usually made of aluminum alloy, its housing is approximately bell-shaped, and it is fastened to the power section by studs and nuts or cap screws. In most cases, this section supports a propeller thrust bearing, a propeller-governor drive shaft, and a propeller reduction-gear assembly if the engine provides for propeller speed reduction. It may also include an oil scavenge pump and a cam-plate or cam-ring mechanism.

This section may also provide the mountings for a propeller-governor control valve, a crankcase breather, an oil sump, magnetos, and magneto distributors. The engines which have magnetos mounted on the nose case are usually of the higher power ranges. The advantage of mounting the magneto on the nose section is in cooling. When the magnetos are on the nose section of the engine, they are exposed to a large volume of ram air; thus, they are kept much cooler than is the case when they are mounted on the accessory section.

The **main,** or **power, section** may be of one-piece or two-piece construction and usually consists of one, two, or possibly three pieces of high-strength heat-treated aluminum-alloy or steel forging, bolted together if there is more than one piece. The use of a two-part main power section for a radial engine makes it possible to add strength to this highly stressed section of the engine. The cam-operating mechanism is usually housed and supported by the main crankcase section. At the center of each main crankcase web section are crank-

shaft bearing supports. Cylinder mounting pads are located radially around the outside circumference of the power section. The cylinders are fastened to the pads by means of studs and nuts or cap screws. Oil seals are located between the front crankcase section and the main crankcase. Similar seals are installed between the power section and the fuel-distribution section.

The **fuel-induction and -distribution section** is normally located immediately behind the main power section and may be of either one-piece or two-piece construction. It is sometimes called the **blower section** or the **supercharger section** because its principal function is to house the blower or supercharger impeller and diffuser vanes. There are openings on the outside circumference of the housing for attaching the individual induction pipes, a small opening for the attachment of the manifold pressure line, and internal passages which lead to the supercharger drain valve.

The **accessory section** provides mounting pads for the accessory units, such as the fuel pumps, vacuum pumps, lubricating-oil pumps, tachometer generators, generators, magnetos, starters, two-speed supercharger-control valves, oil filtering screens, Cuno filters, and other items of accessory equipment. In some aircraft powerplants, the cover for the supercharger rear housing is made of an aluminum-alloy or a magnesium-alloy casting in the form of a heavily ribbed plate that provides the mounting pads for the accessory units; but in other powerplants, the housings for the accessory units may be mounted directly on the rear of the crankcase. Regardless of the construction and location of the accessory housing, it contains the gears for driving the accessories which are operated by engine power.

● CYLINDERS

The **cylinder** of an internal-combustion engine converts the chemical heat energy of the fuel to mechanical energy and transmits it through pistons and connecting rods to the rotating crankshaft. In addition to developing the power from the fuel, the cylinder dissipates

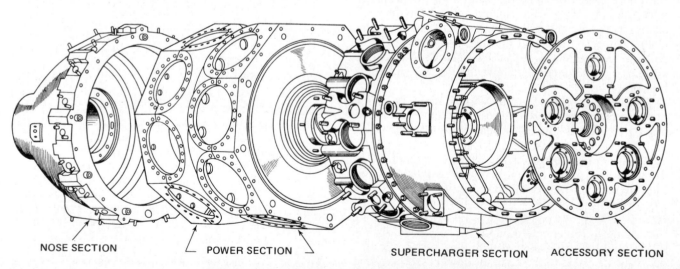

NOSE SECTION   POWER SECTION   SUPERCHARGER SECTION   ACCESSORY SECTION

FIG. 3-12 Crankcase for a twin-row radial engine.

a substantial portion of the heat produced by the combustion of the fuel, houses the piston and connecting-rod assembly, supports the valves and a portion of the valve actuating mechanism, and supports the spark plugs.

The cylinder assembly used for present-day engines usually includes the following components: (1) cylinder barrel with an integral skirt, (2) cylinder head, (3) valve guides, (4) valve rocker-arm supports, (5) valve seats, (6) spark-plug bushings, and (7) cooling fins. The cylinder assemblies—together with the pistons, connecting rods, and crankcase section to which they are attached—may be regarded as the main power section of the engine.

The two major units of the cylinder assembly are the cylinder barrel and the cylinder head. These are shown in Fig. 3-13. The principal requirements for this assembly are (1) the strength to withstand the internal pressures developed during operation at the temperatures which are normally developed when the engine is run at maximum design loads, (2) light weight, (3) the heat-conducting properties to obtain efficient cooling, and (4) a design which makes possible easy and inexpensive manufacture, inspection, and maintenance.

Cylinder Barrel

In general, the barrel in which the piston reciprocates must be made of high-strength steel alloy, be constructed to save weight as much as possible, have

the proper characteristics for operating under high temperatures, be made of a good bearing material, and have high tensile strength. The barrel is usually made of chrome-molybdenum (SEA 4130 or 4140) steel or chrome-nickel-molybdenum steel which is initially forged to provide maximum strength. The forging is machined to design dimensions with external fins and a smooth cylindrical surface inside. After machining, the inside surface is honed to a specific finish to provide the proper bearing surface for the piston rings. The roughness of this surface must be carefully controlled. If it is too smooth, it will not hold sufficient oil for the break-in period, and if it is too rough, it will lead to excessive wear or other damage to both the piston rings and the cylinder wall. The inside of the cylinder barrel may be surface-hardened by means of nitriding, or it may be chrome-plated to provide a long-wearing surface. **Nitriding** is a process whereby the ntirogen from anhydrous ammonia gas is caused to penetrate the surface of the steel by exposing the barrel to the ammonia gas for 40 h or more while the barrel is at a temperature of about 975°F [523°C].

In some cylinders the cylinder is bored with a slight taper. The end of the bore nearest the head is smaller than the skirt end to allow for the expansion caused by the greater operating temperatures near the head. Such a cylinder is said to be **chokebored**; it provides a nearly straight bore at operating temperatures.

The base of the cylinder barrel incorporates (as part of the cylinder) a machined mounting flange by which the cylinder is attached to the crankcase. The flange is drilled to provide holes for the mounting studs or bolts. The holes are reamed for accurate dimensioning. The cylinder **skirt** extends beyond the flange into the crankcase and makes it possible to use a shorter connecting rod. It also makes it possible to reduce the external dimensions of the engine. The cylinders for inverted engines and the lower cylinders of radial engines are provided with extra-long skirts. These skirts keep most of the lubricating oil from draining into the cylinders. This reduces oil consumption and decreases the possibility of **hydraulic lock** (also termed **liquid lock**) which results from oil collected in the cylinder head.

The outer end of the cylinder barrel is usually provided with threads so that it can be screwed and shrunk into the cylinder head, which is also threaded. The cylinder head is heated in an oven to 575 to 600°F [302 to 316°C] and is then screwed onto the cool cylinder barrel.

As mentioned previously, cooling fins are generally machined directly on the outside of the barrel. This method provides the best conduction of heat from the inside of the barrel to the cooling air. On some cylinders, the cooling fins are on aluminum-alloy muffs or sleeves shrunk on the outside of the barrel.

The cylinder barrel for the Continental Tiara engine is shown in Fig. 3-14. The barrel consists of a cylindrical, centrifugally cast sleeve of alloyed grey iron around which is die-cast an aluminum-finned muff. The aluminum muff (jacket) transmits heat from the liner to the cooling fins. It will be noted that the cylinder barrel is attached to the cylinder head by means of through-bolts from the crankcase.

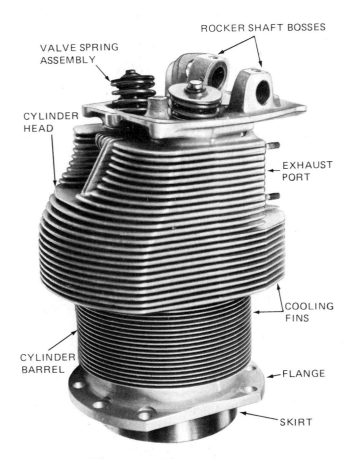

VALVE SPRING ASSEMBLY

ROCKER SHAFT BOSSES

CYLINDER HEAD

EXHAUST PORT

COOLING FINS

CYLINDER BARREL

FLANGE

SKIRT

FIG. 3-13 Cylinder assembly.

FIG. 3-14 Cylinder barrel for the Tiara engine. *(Teledyne Continental)*

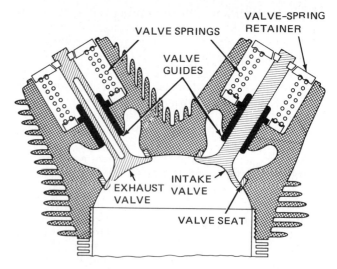

FIG. 3-15 Installation arrangement for valve guides.

Cylinder Heads

The cylinder head encloses the combustion chamber for the fuel-air mixture and contains the intake and exhaust valves, valve guides, and valve seats. The cylinder head also provides the support for the rocker shafts upon which the valve rocker arms are mounted.

The openings into which the spark plugs are inserted are provided in the cylinder head at positions designed to provide the best burning pattern. The spark-plug openings may contain bronze bushings shrunk and staked or pinned into the head, and in some cylinders the threads are reinforced with steel inserts called Heli-Coils. The Heli-Coil inserts make it possible to restore the thread by replacement of the inserts.

Cylinder heads are usually made of cast-aluminum alloy, (AMS 4220 or equivalent) to provide a maximum of strength with minimum weight. One disadvantage of aluminum alloy for cylinder heads is that the coefficient of expansion of aluminum is considerably greater than that of steel. This disadvantage is largely overcome through the method by which the cylinder heads are attached to the cylinder barrels.

The cooling fins are cast or machined on the outside of the cylinder head in a pattern to provide the most efficient cooling and to take advantage of cylinder-head cooling baffles. The area surrounding the intake passage and valve does not usually have cooling fins because the fuel-air mixture entering the cylinder carries the heat away. The intake side of the cylinder head can be quickly identified by noting which side is not finned.

As shown in Fig. 3-15, the valve guides are positioned to support and guide the stems of the valves. The valve guides are shrunk into bored bosses with a 0.001- to 0.0025-in [0.0254- to 0.0635-mm] tight fit. Before the valve guides are installed, the cylinder head is heated to expand the holes into which the guides are to be installed. The guides are then pressed into place

or driven in with a special drift. When the cylinder head cools, the guide is gripped so tightly that it will not become loose even under severe heating conditions. It is common practice when valve guides are replaced to install new guides which are approximately 0.002 in [0.05 mm] larger than the holes in which they are to be installed. Valve guides are made of aluminum bronze, tin bronze, or steel, and in some cylinders the exhaust-valve guide is steel while the intake-valve guide is made of bronze.

Because the aluminum-alloy cylinder-head material does not provide serviceable valve seats, valve-seat inserts are shrunk into place for both the intake and exhaust valves. These inserts are made of forged chrome-molybdenum steel or bronze and are installed in the heated head just before the head is screwed onto the barrel. In some cases the exhaust-valve seat is made of steel with a layer of Stellite (a very hard, heat-resisting alloy) bonded on the seat surface to provide a more durable seat. Valve-seat inserts are replaceable when they have been reground to the extent that they are no longer within approved dimension limits. When it is necessary to replace valve guides and valve seats, the valve guides should be replaced first, because the pilots for the seat tools are centered by the valve guides.

The interior shape of the cylinder head may be flat, peaked, or hemispherical, but the latter shape is preferred because it is more satisfactory for scavenging the exhaust gases rapidly and thoroughly. The cylinder head for the Continental Tiara engine shown in Fig. 3-16 is essentially flat at the top of the combustion chamber.

The three methods used for joining the cylinder barrel to the cylinder head are (1) the threaded joint, (2) the shrink fit, and (3) the stud-and-nut joint. The method most commonly employed for modern engines is the threaded joint.

The threaded joint is accomplished by chilling the cylinder barrel, which has threads at the head end, and heating the cast cylinder head to about 575°F [302°C] as previously explained. The cylinder head is threaded to receive the end of the barrel. A jointing compound

FIG. 3-16 Cylinder head for the Tiara engine. *(Teledyne Continental)*

is placed on the threads to prevent compression leakage, and then the barrel is screwed into the cylinder head. When the cylinder head cools, it contracts and grips the barrel tightly.

The cylinder head is provided with machined surfaces at the intake and exhaust openings for the attachment of the intake and exhaust manifolds. The manifolds are held in place by means of bolted rings which fit against the flanges of the manifold pipes. The intake pipes are usually provided with synthetic-rubber gaskets which seal the joint between the pipe and the cylinder. Exhaust pipes are usually sealed by means of metal or metal and asbestos gaskets. The mounting studs are threaded into the cast cylinder heads and are usually not removed except in case of damage.

The cylinder heads of currently operating radial engines are provided with fittings to accommodate rocker box intercylinder drain lines (hoses) which allow for evening of pressure and oil flow between cylinder heads. If oil flow is excessive in one or more rocker boxes, the excess will be relieved by flowing to other rocker boxes. The intercylinder drain lines also assure adequate lubrication for all rocker boxes. If one or more of the rocker box intercylinder drain lines becomes clogged, it is likely that excessive oil consumption will occur and that the spark plugs in the cylinders adjacent to the clogged lines will become fouled.

Cylinder Finish

Since air-cooled cylinder assemblies are exposed to conditions leading to corrosion, they must be protected against this form of deterioration. One method is to apply a coating of baked, heat-resistant enamel. In the past, this enamel was usually black; however, manufacturers have developed coatings which not only pro-

vide corrosion protection but also change color when over-temperature conditions occur. The Continental Gold developed by Teledyne Continental Motors is normally gold in color but turns pink when subjected to excessive temperatures. The blue-gray enamel used on Lycoming engines also changes color when over-heated, enabling the technician to detect possible heat damage during the inspection of the engine.

In the past, manufacturers *metallized* engine cylinders, particularly for large radial engines, by spraying a thin layer of molten aluminum on the cylinders with a special metallizing gun. The aluminum coating, when properly applied, was effective in providing protection against the corrosive action of saltwater spray, salt air, and the blasting effect of sand and other gritty particles carried by the cooling airstream.

● PISTONS

Construction

The **piston** is a plunger that moves back and forth or up and down within an engine cylinder barrel. It transmits the force of the burning and expanding gases in the cylinder through the connecting rod to the engine crankshaft. As the piston moves down (toward the crankshaft) in the cylinder during the intake stroke, it draws in the fuel-air mixture. As it moves upward (toward the cylinder head), it compresses the charge. Ignition takes place, and the expanding gases cause the piston to move toward the crankshaft. On the next stroke (toward the head), it forces the burned gases out of the combustion chamber.

In order to obtain maximum engine life, the piston must be able to withstand high operating temperatures and pressures; hence, it is usually made of aluminum alloy which may be either forged or cast. Aluminum alloy AMS 4140 is often used for forged pistons. Cast pistons may be made of Alcoa 132 alloy. Aluminum alloy is used because it is light in weight, has a high heat conductivity, and has excellent bearing characteristics.

A cross section of a typical piston is illustrated in Fig. 3-17. The top of the piston is the **head.** The sides form the skirt. The underside of the piston head often contains ribs or other means of presenting maximum surface area for contact with the lubricating oil splashed on it. This oil carries away part of the heat conducted through the piston head.

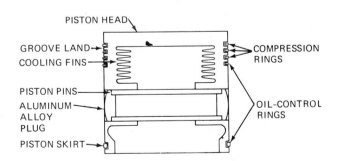

FIG. 3-17 Cross section of an assembled piston.

Some pistons are constructed with a slightly oval cross section. The diameter perpendicular to the piston pin is greater to allow for more wear of the piston due to additional side thrust against the cylinder walls and to provide a better fit at operating temperatures. Such a piston is called **cam ground** and must be installed as indicated by a mark on the top of the piston head.

Grooves are machined around the outer surface of the piston to provide support for the **piston rings.** The metal between the grooves is called a **groove land** or simply a **land.** The grooves must be accurately dimensioned and concentric with the piston.

The piston and ring assembly must form as nearly as possible a perfect seal with the cylinder wall. It must slide along the cylinder wall with very little friction. The engine lubricating oil aids in forming the piston seal and in reducing friction. All the piston assemblies in any one engine must be balanced. This means that each piston must weigh within $\frac{1}{4}$ ounce (oz) [7.09 gram (g)] of each of the others. This balance is most important in order to avoid vibration while the engine is operating. In any case, the weight limitations specified by the manufacturer must be observed.

The piston illustrated in the cross-sectional drawing of Fig. 3-17 has five piston rings; however, some pistons are equipped with four rings and many operate with only three rings.

The parts of a complete piston assembly are shown in Fig. 3-18. This illustration shows the piston, piston pin, pin retainer plugs, oil rings, and compression rings.

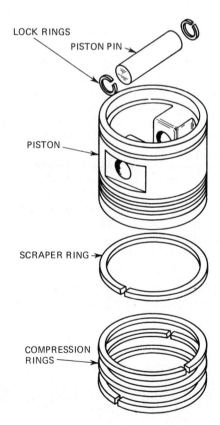

LOCK RINGS
PISTON PIN
PISTON
SCRAPER RING
COMPRESSION RINGS

FIG. 3-18 Components of a complete piston assembly.

Piston Speed

In order to appreciate the loads imposed on a piston and connecting-rod assembly, it is helpful to consider the speed at which the piston must travel in the cylinder. In order to move at high speeds with a minimum of stress the piston must be as light as possible. If an engine operates at 2000 rpm, the piston will start and stop 4000 times in 1 min, and if the piston has a 6-in [15.24-cm] stroke, it may reach a velocity of more than 35 mph [56.32 km/h] at the end of the first quarter of crankshaft rotation and at the beginning of the fourth quarter of rotation.

Piston Temperature and Pressure

The temperature inside the cylinder of an airplane engine may exceed 4000°F [2204°C] and the pressure against the piston during operation may be as high as 500 pounds per square inch (psi) [3447.5 kilopascals (kPa)] or higher. Since aluminum alloy is light and strong and conducts the heat away rapidly, it is generally used in piston construction. The heat in the piston is carried to the cylinder wall through the outside of the piston and is transmitted to the engine oil in the crankcase through ribs or other means on the inside of the piston head. Fins increase the strength of the piston and are more generally used than other methods of cooling.

Piston and Cylinder Wall Clearance

Piston rings are used as seals to prevent the loss of gases between the piston and cylinder wall during all strokes. It would be desirable to eliminate piston rings by having pistons large enough to form a gastight seal with the cylinder wall, but in that case the friction between the piston and the cylinder wall would be too great and there would be no allowance for expansion and contraction of the metals. The piston is actually made a few thousandths of an inch smaller than the cylinder, and the rings are installed in the pistons to seal the space between the piston and cylinder wall. If the clearance between the piston and the cylinder wall became too great, the piston could wobble, causing **piston slap.**

Types of Pistons

Pistons may be classified according to the type of head used. These are flat, recessed, concave, convex, or truncated cone, all of which are illustrated in Fig. 3-19. Pistons in modern engines are usually of the flathead type. The skirt of the piston may be of the trunk type, trunk type relieved at the piston pin, or slipper type. Typical pistons for modern engines are shown in Fig. 3-20. It will be noted that some pistons have the skirt cut out at the bottom to clear the crankshaft counterweights. Slipper-type pistons are no longer used in modern engines because they do not provide adequate strength and wear resistance.

The horsepower ratings of engines of the same basic design are changed merely by the use of different pistons. A domed piston increases the compression ratio

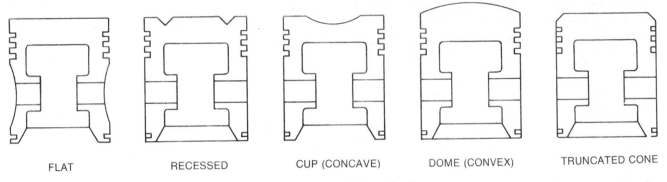

FLAT   RECESSED   CUP (CONCAVE)   DOME (CONVEX)   TRUNCATED CONE

FIG. 3-19 Types of piston heads.

and the bmep when the engine is operating at a given rpm.

Piston-Ring Construction

Piston-rings are usually made of high-grade gray cast iron that provides the spring action necessary to maintain a steady pressure against the cylinder wall, thus retaining the necessary seal. Cast-iron rings do not lose their elasticity even when they are exposed to rather high temperatures. The rings are split so they can be slipped over the outside of the piston and into the ring grooves which are machined on the circumference of the piston. Some compression rings are given a chrome-plated surface on the face of the ring. *This type of ring must never be used in a chrome-plated cylinder.*

As shown in Fig. 3-21, the piston-ring gap may be a plain butt joint, a step joint, or an angle joint. The butt joint is commonly used in modern aircraft engines. When a piston-ring is installed in a cylinder, there must be a specified gap clearance between the ends of the joint to allow for heat expansion during the operation of the engine. This gap dimension is given in the Table of Limits for the engine. If a piston-ring does not have sufficient gap clearance, the ring may sieze against the wall of the cylinder during operation and cause scoring of the cylinder or failure of the engine. The joints of the piston-rings must be staggered around the circumference of the piston in which they are installed at the time that the piston is installed in the cylinder. This is to reduce **blowby,** that is, to reduce the flow of gases

from the combustion chamber by the pistons and into the crankcase. This situation is evidenced by oil vapor and blue smoke being emitted from the engine breather. This same indication may occur as a result of worn piston rings; in this case, it is caused by oil entering the combustion chamber, where it burns and a part of the combustion gases blows by the piston rings into the crankcase and out the breather. The greatest wear in a reciprocating engine usually occurs between the piston rings and the cylinder walls in a reciprocating engine; excessive blue smoke out the exhaust or the engine breather indicates that repairs should be made. The side clearance of piston rings in the piston ring grooves is important to allow for free movement of the rings in the grooves, but the clearance should not be great enough to permit any appreciable leakage of gases. The side clearance for the various rings is specified in the Table of Limits for the engine.

Functions of Piston Rings

The importance of the piston rings in a reciprocating engine cannot be overemphasized. The three principal functions are (1) to provide a seal to hold the pressures in the combustion chamber, (2) to prevent excessive oil from entering the combustion chamber, and (3) to conduct the heat from the piston to the cylinder walls. Worn or otherwise defective piston rings will cause loss of compression and excessive oil consumption. Defective piston rings will usually cause excessively high oil discharge from the crankcase breather and an

FIG. 3-20 Several types of pistons.

FIG. 3-21 Piston-ring joints.

BUTT STEP ANGLE

excessive amount of blue smoke from the exhaust of the engine during normal operation. The smoke is normal when an engine is first started, but it should not continue for more than a few moments.

Types of Piston Rings

Piston rings in general may be of the same thickness throughout the circumference or they may vary, but aircraft engine piston rings are almost always of the same thickness all the way around. Piston rings may be classified according to function as (1) compression rings and (2) oil rings.

The purpose of **compression rings** is to prevent gases from escaping past the piston during engine operation. They are placed in the ring grooves immediately below the piston head. The number of compression rings used on a piston is determined by the designer of the engine, but most aircraft engines have three or four piston rings for each piston.

The cross section of the compression ring may be rectangular, tapered, or wedge-shaped. The rectangular cross section provides a straight bearing edge against the cylinder wall. The tapered and wedge-shaped cross sections present a bearing edge which is supposed to hasten the seating of a new ring against the hardened surface of the cylinder wall. Figure 3-22 illustrates cross sections of rectangular, tapered, and wedge-shaped compression rings.

The principal purpose of **oil rings** is to control the quantity of lubricant supplied to the cylinder walls and to prevent this oil from passing into the combustion chamber. The two types of oil rings are **oil-control rings** and **oil-wiper rings** (sometimes called **oil-scraper rings**).

Oil-control rings are placed in the grooves immediately below the compression rings. There may be only one oil-control ring to a piston, or there may be two

or three. The purpose of the oil-control ring is to control the thickness of the oil film on the cylinder wall. The oil-control-ring groove is often provided with drilled holes to the inside of the piston to permit excess oil to be drained away. The oil flowing through the drilled holes provides additional lubrication for the piston pins.

If too much oil enters the combustion chamber, it will burn and may leave a coating of carbon on the combustion chamber walls, the piston head, and the valve heads. This carbon can cause the valves and piston rings to stick if it enters the valve guides and the ring grooves. In addition, the carbon may cause detonation and preignition. If the operator of an aircraft engine notices increased oil consumption and heavy blue smoke from the exhaust, it is a good indication that the piston rings are worn and not providing the seal necessary for proper operation.

Oil-wiper or -scraper rings are placed on the skirts of the pistons to regulate the amount of oil passing between the piston skirts and the cylinder walls during each of the piston strokes. The cross section is usually beveled, and the beveled edge is installed in either of two positions. If the beveled edge is installed nearest the piston heads, the ring scrapes oil toward the crankcase. If installed with the beveled edge away from the piston head, the ring serves as a pump to keep up the flow of oil between the piston and the cylinder wall. Figure 3-23 shows cross sections of oil-control, oil-wiper, and bottom-oil-ring installations. The engine technician must make sure the piston rings are installed according to the manufacturer's overhaul instructions.

Piston-Ring Cross Section

In addition to the cross sections of compression rings described previously, we must consider the shapes of oil rings. Oil-control rings usually have one of three cross sections: (1) ventilated, (2) oil-wiper (tapered with narrow edge up), or (3) uniflow-effect (tapered with wide edge up). The choice of the cross section to be used is normally determined by the manufacturer. In some cases the upper oil-control ring is made up of two or more parts. The **ventilated-type** oil-control rings are usually of a two-piece construction, made with a number of equally spaced slots around the entire circumference of the ring to allow the oil to drain through to the holes in the ring groove and then into the crankcase. Another ring assembly consists of two thin steel rings, one on each side of a cast-iron ring. The cast-iron ring has cutouts along the sides to permit the flow of oil into the groove and through drilled holes to the inside of the piston. The choice of the particular ring to be used with an engine is normally determined by the manufacturer. Where there is more than one po-

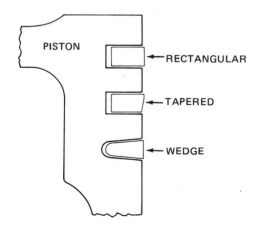

FIG. 3-22 Cross sections of compression rings.

PISTON

RECTANGULAR

TAPERED

WEDGE

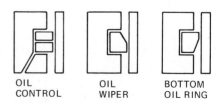

OIL CONTROL   OIL WIPER   BOTTOM OIL RING

FIG. 3-23 Oil-ring installations.

sition in which a ring may be installed, it should be marked to show the correct installation.

Narrow surfaced rings are preferred to wide ones because they adapt themselves better to the wall of the cylinder. Regardless of the cross section or the width, all modern piston rings are constructed to withstand wear and deterioration. Wear is generally caused by fine abrasive particles in the lubricating oil and by the friction existing between the rings, ring grooves, and cylinder walls.

Wedge-shaped piston rings are fitted to beveled-edge grooves to obtain a sliding, self-cleaning action that will prevent sticking between the ring and the groove. Also, the ring lands remain stronger where beveled-edge grooves are between them. Since compression rings operate at the highest temperatures and are a greater distance from the oil source, they receive less lubrication than other rings and have the greatest tendency to stick; therefore, wedge-shaped rings are installed as compression rings; that is, they are placed in the ring grooves immediately below the piston head.

On certain radial-type aircraft engines, the oil-wiper rings on the upper cylinders are faced toward the piston head or dome to carry more oil to the top piston rings; hence, the oil rings above the top ring serve as wiper rings. In that case, the oil-control rings on the lower cylinders are normally faced toward the crankshaft to prevent over-lubrication.

Piston Pins

A **piston pin**, sometimes called a **wrist pin**, is used to attach the piston to the connecting rod. It is made of steel (AMS 6274 or AMS 6322) hollowed for lightness and surface-hardened or through-hardened to resist wear. The pin passes through the piston at right angles to the skirt so that the piston can be anchored to the connecting-rod assembly. A means is provided to prevent the piston pin from moving sideways in the piston and damaging the cylinder wall. The piston pin is mounted in bosses and bears directly on the aluminum alloy of which the pistons are made. When the piston is made of aluminum alloy, however, the bosses may or may not be lined with some nonferrous (no iron or steel) metal, such as bronze. Figure 3-24 shows a piston pin in a piston-pin boss. The piston pin passes through the piston bosses and also through the small end of the connecting rod which rides on the central part of the pin.

Piston pins are usually classified as **stationary** (rigid), **semifloating**, or **full-floating**. The stationary type is not free to move in any direction and is securely fastened in the boss by means of a set screw. The semifloating piston pin is securely held by means of a clamp screw in the end of the connecting rod and a half slot in the pin itself. The full-floating type is free to run or slide in both the connecting rod and the piston and is the most widely used in modern aircraft engines.

Piston-Pin Retainers

The three devices used to prevent contact between the piston-pin ends and the cylinder wall are circlets, spring rings, and nonferrous-metal plugs. Figure 3-25 shows these three retainers.

Circlets resemble piston rings and fit into grooves at the outside end of each piston boss.

Spring rings are circular spring-steel coils which fit into circular grooves cut into the outside end of each piston boss to prevent the movement of the pin against the cylinder wall.

Nonferrous-metal plugs, usually made of aluminum alloy, are called **piston-pin plugs** and are used in most aircraft engines. They are inserted in the open ends of a hollow piston pin to prevent the steel pin end from bearing against the cylinder wall. The comparatively soft piston-pin plugs may bear against the cylinder walls without damage to either the plug or the wall.

Piston pins are fitted into the pistons and the connecting rod with clearances of less than 0.001 in [0.0254 mm]. This is commonly called a "push fit" because the pin can be inserted in the piston boss by pushing with the palm of the hand. The proper clearances for piston pins, bosses, and connecting rods are listed in the Table of Limits for any particular engine.

Since the piston-pin bearing surfaces are not pressure-lubricated, it is a common practice to drill holes through the piston-pin bosses to supply oil to the bearing surfaces.

● CONNECTING-ROD ASSEMBLIES

A variety of connecting-rod assemblies have been designed for the many different types of engines. Some of the arrangements for such assemblies are shown in Fig. 3-26. The **connecting rod** is defined as the link which transmits forces between the piston and the crankshaft of an engine. It furnishes the means of con-

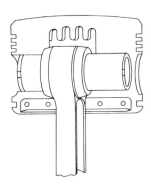

FIG. 3-24 Piston pin in a piston-pin boss.

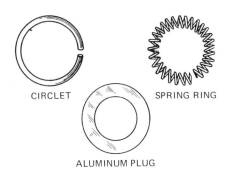

CIRCLET   SPRING RING

ALUMINUM PLUG

FIG. 3-25 Piston-pin retainers.

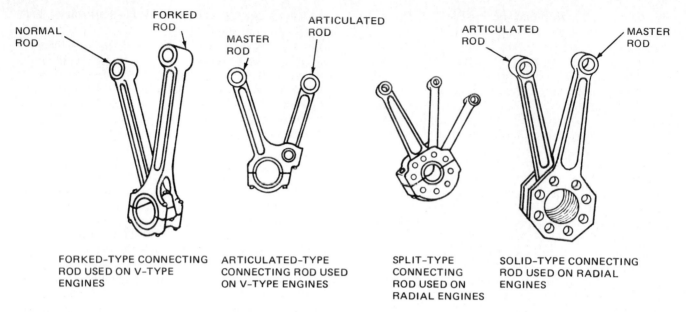

FORKED-TYPE CONNECTING ROD USED ON V-TYPE ENGINES

ARTICULATED-TYPE CONNECTING ROD USED ON V-TYPE ENGINES

SPLIT-TYPE CONNECTING ROD USED ON RADIAL ENGINES

SOLID-TYPE CONNECTING ROD USED ON RADIAL ENGINES

FIG. 3-26   Connecting-rod assemblies.

verting the reciprocating motion of the piston to a rotating movement of the crankshaft in order to drive the propeller.

A tough steel alloy (SAE 4340) is the material used for manufacturing most connecting rods, but aluminum alloy has been used for some low-power engines. The cross-sectional shape of the connecting rod is usually like either the letter H or the letter I, although some have been made with a tubular cross section. The end of the rod which connects to the crankshaft is called the **large end** or **crankpin end,** and the end which connects to the piston pin is called the **small end** or the **piston-pin end.** Connecting rods, other than tubular types, are manufactured by forging in order to provide maximum strength.

Connecting rods stop, change direction, and start at the end of each stroke; hence, they must be light in weight to reduce the inertia forces produced by these changes of velocity and direction. At the same time, they must be strong enough to remain rigid under the severe loads imposed under operating conditions.

There are three principal types of connecting-rod assemblies: (1) the plain type, shown in Fig. 3-27, (2) the fork-and-blade type, shown in Fig. 3-28, and (3) the master-and-articulated type, shown in Fig. 3-29.

Plain Connecting Rod

The plain connecting rod is used on in-line engines and opposed engines. The small end of the rod usually has a bronze bushing to serve as a bearing for the piston

pin. This bushing is pressed into place and then reamed to the proper dimension. The large end of the rod is made with a cap, and a two-piece shell bearing is installed. The bearing is held in place by the cap. The outside of the bearing flange bears against the sides of the crankpin journal when the rod assembly is installed on the crankshaft. The bearing inserts are often made of steel and lined with a nonferrous bearing material, such as lead bronze, copper lead, lead silver, or babbitt (a soft bearing alloy, silver in color and composed of tin, copper, and antimony). Another type of bearing insert is made of bronze and has a lead plating for the bearing surface against the crankpin.

The two-piece bearing shell fits snugly in the large end of the connecting rod and is prevented from turning by dowel pins or by tangs which fit into slots cut into the cap and the connecting rod. The cap is usually secured on the end of the rod by bolts; however, some rods have been manufactured with studs for holding the cap in place.

During inspection, maintenance, repair, and overhaul, the proper fit and balance of connecting rods are obtained by always replacing the connecting rod in the

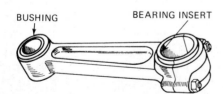

FIG. 3-27   Plain-type connecting rod.

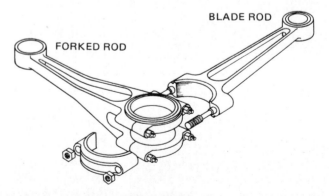

FIG. 3-28   Fork-and-blade connecting rod.

FIG. 3-29 Master-and-articulated connecting-rod assembly.

same cylinder and in the same relative position as it was before removal. The connecting rods and caps are usually stamped with numbers to indicate their position in the engine. The rod assembly for the no. 1 cylinder is marked with a 1, the assembly for the no. 2 cylinder is marked with a 2, and so on.

Fork-and-Blade Connecting-Rod Assembly

The fork-and-blade connecting rod, illustrated in Fig. 3-28, is generally used in V-type engines. The **forked rod** is split on the large end to provide space for the **blade rod** to fit between the prongs.

One two-piece bearing shell is fastened by lugs or dowel pins to the forked rod. Between the prongs of the forked rod, the center area of the outer surface of this bearing shell is coated with a nonferrous bearing metal to act as a journal for the blade rod and cap.

During overhaul or maintenance, the fork-and-blade connecting rods are always replaced on the crankshaft in the same relative positions as they occupied in their original installation. This ensures the proper fit and engine balance. Specific instructions for overhaul operations are given in the manufacturer's overhaul manual, and such instructions must be followed carefully in order to obtain the best results. This applies to all aircraft engines.

Master-and-Articulated Connecting-Rod Assembly

The **master-and-articulated rod assembly** is used primarily for radial engines, although some V-type engines have employed this type of rod assembly. The complete rod assembly for a seven-cylinder radial engine is shown in Fig. 3-29.

The **master rod** in a radial engine is subjected to some stresses not imposed upon the plain connecting rod; hence, its design and construction must be of the highest quality. It is made of an alloy-steel forging, ma-

chined and polished to final dimensions and heat-treated to provide maximum strength and resistance to vibration and other stresses. The surface must be free of nicks, scratches, or other surface damage which may produce a stress concentration and ultimate failure.

The master rod is similar to other connecting rods except that it is constructed to provide for the attachment of the articulated rods (link rods) on the large end. The large end of the master rod may be a two-piece type or a one-piece type, as shown in Fig. 3-30.

If the large end of the master rod is made of two pieces, the crankshaft is one solid piece. If the rod is one piece, then the crankshaft may be of either two-piece or three-piece construction. Regardless of the type of construction, the usual bearing surfaces must be supplied.

Master-rod bearings are generally of the plain type and consist of a split shell or a sleeve, depending upon whether the master rod is of the two-piece type or the one-piece type. The bearing usually has a steel or bronze backing with a softer nonferrous material bonded to the backing to serve as the actual bearing material. In low-power engines, babbitt material was suitable for the bearing surface, but it was found to be lacking in the durability necessary for the higher power engines. For this reason, bronze, leaded bronze, and silver have been used in later engines. The actual bearing surface is usually plated with lead in order to reduce the friction as much as possible. During operation the bearing is cooled and lubricated by a constant flow of lubricating oil.

The **articulated rods** (link rods) are hinged to the master rod flanges by means of steel **knuckle pins.** Each articulated rod has a bushing of nonferrous metal, usually bronze, pressed or shrunk into place to serve as a knuckle-pin bearing. Aluminum-alloy link rods have been used successfully in some lower power radial engines. With these rods, it is not necessary to provide bronze bushings at the ends of the rod because the aluminum alloy furnishes a good bearing surface for the piston pins and knuckle pins.

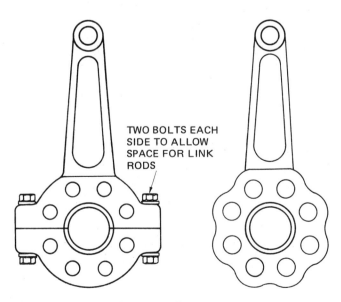

TWO BOLTS EACH SIDE TO ALLOW SPACE FOR LINK RODS

FIG. 3-30 Types of master rods.

Articulated rods, when made of steel, are usually constructed in an I or H cross section. These configurations give the greatest strength and resistance to distortion with the lightest weight.

The knuckle pin resembles a piston pin. It is usually made of nickel steel, hollowed for lightness and for permitting the passage of lubricating oil and surface-hardened to reduce water.

The articulated rod is bored and supplied with bushings at each end. One end receives the piston pin, and the other end receives the knuckle pin. The knuckle-pin bore in the articulated rod includes a bushing of nonferrous metal, which is usually bronze. It is pinned, pressed, shrunk, or spun into place. The bushing must be bored to precise dimension and alignment.

Knuckle pins installed with a loose fit so that they can turn in the master-rod flange holes and also turn in the articulated rod bushings are called **full-floating knuckle pins.** Knuckle pins also may be installed so that they are prevented from turning in the master rod by means of a tight press fit. In either type of installation a lock plate on each side bears against the knuckle pin and prevents it from moving laterally (sideways).

Figure 3-31 shows a knuckle-pin and lock-plate assembly for a full-floating arrangement, and Fig. 3-32 shows a stationary knuckle-pin and lock-plate assembly.

● THE CRANKSHAFT

The **crankshaft** transforms the reciprocating motion of the piston and connecting rod into rotary motion for turning the propeller. It is a shaft composed of one or more cranks located at definite places between the ends. These **cranks,** sometimes called **throws,** are formed by forging offsets into a shaft before it is machined. Since the crankshaft is the backbone of an internal-combustion engine, it is subjected to all the forces developed within the engine and must be of very strong construction. For this reason, it is usually forged from some extremely strong steel alloy, such as chromium-nickel-molybdenum steel (SAE 4340).

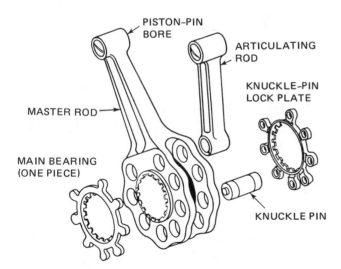

FIG. 3-31 Master rod with full-floating knuckle-pin and lock-plate assembly.

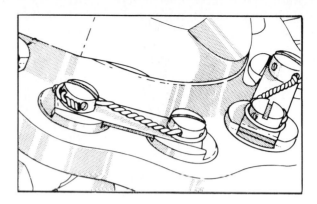

FIG. 3-32 Stationery knuckle-pin and lock-plate assembly.

A crankshaft may be constructed of one or more pieces. Regardless of whether it is of one-piece or multipiece construction, the corresponding parts of all crankshafts have the same names and functions. The parts are (1) the main journal, (2) the crankpin, (3) the crank cheek or crank arm, and (4) the counterweights and dampers. Figure 3-33 shows the nomenclature of a typical crankshaft.

Main Journal

The **main journal** is the part of the crankshaft that is supported by and rotates in a **main bearing.** Because of this it may also properly be called a **main-bearing journal.** This journal is the center of rotation of the crankshaft and serves to keep the crankshaft in alignment under all normal conditions of operation. The main journal is surface-hardened by nitriding for a depth of 0.015 to 0.025 in [0.381 to 0.635 mm] to reduce wear. Every aircraft engine crankshaft has two or more main journals to support the weight and operational loads of the entire rotating and reciprocating assembly in the power section of the engine.

Crankpin

The **crankpin** can also be called a **connecting-rod-bearing journal** simply because it is the journal for a connecting-rod bearing. Since the crankpin is off center from the main journals, it is sometimes called a **throw.** The crankshaft will rotate when a force is applied to the crankpin in any direction other than parallel to a line directly through the center line of the crankshaft.

The crankpin is usually hollow for three reasons: (1) It reduces the total weight of the crankshaft, (2) it

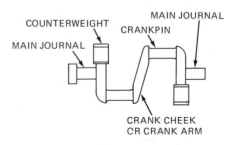

FIG. 3-33 Nomenclature for twin-row radial crankshaft engine.

provides a passage for the lubricating oil, and (3) it serves as a chamber for collecting carbon deposits, sludge, and other foreign substances which are thrown by centrifugal force to the outside of the chamber where they will not reach the connecting-rod-bearing surface. For this reason the chamber is often called the **sludge chamber.** On some engines a drilled passage from the sludge chamber to an opening on the exterior surface of the connecting rod makes it possible to spray clean oil on the cylinder walls.

Lubrication of the crankpin bearings is accomplished by oil taken through drilled passages from the main journals. The oil reaches the main journals through drilled passages in the crankcase and in the crankcase webs which support the main bearings. During overhaul the technician must see that all oil passages and sludge chambers are cleared in accordance with the manufacturer's instructions.

Crank Cheek

The **crank cheek,** sometimes called the **crank arm,** is the part of the crankshaft which connects the crankpin to the main journal. It must be constructed to maintain rigidity between the journal and the crankpin. On many engines, the crank cheek extends beyond the main journal and supports a counterweight used to balance the crankshaft. The crank cheeks are usually provided with drilled oil passages through which lubricating oil passes from the main journals to the crankpins.

Counterweights and Dampers

The purpose of **counterweights** and **dampers** is to relieve the whip and vibration caused by the rotation of the crankshaft. They are suspended from or installed in specified crank cheeks at locations determined by the design engineers. Crankshaft vibrations caused by power impulses may be reduced by placing floating dampers in a counterweight assembly. The need for counterweights and dampers is not confined to aircraft engines. Any machine with rotating parts may reach a speed at which so much vibration occurs in the revolving mass of metal that it must be reduced or the machine will eventually destroy itself.

The purpose of the **counterweight** is to provide static balance for a crankshaft. If a crankshaft has more than two throws, it does not always require counterweights because the throws, being arranged symmetrically opposite each other, balance each other. A single-throw crankshaft, such as that used in a single-row radial engine, must have counterbalances to offset the weight of the single throw and the connecting rod and piston assembly attached to it. This crankshaft is illustrated in Fig. 3-34.

Dampers or **dynamic balances** are required to overcome the forces which tend to cause deflection of the crankshaft and torsional vibration. These forces are generated principally by the power impulses of the pistons. If we compute the force exerted by the piston of an engine near the beginning of the power stroke, we shall find that 8000 to 10,000 lb [35 584 to 44 480 newtons (N)] is applied to the throw of a crankshaft. As the engine runs, this force is applied at regular intervals to the different throws of the crankshaft on an in-line or opposed engine and to the one throw of a single-row radial engine. If the frequency of the power impulses is such that it matches the natural vibration frequency of the crankshaft and propeller as a unit or of any moving part of the engine, then severe vibration will take place. The dynamic balances may be pendulum-type weights mounted in the counterweight (Fig. 3-35) or they may be straddle-mounted on extensions of the crank cheeks. In either case, the weight is free to move in a direction and at a frequency which will dampen out the natural vibration of the crankshaft. Dynamic balances are shown in Fig. 3-36.

The effectiveness of a dynamic damper can be understood by observing the operation of a pendulum. If a simple pendulum is given a series of regular impulses at a speed corresponding to its natural frequency, using a bellows to simulate a modified power impulse in an engine, it will begin swinging or vibrating back and forth from the impulses, as shown in the upper half of Fig. 3-35.

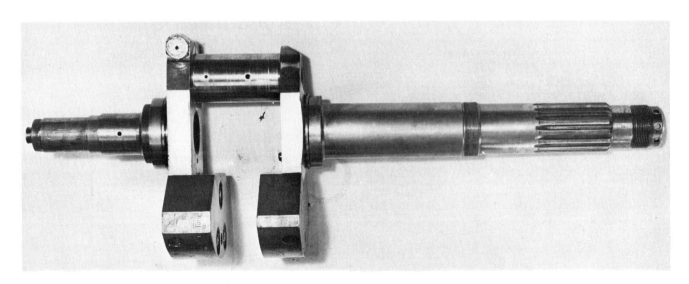

FIG. 3-34 Single-throw crankshaft with counterweights.

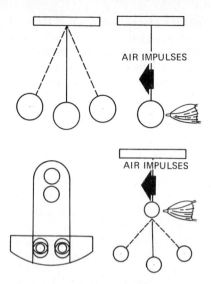

FIG. 3-35 Dynamic balances and principles of operation.

Another pendulum, suspended from the first, will absorb the impulse and swing itself, leaving the first pendulum stationary, as shown in the lower portion of Fig. 3-35. The dynamic damper, then, is a short pendulum hung on the crankshaft and tuned to the frequency of the power impulses to absorb vibration in the same manner as the pendulum illustrated in the lower part of the illustration. A **mode number** is used to indicate the correct type of damper for a specific engine.

Types of Crankshafts

The four principal types of crankshafts are (1) the single-throw, (2) the double-throw, (3) the four-throw, and (4) the six-throw. Figure 3-37 shows the crankshaft for an in-line engine, a single-row radial engine, and a double-row radial engine. Each individual type of crankshaft may have several configurations, depending on the requirements of the particular engine for which it is designed. An engine which operates at a high speed and power output requires a crankshaft more carefully balanced and with greater resistance to wear and distortion than an engine which operates at slower speeds.

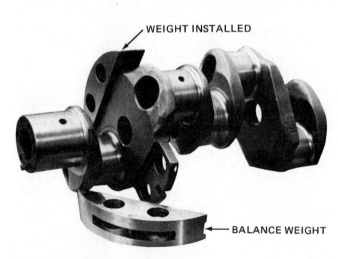

FIG. 3-36 Dynamic balance weights.

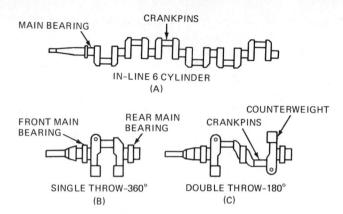

FIG. 3-37 Three types of crankshafts.

Single-Throw Crankshaft. The type of crankshaft and the number of crankpins it contains correspond in every case to the engine cylinder arrangement. The position of a crank on any crankshaft in relation to other cranks on the same shaft is given in degrees.

The single-throw, or 360°, crankshaft is used in single-row radial engines. It may be of single-piece or two-piece construction with two main bearings, one on each end. A single-piece crankshaft is shown in Fig. 3-38. This crankshaft must be used with a master rod which has the large end split.

Two-piece single-throw crankshafts are shown in Fig. 3-39. The first of these (A) is a clamp-type shaft, sometimes referred to as a **split clamp** crankshaft. The front section of this shaft includes the main-bearing journal, the front crank-cheek and counterweight assembly, and the crankpin. The rear section contains the clamp by which the two sections are joined, the rear crank-cheek and counterweight assembly, and the rear main-bearing journal. The spline-type crankshaft (B) has the same parts as the clamp-type, with the exception of the device by which the two sections are joined. In this shaft the crankpin is divided, one part having a female spline and the other having a male spline to match. When the two parts are joined, they are held securely in place by means of an alloy-steel bolt.

Double-Throw Crankshaft. The double-throw, or 180°, crankshaft is generally used in a double-row radial engine. When used in this type of engine, the crankshaft

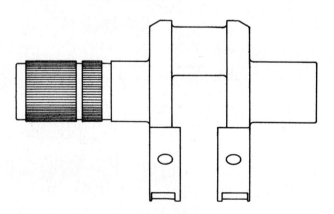

FIG. 3-38 Single-piece crankshaft.

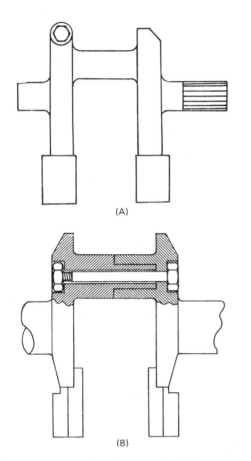

FIG. 3-39 Two-piece single-throw crankshafts.

has one throw for each row of cylinders. The construction may be one-piece or three-piece, and the bearings may be of the ball type or roller type.

Four-Throw Crankshaft. Four-throw crankshafts are used in four-cylinder opposed engines, four-cylinder in-line engines, and V-8 engines. In the four-throw crankshaft for an in-line or opposed engine two throws are placed 180° from the other two throws. There may be three or five crankshaft main journals, depending on the power output and the size of the engine. The bearings for the four-cylinder opposed engine are of the plain, split-shell type. In the four-throw crankshaft, illustrated in Fig. 3-40, lubrication for the crankpin bearings is provided through passages drilled in the

crank cheeks. During operation, oil is brought through passages in the crankcase webs to the main-bearing journals. From the main-bearing journals, the oil flows through the crank-cheek passages to the crankpin journals and the sludge chambers in the journals.

Six-Throw Crankshaft. Six-throw crankshafts are used in six-cylinder in-line engines, twelve-cylinder V-type engines, and six-cylinder opposed engines. Since the in-line and V-type engines are not in general use in the United States, we shall limit our discussion to the type of shaft used in the six-cylinder opposed engine.

A crankshaft for a Continental six-cylinder opposed aircraft engine is shown in Fig. 3-41. This is a one-piece six-throw 60° crankshaft machined from an alloy-steel (SAE 4340) forging. It has four main journals and one double-flanged main-thrust journal. The shaft is heat-treated for high strength and nitrided to a depth of 0.015 to 0.025 in [0.381 to 0.635 mm], except on the propeller splines, for maximum wear. The crankpins and main-bearing journals are ground to close limits of size and surface roughness. After grinding, nitriding, and polishing, the crankshaft is balanced statically and dynamically. Final balance is attained after the assembly of the counterweights and other parts.

As shown in Fig. 3-41, the crankshaft is provided with dynamic counterweights. Since the selection of the counterweights of the correct mode is necessary to preserve the dynamic balance of the complete assembly, they cannot be interchanged on the shaft or between crankshafts. For this reason, neither counterweights nor bare crankshafts are supplied alone.

The crankshaft is line-bored the full length to reduce weight. Splined shafts have a threaded plug installed at the front end. The crankpins are recessed at each end to reduce weight. Steel tubes permanently installed in holes drilled through the crank cheeks provide oil passages across the lightening holes to all crankpin surfaces from the main journals. A U-shaped tube, permanently installed inside the front end of the shaft bore, conducts oil from the second main journal to the front main-thrust journal.

Propeller Shafts

Aircraft engines are equipped with one of three types of propeller mounting shafts: taper shafts, spline shafts, and flange shafts. In the past, low-power engines have

PROPELLER
SHAFT

FIG. 3-40 Four-throw crankshaft.

DYNAMIC COUNTERWEIGHT MOUNTING

FIG. 3-41 Crankshaft for a six-cylinder opposed engine.

often been equipped with **tapered** propeller shafts. The propeller shaft is an integral part of the crankshaft (Fig. 3-40). The tapered end of the shaft forward of the main bearing is milled to receive a key which positions the propeller in the correct location on the shaft. The shaft is threaded at the forward end to receive the propeller retaining nut.

Crankshafts with **spline** propeller shafts are shown in Figs. 3-34 and 3-41. As can be seen in the illustrations, the splines are rectangular grooves machined in the shaft to mate with grooves inside the propeller hub. One spline groove may be blocked with a screw (or otherwise) to assure that the propeller will be installed in the correct position. A wide groove inside the propeller hub receives the blocked, or "blind," spline. The propeller is mounted on the spline shaft with front and rear cones to assure correct position longitudinally and radially. A retaining nut on the threaded front portion of the shaft holds the propeller firmly in place when properly torqued. The installation of propellers is described in other chapters of this text.

It will be noted that the propeller shaft in Fig. 3-34 is threaded about halfway between the forward end and the crank throw. This threaded portion of the shaft is provided to receive the thrust-bearing retaining nut, which holds the thrust bearing in the nose case of the engine. The shaft in Fig. 3-41 is not threaded aft of the splines because the design of the engine case eliminates the need for a thrust nut.

Spline propeller shafts are made in several sizes, depending upon engine horsepower. These numbers are identified as SAE-20, 30, 40, 50, 60, and 70. High-power engines have shafts from SAE-50 to SAE-70 and low-power engines are equipped with shaft sizes from SAE-20 to SAE-40.

A **flange-type** shaft is used with many modern opposed engines with power ratings up to 450 hp [335.57 kW]. Figure 3-42 shows a shaft of this type. A short stub shaft extends forward of the flange to support and center the propeller hub. Six high-strength bolts or studs are used to secure the propeller to the flange. In this type of installation, it is most important that the bolts or studs be tightened in a sequence which will provide a uniform stress. It is also necessary to use a torque wrench and apply torque as specifed in the manufacturer's service manual.

Some aircraft utilize **propeller-shaft extensions** to move the propeller forward, thus permitting a more stream-lined nose design. Special instructions are provided by the aircraft manufacturer for service of such extensions.

Propeller-shaft loads are transmitted to the nose section of the engine by means of thrust bearings and forward main bearings. On some opposed-type engines the forward main bearing is flanged to serve as a thrust bearing as well as a main bearing. The nose section of an engine is either a separate part or is integral with the crankcase. For either type of design, the nose section transmits the propeller shaft loads from the thrust bearing to the crankcase, whence they are applied to the aircraft structure through the engine mounts.

● BEARINGS

A **bearing** is any surface that supports or is supported by another surface. It is a part in which a journal, pivot, pin, shaft, or similar device turns or revolves. The bearings used in aircraft engines are designed to produce a minimum of friction and a maximum of wear resistance.

A good bearing has two broad characteristics: (1) It must be made of a material that is strong enough to withstand the pressure imposed on it and yet permit the other surface to move with a minimum of wear and friction, and (2) the parts must be held in position within very close tolerances to provide quiet and efficient operation and at the same time permit freedom of motion.

Bearings must reduce the friction of moving parts

FIG. 3-42 Flange-type propeller shaft.

and also take thrust loads, radial loads, or a combination of thrust and radial loads. Those which are designed primarily to take thrust loads are called **thrust bearings**.

Plain Bearings

Plain bearings are illustrated in Fig. 3-43. These bearings are usually designed to take radial loads; however, plain bearings with flanges are often used as thrust bearings in opposed aircraft engines. Plain bearings are used for connecting rods, crankshafts, and camshafts of low-power aircraft engines. The metal used for plain bearings may be silver, lead, an alloy (such as bronze or babbitt), or a combination of metals. Bronze withstands high compressive pressure but offers more friction than babbitt. On the other hand, babbitt offers less friction but cannot withstand high compressive pressures as well as bronze. Silver withstands compressive pressures and is an excellent conductor of heat, but its frictional qualities are not dependable.

Plain bearings are made with a variety of metal combinations. Some bearings in common use are steel-backed with silver or silver-bronze on the steel and a thin layer of lead then applied for the actual bearing surface. Other bearings are bronze-backed and have a lead or babbitt surface.

Roller Bearings

The **roller bearings** shown in Fig. 3-44 are one of the two types known as "antifriction" bearings because the rollers eliminate friction to a large extent. These bearings are made in a variety of shapes and sizes and can be adapted to both radial and thrust loads. Straight roller bearings are generally used only for radial loads; however, tapered roller bearings will support both radial and thrust loads.

The bearing **race** is the guide or channel along which the rollers travel. In a roller bearing, the roller is situated between an inner and an outer race, both of which are made of case-hardened steel. When a roller is tapered, it rolls on a cone-shaped race inside an outer race.

Roller bearings are used in high-power aircraft engines as main bearings to support the crankshaft. They are also used in other applications where radial loads are high.

Ball Bearings

Ball bearings provide less rolling friction than any other type. A ball bearing consists of an inner race and

(A)

(B)

FIG. 3-44 Roller bearings. *(Timkin Roller Bearing Co.)*

an outer race, a set of polished steel balls, and a ball retainer. Some ball bearings are made with two rows of balls and two sets of races. The races are designed with grooves to fit the curvature of the balls in order to provide a large contact surface for carrying high radial loads.

A typical ball-bearing assembly used in an aircraft engine is shown in Fig. 3-45. In this assembly, the balls are controlled and held in place by means of the ball retainer. This retainer is necessary to keep the balls properly spaced, thus preventing them from contacting one another.

FIG. 3-43 Plain bearings.

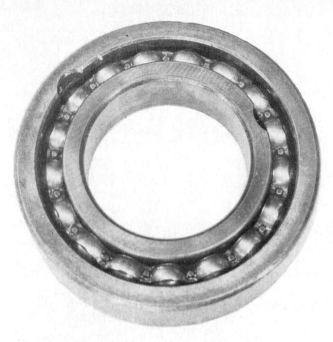

FIG. 3-45 Ball-bearing assembly.

Ball bearings are commonly used for thrust bearings in large radial engines and gas-turbine engines. Because of their construction, they can withstand heavy thrust loads as well as radial or centrifugal loads. They are also subject to gyroscopic loads, but these are not critical. A ball bearing designed especially for thrust loads is made with exceptionally deep grooves for the ball races. Bearings designed to resist thrust in a particular direction will have a heavier race design on the side which takes the thrust. It is important to see that this type of bearing is installed with the correct side toward the thrust load.

In addition to the large ball bearings used as main bearings and thrust bearings, many smaller ball bearings will be found in generators, magnetos, starters, and other accessories used on aircraft engines. For this reason, the engine technician should be thoroughly familiar with the inspection and servicing of such bearings.

Many bearings, particularly for accessories, are prelubricated and sealed. These bearings are designed to function satisfactorily without lubrication service between overhauls. In order to avoid damaging the seals of the bearings, it is essential that the correct bearing pullers and installing tools be employed when removing or installing them.

● PROPELLER REDUCTION GEARS

Reduction gearing between the crankshaft of an engine and the propeller shaft has been in use for many years. The purpose of this gearing is to allow the propeller to rotate at the most efficient speed to absorb the power of the engine while the engine turns at much higher rpm in order to develop full power. As noted in the previous chapter, the power output of an engine is directly proportional to its rpm. It follows, therefore, that an engine will develop twice as much power at 3000 rpm as it will at 1500 rpm. Thus, it is advantageous from a power-weight point of view to operate an engine at as high an rpm as possible so long as such factors as vibration, temperature, and engine wear do not become excessive.

A propeller cannot operate efficiently when the tip speed approaches or exceeds the speed of sound (1116 ft/s [340.16 m/s] at standard sea-level conditions). An 8-ft [2.45-m] propeller tip travels approximately 25 ft [7.62 m] in one revolution; hence, if the propeller is turning at 2400 rpm (40 rev/s), the tip speed is 1000 ft/s [304.8 m/s]. A 10-ft [3.05-m] propeller turning at 2400 rpm would have a tip speed of 1256 ft/s [382.83 m/s], which is well above the speed of sound.

Small engines that drive propellers of no more than 6 ft [1.83 m] in length can operate at speeds of over 3000 rpm without creating serious propeller problems. Larger engines, such as the Avco Lycoming IGSO-480 and the Teledyne Continental Tiara T8-450, are equipped with reduction gears. The IGSO-480 operates at 3400 rpm maximum and this is reduced to 2176 rpm for the propeller by means of the 0.64:1 planetary reduction-gear system. The T8-450 engine operates at a maximum of 4400 rpm and this is reduced to about 2200 rpm for the propeller by means of a ratio 0.5:1 offset spur reduction gear. This ratio could also be expressed as 2:1. It must be remembered that when reduction gears are employed, the propeller always rotates slower than the engine.

Reduction gears are designed as simple spur gears, planetary gears, bevel planetary gears, and combinations of spur and planetary gears. A **spur-gear** arrangement is shown in Fig. 3-46. The driven gear turns in a direction opposite that of the drive gear; hence, the propeller direction will be opposite that of the engine crankshaft. The ratio of engine speed to propeller speed is inversely proportional to the number of teeth on the crankshaft drive gear and the number of teeth on the driven gear.

Arrangements for **planetary gears** are shown in Fig. 3-47. In Fig. 3-47(A) the outer gear, called the **bell gear,**

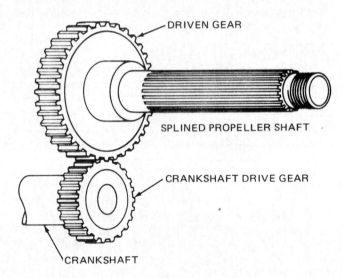

FIG. 3-46 Spur-gear arrangement.

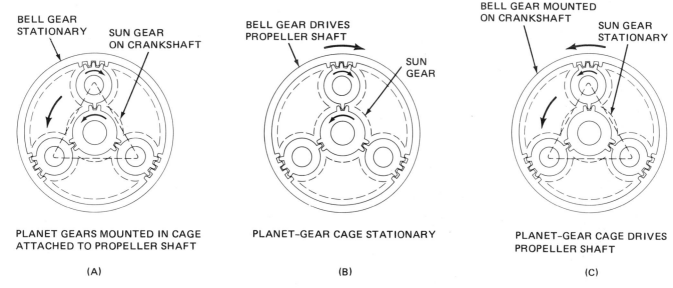

BELL GEAR STATIONARY
SUN GEAR ON CRANKSHAFT

BELL GEAR DRIVES PROPELLER SHAFT
SUN GEAR

BELL GEAR MOUNTED ON CRANKSHAFT
SUN GEAR STATIONARY

PLANET GEARS MOUNTED IN CAGE ATTACHED TO PROPELLER SHAFT

PLANET-GEAR CAGE STATIONARY

PLANET-GEAR CAGE DRIVES PROPELLER SHAFT

(A)   (B)   (C)

FIG. 3-47   Different arrangements for planetary gears.

is stationary and is bolted or otherwise secured to the inside of the engine nose case. The planet gears are mounted on a **carrier ring,** or **cage,** which is attached to the propeller shaft. The **sun gear** is mounted on the forward end of the crankshaft. When the crankshaft turns, the pinion (planet) gears rotate in a direction opposite that of the crankshaft. These gears are meshed with the stationary bell gear—hence, they "walk" around the inside of the gear, carrying their cage with them. Since this assembly is attached to the propeller shaft, the propeller will turn in the same direction as the crankshaft and at a speed determined by the number of teeth on the reduction gears.

In Fig. 3-47(B), the planet gears are mounted on stationary shafts so that they do not rotate as a group around the sun gear. When the crankshaft rotates, the sun gear drives the planet pinions and they, in turn, drive the bell gear in a direction opposite the rotation of the crankshaft.

The arrangement where the sun gear is stationary is shown in Fig. 3-47(C). Here the bell gear is mounted on the crankshaft and the planet gear cage is mounted on the propeller shaft. The planet gears walk around the sun gear as they are rotated by the bell gear in the same direction as the crankshaft.

In a **bevel-planetary-gear** arrangement (Fig. 3-48), the planet gears are mounted in a forged steel cage attached to the propeller shaft. The bevel drive gear (sun gear) is attached to the forward end of the crankshaft, and the stationary bell gear is attached to the engine case. As the crankshaft rotates, the drive gear turns the pinions and causes them to walk around the stationary gear, thus rotating the cage and the propeller shaft. The bevel-gear arrangement makes it possible to use a smaller-diameter reduction-gear assembly, particularly where the reduction-gear ratio is not great.

Reduction-gear systems for gas-turbine engines have a much higher reduction ratio than those for reciprocating engines. For example, the United Aircraft of Canada PT6A-27 engine employs a two-stage 15:1 reduction-gear system. This is necessary because the

power-turbine speed is 33,000 rpm. The Rolls Royce Dart 7 engine has a gear ratio of 10.75:1 and the Avco Lycoming T53-L-13B engine has a gear ratio of 0.31:1 (3.2:1). This engine is used for helicopters; hence, the output shaft 6300 rpm can be reduced further through the power-train gearbox. Reduction-gear systems for turbine engines are described in the sections covering turboprop and turboshaft engines.

The gears utilized in engine reduction gear systems are subjected to very high stresses and are therefore machined from high-quality alloy-steel forgings. The larger shafts are supported by ball bearings designed to absorb and transmit to the engine case all loads imposed upon them.

Since reduction-gear systems are of critical importance in the reliability of engines, the overhaul of engines having reduction-gear systems, with the exception of spur-gear systems, is classified as a major repair. The overhaul and return to service of such engines must be under the direction of persons suitably certificated by the Federal Aviation Administration.

FIG. 3-48   Bevel-planetary-gear arrangement.

● VALVES AND ASSOCIATED PARTS

Definition and Purpose

In general, a **valve** is any device for regulating or determining the direction of flow of a liquid, gas, etc., by a movable part which opens or closes a passage. The word "valve" is also applied to the movable part itself.

The main purpose of valves in an internal-combustion engine is to open and close **ports,** which are openings into the combustion chamber of the engine. One is called the **intake port,** and its function is to allow the fuel-air charge to enter the cylinder. The other is called the **exhaust port** because it provides an opening through which burned gases are expelled from the cylinder.

Each cylinder must have at least one intake port and one exhaust port. On some liquid-cooled engines of high power output, two intake and two exhaust ports are provided for each cylinder. The shape and form of all valves are determined by the design and specifications of the particular engine in which they are installed.

Poppet-Type Valves

The word "poppet" comes from the popping action of the valve. This type of valve is made in four general configurations with respect to the shape of the valve head. These are (1) the flat-headed valve, (2) the semitulip valve, (3) the tulip valve, and (4) the mushroom valve. These are illustrated in Fig. 3-49.

Valves are subjected to high temperatures and a corrosive environment; hence, they must be made of metals which resist these deteriorating influences. Since intake (or inlet) valves operate at lower temperatures than exhaust valves, they may be made of chrome-nickel steel. Exhaust valves, which operate at higher temperatures, are usually made of nichrome, silchrome, or cobalt-chromium steel. Poppet valves are made from these special steels and forged in one piece.

A valve **stem** is surface-hardened to resist wear. Since the **tip** of the valve must resist both wear and pounding, it is made of hardened steel and welded to the end of the stem. There is a machined **groove** on the stem near the tip to receive split-ring stem keys. These stem keys hold a split lock ring to keep the valve-spring retaining washer in place.

The stems of some valves have a narrow groove below the lock-ring groove for the installation of **safety circlets** or **spring rings** which are designed to prevent the valves from falling into the combustion chambers if the tip should break during engine operation or during valve disassembly and assembly. A poppet-valve installation for a radial engine is shown in Fig. 3-50.

Exhaust Valves

Exhaust valves operate at high temperatures, and they do not receive the cooling effect of the fuel-air charge; hence, they must be designed to dissipate heat rapidly. This is accomplished by making the exhaust valve with a hollow stem and, in some cases, with a hollow mushroom head, and by partly filling the hollow portion with metallic sodium. The sodium melts at a little over 200°F [93.3°C], and during operation it flows back and forth in the stem, carrying heat from the head and dissipating it through the valve guides into the cylinder head. The cylinder head is cooled by means of fins, as explained previously. In some engines, exhaust valve stems contain a salt as the cooling agent.

Lower-power engines are not all equipped with sodium-filled exhaust valves. It is important, however, for the technician to determine whether the exhaust-valve stems upon which he may be working are of this type. *Under no circumstances should a sodium valve be cut open, hammered, or otherwise subjected to treatment which may cause it to rupture. Furthermore, sodium valves must always be disposed of in an appropriate manner.*

The faces of high-performance exhaust valves are often made more durable by the application of about

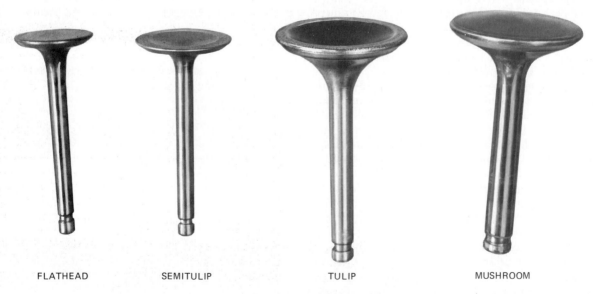

FLATHEAD   SEMITULIP   TULIP   MUSHROOM

FIG. 3-49 Types of poppet valves.

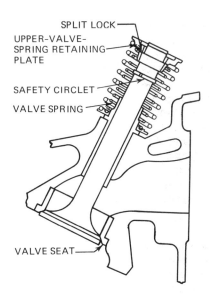

FIG. 3-50 Poppet-valve installation for a radial engine.

$\frac{1}{16}$ in [1.59 mm] of a material called Stellite. This alloy is welded to the face of the valve and then ground to the correct angle. Stellite is resistant to high-temperature corrosion and also withstands the shock and wear associated with valve operation exceptionally well.

The face of the valve is usually ground to an angle of either 30 or 45°. In some engines, the intake-valve face is ground to an angle of 30° and the exhaust-valve face is ground to 45°. The 30° angle provides better airflow and the 45° angle allows increased heat flow from the valve to the valve seat. This is of particular benefit to the exhaust valve.

The tip of the valve stem is often made of high-carbon steel or Stellite so it can be hardened to resist wear. It must be remembered that the tip of the valve stem is continuously receiving the impact of the rocker arm as the rocker arm opens and closes the valve.

Intake Valves

Specially cooled valves are not generally required for the intake port of an engine because the intake valves are cooled by the fuel-air mixture. For this reason, the most commonly used intake valves have solid stems and the head may be flat or of the tulip type. The valve is forged from one piece of alloy steel and then machined to produce a smooth finish. The stem is accurately dimensioned to provide the proper clearance in the valve guide. The intake-valve stem has a hardened tip similar to that of the exhaust valve.

Intake valves for low-power engines usually have flat heads. Tulip-type heads are often used on the intake valves for high-power engines because the tulip shape places the metal of the head more nearly in tension, thus reducing the stresses where the head joins the stem.

Valve Seats

The aluminum alloy used for making engine cylinder heads is not hard enough to withstand the constant hammering produced by the opening and closing of the valves. For this reason, bronze or steel valve seats are shrunk or screwed into the circular edge of the valve openings in the cylinder head. A typical six-cylinder opposed engine has forged aluminum-bronze intake-valve seats and forged chrome-molybdenum steel seats for the exhaust valves. The engines in common use today usually have the valve seats shrunk into the seat recesses, as shown in Fig. 3-51. Before the seat is installed, its outside diameter is from 0.007 to 0.015 in [0.178 to 0.381 mm] larger than the recess in which it is to be installed. In order to install the seats, the cylinder head must be heated to 575°F [301°C] or more. The seat and mandrel are chilled with dry ice and the seat is pressed or drifted into place while the cylinder head is hot. Upon cooling, the head recess shrinks and grips the seat firmly.

It is necessary to replace valve seats only after they have worn beyond the limits specified by the manufacturer. Repeated grinding of the seats eventually makes it necessary to replace them.

Valve seats are ground to the same angle as the face of the valve, or they may have a slightly different angle to provide an "interference fit." In order to obtain good seating at operating temperatures, valve faces are sometimes ground to an angle $\frac{1}{4}$ to 1° less than the angle of the valve seats. This provides a line contact between the valve face and seat and permits a more positive seating of the valve, particularly when the engine is new or freshly overhauled. The importance of proper valve and seat grinding and lapping is discussed at length in the section on cylinder overhaul in Chap. 12.

Valve Springs

Valves are closed by helical-coiled springs. Two or more springs, one inside the other, are installed over the stem of each valve. If only one spring were used on each valve, the valve would surge and bounce because of the natural vibration frequency of the spring. Each spring of a pair of springs is made of round spring-steel wire of a different diameter, and the two coils differ in pitch. Since the springs have different frequencies, the two springs together rapidly damp out all spring-surge vibrations during engine operation. A second reason for the use of two (or more) valve springs on each valve is that it reduces the possibility of failure by breakage from heat and metal fatigue.

The valve springs are held in place by means of steel **valve-spring retainers,** which are special washers shaped

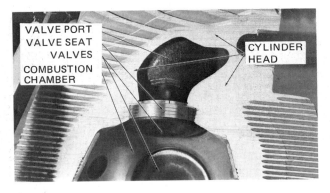

FIG. 3-51 Valve seat in the cylinder head.

to fit the valve springs. The lower retainer seats against the cylinder head, and the upper retainer is provided with a conical recess into which the split stem keys (keepers) fit. The valve-spring retainers are sometimes called the upper and lower **valve-spring seats**.

● VALVE OPERATION AND MECHANISMS

Principles

In order to understand valve operation and timing, it is essential that the fundamental principles of engine operation be kept in mind. It will be remembered that most modern aircraft engines of the piston type operate on the four-stroke-cycle principle. This means that the piston makes four strokes during one cycle of operation. During the four strokes, five events take place: (1) intake, (2) compression, (3) ignition, (4) combustion, and (5) exhaust. During these events the operation of the valves directs the flow of the fuel-air mixture and the burned gases. At the intake stroke the intake valve must be open and the exhaust valve closed. The valves are both closed during the compression, ignition, and combustion events, and then the exhaust valve opens near the end of the power stroke (Fig. 3-52).

During one cycle of the engine's operation, the crankshaft makes two revolutions and the valves each perform one operation. It is therefore apparent that the valve-operating mechanism for an intake valve must make one operation for two turns of the crankshaft.

On an opposed or in-line engine which has single lobes on the camshaft, the camshaft is geared to the crankshaft to produce one revolution of the camshaft for two revolutions of the crankshaft. The cam drive gear on the crankshaft has one-half the number of teeth that the camshaft gear has, thus producing the 1:2 ratio.

On radial engines which utilize cam rings or cam plates to operate the valves, there may be three, four, or five cam lobes on the cam ring. The ratio of crankshaft to cam-ring rotation is then 1:6, 1:8, and 1:10, respectively.

Piston Position

In any complete discussion of valve operation and timing, it is necessary that we take into consideration the position of the piston. A piston has two extreme positions, **top dead center** (TDC) and **bottom dead center** (BDC). These positions are often called top center (TC) and bottom center (BC).

Top dead center occurs when the piston is at the exact top of its stroke, with the crank throw and connecting rod in perfect alignment. The piston is at bottom dead center when the crankshaft has turned 180° from top dead center. At this time the piston is at the bottom of its stroke and the crank throw is aligned with the connecting rod.

During the operation of an engine, as the piston approaches and leaves the top- or bottom-center positions, its linear (up-and-down) motion becomes small in comparison with crankshaft angular travel. This is illustrated in the diagrams of Fig. 3-53.

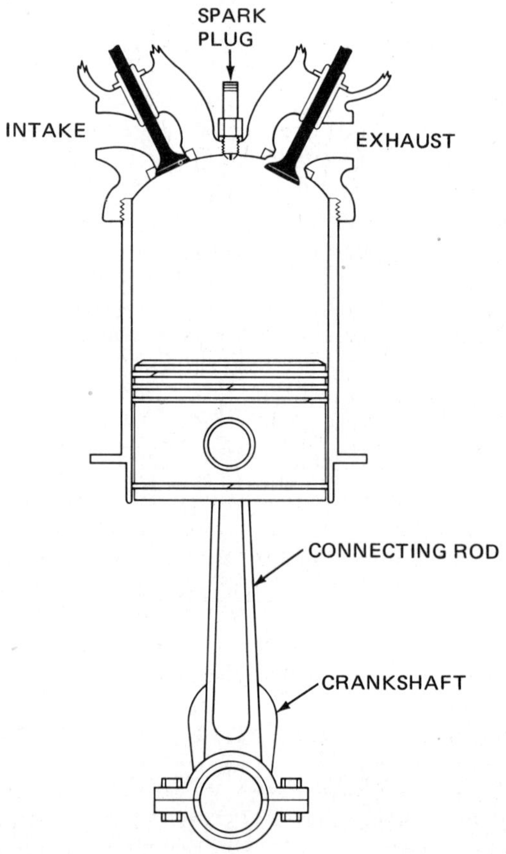

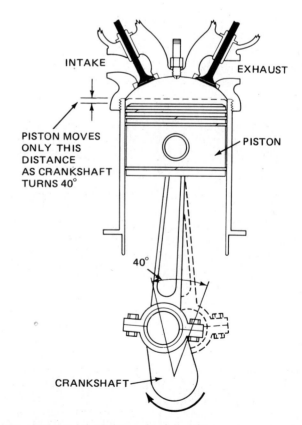

FIG. 3-52 Exhaust valve open.

FIG. 3-53 Crankshaft and piston travel.

56

As the piston leaves the top-center position, it accelerates and attains its maximum speed when the connecting rod and crank throw are at right angles. From this position, the piston loses speed, passes through the halfway position, and reaches bottom center. Leaving the bottom-center position, the piston's travel action is reversed until the piston again reaches top center.

During the first 90° of crankshaft rotation from top center, the piston moves more than half the stroke. Also, when the crankshaft rotates from 90° before top center (BTC) to top center, the piston moves through more than half of its stroke. The piston travel is comparatively slow around top center; hence, all instructions for timing emphasize the importance of locating the top-center position exactly.

Abbreviations for Valve Timing Positions

In a discussion of the timing points for an aircraft engine, it is convenient to use abbreviations. The abbreviations commonly used in describing crankshaft and piston positions for the timing of valve opening and closing are as follows:

After bottom center	ABC	Exhaust closes	EC
After top center	ATC	Exhaust opens	EO
Before bottom center	BBC	Intake closes	IC
Bottom center	BC	Intake opens	IO
Bottom dead center	BDC	Top center	TC
Before top center	BTC	Top dead center	TDC

Engine Timing Diagram

To provide a visual concept of the timing of valves for an aircraft engine, a valve-timing diagram is used. The diagram for the Continental Model E-165 and E-185 engines is shown in Fig. 3-54. A study of this diagram reveals the following specifications for the timing of the engine.

IO	15° BTC	EO	55° BBC
IC	60° ABC	EC	15° ATC

Reason suggests that the intake valve should open at top center and close at bottom center. Likewise, it seems that the exhaust valve should open at bottom center and close at top center. This would be true except for the inertia of the moving gases and the time required for the valves to open fully. Near the end of the exhaust stroke, the gases are still rushing out the exhaust valve. The inertia of the gases causes a low-pressure condition in the cylinder at this time. Opening the intake valve a little before top center takes advantage of the low-pressure condition to start the flow of fuel-air mixture into the cylinder, thus bringing a greater charge into the engine and improving **volumetric efficiency.** If the intake valve should open too early, exhaust gases would flow out into the intake manifold and ignite the incoming fuel-air mixture. The result would be **backfiring.** Backfiring also occurs when an intake valve sticks in the open position. The exhaust valve closes shortly after the piston reaches top center and prevents reversal of the exhaust flow back into the cylinder. The angular distance through which both valves

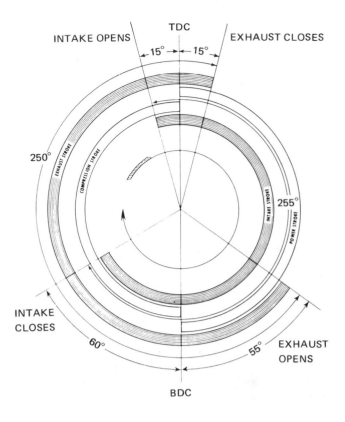

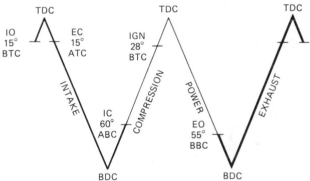

FIG. 3-54 Diagram for valve timing.

are open is called **valve overlap,** or **valve lap.** When the intake valve opens 15° BTC and the exhaust valve closes 15° ATC, the valve overlap is 30°.

Fig. 3-54 shows two diagrams that may be used as guides for valve timing. Either one may be employed to indicate the points in the cycle where each valve opens and closes. In the diagrams of Fig 3-54, the intake valve remains open 60° after bottom center. This is to take advantage of the inertia of the fuel-air mixture rushing into the cylinder, because the mixture will continue to flow into the cylinder for a time after the piston has passed bottom center. The total period during which the intake valve is open is designed to permit the greatest possible charge of fuel-air mixture into the cylinder.

The exhaust valve opens before bottom center for two principal reasons: (1) more thorough scavenging of the cylinder and (2) better cooling of the engine. Most of the energy of the burning fuel is expended by the time the crankshaft has moved 120° past top center on the power stroke and the piston has moved almost to its lowest position. Opening the exhaust valve at this time allows the hot gases to escape early, and less heat is transmitted to the cylinder walls than would be the case if the exhaust valve remained closed until the piston reached bottom center. The exhaust valve is not closed until after top center because the inertia of the gases aids in removing additional exhaust gas after the piston has passed top center.

The opening or closing of the intake or exhaust valves after top or bottom center is called **valve lag.** The opening or closing of the intake or exhaust valves before bottom or top center is called **valve lead.** Both valve lag and valve lead are expressed in degrees of crankshaft travel. For example, if the intake valve opens 15° before top center, the valve lead is 15°.

It will be noted from the diagrams of Fig. 3-54 that the valve lead and valve lag are greater in relation to the bottom-center position than they are to the top-center position. One of the reasons for this is that the piston travel per degree of crankshaft travel is less near bottom center than it is near top center. This is illustrated in Fig. 3-55. In the diagram (A), the circle represents the path of the crank throw and the point *C* represents the center of the crankshaft. *TC* is the position of the piston pin at top center, and *BC* is the position of the piston pin at bottom center. The numbers show the positions of the piston pin and the crank throw at different points through 180° of crankshaft travel. It will be noted that the piston travels much farther during the first 90° of crankshaft travel than it

does during the second 90° and that the piston will be traveling at maximum velocity when the crankthrow has turned 80 to 90° past top center.

Using the valve-timing specifications for the diagrams of Fig. 3-54, it is possible to determine (1) the rotational distance through which the crankshaft travels while each valve is open and (2) the rotational distance of crankshaft travel while both valves are closed. Since the intake valve opens at 15° BTC and closes at 60° ABC, it is seen that the crankshaft rotates 15° from the point where the intake valve opens to reach TC, then 180° to reach BC, and another 60° to the point where the intake valve closes. The total rotational distance of crankshaft travel with the intake valve open is therefore 15° + 180° + 60°, or a total of 255°. Using the same reasoning, it is seen that crankshaft travel while the exhaust valve is open is 55° + 180° + 15°, or a total of 250°. Valve overlap at TC is 15° + 15°, or 30°, and through BC is 55° pl 60°, or 115°. The total rotational distance of the crankshaft while both valves are closed is determined by noting when the intake valve closes on the compression stroke and when the exhaust valve opens on the power stroke. It can be seen from the diagram that the intake valve opens 60° ABC and that the crankshaft must therefore rotate 120° (180° − 60°) from intake-valve opening to TC. Since the exhaust valve opens 55° BBC, the crankshaft rotates 125° (180° − 55°) from TC to the point where the exhaust valve opens. The total rotational distance that the crankshaft must travel from the point where the intake valve closes to the point where the exhaust valve opens is then 120° + 125°, or 245°. The time the valves are off their seat is their **duration.** For example, the duration of the exhaust valve above is 250° of crankshaft travel.

Firing Order

In any discussion of valve or ignition timing, it is essential that we consider the firing order of various engines because all parts associated with the timing of the engine must be designed and timed to comply with the firing order. As the name implies, the **firing order** of an engine is the order in which the cylinders fire.

The firing order of in-line V-type, and opposed engines is designed to provide for balance and to eliminate vibration to the extent that this is possible. The firing order is determined by the relative positions of the throws on the crankshaft and the positions of the lobes on the cam shaft.

The firing order of a single-row radial engine which operates on the four-stroke cycle must always be by alternate cylinders, and the engine must have an odd number of cylinders. Twin-row radial engines are essentially two single-row engines joined together. This means that alternate cylinders in each row must fire in sequence. For example, an 18-cylinder engine consists of two single-row nine-cylinder engines. The rear row of cylinders has the odd numbers 1, 3, 5, 7, 9, 11, 13, 15, and 17. Alternate cylinders in this row are 1, 5, 9, 13, 17, 3, 7, 11, and 15. The front row has the numbers 2, 4, 6, 8, 10, 12, 14, 16, and 18, and the alternate cylinders for this row are 2, 6, 10, 14, 18, 4, 8, 12, and 16. Since the firing of the front and rear rows of cyl-

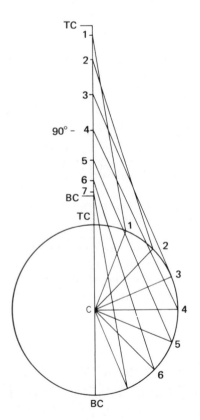

FIG. 3-55   Relation between piston travel and crankshaft travel.

inders is started on opposite sides of the engine, the first cylinder to fire after no. 1 is no. 12. Starting with the no. 12 cylinder, the front-row firing sequence is then 12, 16, 2, 6, 10, 14, 18, 4, and 8. By combining the rear row firing with the front row firing, we obtain the firing order for the complete engine, thus: 1, 12, 5, 16, 9, 2, 13, 6, 17, 10, 3, 14, 7, 18, 11, 4, 15, and 8.

The following table gives the firing orders for the majority of engine types:

TABLE 3-1

Type	Firing order
4-cyl. in-line	1-3-4-2 or 1-2-4-3
6-cyl. in-line	1-5-3-6-2-4
8-cyl. V-type (CW)	1R-4L-2R-3L-4R-1L-3R-2L
12-cyl. V-type (CW)	1L-2R-5L-4R-3L-1R-6L-5R-2L- 3R-4L-6R
4-cyl. opposed	1-3-2-4 or 1-4-2-3
6-cyl. opposed	1-4-5-2-3-6
8-cyl. opposed	1-5-8-3-2-6-7-4
9-cyl. radial	1-3-5-7-9-2-4-6-8
14-cyl. radial	1-10-5-14-9-4-13-8-3-12-7-2-11-6
18-cyl. radial	1-12-5-16-9-2-13-6-17-10-3-14-7-18- 11-4-15-8

As an aid in remembering the firing order of large radial engines, technicians often use "magic" numbers. For a 14-cylinder radial engine, the numbers are +9 and −5 and for an 18-cylinder engine the numbers are +11 and −7. To determine the firing order of a 14-cylinder engine, the technician starts with the number 1, the first cylinder to fire. Adding +9 gives the number 10, the second cylinder to fire. Subtracting 5 from 10 gives 5, the third cylinder to fire. Adding 9 to 5 gives 14, the fourth cylinder to fire. Continuing the same process will give the complete firing order. The same technique is used with an 18-cylinder engine by applying the magic numbers +11 and −7.

Purpose of Valve-Operating Mechanism

The purpose of a valve-operating mechanism in an aircraft engine is to control the timing of the valves of the engine so that each valve will open at the correct time, remain open for the required length of time, and close at the proper time. The mechanism should be simple in design, be ruggedly constructed, and give satisfactory service for a long time with a minimum of inspection and maintenance.

The two types of valve-operating mechanisms most generally used today are the type found in the opposed engine and the type used in a typical radial engine. Since both these engines are equipped with overhead

valves (valves in the cylinder head), the valve-operating mechanisms for each are quite similar.

Valve Mechanism Components

A standard valve-operating mechanism includes certain parts which are found in both opposed and radial engines. These parts may be described briefly as follows:

Cam. A device for actuating the valve-lifting mechanism.

Valve lifter or tappet. A mechanism to transmit the force of the cam to the valve pushrod.

Pushrod. A steel or aluminum-alloy rod or tube situated between the valve lifter and the rocker arm of the valve-operating mechanism to transmit the motion of the valve lifter.

Rocker arm. A pivoted arm mounted on bearings in the cylinder head to open and close the valves. One end of the arm presses on the stem of the valve and the other end receives motion from the pushrod.

The valve-operating cam in an opposed or in-line engine consists of a shaft with a number of cam lobes sufficient to operate all the intake and exhaust valves of the engine. In a typical six-cylinder opposed engine, the cam-shaft has three groups of three cam lobes, as shown in Fig. 3-56. In each group, the center lobe actuates the valve lifters for the two opposite intake valves, whereas the outer lobes of each group actuate the lifters for the exhaust valves.

In a radial engine, the valve-actuating device is a **cam plate** (or **cam ring**) with three or more lobes. In a five-cylinder radial engine, the cam ring usually has three lobes; in a seven-cylinder radial engine, the cam ring has three or four lobes; and in a nine-cylinder radial engine, the cam ring has four or five lobes.

Valve-operating mechanisms for in-line and V-type engines utilize a camshaft similar to the type installed in an opposed engine. The shaft incorporates single cam lobes placed at positions along the shaft which will enable them to actuate the valve lifters or rocker arms at the correct time. Some in-line and V-type engines have overhead camshafts. In such cases the actuating mechanism may be arranged as shown in Fig. 3-57. The camshaft is mounted along the top of the cylinders and is driven by a system of bevel gears through a shaft leading from the crankshaft drive gear.

Valve Mechanism for Opposed Engines

A simplified drawing of a valve-operating mechanism is shown in Fig. 3-58. The valve action starts with the crankshaft timing gear, which meshes with the cam-

FIG. 3-56 Camshaft for a six-cylinder opposed engine.

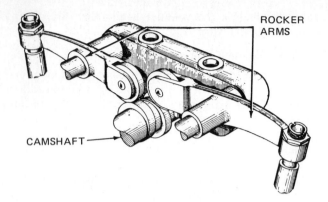

FIG. 3-57 Overhead cam mechanism.

shaft gear. As the crankshaft turns, the camshaft also turns, but at one-half the rpm of the crankshaft. This is because a valve operates only once during each cycle and the crankshaft makes two revolutions per cycle. A cam lobe on the camshaft raises the cam roller and hence the pushrod to which the cam roller is attached. The ramp on each side of the cam lobe is designed to reduce opening and closing shock through the valve operating mechanism. In opposed engines, a cam roller is not employed, and in its place is a tappet or a hydraulic lifter. The pushrod raises one end of the rocker arm and lowers the other end, thus depressing the valve,

working against the tension of the valve spring which normally holds the valve closed. When the cam lobe has passed by the valve lifter, the valve will close by the action of the valve spring or springs.

The valve-operating mechanism for an opposed-type aircraft engine is shown in Fig. 3-59. The valve-actuating mechanism starts with the drive gear on the crankshaft. This gear may be called the **crankshaft timing gear** or the **accessory drive gear.** Mounted on the end of the camshaft is the **camshaft gear,** which has twice as many teeth as the crankshaft gear. In some engines the mounting holes in the camshaft gear are spaced in such a manner that they will line up with the holes in the camshaft flange in only one position. In other engines, a dowel pin in the end of the crankshaft mates with a hole in the crankshaft gear to assure correct position. Thus, when the timing marks on the camshaft and the crankshaft gear are aligned, the camshaft will be properly timed with the crankshaft.

Adjacent to each cam lobe is the **cam follower face** which forms the base of the **hydraulic valve lifter** or tappet assembly. The outer cylinder of the assembly is called the **lifter body.** Inside the lifter body is the **hydraulic unit assembly** consisting of the following parts: **cylinder, plunger, plunger springs, ball check valve,** and **oil inlet tube.** Figure 3-60 is an illustration of the complete lifter assembly. During operation, engine oil under pressure is supplied to the oil reservoir in the lifter body through an inlet hole in the side. Since this oil is under pressure directly from the main oil gallery of the engine, it flows into the oil inlet tube, through the ball check valve, and into the cylinder. The pressure of the oil forces the plunger against the **pushrod socket** and takes up all the clearances in the valve-operating mechanism during operation. For this reason, a lifter of this type has been called a "zero-lash lifter." When the cam is applying force to the cam follower face, the oil in the cylinder tends to flow back into the oil reservoir, but this is prevented by the ball check valve.

During overhaul of the engine, the hydraulic valve lifter assembly must be very carefully inspected. All

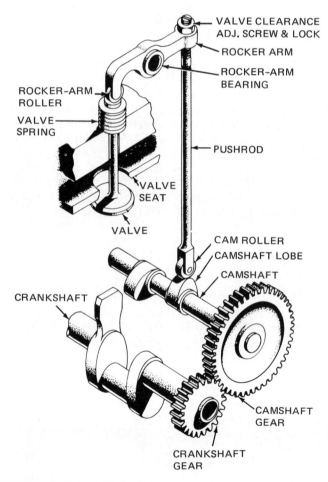

FIG. 3-58 Valve-operating mechanism.

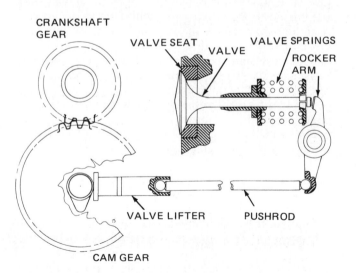

FIG. 3-59 Valve-operating mechanism for an opposed engine.

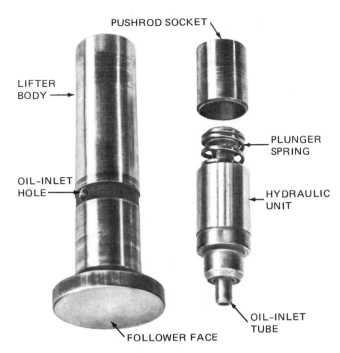

FIG. 3-60 Hydraulic valve lifter assembly.

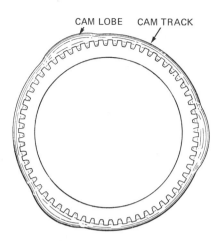

FIG. 3-61 Cam ring.

the parts of one assembly must be reassembled together in order, so as to assure proper operation.

The ball end of the hollow valve pushrod fits into the pushrod socket, or cup, which bears against the plunger in the lifter. Both the socket and the ball end of the pushrod are drilled to provide a passage for oil to flow into the pushrod. This oil flows through the pushrod and out a hole at the end (the hole fits the pushrod socket in the **rocker arm**), thus providing lubrication for the rocker-arm bearing (bushing) and valves. The rocker arm is drilled to permit oil flow to the bearing and valve mechanism.

The rocker arm is mounted on a steel shaft which, in turn, is mounted in rocker-shaft bosses in the cylinder head. The rocker-shaft bosses are cast integrally with the cylinder head and then are machined to the correct dimension and finish for the installation of the rocker shafts. The rocker-shaft dimension provides a push fit in the boss. The shafts are held in place by the rocker box covers or by covers over the holes through which they are installed. The steel rocker arms are fitted with bronze bushings to provide a good bearing surface. These bushings may be replaced at overhaul if they are worn beyond acceptable limits.

One end of each rocker arm bears directly against the hardened tip of the valve stem. When the rocker arm is rotated by the pushrod, the valve is thus depressed against valve-spring pressure. The distance the valve opens and the time it remains open are determined by the height and contour of the cam lobe.

Valve Mechanism for Radial Engines

Depending on the number of rows of cylinders, the valve-operating mechanism of a radial engine is operated by either one or two cam plates (or cam rings). Only one plate (or ring) is used with a single-row radial engine, but a double cam track is required. One cam track operates the intake valves, and the other track operates the exhaust valves. In addition, there are the necessary pushrods, rocker-arm assemblies, and tappet assemblies that make up the complete mechanism.

A **cam ring** (or cam plate), such as the one shown in Fig. 3-61, serves the same purpose in a radial engine as a camshaft serves in other types of engines. The cam ring is a circular piece of steel with a series of cam lobes on the outer surface. Each cam lobe is constructed with a **ramp** on the approach to the lobe to reduce the shock that would occur if the lobe rise were too abrupt. The **cam track** includes both the lobes and the surface between the lobes. The **cam rollers** ride on the cam track.

Figure 3-62 illustrates the gear arrangement for driving a cam plate (or ring). This cam plate has four lobes on each track; hence, it will be rotated at one-eighth crankshaft speed. Remember that a valve operates only once during each cycle and that the crankshaft makes two revolutions for each cycle. Since there are four lobes on each cam track, the valve operated by one set of lobes will open and close four times for each revolution of the cam plate. This means that the cylinder has completed four cycles of operation and that the crankshaft has made eight revolutions.

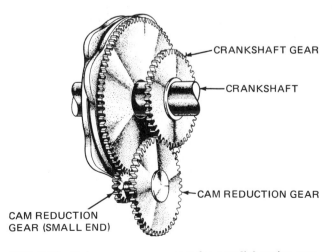

FIG. 3-62 Drive-gear arrangement for a radial-engine cam.

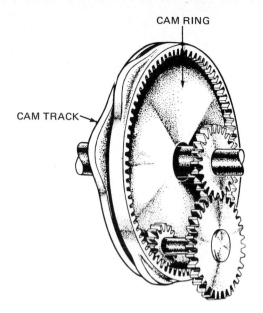

FIG. 3-63 Cam plate driven by an internal gear.

In the illustration it will be noted that the crankshaft gear and the large cam reduction gear are the same size; hence, the cam reduction gear will turn at the same rpm as the crankshaft. The small cam reduction gear is only one-eighth the diameter of the cam plate gear, and this provides the reduction to make the cam

plate turn at one-eighth crankshaft speed. The rule for cam-plate speed with respect to crankshaft speed may be given as a formula thus:

$$\text{cam-plate speed} = \frac{1}{\text{no. of lobes} \times 2}$$

Figure 3-63 illustrates a cam plate driven by an internal gear. In an arrangement of this type the cam plate turns opposite the direction of engine rotation. A study of the operation of the cam will lead us to the conclusion that a four-lobe cam turning in the opposite direction from the crankshaft will be used in a nine-cylinder radial engine. In the diagrams of Fig. 3-64 the numbers on the large outer ring represent the cylinders of a nine-cylinder radial engine. The firing order of such an engine is always 1-3-5-7-9-2-4-6-8. The small ring in the center represents the cam ring. In the first diagram we note that the no. 1 cam is opposite the no. 1 cylinder. We may assume that the cam is operating the no. 1 intake valve. In moving from the no. 1 cylinder to the no. 3 cylinder, the next cylinder in the firing order, we see that the crankshaft must turn 80° in the direction shown. Since the cam is turning at one-eighth crankshaft speed, a lobe on the cam will move 10° while the crankshaft is turning 80°. Thus, we see that the no. 2 cam lobe will be opposite the no. 3 cylinder as shown by the second diagram. When the crankshaft has turned

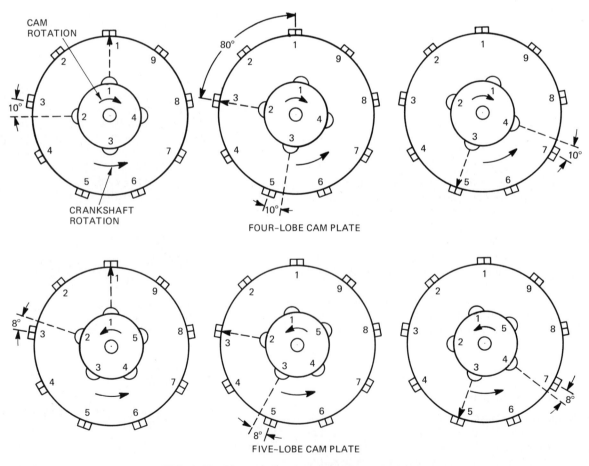

FIG. 3-64 Diagrams to show cam-plate operation.

another 80° to the intake operation of the no. 5 cylinder, the no. 3 cam lobe is opposite the no. 5 cylinder.

If we draw a similar diagram for a nine-cylinder radial engine with a five-lobe cam, we will note that the cam must travel in the same direction as the crankshaft. This is because there will be 72° between the centers of the cam lobes and 80° between the cylinders firing in sequence. The cam plate will turn one-tenth crankshaft rpm; hence, as the crankshaft turns 80°, the cam plate will turn 8°, and this will align the next operating cam with the proper valve mechanism.

The valve-operating mechanism for a radial engine is shown in Fig. 3-65. All the main parts are labeled and should be carefully studied. Since the cam in this illustration has three lobes and is turning opposite the direction of the crankshaft, we can determine that the valve mechanism must be designed for a seven-cylinder radial engine.

The valve **tappet** in this mechanism is spring-loaded to reduce shock and is provided with a cam roller to bear against the cam track. The tappet is enclosed in a tube called the **valve tappet guide.** The valve tappet is drilled to permit the passage of lubricating oil into the hollow **pushrod** and up to the rocker-arm assembly. The rocker arm is provided with a **clearance-adjusting screw** so proper clearance can be obtained between the rocker arm and valve tip. This clearance is very important because it determines when the valve will start to open, how far it will open, and how long it will stay open.

The pushrod transmits the lifting force from the valve tappet to the rocker arm in the same manner as that described for the opposed-type engines. The rod may be made of steel or aluminum alloy. Although it is called a rod, it is actually a tube with steel balls pressed into the ends. The length of the pushrod depends upon the distance between the tappet and the rocker-arm sockets.

An aluminum-alloy tube, called a **pushrod housing,** surrounding each pushrod provides a passage through which the lubricating oil can return to the crankcase, keeps dirt away from the valve-operating mechanism, and otherwise provides protection for the pushrod.

The rocker arm in the radial engine serves the same purpose as it does in the opposed engine. Rocker-arm assemblies are usually made of forged steel and are supported by a bearing which serves as a pivot. This bearing may be a plain, roller, or ball type. One end of the arm bears against the pushrod, and the other end bears on the valve stem. The end of the rocker arm bearing against the valve stem may be plain, or it may be slotted to receive a steel **rocker-arm roller.** The other end of the rocker arm may have either a threaded split clamp and locking bolt or a tapped hole in which is mounted the valve clearance-adjusting screw. The adjusting ball socket is often drilled to permit the flow of lubricating oil.

Typical rocker arms are illustrated in Fig. 3-66. Rocker arms shown at (B) and (C) are designed for opposed-type engines. The rocker arm at (A) is used in a Pratt & Whitney R-985 radial engine.

Valve Clearance

Every engine must have a slight clearance between the rocker arm and the valve stem. When there is no clearance, the valve may be held off its seat when it should be seated (closed). It is apparent that this will cause the engine to operate erratically and eventually the valve will be damaged. If, however, an engine is equipped with hydraulic valve lifters, there will be no apparent clearance at the valve stem during engine operation.

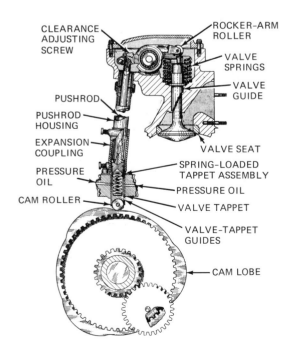

FIG. 3-65   Valve-operating mechanism for a radial engine.

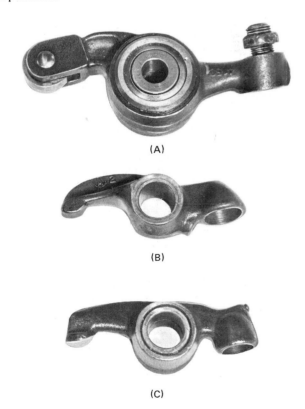

(A)

(B)

(C)

FIG. 3-66   Typical rocker arms.

The **cold clearance** for the valves on an engine is usually much less than the "hot" (or operating) clearance. This is true except when the engine is equipped with an overhead cam. The reason for the difference in hot and cold clearances is that the cylinder on an engine becomes much hotter than the pushrod and therefore expands more than the pushrod. In effect, this shortens the pushrod and leaves a gap between the pushrod and the rocker arm or between the rocker arm and the valve stem. The hot valve clearance of an engine can be as much as 0.070 in [1.778 mm], while the cold clearance may be 0.010 in [0.254 mm].

In adjusting the valve clearance of an engine, the technician must make sure that the cam is turned to a position where it is not applying any pressure to the pushrod. For any particular cylinder, it is good practice to place the piston in the position for the beginning of the power stroke. At this point both cams are well away from the valve tappets for the valves being adjusted.

On an adjustable rocker arm, the locknut is loosened and a feeler gage of the correct thickness is inserted between the rocker arm and the valve stem. The adjusting screw is turned to a point where a slight drag is felt on the feeler gage. The lock screw or nut is then tightened to the proper torque while the adjusting screw is held in place. After the adjusting screw has been locked, a feeler gage 0.001 in [0.0254 mm] thicker than the gage used for adjusting the clearance cannot be inserted in the gap if the clearance is correct.

It must be emphasized at this point that valve timing and adjustment, particularly that of the exhaust valve, have an important effect on the **heat-rejection** (cooling) of the engine. If the exhaust valve does not open at precisely the right moment, the exhaust gases will not leave the cylinder head when they should, and heat will continue to be transferred to the walls of the combustion chamber and cylinder. On the other hand, the exhaust valve must be seated long enough to transfer the heat of the valve head to the valve seat; otherwise, the valve may overheat and warp or burn. Inadequate valve clearance may prevent the valves from seating positively during start and warmup; if the valve clearance is excessive, the valve-open time and the valve overlap will be reduced.

When it is necessary to adjust the valves for an engine which is designed with a floating cam ring, special procedures must be followed. The floating cam ring for an R-2800 engine may have a clearance at the bearing of 0.013 to 0.020 in [0.330 to 0.508 mm], and this will affect the valve adjustment if it is not eliminated at the point where the valves are being adjusted. The clearance is called **cam float** and is eliminated by depressing certain valves while others are being adjusted.

Each valve tappet which is riding on a cam lobe applies pressure to the cam ring because of the valve springs. Therefore, if the pressure of the valves on one side of the cam ring is released, the ring will tend to move away from the tappets which are applying pressure. This will eliminate the cam float on that side of the cam ring. The valves whose tappets are resting on the cam ring at or near the point where there is no clearance between the ring and the bearing surface,

and which are between lobes, are adjusted, and then the crankshaft is turned to the next position. Certain valves are depressed, and other valves on the opposite side of the engine are adjusted. Table 3-2 is a chart showing the proper combinations for adjusting the valves on an R-2800 engine.

According to the chart, the valve adjustment begins with the no. 1 inlet and the no. 3 exhaust valves. The number 11 piston is placed at top center on its exhaust stroke. In this crankshaft position, the no. 15 exhaust tappet and the no. 7 inlet tappet are riding on top of cam lobes and applying pressure to the cam ring. When these two valves are depressed, the pressure is released from this side of the cam ring and the pressure of the tappets on the opposite side of the ring eliminates the cam ring float. The no. 1 inlet and the no. 3 exhaust valves are then adjusted for proper clearance. The adjustment is made only when the engine is cold.

Care must be exercised when depressing the valves on the engine. If a valve which is closed is completely depressed, the ball end of the pushrod may fall out of its socket. If the valve-adjusting chart is followed closely, only the valves which are open will be depressed.

On modern opposed-type engines the rocker arm is not adjustable and the valve clearance is adjusted by changing the pushrod. If the clearance is too great, a longer pushrod is used. When the clearance is too small, a shorter pushrod is installed. A wide range of clearances is allowable because the hydraulic valve lifters take up the clearance when the engine is operating. Valve clearance in these engines is normally checked only at overhaul.

Valve Timing

The manufacturer of an engine specifies in the maintenance and overhaul instructions the exact timing of the valves that will obtain the best possible performance. These instructions must be carefully followed and can be disregarded only when such deviation from the manufacturer's instructions is specifically ap-

TABLE 3-2 Valve-adjusting chart

Set piston at top center of its exhaust stroke	Depress rockers		Adjust valve clearances	
	Inlet	Exhaust	Inlet	Exhaust
11	7	15	1	3
4	18	8	12	14
15	11	1	5	7
8	4	12	16	18
1	15	5	9	11
12	8	16	2	4
5	1	9	13	15
16	12	2	6	8
9	5	13	17	1
2	16	6	10	12
13	9	17	3	5
6	2	10	14	16
17	13	3	7	9
10	6	14	18	2
3	17	7	11	13
14	10	18	4	6
7	3	11	15	17
18	14	4	8	10

proved by manufacturers' bulletins, FAA airworthiness directives, or a similar authority.

When a new engine is designed, there are many theoretical calculations before even the first engine of the new model is constructed. When the prototype engine leaves the factory, it is placed on a test block and operated at various speeds, with different grades of fuel, at different conditions of temperature, pressure, and humidity, and with various adjustments of valve and ignition timing. The horsepower developed by the engine under every conceivable set of conditions is accurately indicated and recorded. Finally, the manufacturer is able to state positively what adjustments will be permitted in the valve and ignition timing.

The power delivered by an engine at a given speed depends to a great extent upon the valve timing. If the best performance is obtained at high speeds, the performance is less efficient at low speeds. The reverse is also true. For this reason, valve timing is usually specified to obtain the best average results throughout the usual speed range of the airplane in which the engine is installed. Generally, it may be expected that the engine is timed to give the most efficient performance at or near normal cruising speed.

The valve timing on modern aircraft engines is designed to remain the same under all conditions of operation after it is originally assembled with correct timing.

The basic principle of valve timing is to make sure that one of the valves is opening at the correct time. Since all cams are normally on one shaft or one cam plate, if one valve is in time, the others must be in time also. In a V-type engine, two cams must be timed. In an opposed engine, the crankshaft gear can be installed in only one position and the camshaft gear can be installed on the camshaft in only one position. If the two gears are meshed with the marked gear teeth together, than the camshaft is properly timed. This is illustrated in Fig. 3-67.

The valve timing of a V-type engine with an overhead cam is accomplished by adjusting the position of a vernier gear until the camshaft is in the correct position. The no. 1 piston is placed in the position where the intake valve is required to begin opening. The camshaft is turned in the normal direction of rotation until the cam lobe for the intake valve of the no. 1 cylinder is starting to apply pressure to the rocker arm and valve. At this time the gears are meshed with all clearance or backlash taken up. When this is accomplished, the camshaft is timed. On a V-type engine it is necessary to time each camshaft separately. In all cases when checking valve timing, the valve clearances on the no. 1 cylinder of the engine should be set to the hot (operating) clearance.

In order to determine the position of a piston for valve timing, certain special tools and instruments are used. One of these is the **top-center indicator,** which determines the top-dead-center position of the piston in the cylinder. Another is the **timing disk,** which is used to measure the crankshaft rotation in degrees and to aid in determining when the crankshaft is in the correct position for timing the valves. Figure 3-68 shows how these tools are used.

The top-center indicator is a hinged lever mounted in an adapter which fits the spark-plug opening. One end of the lever extends into the cylinder where it may bear against the top of the piston, and the other end of the lever is a pointer which moves along a scale. When the piston approaches top center, it presses upward on the end of the indicator lever and moves the pointer along the scale.

The timing-disk indicator consists of two parts—the disk itself and a pointer. The timing disk is fastened to the nose case of the engine in such a manner that the propeller shaft extends through the center of the disk. The pointer is clamped or otherwise fastened to the propeller shaft. A timing handle (propeller shaft wrench) is made to fit the splines of the shaft and is used to turn the crankshaft.

To locate the top dead center of a cylinder, the tools are installed as shown in the drawing. The crankshaft is then turned in the normal direction of rotation until the top-center indicator starts to move. At this point, a mark is placed at the pointer position on the scale of the top-center indicator and also at the position of the pointer on the timing disk. The crankshaft is turned farther in the direction of rotation until the top-center indicator pointer has moved to its limit and back to the mark previously placed on the scale. The turning is stopped at this point, and another mark is placed on the timing disk. The crankshaft is then turned in reverse to a point where the piston has moved away from the arm of the top-center indicator. The top-center position of the piston is found by turning the crankshaft again in the direction of normal rotation until the pointer on the crankshaft is exactly halfway between the marks on the timing disk. The engine is turned in reverse and then turned in the direction of normal rotation to take up any clearance or slack (backlash) that may exist in the various parts in order to ensure accurate timing.

It is possible to find the approximate top center by turning the crankshaft until the pointer of the top-center indicator moves to its extreme position. It should be noted, however, that the crankshaft can be moved back and forth a small amount when the piston is at top center without showing appreciable movement on the top-center indicator. This means that the top-center position cannot be located precisely in this manner.

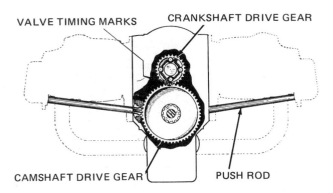

FIG. 3-67 Meshing of marked teeth on cam drive gears.

VALVE TIMING MARKS CRANKSHAFT DRIVE GEAR

CAMSHAFT DRIVE GEAR PUSH ROD

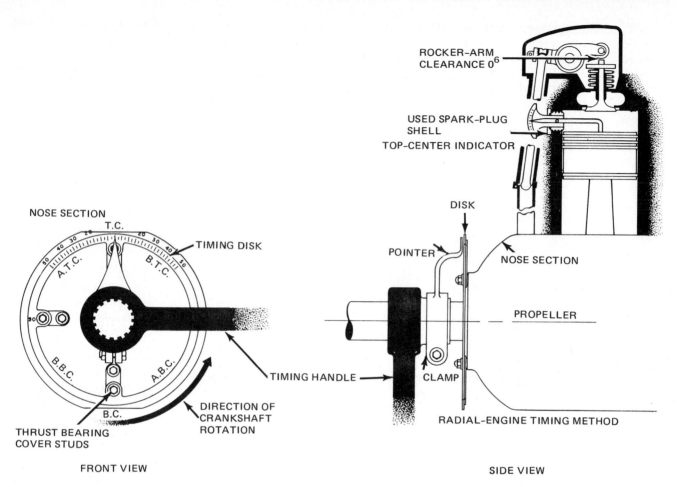

FRONT VIEW

SIDE VIEW

FIG. 3-68   Use of timing disk and pointer.

When the method described with the timing disk is used, the top-center position is located very accurately.

After the position of the top dead center is located, the timing disk may be adjusted to show this position as 0°, as indicated in the drawing. Thereafter, the indication on the timing disk may be used to show the number of degrees the crankshaft is turned before or after top center. When the crankshaft is placed in any given position, it should be turned in reverse first and then brought to the desired position by being turned in the normal direction of rotation. As explained above, this will eliminate any backlash that may exist in the mechanism. This is particularly important when timing the ignition.

Time-Rite Piston-Position Indicator

A very useful and popular timing device is called the Time-Rite piston-position indicator. This device is illustrated in Fig. 3-69. The Time-Rite instrument is designed to afford precision timing for all reciprocating aircraft engines by direct measurement of piston travel.

Description. The Time-Rite consists principally of a body which screws into the spark-plug hole, a pivot arm which contacts the head of the piston, an automatically referenced slide pointer, and an adjustable calibrated scale. These design features eliminate the

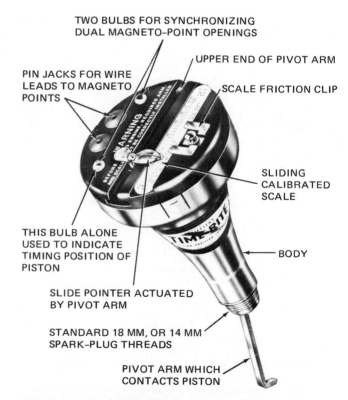

FIG. 3-69   Time-Rite piston-position indicator.

66

need for finding top dead center and compensate for the variables involved in accurate piston positioning.

Calibrated scales are available for all types of engines, and all scale calibrations are obtained in cooperation with and are approved by the engine manufacturers.

Because of the difference in spark-plug locations and piston-dome shapes, several different pivot arms are available to adapt the Time-Rite to all aircraft engines.

All arms are easily interchanged in accordance with instruments furnished with each instrument.

Use of the Time-Rite. The proper use of the Time-Rite piston-position indicator is best described in steps, as illustrated in Fig. 3-70.

Step 1. Remove the front spark plug and gasket from cylinder no. 1. Screw the Time-Rite into the spark-plug

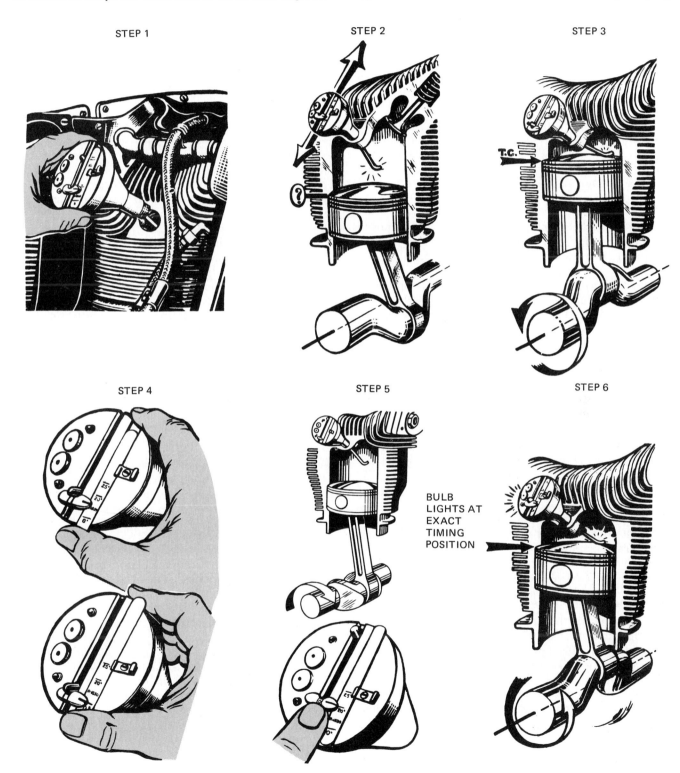

FIG. 3-70 Steps in the use of the Time-Rite indicator.

bushing after making sure that the piston is not near top center. The instrument cannot be installed readily with the piston at or near the top of the stroke.

With some engines it may not be possible to screw the Time-Rite into the spark-plug bushing when using a hooked pivot arm, since the arm hits either the cylinder wall or the top of the cylinder head. In this case simply hold the cap so that it will not rotate, and screw the body into the bushing. Be sure that the correct arm and scale are installed properly before timing the engine.

Step 2. Turn the cap so that the slot is parallel with the vertical axis of the cylinder. The scale should be to the right of the slot when the indicator is installed in radial engines.

Step 3. Turn the engine in the direction of rotation so that the piston goes through the top-center position. This will leave the slide pointer at the highest point of piston travel. This operation takes the place of finding top dead center, as is required with other methods of engine timing.

Step 4. Set the zero position of the scale opposite the slide-pointer reference mark. Be sure that the correct scale is used for the engine being timed.

Step 5. Turn the engine back through the top-center position so that the piston has reached some point before the desired timing position. Set the slide pointer opposite the desired timing position on the scale.

Step 6. Turn the engine in the direction of rotation until the pivot arm just touches the slide pointer. The bulb will light to indicate that the piston is at the exact timing position.

As with many instruments, there are variations in the methods used to obtain the same results. For example, after the zero position of the scale is located as in step 4, the timing position of the engine can be found by using the slide pointer. In step 5, instead of the pointer being placed opposite the desired timing position on the scale, it is placed well above any of the scale marks. Then the engine is turned slowly in the normal direction until the pivot arm moves the pointer to the timing position on the scale. This method must be employed if the light is inoperative.

Checking Piston Position with a Timing Plug

A common method to determine piston position in an opposed engine is to use a timing plug inserted in the top spark-plug hole of the no. 1 cylinder and a protractor, or "degree wheel," mounted with an adapter on the propeller spinner. The **timing plug** consists of a knurled knob with a threaded section sized to fit either a 14-mm or 18-mm spark-plug opening and a rod from $2\frac{1}{2}$ to 3 in [6.35 to 7.62 cm] long. When the plug is installed in the spark-plug hole, the rod will extend into the upper portion of the cylinder where it can contact the top of the piston as it nears top-center position.

The adapter for the 360° protractor is called a "hat"

or "funnel" and is installed over the propeller spinner with rubber bands or some other form of attachment. The protractor disk has a weighted pointer provided at the center of the protractor. Thus, when the protractor is mounted on the spinner of the engine, the pointer remains in a vertical position, pointing downward even though the portractor is rotated.

The procedure for use of the devices just described for determining piston position is as follows:

1. Install the adapter (timing hat) on the propeller spinner with the protractor and pointer (timing disk).

2. Open the cowling and remove the top spark plugs from all the cylinders.

3. Locate the compression stroke of the no. 1 cylinder by placing the thumb over the spark-plug hole and turning the propeller in the normal direction of rotation. Pressure indicates the compression stroke.

4. Insert the eraser end of a lead pencil into the spark-plug hole of the no. 1 cylinder and slowly rotate the engine in the normal direction until the top of the piston touches the pencil as it comes up on the compression stroke. At this time the top of the piston should be from $2\frac{1}{2}$ to 3 in [6.35 to 7.62 cm] from the top of the cylinder.

5. Screw the timing plug into the no. 1 spark-plug hole finger tight. There should be no play or movement of the plug with respect to the cylinder head.

6. Turn the propeller in the normal direction of rotation very slowly and carefully until the top of the piston contacts the end of the timing plug. Do not use excessive force in moving the propeller.

7. Rotate the protractor until the 0°/360° mark is directly under the pointer.

8. Turn the propeller in a direction opposite from normal rotation until the piston touches the timing plug again. Note the number of degrees indicated on the protractor.

9. While the piston in the no. 1 cylinder is still touching the timing plug firmly, rotate the protractor (degree wheel) a distance one-half the number of degrees noted in step 8. This rotation should be in the direction of normal engine rotation.

10. Turn the propeller in the direction of normal engine rotation until the pointer at the protractor indicates the 0°/360° position. This is bottom dead center (BDC). Top dead center (TDC) is 180° from this point in the direction of rotation. Remove the timing plug before turning the engine to the TDC position.

Having located the TDC of the no. 1 cylinder, the technician can mark the points of ignition firing and of valve opening and closing on the protractor (degree wheel or timing disk). These points can then be utilized for valve or ignition timing checks. Any position for valve operation or ignition timing must be approached in the direction of normal engine rotation in order to take up any backlash in the operating mechanisms.

When using a timing disk with any engine having propeller reduction gears, the ratio of gear reduction must be applied. If the reduction gearing has a 2:1 ratio, the crankshaft of the engine will be turning twice as fast as the propeller. In this case, if the timing disk shows a rotation of 110°, the engine will have turned

220°. The technician checking timing of any kind must make certain that she or he knows whether the engine is geared and what the ratio is.

Checking Valve Timing

The correct procedure for checking or adjusting the valve timing on a particular engine is given in the manufacturer's overhaul and maintenance manual. For best results, the manufacturer's recommendations should always be followed. The proper timing for the valves of an engine is often shown on the engine data plate. For example, the indication may read: *I.O. 15° BTC, E.C. 20° ATC*. These instructions mean that the intake valve opens 15° before top center and that the exhaust valve closes 20° after top center. To check the timing of the valves it is merely necessary to place the piston in the position called for and then to see if the required valve action is taking place.

It must be remembered that the piston makes four strokes per cycle and must be on the correct stroke when valve timing is being checked. Since the intake valve must open at or near the end of the exhaust stroke, the proper stroke of the piston can be determined by watching the valve action. By rotating the engine in the normal direction of rotation and noting when the exhaust valve begins to close, the technician will at the same time see that the intake valve is beginning to open. At this time the piston is near top center at the end of the exhaust stroke and the beginning of the intake stroke. If it is desired to find top center at the end of the compression stroke, the engine should be turned to a position near midway between the point where the intake valve closes and the exhaust valve opens. Top center is not usually exactly halfway between these points, but this will serve to place the piston on the correct stroke for compression.

● THE ACCESSORY SECTION

The **accessory section** of an engine provides mounting pads for the accessory units, such as the fuel pressure pumps, fuel-injector pumps, vacuum pumps, oil pumps, tachometer generators, electric generators, magnetos, starters, two-speed supercharger-control valves, oil screens, hydraulic pumps, and other units. Regardless of the construction and location of the accessory housing, it contains and supports the gears for driving those accessories which are operated by engine power.

Accessory sections for aircraft engines vary widely in shape and design because of the engine and aircraft requirements. The illustrations in the accompanying text show accessory sections designed for a radial engine and an opposed engine.

Accessory Section for a Radial Engine

The accessory section which is shown in Fig. 3-71 is designed for the Pratt & Whitney R-985 Junior Wasp radial engine and is called the **rear case**. This case section attaches to the rear of the supercharger case and supports the accessories and accessory drives. The front face incorporates a vaned diffuser, and the rear face contains an intake duct with three vanes in the elbow. The case also includes an oil pressure chamber containing an oil strainer and check valve, a three-

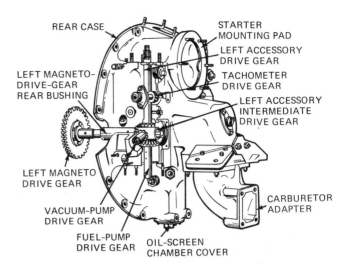

FIG. 3-71 Accessory section for the Pratt & Whitney R-985 radial engine.

section oil pump, and an oil pressure-relief valve. Mounting pads are provided for the carburetor adapter, two magnetos, a fuel pump, the starter vacuum-pump adapter, a tachometer drive, and the generator. The accessories are driven by three shafts which extend entirely through the supercharger and rear sections. Each shaft, at its forward end, carries a spur gear which meshes with a gear coupled to the rear of the crankshaft. The upper shaft provides a drive for the starter and for the generator. Each of the two lower shafts drives a magneto through an adjustable, flexible coupling. Four vertical drives are provided for by a bevel gear keyed to each magneto drive shaft. Two vertical drive shafts are for operating accessories, and two tachometers are driven from the upper side of the bevel gears. The undersides of the bevel gears drive an oil pump on the right side and a fuel pump on the left. An additional drive, for a vacuum pump, is located at the lower left of the left magneto drive.

Accessory Section for an Opposed Engine

The **accessory case** for a Continental six-cylinder opposed engine is shown in Fig. 3-72. This case is constructed of magnesium and is secured to the crankcase rear flange by 12 hex-head screws.

The accessory case conforms to the shape of the crankcase's rear flange and is open at its front side within the height of the crankcase. The accessory case extends below the crankcase, forming a closed compartment which serves as the oil sump in dry-sump engines.

The rear surface of the accessory case is provided with raised, machined pads for mounting of the starter, the generator, magnetos, left- and right-side accessory drive adapters, oil screen housing, the tachometer drive housing, and the accessory drive's idler-gear shaft. Tapped holes and studs provide attachments for adapters, housings, and accessories.

The pressure oil screen assembly is screwed into a housing which is attached by five hex-head screws to the rear surface of the accessory case below the right magneto. The open front end of the tubular screen assembly fits closely into a counterbore in the cases

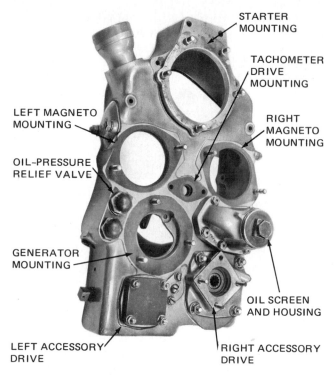

STARTER MOUNTING

TACHOMETER DRIVE MOUNTING

LEFT MAGNETO MOUNTING

RIGHT MAGNETO MOUNTING

OIL-PRESSURE RELIEF VALVE

GENERATOR MOUNTING

OIL SCREEN AND HOUSING

LEFT ACCESSORY DRIVE

RIGHT ACCESSORY DRIVE

FIG. 3-72   Accessory case for a six-cylinder opposed engine.

of wet-sump engines, while the cases of dry-sump models are equipped with a check-valve assembly which fits in the same counterbore and receives a slightly different screen over its rear shoulder. The check valve offers no resistance to oil leaving the front end of the screen, but it prevents return flow during periods of idleness. The cavity surrounding the oil screen is connected by a passage to the pressure-oil-pump outlet port. The pressure-oil-screen exit cavity of cases for wet-sump engines is drilled and tapped through from the right side of the case for an outlet-oil-line elbow fitting. In dry-sump engines the screen exit is connected to the rear end of the crankcase's right oil gallery.

The housing for a gear-type pressure oil pump is attached to the front surface of the accessory case's rear wall by five screws. One screw is installed from the rear and lies within the generator pilot counterbore of the case's rear surface. Two of the pressure-oil-screen housing's attaching screws pass through the case's rear wall and into the pump housing's tapped holes. The other two points of pump housing attachment are the tachometer-drive housing's two attaching screws. The lower, or driving, impeller of the pressure pump has a squared hole in the center of the camshaft gear web. A rearward extension of the upper, or driven, pump impeller passes through the rear wall of the case and through a small cast housing provided with a shaft oil seal and threaded to receive the tachometer-drive conduit nut. The tachometer-drive cable end enters and is driven by a slotted hole in the impeller shaft.

Accessory cases for wet-sump engines of this model are equipped with an oil suction tube attached to the inlet port of the pressure oil pump and extending down-

ward through a hole in the bottom case surface. The oil-sump bypass tube is installed in the case hole surrounding the suction tube and is sealed to the case and to the oil-sump inlet opening by two hydraulic O rings installed in grooves in the outer surface of the tube. The suction tube is attached to the left side wall of the accessory case by a clip, a speed nut, and a roundhead screw. Accessory cases for dry-sump engines do not have the large bottom hole for the bypass tube, but they are tapped to receive a $\frac{3}{8}$-18 drain plug. Instead of a suction tube, the cases of dry-sump engines are equipped with an oil inlet tube connected to the pressure pump inlet port and extending through the left side of the case for connection of the oil inlet hose.

The generator is mounted on a pad below the left magneto and is centered by a pilot which fits in a case counterbore. Three studs attach the assembly. Below the generator, a pad is provided for mounting the left-side accessory drive adapter. The adapter is bored lengthwise for a shaftgear bearing. The rear end of the left-side accessory drive shaftgear is splined to receive and drive the accessory drive shaft. A fuel pump is normally mounted on this drive. The left-side accessory drive shaftgear installed in all dry-sump engines of this model has a forward projection through the front wall of the case which drives a gear-type oil scavenge pump. The scavenge pump is mounted on a pad machined on the front side of the case at the lower left corner. It is retained by six hex-head screws. The left-side accessory drive shaftgear is driven by an idler gear which rotates on a shaft installed from the rear side of the case and retained by two screws. The right-side accessory drive, consisting of a shaftgear and an adapter, is installed on a mount pad below the oil-pressure screen housing.

Magnetos are mounted on pads at the left and right sides of the accessory case's rear surface and are attached by two studs and nuts each. Magneto drive gears are independently mounted on supports attached to the rear of the crankcase and are engaged to the magnetos through rubber-padded couplings.

The starter mounting pad is at the upper rear of the accessory case. The adaptors used depend upon the type of starter to be installed on the engine.

A vertical oil passage is drilled from the right side of the accessory case's bottom surface to the outlet of the pressure oil screen. It is plugged at the case's bottom surface. An intersecting diagonal passage is drilled from the right side of the case through a rib along the front case wall to outlets in the front bearings of the idler gear shaft and the left-side accessory drive shaftgear. For use with dry-sump engines, this passage is always plugged with a plain $\frac{1}{8}$-in [3.18-mm] pipe plug at the right-side surface of the case. For wet-sump engines which are not equipped with a left-side accessory drive, the diagonal passage must be closed with a special extention plug which fits closely in the drilled hole to prevent escape of oil through the open shaft bearings. A small hole drilled from the right-side accessory drive mounting pad into the vertical oil passage registers with an oil hole in the drive adapter leading to the shaftgear bearing.

Of the two bronze acorn caps below the left magneto,

the upper one guides and covers the oil pressure-relief valve and spring. The relief-valve seat registers with the rear end of the left oil gallery.

● REVIEW QUESTIONS

1. Name four of the most common engine classifications by cylinder arrangement.
2. What types of engines provide the best power-weight ratio?
3. What types of engines provide the least drag in flight?
4. Discuss the comparative advantages and disadvantages of air cooling and liquid cooling for aircraft engines.
5. Describe some of the effects of excessive heat in an aircraft engine.
6. Describe how cylinders are designed for air cooling.
7. Explain the use of pressure baffles in the air cooling of a reciprocating engine.
8. Of what material are baffles usually made?
9. What may happen if baffles are not properly installed?
10. What precautions should be observed in removing and installing baffles?
11. How are baffles attached to the engine?
12. How is a reciprocating engine in a helicopter provided with adequate cooling air?
13. What liquid is used as a coolant in a liquid-cooled engine?
14. Describe the functions of a crankcase.
15. Describe the construction of a crankcase for a six-cylinder opposed engine.
16. Of what material is a crankcase usually made?
17. Describe the arrangement of the fuel-induction system.
18. Name the principal sections of the crankcase for a radial engine. Describe the function of each.
19. What accessories may be found mounted on the nose section of some radial engines?
20. What principal parts are supported by the power section?
21. Name the accessories likely to be mounted on the accessory section of a radial engine.
22. Name the parts of a cylinder assembly.
23. What are the principal requirements of a cylinder assembly?
24. What is meant by a *chokebored cylinder*?
25. What is the reason for chokeboring a cylinder?
26. Which dimension is the smaller in a chokebored cylinder?
27. Of what material is a cylinder barrel constructed and by what processes is it manufactured?
28. Why is a cylinder barrel manufactured with an extended skirt?
29. What is meant by *nitriding*?
30. By what two methods are cooling fins provided for cylinder barrels?
31. Describe the construction of a cylinder head.
32. Give one disadvantage of employing aluminum alloy in the construction of engine cylinder heads.
33. Why are valve guides installed in a cylinder head?
34. What is the purpose of a Heli-Coil insert?
35. How is the cylinder head attached to a cylinder barrel?
36. What type of fit is required for the installation of valve guides in a cylinder head?
37. How can the intake-valve side of a cylinder head be distinguished from the exhaust-valve side?
38. Explain the reason for valve seats in a cylinder head and describe the method of installation.
39. When the technician must replace both a valve guide and a valve seat for the same valve, what is the correct sequence of operations? Why?
40. What is the principal function of a piston?
41. Why are pistons made of aluminum alloy?
42. How is a piston cooled?
43. What is the weight limitation with respect to pistons installed in the same engine?
44. Discuss the requirements of piston rings and their installation on the piston.
45. Why should the piston rings near the top of a piston have more side clearance than those near the bottom of the piston?
46. Why is it necessary to have comparatively large clearances between the pistons and cylinder walls?
47. What is the purpose of the holes drilled through the piston wall in the oil-ring grooves?
48. Why is the *piston-ring gap* important?
49. Describe the differences in the construction of different types of piston rings and explain the function of each.
50. At what location on the piston are the *compression rings* installed?
51. What types of piston rings would normally be installed in the top four grooves of a five-ring piston?
52. What will occur if an *oil-wiper* (-scraper) *ring* in the top part of a piston is installed with the bevel side toward the bottom of the piston?
53. What is a *piston-ring compressor* and how is it used?
54. Describe the construction of a *piston pin*.
55. What is a full-floating piston pin?
56. Where does the greatest wear normally occur in a reciprocating aircraft engine?
57. What precautions apply to the use of chrome-plated piston rings?
58. What means are employed to prevent the ends of the piston pins from contacting and scoring the cylinder walls?
59. What is the function of a *connecting rod*?
60. Describe the construction of a plain connecting rod.
61. What are the three principal types of connecting-rod assemblies?
62. In what type of engine would a *fork-and-blade* connecting-rod assembly be employed?
63. Describe the *master-and-articulated* rod assembly.
64. What type of crankshaft is required in a radial engine using a single-piece master rod?
65. Describe the construction of a master-rod bearing.
66. What device is used to hold *knuckle pins* in place?

67. Name the parts of the crankshaft.
68. Describe the oil passages and sludge chambers in a crankshaft.
69. Why are *counterweights* needed on many crankshafts?
70. Explain the operation of *dynamic dampers*.
71. Why is a plain bearing faced with a soft metal such as lead, silver or babbitt?
72. Describe two types of antifriction bearings.
73. What type of bearing produces the least rolling friction?
74. What type of loads are normally applied to plain bearings? To ball bearings?
75. Where are straight roller bearings used in high-power radial engines?
76. What type of bearing is used as a thrust bearing in most radial engines?
77. Why is a ball bearing particularly suited for thrust loads?
78. What type of bearing is normally used for the master-rod bearing in a radial engine?
79. What are the purposes of propeller reduction gears?
80. Describe three types of reduction-gear arrangements.
81. If an engine has a propeller reduction-gear ratio of 0.750:1, what is the propeller speed when the engine is turning at 2800 rpm?
82. If a turboprop engine is turning at 24,000 rpm, what propeller gear ratio is required to obtain a propeller rpm of 1800?
83. What type of construction contributes to the cooling of exhaust valves?
84. Describe the use of *Stellite* in the construction of exhaust valves.
85. Why are two coiled springs used to hold valves in the closed position?
86. Describe the valve operating mechanism for an opposed engine.
87. How is *valve stretch* measured?
88. Why is the angle of the valve face important?
89. When grinding a valve face and seat, what is meant by *interference fit*? What benefit is derived from this fit?
90. Explain *valve lap*, *valve lead*, and *valve lag*.
91. Why is valve overlap (lap) incorporated in the design of an aircraft engine?
92. An aircraft engine has the following valve-timing specifications: IO 15° BTC, EO 50° BBC, IC 55°ABC, EC 14°ATC. Through what distance of crankshaft rotation is the intake valve open?
93. With the above specifications, through what distance of crankshaft rotation is the exhaust valve open? Closed?
94. If the intake valve in one cylinder of a four-stroke-cycle engine has just closed, on what stroke is the piston in that cylinder?
95. If the exhaust valve in one cylinder of a four-stroke-cycle engine has just opened, on what stroke is the piston in that cylinder?
96. Relative to the crankshaft speed, what is the speed of rotation for the camshaft in an opposed engine?
97. During operation, at what points in an engine cycle will a piston be moving at maximum velocity?
98. How is correct valve timing assured when an opposed-type engine is assembled?
99. Discuss valve timing for a V-type engine.
100. What is meant by *cam float* in a radial engine?
101. What precaution must be taken with respect to cam float when adjusting the valves of certain radial engines?
102. Name the parts of a typical hydraulic valve lifter.
103. Explain the term *zero lash* when referring to a valve lifter.
104. Describe the operation of a hydraulic valve lifter.
105. What is the operational valve clearance in an engine employing hydraulic (zero lash) valve lifters?
106. Give the rotational speed and the direction of rotation for a four-lobe cam plate with respect to crankshaft rotation in a seven-cylinder radial engine.
107. What is the purpose of the ramp on each side of the lobes on a cam ring?
108. Explain the difference between hot and cold valve clearances.
109. What causes the hot clearance of a valve in a typical aircraft engine to be greater than its cold clearance?
110. What valve clearance should be used when checking the valve timing of an engine?
111. If valve clearance is excessive, what effect does this have on valve timing?
112. What would be the result if engine backlash were not eliminated before one checked or adjusted the valve timing?
113. Describe the procedure for locating the top-dead-center position of a particular piston.
114. Describe the use of a timing disk and pointer.
115. Give the steps employed when setting timing with a Time-Rite piston-position indicator.
116. Explain the use of a top-dead-center indicator.
117. Describe the procedure for using a timing plug to locate the top dead center of a piston.
118. Why is it necessary to consider the gear ratio of propeller reduction gears when locating piston position with a disk attached to the propeller spinner?
119. Describe the accessory section of an engine and name the accessories which are generally mounted on this section.

4 BASIC FUEL SYSTEMS AND FLOAT-TYPE CARBURETORS

This chapter primarily explains basic fuel systems and their relation to the engine and to the supply of fuel to the engine through the carburetor. The theory, operation, construction, and maintenance of float-type carburetors used on small- and medium-sized reciprocating engines is covered in detail to give the technician a good working knowledge. Other types of fuel metering and fuel control devices are described in later chapters. Additional information on aircraft and engine fuel systems is provided in a related text of this series, *Aircraft Maintenance and Repair*. Requirements for aircraft fuel systems are set forth in Federal Aviation Regulations (FAR), parts 23 and 25.

● FUEL SYSTEMS

The complete fuel system of an airplane can be divided into two principal sections: the aircraft fuel system and the engine fuel system. The **aircraft fuel system** consists of the fuel tank or tanks, the fuel boost pump, the tank strainer (also called a finger strainer), fuel tank vents, the fuel lines (tubing and hoses), fuel-control or -selector valves, the main (or master) strainer, fuel-flow and pressure gages, and fuel drain valves. Fuel systems for different aircraft will vary in complexity and may or may not include all of the foregoing components. The **engine fuel system** begins where the fuel is delivered to the engine-driven pump and includes all of the fuel controlling units from this pump through the carburetor or other fuel metering device.

Requirements for Fuel Systems

The complete fuel system of an aircraft must be capable of delivering a continuous flow of clean fuel under positive pressure from the fuel tank or tanks to the engine under all conditions of engine power, altitude, and aircraft attitude, and throughout all types of flight maneuvers for which the aircraft is certificated or approved. In order to accomplish this and to provide for maximum operational safety, certain conditions must be met:

1. Gravity systems must be designed with the fuel tank placed a sufficient distance above the carburetor to provide such fuel pressure that the fuel flow can be 150 percent of the fuel flow required for takeoff operation.

2. A pressure, or pump, system must be designed so that the system can provide 0.9 lb/h [0.41 kg/h] of fuel flow for each takeoff horsepower delivered by the engine, or 125 percent of the actual takeoff fuel flow of the engine, at the maximum power approved for takeoff.

3. A **boost pump**, usually located at the lowest point in the fuel tank, must be available for engine starting, for takeoff, for landing, and for use at high altitudes. It must have sufficient capacity to substitute for the engine-driven fuel pump at any time that the engine-driven pump should fail.

4. Fuel systems must be provided with valves so that fuel can be shut off and prevented from flowing to any engine. Such valves must be accessible to the pilot and flight engineer.

5. In the case of systems in which outlets are interconnected, it should not be possible for fuel to flow between tanks in quantities sufficient to cause an overflow from the tank vent when the airplane is operated in the condition most apt to cause such overflow when the tanks are full.

6. Multiengine airplane fuel systems should be designed so that each engine is supplied from its own tank, lines, and fuel pumps; however, means may be provided to transfer fuel from one tank to another or to run two engines from one tank in an emergency. This is accomplished by a **cross-flow system** and valves.

7. A gravity-feed system should not supply fuel to any one engine from more than one tank unless the tank airspaces are interconnected in such a manner as to ensure equal fuel feed.

8. Fuel lines should be of a size to carry the maximum required fuel flow under all conditions of operation and should have no sharp bends or rapid rises which would tend to cause vapor accumulation and subsequent vapor lock. Fuel lines must be kept away from hot parts of the engine insofar as possible.

9. Fuel tanks should be provided with drains and sumps to permit the removal of the water and dirt which usually accumulate in the bottom of the tank. Tanks must also be vented with a positive-pressure venting system to prevent the development of a low pressure which will restrict the flow of fuel and cause the engine to stop. Fuel tanks must be able to withstand, without failure, all loads to which they may be subjected during operation.

10. Fuel tanks must be provided with baffles if the tank design is such that a shift in fuel position will cause an appreciable change in the balance of the aircraft. This applies chiefly to wing tanks, where a sudden shift of fuel weight can cause loss of aircraft control. Baffles also aid in preventing fuel sloshing, which can contribute to vapor lock.

Gravity-Feed Fuel Systems

A gravity-feed fuel system is one in which the fuel is delivered to the engine solely by gravity. (See items 1 and 7 in the foregoing list.) The gravity system does not require a boost pump because the fuel is always under positive pressure to the carburetor. A fuel quantity gage must be provided to show the pilot the quantity of fuel in the tanks at all times. The system includes fuel tanks, fuel lines, a strainer and sump, a fuel cock or shutoff valve, a priming system (optional), and a fuel quantity gage. The carburetor may also be considered to be part of the system. A gravity-feed fuel system is shown in Fig. 4-1.

Pressure Systems

For aircraft in which it is not possible to place the fuel tanks at the required distance above the carburetors or other fuel metering devices and in which a greater pressure is required than can be provided by gravity, it is necessary to utilize fuel boost pumps and engine-driven fuel pumps.

For a fuel system which relies entirely on pump pressure, the **fuel boost pump** is located in the bottom of the fuel tank and may be either inside or outside the tank. In many systems the boost pump is submerged in the fuel at the bottom of the tank. In some systems (Fig. 4-2) gravity feeds the fuel to reservoir tanks and then through the fuel-selector valve to the auxiliary (boost) fuel pump. This system utilizes a fuel-injection metering unit and requires more pressure than can be supplied by gravity alone.

In a pressure system the **engine-driven fuel pump** is in series with the boost pump and the fuel must flow through the engine-driven pump to reach the fuel metering unit. The pump must therefore be designed so that fuel can bypass it when the engine is not running. This is normally accomplished by means of a **bypass valve**. The pump must also include a **relief valve** or similar unit to permit excess fuel to return to the inlet side of the pump. The engine-driven pump must be capable of delivering more fuel to the engine than is required for any mode of operation.

The fuel boost pump supplies fuel for starting the engine, and the engine pump supplies the fuel pressure necessary for normal operation. During high-altitude operation, takeoff, and landing, the boost pump is operated to ensure adequate fuel pressure. This is particularly important during landing and takeoff in case of engine-pump failure.

Vapor Lock

The condition known as **vapor lock** is caused by fuel vapor and air collecting in various sections of the fuel system. The fuel system is designed to handle liquid fuel rather than a gaseous mixture. When a substantial amount of vapor collects, it interferes with the operation of pumps, valves, and the fuel metering section of the carburetor. Vapor is caused to form by the low atmospheric pressure of high altitude, by excessive fuel temperature, and by turbulence (or sloshing) of the fuel.

The best solution to the vapor-lock problem is the use of a boost pump, which is why the boost pump is operated at high altitudes. The boost pump applies positive pressure to the fuel in the lines, reducing the tendency of the fuel to vaporize and forcing vapor bubbles through the system and out through the venting devices. Since the boost pump is located in the bottom of the fuel tank or below the tank, it will always have a good supply of fuel and will continue to force fuel through the supply lines, even though vapor bubbles may be entrained in the fuel.

Fuel pumps and carburetors are often equipped with vapor separating devices (which will be discussed later, in relation to various types of fuel metering units). **Vapor separators** are chambers provided with float valves or other types of valves which open when a certain amount of vapor accumulates. When the valve opens, the vapor is vented through a line to the fuel tank. The vapor is thereby prevented from entering the fuel metering system and interfering with normal operation.

The design of a fuel system must be such that an accumulation of vapor is not likely to occur. Fuel lines must not be bent into sharp curves; neither should there be sharp rises or falls in the line. If a fuel line rises and then falls, vapor can collect in the high point of the resulting curve. It is therefore desirable that the fuel line have a continuous slope upward or downward from the tank to the boost pump and a continuous slope upward or downward from the boost pump to the engine-driven pump to prevent vapor from collecting anywhere in the line.

Fuel Strainers and Filters

We have mentioned that all aircraft fuel systems must be equipped with strainers and/or filters to remove dirt particles from the fuel. Strainers are often installed in the fuel tank (cell) outlets or they may be integral with the fuel boost pump assembly. Fuel tank strainers have a comparatively coarse mesh, some being as coarse as eight mesh to 1 in [2.54 cm]. **Fuel sump strainers**, also called **main strainers** or **master strainers**, are located at the lowest point in the fuel system be-

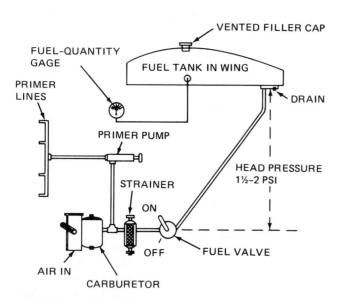

FIG. 4-1   Gravity-feed fuel system.

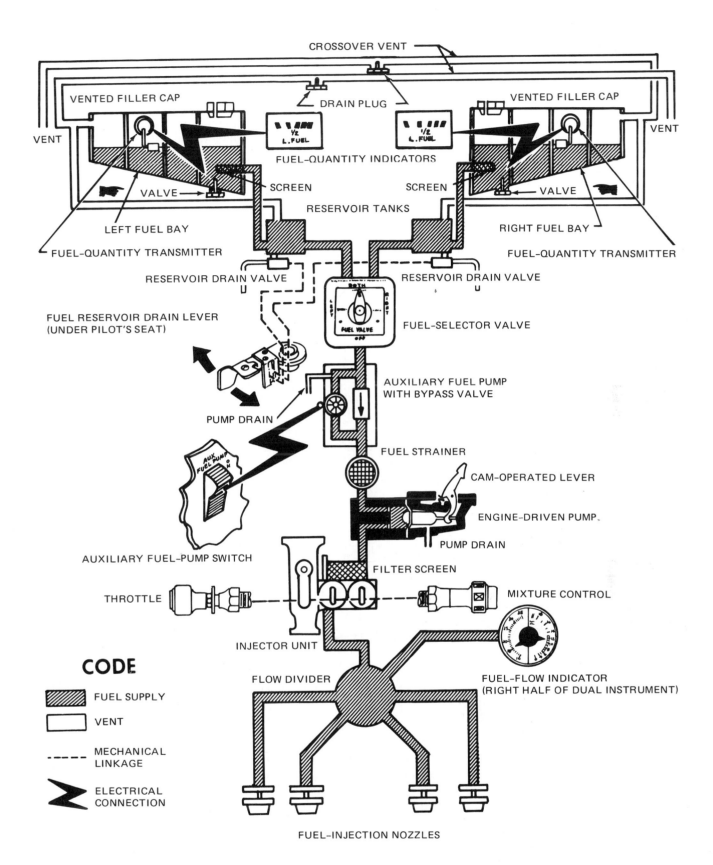

TO ENSURE DESIRED FUEL CAPACITY WHEN REFUELING, PLACE THE
FUEL-SELECTOR VALVE IN EITHER LEFT OR RIGHT POSITION TO PRE-
VENT CROSS-FEEDING.

FIG. 4-2   Pressure fuel system.   *(Cessna Aircraft Co.)*

tween the fuel tank and the engine and are much finer in mesh size, usually being 40 or more mesh per inch. The fuel filters installed in carburetors and other fuel metering units may be screens or sintered (heat-bonded) metal filters. Many of these are designed to remove all particles larger than 40 microns. A **micron** is one one-thousandth of a millimeter.

Fuel strainers and filters should be checked and cleaned as set forth in the aircraft service manual, which in many cases means as often as after every 25 h of operation.

Fuel System Precautions

In servicing fuel systems, it must be remembered that fuel is flammable and that the danger of fire or explosion always exists. The following precautions should be taken:

1. Aircraft being serviced or having the fuel system repaired should be properly grounded.
2. Spilled fuel should be neutralized or removed as quickly as possible.
3. Open fuel lines should be capped.
4. Fire extinguishing equipment should always be available.
5. Metal fuel tanks must not be welded or soldered unless they have been adequately purged of fuel fumes. Keeping a tank or cell filled with carbon dioxide will prevent explosion of fuel fumes.

● PRINCIPLES OF CARBURETION

In the discussion of heat-engine principles, it was explained that a heat engine converts a portion of the heat of a burning fuel to mechanical work. In order to obtain heat from fuel it is necessary that the fuel be burned, and the burning of fuel requires a combustible mixture. The purpose of **carburetion,** or **fuel metering,** is to provide the combustible mixture of fuel and air necessary for the operation of an engine.

Since gasoline and other petroleum fuels consist of carbon (C) and hydrogen (H) chemically combined to form hydrocarbon molecules (CH), it is possible to burn these fuels by adding oxygen (O) to form a gaseous mixture. The carburetor mixes the fuel with the oxygen of the air to provide a combustible mixture which is supplied to the engine through the induction system. The mixture is ignited in the cylinder, with the result that the heat energy of the fuel is released and the fuel-air mixture is converted to carbon dioxide (CO_2), water (H_2O), and possibly some carbon monoxide (CO).

The carburetors used on aircraft engines are comparatively complicated because they play an extremely important part in engine performance, mechanical life, and the general efficiency of the airplane. This is caused by the widely diverse conditions under which airplane engines are operated. The carburetor must deliver an accurately metered fuel-air mixture for engine loads and speeds between wide limits and must provide for automatic or manual mixture correction under changing conditions of temperature and altitude, the while being subjected to a continuous vibration that tends to upset the calibration and adjustment. For these rea-sons, many accurately constructed and precisely adjusted parts are included in an aircraft carburetor assembly. Sturdy construction is essential for all parts of an aircraft carburetor, to provide durability and resistance to the effects of vibration. A knowledge of the functions of these parts is essential to the understanding of carburetor operation.

Fluid Pressure

In the carburetor system of an internal-combustion engine, liquids and gases, collectively called **fluids,** flow through various passages and orifices (holes). The volume and density of liquids remain fairly constant, but gases expand and contract as a result of surrounding conditions.

The atmosphere surrounding the earth is like a great pile of blankets pressing down on the earth's surface. **Pressure** may be defined as force acting upon an area. It is commonly measured in pounds per square inch (psi), inches of mercury (inHg), centimeters of mercury (cmHg) or kilopascals (kPa). The **atmospheric pressure** at any place is equal to the weight of a column of water or mercury a certain number of inches, centimeters, or millimeters in height. For example, if the cube-shaped box shown in Fig. 4-3, each side of which is 1 in² [6.45 cm²] in area, is filled with mercury, that quantity of mercury will weigh 0.491 lb [2.18 N]; hence, a force of 0.491 lb is acting on the bottom square inch of the box. If the same box were 4 in [10.16 cm] high, the weight of the mercury would be 4 × 0.491, or 1.964 lb [8.74 N]; hence, the downward force on the bottom of the box would be 1.964 lb. Therefore, each inch of height of a column of mercury represents 0.491 psi [3.385 kPa] pressure. To change inches of mercury to pounds per square inch, simply multiply by 0.491. For example, if the height of a column of mercury is 29.92 inHg [101.34 kPa], multiply 29.92 by 0.491 and the product is 14.69, which is standard atmospheric pressure at sea level.

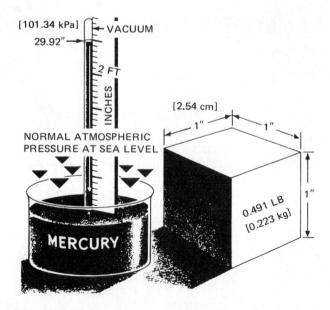

FIG. 4-3 Measuring atmospheric pressure.

Referring again to Fig. 4-3, a glass tube about 36 in [91.44 cm] long, with one end sealed and the other end open, is filled completely with mercury. The tube is then placed in a vertical position with the open end submerged in a small container partly filled with mercury. If this experiment is performed at sea level under standard conditions, the mercury will sink and come to rest at a level of 29.92 in [76 cm] above the mercury in the container. There is then a vacuum above the mercury in the tube; hence, there is no atmospheric pressure above the mercury in the tube.

The atmospheric pressure acts on the surface of the mercury in the container of Fig. 4-3. The weight of the mercury column above the surface of the mercury in the container must therefore equal the weight of the air column above the same surface. The length of the mercury column in the tube indicates the atmospheric pressure and is measured by means of a scale placed beside the tube or marked on its surface. Atmospheric pressure is expressed in *pounds per square inch, inches of mercury, kilopascals,* or *millibars* (mb).

Standard sea-level pressure is 14.7 psi, 29.92 inHg, 101.34 kPa, or 1013 mb. It is well to remember these quantities because they may be encountered often in discussions of pressures in general or especially atmospheric pressures.

From observations such as those described above, the National Aeronautics and Space Administration (NASA) and the International Committee for Aeronautical Operations (ICAO) have established a **standard atmosphere** for comparison purposes. Standard atmosphere is defined as a pressure of 29.92 inHg at sea level with a temperature of 15°C (59°F) when the air is perfectly dry at latitude 40°N. This is a purely fictitious and arbitrary standard, but it has been accepted and should be known to pilots, technicians, and others engaged in aircraft work.

The pressure of the atmosphere varies with the altitude.

● At 5000 ft [1524 m] it is 24.89 inHg [84 kPa]
● At 10,000 ft [3046 m] it is 20.58 inHg [69.7 kPa]
● At 20,000 ft [6096 m] it is 13.75 inHg [46.57 kPa]
● At 30,000 ft [9144 m] it is 8.88 inHg [30 kPa]
● At 50,00 ft [15,240 m] it is only 3.346 inHg [11.64 kPa].

Expressing the same principle in different terms, the pressure of the atmosphere at sea level is 14.7 psi [101.34 kPa], but at an altitude of 20,000 ft [6096 m] the pressure is only about 6.74 psi [46.47 kPa].

The pressure exerted on the surface of the earth by the weight of the atmosphere is called **absolute pressure** and can be measured by a barometer. A **relative pressure** assumes that the atmospheric pressure is zero. Relative or differential pressures are usually measured by fuel pressure gages, steam gages, etc. This means that, when a pressure is indicated by such a gage, the pressure actually shown is so many pounds per square inch above atmospheric pressure. This pressure is often indicated as **psig**, meaning psi gage. When absolute pressure is indicated, it is shown as **psia**, meaning psi absolute.

The effect of atmospheric pressure is important to the understanding of all aircraft fluids, including fuel, oil, water, hydraulic fluid, etc. The effect of atmospheric pressure on liquids can be demonstrated by a simple experiment. Place a tube in a glass of water, place your finger over the open end of the tube, and take the tube out of the water, retaining your finger at the end. The water will not run out of the tube until your finger is removed. This shows the importance of providing and maintaining open vents to the outside atmosphere for tanks, carburetor chambers, and other parts or units which depend on atmospheric venting for their operation.

The Venturi Tube

Figure 4-4 shows the operation of a **venturi tube,** which was originally used for the measurement of the flow of water in pipes. This device consists of a conical, nozzlelike reducer through which the air enters, a narrow section called the **throat,** and a conical enlargement for the outlet, which attains the same size as the inlet but more gradually.

The quantity of air drawn through the inlet will be discharged through the same size opening at the outlet. The velocity of the fluid must therefore increase as it passes through the inlet cone, attain a maximum value in the throat, and thereafter gradually slow down to its initial value at the outlet. The pressure at the throat is consequently less than that at either the entrance or the exit.

Figure 4-4 shows manometers connected to the venturi. These are gages for measuring pressure and are similar in principal to barometers.

The operation of the venturi is based upon **Bernoulli's principle,** which states that the total energy of a particle in motion is constant at all points on its path in a steady flow; therefore, at a higher velocity the pressure must decrease. The pressure in the throat of the venturi tube is less than the pressure in either end of the tube because of the increased velocity in the constricted portion. This is explained by the fact that the same amount of air passes all points in the tube in a given time.

An analogy can be made with men marching down a street. If the street is wide enough for the men to march 20 abreast and then narrows to an alley where they can march only 10 abreast, the men marching through the alley must walk twice as fast as they did

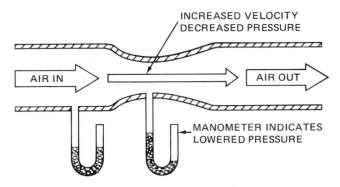

FIG. 4-4 Operation of a venturi tube.

in the street for the same number of men to pass a point in the alley in a given period of time. If they did not increase their marching speed in the alley, there would soon be a traffic jam at the entrance to the alley.

The venturi illustrates the relation existing between pressure (force per unit area) and velocity in a moving column of air. In equal periods of time, equal amounts of air flow through the inlet, which has a large area; through the throat, which has a small cross-sectional area; and then out through the outlet, which also has a large area.

If any body, fluid or solid, is at rest, force must be applied in order to set it in motion. If the body is already in motion, force must be applied to increase its velocity. If a body in motion is to have its velocity decreased, or if the body is to be brought to a state of rest, an opposing force must be applied.

In Fig. 4-4, if the cross-sectional area of the inlet is twice that of the throat, the air will move twice as fast in the throat as it does in the inlet and outlet. Since we have agreed that the velocity of any moving object cannot be decreased without applying an opposing force, the pressure of the air in the outlet portion of the tube must be greater than it is in the throat. From this it can be understood that the pressure in the throat must be less than it is at either end of the tube.

The Venturi in a Carburetor

Figure 4-5 shows the venturi principle applied in a simplified carburetor. The amount of fluid which flows through a given passage in any unit of time is directly proportional to the velocity at which it is moving. The velocity is directly proportional to the difference in applied forces. If a fuel discharge nozzle is placed in the venturi throat of a carburetor, the effective force applied to the fuel will depend on the velocity of air going through the venturi. The rate of flow of fuel through the discharge nozzle will be proportional to the amount of air passing through the venturi, and this will determine the supply of the required fuel-air mixture delivered to the engine. The ratio of fuel to air should be varied within certain limits; hence, a mixture-control system is provided for the venturi-type carburetor.

Review of the Engine Cycle

The conventional aircraft internal-combustion engine is a form of heat engine in which the burning of

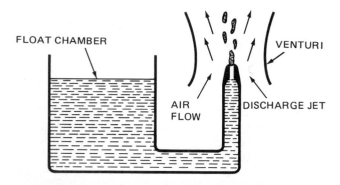

FIG. 4-5 Venturi principle applied to a carburetor.

the fuel-air mixture occurs inside a closed cylinder and in which the heat energy of the fuel is converted into mechanical work to drive the propeller.

The **engine cycle** (Otto cycle) must be understood and remembered in order to learn the process of carburetion. The fuel and air must be mixed and inducted into the cylinder during the **intake stroke;** the fuel-air charge must be compressed during the **compression stroke;** the charge must be ignited and must burn and expand in order to drive the piston downward and cause the crankshaft to revolve during the **power stroke;** and finally, the burned gases must be exhausted (or scavenged) during the **exhaust stroke.**

The quantity and the nature of the charge of fuel and air inducted into the engine cylinder must be given considerable attention because the power, speed, and operating efficiency of the engine are governed largely by this charge.

Fuel-Air Mixtures

Gasoline and other liquid fuels will not burn in the liquid state, but must be vaporized and combined with correct amounts of oxygen in order to form a combustible mixture. The mixture of fuel and air is described as **chemically correct** when there is just enough oxygen present in the mixture to burn the fuel completely. If there is not quite enough air, combustion may occur but it will not be complete. If there is either too much or too little air, the mixture will not burn.

As mentioned previously, burning is a chemical process. Gasoline is composed of carbon and hydrogen, and a gasoline called isooctane has the formula C_8H_{18}. During the burning process this molecule must be combined with oxygen to form carbon dioxide (CO_2) and water (H_2O). The equation for the process may be

$$2C_8H_{18} \times 25O_2 = 16CO_2 + 18H_2O$$

Thus, we see that 2 molecules of this particular gasoline require 50 atoms of oxygen for complete combustion. The burning of fuel is seldom as complete as this, and the resulting gases would likely contain carbon monoxide (CO). In such a case the equation could be

$$C_8H_{18} + 12O_2 = 7CO_2 + 9H_2O + CO$$

Air is a mechanical mixture containing by weight about 75.3 percent nitrogen, 23.15 percent oxygen, and a small percentage of other gases. The nitrogen is a relatively inert gas which has no chemical effect on combustion. The oxygen is the only gas in the mixture which serves any useful purpose as far as the combustion of fuel is concerned.

Gasoline will burn in a cylinder if mixed with air in a ratio ranging between 8 parts of air to 1 part of fuel and 18 parts of air to 1 part of fuel (by weight). That is, the air-fuel (A/F) ratio would be from 8:1 to 18:1 for combustion. This means that the air-fuel mixture can be ignited in a cylinder when the ratio is anywhere from being as *rich* as 8 parts of air by weight to 1 part of fuel by weight to being as *lean* as 18 parts of air by weight to 1 part of fuel by weight. In fuel-air mixtures, the proportions are expressed on the basis of weight, because a ratio based on volumes would be subject to inaccuracies resulting from variations of temperature and pressure.

The proportions of fuel and air in a mixture may be expressed as a ratio (such as 1:12) or as a decimal fraction. The ratio 1:12 becomes 0.083 (which is derived by dividing 1 by 12). The decimal proportion is generally employed in charts and graphs to indicate fuel-air (F/A) mixtures.

Fuel-air mixtures employed in the operation of aircraft engines are described as **best power mixture, lean best power mixture, rich best power mixture,** and **best economy mixture.** The graph of Fig. 4-6 illustrates the effects of changes in F/A ratio for a given engine operating at a given rpm. The mixture at *A* is the lean best power mixture and is the point below which any further leaning of the mixture will rapidly reduce engine power; in other words, the lean best power mixture is the leanest mixture that can be used and still obtain maximum power from the engine. The mixture indicated by *B* in Fig. 4-6 is the rich best power mixture and is the richest mixture that can be used and still maintain maximum power from the engine. The mixtures from *A* to *B* in the graph therefore represent the best power range for the engine. Points *C* and *D* in Fig. 4-6 are the limits of flammability for the F/A mixture; that is, the F/A mixture will not burn at any point richer than that represented by point *C* or at any point leaner than that represented by point *D*.

The chart of Fig. 4-7 illustrates how the best power mixture will vary for different power settings. It will be observed that there is a very narrow range of fuel-air ratios for the best power mixture. For example, the setting for 2900 rpm is 0.077, for 3000 rpm is 0.082, and for 3150 rpm is 0.091. Any other fuel-air ratios than those given will result in a rapid falling off of power. Internal-combustion engines are so sensitive to the proportioning of the fuel and air that the mixture ratios must be maintained within a definite range for any given engine. A perfectly balanced fuel-air mixture is approximately 15:1, or 0.067. This is called a **stoichiometric mixture;** it is one in which all of the fuel and oxygen in the mixture can be combined in the burning process. For a variety of reasons, the stoichiometric mixture is not usually the best to employ.

Specific fuel consumption (sfc) is the term used to indicate the economical operation of an engine. Brake specific fuel consumption (bsfc) is a ratio which shows the amount of fuel consumed by an engine in lb/h for each brake horsepower developed. For example, if an engine is producing 147 hp [109.62 kW] and burns 10.78 gal/h [40.807 L/h] of fuel, the fuel weight being 6 lb/gal [0.719 kg/L], the specific fuel consumption would be 0.44 lb/hp/h [0.15 kg/kW/h].

The chart of Fig. 4-8 was derived from a test run to determine the effects of fuel-air ratios. It should be noted that the lowest specific fuel consumption in this case occurred with a fuel-air ratio of approximately 0.067 and that maximum power was developed at F/A ratios between 0.074 and 0.087. For this particular engine, we may say that lean best power is at point *A* on the chart (F/A ratio 0.074) and that rich best power is at point *B* (F/A ratio 0.087).

It should be noted further that specific fuel consumption increases substantially as the mixture is leaned or enriched from the point of lowest specific fuel consumption. From this observation it is quite apparent that excessive leaning of the mixture in flight will not produce maximum economy. Furthermore, excessive leaning of the mixture is likely to cause detonation, as mentioned previously.

If detonation is allowed to continue, the result will be mechanical damage or failure of the top of the pistons and rings. In severe cases, cylinder heads may be fractured. It is therefore important to follow the engine operating instructions regarding mixture-control settings, thereby avoiding detonation and its unfavorable consequences. A careful observance of cylinder-head temperature and/or exhaust gas temperature (EGT) will in most cases enable the pilot to take corrective action before damage occurs. A reduction of power and an enrichment of the mixture will usually suffice to eliminate detonation.

We have already stated that if there is too much air or too much fuel, the mixture will not burn. In other

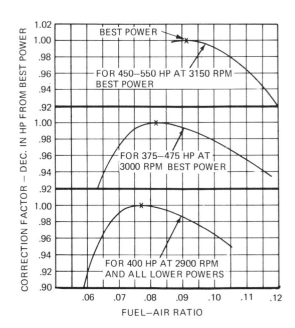

FIG. 4-7 Chart to show best power mixture for different power settings.

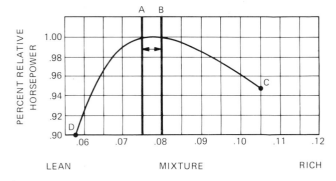

FIG. 4-6 Effects of fuel-air ratios on power at a constant rpm.

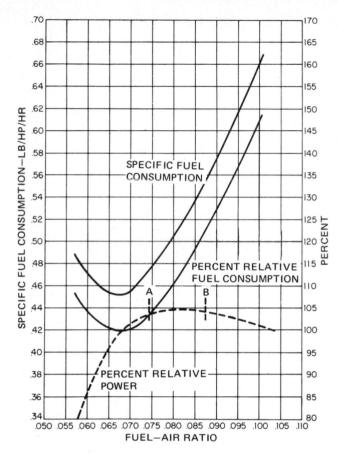

FIG. 4-8 Effects of fuel-air ratios and power settings on fuel consumption.

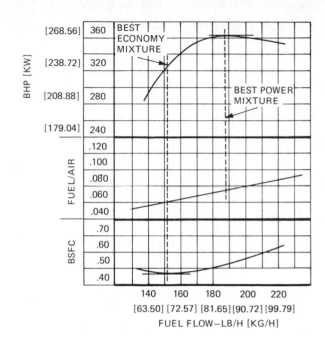

FIG. 4-9 Best economy mixture and best power mixture at constant throttle and constant rpm.

words, when the mixture is excessively rich or excessively lean, it approaches the limit of flammability; as it approaches this limit, the rate of burning decreases until it finally reaches zero. This is much more pronounced on the lean side than it is on the rich side of the correct proportion of fuel and air.

It has been noted that the best power mixtures of fuel and air for the operation of an engine are those which enable the engine to develop maximum power. There is, however, a mixture of fuel and air which will produce the greatest amount of power for a given consumption of fuel. This is called the **best economy mixture,** attained by leaning the mixture below the lean best power mixture. As the mixture is leaned, both power and fuel consumption drop, but fuel consumption decreases more rapidly than engine power until the best economy mixture is reached. The point is reached with a mixture somewhere between 0.055 and 0.065, depending on the particular engine and the operating conditions. It must be emphasized that leaning of the mixture below the best power mixture is not practiced with the engine developing its maximum power. Usually the engine would be operating at less than 75 percent power. The operator's manual should be consulted for specific information on the operation of a particular powerplant and aircraft combination.

The chart of Fig. 4-9 provides a graphic illustration of the difference between best economy mixture and best power mixture. This chart is based upon a con-stant-throttle position with a constant rpm. The only variable is the fuel-air ratio. With a very lean fuel-air ratio of about 0.055, the engine delivers 292 bhp [217.7 kW] with a fuel flow of 140 lb/h [63.50 kg/h] and the bsfc is about 0.48 lb/hp/h [0.218 kg/kW/h]. The best economy mixture occurs when the fuel-air ratio is approximately 0.062. At this point the bhp is 324 and the fuel flow is 152 lb/h [68.95 kg/h]. The bsfc is then 0.469 lb/hp/hr [0.213 kg/kW/h]. As the strength of the fuel-air mixture is increased, a point is reached where the engine power has reached a peak and will begin to fall off. This is the best power mixture, and it is shown to be approximately 0.075, or 1:13.3. At this point the bhp is 364 and the bsfc is 0.514 with a fuel flow of 187 lb/h [84.82 kg/h].

We have now established the effects of fuel-air ratio when other factors are constant, and we can see that the mixture in the operation of the engine will have a profound effect on the performance. We must, however, explore the matter further because engine operating temperature must be considered. If an engine is operated at full power and at the best power mixture, as shown in the upper curve of the chart in Fig. 4-7, it is likely that the cylinder-head temperature will become excessive and detonation will result. For this reason, at full power settings the mixture will be enriched beyond the best power mixture. This is the function of the **economizer,** or **enrichment valve,** in the carburetor or fuel control as will be explained later. The extra fuel will not burn but will vaporize and absorb some of the heat developed in the combustion chamber. At this time the manual mixture control is placed in the **full rich** or **auto rich** position and the fuel-air ratio will be at or above the rich best power mixture.

When operating under cruising conditions of rpm and manifold pressure, it is possible to set the mixture at the **lean best power** value in order to save fuel and still obtain a maximum value of cruising power from the engine. If it is desired to obtain maximum fuel economy at a particular cruise setting, the manual mixture control will be used to lean the mixture to the best economy F/A ratio. This will save fuel but will result in a power reduction of as much as 15 percent.

The chart of Fig. 4-10 illustrates graphically the requirements of an aircraft engine with respect to F/A ratio and power output. As shown in the illustration, a rich mixture is required for very low power settings and for high power settings. When the power is in the 60 to 75 percent range, the F/A ratio can be set for lean best power or for best economy. The curve shown in Fig. 4-10 will vary for different engines.

The effect of the F/A mixture on cylinder-head and exhaust-gas temperatures is illustrated in Fig. 4-11. It can be seen that the temperatures rise as the mixture is leaned to a certain point; however, continued leaning leads to a drop in the temperatures. This is true if the engine is not operating at high power settings.

An excessively lean mixture may cause an engine to **backfire** through the induction system or to stop completely. A backfire is caused by slow flame propagation. This happens because the fuel-air mixture is still burning when the engine cycle is completed. The burning mixture ignites the fresh charge when the intake valve opens, the flame travels back through the induction system, the combustible charge is burned, and often any gasoline that has accumulated near the carburetor is burned. This occurs because flame propagation speed decreases as the mixture is leaned. Thus, a mixture which is lean enough will still be burning when the intake valve opens.

The **flame propagation** in an engine cylinder is the rate at which the flame front moves through the mixture of fuel and air. The flame propagation is most rapid at the best power setting and falls off substantially on either side of this setting. If the mixture is too lean, the flame propagation will be so slow that the mixture will still be burning when the intake valve opens, thus igniting the mixture in the intake manifold and causing a backfire. The effect of the F/A ratio on flame propagation is illustrated in Fig. 4-12.

Backfiring is not the same as **kickback,** which occurs when the ignition is advanced too far at the time that the engine is to be started. If the mixture is ignited before the piston reaches top center, the combustion

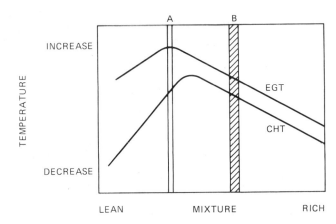

FIG. 4-11   Effect of fuel-air mixture on cylinder-head and exhaust-gas temperatures.

pressure may cause the piston to reverse its direction and turn the crankshaft against the normal direction of rotation.

Afterfiring is caused when raw fuel is permitted to flow through the intake valve into the cylinder head, then out the exhaust valve into the exhaust stack, manifold and muffler, and heater muff. The fuel can cause a fire or explosion that can be very damaging to the exhaust and cabin-heating systems. During starting, engines equipped with continuous fuel injection *must not* have the mixture control advanced out of IDLE CUTOFF until the engine is rotating.

Effects of Air Density

Density may be defined simply as the *weight per unit volume of a substance.* The weight of 1 cubic foot (ft^3) [28.32 L] of dry air at standard sea-level conditions is 0.076475 lb [0.0347 kg]. A pound [0.4536 kg] of air under standard conditions occupies approximately 13 ft^3 [368.16 L].

The density of air is affected by pressure, temperature, and humidity. An increase in pressure will *increase* the density of air, an increase in temperature will *decrease* the density, and an increase in humidity will *decrease* the density. It is evident, therefore, that the fuel-air ratio is affected by air density. For example, an aircraft engine will have less oxygen to burn with the fuel on a warm day than it will on a cold day at the same location; that is, the mixture will be richer when the temperature is high and the engine cannot

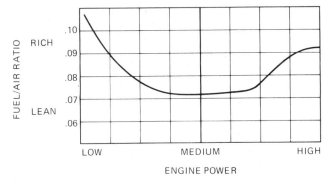

FIG. 4-10   Fuel-air ratios required for different power settings.

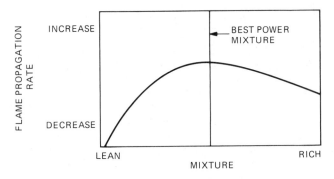

FIG. 4-12   Effect of fuel-air ratio on flame propagation.

produce as much power as when the air is cool. The same is true at high altitudes, where air pressure decreases and density decreases. For this reason, pilots usually "lean out" the mixture at higher altitudes to avoid an overrich mixture and waste of fuel.

Water vapor in the air (humidity) decreases the density because a molecule of water weighs less than a molecule of oxygen or a molecule of nitrogen. Therefore, a pilot must realize that an engine will not develop as much power on a warm, humid day as it will on a cold, dry day. This is because the less-dense air provides less oxygen for fuel combustion in the engine.

Effect of Pressure Differential in a U-Shaped Tube

Figure 4-13 shows two cross-sectional views of a U-shaped glass tube. In the upper view, the liquid surfaces in the two arms of the tube are even because the pressures above them are equal. In the lower view, the pressure in the right arm of the tube is reduced while the pressure in the left arm of the tube remains the same as it was before. This causes the liquid in the left arm to be pushed down while the liquid in the right arm is raised, until the differences in the weights of the liquid in the two arms are exactly proportional to the difference in the forces applied on the two surfaces.

Pressure Differential in a Simple Carburetor

The principle explained in the preceding paragraph is applied in a simple carburetor, such as the one shown in Fig. 4-5. The rapid flow of air through the venturi reduces the pressure at the discharge nozzle so that the pressure in the fuel chamber can force the fuel out into the airstream. Since the airspeed in the tube is comparatively high and there is a relatively great reduction in pressure at the nozzle during medium and high engine speeds, there is a reasonably uniform fuel supply at such speeds.

When the engine speed is low and the pressure drop in the venturi tube is slight, the situation is different. This simple nozzle, otherwise known as a **fuel discharge nozzle,** in a carburetor of fixed size does not deliver a continuously richer mixture as the engine suction and airflow increase. Instead, a plain discharge nozzle will give a fairly uniform mixture at medium

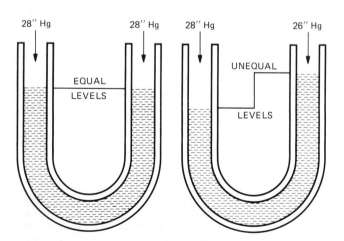

FIG. 4-13   Pressure effects on fluid in a U-shaped tube.

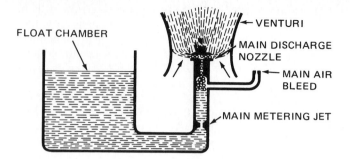

FIG. 4-14   Basic venturi-type carburetor.

and high speeds; but at low speeds and low suction, the delivery falls off greatly in relation to the airflow.

This is partly because some of the suction force is consumed in raising the fuel from the float level to the nozzle outlet, which is slightly higher than the fuel level in the fuel chamber to prevent the fuel from overflowing when the engine is not operating. It is also caused by the tendency of the fuel to adhere to the metal of the discharge nozzle and break off intermittently in large drops instead of forming a fine spray. The discharge from the plain fuel nozzle is, therefore, retarded by an almost constant force, which is not important at high speeds with high suction but which definitely reduces the flow when the suction is low because of reduced speed.

Figure 4-14 shows how the problem is overcome in the design and construction of the venturi-type carburetor. Air is **bled** from behind the venturi and passed into the **main discharge nozzle** at a point slightly below the level of the fluid, causing the formation of a finely divided fuel-air mixture which is fed into the airstream at the venturi. A metering jet between the fuel chamber and the main discharge nozzle controls the amount of fuel supplied to the nozzle. A **metering jet** is an orifice, or opening, which is carefully dimensioned to meter (measure) fuel flow accurately in accordance with the pressure differential existing between the float chamber and the discharge nozzle. The metering jet is an essential part of the main metering system.

The Air Bleed

The **air bleed** in a carburetor lifts an emulsion of air and liquid to a level higher above the liquid level in the float chamber than would be possible with unmixed fuel. Figure 4-15 shows a person sucking on a straw placed in a glass of water. The suction is great enough to lift the water above the level in the glass without drawing any into the mouth. In Fig. 4-16 a tiny hole has been pricked in the side of the straw above the surface of the water in the glass and the same suction is applied as before. The hole causes bubbles of air to enter the straw, and the liquid is drawn up in a series of small drops or slugs.

In Fig. 4-17 the air is taken into the main tube through a smaller tube which enters the main tube below the level of the water. There is a restricting orifice at the bottom of the main tube; that is, the size of the main tube is reduced at the bottom. Instead of a continuous series of small drops or slugs being drawn up through

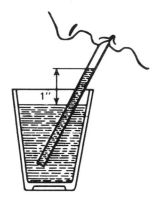

FIG. 4-15   Suction lifting a liquid.

FIG. 4-17   Air bleed breaking up a liquid.

the tube when the person sucks on it, there is a finely divided mixture of air and water formed in the tube.

Since there is a distance through which the water must be lifted from its level in the glass before the air begins to pick it up, the free opening of the main tube at the bottom prevents a very great suction from being exerted on the air-bleed hole or vent. If the air openings were too large in proportion to the size of the main tube, the suction available to lift the water would be reduced.

In Fig. 4-17 the ratio of water to air could be modified for high and low airspeeds (produced by sucking on the main tube) by changing the dimensions of the air bleed, the main tube, and the opening at the bottom of the main tube.

In Fig. 4-14 the carburetor nozzle has an air bleed, as explained previously. We can summarize our discussion by stating that the purpose of this air bleed in the discharge nozzle is to assist in the production of a more uniform mixture of fuel and air throughout all operating speeds of the engine.

Vaporization of Fuel

The fuel leaves the discharge nozzle of the carburetor in a stream which breaks up into various sizes of drops suspended in the airstream, where they become even more finely divided. Vaporization occurs on the surfaces of each drop, causing the very fine particles to disappear and the large particles to decrease in size.

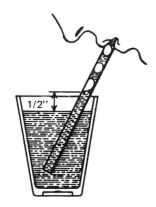

FIG. 4-16   Effect of air bleed.

The problem of properly distributing the particles would be simple if all the particles in each drop vaporized completely before the mixture left the intake pipe, but some particles of the fuel enter the engine cylinders while they are still in a liquid state and thus must be vaporized and mixed in the cylinder during the compression stroke.

The completeness of vaporization depends upon the volatility of the fuel, the temperature of the air, and the degree of atomization. **Volatile** means readily vaporized; therefore, the more volatile fuels evaporate more readily. Higher temperatures increase the rate of vaporization; hence, carburetor air-intake heaters are sometimes provided. Some engines are equipped with "hot-spot" heaters which utilize the heat of exhaust gases to heat the intake manifold between the carburetor and the cylinders. This is usually accomplished by routing a portion of the engine exhaust through a jacket surrounding the intake manifold. In another type of hot-spot heater, the intake manifold is passed through the oil reservoir of the engine. The hot oil supplies heat to the intake manifold walls, and the heat is transferred to the fuel-air mixture.

The degree of atomization is the extent to which fine spray is produced; the more fully the mixture is reduced to fine spray and vaporized, the greater is the efficiency of the combustion process. The air bleed in the main discharge-nozzle passage aids in the atomization and vaporization of the fuel. If the fuel is not fully vaporized, the mixture may run lean even though there is an abundance of fuel present.

The Throttle Valve

A **throttle valve,** usually a **butterfly-type** valve, is incorporated in the fuel-air duct to regulate the fuel-air output. The throttle valve is usually an oval-shaped metal disk mounted on the throttle shaft in such a manner that it can completely close the throttle bore. In the closed position, the plane of the disk makes an angle of about 70° with the axis of the throttle bore. The edges of the throttle disk are shaped to fit closely against the sides of the fuel-air passage. The arrangement of such a valve is shown in Fig. 4-18. The amount of air flowing through the venturi tube is reduced when the valve is turned toward its closed position. This reduces the suction in the venturi tube, so that less

FIG. 4-18 Throttle valve.

fuel is delivered to the engine. When the throttle valve is opened, the flow of the fuel-air mixture to the engine is increased. Opening or closing the throttle valve thus regulates the power output of the engine. In Fig. 4-19 the throttle valve is shown in the open position.

● FLOAT-TYPE CARBURETORS

Essential Parts of a Carburetor

The carburetor consists essentially of a main air passage through which the engine draws its supply of air, mechanisms to control the quantity of fuel discharged in relation to the flow of air, and a means for regulating the quantity of fuel-air mixture delivered to the engine cylinders.

The essential parts of a float-type carburetor are (1) the float mechanism and its chamber, (2) the strainer, (3) the main metering system, (4) the idling system, (5) the economizer (or power enrichment) system, (6) the accelerating system, and (7) the mixture-control system.

In the float-type carburetor, atmospheric pressure in the fuel chamber forces fuel from the discharge nozzle when the pressure is reduced at the venturi tube. The intake stroke of the piston reduces the pressure in the engine cylinder, thus causing air to flow through the intake manifold to the cylinder. This flow of air passes through the venturi of the carburetor and causes the reduction of pressure in the venturi which, in turn, causes the fuel to be sprayed from the discharge nozzle.

The Float Mechanism. As previously explained, the float in a carburetor is designed to control the level of fuel in the float chamber. This fuel level must be maintained slightly below the discharge-nozzle outlet holes in order to provide the correct amount of fuel flow and to prevent leakage of fuel from the nozzle when the engine is not running. The arrangement of a float mechanism in relation to the discharge nozzle is shown in the diagram of Fig. 4-20. In the diagram it will be noted that the float is attached to a lever which is pivoted and that one end of the lever is engaged with the float needle valve. When the float rises, the needle valve closes and stops the flow of fuel into the chamber. At this point, the level of the fuel is correct for proper operation of the carburetor, provided that the needle-valve seat is at the correct level.

As shown in Fig. 4-20, the float-valve mechanism includes a needle and a seat. The needle valve is constructed of hardened steel, or it may have a synthetic rubber section which fits the seat. The needle seat is usually made of bronze. There must be a good fit between the needle and seat to prevent fuel leakage and overflow from the discharge nozzle.

During the operation of the carburetor, the float assumes a position slightly below its highest level to allow a valve opening sufficient for replacement of the fuel as it is drawn out through the discharge nozzle. If the fuel level in the float chamber is too high, the mixture will be rich, and if the fuel is too low, the mixture will be lean. In order to adjust the fuel level for the carburetor shown in Fig. 4-20, washers are placed under the float needle seat. If the fuel level (float level) needs to be raised, washers are removed from under the seat. If the level needs to be lowered, washers are added. The specifications for the float level are given in the manufacturer's overhaul manual.

For some carburetors, the float level is adjusted by bending the float arm. This is true of many automobile carburetors as well as of some aircraft carburetors.

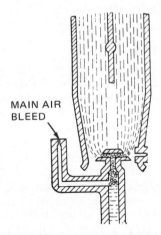

MAIN AIR BLEED

FIG. 4-19 Throttle valve in open position.

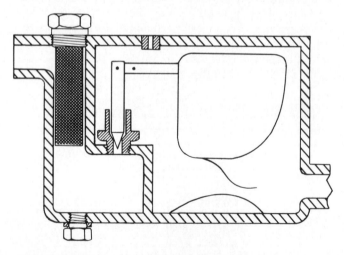

FIG. 4-20 Float and needle-valve mechanism in a carburetor.

Figure 4-21 shows two additional types of float mechanisms. The upper drawing illustrates the **concentric** (having a common center) float and valve, while the lower drawing illustrates an **eccentric** (off-center) float and valve.

The Fuel Strainer. In most carburetors, the fuel supply must first enter a strainer chamber, where it passes through a strainer screen. The **strainer** consists of a fine wire mesh or other type of filtering device, cone-shaped or cylindrically shaped, located so that it will intercept any dirt particles which might clog the needle-valve opening or, later, the metering jets. The strainer is usually removable so that it can be taken out and thoroughly drained and flushed. It is retained by a strainer plug or a compression spring. A typical strainer is shown in Fig. 4-22.

The Main Metering System. The **main metering system** controls the fuel feed in the upper half of the engine speed range as used for cruising and full-throttle operations. It consists of three principal divisions, or units: (1) the **main metering jet** through which fuel is drawn from the float chamber; (2) the **main discharge nozzle,** which may be any one of several types; and (3) the **passage leading to the idling system.**

Although the previous statement is correct, it should be understood that some authorities state that the purpose of the main metering system is to maintain a constant fuel-air mixture at all throttle openings throughout the power range of engine operation. The same authorities divide the main metering system into four parts: (1) the venturi, (2) a metering jet which measures the fuel drawn from the float chamber, (3) a main discharge nozzle, including the main air bleed, and (4) a passage leading to the idling system. It is apparent that these are merely two different approaches to the same thing.

The three functions of the main metering system are (1) to proportion the fuel-air mixture, (2) to decrease the pressure at the discharge nozzle, and (3) to control the airflow at full throttle.

FIG. 4-22 Carburetor fuel strainer.

The airflow through an opening of fixed size and the fuel flow through an air-bleed jet system respond to variations of pressure in approximately equal proportions. If the discharge nozzle of the air-bleed system is located in the center of the venturi so that both the air-bleed nozzle and the venturi are exposed to the suction of the engine in the same degree, it is possible to maintain an approximately uniform mixture of fuel and air throughout the power range of engine operations. This is illustrated in Fig. 4-23, which shows the air-bleed principle and the fuel level of the float chamber in a typical carburetor. If the main air bleed of a carburetor should become restricted or clogged, the F/A mixture would be excessively rich because more of the available suction acts upon the fuel in the discharge nozzle and less air is introduced with the fuel.

The full power output from the engine makes it necessary to have above the throttle valve, a manifold suction (reduced pressure, or partial vacuum) which is between 0.4 and 0.8 psi [2.8 and 5.5 kPa] at full engine speed. However, more suction is desired for metering and spraying the fuel, and this is obtained from the venturi. When a discharge nozzle is located in the central portion of the venturi, the suction obtained is several times as great as the suction found in the intake manifold.

Thus, it is possible to maintain a relatively low manifold vacuum (high manifold pressure). This results in

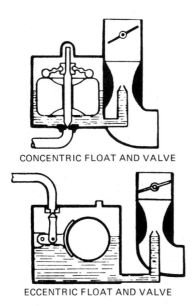

CONCENTRIC FLOAT AND VALVE

ECCENTRIC FLOAT AND VALVE

FIG. 4-21 Concentric and eccentric float mechanisms.

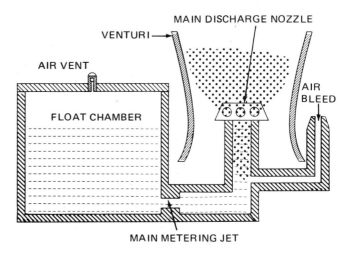

FIG. 4-23 Location of air-bleed system and main discharge nozzle.

high volumetric efficiencies. In contrast, high manifold vacuums result in low volumetric efficiencies.

We have stated previously that the venturi tube affects the air capacity of the carburetor. Hence, it is apparent that the tube should be obtainable in various sizes, so that it can be selected according to the requirements of the particular engine for which the carburetor is designed.

By itself, the main metering system does not accomplish all its functions unaided, but when the other essential parts of a carburetor are examined, the whole system of carburetion becomes apparent. The main metering system for a typical carburetor is shown in Fig. 4-24.

Idling System. At idling speeds, the airflow through the venturi of the carburetor is too low to draw sufficient fuel from the discharge nozzle, so the carburetor cannot deliver enough fuel to the engine to keep it running. At the same time, with the throttle nearly closed, the air velocity is high and the pressure is low between the edges of the throttle valve and the walls of the air passages. Furthermore, there is very high suction on the intake side of the throttle valve. Because of this situation, an idling system with an outlet at the throttle valve is added. This idling system delivers fuel only when the throttle valve is nearly closed and the engine is running slowly. An **idle cutoff** valve stops the flow of fuel through this idling system on some carburetors, and this is used for stopping the engine. An increased amount of fuel (richer mixture) is used in the idle range because at idling speeds the engine may not have enough air flowing around its cylinders to provide proper cooling.

Figure 4-25 shows a three-piece main discharge assembly, with a main discharge nozzle, main air bleed, main discharge-nozzle stud, idle feed passage, main metering jet, and accelerating well screw. This is one of the two types of main discharge-nozzle assemblies used in updraft, float-type carburetors. An **updraft** carburetor is one in which the air flows upward through the carburetor to the engine. The other type has the main discharge nozzle and the main discharge-nozzle stud combined in one piece screwed directly into the

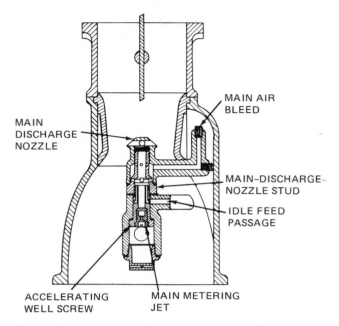

FIG. 4-25 Three-piece main discharge assembly.

discharge-nozzle boss, which is part of the main body casting, thereby eliminating the necessity of having a discharge-nozzle screw.

Figure 4-26 is a drawing of a conventional idle system, showing the idling discharge nozzles, the mixture adjustment, the idle air bleed, the idle metering jet, and the idle tube. Note that the fuel for the idling system is taken from the fuel passage for the main discharge nozzle and that the idle air-bleed air is taken from a chamber outside the venturi section. Thus the idle air is at air-inlet pressure. The idle discharge is divided between two discharge nozzles, and the relative quantities of fuel flowing through these nozzles are dependent on the position of the throttle valve. At very low idle, all the fuel passes through the upper orifice, since the throttle valve covers the lower orifice.

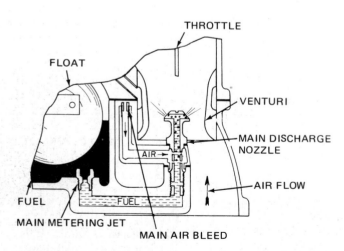

FIG. 4-24 Main metering system in a carburetor. *(Energy Controls Div., Bendix Corp.)*

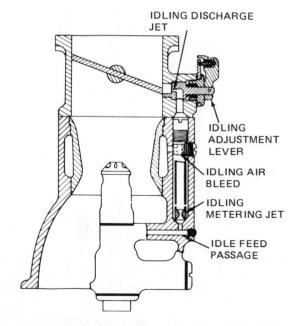

FIG. 4-26 Conventional idle system.

In this case, the lower orifice acts as an additional air bleed for the upper orifice. As the throttle is opened further, exposing the lower orifice, additional fuel passes through this opening.

Since the idle mixture requirements vary with climatic conditions and altitude, a needle-valve-type mixture adjustment is provided to vary the orifice in the upper idle discharge hole. Moving this needle in or out of the orifice varies the idle fuel flow accordingly to supply the correct fuel-air ratio to the engine.

The idling system described above is used in the Bendix-Stromberg NA-S3A1 carburetor and will not necessarily be employed in other carburetors. The principles involved are similar in all carburetors, however.

Figure 4-27 shows a typical float-type carburetor at (A) idling speed, (B) medium speed, and (C) full speed. The greatest suction (pressure reduction) in the intake manifold above the throttle is at the lowest speeds, when the smallest amount of air is received, which is also the condition requiring the smallest amount of fuel. When the engine speed increases, more fuel is needed but the suction in the manifold decreases. For this reason, the metering of the idling system is not accomplished by the suction existing in the intake manifold. Instead, it is controlled by the suction existing in a tiny intermediate chamber, or slot, formed by the

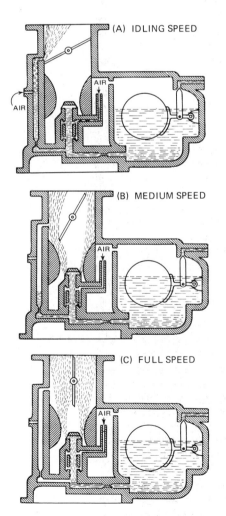

FIG. 4-27 Float-type carburetor at different engine speeds.

idling discharge nozzle and the wall of the carburetor at the edge of the throttle valve. This chamber has openings into the barrel of the carburetor, both above and below the throttle.

In the drawings of Figs. 4-26 and 4-27, note that there is a small chamber surrounding the main discharge-nozzle passage just below the main air-bleed inlet. This chamber serves as an **accelerating well** to store extra fuel that is drawn out when the throttle is suddenly opened. If this extra supply of fuel were not immediately available, the fuel flow from the discharge nozzle would be momentarily decreased and the mixture to the combustion chamber would be too lean, thereby causing the engine to hesitate or misfire. In many carburetors an **accelerating pump** is used to force an extra supply of fuel from the discharge nozzle when the throttle is opened quickly. The operation of the accelerating pump is explained later in this chapter.

In Fig. 4-27, note that when the engine is operating at intermediate speed, the accelerating well still holds some fuel. However, when the throttle is wide open, all the fuel from the well is drawn out. At full power, all fuel is supplied through the main discharge and economizer system, and the idling system then acts as an auxiliary air bleed to the main metering system. The main metering jet provides an approximately constant mixture ratio for all speeds above idling, but it has no effect during idling. Remember that the purpose of the accelerating well is to prevent a power lag when the throttle is opened suddenly.

The Economizer System. An **economizer**, or **power enrichment, system,** is essentially a valve which is closed at low engine and cruising speeds but is opened at high speeds to provide an enriched mixture to reduce burning temperatures and prevent detonation. In other words, it supplies and regulates the additional fuel required for all speeds above the cruising range. An economizer is also a device for enriching the mixture at increased throttle settings. It is important, however, that the economizer close properly at cruising speed; otherwise, the engine may operate satisfactorily at full throttle but will "load up" at cruising speed and below because of the extra fuel being fed into the system. The extra-rich condition will be indicated by rough running and by black smoke emanating from the exhaust.

The economizer gets its name from the fact that it enables the pilot to obtain maximum economy in fuel consumption by providing for a lean mixture during cruising operation and a rich mixture for full-power settings. Most economizers in their modern form are merely enriching devices. The carburetors equipped with economizers are normally set for their leanest practical mixture delivery at cruising speeds, and enrichment takes place as required for higher power settings.

Three types of economizers for float-type carburetors are (1) the needle-valve type, (2) the piston type, and (3) the manifold-pressure-operated type. Figure 4-28 illustrates the **needle-valve type** economizer. This mechanism utilizes a needle valve which is opened by the throttle linkage at a predetermined throttle position. This permits a quantity of fuel, in addition to the

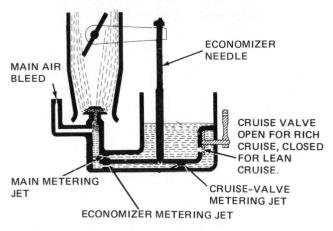

FIG. 4-28 Needle-valve-type economizer.

fuel from the main metering jet, to enter the discharge-nozzle passage. As shown in the diagram, the economizer needle valve permits fuel to bypass the cruise-valve metering jet.

The **piston-type economizer,** illustrated in Fig. 4-29

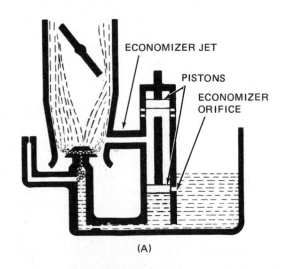

(A)

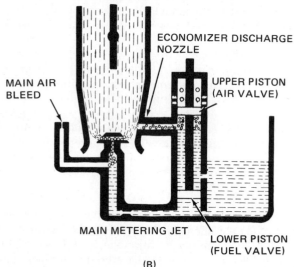

(B)

FIG. 4-29 Piston-type economizer.

is also operated by the throttle. The lower piston serves as a fuel valve, preventing any flow of fuel through the system at cruising speeds. (See view A.) The upper piston functions as an air valve, allowing air to flow through the separate economizer discharge nozzle at part throttle. As the throttle is opened to higher power positions, the lower piston uncovers the fuel port leading from the economizer metering valve and the upper piston closes the air ports (view B). Fuel fills the economizer well and is discharged into the carburetor venturi where it adds to the fuel from the main discharge nozzle. The upper piston of the economizer permits a small amount of air to bleed into the fuel, thus assisting in the atomization of the fuel from the economizer system. The space below the lower piston of the economizer acts as an accelerating well when the throttle is opened.

The **manifold-pressure-operated economizer,** illustrated in Fig. 4-30 has a bellows which is compressed when the pressure from the engine blower rim produces a force greater than the compression spring in the bellows chamber. As engine speed increases, the blower pressure will also increase. This pressure collapses the bellows and causes the economizer valve to open. Fuel then flows through the economizer metering jet to the main discharge system. The operation of the bellows and spring is stabilized by means of a dashpot, as shown in the drawing.

The Accelerating System. When the throttle controlling an engine is suddenly opened, there is a corresponding increase in the airflow; but because of the inertia of the fuel, the fuel flow does not accelerate in proportion to the airflow increase. Instead, the fuel lags behind and causes a temporary lean mixture. This, in turn, may cause the engine to miss or backfire, and it is certain to cause a temporary reduction in power. To prevent this condition, all carburetors are equipped with an **accelerating system.** This is either an accelerating pump or an accelerating well, which has been mentioned previously. The function of the accelerating system is to discharge an additional quantity of fuel into the carburetor airstream when the throttle is sud-

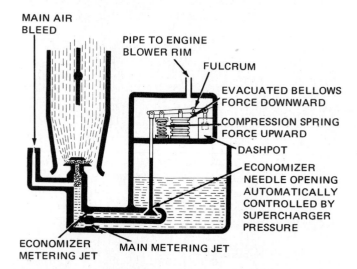

FIG. 4-30 Manifold-pressure-operated economizer.

denly opened, thus causing a temporary enrichment of the mixture and producing a smooth and positive acceleration of the engine.

It has been explained that the accelerating well is a space around the discharge nozzle and is connected by holes to the fuel passage leading to the discharge nozzle. The upper holes are located near the fuel level and are uncovered at the lowest pressure that will draw fuel from the main discharge nozzle; hence, they receive air during the entire time that the main discharge nozzle operates.

Very little throttle opening is required at idling speeds. When the throttle is suddenly opened, air is drawn in to fill the intake manifold and whichever cylinder is on the intake stroke. This sudden rush of air temporarily creates a high suction at the main discharge nozzle, brings into operation the main metering system, and draws additional fuel from the accelerating well. Because of the throttle opening, the engine speed increases and the main metering system continues to function.

The **accelerating pump,** illustrated in Fig. 4-31, is a sleeve-type piston pump operated by the throttle. The piston is mounted on a stationary hollow stem screwed into the body of the carburetor. The hollow stem opens into the main fuel passage leading to the discharge nozzle. Mounted over the stem and piston is a movable cylinder, or sleeve, which is connected by the pump shaft to the throttle linkage. When the throttle is closed, the cylinder is raised and the space within the cylinder fills with fuel through the clearance between the piston and the cylinder. If the throttle is quickly moved to the open position, the cylinder is forced down, as shown in Fig. 4-32, and the increased fuel pressure also forces the piston partway down along the stem. As the piston moves down, it opens the pump valve and permits the fuel to flow through the hollow stem into the main fuel passage. With the throttle fully open and the accelerating pump cylinder all the way down, the spring pushes the piston up and forces most of the fuel out of the cylinder. When the piston reaches its highest position, it closes the valve and no more fuel flows toward the main passage.

There are several types of accelerating pumps, but each serves the purpose of providing extra fuel during

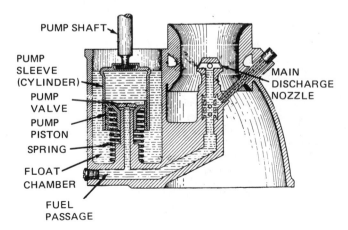

FIG. 4-31 **Movable-piston-type accelerating pump.**

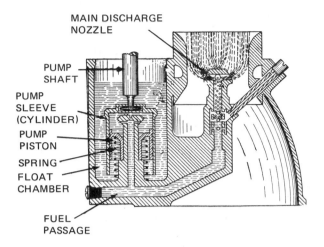

FIG. 4-32 **Accelerating pump in operation.**

rapid throttle opening and acceleration of the engine. When a throttle is moved slowly toward the OPEN position, the accelerating pump does not force extra fuel into the discharge system. This is because the spring in the pump holds the valve closed unless the fuel pressure is great enough to overcome the spring pressure. When the throttle is moved slowly, the trapped fuel seeps out through the clearance between the piston and the cylinder and the pressure does not build up enough to open the valve.

Mixture-Control System. At higher altitudes, the air has less pressure, density, and temperature. The weight of the air taken into an unsupercharged (naturally aspirated) engine decreases with the decrease in air density, and the power is reduced in approximately the same proportion. Since the quantity of oxygen taken into the engine decreases, the fuel-air mixture becomes too rich for normal operations. The mixture proportion delivered by the carburetor becomes richer at a rate inversely proportional to the square root of the increase in density of the air.

It must be remembered that the density of the air changes with temperature and pressure. If the pressure remains constant, the density of the air will vary according to temperature, increasing as the temperature drops. This will cause a leaning of the fuel-air mixture in the carburetor because the denser air contains more oxygen. The change in air pressure due to altitude is considerably more of a problem than the change in density due to temperature changes. At 18,000 ft [5486.4 m] altitude, the air pressure is approximately one-half the pressure at sea level. Hence, in order to provide a correct mixture, the fuel flow must be reduced to almost one-half what it would be at sea level. The adjustment of fuel flow to compensate for changes in air pressure and temperature is a principal function of the mixture control.

Briefly, a **mixture-control system** can be described as a mechanism or device by means of which the richness of the mixture entering the engine during flight can be controlled to a reasonable extent. This control should exist through all normal altitudes of operation.

The functions of the mixture-control system are (1) to prevent the mixture from becoming too rich at high

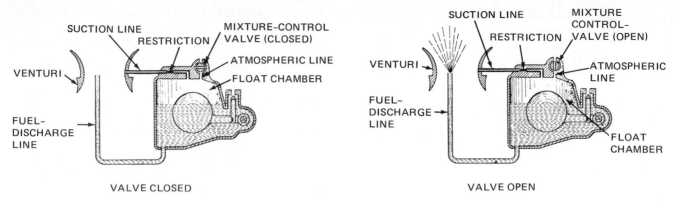

FIG. 4-33   Back-suction-type mixture control.

altitudes and (2) to economize on fuel during engine operation in the low-power range where cylinder temperature will not become excessive with the use of the leaner mixture.

Mixture-control systems may be classified according to their principles of operation as (1) the **back-suction** type, which reduces the effective suction on the metering system; (2) the **needle** type, which restricts the flow of fuel through the metering system; and (3) the **air-port** type, which allows additional air to enter the carburetor between the main discharge nozzle and the throttle valve. Figure 4-33 shows two views of a back-suction mixture-control system. The left view shows the mixture-control valve in the closed position. This cuts off the atmospheric pressure from the space above the fuel in the fuel chamber. Since the float chamber is connected to the low-pressure area in the venturi of the carburetor, the pressure above the fuel in the float chamber will be reduced until fuel is no longer delivered from the discharge nozzle. This acts as an **idle cutoff** and stops the engine. In some carburetors, the end of the back-suction tube is located where the pressure is somewhat higher than that at the nozzle, thus making it possible for the mixture-control valve to be completely closed without stopping the flow of fuel. The fuel flow is varied by adjusting the opening of the mixture-control valve. To lean the mixture, the valve is moved toward the CLOSED position, and to enrich the mixture, the valve is moved toward the OPEN position. The right-hand drawing in Fig. 4-33 shows the valve in the FULL RICH position.

In order to reduce the sensitivity of the back-suction mixture control, a disk-type valve is sometimes used. This valve is constructed so that a portion of the valve opening can be closed rapidly at first and the balance of the opening can be closed gradually. The disk-type valve is shown in Fig. 4-34. This assembly is called an **altitude-control-valve disk and plate.** The arrangement of the mixture control for an NA-S3A1 carburetor is shown in Fig. 4-35.

A needle-type mixture control is shown in Fig. 4-36. In this control, the needle is used to restrict the fuel passage through the main metering jet. When the mixture control is in the FULL RICH position, the needle is in the fully raised position and the fuel is accurately measured by the main metering jet. The needle valve is lowered into the needle-valve seat to lean the mixture, thus reducing the supply of fuel to the main discharge nozzle. Even though the needle valve is completely closed, a small bypass hole from the float chamber to the fuel passage allows some fuel to flow; hence, the size of this bypass hole determines the control range.

The air-port type of mixture control, illustrated in Fig. 4-37 has an air passage leading from the region between the venturi tube and the throttle valve to atmospheric pressure. In the air passage is a butterfly valve which is manually controlled by the pilot in the cockpit. It is apparent that when the pilot opens the butterfly valve in the air passage, air which has not been mixed with fuel will be injected into the fuel-air mixture. At the same time, the suction in the intake manifold will be reduced, thereby reducing the velocity of the air coming through the venturi tube. This will

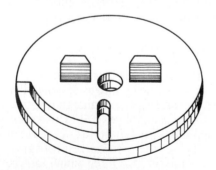

FIG. 4-34   Disk-type mixture control.

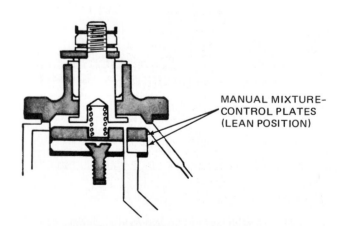

FIG. 4-35   Mixture control for NA-S3A1 carburetor.

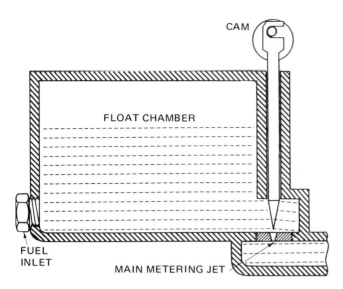

FIG. 4-36 Needle-type mixture control.

further reduce the amount of fuel being drawn into the intake manifold.

Idle Cutoff

The term *idle cutoff* has been mentioned previously; it describes the position of certain mixture controls in which the control is enabled to stop the flow of fuel into the intake airstream. Some float-type carburetors and the majority of pressure-type carburetors incorporate the IDLE CUTOFF position in the mixture-control system.

Essentially, the idle cutoff system stops the flow of fuel from the discharge nozzle and is hence used to stop the engine. This provides an important safety factor, because it eliminates the combustible mixture in the intake manifold and prevents the engine from firing as a result of a hot spot in one or more of the cylinders. In some cases, engines which have been stopped by turning off the ignition switch have kicked over after stopping, thus creating a hazard to someone who may move the propeller. The engine ignition switch is turned off *after* the engine is stopped by means of the idle cutoff. This procedure also eliminates the possibility of unburned fuel entering the cylinder and washing the oil film from the cylinder walls.

In general, when an aircraft carburetor or fuel control unit is equipped with the idle cutoff on the mixture control, the engine may be started with the mixture control in the IDLE CUTOFF position. The fuel for starting is supplied through the priming system. This procedure aids in preventing backfiring when starting because there is no fuel in the air between the carburetor and the cylinder.

When starting an engine with this procedure, the mixture control is placed in the FULL RICH position as soon as the engine fires and begins to run. In any event, the instructions in the operator's manual should be followed.

Automatic Mixture Control

Originally, the mixture control was manually operated on all airplane engines, but the more recent aircraft carburetors are often equipped with a device for automatically controlling the mixture as altitude changes. **Automatic mixture-control systems** may be operated on the back-suction principle and the needle-valve principle or by throttling the air intake to the carburetor. In the last type of automatic mixture control, the control regulates power output within certain limits in addition to exercising its function as a mixture control.

In automatic mixture-control systems operating on the back-suction and needle-valve principles, the control may be directly operated by the expansion and contraction of a pressure-sensitive evacuated bellows through a system of mechanical linkage. This is the simplest form of automatic mixture control and is generally found to be accurate, reliable, and easy to maintain. Some mixture-control valves, such as the one illustrated in Fig. 4-38 are operated by bellows vented to the atmosphere; hence, the fuel flow is proportional to the atmospheric pressure. Fig. 4-39 shows the bellows type of mixture-control valve installed on a carburetor as a back-suction control device. As atmospheric pressure decreases, the bellows will expand and begin to close the opening into the fuel chamber. This

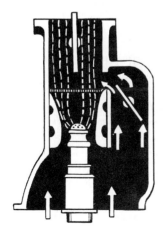

FIG. 4-37 Air-port-type mixture control.

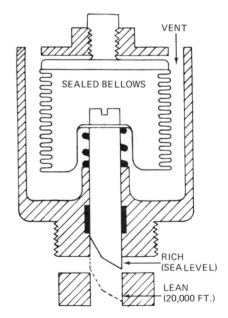

FIG. 4-38 Automatic mixture-control mechanism.

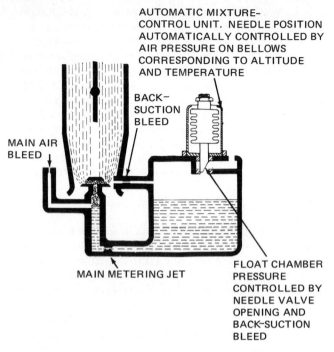

AUTOMATIC MIXTURE-
CONTROL UNIT. NEEDLE POSITION
AUTOMATICALLY CONTROLLED BY
AIR PRESSURE ON BELLOWS
CORRESPONDING TO ALTITUDE
AND TEMPERATURE

BACK-
SUCTION
BLEED

MAIN AIR
BLEED

MAIN METERING JET

FLOAT CHAMBER
PRESSURE
CONTROLLED BY
NEEDLE VALVE
OPENING AND
BACK-SUCTION
BLEED

FIG. 4-39 Automatic mixture control in operation.

will cause a reduction of pressure in the chamber, resulting in a decreased flow of fuel from the discharge nozzle. In some systems having external superchargers (not illustrated here), both the fuel chamber and the bellows may be vented to the carburetor intake to obtain the correct mixtures of fuel and air.

Automatic controls often have more than one setting in order to obtain the correct mixtures for cruising and high-speed operation. In addition to the automatic con-

trol feature, there is usually a provision for manual control if the automatic control fails.

When the engine is equipped with a fixed-pitch propeller which allows the engine speed to change as the mixture changes, a manually operated mixture control can be adjusted by observing the change in engine rpm as the control is moved. Obviously, this will not succeed with a constant-speed propeller.

If a constant-speed propeller cannot be locked into fixed-pitch position, and if the extreme pitch positions cause engine speeds outside the normal operating range in flight, it is necessary to have an instrument of some type to indicate fuel-air ratio or power output.

If the propeller can be locked in a fixed-pitch position, and if this does not lead to engine speeds outside the normal flight operating range, the following expressions may be employed to describe the manual adjustments of the mixture control.

Full rich The mixture-control setting in the position for maximum fuel flow.

Rich best power The mixture-control setting which, at a given throttle setting, permits maximum engine rpm with the mixture control as far toward full rich as possible without reducing rpm.

Lean best power The mixture-control setting which, at a given throttle setting, permits maximum engine rpm with the mixture control as far toward lean as possible without reducing rpm.

Downdraft Carburetors

So far we have principally dealt with **updraft** carburetors. **Updraft** means that the air through the carburetor is flowing upward. A **downdraft carburetor** (Fig. 4-40) takes air from above an engine and causes it to

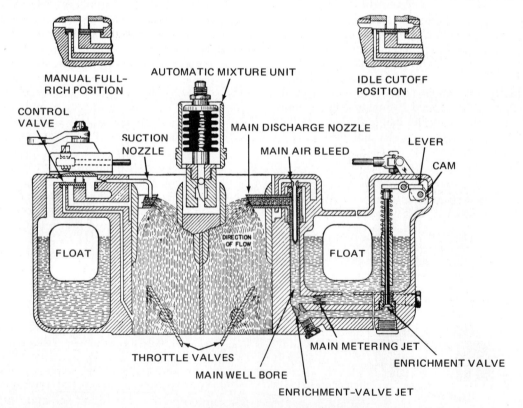

MANUAL FULL-
RICH POSITION

AUTOMATIC MIXTURE UNIT

IDLE CUTOFF
POSITION

CONTROL
VALVE

SUCTION
NOZZLE

MAIN DISCHARGE NOZZLE

MAIN AIR BLEED

LEVER

CAM

FLOAT

DIRECTION
OF FLOW

FLOAT

THROTTLE VALVES

MAIN WELL BORE

ENRICHMENT-VALVE JET

MAIN METERING JET

ENRICHMENT VALVE

FIG. 4-40 Downdraft carburetor.

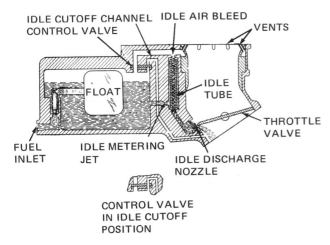

IDLE CUTOFF CHANNEL CONTROL VALVE
IDLE AIR BLEED
VENTS
FLOAT
IDLE TUBE
THROTTLE VALVE
FUEL INLET
IDLE METERING JET
IDLE DISCHARGE NOZZLE
CONTROL VALVE IN IDLE CUTOFF POSITION

FIG. 4-41   Downdraft carburetor, showing its idling system.

flow down through the carburetor. Those who favor downdraft carburetors claim that this type reduces fire hazard, provides a better distribution of mixture to the cylinders of an upright engine, and has less tendency to pick up sand and dirt from the ground.

Downdraft carburetors are very similar in function and systems to the updraft types. Figure 4-40 illustrates one of the several types of downdraft carburetors used for aircraft. This particular model has two float chambers, two throttle valves, and an automatic mixture-control unit.

Figure 4-41 illustrates a portion of a downdraft carburetor and its idling system, emphasizing the path of the fuel leaving the float chamber and the position of the idle air bleed. When the engine is not operating, this air bleed, in addition to its other functions, prevents the siphoning of fuel. An average intake pressure to the mixture-control system is supplied by the series of vents at the entrance to the venturi.

Model Designation

All Bendix-Stromberg float-type aircraft carburetors carry the general model designation NA followed by a hyphen; the next letter indicates the type, as shown in Table 4-1.

The final numeral indicates the nominal rated size of the carburetor, the size starting from 1 in [2.54 cm], which is no. 1, and increasing in $\frac{1}{4}$-in [6.35-mm] steps. For example, a 2-in [5.08 cm] carburetor is no. 5. The actual diameter of the carburetor barrel opening is

TABLE 4-1

Type letter	Type description
S and R	Single barrel
D	Double barrel, float chamber to rear (obsolescent)
U	Double barrel, float chamber between barrels
Y	Double barrel, double chamber fore and aft of barrels
T	Triple barrel, double float chamber fore and aft of barrels
F	Four barrel, two separate float chambers

$\frac{3}{16}$ in [4.76 mm] greater than the nominal rated size in accordance with the standards of the Society of Automotive Engineers. A final letter is used to designate various models of a given type. This system of model designation applies to inverted or downdraft as well as to updraft carburetors. The model designation and serial number are found on an aluminum tag riveted to the carburetor. There are so many Bendix-Stromberg carburetors that it is necessary to consult the publications of the manufacturer to learn the details of the designation system, but the explanation above is ample for ordinary purposes.

Typical Float-Type Carburetors

A comparatively simple float-type carburetor is illustrated in the drawing of Fig. 4-42. This is the Bendix-Stromberg NA-S3A1 carburetor used on a number of small aircraft engines. The NA-S3A1 carburetor is a single-barrel updraft model with a single hinge-type float, a main metering system of the plain tube type with an air bleed to the main discharge nozzle, an idling system, and a back-suction-type mixture control, manually operated. An external view of the carburetor is shown in Fig. 4-43.

A careful examination of the drawing of Fig. 4-42 will show how the principles explained in previous sections are utilized in this carburetor to control fuel flow and to provide a suitable mixture for engine operation.

Another carburetor commonly used for light aircraft engines is the Marvel-Schebler MA-3, shown in Fig. 4-44. This carburetor is somewhat more complex than the Bendix-Stromberg NA-S3A1 mentioned previously. The MA-3 carburetor has a double float assembly hinged to the upper part of the carburetor. This upper part is called the **throttle body** because it contains the throttle assembly. The fuel inlet, the float needle valve, and two venturis are also contained in the throttle body.

The carburetor body-and-bowl assembly contains a crescent-shaped fuel chamber surrounding the main air passage. The main fuel discharge nozzle is installed at an angle in the air passage with the lower end leading into the main fuel passage. An accelerating pump is incorporated in one side of the body-and-bowl assembly. The pump receives fuel from the fuel chamber and discharges accelerating fuel through a special accelerating pump discharge tube into the carburetor bore adjacent to the main discharge nozzle. The carburetor also includes an altitude mixture-control unit.

In addition to the MA-3 carburetor, the Marvel-Schebler/Tillotson Division of the Borg-Warner Company manufactures the MA3-SPA, MA4-SPA, MA4-5, MA4-5AA, MA-5, MA6-AA, and HA6 carburetors. These models are essentially the same as the MA-3 in design and operation, but they are designed for larger engines. The HA6 carburetor employs the same float and control mechanism; however, the airflow is horizontal rather than updraft. The essential features of the carburetors are the same, but size and minor variations account for the different model numbers.

The HA6 carburetor, having the horizontal airflow, is mounted on the rear of the oil sump of the engine for which it is designed. The air-fuel mixture then passes

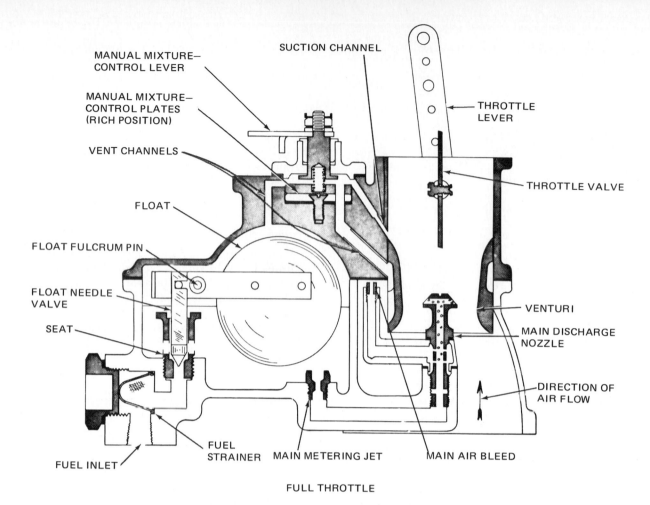

MANUAL MIXTURE—CONTROL LEVER

MANUAL MIXTURE—CONTROL PLATES (RICH POSITION)

VENT CHANNELS

FLOAT

FLOAT FULCRUM PIN

FLOAT NEEDLE VALVE

SEAT

FUEL INLET

FUEL STRAINER

MAIN METERING JET

MAIN AIR BLEED

SUCTION CHANNEL

THROTTLE LEVER

THROTTLE VALVE

VENTURI

MAIN DISCHARGE NOZZLE

DIRECTION OF AIR FLOW

FULL THROTTLE

FIG. 4-42   Drawing of Bendix-Stromberg NA-S3A1 carburetor.

through the oil sump where it picks up heat from the oil and thus provides better vaporization of the fuel.

A simplified drawing of the arrangement and operation of an MA-type carburetor is given in Fig. 4-45.

The principal features of a more complex carburetor

FIG. 4-43   Photograph of Bendix-Stromberg NA-S3A1 carburetor.

FIG. 4-44   Marvel-Schebler MA-3 carburetor.

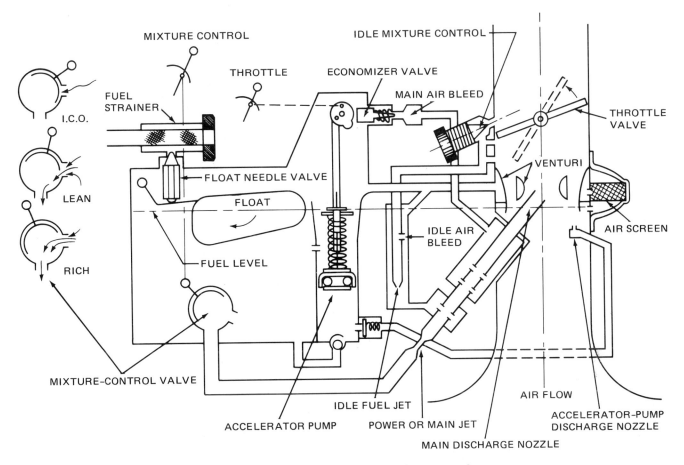

FIG. 4-45 Simplified drawing of Marvel-Schebler carburetor.

are illustrated in Fig. 4-46. This is a Bendix-Stromberg NAR-type carburetor, which includes all the systems explained previously. The operation of these systems can be easily understood through a careful study of the drawing. Observe particularly the float and needle valve, the needle-type mixture control, the economizer system, the accelerating pump, the idle system, the main metering system, and the air bleeds. The names of all the principal parts are included in the illustration, and these should be memorized by the student.

Disadvantages of Float-Type Carburetors

Float-type carburetors have been improved steadily by their manufacturers, but they have two important disadvantages or limitations: (1) The fuel-flow disturbances in maneuvers may interfere with the functions of the float mechanism, resulting in erratic fuel delivery, sometimes causing engine failure, and (2) when icing conditions are present, the discharge of fuel into the airstream ahead of the throttle causes a drop of temperature and a resulting formation of ice at the throttle valve.

● CARBURETOR ICING

Water Vapor in Air

In addition to gases, the air always contains some water vapor, but there is an upper limit to the amount of water vapor (as an invisible gas) that can be con-

tained in air at a given temperature. The capacity of air to hold water increases with the temperature. When air contains the maximum possible amount of moisture at a given temperature in a given space, the pressure exerted by the water vapor is also at a maximum and the space is then said to be **saturated.**

Humidity

Humidity, in simple terms, is moisture or dampness. The **relative humidity** of air is the ratio of the amount of moisture which the air does have to what it could contain, usually expressed as percent relative humidity. For example, if we have saturated air at 20°F [−6.67°C] and the temperature is increased to 40°F [4.44°C], the relative humidity will drop to 43 percent if the barometric pressure has remained unchanged. If this same air is heated further without removing or adding moisture, its capacity for holding water vapor will increase and its relative humidity will be less.

Lowering the temperature of air reduces its capacity to hold moisture. For example, if air at 80°F [26.67°C] has a relative humidity of 49 percent and it is suddenly cooled to 62°F [16.67°C], it will be found that the relative humidity is then 100 percent because cooling the air has increased its relative humidity.

In the case just mentioned, a relative humidity of 100 percent means that the air is saturated; that is, it contains all the moisture that it can hold. If it should be cooled still more and its moisture capacity decreases, some of the water vapor will condense. The

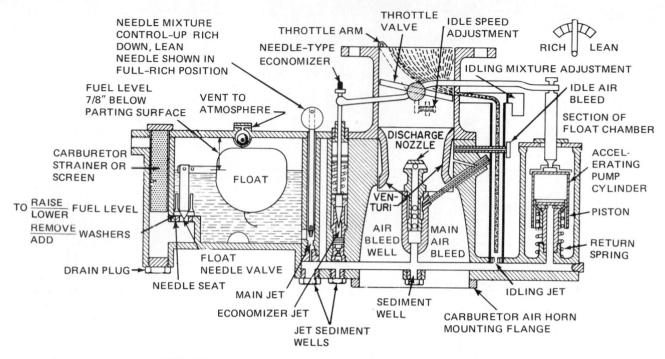

FIG. 4-46 Drawing of Bendix-Stromberg series NAR carburetor.

temperature at which the moisture in the air begins to condense is called the **dew point.**

Vaporization

The addition of heat changes a solid into a liquid, and it can change a liquid into a gas or vapor. The process of converting a liquid into a vapor is called **vaporization.** As the liquid is heated, the more rapidly moving molecules escape from the surface in a process called **evaporation.** Thus, when a pan of water is put on a hot stove, bubbles of water vapor begin to form at the bottom of the pan and rise through the cooler water above them and then collapse, causing the "singing" noise associated with boiling liquids. When all the water in the pan is hot enough for the bubbles to reach the surface easily, vaporization takes place throughout the water, accompanied by a violent disturbance. This vaporization is commonly called **boiling.**

Latent Heat of Vaporization

When a liquid evaporates, it uses heat. If one gram of water is raised in temperature from 0°C (32°F) to 100°C (212°F), 100 calories are required. (A **calorie** is defined as the quantity of heat required to raise the temperature of one gram of water from 14.5 to 15.5°C.)

When more heat is applied, the liquid water at 100°C is changed to water vapor at 100°C. This evaporation requires 539 calories (cal). The temperature of the water does not rise during this process of changing from a liquid to a gaseous state; hence, it is evident that 539 cal are required for the change of state. The reverse is also true; that is, when 1 g of water vapor is condensed to liquid water, 539 cal are given off. This energy, (539 cal) required for the change of state from liquid to vapor or from vapor to liquid is called the **latent heat of vaporization or condensation.**

Laws of Evaporation

There are six **laws of evaporation** which have a bearing on the formation of ice in carburetors:

1. The rate of evaporation increases as the temperature increases. Heat increases the rate at which molecules move; hence, hot water evaporates faster than cold water.

2. The rate of evaporation increases with an increase of the surface area of the liquid. More molecules can escape in a given time from a large surface than from a small surface; hence, water in a big pan will evaporate faster than it will in a small vase.

3. The rate of evaporation increases when the atmospheric pressure decreases. As the weight of the air above a body of water decreases, the escaping water molecules encounter less resistance. In other words, evaporation is faster under low pressure than it is under high pressure. For example, it is possible to freeze water by the cooling effect of evaporation if a stream of dry air is passed over water in a partial vacuum.

4. The rate of evaporation varies with the nature of the exposed liquid. For example, alcohol evaporates faster than water.

5. The rate of evaporation of water is decreased when the humidity of the air is increased. The escaping molecules of water can get away more easily if there are only a few molecules of water already in the air above the water. Conversely, if the humidity is high, the escaping molecules encounter more resistance to their departure from the water.

6. The rate of evaporation increases with the rate of change of the air in contact with the surface of the liquid. Wet clothes dry faster on a windy day than on a day when the air is calm.

The Cooling Effect of Evaporation

Every gram of water that evaporates from the skin takes heat from the skin and the body is cooled. That is the reason for human perspiration. If a person stands in a breeze, the perspiration evaporates more rapidly and the body is cooled faster. If the relative humidity is great, a person suffers from the heat on a hot day because the perspiration does not evaporate fast enough to have a noticeable cooling effect. Even the commercial production of ice depends upon the cooling effect of evaporation.

Carburetor Ice Formation

When fuel is discharged into the low-pressure area in the carburetor venturi, it evaporates rapidly. This evaporation of the fuel cools the air, the walls, and the water vapor. If the humidity (moisture content) of the air is high and the metal of the carburetor is cooled below 32°F [0°C], ice forms and interferes with the operation of the engine. The fuel-air passages are clogged, the mixture flow is reduced, and the power output drops. Eventually, if the condition is not corrected, the drop in power output may cause engine failure. The formation of ice in the carburetor may be indicated by a gradual loss of engine speed, a loss of manifold pressure, or both, without change in the throttle position.

It is of extreme importance that a pilot recognize the symptoms of carburetor icing and the weather conditions which may be conducive to icing. The principal effects of icing are the loss of power (drop in manifold pressure without a change in throttle position), engine roughness, and backfiring. The backfiring is caused when the discharge nozzle is partially blocked, which causes a leaning of the mixture.

Standard safety procedures with respect to icing are (1) checking carburetor heater operation before take-off, (2) turning on the carburetor heater when reducing power for gliding or landing, and (3) using carburetor heat whenever icing conditions are believed to exist.

If the mixture temperature is slightly above freezing, there is little danger of carburetor icing. For this reason, a **mixture thermometer** (carburetor air temperature [CAT] gage) is sometimes installed between the carburetor and the intake valve. This instrument not only indicates low temperatures in the carburetor venturi, which would lead to icing conditions, but also indicates high temperatures when a carburetor air-intake heater is used in order to avoid excessively high temperatures, which would cause detonation, preignition, and loss of power.

In general, ice may form in a carburetor system by any one of three processes: (1) The cooling effect of the evaporation of the fuel after being introduced into the airstream may produce what is called **fuel ice** or **fuel evaporation ice;** (2) water in suspension in the atmosphere coming in contact with engine parts at a temperature below 32°F [0°C] may produce what is called **impact ice** or **atmospheric ice;** (3) the freezing of the condensed water vapor of the air at or near the throttle forms what is known as **throttle ice** or **expansion ice.** The classification of carburetor ice in three groups is purely arbitrary. Throttle ice is most likely to form when the throttle is partially closed, as when letting down for a landing. The air pressure is decreased and the velocity is increased as the air passes the throttle. These changes cause a rapid decrease in air temperature, with a resultant formation of ice on the throttle if the humidity is high and the temperature of the air is a few degrees above freezing. The pilot bothered by carburetor icing is not particularly interested in the name to give the ice—only in wanting to know how to get rid of it, avoid its repetition, and take measures on the ground to remedy the condition producing it.

To get rid of carburetor icing, the pilot can fly slower, fly at an altitude where the air is warmer, turn on the carburetor air heater, and do other things beyond the scope of this discussion. However, it is better to avoid icing conditions in the first place.

Some pilots erroneously believe that carburetor icing does not take place when the free-air temperature is above the freezing point. Ice can be formed when the inlet-air temperature is above the freezing point and when the relative humidity is below 100 percent. Water condensation takes place, and since heat is absorbed from the air-vapor mixture by the evaporating fuel, the mixture drops below 32°F [0°C], thus freezing the condensed water vapor. The most severe carburetor icing conditions may occur with the temperature between 50 and 60°F [10 and 15.56°C] with a humidity above 60 percent. Under these conditions, the moisture in the air is frozen and deposited in the carburetor, where the ice deposit continues to grow in size until it may lock the throttle valve or restrict the amount of air entering the system to a degree that will cause engine failure.

When the fuel is vaporized in the carburetor as the fuel sprays out of the nozzle, the temperature of the incoming air usually drops at least 30°F [−1.11°C] and may drop as much as 70°F [21.11°C], depending on several factors. As the throttle is opened, the temperature of the incoming air drops even further.

The variable-venturi carburetor and the pressure-injection carburetors are relatively free from carburetor icing troubles. But even in these carburetors, ice or snow already formed in the atmosphere may get into the carburetor induction system unless the powerplant is provided with an air intake or scoop located where it can receive air free from ice or snow.

Carburetor Air-Intake Heaters

The exhaust-type carburetor air-intake heater is essentially a jacket or tube through which the hot exhaust gases of the engines are passed to warm the air flowing over the heated surface before the air enters the carburetor system. The principal value of carburetor air heat is to eliminate or prevent carburetor icing. The amount of warm air entering the system can be controlled by an adjustable valve.

The **alternate air-inlet** heating system has a two-position valve and an air scoop. When the passage from the scoop is closed, warm air from the engine compartment is admitted to the carburetor system. When the passage from the scoop is open, cold air comes from the scoop. Since the heating of the air for the carburetor depends on the free-air temperature, the engine temperature, the cowl-flap position, and other variable factors, this system is not always very dependable.

In a third type of carburetor air-intake heater, the air is heated by the compression which occurs in the external supercharger, but it becomes so hot that it is passed through an **intercooler** to reduce its temperature before entering the carburetor. Shutters at the rear of the intercooler can be opened or closed to regulate the degree of cooling to which the air warmed by the supercharger is subjected. Air entering the engine at too high a temperature can lead to detonation.

Excessively High Carburetor Air Temperatures

At first thought, it seems foolish to heat the air first and then cool it when the purpose is to raise its temperature to avoid icing the carburetor. However, it has been stated before in this text that excessively high carburetor air temperatures are not wanted. Air expands when heated, and its density is reduced. Lowering the density of the air reduces the mass; that is, it cuts down the *weight* of the fuel-air charge in the engine cylinder, thus reducing volumetric efficiency. This results in a loss of power, because power depends upon the weight of fuel-air mixture burned in the engine. Note, also, that since the weight of the air is decreased while the fuel weight remains essentially unchanged, the mixture is enriched and power is decreased when carburetor heat is turned on.

Another danger inherent in a high fuel-air temperature is detonation. If the air temperature is such that further compression in the cylinder raises the temperature to the combustion level of the fuel, detonation will occur. As explained previously, detonation causes excessive cylinder-head temperatures and may lead to piston damage and engine failure. This is one of the reasons why it is necessary to employ an intercooler with a high-pressure supercharger, especially at low altitudes where the air is dense.

● INSPECTION AND OVERHAUL OF FLOAT-TYPE CARBURETORS

Inspection in Airplane

Remove the carburetor strainer and clean it frequently. Flush the strainer chamber with gasoline to remove any foreign matter or water. Inspect the fuel lines to make certain that they are tight and in good condition. Inspect the carburetor to be sure that all safety wires, cotter pins, etc., are in place and that all parts are tight. On those models having economizer or accelerating pumps, clean the operating mechanism frequently and put a small quantity of oil on the moving parts.

When inspecting the carburetor and associated parts, it is particularly important to examine the mounting flange closely for cracks or other damage. The mounting studs and safety devices should be checked carefully for security. If there is an air leak between the mounting surfaces or an air leak because of a crack, the fuel-air mixture may become so lean that the engine

will fail. A very small leak can cause overheating of the cylinders and power loss.

Disassembly

Great care must be taken when disassembling a float-type carburetor to make sure that parts are not damaged. The sequence described in the manufacturer's overhaul manual should be used if a manual is available. This sequence is designed to ensure the disassembly of the carburetor in such a way that parts still on the assembly will not interfere with parts to be removed. As parts are removed, they should be placed in a tray with compartments to keep the components of each assembly together. When this is done, there is much less likelihood of installing parts in the wrong position when the carburetor is reassembled.

The tools used in the disassembly of a carburetor should be of the proper type. Screwdrivers should have the blades properly ground to avoid slipping in screw slots and damaging the screw heads or gouging the aluminum body of the carburetor. Metering jets and other specially shaped parts within the carburetor should be removed with the tools designed for the operation. A screwdriver should not be used for prying, except where specific instructions are given that this should be done.

Cleaning

The cleaning of carburetor parts is usually described in the manufacturer's overhaul manual, but there are certain general principles which may be followed. The first step is to remove oil and grease by using a standard petroleum solvent, such as Stoddard Solvent (Federal Specification P-S-661 or the equivalent). The parts to be cleaned should be immersed in the solvent for 10 to 15 min, rinsed in the solvent, and then dried.

To remove carbon and gum from the carburetor parts, a suitable carbon remover should be employed. Carbon remover MIL-C-5546A or the equivalent may be used. The remover should be heated to about 140°F [60°C] and the parts immersed in it for 30 min. The parts should then be rinsed thoroughly in hot water (about 176°F [80°C]) and dried with clean, dry compressed air, with particular attention paid to internal passages and recesses.

Wiping cloths or rags should never be used to dry carburetor parts because of the lint which will be deposited on the parts. Small particles of lint can obstruct jets, jam close-fitting parts, and cause valves to leak.

If aluminum parts are not corroded but still have some deposits of carbon, these deposits may be removed with No. 600 wet-or-dry paper used with water. After this the parts should be rinsed with hot water and dried.

Aluminum parts that are corroded can be cleaned by immersion in an alkaline cleaner Formula T or an equivalent agent inhibited against attack on aluminum. The parts should be immersed for 10 to 15 min with the cleaner temperature at 190 to 212°F [88 to 100°C]. The cleaning solution can be made by mixing ingredients as follows:

Sodium phosphate dibasic	2 lb [0.907 kg]
Sodium metasilicate	1 lb [0.454 kg]

Soap (Fed. Spec. P-S-598)	0.8 lb [0.363 kg]
Sodium dichromate	0.2 lb [0.091 kg]
Water (near boiling temperature)	10 gal [37.85 L]

After immersion in the cleaning solution, the parts should be rinsed in hot water and then in cold water. Following the rinse, the parts should be immersed for 5 to 10 min in a chromic phosphoric acid solution consisting of 3.5 pints (pt) [1.85 L] of 75 percent phosphoric acid and 1.75 lb [0.794 kg] of chromic acid to 10 gal [37.85 L] of water. The temperature of the solution should be 181.4 to 199.4°F [83 to 93°C] when the parts are immersed. This process will remove corrosion, paint, and anodic coating.

Finally, it is necessary to rinse the parts thoroughly with cold water followed by hot water. Corroded areas can be smoothed with No. 600 wet-or-dry paper and water. This should be followed by another hot-water rinse, after which the parts must be dried carefully. Internal passages and recesses should be checked to see that they are clear and that no moisture is retained.

The foregoing process is one of several which may be used for cleaning carburetor parts. Cleaning solutions suitable for aluminum and steel may be obtained from any chemical manufacturer who specializes in industrial chemicals and cleaning solutions. In every case, the manufacturer's directions for use should be carefully followed. It is particularly important that the operator use caution to avoid injury, because many cleaning chemicals will burn the skin and may cause blindness if splashed into the eyes. Protective clothing, gloves, and goggles should always be worn when dangerous chemicals are used.

After the cast aluminum parts of a carburetor have been cleaned and stripped, it is essential that they be treated with some type of coating for protection and prevention of corrosion. It is common practice to use chemical coatings, such as the Alodyne 1200 conversion or Iridite Conversion. These coatings may be effectively applied by following the manufacturer's instructions carefully.

Inspection of Parts

Before assembly of the carburetor, all parts should be inspected for damage and wear. Inspections for a typical carburetor are as follows:

1. Check all parts for bends, breaks, cracks, or crossed threads.
2. Inspect the fuel strainer assembly for foreign matter or a broken screen.
3. Inspect the float needle and seat for excessive wear, dents, scratches, or pits.
4. Inspect the mixture-control plates for scoring or improper seating.
5. Inspect the float assembly for leaks by immersing in hot water. Bubbles will issue from a point of leakage.
6. Inspect the throttle shaft's end clearance and the play in the shaft bushings.

In addition to the foregoing inspections, certain assemblies must be checked for fits and clearances. Among these are the fulcrum bushing in the float, the fulcrum pin, the slot in the float needle, the pin in the float assembly, the bushing in the cover assembly, the mixture-control stem, and the throttle shaft and bushings. The limits for these assemblies are given in the Table of Tolerance Values in the manufacturer's overhaul manual.

The foregoing inspections are those specified for the Bendix-Stromberg NA-S3A1 carburetor. Other carburetors, such as the Marvel-Schebler MA-3, MA-4, and MA-6 series, will have additional inspections specified in the overhaul manual. In each case the manufacturer's instructions should be followed carefully.

Inspection of Metering Jets

The sizes of the metering jets in a carburetor are usually correct because these sizes are established by the manufacturer. Sometimes a jet may be changed or drilled to increase the size, so it is always well to check the sizes when the carburetor is overhauled. The correct sizes are given in the specifications of the manufacturer in the overhaul manual, and the size numbers are usually stamped on the jet. The number on the jet corresponds to a numbered drill shank; hence, it is possible to check the size of the jet by inserting the *shank* of a numbered drill into the jet as shown in Fig. 4-47. If the drill shank fits the jet without excessive play, the jet size is correct. The number of the jet should also be checked against the specifications in the overhaul manual to see that the correct jet is installed. Metering jets should also be examined closely to see that there are no scratches, burrs, or other obstructions in the jet passage because these will cause local turbulence, which interferes with normal fuel flow. If a metering jet is defective in any way, it should be replaced by a new one of the correct size.

Repair and Replacement

The repair and replacement of parts for a carburetor depend on the make and model being overhauled and

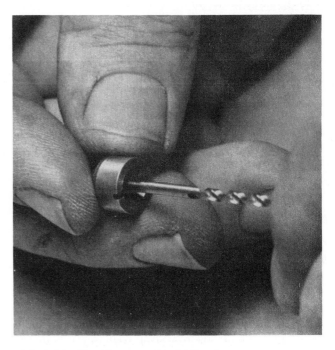

FIG. 4-47 Checking metering jet for size.

should be performed in accordance with the manufacturer's instructions. It is always proper to replace gaskets and fiber washers and any other part which shows substantial signs of wear or damage. When clearances and other dimensions are not within the specified limits, the parts involved must be repaired or replaced.

A carburetor float is usually made of formed brass sheet and can be checked for leaks by immersing it in hot water. The heat will cause the air and any fuel fumes in the float to expand, thus making a stream of bubbles emerge from the leak.

If the float is found to have leaks, the leaks should be marked with a pencil or other means which will not cause damage. A small hole may then be drilled in the float to permit the removal of any fuel which may have been trapped inside. After the hole is drilled, the fuel should be drained and the float then immersed in boiling water until all fuel fumes are evaporated from the inside. This will permit soldering of the leaks without the danger of explosion. As a further precaution, the float should never be soldered with an open flame. The small leaks should be soldered before the drilled hole is sealed. Care must be taken to apply only a minimum of solder to the float, because the weight must not be increased more than necessary. An increase in the weight of the float will cause an increase in the fuel level which, in turn, will increase fuel consumption.

After the float is repaired, it should be immersed in hot water in order to determine that all leaks have been sealed properly.

Checking the Float Level

As explained previously, the fuel in the float chamber of a carburetor must be maintained at a level which will establish the correct fuel flow from the main discharge nozzle when the carburetor is in operation. The fuel level in the discharge nozzle is usually between $\frac{3}{16}$ and $\frac{1}{8}$ in [4.76 and 3.18 mm] below the opening in the nozzle.

After the carburetor is partially assembled according to the manufacturer's instructions, the float level may be checked. In a Bendix-Stromberg carburetor where the float and needle seat is in the lower part of the carburetor, the float level may be tested as follows:

1. Mount the assembled main body in a suitable fixture so that it is level when checked with a small spirit level.

2. Connect a fuel supply line to the fuel inlet in the main body, and regulate the fuel pressure to the value given on the applicable specification sheet. This pressure is $\frac{1}{2}$ psi [3.45 kPa] for a NA-S3A1 carburetor used in a gravity-feed fuel system. When the fuel supply is turned on, the float chamber will begin to fill with fuel and the flow will continue until it is stopped by the float needle on its seat.

3. Using a depth gage, measure the distance from the parting surface of the main body to the level of the fuel in the float chamber approximately $\frac{1}{2}$ in [12.7 mm] from the side wall of the chamber, as shown in Fig. 4-48. If the measurement is taken adjacent to the side of the float chamber, a false reading will be obtained. The fuel level for the NA-S3A1 carburetor should be $\frac{13}{32} \pm \frac{1}{64}$ in [10.32 ± 0.397 mm] from the parting surface.

FLOAT

FIG. 4-48 Checking float level.

4. If the level of the float is not correct, remove the needle and seat and install a thicker washer under the seat to lower the level, or a thinner washer to raise the level. A change in washer thickness of $\frac{1}{64}$ in [0.397 mm] will change the level approximately $\frac{5}{64}$ in [1.98 mm] for a NA-S3A1 carburetor.

Two different test procedures are used to establish the correct float level and the float-valve operation for the Marvel-Schebler MA series carburetors. The first of these is carried out during assembly after the float and lever assembly is installed. The throttle body is placed in an upside-down position as shown in Fig. 4-49. The height of the lower surface of each float above the gasket and screen assembly is then measured. For the MA-3 and MA-4 carburetors, this distance should be $\frac{7}{32}$ in [5.56 mm]. For the MA-4-5 carburetor, the distance is $\frac{13}{64}$ in [5.16 mm]. When the throttle body is placed in the upside-down position, the float needle is bearing against the float valve and holding it in the closed position. This is the same position taken by the float when the carburetor is in the normal operating position and the float chamber is filled with fuel.

The method for testing the float-valve operation is illustrated in Fig. 4-50 and is performed after the carburetor is completely assembled. The procedure is as follows:

1. Connect the inlet fitting of the carburetor to a fuel pressure supply of 0.4 psi [2.76 kPa].

2. Remove the bowl drain plug, and connect a glass tube to the carburetor drain connection with a piece of rubber hose. The glass tubing should be positioned vertically beside the carburetor.

3. Allow the fuel pressure at 0.4 psi to remain for a period of at least 15 min, and then raise the fuel pressure to 6.0 psi [41.37 kPa]. (There will be a slight rise

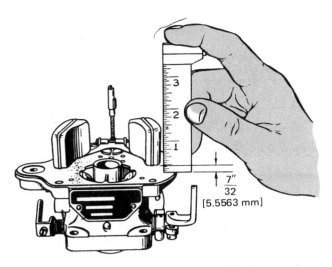

FIG. 4-49 Measuring float distance on MA-3 carburetor.

in the fuel level as the pressure is increased.) Allow the 6.0 psi pressure to remain for at least 5 min after the fuel level has stabilized.

4. If the fuel does not rise to the level of the parting surface of the castings or run out of the nozzle, which can be observed through the throttle bore, the float valve and seat are satisfactory. If fuel is observed running out the nozzle, the bowl and throttle body must be separated and the float valve and seat cleaned or replaced.

In Fig. 4-50 the fuel level, shown as DISTANCE "A," will automatically be correct if the float height is correct and the float valve does not leak.

The foregoing procedures are given as typical operations in the inspection and overhaul of float-type carburetors; however, it is not possible in this text to give complete, detailed overhaul operations, for specific float-type carburetors. When faced with the necessity of overhauling a particular carburetor, the technician should obtain the correct manufacturer's manual and all special bulletins pertaining to the carburetor and should check the FAA Engine Specification for the particular engine upon which the carburetor is to be used to see if any parts changes or modifications are called for. The overhaul procedure should then be carried out according to the applicable instructions.

Troubleshooting

The troubleshooting chart on page 102 provides some typical procedures for determining and correcting float-type carburetor malfunctions; however, it is not intended to cover all possible problems which may occur with a carburetor system. It must be understood that there are many types of carburetors and that numerous variations exist in their operational characteristics.

Adjusting Idle Speed and Idle Mixture

The correct idle speed and idle mixture are essential for the most efficient operation of an engine, particularly on the ground. The idle speed is established by the manufacturer at a level to keep the engine running smoothly, reduce overheating, and avoid spark-plug fouling. A typical idle speed is 600 ± 25 rpm; however, this will vary somewhat with different aircraft.

The idle speed for an engine with a float-type carburetor is adjusted by turning the screw which bears against the throttle stop. Thus, the idle speed is established by varying the degree of throttle opening when the throttle lever is completely retarded. Usually, turning the screw to the right will increase the idle speed.

To adjust idle mixture, the following steps should be taken:

1. Run the engine until it is operating at normal operating temperature.
2. Operate the engine at IDLE and adjust for the correct idle speed.
3. Turn the idle mixture adjustment toward LEAN until the engine begins to run rough.
4. Turn the mixture adjustment toward RICH until the engine is operating smoothly and the rpm has dropped slightly from its peak value.
5. Using the manual mixture control in the cockpit, move the control slightly toward LEAN. The rpm should

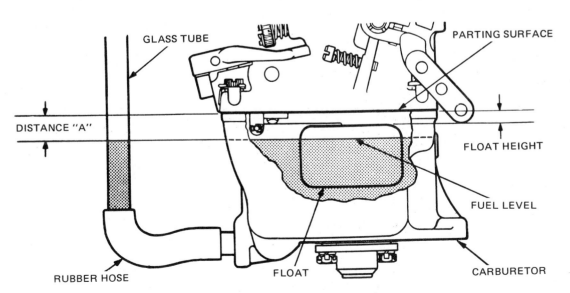

FIG. 4-50 Testing float-valve operation.

Trouble	Cause	Correction
Carburetor leaks when engine is stopped.	Float needle valve not seated properly because of dirt on seat.	Tap carburetor body with soft mallet while engine is running. Remove and clean carburetor. Check float level.
	Float needle valve worn.	Replace float needle valve.
Mixture too lean at idle.	Fuel pressure too low.	Adjust fuel pressure to correct level.
	Idle mixture control out of adjustment.	Adjust idle mixture control.
	Obstruction in idle metering jet.	Disassemble and clean carburetor.
	Air leak in the intake manifold.	Check intake manifold for tightness at all joints. Tighten assembly bolts.
Mixture too lean at cruising speed.	Air leak in the intake manifold.	Check intake manifold for tightness at all joints. Tighten assembly bolts.
	Automatic mixture control out of adjustment.	Adjust automatic mixture control.
	Float level too low.	Check and correct float level.
	Manual mixture control not set correctly.	Check setting of manual mixture control. Adjust linkage if necessary.
	Fuel strainer clogged.	Clean fuel strainer.
	Fuel pressure too low.	Adjust fuel-pump relief valve.
	Obstruction in fuel line.	Check fuel flow and clear any obstructions.
Mixture too lean at full-power setting.	Same causes as those for lean cruise.	Make corrections the same as those for lean cruise.
	Economizer not operating correctly.	Check economizer system for operation. Adjust or repair as required.
Mixture too rich at idle.	Fuel pressure too high.	Adjust fuel pressure to correct level.
	Idle mixture control out of adjustment.	Adjust idle mixture.
	Primer line open.	See that primer system is not feeding fuel to engine.
Mixture too rich at cruising speed.	Automatic mixture control out of adjustment.	Adjust automatic mixture control.
	Float level too high.	Adjust float level.
	Manual mixture control not set correctly.	Check setting of manual mixture control. Adjust linkage if necessary.
	Fuel pressure too high.	Adjust fuel pump relief valve for correct pressure.
	Economizer valve open.	Check economizer for correct operation. Quick acceleration may clear.
	Accelerating pump stuck open.	Quick acceleration of engine may remove foreign material from seat.
	Main air bleed clogged.	Disassemble carburetor and clean air bleed.
Poor acceleration. Engine backfires or misses when throttle is advanced.	Accelerating pump not operating properly.	Check accelerating pump linkage. Remove carburetor, disassemble, and repair accelerating pump.

increase slightly (about 20 rpm) before it begins to fall off and the engine starts to misfire. Returning the mixture control to FULL RICH should cause the engine operation to become smooth.

● REVIEW QUESTIONS

1. Into what two principal sections may the fuel system of an aircraft be divided?
2. Name the principal components of a basic fuel system.
3. What is the minimum flow rate for a gravity fuel system? A pressure fuel system?
4. Under what conditions is the fuel boost pump usually operated?
5. What are the requirements for fuel-control valves?
6. What factor determines the size of fuel lines?
7. Discuss the configuration and location of fuel lines installed in an aircraft.
8. What provision is made to remove dirt and water from a fuel tank?
9. What is the purpose of baffles in a fuel tank?
10. Why does an engine-driven fuel pump require a bypass system and a relief valve?
11. Explain *vapor lock* and list three primary causes.
12. Describe methods whereby the tendency toward

the accumulation of vapor in the fuel system may be reduced or eliminated.

13. What is a vapor separator?
14. How is the bending and routing of fuel lines related to vapor formation and accumulation?
15. Discuss various types of fuel strainers or filters and tell where they are located in the system.
16. Discuss the effectiveness of filters installed in carburetors and other fuel metering devices.
17. Give some precautions to be observed in connection with servicing or repairing fuel systems.
18. What is the purpose of carburetion or fuel metering?
19. Of what chemical elements are gasoline and other petroleum fuels composed?
20. What is *standard sea-level pressure*?
21. What are the other elements of a standard atmosphere?
22. Differentiate between absolute pressure and gage pressure.
23. Explain the operation of a venturi tube.
24. How is the venturi tube utilized in a carburetor?
25. What chemical action takes place when gasoline burns in an engine?
26. What fuel-air ratios are considered the best power range?
27. Express in decimals the approximate lean best power ratio for fuel and air in a reciprocating engine.
28. Explain the best economy mixture of fuel and air.
29. What is meant by *specific fuel consumption* (sfc) of an engine?
30. If a fuel-air mixture is excessively lean, what effect does this condition have on the rate of burning of the mixture?
31. How is the tendency toward detonation affected by the fuel-air ratio?
32. Why does an engine often backfire when the fuel-air mixture is too lean?
33. Define *flame propagation*.
34. How does the fuel-air ratio affect flame propagation?
35. What is the cause of kickback when starting an engine?
36. Why is a rich mixture used during takeoff?
37. What is meant by *density* of the air?
38. What atmospheric conditions affect air density?
39. Why does a pilot lean the fuel-air mixture when flying at high altitudes?
40. What precautions should a pilot consider when taking off on a hot, humid day?
41. If an airplane is flying at a constant altitude and moves from a cold area into a warmer area, what is the effect on the fuel-air mixture?
42. When carburetor heat is applied, what effect will it have on the density of the air entering the engine? How does this affect the fuel-air ratio?
43. What effect does the application of carburetor heat have on the engine power output?
44. How is the fuel discharge nozzle in a float-type carburetor located with respect to the fuel level in the float chamber?
45. Explain the use of an air bleed in a float-type carburetor.

46. What effect would a clogged main air bleed have on the fuel-air ratio?
47. What is the function of the main metering jet in a carburetor?
48. How and why is the fuel-air mixture sometimes heated before it enters the combustion chambers of an engine?
49. Describe a throttle valve and its operation.
50. What are the essential parts of a float-type carburetor?
51. Explain the importance of float level in a carburetor.
52. How is the float level usually adjusted?
53. Describe the main metering system in a carburetor.
54. When an engine is operating at full power, what is the approximate amount of suction (vacuum) in the intake manifold on the engine side of the throttle?
55. Describe the idling system of a typical float-type carburetor, including the idle air bleed and idle mixture.
56. Why is the fuel-air ratio enriched during the idle operation?
57. What is the idle cutoff?
58. Describe the operation of the accelerating well in a Bendix NA-S3A1 carburetor.
59. What is the purpose of the economizer valve in a carburetor?
60. What would be the effect on engine operation if the economizer valve stuck open?
61. What would happen if the economizer valve stuck in the closed position?
62. Describe the operation of the needle-type economizer valve.
63. Explain the operation of the piston-type accelerating system.
64. Why is an accelerating system needed?
65. During the operation of an aircraft engine, what would indicate that the accelerating system was not operating?
66. What are the principal functions of the mixture control system?
67. Explain the operation of the back-suction type of mixture control.
68. Why is it desirable to employ the idle cutoff for stopping an engine?
69. On an aircraft with a mixture control having an idle cutoff position, in what position should the mixture control be placed when starting the engine? Why?
70. Describe the operation of an automatic mixture control.
71. What would be the effect on engine operation if the automatic mixture control should stick in the sea-level position?
72. Why is it necessary to have a fixed-pitch propeller position when adjusting the mixture?
73. Differentiate between updraft and downdraft carburetors.
74. What is the diameter of the carburetor barrel opening in an NA-R9B carburetor?
75. Describe the float and needle-valve assembly of the Marvel MA-3 carburetor.

76. What are the disadvantages of a float-type carburetor?
77. Describe some of the conditions under which carburetor icing may occur.
78. What temperature change occurs when water vaporizes?
79. Define *latent heat of vaporization*.
80. Give six laws of vaporization.
81. What are the symptoms of carburetor icing during the operation of an aircraft engine?
82. What precaution is taken to prevent carburetor icing?
83. Describe fuel ice, impact ice, and throttle ice.
84. How is it possible for carburetor icing to take place when the free-air temperature is above the freezing point?
85. Describe methods for heating the carburetor intake air.
86. What are the effects of excessively high carburetor air temperature?
87. How is engine operation affected when there is an air leak at the carburetor mounting flange?
88. What items should be checked during the inspection of a carburetor when the engine is mounted in the airplane?
89. Describe the cleaning of carburetor parts.
90. What inspections should be made of the parts of a float-type carburetor?
91. What would be the effect of a worn throttle-shaft bearing or bushing on the operation of the engine?
92. How is the size of a metering jet checked?
93. Describe the repair of a leaking float.
94. Give the procedure for checking the float level of a Bendix NA-S3A1 carburetor.
95. How is the float level of a Marvel MA-3 carburetor checked?
96. If fuel drips from the carburetor when the airplane is not operating, what is the likely cause?
97. What is the effect on engine operation when the float level is too high?
98. What are the symptoms of an excessively rich mixture?
99. If the engine backfires or cuts out momentarily when the throttle is moved forward, what is the likely cause?

5 PRESSURE-INJECTION CARBURETORS

In Chap. 4 we examined the operation of float-type carburetors for the control of fuel to a piston engine. Originally, float-type carburetors were used for almost all types of piston engines, but today pressure-injection carburetors are used with many engines and continuous-flow fuel injection is used with many others. The purpose of this chapter is to explain the principles and operation of typical **pressure-injection** carburetors, also called **pressure discharge** carburetors.

● PRINCIPLES OF PRESSURE INJECTION

Introduction

The pressure-injection carburetor is a radical departure from float-type carburetor designs and presents an entirely different approach to the problem of aircraft engine fuel feed. It employs the simple method of metering the fuel through fixed orifices according to air venturi suction and air-impact pressure—combined with the new function of atomizing the fuel spray under positive pump pressure.

Advantages

Among the more important advantages obtained from the use of the pressure-injection carburetor are the following:

1. Ice does not form from the vaporization of fuel in the throttle body of the carburetor.
2. Since the system is entirely closed, it operates normally during all types of flight maneuvers. Gravity and inertia have very little effect.
3. The fuel is accurately and automatically metered at all engine speeds and loads regardless of changes in altitude, propeller pitch, or throttle position.
4. Atomizing the fuel under pressure results in smoothness, flexibility, and economy of powerplant operation.
5. The settings are simple and uniform.
6. Protection against fuel boiling and vapor lock is provided.

Principles of Operation

The basic principle of the pressure-injection carburetor can be explained briefly by stating that mass airflow is utilized to regulate the pressure of fuel to a metering system which governs the flow of fuel according to the pressure applied. The carburetor therefore increases fuel flow in proportion to mass airflow and maintains a correct fuel-air ratio in accordance with the throttle and mixture settings of the carburetor.

The fundamental operation of a pressure-injection carburetor may be shown with the simplified diagram of Fig. 5-1. Shown in this diagram are four of the main parts of a pressure carburetor system: (1) the **throttle unit,** (2) the **regulator unit,** (3) the **fuel-control unit,** and (4) the **discharge nozzle.**

When the carburetor is operating, the air flows through the throttle unit in an amount governed by the opening of the throttle. At the entrance to the air passage are impact tubes which develop a pressure proportional to the velocity of the incoming air. This pressure is applied to chamber A in the regulator unit. As the air flows through the venturi, a reduced pressure is developed in accordance with the velocity of the airflow. This reduced pressure is applied to chamber B in the regulator unit. It is readily seen that the comparatively high pressure in chamber A and the low pressure in chamber B will create a differential of pressure across the diaphragm between the two chambers. The force of this pressure differential is called the **air metering force,** and as it increases, it opens the poppet valve and allows fuel under pressure from the fuel pump to flow into chamber D. This unmetered fuel exerts force upon the diaphragm between chamber D and chamber C, and thus tends to close the poppet valve. The fuel flows through one or more metering jets in the fuel-control unit and thence to the discharge nozzle. Chamber C of the regulator unit is connected to the output of the fuel-control unit to provide **metered fuel** pressure to act against the diaphragm between chambers C and D. Thus, unmetered fuel pressure acts against the D side of the diaphragm and metered fuel pressure acts against the C side. The fuel pressure differential produces a force called the **fuel metering force.**

When the throttle opening is increased, the airflow through the carburetor is increased and the pressure in the venturi is decreased. Thus, the pressure in chamber B is lowered, the impact pressure to chamber A is increased, and the diaphragm between chambers A and B moves to the right because of the differential of pressure (air metering force). This movement opens the poppet valve and allows more fuel to flow into chamber D. This increases the pressure in chamber D and tends to move the diaphragm and the poppet valve to the left against the air metering force; however, this movement is modified by the pressure of metered fuel in chamber C. The pressure differential between chambers C and D (fuel metering force) is balanced against the air metering force at all times when the engine is operating

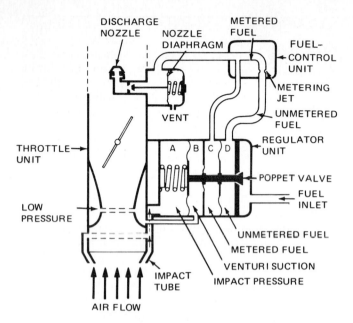

FIG. 5-1 Simplified diagram of a pressure-injection carburetor.

at a given setting. Chamber C pressure is established at approximately 5 psi [34.48 kPa] by the spring-loaded, diaphragm-operated main discharge-nozzle valve. This valve prevents leakage from the nozzle when the engine is not operating.

When the throttle opening is reduced, air metering force decreases and the fuel metering force starts to close the poppet valve. This causes a decrease in fuel metering force until it is again balanced by the air metering force.

Note particularly that an increase in airflow through the carburetor results in an increase in the fuel metering pressure across the metering jets in the fuel-control section, and this increase causes a greater flow of fuel to the discharge nozzle. A decrease in airflow has the converse effect.

It must be understood that the regulator section of a pressure-injection carburetor cannot regulate fuel pressure accurately at idling speeds because the venturi suction and the air-impact pressure are not effective at low values. Therefore, it is necessary to provide **idling valves** which are operated by the throttle linkage in order to meter fuel in the idling range, and which have springs in the pressure regulators in order to keep the poppet valve from closing completely. These idling-control valves and springs are discussed in the sections dealing with particular carburetors.

To summarize the operation of the regulator section of a pressure-injection carburetor, the functions of the regulator's air and fuel chambers may be described as follows:

Chamber A. This chamber receives ambient or impact air pressure, depending upon the type of carburetor. This pressure tends to open the main fuel poppet valve to allow fuel to flow into the regulator section.

Chamber B. Chamber B receives venturi suction, either from the main venturi of the carburetor or from a boost venturi. This helps to move the diaphragm between chambers A and B in a direction to open the main fuel poppet valve. The force developed by chamber A and the force developed by chamber B work together to open the poppet valve. If the diaphragm between the two chambers became ruptured to an appreciable degree, the poppet valve would close sufficiently to stop or greatly reduce the power of the engine.

Chamber C. Metered fuel enters chamber C at a pressure slightly lower (approximately $\frac{1}{4}$ psi [1.72 kPa]) than the pressure in chamber D. This prevents the higher fuel pressure in chamber D from closing the poppet valve and maintains a small pressure differential across the diaphragm between chambers C and D.

Chamber D. Unmetered fuel from the system's fuel pump enters chamber D. The pressure of this fuel acts against the fuel diaphragm and would close the poppet valve except for the balancing forces provided by other chambers.

Types of Pressure-Injection Carburetors

A number of different types of pressure-injection carburetors have been manufactured, the majority having been developed by the Energy Controls Division of the Bendix Corporation. Models have been manufactured for almost all sizes of reciprocating engines.

The carburetor for small engines has a single venturi in a single barrel and is designated by the letters PS, meaning, a pressure-type single-barrel carburetor.

A pressure carburetor for larger engines has a double barrel with boost venturis and is designated by the letters PD (pressure double). The triple-barrel carburetor is designated by the letters PT, and the rectangular-barrel carburetor is designated by PR.

The numbers following the letter designation generally indicate the bore size of a carburetor or injection unit. Nominal bore sizes are designated in increments of $\frac{1}{4}$ in [6.35 mm], beginning with no. 1 as a 1-in [2.54-cm] nominal bore size. The actual bore diameter is $\frac{3}{16}$ in [4.7625 mm] larger than the nominal size. For example, the diameter of the no. 10 is 9 times $\frac{1}{4}$ in larger than 1 in, for a nominal diameter of 3.25 in [8.26 cm]. The actual bore diameter is $\frac{3}{16}$ in larger than the nominal diameter, or 3.4375 in [8.7313 cm].

Identification Plates

Each new carburetor is identified with a **specification plate** or **production identification plate** attached to the main body. The plate identifies the manufacturer, unit serial number, model designation, and parts list and issue numbers to which the unit was manufactured.

Carburetors which are overhauled or modified are identified by means of a **service replacement identification plate**. A production identification plate and a service replacement identification plate are shown in Fig. 5-2.

Operating Units

The corresponding operating units comprising the various pressure-injection carburetors are similar in many respects. Each carburetor or injection system

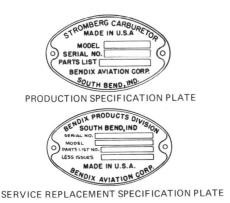

PRODUCTION SPECIFICATION PLATE

SERVICE REPLACEMENT SPECIFICATION PLATE

FIG. 5-2 Identification plates.

includes four to six principal units serving similar functions in the different carburetors.

The **throttle unit,** or **throttle body unit,** controls and measures the mass airflow through the carburetor to the engine. It contains the throttle and the venturi or venturis. By sensing the impact force and the velocity of the air, this unit provides the regulating forces used to influence the fuel pressure and the fuel-air mixture ratios.

The **regulator unit,** or **pressure-regulator unit,** automatically regulates the fuel pressure applied to the metering elements in the fuel-control unit. This regulation is accomplished by means of diaphragms responding to mass airflow and fuel flow, as explained previously.

The **fuel-control unit** receives fuel under varying pressures from the regulator unit and meters it to the discharge nozzle. The fuel-control unit contains one or more metering jets and may also include the manual mixture control. If a manual mixture control is included in the unit, the unit will usually contain jets (metering orifices) for auto-lean, auto-rich, and sometimes full-rich operating conditions. It will also include a means whereby all fuel flow can be stopped to provide idle cutoff for stopping the engine. The fuel-control section often contains the idle valve, by which the engine is controlled in the idling range.

The **discharge nozzle** for a pressure-injection carburetor may be somewhat complex, or it may be merely a fuel passage leading into the airstream on the engine side of the throttle, with openings through which the fuel is sprayed into the airstream. Discharge nozzles are usually equipped with a valve to prevent fuel flow at very low pressures. Details of discharge nozzles will be explained for the specific carburetors described in this text.

An **accelerating pump** or similar device is usually required for a pressure-injection carburetor, even as it is for a float-type carburetor. Some models of pressure carburetors employ diaphragm-type accelerating pumps, and others use throttle-operated piston pumps. A typical diaphragm pump consists of a diaphragm which separates a fuel chamber from an air chamber. The air chamber is connected to the throttle bore on the engine side of the throttle. During low-power operation, the air chamber of the pump has low pressure because of

the restricted airflow past the nearly closed throttle. This low pressure on one side of the diaphragm, plus the fuel pressure on the other side, moves the diaphragm in a direction to completely fill the fuel chamber with fuel. When the throttle is opened suddenly, the pressure on the engine side of the throttle increases. This pressure is transmitted through the vacuum passage to the air chamber of the pump. The increased air pressure, together with the pump-spring force, moves the diaphragm against the fuel chamber and forces fuel out through the discharge nozzle, providing an extra supply of fuel for the acceleration period. Figure 5-3 shows a typical pressure discharge nozzle with a diaphragm-type accelerating pump.

An **enrichment valve,** commonly called a **power enrichment valve** or **fuel-head enrichment valve,** depending on how it is actuated, takes the place of the economizer valve in the float-type carburetor. At high-power settings it is necessary that the fuel-air ratio be increased to provide the extra cooling necessary. Pressure-injection carburetors are usually designed so that the enrichment valve opens automatically when the throttle is moved to the higher power settings.

A number of pressure-injection carburetors make use of **automatic mixture-control** (AMC) **systems** which adjust the F/A mixture to compensate for changes in altitude. As previously explained, the pressure of the air decreases as altitude increases and this results in a decrease of available oxygen for a given volume of air. Thus, the mixture will become richer as altitude increases unless the fuel flow is reduced in proportion to the decrease in air pressure. An automatic mixture-control unit is illustrated in Fig. 5-4. The bellows is filled with a measured amount of nitrogen and an inert oil to make it sensitive to temperature and pressure changes. As the bellows expands and contracts, it adjusts a valve which decreases or increases the air flow from the impact tubes to chamber A of the regulator unit. This flow exists because there is a small air bleed

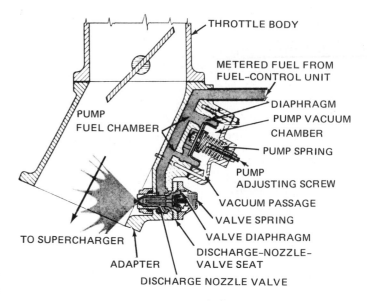

FIG. 5-3 Discharge nozzle with diaphragm-type accelerating pump.

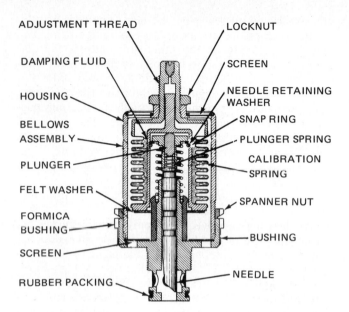

FIG. 5-4   Automatic mixture-control unit.

Labels for Fig. 5-4:
ADJUSTMENT THREAD
DAMPING FLUID
HOUSING
BELLOWS ASSEMBLY
PLUNGER
FELT WASHER
FORMICA BUSHING
SCREEN
RUBBER PACKING
LOCKNUT
SCREEN
NEEDLE RETAINING WASHER
SNAP RING
PLUNGER SPRING
CALIBRATION SPRING
SPANNER NUT
BUSHING
NEEDLE

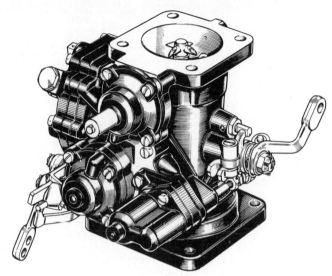

FIG. 5-5   Bendix PS-5C carburetor.   (*Energy Controls Div., Bendix Corp.*)

in the diaphragm between chambers *A* and *B*. If the airflow is reduced by the action of the automatic mixture-control valve, the pressure in chamber *A* will be reduced and the air diaphragm will move in a direction to close the fuel poppet valve. Thus, when the bellows of the AMC expands the airflow and air pressure for chamber *A* decrease and the air diaphragm moves to reduce the opening of the fuel poppet valve, thereby reducing the pressure of the fuel applied to the metering jets. This, of course, reduces the fuel flow from the discharge nozzle and leans the fuel-air mixture.

It must be pointed out that the AMC cannot effectively modify chamber *A* pressure except in conjunction with the automatic mixture-control air bleeds located between chambers *A* and *B*. These bleeds tend to neutralize the pressure between *A* and *B*; however, the continuous flow of automatically controlled air to chamber *A* maintains the correct pressure differential.

● PRESSURE CARBURETOR FOR SMALL ENGINES

In order to provide the benefits of pressure-injection carburetion for small aircraft engines, the Bendix Products Division (now Energy Controls Division) developed the PS series of carburetors. Figure 5-5 illustrates the PS-5C carburetor used on the Continental 0-470 series engines.

The PS-5C carburetor utilizes the principles explained previously in this chapter. It includes a throttle unit, regulator section, fuel-control unit, discharge nozzle, manual mixture control, acceleration pump, and idle system.

General Description

The PS-5C injection carburetor is a single-barrel updraft unit that provides a closed fuel system from the engine fuel pump to the carburetor discharge nozzle. Its function is to meter fuel through a fixed jet to the

engine in proportion to mass airflow. The discharge nozzle is located downstream of the throttle valve to prevent ice from forming in the carburetor. This carburetor provides positive fuel delivery regardless of aircraft altitude or attitude and maintains proper fuel-air ratios regardless of engine speed, propeller load, or throttle-lever position.

Basic Operation

The drawing of Fig.5-6 should be studied to obtain a visual idea of the operating principles as they are described.

Air enters the carburetor throttle section through the air intake and passes through the venturi tube past the throttle valve and into the intake manifold of the engine. The flow of air is controlled by a conventional butterfly-type throttle valve. Intake air (impact pressure) also enters the annular (ringlike) space between the outside diameter of the venturi tube and the flange of the carburetor main body and flows through internal channels to chamber *A* of the regulator section and to the discharge-nozzle air bleed. It also flows to the high-pressure side of the manual mixture-control needle valve.

The velocity of the air flowing through the venturi creates a low-pressure area at the throat of the venturi tube. This low pressure is transmitted through internal channels to chamber *B* of the regulator and is known as **regulated venturi suction**. This pressure is also transmitted to the low-pressure side of the control diaphragm of the discharge nozzle and to the needle valve of the manual mixture control.

Since the intake air pressure in chamber *A* is greater than the regulated venturi suction in chamber *B*, a pressure differential is created. This pressure differential acts upon the air diaphragm which separates the two chambers. The differential force (air metering force) acting on the air diaphragm increases or decreases with changes of airflow through the carburetor. Further control of the air metering force is provided by the manual mixture control.

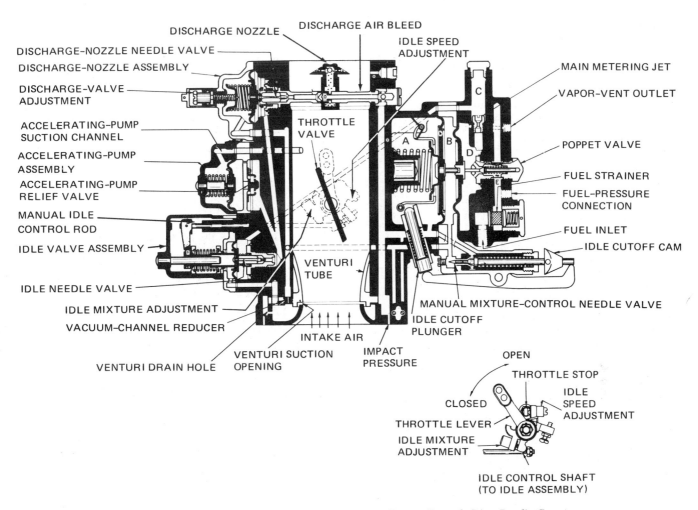

FIG. 5-6 Drawing of the PS-5C carburetor. *(Energy Controls Div., Bendix Corp.)*

Movement of the air diaphragm in response to the air metering force is applied to the regulator poppet valve through a stem arrangement. The degree of opening of this poppet valve determines the pressure of the unmetered fuel that is applied to the metering jet.

Fuel at engine pump pressure flows through the strainer into chamber E and past the poppet valve into chamber D of the regulator. The pressure of the fuel in chamber D is somewhat lower than the pressure in chamber E because of the pressure drop across the poppet valve. Fuel at this pressure is termed **unmetered fuel.** The pressure of this fuel is regulated by the degree of opening of the poppet valve. The poppet-valve movement is controlled by the forces acting on the air and fuel diaphragms. Chamber A pressure minus chamber B pressure is the force acting on the **air diaphragm**, while chamber D pressure minus chamber B pressure is the force acting on the **fuel diaphragm**.

Fuel from chamber D flows through the **main metering jet** to chamber C. This jet meters fuel to the discharge nozzle during engine speeds above the idle range.

Metered fuel from chamber C flows through the idle needle valve to the fuel side of the discharge-nozzle diaphragm. The opposite side of the diaphragm is exposed to regulated venturi suction and receives the force of an adjustable spring. When metered fuel pressure on the fuel side of the diaphragm overcomes the spring pressure, the needle valve will open and allow fuel to be discharged through the nozzle seat into the discharge-nozzle assembly. A fuel pressure drop is established as the fuel passes the discharge-nozzle needle valve. A supply of air from the impact pressure channel is directed to the discharge nozzle to help atomize the fuel as it is discharged. This serves the same purpose as the main air bleed in a float-type carburetor.

As the throttle is opened, airflow through the carburetor will increase, thus causing the air metering force to increase. Since the air metering force is stronger than the differential pressure across the fuel diaphragm (fuel metering force), the regulator poppet valve will open farther. This will permit a greater volume of fuel to flow into chamber D, thus increasing the pressure of the unmetered fuel pressure applied to the main metering jet. When fuel pressure in chamber D is increased, the differential across the fuel diaphragm is increased and the poppet-valve movement is stopped. At this time the air metering force is equal to the differential across the fuel diaphragm and the poppet valve will hold its position until the air metering force increases or decreases. The fuel leaving chamber D is now metered by the main metering jet and not by the idle needle valve as it is during the idling range of engine speeds.

Idling System

In the idle range at low airflows, the differential pressure (unmetered fuel pressure minus regulated venturi suction) across the large fuel diaphragm, plus the poppet-valve spring force, will try to move the poppet valve to the closed position. However, this force is opposed by the regulator spring force in chamber A, plus a small amount of air metering force. The two opposing sets of forces come into balance, holding the poppet valve open sufficiently to allow an ample amount of fuel to pass for idling purposes.

After passing through the main metering jet, the fuel is exposed to the idle needle valve and its diaphragm. The fuel pressure on one side of the diaphragm and the unregulated venturi suction on the air side of the diaphragm establish a differential pressure which tends to open the idle needle valve. The movement of the valve, however, is restricted and controlled by a fork on the throttle lever. At this point in the operating range, the actual metering of fuel is accomplished by the idle needle valve, because the orifice created by the needle valve and its seat is smaller than the main metering jet. As the throttle lever is opened, the fork on the idle lever moves out of contact with the idle pushrod and allows the differential pressure acting on the diaphragm to hold the idle needle valve open at or above the high cruise throttle-lever position.

The idle needle-valve assembly also serves as a power enrichment valve for high engine power settings. When the throttle reaches a position within 28° of wide open, an arm on the throttle shaft engages the manual idle-control rod and releases the pressure of the discharge diaphragm spring, thus allowing the idle needle valve to move away from the needle-valve seat. This increases the orifice area to permit an increased fuel flow to the discharge nozzle.

Accelerating Pump

A single-diaphragm vacuum-operated accelerating pump, which compensates for the lag in fuel flow that occurs when the throttle is opened rapidly, is incorporated in the Model PS-5C injection carburetor. The pump is composed of two chambers separated by a spring-loaded diaphragm. One side of the diaphragm is open to pressure above the throttle, while metered fuel pressure is applied to the other side of the diaphragm. This pressure differential causes the diaphragm to move in such a direction that it compresses the spring. When the throttle is opened rapidly, pressure above the throttle increases, causing a corresponding increase in the pressure on the spring side of the diaphragm. This increase of pressure, plus the force of the spring moves the diaphragm in a direction that displaces the fuel on the opposite side of the diaphragm. The displaced fuel temporarily increases the discharge-nozzle pressure and causes the nozzle to open farther, thus providing a momentary rich mixture for acceleration. The accelerating pump is shown in the diagram of Fig. 5-6. Note that this pump operates in a manner similar to the pump described previously in this chapter.

Manual Mixture Control

A **manual mixture-control valve** provides a means of correcting for natural enrichment at altitude and for adjusting mixture at other times to obtain the most satisfactory operating conditions. The mixture control consists of a needle valve positioned in its seat by a pilot-controlled lever or knob. On one side of the seat is chamber B pressure, and on the other side is chamber A pressure. As the needle is moved out off its seat, chamber A pressure bleeds into chamber B, thus decreasing the pressure differential across the air diaphragm. This causes the poppet valve to close partially and thus reduces the fuel pressure in chamber D, which results in a leaner mixture. The reverse action takes place when the needle is moved back in its seat.

When the **manual mixture-control lever** is moved to the IDLE CUTOFF position, the **idle cuttoff cam** on the linkage actuates a rocker arm which causes the plunger to move inward against the **release lever** in chamber A. The release lever compresses the regulator diaphragm. While this is taking place, the manual mixture-control needle is being pulled out of its seat, thus reducing the air metering force. With the air metering force reduced, the differential pressure across the large fuel diaphragm, plus the spring force on the regulator poppet valve, closes the poppet valve, thus shutting off the fuel through the carburetor and to the engine.

Idle Speed and Mixture Adjustment

Idle speed and mixture adjustment are accomplished in much the same way for pressure-type carburetors as for float types. The objective is to adjust the idle mixture so that it is a slightly richer lean best power mixture. The procedures are set forth in the manufacturer's maintenance manual for the particular carburetor being adjusted.

● PRESSURE CARBURETORS FOR LARGE ENGINES

A number of different models of pressure-injection carburetors have been designed for large engines; however, they all utilize the principles explained previously. These carburetors vary in size, type of mixture control, type of enrichment valves, shape of throttle body, and type of discharge nozzle. In this section we shall describe a typical model which illustrates the most important principles of operation.

It is important for technicians who may be called upon to work on older aircraft, such as the DC-3, DC-6, DC-7, and others, to have an understanding of the principles of pressure carburetion. It should be noted that the principles are similar to those previously discussed for smaller aircraft.

Bendix-Stromberg PR-Type Carburetor

For the purpose of this discussion, a Bendix PR-58-type carburetor is used because it possesses most of the characteristics of typical pressure-injection carburetors for large engines, including both the PD and PT types. A drawing of this carburetor system is shown in Fig. 5-7.

Principal Units

To understand the operation of the complete carburetor, one should note the construction and operation of each unit and its function in relation to the other units. There are four basic units of the carburetor: (1) the **throttle unit,** (2) the **pressure-regulator unit,** (3) the **fuel-control unit,** and (4) the **automatic mixture-control unit.** In addition, this particular model of the PR-58 carburetor is equipped to operate with a water-alcohol ADI (antidetonant-injection) system which includes the derichment valve in the fuel-control unit and a separate water-alcohol (W/A) regulator. The W/A regulator is shown schematically in Fig. 5-11.

The Throttle Unit. The **throttle unit,** illustrated in Fig. 5-7, is the foundation of the carburetor and contains the main venturi plates, boost venturi tubes, impact tubes, and throttle valves. The principal functions of the throttle unit are to control and provide a means

of measuring the mass airflow to the engine, thus controlling engine power output. Provisions are made on the throttle unit for mounting the automatic mixture control, the pressure regulator unit, a mechanically operated accelerating pump, and the adapter for the fuel feed-valve assembly. The **fuel feed valve** has provision for mixing the ADI fluid with the fuel, and the mixture of fuel and fluid is then sprayed into the entering airstream through a slinger ring at the entrance to the internal blower of the engine.

The accelerating pump, shown attached to the throttle body in Fig. 5-7, is throttle-actuated. It consists of a piston, a fill check valve, a relief check valve, and fuel chambers. Fuel from chamber D flows into the cylinder of the pump through the fill check valve. When the throttle is advanced, the piston forces the fuel out through a passage to the **balance chamber** at the end of the poppet-valve stem. The additional fuel pressure in the balance chamber momentarily unbalances the balance diaphragm force and causes the poppet valve to open to a greater extent than would occur through the action of the air diaphragm. Additional fuel is thus delivered through the fuel-control unit to the fuel feed valve. Note particularly that the force of the balance

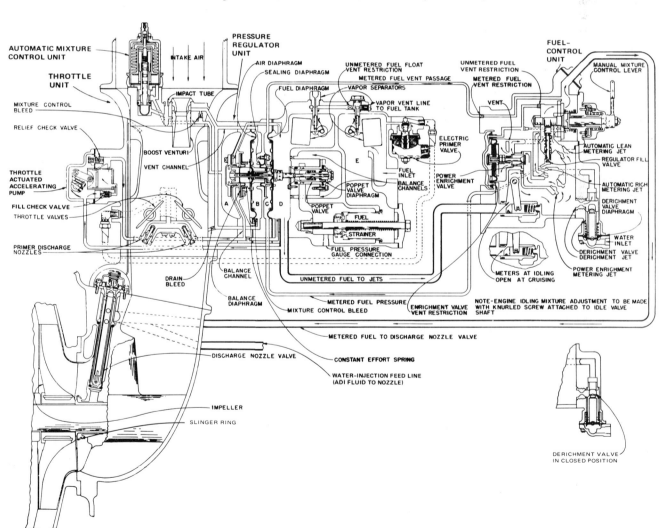

FIG. 5-7   Drawing of a Bendix PR-58-type carburetor system.   *(Energy Controls Div., Bendix Corp.)*

diaphragm opens the main poppet valve when pressure is increased in the balance chamber. The functions of the balance chamber and diaphragm are explained in the next section of this chapter.

The Pressure-Regulator Unit. The **pressure-regulator unit** for the PR-type carburetor is essentially the same as described previously. It does, however, include an additional chamber, **chamber E**, to which the fuel inlet line from the main fuel system pump is connected. The pressure in chamber E is therefore the same as the fuel pump pressure. Before entering chamber E, the fuel passes through the carburetor fuel strainer.

The functions of the pressure regulator's other four chambers may be described as follows:

Chamber A is subjected to the pressure from the carburetor impact tubes, so that the pressure in the chamber increases as air mass and velocity increase. Chamber A pressure acts against the **air diaphragm** through the poppet-valve stem to open the poppet valve.

Chamber B is subjected to the reduced air pressure created by the **boost venturi** in the throat of the throttle section. Since chamber B is on the opposite side of the air diaphragm from chamber A, the reduced pressure in chamber B adds to the poppet-valve opening force produced by chamber A. Thus, both chambers A and B increase poppet-valve opening force as mass airflow increases through the throttle as a result of opening the throttle.

Chamber C is subjected to metered fuel pressure fed from the regulator fill valve in the fuel-control unit. This pressure is established at approximately 5 psi [34.48 kPa] by the spring control in the **fuel feed valve** (discharge nozzle) and remains essentially constant because the fuel feed valve opens and permits fuel to flow whenever the pressure exceeds 5 psi.

Chamber D is subjected to unmetered fuel pressure, which balances against the metered fuel pressure in chamber C. The differential between the two pressures produces a force across the **fuel diaphragm** to balance the effect of the force produced by the **air diaphragm**. The main fuel poppet valve is therefore opened to the extent necessary to allow the forces to balance. Both the fuel diaphragm and the air diaphragm are connected to the poppet-valve shaft; thus, when unmetered fuel pressure in chamber D increases, the force produced tends to close the poppet valve. This reduces the pressure until the fuel diaphgram's force is equal to the air diaphragm's force. The result of the actions described is that the poppet valve will increasingly open as mass airflow through the carburetor increases and will increasingly close as airflow decreases.

At idling speeds, however, the forces acting on the air diaphragm are not sufficient to keep the poppet valve open. For this reason a spring is incorporated in the regulator to prevent the poppet valve from closing. Fuel can then continue to flow through chamber D to the fuel-control unit, where it is metered by the **idle valve**. The idle valve is operated by the throttle and is effective for metering in the first 10° of throttle travel. Beyond that point, it has no metering effect.

Between the fuel chambers ("wet side") and the air chambers ("dry side") of the pressure-regulator unit is a solid, metal partition through which the poppet-valve stem extends. In order to ensure that no fuel can pass by the stem into chamber B, a **sealing diaphragm** is installed around the valve stem where it passes through the partition. This sealing diaphragm provides an effective seal between chambers B and C; however, it has an effect on the poppet valve for which a correction must be made.

● The regulated fuel pressure in chamber C applies force to the sealing diaphragm, and this force tends to close the poppet valve.

● In addition, the low pressure in chamber B applies force to the sealing diaphragm, also in a direction tending to close the poppet valve.

It is therefore necessary to apply a balancing force, which is accomplished by means of the **balance chamber and diaphragm** located in chamber A at the end of the poppet-valve stem. The balance diaphragm is exposed to chamber D fuel pressure on one side and chamber A pressure on the other. This provides an equalizing force against the chamber C and chamber B pressures that are applied to the opposite sides of the sealing diaphragm. The net result is that equal forces are exerted in opposite directions, thus canceling each other; consequently no force is applied to the poppet valve except that applied through the air diaphragm and the fuel diaphragm. The end result is that the poppet valve opens in a degree established by the mass volume of airflow through the carburetor.

Included in the pressure-regulator unit are two **vapor separators.** As the name implies, these units are designed to remove the fuel vapor and air which would otherwise collect in the regulator unit and interfere with proper operation. Figure 5-8 illustrates the principle of operation. As vapor collects in either chambers E or D, the level of fuel drops and the float is lowered. A valve connected to the arm of the float is withdrawn from its seat and the vapor is permitted to escape. During normal operation, the float assumes a level such that vapor is released as it accumulates; that is, the valve does not continuously open and close.

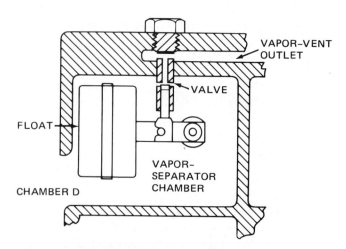

FIG. 5-8 Operating principle of a vapor separator.

If the float should stick closed, vapor would accumulate in the fuel chamber. Vapor bubbles would then be passed along the fuel passages, causing erratic operation of the system plus a leaning of the mixture. If the float should stick open or if the float should sink because of a leak in the float, fuel would be vented back to the fuel tank or overboard.

A vapor separator can be tested by disconnecting the vent line where it connects to the carburetor, placing the manual mixture control in AUTO RICH, and turning on the fuel boost pump. There should be an immediate discharge of a small amount of air and fuel from the vent connection; the discharge should then stop, except perhaps for a slow drip. If the valve is stuck open, there will be a continuous flow of fuel; and if the valve is stuck closed, there will be no flow.

In performing this test, the technician should be cautious to avoid any spillage of fuel. A suitable fire extinguisher should always be at hand in case of accidental ignition of fuel.

The Fuel-Control Unit. The **fuel-control unit** is attached to the regulator unit. It incorporates a series of fixed metering jets, a manual mixture-control valve, an idle valve, a fuel-head enrichment valve, a regulator fill valve, and, in this particular model, a derichment valve. The fuel metering jets in this unit are the auto-rich metering jet, the power-enrichment metering jet, the auto-lean metering jet, and the derichment jet.

The **manual mixture-control valve** has three different positions: (1) IDLE CUTOFF, (2) AUTO LEAN, and (3) AUTO RICH. Some units include a position called FULL RICH. The manual mixture control consists of a disk, or plate, with apertures which allow fuel to flow through channels from the various metering jets, depending on the position of the control. Figure 5-9 shows the positions of the plate for the different settings.

The **fuel-head power enrichment valve** is operated by a diaphragm (see Fig. 5-7). Metered fuel pressure is applied to one side of the diaphragm and, together with the force of the diaphragm spring, tends to keep the valve closed. Unmetered fuel pressure from chamber D is applied to the opposite side of the diaphragm. At cruise power settings, the metered fuel pressure and spring force keep the valve closed because the force of the unmetered pressure applied to the diaphragm is less than the combined forces acting against the opposite side of the diaphragm. At high-power settings, the fuel pressure from chamber D increases to a level greater than that of the metered fuel pressure plus spring force, and the enrichment valve is opened to supply the extra fuel needed for cooling.

The **idle valve** is located in the main unmetered fuel passage and is actuated by the throttle. It is designed to provide metering only during the first 10° of throttle travel. Thereafter, it is completely open and has no effect on fuel flow.

The **derichment valve** is designed to cause a leaning of the fuel-air mixture when the ADI fluid (water-alcohol) is being injected into the fuel-air mixture at the fuel discharge nozzle. The valve is diaphragm-operated, with metered fuel pressure being applied to one side of the diaphragm and water-alcohol pressure to the other side. When the ADI system is operating, the water-alcohol pressure is great enough to close the derichment valve, thus reducing the amount of fuel flowing to the fuel feed valve. Figure 5-7 shows how the derichment valve is located with respect to the other elements of the fuel-control unit.

The **regulator fill valve** is located in the passage which supplies regulated fuel pressure to chamber C in the pressure-regulator unit. The valve is a poppet type and is operated by means of the manual mixture-control lever. It is open during all operating positions of the mixture control; however, it is closed when the control is placed in IDLE CUTOFF to provide for a complete cutoff of all fuel flow.

Automatic Mixture Control. The **automatic mixture-control (AMC) unit,** illustrated as part of Fig. 5-7, is mounted in the throttle unit at the air inlet where it senses the pressure of the air entering the carburetor. The unit illustrated is similar to that described previously; it consists of a sealed metallic bellows which operates a contoured needle valve located in the passage that carries impact air pressure to chamber A of the pressure-regulator section. The **automatic mixture-control bleed** permits a small flow of air between chambers A and B; this airflow is drawn out of chamber B by boost venturi suction and returned to the airstream entering the carburetor. The air pressure in chamber A is therefore dependent on continuous replenishment through the passage controlled by the automatic mixture-control valve. When incoming air pressure increases, the mixture-control bellows contracts and moves the needle valve to a more open position. This permits greater airflow and higher pressure in chamber A, which, in turn, causes the main fuel poppet valve to open further and increase the fuel pressure to the discharge nozzle. The reverse action takes place as incoming air pressure decreases with higher altitude.

The sealed bellows of the AMC contains nitrogen at approximately sea-level pressure and an inert oil which acts as a dampening fluid.

Summary of Operating Principles

Air enters the carburetor at the air intake and passes downward through the main venturi plates and the boost venturi tubes past the throttle valves and into the internal supercharger. The flow of air is controlled by

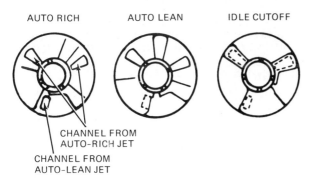

AUTO RICH   AUTO LEAN   IDLE CUTOFF

CHANNEL FROM
AUTO-RICH JET

CHANNEL FROM
AUTO-LEAN JET

FIG. 5-9 Positions of the manual mixture-control-valve plate.

the throttle valves. Referring to Fig. 5-7, we see that the boost venturi tubes in conjunction with the main venturi plates create a suction which is transmitted from the throats of the boost venturi tubes to chamber *B* of the regulator. A portion of air flows into the impact tubes through an internal channel past the automatic mixture-control needle where the flow is restricted in accordance with air pressure. Low pressure increases the restriction, and high pressure decreases it. The airflow, thus restricted, is termed **regulated impact pressure.** This regulated impact pressure is transmitted to the chamber in the regulator labeled *A*, and although it is lower than the original impact pressure, it is greater than the venturi suction in chamber *B*. This pressure differential between chambers *A* and *B*, acting on the air diaphragm between these two chambers, produces a force to the right, which tends to open the poppet valve. This force, termed the **air metering force,** increases or decreases as the throttle valves are opened or closed to permit more or less air to flow through the throttle unit.

Fuel enters the regulator at the fuel inlet and passes through the fuel strainer into chamber *E* of the regulator. The pressure of the fuel in this chamber is the same as the pressure delivered by the engine-driven fuel pump and is the highest pressure in the carburetor. Fuel vapor or air is eliminated from chamber *E* by the action of the **vapor separator.** The valve of the separator opens when vapor or air accumulates in chamber *E* and allows the vapor or air to escape through a line to the fuel tank. An identical system is used to vent the unmetered fuel chamber *D*. These two venting systems are connected by an internal channel.

In reference again to Fig. 5-7, note that the liquid fuel flows past the poppet valve into chamber *D* of the regulator. The pressure of the fuel in chamber *D* is lowered, owing to the restriction of the poppet valve. The fuel at this reduced pressure is termed **unmetered fuel.**

The fuel flows from chamber *D* of the regulator through an internal channel into the fuel-control unit, past the idle valve, and through the metering jets. As the fuel passes through the jets, the fuel pressure is further reduced and remains at approximately 5 psi [34.48 kPa], as determined by the spring force in the control valve of the fuel feed valve (discharge nozzle). The fuel pressure between the fuel-control jets and the fuel feed valve is termed **metered fuel pressure.** A portion of the metered fuel is directed through the regulator fill valve to a passage leading to chamber *C* in the regulator. This fuel pressure in chamber *C* is applied against the fuel diaphragm, opposing the force from chamber *D*. The pressure differential between chambers *C* and *D* acts on the fuel diaphragm and creates a force to the left, which tends to close the main fuel poppet valve. This force is termed the **fuel metering force.**

The **air metering force** (pressure in chamber *A* combined with the negative pressure in chamber *B*) acting on the air diaphragm causes the diaphragm to move to the right, thus opening the poppet valve. As the opening of the poppet valve increases, the pressure of the unmetered fuel in chamber *D* increases, thereby increasing fuel flow through the metering jets. Note that fuel pressure established by the position of fuel and

air diaphragms is called fuel metering pressure and is the result of the fuel metering force created by the pressure differential between chambers *C* and *D*. The fuel metering force is regulated by and equal to the air metering force, except in the idle range.

The **idle spring** in the regulator holds the poppet valve open in the idle range. This is necessary because the air metering force is not great enough to open the poppet valve in this range, and thus the fuel metering force is slightly higher than the air metering force; without the idle spring, the poppet valve would close and the engine would stop. The idle spring applies sufficient force to overcome the fuel metering force—hence the poppet valve remains open. To provide fuel metering in the idle range, the idle valve in the fuel-control unit is adjusted by the throttle in the first 10° of travel. The proper amount of fuel is thus delivered to the fuel feed valve to maintain the correct mixture in the idle range.

The fuel which has passed through the metering jets in the fuel-control unit is **metered fuel.** When the airplane is cruising with the manual mixture control in the AUTO LEAN position, the fuel flow is limited by the size of the **auto lean** jet. This jet is selected to give the maximum fuel economy. Additional fuel for best power when cruising is added by moving the manual mixture control to the AUTO RICH position, whereupon fuel also flows through the **auto-rich** jet.

For takeoff and emergency operation, a diaphragm-operated **power enrichment valve** is incorporated in the fuel-control unit. When the force of the unmetered fuel pressure on the enrichment-valve diaphragm is sufficient to overcome the combined force of the metered fuel pressure and the enrichment-valve spring, the valve opens and allows additional fuel to flow through the open valve. Since the power enrichment jet is larger than the auto-rich jet, a richer mixture will be obtained whenever the enrichment valve is open. When the power enrichment valve is wide open, the metering of the fuel is accomplished by the auto-lean and power enrichment jets.

From the facts given above it is apparent that this carburetor is fully automatic. The correct fuel-air ratio is maintained for any power or mixture selection, according to the engine manufacturer's requirements, without manual adjustment in flight.

Typical Fuel-Air Ratio Curve

Figure 5-10 is a typical fuel-air ratio curve obtained with a pressure-injection carburetor. The cruise and takeoff values are maintained under altitude and temperature change, independently of propeller pitch, engine speed, or throttle position.

The idling range is from 0 to 1900 lb/h [861.83 kg/h] of air, and the fuel flow is controlled by the position of the idle valve, which is controlled by the throttle linkage.

On the fuel-air ratio curve of Fig. 5-10, the **cruising range** is from 1900 to 5300 lb/h [861.83 to 2404.04 kg/h] of air. After about 10° of throttle movement, the idle needle valve is drawn out to its open position. The metering then shifts from the orifice formed between the idle needle valve and seat to the auto-lean metering

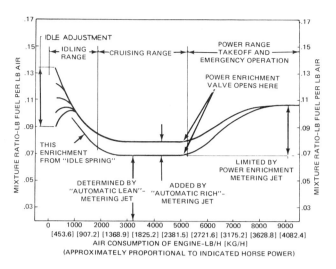

FIG. 5-10 Fuel-air ratio curve (pressure-injection carburetor).

jet and the auto-rich metering jet, provided that the manual selector valve is in the RICH position.

If the selector valve is in the LEAN position, the auto-rich metering jet passage is closed. Rich-automatic cruise is determined by the capacity of the auto-lean and the auto-rich metering jets. Lean-automatic cruise is obtained when the manual control is placed in such a position that the channel through the valve plates from the auto-rich metering jet is closed, thus reducing the fuel flow. (The terms ''rich automatic'' and ''lean automatic'' refer to the cruise conditions of the aircraft engine rather than to elements of the carburetor.)

The **power range,** as shown on the curve, is from the upper limit of cruising, 5300 to 9500 lb/h [2404.04 to 4309.13 kg/h] of air. Gradually increased fuel-air ratios are required from either the RICH CRUISE or LEAN CRUISE positions up to 9500 lb/h of air, or maximum air consumption.

● WATER INJECTION

We have discussed water injection briefly in connection with the PR58-type carburetor because this particular unit and system is designed to accommodate water injection as a means of providing maximum engine power at takeoff.

Water injection, also called **antidetonant injection** (ADI), is the use of water with the fuel-air mixture to provide cooling for the mixture and the cylinders so that additional power can be drawn from the engine without danger of detonation.

Instead of using pure water for the ADI system, it is necessary to use a water-alcohol mixture (methanol) with a small amount of water-soluble oil added. The alcohol prevents freezing of the water during cold weather and at high altitudes. The water-soluble oil is added to prevent the corrosion which would occur in the units of the system if they lacked oil. The water-alcohol-oil mixture is called **antidetonant fluid,** or simply **ADI fluid.** In servicing the ADI system, the technician must be sure that the correct mixture of fluid components is used. This information is contained in the manufacturer's service instructions.

Advantages

It is often necessary to use the maximum power which an engine can produce, such as for taking off from short fields, for military emergencies, and for times when it is necessary to ''go around'' after making a landing approach. A ''dry'' engine—that is, one without water injection—is limited in its power output by the detonation which results when operating limits are exceeded. The injection of water into the fuel-air mixture has the same effect as the addition of antiknock compounds in that it permits the engine to deliver greater power without the danger of detonation.

The average engine operating without water injection requires a rich mixture with a ratio of approximately 10 parts air to 1 part fuel by weight (F/A ratio 0.10). With this mixture a portion of the fuel is unburned and acts as a cooling agent. The additional unburned fuel subtracts from the power of the engine. But when water is added to the fuel-air mixture in proper quantities, the power of the engine can be increased. The water cools the fuel-air mixture, thus permitting a higher manifold pressure to be used. In addition, the fuel-air ratio can be reduced to the rich best power mixture, thus deriving greater power from the fuel consumed. When water injection is employed, the fuel-air ratio can be reduced to approximately 0.08, which is a much more efficient mixture than the 0.10 ratio required otherwise.

The use of water injection permits an increase of 8 to 15 percent in takeoff horsepower. During World War II this extra power contributed greatly to the superiority of the United States air forces.

The equipment required for water injection includes a storage tank, a pump, a water regulator, a derichment valve, and necessary circuits and controls.

Principles of Operation

The **water-alcohol (W/A) regulator** is the unit which makes possible the injection of ADI fluid into the fuel at the fuel feed valve in a quantity which ensures a correct volume of the water-alcohol mixture. If too much of the ADI fluid were injected, the cooling effect would reduce the power of the engine. Note that the expansion of gases due to heat enables an engine to develop power. If insufficient ADI fluid were injected, the engine would overheat and detonation would occur.

Note in Fig. 5-11 that this diagram represents one particular type of W/A regulator, one which is similar to that used in the PR58-type carburetor on a P&W R2800 engine. This regulator includes three diaphragm-operated valves. These are the **metering-pressure control valve,** the **check valve,** and the **W/A enrichment valve.**

The **metering pressure-control valve** is operated by chamber *D* (unmetered) fuel pressure from the carburetor applied to one side of the diaphragm and W/A pump pressure applied to the opposite side. When the pilot turns on the ADI switch in the cockpit, ADI fluid flows from the pump to the regulator. The metering pressure-control valve will be open because of the unmetered fuel pressure applied to the valve diaphragm. When pump pressure builds up to the level of unmetered fuel pressure, the valve will begin to close.

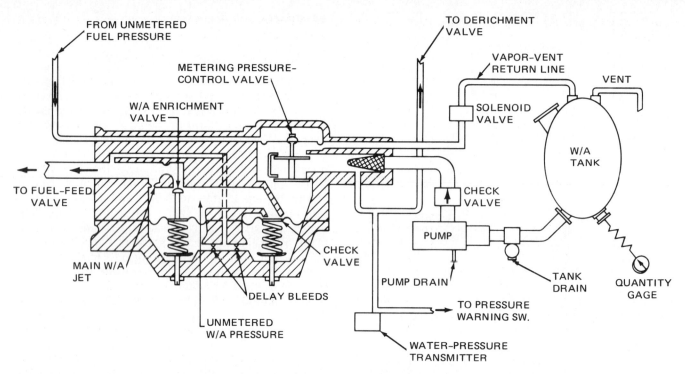

FIG. 5-11 Schematic diagram of a water-alcohol (ADI) regulator.

The **check valve** would be normally closed when the system is not operating; however, it will begin to open slowly when pump pressure is applied to the diaphragm. The valve cannot open immediately because of the delay bleed. The **delay bleed** provides time for the derichment valve to close before the ADI fluid starts to discharge into the fuel feed valve.

When the system is *not* operating, the fuel backs up into the W/A feed line. Hence, when the system is turned on, fuel will be the first substance injected into the fuel feed valve. If the carburetor is set for takeoff at FULL RICH or EMERGENCY RICH and additional fuel is injected from the W/A line, the overrich mixture will cause the engine to lose power and there will be a definite hesitation in the operation of the engine. The use of delay bleeds with the check valve prevents this situation because the derichment valve closes and leans the mixture before the extra fuel is injected from the W/A feed line. Because of the leaner mixture caused by the derichment valve, the extra fuel injected from the W/A line does not enrich the fuel-air mixture sufficiently to cause engine hesitation.

The **W/A enrichment valve** modifies the flow of ADI fluid in connection with the main W/A jet. This valve is closed when the system is not operating. The operation of the ADI system is initiated when the pilot turns on the ADI-control switch in the cockpit. This switch is in series with a pressure switch operated by engine oil pressure or manifold pressure; the engine must therefore be operating at a comparatively high power setting before the electrical power can be directed to the ADI pump. When the system is operating, a **water pressure transmitter** connected to the regulator sends electric signals to the **water pressure indicator** in the cockpit. Water pressure from the regulator is also directed to a pressure warning switch which controls the **water pressure warning light** in the cockpit. The

light is on while the system is operating. If the ADI fluid supply should become exhausted while the system is operating, the pressure switch will open and the warning light in the cockpit will turn off. At the same time, the derichment valve will open and permit enrichment fuel to flow through the fuel-control unit, thus providing the necessary cooling to avoid detonation. Some systems utilize a float-operated switch to turn off the fluid pump before the fluid supply is exhausted. This is particularly important if a vane-type pump is employed. Such a pump can maintain pressure on the derichment valve even though the fluid supply is exhausted. In such a case, the valve would not open to allow enrichment fuel to flow, and the engine would still be operating on the best power mixture. This would allow detonation to occur.

The ADI system is particularly advantageous under conditions of high humidity. The water vapor in humid air displaces oxygen, so that a particular fuel-air ratio will increase in richness as humidity increases. Therefore, when an aircraft is taking off using the emergency rich mixture, high humidity will further enrich the mixture and cause a substantial loss of power. When an aircraft takes off wet (with ADI), the fuel-air ratio is set for best power and the enrichment caused by high humidity is not great enough to cause an appreciable loss of power. The water injected in the fuel does not have an appreciable effect on the F/A ratio because it does not displace the oxygen in the air.

● **INSPECTIONS AND MAINTENANCE**

Inspection of Engine Fuel Systems and Carburetors

If possible, inspections should be carried out in accordance with manufacturer's or operator's instructions as set forth in appropriate manuals. If these are not available, the following general practices can be followed:

1. Remove cowling as necessary to gain access to the items to be inspected. Place cowling sections in suitable racks to avoid damage.

2. Examine all fuel-line connections and fittings for signs of leakage. If fuel leakage is discovered, correct by tightening or replacing the fitting. If a leak cannot be stopped by applying specified torque, the fitting or tube end should be replaced. Tubing fittings must not be overtorqued because of the danger of crushing the metal of the tubing and causing irreparable damage.

3. Observe the condition of the hoses. The outer surface should be smooth and firm and free of blisters, bulges, collapsed bends, or deep cracks. Small blisters can be accepted, provided that there is no fuel leakage when the blister is punctured with a pin and that there is no leakage when the hose is tested at $1\frac{1}{2}$ times working pressure. Appreciable bulging at the hose fittings or clamps requires that the hose be replaced. Fine cracks which do not penetrate to the first fabric layer are acceptable.

4. Carefully examine the condition of the hoses and tubing at the clamps or brackets used for mounting. Both the mounting and the line should be checked for wear and looseness. A loose mounting will cause wear.

5. Check the metal fuel lines for wear, nicks, cuts, dents, and collapsed bends. Small nicks and cuts which do not extend deeper than 10 percent of the wall thickness and are not in the heel of a bend may be repaired by stoning and polishing with crocus cloth. Dents which are not deeper than 20 percent of the tubing diameter and are not in the heel of a bend are acceptable. Dents can be removed by drawing a steel bullet through the tubing with an attached steel cable. Tubing which is not repairable must be replaced.

6. Remove the drain plugs in the carburetor and sumps to eliminate water and sediment. See that the plugs are reinstalled with proper torque and safetying. Install new washers with the plugs where required.

7. Remove all fuel screens and filters to clean them and to check their condition. Collapsed screens, and filters which do not provide free fuel flow, must be replaced. Main-line fuel screen sumps and tank drains should be opened briefly at preflight inspections to remove water and sediment.

8. If the fuel system includes an engine-driven fuel pump, check the pump for security, oil leakage from the mounting, and proper safetying of mounting nuts, bolts, or screws. Check the electric fuel boost pump for the operation and security of the fuel connections and electric connections. The brushes of the pump motor should be replaced in accordance with the schedule set forth in the service manual.

9. Check the carburetor for security of mounting, fuel leakage, and proper safetying. Check the gasket at the mounting flange or base to determine if there is a possibility of air leakage. Examine the throttle shaft bearings and the control arms for the throttle and mixture control for excessive play. Remember that excessive clearance at the throttle shaft can allow air to enter the carburetor and lean the mixture. Apply lubricant to the bearings and moving joints in accordance

with the service instructions or the approved lubrication chart.

Removal and Installation of Carburetors

A carburetor is an engine accessory, and the same principles of good mechanical practice apply to it as with any other accessory. Approved procedures for the removal and installation of carburetors are contained in the maintenance manual for each type of engine installation. The maintenance manual instructions should be followed whenever a manual is available; however, when a service or maintenance manual is not available, certain generally approved practices can be followed:

1. See that a correct type of fire extinguisher is at hand before starting work on any part of the fuel system, including the carburetor.

2. Remove cowling as necessary to gain access to the carburetor, and place the cowling in a rack to prevent damage.

3. Shut off the main fuel valve to avoid fuel spillage.

4. Disconnect and cap or plug the fuel and vent lines. Use the proper types of wrenches to loosen the fittings.

5. Disconnect the throttle and mixture-control cables or rods. Secure the throttle in the closed position. Retain the clevis pins or volts in the ends of the control fittings, and secure them with cotter pins or safety wire. Tag the control fittings for identification if necessary.

6. Cut the safety wire from the air scoop or duct attachment. Do not allow bits of safety wire to fall into the engine nacelle. Scraps of wire, tape, and other debris should be placed in a trash container.

7. Remove the carburetor attachment bolts or nuts and examine their condition. Discard damaged nuts, bolts, or washers.

8. Great care must be taken during removal and replacement of the carburetor. After the linkages and hose attachments have been disconnected, all loose nuts, bolts, cotter pins, and other parts must be cleared away before the carburetor is lifted from its mounting.

9. Remove the carburetor carefully from the mounting to avoid damaging the threads of the mounting studs. If the carburetor is a downdraft type, install a cover plate over the carburetor opening immediately and secure it with nuts or bolts. Material falling into the carburetor opening of the engine could make it necessary to remove the engine from the aircraft. The cover plate must remain in place until the carburetor is reinstalled.

10. The installation procedure for a carburetor is generally the reverse of the removal procedure. The same care must be taken to avoid damage and to prevent foreign material from entering the engine or fuel lines.

11. The replacement carburetor, particularly if it is a downdraft type, must have the throttle secured in the closed position. Place a new carburetor gasket on the mounting or carburetor deck. If there are matching bleed holes on the carburetor and mounting, the gasket must have a hole located to fit the bleed holes. The gasket must be installed so that the hole in the gasket

matches the holes in the carburetor and mounting. In some cases a carburetor deck screen takes the place of the gasket. This screen prevents larger bits of material, such as nuts and washers, from falling into the engine during inspection and service, and it also stops other material which may be drawn into the engine through the air-inlet duct.

12. Mount the carburetor in the correct position on the engine. Make sure that the correct types of nuts, bolts, and washers are installed. The tightening of the nuts or other fasteners must be done progressively and alternately on opposite sides of the mounting. When the nuts or bolts have been tightened to a firm position, they should be torqued with a suitable torque wrench to the value given for the type and size of fastener employed.

13. Connect the fuel and vent lines. Do not use any lubricant or thread compound other than clean engine oil.

Preparation of a New or Overhauled Carburetor

Pressure-type carburetors are normally stored with the fuel section filled with a light preservative oil to prevent interior corrosion and to keep the diaphragms flexible. When a new or overhauled carburetor is removed from extended storage, it should be prepared according to the manufacturer's instructions, before installation and operation. This preparation usually consists of removing the drain plugs and caps from the fittings to permit the complete drainage of preservative fluids from the inside. When this has been accomplished, fill the fuel section of the carburetor with gasoline and replug the drains. The gasoline should be allowed to remain in the carburetor for at least 8 h in order to soak and soften the diaphragms. See that gasoline is not permitted to enter the air section of the regulator unit.

Removal of residual preservative oil from a previously stored pressure-type carburetor can be accomplished after the carburetor is installed on the engine. In this case, remove the shipping plugs and caps from the carburetor and drain the preservative oil insofar as possible before installing the carburetor on the engine. Install the carburetor and connect the fuel and vent lines. Turn on the main fuel valve, move the mixture control to the AUTO RICH position, and turn on the fuel boost pump. Fuel will flow into and through the fuel sections of the carburetor and out the discharge nozzle. Depending on whether the carburetor is a downdraft or updraft type, the fuel will flow out of the supercharger drain at the bottom of the engine or out the bottom of the carburetor and the air-intake drain. When the fuel has cleared the carburetor of oil and oil-free fuel is draining out, turn off the boost pump and place the mixture control in the IDLE CUTOFF position. This closes the fuel passages and retains the fuel in the carburetor. The engine should not be operated for at least 8 h after the fuel is placed in the carburetor.

Adjusting Carburetor Controls

The principal controls for the pressure-discharge carburetor are the throttle and the mixture control. Both are operated through linkages to levers or push-pull knobs in the cockpit. The linkages from the cockpit to the engine nacelle may be flexible wires, cables with pulleys, bellcranks and push-pull rods, or a combination of these. The various types are discussed in detail in Chap. 14.

The satisfactory operation of the engine depends on the correct adjustment of the carburetor controls; the adjustment must be such that the controlled unit in the carburetor is moved through its full range by the cockpit lever or knob. To make sure this takes place, the cockpit controls are rigged with a small amount of **springback** at the end of travel in each direction. For example, when the throttle is completely closed, the throttle control in the cockpit is not quite to the end of its travel in the closed position. If the cockpit throttle control is pulled all the way to the end of its travel and then released, it will spring back about $\frac{1}{8}$ in [3.18]. In practice, the amount of springback is specified in the maintenance manual for the particular installation.

To check the springback on a carburetor control, move the cockpit control to the end of its travel in one direction or the other. Examine the control lever at the carburetor to see whether it is against the stop. Also, note whether the cockpit control has springback. If adjustment is needed, disconnect the adjustable control fitting and either shorten or lengthen it sufficiently to achieve the required result. Check each control in both directions and adjust for equal springback both ways. After adjusting, make sure that the fittings, clevis pins, bolts, and other fasteners are properly safetied.

For controls which utilize detents to establish the positions for particular functions, the control linkage must be adjusted to ensure that the position of the control on the carburetor coincides with the position established by the detent on the cockpit control. A **detent** is a device incorporated in a lever control to give the operator a "feel" that the lever is in a desired position. It often consists of a spring-loaded ball or rounded wedge which slips into a notch when the control reaches an indexed position. For example, the mixture-control lever in the cockpit may be marked for IDLE CUTOFF, AUTO LEAN, and AUTO RICH. The detent will engage and tend to hold the lever in each of these positions. When rigging or adjusting the control, the technician must make sure that the carburetor control is in the position indicated by the cockpit control.

Adjusting Idle Mixture and Idle Speed

The only adjustments that can be made on a pressure-type carburetor while it is installed on the engine are the **idle mixture** and **idle speed**. Usually, the procedures for accomplishing these adjustments are given in the maintenance manual; however, certain general practices are satisfactory for most installations.

The idle mixture *must* be correct. If the idle mixture is too lean, the engine will be difficult to start. The idle mixture must be richer than the best power mixture because when the engine is cold, the fuel does not vaporize fully and the mixture is not as rich as when the engine is warm. If the mixture is too lean at starting, backfiring will occur, sometimes causing an induction system fire or other damage to the system.

If the idle mixture is too rich, the spark plugs will

become fouled during idling because of the unburned carbon deposited on the spark-plug electrodes. In addition, an overrich idle mixture will affect the operation of the engine at operating speeds, reducing power and consuming extra fuel.

To check the idle mixture, run the engine at warm-up speed until it has reached normal operating temperature. If the engine is equipped with a constant-speed or two-position propeller, place the propeller control in the HIGH RPM (low-pitch) position. Adjust the engine rpm to the correct idling speed by retarding the throttle. If the idle speed is above the correct value when the throttle is fully retarded, adjust the screw at the idle speed throttle stop until the rpm can be set at the correct value as specified in the operator's manual.

If the engine is equipped with a manifold pressure gage, use this instrument for checking the idle mixture.

With the engine operating at correct idle speed, slowly move the manual mixture control in the cockpit toward the IDLE CUTOFF position. Watch the manifold pressure gage and note whether there is a slight decrease in manifold pressure (MAP) and an increase in rpm just before the engine rpm drops and the MAP rises. This decrease should be about $\frac{1}{4}$ inHg [0.85 kPa], and the rpm should increase 50 to 70. This small increase in MAP indicates that the mixture is slightly richer than the best power mixture and is passing through the best power ratio as the control moves toward IDLE CUTOFF. To avoid stopping the engine during this check, return the mixture control to AUTO RICH as soon as the engine starts to cut out.

If the drop in MAP is greater than $\frac{1}{4}$ inHg and the rpm increase is more than 100 during leaning of the mixture control, it is an indication that the idle mixture is too rich. Adjust the idle mixture on the carburetor and perform the mixture check again. Between the checks, run the engine at increased speed for a short time to clear any accumulated carbon from the combustion chambers. Remember also that the $\frac{1}{4}$ inHg stated here for this check and the rpm increase may not agree with the service or operator's manual for the particular engine being checked. Always use the values given in the appropriate manual.

If there is no drop in MAP and no increase in rpm during the idle mixture check, it is evident that the mixture is too lean. In this case, adjust the mixture on the carburetor to provide the proper enrichment.

When checking the idle mixture for an installation which does not include a manifold pressure gage, the tachometer alone is used to determine the degree of richness. During the idle mixture check, lean the mixture with the manual mixture control in the cockpit and observe the tachometer. There should be an increase in speed of approximately 20 rpm, or whatever is specified in the appropriate manual, just before the engine begins to cut out. If the change in rpm is not correct, adjust the mixture at the carburetor in the same way as when using the MAP as a guide. If the increase in rpm is greater than that specified, the mixture is too rich. If there is little or no increase in rpm, the mixture is too lean.

Upon completion of the idle mixture adjustment, use the screw on the throttle stop to adjust the idle speed at the fully retarded position. Minimum manifold pressure should be indicated by the MAP gage.

Malfunctions of Pressure-Type Carburetors

Malfunctions of a pressure-type carburetor usually require that the carburetor be removed and checked on a flow bench. Minor adjustments of idling mixture, idle speed, and control linkages can be made in the field, but such problems as plugged passages, ruptured diaphragms, and sticking valves cannot be corrected without at least partial disassembly. This can be done only in an approved carburetor overhaul shop.

The technician determines when it is necessary to remove a carburetor for internal repair. Certain symptoms will give indications of specific carburetor problems.

If the engine stops or will not start, several steps can be taken to determine if the carburetor is at fault:

1. Turn on the fuel boost pump and move the manual mixture control to AUTO RICH. Fuel should flow from the discharge nozzle and out the drain.

2. If there is no fuel flow from a small PS-type carburetor, check to see that the idle cutoff plunger is not stuck in the IDLE CUTOFF position. If it is stuck, a light tap should release it. Lubricate the plunger periodically to prevent sticking.

3. If there is no fuel flow and the idle cutoff plunger is not stuck, check to see if fuel pressure is available at the carburetor fuel inlet. With the fuel boost pump on, loosen the fuel connection at the carburetor. If there is no fuel flow, correct the trouble in the fuel system

4. If fuel pressure is available at the carburetor fuel inlet and fuel will not flow through the carburetor when the mixture control is placed in the AUTO RICH position, this indicates that the poppet valve is not opening or that the carburetor fuel screen is completely plugged. If the fuel screen is checked and found clear, then the idle spring in the fuel regulator unit is probably broken. Remove and replace the carburetor.

If an engine equipped with a PR-type carburetor will start and run at idle, but cuts out at higher speeds, this indicates that the air diaphragm between the *A* and *B* chambers is ruptured and therefore applies no force to open the poppet valve. In this case, the carburetor must be removed and overhauled.

When an engine runs excessively rich at all operating speeds, even when the mixture control is in the AUTO LEAN position, it is likely that the fuel diaphragm is ruptured. In a PD- or PR-type carburetor, a ruptured fuel diaphragm would permit excessive opening of the poppet valve, with resultant excess fuel pressure across the fuel metering jets. With a PS-type carburetor, a leaking or ruptured fuel diaphragm would result in excessive opening of the poppet valve. In addition, fuel would flow through the diaphragm into the *B* chamber of the carburetor from whence it would be sucked out into the venturi section of the carburetor. This, of course, would result in an exceptionally rich mixture.

In any case where a pressure-type carburetor causes the engine to run excessively rich, the carburetor must be removed for repair.

If an engine runs normally at low altitudes but runs rich at high altitudes, this indicates that the automatic mixture control is not functioning or that the AMC air bleed between chambers A and B is plugged. With some carburetors, the AMC unit can be removed for checking and cleaning without disturbing other sections of the carburetor. The manufacturer's manual should be checked to determine whether the AMC can be removed.

A malfunction of the accelerating pump is indicated by the engine cutting out or hesitating when the throttle is advanced. If the accelerating pump is a diaphragm type, the diaphragm may be ruptured, and the carburetor must be removed for repair.

In any case where the performance of a pressure-type carburetor cannot be made satisfactory by adjustment of the idle mixture, idle speed, or controls, the carburetor must be removed for repair. As stated previously, the overhaul and repair of pressure-type carburetors must be done in an approved carburetor repair facility where the overhauled units can be tested on a flow bench to ensure satisfactory operation.

● REVIEW QUESTIONS

1. List the advantages of a pressure-injection carburetor.
2. Name the principal units of a pressure-type carburetor.
3. How does the pressure-type carburetor measure airflow into the engine?
4. Explain how the air metering force is developed.
5. Under what conditions does the air metering force increase?
6. What are the four fuel regulating chambers in the PD carburetor and what pressures are applied to each?
7. Give the principal functions of the fuel-control unit.
8. Why is an idling valve needed in the fuel-control unit?
9. What prevents the poppet valve from closing at idling speeds when the air metering force is not sufficient to keep it open?
10. In simple terms, what is the function of the fuel-regulator section or unit in a pressure-type carburetor?
11. What functional items may be found in the fuel-control unit of a pressure-type carburetor?
12. Why is the discharge nozzle equipped with a valve?
13. Describe the operation of a diaphragm-operated accelerating pump.
14. Explain the purpose and function of an enrichment valve.
15. Describe the construction and operation of an automatic mixture-control unit.
16. In a PS-type carburetor, what pressure is modified by the manual mixture control?
17. To what chamber of the regulator section is unmetered fuel pressure applied in the PS-type carburetor?
18. If the impact pressure openings were clogged in a PS carburetor, what would be the effect on engine operation?
19. Trace the fuel flow from unit to unit through the PS carburetor.
20. What is the effect of a pressure increase in the D chamber of a PS-type carburetor?
21. What arrangement is made in a pressure carburetor to allow fuel flow for idling when the throttle is closed?
22. Describe the idling system of the PS-type carburetor.
23. Explain the function of the idle cutoff cam in a PS-type carburetor.
24. Name the four principal units of a PR-58 carburetor.
25. What are the functions of the throttle unit?
26. Describe the operation of the accelerating pump on a PR-58 carburetor.
27. What would be the effect if the air diaphragm between chambers A and B should become ruptured?
28. What unit establishes the metered fuel pressure (chamber C) in a PR-58 carburetor?
29. To what extent is the idle valve controlled by the throttle?
30. What is the function of the sealing diaphragm in the partition between chambers B and C?
31. Explain the purpose of the balance chamber at the end of the poppet-valve system.
32. What would happen if a vapor-separator valve or vent line became clogged during operation?
33. If a vapor-separator float should leak, what would be the result? How can the vapor separator be tested?
34. Describe the fuel-control unit for a PR-58 carburetor and explain the function of each part which affects fuel flow.
35. What special valve must be included when the carburetor is designed for use with a water-alcohol injection system?
36. Why is a power enrichment valve necessary?
37. What causes the power enrichment valve to open?
38. What is the function of the regulator fill valve?
39. Describe the operation of the automatic mixture control.
40. What would be the effect if the AMC air bleed between chambers A and B should become clogged?
41. In what position is the manual mixture-control lever placed for best fuel economy during operation?
42. Explain the value of a water-injection system.
43. Why is it necessary to mix alcohol with the water for use in the water-injection system? Why is water-soluble oil added?
44. How does the fuel-air ratio when using the ADI system compare with the F/A ratio when the engine is operated without ADI?
45. Give a brief description of the water-alcohol regulator.
46. At what point in the system is the ADI fluid mixed with the fuel?
47. What would happen if the ADI fluid became exhausted while operating the system on takeoff?
48. Why are delay bleeds necessary in the regulator?
49. How is the ADI system controlled by the pilot?
50. What prevents operation of the ADI system at low-power settings?

51. What is the effect of high humidity in the air during takeoff without water injection?
52. Describe a typical inspection of an engine fuel system.
53. What precaution must be observed in tightening fluid-line fittings?
54. What conditions of a fuel hose require the hose to be replaced?
55. What should be done with metal tubing which has been dented 25 percent of its diameter?
56. Describe the inspection of a fuel pump installation.
57. How would an installed carburetor be inspected?
58. What would be the effect of an excessively loose throttle shaft?
59. What precautions should be observed in the removal and replacement of a downdraft carburetor?
60. Describe the removal and installation of a carburetor.
61. Why should the throttle of a downdraft carburetor be secured in the closed position during removal and replacement?
62. What precaution must be observed in the installation of a new gasket for a carburetor?
63. Describe the procedure for tightening nuts or bolts in the installation of a carburetor.

64. What treatment should be given a new or overhauled carburetor which has been in storage before placing it in service?
65. What adjustments may be made to pressure-type carburetors in the field?
66. What is the importance of springback when adjusting carburetor control linkages?
67. What precautions must be taken in adjusting control which incorporates detents?
68. Describe the procedure for setting or adjusting the idle mixture.
69. Why should the idle mixture be somewhat richer than the best power mixture?
70. At what location is adjustment for idle speed made?
71. What malfunctions are indicated if fuel will not flow through a pressure -type carburetor when the fuel boost pump is turned on and the mixture control placed in the AUTO RICH position?
72. What malfunctions are indicated if an engine with a pressure-type carburetor runs excessively rich?
73. What problem is indicated if an engine runs rich at high altitudes but operates normally at low altitudes?
74. How will an engine respond when the accelerating pump is not functioning?

6 FUEL-INJECTION SYSTEMS

● DEFINITION

Fuel injection is the introduction of fuel or a fuel-air mixture into the induction system of an engine or into the combustion chamber of each cylinder by means of a pressure source other than the pressure differential created by airflow through the venturi of a carburetor. The usual pressure source is an injection pump, which may be any one of several types. A **fuel-injection carburetor** discharges the fuel into the airstream at or near the carburetor. A **fuel-injection system** discharges the fuel into the intake port of each cylinder just ahead of the intake valve, or directly into the combustion chamber of each cylinder.

Fuel-injection systems have a number of advantages, among which are the following:

1. Freedom from vaporization icing, thus making it unnecessary to use carburetor heat except under the most severe atmospheric conditions
2. More uniform delivery of the fuel-air mixture to each cylinder
3. Improved control of the fuel-air ratio
4. Reduction of maintenance problems
5. Instant acceleration of the engine after idling, with no tendency to stall
6. Increased engine efficiency

Fuel-injection systems have been designed for all types of reciprocating aircraft engines from the 4-cylinder opposed engines up to the 28-cylinder Pratt & Whitney R-4360 engine. They are presently used on a wide variety of light engine airplanes, large commercial aircraft, helicopters, and military aircraft.

● CONTINENTAL CONTINUOUS-FLOW INJECTION SYSTEM

The Continental fuel-injection system is of the multinozzle, *continuous-flow* type which controls fuel flow to match engine airflow. The fuel is discharged into the intake port of each cylinder. Any change in air throttle position or engine speed, or a combination of both, causes changes in fuel flow in the correct relation to engine airflow. A manual mixture control and a pressure gage, indicating metered fuel pressure, are provided for precise leaning at any combination of altitude and power setting. Since fuel flow is directly proportional to metered fuel pressure, the settings can be predetermined and the fuel consumption can be accurately predicted. The continuous-flow system permits the use of a typical rotary-vane fuel pump in place of a much more complex and expensive plunger-type pump. There is no need for a timing mechanism because each cylinder draws fuel from the discharge nozzle in the intake port as the intake valve opens.

The Continental fuel-injection system consists of four basic units. These are the **fuel-injection pump,** the **fuel-air control unit,** the **fuel manifold valve,** and the **fuel discharge nozzle.** These units are shown in Fig. 6-1.

Fuel-Injection Pump

The **fuel-injection pump,** Fig. 6-2, is a positive-displacement rotary-vane type with a splined shaft for

FUEL–AIR
CONTROL UNIT

INJECTOR
PUMP

FUEL
MANIFOLD
VALVE

NOZZLES

FIG. 6-1 Units of the Continental fuel-injection system.

connection to the accessory drive system of the engine. A spring-loaded diaphragm-type relief valve, which also acts as a pressure regulator, is provided in the body of the pump. Pump outlet fuel pressure passes through a calibrated orifice before entering the relief valve chamber, thus making the pump delivery pressure proportional to engine speed.

Fuel enters the pump assembly at the swirl well of the **vapor separator,** as shown in the drawing. At this point, any vapor in the fuel is forced upward to the top of the chamber, where it is drawn off by means of the **vapor ejector.** The vapor ejector is a small pressure jet of fuel which feeds the vapor into the vapor return line, where it is carried back to the fuel tank. There are no moving parts in the vapor separator, and the only restrictive passage is used in connection with vapor removal; hence, there is no restriction of main fuel flow.

Disregarding the effects of altitude or ambient air conditions, the use of a positive-displacement engine-driven pump means that changes in engine speed affect total pump flow proportionally. The pump provides greater capacity than is required by the engine; hence, a recirculation path is required. When the relief valve and orifice are placed in this path, fuel pressure proportional to engine speed is provided. These provisions assure proper pump pressure and delivery for all engine operating speeds.

A check valve, shown in the drawing, is provided so that boost pressure to the system from an auxiliary pump can bypass the engine-driven pump when starting. This feature is also available to suppress vapor formation under high ambient temperatures of the fuel. Furthermore, this permits use of the auxiliary pump as a source of fuel pressure in the event of failure of the engine-driven pump.

Fuel-Air Control Unit

The **fuel-air control unit,** shown in Fig. 6-3, occupies the position ordinarily used for the carburetor at the intake manifold inlet. The unit includes three control elements, one for air in the **air throttle assembly** and two for fuel in the **fuel-control assembly,** which is mounted on the side of the air throttle assembly.

The air throttle assembly is an aluminum casting which contains the shaft and butterfly valve. The casting bore size is tailored to the engine size, and no venturi or other restriction is employed. Large shaft bosses provide an adequate bearing area for the throttle shaft so that there is a minimum of wear at the shaft bearings. Wave washers are used to provide protection against vibration. A conventional idle speed adjusting screw is mounted in the air throttle shaft lever and bears against a stop pin in the casting. The air throttle assembly is shown in Fig. 6-3.

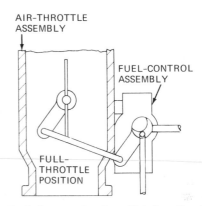

FIG. 6-3 Fuel-air control unit. *(Teledyne Continental)*

The fuel-control unit, shown in Figs. 6-3 and 6-4, is made of bronze for best bearing action with the stainless-steel valves. The central bore contains a metering valve at one end and a mixture-control valve at the other end. These rotary valves are carried in oil-impregnated bushings and are sealed against leakage by O rings. Loading springs are installed in the center bore of the unit between the end bushings and the large end of each control shaft to force the valve ends against a fixed plug installed in the middle of the bore. This arrangement ensures a close contact between the valve faces and the metering plug. The bronze metering plug has one passage that mates with the fuel return port and one through passage that connects the mixture-control-valve chamber with the metering-valve chamber. O rings are used to seal the metering plug in the body.

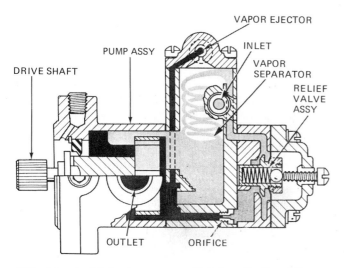

FIG. 6-2 Fuel-injection pump. *(Teledyne Continental)*

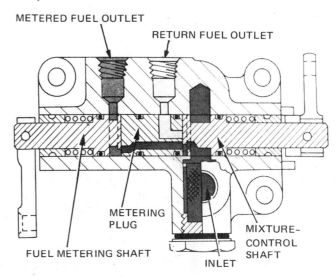

FIG. 6-4 Fuel-control unit. *(Teledyne Continental)*

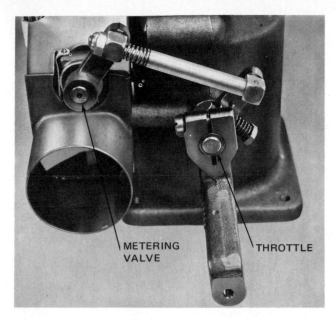

FIG. 6-5 Linkage from throttle to fuel metering valve.

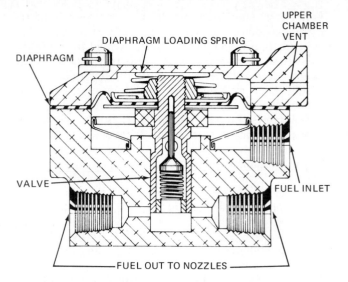

FIG. 6-6 Fuel manifold valve. (*Teledyne Continental*)

Each stainless-steel rotary valve includes a groove which forms a fuel chamber. A contoured end face of the mixture-control valve aligns with the passages in the metering plug to regulate the fuel flow from the fuel chamber to the metering valve or to the return fuel outlet. A control lever is mounted on the mixture-control-valve shaft for connection to the cockpit mixture control. If the mixture control is moved toward the lean position, the mixture-control valve in the fuel-control unit causes additional fuel to flow through the return line to the fuel pump. This, of course, reduces the fuel flow through the metering plug to the metering valve. Rotation of the fuel-control valve toward the RICH position causes more fuel to be delivered to the metering valve and less to the return fuel outlet.

On the metering valve, a cam-shaped cut is made on one outer part of the end face which bears against the metering plug. As the valve is rotated, the passage from the metering plug is opened or closed in accordance with the movement of the throttle lever. As the throttle is opened, the fuel flow to the metered fuel outlet is increased. Thus, fuel is measured to provide the correct amount for the proper fuel-air ratio. The linkage from the throttle to the fuel metering valve is shown in Fig. 6-5.

Fuel Manifold Valve

The **fuel manifold valve** is illustrated in Fig. 6-6. The fuel manifold valve body contains a fuel inlet, a diaphragm chamber, a valve assembly, and outlet ports for the lines to the individual fuel nozzles. The spring-loaded diaphragm carries the valve plunger in the central bore of the body. The diaphragm is enclosed by a cover which retains the diaphragm loading spring.

When the engine is not running, there is no pressure on the diaphragm to oppose the spring pressure; hence, the valve will be closed to seal off the outlet ports. Furthermore, the valve in the center bore of the valve plunger will be held on its seat to close the passage through the plunger. When fuel pressure is applied to the fuel inlet and into the chamber below the dia-

phragm, the diaphragm will be deflected and will raise the plunger from its seat. The pressure will also open the valve inside the plunger and allow fuel to pass through to the outlet ports.

A fine screen is installed in the diaphragm chamber so that all fuel entering the chamber must pass through the screen to filter out any foreign particles. During inspection or troubleshooting of the system, this screen should be cleaned.

Fuel Discharge Nozzle

The **fuel discharge nozzle,** shown in Fig. 6-7, is mounted in the cylinder head of the engine with its outlet di-

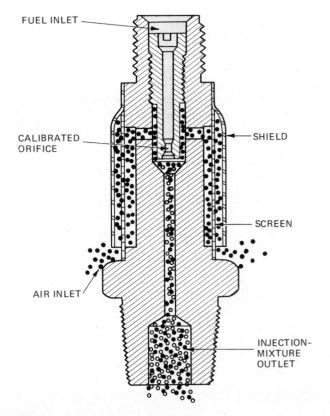

FIG. 6-7 Fuel discharge nozzle.

rected into the intake port. The nozzle body contains a drilled central passage with a counterbore at each end. The lower end of the nozzle is used for a fuel-air mixing chamber before the spray leaves the nozzle. The upper bore contains a removable orifice for calibrating the nozzles.

Near the top of the nozzle body, radial holes connect the upper counterbore with the outside of the nozzle body to provide for air bleed. These holes enter the counterbore above the orifice and draw outside air through a cylindrical screen fitted over the nozzle body. The screen keeps dirt and other foreign material out of the interior of the nozzle. A press-fitted shield is mounted on the nozzle body and extends over the greater part of the filter screen, leaving an opening near the bottom. This provides both mechanical protection and an air path of abrupt change of direction as an aid to cleanliness. Nozzles are calibrated in several ranges, and all nozzles furnished for one engine are of the same range. The range is identified by a letter stamped on the hex of the nozzle body.

Complete System

The complete Continental fuel-injection system installed on an engine is shown in Fig. 6-8. This diagram shows the fuel-air control unit installed at the usual location of the carburetor, the pump installed on the accessory section, the fuel manifold valve installed on the top of the engine, and the nozzles installed in the cylinders at the intake ports. The simplicity of the system, which contributes to ease of maintenance and economy of operation, is clearly apparent. A diagram of the complete system is shown in Fig. 6-9.

To summarize, the Continental fuel-injection system utilizes fuel pressure (established by engine rpm and the relief valve) and a variable orifice (controlled by throttle position) to meter the correct volume and pressure of fuel for all power settings. The mixture is controlled by adjusting the quantity of fuel returned to the pump inlet.

Installations for Turbocharged Engines

To permit the Continental fuel-injection system to operate at the high altitudes encountered by turbocharged engines, the fuel nozzles are modified by incorporating a shroud to direct pressurized air to the air bleeds. The low air pressure at high altitudes is not sufficient for air to enter the bleeds and mix with the fuel prior to injection. Ram air is therefore applied to the nozzle shrouds, which direct it to the air bleeds. A shrouded fuel nozzle is illustrated in Fig. 6-10.

Another modification required for high-altitude operation is an **altitude compensating valve** in the fuel pump. This valve is operated by an aneroid bellows which moves a tapered plunger in an orifice. The plunger's movement varies the amount of fuel bypassed through the relief valve and thus compensates for the effect of lower pressures at high altitude. A pump which incorporates the aneroid valve is shown in Fig. 6-11.

Adjustments

The **idle speed adjustment** for the Continental fuel-injection system is made with a conventional spring-loaded screw (Fig. 6-12) on the throttle lever of the fuel-control unit. The screw is turned to the right to increase and to the left to decrease idle speed.

The **idle mixture adjustment** (Fig. 6-12) is accomplished by means of the locknut at the metering-valve end of the linkage between the metering valve and the throttle lever. Tightening the nut to shorten the linkage provides a richer mixture, and backing off the nut leans the mixture. The mixture should be adjusted slightly richer than best power, as is the case with all systems. This is checked after adjustment by running the engine at idle and slowly moving the mixture control toward IDLE CUTOFF. Idle speed should increase slightly just before the engine begins to stop. When the manifold pressure gage is used for the check, the MAP should be seen to decrease slightly as the mixture is leaned with the manual mixture control.

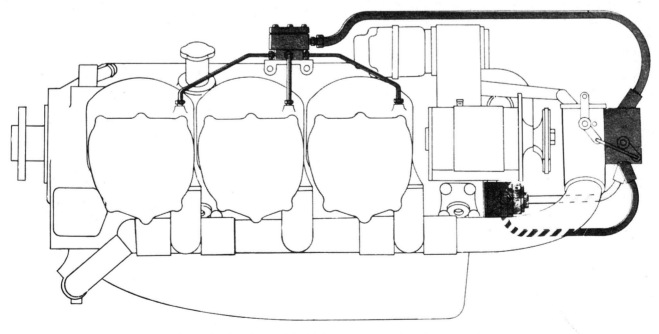

FIG. 6-8 Continental fuel-injection system installed on engine.

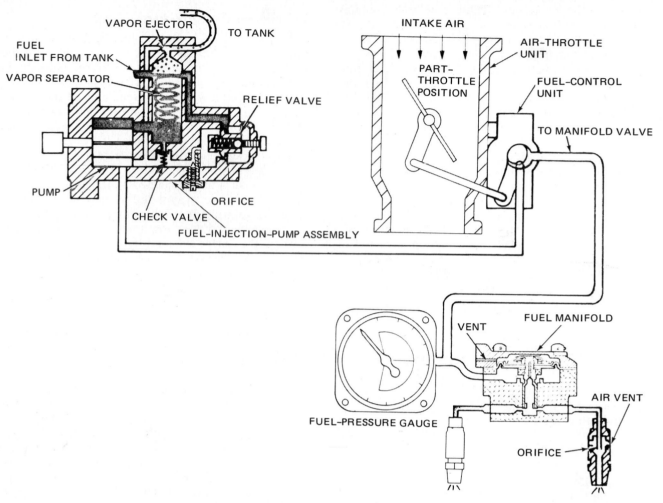

FIG. 6-9 Schematic diagram of the Continental fuel-injection system. *(Teledyne Continental)*

Pump Pressure for the Injection System

Since the flow of a liquid through a given orifice will increase as the pressure increases, the fuel pressure delivered by the engine-driven fuel pump must be correct if the flow through the metering unit and the fuel nozzles is to be correct. The fuel pump adjustment is to be accomplished as directed by the manufacturer in service manuals or bulletins.

The fuel pump pressure is adjusted at both idle speed and at full-power rpm. A pressure gage is connected to the fuel pump outlet line or the metering unit inlet

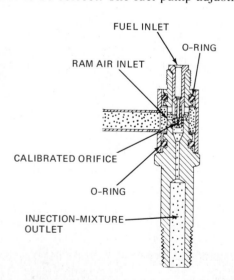

FIG. 6-10 Shrouded fuel nozzle. *(Teledyne Continental)*

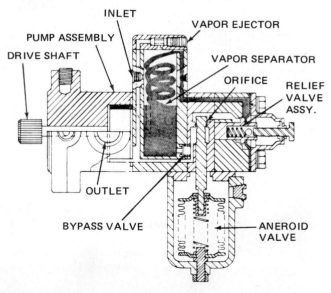

FIG. 6-11 Fuel pump incorporating the aneroid valve.

126

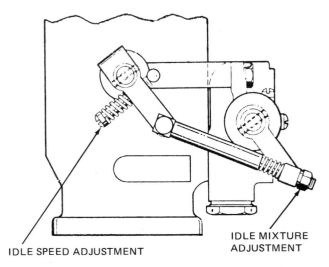

IDLE SPEED ADJUSTMENT   IDLE MIXTURE ADJUSTMENT

FIG. 6-12 Idle speed and mixture adjustments.

line by means of a tee fitting. The technician can then make the fuel adjustment in accordance with the instructions for the system being tested.

To illustrate the fuel pressure requirements for typical engines, the following information is quoted from the *Teledyne Continental Service Bulletin*:

Engine Model—IO-346-A,B
600 RPM
Pump pressure	7–9 psi	48–62 kPa
Nozzle pressure	2.0–2.5 psi	13.8–17.25 kPa

2700 RPM
Pump pressure	19–21 psi	131–145 kPa
Nozzle pressure	12.5–14.0 psi	86–96.5 kPa
Fuel flow	78–85 lb/hr	13–14 gal/hr 35.4–38.5 kg/hr

Engine Model—GTSIO-520-C
RPM 450
Pump pressure	5.5–6.5 psi	38–45 kPa
Nozzle pressure	3.5–4.0 psi	24–28 kPa

RPM 2400
Pump pressure	30–33 psi	207–228 kPa
Nozzle pressure	16.5–17.5 psi	114–121 kPa
Fuel flow	215–225 lb/hr	36–38 gal/hr 98–102 kg/hr

Inspections

In order to avoid any difficulty with the fuel-injection system, it is well to perform certain inspections and checks, even though no operating discrepancies have been noted. The following inspections are recommended:

1. Check all attaching parts for tightness. Check all safetying devices.
2. Check all fuel lines for leaks and for evidence of damage, such as sharp bends, flattened tubes, or chafing from metal-to-metal contact.
3. Check the control connections, levers, and linkages for tight attaching parts, for safetying, and for lost motion owing to wear.
4. Inspect nozzles for cleanliness, with particular attention to air screens and orifices. Use a standard ½-in [12.70-mm] spark-plug wrench (deep socket) to remove the nozzles. Do not remove the shields to clean

the air screen in the nozzles. Do not use wire or other objects to clean the orifices. To clean the nozzles, remove them from the engine and immerse them in fresh cleaning solvent. Use compressed air to dry.

5. Unscrew the strainer plug from the fuel injection-control valve, and clean the screen in fresh cleaning solvent. Reinstall, safety, and check for leaks.

6. In periodic lubrication, apply a drop of engine oil on each end of the air throttle shaft and at each end of the linkage between the air throttle and the fuel metering valve. No other lubrication is required.

7. In the event that a line fitting in any part of the injection system must be replaced, only a fuel-soluble lubricant (such as engine oil) is authorized on the fitting threads as installation. Do not use any other form of thread compound.

8. If a nozzle is damaged and requires replacement, it is not necessary to replace the entire set. Each replacement nozzle must match the one removed, as marked.

Operation

For the operation of an aircraft engine equipped with the Continental fuel-injection system, certain facts must be remembered by the technician or pilot:

1. When starting the engine, it is easy to flood the system if the timing of the starting events is not correct. When the mixture control is in any position other than IDLE CUTOFF, when the throttle is open (even slightly), and when the auxiliary fuel pump is operating, fuel will be flowing into the intake ports of the cylinders. Therefore, the engine should be started within a few seconds after the auxiliary fuel pump is turned on.
2. The engine cannot be started without the auxiliary fuel pump because the engine-driven pump will not supply adequate pressure until the engine is running.
3. The auxiliary fuel pump should be turned off during flight. It may be left on during takeoff as a safety measure.
4. For takeoff, the throttle should be fully advanced and the mixture control should be set at FULL RICH.
5. For cruising, the engine rpm should be set according to instructions in the operator's handbook. The mixture control may be set for best power or for an economy cruising condition, depending on the desire of the operator. Care must be exercised to make sure that the mixture is not leaned too much.
6. Before reducing power for descent, set the mixture to best power. Once the traffic pattern is entered, the mixture control must be set to FULL RICH and kept in this position until after landing.
7. The engine is stopped by moving the mixture control to IDLE CUTOFF after the engine has been idled for a short time. All switches should be turned off immediately after the engine is stopped.

Troubleshooting

Before assuming that the fuel-control system for an engine is at fault when the engine is not operating properly, the technician should make sure that other factors are not involved. A rough-running engine may have ignition problems which should be checked before dis-

Trouble	Probable cause	Corrective action
Engine will not start.	No fuel to engine.	Check mixture control for proper position, auxiliary pump ON and operating, feed valves open, fuel filters open, tank fuel level.
No gage pressure.	Engine flooded.	Reset throttle, clear engine of excess fuel, and try another start.
With gage pressure.	No fuel to engine.	Loosen one line at nozzle. If no fuel flow shows with metered fuel pressure on gage, replace the fuel manifold valve.
Rough idle.	Nozzle air screens restricted. Improper idle mixture adjustment.	Remove nozzles and clean. Readjust as described under *Adjustments*, page 125.
Poor acceleration.	Idle mixture too lean. Linkage worn.	Readjust as described under *Adjustments*, page 125. Replace worn elements of linkage.
Engine runs rough.	Restricted nozzle. Improper mixture.	Remove and clean all nozzles. Check mixture-control setting. Improper pump pressure; replace, or call an authorized representative to adjust.
Low gage pressure.	Restricted flow to metering valve. Inadequate flow from pump. Mixture-control-level interference.	Check mixture control for full travel. Check for clogged fuel filters. May be worn fuel pump or sticking relief valve. Replace the fuel pump assembly. Check for possible contact with cooling shroud.
High gage pressure.	Restricted flow beyond metering valve. Restricted recirculation passage in pump.	Check for restricted nozzles or fuel manifold valve. Clean or replace as required. Replace pump assembly.
Fluctuating gage pressure.	Vapor in system; excess fuel temperature. Fuel in gage line. Leak at gage connection.	If not cleared with auxiliary pump, check for clogged ejector jet in the vapor-separator cover. Clean only with solvent, no wires. Drain the gage line and repair the leak.
Poor idle cutoff.	Engine getting fuel.	Check mixture control to be in full IDLE CUTOFF. Check auxiliary pump to be OFF. Clean nozzle assemblies (screens) or replace. Replace manifold valve.

mantling the fuel-control system. The troubleshooting chart provided here is a guide for determining what may be at fault. The items in the chart are arranged in sequence of the approximate ease of checking and not necessarily in the order of probability.

● BENDIX RSA FUEL-INJECTION SYSTEM

The Bendix RSA-type fuel-injection system is a continuous-flow type and is based on the same principles as the pressure-type carburetors described in Chap. 5. The principal units of the system are shown in Fig. 6-13. A **fuel regulator section,** consisting of two diaphragms and four chambers, serves to regulate the fuel pressure to the flow divider, which distributes the fuel to the fuel nozzles. The quantity of air consumed by the engine is measured by a venturi and by impact tubes in the **airflow section** of the system. The reduced pressure created by the venturi is applied to one side of the **air diaphragm** in the regulator section, and the increased air pressure from the impact tubes is applied

to the opposite side of the diaphragm. The air diaphragm acts to open the ball valve, permitting fuel to flow to the flow divider and thence to the fuel nozzles. An increase of airflow through the airflow section thus causes an increase in fuel flow to the engine.

The fuel pressure from the fuel pump is applied to one side of the **fuel diaphragm** in the regulator section, and metered fuel is applied to the opposite side of the fuel diaphragm. The fuel diaphragm tends to close the ball valve, depending on the differential fuel pressure across the diaphragm. Thus, the fuel pressures and air pressures are balanced to provide correct fuel flow and pressure to the fuel nozzles.

Airflow and Regulator Sections

The airflow and regulator sections of the RSA fuel-injection system are shown in Fig. 6-13. An external view of the complete **injector unit** is shown in Fig. 6-14. The injector includes the airflow section, the regulator section, and the fuel metering section.

Air entering the airflow section produces ram pressure on the impact tubes (shown in Fig. 6-13) and this

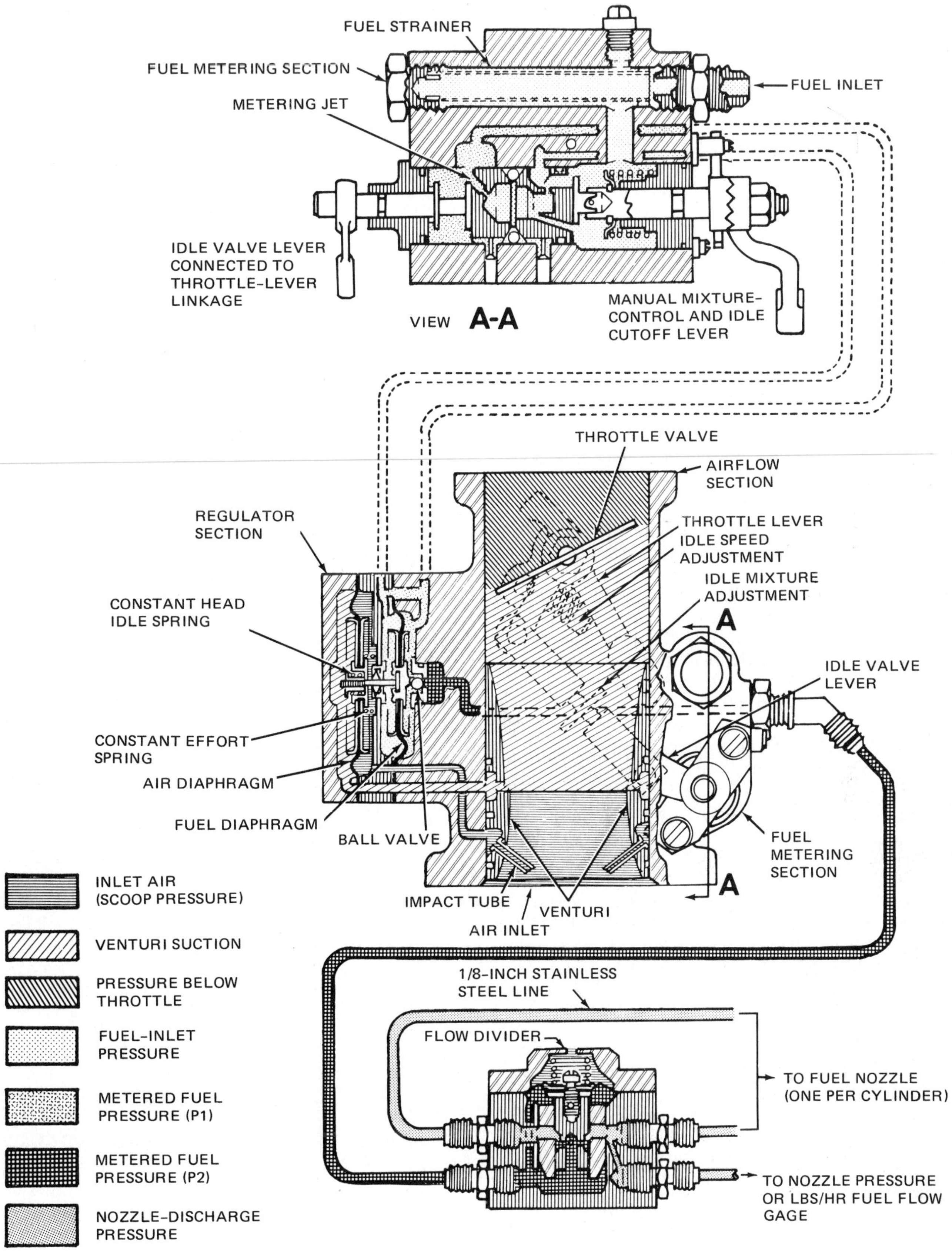
FUEL STRAINER
FUEL METERING SECTION
METERING JET
FUEL INLET
IDLE VALVE LEVER CONNECTED TO THROTTLE-LEVER LINKAGE
MANUAL MIXTURE-CONTROL AND IDLE CUTOFF LEVER
VIEW A-A
THROTTLE VALVE
AIRFLOW SECTION
REGULATOR SECTION
THROTTLE LEVER
IDLE SPEED ADJUSTMENT
IDLE MIXTURE ADJUSTMENT
A
CONSTANT HEAD IDLE SPRING
IDLE VALVE LEVER
CONSTANT EFFORT SPRING
AIR DIAPHRAGM
FUEL DIAPHRAGM
BALL VALVE
FUEL METERING SECTION
A
IMPACT TUBE
VENTURI
AIR INLET
INLET AIR (SCOOP PRESSURE)
VENTURI SUCTION
PRESSURE BELOW THROTTLE
FUEL-INLET PRESSURE
METERED FUEL PRESSURE (P1)
METERED FUEL PRESSURE (P2)
NOZZLE-DISCHARGE PRESSURE
1/8-INCH STAINLESS STEEL LINE
FLOW DIVIDER
TO FUEL NOZZLE (ONE PER CYLINDER)
TO NOZZLE PRESSURE OR LBS/HR FUEL FLOW GAGE

FIG. 6-13 Bendix RSA-type fuel-injection system. *(Cessna Aircraft Co.)*

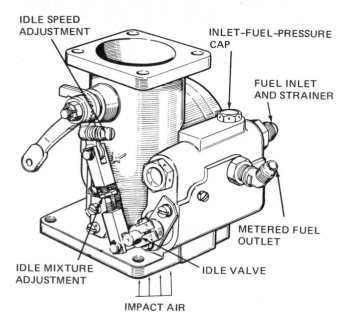

IDLE SPEED ADJUSTMENT

INLET-FUEL-PRESSURE CAP

FUEL INLET AND STRAINER

METERED FUEL OUTLET

IDLE MIXTURE ADJUSTMENT

IDLE VALVE

IMPACT AIR

FIG. 6-14 Bendix RSA injector unit. *(Energy Controls Div., Bendix Corp.)*

pressure is directed to the right-hand side of the air diaphragm in the regulator section, thus causing the diaphragm to move to the left. Air flowing through the venturi causes a decrease in pressure, which is directed to the left-hand side of the air diaphragm, thereby producing additional force to move the diaphragm to the left. A study of the drawing will reveal that the movement of the air diaphragm to the left will open the ball valve and permit fuel to flow to the flow divider. The force produced by the air pressure acting on the air diaphragm is called the **air metering force.** It should be noted that the movement of the air diaphragm will also move the fuel diaphragm, because the two diaphragms are connected by means of the **regulator stem.**

As shown in Fig. 6-13, two different fuel pressures are applied to the fuel diaphragm. **Inlet fuel pressure,** also called **unmetered fuel pressure,** is applied to one side of the fuel diaphragm and **metered fuel pressure** is applied to the other side. Inlet fuel pressure is applied to the left side of the diaphragm and metered fuel pressure is applied to the ball valve side of the diaphragm. Metered fuel pressure is the pressure of the fuel after it has passed through the fuel strainer, the manual mixture-control rotary plate, the main metering jet, and the rotary idle plate in the fuel metering section. Unmetered (inlet) fuel pressure is the pressure of the fuel not only at the inlet of the fuel metering section but also after the fuel has passed through the fuel strainer. The force produced by the fuel pressures acting on the fuel diaphragm is called **fuel metering force.**

Operation of the Regulator Section

At any given engine rpm and throttle setting, the impact air pressure, venturi air pressure, and inlet fuel pressure will all be constant. The metered fuel pressure is therefore controlled by the ball valve to provide constant fuel pressure to the flow divider. As the ball valve opens, metered fuel pressure tends to decrease on the right side of the fuel diaphragm. This causes an

immediate movement of the regulator diaphragms to close the ball valve, because the differential pressure across the fuel diaphragm increases as metered fuel pressure decreases. As the ball valve moves toward the closed position, the metered fuel pressure increases, thus decreasing the differential pressure across the fuel diaphragm. This reduces the force tending to close the valve, and the valve remains open sufficiently to produce a balanced, or stable, condition.

Since the air differential pressure is a function of the airflow and the fuel differential pressure is a function of the fuel flow, the correct fuel-air ratio is always maintained, regardless of the quantity of air being consumed by the engine. As explained previously—as airflow through the airflow section increases, the air diaphragm applies a greater force, tending to open the ball valve.

Because of the low level of air forces during very low or idling engine speeds, it is necessary to provide a means to allow adequate fuel flow for these conditions. This is accomplished by means of the **constant-head idle spring** and the **constant-effort spring** installed on opposite sides of the air diaphragm in the regulator section. These springs are balanced in such a manner that they provide a constant fuel differential pressure for the idling range of engine operation. They do not permit the ball valve to close and completely shut off fuel flow. As airflow increases and the air diaphragm compresses the constant head idle spring, the spring retainer eventually touches the air diaphragm and the spring then acts as a solid member.

Fuel Metering Section

A cutaway drawing of the fuel metering section for the RSA fuel-injection system is shown in Fig. 6-15. This section includes the fuel strainer, the main metering jet, a manual mixture-control valve, and an idle valve. As shown in Fig. 6-14, the idle valve is linked to the throttle and an idle mixture adjustment is incorporated in the linkage. The main metering jet is a precision-sized passage between inlet fuel pressure and metered fuel pressure. In some models a fuel enrich-

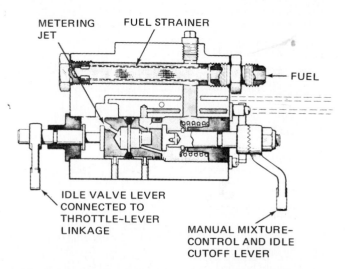

METERING JET

FUEL STRAINER

FUEL

IDLE VALVE LEVER CONNECTED TO THROTTLE-LEVER LINKAGE

MANUAL MIXTURE-CONTROL AND IDLE CUTOFF LEVER

FIG. 6-15 Fuel metering section of the Bendix RSA fuel-injection system.

ment jet is included in parallel with the main metering jet. As explained previously, the fuel enrichment jet serves to provide an enriched fuel mixture for high-power settings, thus preventing detonation and over-heating. This jet is uncovered by the idle mixture rotary plate at advanced throttle settings.

The manual mixture-control valve is actuated by the mixture-control lever or knob in the cockpit. The range of operation is from FULL RICH to IDLE CUTOFF(ICO). During normal operation, the mixture control is set to provide the best fuel-air ratio for the conditions of flight. The operator's handbook provided by the manufacturer of the aircraft describes how to set the fuel mixture for the most satisfactory operation.

As shown in Fig. 6-14, the idle speed is adjusted externally by a screw on the throttle linkage to set the closed throttle position for the desired engine idle speed. The idle mixture is externally adjusted by turning the threaded wheel, which changes the relation between the idle valve position and the throttle position.

Flow Divider

The flow divider (Fig. 6-16) serves to distribute metered fuel from the injector to the fuel nozzles, one of which is located in each cylinder head of the engine. The external appearance of the flow divider is shown in Fig. 6-17. The flow divider keeps metered fuel under pressure during normal operation and shuts the fuel flow off when the mixture control is placed in IDLE CUTOFF.

The operation of the flow divider can be understood by studying Fig. 6-16. The **flow divider needle** is attached to a diaphragm and spring, as seen in the top of the drawing. When there is no fuel pressure, as at IDLE CUTOFF, the spring holds the needle down and fuel cannot flow through the passage in the center of the needle because the annulus (ring) around the outside diameter of the needle is blocked off. As fuel pressure builds up against the diaphragm and the bottom of the needle, the spring pressure is overcome and the needle rises. This opens the annulus to the fuel passages, and fuel can then flow to the nozzles.

Because the regulator meters a fixed amount of fuel to the flow divider, the valve opens only as far as necessary to deliver the proper amount of fuel to the nozzles. At idle speeds, the required opening of the valve is very small and the discharge-nozzle pressure is low. The fuel is therefore divided by the flow divider at low engine speeds. At the point where the fuel flow

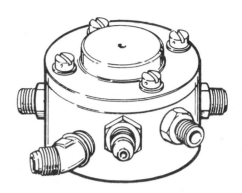

FIG. 6-17 External view of the flow divider.

and pressure are high enough to open the valve completely, fuel pressure is controlled by the nozzles.

Air-Bleed Fuel Nozzles

The fuel nozzles for the individual cylinders are constructed with air bleeds to provide for atomization and improved vaporization of the fuel. A diagram of a nozzle is given in Fig. 6-18. Each nozzle incorporates a calibrated jet, the size of which is determined both by the fuel inlet pressure available and by the maximum fuel flow required by the engine. All nozzles are calibrated to flow the same amount of fuel within 2 percent and are interchangeable between engines and cylinders.

The fuel is discharged through the nozzle jet into an ambient air pressure chamber within the nozzle. The fuel is then drawn into the individual intake-valve chambers and through the intake valves into the combustion chambers.

Because the fuel pressure at the nozzles is directly proportional to the fuel flow, a simple pressure gage (Fig. 6-19) is commonly used as a fuel flowmeter. The gage can be calibrated in psi, gallons per hour, or pounds per hour, and the face is marked accordingly.

The shrouded type of fuel discharge nozzle (Fig. 6-20) is usually used with turbocharged engines. The purpose of the shroud is to deliver compressor discharge air pressure to the air bleeds of the nozzles; otherwise, the fuel pressure would be considerably higher than the ambient air pressure, and raw fuel would pass through the air bleeds and onto the cylinder heads.

Automatic Mixture Control

On the model RSA-5AB1 fuel-injection system, an **automatic mixture control** is included in the injector unit. An external view of the injector is given in Fig.

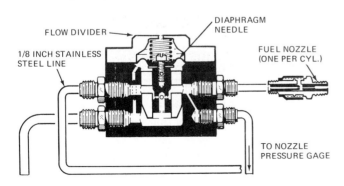

FIG. 6-16 Cutaway view of the flow divider.

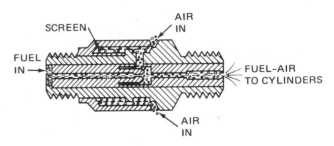

FIG. 6-18 Schematic diagram of an air-bleed fuel nozzle.

FIG. 6-19   Fuel pressure gage (flow indicator).

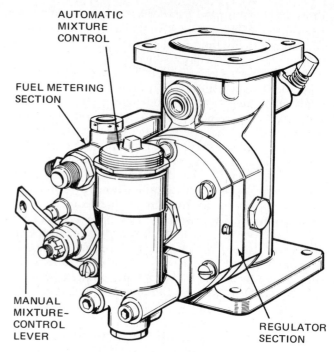

FIG. 6-21   Bendix RSA-5AB1 fuel injector with automatic mixture control.

6-21. This mixture control functions in much the same manner as that described in Chap. 5.

The automatic mixture control, illustrated in Fig. 6-22, adjusts what may be termed a *bypass* valve between the air chambers in the regulator section. The effect reduces the impact air pressure in the regulator when the air becomes less dense, thus reducing the amount of fuel flowing to the engine. As aircraft altitude increases, air density decreases and engine power decreases. When the pilot advances the throttle to produce more power, the regulator section increases the fuel flow, but the actual quantity of air entering the engine does not increase at the same rate because of its lower density. This causes a rich condition and requires adjustment of the fuel-air mixture. The automatic mixture control accomplishes the adjustment without manual control.

The automatic mixture control consists of a sealed bellows (filled with helium gas and a small amount of an inert oil) which controls a contoured needle that is moved in and out of a fixed orifice. The orifice is in a passage between the two air chambers in the regulator section of the injector. When air density decreases, the bellows expands and the orifice size effectively in-

creases. This reduces the effect of the impact air pressure on the air diaphragm, results in less air metering force, and therefore tends to open the ball valve. The ball valve moves toward the closed condition, reducing fuel flow to the flow divider.

When an aircraft equipped with an injection system has no automatic mixture control, the pilot must adjust the mixture manually as the airplane gains altitude;

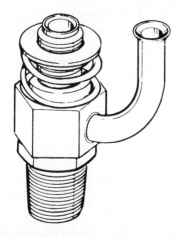

FIG. 6-20   Shrouded type of fuel-discharge nozzle used with turbocharged engines.

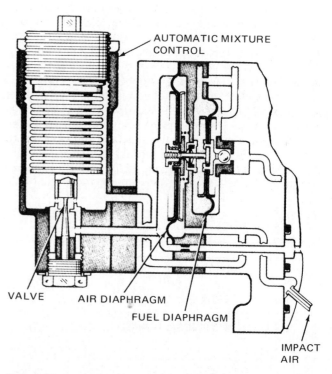

FIG. 6-22   Cutaway view of the Bendix RSA-5AB1 fuel injector, showing the operation of the automatic mixture control.

otherwise, the fuel-air mixture will increase in richness as the altitude increases and the engine will not operate efficiently.

Installation of the Injection System

The following information with respect to the installation of the RSA fuel-injection system was provided by the Energy Controls Division of the Bendix Corporation and is given here to serve as an example of the information supplied to aircraft operators and maintenance personnel:

1. The injector can be mounted on the inlet flange of the engine's intake manifold at any attitude that facilitates engine-to-airframe combination installation, taking into consideration that the throttle linkage and the manual mixture-control linkage must be attached to the unit.

2. An allowance should be made for adequate ventilation to the injector because of possibly high ambient temperatures within the engine nacelle.

3. The flow divider can be mounted at an optimum location with a predetermined bracket configuration; however, it must be mounted with the nozzle-line fittings in a horizontal plane.

4. On engines where the nozzle is installed horizontally, particular attention should be paid to the identification mark stamped on one of the hex flats on the nozzle body. This mark is located 180° from the air-bleed hole and must appear on the lower side of the nozzle. This assures that the air-bleed hole is on top—to reduce fuel bleeding from this opening just after shutdown.

5. A flexible hose is used from the engine-driven fuel pump to the injector fuel inlet. This hose size may differ according to installations.

6. Fuel strainer configuration may differ according to installation requirements. In most cases a 74-micron screen is used.

7. In most installations a No. 4 flexible hose is used from the injector outlet to the flow divider. Later model injectors have an alternate fuel outlet 180° from the standard outlet.

8. A $\frac{1}{8}$-inch (outside diameter) stainless steel tube is routed from a restricted fitting (marked GAGE) on the flow divider to the firewall. A No. 3 low-pressure hose is usually used from the firewall to the gage. In all cases, the hose volume should be held to a minimum.

9. In installations where an inlet fuel pressure gage is used, a No. 4 flexible hose is connected to the fuel-pressure takeoff fitting on the fuel metering system.

10. The nozzle line length will depend on the engine's installation and the location of the flow divider. The nozzle lines are formed from 0.085–0.090-in inside diameter (ID) $\times$ $\frac{1}{8}$-in outside diameter (OD) stainless steel tubing, with suitable fittings to connect to the top of the nozzle and to the flow divider. The lines are clamped at suitable locations to reduce line vibration.

Troubleshooting

The accompanying chart may be used to troubleshoot the Bendix RSA fuel-injection system. It is important, however, to keep in mind any modifications of the units that may change the troubleshooting procedures.

Operation of the Bendix RSA Fuel-Injection System

A principal consideration in the ground operation of an engine equipped with an RSA fuel-injection system is the temperature within the engine nacelle. In flight operation, the engine and nacelle are adequately cooled by the rush of air that results both from the ram effect and from the volume of air moved by the propeller. On the ground, however, there is little or no ram effect, and the volume of air from the propeller is not adequate for cooling at idling speeds during hot weather conditions.

High temperatures in the engine nacelle cause fuel vaporization that, in turn, affects the way the engine responds in its starting, idling, and shutdown. The individual operating the engine should be aware of the effects of high temperatures and should adjust the operating procedures accordingly.

In hot weather, after the engine is shut down, the high temperatures in the engine nacelle cause the fuel in the nozzle feed lines to vaporize and escape through the nozzles into the intake manifold. For this reason, it is not necessary to prime the engine if it is restarted within 20 to 30 min after shutdown. The engine is started with the mixture control in the IDLE CUTOFF position; as soon as the engine fires, the control is advanced to the FULL RICH position. The fuel feed lines quickly fill and supply fuel at the nozzles before the engine can stop. If an engine has been stopped for more than 30 min, it is likely that the fuel vapor in the manifold has dissipated and that some priming may be needed to start the engine.

The starting procedure for an engine equipped with an RSA fuel-injection system has been designed to avoid flooding of the engine and to provide positive results. The normal steps for a cold start are as follows:

1. See that the mixture control is in the IDLE CUTOFF position.

2. Adjust the throttle to $\frac{1}{8}$ open.

3. Turn the master switch ON.

4. Turn the fuel boost pump switch ON.

5. Move the mixture control to FULL RICH until the fuel flow gage reads 4 to 6 gal/h [15.14 to 22.71 L/h], then immediately return the control to IDLE CUTOFF. If the aircraft does not have a fuel flow indicator, place the mixture control in the FULL RICH position for 4 to 5 s and return it to IDLE CUTOFF. Placing the mixture control in the FULL RICH position allows fuel to flow through the nozzles into the intake manifold to prime the engine.

6. Engage the starter.

7. As soon as the engine starts, move the mixture control to FULL RICH.

When starting a hot or warm engine, it is probably not necessary to prime the engine; otherwise, the procedure is the same as for a cold engine start.

The foregoing procedure has proved generally successful; however, for any particular aircraft, the instructions given in the operator's manual should be followed.

TROUBLESHOOTING CHART

Problem	Probable cause	Remedy
Hard starting.	Technique.	Refer to aircraft manufacturer's recommended starting procedure.
	Flooded.	Clear engine by cranking with throttle open and mixture control in ICO.
	Throttle valve opened too far.	Open throttle to a position at approximately 800 rpm.
	Insufficient prime (usually accompanied by a backfire).	Increase amount of priming.
Rough idle.	Mixture too rich or too lean.	Confirm with mixture control. A too-rich mixture will be corrected and roughness decreased during lean-out, while a too-lean mixture will be aggravated and roughness increased. Adjust idle to give a 25 to 50 rpm rise at 700 rpm.
	Plugged nozzle(s) (usually accompanied by high takeoff fuel-flow readings).	Clean nozzles.
	Slight air leak into induction system through manifold drain check valve. (Usually able to adjust initial idle, but rough in 1000 to 1500 rpm range.)	Confirm by temporarily plugging drain line. Replace check valves as necessary.
	Slight air leak into induction system through loose intake pipes or damaged O rings. (Usually able to adjust initial idle, but rough in 1000 to 1500 rpm range.)	Repair as necessary.
	Large air leak into induction system. Several cases of $\frac{1}{8}$-in [3.175-mm] pipe plugs dropping out. (Usually unable to throttle engine down below 800 to 900 rpm.)	Repair as necessary.
	Internal leak in injector. (Usually unable to lean out idle range.)	Replace injector.
	Unable to set and maintain idle.	Replace injector
	Fuel vaporizing in fuel lines or distributor. (Encountered only under high ambient temperature conditions or following prolonged operation at low idle rpms.)	Cool engine as much as possible before shutoff. Keep cowl flaps open when taxiing or when the engine is idling. Use fast idle to aid cooling.
Low takeoff fuel flow.	Strainer plugged.	Remove strainer and clean in a suitable solvent. Acetone or MEK is recommended.
	Injector out of adjustment.	Replace injector.
	Faulty gage.	In a twin-engine installation, crisscross the gages. Replace as necessary. In single-engine, change the gage.
	Sticky flow divider valve.	Clean the flow divider valve.
High fuel flow indicator reading.	Plugged nozzle if high fuel flow is accompanied by loss of power and roughness.	Remove and clean.
	Faulty gage.	Crisscross the gages and replace if necessary.
	Injector out of adjustment.	Replace injector.
Staggered mixture-control levers.	If takeoff is satisfactory, do not be too concerned about staggered mixture-control levers because some misalignment is normal with twin-engine installation.	Check rigging.
Poor cutoff.	Improper rigging of aircraft linkage to mixture control.	Adjust.
	Mixture-control valve scored or not seating properly.	Eliminate cause of scoring (usually burr or dirt) and lap mixture-control valve and plug on surface plate.
	Vapor in lines.	Cool engine as much as possible before shutoff. Keep cowl flaps open when taxiing or when the engine is idling. Use fast idle to aid cooling.
Rough engine (turbocharged) and poor cutoff.	Air-bleed hole(s) clogged.	Clean or replace nozzles.

During ground operations, particularly in hot weather, every effort should be made to keep the engine and nacelle temperatures as low as possible. To do this, keep ground operations at a minimum, keep the engine rpm as high as practical, and keep the cowl flaps open. Upon restarting a hot engine, operate the engine at 1200 to 1500 rpm for a few minutes to dissipate residual heat. Higher rpm also aids in cooling the fuel lines by increasing fuel pressure and flow.

The idle speed and mixture should be adjusted to affect a compromise which will best serve the engine's operating requirements for both cool and hot weather. A comparatively high idling speed (700 to 750 rpm) is best for hot weather; however, it should not be so high that it makes the aircraft difficult to operate on the ground.

The idle mixture for hot weather operation should be set to provide a 50-rpm rise when moving the mixture control to IDLE CUTOFF. The richer setting increases pressure and helps to dissipate any vapor which may form in the system.

Before stopping a hot engine, it should be run at increased rpm for a few minutes to eliminate as much heat as possible. The mixture control is then moved slowly to IDLE CUTOFF. When the engine has stopped, the ignition switch is turned OFF. If the engine is quite warm, it is likely that it will idle roughly for several seconds before stopping because of the vaporized fuel feeding from the fuel nozzles. This is likely even though the idle cutoff will completely stop the flow of fuel to the flow divider.

The operator is cautioned that great care must be taken in moving the propeller on an engine that has just been shut down. Sometimes hot spots may exist in one or more combustion chambers and fuel vapor can be ignited and cause the engine to kick over. It is best to avoid moving the propeller until the engine has cooled for several minutes.

Field Adjustments

As with the pressure discharge carburetor, field adjustments for the fuel-injection system are generally limited to idle speed and idle mixture. The procedure for making these adjustments is given below; however, the procedure given in the airplane operator's manual should be followed for a particular type and model of aircraft.

1. Perform a magneto check in accordance with instructions. If the rpm drop for each magneto is satisfactory, proceed with the idle mixture adjustment.

2. Retard the throttle to the IDLE position. If the idle speed is not in the recommended range, adjust the rpm with the idle speed adjusting screw adjacent to the throttle lever. See Fig. 6-14.

3. When the engine is idling satisfactorily, slowly move the mixture-control lever in the cockpit toward the IDLE CUTOFF position. Observe the tachometer or manifold pressure (MAP) gage. If the engine rpm increases slightly or the manifold pressure decreases, this indicates that the mixture is on the rich side of best power. An immediate decrease in rpm or increase

in manifold pressure indicates that the idle mixture is on the lean side of best power. An increase of 25 to 50 rpm or a decrease in MAP of $\frac{1}{4}$ in Hg [0.85 kPa] should provide the mixture that is rich enough to provide satisfactory acceleration under all conditions and lean enough to prevent spark-plug fouling or rough operation.

4. If the idle mixture is not correct, turn the idle adjustment (see Fig. 6-14) one or two notches in the direction required for correction and recheck, as explained previously. Make additional adjustments if necessary until the mixture is satisfactory.

5. Between adjustments, clear the engine by running it up to 2000 rpm before making the mixture check.

6. The mixture adjustment is made by lengthening the linkage between the throttle lever and idle valve to enrich the mixture and by shortening the linkage to lean the mixture. The center screw has right-hand threads on both ends, but one end has a no. 10-24 thread and the other has a no. 10-32 thread. For easy reference, consider only the coarse-threaded end. When this end of the link is backed out of its block, the mixture is enriched. Leaning is accomplished by screwing the coarse-threaded end of the link into the block.

7. If the center screw bottoms out in one of the blocks before a satisfactory mixture is achieved, an additional adjustment must be made. First, measure the distance between the blocks. Then disconnect one of the blocks from its lever by removing the link pin. Turn the block and adjustment screw until the adjusting wheel is centered and the distance between the blocks is the same as previously measured. There is now additional adjustment range and the reference point is retained.

8. Make the final idle speed adjustment to obtain the desired idling rpm with the throttle closed.

9. If the setting does not remain stable, check the idle linkage for looseness. In all cases, allowance should be made for the effect of weather conditions. The prevailing wind can add to or reduce the propeller load and affect the engine rpm. During the idle mixture and rpm checks, the airplane should be placed crosswind.

Inspection and Maintenance

The routine inspection and maintenance of an engine fuel system that includes an RSA fuel-injection system are similar to the inspection and maintenance of carburetors and fuel systems. The principal items to note are tightness and safetying of nuts and bolts, leakage from lines and fittings, and looseness in control linkages. Fuel strainers should be removed and cleaned as specified. Minor fuel stains at the fuel nozzles are normal and not a cause for repair.

When a new injector unit is installed, the injector inlet strainer should be removed and cleaned after 25 h of operation. Thereafter, the strainer should be cleaned at each 50-h inspection.

If an aircraft engine is equipped with a fuel injector that includes an automatic mixture control (AMC), the operator should be alert for signs of problems with the unit. Dirt can build up on the needle and cause rich operation and possible sticking of the needle, with resultant loss of altitude compensation. The following instructions are provided for cleaning the unit:

1. Carefully remove the AMC unit. If the gasket is damaged, replace it with a new gasket with the appropriate Bendix part number.

2. Remove the 9/16-24 plug and immerse the unit in clean naphtha or other approved petroleum solvent. Invert the unit so that it will fill completely with the solvent. Exercise the AMC needle with a hardwood or plastic rod to facilitate cleaning. Shake the unit vigorously while allowing the solvent to drain. Repeat several times to wash out all traces of contaminants.

3. Drain the unit and allow the solvent to evaporate completely. Do not dry with compressed air.

4. Replace the plug and install the unit on the injector. Torque to 50 to 60 pound inches (lb·in) [5.65 to 6.78 newton-meters (N·m)].

Lubrication of the injector should be accomplished in accordance with the approved lubrication chart for the particular installation. The clevis pins used in connection with the throttle and the manual mixture control should be checked for freedom of movement and lubricated if necessary.

Lubricate the throttle shaft bushings by placing a drop of engine oil on each end of the throttle shaft so that it can work into the bushings.

Use care in cleaning and oiling the air filter element. If the element is replaced with excessive oil clinging to it, some of the oil will be drawn into the injector and will settle on the venturi. This can greatly affect the metering characteristics of the injector.

Preparation for Storage

The storage and preservation of an injector unit is accomplished in a manner similar to that employed for carburetors and other fuel-control units. If a unit is to be taken out of service for less than 28 days, it may be left filled with clean fuel or calibration fluid. Any unit taken out of service for a period of 28 days or longer—or being returned to overhaul—must be flushed with preserving oil. Any good grade of clean no. 10 nondetergent oil is satisfactory.

Caution must be exercised when handling or working with the fuel injector to prevent oil or fuel from entering the air section of the unit. Fluid can easily enter the air section through the impact tubes or the suction passages of the venturi. Although no permanent damage will result to the parts in the regulator if oil or fuel gets into them for short periods of time, it may affect the air signals from the venturi to the regulator by blocking the air passages. The effect will be more noticeable in extremely cold environments.

The following instructions are given by the manufacturer for the preservation of an RSA injector unit:

1. Remove the plugs and/or caps from the fuel ports and drain all residual fuel from the unit.

2. Replace all plugs and caps except those for the fuel inlet and outlet.

3. Introduce oil from a filtered source (10 micron) into the fuel inlet by gravity pressure only until oil flows from the outlet port.

4. Dump the oil from the unit. A film of oil on internal parts of the fuel section is sufficient to preserve the control.

5. Replace the caps and/or plugs in the fuel inlet and outlet.

After the injector unit is flushed with preserving oil, it should be protected from dust and dirt and given such protection against moisture as climatic conditions at the point of storage require. If the unit is to be stored near or shipped over saltwater, the following additional steps must be taken:

1. Spray the exterior of the injector with an approved preservative oil.

2. Pack the injector in a dustproof container, wrap the container with moistureproof and vaporproof material, and seal. Pack the wrapped unit in a suitable shipping case. Pack a $\frac{1}{2}$-lb [0.23-kg] bag of silica gel crystals in the dustproof container with the injector, but do not allow the bag to touch the injector.

● FUEL INJECTION FOR LARGE ENGINES

Direct fuel-injection systems for large engines differ from the systems previously described in that the fuel is injected directly into the combustion chamber of the large engine. The fuel is injected by means of a piston pump regulated to deliver accurately measured quantities of fuel according to the quantity of air entering the cylinders through the intake valves from the intake manifold.

The **master control unit** of a direct fuel-injection system is identical in most respects to the pressure-injection carburetors previously described in this chapter. Figure 6-23 is an illustration of a master control unit

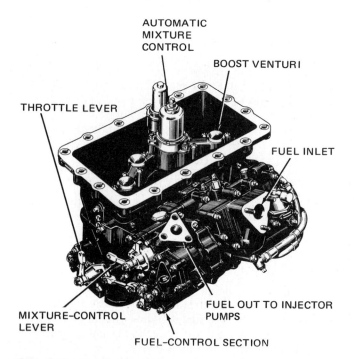

FIG. 6-23 Bendix master control unit. *(Energy Controls Div., Bendix Corp.)*

manufactured by the Bendix Products Division (now the Energy Controls Division) of Bendix Corporation. Figure 6-24 is a schematic diagram of the complete fuel system, showing how the various units are arranged. The upper center drawing in Fig. 6-24 illustrates the airflow and regulating section of the master control.

The function of the master control unit is to meter fuel to the injection pump. The quantity of fuel is automatically controlled in proportion to the air entering the engine to produce the most efficient fuel-air ratios for various engine operating conditions of load, speed, altitude, and temperature.

The air flowing through the compound venturi of the control unit creates a partial vacuum, or suction. This low pressure is transmitted to one side of an air-metering-force diaphragm, chamber B. Air-scoop impact pressure is transmitted from the impact tubes to the opposite side of this diaphragm, chamber A.

The **automatic mixture-control unit** corrects the air metering pressures to compensate for variations in atmospheric pressure and temperature. Thus, the resultant air metering force is a measure of the mass, or weight, of the air entering the engine.

The diaphragm shown between chambers C and D in Fig. 6-24 is balanced between unmetered fuel on one side and metered fuel on the opposite side. When these two pressures are equal, no fuel flows through the metering jet. The air metering force, which is the difference of pressure between chambers A and B, acts upon

the air diaphragm and tends to upset the fuel pressure balance by opening the poppet valve and allowing fuel to flow into chamber D. This causes a pressure differential across the metering jet which is exactly proportional to the mass airflow through the venturi, since the air and fuel diaphragms are connected. The metered fuel flows from the control unit to the injection pump, as shown in the diagram.

● BENDIX PR-58S2 INJECTION SYSTEM

Master Control

The master control for the Bendix PR-58S2 injection system designed for the Wright R-3350 engine is illustrated in Fig. 6-25. It will be recognized that this control system is similar to the pressure-injection carburetor for large engines. The A and B chambers in the regulator unit provide the **air metering force** across the air diaphragm, and the C and D chambers provide the **fuel metering head**, or force. The main poppet-valve opening is controlled by the balance between these two forces.

The metered fuel, instead of being routed to a discharge nozzle, is directed to the fuel-injection pumps. The output of the pumps is controlled by a diaphragm balanced between boost venturi suction and metered fuel pressure.

Chamber E of the regulator unit contains a fuel strainer, a vapor separator, and fuel at engine pump

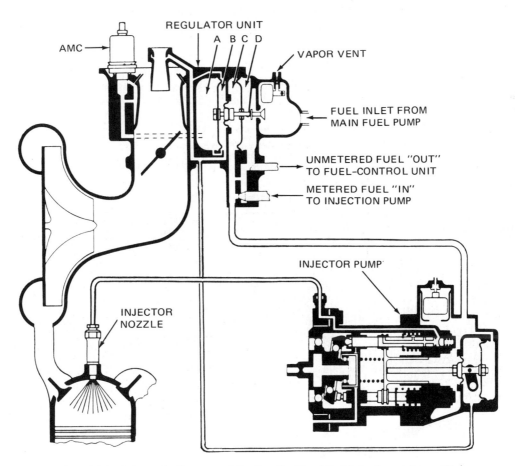

FIG. 6-24 Schematic diagram of the Bendix direct fuel-injection system *(Energy Controls Div., Bendix Corp.)*

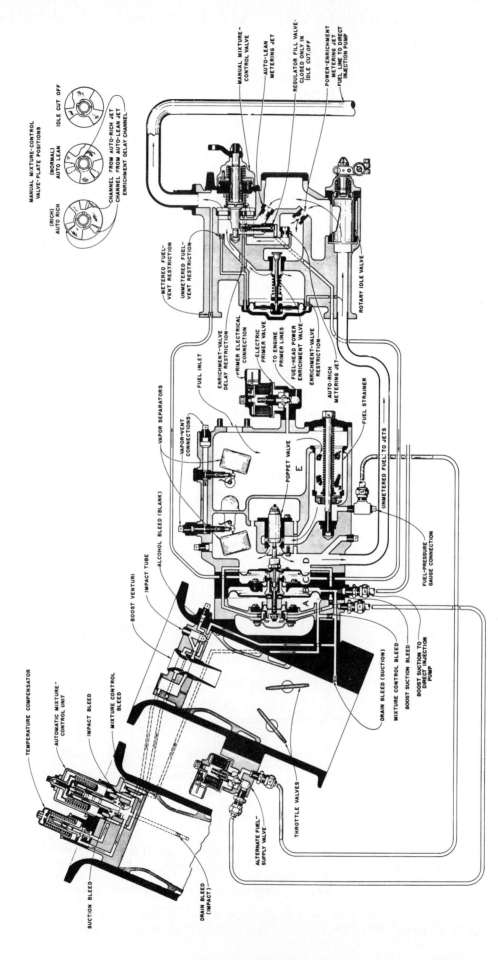

MANUAL MIXTURE-CONTROL VALVE-PLATE POSITIONS

(RICH) AUTO RICH

(NORMAL) AUTO LEAN

IDLE CUT OFF

CHANNEL FROM AUTO-RICH JET
CHANNEL FROM AUTO-LEAN JET
ENRICHMENT DELAY CHANNEL

MANUAL MIXTURE-CONTROL VALVE
AUTO-LEAN METERING JET
AUTO-LEAN METERING JET
REGULATOR FILL VALVE-CLOSED ONLY IN IDLE CUT-OFF
POWER-ENRICHMENT METERING JET
FUEL LINE TO DIRECT INJECTION PUMP

METERED FUEL-VENT RESTRICTION
UNMETERED FUEL-VENT RESTRICTION

ROTARY IDLE VALVE

ENRICHMENT-VALVE DELAY RESTRICTION
PRIMER ELECTRICAL CONNECTION
ELECTRIC PRIMER VALVE
TO ENGINE PRIMER LINES
FUEL-HEAD POWER ENRICHMENT-VALVE RESTRICTION
AUTO-RICH METERING JET
FUEL STRAINER

FUEL INLET

VAPOR SEPARATORS
VAPOR-VENT CONNECTIONS

ALCOHOL BLEED (BLANK)

POPPET VALVE

E

D

UNMETERED FUEL, TO JETS

FUEL-PRESSURE GAUGE CONNECTION

BOOST VENTURI
IMPACT TUBE

A B C

DRAIN BLEED (SUCTION)
MIXTURE CONTROL BLEED
BOOST SUCTION BLEED
BOOST SUCTION TO DIRECT INJECTION PUMP

TEMPERATURE COMPENSATOR
AUTOMATIC MIXTURE-CONTROL UNIT
IMPACT BLEED
MIXTURE CONTROL BLEED

SUCTION BLEED

DRAIN BLEED (IMPACT)

ALTERNATE FUEL-SUPPLY VALVE

THROTTLE VALVES

FIG. 6-25 Schematic diagram of the Bendix PR-58S2 master control. (*Bendix Corp.*)

pressure. The poppet valve is located between chambers *D* and *E*. When chamber *D* fills with fuel, the vapor separator in chamber *D* eliminates any vapor not discharged from the vapor separator in chamber *E*. Fuel flow from chamber *D* (unmetered fuel pressure) through the metering jets in the fuel-control unit is accompanied by a pressure drop in chamber *C* (metered fuel pressure). The direct fuel-injection pump provides chamber *C* with a constant back pressure (about 6 or 7 psi [41.37 or 48.27 kPa]). This metered fuel pressure does not vary any appreciable amount during engine operation, although the fuel flow increases proportionally to engine demand. During operation, *B* chamber pressure (boost venturi suction) is approximately one-third of the pressure in chamber *A*.

Electric Primer Valves

The PR-58S2 master control uses two electric primer-valve assemblies. One primer supplies unmetered fuel to the engine priming nozzles, and the other provides an alternate fuel-supply system. The electric primer-valve position is normally closed by spring action. When energized, the coil overcomes spring force and opens the valve. The coil's design permits continuous operation without an excessive temperature rise. The installation of the primer to the master control does not provide a positive ground. For this reason, a shielded cable or similar means must be used for direct connections to the airframe to provide satisfactory grounding for the solenoid coil.

The alternate fuel-supply system was designed to provide a source of fuel when icing occurs in the induction system. Such icing causes a drop in metering suction, resulting in an undesirable drop in fuel flow.

The alternate fuel-supply system, by using an electric primer, supplies fuel at engine pump pressure to the balance diaphragm located in chamber *A* of the regulator unit. This pressure compensates for the loss of metering suction caused by ice and again establishes a metering head that provides an adequate fuel flow. The throttle position must be adjusted to provide the required fuel-air ratio until such time as the ice condition has been corrected. Extreme altitude may cause ice crystals to form in the inlet fuel filter; however, the bypass-valve-type strainer will supply adequate fuel in the event that the screen becomes clogged.

Operating Characteristics

The fuel-air ratio curve in Fig. 6-26 illustrates graphically a fuel-airflow schedule for the average PR-58S2 master control. Following the curve from the idle range through takeoff, the regulator unit maintains the correct metering pressure and supplies fuel ready for metering to the fuel control at all times. The cruise and takeoff fuel-air ratios are maintained with respect to altitude change, independently of propeller pitch, engine speed, or throttle position; however, the volume of fuel delivered will vary in proportion to airflow.

Variations of the curve in the idle range indicate different idle adjustments, and as the airflow increases, the fuel-air ratio becomes leaner. At about 4000 lb/h [1814.37 kg/h] airflow, metering suction has overtaken constant-head idle-spring tension and the poppet-valve stem is functioning as an integral unit; hence, the fuel-air ratio curve follows a single line. The rotary idle valve, past its metering phase, supplies fuel to the metering jets. Fuel metering through the cruise range is across the auto-lean metering jet. With the manual mix-

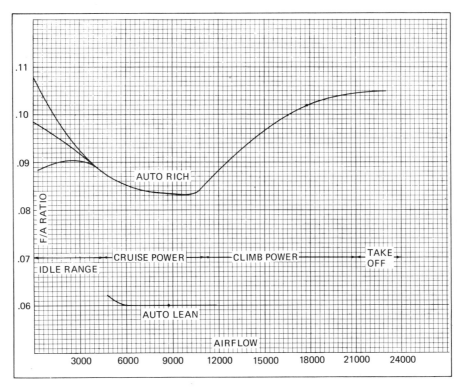

FIG. 6-26 Fuel-air ratio curve for the Bendix PR-58S2 direct fuel-injection system.

ture control in the AUTO LEAN position, the fuel supplied across the auto rich jet is blanked off; otherwise, the fuel flows are the same as in AUTO RICH. From the fuel curves illustrated, the fuel-air ratio through the cruise range in AUTO RICH is about 1 part fuel to 12 parts air. In AUTO LEAN the ratio is about 1 part fuel to 16 parts air. During advanced cruise, the power enrichment valve opens and increasing fuel flow is indicated by the upswing on the curve. The power enrichment valve meters fuel in conjunction with the auto-lean and auto-rich metering jets until the combined metering area of the power enrichment valve and the auto-rich metering jet is greater than the area of the power enrichment metering jet. At this time the auto-lean and power enrichment jets assume the metering and the curve tends to level off, adequately supplying the engine with the proper fuel-air mixture to perform safely through the power range.

Fuel-Injection Pump

The **fuel-injection pump,** illustrated in Fig. 6-27, directs the fuel under a relatively high pressure to the discharge nozzles in the cylinders. The pump illustrated provides for nine cylinders. Two such pumps are used for 18-cylinder engines and are synchronized by means of an interconnecting control rod. The right-hand pump supplies fuel to the front cylinders of the engine and the left-hand pump supplies fuel to the rear cylinders.

The injection pump automatically divides the metered fuel into exactly equal charges for each engine cylinder and forces one such measured charge of fuel directly into each cylinder, timed for injection during the intake stroke.

A **wobble plate** within the pump housing is driven from the engine at one-half engine speed. This imparts a continuous reciprocating motion to the pump pistons, or plungers. The effective stroke of the plungers is controlled by the position of the bypass sleeves which surround the plungers.

The **control diaphragm** shown at the extreme right of the pump drawing serves to control the axial position of the bypass sleeves. The pressure of the fuel from the control unit acts upon this spring-loaded dia-

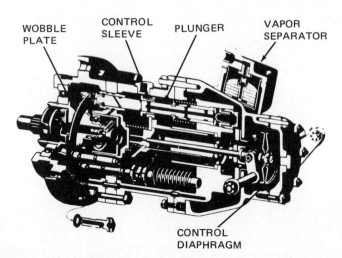

FIG. 6-27 **Direct fuel-injection pump.** *(Bendix Corp.)*

phragm, causing it to move inward as the pressure increases. The air chamber at the inner side of the diaphragm is connected to the low-pressure air from the boost venturi of the control unit. The velocity of the air through the control unit determines the amount of suction to the air side of the fuel-pump-control diaphragm. As the velocity increases, the suction increases, thus tending to aid the fuel pressure in moving the diaphragm inward to increase the effective stroke of the plungers. The air and fuel forces applied to the pump diaphragm are opposed by a heavy spring in the center of the pump. This spring moves the diaphragm outward as fuel pressure and air suction decrease.

The operation of the direct fuel-injection system is such that, as the throttle is opened, the airflow through the control unit increases. The quantity of air acting through the control unit and pump diaphragm determines the quantity of fuel delivered by the pump, thus maintaining the correct ratio of fuel and air to the engine.

Discharge Nozzle

The **discharge nozzle** used with the direct fuel-injection system is a very important unit of the system. Essentially, it is a spring-loaded control valve designed to keep the fuel inlet closed until a pressure of approximately 500 psi [3447.5 kPa] is applied. The valve opens instantly, giving a sharp start to the fuel injection, and closes instantly when the pressure is released. The fuel is injected into the cylinder in a fine spray at a high velocity, which causes it to become completely vaporized and well mixed with the air in the cylinder.

This fuel discharge nozzle is known as a **pressure discharge nozzle** because a pressure of 500 psi is required before the nozzle will open and permit fuel flow. The continuous-flow nozzle described previously does not incorporate a valve, and fuel flows whenever fuel pressure exists at the nozzle. The amount of flow is determined by throttle position and engine requirements.

The spray pattern of a discharge nozzle is important because it determines the degree of fuel atomization accomplished. Spray patterns are tested on a flow bench which provides the correct fuel pressure for operation of the nozzle. The nozzle is mounted in a chamber with a transparent viewing window to permit observation of the spray pattern. The flow bench is also used to determine whether the nozzle is adjusted properly for valve opening pressure on the pressure discharge nozzle.

Note that pressure discharge and pressure-injection fuel-control units cannot be repaired in the field. Adjustments for throttle position, mixture-control position, idle speed, and idle mixture can be made in the field, and dual injection pumps can be synchronized. But for repairs and major adjustments, the units must be removed from the engine and serviced in a properly equipped shop.

Synchronization

The two injection pumps mounted on the 18-cylinder R-3350 engine must be accurately synchronized. Since each pump supplies fuel for a separate group of nine

FIG. 6-28 Injection pumps, showing the use of the synchronization bar.

cylinders, lack of synchronization will cause one group of nine cylinders to receive a greater quantity of fuel than the other group. This will result in rough engine operation.

Synchronization of the injection pumps is ensured by means of the **synchronization bar,** which connects the synchronizing arms of the pumps together. Thus, the movement of the control mechanism (bypass-control plate) of one pump must coincide with the identical movement of the control mechanism of the other pump. The bypass sleeves in both pumps will be in the same position relative to the pump plungers, and the plungers in both pumps will have the same effective stroke. At inspections, the synchronization bar should be checked for looseness and, if necessary, should be adjusted. The arrangement of the synchronization bar is shown in Fig. 6-28.

Inspection and Maintenance

Inspection and maintenance procedures for the direct fuel-injection system described here are similar in most respects to those for carburetors and fuel-injection systems. In addition to the standard inspections for leakage, condition of hoses and lines, tightness and safetying of nuts and bolts, and operation of control linkages, special instructions for the particular system should be followed.

Field adjustments can be made for idle speed and idle mixture, and synchronization can be adjusted by means of the adjustable fittings at the ends of the synchronizing bar.

Fuel-injection nozzles should be removed periodically according to the approved inspection schedule for inspection and cleaning.

The master control unit is similar to the pressure discharge carburetor described in Chap. 5. The same precautions regarding removal, installation, and storage should be followed for the master control.

● REVIEW QUESTIONS

1. What is meant by *fuel injection*?
2. List the advantages of a fuel-injection system.
3. What is meant by a *continuous-flow* fuel-injection system?
4. What are the four basic units of the Continental continuous-flow fuel-injection system?
5. Describe the operation of the vapor separator and of the vapor ejector.
6. What is the function of the air throttle assembly?
7. How many fuel-control units are included in the fuel-control assembly? What are they?
8. Describe the operation of the fuel manifold valve for the Continental fuel-injection system.
9. At what locations on the engine are the fuel discharge nozzles installed?
10. Describe the construction of a fuel discharge nozzle for the Continental fuel-injection system.
11. What adjustments may be made in the field for the operation of the Continental fuel-injection system?
12. What special feature is required for the fuel discharge nozzles installed on a turbocharged engine?
13. Explain the operation of the altitude compensating valve.
14. How would you check the idle mixture for the Continental fuel-injection system?
15. List typical inspections which should be made pe-

riodically for the Continental fuel-injection system.

16. What precautions must be taken to avoid flooding the engine when starting an engine equipped with the Continental system?

17. When troubleshooting a rough-running engine, what checks should be made before checking the fuel-injection system?

18. List the probable causes for low fuel gage pressure and give the procedures for correction.

19. List the probable causes for fluctuating fuel gage pressure and give the procedures for correction.

20. Describe the operation of the fuel-regulator section for the Bendix RSA fuel-injection system.

21. What would be the effect on the operation of the RSA fuel-injection system if the impact tubes became clogged?

22. Explain the function and describe the operation of the fuel metering section of the RSA injector.

23. What factors produce an air metering force in the Bendix RSA fuel-injection system?

24. As air metering force increases in the Bendix RSA fuel-injection system, what is the effect on fuel flow to the fuel nozzles?

25. What means is provided in the RSA system to provide for adequate fuel flow at idling engine speeds?

26. How is idle speed adjusted on the RSA injector?

27. What active components are included in the fuel metering section?

28. Describe the linkage between the throttle and the idle valve on the RSA injector.

29. Describe the operation of the flow divider.

30. What factors determine the size of the calibrated jet in the fuel nozzle?

31. Why is it possible to use a fuel pressure gage to indicate fuel flow through the RSA system?

32. Describe the operation of the automatic mixture control on the RSA-5AB1 injector.

33. In what position must the flow divider valve be installed for the RSA system?

34. Why must horizontally installed fuel nozzles be installed with the identification mark on the lower side?

35. What type of tubing is used for nozzle fuel lines?

36. List the causes and remedies for a rough-idling engine equipped with an RSA system.

37. Explain the need for care in the operation of an engine with fuel injection during hot weather.

38. What methods are employed to keep engine heat at a minimum during ground operation.

39. Give the procedure for starting an engine equipped with an RSA fuel-injection system.

40. What is the cause when an engine continues to idle for a few seconds after the mixture control is placed in IDLE CUTOFF?

41. Describe the procedure for checking and adjusting the idle mixture for an RSA fuel-injection system.

42. List the items to be checked and serviced during a periodic inspection of the RSA system.

43. Describe what you would do to prepare an RSA injector for long-term storage.

44. What is the difference between direct fuel injection and continuous fuel injection?

45. Compare the master control for a Bendix direct fuel-injection system with a pressure discharge carburetor.

46. What two forces are employed to control the output of the direct fuel-injection pumps?

47. What are the functions of the two electric primer valves in the PR-58S2 system?

48. What is the purpose of the alternate fuel-supply system in the PR-58S2 master control unit?

49. Discuss the operating characteristics of an engine utilizing a direct fuel-injection system.

50. What is the function of the wobble plate in the injection pump?

51. What fuel pressure is required to open the fuel discharge nozzles?

52. What is the specific function of the synchronization bar?

53. What is the effect on engine operation when the synchronization bar is out of adjustment? Why?

54. What field adjustments and maintenance can be performed on the PR-58S2 system?

7 INDUCTION SYSTEMS, SUPER-CHARGERS, TURBOCHARGERS, AND EXHAUST SYSTEMS

● GENERAL DESCRIPTION

The complete induction system for an aircraft engine includes three principal sections: (1) the air scoop and ducting leading to the carburetor, (2) the carburetor, or air-control section, of an injection system, and (3) the intake manifold and pipes. These sections constitute the passages and controlling elements for all the air which must be supplied to the engine. An induction system is shown in Fig. 7-1.

Air Scoop and Ducting

The ducting system for a nonsupercharged (naturally aspirated) engine comprises four principal parts: (1) the air scoop, (2) air filter, (3) alternate-air valve, and (4) carburetor air heater, or heater muff.

A typical **air scoop** is simply an opening facing into the airstream. This scoop receives ram air, usually augmented by the propeller slipstream. The effect of the air velocity is to "supercharge" the air a small amount, thus adding to the total weight of air received by the engine. The power increase thus afforded may be as much as 5 percent. The design of the air scoop has a substantial effect on the amount of increased power provided by ram air pressure.

The **air filter** is installed at or near the air scoop to remove dust, sand, and larger foreign particles from the air before it is carried to the engine. The filter usually consists of a mat of metal filaments encased in a frame and dipped in oil. The oil film on the metal filaments catches and holds dust and sand particles. Normally, the air filter should be removed, cleaned, and reoiled after every 25 h of engine operation. If the airplane is operating in a particularly dusty or sandy area, the filter should be serviced at more frequent intervals. Many modern aircraft engines are equipped with nonmetallic air filters that are cleaned by washing in a detergent solution and rinsing in water; they are then blown dry with compressed air at not over 100 psi [689.5 kPa]. These filters can also be cleaned on a frequent basis merely by blowing air through them from the inside out; this would be done in case of operating in extremely dusty conditions.

The **alternate-air valve** is operated by means of the carburetor heat control in the cockpit. The valve is simply a gate which closes the main air duct and opens the duct to the heater muff when the control is *on*. During normal operation, the gate closes the passage to the heater muff and opens the main air duct. The gate is often provided with a spring which tends to keep it in the normal position.

The **heater muff** is a shroud placed around a section of the exhaust pipe. The shroud is open at the ends to permit air to flow into the space between the exhaust pipe and the wall of the shroud. A duct is connected from the muff to the main air duct. During operation of the carburetor air heater system, **protected air** within the engine compartment flows into the space around the exhaust pipe where it is heated before being carried to the main air duct. It should be noted that carburetor air heat should be applied only if necessary to prevent ice formation and to keep rain out of the carburetor. The use of heated air during periods of high-power operation is likely to cause detonation and will definitely cause a reduction in engine power output.

Some induction systems are designed to permit direct ram air to enter the carburetor without first passing through an air filter. In such a case the air filter is installed in an alternate duct. Thus, when the airplane is operating on the ground in sandy or dusty conditions, the direct air duct is closed by means of a gate valve and air is drawn into the carburetor through the air filter; after takeoff, when the airplane is flying in clean air, the air intake is shifted back to the direct duct. The alternate air source is also useful when flying through heavy rain; the protected air from within the nacelle, being free of rain, enables the engine to continue operation in a normal manner. However, it must be remembered that the air filter reduces air pressure to the carburetor to some extent, thus reducing the power output. Since it is less dense than cool air, heated air results in a loss of power. For maximum power, therefore, it is desirable that a free flow of unheated air be provided for the engine.

Intake Manifolds

The typical **opposed-type** (or **flat-type**) aircraft engine has an induction system with an individual pipe leading to each cylinder. On some models of this type, one end of each pipe is bolted to the cylinder by means of a flange and the other end fits into a slip joint in the manifold. On other models of this type, the pipes are connected to the manifold by short sections of rubber (or synthetic rubber) hose held by clamps. In still other models of this type, the carburetor is mounted on the oil sump and the fuel-air mixture flows from the carburetor through passages in the oil sump and then out through each of the individual pipes leading to the engine cylinders. As the mixture of fuel and air flows through the passages in the oil sump, heat is transferred from the oil to the fuel-air mixture. This arrangement accomplishes two purposes to a certain extent: (1) It cools the oil slightly, and (2) it increases the temperature of the fuel-air mixture slightly for a better va-

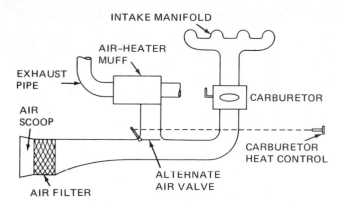

FIG. 7-1 Diagram of a simple induction system.

porization of the fuel. An arrangement whereby heat is applied to the fuel-air mixture by means of heated oil or by proximity to the exhaust manifold is usually termed a **hot spot.** The intake manifold and pipes for an opposed engine are shown in Fig. 7-2.

The type of induction system used on a radial-type engine principally depends on the horsepower output desired from the engine. On a small radial engine of low output, the air is drawn through the carburetor, mixed with fuel in the carburetor, and then carried to the cylinders through individual intake pipes. In some engines, an intake manifold section is made a part of the main engine structure. The fuel-air mixture is carried from the outer edge of the manifold section to the separate engine cylinders by individual pipes, which are connected to the engine by means of a slip joint. The purpose of the slip joint is to prevent the damage which would otherwise occur from the expansion and contraction caused by changes in temperature.

In a typical **high-output radial engine,** an internal blower or a supercharger is located in the rear section of the engine. The fuel-air mixture passes from the carburetor through the supercharger or blower and then flows out through the diffuser section and individual intake pipes to the engine cylinders. An arrangement of this type is shown in Fig. 7-3. In addition to providing some supercharging, the internal impeller (blower) helps to atomize and vaporize the fuel and to ensure an equal distribution of the fuel-air mixture to all cylinders.

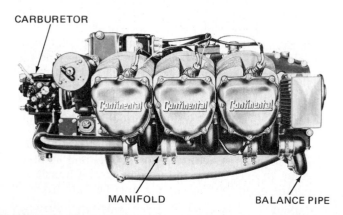

FIG. 7-2 Intake manifold and pipes for an opposed engine.

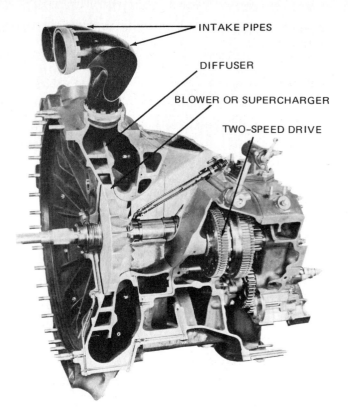

FIG. 7-3 Supercharger section and intake pipes in a radial engine.

Importance of Gastight Seal

The portion of the intake system of an engine between the carburetor and the cylinders must be installed gastight to provide proper engine operation. When the manifold pressure is below atmospheric pressure, which is always the case with unsupercharged (naturally aspirated) engines, an air leak in the manifold system will allow air to enter and lean out the fuel-air mixture. This can cause overheating of the engine, detonation, backfiring, or complete stoppage. Small induction system leaks will have the most noticeable effect at low rpm because the pressure differential between the atmosphere and the inside of the intake manifold increases as rpm decreases.

In a supercharged engine, a portion of the fuel-air mixture will be lost if leakage occurs in the intake manifold or pipes. This, of course, will cause a reduction in power and a waste of fuel.

One method of forming a gastight connection for intake pipes is to provide a synthetic rubber packing ring and a packing retaining nut to form a slip-joint seal at the distribution chamber, thus allowing the intake pipes to slide in and out of the distribution-chamber opening while the metal of the engine cylinder is expanding and contracting from changes of temperature. Obviously, it is necessary to place a gasket at the cylinder intake port between the pipe flange and the cylinder port and to secure the flange rigidly with bolts and nuts.

Another method of forming a gastight connection for intake pipes is to use a packing ring and a packing retaining nut which screws into or over the intake-port opening. Still another method is to have short stacks protruding from the intake ports, using rubber couplings to connect the pipes to these protruding stacks.

Induction System for Six-Cylinder Opposed Engine

The principal assemblies of an induction system for the Cessna Model 310 airplane equipped with a Continental IO-470-D engine are shown in Fig. 7-4. With this system, ram air enters the air box at the left rear engine baffle and is ducted to the rear, where it passes through an air filter before entering the fuel-air control unit. Between the filter and the fuel-air control unit is an induction air door which serves as a gate to close the heater duct or the main air duct, depending on the position of the control. In case the air filter becomes clogged, the door will open automatically to allow air to enter from the heater duct.

From the fuel-air control unit, air is supplied to the cylinders through intake manifold piping. This piping is arranged in jointed sections along the lower side of the cylinders on each side of the engine to form two manifolds leading from the Y fitting at the fuel-air control unit. A part of this arrangement is shown in Fig. 7-5. A balance pipe is connected between the two manifolds at the front end. This pipe equalizes the pressure in the manifolds, thus providing for a more uniform

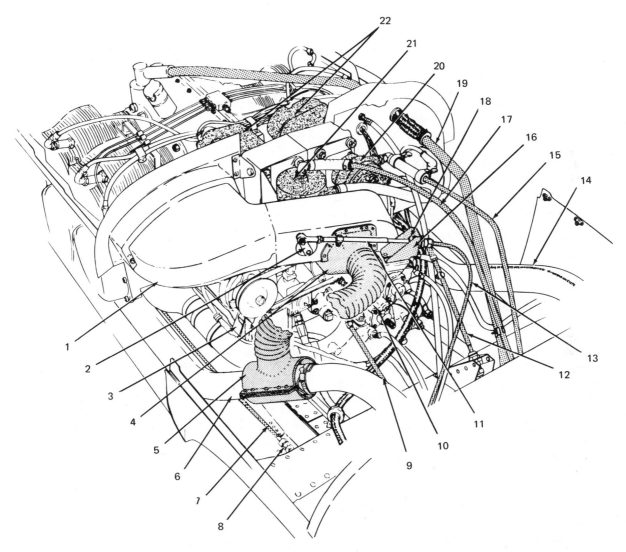

1. Carburetor air box assembly
2. Alternate air actuating arm
3. Engine mount
4. Carburetor heat adapter
5. Carburetor heat shroud
6. Left exhaust stack
7. Propeller control conduit
8. Conduit connector
9. Throttle control sliding end
10. Carburetor
11. Alternate air control sliding end
12. Mixture control sliding end
13. Alternate air control conduit
14. Right exhaust stack
15. Vacuum line
16. Control mounting bracket
17. Air-oil separator line
18. Alternate air connector
19. Crankcase breather line
20. Vacuum pump
21. Air-oil separator
22. Magnetos

FIG. 7-4 Induction system for a six-cylinder opposed engine. *(Cessna Aircraft Co.)*

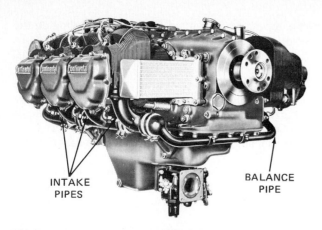

FIG. 7-5 Intake pipes and balance pipe. *(Teledyne Continental)*

airflow to the cylinders. Short sections of individual piping lead from the manifold to each cylinder intake port. Fuel is injected continually into the intake ports of each cylinder while the engine is operating.

It must be emphasized that induction systems are designed in many ways to suit the operation of various engine and aircraft combinations. Most systems, however, include the elements described in this section, and each is designed to provide the engine with adequate air for the most efficient operation. The characteristics of induction systems utilized with superchargers are described in following sections.

Induction System Icing

In Chap. 4 we discussed carburetor icing and the use of the carburetor air heater. It was explained how impact ice, fuel evaporation ice, and throttle ice may occur in the carburetor. In this section, we shall discuss further the formation of ice with relation to the induction system.

Ice can form anywhere from the inlet of the induction system (air scoop) to the intake manifold between the carburetor and the intake port of the cylinder. The nature of the ice formation depends on atmospheric temperature and humidity and the operating conditions of the engine. As explained previously, if the air scoop and ducting leading to the carburetor are at a temperature the freezing point of water, impact ice will form when particles of water in the air strike the cold surfaces, particularly at the air screen and at turns in the intake duct. Any small protrusions in the duct are also likely to start the formation of ice. Icing can be detected by a reduction in engine power when the throttle position remains fixed. If the aircraft is equipped with a fixed-pitch propeller, the engine rpm will decrease. With a constant-speed propeller, the manifold pressure will decrease and the engine power will drop, even though the engine rpm remains constant.

An aircraft which may be operated in icing conditions should be equipped with a **carburetor air temperature gage.** This instrument reads the temperature of the air as it enters the carburetor and makes it possible to detect the existence of icing conditions. If the carburetor air temperature is below 32°F [0°C] and there is loss of engine power, it can be assumed that icing exists and that carburetor heat should be applied.

The formation of ice in an induction system is prevented by the use of carburetor heat. (With some older installations for large engines, alcohol was sprayed into the air-inlet duct to reduce the formations of ice; the system consisted of an alcohol reservoir, an electric pump, a spray nozzle, and controls arranged in a system to be used by the pilot as needed.) For small aircraft, the air is often heated by means of a muff around the exhaust manifold; the heat of the exhaust raises the temperature of the air before it flows to the carburetor. For altitude engines equipped with carburetors, the air heater must be able to produce a heat rise of 120°F [67°C] when operating in air free of visible moisture at a temperature of 30°F [17°C] with engine power at 60 percent of the maximum continuous rating. An engine equipped with a pressure carburetor requires a preheater which can produce a heat rise of 100°F [56°C] under the same conditions.

A muff-type heater is shown in Fig. 7-1. During inspections, it is important that the muff be removed so that the exhaust pipe can be examined for cracks or holes caused by heat and corrosion. If exhaust gas leaks into the muff, it causes a reduction of the oxygen available for fuel combustion and hence a loss of power. Exhaust gas leaking from the exhaust pipes might also enter the cockpit and endanger personnel.

The carburetor heat control must be rigged with **springback** to ensure that the air valve is completely open or closed when the control is in the HEAT OFF (cold) or the HEAT ON (hot) position. With springback, the valve is completely closed or completely open before the control reaches the limit of its travel.

● PRINCIPLES OF SUPERCHARGING

The purpose of supercharging and of turbocharging is to allow an engine to develop a maximum of power when operating at high altitudes. At altitude, an unsupercharged (naturally aspirated) engine will lose power because of the reduced density of the air entering the engine.

The value of supercharging for reciprocating aircraft engines was clearly demonstrated during World War II. The **turbosupercharger** (turbine-driven supercharger) was of particular importance because it permitted our bombers to operate at extreme altitudes, from which vantage point they could bomb enemy targets with much less danger of attack than at lower altitudes. After the war, engine-driven and turbine-driven superchargers were used on commercial airliners to permit high-altitude flight.

Supercharging is the process of increasing the manifold pressure (MAP) of an engine to a level above 30 inHg [101.61 kPa]. **Turbocharging** is used to increase MAP up to 30 inHg at altitude with a turbine-driven blower or compressor. This permits the development of sea-level power and makes it possible for aircraft to fly at altitudes of 20,000 to 30,000 ft [6096 to 9144 m].

The Properties of Gases as Related to Supercharging

Understanding the principles of supercharging requires a knowledge of mass, volume, and density as applied to the properties of gases. All matter can be classified as solids and fluids. In turn, fluids include both liquids and gases. Solids, liquids, and gases all have weight, but the weight of gases is by no means a constant value under all conditions. For example, at sea-level pressure about 13 ft³ [368.16L] of air weighs 1 lb [0.45 kg], but at greater pressures the same volume would weigh more, and at lower pressures it would weigh less.

Mass is not the same as weight, but for an ordinary layperson's discussion it is often used as if it were the same. Mass should not be confused with volume either, because volume designates merely the space occupied by an object and does not take density or pressure into consideration. The relations among these various factors is explained by the various laws pertaining to the behavior of gases.

From experience it is known that any gas can be compressed to some extent, and that this compression is accomplished by exerting some force on the gas, that is, by increasing its pressure. Boyle's law expresses the relationship between pressure and volume as follows: *In any sample of gas, the volume is inversely proportional to the absolute pressure if the temperature is kept constant.*

With reference to the quantity of gas (air) in a closed cylinder fitted with a movable piston (as shown in Fig. 7-6), if it is assumed that the temperature is constant and that there is no leakage past the piston, then the volume and pressure are inversely related—in accordance with Boyle's law.

At the left in Fig. 7-6, 15 ft³ [424.8 L] of gas weighs 1 lb at 10 psia [68.95 kPa] pressure. In the picture at the right, 15 ft³ of gas weighs 2 lb [0.91 kg] at a pressure of 20 psia [137.9 kPa] because more gas is packed into the cylinders. The volume is the same in both pictures, but the pressure is changed, and hence the mass (quantity) of air below the piston is changed.

In Fig. 7-7, cylinders are shown with their pistons. The one at the left has a volume of 10 in³ [0.17 L] under the piston, the one in the middle has a volume of 5 in³ [0.08 L] under the piston, and the one at the right has a volume of 20 in³ [0.33 L] under the piston.

If the density of the air in the cylinder at the left is accepted as standard, the air in the middle cylinder has a density of 2 and the air in the cylinder at the right has a density of only ½.

The weight of the air surrounding the earth is sufficient to exert considerable pressure on objects at sea

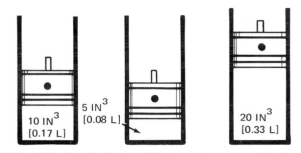

15 PSIA [103.43 kPa]   30 PSIA [208.85 kPa]   7.5 PSIA [52.71 kPa]

FIG. 7-7   Relative volumes and pressures.

level. At altitudes above sea level the pressures, density, and temperatures of the air are lower. At an altitude of 20,000 ft [6096 ml] the pressure and density of the atmosphere are only about one-half their value at sea level.

Unless something is done to offset the decreasing density, at sea level the engine would receive the full weight of air, at 10,000 ft [3048 m] it would receive only three-fourths as much air by weight as at sea level, at 20,000 ft [6096 m] it would get only one-half as much air by weight as at sea level, at 30,000 ft [9144 m] it would obtain only one-third as much air by weight as at sea level, and at 40,000 ft [12 192 m] it would receive only one-fourth as much air by weight as at sea level. These relations are illustrated in Fig. 7-8.

These relations are important in any discussion of

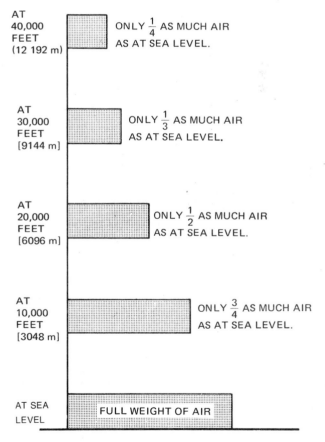

FIGURES ARE APPROXIMATE AND MEAN WEIGHT OF AIR.

FIG. 7-8   Effect of altitude on the density of air.

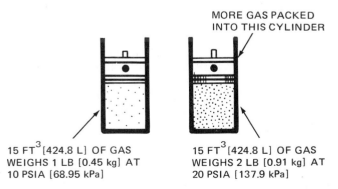

15 FT³ [424.8 L] OF GAS WEIGHS 1 LB [0.45 kg] AT 10 PSIA [68.95 kPa]

15 FT³ [424.8 L] OF GAS WEIGHS 2 LB [0.91 kg] AT 20 PSIA [137.9 kPa]

FIG. 7-6   Quantity of gas (air) charge.

an internal-combustion engine because the power developed by the engine depends principally on the mass of the induced charge. An engine that is not supercharged can induce only a definite value according to its volumetric efficiency and piston displacement. In order to increase the mass of the charge, it is necessary to increase the pressure and the density of the incoming charge by means of a supercharger or turbocharger. Therefore, it is possible to say that the function of a supercharger or turbocharger is to increase the quantity of air (or fuel-air mixture) entering the engine cylinders.

Superchargers were developed originally for the sole purpose of increasing the density of the air taken into the engine cylinders at high altitudes in order to obtain the maximum power output. However, with improvements in the production of fuels and in the design of engines, it became possible to operate a supercharger at low altitudes to increase the induction system pressure (thus increasing the charge density) *above* the normal value of atmospheric pressure.

Figure 7-9 illustrates the effect of temperature on gas volume. The elastic property of gases is demonstrated when any change occurs in the temperature. If the temperature of a given quantity of any gas is raised and the pressure is held constant, the gas will expand in proportion to absolute temperature. This is expressed in the following equation:

$$\frac{V_1}{V_2} = \frac{T_1}{T_2} \quad \text{pressure constant}$$

The foregoing equation is known as Charles's law, attributed to the French mathematician and physicist Jacques A. C. Charles (1746–1823).

In Fig. 7-9 the temperature of the gas in the first cylinder is at 0°C, or 273 K (kelvin). When the gas temperature is raised to 273°C, the absolute temperature is doubled; hence, the volume is doubled. To illustrate further, if a quantity of gas has a volume of 10 ft³ [283.2 L] at 10°C and we wish to find the volume at 100°C, we must convert the temperature indications to absolute (kelvin) values. To do this, we merely add 273 to the celsius value. Then, 10°C becomes 283 K and 100°C becomes 373 K. Applying the formula,

$$\frac{10}{V_2} = \frac{283}{373} \quad \text{pressure constant}$$

$$283V_2 = 3730$$

$$V_2 = \frac{3730}{283} = 13.18 \text{ ft}^3 \text{ [373.26 L]}$$

Thus, we see that the increase in temperature has increased the volume of the gas from 10 to 13.18 ft³ [373.26 L], the pressure remaining constant.

The pressure of a gas varies in proportion to the absolute temperature if the volume is held constant. This is expressed by the equation

$$\frac{P_1}{P_2} = \frac{T_1}{T_2} \quad \text{volume constant}$$

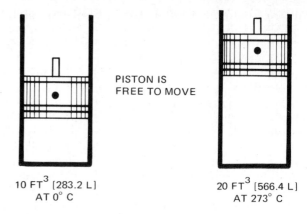

10 FT³ [283.2 L]
AT 0° C
20 FT³ [566.4 L]
AT 273° C

FIG. 7-9   Effect of temperature on gas volume.

This principle is illustrated in Fig. 7-10 where the gas temperature has been doubled; that is, it has been raised from 0°C (273 K) to 273°C (546 K). Since the absolute temperature has been doubled with the volume held constant, the pressure has doubled.

Manifold Pressure

Manifold pressure (MAP) is the pressure in the intake manifold of the engine. The weight of the fuel-air mixture entering the engine cylinders is measured by the manifold pressure and the temperature of the mixture. In an ordinary automobile engine the manifold pressure is less than outside atmospheric pressure because of the air-friction losses in the air-induction system. In a supercharged engine, however, the manifold pressure may be higher than the pressure of the atmosphere. When the supercharger is operating, the manifold pressure may be greater or less than the atmospheric pressure, depending on the settings of the supercharger control and throttle.

It must be pointed out that manifold pressure is of prime importance on high-performance engines equipped with constant-speed propellers. If the manifold pressure is too high, detonation and overheating will occur. These conditions will damage the engine and cause engine failure if permitted to exist for an appreciable length of time. During the operation of a supercharged engine, it is particularly important that the pilot or flight engineer pay strict attention to the power settings (rpm and manifold pressure) of the engines.

If an engine has a particularly high compression ratio, it is quite likely that the supercharger cannot be used at all until an altitude of 5000 ft [1524 m] or more

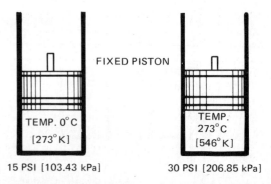

15 PSI [103.43 kPa]
30 PSI [206.85 kPa]

FIG. 7-10   Effect of temperature on gas pressures.

is reached. If the supercharger on such an engine is operated at low altitudes, the pressures and temperatures in the combustion chambers will cause detonation and preignition.

Another factor governing the manifold pressure to be used with an engine is the octane rating or performance number of the fuel. If a fuel has very high anti-detonation characteristics, the maximum manifold pressure may be higher than it would be with a fuel having a lower anti-knock rating. For this reason, it is essential that the person servicing an airplane note the fuel rating marked on the fuel tank cover. If it becomes necessary in an emergency to use a fuel with a rating lower than that specified for the airplane, the pilot can avoid trouble by operating the engine at lower manifold pressures than normal.

Purposes of Supercharging

The main purpose of supercharging an aircraft engine is to increase the manifold pressure above the pressure of the atmosphere in order to provide high power output for takeoff and to sustain the maximum power at high altitudes.

Increased manifold pressure increases the power output in two ways:

1. It increases the weight of the fuel-air mixture (charge) delivered to the cylinders of the engine. At a constant temperature, the weight of the fuel-air mixture that can be contained in a given volume of space is dependent on the pressure of the mixture. If the pressure on any given volume of gas is increased, the weight of that gas is increased because the density is increased.

2. It increases the compression pressure. The compression ratio for any given engine is constant; hence, the greater the pressure of the fuel-air mixture at the beginning of the compression stroke, the greater will be the **compression pressure,** the latter being the pressure of the mixture at the end of the compression stroke. Higher compression pressure causes a higher mean effective pressure and consequently a higher engine output.

Figure 7-11 illustrates the increase of the compression pressure. In the cylinder at the extreme left, the pressure is only 36 inHg [121.93 kPa] at the beginning of the compression stroke. At the end of the compression stroke the compression pressure in the second

cylinder from the left is 270 inHg [914.49 kPa]. In this case, the intake pressure is comparatively low.

The pressure is 45 inHg [154.42 kPa] in the third cylinder from the left in Fig. 7-11, and the compression pressure in the cylinder at the extreme right is 405 inHg [1371.74 kPa]. Since the pressure is relatively high at the beginning of the compression stroke, it is still higher at the end. In this case the intake pressure is high in comparison with that represented by the two drawings to the left. Increased temperature as a result of the compression also adds to the pressure.

Relation between Horsepower and Manifold Pressure

The relation between the manifold pressure and the engine power output for a certain engine at maximum rpm is shown on the chart in Fig. 7-12, where the manifold pressure in inches of mercury is plotted horizontally and the horsepower is plotted vertically. By referring to a manifold pressure of 30 and then following the vertical line upward until it intersects the curve, it is found that the curve intersects the 30 inHg [101.61 kPa] line at a horsepower of about 550 [410.14 kW]. This means that, when the engine is not supercharged, the theoretical maximum pressure in the intake manifold is assumed to be almost 30 inHg, which is the atmospheric pressure at sea level, and the power developed by the engine is about 550 hp. However, in actual practice it is impossible in an unsupercharged engine to obtain an intake manifold pressure of 30 in because of friction losses in the manifold.

Refer to a manifold pressure reading of 45, and follow the vertical line upward until it intersects the curve. This shows that if the manifold pressure is increased to 45 in by supercharging, the engine output is then about 1050 hp [782.99 kW].

It may occur to a student that the manifold pressure could be increased indefinitely to obtain more power, but it must be strongly emphasized that this is not true. Excessive manifold pressure always adversely affects the operation of the engine and ultimately damages it permanently as a result of high stresses, detonation, and high temperatures.

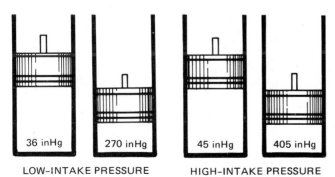

LOW-INTAKE PRESSURE   HIGH-INTAKE PRESSURE

FIG. 7-11   Effects of air pressure entering cylinder.

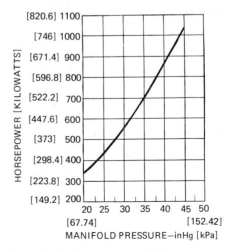

FIG. 7-12   Relation between manifold pressure and horsepower.

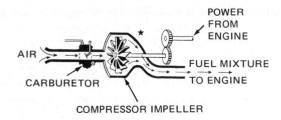

FIG. 7-13 Sea-level supercharger.

The Sea-Level Supercharger and the Sea-Level Engine

The **sea-level supercharger,** illustrated in Fig. 7-13, is sometimes called a **ground boost blower.** It is not a true supercharger unless it raises the MAP above 30 inHg. This system consists of a centrifugal compressor connected to the air intake, together with the necessary mechanism for driving it from the engine crankshaft through a gear train. The pumping capacity depends on the size of the impeller and the speed at which it is rotated.

If other factors remained the same, the gain in power output from the engine would be in proportion to the increase of pressure. However, the other factors do not remain constant, and there are limits beyond which safe operation is not possible. One of these factors is temperature. When the air is compressed, its temperature is raised. This reduces the efficiency of the supercharger because heated air expands and increases the amount of power required to compress it and push it into the cylinders. It also reduces the efficiency of the engine because any gas engine operates better if the intake mixture of fuel and air is kept cool. When the fuel-air mixture reaches an excessively high temperature, preignition and detonation may take place, resulting in a loss of power and often in a complete mechanical failure of the powerplant.

The rise in temperature resulting from the supercharging is in addition to the heat generated by the compression within the engine cylinders. For this reason, the combined compression of the supercharger and of the cylinders must be kept within the correct limits determined by the antiknock qualities or octane ratings of the fuel used in the engine.

If a supercharger were designed with enough capacity to raise the pressure of air at sea level from 14.7 to 20 psi [101.36 to 137.9 kPa], it would be possible to obtain about 40 percent more power than would be generated if there were no supercharging to increase the air pressure. If this supercharger were installed with a 1000-hp [745.7-kW] engine, the piston displacement of the 1000-hp engine would not need to be any greater than the piston displacement of a 710-hp [529.48-kW] engine that was not supercharged.

However, it must be understood that an engine which is to be provided with a supercharger must be designed to withstand the higher stresses developed by the increased power. It is not simply a matter of adding a supercharger to a 710-hp engine in order to obtain a 1000-hp output.

The difference between 710 and 1000 hp is obviously 290 hp [216.25 kW], but it requires about 70 hp [52.22 kW] to operate the supercharger for this imaginary engine; hence, not merely 290 hp but 290 plus 70, or 360 hp [268.56 kW], must be developed within the engine in order to obtain the 1000-hp output. This means that the supercharger must account for the development of an additional 360 hp when added to the 710-hp engine in order actually to obtain the 1000 hp desired.

Engines are no longer being manufactured with sea-level superchargers, although many of these "altitude" engines are still in operation. Engines currently being manufactured employ **turbochargers** to **turbonormalize** the engine. The turbocharger provides the intake air compression that makes it possible to maintain sea-level pressure at high altitudes.

Internal and External Superchargers

Most superchargers used on conventional airplanes are alike in that an impeller (blower) rotating at high speed is used to compress either the air before it is mixed with the fuel in the carburetor or the fuel-air mixture which leaves the carburetor. It is therefore possible to classify superchargers according to their *location* in the induction system of the airplane as either an internal type or an external type.

When the supercharger is located between the carburetor and the cylinder intake ports, it is an **internal type**, as shown in Fig. 7-14. Air enters the carburetor at atmospheric pressure and is mixed in the carburetor with the fuel. The fuel-air mixture leaves the carburetor at near-atmospheric pressure, is compressed in the supercharger to a pressure greater than atmospheric, and then enters the engine cylinders. The power required to drive the supercharger impeller is transmitted from the engine crankshaft by means of a gear train. Because of the high gear ratio, the impeller rotates much faster than the crankshaft. If the gear ratio is adjustable for two different speeds, the supercharger is described as a **two-speed supercharger.** In general, the internal-type supercharger may be used with an engine which is not expected to operate at very high altitudes or, in any event, where it is not necessary for air to be delivered under pressure to the carburetor intake.

An **external-type** supercharger delivers compressed air to the carburetor intake, as shown in Fig. 7-15. The air is compressed in the supercharger and then delivered through an air cooler to the carburetor, where it

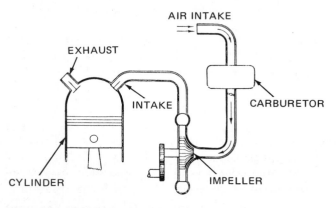

FIG. 7-14 Location of an internal-type supercharger.

150

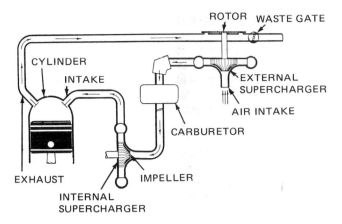

FIG. 7-15 Location of an external-type supercharger.

is mixed with the fuel. Since the power required to drive the ordinary type of external supercharger is obtained from the action of the engine exhaust gases against a bucket wheel, or turbine, the external type is also called a **turbosupercharger** or **turbocharger,** depending on whether it supercharges the air or merely maintains sea-level pressure. The speed of the impeller depends only on the quantity and pressure of the exhaust gases directed against the bucket wheel; hence, the turbosupercharger is also a **multispeed** supercharger. The volume of exhaust directed through the turbine is determined by the position of the **waste gate.** The waste gate is operated by means of a control in the cockpit.

Stages

A **stage** is an increase in pressure. Superchargers can be classified as single-stage, two-stage, or multiple-stage, according to the number of times an increase of pressure is accomplished. The single-stage supercharging system may include either a single-speed or a two-speed internal supercharger. Even though the two-speed internal supercharger has two definite speeds at which the impeller can rotate, only one stage (boost in pressure) can be accomplished at any time; hence, it is still a single-stage system. A single-stage supercharger system is shown in Fig. 7-14.

The Effect of Altitude on Engine Power Output

When an airplane climbs above the surface of the earth, the pressure of the atmosphere decreases and the density of the air also decreases. The less dense air, described as "thin air" by laypeople, offers less resistance to the flight of the airplane than the denser air near the earth's surface for the same reason that it is easier to swim in water than it would be to swim in lubricating oil, molasses, or some other fluid having a density greater than water. In addition, the back pressure on the engine exhaust gases is reduced, and the air at the higher altitudes is colder than that near the surface of the earth. All of these factors tend to increase the effectiveness of an engine.

Associated with these advantages, there are several disadvantages to increased altitude. The air at higher altitudes weighs less per cubic foot and there is less air pressure available for pushing it into the cylinders of the engine; hence, the power output of an engine decreases as the altitude increases. This can be compared with driving an automobile up a mountain road. The motorist finds that there is a loss of engine power output when climbing toward the top of the mountain. The automobile engine receives the same *volume* of air, but the *weight* of the air is less; hence, there is not enough oxygen delivered to the engine cylinders to provide efficient combustion for the fuel-air mixture.

There are certain minor problems associated with engine operation at increased altitude, whether the engine is in an automobile or in an airplane. For example, the reduced atmospheric pressure lowers the boiling point of gasoline so that the fuel system is in danger of vapor lock unless it has been especially designed to avoid this problem. Also, the reduced density of the atmosphere at higher altitudes causes the air to have less resistance to the passage of electricity. This tends to allow the electric current to "leak out" of the ignition system before it produces the sparks in the spark plugs.

Returning to the major problems of engine power output at increased altitudes, we find that an automobile engine which delivers 100 hp [74.57 kW] at sea level can deliver only about 60 hp [44.74 kW] at the top of Pikes Peak, which is about 14,000 ft [4267.20 m] above sea level. Since comparatively few automobiles are driven to the top of Pikes Peak, this loss of power with increased altitude does not have much effect on the work of the automobile engineers who design engines, but it is a very real problem to the designers of powerplants for airplanes.

Many airplanes are required to climb to altitudes of more than 20,000 ft [6096 m]. At these high altitudes a cubic foot of air weighs only about one-half of what it weighs at sea level. In order to have the same *weight* of air at 20,000 ft [6096 m] as at sea level, there must be twice as much *volume* of the thinner air. Likewise, at 40,000 ft [12 192 m] the volume of the thinner air must be four times as great because 1 ft^3 [28.32 L] of air at that altitude weighs only about one-fourth of what it weighs at sea level.

The actual air pressure at 20,000 ft [6096 m] altitude is less than one-half the air pressure at sea level, being about 13.75 inHg [46.61 kPa] at 20,000 ft and 29.92 inHg [101.34 kPa] at sea level. The density of the air, however, does not decrease as much as the pressure decreases; this is because of the decrease in temperature. The standard temperature at sea level is 59°F [15°C] and at 20,000 ft is −12.3°F [−24.61°C]; hence, the actual weight of a given volume of air is slightly more than one-half the weight that the same volume of air would have at standard sea-level conditions. At 40,000 ft [12 192 m] the **pressure** of the air is less than one-fifth the sea-level pressure, but because of the temperature decrease, the **weight** of a given volume of air is about one-fourth the weight of the same volume at sea level.

A sea-level supercharger or turbocharger provides an effective means for increasing the **pumping capacity** of the engine with a minimum increase in weight, but a powerplant equipped with the sea-level supercharger is affected by changes in altitude in the same manner as an unsupercharged engine, as shown in Fig. 7-16.

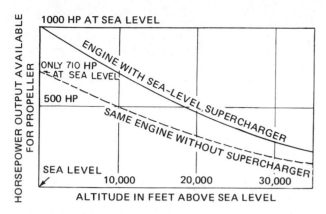

FIG. 7-16 Effect of altitude with a sea-level supercharger.

The human body has a supercharging system of its own. When climbing a mountain or riding in an airplane to higher altitudes, a person's lungs attempt to compensate for the thinness of the air by pumping in more air, thereby trying to obtain the quantity of oxygen needed for the combustion process of the body. In itself, the gasoline engine has no provision for such a process. If it is to provide adequate power at higher altitudes, the engine must be provided with special supercharging equipment to compensate for the thinness of the atmosphere.

It may appear at this point that the solution to the problem is simply to install a blower of enough capacity to take care of the highest altitudes at which the airplane is expected to fly. Unfortunately, it is necessary to provide enough power at increased altitudes and also *to avoid excessive power at the lower altitudes*. If the dense air at or near sea level were compressed as much as it is necessary to compress the thin air at high altitudes, there would be several unsatisfactory conditions. First, the engine would be overloaded and would soon fail because of the excessively high temperatures and pressures of the air delivered to the cylinders. Second, the extra power required to drive such a supercharger at or near sea level would be a waste of energy.

It is obvious that the greater the capacity of the supercharging equipment, the more efficient will be the performance at high altitudes, but it is also apparent that the airplane has to take off from the surface of the earth and climb through low altitudes, where the air is dense, in order to reach the higher altitudes where the supercharging equipment is needed. Therefore, some provision must be made for slowing down the supercharging process or otherwise reducing its effect while it is operating at or near sea level.

The foregoing facts account for the basic difference between a sea-level supercharger and an altitude supercharger. The capacity of the sea-level supercharger is determined by the operating conditions which the engine can safely experience at or near sea level and the supercharging required at higher altitudes. It has no special controls or regulating devices, but there must be some provision for protecting it against the adverse effects of excessive temperatures and pressures at the lower levels.

In theory, the ideal altitude supercharger would provide the engine with the same weight of air, thus enabling it to deliver full power regardless of the altitude. Some superchargers have been designed and constructed according to this theory, but in practice a compromise is required—one taking into consideration not only the engine efficiency but the overall efficiency of the airplane and its effectiveness in performing the task for which it was designed. Some turbochargers currently manufactured are controlled by systems which effectively maintain sea-level MAP to the altitude established for the particular aircraft.

Factors Considered in Designing Altitude Superchargers

If the power to run the supercharger is taken from the engine crankshaft, the net gain in horsepower obtained by supercharging is reduced.

The net gain in horsepower obtained from supercharging is not fully reflected in overall airplane performance because the supercharging equipment requires additional space and adds to the weight of the airplane.

The speed changing devices or multiple-stage compressors, or both, required to make a supercharger effective at different altitudes need additional space, add to the weight of the airplane, and complicate the operation, inspection, and maintenance procedures.

The degree of supercharging must be restricted within definite limits to avoid the dangers of preignition and detonation which result from excessive temperatures and pressures.

A special cooling apparatus must be used to reduce the temperature of the fuel-air mixture because of the excessive heat resulting from the extra compression required at extreme altitudes. This apparatus requires space, adds to the weight of the airplane, and complicates operation, inspection, and maintenance. The special radiators used for this cooling are called **intercoolers** or **aftercoolers,** depending on their location with reference to the carburetor.

● INTERNAL SINGLE-SPEED SUPERCHARGER

System for Six-Cylinder Opposed Engine

An internal, single-speed supercharger consists of a gear-driven impeller placed between the carburetor and the intake ports of the cylinders. Figure 7-17 shows the supercharger installation for a Lycoming series GSO-480 engine. The supercharger is mounted on the rear end of the engine and is driven at 11.27 times crankshaft speed. At this speed the supercharger is capable of delivering a maximum manifold pressure of 48 inHg [162.8 kPa].

An exploded view to show the supercharger components is provided in Fig. 7-18. From the arrangement and design of the parts, it is seen that the fuel-air mixture from the carburetor is drawn into the center of the impeller, whence it is thrown outward (by centrifugal force) through the diffuser and into the supercharger housing. From here it flows into the portion of the induction system surrounded by the oil sump and then into the individual intake pipes to each cylinder.

FIG. 7-17 Supercharger installation for a light opposed engine. *(Lycoming Div. Avco Corp.)*

Engine Performance with Supercharger

The effects of supercharging on the performance of a particular engine can be seen by examination of the fuel-consumption and power curves in Fig. 7-19. The chart on the left shows the performance of the Lycoming series GO-480-D engines, and the chart on the right shows the performance of the series GSO-480 engines. Note that the engines of both series have a maximum rpm of 3400 rpm and that the compression of both engines is 7.30:1. The maximum power available from the GO-480-D engine is 275 hp [205.07 kW], whereas the maximum power available from the GSO-480 engine is 340 hp [253.54 kW].

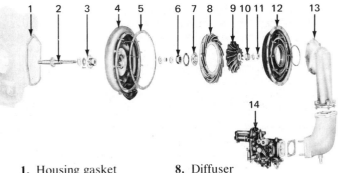

1. Housing gasket
2. Impeller shaftgear
3. Shaft bearing
4. Housing assembly
5. Housing gasket
6. Driveshaft seal
7. Driveshaft oil seal retainer
8. Diffuser
9. Impeller
10. Impeller nut spacer
11. Impeller locknut
12. Air inlet adapter assembly
13. Air inlet housing assembly
14. Carburetor

FIG. 7-18 Exploded view of a supercharger.

Because of the higher manifold pressures used with the supercharged engine, it is necessary to employ fuel with a comparatively high performance number. On the chart for this engine, the performance rating of the fuel is shown to be 100/130. The technician who services an airplane having a supercharged engine must make certain that the fuel tanks are filled with the correct grade of fuel.

Service and Maintenance

No service or maintenance is required with the supercharger just described other than normal inspections. At the time of engine overhaul the supercharger is inspected and overhauled in accordance with the instructions given in the manufacturer's engine overhaul manual.

Single-Speed Supercharger for a Radial Engine

The principal components of the supercharger section for the Pratt & Whitney R-985 engine are shown in Fig. 7-20. It will be observed that the supercharger consists of an impeller mounted in the supercharger (blower) case of the engine, immediately forward of the rear case. The rear case contains the **diffuser vanes** which distribute the fuel-air mixture evenly to the nine intake pipes attached by means of seals and packing nuts to the openings around the outside of the case.

The impeller is driven by the engine at a speed of ten times the crankshaft speed. This provides for a maximum manifold pressure of 37.5 inHg [127.01 kPa] and a power output of 450 hp [335.57 kW].

● TWO-SPEED INTERNAL SUPERCHARGER

A two-speed internal supercharger is designed to permit a certain degree of supercharging at sea level and additional supercharging for altitude operation. Typical of a two-speed system is that installed on the Pratt & Whitney R-2800 Double Wasp series CB engines. The impeller for this supercharger is driven at a ratio of 7.29:1 by the low-ratio clutch and at 8.5:1 by the high-ratio clutch. In some models of the engine the high gear ratio may be as much as 9.45:1.

Supercharger Case

The supercharger case for the R-2800 engine is a magnesium-alloy casting attached to the crankcase's rear section. To the rear of the supercharger case is the intermediate rear case. The impeller of the supercharger is housed between the two sections. The front face of the case incorporates a steel liner to accommodate the impeller shaft front oil-seal rings and a bushing to support the rear end of the rear counterweight intermediate-drive-gear shaft. A vaned diffuser is secured to the rear face of the case. Located around the periphery of the case are nine fuel-air mixture outlet ports. Attached to each port is a V-shaped intake pipe through which the fuel-air mixture is carried to one front and one rear cylinder.

Intermediate Rear Case

The intermediate rear case (attached to the rear of the supercharger case) houses the impeller and acces-

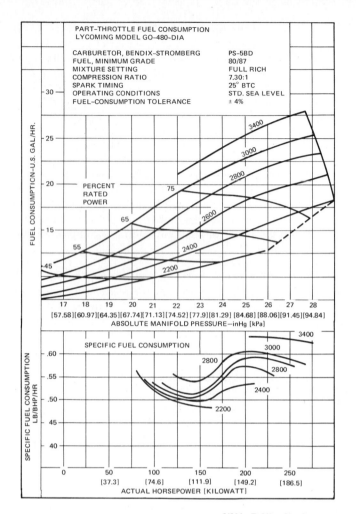

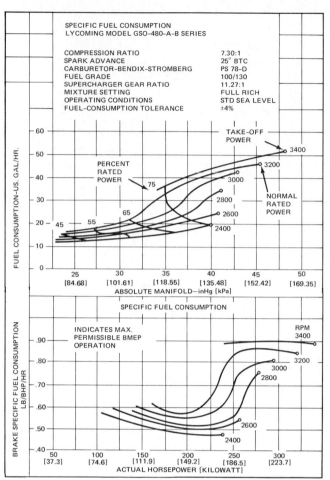

FIG. 7-19 Fuel-consumption and power curves.

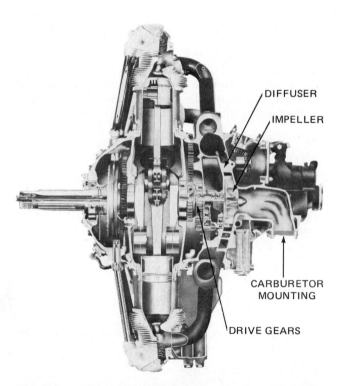

FIG. 7-20 Single-speed internal supercharger for a radial engine.

sory drive shafts and provides support for the front end of the shafts in the various accessory gear trains. A supercharger intake insert, which forms a throat for the impeller, is located in the front face of the case. The web of the insert incorporates a steel liner which serves the dual purpose of accommodating the impeller shaft center oil seal rings and providing a fuel distributor for the fuel slinger.

The rear face of the case incorporates a steel liner to accommodate the impeller shaft rear oil-seal rings and bushings to support the front ends of two clutch shafts or two impeller intermediate-drive-gear shafts, a generator drive-gear shaft, and a vacuum pump shaft.

Impeller and Impeller Shaft

The impeller is splined to the impeller shaft, and the assembly is supported on the accessory drive shaft on bronze bearings, which are installed in the inside diameter of the impeller shaft. The impeller is driven by a train of gears, starting with the accessory drive shaft adapter, which is splined to the front end of the accessory drive shaft and the rear of the crankshaft. The drive continues through the accessory spring drive gear (which is splined to the rear end of the accessory drive shaft) to the clutch shaft-drive pinion, to the clutch gears, and to the integral spur gears on the rear of the impeller shaft.

Clutches

A desludger-type dual clutch is mounted on each side of and parallel to the impeller shaft. Each clutch shaft is supported at its front end by a bushing in the rear of the intermediate rear case and at its rear end by a bushing in the rear case. The **cones** of the clutches are splined to the clutch shaft and engage the clutch gears from the inside through a cone-shaped facing (lining). This arrangement can be seen in the drawings of Fig. 7-21. Observe that the high-ratio gear is larger in diameter than the low-ratio gear. This accounts for the difference in drive ratio.

A narrow ring gear with a drilled vent hole rides on each clutch-gear support. These ring gears turn at a different rate from their adjacent clutch gears, so that the single vent holes will line up at intervals with one of several corresponding holes drilled in the clutch-gear supports. As these holes line up, a spurt of oil is allowed to escape from behind the cones, thus preventing the accumulation of sludge.

A study of Fig. 7-21 will show how the clutch mechanism is operated. In the drawing on the left, the high-ratio clutch is engaged because of the oil pressure applied to the space between the support assembly and the high-ratio cone. The force presses the cone facing against the inside of the gear, thus causing the gear to turn with the shaft and cone. In the drawing at the right, the pressure oil is directed to the space between the support for the low-ratio gear and the low-ratio cone, thus engaging the low-ratio clutch. An exploded view of the clutch and gear assembly is shown in Fig. 7-22.

An understanding of the basic arrangement for the two-speed gear drive can be obtained by studying the simplified drawing of Fig. 7-23. In this drawing, gear 5 is the spur gear driven by the accessory drive gear. Gear 2 is the low-ratio drive gear, and gear 1 is the high-ratio drive gear. If the clutch inside gear 2 is engaged, gear 2 will be turning with the shaft and will be driving gear 4, which is integral with the impeller shaft. It is apparent that the impeller shaft will be turning about $1\frac{1}{2}$ times the speed of the clutch drive shaft. This is the low-ratio drive. Remember that the clutch in gear 1 is not engaged and gear 1 is idling.

If the clutch in gear 1 is engaged, then gear 1 will

① IMPELLER RATIO SELECTOR VALVE
② IMPELLER DRIVE SHAFT
③ ACCESSORY DRIVE SHAFT
④ LOW-RATIO CLUTCH

⑤ HIGH-RATIO CLUTCH
⑥ STARTER JAW
⑦ ACCESSORY DRIVE GEAR
⑧ CLUTCH SHAFT PINION

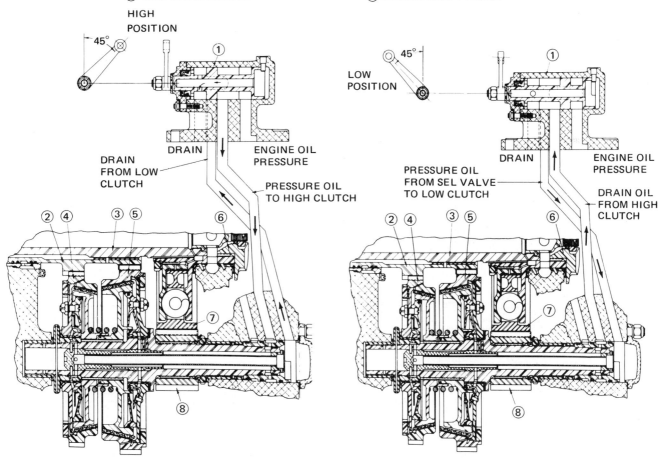

SELECTOR VALVE IN HIGH POSITION
WITH HIGH-CLUTCH ENGAGED

SELECTOR VALVE IN LOW POSITION
WITH LOW-CLUTCH ENGAGED

FIG. 7-21 Clutch operation in a two-speed internal supercharger.

155

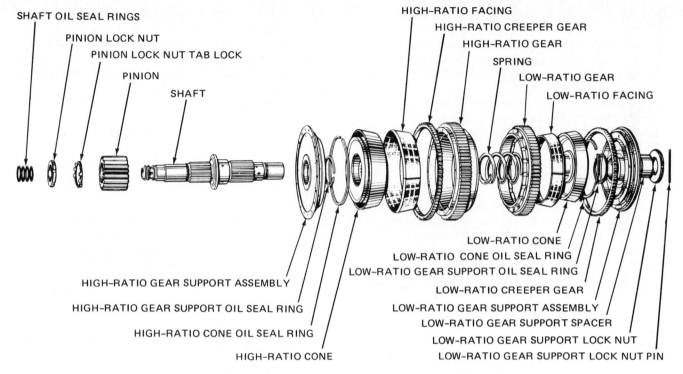

FIG. 7-22 Exploded view of clutch and gear assembly.

turn at the same rate as the clutch drive shaft and will drive gear 3 at about 2½ times the speed of the clutch drive shaft. This is the high-ratio drive. At this time, gear 2 will be idling. The clutches for both gears cannot be engaged at the same time.

Selector Valve

A manually operated, two-position impeller ratio-selector valve is mounted on the top of the engine rear case. If the valve is moved into the HIGH position, pressure will be directed to the chamber between the cone and the high-ratio clutch gear support. If the valve is moved into the LOW position, pressure oil will be similarly directed to each low-ratio clutch. The oil pressure causes the cones to engage the gears, which operate in parallel to drive the impeller. The selector valve is shown in each of the drawings of Fig. 7-21.

THE TURBOSUPERCHARGER

A **turbosupercharger** or **turbocharger** is an external supercharger designed to be driven by means of a turbine wheel which receives its power from the engine exhaust. Ram air pressure or alternate air pressure is applied to the inlet side of the turbocompressor (blower), and the output is supplied to the inlet side of the carburetor or fuel injector. If a high degree of air compression is done by the compressor, it may be necessary to pass the compressed air through an intercooler to reduce the air temperature. As mentioned previously, if the carburetor air temperature is too high, detonation will occur.

The exhaust gases are usually diverted from the main exhaust stack by means of a **waste gate.** On large en-

FIG. 7-23 Simplified illustration of gear operation in a two-speed supercharger.

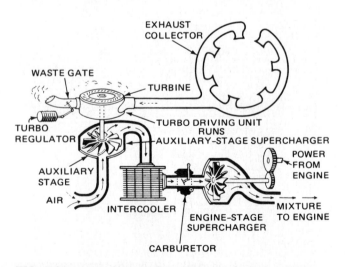

FIG. 7-24 Arrangement of a turbosupercharger system.

gines, the position of the waste gate is automatically controlled by reference to selected manifold pressure. The waste gate is closed to direct the exhaust gases through the turbine. The degree of closing determines the amount of air pressure boost obtained from the supercharger.

The general arrangement of a turbosupercharger system is shown in the schematic diagram of Fig. 7-24.

A turbosupercharger system can be used to maintain a given manifold pressure to the design altitude of the system. Above this altitude, which is called the **critical altitude,** the manifold pressure will begin to fall off as altitude is increased. We may therefore define critical altitude as the altitude above which a particular engine-supercharger combination will no longer deliver full power.

● TURBOCHARGERS FOR LIGHT-AIRCRAFT ENGINES

A turobcharger installation for a light airplane engine is shown in Fig. 7-25. This is the Rajay Turbo 200 system, using a turbocharger manufactured by Rajay Industries, Inc., of Long Beach, California, for installation on the Piper PA-23 150 and 160 airplanes.

It will be observed from the photograph that the exhaust pipes from each cylinder are all connected into one main exhaust stack. A waste gate is placed near the outlet of the stack to block the exit of the exhaust gases and direct them through a duct into the turbine. The turbine drives the compressor, which increases the air pressure to the carburetor inlet.

This particular turbocharger is designed for use at altitudes above 5000 ft [1524 m] because maximum engine power is available without supercharging below that altitude.

Description

The Rajay 200 turbocharger is a 12.5-lb [5.67-kg] unit of high-speed turbine equipment designed primarily for use on small high-performance diesel engines. The basic design has been modified to be compatible with the aircraft powerplant application.

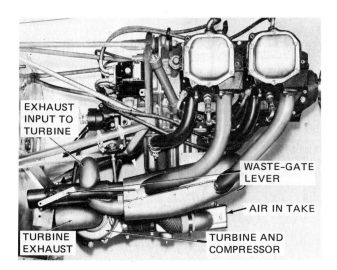

FIG. 7-25 **Turbocharger installation for a light opposed engine.** *(Rajay Industries)*

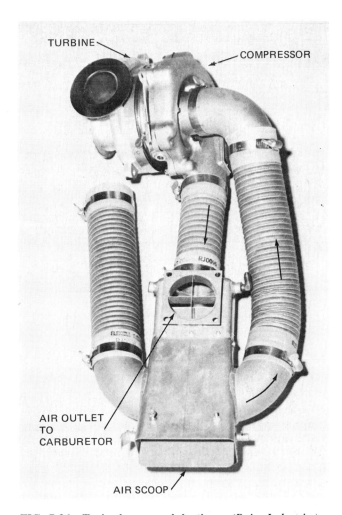

FIG. 7-26 **Turbocharger and ducting.** *(Rajay Industries)*

The unit consists of a precision-balanced rotating shaft with a radial inflow turbine wheel on one end and a centrifugal compressor impeller on the other end, each with its own housing. The turbine, driven by engine exhaust gases, powers the impeller, which supplies air under pressure to the carburetor inlet. This higher pressure supplies more air by weight to the engine with the advantage of a proportionally higher power output and a minimum increase in weight.

A photograph of the turbocharger and its ducting to the carburetor air box is shown in Fig. 7-26. This photograph may be compared with the diagram of Fig. 7-27 to gain a clear understanding of the operating principles.

In the carburetor air box is a **swing check valve** which is open during **naturally aspirated** operation, that is, when the turbocharger is not in operation. The check-valve operation is automatic, and the valve will close when turbo boost pressure is greater than ram air pressure. When the waste gate in the exhaust stack outlet is closed or partially closed, exhaust gases are directed to the turbine. The rotation of the turbine causes the compressor to draw air from the air box and deliver it under pressure through the duct forward of the carburetor.

It will be observed in the diagram that carburetor heat may be obtained through the alternate air duct,

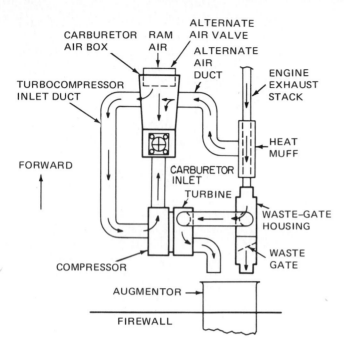

CARBURETOR AIR BOX — RAM AIR — ALTERNATE AIR VALVE — ALTERNATE AIR DUCT — ENGINE EXHAUST STACK — TURBOCOMPRESSOR INLET DUCT — HEAT MUFF — FORWARD — CARBURETOR INLET — TURBINE — WASTE–GATE HOUSING — WASTE GATE — COMPRESSOR — AUGMENTOR — FIREWALL

FIG. 7-27 Turbocharger system.

regardless of whether or not the supercharger is operating. If carburetor heat is desired, the alternate air valve is closed, shutting off ram air and allowing air to be drawn through the heater muff. With this installation, carburetor heat should not be used when the turbo boost is more than 5 inHg [16.94 kPa].

The bearings of the turbocharger are of the sleeve-journal type and utilize pressure lubrication from the engine. This type of bearing is low in cost and has a high degree of reliability.

The turbine and turbine housing are cast of high-temperature alloys, with the central main housing, compressor housing, and impeller cast of aluminum alloy for light weight and good thermal characteristics. The design and construction result in a unit which is completely air-cooled.

Lubrication System

Lubricant for the turbocharger is supplied by a line connected to a fitting on the engine-governor, fuel-pump, dual-drive pad. A fitting included in this lubricant supply line incorporates a pressure-regulator poppet valve to reduce engine gallery oil pressure from the normal 60- to 80-psi [414- to 552-kPa] range to a pressure of 30 to 50 psi [207 to 345 kPa], which is required for the turbocharger. At this pressure range, between 1 and 2 qt/min [0.95 and 1.89 L/min] of lubricating oil will be supplied to the unit. This quantity of oil is but a small percentage of the total oil pump capacity. The oil supplied to the turbocharger is normally returned to the engine sump by way of the bypass pressure-relief valve.

Incorporated in the turbo lubricant supply line is a pressure switch which will activate a red warning light in the event that turbocharger oil pressure is below 27 to 30 psi [186 to 207 kPa]. If the oil pressure is lost, the pilot simply removes the turbocharger from service by opening the waste gate and returns the engine to naturally aspirated operation to save the turbocharger

bearings. The turbocharger lubricating-oil sump is scavenged by means of the fuel-pump drive gears contained in the dual drive unit for the pump.

Controls

As previously explained, the principal factor in turbocharger operation is the degree of waste gate closure. This determines the amount of the total engine exhaust-gas flow through the turbine and the resulting level of boost. A separate push-pull control with a precise vernier adjustment is installed for actuation of the waste gate. This permits convenient, exact matching of manifold pressures for both engines on the airplane.

Operation

Since the Rajay 200 turbocharger is designed for operation at altitudes above 4000 to 6000 ft [1219 to 1829 m], the ground operation of the engine is the same as that required for an unsupercharged engine. During climb, when the airplane reaches an altitude where power begins to fall off, the pilot places the throttle in the wide open position and begins to close the waste gate with the separate control located on the fuel-valve-selector console. The operation of the turbocharger makes it possible to maintain a sea-level manifold pressure of 28 inHg [94.83 kPa] absolute during the climb of the aircraft to an altitude of 20,000 ft [6096 m].

Typical engine operating conditions for climb are as follows:

1. 2400 rpm maximum and 2200 rpm minimum with turbo operative.

2. 25 to 28 inHg [84.67 to 94.83 kPa] maximum manifold pressure.

3. 400 to 475°F [204.24 to 245.87°C] maximum cylinder-head temperature with turbo operating. A cylinder-head temperature gage is required.

4. Carburetor inlet air temperature 100 to 160°F [37.78 to 71.12°C], depending upon the power setting and the temperature of outside air. A performance number for the fuel will have been specified to preclude any possibility of detonation.

After the airplane has attained the desired cruising altitude, power should be reduced to 23 to 25 inHg [77.9 to 84.67 kPa] MAP and the rpm to the 2200 to 2300 cruising range. The aircraft is then trimmed for cruising speed, and the fuel-air mixture adjusted for best economy. When the engine is operating below 75 percent power, leaning can be accomplished by pulling the mixture control back slowly until there is a slight drop in MAP. The mixture control is then moved forward until smooth, steady engine operation is attained. It should be emphasized that manual leaning must not be done when an engine is operating at more than 75 percent of power or when the MAP exceeds a certain value specified in the operator's handbook.

When it is desired to let down and prepare for landing, the turbocharger should be shut off by opening the waste gate and carburetor heat should be turned on. The throttle should not be closed suddenly because rapid cooling of the cylinder heads may cause cracking or other damage. During letdown the throttle should not be entirely closed and should occasionally be opened sufficiently to clear the engine.

The Rajay 260 Turbocharger

Another good exmaple of a turbocharger system for light aircraft is the Rajay 260, designed for installation on the Piper Cherokee airplane. The system operates on the same principles described for the Rajay 200.

The principal components of the turbocharger for the Rajay system are illustrated in Fig. 7-28. The housings are machined castings, whereas the turbine and compressor impeller are formed by investment casting, machining, and grinding.

The Rajay 260 system as installed on the Cherokee aircraft is shown in Fig. 7-29. Intake air is **turbonormalized** by controlling the waste gate from the cockpit. Turbonormalizing merely means increasing intake air pressure to approximately normal sea-level value.

When the waste gate is closed or partially closed, exhaust gas is directed to the turbine. The pilot determines how much the waste gate should be closed by observing the manifold pressure gage. An automatic flap valve in the air box directs either turbocharger air or unpressurized air to the carburetor, depending on which is greater. When turbocharger air pressure is greater than the unpressurized air pressure from the air filter, the airbox valve closes the unpressurized air inlet and directs turbocharger air to the carburetor.

● AVCO LYCOMING AUTOMATIC TURBOCHARGER

In this section we shall describe the operation of a typical automatic turbocharger manufactured by the AiResearch Manufacturing Co., a division of the Garrett Corp., and used on the Avco Lycoming TIO-540-A2A engine which powers the Piper Navajo airplane. This is a **sea-level boosted** engine; that is, the turbocharger can operate at sea level and above to produce the selected power from the engine until an altitude of approximately 19,000 ft [5791.20 m] is attained. Thus, the engine can produce sea-level power up to the critical altitude of approximately 19,000 ft.

At sea level, without turbocharging, the engine named above can produce 290 hp [216.25 kW]. With turbocharging, the same engine will have manifold pressure boosted by 10 inHg [33.87 kPa] and the horsepower will be 310 [231.17 kW]. In addition to the increased horsepower at sea level, the engine can develop rated power up to approximately 19,000 ft altitude as stated above. At this altitude, a **normally,** or **naturally, aspirated engine** (one without supercharging) can develop only about one-half the power that can be developed at sea level. This is because of the lower density of the air at higher altitudes.

The turbocharger is so named because it is a supercharger driven by a turbine. The power to drive the turbine is furnished by the high-velocity gases from the engine exhaust, as explained in the previous section.

The turbocharger under discussion utilizes a centrifugal compressor attached to a common shaft with the turbine. This arrangement is illustrated in Fig. 7-30. This turbocharger has three control components, which enables it to provide automatically the power that the pilot has selected for operation. These are the **density controller,** the **differential pressure controller,** and the **exhaust bypass-valve assembly** (waste gate).

Density Controller

The density controller (Fig. 7-31) is designed to sense the *density* of the air after it has passed through the compressor, that is, between the compressor and the

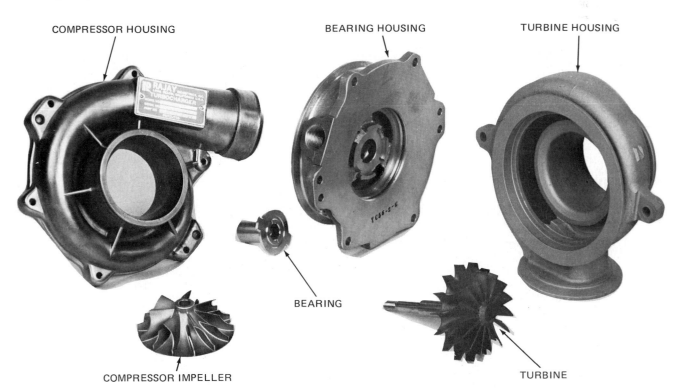

COMPRESSOR HOUSING   BEARING HOUSING   TURBINE HOUSING

BEARING

COMPRESSOR IMPELLER

TURBINE

FIG. 7-28   Components of a Rajay turbocharger. *(Rajay Industries)*

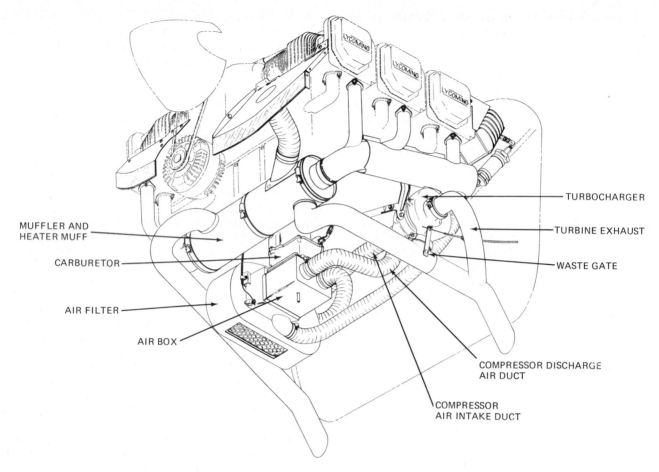

FIG. 7-29   Installation of the Rajay 260 turbocharger system.   *(Rajay Industries)*

throttle valve. The air pressure in this area is called the **deck pressure.** Density is determined by pressure and temperature, so that the density controller is equipped with a bellows assembly which contains dry nitrogen for temperature sensitivity, a valve, springs, and a housing with oil inlet and outlet ports. As tem-

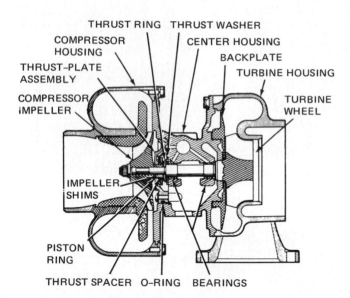

FIG. 7-30   Arrangement of the turbocharger turbine and compressor on a single shaft.   *(Lycoming Div., Avco Corp.)*

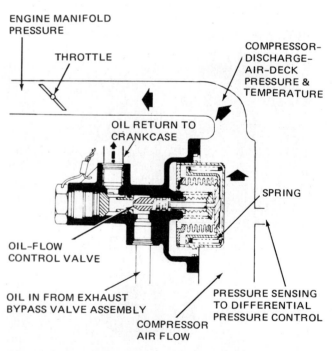

FIG. 7-31   Density controller.   *(Lycoming Div., Avco Corp.)*

perature increases, the bellows tends to expand; and as pressure increases, the bellows tends to contract.

The bellows assembly extends into the deck pressure airstream where it is exposed to the air entering the engine. If the density is not equal to that required for full power operation, the controller will call for additional supercharging.

The bellows controls a metering valve which permits oil to escape back into the engine or restrains the oil and causes a pressure buildup to the exhaust bypass valve. When oil pressure builds up, the bypass valve closes and directs exhaust gases to the turbine, thus increasing compressor rpm which, in turn, increases compression of the incoming air. Pressurized oil in the exhaust bypass-valve actuator causes the butterfly valve to close an amount proportional to the oil pressure.

The density controller operates to restrain oil only during full-power operation. At part-throttle settings, the control of the turbocharger is assumed by the differential pressure controller.

It must be remembered that the density controller is designed to maintain deck pressure and density at a level required for full-power operation of the engine. If the power selected is less than full power, the deck pressure will be less than required for full power and the controller will not permit any oil to bypass back to the engine. The control of the oil pressure to the exhaust bypass valve will therefore be accomplished by the differential pressure controller.

Differential Pressure Controller

The differential pressure controller is illustrated in Fig. 7-32. Note that the unit incorporates a diaphragm which is exposed on one side to deck pressure and on the other side to inlet manifold pressure. The diaphragm is connected to a valve and is subjected to spring pressure, which holds the valve closed until the differential pressure across the diaphragm is sufficient to open the valve. The unit is set to provide approximately 2 to 4 inHg [6.77 to 13.55 kPa] pressure drop across the throttle at a specified manifold pressure; how-

ever, the usual differential is approximately 2 inHg. If the required pressure differential is exceeded, the oil valve will open sufficiently to reduce supercharging until the desired pressure differential is restored.

Exhaust Bypass-Valve Assembly

The exhaust bypass-valve assembly is illustrated in Fig. 7-33. When this valve is open to its maximum design limit, almost all the exhaust gas is dumped overboard and not directed to the turbine. Hence, minimum supercharging is applied. When the valve is in the fully closed position, there is still a clearance of 0.023 to 0.037 in [0.58 to 0.94 mm] shown as dimension A in the drawing. By keeping the butterfly from completely closing and touching the walls of the valve housing, the danger of the valve's sticking is eliminated. Furthermore, the possibility of turbocharger overspeeding is reduced when all the exhaust gas is supplied to the turbine at high altitude. When the butterfly is fully open, there is a clearance of from 0.73 to 0.75 in [18.54 to 19.05 mm] between the edge of the butterfly and the wall of the valve housing. This is shown as dimension B in the drawing. It must be remembered that the dimensions given here are for one model of valve only; the others will have different dimensions. The manufacturer's manual should be consulted to ensure that the correct dimension is established when overhauling or adjusting the valve.

The butterfly moves in proportion to the amount of oil pressure built up by the controller. Therefore, the butterfly can be at any position between fully open and fully closed. Thus, the engine can deliver constant power for any power setting selected by the pilot as long as the aircraft is below the critical altitude for the selected power setting. The critical altitude is the point at which the butterfly is fully closed and the engine is delivering full rated power. At any altitude above the critical altitude the engine cannot deliver full rated power because the air density decreases and the amount of air entering the engine is not sufficient to produce full power. During operations below the critical altitude, such as cruise conditions and power-on letdown, the

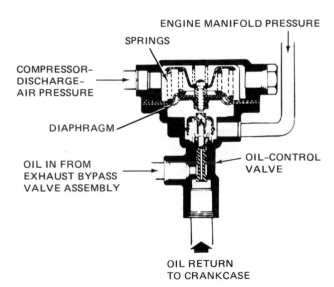

FIG. 7-32 Differential pressure controller. *(Lycoming Div., Avco Corp.)*

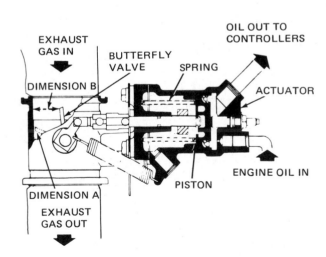

FIG. 7-33 Exhaust bypass-valve assembly. *(Lycoming Div., Avco Corp.)*

butterfly will be positioned somewhere between fully open and fully closed.

Operation of the Turbocharger-Control System

A diagram of the automatic turbocharger-control system under discussion is given in Fig. 7-34. Note that air enters the turbocharger compressor through the air filter and inlet duct. From the compressor the air passes the density controller (where the density is sensed) and then continues through the throttle valve to the intake valve of the engine. From the exhaust valve the exhaust gases enter the exhaust manifold and flow to either the exhaust bypass valve or the turbine, depending on the position of the butterfly valve. If the system is calling for maximum supercharging, the major portion of the exhaust gases must pass through the turbine because the butterfly valve is closed. If no supercharging is required, the butterfly valve will be open and the gases pass out through the exhaust stacks.

Remember, the position of the butterfly valve of the exhaust bypass depends on the oil pressure in the actuator. Engine oil pressure is fed to the area above the piston in the actuator and an oil outlet is connected to the same area. If oil can flow freely from the cylinder through the return line, the piston will not be moved and the butterfly valve will remain open. The oil flow out of the oil-pressure line is controlled by the density controller and the differential pressure control. If one or the other of these units permits the oil to return to the crankcase freely, pressure will not build up in the exhaust bypass actuating cylinder and the butterfly valve

will remain open. If one of the controlling units is restraining the oil return, the bypass valve in the other unit will be completely closed and pressure will be built up in the actuating cylinder of the exhaust bypass valve. The piston will then move to close the butterfly valve and exhaust gases will be directed to the turbine. This will increase the rpm of the turbine and compressor and the incoming air will be supercharged or pressurized.

Since the exhaust bypass-valve assembly contains O rings, seals, springs, and a piston, there is a certain amount of friction, which may result in slightly different valve positions for the same oil pressure, dependent on whether the pressure is increasing or decreasing at the time the setting is made. Engine oil under pressure is fed through an orifice to the top of the piston and is bled back to the engine through the controllers, as previously explained. In all cases, the controllers will adjust the butterfly valve to establish the air density or differential pressure required to produce the power called for by the pilot.

The controllers act independently to regulate the pressure on the exhaust-bypass-valve actuator piston. Since they serve two different functions, the controllers can be analyzed separately. It must be remembered that only one of the controllers will be operating at any given time. At full throttle, the density controller regulates the amount of supercharging; at part throttle, the differential pressure controller performs this function.

The density controller is designed to hold the air

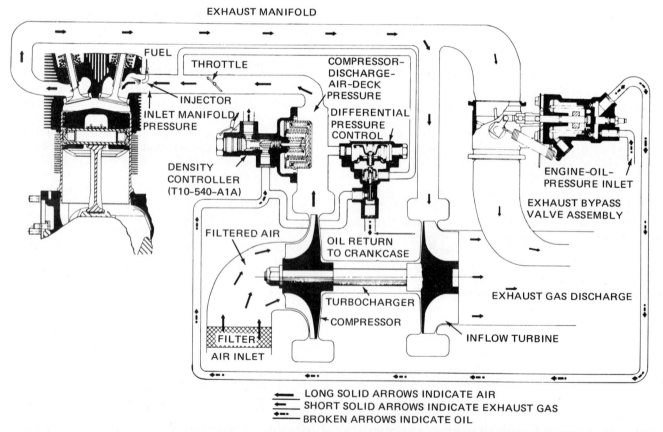

FIG. 7-34 Schematic diagram of the automatic turbocharger-control system. *(Lycoming Div., Avco Corp.)*

density constant at the injector entrance (intake area in the cylinder head). As the air temperature is increased, a higher manifold pressure is required and this results in a greater temperature rise across the compressor. In turn, wide-open-throttle manifold pressure increases with either altitude or outside air temperature. In a full-throttle climb, the gain is 3 to 4 inHg [10.16 to 13.55 kPa] manifold pressure between sea level and critical altitude.

As deck pressure and temperature change, the bellows in the density controller is either expanded or contracted which, in turn, repositions the oil-bleed valve, changes the amount of bleed oil, establishes a new oil pressure on the exhaust bypass-valve actuator piston, and repositions the butterfly valve. Controllers, engines, and turbochargers have individual differences; and with the length of time required to establish temperature equilibriums, two powerplant packages will show somewhat different manifold pressures.

The density controller can be adjusted, but it must be done under controlled conditions with ample stabilization time. Since this unit regulates wide-open throttle power and manifold pressure only, adjustments should not be made to correct any part-throttle discrepancies. The controller should be adjusted by authorized personnel and must be adjusted to the curve found in the engine operation manual.

If the differential pressure controller were not used, the density controller would attempt to position the exhaust bypass valve so that the air density at the injector entrance would always be that which is required for maximum power. Since this high air density is not required during part-throttle operation, the differential pressure controller reduces the air pressure to the correct level and causes the exhaust bypass valve to modulate over as high an operating range as possible.

As explained previously, the differential pressure controller contains a diaphragm connected to a bleed valve in the oil line between the exhaust bypass valve and the engine crankcase. One side of the diaphragm senses air pressure before the throttle valve and the other side senses air pressure after the throttle valve. When operating at wide-open throttle, the air-pressure drop across the throttle (and across the diaphragm) is at a minimum and the spring in the differential controller holds the oil-bleed valve in the closed position. Therefore, at wide-open throttle the differential pressure controller is not working and the density controller determines the position of the exhaust bypass valve.

As the throttle is partially closed, the air pressure drop across the throttle valve and across the diaphragm is increased. This pressure differential on the diaphragm opens the bleed valve and allows oil to bleed back to the crankcase. This establishes a different oil pressure on the exhaust bypass-valve actuator piston and changes the position of the exhaust bypass valve. This cycle repeats itself until equilibrium is established.

The differential pressure controller regulates the amount of supercharging to a point just slightly higher than that required for part-throttle operation and therefore reduces the time required for the system to seek and find equilibrium. The unit performs a necessary function, but it has a side effect that must be understood. When the pilot changes the position of the throttle, a long chain of events is triggered. Since heat transfers and turbocharger inertia are involved, many cycles may be required to reach a new equilibrium.

The sequence of events which occur as the result of a change in throttle position are as follows:

1. The pilot moves the throttle and establishes a different pressure drop across the throttle valve.
2. The differential controller diaphragm senses the change and repositions the bleed valve.
3. The new bleed-valve setting changes the oil flow, which establishes a new pressure on the exhaust bypass-valve actuating cylinder.
4. The changed pressure on the actuating cylinder piston causes the piston to reposition the butterfly valve.
5. The new butterfly-valve position changes the amount of exhaust gas flow to the turbine.
6. This changes the amount of supercharging which, in turn, changes the air pressure at the injector entrance.
7. The new pressure then changes the pressure drop across the throttle valve and the sequence returns to step 2 and repeats until equilibrium is established.

The net result of these events is an effect called **throttle sensitivity.** When comparing the operation with that of naturally aspirated engines, the supercharged engine's manifold pressure setting will require frequent resetting if the pilot does not move the throttle controls slowly and wait for the system to seek its stabilization point before making corrective throttle settings. Variations of temperature can also cause the turbocharger system to fluctuate. For instance, the bellows in the density controller responds relatively slowly to temperature changes. Changes in oil temperature have some effect on the system, too, because of corresponding changes in oil viscosity. A "thin" oil requires greater restriction than a "thick" oil to maintain the same back pressure on the exhaust bypass-valve actuating piston.

During acceleration, the supercharging system discussed in this section is noticeably subject to variations because all the components of the control system are continually resetting themselves and little time is available for stabilization. An **overboost** condition occurs when the manifold pressure exceeds the limits at which the engine was tested and certified by the FAA. This can be detrimental to the life and performance of the engine. Overboost can be caused by malfunctioning controllers or by an improperly operating exhaust bypass valve. The control system on the engine is designed to prevent overboost, but owing to the many components and the response time involved, another related condition known as **overshoot** must be considered. Overshoot is a condition of the automatic controls not having the ability to respond quickly enough to check the inertia of the turbocharger speed increase with rapid engine throttle advance.

Overshoot differs from overboost in that the high manifold pressures last for only a few seconds. This condition can usually be overcome by smooth throttle operation. With unstabilized temperatures and during transient conditions, a certain amount of overshoot must be expected. But overshoot can be held to an acceptable level by avoiding a too violent opening of the throttles and by changing the oil and filter at regular intervals specified by the manufacturer. It may also be necessary to disassemble and clean certain assemblies in accordance with manufacturers' instructions.

● INSPECTION, MAINTENANCE, AND REPAIR OF SUPERCHARGERS AND TURBOCHARGERS

The same principles of inspection, maintenance, and repair that apply to other sections of the powerplant system apply to superchargers and turbochargers. Visual inspection of all visible parts should be accomplished daily to observe oil leaks, exhaust leaks, cracks in the metal of "hot sections," loose or insecure units, and other unacceptable conditions. Note that exhaust ducts, waste gates, nozzle boxes, and turbines are subjected to extremely high temperatures; thus, cracks develop because of the continued expansion and contraction of the metal as temperature changes occur.

The manufacturer's manual will specify the most important inspections to be accomplished and the service time established for periodic inspections. An inspection of a complete system should include the following, in addition to any other inspections specifically required by the company's operation manual or the manufacturer's maintenance manual:

1. Mounting of all units.
2. Oil leaks or dripping from any unit.
3. Security of oil lines.
4. Security and condition of electric wiring.
5. Cracks in ducting and other metal parts, including the turbine and housing.
6. Warping of metal ducts.
7. Operation of the complete system to determine performance, to discover undesirable sounds, and to note evidence of vibration. Unusual sounds and appreciable vibration require removal and replacement of the turbocharger to correct the faulty condition.

Improper lubrication or the use of an incorrect lubricant can cause serious malfunctions and the failure of units. Because of the high temperatures to which a turbine wheel is exposed, the turbine shaft is also subject to high temperatures. This can cause "coking" of the lubricant, with a subsequent buildup of carbon (coke) at turbine shaft seals and bearings. An appreciable amount of coking can cause failure of both turbine shaft seals and bearings. Leaking shaft seals permit hot exhaust gases to reach the shaft bearings, where additional coking is likely to occur. Coking of the bearings is likely to limit the rpm that the turbine and compressor assembly can attain. In this case, the turbocharger will require removal and replacement. Because of the problems caused by coking in the turbine area, it is most essential that the proper type of lubricant be employed. The service manual for the aircraft will provide this information.

All overhauling and testing of superchargers and tur-

bochargers should be accomplished at authorized service stations. This is particularly important because of the need to balance the turbine and compressor assembly accurately. These units rotate at speeds of up to 70,000 rpm; hence, a slight unbalance can cause severe vibration and ultimate disintegration and failure. Another important reason that an authorized station should perform the overhaul of turbosuperchargers and turbochargers is that such a station is equipped with all necessary special tools and keeps up to date on any bulletins or special information needed for the proper repair of all units.

● THE TURBOCOMPOUND ENGINE

Introduction

The application of gas turbines to aircraft powerplants has advanced through various stages and now come to the point that the gas turbine is the prime source of power for large aircraft and is being rapidly adapted to lighter aircraft.

One of the first ways of utilizing the gas turbine to increase engine power is found in the turbosupercharger. As previously discussed, the turbosupercharger employs an exhaust-driven turbine to drive a compressor. The compressor increases the intake air pressure of a conventional reciprocating engine and enables the engine to burn a greater mass of fuel and air than would be otherwise possible. In the turbosupercharger the turbine is used *indirectly* to increase engine power.

A *direct* method for increasing power by means of the gas turbine is employed in the **turbocompound** engine. This application of the gas turbine was found to be quite effective in increasing engine power; however, the need for such a device has been obviated by the advent of the gas-turbine engine.

Since aircraft with turbocompound engines are still in operation in various parts of the world, it is believed useful to provide a brief description of such a system. Two of the large aircraft employing turbocompound engines are the Douglas DC-7 and the Lockheed Super Constellation

Description

The basic model Wright turbocompound engine, manufactured by the Wright Aeronautical Division of the Curtiss-Wright Corporation, is a conventional aircooled reciprocating engine having 18 cylinders in a two-row configuration. It is equipped with a two-speed supercharger, direct fuel injection, a low-tension ignition system with automatic spark advance, and a low-flow torquemeter. Fig. 7-35 is an illustration of the engine, showing many of the external details, including the blow-down turbines.

Blow-down turbines, also called **power-recovery turbines** (PRT), are the *velocity* type—which means that the velocity of the exhaust gases rather than the pressure of the gases is used to produce the power. With this type of design, the turbines create a minimum of back pressure on the exhaust system. Each of the three PRTs is powered by the exhaust gases from six of the cylinders. The turbines are cooled by ram air ducted

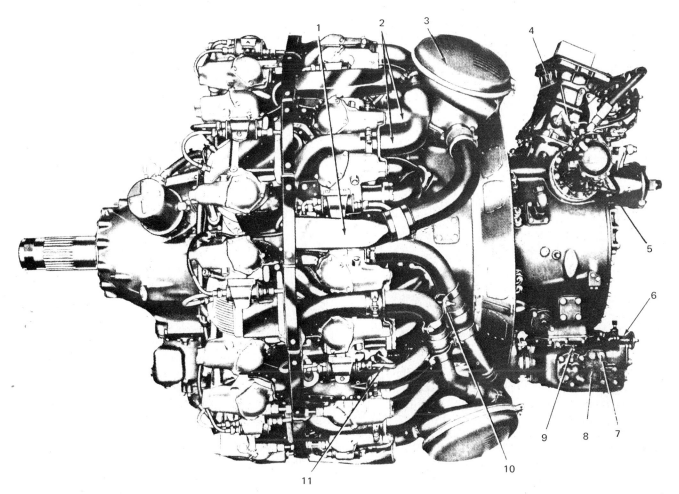

1. Power recovery turbine cooling air scoop
2. Exhaust pipes to turbine
3. Power recovery turbine and cover
4. Master control for direct fuel injection
5. Fuel injection pump
6. Supercharger clutch control valve
7. Oil temperature gage connection
8. Pre-oiling connection
9. Pre-oiling vent connection
10. Exhaust pipe supporting clamp
11. Ignition lead

FIG. 7-35 Wright turbocompound engine.

from the intercylinder baffles. The turbine to crankshaft ratio is 9.7:1, which does not always exist during operation because of the fluid clutch between the turbine drive and the crankshaft gear.

Figure 7-36 shows the method by which the PRTs are coupled to the crankshaft of the engine. The turbine wheel is connected by a hollow shaft and a coupling to a bevel drive gear. The vibration of the shaft is absorbed by a vibration damper consisting of spring-loaded plates and disks. The bevel drive gear engages another bevel gear which is connected by a shaft to the fluid coupling impeller. The rear half of the fluid coupling is connected by a splined shaft to a pinion which meshes with the crankshaft drive gear, thus delivering the turbine power to the crankshaft.

During operation of the turbocompound engine, the PRTs can rotate at a speed of more than 28,000 rpm. It can be seen, then, that great centrifugal stresses are imposed upon the turbines and turbine buckets. For this reason, regular inspections must be accomplished on the PRTs to detect cracks, burn damage, or other

conditions which could lead to disintegration. Operation of the engines must be accomplished with care and in accordance with the manufacturer's instructions to prevent the rapid heating and cooling which leads to warpage and cracking of the components.

Performance

The Wright model TC19DA turbocompound engine is normally rated at 3250 hp [2423.53 kW]; however, it may operate at 3700 hp [2759.09 kW] for takeoff. This compares with other R-3350 engines—without blow-down turbines—which deliver 2700 hp [2013.39 kW] at 2900 rpm. The recommended cruising performance of the engine is 1800 hp [1342.8 kW] at 2200 rpm. At this speed, the specific fuel consumption is 0.37 lb/hp/h [.23 kg/kW/h]. Since the weight of the engine is 3240 lb [1470 kg], the specific weight is 0.92 lb/hp [0.56 kg/kW].

It will be noted from these figures that the performance of the turbocompound engine is substantially better than that of the conventional reciprocating engines.

165

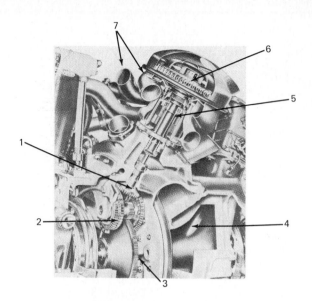

1. Bevel drive gear
2. Fluid coupling
3. Crankshaft gear
4. Diffuser section
5. Vibration damper
6. Turbine wheel
7. Exhaust pipes to turbine

FIG. 7-36 Power-recovery turbine and coupling units.

● RECIPROCATING ENGINE EXHAUST SYSTEMS

One of the most critical systems employed in the engine's operation is the **exhaust system,** which removes the products of combustion from the engine safely and effectively. Since exhaust gases are both toxic and very hot, considerable care must be exercised in the design, construction, and maintenance of the exhaust system.

The maintenance and inspections of the engine exhaust system must be accomplished on a regular basis and in accordance with the manufacturer's instructions. Poor maintenance and lack of inspections can lead to a nacelle fire, toxic gases entering the cockpit and cabin, damage to parts and structure in the nacelle, and poor engine performance.

Development of Exhaust Systems

Exhaust systems for the early aircraft engines were very simple. The engine exhaust was expelled from the exhaust port of each cylinder separately through short steel stacks attached to the exhaust ports. These systems were noisy and often permitted the exhaust gases to flow into the open cockpits of the aircraft. Pilots flying at night were often able to troubleshoot their engines by observing the color of the exhaust flame. A short, light-blue flame usually indicated that the mixture was correct and that the engine was operating satisfactorily. If the flame was shorter than normal for the engine, a lean mixture was indicated. When the flame was white or reddish in color, the mixture was excessively rich. If only one cylinder produced a white or red flame, valve trouble or worn piston rings were likely causes.

The next step in the development of exhaust systems was the installation of **exhaust manifolds** for in-line and opposed engines and collector rings for radial engines.

Through these devices, the exhaust gases were directed outward and down, thus reducing the likelihood of gases entering the cockpit or cabin. The use of manifolds and exhaust pipes led to the design of muffs and other heat exchanging equipment whereby a portion of the heat of the exhaust could be collected and employed for cabin heating, carburetor anti-icing, and windshield defrosting.

Modern aircraft are equipped with exhaust manifolds, heat exchangers, and mufflers. In addition, some exhaust systems include turbochargers, augmenters, and other devices. Inconel or other heat- and corrosion-resistant alloys are used in the manufacture of most exhaust systems.

Exhaust Systems for Opposed Engines

While many different types of exhaust systems have been designed and built for opposed aircraft engines, all include the essential features required for an effective exhaust system. Figure 7-37 shows a relatively simple exhaust system for a four-cylinder opposed aircraft engine. The exhaust manifold consists of risers leading from the muffler to the exhaust ports. The risers are attached by means of flanges, studs, and heat-resistant brass nuts. A copper-asbestos gasket is fitted between the flange and the cylinder to seal the installation and prevent the escape of exhaust gases. The tubing used in the construction of the muffler, shroud, and risers is corrosion-resistant steel.

The exhaust system for a four-cylinder opposed engine installed in an experimental helicopter is shown in Fig. 7-38. This view clearly shows the clamp-tube expansion joints which are essential to permit the tubing to expand and contract without generating cracks. The risers are attached to the Y-type manifold which passes through the muffler (heat exchanger).

The arrangement of the exhaust manifold system for a six-cylinder opposed engine with a turbocharger is shown in Fig. 7-39. The corrosion-resistant steel risers are attached to the exhaust ports on the lower sides of the cylinders by means of studs and heat-resistant nuts. The risers are curved to direct the exhaust flow toward the rear in the manifold. The opposite side of the engine has a similar arrangement and the exhaust flow from

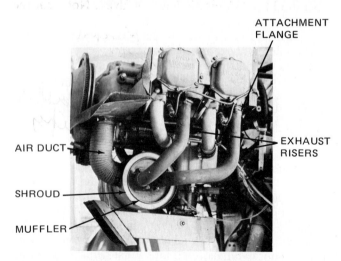

FIG. 7-37 Exhaust system for a four-cylinder opposed engine.

FIG. 7-38 Exhaust system on a four-cylinder helicopter engine.

both sides of the engine passes through exhaust pipes to the waste gate at the rear of the engine. When the engine is operated with turbocharging, the waste gate directs a portion of all of the exhaust through the turbocharger turbine. Expansion joints are provided in the system to allow for uneven expansion and contraction due to changes in temperature.

An exhaust system for a light airplane, including the cabin heating system, is shown in Fig. 7-40. In this system, the exhaust is directed through the mufflers and out below the engine through a common tailpipe. A crossover pipe carries the exhaust from the left-hand cylinders to the tailpipe of the right-hand muffler. The exhaust risers are attached to the mufflers by means of clamps which allow for expansion and contraction. Stainless-steel shells, or **shrouds,** are placed around the mufflers to capture the heat from the mufflers and direct it to the heater hoses. Shrouds of both mufflers are connected to a flexible duct system which routes outside air into the space between the mufflers and their shrouds. Heated air from the shrouds is carried through flexible ducts to plenum chambers in which the heat-control valves are located. Figure 7-41 shows the mufflers with the shrouds removed. Note that the

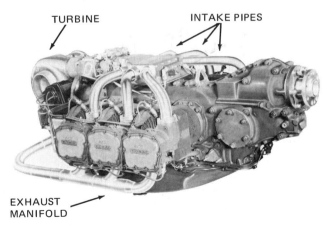

TURBINE   INTAKE PIPES

EXHAUST MANIFOLD

FIG. 7-39 Exhaust manifold system for a six-cylinder opposed engine with a turbocharger. *(Lycoming Div., Avco Corp.)*

shrouds are clamped around the mufflers by means of flanges joined by screws. This construction permits regular removal of the shrouds to inspect the mufflers for cracks or other signs of deterioration.

The exhaust system employed on a light twin-engine airplane is shown in Fig. 7-42. A combustion-type heating system is installed in this aircraft, so there is no need for a heat-exchanging system associated with the exhaust. The system on each side of the engine consists of three risers, a flexible joint between the aft riser and the muffler, and the tailpipe attached to the structure by means of a hanger with a spring-type **isolator** to allow for expansion and flexibility. Figure 7-43 illustrates the construction of the flexible joint between the aft riser and the muffler. Between the risers are slip joints which aid in alignment and allow for expansion.

Exhaust Systems for Radial Engines

The original radial engines usually disposed of gases through short stacks. The stacks served to prevent extreme, rapid temperature changes for the exhaust valves and the exhaust area of the cylinder head. They also carried the hot exhaust gases away from the immediate area of the cylinder head. Experience revealed that for most installations it was desirable to collect the exhaust in a single manifold and discharge it at a point where the heat could not affect the aircraft structure and the gases would not be ingested into the engine intake or drawn into the cockpit or cabin.

The exhaust collector ring for a nine-cylinder R-985 engine, shown in Fig. 7-44, is fabricated in sections with short stacks leading to the exhaust port of each cylinder. The construction of the collector ring is such that individual sections can be removed for maintenance and repair. The separate sections also permit expansion and contraction without causing dangerous stresses and warpage.

The collector ring for a 14-cylinder twin-row radial engine, shown in Fig. 7-45, is made up of seven sections, each with two exhaust inlets. The section of the ring opposite the exhaust outlet is the smallest because it carries the exhaust from only two cylinders. From that point to the outlet side, the sections are increased in diameter to provide for the additional gases they must carry. The longer exhaust stacks connected to the ports of the collector ring reach forward to connect with the front cylinders. The exhaust stacks from the cylinders are joined to the collector-ring ports by means of sleeve connections to allow for expansion and flexibility. Each section of the collector ring is attached to the blower section of the engine by means of a bolted bracket.

Exhaust Augmentors

Some aircraft exhaust systems include **exhaust augmentors,** also referred to as **ejectors.** For installation of augmentors with an 18-cylinder engine such as the R-2800, the exhaust is collected from the right side of the engine and discharged into the bellmouth of the right-hand augmentor and from the left side of the engine to be discharged into the left augmentor. Four of the exhaust pipes on each side of the engine handle the exhaust from two cylinders each. The firing of the two cylinders feeding into each of the exhaust stacks

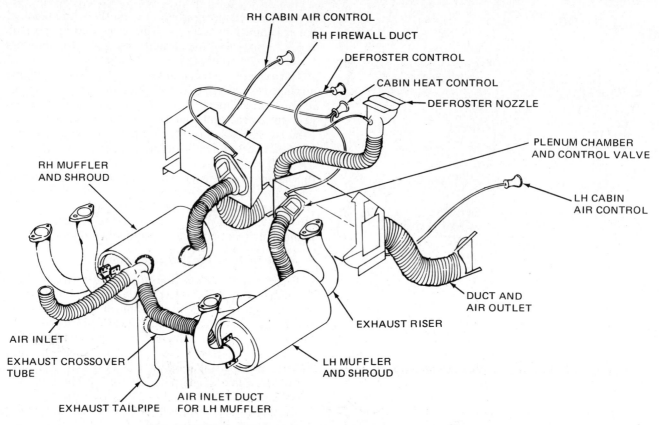

FIG. 7-40 Exhaust system for a light airplane. *(Cessna Aircraft Co.)*

Labels in figure:
RH CABIN AIR CONTROL
RH FIREWALL DUCT
DEFROSTER CONTROL
CABIN HEAT CONTROL
DEFROSTER NOZZLE
PLENUM CHAMBER AND CONTROL VALVE
LH CABIN AIR CONTROL
RH MUFFLER AND SHROUD
DUCT AND AIR OUTLET
EXHAUST RISER
LH MUFFLER AND SHROUD
AIR INLET
EXHAUST CROSSOVER TUBE
EXHAUST TAILPIPE
AIR INLET DUCT FOR LH MUFFLER

is separated as much as possible to provide for maximum exhaust flow without excessive back pressure.

A simplified drawing of an augmentor is shown in Fig. 7-46. It should be noted that the augmentor produces a venturi effect which increases the flow of air from the engine nacelle. This increased flow through the nacelle aids in cooling the engine and provides a small amount of "jet" thrust. Augmentor tubes must be in perfect alignment with the exhaust flow to produce maximum effect.

The augmentor tubes are constructed of corrosion-resistant steel and sometimes contain an adjustable vane which can be controlled from the cockpit. In case the engine is operating at a temperature below that desired, the pilot can close the vane to reduce the cross-sectional area of the augmentor as much as 45 percent to increase the operating temperature of the engine.

Inspection and Maintenance of Exhaust Systems

The importance of proper inspection and maintenance of exhaust systems cannot be overemphasized. Defective systems can lead to (1) engine fire, (2) engine failure, (3) structural failure, or (4) carbon monoxide poisoning of the passengers and flight crew.

The exhaust system components are subjected to extreme temperatures, and the resulting expansion and contraction produce stresses which often lead to cracks and distortion due to warpage. It is therefore essential that the exhaust system of every engine be thoroughly inspected at regular intervals.

Precautions: In the inspection and service of exhaust systems, certain precautions must be observed. Failure to employ adequate care in working with exhaust systems can result in their damage and deterioration.

Corrosion-resistant exhaust system parts must be protected against contact with zinc-coated (galvanized) tools or any zinc-coated metal parts. Furthermore, lead pencils must not be used to mark exhaust-system parts. At high temperatures, the metal of the exhaust system will absorb the zinc or lead of the lead pencil, and this will materially affect the molecular structure of the metal. Because of this, the softened metal will likely be subject to the development of cracks in the marked areas.

The cleaning of exhaust system parts must be done with care. This is particularly true of those parts which are ceramic-coated. Such parts must not be cleaned with harsh alkaline cleaners or by sandblasting. Degreasing with a suitable solvent will usually suffice. For a particular make and model of aircraft, the instructions of the manufacturer should be followed.

The assembly of an exhaust system after inspection and repair is most critical. After the exhaust stacks or risers are secured to the cylinders, all other parts should be installed so that joints and other connections are in proper alignment to prevent exhaust leakage. Nuts, bolts, and clamp screws must be tightened to the correct torque. Overtorquing will probably result in failure.

Procedures: The procedures for performing exhaust system inspections are given in the manufacturer's

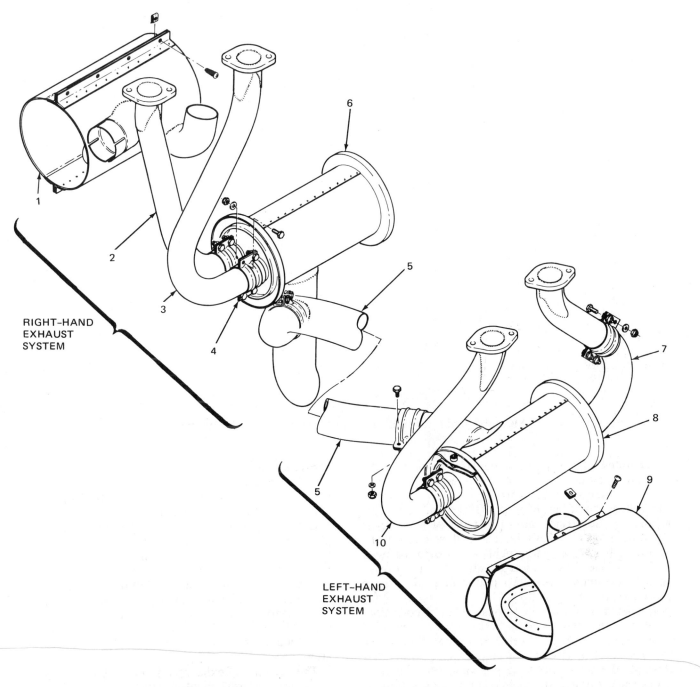

RIGHT-HAND EXHAUST SYSTEM

LEFT-HAND EXHAUST SYSTEM

1. Right-hand shroud-half assembly
2. Right-hand outboard-riser assembly
3. Right-hand inboard-riser assembly
4. Clamp half
5. Crossover-tube assembly
6. Right-hand muffler-weld assembly
7. Left-hand aft-riser assembly
8. Left-hand muffler-weld assembly
9. Left-hand muffler-shroud assembly
10. Left-hand fwd-riser assembly

FIG. 7-41 Exhaust system with muffler shrouds removed. *(Cessna Aircraft Co.)*

maintenance manual for the aircraft. In general, the required inspections are similar in type and scope for most aircraft. The following steps are typical:

1. Remove the engine cowling sufficiently to see all parts of the exhaust system.

2. Examine all parts for cracks, wear, looseness, dents, corrosion, and any other apparent deterioration. Pay particular attention to attaching flanges, welded

joints, slip joints, muffler shrouds, clamps, and attachment devices.

3. Check all joints for signs of exhaust leakage. Leakage can cause hot spots, in addition to being hazardous to passengers and crew. With supercharged engines operating at high altitudes, exhaust leaks assume the nature of blow torches because of the sea-level pressures maintained in the system. These leaks are a fire hazard as well as being damaging to parts and

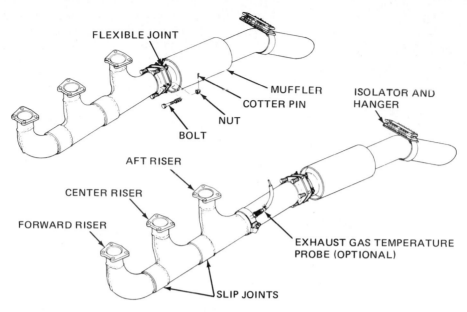

FIG. 7-42 **Exhaust system components for a six-cylinder opposed engine.** *(Cessna Aircraft Co.)*

structure. Evidence of leakage is a light-gray or sooty spot at any slip joint or at any other point where pipes are joined. Leakage spots will also reveal cracks in the system.

4. After a thorough visual inspection, the exhaust system should be pressure-checked. Attach the pressure side of an industrial vacuum cleaner to the tailpipe opening, using a suitable rubber plug to provide a seal. With the vacuum cleaner operating, check the entire system by feel or with a soap solution to reveal leaks. After the pressure test, remove the soap suds with water and dry the system components with compressed air.

5. For a complete inspection of the exhaust system it is often necessary to disassemble the system and check individual components. Disassemble the system according to the manufacturer's instructions, being careful to examine all attaching parts, such as clamps, brackets, bolts, nuts, and washers.

6. Remove the shrouds from the mufflers. Use rubber plugs to seal the openings and apply $2\frac{1}{2}$ psi [13.79 kPa] air pressure while the muffler is submerged in water. Seal the exhaust stacks and pipes and test in the same manner, using $5\frac{1}{2}$ psi [34.48 kPa]. Pressures

used for testing may vary, but whatever the manufacturer recommends should be used.

7. After all components of the exhaust system are examined and found satisfactory, reassemble the system on the engine loosely to allow for adjustment and proper alignment. Tighten the stack attachments to the cylinder exhaust ports first, using a torque wrench. Be sure that the proper type of heat-resistant nut is used and that new gaskets have been installed. Next, tighten all other joints and attachments, making sure that all parts are in correct alignment.

8. After the exhaust system has been installed, run the engine long enough to bring it up to normal operating temperature. Shut the engine down and remove

0.61 $^{+0.00}_{-0.02}$
TYPICAL

FIG. 7-43 **Flexible joint construction.** *(Cessna Aircraft Co.)*

FIG. 7-44 **Exhaust collector ring for a nine-cylinder radial engine.**

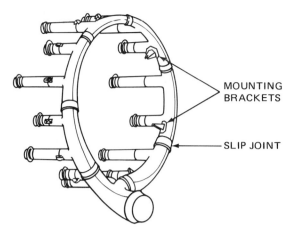

FIG. 7-45 Exhaust collector ring and stacks for a 14-cylinder twin-row radial engine.

MOUNTING BRACKETS

SLIP JOINT

the cowling. Inspect each exhaust port and all joints where components are attached to one another. Look for signs of exhaust leaks, such as a light-gray or sooty deposit. If a leak is found, loosen the connection and realign it.

9. If an exhaust system includes augmentors, inspect the augmentors in the same manner as for the other components. Leaks in the augmentors can cause fires and the escape of gases into the cockpit or cabin. The alignment of the augmentors is particularly critical. The manufacturer's specifications must be followed precisely.

10. On a system which includes a turbocharger, special inspections of the turbine and compressor assemblies must be made. Inspect the interior of all units for the buildup of coke deposits. These deposits can cause the waste gate to stick, causing excessive boost. In the turbine, carbon deposits will cause a gradual lessening of turbine efficiency, with a resulting decrease in engine power. Wherever coke buildup is found in any unit, remove it in accordance with the manufacturer's instruction.

Repairs: Exhaust system components which have become burned, cracked, warped, or worn to the extent that leakage occurs should usually be replaced with new parts. In certain instances, cracks can be repaired by heliarc (inert gas) welding with the proper

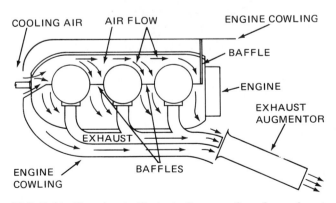

FIG. 7-46 **Drawing to illustrate the operation of an exhaust augmentor.**

COOLING AIR   AIR FLOW   ENGINE COWLING

BAFFLE

ENGINE

EXHAUST AUGMENTOR

EXHAUST

BAFFLES

ENGINE COWLING

type of welding rod. Care must be taken to avoid any repair which will cause a rough spot or protrusion inside an exhaust pipe or muffler. Any such area will create a hot spot and eventual burn-through.

Dents can sometimes be removed, provided that the dent has not caused a thin spot due to internal erosion and burning. Dents are removed by placing the exhaust pipe over a suitable mandrel and working out the dent with a soft hammer.

After any repair to a component of the exhaust system, a pressure test should be made as previously described.

● REVIEW QUESTIONS

1. Name the principal sections of an induction system.
2. What is the function of an alternate air valve?
3. How would you service the different types of air filters?
4. Describe the carburetor heat system.
5. Describe the function and construction of the heater muff.
6. What means can be used to prevent the entrance of rain into the induction system?
7. How does an air filter affect engine power?
8. What is meant by a *hot spot?*
9. What would be the effect if carburetor heat were turned on during full-power operation?
10. What is the effect of an air leak in the manifold between the carburetor and the cylinder for an unsupercharged (naturally aspirated) engine?
11. What is the function of the balance pipe connected between the forward ends of the intake manifolds on some engines?
12. Where is impact ice likely to form in the induction system of an airplane?
13. What is the first indication of carburetor ice in an engine equipped with a fixed-pitch propeller?
14. What instrument indication will be observed when carburetor ice forms in an engine equipped with a constant-speed propeller?
15. What instrument is useful in determining whether icing may occur in the carburetor?
16. What is the temperature rise required for a carburetor air-heating system?
17. Discuss the need for springback in the adjustment of the carburetor heat control.
18. Why is it necessary to remove the heater muff when inspecting a carburetor air-heating system?
19. What are the advantages of supercharging?
20. What is the effect of supercharging on the density of air? How does it affect the air temperature?
21. Compare the density of the atmosphere at sea level with the density at 20,000 ft altitude.
22. Explain the significance of manifold pressure in the operation of an aircraft engine.
23. What is the effect of excessive manifold pressure?
24. How does a supercharger affect the mean effective pressure in the cylinders of an engine?
25. What is an internal supercharger?
26. What engine instruments are particularly important in the operation of an aircraft engine?

27. What is meant by the term *sea-level supercharger?*
28. When an engine is equipped with a direct-drive sea-level supercharger, what effect does altitude have on the power output?
29. What means is used to prevent excessive intake air temperature when a supercharger is in operation?
30. Describe an internal supercharger for a six-cylinder opposed engine.
31. Explain the operation of a two-speed internal supercharger.
32. How is engine power delivered to a mechanically driven supercharger?
33. What device is used to change the speed of a two-speed supercharger?
34. What is the energy source for a turbocharger?
35. Describe the function of the waste gate in a turbocharger system.
36. Describe the operation of a manually controlled turbocharger.
37. What engine pressure is sensed in an automatic turbocharger-control system to prevent overboost?
38. What is meant by *critical altitude?*
39. Describe the operation of a density controller.
40. How does the differential pressure controller maintain the correct boost at part-throttle settings?
41. Explain the operation of the exhaust bypass-valve assembly.
42. Describe the turbosupercharger system for a large engine.
43. Give a brief description of the electronic turbo-supercharger-control system.
44. What is the cause of cracks in turbines, nozzles, and diffusers in a turbosupercharger?
45. Explain the importance of proper lubrication for a turbosupercharger.
46. What is coking, and what problems may it cause in a turbosupercharger?
47. What is the function of the exhaust shroud used in connection with the turbosupercharger?
48. Explain the function of an intercooler.
49. How is overspeed prevented in the electronic turbosupercharger-control system?
50. What inspections should be made regularly for a turbosupercharger installation?
51. What is the purpose of the turbines in a turbocompound engine?
52. Explain the operating principles of the power-recovery turbines.
53. Describe the coupling of the power-recovery turbines to the crankshaft of the engine.
54. What are the principal hazards associated with a defective exhaust system for reciprocating engines?
55. Describe the exhaust system for an opposed-type engine.
56. How are risers or exhaust stacks attached to the exhaust ports of cylinders?
57. What provisions are made to allow for the expansion and contraction of exhaust system components?
58. Of what material are exhaust system components usually made?
59. Explain the operation of the device which extracts heat from the exhaust system for use in cabin heating and for carburetor air heat.
60. Describe the exhaust system for a radial engine.
61. Why is the collector ring constructed in sections?
62. Describe the operation and purpose of augmentors.
63. Why is it important that the augmentor tubes be in perfect alignment?
64. Of what material are augmentor tubes constructed?
65. Describe the arrangement of the exhaust pipes for an 18-cylinder engine where the exhaust system includes augmentor tubes.
66. What are the principal causes of deterioration in an exhaust system?
67. What precautions must be observed with respect to other metals or marking materials contacting the material of the exhaust system?
68. Discuss the cleaning of exhaust system parts. What precaution must be observed with respect to ceramic-coated exhaust manifolds?
69. List the procedures for a typical exhaust system inspection.
70. What is an accepted procedure for testing mufflers and exhaust pipes?
71. If an exhaust system includes a turbocharger, what inspections should be made with respect to the turbine?
72. What precautions must be observed when it is decided that exhaust system components can be repaired?
73. Describe the correct procedure for assembling an exhaust system after inspection and repairs.
74. What tests should be made after an exhaust system has been assembled?

8 LUBRICANTS AND LUBRICATING SYSTEMS

● CLASSIFICATION OF LUBRICANTS

A **lubricant** is any natural or artificial substance having greasy or oily properties which can be used to reduce friction between moving parts or to prevent rust and corrosion on metallic surfaces. Lubricants may be classified according to their origins as animal, vegetable, mineral, or synthetic.

Animal Lubricants

Examples of lubricants having an animal origin are tallow, tallow oil, lard oil, neat's-foot oil, sperm oil, and porpoise-jaw oil. These are highly stable at normal temperatures, so they can be used to lubricate firearms, sewing machines, clocks, and other light machinery and devices. Porpoise-jaw oil, for example, is used to lubricate expensive watches and very delicate instruments. However, animal lubricants cannot be used for internal-combustion engines because they produce fatty acids at high temperatures.

Vegetable Lubricants

Examples of vegetable lubricants are castor oil, olive oil, rape oil, and cottonseed oil. These oils tend to oxidize when exposed to air. Vegetable and animal oils have a lower coefficient of friction than do most mineral oils, but they wear away steel rapidly because of their ability to loosen the bonds of iron on the surface.

Castor oil, like other vegetable oils, will not dissolve in gasoline. For this reason it was used in rotary engines where the crankcase was used as a part of the induction system. It oxidizes easily and causes gummy conditions in an engine.

Mineral Lubricants

Mineral lubricants are used to a large extent in the lubrication of aircraft internal-combustion engines. They may be classified as solids, semisolids, and fluids.

Solid Lubricants. Solid lubricants, such as mica, soapstone, and graphite, do not dissipate heat rapidly enough for high-speed machines, but they are fairly satisfactory in a finely powdered form on slow-speed machines. They fill the low spots in the metal on a typical bearing surface to form a perfectly smooth surface, and at the same time they provide a slippery film that reduces friction. When a solid lubricant is finely powdered and is not too hard, it may be used as a mild abrasive to smooth the surface previously roughened by excessive wear or by machine operations in a factory. Some solid lubricants can carry heavy loads, and hence they are mixed with certain fluid lubricants to reduce the wear between adjacent surfaces subjected to high unit pressures. Powdered graphite is used instead of oils and greases to lubricate firearms in extremely cold weather, because oils and greases become thick and gummy, rendering the firearms inoperative.

Semisolid Lubricants. Extremely heavy oils and greases are examples of semisolid lubricants. Grease is a mixture of oil and soap. It gives good service when applied periodically to certain units, but its consistency is such that it is not suitable for circulating or continuous-operating lubrication systems. In general, sodium soap is mixed with oil to make grease for gears and hot-running equipment, calcium soap is mixed with oil to make cup grease, and aluminum soap is mixed with oil to make grease for ball-bearing and high-pressure applications.

Fluid Lubricants (Oils). Fluid lubricants (oils) are used as the principal lubricant in all types of internal-combustion engines because they can be pumped easily and sprayed readily and because they absorb and dissipate heat quickly and provide a good cushioning effect.

Summary of Advantages of Mineral-Base Lubricants. In general, lubricants of animal and vegetable origin are chemically unstable at high temperatures, often perform poorly at low temperatures, and are unsuited for aircraft engine lubrication. On the other hand, lubricants having a mineral base are chemically stable at moderately high temperatures, perform well at low temperatures, and are widely used for aircraft engine lubrication.

Synthetic Lubricants

Because of the high temperature required in the operation of gas-turbine engines, it became necessary for the industry to develop lubricants which would retain their characteristics at temperatures that cause petroleum lubricants to evaporate and break down into heavy hydrocarbons. Synthetic lubricants do not evaporate or break down easily and do not produce coke or other deposits. These new lubricants are called synthetics because they are not made from natural crude oils. Typical synthetic lubricants are the Type I, alkyl diester oils (MIL-L-L7808); and the Type II, polyol ester oils (MIL-L-23699).

The Source of Engine Lubricating Oils

Petroleum, which is the source of volatile fuel gasoline, is also the source of engine lubricating oil. Crude petroleum is refined by the processes of distillation, dewaxing, chemical refining, and filtration.

173

In the process of distillation, crude petroleum is separated into a series of products varying from gasoline to the heaviest lubricating oils according to the boiling point of each. The dewaxing process essentially consists in chilling the waxy oil to low temperatures and allowing the waxy constituents to crystallize, after which the solid wax can be separated from the oil by filtration. After the removal of the wax, resinous and asphaltic materials are removed from the lubricating oil by chemical refining. The oil is then treated with an absorbent which removes the last traces of the chemical refining agents previously used, improves the color, and generally prepares it for shipment and use.

Base Crudes and Lubricating-Oil Production

There are two **base crudes: naphthenic** and **paraffinic**. In the United States, the true **paraffinic-base crudes** are found in the Pennsylvania or Allegheny region. These are regarded as the oldest crudes ever discovered and are believed to have originated about 100 million years ago. The **naphthenic-base crudes** are found principally in southern Texas and California. These crudes are probably about 60 million years old. Lubricants made from naphthenic-base crudes are sometimes called **coastal** or **asphaltic-base** oils.

The midcontinent field is another source of supply for the manufacture of lubricating oils. The base crude found in the midcontinent area is equivalent to a natural blend of paraffinic- and naphthenic-base crudes and is obtained principally in east Texas, Oklahoma, Kansas, and Illinois. This crude is believed to be about 80 million years old.

Figure 8-1 is a map prepared by the Sinclair Refining Company showing the principal crude areas in the United States. Notice that the paraffinic-base crudes are found not only in western Pennsylvania but also in several other states which can be regarded as located in the Appalachian area. The naphthenic-base crudes are shown as found along the Gulf Coast and in California. In addition to the midcontinent region already mentioned, the maps show that mixed-base crudes are found in the Rocky Mountain area and also in several states to the west of the Appalachian area, such as Michigan, Indiana, and Illinois.

Table 8-1 shows the natural characteristics of oils

FIG. 8-1 Map showing the principal crude-oil areas in the United States.

made from the three crudes without special refining or treating. The terms such as API gravity, flash point, etc., given in the vertical column at the left of the table are all explained in this chapter. In studying this table, it should be understood that special refining processes make it possible for the refiner to alter the natural characteristics. For example, it is possible to make oils having the same specifications as Pennsylvania oils from other crudes, such as those found in the midcontinent areas.

● LUBRICATING-OIL PROPERTIES

The most important properties of an aircraft engine oil are its flash point, viscosity, pour point, and chemical stability. There are various tests for these properties which can be made at the refinery and in the field. In addition, there are tests which are of interest principally to the petroleum engineers at the refinery, although all personnel interested in aircraft engine lubrication should have some familiarity with such tests so that they can intelligently read reports and specifications pertaining to petroleum products.

Some of the properties tested at the refinery are the gravity, color, cloud point, carbon residue, ash residue, oxidation, precipitation, corrosion, neutralization, and oiliness.

Gravity

The **gravity** of a petroleum oil is a numerical value which serves as an index of the weight of a measured volume of the product. There are two scales generally used by petroleum engineers. One is the specific-gravity scale, and the other is the American Petroleum Institute (API) gravity scale. *Gravity is not an index to quality.* It is a *property*, of importance only to those operating the refinery—but it is a convenient *term* for use in figuring the weights and measures and also in the distribution of lubricants.

Specific gravity is the weight of any substance compared with the weight of an equal volume of a standard substance. When water is used as a standard, the specific gravity is the weight of a substance compared with the weight of an equal volume of distilled water.

A hydrometer is used to measure specific gravity, and it is also used in conducting the API gravity test.

Formerly, the petroleum industry used the Baumé scale, but it has been superseded by the API gravity scale which magnifies that portion of the specific-gravity scale which is of greatest interest for testing petroleum products. The test is usually performed with a hydrometer, using a thermometer and a conversion scale for temperature correction to the standard temperature of 60°F [15.56°C], as shown in Fig. 8-2.

Water has a specific gravity of 1.000, weighs 8.328 lb/gal [3.78 kg/gal], and has an API gravity reading of 10 under standard conditions for the test. An aircraft lubricating oil, which has a specific gravity of 0.9340 and weighs 7.778 lb/gal [3.53 kg/gal], has an API gravity reading of 20 under standard conditions. If an aircraft lubricating oil has a specific gravity of 0.9042 and weighs 7.529 lb/gal [3.42 kg/gal] its API gravity reading is 24. These are merely examples of the relation between specific-gravity figures and API gravity readings. In

	Pennsylvania Crude	Midcontinent Crude	Coastal Crude
API gravity	High: Light oils 27–32° Heavy oils 25–27°	Intermediate: Light oils 24–27° Heavy oils 22–24°	Low: Light oils 19–24° Heavy oils 17–22°
Flash point	High: Light oils 390–450°F [199–232°C] Heavy oils 450–550°F [232–288°C]	Intermediate: Light oils 330–420°F [166–216°C] Heavy oils 410–525°F [210–274°C]	Low: Light oils 300–385°F [149–196°C] Heavy oils 370–480°F [188–249°C]
Fire point	High: Light oils 450–510°F [232–266°C] Heavy oils 510–620°F [266–327°C]	Intermediate: Light oils 375–480°F [191–249°C] Heavy oils 450–600°F [232–316°C]	Low: Light oils 350–425°F [177–218°C] Heavy oils 425–535°F [218–279°C]
Viscosity	For any common viscosity at a specific high temperature, oils in this group will be lighter than midcontinent or coastal at a corresponding low temperature	For any common viscosity at a specific high temperature, oils in this group will be lighter than coastal and heavier than Pennsylvania at a corresponding low temperature	For any common viscosity at a specific high temperature, oils in this group will be heavier than Pennsylvania or midcontinent at a corresponding low temperature
SAE 50 or aviation grade 100	Example: Viscosity: At 210°F [99°C], 100 s At 100°F [38°C], 1,200 s At 0°F [18°C], 110,000 s Viscosity index 99.3	Example: Viscosity: At 210°F, 100 s At 100°F, 1,600 s At 0°F, 600,000 s Viscosity index 71.8	Example: Viscosity: At 210°F, 100 s At 100°F, 2200 s At 0°F, 2,000,000 + s Viscosity index 30.6
Pour point	High due to paraffinicity	Intermediate	Low but extremely viscous
Carbon (amount)	High but stable	Medium	Low
Carbon (character)	Hard	Medium	Soft

actual practice, when the specific gravity and the weight in pounds per gallon at standard temperature are known, the API reading can be obtained from a table prepared for this purpose.

Flash Point

The **flash point** of an oil is the temperature to which the oil must be heated in order to give off enough vapor to form a combustible mixture above the surface that will momentarily flash or burn when the vapor is brought into contact with a very small flame. Because of the high temperatures at which aircraft engines operate, it is important that the oil used in such engines have a high flash point. The rate at which oil vaporizes in an engine depends on the temperature of the engine and the grade of the oil. If the vaporized oil burns, the engine is not properly lubricated. The operating temperature of any particular engine determines the grade of oil which should be used.

The **fire point** is the temperature to which any substance must be heated in order to give off enough vapor to burn continuously when the flammable air-vapor mixture is ignited by a small flame. The fire-point test is mentioned occasionally in reports on lubricants, but it is not used as much as the flash-point test and should not be confused with it.

Lubricating oils can be tested by means of the **Cleveland open cup** in accordance with the recommendations of the American Society for Testing and Materials (ASTM). This apparatus, shown in Fig. 8-3, is simple and adaptable to a wide range of products. It can be used for both flash-point and fire-point tests. When a test is made, the amount of oil, the rate of heating, the size of the igniting flame, and the time of exposure are all specified and must be carefully controlled to obtain accurate results from the test.

In a test of stable lubricating oils, the fire point is usually about 50 or 60°F [28 to 33°C] higher than the flash point. It should be understood that the determina-

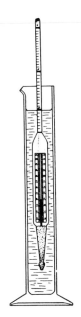

FIG. 8-2   Hydrometer for determining API gravity.

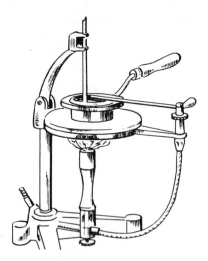

FIG. 8-3 Cleveland open cup tester.

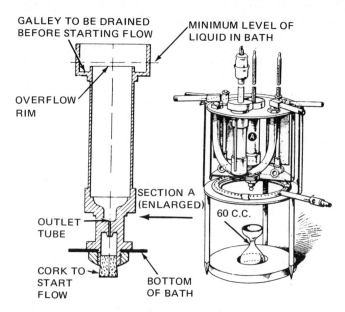

FIG. 8-4 Saybolt Universal viscosimeter.

tion of the fire point does not add much to a test, but the flash point of oil gives a rough indication of its tendency to vaporize or to contain light volatile material.

If there is any foaming during the flash-point test, this indicates the presence of moisture in the sample of oil being tested. If there are a few popping noises during the test, these indicate the presence of a very small quantity of moisture, or what the scientists refer to as a **trace** of moisture.

In a comparison of oils, if one has a higher or lower flash or fire point, this does not necessarily reflect on the quality of the oil—not unless the fire point or flash point is exceptionally low in comparison with the fire or flash point of similar conventional oils.

If oil which has been used in an aircraft engine is tested and found to have a very low flash point, this indicates that the oil has been diluted by engine fuel. If the oil has been diluted only slightly with aviation-grade gasoline, the fire point is not lowered much because the gasoline in the oil ordinarily evaporates before the temperature of the fire point is reached. If the oil has been greatly diluted by gasoline, the fire point will be very low.

In testing oil which has been used in an engine, it is possible to obtain more accurate results from the flash-point and fire-point tests if the sample of oil is obtained from the engine while both the engine and the oil are still hot.

Viscosity

Viscosity is technically defined as the fluid friction, (or the body) of an oil. In simple terms, viscosity may be regarded as the resistance an oil offers to flowing. A heavy-bodied oil is high in viscosity and pours or flows slowly; it may be described as a **viscous** oil. The lower the viscosity, the more freely an oil pours or flows at temperatures above the pour point. Oil that flows readily is described as having a low viscosity. The amount of fluid friction exhibited by the oil in motion is a measure of its viscosity.

The **Saybolt Universal viscosimeter,** illustrated in Fig. 8-4, is a standard instrument for testing petroleum products and lubricants. The tests are usually made at temperatures of 100, 130, and 210°F [38, 54, and 99°C]. This instrument has a tube in which a specific quan-

tity of oil is brought to the desired temperature by a surrounding liquid bath. The time in seconds required for exactly 60 cm³ of the oil to flow through an accurately calibrated outlet orifice is recorded as **seconds Saybolt Universal viscosity.**

Commercial aviation oils are generally classified by symbols such as 80, 100, 120, and 140, which approximate the seconds Saybolt Universal viscosity at 210°F. Their relation to Society of Automotive Engineers (SAE) numbers are given in Table 8-2.

Engineers use viscosity-temperature charts, published by the ASTM, in order to find quickly the variation of viscosity with the temperature of petroleum oils when the viscosities and any two temperatures are known. The two known temperatures are plotted on a chart, and a straight line is drawn between these points. When the straight line is extended beyond the two known points, the viscosities at other temperatures can be read on the chart from that line.

The **viscosity index,** abbreviated VI, is an arbitrary method of stating the rate of change in viscosity of an oil with changes of temperature. The viscosity index of any specific oil is based on a comparative evaluation with two series of standardized oils: one having an assigned viscosity-index value of 100, which is somewhat typical of a conventionally refined Pennsylvania oil, and the other series having an assigned viscosity-index rating of 0, which is typical of certain conventionally refined naphthenic-base oils. The viscosity characteristics of these two series of standardized oils have been arbitrarily chosen and adopted by the ASTM.

The average viscosity-index values of some of the

TABLE 8-2 Grade Designations for Aviation Oils

Commercial aviation no.	Commercial SAE no.	AN specification no.
65	30	1065
80	40	1080
100	50	1100
120	60	1120
140	70	

better known lubricating oils, depending upon their geographical source, are as follows: (1) Pennsylvania, 95 to 100; (2) midcontinent, 62 to 75; (3) east Texas, 50 to 60; (4) California, 0 to 35; Gulf Coast, 10 to 30; Republic of Colombia, 30 to 40; Peru, 15 to 25. These values are for the regular, conventional refining processes.

When a method of refining known as **solvent treat** is used, the average viscosity-index values may be raised, depending on the geographical source, as follows: (1) Pennsylvania, 100 to 110; (2) midcontinent, 80 to 100; (3) east Texas, 80 to 95; and (4) California, 40 to 65. It should be understood that an assigned viscosity-index value of 100 is typical of the conventionally refined Pennsylvania oil; hence, the use of a refining process for raising the viscosity-index value can increase the value above 100. However, when the solvent-treat process of refining is used, close control must be exercised during the refining or the oil will lack some of its natural lubricating and corrosion inhibiting properties.

Certain compounds can be added to the oil at the plant to raise the viscosity-index value above what it would be by any normal refining process, but this should not be interpreted by the reader to mean that it is safe to purchase compounds and dump them into the oil after it is received from the refinery or one of its agents.

The viscosity-index value is not fixed for all time when the oil is sold by the refinery or its distributors. If a lubricating oil is subjected to high pressure without any change in the temperature, the viscosity increases. Naphthenic oils of high viscosity vary more with pressure than do paraffinic oils. Those oils known as **fixed oils** vary less in viscosity than do either the naphthenic or the paraffinic oils.

The general rule is that oils of lower viscosity are used in colder weather and that oils of higher viscosity are used in warmer weather. But it is also important to choose an oil which has the lowest possible viscosity in order to provide an unbroken film of oil while the engine is operating at its maximum temperature, thus reducing friction to the minimum when the engine is cold. The type and grade of oil to be used in an engine is specified in the operator's manual.

No table of recommended operating ranges for various grades of lubricating oil can have more than a broad, general application because the oil must be especially selected for each particular make, model, type, and installation of engine, not forgetting the operating conditions of both the engine and the airplane in which it is installed. However, a few recommendations may provide a starting point from which those selecting oil can proceed.

The grade of lubricating oil to be used in an aircraft engine is determined by the operating temperature of the engine and by the operating speeds of bearings and other moving parts. Commercial aviation grade no. 65 (SAE 30 or AN 1065) may be used at ground air temperatures of 4°C (40°F) and below. The **oil-in** temperature is the temperature of the oil before it enters the engine, as indicated by a thermometer bulb or other temperature measuring device located in the oil system near the engine oil pump. The safe oil-in temperature for this grade is from 20°C (68°F) to 95°C (203°F).

Commercial grade no. 100 (SAE 50 or AN 1100) may be used in some engines for all temperatures, as directed by the manufacturer's instructions. For example, the Beechcraft G-18S airplane manual specifies MIL-L-6082, grade 1100 (SAE 50), for both summer and winter operation.

It will be noted that lubricating oils used in aircraft engines have a higher viscosity than those used in automobile engines. This is because aircraft engines are designed with greater operating clearances and operate at higher temperatures. Some manufacturers specify different grades of oil, depending on outside air temperatures. For the Continental IO-470D engines installed on the Cessna 310 airplane, grade 1100 oil is recommended when temperatures are generally above 40°F [4°C] and grade 1065 oil when the temperatures are below 40°F.

When servicing the oil tank for an engine installed in an airplane, the operator can find the proper grade of oil marked on or near the filler cap. If this marking should become obliterated, the operator's manual for the airplane will contain the information.

Viscosity of Synthetic Turbine-Engine Oils. The viscosity of the synthetic oils used in gas-turbine engines is generally expressed in units of the centimeter-gram-second (cgs) system. Under this system, the basic unit for the **coefficient of absolute viscosity** is the **poise** (P), named for the French physiologist Jean L. M. Poisuille (1799–1869). If we imagine a flat plate being drawn across the surface of a layer of oil, the force necessary to move the plate at a given velocity is a measure of the viscosity of the oil. If the layer of oil is 1 cm thick and the plate is moved at the rate of 1 cm/s, the total number of **dynes** of force required to move the plate, divided by the area of the plate in cm², will equal the coefficient of viscosity in poises. To express this in different terms, *when 1 dyne will move a plate 1 cm square at a rate of 1 cm/s across the surface of a liquid with a thickness of 1 cm, the coefficient of absolute viscosity is 1 P.*

The viscosity of turbine engine oil is considerably less than 1 P; hence, the **centipoise** (cP, 0.01 P) is used to express the viscosity.

Because the density of an oil is an important factor, it is common practice to employ the unit for **kinematic viscosity** in establishing the characteristics of gas-turbine lubricants. The unit for kinematic viscosity is the same as the poise when the density of a liquid is 1 gm/cm³. Kinematic viscosity is expressed in **stokes** (St) [m_2/s × 10^{-4}] or **centistokes** (cSt), 1 cSt being equal to 0.01 St. Kinematic viscosity in stokes is equal to absolute viscosity in poises divided by the density of the liquid in grams per cubic centimeters. The Saybolt Universal viscosity of an oil having a kinematic viscosity of 5 cSt is approximately 42.6. This is roughly equivalent to what we know as 20-weight lubricating oil.

Viscosity and Cold-Weather Starting. The **pour point** of an oil indicates how fluid it is at low temperatures under laboratory conditions, but it does not necessarily measure how pumpable it is under field conditions. The **viscosity** is a far better indication of whether or not the oil will make it possible to start at low temperatures and how well the oil can be pumped. At low temperatures it is desirable to have a combination of a low pour point and low viscosity if the proper viscosity for operating temperatures is still retained.

In order to thin the lubricating oil for starting engines in cold weather, engine gasoline may be added directly to the oil if provisions are made for this in the powerplant of the airplane. The cold oil, diluted with gasoline, circulates easily and provides the necessary lubrication. Then when the engine reaches its normal operating temperature, the gasoline evaporates and leaves the oil as it was before it was diluted.

When **oil dilution** is practiced, less power is needed for starting and the starting process is completed more quickly. The only important disadvantage is that the presence of ethyl gasoline in the oil may cause a slight corrosion of the engine parts, but this objection is outweighed by the advantages.

The grade of oil to be used for oil dilution in starting depends on various factors, but two general rules may prove useful in the absence of specific instructions for the particular powerplant installed on the airplane under operation. These broad rules, subject to exceptions, of course, are as follows:

1. Never use an oil lighter than commercial aviation 100 (SAE 50 or AN 1100) for dilution.

2. Select the oil to be used for diluting with engine fuel according to the ground temperature, as previously explained.

Color

The **color** of a lubricating oil is obtained by reference to transmitted light; that is, the oil is placed in a glass vessel and held in front of a source of light. The intensity of the transmitted light must be known in conducting a test because it may cause the color to vary.

The apparatus used for a color test is that approved by the American Society for Testing and Materials and is called an **ASTM Union Colorimeter.** Colors are assigned numbers ranging from 1 (lily white) to 8 (darker than claret red). Oils darker than no. 8 color value are diluted with kerosene (85 percent kerosene by volume and 15 percent lubricating oil by volume) and then observed in the same manner as oils having color values from 1 to 8.

When reflected light (as distinguished from direct light) is used in a color test, the color is called the **bloom** and is used, among other things, to indicate the origin and refining method of the oil.

Cloud Point

The **cloud point** is the temperature at which the separation of wax becomes visible in certain oils under prescribed testing conditions. When such oils are tested, the cloud point is a temperature slightly above the solidification point. If the wax does not separate before solidification, or if the separation is not visible, the cloud point cannot be determined.

Pour Point

The **pour point** of an oil is the temperature at which the oil will just flow without disturbance when chilled. In practice, the pour point is the lowest temperature

at which an oil will flow (without any disturbing force) to the pump intake. The fluidity of the oil is a factor of both pour test and viscosity. If the fluidity is good, the oil will immediately circulate when engines are started in cold weather. Petroleum oils, when cooled sufficiently, may become plastic solids as a result either of the partial separation of the wax or of the congealing of the hydrocarbons composing the oil. To lower the pour point, **pour-point depressants** are sometimes added to oils which contain substantial quantities of wax.

Figure 8-5 shows an apparatus used for conducting the pour-point and cloud-point tests. The parts to be especially observed are the thermometers, the glass test jar, the jacket, the ring gasket, the cooling bath, and the cork or felt disk.

The general statement is sometimes made that the pour point should be within 5°F [3°C] of the average starting temperature of the engine, but this should be considered in connection with the viscosity of the oil, since the oil must be viscous enough to provide an adequate oil film at engine operating temperatures. Therefore, for cold-weather starting, the oil should be selected in accordance with the operating instructions for the particular engine, considering both the pour point and the viscosity.

Carbon-Residue Test

The purpose of the **carbon-residue test** is to study the carbon-forming properties of a lubricating oil. There are two methods: (1) the Ramsbottom carbon-residue test and (2) the Conradson test. The Ramsbottom test is widely used in Great Britain and is now preferred by many American petroleum engineers because it seems to yield more practical results than the Conradson test, which was formerly more popular in the United States.

The apparatus for the **Ramsbottom test** is shown in Fig. 8-6. A specific amount of oil is placed either in a heat-treated glass bulb well or in a stainless-steel bulb used for the same purpose. The oil is then heated to a high temperature by a surrounding molten-metal bath for a prescribed time. The bulb is weighed before and after the test. The difference in weight is divided by the weight of the oil sample and multiplied by 100 to obtain the percentage of carbon residue in the sample.

The apparatus for the **Conradson test** is shown in Fig. 8-7. Oil is evaporated under specified conditions for conducting the test. The parts to be observed in

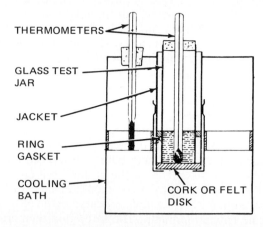

FIG. 8-5 Apparatus for cloud-point and pour-point tests.

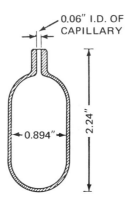

FIG. 8-6 Bulb for Ramsbottom carbon-residue test.

the illustration are the burner, the porcelain crucible, the wire gauze, the spun sheet-iron crucible, the Skidmore iron crucible, the asbestos block, the hood, and the flame gage. The carbon residue from the Conradson test should not be compared directly with the carbon residue from the Ramsbottom test, since the residues are obtained under different test conditions, but tables have been prepared by engineers which give the average relation between the results of tests performed by the two methods.

Petroleum engineers advise those who are not in their field to be cautious in evaluating carbon-residue tests, since the carbon deposits from oil vary with the type and mechanical condition of the engine, the service conditions, the cycle of operation, the other characteristics of the oil, and the method of carbureting the fuel. In the early days of internal-combustion engines, carbon-residue tests were more important as an indication of the carbon-forming properties of lubricating oil than they are today. The methods now used to refine petroleum products tend to make the carbon-residue tests less useful than they were originally.

The Ash Test

The **ash test** is an extension of the carbon-residue test. If an unused (new) oil leaves almost no ash, it is regarded as pure. The **ash content** is a percentage (by weight) of the residue after all carbon and carbonaceous matter have been evaporated and burned.

In a test of used lubricating oil, the ash is analyzed chemically to determine the content of iron, which shows the rate of wear; sand or grit, which comes from the atmosphere; lead compounds, which come from leaded gasoline; and other metals and nonvolatile materials. The ash analysis tells something about the per-

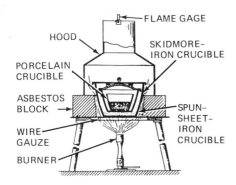

FIG. 8-7 Apparatus for Conradson carbon test.

formance of the engine lubricating oil, but it is only one of many tests which are used to promote efficiency.

Oxidation Tests

Aircraft engine lubricating oils may be subjected to relatively high temperatures in the presence of both air and what the engineers call catalytically active metals or metallic compounds. This causes the oil to oxidize. It increases the viscosity, and it also forms sludge, carbon residues, lacquers or varnishes (asphaltines), and sometimes inorganic acids.

There are several methods of testing for oxidation, the details of which do not interest most people outside the research laboratories, although the conclusions are important to aircraft engine personnel in general. The U.S. Air Force has its own oxidation test, the U.S. Navy has its work factor test, and engine manufacturers often have their own tests.

It has been found that, when the carbon residue of engine oils is lowered below certain limits, the products of oxidation are soluble in hot oil. Deposits of lacquer form on the metallic surfaces, such as on the pistons, in the ring grooves, and on valve guides and stems, and anywhere that the oil flows comparatively slowly in the engine. In addition, a sludge of carbonlike substance forms in various places. To overcome this situation, certain compounds known as **antioxidant** and **anticorrosion** agents have been used to treat lubricating oils before they are sold to the public.

Precipitation Number

The **precipitation number** recommended by the ASTM is the number of milliliters of precipitate formed when 10 ml of lubricating oil are mixed with 90 ml of petroleum naphtha under specified conditions and then centrifuged (subjected to centrifugal force) under prescribed conditions. The volume of sediment at the bottom of the centrifuge tube (container) is then taken as the ASTM precipitation number.

Corrosion and Neutralization Number

Lubricating oils may contain acids. The **neutralization number** recommended by the ASTM is the weight in milligrams of potassium hydroxide required to neutralize 1 g of oil. A full explanation of this topic belongs in the field of elementary chemistry and is beyond the scope of this text.

The neutralization number does *not* indicate the corrosive action of the used oil that is in service. For example, it has been found that in certain cases an oil having a neutralization number of 0.2 might have high corrosive tendencies in a short operating period, whereas another oil having a neutralization number of 1.0 might have no corrosive action on bearing metals.

Oiliness

Oiliness is the property that makes a difference in reducing the friction when lubricants having the same viscosity but different oiliness characteristics are compared under the same conditions of temperature and film pressure. Oiliness, contrary to what might be expected, depends not only on the lubricant but also on the surface to which it is applied. Oiliness has been compared with metal wetting, but oiliness is a wetting effect that reduces friction, drag, and wear. It is especially important when the film of oil separating rubbing surfaces is very thin, when the lubricated parts are very hot, or when the texture (grain) and finish of the metal are exceedingly fine. When some oil films are formed, there may be almost no viscosity effects, and then the property of oiliness is the chief source of lubrication.

Extreme-Pressure (Hypoid) Lubricants

When certain types of gearing are used, such as spur-type gearing and hypoid-type gearing, there are high tooth pressures and high rubbing velocities that require the use of a class of lubricants called **extreme-pressure (EP) lubricants,** or **hypoid lubricants.** Most of these special lubricants are mineral oils containing loosely held sulfur or chlorine or some highly reactive material. If ordinary mineral oils were used by themselves, any metal-to-metal contact in the gearing would usually cause scoring, galling, and the local seizure of mating surfaces.

The characteristics of these EP lubricants are as follows:

1. They prevent the galling or scoring of mating surfaces.
2. They reduce wear by conditioning the mating surfaces.
3. They make possible a low degree of friction.

Chemical and Physical Stability

An aircraft engine oil must have **chemical stability** against oxidation, thermal cracking, and coking. It must have **physical stability** with regard to pressure and temperature.

Some of the properties discussed under other topic headings in this chapter are closely related to both chemical and physical stability. The oil must have resistance to emulsion; this characteristic is termed **demulsibility** and is a measure of the oil's ability to separate from water. Oil that is emulsified with water does not provide a high film strength or adequate protection against corrosion. Aircraft engine oil should also be nonvolatile, and there should be no objectionable compounds of decomposition with fuel by-products. The viscosity characteristics should be correct, as we have explained in detail. If anything is added during the refining process, the resultant should be uniform in quality and pure.

When all the other factors are favorable, the oil should have a minimum coefficient of friction, maximum adhesion to the surfaces to be lubricated, oiliness characteristics, and adequate film strength.

● LUBRICANT REQUIREMENTS AND FUNCTIONS

Characteristics of Aircraft Lubricating Oil

The proper lubrication of aircraft engines requires the use of a lubricating oil which has the following characteristics:

1. It should have the proper **body (viscosity)** at the engine operating temperatures usually encountered by the airplane engine in which it is used; it should be distributed readily to the lubricated parts; and it must resist the pressures between the various lubricated surfaces.

2. It should have **high antifriction characteristics** to reduce the frictional resistance of the moving parts when separated only by boundary films. An ideal fluid lubricant provides a strong oil film to prevent metallic friction and to create a minimum amount of oil friction, or oil drag.

3. It should have **maximum fluidity at low temperatures** to ensure a ready flow and distribution when starting at low temperatures. Some grades of oil become practically solid in cold weather, causing high oil drag and impaired circulation. Thus, the ideal oil should be as thin as possible and yet stay in place and maintain an adequate film strength at operating temperatures.

4. It should have **minimum changes in viscosity with changes in temperature** to provide uniform protection where atmospheric temperatures vary widely. The viscosity of oils is greatly affected by temperature changes. For example, at high operating temperatures, the oil may be so thin that the oil film is broken and the moving parts wear rapidly.

5. It should have **high antiwear properties** to resist the wiping action that occurs wherever microscopic boundary films are used to prevent metallic contact. The theory of fluid lubrication is based on the actual separation of the metallic surfaces by means of an oil film. As long as the oil film is not broken, the internal friction (fluid friction) of the lubricant takes the place of the metallic sliding friction which otherwise would exist.

6. It should have **maximum cooling ability** to absorb as much heat as possible from all lubricated surfaces—and especially from the piston head and skirt. One of the reasons for using liquid lubricants is that they are effective in absorbing and dissipating heat. Another reason is that liquid lubricants can be readily pumped or sprayed. Many engine parts, especially those carrying heavy loads at high rubbing velocities, are lubricated by oil under direct pressure. Where direct-pressure lubrication is not practical, a spray of mist of oil provides the required protection. Regardless of the method of application, the oil absorbs the heat and later dissipates it through coolers or heat exchangers.

7. It should offer the **maximum resistance to oxidation,** thus minimizing harmful deposits on the metal parts.

8. It should be **noncorrosive** to the metals in the lubricated parts.

The Functions of Engine Oil

Engine oil performs these functions:

1. It lubricates, thus reducing the friction between moving parts.
2. It cools the various parts of the engine.
3. It tends to seal the combustion chamber by filling the spaces between the cylinder walls and piston rings, thus preventing the flow of combustion gases past the rings.
4. It cleans the engine by carrying sludge and other

residues away from the moving engine parts and depositing them in the engine oil filter.

5. It aids in preventing corrosion by protecting the metal from oxygen, water, and other corrosive agents.

6. It serves as a cushion between parts where impact loads are involved.

Lubricating Oils for Gas-Turbine Engines

It has been mentioned that lubricating oils for gas-turbine engines are usually of the *synthetic* type. This means that the oils are not manufactured in the conventional manner from petroleum crude oils. Petroleum lubricants are not suitable for modern gas-turbine engines because of the high temperatures encountered during operation. These temperatures often exceed 500°F [260°C], and at temperatures of this level, petroleum oils tend to break down. The lighter fractions of the oil evaporate, thus leaving carbon and gum deposits; the lubricating characteristics of the oil rapidly deteriorate, too.

Synthetic oils are designed to withstand high temperatures and still provide good lubrication. The first generally acceptable lubricating oil conformed to MIL-L-7808 and is known as type I, an aklyl diester oil. During recent years, the type II oil, a polyol ester lubricant, has been found most satisfactory. This oil meets or exceeds the requirements of MIL-L-23699.

The type II synthetic lubricant is also described as a *5-centistoke* oil. This means that the oil must have a minimum kinematic viscosity of 5 cSt at a temperature of 210°F[99°C]. This specification is necessary because the oil must maintain sufficient body to carry all applied loads at operating temperatures.

Lubricants for gas-turbine engines must pass a variety of exacting tests to ensure that they have the characteristics required for satisfactory performance. Among the characteristics tested are specific gravity, acid-forming tendencies, metal corrosion, oxidation stability, vaporphase coking, gear scuffing, effect on elastomers, and bearing performance. These tests are designed to provide indications that the oil will supply the needed lubrication under all conditions of operation.

Synthetic oils are likely to dissolve paints and therefore should be removed immediately if spilled on a painted surface. They also may tend to soften certain types of materials used for seals; for this reason, great care must be taken if a system is to be converted from a petroleum lubricant to a synthetic lubricant. It must be determined that the gaskets, seals, and hoses are of a material designed for synthetic oils.

● AVIATION GREASES

Oil and Grease Compared for Aviation Purposes

Oil is the ideal lubricant under many conditions, especially when the design and operation of the equipment make it possible to apply the lubricant as a fluid. However, in many aircraft engine accessories, airplane control mechanisms, linkage bearings, and landing wheels, fluid oil cannot be used. There are several reasons why this is true: (1) The inaccessibility of the parts or the physical difficulty of servicing the parts may make it impractical to lubricate regularly with oil;

(2) the oil may reach sensitive parts where it will cause damage, such as the electric fields of instruments, motors, generators, magnetos, etc.; and (3) under many conditions, oil leaks too rapidly to maintain good lubrication.

Factors Considered in Selecting Greases

The many different applications or uses of grease lubricants have made it necessary to develop greases to serve a wide variety of purposes. In selection of a grease, the *type of friction* encountered must be considered. We must consider *rolling friction, sliding friction,* and *wiping friction.* Another most important factor is the *temperature* under which the grease must work. If the temperature is high, a high-temperature grease must be employed to avoid losing the grease by its melting and flowing out of the bearing. A third principal factor to be considered is the **speed of relative motion between the bearing surfaces.** The characteristics of a grease used for high-speed bearings must be different from those of a grease used in low-speed applications. **Bearing load** is important. If the grease is too light and not sufficiently cohesive, it may be pressed out to the extent that metal surfaces come in contact.

Grease may be either semifluid or nonfluid. It is usually made by combining oil with one or more types of soap, polyurea, or other stabilizing agent. It is apparent that the quality and the grade of the oil combined with the thickener affect the quality of the product. In addition, the quantity and type of thickener base, the actual composition of the product, the manufacturing processes, and the uniformity of production are factors which determine the character and quality of a grease lubricant.

Consistency or Penetration

The **consistency** or **penetration,** the most important characteristic of a grease, is measured by an apparatus called a **penetrometer** (Fig. 8-8). A container full of grease to be tested is placed on a table. A cone is lowered until the tip barely touches the top surface of the grease sample. An actuating plunger is released to permit the cone to rest free on the grease for 5 s. The depth to which the cone sinks into the grease is indicated on a scale calibrated to measure the depth of penetration in tenths of a millimeter.

If the grease is soft, the cone will penetrate deeper and give a higher reading than it will if the grease is

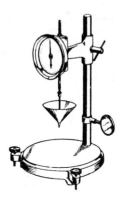

FIG. 8-8 Penetrometer for testing grease consistency.

hard. The standard temperature for this test is 77°F [25°C]. Other variables, such as the presence of air pockets in the grease, are rigidly controlled to make all tests as uniform as possible. "Unworked" samples are sometimes tested in their original containers. Some types of greases are "worked" before the test to give them uniform consistency. In the report on the test it is stated whether or not the grease was worked before the test.

Gear lubricants and certain other greases, especially semifluid greases, are too soft to obtain accurate readings with a penetrometer. Other tests have been developed to indicate the flow characteristics of these greases.

Grease Numbers

Once greases were described as soft, medium, or hard, but the modern method is to follow a procedure recommended by the National Lubricating Grease Institute (NLGI). This organization assigns what it calls NLGI numbers that represent worked penetration limits. For example, a grease commonly described as soft is given an NLGI number of 0 if the penetration reading from the penetrometer test is from 355 to 385; if it is from 310 to 340, the NLGI number is 1. Likewise, if the grease is commonly described as medium it is given an NLGI of 2 or 3, depending on the penetrometer test. In a similar manner, if the grease is commonly described as hard, it receives an NLGI number of 4 or 5, again depending on the penetrometer test.

NLGI numbers may be compared with the SAE numbers for oils. They designate the grade but not the quality. They are important to anyone who wishes to know for sure exactly what grade of grease is being obtained. Before these numbers were established, manufacturers described their products according to their own whims.

Dropping Point

The **dropping point** of a grease is the temperature at which a sample of the grease passes from a semisolid to a fluid state under conditions of rest. It is sometimes erroneously called the "melting point," but this term should be avoided.

Figure 8-9 illustrates the apparatus for making the dropping-point test. A small sample is placed in a small metal cup which has an orifice in the bottom. The cup is placed in a special test tube. A thermometer is ad-

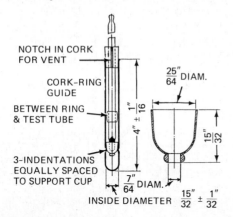

FIG. 8-9 Apparatus for dropping-point test.

justed to read the temperature accurately. The equipment is then placed in an oil bath and heated at a specified rate. Grease gradually protrudes through the orifice as the temperature increases. When a drop falls or reaches the bottom of the test tube, the temperature indicated by the thermometer at that instant is recorded as the dropping point of the grease.

The ASTM reports that the dropping point does not relate *directly* to the performance of a grease in an airplane under field conditions. The design of a lubricated bearing, the rotational speeds, and other factors are more important than the dropping-point characteristic.

Special Tests

There are many tests for greases which are beyond the scope of this text. Greases are tested for cohesion, bleeding, congealing, film strength, water resistance, storage ability, pumpability, adhesion, low-temperature torque, and high-temperature oxidation. The procedures for performing these tests vary widely according to the special requirements or specifications for which the test is made.

Requirements for Aircraft Lubricating Greases

The characteristics required of aircraft engine lubricants apply in general to aircraft lubricating greases, but those which are especially important in judging greases are as follows:

1. The grease should have the **proper consistency** for the method of application employed and for complete distribution to the surfaces to be lubricated.

2. The grease should have **high antiwear properties** to prevent metallic contact where pressures are applied suddenly, and especially where boundary films are expected to protect the surfaces.

3. The grease should have **adequate plasticity at low temperatures** to prevent sluggish action and the sticking of moving parts. **Plasticity** is a state or condition of being easily subjected to molding into any form under pressure. In this case, it merely means that the grease should not become gummy or hard at low temperatures but should flow easily.

4. The grease should have **high antifriction characteristics** to reduce the frictional resistance of the moving parts.

5. The grease should offer **resistance to separation** at high temperatures or when subjected to heavy pressures. This means that it should not break down into its chemical components, thus losing its lubricating properties. In other words, it should be chemically stable.

Types of Greases

Most grease lubricants can be divided into six groups—calcium-soap greases, sodium-soap greases, aluminum-soap greases, lithium-soap greases, special greases, and mixed-base greases.

Calcium-base greases. These greases are a combination of light to medium weights or grades of oil with calcium soaps, examples being cup greases and pressure system greases. The texture is smooth, and there is no tackiness at normal temperatures. These lubricants are rarely used at operating temperatures above 175°F [79°C], where high centrifugal speeds may throw out the lubricant, or where the lubricated part is subject to shock loading. These greases are more suitable at low temperatures. Also, they are very water repellent.

Sodium-base greases. These greases are a combination of medium to heavy weights or grades of oil with sodium soaps. They are generally used to lubricate antifriction ball, roller, and needle-type bearings that are subjected to shock loads or high temperatures, and they are used when there is need to retain the lubricant under conditions of high rotation or centrifugal motion. They are very susceptible to moisture and are very adhesive and cohesive. The texture may be either smooth or fibrous, depending on the amount of milling or working they receive during the manufacturing process. They have a wide range of adaptability to temperature and rotating-speed changes. Common aircraft applications are the lubrication of engine accessories and wheel bearings, but it should be understood that sodium-base greases can be manufactured to meet a very great variety of aircraft engine and general aircraft requirements.

Aluminum-base greases. These greases are suitable for plain bearings when the speed of rotation is slow, the part is subjected to shock loads, and the part or unit is exposed to water frequently. They are water repellent, and resist thinning at high temperatures and shock loading. Under certain conditions, these greases may be tacky, stringy, or very adhesive and cohesive. They are widely used on airplanes to lubricate Lord mountings, landing-gear joints, and similar parts or units.

Lithium-base greases. These greases are used for extremely low temperature applications. One such grease conforms to specification MIL-G-7421B. This is a smooth, brown, buttery grease consisting of a synthetic-oil base and a lithium-soap thickener. Additional materials are added to provide resistance to oxidation and corrosion. This grease is designed to operate over the temperature range of −100 to 250°F [−73 to 121°C]. It is intended for use in antifriction bearings, gears, rolling and sliding surfaces requiring small breakaway torques, electrical equipment, small actuators, and other applications where low temperatures are likely to be encountered.

Since this grease is made with a synthetic-oil base, it is likely to soften natural rubber and some paints. For this reason it must be used only where the associated materials will not be affected.

Figure 8-10 is a chart showing the uses of various greases. These are typical of the many types which have been developed for aircraft use.

Special greases. These greases are usually made of soap bases other than calcium, sodium, or aluminum, examples of the soaps used being barium, lithium, and lead. Greases made from barium soap are used for the same purposes as aluminum-soap lubricants. Greases made from lead soaps are used where the rubbing conditions or the high loading make it necessary to obtain a very tough lubricating film. Lithium soap is used to make greases that are highly water repellent, and for

SPECIFICATION	TEMPER-ATURE RANGE, °F	TYPICAL USES
MIL-G-7711A	−40 to 300	Extreme General purpose
MIL-G-21164C	−100 to 275	Extreme pressure applications at both low and high temperatures. This grease has a synthetic base.
MIL-G-3545C	−20 to 350	Ball and roller bearings, engine accessories and wheel bearings.
MIL-G-23827A	−100 to 275	Low-temperature grease for general airframe lubrication including instruments.
MIL-G-25013D	−100 to 450	For extreme low and high temperature applications for all antifriction bearings.

FIG. 8-10 Chart showing various greases and their applications.

purposes where there must be a minimum drag at very low temperatures. In this connection, it should be understood that when low-temperature greases are desired, the oil mixed with the soap is usually very light, with a low pour-test rating and a high viscosity index.

Mixed-base greases. These greases are made up of a combination of several soap bases. Usually, if one soap base predominates in the combination, the grease is classified under the heading of that base.

Additives

Graphite is usually added to the conventionally made calcium- or sodium-soap-base greases according to the application for which the grease is intended, especially for sliding surfaces that are not easily reached for frequent lubrication. Graphite is especially valuable in a grease used as an *antiseize thread* lubricant.

Talc is sometimes used instead of graphite for inaccessible sliding surfaces.

Mica may be added where high temperatures are frequently encountered. It is also valuable when added to grease prepared as an antiseize lubricant. Spark-plug thread lubricants usually contain some mica.

Zinc, lead, and other metal oxides are either used as general lubricants or added to grease for antiseize purposes. They produce greases that are approximately white in color.

Special Engine Lubricants

Oil is the principal aircraft engine lubricant, but **special lubricants,** most of which are greases, are used for special purposes and given names that usually indicate their use. Examples are thread lubricant, spark-plug lubricant, high-melting-point grease, petrolatum, and rust-preventive compound.

Petrolatum is a pure, refined mineral oil which can be used as a light petroleum grease or as a combined rust preventive and lubricant. It is commonly used on aircraft battery terminals.

Thread lubricant is composed of about 75 percent heavy mineral oil and about 25 percent lead soap. It is a semifluid composition used on the threaded parts of aircraft engines to prevent oxidation or corrosion.

Spark-plug lubricant is about 60 percent heavy mineral oil and about 40 percent finely ground mica. It is used on the threads of spark plugs to make it easier to remove them from the engine cylinders and replace them.

High-melting-point grease, also called **fiber grease,** is composed of sodium soap and mineral oil. It may vary in consistency from soft to hard. It is used to lubricate "hot-running" equipment, particularly electrical units.

Rust-preventive compounds are used to prevent the corrosion of the unpainted surfaces of aircraft engines in storage. They are available in two grades. The light grade is composed of 90 percent vegetable oil, 7 percent alcohol, and 3 percent triethanolamine and is especially effective in preventing the corrosion of engine cylinder barrels, valves, and valve mechanisms exposed to the action of ethyl fuels. This light grade is used on the internal surfaces of engines because it is not necessary to remove it when the engine is prepared for operation after being in storage. The heavier grade is applied to the external parts of engines and can be removed by using kerosene as a solvent.

A suitable corrosion-preventive oil complies with the specification MIL-C-6529C. This preservative oil is made up of aircraft lubricating oil, grade 1100, and corrosion-preventive additives. The material may be used full strength, or it may be mixed with 75 percent lubricating oil, and in this diluted condition it can be used for regular engine operation.

● THE NEED FOR LUBRICATION

There are many moving parts in an aircraft engine. Some reciprocate and some rotate, but regardless of the motion, each moving part must be guided in its motion or held in a given position during motion. The contact between surfaces moving in relation to each other produces friction, which consumes energy. This energy is transformed into heat at comparatively low temperatures and therefore reduces the power output of the engine. Furthermore, the friction between moving metallic parts causes wear. If lubricants are used, a film of lubricant is applied between the moving surfaces to reduce wear and to lower the power loss.

Sliding Friction

When one surface slides over the other, the interlocking particles of metal on each surface offer a resistance to motion known as **sliding friction.** If any supposedly smooth surface is examined under the microscope, it will be seen that it has hills and valleys. The smoothest possible surface is only relatively smooth. No matter how smooth the surfaces of two objects may appear to be, when they slide over each other, the hills in one catch in the valleys of the other.

Rolling Friction

When a cylinder or sphere rolls over the surface of a plane object, the resistance to motion offered by the surfaces to each other is known as **rolling friction.** In addition to the interlocking of the surface particles which occurs when two plane objects slide over each other, there is a certain amount of deformation of both the cylinder or sphere and the plane surface over which it rolls. There is less rolling friction when ball bearings are used than there is when roller bearings are employed. At first thought, this may not seem reasonable, but actually rolling friction is less than sliding friction and is always preferred by mechanical designers when the surface permits what they call **line** or **point contact.** A simple explanation of the reduction of friction obtained with a rolling contact is that the interlocking of surface particles is considerably less than in the case of sliding friction. Therefore, even when the deformation is added, the total friction by rolling contact is less than it is by sliding contact.

Wiping Friction

Wiping friction occurs particularly between gear teeth. Some gear designs, such as the hypoid gears and worm gears, have greater friction of the wiping type than do other designs, such as the simple spur gear. Wiping friction involves a continually changing load on the contacting surfaces, both in intensity and in direction of movement, and it usually results in extreme pressure, for which special lubricants are required. Lubricants for this purpose are called extreme-pressure (EP) lubricants.

Factors Determining the Amount of Friction

The amount of friction between two solid surfaces depends largely on the rubbing of one surface against the other, the condition and material of which the surfaces are made, the nature of contact movement, and the load carried by the surfaces. The friction usually decreases at high speeds. When a soft bearing material is used in conjunction with hard metals, the softer metal can mold itself to the form of the harder metal, thus reducing friction. Increasing the load increases the friction.

The introduction of lubricant between two moving metallic surfaces produces a film which adheres to both surfaces. The movement of the surfaces causes a shearing action in the lubricant. In this manner the metallic friction between surfaces in contact is replaced by the smaller internal friction within the lubricant. Only fluid lubricants with a great tendency to adhere to metal are able to accomplish this purpose, since they enter where the contact between the surfaces is closest and where the friction would be the greatest if there were no lubrication. The adhesive quality of the lubricant tends to prevent actual metallic contact. The viscosity tends to keep the lubricant from being squeezed out by the pressure on the bearing surfaces.

Although the amount of friction between two solid surfaces depends on the load carried, the rubbing speed, and the condition and material of which the surfaces are made, the fluid friction of a lubricant is not affected in the same manner. The **internal friction of the lubricant** that replaces the metallic friction between moving parts is determined by the rubbing speed, the area of the surfaces in contact, and the viscosity of the lubricant. It is not determined by the load, by the condition of the surfaces, or by the materials of which they are made.

CHARACTERISTICS AND COMPONENTS OF LUBRICATION SYSTEMS

The purpose of a lubrication system is to supply oil to the engine at the correct pressure and volume to provide adequate lubrication and cooling for all parts of the engine which are subject to the effects of friction. The oil tank must have ample capacity, the oil pump volume and pressure must be adequate, and the cooling facilities for the oil must be such that the oil temperature is maintained at the proper level to keep the engine cool. Several typical systems are described in this section.

The lubrication is distributed to the various moving parts of a typical internal-combustion engine by pressure, splash, and spray.

Pressure Lubrication

In a typical pressure lubrication system, a mechanical pump supplies oil under pressure to the bearings. The oil flows into the inlet (or suction) side of the pump, which is usually located higher than the bottom of the oil sump so that sediment which falls into the sump will not be drawn into the pump. the pump may be either the eccentric-vane type or the gear type, but the gear type is more commonly used. It forces oil into an oil manifold, which distributes the oil to the crankshaft bearings. A pressure-relief valve is usually located near the outlet side of the pump.

Oil flows from the main bearings through holes drilled in the crankshaft to the lower connecting-rod bearings. Each of these holes through which the oil is fed is located so that the bearing pressure at that point will be as low as possible.

Oil reaches a hollow camshaft through a connection with the end bearing or the main oil manifold and then flows out of the hollow camshaft to the various camshaft bearings and cams.

Lubrication for overhead valve mechanisms on reciprocating engines, both in conventional airplanes and in helicopters, is supplied by the pressure system through the valve pushrods. Oil is fed from the valve tappets (lifters) into the pushrods. From there, it flows under pressure to the rocker arms, rocker-arm bearings, and valve stems.

The engine cylinder surfaces and piston pins receive oil sprayed from the crankshaft and also from the crankpin bearings. Since oil seeps slowly through the small crankpin clearances before it is sprayed on the cylinder walls, considerable time is required for enough oil to reach the cylinder walls, especially on a cold day when the oil flow is more sluggish. This situation is one of the chief reasons for diluting the engine oil with engine fuel for starting in very cold weather.

Splash Lubrication and Combination Systems

Pressure lubrication is the principal method of lubrication used on all aircraft engines. Splash lubrication may be used in addition to pressure lubrication on aircraft engines, but it is never used by itself. Hence, aircraft engine lubrication systems are always of either the pressure type or the combination pressure, splash, and spray type, usually the latter. For this reason, this text discusses the pressure type of lubrication system but calls attention to those units or parts which are splash-lubricated or spray-lubricated. The bearings of gas-turbine engines are usually lubricated by means of oil jets that spray the oil under pressure into the bearing cavities. Further information on the lubrication systems for gas-turbine engines is provided in the sections of this text that describe such engines.

Principal Components of a Lubrication System

An aircraft engine lubrication system includes a pressure oil pump, an oil pressure-relief valve, an oil reservoir (either as a part of the engine or separate from the engine), an oil pressure gage, an oil temperature gage, an oil filter, and the necessary piping and connections. In addition, many lubrication systems include oil coolers and/or temperature regulating devices. Oil-dilution systems are included when they are deemed necessary for cold-weather starting. The block diagram of Fig. 8-11 shows the principal components of a lubrication system for a reciprocating engine and where the components are located in the system. The system illustrated is called a **dry-sump** system because the oil is pumped out of the engine into an external oil tank.

In the system illustrated in Fig. 8-11, oil flows from the oil tank to the engine-driven pressure pump. The oil temperature is sensed before it enters the engine; that is, the temperature of the oil in the oil-in line is sensed and the information is displayed by the engine oil temperature gage. The pressure pump has greater capacity than is required by the engine; hence, a pressure-relief valve is incorporated to bypass excess oil back to the inlet side of the pump. A pressure gage connection, or sensor, is located on the pressure side of the pressure pump to actuate the oil pressure gage. The oil screen (oil filter) is usually located between the pressure pump and the engine system; oil screens are provided with bypass features to permit unfiltered oil to flow to the engine in case the screen becomes clogged, since unfiltered oil is better than no oil. After the oil has flowed through the engine system, it is picked up by the scavenge pump and returned through the oil cooler to the oil tank. The scavenge pump has a capacity much greater than that of the pressure pump, because the oil volume it must handle is increased as a result of the air bubbles and foam that are entrained during engine operation. The oil cooler usually incorporates a thermostatic control valve that bypasses the oil around the cooler until the oil temperature reaches a proper value. To prevent pressure buildup in the oil tank, a vent line is connected from the tank to the crankcase of the engine. This permits the oil tank to vent through the engine venting system.

Oil systems often include units or features not shown in the diagram of Fig. 8-11. Some systems include oil-dilution capabilities by which gasoline is introduced into the oil system through the in line between the oil tank and the engine just prior to engine shutdown. Check valves are employed in some systems to prevent oil from flowing by gravity to the engine when the

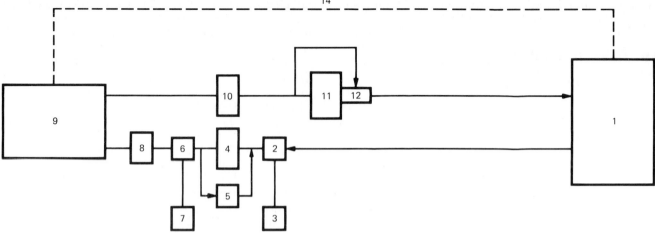

FIG. 8-11   Block diagram to show relative locations of lubricating system components.

engine is inoperative. Details of typical systems are explained and illustrated in this chapter.

Oil Capacity

The capacity of the lubrication system must be sufficient to supply the engine with an adequate amount of oil at a temperature not in excess of the maximum which has been established as safe for the engine. On a multiengine airplane, the lubrication system for each engine must be independent of the systems for the other engines.

The usable tank capacity must not be less than the product of the endurance of the airplane under critical operating conditions and the maximum oil consumption of the engine under the same conditions, plus an adequate margin to ensure satisfactory circulation, lubrication, and cooling. In lieu of a rational analysis of airplane range, a fuel-oil ratio of 30:1 by volume is acceptable for airplanes not provided with a reserve or transfer system. If a transfer system is provided, a fuel-oil ratio of 40:1 is considered satisfactory.

Oil Tanks

Dry-sump engine lubrication systems require a separate oil tank for each engine system. These tanks are constructed of welded sheet aluminum, riveted aluminum, or stainless steel. Some aircraft are equipped with synthetic rubber tanks similar to fuel cells.

An outlet for the tank is normally located at the lowest section of the tank to permit complete drainage, either in the ground position or in a normal flight attitude. If the airplane is equipped with a propeller feathering system, a reserve of oil must be provided for feathering, either in the main tank or in a separate reservoir. If the reserve oil supply is in the main tank, the normal outlet must be arranged so the reserve oil supply cannot be drawn out of the tank except when it is necessary to feather the propeller.

Some oil tanks, particularly those that carry a large supply of oil, are designed with a **hopper** that partially isolates a portion of the oil in the tank from the main body of the oil. Oil to the engine is drawn from the bottom of the hopper, and return oil feeds into the top of the hopper. This permits a small portion of the oil in the tank to be circulated through the engine, thus

permitting a rapid warm-up of the engine. The hopper is open to the main body of the oil by means of holes near the bottom, and in some cases the hopper is equipped with a flapper valve that allows oil to flow into the hopper as the oil in the hopper is consumed. The hopper is also termed a **temperature accelerating well.**

Provision must be made to prevent the entrance into the tank itself or into the tank outlet of any foreign object or material which might obstruct the flow of oil through the system. The oil tank outlet must not be enclosed by any screen or guard which would reduce the flow of oil below a safe value at any operating temperature condition. The diameter of the oil outlet must not be less than the diameter of the inlet to the oil pump. That is, the pump must not have greater capacity than the outlet of the tank.

Oil tanks must be constructed with an expansion space of 10 percent of the tank capacity or ½ gal [1.89 L], whichever is greater. It must not be possible to fill the expansion space when refilling the tank. This provision is accomplished by locating the filler neck or opening in such a position that it is below the expansion space in the tank. Reserve oil tanks which have no direct connection to any engine must have an expansion space which is not less than 2 percent of the tank capacity.

Oil tanks designed for use with reciprocating engines must be vented from the top of the expansion space to the crankcase of the engine. The vent opening in the tank must be so located that it cannot be covered by oil under any normal flight condition. The vent must be so designed that condensed water vapor which might freeze and obstruct the line cannot accumulate at any point. Oil tanks for acrobatic aircraft must be designed to prevent hazardous loss of oil during acrobatic maneuvers, including short periods of inverted flight.

If the filler opening for an oil tank is recessed in such a manner that oil may be retained in the recessed area, a drain line must be provided which will drain such retained oil to a point clear of the airplane. The filler cap must provide a tight seal and must be marked with the word "oil" and the capacity of the tank.

The strength of the oil tank must be such that it can withstand a test pressure of 5 psi [34.48 kPa] and can support without failure all vibration, inertia, and fluid loads which will be imposed during operation.

The quantity of oil in the oil tank or in the wet sump of an engine can usually be determined by means of a dipstick. In some cases the dipstick is attached to the filler neck cap. Before any flight, it is standard practice to check visually the quantity of oil.

Plumbing for the Lubrication System

The plumbing for the oil system is essentially the same as that required for fuel systems or hydraulic systems. Where lines are not subject to vibration, they are constructed of aluminum-alloy tubing and connections are made with approved tubing fittings, AN or MS type. In areas near the engine or between the engine and the fire wall where the lines are subject to vibration, synthetic hose of an approved type is used. The hose connections are made with approved hose fittings which are securely attached to the hose ends. Fittings of this type are described in another text of this series.

Hose employed in the engine compartment of an airplane should be of a fire-resistant type to minimize the possibility of hot oil being discharged into the engine area if a fire occurs.

The size of oil lines must be such that they will permit flow of the lubricant in the volume required without restriction. The size for any particular installation is specified by the manufacturer of the engine.

Oil Temperature Regulator

As indicated by its name, the oil temperature regulator is designed to maintain the temperature of the oil for an operating engine at the correct level. Such regulators are also termed **oil coolers** because cooling of the engine oil is one of the principal functions of such units. Oil temperature regulators are manufactured in a number of different designs, but their basic functions remain essentially the same.

One type of oil temperature regulator is illustrated in Fig. 8-12. The outer cylinder of this particular unit

is about 1 in (2.54 cm) larger in diameter than the inner cylinder. This provides an oil passage between the cylinders and enables the oil to bypass the core either when the oil is at the correct operating temperature or when the oil is too cold. When the oil from the engine is too hot for proper engine operation, the oil is routed through the cooling tubes by the viscosity valve. Note that the oil which passes through the core is guided by baffles which force it to flow around these tubes and thus to flow through the length of the core several times. The ends of the tubes are hexagonal in shape and form "honeycombs" at both ends of the unit.

The oil flow through the cooling portion of the oil temperature regulator is controlled by some type of thermostatic valve. This valve may be called a **thermostatic control valve** or simply the **oil cooler valve.** This valve is so designed that the temperature of the oil causes it to open or close, routing the oil for little or no cooling when the oil is cold and for maximum cooling when the oil is hot. If the control valve should become inoperative or otherwise fail, the oil will still flow through or around the cooling portion of the unit.

The installation of this unit is simple. It is placed in position, mounting straps are installed, and then the inlet and outlet fittings are attached. Its removal is accomplished in the reverse order.

Oil Viscosity Valve

The **oil viscosity valve,** illustrated in Fig. 8-13, is generally considered part of the oil temperature-regulator unit and is employed in some oil systems. The viscosity valve consists essentially of an aluminum-alloy housing and a thermostatic control element. The valve is attached to the oil cooler valve. Together, the oil cooler valve and the oil viscosity valve, which form the **oil temperature-regulator unit,** have the twofold duty of maintaining a desired temperature and of keeping the viscosity within required limits by controlling the passage of the oil through the unit.

Through its thermostatic control, the viscosity valve routes the oil through the cooling core of the coil cooler when the oil is hot and causes the oil to bypass the core when the oil is not warm enough for correct engine lubrication.

In Fig. 8-13, when the oil is cold, the valve will be off its seat and oil can flow through the opening on the left. This passage permits the oil to flow from the area around the outside of the cooler; hence, it does not become cooled. As the oil becomes heated, the valve

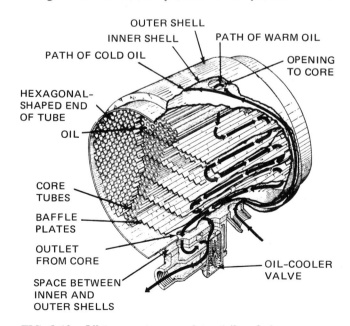

FIG. 8-12 Oil temperature regulator (oil cooler).

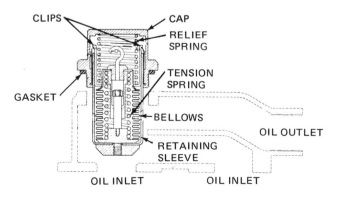

FIG. 8-13 Oil viscosity valve.

closes, thus forcing the oil to flow through the opening on the right, which leads from the radiator section of the cooler. This, of course, exposes the oil to the cooling action of the radiator section.

Oil Pressure-Relief Valves

The pressure of the oil must be great enough to lubricate the engine and its accessories adequately under all operating conditions. If the pressure becomes excessive, the oil system may be damaged and there may be leakage.

The purpose of an **oil pressure-relief valve** is to control and limit the lubricating-oil pressure, to prevent damage to the lubrication system itself, and to ensure that the engine parts are not deprived of adequate lubrication because of a system failure. As noted previously and illustrated in Fig. 8-11, the oil pressure-relief valve is located in the area between the pressure pump and the internal oil system of the engine; it is usually built into the engine. There are several types of oil pressure-relief valves, a few of which are described below.

Single Pressure-Relief Valve. The typical single pressure-relief valve, illustrated in Fig. 8-14, has a spring-loaded plunger, which has a tapered valve at one end, an adjusting screw for varying the spring tension, a locknut to keep the adjusting screw tight, a passage from the pump, a passage to the engine, and a passage to the inlet side of the pump. Normally, the valve is held against its seat by spring tension, but when the pressure from the pump to the engine becomes excessive, the increased pressure pushes the valve off its seat. The oil then flows past the valve and its spring mechanism and is thus bypassed to the inlet side of the oil pump.

Compensating Oil Pressure-Relief Valve and Its Thermostatic Control Valve. The **compensating oil pressure-relief valve,** working with its **thermostatic control valve,** automatically regulates the pressure of the oil in the lubrication system. These units are illustrated in Fig. 8-15, which shows the passage A for high-pressure oil arriving from the pump, passage B to the engine, the low-pressure oil passage C, the route for the return of oil to the pump, the metered (low-pressure) oil chamber, and the two units already mentioned.

The oil pump sends oil under pressure through the strainer to passage A. From passage A, most of the oil flows through passage B to the engine. Referring to the illustration, some of the oil flows through a restriction that reduces its pressure and then circulates it over

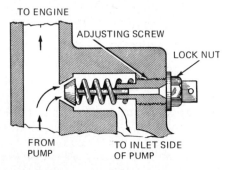

FIG. 8-14 Single pressure-relief valve.

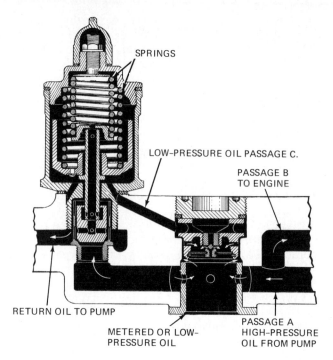

FIG. 8-15 Compensating oil pressure-relief valve.

and around the thermostatic control valve. From this valve, low-pressure oil flows through a hollow shaft to lubricate units other than those lubricated by the major portion of the oil, which leaves through passage B to the engine.

The thermostatic control valve is designed and adjusted to open at a prescribed critical temperature, usually 40°C (104°F). When the temperature is below 40°C, the thermostatic control valve is closed and the oil in passage C, which leads to the main relief valve, is not under pressure. When the temperature of the oil reaches 40°C, the valve opens and allows low-pressure oil to flow through passage C to the main relief valve, where it acts on the compensating piston and overcomes the force of the outer spring in that unit.

When the thermostatic control valve is closed, the pressure of the oil in passage A must be great enough to overcome the tension of two springs before the main relief valve can be forced off its seat, thus permitting excess oil to return to the inlet side of the oil pump. When this happens, the oil in both passage A and passage B must be under very high pressure.

When the thermostatic control valve is open, the pressure in passage A need not be as great as it must be when the thermostatic control valve is closed (in order to allow excess oil to return to the inlet side of the pump). The reason is that the pressure in passage A needs to overcome the tension of only one spring in order that the oil pressure may be restored to its normal operating value.

The usual relief valve of this type can be adjusted for the pressure setting, or "kick-out" pressure, by removing a cap, loosening a locknut, and turning a screw. Generally, turning the pressure adjusting screw to the right, in a clockwise direction, raises the pressure setting required to kick out; and turning the screw to the left, that is, in a counterclockwise direction, lowers the kick-out pressure.

General Design of Relief Valves. Oil pressure-relief valves utilized in modern light airplane engines are comparatively simple in design and construction and usually operate according to the principle illustrated in Fig. 8-16. The relief-valve assembly consists of a plunger and a spring mounted in a passage of the oil pump housing. When oil pressure becomes too high, the pressure moves the plunger against the force of the spring to open a passage, allowing oil to return to the inlet side of the pump.

Oil pressure-relief valves for large reciprocating engines are usually of the compensating type previously described. This type of relief valve ensures adequate lubrication to the engine when the engine is first started by maintaining a high pressure until the oil has warmed sufficiently to flow freely at a lower pressure. The relief-valve setting can usually be adjusted by means of a screw which changes the pressure on the spring or springs controlling the valve. Some of the simpler types of relief valves do not have an adjusting screw; and in these cases, if the relief-valve pressure setting is not correct, it is necessary to change the spring or insert one or more washers behind the spring. The initial, or basic, oil pressure adjustment for an engine is made at the factory or engine overhaul shop.

Strainer-Type Filters

The purpose of any filter is to remove solid particles of foreign matter from the oil before it enters the engine. The **strainer-type oil filter** is simply a tubular screen. Some of these filters are designed so that they will collapse if they become clogged, thus permitting the continuation of normal oil flow. Other screens or filters are designed with relief valves which open if they become clogged (Fig. 8-17).

Disposable Filter Cartridge

Many modern engines for light aircraft incorporate oil systems which utilize external oil filters containing disposable filter cartridges in filter cannisters. An exploded view of such a filter assembly is shown in Fig. 8-18. Note that the filter assembly includes an adapter

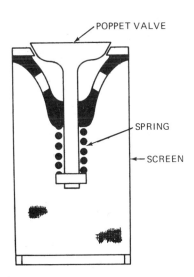

FIG. 8-17   Oil screen with relief (bypass) valve.

by which the unit is mounted on the engine. The adapter includes a receptacle and fittings for the installation of the oil temperature bulb. Also included in the filter adapter is the **thermostatic valve** by which oil is by-passed around the oil cooler until the temperature is at an acceptable level.

In servicing the oil filter, the disposable cartridge is removed and a new element installed. Before the used element is discarded, the outer paper covering is removed and the filter element is opened up and spread out to permit examination for metal particles. The inside of the cannister is also examined for metal particles, which may indicate failure of some part of the engine.

Cuno Oil Filter

The Cuno oil filter has a series of laminated plates, or disks, with one set of disks rotating in the spaces between the other disks. The oil is forced through the spaces between the disks, flowing from the outside of the disks, between the disks and spacers to the inside passage, and thence to the engine. Foreign-matter particles are stopped at the outer diameter of the disks. The minimum size of the particles filtered from the oil is determined by the thickness of the spacers between the disks. The accumulation of matter collected at the outer diameter of the disks is removed by rotating the movable disks, which is accomplished by means of a handle outside the filter case. After long periods of time, the filter case is opened and the sludge removed. At this time, also, the entire filter assembly can be thoroughly inspected and cleaned and the sludge be examined for metal particles.

Inspection of Oil Filter

The oil filter provides an excellent method for discovering internal engine damage. During the inspection of the engine oil filter, the residue on the screens, disks, or disposable filter cartridge and the residue in the filter housing are carefully examined for metal particles. A new engine or a newly overhauled engine will often have a small amount of fine metal particles in the screen, or filter, but this is not considered abnormal. After the engine has been operated for a time and the oil has

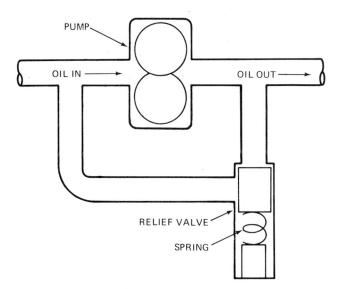

FIG. 8-16   Simplified drawing of oil pressure-relief valve and oil pump.

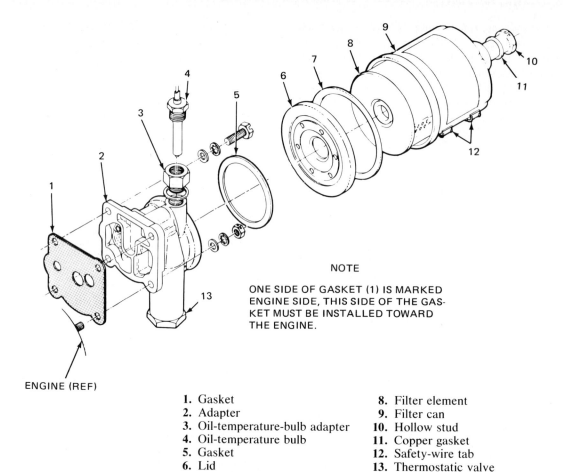

NOTE

ONE SIDE OF GASKET (1) IS MARKED
ENGINE SIDE, THIS SIDE OF THE GAS-
KET MUST BE INSTALLED TOWARD
THE ENGINE.

ENGINE (REF)

1. Gasket	**8.** Filter element
2. Adapter	**9.** Filter can
3. Oil-temperature-bulb adapter	**10.** Hollow stud
4. Oil-temperature bulb	**11.** Copper gasket
5. Gasket	**12.** Safety-wire tab
6. Lid	**13.** Thermostatic valve
7. Gasket	

FIG. 8-18   External oil filter with disposable cartridge.

been changed one or more times, there should not be an appreciable amount of metal particles in the oil screen. If an unusual residue of metal particles is found in the oil screen, the engine should be taken out of service and disassembled to determine the source of the particles. This precaution will often result in preventing a disastrous engine failure in flight.

Oil Filter for Turbojet Engine

The main oil filter for a turbojet engine is illustrated in Fig. 8-19. The oil enters the inlet port, surrounds the filter cartridge, and flows through the cartridge to the inner oil chamber and out to the engine. If the filter becomes clogged, the oil is bypassed through the pressure-relief valve to the discharge port. A differential pressure of 14 to 16 psi [96.53 to 110.32 kPa] is required to unseat the relief valve.

The filter cartridge is composed of a stack of pancake elements capable of filtering out particles larger than 46 micrometers (μm) in size. With the oil temperature at 150°F [65.56°C] and a flow of approximately 15 gal/min, the pressure drop across a clean filter is about 6 psi. The estimated maximum pressure drop across a clogged filter is 23 psi.

A magnetic detector is installed in the side of the filter case to indicate the presence of metal contamination without opening the filter. When the magnetic detector picks up ferrous-metal particles, the center plug will become grounded to the case. If a "hot" test

light is connected between the center terminal of the detector and ground, the light will burn and indicate metal particles on the detector.

Oil Separator

In any air system where the presence of oil or oil mist may exist, it is often necessary to utilize a device called an oil separator. This device is usually placed in the discharge line from a vacuum pump or air pump,

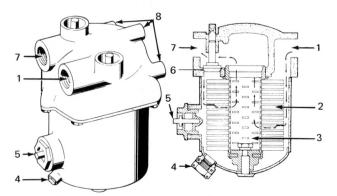

1. Inlet port	**5.** Magnetic chip detector
2. Filter disk	**6.** Pressure-relief valve
3. Inner oil chamber	**7.** Discharge port
4. Drain plug	**8.** Mounting bosses

FIG. 8-19   Main oil filter for a turbojet engine. *(General Electric Co.)*

190

and its function is to remove the oil from the discharge air. The oil separator contains baffle plates which cause the air to swirl around and deposit any oil on the baffles and on the sides of the separator. The oil then drains back to the engine through the oil outlet. The separator must be mounted at an angle of about 20° to the horizontal with the oil drain outlet at the lowest point. By eliminating oil from the air, the oil separator prevents the deterioration of rubber components in the system. This is particularly important in the case of deicer systems where rubber boots on the wings' leading edges are inflated with air from the vacuum pump.

Oil Pressure Gage

An **oil pressure gage** is an essential component of any engine oil system. These gages are usually of the bourdon-tube type and are designed to measure a wide range of pressures, from no pressure up to above the maximum pressure which may be produced in the system. The oil gage line, which is connected to the system near the outlet of the engine pressure pump, is filled with low-viscosity oil in cold weather to obtain a true indication of the oil pressure during the warm-up of the engine. A restricting orifice is placed in the oil gage line to retain the low-viscosity oil and to prevent damage from pressure surges. If high-viscosity oil is used in cold weather, the oil pressure reading will lag behind the actual pressure developed in the system.

Oil Temperature Gage

The temperature probe for the oil temperature gage is located in the oil inlet line or passage between the pressure pump and the engine system. On some engines the temperature probe (sensor) is installed in the oil filter housing. Temperature instruments are usually of the electrical or electronic type. These are described in Chap. 13.

Oil Pressure Pumps

Oil pressure pumps may be of either the gear type or the vane type. A gear-type pump usually consists of two specially designed, close-fitting gears rotating in a case which is accurately machined to provide minimum space between the gear teeth and the case walls. The operation of a typical gear pump is shown in Fig. 8-20. The gear-type pump is utilized in the majority of reciprocating engines.

The capacity of all engine oil pressure pumps is greater than the engine requires, and excess oil is returned to the inlet side of the pump through the pressure-relief valve. This makes it possible for the pump to increase its oil delivery to the engine as the engine wears and clearances become greater.

If an engine pressure pump does not produce oil pressure within 30 s after the engine is started, this is an indication that the pump has lost its prime, probably due to wear. When the side clearance of the gears in the pump becomes too great, oil bypasses around the gears and pressure cannot be developed. In this case the pump must be replaced.

Scavenge Pump

The scavenge pump or pumps for a lubrication system are designed with a greater capacity than the pres-

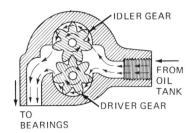

FIG. 8-20 Gear-type oil pump.

sure pump. In a typical engine, the gear-type scavenge pump is driven by the same shaft as the pressure pump, but the depth of the scavenge pump gears is twice that of the pressure pump gears. This gives the scavenge pump twice the capacity of the pressure pump. The reason for a higher capacity scavenge pump is that the oil which flows to the sump in the engine is somewhat foamy and therefore has a much greater volume than the air-free oil which enters the engine via the pressure pump. In order to keep the oil sump drained, the scavenge pump must handle a much greater volume of oil than the pressure pump.

Oil-Dilution System

Fig. 8-21 is a schematic diagram showing how the oil-dilution system is connected between the fuel system and the oil system. In the diagram one can see that a line is connected to the fuel system on the pressure side of the fuel pump. This line leads to the oil-dilution solenoid valve; and from the solenoid valve the line leads to the Y drain, which is in the engine inlet line of the oil system. If the system does not include a Y drain, the oil-dilution line may be connected at some other point in the engine inlet line before the pressure pump inlet. The oil-dilution solenoid valve is connected to a switch in the cockpit so that the pilot can dilute the oil after flight and before shutting down the engine. A cutaway drawing of the solenoid valve is shown in Fig. 8-22.

If an oil-dilution system's control valve becomes defective and leaks or remains open, gasoline will continue to be introduced into the engine oil during operation. This will result in low oil pressure, high oil temperature, foaming of the oil, high fuel consumption, and excessive oil fumes emitted from the engine breather.

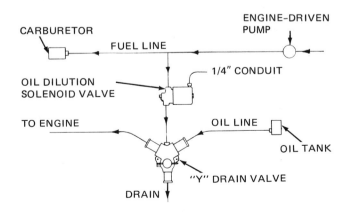

FIG. 8-21 Diagram of an oil-dilution system.

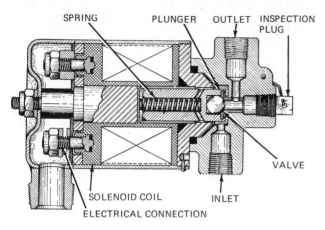

FIG. 8-22   Solenoid valve.

ENGINE DESIGN FEATURES RELATED TO LUBRICATION

We have discussed the design of reciprocating engines and engine parts in general; however, at this point it is important to emphasize certain features directly related to lubrication.

Sludge Chambers

In some engines, the crankshaft is designed with chambers in the hollow connecting-rod journals by which carbon sludge and dirt particles are collected and stored. The chambers may be made by means of metal spools inserted in the hollow crankpins (journals) or by plugs at each end of the hollow journals. At overhaul it is necessary to disassemble or remove the chambers and remove the sludge. Great care must be taken to ensure that the chambers are properly reassembled so that oil passages are not covered or plugged in any way. During the overhaul of crankshafts, all oil passages must be cleaned.

Intercylinder Drains

To provide lubrication for the valve-operating mechanisms in many radial engines, the valve rocker box cavities are interconnected with oil tubes called **intercylinder drains.** The drains ensure adequate lubrication for the valve mechanisms and provide a means whereby the oil can circulate and return to the sump. The drain tubes must be kept clear and free of sludge. If the drain tubes become partially or completely plugged, excess oil will build up in the rocker box area and some of this oil will be drawn into the cylinder through the valve guide during the intake stroke. This will, of course, cause fouling of the spark plugs, particularly in the lower cylinders, and result in the improper lubrication and cooling of the valve mechanism.

Oil Control in Inverted and Radial Engines

Some of the cylinders in a radial engine and all of the cylinders in an inverted engine are located at the bottom of the engine. It is necessary, therefore, to incorporate features to prevent these cylinders from being flooded with oil. This is accomplished by means of long skirts on the cylinders and an effective scavenging system. During the operation of these engines, oil which falls into the lower cylinders is immediately thrown back out into the crankcase. The oil then drains

downward and collects in the crankcase outside the cylinder skirts. From this point the oil drains into the sump, so that there is no buildup in the crankcase. The oil in the crankcase during operation is primarily in the form of a mist or spray. This oil lubricates the cylinder walls, pistons, and piston pins.

Excessive oil is prevented from entering the cylinder heads by means of oil-control rings on the pistons, as explained previously. Some pistons incorporate drain holes under the oil-control rings. Oil from the cylinder walls passes through these drain holes to the inside of the piston and is then thrown out into the crankcase.

TYPICAL LUBRICATION SYSTEMS

Oil System for Wet-Sump Engine

The lubrication system for the Continental IO-470-D engine installed on the Cessna 310 aircraft is shown in Fig. 8-23. Lubricating oil for the engine is stored in the sump, which is attached to the lower side of the engine. Oil is drawn from the sump through the suction oil screen, which is positioned in the bottom of the sump. After passing through the gear-type oil pump, the oil is directed through the oil filter screen and thence along an internal gallery to the forward part of the engine where the oil cooler is located. A bypass check valve is placed in the bypass line around the filter screen to provide for oil flow in case the screen becomes clogged. A nonadjustable pressure-relief valve permits excess pressure to return to the inlet side of the pump.

Oil temperature is controlled by a thermally operated valve which either bypasses the oil around the externally mounted cooler or routes it through the cooler

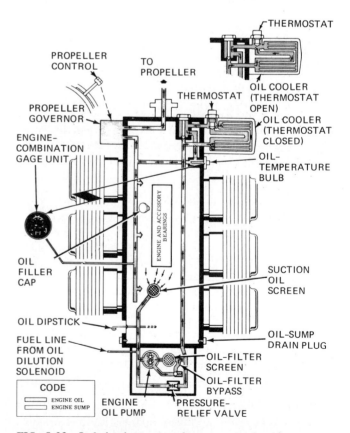

FIG. 8-23   Lubrication system for an opposed engine.

passages. Drilled and cored passages carry the oil from the oil cooler to all parts of the engine which require lubrication. Oil from the system is also routed through the propeller governor to the crankshaft and thence to the propeller for control of pitch and engine rpm.

The oil temperature bulb is located at a point in the system where it senses oil temperature after the oil has passed through the cooler. Thus, the temperature gage indicates the temperature of the oil *before* it passes through the hot sections of the engine.

The oil-pressure indicating system consists of plumbing that attaches to a fitting on the lower left portion of the crankcases between the no. 2 and no. 4 cylinders. The plumbing is routed through the wings, into the cabin, and to the forward side of the instrument panel. Here it connects to a separate engine gage unit for each engine. A restrictor is incorporated in the elbow of the engine fitting to protect the gage from pressure surges and to limit the loss of engine oil in case of a plumbing failure. This restriction also aids in retaining the light oil which may be placed in the gage line for cold-weather operation.

This lubrication system may be equipped with provision for oil dilution. A fuel line is connected from the main fuel strainer case to an oil-dilution solenoid valve mounted on the engine fire wall. From the solenoid valve a fuel line is routed to a fitting on the engine which connects with the suction side of the engine oil pump. When the oil-dilution switch is closed,

fuel flows from the fuel strainer to the inlet side of the oil pump. A total of 4 quarts (qt) [3.79 L] of fuel is required for dilution in this particular engine.

Oil System for Dry-Sump Engine

Typical of lubrication systems for large aircraft equipped with dry-sump reciprocating engines is the system illustrated in Fig. 8-24. This system is installed in the Convair 340 airliner, which is equipped with Pratt & Whitney R-2800 engines. The oil for this system is stored in the tank shown in the upper rear portion of the nacelle. Oil is drawn from the tank through a firewall shutoff valve to the engine inlet, where it enters the engine-driven oil pump, which circulates the oil under pressure through the engine. The oil is then picked up by the scavenge pump and forced through the engine outlet to the cooler and back to the supply tank where it started.

The oil system includes, in the oil return line, a thermostat which controls the cooler flap mechanism, thus maintaining the oil temperature within the required limits. If the oil is cool, the cooler flap is closed; but as the oil temperature increases, the flap opens and allows cooling air to flow through the cooler in sufficient quantity to maintain the desired temperature.

Lubrication System for a Gas-Turbine Engine

Gas-turbine engines have been designed and manufactured in many different configurations; thus, there

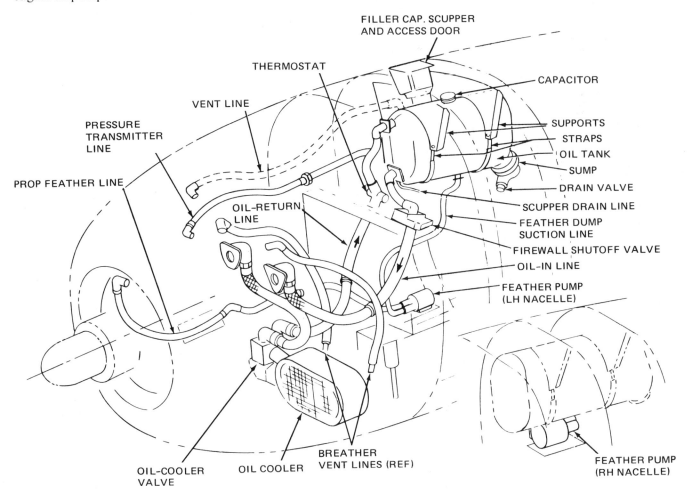

FIG. 8-24 Lubrication system for the Convair 340 airliner.

are correspondingly different designs for the lubrication systems of such engines. Some systems employ fuel-oil oil coolers, others utilize ram air for cooling the oil, and some do not employ oil coolers. The latter systems are referred to as "hot-tank" systems because the oil returning to the oil tank is quite hot. Gas-turbine lubrication systems are usually of the dry-sump type, in that the oil is scavenged from the engine and stored in an oil tank.

A gas-turbine engine lubrication system that may be considered typical is that of the Pratt & Whitney JT8D, which powers the Boeing 727 and 737 aircraft, the Douglas DC-9, the S.N.I.A.S. Super Caravelle, and the Dassault Mercure 2. The system for the JT8D engine is shown in Fig. 8-25.

For readers not familiar with the configuration of turbine engines at this point, reference to the chapters describing turbine engines will be helpful for interpreting the drawing of Fig. 8-25 and the description of the system.

The JT8D lubrication system is of a self-contained, high-pressure design consisting of a pressure system which supplies lubrication to the main engine bearings and to the accessory drives, and a scavenge system by which oil is withdrawn from the bearing compartments and from the accessories and then returned to the oil tank. A breather system connecting the individual bearing compartments and the oil tank completes the lubrication system.

Oil is gravity fed from the main oil tank into the main oil pump *A* in the gearbox. The pressure section of the main oil pump forces oil through the main oil strainer

C located immediately downstream of the pump discharge. The main oil-strainer filter element is a stacked type, a reusable filter element, or a disposable filter element. It is designed to remove all particles in the oil having a diameter greater than 40 μm. If the filter element becomes clogged, the bypass valve *D* will move off its seat and the oil will bypass through the center of the filter.

Proper distribution of the total oil flow to the various locations is maintained by metering orifices and clearances. The main oil pump is regulated by a valve to maintain a specified pressure and flow. This valve is shown at *B* in Fig. 8-25. Note that the valve is located such that when oil pressure becomes too high, the valve will open and return a portion of the oil to the inlet side of the pump. Oil pressure, relative to internal engine breather pressure (tank pressure), and oil flow are essentially constant with changes in altitude and engine speed.

Oil leaves the gearbox and flows to the **fuel-coolant oil cooler,** where a portion of the heat of the oil is transferred to the fuel flowing to the fuel-control unit. If the cooler is blocked, an oil cooler bypass valve *F* opens to permit the continuous flow of oil. Oil leaves the cooler and flows into the oil pressure tubing to the main bearing compartments. The **pressure sense line** maintains a constant oil pressure at the bearing jets, regardless of the pressure drop of the oil at the fuel-oil cooler. The pressure sense line can be seen in the drawing of Fig. 8-25 leading from the outlet of the fuel-oil cooler back to the pressure regulating valve *B*.

Oil for the no. 1 bearing enters the inlet case through

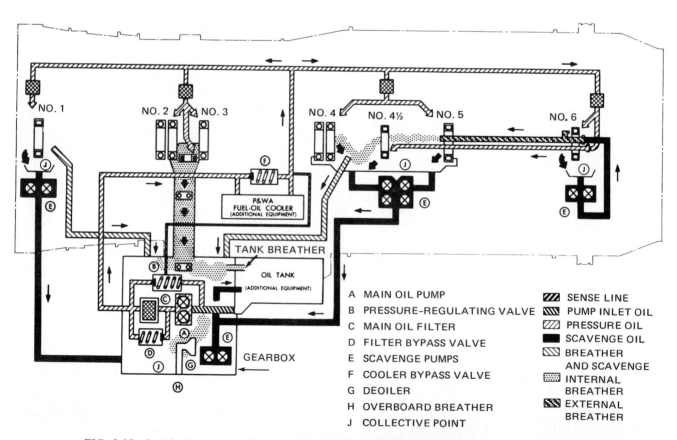

A	MAIN OIL PUMP		▨	SENSE LINE
B	PRESSURE-REGULATING VALVE		▨	PUMP INLET OIL
C	MAIN OIL FILTER		▨	PRESSURE OIL
D	FILTER BYPASS VALVE		■	SCAVENGE OIL
E	SCAVENGE PUMPS		▨	BREATHER AND SCAVENGE
F	COOLER BYPASS VALVE		▨	INTERNAL BREATHER
G	DEOILER		▨	EXTERNAL BREATHER
H	OVERBOARD BREATHER			
J	COLLECTIVE POINT			

FIG. 8-25 Lubrication system for the Pratt & Whitney JT8D turbofan engine. *(Pratt & Whitney)*

a tube in the bottom inlet guide vane. For engines equipped with oil-dampened no. 1 bearings, a transfer tube from the front accessory support leads back into the bearing support to supply oil to a cavity around the bearing's outer race. The remainder of the oil moves up the tube and is then routed through a small strainer in the front accessory drive support into the accessory drive gearshaft. It moves to the outer wall holes in the front hub and in the inner race retaining nut to the front of the no. 1 bearing.

Oil enters the no. 2 and no. 3 bearing compartment through a small strainer and is sprayed onto the bearings through a three-legged oil-nozzle assembly. A **front leg,** or nozzle, directs oil toward the no. 2 bearing, a second directs oil toward the no. 3 bearing, and a third directs oil toward the gearbox driveshaft upper bearing. Oil flows through holes in the rear hub to the inside diameter (ID) of the no. 2 bearing. Flow through the gearbox drive bevel-gear holes carries oil to the ID of the no. 3 bearing.

Pressure oil for the no. 4 and no. 5 bearing locations flows into the engine through a tube at the eight o'clock location at the left side of the fan discharge diffuser's outer duct. It then flows upward around the diffuser case to the ten o'clock position and inward through the inner passage of the dual concentric pressure and breather tubing to the no. 4 bearing support. Here it is directed rearward through an elbow and flows into the multipassage no. 4 bearing oil-nozzle assembly.

The no. 4 bearing oil-nozzle assembly has an inlet passage, outlet holes at the bottom directing oil toward the no. 4 bearing, and an outlet passage toward the rear. An oil strainer is positioned inside the inlet passage. The outlet passage toward the rear accommodates the long oil tube of the no. 5 bearing oil-nozzle assembly. Oil passes rearward through this tube and is then directed through the no. 5 bearing oil-nozzle assembly. From the oil nozzle it passes under the bearing race and through the seal plate to the no. 5 bearing compartment.

Oil flows to the no. 6 bearing area through a tube located in the upper turbine exhaust strut and down into the scavenge pump housing of the no. 6 bearing. In the scavenge pump housing it passes through a small strainer and then down into the outer passage of the no. 6 bearing oil-nozzle assembly.

For engines having *oil-damped* no. 6 bearings, oil flows from the oil scavenge pump through a tube to the no. 6 bearing housing. The oil is then distributed to a cavity formed between the housing and the bearing's outer race. Seal rings around the bearing's outer race help contain oil in the cavity.

The oil flows forward in the oil nozzle's outer passage and divides into two streams. One stream flows outward through small holes on the outside diameter (OD) of the nozzle's outer passage tube to lubricate the no. 6 bearing area. From the same nozzle's outer passage tube, the other stream continues forward through holes on the nozzle's outer front face and into the outer passage of the turbine bearings' oil pressure and scavenge tubes assembly **(trumpet)** inside the front-compressor drive-turbine rotor. The oil continues forward through the single (short) pressure tube in the oil trumpet to the no. $4\frac{1}{2}$ bearing area.

As previously mentioned, the outward stream for the no. 6 bearing area flows into the rear hub of the front-compressor drive-turbine rotor. Through two sets of holes in the hub, it flows to the no. 6 bearing seals and to the no. 6 bearing's inner race.

The pressure oil in the oil trumpet flows forward and out through an oil baffle and then through holes in the long turbine shaft to cool the no. $4\frac{1}{2}$ bearing seal spacers and to lubricate the no. $4\frac{1}{2}$ bearing.

The scavenge oil system of the engine includes four gear-type pumps E with five pump stages. The pumps scavenge the main bearing compartments and deliver the scavenged oil to the engine oil tank. It will be noted that the scavenge pump for the no. 4 and no. 5 bearing area has two pump stages.

The single-stage scavenge pump for the no. 1 bearing compartment is located in the cavity of the front accessory drive housing. The pump is driven by the front accessory drive gearshaft located in the front hub of the front-compressor rotor. The pump picks up the oil, sends it outward through a passage in the housing, and then sends it down a tube located in the bottom vane of the inlet case.

The second scavenge pump is located in the scavenge stage of the main oil pump assembly in the accessory drive gearbox. Scavenge oil from the gearbox driveshaft bearings and from the no. 2 and no. 3 bearings is pumped from its collection point in the gearbox. It will be noted that the scavenge oil from the no. 2 and no. 3 bearings drains down the outside of the accessory driveshaft to the gearbox. One bevel gear drives both the pressure and scavenge stages of the pump.

The third pump, with two stages driven by the same gear, is the no. 4 and no. 5 bearings' oil scavenge pump assembly located inside the diffuser case. Together, the two stages of the pump scavenge oil from the no. 4 and no. 5 bearing areas. In addition, scavenge oil from the no. 6 bearing areas, after flowing forward through the two long scavenge tubes in the oil trumpet, flows into this compartment. A tube in the combustion-chamber heat shield allows passage of the oil forward from the no. 5 bearing cavity.

The discharge from the pump is carried forward into the scavenge adapter at just below the nine o'clock position in the no. 4 bearing support and then outboard. It flows through the inner tube of the dual concentric tubing to the outside of the diffuser case, then downward to the eight o'clock position where it is routed through the fairing to the outside of the diffuser's outer duct.

The fourth scavenge pump is located in the no. 6 bearing's scavenge pump housing where it is driven by a gear-shaft bolted to the rear of the turbine rotor's fourth-stage rear hub. It scavenges oil from the no. 6 bearing compartment and pumps it upward into the inner passage of the no. 6 bearing oil-nozzle assembly.

The oil flows forward in the oil nozzle's inner passage and is discharged through the center hole in the front of the nozzle. It passes forward into the inner passage of the oil trumpet and continues forward through the two long scavenge tubes, as previously mentioned. At the front of the trumpet, the oil flows outward through holes in the front-compressor drive turbine shaft (inner shaft) and in the front of the no. $4\frac{1}{2}$ bearing's inner race retaining nut. The oil is then spun outward through holes in the rear-compressor drive-turbine shaft (outer shaft) and into the no. 4 bearing cavity.

The return oil passed forward by the two rearmost pumps, as well as that from the front oil suction pump, is directed into the gearbox cavity. From here the oil is pumped by the scavenge stage of the pump to the oil tank. Within the tank the oil passes through a de-areator where the major part of the entrapped air is removed.

To ensure proper oil flow and to maintain satisfactory scavenge pump performance during operation, the pressure in the bearing cavities is controlled by the **breather system.** The atmosphere of the no. 2 and no. 3 bearing cavity vents into the accessory gearbox. Breather tubes in the compressor inlet case and diffuser case discharge through external tubing into the accessory drive gearbox. Breather air from the no. 6 bearing compartment comes forward through the oil pressure and scavenge tubes assembly (oil trumpet), along with the scavenge oil from that compartment, to the diffuser-case cavity.

In the gearbox, vapor-laden atmosphere passes through rotary breather impellers (mounted on the starter drive gear shaft) where the oil is removed. The relatively oil-free air reaching the center of the gearshaft is conducted overboard.

● INSPECTION AND MAINTENANCE

Many of the important points to be observed in the inspection and maintenance of an aircraft engine lubrication system have been explained in the preceding pages. It is also obvious that before an airplane is serviced, the handbooks and manuals issued by the manufacturers of the equipment for that particular make, model, and type of airplane should be consulted. The following general instructions apply broadly to all airplane lubrication systems.

Filling the Oil Tank

Do not try to fill the oil-supply tank to the top. There must be an air space left for the expansion of oil, and the tank is usually constructed so the expansion space cannot be filled. Make sure that the correct type and grade of oil is placed in the tank. Note the markings on the filler cap or consult the operator's manual. The word "oil," the capacity of the tank, the type of oil, and the grade of oil should be marked on or near the filler cap. Care must be taken to avoid mixing synthetic oil with petroleum-base oils, because the oils are not compatible, and damage to gaskets and seals may occur.

Inspection of Tanks

The oil-supply tank is inspected periodically for the general condition of the attaching straps or other devices used to secure the tank in place, the condition of the seams and walls of the tank, signs of leakage, and the security of the attached plumbing. Self-sealing oil tanks are inspected and repaired in the same manner as self-sealing fuel tanks.

To remove an oil tank for replacement or repair, the oil is first drained from the lubrication system by means of the Y drain or other main drain fitting. In some cases a special drain plug is located in the bottom of the oil tank to drain oil from the propeller feathering reserve space. The vent lines, oil inlet and outlet lines, scupper overflow tube, and all other attached plumbing must be removed. The attachment bolts, clamps, straps, and any other attaching fittings are removed or disconnected. When the tank is completely clear of all attachments, it is lifted from the airplane. The installation of an oil tank can be considered the reverse of removal. After all checks have been made and the oil tank mounting has been inspected, the oil tank is lifted into place and secured with the tank straps or other means of attachment. Lines and fittings are then connected as in the original installation. Fittings are torqued to the correct values by means of a special torque wrench.

Inspection of Plumbing

The inspection of the plumbing for an oil system is similar to the inspection of any other plumbing. The tubing, hose, tubing fittings, hose fittings, hose clamps, and all other components of the system are inspected for cracks, holes, dents, bulges, and other signs of damage that might restrict the flow or cause a leak. All lines are inspected to see that they are properly supported and are not rubbing against a structure. Fittings should be checked for signs of improper installation, overtorquing, excessive tension, or other conditions which may lead to failure.

Inspection of Screens and Filters

When screen-type filters are used, they should be removed, cleaned, and inspected for breaks whenever the lubrication system is given a major inspection. At this time, the residue in the screen and in the screen housing should be checked for metal particles or chips which would indicate internal failure in the engine.

Special types of filters, such as the stacked-disk type, should be serviced according to the manufacturer's instructions. Such filters are so constructed that improper treatment may render the filter unserviceable.

Lubricant Service for Gas-Turbine Engines

In the past it has generally been the practice to drain and replace the lubricating oil in both reciprocating and gas-turbine engines at specified intervals. At these times the oil filters or screens were examined for residues of metal to detect incipient failures in the engine. But experience and extensive investigations have shown that the periodic changes of oil are not necessary for gas-turbine engines using synthetic lubricants. Airline engineers have cooperated with oil companies and engine manufacturers in studies to determine the most effective and economical procedures with respect to the use of lubricants. The result is that engine oil is seldom if ever changed between engine overhaul periods.

In order to ensure continued effectiveness of the engine lubricant, two procedures are employed; filtering of the oil and oil analysis.

A typical procedure employed with an engine such as the Pratt & Whitney JT8D turbofan engine is the **filtering** of the oil every 250 h of operation. This is accomplished by removing the main oil screen (or filter) and installing a filter adapter. The filter adapter provides inlet and outlet connections by which a 15-μm filter is connected externally to the engine. When the filter has been connected, the engine is operated at medium speed for about 5 min. This causes all the engine oil to be passed through the filter, thereby removing even the very smallest particles of suspended material.

This filtering method provides for thorough internal engine cleaning because the hot circulating oil flushes the inside of the engine while the engine is running. If the oil were drained and filtered, pockets or pools of unfiltered oil would remain in the engine along with residues which would normally settle in such pockets or pools.

After the filtering operation is completed, the 40-μm engine filter, having been examined for residues and cleaned, is reinstalled in the engine. Residues from the special 15-μm filter and from the engine oil filter are examined for metal particles. The metals are identified by means of a magnet and by various chemicals to determine their type in order to detect the possibility of failure in the engine. Methods for identification of metal particles are described in another section of this text.

To determine whether the lubricant in a turbine engine has the required characteristics and to discover the possibility of incipient failure in the engine, it is common practice to perform an **oil analysis** of an engine oil sample at intervals determined by service experience. The process includes a spectrographic analysis of the oil and a test for acidity. The spectrographic analysis reveals the type and amount of metals suspended in the oil, and this information is used to detect the possible failure of bearings or other parts in the engine. When the analysis shows that the quantity of one or more metals in the oil is excessive, the cause is determined and the part (or parts) involved is replaced as necessary. Many potential engine failures have been prevented in this manner. The oil analysis procedure may be compared with the analysis of body fluids in the medical profession to diagnose physical illnesses.

When a new model of engine is placed in service, oil analysis is usually performed more often than for an engine which has been proved by many hours of service experience. Oil from the newer model engine may be analyzed every 100 h of operation, whereas the oil from an engine with many hours of service experience may require analysis no more often than every 200 h of operation.

The filtering and analysis of gas-turbine lubricants makes it unnecessary to change the oil between engine overhaul periods. It is only necessary to add oil from time to time as it is used by the engine. The oil consumption for very large gas-turbine engines, including turbofan types, is usually less than 3 lb/h [1.36 kg/h].

Care in Handling Synthetic Lubricants

It must be emphasized that the handling of synthetic lubricants requires precautions not needed for petroleum lubricants. Synthetic lubricants have a high solvent characteristic which causes them to penetrate and dissolve paints, enamels, and other materials. In addition, when synthetic oils are permitted to touch or remain on the skin, physical injury can result. It is therefore essential that the technician handling synthetic lubricants use every care to ensure that the lubricants are not spilled or allowed to be in contact with the skin. If a synthetic lubricant is spilled, it should be cleaned up immediately by wiping up, washing, or handling with a suitable cleaning agent. Safety precautions established by the aircraft operator should be observed carefully.

● REVIEW QUESTIONS

1. Describe a lubricant.
2. Why is a mineral lubricant more satisfactory for aircraft engines than either a vegetable or an animal lubricant?
3. Why are fluid lubricants used in aircraft engines?
4. What are the principal properties of a lubricant?
5. What are the sources of mineral lubricants?
6. What is the significance of API Gravity?
7. Explain the importance of flash point.
8. Define *viscosity*.
9. Why is the viscosity of an engine oil important?
10. What is the viscosity index of an engine oil?
11. Compare viscosity designations for the commercial aviation number, the commercial SAE number, and the AN specification number.
12. Why is an engine oil with a comparatively high viscosity required for aircraft engines?
13. What is the principal determining factor in selecting a grade of lubricating oil to be used in an aircraft engine?
14. For what purpose is the *centistoke* unit employed?
15. Under what conditions would it be advisable to select a lubricating oil with a very low pour point?
16. What is meant by *demulsibility?*
17. Give characteristics which should be possessed by a good engine lubricating oil.
18. What are the principal functions of the lubricating oil in an engine?
19. How does an engine lubricating oil reduce the possibility of corrosion in an engine?
20. Why was it necessary to develop synthetic lubricants for gas-turbine engines?
21. What types of synthetic lubricants are most commonly used for gas-turbine engines?
22. What would happen if a petroleum lubricant were used in a gas-turbine engine that required a synthetic lubricant?
23. Of what materials are aviation greases made?
24. What are the principal requirements for aircraft greases?
25. What factors are considered in selecting a grease for a particular application?
26. When would you use a corrosion-preventive oil?
27. Describe three types of friction.
28. What is meant by *pressure* lubrication?
29. Name the principal components of a dry-sump engine lubrication system.
30. By what means is lubricating oil distributed to the various parts of an engine?
31. How is lubricating oil for the overhead valve com-

ponents in an opposed engine distributed to these parts?

32. How are the bearings of a gas-turbine engine lubricated?

33. In a basic dry-sump lubrication system, how are the principal units arranged in sequence?

34. Why are check values sometimes installed in the engine oil-in line?

35. What is the required capacity of an engine oil system?

36. What is the purpose of a hopper in an engine oil tank?

37. Describe the principal features of an engine oil tank.

38. What internal pressure should an engine oil tank be able to withstand?

39. Compare the plumbing for a lubricating system with that for a fuel system.

40. Describe the expansion space required for an oil tank.

41. What prevents oil from filling the expansion space of an oil tank at the time the tank is filled?

42. How is an engine oil tank vented to prevent pressure buildup?

43. Explain the size requirement for the oil line leading from the oil tank to the engine pressure pump.

44. Describe the operation of an oil temperature regulator.

45. What type of device enables the oil temperature regulator to control oil temperature?

46. What would happen if the thermostatic control valve for an oil temperature regulator should become inoperative?

47. Describe the operation of a viscosity valve.

48. What is the function of an oil-system relief valve?

49. At what point in an oil system is the relief valve located?

50. Describe the operation of a compensating oil pressure-relief valve.

51. Describe two different methods by which relief valves may be adjusted.

52. Why is the capacity of an oil pressure pump greater than the engine requires?

53. What happens to the excess oil supplied by the pressure pump?

54. Describe four types of oil filters.

55. What determines the minimum size of particles removed from the oil by a Cuno filter?

56. How is a disk-type filter such as the Cuno serviced?

57. What inspection should be made with respect to the engine at the time an oil filter is serviced?

58. What design feature is usually incorporated to prevent a clogged filter from restricting oil flow?

59. Describe the stacked-disk oil filter for a gas-turbine engine.

60. What is the purpose of an oil separator?

61. From what point in an engine lubrication system is the oil pressure taken?

62. Why is a low-viscosity oil placed in the oil gage line?

63. Where is the oil temperature sensor placed in an engine lubrication system?

64. What will occur when the side clearance of the gears in a gear-type oil pressure pump becomes too great?

65. Why is the capacity of the scavenge pump in a dry-sump oil system greater than the capacity of the pressure pump?

66. Describe the operation of an oil-dilution system.

67. What is the purpose of oil dilution?

68. After the engine is operating again, what happens to the gasoline that was mixed with the oil during oil dilution?

69. What will occur if the oil-dilution solenoid valve leaks or remains open?

70. Describe the purpose and location of the sludge chambers in a crankshaft.

71. What precaution must be observed in assembling sludge chambers during engine overhaul?

72. What would happen if the intercylinder oil drains in a radial engine became clogged?

73. What design features prevent the lubricating oil from flooding the cylinders in an inverted or radial engine?

74. Describe the principal features of the oil system for the Teledyne Continental IO-470-D engine.

75. Name the principal components for the dry-sump oil system of the R-2800 engine in a Convair 340 airplane.

76. What is meant by a hot-tank oil system on a gas-turbine engine?

77. How many scavenge and pressure pump units are used in the Pratt & Whitney JT8D engine lubricating system?

78. What type of oil cooler is used for the JT8D engine?

79. What function does the oil cooler for the JT8D engine perform other than cooling the oil?

80. What are the important points to remember when filling an oil tank?

81. Why is it necessary for an engine oil tank to be vented?

82. Where may the technician find the correct type and grade specifications required for the oil to fill a particular oil tank?

83. Why should a reciprocating engine be warmed up before one attempts to fly the airplane?

84. Describe the inspection of an oil system.

85. What should be done if synthetic lubricant is spilled during the servicing of an aircraft?

86. What safety precautions should be observed with respect to handling synthetic lubricants?

⑨ PRINCIPLES OF ELECTRICITY

The maintenance, service, and troubleshooting of aircraft engines requires a thorough understanding of ignition systems on the part of the maintenance technician. The service and maintenance of ignition systems, generator systems, starters, and other electrical devices that are or may be associated with engines requires knowledge of the principles of electricity and magnetism. It is not the purpose of this chapter to provide a complete course in electrical theory and practice; however, the basic principles are examined to the extent necessary for the powerplant technician to perform all the normal duties with respect to ignition systems and the other electrical systems related to the powerplant. A complete coverage of electrical and electronic theory and practice is provided in the associated text, *Aircraft Electricity and Electronics*.

● THE ELECTRIC CURRENT

Principles

Many persons believe that the electric current is a mysterious force which cannot be comprehended by the ordinary individual and that anyone who does understand the nature of electricity and the operation of electrical equipment must be a special type of genius. This idea, of course, has no foundation in fact. Anyone of reasonable intelligence can with deligence learn to understand electricity in a comparatively short time.

The first requirement for an understanding of electricity is to know what electricity consists of and where it originates. We must therefore consider the nature of matter itself. From the point of view presented here, **matter** is anything which has substance and occupies space. Air, liquids, metals, wood, plastics, and gases all represent various forms of matter.

At the present time, matter in its many forms is known to consist of more than 100 elements. An element is a basic substance, unique in its nature and character and unlike any other element. The element is composed of **atoms,** which are the smallest particles of a substance that can exist without changing the nature of the substance.

Some substances are **compounds,** which are made up of a chemical combination of elements. Componds are composed of **molecules,** the molecule being a chemical combination of two or more atoms. Water is a compound because it is a chemical combination of the elements hydrogen and oxygen. Other componds are sugar, salt, sulfuric acid, cellulose, etc. Many of the common substances with which we are familiar are composed of mixtures of elements or mixtures of compounds.

The basic component of electricity is the **electron,** which is found in the atom. Every atom consists of a **nucleus** with one or more orbiting electrons. The atom of hydrogen has a single particle as a nucleus and a single electron orbiting this particle. This is the simplest atom and may be represented by the diagram of Fig. 9-1. The nucleus of the hydrogen atom, called a **proton,** carries a **positive** electric charge, and the electron carries a **negative** electric charge. The positive charge of the proton is equal and opposite to the negative charge of the electron. Most of the elements have a comparatively large number of protons and electrons, and some have electrons which are easily dislodged from the outer orbit and shifted to other atoms.

It is sufficient to state here that the electrons around the nucleus of an atom are arranged in levels called "shells," as indicated in Fig. 9-2, and that the electrons which make up the electric current come from the outer shells of the atoms. These electrons which move from one atom to another are called **free electrons.** A substance which contains many free electrons is a good **conductor** of electric current, and a substance with very few free electrons is an **insulator;** that is, it will not carry an electric current easily. Most metals have many free electrons and are therefore good conductors. The best conductors are silver and copper. Aluminum is also a very good conductor and is often used in place of copper because the weight of aluminum is about one-third that of copper.

Electric Charges

Before we pursue the study of electric current, we must know something of electric charges. An electric charge may be negative or it may be positive. Since the basic component of electricity is the electron and the electron carries a negative charge, a body which possesses a **negative charge** has an excess of electrons. This means that there are more electrons than there are protons in the body. If the body has a deficiency of electrons, it carries a **positive charge.** When a negatively charged body is connected to a positively charged body with a conductor, the excess electrons of the negative body will flow to the positive body until the charges are equalized. We find, therefore, that the flow of an electric current is normally from negative to positive.

Any person who has ridden in an automobile many times will have experienced an electric shock as the result of electric charges. These charges are developed when a person's clothing slides along the plastic covering of the seat. If the person touches the metal handle of the car door after the charge is developed, a small

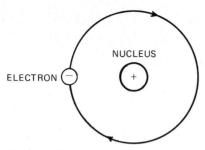

FIG. 9-1   Diagram of a hydrogen atom.

spark will jump between the door handle and the hand or other part of the body which touches the door handle.

Any two dissimilar dry substances when rubbed together will be likely to generate electric charges, one substance becoming negative and the other becoming positive. If a piece of fur is used to rub a glass rod, the fur will acquire a positive charge and the glass rod will have a negative charge.

A simple experiment will serve to demonstrate the effects of small electric charges. With some dry tissue paper, rub a fountain pen or other plastic object. Now hold the pen close to small bits of the tissue paper. The paper will be attracted to the pen and will adhere to it. The paper is attracted to the pen because the pen and paper have unlike charges.

If two objects have like charges, they will repel each other. This is demonstrated when a person combs dry hair with a plastic comb. After vigorous combing, the hairs repel one another and stand out from the head. The repulsion of like charges can also be demonstrated by using two small pith balls suspended together, as shown in Fig. 9-3. If a glass rod is charged by rubbing it with fur and then the glass rod is touched simultaneously to both balls, the balls will spring apart.

Electric charges are generated by friction because one substance gives up electrons more easily than the other. When the two are rubbed together, electrons from one substance move to the other substance, thus leaving one negatively charged and the other positively charged.

The **static electric charges** described above often create problems in the operation of aircraft. They cause radio interference, they may start fires when sparks occur in a flammable atmosphere, and they build up on aircraft in flight to the extent that a very strong spark discharge could take place when the aircraft lands.

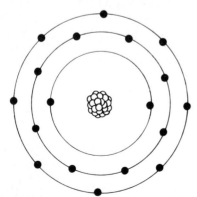

FIG. 9-2   Arrangement of electrons around the nucleus of an atom.

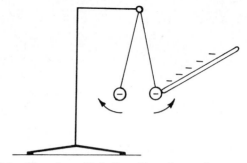

FIG. 9-3   Charging pith balls with a glass rod.

The charge on an airplane is usually dissipated by means of static discharge devices and static conducting tires, which are treated to make them reasonably good conductors of electricity.

When a static charge exists on a body, an **electrostatic field** exists around the body. This field consists of invisible lines of force which have a definite direction. When two bodies having unlike charges are brought near each other, the electrostatic field is comparatively strong. If the charges are sufficiently great, they will cause a spark discharge between the two bodies. Such a discharge can be observed in the familiar flash of lightening, when the charge on a cloud becomes so great that it discharges to the ground or to another cloud. Actually, the flash of lightening may jump from the cloud to the earth or from the earth to the cloud, depending on the nature of the charges.

Voltage, Amperage, and Resistance

When we speak of the electricity used in an electric circuit, we must be able to express certain values with respect to the electricity if we are to know what it will do in the circuit. The first value we must consider is the force which causes the electricity to flow in the circuit. This force is measured in **volts** (V), and it may be considered **electrical pressure.** It is called **electromotive force** (emf). Most of us are familiar with the effects of pressure in a water system, and we know that it causes water to flow out of a faucet or hose nozzle when the valve controlling it is opened. If the pressure is high, there is a comparatively great flow of water, and if the pressure is low, the flow is small. In a given electrical circuit, an increase of voltage (electrical pressure) will cause an increase in the flow of electric current.

The electric current is a movement of electrons through a conductor, and it may be compared with a flow of water through a pipe. There is a difference, however, which must be well understood. An electric circuit must normally form a complete loop or path. In effect, we may say that the electricity must start at a point and return to that point. An electric current will not flow out the end of a conductor in the same manner that water fill flow out of a hose. In an electric circuit, the electricity must start at the generator or battery and then must have a complete path back to the generator or battery.

The flow of electricity through a conductor is measured in **amperes** (A). The ampere represents a flow of one **coulomb** of electricity past a certain point in a circuit in one second. The **coulomb** (C) is said to be equal to approximately 6.24×10^{18} electrons. Observe carefully that the ampere is a **rate of flow** and not a quantity. If we consider the flow of a liquid, we may say that it is flowing at a rate of so many **gallons per second** (gal/s). The gallon is the *quantity* and the gallons per second is the *rate of flow*. In like manner, the coulomb is the *quantity* and the ampere is the *rate of flow*.

Another factor which is most important in the operation of electric circuits is **resistance,** because resistance is found in every circuit . In simple terms, resistance may be called "electrical friction" and compared with mechanical friction. Resistance may be defined as the property of a substance which opposes the flow of an electric current. The unit of resistance is the **ohm** (Ω).

The relation between voltage, amperage, and resistance is explained by **Ohm's law.** Ohm's law may be stated in several ways; however, the following is considered suitable for our purpose: *The current flow in a given circuit is directly proportional to the voltage and inversely proportional to the resistance.* Another statement which may be used to explain Ohm's law is: *One volt of electromotive force will cause one ampere to flow through a circuit having one ohm of resistance.*

From Ohm's law we see that current flow will increase as voltage increases in a given circuit and will decrease as resistance increases. Any one of the values in a circuit can be determined through the use of the formula for Ohm's law when the other two values are known. The formula is

$$I = \frac{E}{R} \qquad R = \frac{E}{I} \qquad E = IR$$

where I = current, A
E = electromotive force, V
R = resistance, Ω

Each of the foregoing equations is a different arrangement of the same formula.

The relative effects of voltage, amperage, and resistance can be understood by considering the simple water system shown in Fig. 9-4. The pump and pressure regulator represent a generator and voltage regulator. The valve represents a variable resistance. If we increase the pressure (voltage) by adjusting the pressure regulator, the rate of flow (amperage) will increase, provided that the valve opening is not changed. If we decrease the opening of the valve (increase resistance) while the pressure remains constant, the rate

of flow will decrease. Thus we see that we can change the amperage in an electric circuit by changing either the voltage or resistance. An increase in voltage increases the amperage, while an increase in resistance decreases the amperage.

A thorough understanding of Ohm's law is important to the aviation maintenance technician because it is often necessary to install electric circuits and components or to replace portions of electric circuits. If the voltage used for a given circuit and the current flow which is likely to exist in the circuit under operating conditions are known, the technician can easily determine the correct size of wire and values of circuit breakers to be used in the circuit.

Electric Power

Power is the rate of doing work. As previously explained, 1 hp is equal to 33,000 ft·lb/min or 550 ft·lb/s. In electricity, the unit of power is the **watt** (W), and it is equal to about 0.00134 hp. We may also state that 746 W is equal to approximately 1 hp. In some instances, the capital letter P may be used in place of W to indicate power.

Using the electrical units already defined, we can determine the power consumed in a circuit with the following formula:

$$W = IE$$

where W = power, W
I = current, A
E = electromotive force, V

If a current of 20 A is flowing in a circuit in which the voltage is 110, the power being consumed in the circuit is 2200 W. To obtain horsepower we divide 2200 by 746 to obtain about 2.95 hp.

Electric Circuits

A simple electric circuit, shown in Fig. 9-5, consists of a source of electric power, a load, and conductors to carry the current from the power source to the load and back to the power source. The circle at the left represents a generator, and the circle at the right represents a motor, which is the load. The arrows indicate the direction of the current through the conductors. Observe that one conductor carries the current to the load and another conductor carries the current from the load back to the generator.

Another simple electric circuit is shown in Fig. 9-6. In this circuit the load is an electric lamp and the circuit is controlled by means of a switch. When the switch is closed, the current can flow and the lamp will op-

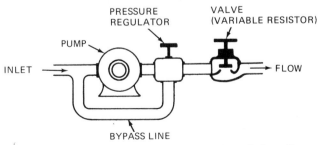

FIG. 9-4 Water analogy to demonstrate the relative effects of voltage, amperage, and resistance.

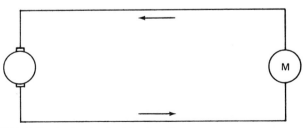

FIG. 9-5 A simple electric circuit.

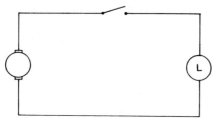

FIG. 9-6   Electric circuit with a switch.

erate. When the switch is off (open), the circuit is "open" and no current can flow.

On an airplane it is common practice to utilize "grounded" circuits in order to save wire and reduce weight. In a grounded circuit the metal structure of the airplane is used as one of the conductors. This circuit, shown in Fig. 9-7, is also called a single-wire circuit because a single wire is used to carry the current in one direction and a switch in the single wire is used for control. A system made up of single-wire circuits is called a **single-wire** system.

There are several different types of circuits, named according to the arrangement of the components in the circuits. The first of these circuits which we shall consider is the **series circuit**. In a series circuit the units are arranged so that the current flowing through one unit must flow through all the other units in the circuit. The simple circuits previously described may be considered as series circuits because all the current must flow through each of the units in the circuit. In Fig. 9-6 the current flows from the generator, through the switch, through the lamp, and back to the generator. In Fig. 9-8 is a series circuit having four lamps in series. All the current in the circuit must flow through each of the lamps. If one lamp should burn out, the current flow would stop and all the lamps would go out. This type of circuit is familiar to most persons because Christmas-tree lights are often connected in series.

A **parallel circuit** is commonly used for lighting and power purposes and is arranged so that each load unit can operate independently of the others. Figure 9-9 illustrates the circuit arrangement for a parallel circuit. In this diagram it can be seen that each of the lamps is connected separately between the power leads and that one lamp is not dependent on another. Therefore, if one lamp burns out, the others will continue to operate. If the switch is open, all the lights will go out because the path to the power source is broken. In this circuit a battery is used as a power source. The battery symbol is shown at the left of the figure.

Still another combination of circuitry is the **series-parallel** circuit. In this arrangement, some of the load units are connected in series and others are connected

in parallel. In this circuit, shown in Fig. 9-10, it will be observed that three of the lamps are in series with one another but that the three lamps as a unit are in parallel with the two other lamps.

Solution of Circuit Values

The technician should be able to determine the values in standard circuits through the use of Ohm's law. In a circuit such as that shown in Fig. 9-11, if the generator voltage and the resistance of the lamp are known, it is easy to compute the current flow in the circuit. Assume that the voltage of the generator is 28 and that the resistance of the lamp is 7Ω. The amperage (current flow) can then be determined as follows:

$$I = \frac{E}{R} = \frac{28}{7} = 4A$$

When several different units are connected in series or parallel or both, the total resistance of the circuits can be determined by the use of standard formulas. When resistances (loads) are connected in series, the total resistance is the sum of the resistances. This is shown by the formula

$$R_t = R_1 + R_2 + R_3 + \ldots$$

In the circuit of Fig. 9-12, the total resistance is the sum of the resistances in the circuit. Since the values for the resistances are given, the formula will appear thus:

$$R_t = 2 + 5 + 3 + 6 = 16 \ \Omega$$

The symbol for a resistance (resistor) in a circuit is the zigzag line shown in the diagrams.

In a parallel circuit the total resistance of a group of resistances is equal to *the reciprocal of the sum of the reciprocals of the resistances*. The **reciprocal** of a number is that number divided into the number 1. For example, the reciprocal of 2 is $\frac{1}{2}$ and the reciprocal of 50 is $\frac{1}{50}$. When a number is multiplied by its reciprocal, the product is 1.

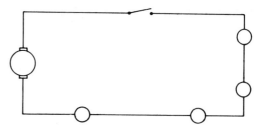

FIG. 9-8   Series circuit.

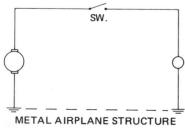

FIG. 9-7   Single-wire circuit.

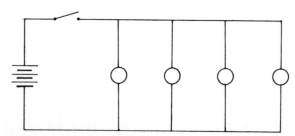

FIG. 9-9   Parallel circuit.

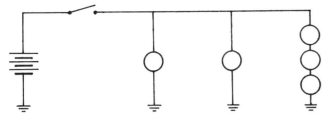

FIG. 9-10   Series-parallel circuit.

The formula for resistances connected in parallel is

$$\frac{1}{R_t} = \frac{1}{R_1} + \frac{1}{R_2} + \frac{1}{R_3} + \cdots$$

or

$$R_t = \frac{1}{1/R_1 + 1/R_2 + 1/R_3 + \cdots}$$

The correctness of the foregoing formula can be easily demonstrated by using a typical example. In the circuit of Fig. 9-13, if $R_1 = 6\ \Omega$, $R_2 = 12\ \Omega$, and $R_3 = 8\ \Omega$, with a battery voltage of 24, by Ohm's law we can determine that the current through R_1 is 4 A, through R_2 it is 2 A, and through R_3 it is 3 A. Then the total current flowing from the battery is $4 + 2 + 3 = 9$ A. Again, by Ohm's law, since $E = 24$ and $I = 9$, $R = \frac{24}{9} = 2.667\ \Omega$.

From the solution we see that the total resistance value is the same, regardless of how we work the problem. Many students have difficulty in understanding why the total resistance in a circuit decreases when more resistances are connected in parallel. We must remember that each resistance connected in parallel provides an additional path for the current flow; hence, more current can flow with a greater number of paths. When more current can flow with a given voltage, it is obvious that there is less total resistance in the circuit.

The principal reason why it is important to know the current value in an electric circuit is so that the proper size of conductor (electrical cable) will be installed to carry the current in a particular circuit. Generally speaking, we can use copper electrical cable of the sizes shown in the following table:

TABLE 9-1

Load, A	AWG (American Wire Gage) Size
Less than 10	18
10–15	14
15–20	12
20–30	10
30–40	8
40–50	6
60–80	4
80–100	2

FIG. 9-11   Circuit to illustrate Ohm's law.

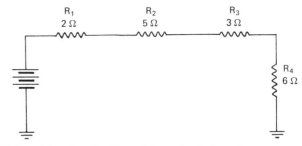

FIG. 9-12   Circuit with resistance loads in series.

For the very heavy, intermittent loads such as those in a start system, it is usually wise to install cable of 0 size upward to 2/0 or 3/0, depending on the value of the load. A safe practice is to install wire at least as large as that originally furnished by the manufacturer.

Alternating Current

Thus far in this section we have dealt principally with **direct current** (DC), that is, current which always flows the same direction in the circuit. It is well, however, for the powerplant technician to understand alternating current (AC), because this type of current will be encountered from time to time in work with powerplants and their associated circuits.

Alternating current periodically changes direction and continuously changes in magnitude. To illustrate alternating current we may use the **sine** curve shown in Fig. 9-14. The sine curve is developed as shown in the illustration and is so named because it represents the values of the sines of all the angles through a complete circle. In trigonometry, the **sine** of an angle is the ratio of the side of a right triangle opposite the angle to the hypotenuse of the triangle. In the diagram, the sine of the 35° angle AOB is AB/AO. If the radius of the circle has a value of 1, the sine of the angle is then merely AB. If we plot the values of the sines for the various angles from 0 to 180° as shown in the drawing, we shall produce a sine curve. Alternating-current values usually follow a sine curve when plotted with a time base such as BCD in the illustration. This means, of course, that the value of an alternating current will be at zero, then increase to a maximum at 90°, back to zero at 180°, to a maximum in the opposite direction at 270°, and again to zero at 360°. The changes in value from 0 to 360° represent 1 **cycle** of alternating current.

The number of cycles which occur per second in an alternating current is the **frequency** of the current. One cycle per second is 1 hertz (Hz). The frequency of the current used in most city power systems is 60 Hz. One-half of a cycle is one **alternation**. In the diagram of Fig. 9-14 the curve from B to C is one alternation and the curve from C to D is one alternation. During the second alternation the current flow is in a direction opposite that of the current for the first alternation.

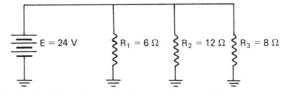

FIG. 9-13   Circuit with resistances in parallel.

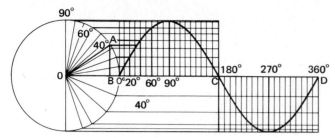

FIG. 9-14 Development of a sine curve.

The sine curve is used to represent either current or voltage in an AC circuit.

One of the principal values of alternating current lies in the fact that the voltage can be stepped up or stepped down by means of a transformer. Thus it is possible to transmit the electric power at a high voltage and low amperage through comparatively small conductors. At the point where the power is to be used, the voltage is reduced to the desired level with a transformer before being delivered to the load. It must be remembered that the voltage drop in a conductor increases in proportion to the current flow. For this reason, there will be a smaller voltage drop in a circuit where the volttage is high and the current low than there will be in a similar circuit where the voltage is low and the current is high.

● MAGNETISM

Magnetism is one of the oldest electrical phenomena known to man. It is believed that Chinese sailors used the **lodestone**, a magnetic oxide of iron, as early as the tenth century **A.D.** for navigational purposes. The lodestone when freely suspended will align itself with the earth's magnetic field and will therefore assume a position pointing generally north and south. Magnetism may be defined as the property of a substance which causes it to attract other magnetic substances.

The Nature of Magnetism

A magnetized substance is surrounded by a **field** of force which has both direction and magnitude. If a magnet is in the shape of a bar as shown in Fig. 9-15, the magnetic field at one end of the bar will be in a direction opposite that of the other end. In order to establish laws or rules for magnetism, the magnet is said to have **poles,** and these poles are named according to the direction of the magnetic field at the pole. When a bar magnet is freely suspended, one end of the bar will point toward the north magnetic pole of the earth and the other end will point toward the south magnetic pole. This is because the earth is a large magnet and

the bar magnet aligns itself with the earth's magnetic field.

The end of a bar magnet which points toward the north pole of the earth is called a "north-seeking" pole or the "north" pole, and the other end is called the "south-seeking" or "south" pole. Although the actual direction of the magnetic field is not known, magnetic force is said, by definition, to travel from the north pole of a magnet to the south pole. For this reason it is often called **magnetic flux,**

When the north pole of one magnet is placed near the south pole of another magnet, there is a very strong attraction between the two poles. On the other hand, when the north pole of one magnet is placed near the north pole of another magnet, there is a repulsion between them. We may therefore state that *opposite poles attract and like poles repel.* Since the north pole of a bar magnet is attracted by the earth's magnetic pole, we know that the magnetic pole near the north geographic pole of the earth is actually a south magnetic pole. The magnetic field of the earth will therefore be in a direction from the geographic south pole to the geographic north pole.

The theory of magnetism can be explained when we consider that atoms may have magnetic polarity as the result of the electrons in orbit about them and that the atoms thus give polarity to the molecules of the substance. If the molecules are aligned in such a manner that their magnetic forces combine, the substance which they form will have magnetic polarity.

It is also said that a group of billions of molecules may make up a **domain** which has magnetic polarity. In a magnetized substance, such as a permanent magnet, the polarized domains are arranged in such a manner that their individual magnetic fields are aligned to reinforce each other. In a nonmagnetized substance the domains are arranged at random and their magnetic fields cancel each other.

When we speak of magnetic fields, we often refer to "magnetic lines of force." Actually, there are no individual lines of force, but we can see what are apparent lines when we perform the experiment shown in Fig. 9-16. When a piece of paper is placed over a horseshoe magnet and iron filings are sprinkled on the paper, the pattern of the magnetic field can be made to appear. The iron filings take positions in lines, so it is natural to refer to the field as being composed of

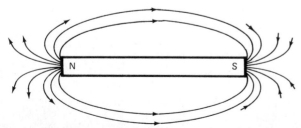

FIG. 9-15 Bar magnet and its field.

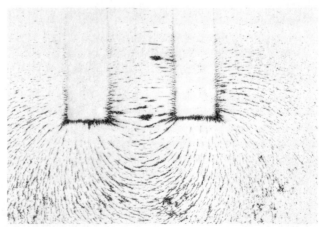

FIG. 9-16 Demonstrating a magnetic field with iron filings.

lines of force. As stated before, we assume that these lines of force are in a direction from north to south. The strength of the magnetic field may also be referred to as the "amount of magnetic flux."

Magnetic fields are encountered wherever a flow of electric current exists, and for this reason (among others) the technician must have an understanding of their nature and their effects. The strength of a magnetic field is determined by its intensity, and the intensity is given in **gauss** (G). One gauss intensity is one line of force per square centimeter.

Magnets

Magnets today are of many sizes and shapes and are made of a variety of materials. One of the most common metals used for permanent magnets is hardened steel. A permanent magnet is one which retains its magnetism indefinitely. Exceptionally strong permanent magnets are made from special alloys, such as alnico and Permalloy. Some of the metals which are used in the alloys are nonmagnetic in the pure state. Permanent magnets are used to provide the fields for magnetos.

Temporary magnets are usually made of soft iron, and they retain their magnetic properties only while within a magnetic field. The field may be produced by an electric coil or by a permanent magnet. In our present study, temporary magnets are used chiefly for solenoids and relays and to produce generator fields. These will be explained as we proceed with discussions of the accessories with which they are associated.

Electromagnets

An **electromagnet** is a magnet produced by means of an electric current. As mentioned previously, whenever there is a flow of an electric current, a magnetic field is produced. In Fig. 9-17, if a current is flowing through a wire as shown, magnetic lines of force will surround the wire in the direction indicated. Actually, the field around the wire is much more extensive than is shown in the diagram. The field around a current-carrying wire extends indefinitely, becoming weaker as the distance from the wire increases. The strength of the field is inversely proportional to the square of the distance from the wire.

The direction of the magnetic field around a current-carrying wire can be determined by the **left-hand rule for conductors.** If the conductor is grasped in the left hand with the thumb pointing in the direction of current flow as shown in Fig. 9-18, the fingers will be encircling the conductor in the direction of the magnetic field. This can be verified by using a compass needle as shown in Fig. 9-19. If a small compass is placed over the current-carrying wire, the needle will point in the direction of the field above the wire.

The effect of the magnetic field around a current-carrying conductor may be multiplied many times by winding the conductor into a coil. When this is done,

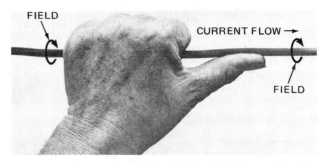

FIG. 9-18   Demonstrating the left-hand rule for conductors.

the magnetic fields of the separate turns are added together and the combined magnetic field passes through the coil as shown in Fig. 9-20. Under these conditions the coil will have the properties of a magnet and will draw magnetic substances toward the center of the coil. When we place a bar of soft iron inside the coil, the effect of the coil is greatly increased because magnetic lines of force pass through soft iron much more easily than they do through the air. The coil with a soft iron core is called an electromagnet. When current is flowing in the coil, the core is magnetic, and when the current is shut off, the core loses its magnetism.

The strength of an electromagnet is determined by the number of turns of insulated wire in the coil and the current flow through the wire. The product of these two quantities is called **ampere-turns.** If the coil of an electromagnet has 200 turns of wire and a current of 5 A is flowing in the coil, the magnetizing force (**magnetomotive force**) is 1000 amp-turns.

The polarity of an electromagnet can be determined by means of the **left-hand rule for coils.** If the left hand is closed and held so that the fingers point in the direction of current flow in the coil, the extended thumb will point in the direction of the field of the electromagnet; that is, the thumb will be pointing toward the north pole. This is illustrated in Fig. 9-21.

It must be remembered that only a few of the metals are magnetic. In general, it is safe to say that most ferrous (iron-base) metals are magnetic; however, there are many metal alloys which contain iron and are not magnetic. Typical of these are the stainless steels. Such metals as aluminum, brass, copper, zinc, silver, and gold are completely nonmagnetic.

The ability of a substance to carry magnetic lines of force is called **permeability.** The permeability of air and other materials which are not magnetic is given a value of approximately 1. Pure annealed iron has a permeability of 6000 to 8000. This means that it will carry magnetic lines of force 6000 to 8000 times as easily as will air. An alloy of nickel and iron, called **Permalloy,** has a permeability of more than 80,000. The opposition of a substance to magnetic lines of force is called **reluctance.** This may be compared with resistance in an electric circuit. If we wish to state a formula for the values in a magnetic circuit, it will appear thus:

$$\text{Magnetic flux} = \frac{\text{magnetomotive force}}{\text{reluctance}}$$

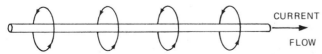

FIG. 9-17   Field around a current-carrying conductor.

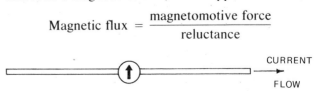

FIG. 9-19   Use of a compass to show field direction.

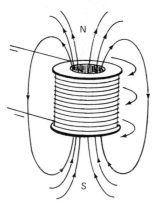

FIG. 9-20   Magnetic field around a coil.

It is not intended that the foregoing formula be used to compute magnetic strength for the purpose of this text because it is not required that the powerplant technician perform this type of problem. If students desire to pursue the study of magnetism further, they may find additional information in a good college physics text. It is sufficient to know that an increase in the number of turns of wire in a coil or in the current through the coil will cause an increase in the magnetic effect of the coil.

● ELECTROMAGNETIC INDUCTION

General Principles

The transfer of electric energy from one circuit to another without the aid of conductors is called **induction.** The transfer of electric energy from one circuit to another by means of a magnetic field is called **electromagnetic induction.** It has been explained that whenever a current flows through a conductor, a magnetic field exists around the conductor. It is also true that when a conductor is "cut" by a changing magnetic field, a voltage will be produced in the conductor.

Purely **magnetic induction** occurs when an unmagnetized magnetic substance is placed within a magnetic field. For example, if a piece of soft iron is placed near a permanent magnet, as shown in Fig. 9-22, the magnetic lines of force will pass through the iron and the iron will become magnetic. As shown in the illustration, the iron will then have the power to attract other magnetic substances.

In Fig. 9-23, the conductor is moved upward through the magnetic field. As it moves, it cuts across the magnetic lines of force and a voltage is induced which causes current to flow in the direction shown by the arrows. If the conductor were moved downward, the

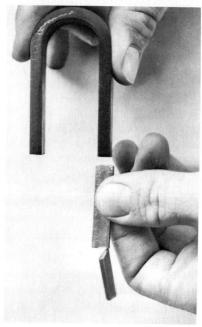

FIG. 9-22   Demonstrating magnetic induction.

current flow would be in the opposite direction. The direction of current flow in a conductor moving across a magnetic field can be determined by use of the **left-hand rule for current flow.** When the left hand is held so that the thumb, forefinger, and middle finger are at right angles to each other as shown in Fig. 9-23, with the thumb pointing in the direction of conductor movement and the forefinger pointing in the direction of the magnetic field, then the middle finger will be pointing in the direction of current flow.

Another method for finding the direction of induced current flow is illustrated in Fig. 9-24. If we assume that a conductor is moving upward through a magnetic field and that the lines of force are bent around the conductor as shown, we can hold the left hand so that the fingers are pointed in the direction of the curved lines of force around the conductor and the thumb will be pointing in the direction of the induced current flow.

Electromagnetic induction takes place whenever there

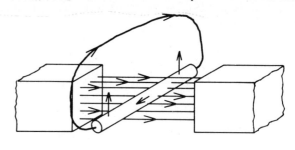

FIG. 9-21   Left-hand rule for coils.

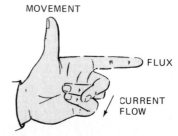

MOVEMENT

FLUX

CURRENT FLOW

FIG. 9-23   Determining the direction of an induced current.

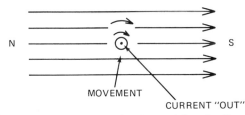

FIG. 9-24 Another method for determining the direction of an induced current.

is a relative movement between a magnetic field and a conductor or conductors such that the conductors cut across (link with) the magnetic lines of force. If the conductors move parallel to the lines of force, there will be no induction. The voltage produced by electromagnetic induction is proportional to the number of lines of force cut per second. When lines of force are being cut at the rate of 10^8 per second, the induced emf will be 1 V.

Lenz's Law

An important principle governing electromagnetic induction is called Lenz's law, named for a German physicist of the nineteenth century, H. F. Emil Lenz. This law may be stated thus: *An induced current is in a direction which produces a magnetic field opposing the change in the field which causes it.* This may be illustrated by the diagrams of Fig. 9-25. In diagram *A* a bar magnet is being moved toward a coil of wire. The field of the bar magnet induces a current in the coil as shown by the meter *M*, and the field produced by this current is such that the north end of the coil field is toward the north end of the bar magnet. Since like poles repel, the field of the coil is opposing the movement of the magnet toward it. When the bar magnet is moved away from the coil, the direction of current flow in the coil and the field of the coil are reversed, as shown in diagram *B*.

One of the methods by which electromagnetic induction takes place has been illustrated in Fig. 9-25. This is the case—called "transformer action"—where the field is moving and the coil is stationary. When the field is stationary and the coil (conductor) is moving, the induction is called "generator action." In a con-

ventional generator the conductors rotate with a drum called the **armature** and the field is produced by stationary electromagnets.

Transformers

In a simple **transformer,** separate coils of wire are wound on a common iron core or on cores arranged so that the magnetic field traverses the separate cores. Two different arrangements for transformers are shown in Fig. 9-26. In 9-26(A) two windings called the primary and the secondary are placed on one common core. In 9-26(B) the primary and secondary windings are placed on separate sections of a core, which is in a square configuration.

In a transformer, the winding into which the current is fed is called the **primary winding** and the winding from which the output is taken is called the **secondary winding.** The output voltage is proportional to the turns ratio of the primary and secondary. This is shown by the formula

$$\frac{E_p}{E_s} = \frac{N_p}{N_s}$$

where E_p = primary voltage
E_s = secondary voltage
N_p = number of turns in primary winding
N_s = number of turns in secondary winding

Example: An alternating current with an emf of 12 V is fed to a transformer having 200 turns in the primary winding and 2000 turns in the secondary winding. What is the output voltage? By the foregoing formula,

$$\frac{12}{E_s} = \frac{200}{2000}$$

Then $200E_s = 24,000$

$$E_s = \frac{24,000}{200} = 120 \ V$$

From this example, we note that the secondary winding has 10 times as many turns as the primary. The output voltage is therefore 10 times the input voltage.

The transformer used in the example is called a **step-up** transformer because the output voltage is higher

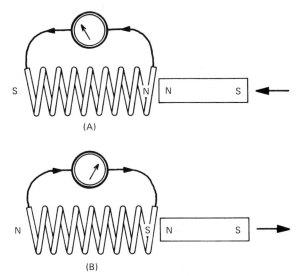

FIG. 9-25 Demonstration of Lenz's law.

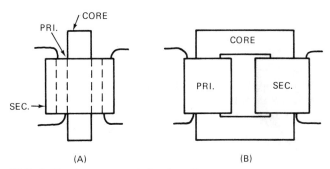

FIG. 9-26 Arrangements for transformers.

than the input voltage; that is, the transformer steps the voltage up. If the secondary winding has fewer turns than the primary winding, the transformer is a **step-down** transformer because the output voltage is less than the input voltage.

Because the total power output of a transformer cannot be greater than the total power input, there must be a decrease in output current when there is an increased output voltage. If a transformer could be 100 percent efficient, we could use the following formula:

$$E_p I_p = E_s I_s$$

Remember that the power in an electric circuit is equal to the voltage times the current. By the formula above, we see that the power in the primary circuit is equal to the power in the secondary circuit when efficiency is 100 percent. Commercial transformers have efficiencies of 95 to 99 percent.

As mentioned previously, a transformer must be operated on an alternating current in order to have a moving magnetic field. If a direct current is connected to a transformer, there will be a momentary surge of voltage at the output terminals during the time that the magnetic field is being built up. Since the direct current is steady, there will be no further change in the field until the direct current is disconnected. At this time there will be a surge of voltage in the reverse direction. In some cases where no alternating current is available, a direct current is rapidly interrupted by vibrating contacts and then fed to the primary of a transformer. The output of the transformer will then be an alternating current having the frequency of the vibrating contact points. This principle has been used for many years to provide the plate voltage for radio sets which must operate from a DC power source.

An important feature of the transformer is the core material. The core must be made of a special soft iron which has low reluctance so that the magnetic field can change rapidly. Furthermore, it must be made of thin laminations insulated from each other to prevent the flow of **eddy currents** which would otherwise develop and cause the core to overheat. In a solid core, the changing magnetic field induces eddy currents in the core material and these currents cause a substantial loss of power as well as heating.

A most important feature of a standard transformer is the fact that it can remain connected to a power source, and if no current is being drawn from the secondary, there will be practically no current flow in the primary. This occurs because the current flow in the primary induces a back voltage in the primary almost equal to the input voltage. This **back emf** prevents any appreciable flow of current in the primary. When a load is connected to the secondary winding, the current flow in the secondary opposes the magnetic field caused by the primary current, and the back voltage in the primary therefore decreases. This allows enough current flow in the primary to supply the required output power. Thus, we see that the transformer is self-regulating. If an excessive load is placed on the transformer, the primary current may be so great that the windings will burn out.

Transformers are manufactured in many sizes and types, and for any particular application, the transformer must have the capacity to handle the power applied to it, it must have the proper turns ratio to provide the amount of step up or step down required, and it must be sufficiently well insulated to withstand the voltages applied to it. It is the responsibility of the technician to see that the transformer selected for a particular job has the proper characteristics to fulfill its purpose.

● CAPACITANCE, INDUCTANCE, AND REACTANCE

Capacitors

It has been pointed out that various substances can be given a negative or positive charge by adding or removing electrons from them. A **capacitor (condenser)** is a device designed to store electrical charges. When two metal plates are placed adjacent to each other with an air gap between them as shown in Fig. 9-27, a simple capacitor is formed. If one plate is connected to the negative terminal of a battery and the other plate is connected through a switch to the positive terminal of the battery, the capacitor will be charged. During the charging process, electrons flow from the negative terminal of the battery to one plate. The excess of electrons on this plate repels electrons from the other plate, and these repelled electrons will flow to the positive terminal of the battery. Now, if the switch is opened, the capacitor will remain charged for a long period of time unless an electrical connection is made between the plates. This is because the positive charge on one plate holds the electrons on the other plate. Between the plates is an **electrostatic field** which may be considered a field of force. The electrostatic field exists between the plates of any charged capacitor.

When the capacitor is disconnected from the battery and the plates are connected by a conductor, the electrons from the negative plate will flow to the positive plate until the charges are equalized. This action is illustrated in Fig. 9-28.

The insulating material between the plates of a capacitor is called the **dielectric**. It may consist of air, waxed paper, mica, glass, Mylar, or a variety of other insulating materials. The most important requirement for a dielectric is that it be a good insulator and be able to withstand the **dielectric stresses** imposed by the ap-

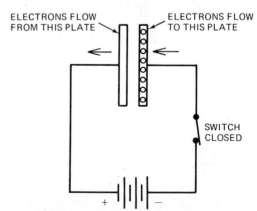

FIG. 9-27 Diagram of a simple capacitor.

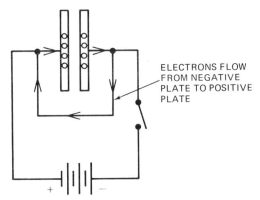

FIG. 9-28 Action of a capacitor.

plied voltages. If the dielectric breaks down, an electrical discharge will take place through it and the capacitor will become useless.

The **capacity** of a capacitor is normally measured in **microfarads.** A microfarad (μF) is one-millionth of a **farad** (F). The capacity is 1 farad when the capacitor will store 1 coulomb of electricity under an emf of 1 volt. One farad is a very high value of capacitance; hence, capacitors rated in microfarads are used for most purposes in a standard electrical system.

The capacitors most often encountered by the powerplant technician are the Mylar-type, the mica-type, and the rolled-paper type. Mylar capacitors can be constructed with alternate layers of metal foil and sheet Mylar. The capacitor can be a flat plate arrangement or it can be two strips of foil separated by a Mylar sheet and rolled into cylindrical form. The mica capacitor is constructed by stacking very thin plates of metal (foil) with thin sheets of mica dielectric between each pair of plates. This construction is illustrated in Fig. 9-29. The paper capacitor consists of two long strips of foil separated by waxed paper or a similar material and rolled into a cylindrical shape. The foil is usually positioned so that the edge of one strip can be contacted at one end of the roll and the edge of the other strip contacted at the opposite end. The roll is then sealed in a paper or metal case for protection.

Variable capacitors are constructed so that one set of plates can be inserted into the other set to any degree desirable by means of a control knob. The plates are spaced so that they do not touch each other, and the air between the plates serves as the dielectric. Variable capacitors are commonly used in electronic circuits.

The capacity, or **capacitance,** of a capacitor is determined by the area of the plates, the distance between the plates, and the **dielectric coefficient** of the insulating material between the plates. In simple terms, the dielectric coefficient of a material is a measure of its ability to convey the effect of electric charges through itself.

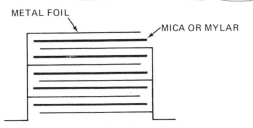

FIG. 9-29 Construction of a mica capacitor.

The dielectric coefficient of a vacuum is 1, and that of other substances is more than 1.

To provide an example of the nature of a capacitor, we may consider a sphere separated into two sections by an elastic diaphragm, as shown in Fig. 9-30. If the two chambers thus formed are connected by pipe to a piston pump as shown and the system is filled with fluid, the sphere may be "charged" first in one direction and then the other. When the pump handle is moved to the left, the chamber A will be filled with fluid and the fluid will be removed from chamber B. This may be compared with the addition of electrons to one plate of a capacitor and the removal of electrons from the other plate. The pump shown in the drawing serves as the battery or generator for this illustration. If the valve is closed after the sphere is charged, the charge will hold. When the valve is opened, the pressure of the diaphragm will drive the fluid from chamber A and fluid will flow back into chamber B, provided that the pump is free to move back to the neutral position.

In an AC system, it appears that a capacitor allows current to flow through it. It is true that the capacitor allows the current to flow, but it does not flow through the capacitor. If the handle of the pump in Fig. 9-30 is moved back and forth, we can see that the fluid will move back and forth in the system. No fluid will actually flow through the sphere. Fluid will flow into one chamber and out of the other chamber, and then the flow will reverse. A similar effect takes place in an AC circuit with a capacitor. The electrons flow *in* to one plate and *out* from the other plate. When the current reverses, the electrons will flow momentarily until the capacitor is charged, and then no more flow can take place because the pressure (voltage) is always in the same direction.

The total capacitance in a circuit where two or more capacitors are connected can be determined by simple computations. If two capacitors are connected in **parallel** as shown in Fig. 9-31, it is merely necessary to add the individual capacitances to determine the total. That is,

$$C_t = C_1 + C_2 + C_3 + \ldots$$

When capacitors are connected in series, the total capacitance is reduced because the effective distance

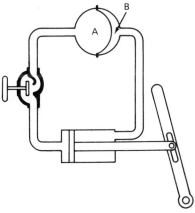

FIG. 9-30 Drawing to illustrate the operation of a capacitor.

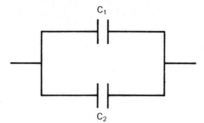

FIG. 9-31 Capacitors in parallel.

between the plates is increased. If two capacitors of equal value are connected in **series** as shown in Fig. 9-32, the total capacitance is one-half the value of one of the capacitors. If the distance between the plates of one of the capacitors is D, the effective distance is increased to $2D$ when the two capacitors are connected in series. The formula for computing total capacitance when capacitors are connected in series is

$$C_t = \frac{1}{1/C_1 + 1/C_2 + 1/C_3}$$

According to the formula, the total capacitance in a circuit where capacitors are connected in series is equal to the reciprocal of the sum of the reciprocals of the individual capacitances.

Example: Three capacitors having values of 2, 4, and 6 μF, respectively, are connected in series. What is the total capacitance? By the formula,

$$\begin{aligned}
C_t &= \frac{1}{\frac{1}{2} + \frac{1}{4} + \frac{1}{6}} \\
&= \frac{1}{\frac{6}{12} + \frac{3}{12} + \frac{2}{12}} \\
&= \frac{1}{\frac{11}{12}} = \frac{12}{11} = 1\frac{1}{11}\mu f
\end{aligned}$$

The principal value of a capacitor (condenser) in a powerplant circuit is to store surges of current and then to release them when the voltage drops. Capacitors are essential to the operation of ignition systems, as will be explained in Chap. 10.

In order to prevent radio interference caused by surging voltages and currents, filter capacitors are used. The effect of these capacitors is to reduce the peaks of voltage and to fill in the low points. This smooths the current flow and reduces the emanation of electromagnetic waves, which cause radio interference.

Inductors

An **inductor,** or **inductance** coil, is simply a coil of wire whose turns are insulated from one another. The coil may have a magnetic core or an air core, depending on the amount of inductance needed. As explained under electromagnetic induction, the moving of mag-

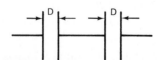

FIG. 9-32 Capacitors in series.

netic lines of force across a conductor will induce a voltage in the conductor. Self-induction occurs when the magnetic field of a conductor induces a voltage in the conductor. Every conductor has a certain amount of inductance, and this inductance is increased by winding the inductor into a coil. An inductance coil is shown in Fig. 9-33.

The effect of an inductance coil is to resist changes in current flow. If the current increases in an inductor, the magnetic field produced by the inductor is increased in strength. This increasing field induces a back voltage in the coil which opposes the increase in current flow. If the current flow is reduced, the change in the magnetic field will produce a voltage which tends to maintain the current flow. This is in accordance with Lenz's law. The value of inductance L is expressed in a unit called the **henry** (H). One henry is the amount of inductance possessed by a coil when a change in current flow of one ampere per second will induce a back emf of one volt. Since the henry is a rather large value of inductance, small inductance coils are rated in **millihenrys** (mH). One millihenry is one-thousandth of a henry.

Reactance

The effect of a capacitor in an AC circuit is called **capacitive reactance,** (X_C) which is measured in ohms because it opposes the flow of current. The effect of an inductor is called **inductive reactance** (X_L), and it is also measured in ohms. Capacitive reactance in an AC system causes the current to lead the voltage. This is illustrated in Fig. 9-34. While the capacitor is charging, the voltage is held down but the current is flowing. As the capacitor approaches the charged condition, the voltage increases and when the voltage of the capacitor has reached maximum, no more current can flow into the capacitor. This means that the current flows out of the capacitor; that is, it is flowing in a direction opposite to that in which it flowed previously. The current flow reaches a peak value at B, and as the voltage increases in the opposite direction, the current flow again decreases and returns to zero as the voltage reaches a peak at C. If an AC circuit contained only capacitance, the current would lead the voltage by 180°. Since every circuit contains some resistance, the lag between voltage and current cannot be as much as 180°.

Inductance in an AC circuit has an effect opposite that of capacitance. That is, inductive reactance cancels capacitive reactance. If an AC circuit has 100 Ω of capacitive reactance and 100 Ω of inductive reactance, the net reactance will be zero. When the reactances are equal, the circuit is said to be **resonant.**

Inductive reactance in an AC circuit causes the current to lag the voltage. In a purely inductive circuit the voltage would lead the current by 180°, as shown in Fig. 9-35. As the voltage rises, the current is held back by the induced back voltage. When the applied voltage

FIG. 9-33 Inductance coil.

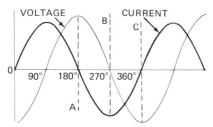

FIG. 9-34　Current leading the voltage.

reaches a peak, the current flow can increase because the field is not inducing a back voltage.

A further understanding of inductance in a circuit may be gained by considering Fig. 9-36. A long spring handle is attached to the shaft supporting a heavy wheel. If the handle is moved quickly, the movement of the wheel will not follow the handle immediately but will lag. After the handle is moved a certain distance and then moved back in the opposite direction, there will be a period when the handle is moving in one direction and the wheel is moving in the opposite direction. This same condition can be seen in Fig. 9-35, where current is flowing in one direction and voltage force is in the opposite direction.

The formulas used for determining capacitive reactance, (X_C) inductive reactance (X_L), and the frequency of resonance are as follows:

$$X_C = \frac{1}{2\pi f C}$$

$$X_L = 2\pi f L$$

$$f = \frac{1}{2\pi\sqrt{LC}}$$

where X_C = capacitive reactance, Ω
X_L = inductive reactance, Ω
f　= frequency, Hz
C　= capacitance, F
L　= inductance, H

From the foregoing formulas it is apparent that the reactance in an AC circuit is dependent on the frequency. As frequency increases in a particular AC circuit, capacitive reactance will decrease and inductive reactance will increase. From this we can see that a capacitor will allow more current flow when the frequency is high and that an inductor will allow more current when the frequency is low.

The use of the foregoing formulas can be demonstrated by solving typical problems. Assume that we wish to determine the capacitive reactance in a circuit

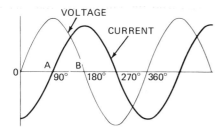

FIG. 9-35　Voltage leading the current.

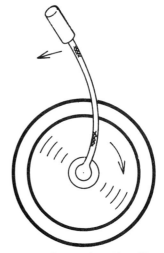

FIG. 9-36　Demonstration to show the effect of inductance.

when the frequency is 400 Hz and the capacitance in the circuit is 0.5 μF. We apply the formula as follows:

$$
\begin{aligned}
X_C &= \frac{1}{2\pi f C} \\
&= \frac{1}{6.28 \times 400 \times 0.0000005} \\
&= \frac{1}{0.001256} \\
&= 796.18\Omega
\end{aligned}
$$

In the foregoing problem it must be noted that capacitance is given in microfarads. Then, in the formula, 0.5 μF is equal to 0.0000005 F or 5×10^{-7} F.

To solve the problem in inductance, we shall assume that a circuit contains 10 mH of inductance and that the frequency is 400 Hz. Substituting in the formula for inductive reactance,

$$
\begin{aligned}
X_L &= 6.28 \times 400 \times 0.010 \\
&= 25.12\Omega
\end{aligned}
$$

With any combination of inductance and capacitance in an AC circuit, there is always one frequency where the capacitive reactance is equal to the inductive reactance. As explained previously, this condition is called **resonance**. To find the frequency where a circuit is resonant when the capacitance is 0.5 μF and the inductance is 10.0 H, we apply the formula for the frequency of resonance:

$$
\begin{aligned}
f &= \frac{1}{2\pi\sqrt{LC}} \\
&= \frac{1}{6.28 \times \sqrt{0.010 \times 0.0000005}} \\
&= \frac{1}{6.28 \times \sqrt{0.000000005}} \\
&= \frac{1}{6.28 \times 0.000001 \times \sqrt{50}}
\end{aligned}
$$

$$= \frac{1}{0.00001 \times 44.4}$$

$$= \frac{1}{0.000444} = 2252.2 \text{ Hz}$$

In the foregoing problems, the decimal figures can be expressed in terms of powers of 10. For example, 0.5 μF can be expressed as 5×10^{-7} F because 10^{-7} = one ten-millionth or 0.0000001. Whenever it becomes necessary to work problems involving long decimals, it is often more convenient to use the powers of 10 for computation when one is familiar with the process.

● MULTIPHASE AC SYSTEMS

It is beyond the scope of this text to give a detailed explanation of multiphase systems; however, we shall give definitions and explanations sufficient for the student to have some understanding of the principles involved. A **multiphase** system has two or more interrelated systems of alternating currents and voltages operating together in three or more conductors.

Values in Three-Phase System

The curves of Fig. 9-37 illustrate the values in a three-phase system and the schematic wiring diagram for a y-wound alternator and motor. If we assume that phase 1 is at a potential of 0 V beginning at 0°, we find that phase 2 is at 86.6 percent of maximum voltage negative and that phase 3 is at 86.6 percent of maximum voltage positive. When the rotor of the alternator has turned 30°, phases 1 and 3 are both at 50 percent of maximum voltage positive and phase 2 is at 100 percent of maximum voltage negative. At a rotation of 60°, phase 1 is at 86.6 percent of maximum voltage positive, phase 2 is again at 86.6 percent of maximum voltage negative, and phase 3 is at 0 voltage. As we continue to note the values at various degrees of rotation, we

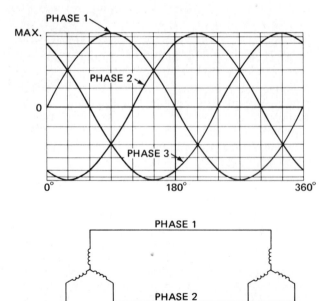

FIG. 9-37   Three-phase electric circuit and resulting curves.

can see that the voltages change steadily in a uniform sequence such that the sum of the negative voltages is always equal to the sum of the positive voltages.

Even though we have used voltage values in the foregoing example, we could also use current values and get the same results. It is true, however, that the current may lead or lag the voltage an amount depending on how much capacitance or inductance is in the circuit.

Alternators and Motors

A three-phase system provides the most efficient distribution of power for motors. In modern aircraft systems, both the alternators and the motors are constructed without brushes or slip rings (sliding contacts), thus greatly reducing the problems that often exist with DC equipment.

A three-phase alternator consists of a stator (stationary winding) in the form of a drum or cylinder with a magnetic rotor. As the rotor turns, the magnetic field of the rotor induces voltages in the windings of the stator, as previously explained. If the output of the alternator is connected to a three-phase stator of an AC motor, the current flow in the windings of the stator will set up a rotating field which turns at the same rate as the rotor of the alternator. This rotating field is employed in the AC motor to develop torque forces in the rotor, thus causing the motor to run.

Use of AC Power

Alternating-current power has a number of advantages over DC power. The principal advantage is that AC power can be transmitted at a higher voltage and lower current, thus making it possible to use smaller diameter conductors. This principle can be illustrated by the following example, where it is desired to transmit 2000 W of power. If a DC voltage of 25 V is employed, it will require 80 A to provide the desired power. When an AC voltage of 250 is used, only 8 A is needed. This means that the AC power will require a conductor with only one-tenth the cross-sectional area required for the DC power.

Alternating current can be reduced or increased instantly and efficiently by means of transformers, thus making it possible to provide the various voltages required for different types of equipment. Furthermore, when a DC voltage is needed, the alternating current can be changed to direct current by means of diode rectifiers and filters. Direct current cannot be increased or decreased by means of transformers. Hence, in a DC system, if higher or lower voltages are required, it is necessary to employ dynamotors or inverters, which are very inefficient and expensive.

When it is desired to operate single-phase equipment from a three-phase source of power, it is merely necessary to connect the load across two of the three phases. In practice, one attempts to divide single-phase loads among the three phases so that the loads on the different legs of the system are as nearly equal as possible. If the loads are greatly out of balance, the efficiency of the system decreases.

When direct current is needed for certain types of equipment, all three phases of the AC system may be connected to a three-phase full-wave rectifier. The cir-

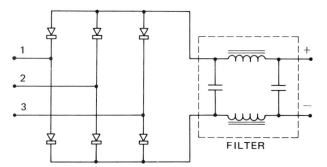

FIG. 9-38 Three-phase full-wave rectifier circuit.

cuit for such a rectifier is shown in Fig. 9-38. Note that the six-diode rectifier units will provide for a continuous flow of current from the three legs of the AC system; hence, a maximum of available power will be supplied to the DC circuit. The ripples which would otherwise be present in the direct current are removed by the filter.

☛ DIRECT-CURRENT GENERATORS AND CONTROLS

Generator Theory

A generator, either for direct current or for alternating current, operates on the principle of electromagnetic induction. The conventional design for a DC generator provides that a rotating armature be placed in a stationary magnetic field, the field being produced, usually, by field windings on field poles. Fig. 9-39 is a simplified drawing of a DC generator. As the armature rotates in the field, the conductors of the armature winding cut across the lines of magnetic flux, thus causing a voltage to be induced in the armature winding. The windings of the armature are connected to the commutator, from which the current is taken by means of brushes.

The direction of the induced voltage in the armature windings depends on the direction of the magnetic field and the direction in which the moving conductor is moving. In Fig. 9-40 a conductor is moving downward as it cuts across magnetic lines of force. The direction of the lines of force is from left to right. We may apply the left-hand rule to determine the direction of voltage or current flow. When the thumb, index finger, and middle finger of the left hand are placed at right angles

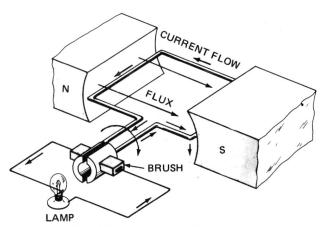

FIG. 9-39 Simplified drawing of a DC generator.

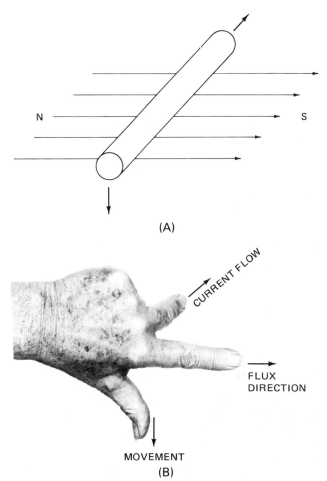

FIG. 9-40 Demonstrating the left-hand rule for generators.

as shown in the illustration, with the thumb pointing in the direction of conductor movement relative to the field and the index finger pointing in the direction of the magnetic flux, the middle finger will point in the direction of the induced voltage. This principle is called the **left-hand rule for generators.** In using the left-hand rule, we assume that the electric current (electrons) moves from negative to positive.

The strength of the voltage produced in the armature of a generator depends on three factors: (1) the strength of the magnetic field, (2) the number of turns of wire on the armature, and (3) the rate at which the armature is turning. Assuming a standard value for each "line of magnetic force," an electromotive force of 1 V will be induced when 100 million lines of force are being cut per second.

Construction of a Generator

The construction of a DC generator is illustrated in Fig. 9-41. The essential electrical parts of this generator are the field poles, armature, commutator, and brushes. The **field poles** are electromagnets, each having a soft iron core surrounded by a **field coil.** Their function is to provide the varying magnetic field necessary for the induction of a voltage in the armature windings. The **armature** consists of a laminated, soft iron core with longitudinal slots which carry the field windings (see Fig. 9-42). As the armature rotates in the magnetic field between the field poles, voltages are induced in its windings. The **commutator** is composed of a series of

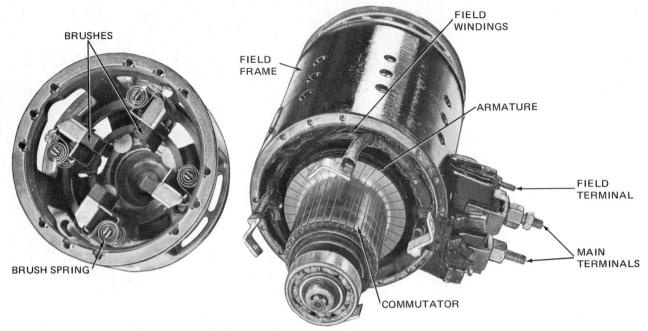

FIG. 9-41 Construction of a DC generator.

copper bars, insulated from one another and mounted in a cylindrical arrangement at the end of the armature. The copper bars are connected to the ends of the armature windings and serve as the electrical contacts through which the current from the armature is delivered to the brushes. The **generator brushes** are made of hard carbon or a composition of carbon and other materials which improve the conducting and wearing qualities of the brushes. The brushes are mounted to make contact with the commutator at points where the voltage differential is the greatest, as shown in Fig. 9-43. The plane cutting through these points is called the **neutral plane,** because at these points there is very little potential difference between adjacent bars of the commutator. If the brushes are not located on the neutral plane, adjacent bars of the commutator under the brushes will have a potential difference and will be short-circuited by the brushes. This will result in arcing of the brushes, loss of power, pitting of the commutator, and rapid wear of the brushes.

The principal structural parts of a generator are the field frame, drive end frame, commutator end frame, brush rigging, bearings, and possibly other parts. The **field frame** is usually an iron cylinder inside which the field pole pieces are mounted and held in place by large screws. The field frame provides a portion of the magnetic circuit between the field poles. The two **end frames** are attached to the field frame, either by screws or by long bolts which extend from one end of the generator to the other. A recess is usually machined in each end

frame to hold the bearing in which the armature shaft is supported.

The purpose of the brush rigging is to provide a mounting for the brush holders and springs. The *negative* brush holder is usually attached directly to the rigging with metal-to-metal contact for grounding purposes. The *positive* brush holder must be insulated from the generator frame to prevent short-circuiting. The positive brush is connected to the positive output terminal, which is also insulated from the frame by means of a phenolic bushing and washers. So that they may always maintain a good contact with the commutator, it is necessary that the brushes be free to move in the brush holders, with the brush springs providing constant pressure.

The electric circuit for a **shunt-type** generator is shown in Fig. 9-44. A shunt-type generator has the field windings connected in parallel with (shunted across) the armature. A variable resistor (voltage regulator) is con-

COMMUTATOR

FIG. 9-42 Armature for a DC generator.

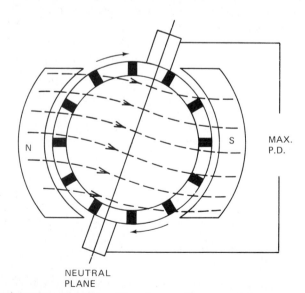

FIG. 9-43 Location of brushes on the commutator.

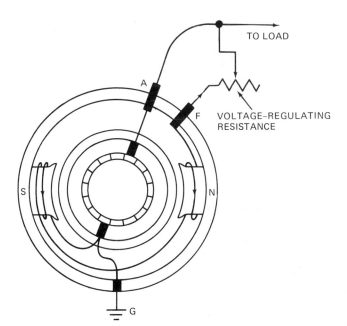

FIG. 9-44   Circuit for a shunt-wound generator.

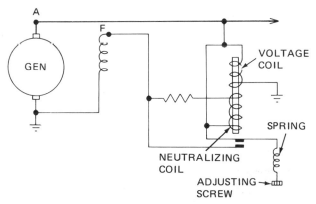

FIG. 9-45   Vibrator-type voltage-control unit.

nected in series with the field windings to control the field current. It will be observed in the drawing that the current taken from the armature flows from one brush to the other through the field winding to provide the magnetic field.

Operation of the Generator

Since the production of voltage in the armature of the generator is dependent on the magnetic field and the magnetic field is dependent on current from the armature, we may ask where the process of electrical generation begins. What actually starts the process? Fortunately, the iron from which the field poles are made has the capability of holding a small amount of magnetism indefinitely, even though there is no current flow through the field windings. This is called **residual magnetism,** and it provides the starting energy for the generator. As the armature starts to rotate, it cuts across the few lines of magnetic force produced by the residual magnetism of the field poles. This is sufficient to induce a voltage in the armature; hence, a current starts to flow from the regulator. The current through the field windings quickly builds up, or "excites," the magnetic field, and the generator voltage increases until limited by the voltage regulator.

Voltage-Control Systems

As mentioned previously, the voltage of a generator depends on the speed of the armature, the number of windings in the armature, and the strength of the field. In order to regulate the voltage to compensate for changes in load and variation of engine speed, it is necessary that one of the voltage controlling factors be changed. The only one of these factors which can be changed conveniently is the strength of the field, and this is accomplished by changing the current flow through the field windings by means of an automatically variable resistance.

For many of the light-airplane systems, the voltage-control unit is of the vibrator type, similar to those used in conventional automobile systems. A schematic

diagram of such a system is shown in Fig. 9-45. This diagram is simplified to show only the voltage control. This particular voltage control is designed to regulate the current in the positive path of the field circuit. The voltage coil is connected directly across the generator output and continually senses generator voltage. If the voltage becomes greater than the value for which the voltage regulator is adjusted, the strength of the magnetic field produced by the voltage coil in the regulator will overcome the spring pressure which holds the contact points together and the points will open. This will break the direct path to the field of the generator and will cause a resistance to be inserted in the circuit. The current flow through the field windings will be reduced, thus weakening the generator field strength. When this occurs, the generator voltage must decrease. It will be noted that when the contact points in the regulator are open and current is flowing through the field, the current must also flow through the neutralizing coil. Since this coil is wound in a direction opposite that of the voltage coil, the magnetic effect of this coil is to reduce the magnetism of the electromagnet. At the same time, since the generator voltage is dropping, the magnetic field produced by the voltage coil is reduced. It is obvious, then, that the total magnetic strength will be quickly reduced and the contact points will be closed by the spring.

The cycle just described occurs many times a second, the actual number being determined by the total load on the circuit and the rpm of the generator. In some regulators the vibration rate is from 50 to 200 Hz.

Another commonly employed vibrator-type regulator is used in the ground circuit of the field windings. A schematic diagram of this regulator is shown in Fig. 9-46. In this diagram it can be seen that the field circuit is complete to ground through the series winding when the contact points are closed. At this time, maximum current is flowing through the field coils of the generator. When the generator voltage exceeds the value for which the regulator is adjusted, the current flowing through the two coils in the regulator produces sufficient magnetic strength to open the contact points. When this occurs, the direct path from the ground to the generator field winding is broken and the field current is required to flow through the resistor. This reduces the field current, thus reducing the generator field strength and the output voltage. When the contact points open, current flow through the series winding

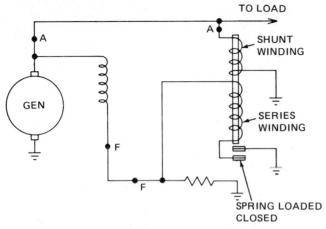

FIG. 9-46 Regulator to control ground circuit of field.

of the regulator is stopped and the magnetic strength of the regulator electromagnet is greatly reduced. As generator voltage drops, the current flow through the shunt winding drops proportionately, thus further reducing the electromagnet strength. All this occurs in a very small fraction of a second, and the contact points close again. This cycle is repeated many times per second, as explained in the previous description.

Reverse-Current Cutout

Every generator system must have an automatic means for disconnecting the generator from the battery when the generator voltage becomes less than the battery voltage. This condition occurs sometimes when an engine is idling, and it always occurs when the engine is stopped. If the battery were left connected to the generator, current from the battery would flow through the generator and probably burn out the armature. If the armature did not burn out, the battery would discharge through the generator.

The **reverse-current cutout** serves as an automatic switch to open the main generator circuit whenever generator voltage drops slightly below battery voltage. A schematic diagram of the cutout relay is shown in Fig. 9-47. In this drawing it will be noted that the electromagnet of the relay has two windings. The series winding consists of a few turns of heavy wire so it can carry the entire output current of the generator. This winding is connected to the contact points through which the current flows to the battery and the electrical load of the system. These contact points are spring-loaded in the normally open position. The shunt wind-

ing, also called the voltage winding, consists of many turns of fine wire, thus providing sufficient resistance to limit the current flow to a fraction of 1 A. This low current draw is necessary because the shunt winding is continuously connected across the generator output when the generator is operating.

When the generator first starts turning, the reverse-current cutout relay points are open. As voltage builds up to the value determined by the voltage-regulator setting, the shunt winding of the relay produces a magnetic field strong enough to close the relay points. This connects the generator to the system. As long as the generator voltage remains above that of the battery, the relay remains closed and the generator supplies current to charge the battery and for all the system loads. When the engine driving the generator is stopped, the generator voltage falls below battery voltage and the battery voltage causes current to feed backward through the cutout relay series winding. Since this current is opposite in direction to its normal flow, it produces a magnetic field which opposes the field of the shunt winding. This reduces the total magnetic strength and allows the spring to open the contact points. It must be explained that the shunt winding will continue to produce a magnetic field when the generator voltage drops because it will be supplied by the battery as long as the relay contact points are closed. The series winding is therefore necessary to produce a field opposing the field of the shunt winding.

Current Regulator

The three-unit regulator unit used with many generator systems includes a **current regulator** to prevent the generator current output from becoming excessive. This condition can occur when excessively heavy loads are applied to the system. The current regulator is shown in Fig. 9-48, which illustrates the complete three-unit regulator arrangement. A study of this diagram will reveal that the current-regulator section of the generator control unit consists of an electromagnet with a series winding through which the entire load current flows and a set of contact points which are connected in series with the voltage-regulator-control circuit. When generator output current exceeds the preset value, the

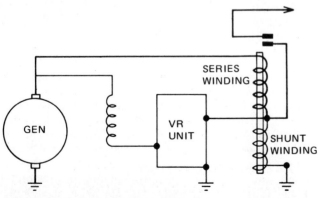

FIG. 9-47 Reverse-current cutout relay.

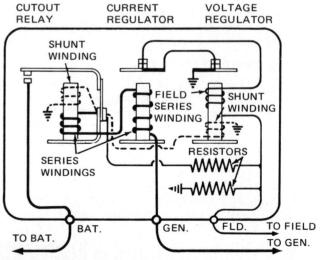

FIG. 9-48 Three-unit regulator.

magnetic field of the current regulator will become strong enough to open the contact points. This will break the direct ground circuit of the generator field winding and cause the field current to flow through the ground resistor. When this occurs, generator field current will decrease substantially and reduce generator voltage. The reduction in voltage will cause a drop in generator output current and a weakening of the magnetic field produced by the current regulator coil. The current regulator contact points will close, and the cycle will repeat in a manner similar to that of the voltage regulator.

Carbon-Pile Voltage Regulator

The most commonly employed type of voltage regulator for DC electrical systems on larger aircraft with DC power systems is called the **carbon-pile** type because it utilizes a stack of carbon disks as a variable resistance element for controlling generator field current. A photograph of such a regulator is shown in Fig. 9-49, and a schematic diagram of the circuit is given in Fig. 9-50.

In this regulator the variable resistance element is connected in series with the field circuit so that any change in the resistance will be accompanied by a change in field current and a corresponding change in generator voltage. The carbon pile consists of alternate disks of hard and soft carbon pressed together by means of a radial leaf spring at one end of the stack. When the disks are pressed together with maximum spring force, the resistance of the pile is low and current flow through the stack will be correspondingly high. At the spring end of the carbon pile is an electromagnet which is shunted across the generator to sense generator voltage. If the voltage becomes excessive, the strength of the electromagnet will exceed the strength of the spring compressing the carbon disks and the pressure will be released. This results in an immediate increase in the resistance of the carbon pile and a reduction of field current. This, of course, reduces the generator voltage. During normal operation of the carbon-pile regulator, when the electric load is steady, a balanced condition will exist in which the strength of the electromagnet in the regulator will equal the strength of the carbon-pile spring. Any change in electric load or generator

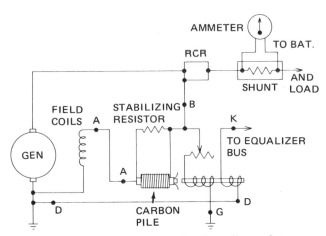

FIG. 9-50 Schematic diagram of carbon-pile regulator circuit.

speed will result in quick adjustment of the regulator to accommodate the new conditions.

Figure 9-50 shows a variable resistor connected in series with the voltage coil of the electromagnet. This variable resistor, also called a rheostat, is used to make fine adjustments in the voltage-regulator setting when the system is in operation. Normally the regulator is set, prior to installation, for a voltage value of 28.5 V. The adjusting rheostat is in the center position when this value is established. When the regulator is installed and in operation, the generator voltage can then be adjusted to balance with other generators connected in the same system.

The stabilizing resistor connected across the carbon pile in the regulator is provided to reduce the surge effect of sudden changes in load and to help reduce arcing between the carbon disks.

An equalizing circuit is usually provided when two or more generators are connected in the same system. This circuit utilizes a winding in each regulator by which the regulator voltage setting will be slightly increased or decreased automatically if the generator is not in balance with the other generators in the system. For example, if a generator is carrying more than its share of the load, current will flow in the equalizer winding in a direction which will strengthen the magnetic field of the voltage coil. This added magnetic strength will reduce the pressure of the spring on the carbon pile and increase the resistance of the coil circuit. When this occurs, the field current of the generator is reduced and the voltage must decrease. The reduced voltage will cause the generator to drop a part of the load, and the other generators in the system will assume the part of the load dropped by the generator which had previously carried too much.

Electronic Voltage Controls

Electronic voltage controls are now in common use on many modern aircraft electrical systems. These may be described as **transistor voltage regulators** or **transistorized voltage regulators**. These devices are described in the associated text, *Aircraft Electricity and Electronics*.

Differential Reverse-Current Relay

Typical of reverse-current relays utlized in 24-V DC systems is the **differential reverse-current relay.** A

FIG. 9-49 Photograph of carbon-pile voltage regulator.

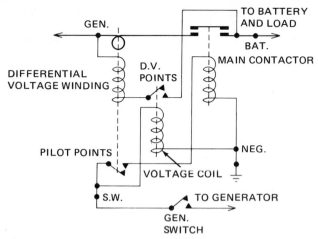

FIG. 9-51 Differential reverse-current relay.

schematic diagram of the circuit for such a unit is shown in Fig. 9-51. This relay acts as both a master switching device and a reverse-current relay. When the generator is operating, there is a differential voltage between the GEN terminal and the BAT terminal of the relay. If the contact points in the differential voltage (D.V.) coil circuit are closed, a current will flow in such a direction that the **pilot points** will close. When the generator switch is turned on, the D.V. points close, thus completing the D.V. coil circuit. The current flowing in the D.V. coil closes the polarized pilot points. When contact points are *polarized*, they are magnetized so that they will close when subjected to a field of one polarity and will open when a field of the opposite polarity is applied. The pilot points in the diagram are polarized to close when the generator voltage is higher than the battery voltage and to open when the differential voltage is in the opposite direction.

In Fig. 9-51, it can be seen that the generator switch must be turned on to close the D.V. points. Then, if the generator voltage is higher than battery voltage, the D.V. coil will cause the pilot points to close. At this time, current can flow through the main contactor coil, thus causing the main contact points to close and connect the generator to the battery.

If the generator speed is reduced to a point where the battery voltage is higher than the generator voltage, there will be a current flow through the D.V. coil and the reverse-current coil in a direction which will cause the pilot points to open. As soon as the pilot points open, the main contactor coil is deenergized and the main contact points open. This disconnects the generator from the battery.

The purpose of the D.V. points in the D.V. coil circuit is to disconnect the circuit completely so that battery voltage will not bleed back through the D.V. circuit when the generator is not operating. The D.V. points are always open when the generator switch is turned off.

Maintenance and Adjustment of Generator-Control Units

It must be emphasized that some adjustments of the control units described in this section must not be made while the unit is mounted on the aircraft. Other ad-justments can be made in accordance with the manu-facturer's service instructions.

The operation of the vibrator-type voltage and current regulators depends on the spring tension which normally holds the contact points closed and on the air gap between the armature and magnet. If the generator voltage is too low, it can be increased by increasing the spring tension on the voltage points. This can usually be accomplished by bending the tabs to which the armature springs are attached or by adjusting the screw which controls the spring tension. If the air gap is not correct, it can be adjusted by loosening the screws and adjusting the contact bracket position.

The cutout relay may require three adjustments. These are the air gap, the point opening distance, and the point closing voltage. The air gap and the point opening adjustments must be made with the battery disconnected. The air gap is adjusted by loosening the screws which hold the armature bracket and raising or lowering the armature. The point opening distance is checked by placing a feeler gage between the contact points when they are open. The opening can be adjusted by bending the upper armature stop or by some other method recommended by the manufacturer.

The closing voltage is checked by connecting a precision voltmeter between the generator terminal of the control unit and ground. The engine speed is then increased until the relay closes. The reading of the voltmeter at this time will give the closing voltage. The closing voltage is adjusted by means of a screw which controls the spring tension on the armature.

A carbon-pile voltage regulator should not be adjusted on the aircraft except with the voltage rheostat. If the generator voltage cannot be adjusted satisfactorily by this means, the regulator should be removed and taken to an electrical shop, where the pile screw and air gap adjustments can be made in accordance with the indications of precision measuring instruments.

Polarizing the Generator Field ("Flashing the Field")

It has been explained that voltage is induced in the armature windings of a generator when the windings are cutting across lines of magnetic force, that is, a magnetic field, as the armature rotates. It has been further pointed out that the magnetic field is produced as the result of current flow through the field windings. Since the direction (polarity) of the generator output is dependent on the direction (polarity) of the magnetic field, we can see that the field must be established in the proper direction if the generator is to function correctly.

Occasionally the field poles of a generator will lose their residual magnetism or will have their polarity reversed as the result of an accidental reversal of current through the field windings. In such cases it is necessary to reestablish the correct polarity by flashing the field. This can be done by various means, but the principle is the same in each case. It is merely necessary to pass a current through the field windings in the proper direction.

For a generator system which completes the ground

circuit of the field through the voltage regulator, it is merely necessary to connect a jumper momentarily from the BAT terminal of the regulator to the GEN terminal. This will cause a current to flow through the generator and field in the correct direction to reestablish the field polarity.

If the generator has the field grounded within the generator case, a positive voltage must be applied to the external field terminal. This is accomplished by connecting a jumper momentarily from the BAT terminal to the FLD terminal of the regulator.

When it is necessary to repolarize the field of a generator in a system utilizing a carbon-pile voltage regulator, the regulator should be removed before a positive voltage is applied to the field terminal. The positive voltage can be obtained from the BAT terminal of the reverse-current relay or from any other terminal which has direct connection to the positive terminal of the battery. A jumper should be connected from the positive battery source to the A (field) terminal of the regulator base or to the field terminal of the generator. If the system incorporates a field switch or a field relay in the field circuit, it will be necessary to close the switch or relay contact to allow current to reach the field terminal of the generator when the jumper is connected from the positive battery source to the A terminal of the regulator base. In all cases, the jumper should be connected for only a fraction of a second.

Generator Troubleshooting

Generator troubleshooting by the technician depends on a thorough understanding of the principles explained in this chapter. We may summarize these principles as follows:

1. Generator voltage and current are produced according to the principles of electromagnetism; that is, when a conductor cuts across (links with) a magnetic field, a voltage is induced in the conductor.

2. The voltage produced by a generator is dependent on the strength of the field, the speed of rotation, and the number of conductors in the armature.

3. The magnetic field for a generator is developed (excited) as a result of current flow through the field windings of the generator.

4. Generator voltage is usually governed by means of a voltage regulator which controls the amount of current flowing in the field windings.

5. The reverse-current cutout relay is necessary to disconnect the generator from the battery when the generator voltage is lower than that of the battery.

6. The reverse-current relay closes when the generator voltage reaches a predetermined value and opens when reverse current flows from the battery to the generator. The reverse current value is preset by adjustment of the relay and is usually between 10 and 20 A.

7. The commutator of a generator acts as a switching device to carry current from the rotating armature to the brushes and thence to the external circuit. The commutator also acts as rectifier to change the alternating current in each winding of the armature to a direct current in the external circuit.

8. The brushes are made of carbon or a carbon compound and maintain constant contact with the commutator under pressure of brush springs. The brushes must be located on the neutral plane of the commutator to prevent short-circuiting between commutator bars and arcing at the brush contact.

When generator trouble exists, the technician should think through the possible causes and then eliminate the causes one by one until the difficulty is located. As an example of a trouble which may be encountered, the generator voltage for a particular system may reach a maximum value of 3 V, regardless of the engine speed. Since a small voltage is being produced, the technician will know that the generator is operating on residual magnetism only and that there is no current flowing through the field circuit. The first unit to check is the voltage regulator because it controls the current through the field. The technician may find that a wire to the regulator is broken or that the regulator is loose on its base and is not making contact at all points. If the regulator and its connections are found to be satisfactory, then it will be necessary to trace the field circuit from the regulator to the generator. If this circuit is intact, it will probably be necessary to trace the circuit within the generator. This can be done with a continuity tester and an ohmmeter or test light. As a last resort, the generator must be removed and repaired.

For troubleshooting generator systems it is useful to employ a troubleshooting chart. The accompanying chart on page 220 is typical of troubleshooting charts for DC generator systems.

Generator Inspections, Maintenance, and Repairs

Between engine overhaul periods the generators should require a minimum of service. Normal service usually includes lubrication of bearings (except for generators with sealed bearings), periodic inspections of circuitry, and checking of output. If inspections reveal that brushes are worn beyond acceptable limits, the brushes should be replaced. If the commutator shows signs of arcing or pitting, the cause should be determined and the commutator resurfaced.

A commutator which is slightly pitted or dirty may be resurfaced by the use of fine sandpaper or an abrasive stick designed for this purpose. The sandpaper is placed over the square end of a small wooden stick and then applied to the commutator while the generator is running. The abrasive will clean the surface of the commutator and will also be carried under the brushes, thus grinding the faces of the brushes to fit the contour of the commutator.The abrasive stick will serve the same purpose. Emery cloth must not be used for this purpose because the emery grit is conductive and will lodge between the commutator bars, thus short-circuiting the bars and causing arcing.

The brush springs should be checked with a brush-spring scale to determine whether the spring tension is correct. The hook end of the scale is placed under the brush spring, and the scale is lifted until the pressure is completely off the brush. The scale is read at this time to find the brush-spring tension. If the tension

TROUBLESHOOTING CHART

Trouble	Probable cause	Remedy
Generator produces voltage, but ammeter reads zero when a load is turned on.	Generator switch not turned on.	Turn on the switch.
	Main generator circuit breaker open.	Reset circuit breaker.
	Defective or improperly connected ammeter.	Check wiring to ammeter. Replace ammeter if necessary
	Defective reverse-current relay.	Repair or replace reverse-current relay.
No voltage or amperage from generator.	Polarity of field reversed or lost.	Repolarize the field.
	High resistance between the brushes and commutator.	Repolarize the field. Clean the commutator.
	Generator armature burned out.	Replace generator.
	Generator drive shaft broken.	Replace generator.
	Armature grounded.	Replace generator.
	Terminal connections faulty.	Correct faulty condition.
	Brushes binding.	Remove and clean brushes.
	Brush spring tension too low.	Adjust or replace brush springs.
Residual voltage only.	Faulty connections at voltage regulator.	Correct faulty connections.
	Open circuit in voltage regulator.	Replace voltage regulator.
	Open field circuit or open field in generator.	Check field circuit and continuity of generator field.
	Generator field circuit grounded.	Replace generator or correct grounded circuit.
Voltage too high.	Voltage regulator not properly adjusted.	Adjust voltage regulator.
	Contact points in voltage regulator stuck.	Repair or replace points.
	Open circuit to voltage coil in regulator.	Replace regulator.
	Open circuit in ground lead of regulator (for generator with positive field terminal).	Repair regulator ground.
Generator voltage too low.	Regulator not correctly adjusted.	Adjust voltage regulator.
	Faulty regulator.	Replace regulatory.
	Faulty voltmeter.	Replace voltmeter.
	Defective armature.	Replace generator.
Voltage fluctuates.	Loose or dirty connections in field circuit.	Clean and tighten connections.
	Voltage-regulator contact points dirty or pitted.	Repair contact points.
	Regulator base contacts loose or dirty.	Clean and tighten base contacts.
	Generator brushes worn or binding.	Replace or clean brushes.
	Commutator dirty or pitted.	Clean or repair commutator.
Excessive arcing at brushes.	Worn or binding brushes.	Clean or replace brushes.
	Brushes not correctly located.	Adjust location of brushes on commutator.
	Dirty, rough, or eccentric commutator.	Clean or resurface commutator.
	Brush spring tension too low.	Adjust or replace brush springs.
Generator burned out after operation.	Main contact points in reverse-current relay stuck closed.	Replace main contact points and replace generator.
	Generator load too high.	Install generator with capacity for load.
Battery has low charge.	Generator capacity too low.	Install generator with capacity for load.
Improper division of load in a parallel system.	Generator switch not turned on.	Turn on all generator switches.
	Voltage adjustments not properly balanced.	Adjust regulators to balance load.
	Equalizer circuit defective.	Correct faulty conditions in equalizer circuit.

is not sufficient, the springs should be adjusted or replaced.

For a particular make or model of generator, the technician should refer to the manufacturer's maintenance manual. Generator information is included in the aircraft maintenance manual.

● ALTERNATING-CURRENT GENERATOR SYSTEMS

In the majority of modern aircraft, alternating-current generators (alternators) are used to produce the electric power necessary for operating the various electric and electronic units in the aircraft. In light aircraft, the three-phase alternating current produced by the alternator is usually converted to direct current by means of a full-wave rectifier (see Fig. 9-38). The direct current is then utilized as in the standard DC system.

In large aircraft, the alternating current is distributed in a three-phase electric system to drive motors and operate other equipment as required. Where DC power is needed, the AC is rectified and the resulting DC is distributed to the appropriate units.

Transistorized and Transistor Voltage Regulators

With alternators in aircraft electric systems, it is common practice to utilize **transistorized** or **transistor** voltage regulators. In these regulators, transistors are used as the controlling elements for field current. In a vibrator-type voltage regulator, all the field (exciting) current for the alternator must pass through the vibrator contact points. The arcing which occurs as the points open and close causes the contact points to wear away. In the *transistorized* regulator, the field current passes through the transistor and only a small control current passes through the vibrator points. For this reason, the points remain in good condition much longer than they would in the vibrator regulator.

In the *transistor* voltage regulator, the control current and the field current for the alternator are both carried by transistors. Thus, there are no contact points to burn or wear and the service life of the regulator is much greater than it is for the other types of regulators.

For a detailed description of transistorized and transistor voltage regulators, the student should refer to the related text, *Aircraft Electricity and Electronics*.

The use of alternators in aircraft electric systems eliminates the need for the reverse-current relays described previously. The rectifiers and diodes used in the AC systems automatically prevent the reverse flow of current which can occur in the DC systems.

Constant-Speed Drive

Large multiengine aircraft utilize a constant-speed drive (CSD) on the accessory section to drive the AC generators. The CSDs are required in order to synchronize the generators and permit them to operate in parallel. A typical CSD unit and system include a governor, hydraulic pumps, hydraulic motors, a differential transmission, an oil supply, and associated control circuitry.

The CSD operates in **overdrive** when it must increase the rpm of the generator, in **straight through** when the engine drive speed is equal to the required generator speed, and in **underdrive** when the engine drive speed is greater than that required for the generator. The governor continually senses the generator speed and signals the CSD unit to make appropriate corrections. The generator speed must be such that its output frequency is 400 Hz.

Additional information on CSD units is given in *Aircraft Electricity and Electronics*.

● THE ELECTRIC MOTOR

Principles of the DC Motor

A typical DC motor is constructed in much the same manner as a DC generator. In fact, some DC motors can be used either as generators or as motors.

The basic principle of a DC motor is illustrated in Fig. 9-52. In the first diagram the current flow is such that the armature has a north polarity at the top and a south polarity at the bottom. Since like poles repel and unlike poles attract, it is apparent that the armature will rotate in a clockwise direction. As it turns to a point almost in alignment with the field poles, the commutator switches the connections to the power source and the polarity of the armature is reversed. The like poles are repelled again, and the armature continues to rotate in a clockwise direction.

The armature of a conventional DC motor is an iron cylinder made up of a stack of laminations. The laminations prevent the flow of eddy currents in the armature, thus preventing overheating and loss of power. The windings of the armature are placed in slots on the surface in the manner described for a generator armature. The ends of the windings are connected to commutator bars so that the windings are in series with one another. When electric power is connected to the commutator on opposite sides, the current flow through the commutator windings will be such that one side of the armature will have a *north* polarity and the other side will have a *south* polarity. This is illustrated in Fig. 9-53. When this condition exists, the side of the armature with the north pole will be repelled from the north field pole and attracted to the south field pole. The same set of conditions exists with respect to the opposite side of the armature. Since the connections to the armature are continually changing as the commutator turns, the top of the armature in the drawing will always have a north polarity and the bottom will always have a south polarity. Thus, the motor will continue to rotate as long as power is applied.

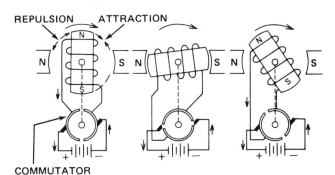

FIG. 9-52 Principle of the DC motor.

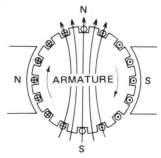

FIG. 9-53 Polarity of a motor armature.

Direct-current motors are often constructed with four field poles. In a motor of this type the armature must be wound in such a manner that four poles will be produced when electric current is flowing through the windings. A schematic diagram illustrating the field of a four-pole armature is shown in Fig. 9-54. This armature is wound so that the two sides of each armature coil are 90° apart. Hence, current flowing into the drawing in one winding will be coming out at a point 90° from the point of entrance. The location of the four brushes on the armature is such that the armature poles will be produced at a position with respect to the field poles where the greatest forces of repulsion and attraction are developed. The brushes are alternately negative and positive; that is, brushes 180° apart are of the same polarity and are connected together.

Types of Motors

Electric motors are generally designed in three different types. These are shunt-wound, series-wound, and compound-wound. The electrical arrangement of each of these types is shown in Fig. 9-55.

The **shunt-wound** motor is also referred to as a **constant-speed** motor because it will always operate at nearly the same rpm, even under a wide variety of loads. When the motor is first started, it will draw a comparatively large surge of current because the armature windings provide a low resistance. However, a voltage will be induced in the armature windings as it rotates because the windings are cutting across the magnetic field produced by the field poles. This induced voltage in the armature is opposite to the voltage being applied from the outside power source. Hence,

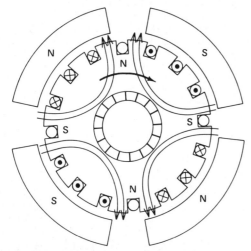

FIG. 9-54 Field of a four-pole motor armature.

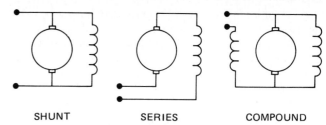

SHUNT   SERIES   COMPOUND

FIG. 9-55 Circuits for different types of electric motors.

as the armature speed increases, the induced back voltage (back emf) in the armature will also increase. When the back emf is approximately equal to the applied voltage, the armature speed will no longer increase because the armature field has decreased almost to zero. If a mechanical load is applied to the motor, the armature will tend to slow down and the back emf will decrease. This will allow more current to flow through the armature from the outside power supply, and the armature magnetic field will increase. This, of course, will increase the torque of the armature, and the motor will drive the load if the load is not beyond the capacity of the motor. As the mechanical load to the motor is increased, the current flow through the motor will also increase.

The shunt-wound motor produces a comparatively low starting torque because the current flow through the field windings is low when the motor is first started. This is because the armature windings take the major portion of the current. As the motor speed increases, the armature current decreases and the field current increases, thus increasing the torque. Because of the low starting torque, the shunt-wound motor is not used where the mechanical load is high at the time the motor is started.

The **series-wound** motor produces a high starting torque because all the current must pass through both the field and the armature windings. For this reason the series motor is used for starting engines and in other applications where a high starting torque is required.

In general, a series motor should not be allowed to operate without a mechanical load applied. If the motor is not driving a load, the rpm may continue to increase until the centrifugal force in the armature causes it to fly apart. This is because the back emf generated in the armature is not sufficient to balance the applied voltage.

The series motor cannot be used with loads requiring a constant speed because the series motor speed varies according to load. When the mechanical load is high, the motor speed will be low, and vice versa.

The **compound-wound** motor incorporates features of both the shunt and series motors. Since it has a series field, it will produce a good starting torque, and because of the shunt field, it will have a reasonably stable speed characteristic. Compound motors are therefore employed where a substantial starting load exists and a steady rpm is desired.

Inspection and Maintenance of Motors

The inspection and maintenance of electric motors are similar to those required for generators. Very little maintenance is normally required, with the possible exception of bearing lubrication. If the motor has sealed bearings, no lubrication is required. Periodic inspections of the brushes and commutator will enable the technician to correct any discrepancies before failure occurs.

● AIRCRAFT STORAGE BATTERIES

Theory of Electrochemical Action

The production of an electric current through electrochemical action is based upon the fact that dissimilar metals, when placed in a particular chemical solution, will react with the solution and with each other, causing at least one of the metals to be decomposed and free electrons to be released.

If we place a strip of zinc and a strip of copper in a glass container so that the strips of metal do not touch each other and then pour a solution of sulfuric acid in the container, we note that chemical action takes place at the zinc strip while the copper strip is apparently not affected. The action of the zinc strip is shown by the formation of hydrogen bubbles which will break free and rise to the surface of the liquid. These bubbles are the result of the combination of zinc ions with the sulfate ions of the sulfuric acid to form zinc sulfate. As each zinc ion combines with a sulfate ion, two electrons are released at the zinc strip. At the same time, hydrogen is released from the sulfuric acid molecules, and this hydrogen forms in bubbles at the zinc plate. The chemical action which takes place may be expressed as follows:

$$Zn + H_2SO_4 = ZnSO_4 + H_2$$

Since electrons are released on the zinc strip, the strip becomes negatively charged. If we connect the zinc strip and the copper strip to a voltmeter as shown in the drawing of Fig. 9-56, we observe that there is a potential difference of more than 1 V between the two strips. When the two strips are connected together, the electrons will flow from the zinc strip to the copper strip.

All ordinary batteries, including dry cells and storage batteries, operate according to the principle explained above. If the battery elements are consumed, as in the case of the zinc strip, the battery cannot be recharged. On the other hand, some combinations of materials make it possible to reverse the chemical action which takes place when the battery is discharging. A battery or cell of this type is commonly called a **storage** or **secondary** battery.

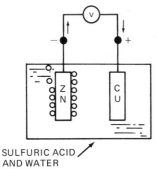

FIG. 9-56 Electrochemical action.

The Lead-Acid Storage Battery

The **lead-acid** storage battery is so named because the active materials are lead, lead peroxide, and sulfuric acid. The positive plates of the battery are composed of a compound of lead peroxide (PbO_2), and the negative plates are composed of pure, spongy lead (Pb). The active materials are held in lead grids, thus forming the plates.

The chemical action which takes place in a storage battery as the battery discharges involves the combining of sulfuric acid with the active material in the plates to form lead sulfate and water. The sulfuric acid (H_2SO_4) breaks up into hydrogen (H_2) ions and sulfate (SO_4) ions. The sulfate ions combine with the lead of the negative plate to form lead sulfate ($PbSO_4$) and with the lead peroxide (PbO_2) of the positive plate to form lead sulfate and oxygen (O_2). The combining of the lead in the negative plate with the SO_4 ions releases electrons on the negative plate. This, of course, is what makes the plate assume a negative charge. The oxygen combines with the hydrogen ions in solution to form water. When the oxygen ions leave the positive plate, they remove electrons from the plate, thus causing it to assume a positive charge.

The potential difference between the plates of a lead-acid storage cell is about 2.2 V when there is no load on the cell. Three cells are connected in series to make a 6-V battery, and six cells in series to make a 12-V battery. Thus, a storage battery of any desired voltage can be produced by connecting the correct number of secondary cells in series.

Construction of Storage Batteries

A typical storage battery is composed of parts such as those illustrated in Fig. 9-57. These are the battery case, made of hard rubber or plastic; the cell containers, which may be integral with or separate from the battery case; the plates joined by plate straps to form plate groups; the plate separators, composed of

FIG. 9-57 Construction of a lead-acid storage battery.

wood, porous rubber, or glass fiber; the terminal posts, attached to the plate straps; the separator protector; the cell covers in which are placed the vent caps; the cell connectors and the electrolyte.

As previously mentioned, the **plates** of the lead-acid cell consist of lead grids in which is held the active material. The active material for the positive plate is lead peroxide, and for the negative plate it is spongy lead. The lead material for the grid includes additional elements to strengthen and improve the conductivity of the plates. Among the elements commonly used in the grids are antimony and silver.

The plates are assembled into plate groups and are connected together by means of the **plate straps.** The negative group includes one plate more than the mating positive group in order to provide a negative plate on the outside of each of the assembled groups. Since the positive plates are softer and less durable than the negative plates, the outer negative plate in each cell groups serves to protect the positive plates.

The **plate groups** are assembled into cell groups with separators between each pair of plates. These **plate separators** are made of wood, or they may include glass-fiber wool mats together with wood or porous rubber separators. The separators must allow a free passage of electrolyte so that they will not interfere with the electrical and chemical action which must take place within the cell. When wood separators are employed, the ribbed side of each separator is placed next to the positive plates to permit the shedding of decomposed active material.

Cell containers are designed to provide support for the assembled plate groups and to hold the electrolyte. These containers are made of hard-rubber composition or plastic. In the bottom of each cell container are four ribs to serve as plate supports. Two of these ribs support the positive plates, and two support the negative plates. This arrangement leaves a space underneath the plates for the accumulation of sediment, which would otherwise tend to short-circuit the plates.

Cell covers are sealed into the tops of the cell containers with a special sealing compound. The compound is applied in a molten condition so that it will completely fill the space between the cover and the container. When the compound cools, it remains sufficiently soft to provide a cushioning effect between the cover and the container and between the separate cells.

The **terminal posts** extend through holes in the cell covers to provide means for connecting the cells to one another. Usually, a lead strap called the cell connector is fused into the terminal posts to connect the negative post of one cell to the positive post of the next, thus providing a series circuit. The spaces around the terminal posts are sealed to prevent the leakage of electrolyte.

The **vent caps,** which are screwed into openings in the cell covers, are designed to prevent the leakage of electrolyte and also to allow gases to escape from the cells. Common types of nonspill vent caps are illustrated in Fig. 9-58. This illustration shows the tubular vent cap used in a battery which has a large space above the plates to hold the electrolyte when the bat-

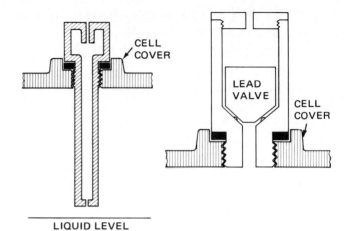

FIG. 9-58 Nonspill vent caps for storage batteries.

tery is on its side or inverted, the open tip of the cap is positioned where it can never be covered by the electrolyte. The other type of cap shown is hollow and contains a lead valve which closes the vent hole whenever the battery is in a position where the electrolyte could flow out the vent. It must be emphasized that each battery cell must be vented to allow the escape of hydrogen gas. During the charging process, especially when the battery is near full charge, a substantial amount of hydrogen gas is released from the electrolyte. This gas must be vented to prevent a pressure explosion of the cell.

A recently designed battery vent cap incorporates a sintered alumina (aluminum oxide) plug in the top of the cap. This plug is sufficiently porous to allow the escape of gas, but it will not allow the passage of electrolyte.

Some aircraft storage batteries are constructed with an integral metal case on the outside of the rubber or plastic case. This metal case, which is usually made of aluminum, protects the battery from damage and also serves as a radio shield. The metal cover of the case provides a sealed compartment which is vented to the outside of the airplane, thus providing for the removal of explosive hydrogen and corrosive acid fumes.

The **main battery terminals** are usually brass studs to which cables can be attached by means of washers and wing nuts. In many instances, smooth prongs are screwed on the terminal studs to provide for quick-disconnect fittings. A quick-disconnect fitting manufactured by the Cannon Electric Company is shown in Fig. 9-59. This plug-type fitting is quickly connected or disconnected merely by turning the large knob on the outside of the fitting.

A more recently manufactured quick-disconnect plug is shown in Fig. 9-60. This is called the Elcon connector and is manufactured by Icore International. The Elcon connector consists of two main assemblies: the terminal assembly attached to the battery to serve as a receptacle, and the connector plug assembly to which the battery cables are connected. The plug assembly is inserted into the receptacle on the battery and is seated firmly by means of the center screw (worm) in the plug. The worm is a cam-locking device that seats securely when the wheel or T-handle is rotated fully to the right.

BASE–ATTACHED TO BATTERY CASE

PLUG ASSEMBLY–ATTACHED TO CABLES

FIG. 9-59 Quick-disconnect plug.

Alkaline Batteries

The alkaline-type secondary (storage) battery provides a very long life with great dependability; however, the cost is greater than that of the lead-acid cell. The nickel-cadmium battery is considered one of the best of the alkaline cells, and it is manufactured in a wide variety of sizes and shapes. The positive plates of the cell are composed of nickel hydroxide mixed with a special type of graphite, and the negative plates are made of a mixture of cadmium oxide and iron oxide. The electrolyte is potassium hydroxide dissolved in water. Cell containers are constructed of nickel-plated steel.

The working voltage of the nickel-cadmium cell is about 1.2 V; hence, a 6-V battery contains five cells. As is true of the lead-acid battery, the capacity of the nickel-cadmium battery is dependent on the total plate area and the thickness of the plates.

One of the principal advantages of the nickel-cadmium battery is its long life. The battery can be left in any state of charge for several years without any appreciable deterioration.

Storage Battery Capacity and Ratings

Storage batteries are rated according to voltage and ampere-hour (Ah) capacity. The ampere-hour rating is usually based upon a discharge period of 5 h; hence, when we state that a particular battery is rated at 34 Ah, we mean that the fully charged battery can furnish current at the rate of approximately 6.8 A for a period of 5 hr before the battery is discharged. The total amount of power a battery can supply is affected by the rate of discharge. If a battery is discharged at a rate greater than the 5-h rate, the total power furnished will be less than it would be at the 5-h rate. On

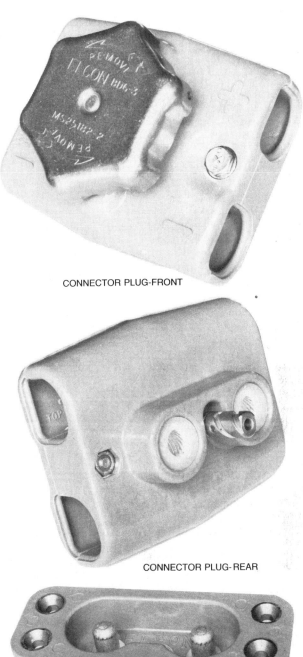

CONNECTOR PLUG-FRONT

CONNECTOR PLUG-REAR

RECEPTACLE-ATTACHED TO BATTERY

FIG. 9-60 Elcon quick-disconnect battery connector. (*Elcon Div., Icore International*)

the other hand, if the battery is discharged at less than the 5-h rate, the total power available will be greater than it would at the 5-h rate.

As previously mentioned, the capacity of a battery depends upon the total area of the plates and the thickness of the plates. As a rule, when the plates are the same size and thickness, the capacity will be proportional to the number of plates.

Testing Lead-Acid Storage Batteries

The most common method for testing the state of charge in a storage battery is to use a hydrometer and test the specific gravity of the electrolyte. The **electrolyte** for a fully charged battery should have a specific gravity of 1.300. As a battery ages, this will slowly decrease, partially because of hard sulfate formation in the plates and partially because a small amount of sulfuric acid is carried away as fumes. It is common practice to consider the battery fully charged if the specific gravity of the electrolyte is between 1.275 and 1.300.

If the specific-gravity reading is less than 1.240 for an aircraft battery, the charge is considered low, because at this level the battery will not carry a heavy, sustained load such as that required for starting the engine for more than a very few minutes. The procedure for testing a storage battery with the hydrometer, illustrated in Fig. 9-61, is as follows:

1. Remove the cell cap, and insert the tip of the hydrometer into the electrolyte.
2. Squeeze the bulb of the hydrometer and release it to draw the electrolyte into the tube.
3. When sufficient electrolyte has entered the tube to float the hydrometer indicator, hold the tube vertically so that the indicator will float free.
4. Read the numerical indication at the liquid level on the float stem.

To test the general condition of a storage battery, a **high-rate discharge tester** is often used. This tester consists of a heavy load resistance and a voltmeter. The prongs of the instrument are pressed against the two terminals of the cell being tested, and the reading of the voltmeter is noted. Normally, for a good cell which is fully charged, the reading may be as low as 1.7 V,

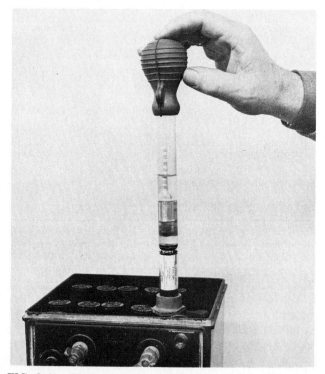

FIG. 9-61   Testing a storage battery with a hydrometer.

and this reading will hold steady for the short time that the tester is connected. If the reading is somewhat lower than 1.7 V and starts to decrease immediately, the cell is either discharged or worn out. If all the cells of a battery give the same reading, the battery is in good condition.

Another simple but effective test is to connect a nominal load (10 to 15 A) to the battery and then read the voltage of each cell with a voltmeter. Cells in poor condition will read lower than the good cells, and a cell which is short-circuited will give a reverse reading.

Charging Storage Batteries

Storage batteries can be charged by either one of two methods. On the aircraft or in any system utilizing a voltage regulator to control the generator voltage, the battery is charged by the **constant-voltage** method. This means that the charging source is adjusted to provide a voltage slightly above that of the battery. In a 24-V system, the battery is nominally rated at 24 V and the generator is set to deliver 28.5 V. The actual voltage of the battery will rise to about 26.4 V as it reaches the fully charged condition; hence, the difference between the battery voltage and the generator voltage will be only 2.1 V when the battery is fully charged. This small voltage is enough to maintain the charge of the battery, but it will not cause it to overcharge or gas excessively.

During the operation of an airplane the electric power system is subjected to a variety of electric loads. As loads, such as radio or lights, are turned on, there is a very slight voltage drop in the generator output. This is quickly sensed by the voltage regulator, which then causes more current to flow through the generator field, thus bringing the generator voltage back to the required level. When this occurs, the generator amperage is increased to take care of the additional electric load, and the battery is not required to supply power for the circuit.

It should be noted that the battery does not supply any power for the aircraft electric system during normal operation of the aircraft. In flight, the battery could be completely disconnected and the aircraft electric system would continue to operate satisfactorily. The battery does serve to stabilize the electric system by absorbing voltage surges and filling in when the voltage drops momentarily as a result of turning on a heavy motor load.

The constant-voltage system may be employed to charge a large number of batteries, provided that the capacity of the source is sufficient to maintain the voltage at a constant level and provided that the batteries all have the same voltage rating. When this system is employed, all the batteries are connected in parallel as shown in Fig. 9-62. The ampere-hour rating of a battery makes little difference when charging by the constant voltage method. The low-rated battery will charge more rapidly than the others, but when it is fully charged, it will draw very little current from the system.

In **constant-current** charging, the batteries must have the same ampere-hour rating because they are connected in series and all receive the same current. If some batteries have a lower capacity than others, the low-capacity batteries will be charged first and will gas

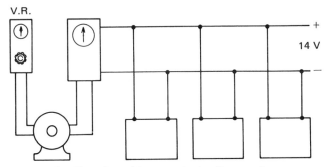

FIG. 9-62 Charging with a constant-voltage system.

furiously as the charging current continues to pass through them.

The usual procedure for constant-current charging is to set the charging source to a given amperage for a specified number of hours and then to reduce the amperage in steps as the batteries near the fully charged state. The batteries must not be charged too rapidly, or they will become overheated and boil. This will cause the loss of electrolyte and may buckle the plates, thus seriously damaging the battery. In general, it is good practice to keep the battery temperature below 110° F [43° C] when charging.

Servicing Nickel-Cadmium Batteries

The servicing of nickel-cadmium batteries varies considerably from that of lead-acid batteries. For example, it is not possible to determine the state of charge of a nickel-cadmium battery either with a hydrometer or with a voltmeter. The specific gravity of the potassium hydroxide electrolyte remains constant regardless of the state of charge. Furthermore, the voltage remains at the same level, either open or under load, until the battery is completely discharged. The only method to determine the state of charge is to charge the battery from a completely discharged condition for a given number of hours at a carefully controlled rate. The time required for charging will indicate the state of charge.

A nickel-cadmium battery can be charged either by the constant-voltage method or the constant-current method. With constant-voltage charging, the battery is charged with the voltage $1\frac{1}{2}$ times the number of cells, regardless of cell capacity. For example, a 19-cell battery would be charged at 28.5 V. Under these conditions, the battery would be charged in 4 h.

When constant-current charging is employed, the battery is commonly charged either at the 5-h rate or the 10-h rate. The current level used depends on the capacity of the battery. For example, charging a 15-Ah battery at the 10-h rate requires a current flow of 1.5 A for 14 h. Charging a 40-Ah battery at the 10-h rate requires 4.0 A for 14 h. It will be noted that the charging rate is $\frac{1}{10}$ the ampere-hour capacity and the charging time is 14 h.

A standard service procedure recommended approximately every 100 h for nickel-cadmium batteries is **capacity reconditioning.** This process involves the complete discharge of the battery and then a recharging to the fully charged state. The purpose of this procedure is to ensure that all cells have approximately the same charge. During the operation of a nickel-cadmium

battery, some cells discharge and charge at slightly different rates than others. Over a period of time, the state of charge among the cells varies considerably and it is necessary to restore cell balance by capacity reconditioning.

During inspections, nickel-cadmium batteries should be checked for water level. Water should be added when it is found that the level is not at the point specified in the battery manual. Care should be taken to avoid the spilling of electrolyte and to prevent the electrolyte from coming in contact with any part of the body. Potassium hydroxide electrolyte is much like household lye and will cause severe burns. Electrolyte spilled in any location should be washed away immediately with water.

For a more detailed discussion of nickel-cadmium batteries, read the section covering such batteries in *Aircraft Electricity and Electronics*.

Safety Precautions

There are three principal sources of danger when handling and servicing storage batteries. These are (1) the presence of hydrogen gas, (2) the corrosive and burning action of the sulfuric acid or potassium hydroxide in the electrolyte, and (3) the possibility of short-circuiting the terminals.

There have been many battery explosions as the result of igniting the hydrogen gas which is constantly present in the space above the electrolyte. This danger is particularly acute in the vicinity of batteries which are on charge or which have recently been on charge. The technician handling batteries should be constantly alert to this danger and make sure that no open flames or sparks are permitted in the vicinity of batteries. It is especially important that all electric loads and power sources be turned off before connecting or disconnecting a storage battery. If a load is on the circuit, or if a charging source is turned on, sparks will be emitted when the battery is connected or disconnected, and these sparks can easily ignite the hydrogen gas.

The handling of sulfuric acid and electrolyte requires great care to avoid the possibility of being burned by the acid. When acid and water are being mixed to provide an electrolyte of the correct specific gravity, the acid should be poured *into* the water carefully and never the reverse. The acid may boil and spit if water is added, and this can result in severe burns. The acid not only will burn the skin but also will eat holes in the clothing and cause severe corrosion of many metals. Therefore, whenever *acid* is spilled, it should be rinsed away with water and neutralized with an alkaline solution such as baking soda and water. Finally, the area should be rinsed with clear water and dried. When the electrolyte from an alkaline battery is spilled, it should be rinsed away with water and neutralized with a mild acid solution. Then the area of the spill should be rinsed with clear water.

A storage battery is capable of delivering a current of several hundred amperes when short-circuited. Care must be taken, therefore, that metal tools, wires, and similar conductive items are not dropped or laid on a battery where they may contact the terminals. If a battery is short-circuited with a pair of pliers or a screwdriver, the arc at the point of contact will melt both the terminal and the tool. A small wire connected between the terminals of a storage battery will immediately become white hot and melt. Short-circuiting a storage battery not only will cause external damage but may also cause the battery plates to warp.

Installation of Storage Batteries

The conditions established by an aircraft manufacturer in the design of the battery installation must comply with the airworthiness requirements set forth by the Federal Aviation Regulations. These include the structural integrity of the aircraft at the battery location and the provision for eliminating explosive and noxious fumes from the vicinity of the battery.

The structure supporting the battery in the aircraft must be such that no damage will occur under any possible operating condition. It must be remembered that a battery represents considerable concentrated weight, and in case of a hard landing, severe loads are placed on the battery mounting structure and adjacent aircraft structure. The means provided for holding the battery in place must also meet the structural requirements. The hold-down bolts and brackets must withstand all possible operating loads.

The battery compartment must be suitably ventilated to remove hydrogen gas and electrolyte fumes. As previously explained, the hydrogen gas is explosive and the acid fumes will corrode all unprotected metal surfaces in the area. If nickel-cadmium batteries are installed in the aircraft, the alkaline electrolyte (potassium hydroxide) will also present corrosion problems; hence, the compartment must be protected against corrosion.

The metal aircraft structure in the vicinity of the battery can be protected by the application of bituminous acid-proof paint. If the battery is located in a sealed compartment, the entire area within the compartment should be coated with acid-proof paint. The compartment should be provided with a venting system to carry away the gases and fumes from the battery. A small ram air scoop on one side can be used to bring air into the compartment, and an outlet vent on the opposite side, headed away from the wind, is satisfactory for the removal of the gases and fumes. The venting system should be designed so that the hydrogen concentration will never exceed 1 percent under any flight conditions.

A battery which is enclosed in a metal case is provided with vent connections to permit the attachment of vent tubes. One tube is connected to a ram air scoop, and the other leads to an outlet which carries the gases outside and clear of the aircraft structure. In some cases a neutralizing sump is placed in the line between the battery case outlet and the tube leading outside the aircraft. This neutralizing sump contains a felt pad saturated with a solution of baking soda to neutralize the acid fumes.

The top of the battery should be kept clean and dry to prevent the slow leakage of current. The battery electrolyte is conductive; hence, current will leak slowly between the battery terminals if electrolyte is permitted to remain on the area between the terminals.

The cable terminals of the battery should be inspected frequently to observe any formation of corrosion. When corrosion is found, the cable should be disconnected from the terminal and all corrosion residue should be removed. After the parts have been thoroughly cleaned and dried, they should be coated with petrolatum and reconnected. The petrolatum will prevent the further formation of corrosion for several weeks.

● REVIEW QUESTIONS

1. Of what is an electric current composed?
2. What is a free electron?
3. What is the difference between a conductor and an insulator?
4. Explain negative and positive charges.
5. What is a static charge?
6. Explain voltage and amperage.
7. Define coulomb.
8. Express Ohm's law in two different ways.
9. If 24 V is applied to a circuit having a resistance of 6 Ω, what will be the current flow?
10. If a current of 5 A is flowing through a resistor with a value of 3 Ω, what will the voltage drop be across the resistor?
11. How much electric power is consumed by a motor which draws a current of 10 A in a 12-V system?
12. What is the total resistance in a circuit when resistors of 3, 6, and 8 Ω are connected in series?
13. Find the total resistance in a circuit having resistors of 4, 8 and 12 Ω connected in parallel.
14. In Fig. 9-63, what is the total resistance?

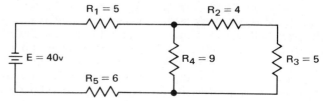

FIG. 9-63

15. In the same circuit, what is the voltage drop across R_3?
16. In the same circuit, what is the current through R_1?
17. What size electric cable would you use in a 28-V circuit for a 1-hp motor which is 75 percent efficient?
18. Define *alternating current*.
19. What is a magnetic substance? A magnetized substance?
20. How can we explain magnetism?
21. What is an electromagnet?
22. How would you determine the magnetic strength of a coil?
23. How would you determine the direction of the magnetic field of a coil?
24. Explain *electromagnetic induction*.
25. State the principle of Lenz's law.
26. Describe a transformer, and tell how the output can be determined.
27. What effect is produced by *capacitance* in an AC circuit?
28. Compare the effect of capacitance with that of *inductance*.

29. How is a capacitor constructed?
30. What is the total capacitance when two 100 μF capacitors are connected in series?
31. Find the capacitive reactance when a 5 μF capacitor and a 10 μF capacitor are connected in parallel in a 400-cyle AC circuit.
32. How much inductance must be placed in the foregoing circuit to produce a condition of *resonance*?
33. What is a three-phase AC circuit?
34. Why is AC power often superior to DC for a power circuit?
35. What electrical principle is used in a generator?
36. Name the principal parts of a generator.
37. What part of a generator circuit is employed for voltage control?
38. Explain the difference between the operation of a vibrator-type voltage regulator and a carbon-pile type.
39. Why is a reverse-current cutout relay required in a generator circuit?
40. Describe the operation of a differential reverse-current relay.
41. How would you polarize, or flash, a generator field?
42. Where would you look for trouble if the generator output was 3 V with the generator operating at normal speed?
43. What trouble would be indicated if you found a generator armature overheated or burned out after normal operation?
44. How do you seat generator brushes?
45. What may be wrong if there is excessive arcing at the brushes?
46. What type of rectifier is used to convert the AC from an alternator to DC for the power circuits?
47. What is the advantage of a transistorized or transistor voltage regulator used with an alternator?
48. Explain the difference between a transistorized and a transistor voltage regulator.
49. Why are reverse-current relays not required for AC electric power systems?
50. Explain what causes the armature of an electric motor to rotate.
51. Why is a series motor used where starting loads are high?
52. What may happen if a series motor is operated with no load?
53. What type of motor is used for constant-speed operations?
54. Explain electrochemical action.
55. What is the open-circuit voltage of a lead-acid cell?
56. Describe the construction of a lead-acid cell.
57. What is the full-charge reading of a hydrometer for a lead-acid cell?
58. What electrolyte is used in an alkaline battery?
59. How is the capacity of a storage battery indicated?
60. What precautions must be taken in the handling of an electrolyte?
61. Give two methods for testing lead-acid cells.
62. What is meant by *constant-voltage* charging?
63. Why is it not possible to determine the state of charge in a nickel-cadmium battery in the same manner employed for lead-acid cells?
64. What voltage is used when charging a nickel-cadmium battery with the constant-voltage method?
65. What is meant by *capacity reconditioning*?
66. What are the requirements for a battery compartment in an airplane?
67. What precautions must be taken in the vicinity of batteries which are being charged?
68. Describe a battery vent system.
69. What treatment is recommended for the prevention of corrosion at the battery terminals?

10 IGNITION SYSTEMS AND COMPONENTS

● PRINCIPLES OF IGNITION

Ignition Event in the Four-Stroke Cycle

During the first event of the four-stroke five-event cycle, the piston moves downward as a charge of combustible fuel and air is admitted into the cylinder. This is the **intake, or admission, stroke.** During the second event, which is the **compression stroke,** the crankshaft continues to rotate and the piston moves upward to compress the fuel-air-mixture.

As the piston approaches the top of its stroke within the cylinder, an electric spark jumps across the points of the spark plugs and ignites the compressed fuel-air mixture. This is the **ignition event,** or the third of the five events. Having been ignited, the fuel-air mixture burns and expands, and the resulting gas pressure drives the piston downward. This causes the crankshaft to revolve. Since it is the only stroke and event that furnishes power to the crankshaft, it is usually called the **power stroke,** and it is the fourth event. The **exhaust stroke** is the fifth event. These facts have been presented before, but they must be reviewed briefly in order to understand aircraft engine ignition.

The electric spark jumps between the electrodes (points) of a spark plug that is installed in the cylinder head or combustion chamber of the engine cylinder. The ignition system furnishes sparks periodically to each cylinder at a certain position of piston and valve travel.

Essential Parts of an Ignition System

The essential parts of an ignition system for a reciprocating engine are a source of high voltage, a timing device to cause the high-voltage source to function at the set position of piston travel, a distributing mechanism to route the high voltage to the various cylinders in the correct sequence, spark plugs to carry the high voltage into the cylinders of the engine and ignite the fuel-air mixture, control switches, and the necessary wiring. The source of the high voltage may be either a **magneto** driven by the engine or an **induction coil** connected to a battery or a generator. In general, it is correct to state that the source of high voltage on most piston-type aircraft powerplants is a magneto, although some have a combination of magneto and battery ignition. Regardless of the source of high voltage, the purpose is to ignite the fuel-air mixture.

Gas-turbine engines utilize ignition systems principally for starting because the fuel-air mixture is self-igniting after it is first started. It is common practice to turn the ignition on during takeoff and sometimes at high altitudes so that the engine will restart immediately if a flameout should occur.

All parts of the aircraft ignition system are enclosed in either flexible or rigid metal covering called **shielding.** This metal covering "receives" and "grounds out" radiations from the ignition system, which would otherwise cause interference (noise) in the radio receiving equipment installed in the airplane.

Magneto Ignition

Magneto ignition is superior to battery ignition because it produces a hotter spark at high engine speeds and it is a self-contained unit, not dependent on any external source of electric energy.

The magneto is a special type of alternating-current generator that produces electric pulsations of high voltage for purposes of ignition. When an aircraft engine is started, the engine turns over too slowly to permit the magneto to operate; hence, *it is necessary to use a booster coil, vibrating interrupter (induction vibrator), or impulse coupling for ignition during starting.*

The **impulse coupling** is designed to give the magneto a momentary high-rotational speed. This coupling is a springlike mechanical linkage between the engine and magneto shaft which "winds up" and "lets go" at the proper moment for spinning the magneto shaft, thus supplying the necessary high voltage for ignition. During the winding-up process, the starter impulse coupling also retards the spark a predetermined amount to prevent backfiring.

When two magnetos fire at the same or approximately the same time through two sets of spark plugs, this is known as a **double,** or **dual, magneto ignition system.** The principal advantages of the dual magneto ignition system are the following: (1) If one magneto or any part of one magneto system fails to operate, the other magneto system will furnish ignition until the disabled system functions again, and (2) two sparks, igniting the fuel-air mixture in each cylinder simultaneously at two different places, give a more complete and quick combustion than a single spark; hence, the power of the engine is increased. The magnetos, which are identical, may be turned on separately (for testing) or both at the same time (during normal operation), by means of an ignition switch. On radial engines, it has been a standard practice to use the right-hand magneto for the front set of spark plugs and the left-hand magneto for the rear set of spark plugs. All certificated reciprocating engines must be equipped with dual ignition.

Dual-ignition spark plugs may be set to fire at the

same instant (synchronized) or at slightly different intervals (staggered). When **staggered** ignition is used, each of the two sparks occurs at a different time. The spark plug on the exhaust side of the cylinder always fires first because the slower rate of burning of the expanded and diluted fuel-air mixture at this point in the cylinder makes it desirable to have an advance in the ignition timing.

The Magnetic Circuit

The magnetic circuit of the magneto may be designed in different ways. One type of design uses **rotating permanent magnets** having two, four, and even eight magnetic poles. These magnets are often made of **alnico,** an alloy of aluminum, iron, nickel, and cobalt that retains magnetism for an indefinite period of time. The magnets rotate under **pole pieces,** which complete a magnetic circuit through a coil core.

Another type of magneto design makes use of **stationary permanent magnets** and a system of rotating inductors which complete the magnetic circuit through the coil core by two different paths, thereby providing two directions through the coil core for the magnetic flux.

The first type of design is called the **rotating-magnet type,** and the second is called the **inductor-rotor type.** The rotating-magnet type is more widely used in aircraft ignition.

Principles of the Rotating-Magnet Type of Magneto

The properties of the common horseshoe magnet are present in the rotating magnet of the magnetos manufactured by the Engine Products Division of the Bendix Corporation. These magnetos are typical examples of the rotating-magnet type.

The horseshoe-shaped permanent magnet shown in Fig. 10-1 has a magnetic field that is represented by many individual paths of invisible magnetic flux, commonly known as "lines" of flux. Each of these lines of flux within the magnet itself extends from the north pole of the magnet (marked with the letter *N*) through the intervening air space to the south pole (marked with the letter *S*), thereby forming the closed loop indicated in the drawing. These lines of magnetic flux are invisible, but their presence can be verified by placing a sheet of paper over the magnet and sprinkling the paper with iron filings. The iron filings will then arrange themselves in definite positions along the lines of flux which compose the magnetic field represented by the solid lines containing arrows.

The lines of flux repel one another; hence, they tend to spread out in the air space between the poles of the magnet, as shown in Fig. 10-1. They also tend to seek the path of least resistance between the poles of the magnet. A laminated soft-iron bar provides an easier path for the lines of flux flowing between the poles than air; consequently, the lines will crowd together if such a bar (sometimes called a **keeper**) is provided near the magnet. The lines of flux then assume the locations and directions shown in Fig. 10-2, where they are concentrated within the bar instead of being spread out as they were in the air space of Fig. 10-1. This heavy concentration of lines of flux within the bar is described as a condition of high "density."

The direction taken by the lines of flux in the laminated soft-iron bar placed in a magnetic field is determined by the polarity of the permanently magnetized horseshoe magnet. For example, in Figs. 10-1 and 10-2, the direction of flow is from the north pole to the south pole. The direction would be reversed if the north pole were the upper pole in the illustrations and the south pole were the lower pole of the pictures.

If the permanent magnet is made of alnico, Permalloy, or some hardened steel, it can retain a large portion of the magnetism induced in it when it was originally magnetized. Since the laminated iron bar is of magnetically "soft" iron, it does not retain much of the magnetism when magnetic lines of flux pass through it. This makes it possible to change the direction of the lines of flux by turning the magnet over so that the north pole of the magnet in Figs. 10-1 and 10-2 is at the top instead of the bottom of the picture.

Current Induced in a Coil

A horseshoe magnet can be used to show that a current can be generated, or **induced,** in a coil of wire. The coil for demonstration purposes should be made with a few turns of heavy copper wire and connected to a **galvanometer,** an instrument that indicates any flow of current by the deflection of its needle (pointer), as shown in Fig. 10-3.

The lines of flux of the horseshoe magnet pass through, or "link," the turns of wire in the coil when in the position shown in Fig. 10-3. When one line of flux passes through one turn of a coil, it is called one **flux linkage.** If one line of flux passes through five turns of a coil, five flux linkages are produced. If five lines of flux pass through five turns of a coil, there are 25 flux linkages, and so on.

In Fig. 10-3(A), if the horseshoe magnet is brought toward the coil from a remote position to the position shown in the drawing, the number of lines of flux, which are linking the coil, constantly increases during the motion of the magnet. In more technical language, there is a change in flux linkages as the horseshoe magnet moves toward the coil, and this change induces

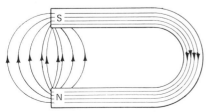

FIG. 10-1 Permanent magnet and field.

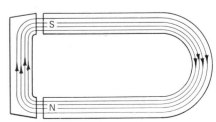

FIG. 10-2 Permanent magnet with flux passing through an iron bar.

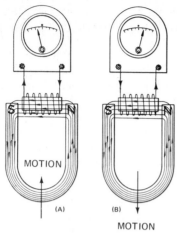

FIG. 10-3 Inducing current with a magnetic field.

a voltage in the coil of wire. This voltage, or electromotive force (emf), causes an electric current to flow around through the circuit, and this is indicated by the deflection of the galvanometer needle.

In Fig. 10-3(B), if the horseshoe magnet is moved back away from the coil, the flux linkages occur in the opposite direction during this movement, and this change in flux linkages induces a current in the coil of wire *in the opposite direction,* as indicated by the movement of the galvanometer needle.

The voltage induced in the coil of wire is proportional to the *rate of change of flux linkages.* The flux linkages can be increased by adding more turns in the coil of wire or by using a stronger magnet having more lines of flux. The **rate** involves an element of time and can be increased by moving the horseshoe magnet near the coil faster, thus increasing the speed of flux change. When any of these methods for increasing the rate of change of flux linkages are tried during an experiment, the galvanometer needle deflection indicates the magnitude of the induced current.

There must be a change in flux linkages to induce a voltage. Voltage is not induced in the coil of wire if the horseshoe magnet is held stationary, regardless of the strength of the magnet or the number of turns of wire in the coil. This principle is applied to the magneto because the lines of flux must have a magnetic path through the coil in the first place and then *there must be a movement of either the coil or the magnet* to produce the change in flux linkages. A voltage in the same proportions could be induced in the coil of wire by holding the horseshoe magnet stationary and moving the coil. This would provide the necessary relative movement to produce the change of flux linkages.

We have previously stated that magnetos can be divided into two types or classes: (1) the rotating-magnet type and (2) the inductor-rotor type. The above explanation of the two methods of changing the flux linkages to induce voltage should clarify this earlier statement.

Whenever there is a current passing through a coil of wire, a magnetic field is established, which has the same properties as the magnetic field of the horseshoe or permanent magnet previously described. An example is the ordinary electromagnet used to operate an electric doorbell.

Lenz's Law

Lenz's law can be stated in terms of induced voltage thus: *An induced voltage, whether caused by self-inductance or mutual inductance, always operates in such a direction as to oppose the source of its creation.*

The same thing can be stated in simpler terms as: When a change in flux linkages produces a voltage which establishes a current in a coil or wire, the direction of the current is always such that its magnetic field opposes the motion or change in flux linkages which produced the current. Lenz's law is of the greatest importance to the operation of the magneto, as explained further in this text.

In Fig. 10-3(A), when the magnet is moved toward the coil, the current flows up the left-hand wire, through the galvanometer, and down the right-hand wire, as shown by the arrows. When the magnet is moved away from the coil, the current flows up the right-hand wire, through the galvanometer, and down the left-hand wire, as shown in Fig. 10-3(B). In this experiment, the coil must be wound exactly as shown in the picture and the poles of the magnet must be as they are shown.

Left-Hand Rule

The polarity of a magnetic field can be determined when the direction of the current and the direction of the winding of a coil are known. If the wire is grasped with the left hand, and if the fingers of the left hand extend around the coil in the direction of the current, the thumb will always point in the direction of the flux, or the north end of the field. This is called the **left-hand rule**.

If the left-hand rule is applied to the current in Fig. 10-3, it will be found that the field which the current establishes opposes the *increase* or change of flux linkages. While the magnet is being moved up toward the coil in Fig. 10-3(A), the usual tendency is to *increase* the flux through the coil core in the direction from right to left of the illustration, as shown by the arrows. However, as soon as the flux starts to increase, a current begins to flow in the coil. It establishes a field of direction from left to right, and this field opposes the increase of magnetic flux and actually exerts a small mechanical force that tends to push the magnet away from the coil.

In Fig. 10-3(B), when the magnet is moving away from the coil, the current flows up the right-hand wire, through the galvanometer, and down the left-hand wire. In accordance with the statement of the left-hand rule, the field of the coil is now helping the field of the magnet. As the magnet is moved away from the coil, the flux linkages occur in the opposite direction and induce a current in the coil which sets up a magnetic field that opposes the change, following the principle of Lenz's law. However, the change is now a decrease; hence, the field of the coil now helps the magnetic field, aiding it in its effort to oppose the change. A small mechanical pull is actually exerted on the magnet by the coil tending to resist the motion of the magnet away from the coil.

If the circuit of the coil is opened, no current can flow in the wire because there is not enough voltage to force the current across the gap where the wire in the circuit is broken or disconnected. To jump the gap requires high voltage, and this would make it necessary to increase greatly the rate of change of flux linkages. We have already discussed the methods of increasing the rate of change of flux linkages. In the case of the simple horseshoe magnet and single coil of Fig. 10-3, if the size of the coil and the size of the magnet were both increased and the rate of movement of the magnet were also increased, the power required to move the magnet rapidly enough to produce the high rate of change of flux linkages would be so great that we would find that our simple experimental device would not be practical. Therefore, the basic design must be changed to provide the compact, efficient source of high voltage required for igniting the fuel-air charge in the engine cylinder. The following pages of this text will tell how this is done.

Rotating Magnet

Figure 10-4 shows a four-pole rotating magnet, similar to those used in some models of Bendix aircraft magnetos. The lines of flux of the rotating magnet, when it is not installed in the magneto, pass from its north pole through the air space to its south pole, in a manner similar to the flow of the lines of flux in the simple horseshoe magnet in Fig. 10-1.

In Fig. 10-5, the **pole shoes** and their extensions are made of soft-iron laminations cast in the magneto housing. The coil core is also made of soft-iron laminations and is mounted on top of the pole-shoe extensions. The pole shoes *D* and their extensions *E*, together with the coil core *C*, form a magnetic path similar to that made by the laminated soft-iron bar (keeper) shown with the ordinary horseshoe magnet in Fig. 10-2. When the magnet is in the position shown in Fig. 10-5, the magnetic path produces a concentration of flux in the core of the coil.

In Fig. 10-6, notice that the rotating magnet has rotated from the position it had in Fig. 10-5. The neutral position of any rotating magnet is that position where one of the pole pieces is centered between the pole shoes in the magneto housing, as shown in Fig. 10-6. When the rotating magnet is in its neutral position, the lines of flux do not pass through the coil core because they are short-circuited by the pole shoes.

Notice especially that primary and secondary windings are *not* shown in the coil core of Figs. 10-5 and 10-6. They are omitted to make it easier to understand the magnetic action. Having learned the action without

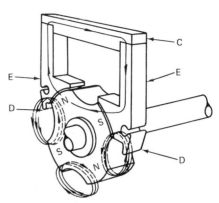

FIG. 10-5 Arrangement of rotating magnet, pole shoes, and core of the coil in a magneto.

windings, the reader will find it easier to understand the functions of the windings explained later in this text.

The curve in Fig. 10-7 shows how the flux in the coil core changes when the magnet is turned with no windings present. This curve is called the **static-flux curve** because it represents the stationary or normal condition of the circuit. If the magnet is turned with no windings on the coil core, the flux will build up through the coil core in first one direction and then the other, as indicated by the curve.

This curve represents both the direction of the flux and its concentration. When the curve is above the horizontal line, the flux is passing through the coil in one direction, and the higher the curve above the line, the greater the number of lines of flux in the core.

When the curve is below the horizontal line, the flux is passing through the coil in the opposite direction, and the lower the curve below the line, the greater the number of lines of flux passing through the core in this other direction.

Whenever the magnet passes through a neutral position, the flux in the coil core falls to zero, and this is shown by the point where the curve touches the horizontal line. Having fallen to zero, the flux then builds up again in the opposite direction, as shown in the curve. Therefore, the greatest change in the flux occurs when the magnet is passing through the neutral position. Note that the curve of Fig. 10-7 has a steep slope at the points corresponding to the neutral positions of the magnet, that is, wherever the curve crosses the horizontal line.

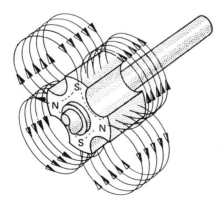

FIG. 10-4 Four-pole rotating magnet.

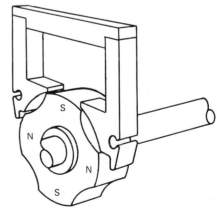

FIG. 10-6 Rotating magnet in neutral position.

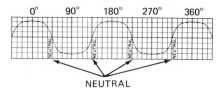

FIG. 10-7 Static-flux curve.

Coil Assembly

The typical coil assembly consists of a laminated soft-iron core around which is placed a **primary** winding and a **secondary** winding. The primary winding consists of a comparatively few turns of insulated copper wire and the secondary winding consists of several thousands of turns of very fine wire. The coil is covered with a case of hard rubber, Bakelite, varnished cambric, or plastic, according to the design requirements of the manufacturer. A primary condenser (capacitor) may be built into the coil between the primary winding and the secondary winding, or it may be connected in the external circuit.

The ends of the coil core extend beyond either end of the coil assembly so that they can be secured to the pole shoe extensions. One end of the primary winding is usually grounded to the core, and the other end is brought out to a terminal connection or to a short length of connecting wire having a terminal on the end. This end of the primary coil is connected in the magnet to the **breaker points,** and provision is also made for a connection to the ignition-switch lead. Note carefully that when the breaker points are closed, current flows from the coil to ground and back from ground to the coil in a complete circuit. This direction of flow alternates with the rotation of the magnet.

One end of the secondary winding is grounded inside the coil, and the other end is brought to the outside of the coil to provide a contact through which the high-tension (high-voltage) current can be carried to the distributor.

This description of a magneto coil does not necessarily apply to all magnetos, but it does give an overall explanation of coil-assembly design and construction. A magneto coil assembly is illustrated in Fig. 10-8.

Breaker Assembly

The **breaker assembly** of a magneto, also referred to as the **contact breaker,** consists of contact points actuated by a rotating cam. Its function is to open and close the circuit of the primary winding as timed to produce a buildup and collapse of the magnetic field. The breaker assembly for one type of magneto is shown as the *contact breaker* in Fig. 10-13 on page 239.

FIG. 10-8 Magneto coil assembly.

Some early magnetos have **lever-type** or **pivot-type** breaker points. These are designed with a movable contact at one end of a lever or arm that is mounted on a pivot. A **cam follower** that rides on the surface of the breaker cam is attached to the breaker arm. Later models of magnetos have pivotless-type breaker assemblies with the movable contact point mounted on a spring-type lever. An additional leaf spring is often used for additional force. Pivotless breaker assemblies are not affected by the wear occurring at the pivot bearings; hence, they remain in adjustment better than pivot types. One type of pivotless breaker assembly is illustrated in the drawing of Fig. 10-9. The breaker contact points for a magneto are made of platinum-iridium alloy or some other heat- and wear-resistant material. The cam and cam follower are lubricated by a felt pad on the cam follower. This pad is saturated with lubricating oil at regular service periods.

The breaker contact points are electrically connected across the primary coil so that there is a complete circuit through the coil when the points are closed and the circuit is broken when the points open. The magneto is timed so that the breaker points close at the position where there is a maximum of magnetic flux through the coil core. At this time there is a minimum of **flux change**.

Primary Condenser

During the operation of the magneto, voltage and current are induced in both the primary and secondary windings of the coil. At the time the breaker points open there is a tendency for the primary current to arc across the points. This results in burning of the contact points and the rate of the collapse of the field is reduced, resulting in a weak spark from the secondary. To overcome this problem, a **primary condenser** (capacitor) is connected across the breaker points. The operation of the primary capacitor is illustrated in Fig. 10-10. The capacitor acts as a storage chamber to absorb the sudden rise of voltage in the primary coil when the breaker points begin to open. The action can be compared to a storage chamber with an elastic diaphragm placed in a hydraulic system. If the fluid valve is suddenly closed, the diaphragm in the chamber stretches to allow fluid to flow into the chamber. This prevents the shock and "banging" that would otherwise occur. In like manner, the primary capacitor prevents arcing between the contact points as they open by absorbing the "inertia" current induced in the pri-

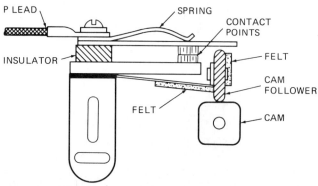

FIG. 10-9 One type of pivotless breaker assembly.

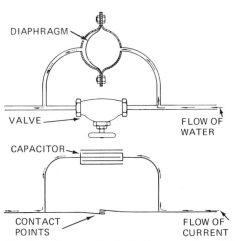

FIG. 10-10 Drawings to illustrate the operation of the primary capacitor.

mary coil by the collapse of the electromagnetic field established by primary current. This field "cuts" the turns of the primary winding as it collapses. The underlying principle is commonly called the **self-inductance of the primary winding.**

The primary capacitor is always connected across the points, but its shape and location vary on different magneto models. It may be round, square, or some other shape. It may be located in the coil housing, in the breaker housing with the breaker points, or on top of the coil. Briefly stated, its function is to absorb self-induced current flowing in the primary circuit.

The primary capacitor must have the correct *capacitance.* Too low a capacitance will permit arcing and burning of the breaker points plus a weakened output from the secondary. If the capacitance is too high, the mismatch between the coil and the capacitor will reduce the voltages developed and result in a weakened spark.

Elements of Magneto Operation

An understanding of the operation of a typical magneto may be enhanced by a study of the several electrical elements or factors that are involved in its operation. The graphic representation in Fig. 10-11 illustrates the strengths and directions of the electrical and magnetic factors in an operating magneto. The factors indicated in the graph are *static flux, breaker timing, primary current, resultant flux, primary voltage* and *secondary voltage.*

The static-flux curve is shown at the top of the graph in Fig. 10-11 and illustrates the nature of the magnetic field (flux) through the core of the coil if there were no windings on the core. As the rotating magnet of the magneto is turned, it can be seen that the flux increases in one direction and then decreases to zero before increasing to a maximum in the opposite direction. The static-flux curve is shown also as a dotted line with the curve for the resultant flux.

When primary and secondary windings arc placed on the coil core and the magneto is rotated, a primary current is induced when the flux begins to reduce from maximum. This current flows during the time that the breaker points are closed as shown on the graph. The direction of the current is such that it produces a magnetic field, as shown in Fig. 10-11, that *opposes* the

change in the magnetic flux produced by the rotating magnet. This is in accordance with Lenz's Law, as explained previously. This effect of the primary current flow produces the resultant flux as shown in the graph. Note that the resultant flux remains very strong in its original direction (above the zero line) until the breaker points open to stop the flow of primary current. Immediately before the breaker points open, it can be seen that the resultant flux is high above the zero line while the static flux is well below the line. This situation causes a high stress in the magnetic circuit so that when the breaker points open there is an extremely rapid change in the flux from high in one direction to high in the opposite direction. The effect of this action may be compared to a spring that is stretched to its limit and then is suddenly released. The rapid change in the magnetic flux through the core of the magneto coil induces the high voltage in the secondary winding of the coil that produces the spark for ignition.

In a magneto the number of degrees of rotation between the neutral position and the position where the contact points open is called the **E-gap angle,** usually shortened to simply **E gap.** The manufacturer of the magneto determines for each model how many degrees beyond the neutral position a pole of the rotor magnet should be in order to obtain the strongest spark at the instant of breaker-point separation. This angular displacement from the neutral position, which is the E-gap angle, varies from 5 to 17°, depending on the make and model; but for a representative type of four-pole magneto, such as the one we are discussing here, the correct E gap is 11°. Notice that the rotating magnet will be in the E-gap position as many times per revolution as there are poles.

When the magnet has reached the position where the contact points are about to open, a few degrees past the neutral position, the primary current is maintaining the original field in the coil core while the magnet has already turned past neutral. The primary current is now attempting to establish a field through the coil core in the opposite direction.

When the contact points are opened, the primary circuit is broken. This interrupts the flow of the primary current and causes an extremely rapid change in flux linkages. In an exceedingly short period of time, the field established by the primary current falls to zero and is replaced by the field of the opposite direction established by the magnet. This process is represented by the almost vertical portion of the resultant-flux curve in Fig. 10-11.

The secondary winding of a magneto coil consists of many turns of fine wire, often over 10,000, wound over the primary on the coil core. The large number of turns in the secondary winding and the very rapid change in flux together cause a high rate of change of flux linkages, which in turn produces the high voltage in the secondary winding.

The extremely rapid flux change represents the dissipation of the energy involved in the stress that existed in the magnetic circuit before the contact points opened. The flux in the coil core would normally change as represented by the static-flux curve in the illustration, but the primary current prevents this change and holds back the flux change while the magnet turns.

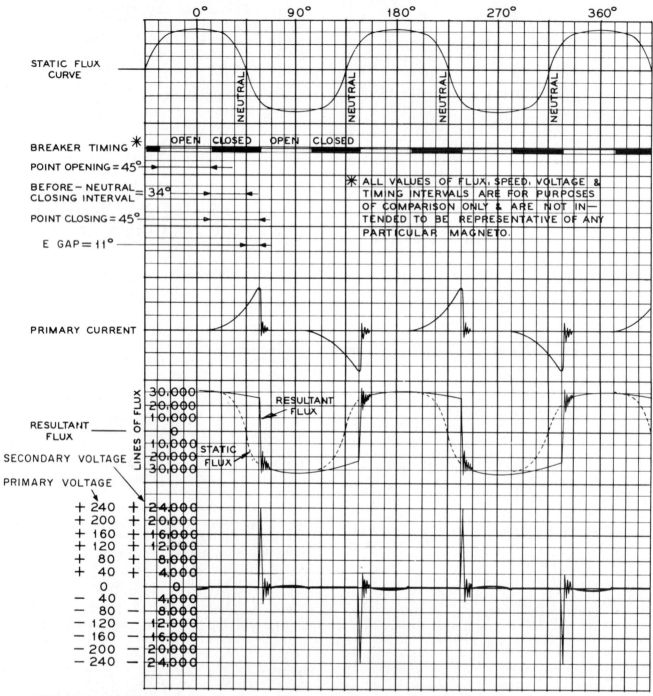

FIG. 10-11 Graphic representation of the electrical and magnetic factors involved in the operation of a magneto, secondary circuit open. *(Bendix Corp.)*

The factors illustrated by the curves in Fig. 10-11 are representative of the conditions that would exist in a rotating magneto when the secondary circuit is open; that is, there is not a complete circuit for the secondary winding. With this condition, both secondary and primary voltages build to a maximum, as shown in the lower part of the graph. It will be noted that secondary voltages are 100 times greater than the primary voltages. This is because the secondary coil has at least 100 times as many windings as the primary coil.

When a magneto is operated with a complete path for secondary current, such as a spark plug or a test gap, the secondary voltage rises only high enough to

cause the current to jump the gap as a spark. The values and directions of the factors of operation in a magneto connected to spark plugs in a normally operating aircraft engine are shown in Fig. 10-12(A). When a magneto is operated on a test stand with open air spark gaps, the values are as shown in Fig. 10-12(B). These values are those that would exist for an eight-pole magneto in which the angular distance through which the breaker points are either open or closed is $22\frac{1}{2}°$.

In Fig. 10-12(A), where the magneto is operating normally on an aircraft engine, the pressure in the cylinder affects the level of secondary voltage necessary to cause the current to jump across the spark-plug gap. In this case secondary emf may reach only 5000 V

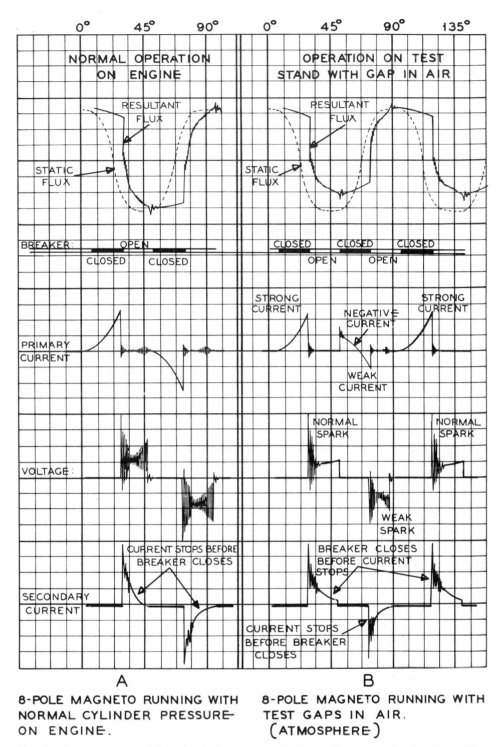

FIG. 10-12 Graphic representation of the electrical and magnetic factors in a magneto operating with a spark gap in the secondary circuit. *(Bendix Corp.)*

before the spark plug fires, and then the voltage diminishes in an oscillating pattern as shown in the graph. The initial oscillations are due to the sudden current load placed on the coil when secondary current starts to flow. The increasing "quench" oscillations are caused by the effect of the turbulence and pressure on the current flowing across the spark-plug gap. The resulting flux change is decreasing at this time and all energy is dissipated just before the breaker points close to begin the next cycle.

When an eight-pole magneto is operated on a test stand with an open-air spark gap, a situation occurs where a negative primary current is developed on each alternate cycle. This is because the secondary current for alternate cycles does not stop flowing before the breaker points close. In this situation the secondary current induces a reverse voltage and current in the primary, which subtracts from the normal buildup of primary current. This causes a weak secondary spark in the ensuing cycle. This weak spark is dissipated before the next closing of the breaker points and the following cycle is again strong. Again, the secondary current does not stop before the breaker points close and another weak spark results.

When eight-pole magnetos were first designed, it was reasoned that the negative current developed on the test stand would affect the operation of the magneto on an engine. However, as pointed out, the pressure in the engine cylinder caused the quenching of the secondary current in every cycle before the closing of the breaker points.

For testing purposes a secondary condenser is incorporated in the secondary circuit of the test stand. This condenser absorbs a portion of the secondary energy and causes the secondary current to stop before the breaker points close. Some of the energy stored in the condenser feeds back into circuit when the breaker points close, and the direction of flow is such that it aids in the buildup of the primary current for the following cycle.

Construction Characteristics of Magnetos

Materials used in the construction of magneto components are selected primarily for their effect in the control of magnetic forces. Strength and durability with respect to mechanical stresses and heat are also considered. Dielectric materials (insulators) must be able to provide adequate insulation to withstand high voltages under all operating conditions.

The **case** of a magneto must be constructed of a nonmagnetic alloy, such as aluminum alloys, so that it will not affect the magnetic circuit. The case supports and protects the operating mechanisms and provides mounting attachments. It completely covers the operating parts of the magneto to prevent the entrance of water, oil, or other contaminating materials. Screened vents are provided to permit ventilation and cooling. Some magnetos are provided with forced air cooling to remove heat from inside the case.

As explained previously, the pole shoes and the coil core of the magneto are constructed of laminated soft iron. Soft iron has high permeability (ability to carry magnetic lines of force) but will not retain magnetism. For this reason, the magnetism in the pole shoes and coil core can change rapidly as required in the operation of the magneto.

The pole shoes and coil core are laminated with insulation between the laminations to reduce the effect of **eddy currents** which develop as a result of the rapidly changing magnetic forces in these parts. Eddy currents interfere with the proper changes in magnetic force and also generate heat.

The magnets employed in magnetos are constructed of very hard alloys, such as alnico or Permalloy. These have proved to be much stronger than hardened steel. The alloys are shaped to meet the requirements of the magneto and are magnetized by means of a strong electromagnetic field. If the design of the magneto requires a rotating magnet, the magnet is mounted on a steel shaft with suitable bearing surfaces for rotation.

The **distributor** of the magneto consists of a rotor made of a durable dielectric material, usually fabricated by molding. The material may be Formica, Bakelite, or some of the more recent thermosetting plastics. In any event, the rotor must be able to withstand high temperatures and high electric stresses. The high-

tension output from the distributor is carried through a distributor cap or blocks in which the high-tension leads for the spark plugs are mounted. Inside the distributor cap or blocks are electrodes which pick up the high-tension current from the rotor electrode. The distributor rotor and cap or blocks are usually coated with a high-temperature wax to prevent moisture absorption and the possibility of high-voltage leakage.

Magneto Speed

Figure 10-13 is a schematic illustration of an aircraft ignition system using a rotating-magnet type of magneto. Notice the **cam** on the end of the magnet shaft. It is *not* a compensated cam; hence, it has as many lobes as there are poles on the magnet; that is, it has four in this case. The number of high-voltage impulses produced per revolution of the magnet is equal, therefore, to the number of poles. The number of cylinder firings per complete revolution of the engine is equal to one-half the number of engine cylinders. Therefore, the ratio of the magneto-shaft speed to that of the engine crankshaft is equal to the number of cylinders divided by twice the number of poles on the rotating magnet. This can be stated in the form of a formula in this manner:

$$\frac{\text{No. of cylinders}}{2 \times \text{no. of poles}} = \frac{\text{magneto shaft speed}}{\text{engine crankshaft speed}}$$

For example, if the uncompensated cam has four lobes, since there are four poles on the magnet, and if the engine has 12 cylinders, then

$$\frac{12 \text{ cylinders}}{2 \times 4 \text{ poles}} = \frac{12}{8} = 1\frac{1}{2}$$

Hence, the magneto speed is $1\frac{1}{2}$ times the engine crankshaft speed.

It should be remembered that a four-stroke-cycle engine fires each cylinder once for each two turns of the crankshaft. Therefore, we know that a 12-cylinder engine will fire six times for each revolution of the crankshaft. Also, a magneto having four lobes on the cam will produce four sparks for each turn of the cam. In the magneto under discussion, the cam is mounted on the end of the magneto shaft, so we know that the magneto produces four sparks for each revolution of the magneto shaft. Then, in order to produce the six sparks needed for each revolution of the crankshaft, the magneto must turn $1\frac{1}{2}$ times.

A nine-cylinder radial engine requires $4\frac{1}{2}$ sparks per revolution and a four-pole magnetic produces 4 sparks per revolution. The ratio of engine speed to magneto speed must therefore be 4 to $4\frac{1}{2}$ or 8 to 9. The engine requires 36 sparks for 8 revolutions and the magneto produces 36 sparks in 9 revolutions.

Polarity or Direction of Sparks

Fundamentally the magneto is a special form of AC generator, modified to enable it to deliver the high voltage required for ignition purposes. In Fig. 10-11, the high rate of change of flux linkages represented by the almost vertical portion of the resultant-flux curve

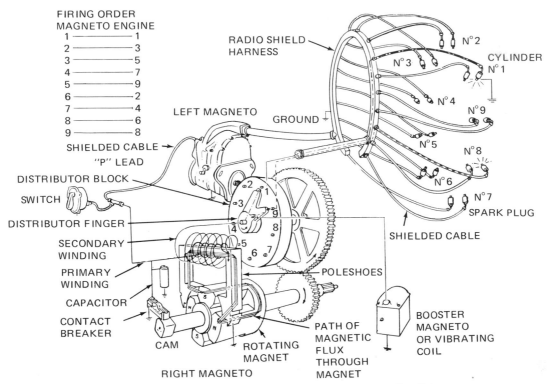

FIRING ORDER
MAGNETO ENGINE

1 —————— 1
2 —————— 3
3 —————— 5
4 —————— 7
5 —————— 9
6 —————— 2
7 —————— 4
8 —————— 6
9 —————— 8

FIG. 10-13 High-tension magneto ignition system. *(Bendix)*

is responsible for the high voltage which produces the secondary spark. From the curves it is apparent that the rapid flux change is downward, then upward, alternating in direction at each opening of the contacts. Since the direction of an induced current depends on the direction of the flux change which produced it, the sparks produced by the magneto are of alternate polarity; that is, they jump one way and then the other, as represented by the secondary current curve in the illustration, which is first above and then below the line, indicating alternate polarity.

● HIGH-TENSION IGNITION SYSTEM

Figure 10-13 illustrates a complete high-tension ignition system for a nine-cylinder radial aircraft engine. This system utilizes a Bendix-Scintilla rotating-magnet type of magneto. It consists of two magnetos, a radio-shield harness, spark plugs, a booster magneto, and a switch. One magneto is shown completely assembled and is illustrated in skeleton form to show the magnetic and electric circuits.

One end of the primary winding is grounded to the magneto, and the other end is connected to the insulated contact point. The other contact point is grounded. The capacitor is connected across the contact points.

The ground terminal on the magneto is electrically connected to the insulated contact point. A wire called the **P lead** connects the ground terminal on each magneto with the switch. When the switch is in the OFF position, this wire provides a direct path to the ground for the primary current; that is, the breaker points are short-circuited. Therefore, when the contact points open, the primary current is not interrupted, thus preventing the production of high voltage in the secondary winding.

One end of the secondary winding is grounded to the magneto, and the other end terminates at the high-tension insert on the coil. The high-tension current produced in the secondary winding is then conducted to the central insert of the distributor finger and across a small air gap to the electrodes of the **distributor block.** High-tension cables in the distributor block then carry it to the spark plugs where the discharge occurs in the engine cylinder.

The Distributor

The distributor finger of Fig. 10-13 is secured to the large distributor gear which is driven by a smaller gear located on the drive shaft of the rotating magnet. The ratio between these gears is always such that the distributor finger is driven at *one-half engine crankshaft speed*. This ratio of the gears ensures the proper distribution of the high-tension current to the spark plugs in accordance with the firing order of the particular engine.

In general, the **distributor rotor** of the typical aircraft magneto is a device that distributes the high-voltage current to the various connections of the distributor block. This rotor may be in the form of a finger, disk, drum, or other shape, depending on the judgment of the magneto manufacturer. In addition, the distributor rotor may be designed with either one or two distributing electrodes. When there are two distributing electrodes, the leading electrode, which obtains high voltage from the magneto secondary, makes its connection with the secondary through the shaft of the rotor, while the trailing electrode obtains a high-tension voltage from the booster by means of a collector ring mounted either on the stationary distributor block or on the rotor itself.

It must be explained that the distributors with the

trailing finger are not employed on late-model aircraft magnetos, although they may be encountered from time to time on magnetos used on older engines. The early systems utilized booster magnetos or high-tension booster coils to provide a strong spark when starting the engine, and the trailing finger of the distributor provided a retarded spark to prevent the engine from kicking back.

Magneto Sparking Order

Almost all piston-type aircraft engines operate on the four-stroke, five-event-cycle principle. For this reason, the number of sparks required for each complete revolution of the engine is equal to one-half the number of cylinders in the engine. The number of sparks produced by each revolution of the rotating magnet is equal to the number of its poles. Therefore, the ratio of the speed at which the rotating magnet is driven to the speed of the engine crankshaft is always one-half the number of cylinders on the engine divided by the number of poles on the rotating magnet, as explained before.

The numbers on the distributor block show the **magneto sparking order** and not the firing order of the engine. The distributor block position marked 1 is connected to the no. 1 cylinder, the distributor block position marked 2 is connected to the second cylinder to be fired, the distributor block position marked 3 is connected to the third cylinder to be fired, and so on.

Some distributor blocks or housings are not numbered for all high-tension leads. In these cases the lead socket for the no. 1 cylinder is marked and the others followed in order according to direction of rotation.

Coming-In Speed of Magneto

To produce sparks, the rotating magnet must be turned at or above a specified number of revolutions per minute, at which speed the rate of change in flux linkages is sufficiently high to induce the required primary current and the resultant high-tension output. This speed is known as the **coming-in speed** of the magneto; it varies for different types of magnetos but averages about 100 to 200 rpm.

Harness Assembly

Note that the harness assembly for the system illustrated in Fig. 10-13 consists of flexible shielding from the magneto housings to the rigid manifold that encircles the crankcase of the radial engine and flexible shielded leads from the manifold to the spark plugs. Thus the complete system is shielded to prevent the emanation of electromagnetic waves which would cause radio interference.

In a system of this type, the lower extremities of the manifold must be provided with drain holes to prevent the accumulation of moisture. In some systems, the manifold is completely filled with a plastic insulating material after the ignition cables are installed. This seals the cables completely away from any moisture.

Occasionally an ignition cable will begin to leak high-tension current, particularly at high altitudes. When the leakage becomes such that it prevents the firing of the spark plug, the cable must be changed, provided

that the manifold is not plastic-filled. To change a cable, a new cable of the proper length is soldered to the end of the cable to be removed. The old cable is then pulled slowly out of the manifold and the new cable is drawn in. The new cable is liberally coated with talc so that it will not tend to stick. If an ignition harness has been in service for an extended period of time and cables begin to break down, it is best to replace the entire set of cables.

Ignition harnesses for opposed engines consist of individual high-tension leads connected to the distributor plate or cap and routed to each spark plug in proper order. These are described later in this chapter.

Ignition Switch and the Primary Circuit

Even though the function of the ignition switch has been explained in this section, the special character of the ignition switch must be emphasized. The usual electric switch is *closed* when it is turned ON. The magneto ignition switch is *closed* when it is turned OFF. This is because the purpose of the switch is to short-circuit the breaker points of the magneto and prevent collapse of the primary circuit required for production of a spark.

After the installation of an ignition switch or after the switch circuit has been rewired, the operation of the circuit must be tested. This can be done with an ohmmeter or a continuity test light. Disconnect the P lead from the magneto and connect it to one terminal of the test unit. Connect the other terminal of the test unit to ground at or near the engine. When the ignition switch is OFF, the test light should burn or the ohmmeter should indicate a complete circuit (little or no resistance). When the ignition switch is turned ON, the test light should go out or the ohmmeter should show infinite resistance (an open circuit).

In working with the magneto system, the technician must always keep in mind that the magneto will be "hot" when the P lead is disconnected or when there is a break in the circuit leading to the ignition switch. If the switch circuit is being repaired, it is important that the primary terminal of the magneto be connected to ground or that the spark plug leads be disconnected.

The ignition switches for modern light aircraft have the appearance of automobile starter switches mounted on the dashboard. The aircraft switch is operated by a key and has positions for OFF, RIGHT, LEFT, BOTH, and START. The switch has a connection for battery power which is used in the START position to actuate the starter contactor or relay. In some cases, the start-ignition switch also includes the master power switch for the aircraft.

Magneto Safety Gap

Magnetos are sometimes equipped with a **safety gap** to provide a return ground when the external secondary circuit is open. One electrode of the safety gap is screwed into the high-tension brush holder, while the grounded electrode is on the safety-gap ground plate. Thus, the safety gap protects against damage from excessively high voltage in case the secondary circuit is accidentally broken and the spark cannot jump between the electrodes of the spark plugs. In such a case, the high-tension spark jumps the safety gap to the ground connection, thereby relieving the voltage in the secondary winding of the magneto.

● TYPES OF MAGNETOS

There are many ways of classifying magnetos. They may be (1) low-tension or high-tension, (2) rotating-magnet or inductor-rotor, (3) single or double, and (4) base-mounted or flange-mounted.

Low-Tension and High-Tension Magnetos

A **low-tension magneto** delivers current at a low voltage by means of the rotation of an armature, wound with only one coil, in the field of a permanent magnet. Its low-voltage current must be transformed into a high-tension (high-voltage) current by means of a transformer.

A **high-tension magneto** delivers a high voltage and has both a primary winding and a secondary winding. An outside induction coil is not needed because the double winding accomplishes the same purpose. The low voltage generated in the primary winding induces a high-voltage current in the secondary winding when the primary circuit is broken.

Rotating-Magnet and Inductor-Rotor Magnetos

In a magneto of the **rotating-magnet** type, the primary and secondary windings are wound upon the same iron core. This core is mounted between two poles, or inductors, which extend to shoes on each side of the rotating magnet. The rotating magnet is usually made with four poles, which are arranged alternately north and south in polarity.

As the magnet rotates, first it sends a magnetic field through the inductors to the core of the coil and back to the opposite pole of the magnet, and then the rotation of the magnet causes the field to reverse. This action was explained in an earlier portion of this chapter and is illustrated in Figs. 10-5 and 10-6. A review of these sections will aid in an understanding of the operation.

The **inductor-rotor** type of magneto has a stationary coil (armature) just as the rotating-magnet type does. The difference lies in the method of inducing a magnetic flux in the core of the coil. The inductor-rotor magneto has a stationary magnet or magnets. As the rotor of the magneto turns, the flux from the magnets is carried through the segments of the rotor to the pole shoes and poles, first in one direction and then in the other.

Single- and Double-Type Magnetos

Two single-type magnetos are commonly used on piston-type engines. The **single-type** magneto is just what its name implies—one magneto.

The **double-type** magneto is generally used on different models of several types of engines. When made for radial engines, it is essentially the same as the magneto made for in-line engines except that two compensated cams are employed. The compensated cam is explained later in this chapter.

The **double-type** magneto is essentially two magnetos having one rotating magnet common to both. It contains two sets of breaker points, and the high voltage is distributed either by two distributors mounted elsewhere on the engine or by distributors forming part of the magneto proper. Since there are two sets of breaker points, an equal number of sparks will be produced by each coil assembly per revolution of the magneto drive shaft.

Base-Mounted and Flange-Mounted Magnetos

A **base-mounted** magneto is attached to a mounting bracket on the engine by means of cap screws, which pass through holes in the bracket and enter tapped holes in the base of the magneto.

A **flange-mounted** magneto is attached to the engine by means of a flange on the end of the magneto. The mounting holes in the flange are not circular; instead, they are slots that permit a slight adjustment, by rotation, in timing the magneto with the engine.

The single-type magneto may be either base-mounted or flange-mounted. The double-type magneto is always flange-mounted.

Symbols Used to Describe Magnetos

Magnetos are technically described by means of letters and figures which indicate make, model, type, etc., as shown in Table 10-1.

Example 1: The type DF18RN is a double-type flange-mounted magneto for use on an 18-cylinder engine designed for clockwise rotation and made by Bendix.

Example 2: The type SF14LU-7 is a single-type flange-mounted magneto for use on a 14-cylinder engine, designed for counterclockwise rotation and made by Bosch, seventh modification.

Note: Each manufacturer uses a dash followed by letters or numerals after the symbol for the product to indicate the series or particular model. The letters F and B are no longer used for civilian magnetos because all are flange-mounted.

TABLE 10-1

Order of Designation	Symbol	Meaning
1	S	Single type
	D	Double type
2	B	Base-mounted
	F	Flange-mounted
4, 6, 7, 9, etc.		Number of distributor electrodes
4	R	Clockwise rotation as viewed from drive-shaft end
	L	Counterclockwise rotation as viewed from drive-shaft end
5	G	General Electric
	N	Bendix
	A	Delco Appliance
	U	Bosch
	C	Delco-Remy (Bosch design)
	D	Edison-Splitdorf

● IGNITION BOOSTERS

If it is impossible under certain conditions to rotate the engine crankshaft fast enough to produce the coming-in speed of the magneto, a source of external high-tension current is required for starting purposes. The various devices used for this purpose are called **ignition boosters.**

An ignition booster may be in the form of a booster magneto, a high-tension coil to which primary current is supplied from a battery, or a vibrator which supplies intermittent direct current from the battery directly to the primary of the magneto. Another device used for increasing the high-tension voltage of the magneto for starting is called an **impulse coupling.** It gives a momentary high rotational speed to the rotor of the magneto during starting.

The Booster Coil

A **booster coil** is a small induction coil. Its function is to provide a shower of sparks to the spark plugs until the magneto fires properly. It is usually connected to the starter switch. When the engine has started, the booster coil and the starter are no longer required; hence, they can be turned off together.

When voltage from a battery is applied to the booster coil, magnetism is developed in the core until the magnetic force on the soft-iron armature mounted on the vibrator overcomes the spring tension and attracts the armature toward the core. When the armature moves toward the core, the contact points and the primary circuit are opened. This demagnetizes the core and permits the spring again to close the contact points and complete the circuit. The armature vibrates back and forth rapidly, making and breaking the primary circuit as long as the voltage from the battery is applied to the booster coil.

The use of booster coils as described here is limited to a few of the older aircraft which are still operating. Most of the modern aircraft employ the **induction vibrator** or an impulse coupling.

The Induction Vibrator

A circuit for an induction vibrators as used with a high-tension magneto is shown in Fig. 10-14. This circuit applies to one engine only, but it is obvious that a similar circuit would be used with each engine of a multiengine airplane. The induction vibrator is energized from the same circuit which energizes the starting solenoid. It is thus energized only during the time that the engines are being started.

One advantage of the induction vibrator is that it reduces the tendency of the magneto to "flash over" at high altitudes, since the booster finger can be eliminated. The function of this induction vibrator is to supply interrupted low voltage for the magneto primary coil, which induces a sufficiently high voltage in the secondary for starting.

The vibrator sends an interrupted battery current through the primary winding of the regular magneto coil. The magneto coil then acts like a battery ignition coil and produces high-tension impulses, which are distributed through the distributor rotor, distributor block, and cables to the spark plugs. These high-

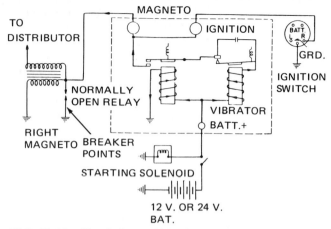

FIG. 10-14   Circuit for an induction vibrator.

tension impulses are produced during the entire time that the magneto contact points are open. When the contact points are closed, sparks cannot be generated, although the vibrator continues to send interrupted current impulses through the magneto contact points without harm to the vibrator or any part of the circuit.

When the ignition switch is in the ON position and the engine starter is engaged, the current from the battery is sent through the coil of a relay which is normally open. The battery current causes the relay points to close, thus completing the circuit to the vibrator coil and causing the vibrator to produce a rapidly interrupted current.

The rapidly interrupted current produced by the vibrator is sent through the primary winding of the magneto coil. By induction, high voltage is created in the secondary winding of the magneto coil, and this high voltage produces high-tension sparks which are delivered to the spark plugs through the magneto distributor block electrodes during the time that the magneto contact points are open.

This process is repeated each time that the magneto contact points are separated, because the interrupted current once more flows through the primary of the magneto coil. The action continues until the engine is firing because of the regular magneto sparks, and the engine starter is released. It should be understood that the vibrator starts to operate automatically when the engine ignition switch is turned to the ON position and the starter is engaged. The vibrator stops when the starter is disengaged.

A schematic diagram of the circuit for an induction vibrator designed for use with light-aircraft engine magnetos is shown in Fig. 10-15. Observe that when the starter switch is closed, battery voltage is applied to the vibrator coil through the vibrator contact points and through the *retard contact points* in the left magneto. As the coil is energized, the breaker points open and interrupt the current flow, thus deenergizing the coil *VC*. The contact points close and again energize the coil, causing the points to open. Thus the contact points of the vibrator continue to make and break contact many times per second, sending an interrupted current through both the main and retard contact points of the magneto. When the magneto turns to its normal firing position, the main contact points open. However, the retard points are still closed and vibrator current

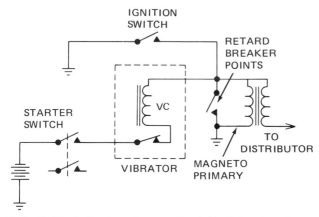

FIG. 10-15   Induction vibrator circuit for light-aircraft engine magneto.

flows to ground through these points. When the magneto has turned to the preset retard position, the retard points open and the vibrator current then flows to ground through the primary winding of the magneto. This induces a high-tension current in the secondary winding, thus providing the necessary retarded spark for starting the engine. A further discussion of this vibrator will be given in the explanation of the complete magneto system.

● BENDIX HIGH-TENSION MAGNETO SYSTEM FOR LIGHT-AIRCRAFT ENGINE

General Description

A typical ignition system for a light-aircraft engine consists of two magnetos, a starter vibrator, combination ignition and starter switch, and harness assembly. These parts are illustrated in Fig. 10-16. This illustration shows the components of the system associated with the Bendix series S-200 magneto. However, it is quite similar to other Bendix systems and the principles are the same.

Magneto

The Bendix S-200 magneto is a completely self-contained unit incorporating a two-pole rotating mag-

net, a coil unit containing the primary and secondary windings, a distributor assembly, main breaker points, retard breaker points, a two-lobe cam, a feed-through-type capacitor, housing sections, and other components necessary for assembly.

The **rotating magnet** turns on two ball bearings, one located at the breaker end and the other at the drive end. A two-lobe cam is secured to the breaker end of the rotating magnet. In a six-cylinder magneto, the rotating magnet turns $1\frac{1}{2}$ times engine speed. Thus, six sparks are produced through 720° of engine rotation, that is, two revolutions of the crankshaft. In a four-cylinder magneto, the rotating magnet turns at engine speed, thus producing four sparks through two revolutions of the crankshaft.

As mentioned previously, the dual breaker magneto incorporates a retard breaker. This breaker is actuated by the same cam as the main breaker and is so positioned that its contacts open a predetermined number of degrees after the main breaker contacts open. A battery-operated starting vibrator used with this magneto provides retarded ignition for starting, regardless of engine cranking speed. The retarded ignition is in the form of a shower of sparks instead of a single spark like that produced by an impulse coupling. It must be remembered that the slow cranking speed of an engine during starting makes it necessary that the ignition be retarded to prevent kickback. At starting speed, if advanced ignition is supplied, the full force of the combustion will be developed before the piston reaches TDC and the piston will be driven back down the cylinder, thus rotating the crankshaft in reverse of the normal direction.

Starting Vibrator

The **starting vibrator assembly** consists of an electromagnet acting upon contact points which make and break the circuit through the electromagnet. This, of course, causes the points to vibrate rapidly, thus developing an interrupted battery current to the primary of the magneto. This interrupted current induces a high voltage in the secondary winding of the magneto coil, thus providing the high-tension current necessary for firing the spark plugs. The starting vibrator provides a

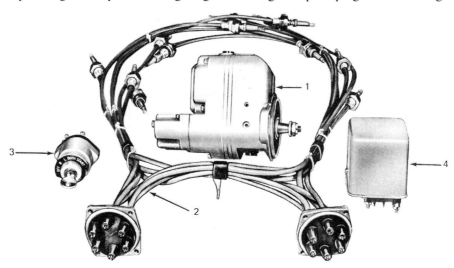

FIG. 10-16   Components of a high-tension ignition system for a light-aircraft engine: (1) magneto, (2) harness assembly, (3) combination switch, (4) vibrator.   *(Bendix)*

shower of sparks with retarded timing and is supplied with battery power only while the starter switch is turned on.

Operation of the System

The operation of the Bendix S-200 magneto system can be understood by studying the diagram of Fig. 10-17. In this circuit the starting vibrator unit includes both a **vibrator** and a **control relay.** The interrupted battery current supplied by the vibrator is controlled by the retard breaker points in the left magneto. The control relay grounds the right magneto during the time that the starter switch is turned on, thus preventing an advanced spark from being applied to the spark plugs.

The vibrator assembly which includes the control relay is used with a *standard* ignition switch. The electrical operation of this switch can be seen in the diagram of Fig. 10-17. In the diagram all the switches and contact points are shown in their normal OFF position.

With the standard ignition switch in its BOTH position and the starter switch turned on, the starter solenoid L3 and the relay coil L1 are energized, thus causing them to close their relay contacts R4, R1, R2, and R3. Relay contact R3 connects the right magneto to ground, rendering it inoperative during starting procedures. Battery current flows through relay contact R1, vibrator points V1, coil L2, through the retard breaker of the left magneto to ground, through the relay contact R2, and through the main breaker to ground. The flow of current through the coil L2 establishes a magnetic field which opens the vibrator points V1 and starts the vibrating cycle. The interrupted battery current thus produced is carried to ground through both sets of breaker points in the left magneto.

When the engine reaches its normal advance firing position, the main breaker opens. However, the current is still carried to ground through the retard breaker, which does not open until the starting *retard position* of the engine is reached. When the retard breaker opens (the main breaker is still open), the vibrator current flows through the primary of transformer T1 (magneto coil), producing a rapidly fluctuating magnetic field around the coil. This causes a high voltage to be induced in the secondary, thus firing the spark plug to which the voltage is directed by the distributor. A shower of sparks is therefore produced at the spark plug owing to the opening and closing of the vibrator points while the main and retard breaker points are open.

When the engine fires and begins to pick up speed, the starter switch is released, thus deenergizing the relay coil L2 and the starter relay L3. This opens the vibrator circuit and the retard breaker circuit, thus rendering them inoperative. The single breaker magneto (right magneto) is no longer grounded; hence, both magnetos are firing in the full advance position.

The schematic diagram of Fig. 10-18 illustrates the operation of a system utilizing a **combination starter-ignition** switch and a starting vibrator which does not include the control relay. When the combination switch is placed in the START position, the right magneto is grounded, the starter solenoid L1 is energized, and current flows through the vibrator L2 to both magneto breaker points and thence to ground if the points are closed. It must be pointed out that all contacts in the combination switch are moved to the **starting** position; hence, they will not be in the position shown in the diagram.

When the engine reaches its normal advance firing position, the main breaker points will open; however, the vibrator current is still carried to ground through the retard breaker, which does not open until the starting retard position of the engine is reached. When the retard breaker opens, the vibrator current flows through the primary of the magneto coil T1, thus inducing a high voltage as explained previously. This voltage provides the retard spark necessary for ignition until the engine speed picks up and the starter switch is released. The combination switch automatically returns to its BOTH position, thus removing the starting vibrator and starter solenoid from the circuit. A study of the switch diagram will also show that the switch circuits of both magnetos are ungrounded; hence, both magnetos will be firing.

The combination switch used with a magneto system has five positions and is actuated by either a switch or a key. The five positions are (1) OFF—both magnetos grounded and not operating; (2) R—right magneto operating, left magneto off; (3) L—left magneto operating, right magneto off; (4) BOTH—both magnetos operating; and (5) START—starter solenoid is operating and the vibrator is energized, causing an intermittent current to flow through the retard breaker on the left magneto while the right magneto is grounded to prevent advanced ignition. The START position on the switch

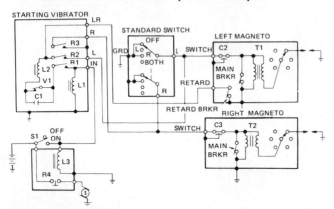

FIG. 10-17 Circuit diagram for Bendix S-200 magneto system. *(Bendix)*

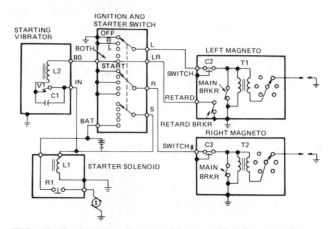

FIG. 10-18 Ignition system with a combination starter-ignition switch.

is a *momentary* contact and is *on* only while being held in this position. When the switch is released, it automatically reverts to the BOTH position.

It will be observed in the diagram that the magnetos are equipped with flow-through-type capacitors C2 and C3, which reduce arcing at the breaker contacts, help to eliminate radio interference, and cause a more rapid collapse of the magnetic field when the breaker points open during normal operation. A capacitor C1 is also necessary in the starter vibrator circuit to produce similar results during the production of the intermittent vibrator current.

Magneto Timing

As we have explained in the general discussion of magnetos, every magneto must be internally timed to produce a spark for ignition at a precise instant. Furthermore, the magneto breaker points must be timed to open when the greatest magnetic field stress exists in the magnetic circuit. This point is called the **E gap,** or **efficiency gap,** and it is measured in degrees past the neutral position of the magnet. The magneto distributor must be timed to deliver the high-tension current to the proper outlet terminal of the distributor block. The internal timing procedure varies to some extent for different types of magnetos; however, the principles are the same in every case.

Distributor timing is usually accomplished during the time the magneto is being assembled. Figure 10-19 shows the matching of the chamfered tooth on the distributor drive gear with the marked tooth on the driven gear. In the illustration the magneto is being assembled for right-hand (clockwise) rotation. The direction of rotation refers to the direction in which the magnet shaft rotates, facing the drive end. When the teeth of the gears are matched as shown, the distributor will be in the correct position with respect to the rotating magnet and breaker points at all times. The large distributor gear also has a marked tooth which can be observed through the timing window on the top of the case to indicate when the distributor is in position for firing no. 1 cylinder. This mark is not sufficiently accurate for timing the opening of the points, but it does show the correct position of the distribution and rotating magnet for timing to no. 1 cylinder.

The following steps are taken to check and adjust the timing of the breaker points for the Bendix S-200 magneto, which does not have timing marks in the breaker compartment:

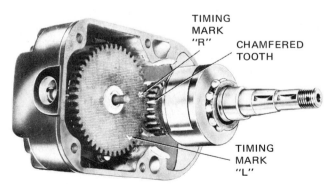

FIG. 10-19 Matching marks on gears for distributor timing.

Within the figure: TIMING MARK "R" CHAMFERED TOOTH TIMING MARK "L"

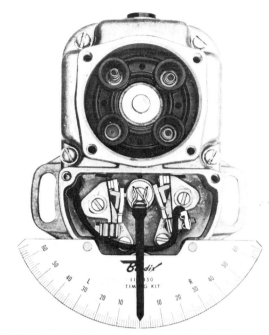

FIG. 10-20 Installation and use of the timing kit.

1. Remove the timing inspection plug from the top of the magneto. Turn the rotating magnet in its normal direction of rotation until the painted, chamfered tooth on the distributor gear is approximately in the center of the inspection window. Then turn it back a few degrees until the magnet is in its neutral position. Because of its magnetism, the rotating magnet will hold itself in the neutral position.

2. Install the timing kit as shown in Fig. 10-20, and place the pointer in the zero position. If the manufacturer's timing kit (Bendix 11-8150) is not available, a substitute can be fabricated using a protractor to provide accurate angular measurement.

3. Connect a suitable timing light across the main breaker points, and turn the magnet in its normal direction of rotation 10° as indicated by the pointer. This is the E-gap position. The main breaker points should be adjusted to open at this point.

4. Turn the rotating magnet until the cam follower is at the high point on the cam lobe, and measure the clearance between the breaker points. This clearance must be 0.018 in ±0.006 in [0.46 mm ±0.15 mm]. If the breaker-point clearance is not within limits, the points must be adjusted for correct setting. It will then be necessary to recheck and readjust the timing for breaker opening. If the breaker points cannot be adjusted to open at the correct time, they should be replaced.

On dual breaker magnetos (those having retard breakers), the retard breaker is adjusted to open a predetermined number of degrees after the main breaker opens, within +2 to 0°. The amount of retard in degrees for any particular magneto is stamped in the bottom of the breaker compartment. In order to set the retard breaker points correctly, it is necessary to add the degrees of retard indicated in the breaker compartment to the reading of the timing pointer when the main breaker points are opening. For example, if the main breaker points open when the timing pointer is at 10° and the required retard is 30°, then the 30° should be added to 10°. The rotating magnet should therefore be turned until the timing indicator reads 40°. The retard breaker points should be adjusted to open at this time.

If an engine is designed for ignition at 20° BTC under normal operating conditions, the retard ignition should be set at least 20° later than the normal ignition. At this time the piston is close enough to TDC that it is not likely to kick back when ignition occurs.

Timing for Magneto with "Cast-In" Timing Marks

Some models of the S-200 magneto and other Bendix magnetos such as the S-1200 have timing marks cast in the breaker compartment. These are illustrated in Fig. 10-21. On each side of the breaker compartment there are timing marks indicating E-gap position and various degrees of retard breaker timing. The marks on the left-hand side, viewed from the breaker compartment, are for clockwise rotating magnetos, and the marks on the right-hand side are for counterclockwise rotating magnetos. The rotation of the magneto is determined by viewing the magneto from the drive end.

The point in the center of the E-gap boss, shown at E in the drawing, indicates the exact E-gap position if the indicator is first set to zero with the magnet in the neutral position. The width of the boss on either side of the point is the allowable tolerance of ±4°. In addition to these marks, the cam has an indented line across its end for locating the E-gap position of the rotating magnet. This position is indicated when the mark on the cam is aligned with the mark at the top of the breaker housing.

The procedure for checking the timing of the magneto is as follows:

1. Disconnect the harness assembly from the magneto and remove the timing window plug.
2. Turn the engine crankshaft in the direction of normal rotation until the painted chamfered tooth of the distributor gear is just becoming visible in the timing window. Continue turning the crankshaft until the line on the end of the cam is aligned with the mark at the top of the breaker housing.
3. Install a suitable point (Bendix 11-8149) on the cam screw so that it indexes with the center of the E-gap mark. The magnet is now at the E-gap position. The position of the cam and pointer should be as shown in Fig. 10-21.
4. Connect a timing light across the main breaker and adjust the breaker to open at this point. Now turn the engine crankshaft until the cam follower is on the

high point of the cam lobe. With a thickness gage, check the clearance between the breaker points to see that the clearance is 0.018 in ±0.006 in [0.46 mm ±0.15 mm]. If the clearance is not correct, the points must be adjusted and then rechecked for correct opening.

5. To set the retard breaker points, first note the number of degrees required for retard breaker opening. This number is marked in the breaker compartment as shown. Turn the engine back to the position where the main breaker points open. Position the pointer at the 0° mark. Now turn the engine in the normal direction of rotation until the pointer is over the required retard degree mark. Using the timing light, adjust the retard breaker contacts to open at this point. Continue turning the crankshaft until the cam follower is on the high point of the cam lobe and measure the breaker-point clearance for 0.018 in ±0.006 in.

Installation of S-200 Magneto

When the S-200 magneto is installed on an engine for the first time, it is necessary to prepare and install a switch terminal connection for the timing light in the following manner:

1. Strip ⅛ in [3.18 mm] of insulation from the end of the shielded wire to be connected to the terminal. Strip the metallic shielding back about 0.672 in [17.07 mm] from the end of the insulation as shown in the drawing of Fig. 10-22.
2. Using Bendix Kit 10-157209, slide the coupling nut, shouldered bushing, insulating bushing, and flat washer over the end of the stripped wire.
3. Fan the strands of wire over the washer and solder, using 50/50 solder. The washer and solder should not exceed a thickness of 0.047 in [1.19 mm]. The fitting is now ready for connecting to the switch terminal of the magneto.

Before the magneto is installed on the engine, it should be checked to make sure that it has the correct direction of rotation. Then proceed as follows:

1. Remove the timing inspection plug from the top of the magneto and turn the magneto in the normal direction of rotation until the painted chamfered tooth

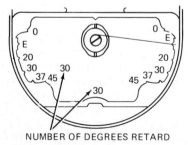

NUMBER OF DEGREES RETARD

FIG. 10-21 Timing marks in the breaker compartment.

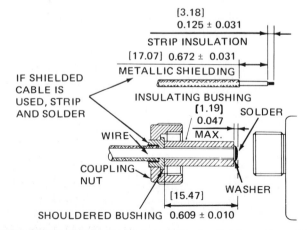

FIG. 10-22 Preparing the connector for the magneto-switch terminal.

of the distributor gear is approximately in the center of the window. The magneto is now in the correct advance position for firing the no. 1 cylinder.

2. Turn the engine to the no. 1 cylinder full-advance firing position (compression stroke) with the use of a suitable piston-position indicator or with a timing disk and top-center indicator.

3. Install the magneto on the engine, and tighten the mounting bolts sufficiently to hold the magneto in position, but loosely enough that it can be rotated.

4. Connect the timing light to the magneto switch terminal, using the previously prepared terminal connection. When using a direct-current continuity light for checking breaker-point opening time, the primary lead from the coil should be disconnected from the breaker points. This will prevent the flow of current from the battery in the tester through the primary winding.

5. If the timing light is out, rotate the magneto housing in the direction of its magnet rotation a few degrees beyond the point where the light comes on. Then slowly turn the magneto in the opposite direction until the light just goes out. Secure the magneto to the engine in this position by tightening the mounting bolts. Recheck the timing of the breaker points by turning the engine in reverse and then rotating it forward until the light goes out. The light should go out when the engine reaches the advance firing position as shown on the timing disk.

6. Remove the timing-light connection and install the switch wire connection to the switch terminal of the magneto.

WARNING: It is most important to note that the magneto is in the "switch on" condition whenever the switch wire is disconnected. It is therefore necessary to have the spark-plug wires disconnected when timing the magneto to the engine; otherwise, the engine could fire and cause injury to personnel.

Alternate Timing Method

Some opposed engines are designed with timing marks on the ring gear inside the crankcase. These engines can have magnetos timed either by using the ring-gear markings or by locating the TDC by means of a piston-position indicator as explained previously. If an engine has the markings on the ring gear, the following procedure is recommended:

1. Remove the magneto timing inspection hole plug and rotate the magneto shaft until the timing pointer inside the magneto gear case is aligned with the marked gear tooth.

2. Remove a spark plug from the no. 1 cylinder.

3. Remove the timing inspection plug located on the side of the crankcase forward of no. 6 cylinder.

4. Rotate the engine in the normal direction of rotation until the no. 1 cylinder is on the compression stroke as indicated by pressure on the thumb held against the spark plug hole. Continue turning the engine until the marked tooth of the gear seen through the timing hole shows that the no. 1 cylinder is in the advanced timing position.

5. With the magneto set as above, place it on the accessory mounting pad.

6. Connect a timing light to the magneto in accordance with instructions.

7. If the timing light is out, rotate the magneto a few degrees in its direction of rotation until the light comes on. Slowly reverse the direction of magneto rotation until the light goes out. Secure the magneto in this position.

Installation of High-Tension Harness

The high-tension spark-plug leads are secured to the proper outlets in the magneto by means of the *high-tension outlet plate* and a rubber *grommet* or *terminal block*. The shielding of the cables is secured in the outlet plate by means of a ferrule, sleeve, and coupling nut. The ferrule and sleeve are crimped on the end of the shielding to form a permanent coupling fitting.

The high-tension cables are inserted through the outlet plate and into the grommet after the insulation has been stripped for about $\frac{1}{2}$ in [12.7 mm] back from the end of the wire. The bare wires are extended through the grommet and secured by means of a small brass washer as shown in Fig. 10-23. A suitable method for

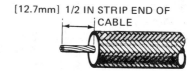

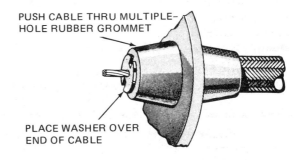

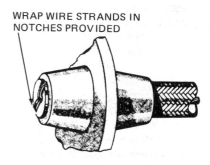

FIG. 10-23   Connecting high-tension leads to the magneto.

CUT OFF CABLE *SQUARELY* AND INSERT
FIRMLY INTO SOCKET OF GROMMET.

INSERT SCREW THROUGH WASHER AND SECURE INTO CABLE.
TIGHTENING FIRMLY BUT NOT EXCESSIVELY.

FIG. 10-24 Use of screws for attaching copper high-tension cable.

securing copper high-tension cable is illustrated in Fig. 10-24. In this method, the wires are cut off even with the insulation and the cable is then inserted into the grommet. A metal-piercing screw is used with a washer to hold the cables in place. The screws penetrate the ends of the stranded copper cable and form threads. The screws must not be turned too tight or they will strip out. The arrangement of the harness assembly and distributor blocks is shown in Fig. 10-25.

Methods for securing ignition leads in distributor caps and for attaching spark-plug terminals are described in maintenance manuals for specific magnetos. In all cases, the manufacturer's instructions should be followed.

During the assembly of the high-tension harness, it is essential to note that the high-tension leads are installed in the outlet plate in the order of engine firing. The order of magneto firing for different magnetos is shown in Fig. 10-26; however, the spark-plug leads must not be connected in the same order. Since the firing order of a typical six-cylinder opposed engine is 1-4-5-2-3-6, the magneto outlets must be connected to spark-plug leads as shown in Table 10-2.

In the practice of connecting the leads for dual magneto systems, the right magneto fires the top spark plugs on the right-hand side of the engine and the bottom spark plugs on the left-hand side of the engine. The left magneto is connected to fire the top spark plugs on the left side of the engine and the bottom spark plugs on the right side of the engine. A circuit diagram for this arrangement is shown in Fig. 10-27. This is the ignition circuit for a Continental 0-470 engine.

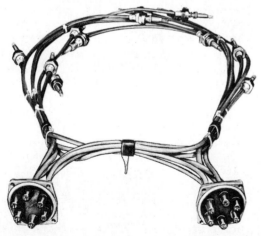

FIG. 10-25 Harness assembly with distributor blocks.

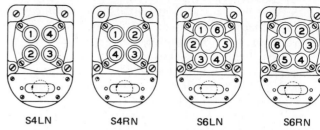

| S4LN | S4RN | S6LN | S6RN |

FIG. 10-26 Magneto firing order. *(Bendix)*

Maintenance

It is recommended that S-200 magnetos be inspected after the first 25 h of operation and every 50 h thereafter. A typical inspection and check are performed as follows:

1. Remove the screws which hold the breaker cover and loosen the cover sufficiently to allow removal of the feed-through capacitor and retard lead terminals from the breakers. The feed-through capacitor and retard leads will remain in the breaker cover when the cover is removed from the magneto.

2. Examine the breaker contact points for excessive wear or burning. Points which have deep pits or excessively burned areas should be discarded. Examine the cam follower felt for proper lubrication. If the felt is dry, apply 2 or 3 drops of Scintilla 10-86527 lubricant or the equivalent. Blot off any excess oil. Clean the breaker compartment with a clean, dry cloth.

3. Check the depth of the spring contact in the switch and retard terminals. The spring depth from the outlet face should not be more than $\frac{1}{2}$ in [12.7 mm].

4. Visually check the breakers to see that the cam follower is securely riveted to its spring. Check the screw that holds the assembled breaker parts together for tightness.

5. Check the capacitor mounting bracket for cracks or looseness. Test the capacitor for a minimum capacitance of 0.30 μF with a suitable capacitor tester.

6. Remove the harness outlet plate from the magneto, and inspect the rubber grommet and distributor block. If moisture is present, dry the block with a soft, dry, clean, lint-free cloth. Do not use gasoline or any solvent for cleaning the block. The solvent will remove the wax coating and possibly cause electrical leakage.

7. Reassemble all parts carefully.

The foregoing directions are indicative of the checks and inspections that should be made, especially if there is any indication of magneto trouble. If possible, the manufacturer's manual should be employed to make sure that no important details are omitted.

TABLE 10-2

Magneto Outlet	Spark-Plug Lead
1	1
2	4
3	5
4	2
5	3
6	6

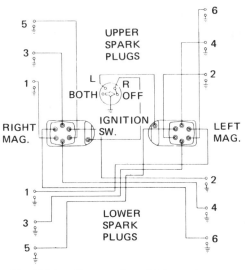

FIG. 10-27 Wiring diagram for a high-tension magneto system on a six-cylinder opposed engine.

FIG. 10-28 Bendix D-3000 magneto. *(Engine Products Div., Bendix Corp.)*

● BENDIX DUAL MAGNETO IGNITION SYSTEMS

The Bendix D-2000 and D-3000 magneto ignition systems were designed to provide dual ignition for aircraft engines with only one magneto. These systems are available for four-, six-, and eight-cylinder engines. A complete system includes the dual magneto, harness assembly, starting vibrator (for the D-2200 system), and an ignition switch. An improved model magneto, the series D-3000, has recently been developed. The D-3000 magneto is identical to the series D-2000 with the exception of a few structural changes.

Magneto

The dual magneto consists of a single driveshaft and rotating magnet that supplies the magnetic flux for two electrically independent ignition circuits. Each ignition circuit includes pole shoes, primary winding, secondary winding, primary capacitor, breaker points, and distributor. Fig. 10-28 is a photograph of a D-3000 magneto showing the block and bearing assembly.

The complete D-3000 dual magneto ignition system is shown in Fig. 10-29. This system utilizes a starting vibrator to provide adequate starting ignition that is controlled by the retard breaker points for the left magneto. The D-2000 dual magneto ignition system employs an impulse coupling to provide the retarded starting ignition; the impulse coupling is described later in this chapter.

The D-3000 magneto shown in Fig. 10-29 has two separate breaker cams mounted on the same shaft. The lower cam operates the main breaker points for both magneto circuits, while the upper cam operates the left magneto retard breaker and the tachometer breaker. The tachometer breaker provides electrical impulses to operate the type of tachometer that utilizes such impulses for rpm indication. If the airplane is equipped with any other type of tachometer, the tachometer breaker is not used. The tachometer breaker is located above the right main breaker and the retard breaker is above the left main breaker. On the D-3000 magneto

that employs an impulse coupling, the upper cam is installed only if a tachometer breaker is required.

The primary capacitors are feed-through types and are mounted in the magneto cover, which is a part of the harness assembly. When the harness is assembled to the magneto, the capacitor leads are attached to the breaker point tabs before the cover is installed over the distributor block.

Harness

As mentioned previously, the harness assembly shown in Fig. 10-29 includes the magneto cover. Each harness is designed for a particular make and model of engine. The assembly is fully shielded to prevent electromagnetic emanations that would interfere with radio and other electronic equipment. The harness shielding consists of tinned copper braid that is impregnated with a silicone-base material. The harness is designed so that any part can be replaced in the field.

Starting Vibrator

The starting vibrator operates on the same principle as the induction vibrator described previously in this chapter. Two types of starting vibrators are used with D-2200 and D-3200 ignition systems, depending upon the type of starter and ignition switches that are employed in the system. One type of starting vibrator includes a relay to ground out the right magneto primary circuit during starting. If this were not done, the right magneto would produce an advanced spark and this would cause the engine to kick back. When the ignition switch incorporates the starter switch, the starting vibrator does not require a relay because the combination switch includes that ground out the right magneto when the switch is in the START position. A photograph of a typical starting vibrator for the D-2200 and D-3200 systems is shown in Fig. 10-30. A D-3000 magneto with a harness for a four-cylinder engine is shown in Fig. 10-31.

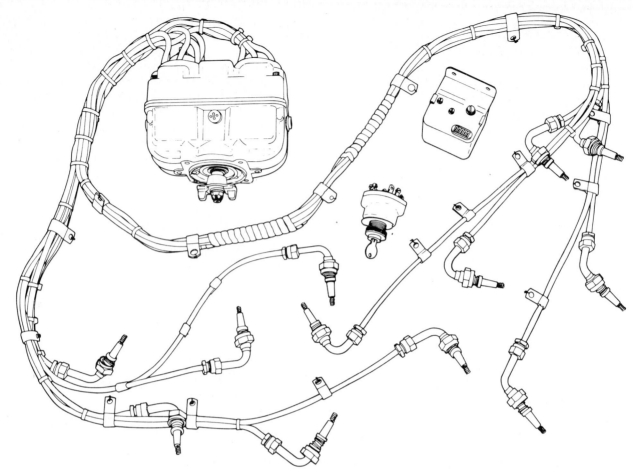

FIG. 10-29 Complete Bendix D-3000 dual magneto ignition system. *(Bendix)*

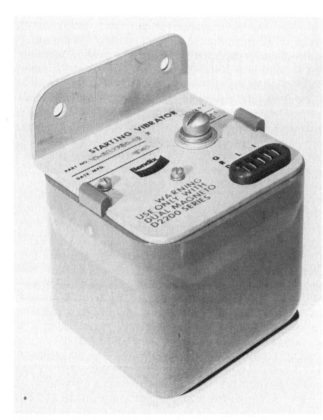

FIG. 10-30 Starting vibrator for the Bendix D-2200 magneto. *(Bendix)*

● THE SLICK MAGNETO FOR LIGHT-AIRCRAFT ENGINES

The Slick series 600 magneto is manufactured for use with six-cylinder opposed aircraft engines and is quite similar in operation to the Bendix magneto previously described. This magneto, however, is not designed with retard breaker points for use with a starting vibrator but is normally equipped with an **impulse coupling** which enables it to produce a retarded spark for starting purposes. Figure 10-32 is a photograph of the magneto with the impulse coupling mounted on the drive shaft.

General Description

The Slick series 600 magneto incorporates a two-hole rotating magnet, a coil which includes both primary and secondary windings, nylon distributor gears, a single set of breaker points, a coaxial capacitor, and other parts necessary for the production and distribution of high-tension current for ignition. The principle of operation is the same as for other similar magnetos, and the timing is arranged to provide for breaker opening at the E-gap position in order to produce the greatest electrical effect.

Timing the Magneto

Figure 10-33 shows the alignment of the drive gear and breaker cam with the magnet shaft. Note that the drive gear must be aligned with the mark on the shaft. This is the first step in internal timing. The main shaft

FIG. 10-31 Bendix dual magneto system for a four-cylinder engine.

should now be turned to a position where a timing pin can be inserted in the small hole in the bottom of the main frame near the drive end. The timing pin engages the rotating magnet shaft to hold the magnet in the E-gap position. The breaker assembly is now installed and adjusted so that the points are just beginning to open. This is illustrated in Fig. 10-34.

The distributor gear is installed with the arrow mark for the proper direction of rotation aligned with the marked tooth on the distributor gear. Direction of rotation is determined from the drive end of the magneto. When the drive turns clockwise, the rotation is *right hand*.

The positioning of the distributor gear is illustrated in Fig. 10-35. The gear in the illustration is installed for right-hand rotation.

Since both gears in the magneto are made of nylon, there is a substantial backlash in the gears at room temperature. After the magneto has warmed to operating temperature, the expansion of the gears reduces the clearance to the proper amount.

The cam oiler pad should be saturated with heavy (SAE 70) oil when the assembly is installed. Two or three drops of the oil are sufficient to provide adequate lubrication. The pad touches the cam lightly at the corner to transfer the lubricant. If excessive oil is applied to the oiling pad, some may flow to the breaker points and cause a poor contact. For this reason, all excess oil must be wiped off the mechanism.

The Impulse Coupling

The impulse coupling installed on the drive shaft of the Slick magneto or the Bendix magneto is designed to provide a retarded spark for starting the engine. The coupling consists of a *shell*, *spring*, and *hub*. The hub is provided with weighted dogs which enable the assembly to accomplish its purpose. These are illustrated in Fig. 10-36. In some manuals the shell is referred to as the **body** and the hub is called the **cam**.

When the impulse coupling is installed on the drive

FIG. 10-32 Magneto with an impulse coupling. *(Slick Electro)*

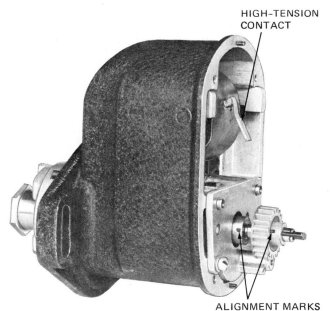

FIG. 10-33 Alignment of breaker cam and drive gear.

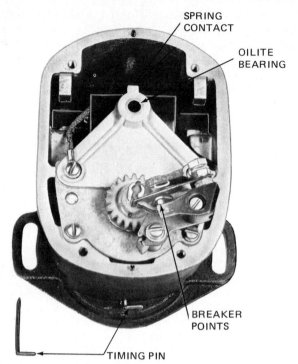

FIG. 10-34　Installation of breaker points.

shaft of the magneto, the shell of the coupling may be rotated by the engine drive for a substantial portion of one revolution while the rotating magnet remains stationary. While this is taking place, the spring in the coupling is being wound up. At the points where the magneto must fire, the dogs are released and the spring unwinds to give the rotating magnet a rapid rotation in the normal direction. This, of course, causes the magneto to produce a strong spark at the spark plug. As soon as the engine begins to run, the weighted dogs are held in the released position by centrifugal force and the magneto fires in its normal advanced position. During starting, the retard spark is produced when the magneto rotation is held back by the impulse coupling.

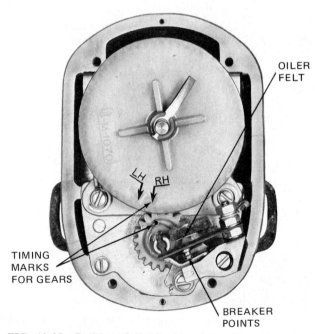

FIG. 10-35　Position of distributor gear.

FIG. 10-36　Components of an impulse coupling.

During normal operation the impulse coupling spring holds the magneto in the advance spark position. If the spring should break, the magneto would continue to rotate but would be in the retard spark position. Hence, the spark plugs fired by this particular magneto would be firing late.

Harness Installation

The high-tension leads to the spark plugs are installed in a manner similar to that described for the Bendix magneto. The leads are inserted through the distributor block housing to a position where they contact the terminals in the distributor block. The leads are held in place be means of coupling nuts. The arrangement of the distributor block in the housing is shown in Fig. 10-37.

Installation on the Engine

The installation of the Slick magneto on the engine is similar to the installation of any other magneto. First, the magneto is placed in the position for firing the no. 1 cylinder. This is accomplished by rotating the magneto shaft until the timing lines are in position, as shown in Fig. 10-38. The timing pin, previously mentioned, is then inserted through the hole in the bottom of the case near the drive end. The engine is placed in the firing position for no. 1 cylinder as explained in the previous section, and the magneto is installed. Slotted holes in the mounting flange permit final adjustment at installation.

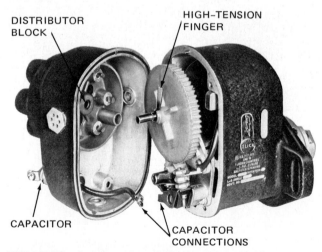

FIG. 10-37　Arrangement of the distributor block.

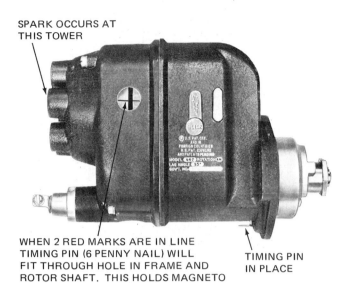

SPARK OCCURS AT THIS TOWER

WHEN 2 RED MARKS ARE IN LINE TIMING PIN (6 PENNY NAIL) WILL FIT THROUGH HOLE IN FRAME AND ROTOR SHAFT. THIS HOLDS MAGNETO FOR TIMING TO ENGINE'

TIMING PIN IN PLACE

FIG. 10-38 Timing lines in position for the no. 1 cylinder.

● SLICK SERIES 4200/6200 MAGNETOS

Slick 4200 and 6200 magnetos are improved designs manufactured to provide ignition for four- and six-cylinder aircraft engines. They can be overhauled in the field according to the manufacturer's instructions.

The 4200 and 6200 magnetos have a common frame and rotor assembly, but the distributor housing, block, and gear must differ between the two models simply because the 4200 model is designed for four-cylinder engines and the 6200 model is designed for six-cylinder engines. The 4200 and 6200 model magnetos are shown in Fig. 10-39.

These magnetos utilize a two-pole magnetic rotor that revolves on two ball bearings located on opposite sides of the rotating magnet. The rotor-and-bearing assembly is contained within the drive and frame.

Bearing preloading is provided by means of a loading spring, thus eliminating the need for selective shimming. The other components that are contained within the drive end frame are a high-tension coil that is retained by wedge-shaped keys in the contact breaker assembly that is secured to the inboard bearing plate with two screws. The contact breaker is actuated by a two-lobe cam at the end of the rotor shaft. The cam also serves to key the rotor gear to the shaft. It will be noted that this construction is similar to that of the Slick series 600 magneto.

To provide a retarded spark for engine starting, the series 4200 and 6200 magnetos employ an impulse coupling. This component is essentially the same as that described for the series 600 magneto.

● OTHER HIGH-TENSION MAGNETOS

Numerous types of high-tension magnetos have been designed for use on aircraft engines; however, it is not deemed essential that all types be described. The basic principles are the same for all such magnetos, and it is only necessary to determine how each is timed internally and timed to the engine. With a good understanding of the principles of operation and timing, the technician can usually adjust any magneto for satisfactory operation. If there is any question concerning a particular magneto and its installation, the manufacturer's manuals for the magneto and the engine should be consulted.

Among popular magnetos not described in this text are the Bendix SF6LN-21, the Bendix 1200, and the Eisemann LA-6 and LA-4 types. The harness installation is essentially the same for most types. The Eisemann magnetos employ a different method for securing the high-tension spark-plug cables from those previously described. The distributor cap is designed for screw connections, and eyelets are installed on the

FIG. 10-39 Slick 4200 and 6200 magnetos. *(Slick Electro)*

ends of the cables. Thus, the cable ends are attached to the proper terminals of the distributor cap by means of screws through the cable eyelets.

● LOW-TENSION IGNITION

Reasons for Development of Low-Tension Ignition

There are several very serious problems encountered in the production and distribution of the high-voltage electricity used to fire the spark plugs of an aircraft engine. High-voltage electricity causes corrosion of metals and deterioration of insulating materials. It also has a marked tendency to escape from the routes provided for it by the designer of the engine.

There are four principal causes for the troubles experienced in the use of high-voltage ignition systems: (1) flashover, (2) capacitance, (3) moisture, and (4) high-voltage corona.

Flashover is a term used to describe the jumping of the high voltage inside a distributor when an airplane ascends to a high altitude. The reason for this is that the air is less dense at high altitudes and hence has less dielectric, or insulating strength.

Capacitance is the ability of a conductor to store electrons. In the high-tension ignition system, the capacitance of the high-tension leads from the magneto to the spark plugs causes the leads to store a portion of the electric charge until the voltage is built up sufficiently to cause the spark to jump the gap of a spark plug. When the spark has jumped and established a path across the gap, the energy stored in the leads during the rise of voltage is dissipated in heat at the spark-plug electrodes. Since this discharge of energy is in the form of a relatively low voltage and high current, it causes burning of the electrodes and shortens the life of the spark plug.

Moisture, wherever it exists, increases conductivity. Thus, it may provide new and unforeseen routes for the escape of high-voltage electricity.

High-voltage corona is a phrase often used to describe a condition of stress which exists across any insulator (dielectric) exposed to high voltage. When the high voltage is impressed between the conductor of an insulated lead and any metallic mass near the lead, an electrical stress is set up in the insulation. Repeated application of this stress to the insulation will eventually cause failure.

Low-tension ignition systems are designed in such a manner that the high voltage necessary to fire the spark plugs is confined to a very small portion of the entire circuit. The greater part of the circuit involves the use of low voltage; hence, the term **low-tension ignition** is used to describe the system.

Operation of the Low-Tension Ignition System

The low-tension ignition system consists of (1) a low-tension magneto, (2) a carbon brush distributor, and (3) a transformer for each spark plug. Figure 10-40 is a diagram showing the principal parts of a simple low-tension system. The distributor is not shown because in this diagram only one spark plug is used. Figure 10-41 is a schematic drawing of a low-tension ignition system, showing the location in the system of the carbon brush distributor.

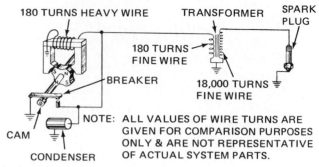

NOTE: ALL VALUES OF WIRE TURNS ARE GIVEN FOR COMPARISON PURPOSES ONLY & ARE NOT REPRESENTATIVE OF ACTUAL SYSTEM PARTS.

FIG. 10-40   Diagram illustrating a low-tension ignition system.

During the operation of the low-tension system, surges of electricity are generated in the magneto generator coil. The peak surge voltage is never in excess of 350 V and probably is nearer 200 V on most installations. This comparatively low voltage is fed through the distributor to the primary of the spark-plug transformer.

While the complete process by which the transformer coil accomplishes its purpose is not a simple one, the following explanation is sufficiently accurate for the technician or other persons who are not responsible for the design of such systems.

At the instant of opening of the breaker contacts which are connected across the magneto generator coil, a rapid flux change takes place in the generator coil core, causing a rapid rise of voltage in this coil. As has already been explained, it is the capacitor connected across the breaker points which actually stops the flow of current when the breaker opens.

The primary capacitor and magneto generator coil of a low-tension system are connected through the distributor directly across the primary winding of the transformer coil, as shown in Fig. 10-41. Therefore, during the time that the voltage across the primary capacitor is rising, as the breaker points open, the natural tendency is for current to start flowing out through the distributor and the primary of the transformer coil.

When this condition has been achieved, there is the situation of a primary capacitor charged to nearly 200 V connected across the primary of the transformer. The result is a very rapid rise of current in the primary, accomplished by a very rapid change in flux linkages (magnetic field) in both coils. The rapid change in the flux linkages in the secondary induces the voltage which fires the spark plug. As soon as the spark gap has been "broken down" (broken through and ionized), current also starts to flow in the secondary circuit.

The **transformer** of the low-tension system is purposely designed to have an appreciable resistance in the primary winding (5 Ω or more). When the second-

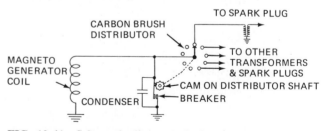

FIG. 10-41   Schematic diagram of a low-tension ignition system with a distributor.

ary current increases, the resistance of the primary coil prevents a corresponding increase of current in that winding, so that as soon as the voltage originally generated in the secondary has been lowered by the flow of secondary current, the spark is extinguished and no further action takes place. From this it will be clear that the duration of the spark in a high-tension system is several times that of a comparable low-tension system. After the secondary current has stopped, the energy stored in the form of a charge in the primary capacitor continues to drain away at a comparatively slow rate through the primary of the transformer coil.

The rather high resistance of the transformer primary winding, which is characteristic of all low-tension transformer coils, helps to bring the primary current to a stop after the spark has been produced. If this were not done, the primary current would continue to flow through the circuit until the distributor finger carbon brush reached the edge of the distributor contact segment, at which time the current would be stopped by the interruption of the contact as the carbon brush moved off the segment. This would cause pitting and burning of the distributor segments.

It should be clear that the spark voltage is produced by the growth of a magnetic field in the transformer core and not by the collapse of the field, as is the case in conventional ignition coils. This fact sometimes raises the question about why the subsequent collapse or decay of the field in the transformer does not produce a second spark at the spark plug. The reason for this is that the rate of decay of the magnetic field in the transformer is determined by the rate of decay of the primary current. It has already been pointed out that the primary current results from the discharge of the primary capacitor and that this current tapers off at a rather slow rate after the secondary current stops. Since the rate of decay of the magnetic field is the same as that of the primary current, it is too slow to produce enough voltage for a second spark at the plug.

The Transformer Coil

Figure 10-42 is a drawing of a typical low-tension transformer coil "telescoped" (that is, pulled out of its case) to show the design. This coil consists of a primary and a secondary winding with a "cigarette" of transformer iron sheet in the center and another cigarette of transformer iron sheet surrounding the primary winding, which is on the outside of the secondary winding. Usually the transformer unit contains two transformers, one for each spark plug in the cylinder.

The complete transformer assembly provides a compact, lightweight unit convenient for installation on the cylinder head near the spark plugs. This permits the use of short high-tension leads from the transformer to the spark plugs, thus reducing to a large extent the opportunities for leakage of high-tension current.

An advantage of the low-tension system is that the failure of the primary or secondary of one transformer will affect only one spark plug. For example, if the primary winding is shorted out, one spark plug will stop firing but the engine will continue to operate well. The "dead" coil and spark plug will be detected when the next magneto check is made on ground run-up.

● LOW-TENSION SYSTEM FOR LIGHT-AIRCRAFT ENGINES

General Description of System

A low-tension system commonly employed on engines for light aircraft is the series S-600 developed by the Bendix Corporation. The model numbers are S6RN-600 for the dual-breaker magneto and S6RN-604 for the single-breaker magneto. These magnetos are also designed for left-hand rotation.

The components of the S-600 low-tension system are shown in Fig. 10-43. This system is designed for use on a six-cylinder opposed engine. Each installation consists of a retard-breaker magneto, single-breaker magneto, starting vibrator, harness assembly, transformer coils, high-tension leads, and either a combination ignition and starter switch or a standard ignition switch.

This system is designed to generate and distribute low-voltage current through low-tension cables to individual high-voltage transformer coils mounted on the engine crankcase. The low voltage is stepped up to a high voltage by the individual transformer coils and then conducted to the spark plug by short lengths of high-tension cable. Both the low-tension and high-

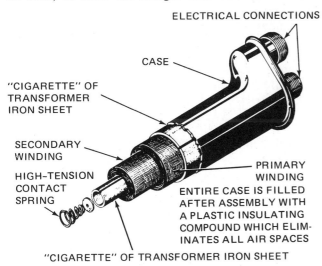

FIG. 10-42 Transformer coil for a low-tension ignition system.

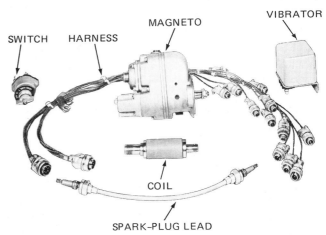

FIG. 10-43 Components of the Bendix S-600 low-tension ignition system. *(Bendix)*

tension cables are shielded to prevent radio interference.

Magneto

The magneto for the S-600 system is almost identical with the S-200 magneto described earlier in this chapter. The principal exceptions in construction are the coil assembly and the distributor assembly. The coil contains one winding only for the generation of the low-tension current. Instead of the high-voltage distributor described for other magnetos, the low-tension magneto employs spring-loaded carbon brushes in the distributor block to bear against metal ring segments on the face of the large distributor gear. The small segment is connected electrically to the carbon brush at the center of the gear which bears against the coil contact.

The magneto has a two-pole rotating magnet and a two-lobe breaker cam on the end of the magnet shaft. The S-600 magneto incorporates a retard breaker in addition to the main breaker. The retard breaker is actuated by the same cam as the main breaker and is so positioned that its contacts open at a predetermined number of degrees after the main breaker contacts open.

A battery-operated starting vibrator used with this magneto provides retarded ignition for starting regardless of engine cranking speed. The retard ignition is in the form of a shower of sparks instead of a single spark obtained by the impulse coupling. This is similar to the operation described for the S-200 high-tension magneto.

A capacitor is required in both the high-tension and low-tension magnetos for the low-voltage circuit. In the magneto under discussion a feed-through-type capacitor is connected across the main breaker points. This capacitor reduces arcing of the points, stores current which feeds to the transformer primary coil shortly after the breaker points open, and aids in reducing radio interference.

Harness Assembly

The harness assembly for the low-tension system is shown in Fig. 10-43. This assembly constitutes two sections, one for each magneto. Each section consists of six shielded low-tension cables which terminate in connectors for the primary end of the transformer coils. The magneto end of each section has one connector plug with six connecting pins. These mate with six-pin sockets in the female section of the connector which is on the magneto.

Transformer Coils

The transformer coils contain primary and secondary windings, with the primary connection on one end of the assembly and the secondary connection on the other end. Both windings are grounded through the metal case. The resistance of the primary winding is between 15 and 25 Ω, and the resistance of the secondary winding is between 5500 and 9000 Ω.

Spark-Plug Leads

The spark-plug leads are short lengths of shielded high-tension cable provided with fittings at the ends by which they are secured to the high-tension ends of the

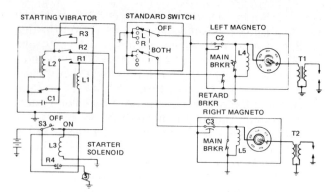

FIG. 10-44 Circuit for a low-tension ignition system.

transformer coils and to the spark plugs. The leads must be of different lengths so that they can reach to the spark-plug locations.

Operation of the Low-Tension System

A schematic diagram of the circuit for a low-tension system is shown in Fig. 10-44. It will be observed that this circuit is very similar to the circuit for the high-tension S-200 system. The system in Fig. 10-44 illustrates the use of the starting vibrator with a control relay and a standard ignition switch. In the diagram all switches and relays are shown in their normal OFF position before operation commences.

When the starter switch S3 is turned on, the starter solenoid L3 and the coil L1 are energized, thus closing contacts R1, R2, R3, and R4. With the magneto switch in the L position, battery current flows through contact R1, the vibrator points, coil L2, contact points R2, and through the main breaker to ground. The magnetic field built up around coil L2 from this current causes the vibrator points to open. Vibration starts and continues as previously explained, and the interrupted current is carried to ground through the main and retard breaker points.

When the engine reaches its normal advance firing position, the main breaker of the magneto opens. However, the vibrator current is still carried to ground through the retard breaker. When the retard position of the engine is reached, the retard breaker opens and the vibrator current flows through coil L4 to ground, thus producing a rapidly changing field around the coil. The buildup and collapse of the field around coil L4 cause a high surge of current to flow through the distributor and the primary winding of the transformer T1. This surge of current in the primary induces a high voltage in the secondary, thus providing the voltage necessary for firing the spark plug.

When the engine fires and begins to pick up speed, the starter switch is released. This deenergizes relay L1 and the starter solenoid L3, thus opening the vibrator circuit and the retard breaker circuit. The switch is then turned to the BOTH position, which permits both magnetos to operate.

Timing the Magneto

For magnetos which do not have timing marks in the breaker compartment, the three raised marks on the drive end of the magneto should be used. These marks are in the recess of the mounting pilot, as shown in Fig. 10-45. The center mark is used for the *neutral position* when timing the magneto. The outer two marks are for the exact E-gap (10°) position, depending on the direction of rotation.

To time the magneto, the rotor is turned in the direction of normal rotation until the painted chamfered tooth is aligned with the center of the inspection window. It is then turned back a few degrees until it locates in its neutral position. Remember, the magnet will tend to place itself in the neutral position when it has been rotated to a position near neutral.

To construct a pointer for timing, a piece of soft wire is bent around the drive shaft, as shown in Fig. 10-45. This wire should be bent to a position where it is aligned with the center timing mark while the magneto is still in the neutral position. The magneto is now turned in the direction of normal rotation until the wire pointer is aligned with the outer timing mark. The magneto is now in the exact E-gap position.

The breaker points should be adjusted to open while the pointer is within $\frac{7}{64}$ in [2.78 mm] of the timing mark. The opening is indicated with a suitable timing light. When the magneto is turned to a position where the cam follower is on a high point of the cam, the breaker-point clearance should be 0.018 in ±0.006 in [0.46 mm ±0.15 mm].

Adjusting Retard Breakers

On magnetos having retard breakers, it is necessary to set the retard breaker to open a predetermined number of degrees after the main breakers open, within +2° to 0°. This can be accomplished by inscribing a line on the mounting flange at the proper distance from the center mark in the pilot recess. On series S-600 magnetos, the line should be scribed $\frac{51}{64}$ in [20.24 mm] from the center line, as shown in Fig. 10-45.

The retard line on the flange can be scribed in the correct position by the use of a divider. Set the divider to the correct dimension ($\frac{51}{64}$ in), and place one point directly over the center mark on the outside diameter (OD) of the pilot. Now scribe a line on the surface of the flange where the other point of the divider touches the OD of the pilot.

After the main breaker has been set to open at E gap, the wire pointer is shifted to index over the center mark without allowing the shaft to move from its position where the main breaker points just open. The shaft is now rotated until the pointer indexes with the scribed line on the flange. By means of the timing light, the retard breaker is adjusted to open at this point.

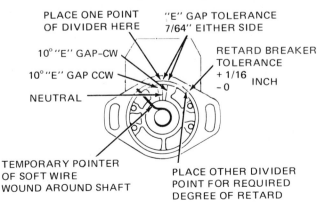

PLACE ONE POINT OF DIVIDER HERE

"E" GAP TOLERANCE 7/64" EITHER SIDE

10° "E" GAP–CW

10° "E" GAP CCW

NEUTRAL

RETARD BREAKER TOLERANCE + 1/16 – 0 INCH

TEMPORARY POINTER OF SOFT WIRE WOUND AROUND SHAFT

PLACE OTHER DIVIDER POINT FOR REQUIRED DEGREE OF RETARD

FIG. 10-45 Installation of the timing pointer.

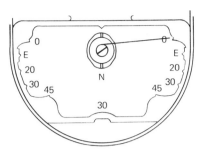

FIG. 10-46 Timing pointer on magneto with timing marks.

Timing Magnetos with Timing Marks in the Breaker Compartments

Magnetos with timing marks in the breaker compartment can be timed as described for the S-200 magneto. Fig. 10-46 illustrates a pointer attached to the cam screw and located with the shaft in the neutral position. The main breaker points should be adjusted to open when the pointer is turned to the E-gap position. The retard breaker should open at 30° after the E-gap position. This is accomplished by resetting the pointer at 0° when the rotor is at E-gap position. Then the rotor is turned until the pointer is at the 30° mark. At this point the retard breaker should open.

Installation

The installation of the S-600 magneto is very similar to the installation of other Bendix magnetos; however, we shall describe the procedure here for emphasis.

1. Check the magneto for direction of rotation to see that the magneto is correct for the engine.

2. Remove the inspection plug from the top of the magneto.

3. Turn the magneto drive shaft in the direction of rotation until the first of the two painted chamfered teeth is aligned approximately with the center of the inspection hole (see Fig. 10-47). The magneto is now in the no. 1 cylinder firing position. (**Note:** The timing mark on the distributor gear is only for timing reference and is not to be used as a reference for adjusting the breaker-point opening.)

4. Turn the engine to the no. 1 cylinder firing position as previously explained.

5. Install the magneto on the engine, and tighten the mounting bolts sufficiently to hold the magneto in position and yet allow it to be rotated.

6. Connect a suitable timing light to the switch (P lead) connection, and connect the common lead of the timing light to a good ground on the engine.

7. If the timing light is out, rotate the magneto hous-

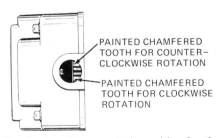

PAINTED CHAMFERED TOOTH FOR COUNTER-CLOCKWISE ROTATION

PAINTED CHAMFERED TOOTH FOR CLOCKWISE ROTATION

FIG. 10-47 Setting the magneto in position for timing to engine.

ing in the direction of its magnet rotation a few degrees beyond the point where the light comes on. Then slowly turn the magneto in the opposite direction until the light just goes out. Secure the magneto housing in this position, and recheck the setting. Connect the switch wire (P lead) to the switch terminal on the breaker cover, and connect the retard wire to the retard connection in the same manner. The fittings for connecting these wires are available through the manufacturer, and they are identified as Kit No. 10-157209 for the switch lead and Kit No. 10-157208 for the retard lead. These are the same kits supplied for the S-200 magneto.

Maintenance for the S-600 Low-Tension System

After the first 25- and 50-h periods of operation and every 50 h thereafter, the breaker assemblies should be checked. Points which have deep pits or excessively burned areas should be discarded.

The oiler felt should be checked for effectiveness. If the fingers are not moistened with oil after squeezing the felt, 2 or 3 drops of oil should be placed on the felt. This oil should be Scintilla 10-186527 or equivalent. After the oil has been allowed to soak in (about 30 min), all excess oil should be blotted off with a clean cloth. Too much oil may foul the breaker contacts and cause excessive burning.

If engine operating troubles develop and they appear to be caused by the ignition system, it is advisable to check the spark plugs, transformer coils, and wiring before working on the magnetos. The transformer coils may be checked with an ohmmeter. The primary resistance should be between 15 and 25 Ω, and the secondary resistance should be between 5500 and 9000 Ω. The resistance is checked between the end terminal of the coil and the metal case.

If the magneto seems to be the source of trouble, a visual inspection may be conducted. Uncouple the magneto connector and separate the harness from the magneto. Remove the four adapter-plate screws, and separate the adapter from the distributor housing. Inspect for the presence of moisture and foreign matter, and remove any that is found. Check for broken leads or damaged insulation. If either is present, remove the magneto and replace it with one known to be in satisfactory condition.

● LOW-TENSION IGNITION SYSTEMS FOR LARGE-AIRCRAFT ENGINES

General Description

In this section we shall describe the low-tension high-altitude ignition system designed by the Scintilla Division (now Engine Products Division) of the Bendix Corporation for use with the Pratt & Whitney R-2800 aircraft engine. The system consists of a magneto, a harness with nine detachable primary leads, two distributors, 18 double high-tension coils, and 36 high-tension leads. As shown in Fig. 10-48, the appearance of these parts on the engine does not differ radically from that of a conventional high-tension system. Electrically, however, the low-tension system is quite different from the high-tension systems, as explained previously.

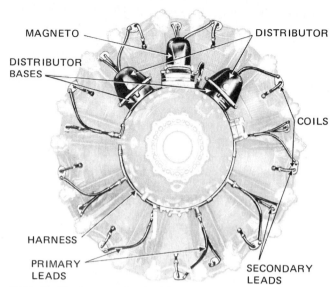

FIG. 10-48   Low-tension ignition system on an R-2800 engine.   *(Bendix)*

Type DLN-10 Magneto

The designation DLN means that the magneto is the dual type, left-hand rotating (counterclockwise), and manufactured by Bendix. It is illustrated in Fig. 10-49. Actually, this magneto can be described as a double-double magneto because it has four coils. Two four-pole magnets are mounted on the magnet shaft, with the poles of one magnet staggered 45° from those of the other. Each magnet operates in conjunction with two pairs of pole shoes molded in the housing. A coil is mounted upon each pair of pole shoes, making four coils in all. Each coil consists of a single primary winding of relatively few turns of heavy wire. There are no secondary windings. One end of each coil winding is grounded to the magneto housing, and the other end is connected through a four-pin electric connector into the harness.

The magneto is mounted on the front of the engine by means of a standard four-bolt flange. It is driven at one and seven-eighths engine crankshaft speed. The magnet shaft is supported on two ball bearings. The housing and cover are of cast-magnesium alloy and are precision machined and secured together with a special clamping ring to provide adequate radio shielding. A

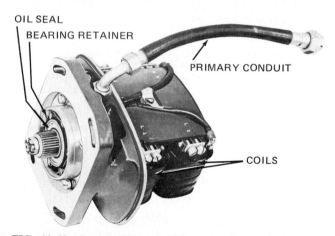

FIG. 10-49   Low-tension magneto.   *(Bendix)*

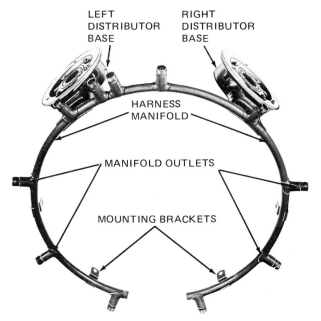

FIG. 10-50 Distributor base and harness assembly. *(Bendix)*

timing plunger is provided which, when depressed, engages a notch cut in the magnet shaft. This facilitates setting of the magneto in its exact firing position when the magneto is timed to the engine.

Harness Assembly and Primary Leads

The harness for this system is a tubular metal manifold. Two permanent distributor bases, nine primary lead outlets, one magneto connection, and one switch connection are integral parts of the manifold. The complete assembly is shown in Fig. 10-50. The outlet connections are four-pin plug-in electrical connectors.

There are 44 low-tension primary wires in the manifold. These function as follows: 4 run from the magneto connector to the distributors (2 to each), 4 run from the switch connector to the distributors (2 to each), and 36 run from the contact segments of the distributors to the primary lead outlets.

Figure 10-51 is a chart of the electrical connections of the system showing the pin-letter designations of the various electrical connectors. All wires within the manifold are No. 18 gage, insulated with rubber and fabric. Since these wires carry only a few hundred volts, the insulation requirements are not critical.

The distributor bases contain the distributor connection plates over which the carbon brushes of the distributor finger travel in making the proper distribution of current to the engine cylinder coils. Short lead wires attached to each segment plate make the electrical connections to the distributors which fit within the distributor bases of the harness.

The detachable primary leads connect the harness outlets to the transformer coils. Each primary lead is

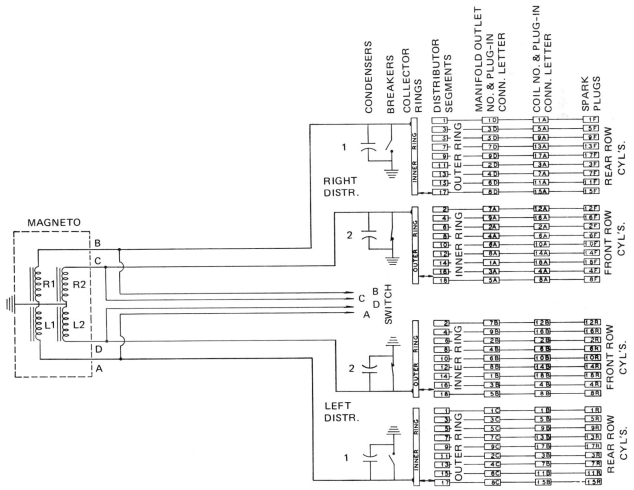

FIG. 10-51 Wiring diagram for a low-tension ignition system on an 18-cylinder radial engine. *(Bendix)*

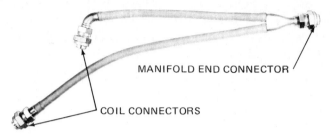

FIG. 10-52 Primary lead assembly.

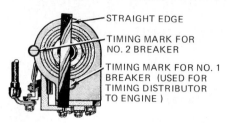

FIG. 10-53 Use of a straightedge with timing step and timing marks.

of forked construction, as shown in Fig. 10-52, having a four-pin electrical connector at each of the branches which lead to the two coils. For purposes of identification, the manifold outlets are numerically designated. The distributor bases are designated as left and right, as viewed from the cockpit or antipropeller end of the engine.

Within each distributor base an oil seal is mounted. The sealing ring of the oil seal assembly mates against a lapped washer on the drive shaft of the distributor unit, thereby preventing the engine oil from entering the distributor.

Distributor Assemblies

The **distributor assemblies** fit within the distributor bases of the harness assembly. They are of clockwise rotation as viewed from the drive end; that is, the rotation is counterclockwise as one looks into the distributor base at the segment plate. Each distributor is secured in position with three screws and is provided with a precision-machined cover. The cover is secured with a clamping ring which ensures an electrically tight seal, preventing radio interference.

The distributor shaft carries two nine-lobe compensated breaker cams, a distributor finger, and the drive coupling. It also carries a lapped washer which mates against the sealing member of the oil seal in the distributor base.

The breakers are mounted adjacent to the cams and are secured to a breaker plate mounted on the distributor housing. A capacitor (condenser) is connected across each breaker and is mounted to the housing with clamps.

The molded **collector plate** is mounted directly over the distributor rotor. The two collector rings in this plate make contact with the two carbon brushes on the upper surface of the distributor rotor. The upper brushes are connected through the rotor to the lower brushes, which make contact with the distributor connection plates in the distributor base section of the harness.

As mentioned previously, the breaker cams are of compensated design. This means that there is a separate lobe for each cylinder of the engine. The no. 1 lobe of each cam is marked with a dot. Each cam is also engraved with an arrow to show the direction of operating rotation. The two cams are mounted on the same shaft with the no. 1 cam and breaker in the top position. The no. 1 breaker of the right distributor fires the front plugs of the rear row of engine cylinders, and the no. 1 breaker of the left distributor fires the rear plugs of the rear row of cylinders. The no. 2 cams fire the front and rear plugs of the front row (even num-

bered) of cylinders. The connections for the spark plugs can be traced in Fig. 10-51.

The distributor shaft turns on two ball bearings and is driven at one-half engine speed. It should be noted that all distributor rotors turn at one-half engine speed for a four-stroke-cycle engine because one complete revolution of the rotor fires all plugs once and this occurs in two revolutions of the crankshaft. Since a compensated cam has one lobe for each cylinder fired, the cam also turns at one-half engine speed and can therefore be driven by the same shaft that drives the distributor rotor.

The drive end of the distributor shaft carries a coupling of standard 17-tooth design. The upper end of the shaft is machined to provide a timing step which is used in conjunction with two timing marks to indicate the opening positions of the two breakers. Figure 10-53 shows the method of locating the opening position of the no. 1 breaker by using a straightedge to align the shaft step with the no. 1 timing mark. The drive coupling of the distributor is provided with vernier ratchets to permit accurate setting of the distributor when it is timed to the engine.

Transformer Coils and High-Tension Leads

Each coil (transformer) assembly consists of two transformer coils permanently assembled into a single case and wired to suitable electrical connections. The coil units are mounted on special brackets secured to the baffle plates of the engine cylinders.

A two-pin electrical connector forms the input connection to the coil assembly. One pin of the connector leads to the primary of each transformer coil. The other end of each primary is grounded to the case. One end of each secondary is also grounded. The other end of each secondary winding terminates at a high-tension contact socket at each end of the coil case.

Each high-tension lead fits directly into one of the high-tension contact sockets of its respective coil. The leads are short, thereby reducing the system's energy losses. Fig. 10-54 shows one of the coil assemblies with its high-tension lead.

Magneto Operation

A simplified schematic diagram of the high-altitude low-tension system for an 18-cylinder radial engine is shown in Fig. 10-55. In the diagram it can be seen that the magneto has two four-pole magnets mounted on

FIG. 10-54 Coil assembly with high-tension lead.

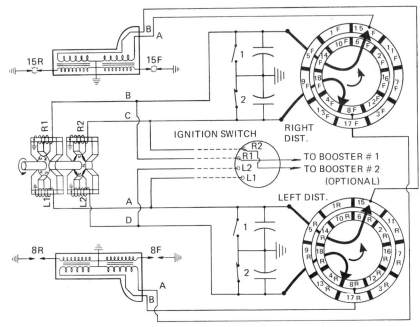

FIG. 10-55 Schematic diagram of a low-tension system.

the same shaft but staggered in position so that the poles of one magnet are aligned with the pole shoes when the poles of the other magnet are in the neutral position. Thus, the magnets are staggered 45°. Also in the magneto are two sets of double-pole shoes and four primary coils. As each magnet rotates one revolution, it produces in each of the two coils on its pole shoes four impulses of current. Since the magnets are staggered 45° from each other, there are eight current impulses per revolution of the shaft. Because each impulse is generated in two coils, each is of a dual nature; that is, they are two separate but simultaneous pulses of low-voltage current. During operation, a pair of pulses will fire the front and rear plugs of one cylinder.

The magnet shaft is driven at one and one-eighth engine speed. This produces nine double spark impulses per engine revolution, which is the requirement for an 18-cylinder four-stroke-cycle engine, such as the Pratt & Whitney R-2800.

Harness and Distributors

The low-voltage impulses from the magneto coils are transmitted through the harness to the distributors. The impulses from the two right-hand coils in the magneto, R1 and R2, are carried to the right-hand distributor which fires the front plugs in each cylinder. The two left-hand coils in the magneto, L1 and L2, are connected to the left-hand distributor and fire the rear plugs of each cylinder. As long as the breaker points are closed, there are two completed circuits in each distributor. One is through the breaker points to ground and back to the grounded coil in the magneto. The other is out through the distributor into the primary winding of the high-tension coil. This second circuit has a relatively high resistance, so most of the current flows through the breaker points to ground. When the magneto low-tension current reaches its peak value, the cam separates the breaker points and interrupts the circuit.

As the breaker points open, the capacitor connected across the points is charged. The capacitor charge plus the remaining current from the magneto coil then surges through the distributor contacts and into the primary coil of the high-tension coil where it induces the high-voltage in the secondary winding. This voltage breaks down the gap between the spark-plug electrodes and produces the spark to ignite the fuel-air mixture. The surge of secondary current is quite short-lived and falls away rapidly. Then the points close again and reconnect the low-resistance circuit to ground. Meanwhile, of course, the rotor of the distributor has made contact with the proper segment of the distributor block so that the surge is directed to the particular spark plug which it is required to fire. As the magnet rotates and another impulse is building up in the magneto circuit, the rotor moves to the next distributor segment, and the next plug in the firing order will receive the coming surge of high voltage. Also, the two distributors are synchronized, and when the right one is firing a front plug, the left is firing the rear plug in the same cylinder. Therefore, the two plugs in any one cylinder are each receiving a high-voltage spark at the same moment, ensuring proper ignition of the fuel-air mixture in the cylinder.

Distribution of the Magneto Output

An examination of Fig. 10-55 will further clarify the action of these circuits at the instant of firing a cylinder. Assume that the no. 15 cylinder is being fired. This cylinder is the seventeenth in the firing order. Coils L1 and R1 are generating impulses in the magneto. The current goes through the harness to the distributors, L1 to the left distributor and R1 to the right distributor. Since the breaker points are opened, the current flows into the distributor rotor, through the carbon brushes and to the no. 17 segment which is connected to the coil for the no. 15 cylinder. It then flows through the harness to the coil for the no. 15 cylinder where it is transformed into high-voltage current and passes through the high-tension lead to the spark plug, causing a spark at the gap. The current from the left distributor is transmitted to the rear plug, and the current from the right distributor is transmitted to the front plug. After this surge of current has ceased, the distributor rotor will progress from the no. 17 segment to the no. 18 segment and coils L2 and R2 in the magneto will be generating impulses which will be transmitted to cylinder no. 8, which is the eighteenth cylinder in the firing order. Note that the firing order for the R-2800 engine is 1-12-5-16-9-2-13-6-17-10-3-14-7-18-11-4-15-8. On an 18-cylinder twin-row radial engine, the even-numbered cylinders are in the front row and the odd-numbered cylinders are in the rear row.

● THE COMPENSATED CAM

Reason for Compensated Cam

In a radial engine, because of the mounting of the link rods on the flanges of the master rod, the travel of the pistons connected to the link rods is not uniform. Normally we would expect that the pistons in a nine-cylinder radial engine would reach top center 40° apart. That is, for each 40° the crankshaft turns, another piston would reach TDC. Since the master rod tips from side to side while it is carried around by the crankshaft, the link rods follow an *elliptical* path instead of the circular path required for uniform movement. For this reason there is less than 40° of crankshaft travel between the TDC positions for some pistons and more than 40° of travel between the TDC positions of other pistons.

In order to obtain ignition at precisely 25° BTC it is necessary to compensate the breaker cam in the magneto by providing a separate cam for each cylinder of the engine.

Design of the Compensated Cam

A compensated cam for a nine-cylinder radial engine is shown in Fig. 10-56. The cam lobe for the no. 1 cylinder is marked with a dot, and the direction of rotation is shown by an arrow. A careful inspection of the cam will reveal that there is a slight difference in the distance between the various lobes. This variation

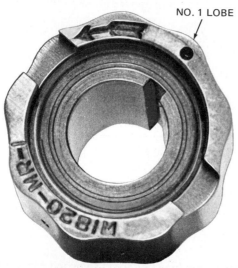

FIG. 10-56 Compensated cam.

is designed into the cam in order to compensate for the nonuniform movement of the pistons. The variation may be as much as 2.5° more or less than 40° for a nine-cylinder radial engine.

The compensated cam turns at one-half crankshaft speed because it produces a spark for each cylinder during each complete revolution. Since the crankshaft must rotate through two turns to fire all the pistons, the cam can turn only one-half crankshaft speed. The compensated cam is normally mounted on the same shaft that drives the distributor because the distributor also can turn only one-half crankshaft speed.

● IGNITION SHIELDING

Since the magneto is a special form of high-frequency generator, it acts like a radio transmitting station while it is in operation. Its oscillations are called **uncontrolled** because they cover a wide range of frequencies. The oscillations of a conventional radio transmitting station are waves of a **controlled** frequency. For this reason the ignition system must be shielded.

Shielding is difficult to define in general terms. Aircraft **radio shielding** is the metallic covering or sheath over all electric wiring and ignition equipment, grounded at close intervals and provided for the purpose of eliminating any interference with radio reception.

If the high-tension cables and switch wiring of the magneto are not shielded, they can serve as antennas from which the uncontrolled frequencies of the magneto oscillations are radiated. The receiving aerial on an airplane is comparatively close to the ignition wiring; hence, the uncontrolled frequencies are picked up by the aerial along with the controlled frequencies from the aircraft radio station, thus causing interference (noise) to be heard in the radio receiver in the airplane.

Design of the Ignition Shielding

The magneto has a metallic cover that is made of a nonmagnetic material. The cover joints are fitted tightly to prevent dirt and moisture from entering. Since it is necessary to cover the cables completely, fittings are provided on the magneto to attach a shielded ignition harness. Provision is made for ventilation to remove condensation and the corrosive gases formed by the arcing of the magneto within the housing.

Shielding of high-tension leads for the ignition system installed on an opposed engine is accomplished by means of a woven wire sheath placed around each spark-plug lead. This sheath is electrically connected to the magneto case and to the spark-plug shells to provide a continuous grounded circuit.

Ignition Wiring System

The **low-tension wiring** on a high-tension magneto consists of a single shielded conductor from the primary coil to the engine ignition switch. Its circuit passes through the fire wall with a connector plug, frequently of a special design, which automatically grounds the magnetos when the plug is disconnected.

High-tension cable differs from low-tension cable in that high-tension cable has a conductor of small cross section and insulation of comparatively large cross section, whereas low-tension cable has a conductor of large cross section and insulation of comparatively small cross section. The reason for this difference is that the capacity to carry current is the primary requisite of low-tension cable, whereas dielectric strength (insulating property) is the most important requirement of high-tension cable.

High-tension cable may consist of several strands of small wire; a layer of rubber, synthetic rubber, or plastic; a glass braid covering; and a neoprene, or plastic, sheath. It is available in several sizes, the most common being 5 and 7 mm.

High-tension wiring is placed in the special conduit arrangement known as the **ignition harness**, mentioned before, or enclosed in a woven wire sheath to provide for radio shielding.

● THE TIMING LIGHT

Timing lights for the synchronizing of magneto breaker points may be simply "hot" test lights, or they may be more complex. If an ordinary hot test light is used for checking the opening of breaker points, it is necessary to disconnect the primary lead from the breaker points, because the light will burn regardless of whether the points are open or closed, owing to the low resistance of the primary windings. There may be a slight but not appreciable dimming when the breaker points open. If the primary circuit is disconnected from the points, the light will go out when the points are open because the current path is open. This is illustrated in Fig. 10-57.

The circuit for another type of test light is shown in Fig. 10-58. In this test unit a vibrator produces an interrupted direct current which is converted to an alternating current by means of a transformer. The alternating current is supplied to the breaker points, and the points are closed because there is a complete circuit through the secondary of the transformer and the lamp. When the breaker points open, the primary of the magneto is in series with the lamp, and the

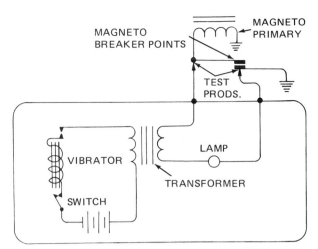

FIG. 10-58 Circuit for a vibrator-type timing light.

inductive reactance of the primary coil will reduce the current, thus causing the lamp to go out or glow very dimly.

Still another commonly used timing light, called the Abbott synchronizer, utilizes a neon lamp. The circuit for this unit is shown in Fig. 10-59. The vibrator converts the direct current into pulsating (square-wave) direct current which is applied to the primary of the synchronizer transformer. When the magneto breaker points are open, a pulsating current flows through the primary of the transformer. This voltage is stepped up about 20 to 1 in the secondary to produce a spike voltage which is sufficient to fire the neon lamp connected across the secondary. When the breaker points are open, some current can flow through the primary of the magneto, but the inductive reactance (X_L) of the magneto primary is great enough to prevent the short-circuiting of the transformer primary. The neon light will therefore glow when the breaker points are open. When the breaker points are closed, the primary of the transformer is short-circuited and no voltage is developed in the secondary of the transformer; hence, the neon lamp will not glow.

When the synchronizer is connected to the magneto primary circuit as shown, the magneto will not be capable of firing because the primary of the magneto is short-circuited through the primary of the transformer. Therefore, it is safe to work on the magneto.

The internal circuit of the Abbott synchronizer is shown in Fig. 10-60. This circuit provides for the timing of two magneto circuits together so that they will fire

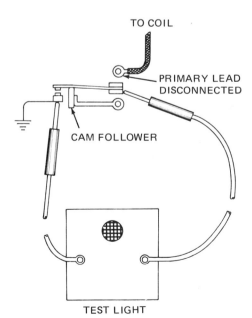

FIG. 10-57 Use of a timing light.

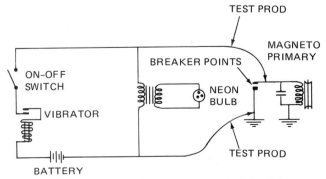

FIG. 10-59 Circuit to demonstrate the principle of the Abbott synchronizer.

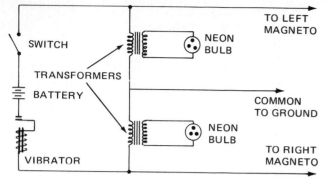

FIG. 10-60 Internal circuit of the Abbott synchronizer.

in unison or synchronization. It will be observed that the circuit shown in Fig. 10-60 is the same as the one shown in Fig. 10-59 except that it utilizes two transformers and two neon lamps to provide for the synchronization of two magnetos on the same engine.

● IGNITION SYSTEMS FOR GAS-TURBINE ENGINES

Ignition systems for gas-turbine engines are required to operate for starting only; their total operating time is therefore almost insignificant compared with an ignition system for a reciprocating engine. For this reason, the gas-turbine ignition system is almost trouble-free.

An important charactertistic of a gas-turbine ignition system is the high-energy discharge at the igniter plug. This is necessary because of the difficulty in igniting the fuel-air mixture under some conditions, particularly at high altitudes when an engine has "flamed out." The high-energy discharge is accomplished by means of a storage capacitor in what is termed a **high-energy capacitor discharge** system. The effect of this system is to produce what appears to be a white-hot ball of fire at the electrodes of the igniter plug. In some designs, the igniter actually "shoots" the electric "flame" several inches.

The technician must use great care when working on gas-turbine ignition systems because of the possibility of a lethal shock.

Ignition System for the Pratt & Whitney JT9D Engine

One of the modern and widely used gas-turbine engines is the Pratt & Whitney JT9D high-bypass fanjet which powers the Boeing 747 aircraft and others. The ignition system for this engine is a good example of modern ignition systems for gas-turbine engines. A drawing of the circuit for this system is shown in Fig. 10-61.

The ignition system for the JT9D engine is the G.L.A. 43035, manufactured by the General Laboratory Associates, Inc. The complete system includes two separate heat-shielded and shock-mounted exciters. Figure 10-61 shows the circuit for one of the exciters. Each exciter supplies ignition energy to a recess-gap igniter plug through a high-tension lead. A small amount of engine fan air is directed to cool the high-tension leads, the exciter boxes, and the igniter plugs.

Input power for operation of the ignition exciters is

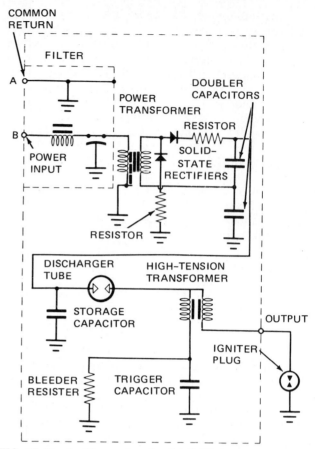

FIG. 10-61 Ignition system for the JT9D turbofan engine.

115 V, 400 Hz AC, with an input current not in excess of 2.5 A to each exciter. The stored energy is 4 J nominal. One joule (J) is the amount of work done in maintaining a current of 1 A against a resistance of 1 Ω for 1 s; this is equivalent to 0.73732 ft·lb.

AC power from the aircraft electric system is applied between the exciter A and B terminals at the input connector. This power is first passed through a **filter circuit** consisting of a reactor and a feed-through capacitor to prevent high-frequency feedback into the aircraft electric system. The reactor also serves as a power choke to limit spark rate variations over the input-voltage frequency range. From the filter, the voltage is applied across the primary of the power transformer.

The high voltage generated in the secondary of the power transformer is rectified in the **doubler circuit** by action of the two solid-state rectifiers and the doubler capacitors so that with each change in polarity a pulse of DC voltage is sent to the storage capacitor. The resistors in the doubler circuit serve to limit the current passing through the rectifiers during those intervals of discharge of the storage capacitor when the voltage has reversed. With successive pulses the storage capacitor assumes a greater and greater charge at increasing voltage.

When the voltage of the storage capacitor reaches the predetermined level for which the spark gap in the discharger tube has been calibrated, the gap breaks down and a portion of the charge accumulated on the storage capacitor flows through the primary of the high-tension transformer and to the trigger capacitor. This flow of current induces in the transformer secondary a voltage high enough to ionize the air gap in the igniter plug. With the gap thus made conductive, the remaining charge on the storage capacitor is delivered to the igniter plug as a high-current, low-voltage spark across the gap. This is the high-energy spark necessary to produce ignition under adverse conditions.

The **bleeder resistor** is provided to dissipate the energy in the circuit if the igniter plug is absent or fails to fire. It also serves to provide a path to ground for any residual charge on the trigger capacitor between cycles. When the storage capacitor has discharged all its accumulated energy, the cycle of operation recommences. Variations in input voltage or frequency will affect the spark repetition rate, but the stored energy will remain virtually constant.

The exciter is reparable using test equipment available in most airline overhaul shops. Positive hermetic seal is ensured through the use of a stainless-steel case weldment. The exciter capability for continuous operation will ensure conformance with potential airline ignition duty cycles. Maximum exciter service life can be attained, however, by following the recommended engine or aircraft operational procedures.

● SPARK PLUGS

The Function of the Spark Plug

The spark plug is the part of the ignition system in which the electric energy of the high-voltage current produced by the magneto or other high-tension device is converted into the heat energy required to ignite the air-fuel mixture in the engine cylinders. The spark plug provides an air gap across which the high voltage of the ignition system produces a spark to ignite the mixture.

Construction of Spark Plugs

An aircraft spark plug fundamentally consists of three major parts: (1) the electrodes, (2) the ceramic insulator, and (3) the metal shell.

Figure 10-62 shows the constructional features of a typical aircraft spark plug manufactured by the Champion Spark Plug Company. In the illustration, the assembly shown in the center of the spark plug is the inner electrode assembly consisting of the terminal contact, spring, resistor, brass cap, and conductor (not labeled in the illustration), and the nickel-clad copper electrode. The insulator, shown between the electrode assembly and the shell, is made in two sections. The main section extends from the terminal contact to a point near the tip of the electrode. The barrel-insulating section extends from near the top of the shielding barrel far enough to overlap the main insulator.

The outer section of the spark plug, illustrated in Fig. 10-62, is a machined steel shell. The shell is often plated to eliminate corrosion and to reduce the possibility of thread seizure. In order to prevent the escape of high-pressure gases from the cylinder of the engine through the spark-plug assembly, internal pressure seals, such as the cement seal and the glass seal, are used between the outer shell and the insulator and also between the insulator and the center electrode assembly.

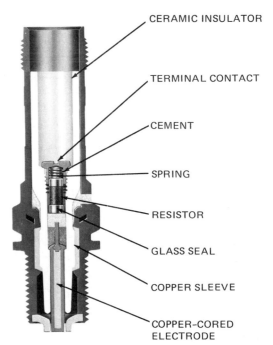

FIG. 10-62 Shielded spark plug. (*Champion Spark Plug Co.*)

The shell of the spark plug includes the radio-shielding barrel. In some spark plugs, the shell and shielding barrel are made in two sections and are screwed together. The two parts should never be disassembled by the technician because during manufacture the correct pressure is applied to provide a gastight seal. Any disturbance of the seal may cause leakage.

The shell and the radio-shielding barrel complete the ground circuit for the radio shielding of the ignition harness. The shell is externally threaded on both ends so that it can be joined to the radio shielding of the ignition harness at the top and can be screwed into the cylinder head at the bottom.

Spark plugs are manufactured with many variations in construction to meet the demands of aircraft engines. **Resistor-type** spark plugs were designed to reduce the burning and erosion of the electrodes in engines having shielded harnesses. The capacitance between the high-tension cable and the shielding is sufficient to store electric energy in quantities which produce a comparatively high-current discharge at the spark-plug electrodes. The energy is considerably greater than is necessary to fire the fuel-air mixture; hence, it can be reduced by means of a resistor in order to provide greater spark-plug life.

Another improvement which leads to greater dependability and longer life is the use of platinum-alloy firing tips and high-temperature alloys in the main center electrode. A spark plug with this type of construction is illustrated in Fig. 10-63.

Unshielded spark plugs are still used in a few light-aircraft engines. An unshielded spark plug is shown in Fig. 10-64.

The construction of spark plugs for aircraft engines is further illustrated in Fig. 10-65. The spark plug on the left is the massive-electrode type, so named because of the size of the center and ground electrodes. This spark plug is a resistor type to reduce electrode

SILVER-CORED
CENTER ELECTRODE

PLATINUM
ELECTRODES

FIG. 10-63   Spark plug with platinum electrodes.   (*Champion Spark Plug Co.*)

erosion. Nickel seals are provided between the insulator and the shell to effectively eliminate gas leakage. The center electrode consists of a copper core with a nickel-alloy sheath. The tip of the insulator is recessed to maintain the proper temperature to prevent fouling and lead buildup. The three ground electrodes are nickel-alloy and are designed to be cleaned with a three-blade vibrator tool. The center electrode is sealed against gas leakage by means of a metal-glass binder.

The spark plug on the right in Fig. 10-65 is a fine-wire type. It is similar in construction to the massive-electrode plug except for the electrodes. The center electrode is of platinum and the two ground electrodes are constructed of either platinum or iridium. The use of platinum and iridium ensures maximum conductivity and minimum wear.

CHAMPION

FIG. 10-64   Unshielded spark plug.

The spark plugs illustrated in Fig. 10-65 are manufactured by the SGL Auburn Spark Plug Company. Formerly these spark plugs were manufactured by AC, a division of General Motors.

Classification of Shell Threads

Shell threads of spark plugs are classified as 14 or 18 mm diameter, **long reach** or **short reach,** thus:

Diameter	Long Reach	Short Reach
14 mm	$\frac{1}{2}$ in [12 mm]	$\frac{3}{8}$ in [9.53 mm]
18 mm	$\frac{13}{16}$ in [20.67 mm]	$\frac{1}{2}$ in [12 mm]

Terminal threads at the top of the radio-shielded spark plugs are either $\frac{5}{8}$-in [15.88-mm] 24 thread or $\frac{3}{4}$-in [19.05-mm] 20 thread. The latter type is particularly suitable for high-altitude flight and for other situations where flashover within the sleeve might be a problem.

The designation numbers for spark plugs provide an indication of the characteristics of the plug. The Champion Spark Plug Company utilizes letters and numbers to indicate whether the spark plug contains a resistor and to indicate the barrel style, mounting thread, reach, hex size, heat rating range, gap, and electrode style. The designations are as follows:

1. No letter or an R. The R indicates a resistor-type plug.

2. No letter, E, or H. No letter—unshielded; E—shielded $\frac{5}{8}$-in 24 thread; H—shielded $\frac{3}{4}$-in 20 thread.

3. Mounting thread, reach, and hex size.

a—18 mm, $\frac{13}{16}$-in reach, $\frac{7}{8}$-in [22.26 mm] stock hex

b—18mm, $\frac{13}{16}$-in reach, $\frac{7}{8}$-in milled hex

d—18 mm, $\frac{1}{2}$-in reach, $\frac{7}{8}$-in stock hex

j—14 mm, $\frac{3}{8}$-in reach, $\frac{13}{16}$-in stock hex

l—14 mm, $\frac{1}{2}$-in reach, $\frac{13}{16}$-in stock hex

m—18 mm, $\frac{1}{2}$-in reach, $\frac{7}{8}$-in milled hex

4. Heat rating range. Numbers from 26 to 50 indicate coldest to hottest heat range. Numbers from 76 to 99 indicate special application aviation plugs.

5. Gap and electrode style. E—two-prong aviation; N—four-prong aviation; P—platinum fine wire; B—two-prong massive, tangent to center; R—push wire, 90° to center.

Spark-Plug Heat Range

The **heat range** of a spark plug is the principal factor governing aircraft performance under various service conditions. The term *heat range* refers to the classification of spark plugs according to their ability to transfer heat from the firing end of the spark plug to the cylinder head.

Spark plugs have been classified as "hot," "normal," and "cold." However, these terms may be misleading because the heat range varies through many degrees of temperature from extremely hot to extremely cold. Thus the word "hot" or "cold" or "normal" in itself does not necessarily tell the whole story.

Since the insulator is designed to be the hottest part of the spark plug, its temperature can be related to the preignition and fouling regions, as shown in Fig. 10-66. Preignition is likely to occur if surface areas in the combustion chamber exceed critical limits or if the spark-plug core nose temperature exceeds 1630°F

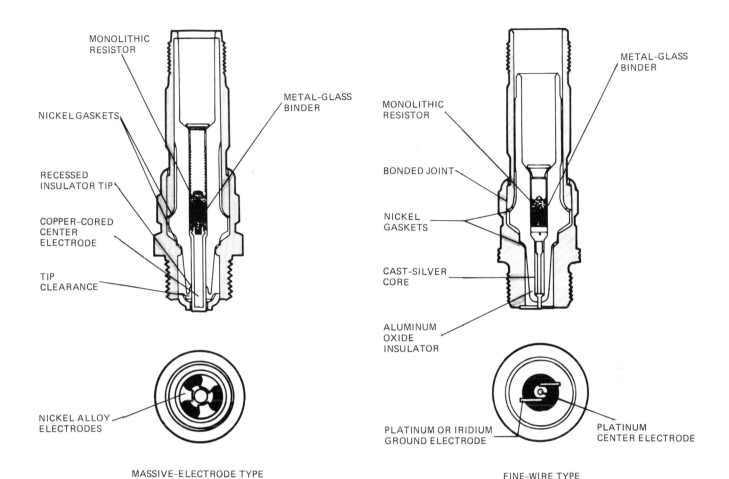

MASSIVE-ELECTRODE TYPE   FINE-WIRE TYPE

FIG. 10-65   Massive-electrode and fine-wire spark plugs.   (*SGL Auburn Spark Plug Co.*)

[888°C]. On the other hand, fouling or shorting of the plug due to carbon deposits is likely to occur if the insulator tip temperature drops below approximately 800°F [427°C]. Thus, spark plugs must operate between fairly well-defined temperature limits, and this requires that plugs be supplied in various heat ranges to meet the requirement of different engines under a variety of operating conditions.

From the engineering standpoint, each individual plug must be designed to offer the widest possible operating range. This means that a given type of spark plug should operate as hot as possible at slow speeds and light load and as cool as possible at cruise and takeoff power. Plug performance therefore depends on the operating temperature of the insulator nose, with the most de-

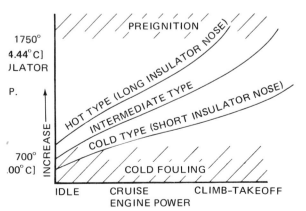

FIG. 10-66   Chart to show spark-plug temperature ranges.

sirable temperature range falling between 1000 and 1250°F [538 and 677°C].

Fundamentally, an engine which runs hot requires a relatively cold spark plug, whereas an engine which runs cool requires a relatively hot spark plug. If a hot spark plug is installed in an engine which runs hot, the tip of the spark plug will be overheated and cause **preignition.** If a cold spark plug is installed in an engine which runs cool, the tip of the spark plug will collect unburned carbon, causing **fouling** of the plug.

A discussion of hot, normal, and cold plugs is technically correct, but it should be emphasized that different heat ranges of aircraft spark plugs cannot be substituted arbitrarily, as is common in automotive practice, because the selection of aircraft spark plugs is governed by the aircraft engine manufacturers' and the Federal Aviation Administration's approvals governing the use of a particular spark plug in any aircraft engine.

The principal factors governing the heat range of aircraft spark plugs are (1) the distance between the copper sleeve around the insulator and the tip of the insulator, (2) the thermal conductivity of the insulating material, (3) the thermal conductivity of the electrode, (4) the rate of heat transfer between the electrode and the insulator, (5) the shape of the insulator tip, (6) the distance between the insulator tip and the shell, and (7) the type of outside gasket used. Hot- and cold-plug construction is illustrated in Fig. 10-67.

Other features of spark-plug construction may affect the heat range to some extent. However, the factors

267

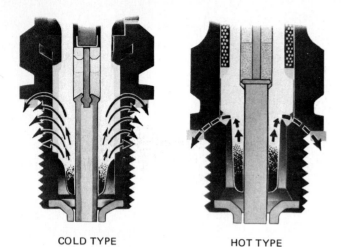

COLD TYPE   HOT TYPE

FIG. 10-67 Constructoin of hot and cold spark plugs.

we have mentioned are those of primary consideration. In all cases, the technician should install the type of spark plug approved for the particular engine being serviced. Some manufacturers use a color-code system on the cylinders to indicate the heat range of spark plugs to be installed.

Servicing Aircraft Spark Plugs

Scheduled servicing intervals are normally determined by the individual aircraft operator. These intervals will vary according to operating conditions, engine models, and spark-plug types. The principal determining factor in the removal and servicing of spark plugs is the width of the spark gap, that is, the distance between the electrodes where the spark is produced. This spark-plug gap increases in width with use until the distance becomes so great that the spark plug must be removed and either regapped or replaced. If the spark-plug gap becomes too wide, a higher secondary voltage must be developed by the ignition system in order to create a spark at the gap. This higher voltage will tend to leak through the insulation of the ignition wiring, thus eventually causing failure of the high-tension leads.

The correct spark-plug gap for a particular spark-plug installation is established by the manufacturer. No spark plug should be operated with a gap greater than that specified in the manufacturer's instructions.

Servicing Procedure

It must be emphasized that an aircraft operator must either follow the manufacturer's instructions for the servicing of aircraft spark plugs or adopt an adequate procedure of his or her own.

In general, the servicing of aircraft spark plugs is accomplished according to the following sequence:

1. **Removal.** Shielded terminal connectors are removed by loosening the elbow nut with the proper size crowfoot or open-end wrench. Care must be taken to avoid damaging the elbow. Terminal sleeve assemblies must be pulled out in a straight line to avoid damaging either the sleeve or the barrel insulator.

The spark plug is loosened from the cylinder bushing or the use of the proper size deep-socket wrench. It is recommended that a six-point wrench be used because

it provides a greater bearing surface than a twelve-point wrench. The socket must be seated securely on the spark-plug hexagon to avoid possible damage to the insulator or connector threads.

As each spark plug is removed, it should be placed in a tray with numbered holes so that the engine cylinder from which the spark plug has been removed can be identified. This is important, because the condition of the spark plug may indicate impending failure of some part of the piston or cylinder assemblies.

If a spark plug is dropped on a hard surface, cracks may occur in the ceramic insulation which are not apparent on visual examination. *Any spark plug which has been dropped should be rejected or returned to the manufacturer for reconditioning.*

In case the threaded portion of a spark plug breaks off in the cylinder, great care must be exercised in the removal of the broken section. Normally it will be necessary to remove the cylinder from the engine. Before an attempt is made to remove the section, a liquid penetrant should be applied around the threads and allowed to stand for at least 30 min. If the broken part is tapped lightly with the end of a punch while the penetrant is working, the vibration will help the liquid to enter the space between the threads. The electrodes of the plug should be bent out of the way with a pin punch or similar tool in order to permit the insertion of a screw extractor (Easyout). A steadily increasing force on the screw extractor should remove the broken part.

If the foregoing process is not successful, it will be necessary to remove the part by cutting with a small metal saw. Such a saw can be made by cutting a hacksaw blade to a size which can be inserted inside the spark-plug section. The saw is carefully manipulated to cut three slots inside the section without cutting deeply enough to touch the threads of the cylinder. The three sections of the part can then be broken out by means of a punch and a light hammer. The blows should be directed toward the center of the spark-plug hole.

After the broken portion of the spark plug has been removed, the threads in the cylinder should be checked carefully for damage. Usually the threads can be cleaned by means of a thread chaser. If there is appreciable damage to the threads, a Heli-Coil insert should be installed. See instructions in the chapter on overhaul practices.

2. **Preliminary inspection.** Immediately after the spark plugs have been removed, they should be given a careful visual inspection and all unserviceable plugs should be discarded. Spark plugs with cracked insulators, badly eroded electrodes, damaged shells, or damaged threads should be rejected.

3. **Degreasing.** All oil and grease should be removed from both the interior and the exterior of the spark plugs according to the degreasing method approved for that particular type of spark plug. Either vapor degreasing or the use of solvents, such as Stoddard solvent, is usually recommended. *Carbon tetrachloride should not be used in cleaning spark plugs* because it is likely to cause corrosion and is poisonous.

4. **Drying.** After they have been degreased, spark plugs should be dried, both inside and out, to remove all traces of solvent. Drying may be accomplished by the use of dry compressed air or by placing the spark plugs in a drying oven.

5. Cleaning. During operation of an aircraft engine, lead and carbon deposits form on the ceramic core, the electrodes, and the inside of the spark-plug shell. These deposits are most readily removed by means of an abrasive blasting machine especially designed for cleaning spark plugs.

The use of an abrasive blast spark-plug cleaner is shown in Fig. 10-68. Instructions for the use of the cleaning unit are as follows:

> *(a)* Install the proper size rubber adapter in the cleaner, and press the firing end of the spark plug into the adapter hole.
> *(b)* Move the control lever to ABRASIVE BLAST, and slowly wobble the spark plug in a circular motion for 3 to 5 s. The wobbling motion angle should be no greater than 20° from vertical to permit the abrasive materials freely to enter the firing end opening and facilitate cleaning.
> *(c)* Continue the wobbling motion, and move the lever to AIR BLAST to remove the abrasive particles from the firing bore.
> *(d)* Remove the plug and examine its firing end. If cleaning is incomplete, repeat the cleaning cycle for 5 to 10 s. If cleaning is still incomplete, check the cleaner and replace the abrasive. For complete service information, refer to the manufacturer's service manual.

Several companies offer spark-plug cleaning machines with the abrasive supported in liquid. This is an excellent cleaning method, but the operator must make sure that the abrasive is of the **aluminum oxide** type. **Silica** abrasive can contribute to later plug fouling. Immediately after cleaning by the wet blast method, the plugs should be oven-dried to prevent rusting and to ensure a satisfactory electrical test.

Excessive use of the abrasive blast is avoided to prevent too much wear of the electrodes and insulators. Spark-plug threads are cleaned by means of a wire wheel with soft bristles. Threads which are slightly nicked may be cleaned by using a chasing die.

The connector seat at the top of the shielding barrel must be cleaned to provide a proper seating surface for the gasket and nut at the end of the ignition harness. If necessary, fine-grained garnet paper or sandpaper may be used to smooth the seat. Emery paper should not be used because the emery compound conducts electricity and may establish a path for leakage of high-voltage current. After the seat is cleaned, the shielding barrel should be thoroughly blown out with an air blast.

6. Regapping. The tools and methods used to set spark-plug gaps will vary with the shape, type, and arrangement of electrodes. In Fig. 10-69, four typical forms of electrode construction are illustrated. These are the single-ground electrode, the two-prong "fine-wire" electrodes, the two-prong ground electrodes, and the four-prong ground electrodes.

The gap in any spark plug is measured by the use of **round wire gages**. These gages are supplied in two sizes for each gap to measure both the minimum and the maximum width for the gap. For example, a spark-plug gap gage will have two wires, one of which is 0.011 in [0.279 mm] and the other 0.015 in [0.381 mm]. When the spark-plug gap is being tested, the smaller-dimension gage must pass through the gap and the larger-dimension gage must be too large to pass through the gap. A gage for checking the gap clearance of a spark plug is shown in Fig. 10-70.

If a spark-plug gap is too large, it is closed by means of a special gap-setting tool supplied by the manufacturer and used according to the manufacturer's instructions. A gap-setting tool is shown in Fig. 10-71.

If the gap of a four-prong or two-prong spark plug has been closed beyond limits, no effort should be made to open the gap. In such a case the plug should be returned to the manufacturer for adjustment.

Single-electrode and two-prong wire electrode plugs are so constructed that the gap can be either opened or closed without danger of cracking the ceramic insulator. A special tool for adjusting the gap of such a plug is shown in Fig. 10-72.

7. Inspection and testing. The final step in preparing a used spark plug for service is the inspection and test. Visual inspection is accomplished by means of a magnifying glass. It is essential that good lighting be provided. The following items are examined: threads, electrodes, shell hexagons ("hex"), ceramic insulation, and the connector seat.

Spark plugs are tested by applying high voltage, equivalent to normal ignition voltage, to the spark plug while the plug is under pressure. This test has commonly been called the "bomb" test. Modern testing devices have been designed by spark-plug manufac-

FIG. 10-68 Use of abrasive blast spark-plug cleaner.

FIG. 10-69 Types of electrode construction.

FIG. 10-70 Spark-plug gap gage.

turers to apply the correct pressure and voltage to the spark plug for this test. Spark plugs are tested under pressure to simulate to some extent the operating conditions of the spark plug. A spark plug that fires satisfactorily under normal atmospheric conditions may fail under pressure because of the increased resistance of the air gap under these conditions. The testing of a spark plug in a pressure tester is shown in Fig. 10-73. Instructions for operation are included with the test unit.

Spark plugs which fail to function properly during the pressure test should be baked in an oven for about 4 h at 225° F [107°C]. This will dry out any moisture within the plug. After it has been baked, the plug should be tested again, and if it fails under the second test, it should be rejected or returned to the manufacturer.

8. Gasket servicing. One of the most important and yet unrecognized essentials of spark-plug installation involves the condition of the solid copper gasket used with the spark plug. When spark plugs are installed, either a new gasket or a reconditioned gasket should be used.

Used spark-plug gaskets should be annealed by being heated to a cherry red and immediately quenched in light motor oil. After the quenching, the oil should be removed with a solvent and the gaskets immersed in a solution of 50 percent water and 50 percent nitric acid to remove oxides. After the acid bath, the gaskets should be carefully rinsed in running water and dried.

Even though good reconditioned spark-plug gaskets may be available, it is recommended that new gaskets be installed with new or reconditioned spark plugs. The additional cost which may be involved is so small that it cannot offset the advantage of new gasket reliability.

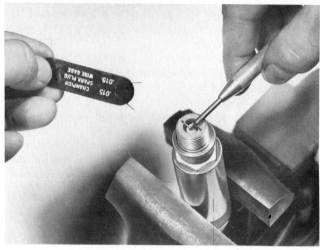

FIG. 10-72 Setting the gap on a fine-wire electrode spark plug.

9. Installation. Before installing spark plugs, the technician must make sure to have the proper type of plug for the engine. Two of the most critical factors are the *heat range* and the *reach*. The effect of improper heat range has been described previously. Briefly, if the plug is too hot, preignition and detonation may occur; if the plug is too cold, it will become fouled.

If a long-reach plug is installed in an engine which is designed for a short-reach plug, some of the threads of the plug will extend into the combustion chamber. In this position, the threads and the end of the plug may become overheated and thus cause preignition. This will result in loss of power, overheating, and possible detonation. The threads will be subject to high-temperature corrosion, and this will cause the plug to stick in the cylinder. The sticking tendency will be aggravated by the formation of carbon and lead residue at the end of the plug. When this condition exists, damage may be caused to the cylinder threads and to the spark plug at the time of removal.

Regardless of the care exerted by technicians in carrying out all the steps previously explained, their best efforts are in vain if they fail to install the spark plugs properly. The first step is to inspect the spark plug and the cylinder bushing. The threads of each should be

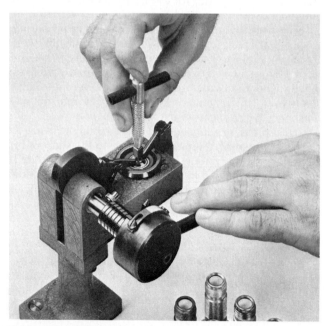

FIG. 10-71 Gap-setting tool.

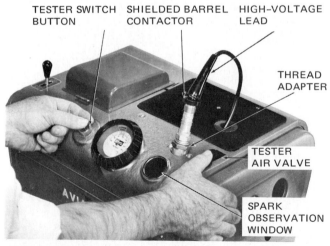

FIG. 10-73 Pressure-testing a spark plug. (*Champion Spark Plug Co.*)

clean and free of damage. A light coating of approved *antiseize compound* should be applied to the threads of the spark plug. It should not, however, be applied to the first two threads because the material may run down on to the electrodes when hot, thus shorting the electrodes of the spark plug.

The spark plug is then installed together with the gasket in the cylinder bushing. It should be possible to screw the spark plug into the cylinder bushing by hand. The spark plug is tightened by means of a deep-socket wrench with a torque handle. It is very important that the spark plug be tightened to the torque specified by the manufacturer. The usual torque for 18-mm spark plugs is 360 to 420 lb·in [40.68 to 47.46 N·m], and the torque for 14-mm plugs is 240 to 300 lb·in [27.12 to 33.9 N·m]. Overtightening a spark plug may damage the threads and make the spark plug difficult to remove or, in extreme cases, change the gap setting. Overtightening will also cause the spark plug to stick in the cylinder.

The ignition lead is connected to the spark plug by inserting the terminal sleeve in a straight line into the shielding barrel and tightening the terminal nut to the top of the shielding barrel. The terminal sleeve must be clean and dry and it must not be touched by the hands, as moisture or acid might eventually cause failure.

The terminal nut at the top of the spark plug should be tightened as far as possible by hand and then turned about one-quarter turn with the wrench. A good snug fit is all that is required. Overtightening may cause damage to the elbow and the threads.

After a complete set of spark plugs has been installed, it is wise to run the engine and perform a magneto check to determine the general condition of the ignition system. A drop in engine rpm beyond the specified limit will require a check of the magneto, ignition cables, or spark plugs.

● TURBINE ENGINE IGNITERS

Although a gas-turbine engine igniter serves the same purpose as a spark plug, the design and configuration of these units are considerably different from the spark plugs for reciprocating engines. Furthermore, the sizes and shapes of igniters have not been standardized appreciably; hence, there are no fixed rules for service and maintenance. In every case it is essential that the manufacturer's recommendations for cleaning and reconditioning be followed.

Fig. 10-74 illustrates different types of igniter plugs. Since the igniters are designed to operate at a lower surrounding pressure than the spark plug, the spark gaps are greater. The power source for the igniter supplies a very high level of energy; hence, the spark produced is of relatively high amperage and resembles a white-hot flame rather than a spark. This spark discharge causes a much more rapid erosion of the electrodes than the spark of a spark plug. However, the igniter operates each time just long enough to start the engine, and the total erosion over a long period of engine operation is not great. In some cases the igniter also operates during takeoff of the airplane in order to ensure a restart if the engine should flame out.

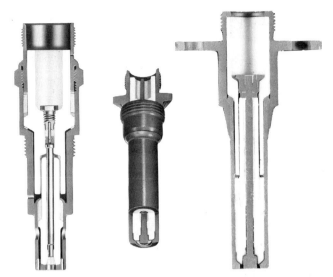

IGNITER AA 63 S
FOR P & W JT4A
ENGINE

IGNITER FHE 151
FOR P & W JT12
ENGINE

IGNITER FHE 19 – 6H
FOR ROLLS ROYCE
DART TURBOPROP

FIG. 10-74 Turbine igniter plugs.

Service and Inspection

The igniter is inspected visually for burning and erosion of the electrode or shell, cracking of the ceramic insulator, and damage to threads, hex, or flange. If damage is apparent, the igniter should be discarded.

The inside of the shell may be cleaned with a swab dampened with a standard petroleum cleaning solvent conforming to PS-661 specification. If the inside of the barrel insulator is stained, a nonconductive abrasive powder such as Bon Ami may be used with the solvent and swab. After cleaning, all traces of the abrasive should be removed with a clean swab dampened in solvent.

The connector seat at the top of the barrel should be cleaned to ensure a good contact between the harness connector and the barrel. If necessary, a finely grained garnet or sandpaper can be used. Care should be taken to see that abrasive particles do not enter the barrel.

The firing end of the igniter can be cleaned only if a cleaning procedure is approved by the manufacturer. Some igniters have semiconductor materials at the firing end, and these can be damaged if normal cleaning procedures are followed.

● SERVICE AND MAINTENANCE OF OF IGNITION SYSTEMS

The procedures and techniques given in this section provide a general guide to ignition system maintenance, but they do not supersede instructions provided in approved maintenance manuals. In all cases, the manuals provided by manufacturers should be consulted. Airline maintenance departments furnish approved manuals for maintenance operations to suit their particular needs. These manuals are developed in cooperation with manufacturers, maintenance experts, and airline engineers to provide the most effective methods and procedures for the maintenance of the equipment operated by the airline.

Magnetos

The inspection and maintenance of a magneto is not difficult, but it requires careful attention to detail. In any case, the procedures should follow instructions given in an approved maintenance manual. A complete inspection of a magneto generally includes the following:

1. Removal of the magneto in accordance with approved instructions. The technician must note that the magneto circuit is likely to be in an ON condition when the P lead is disconnected.

2. Removal of the distributor blocks, high-tension plate, terminal block, or other part which may hold the high-tension leads.

3. Removal of the cover over the breaker-point assembly.

4. Examination and service of the breaker-point assembly and capacitor. If the breaker points have a smooth and frosty appearance, they are in good condition. If they are badly pitted, the cause should be determined and the points replaced. Breaker points which have mild pitting or roughness may be reconditioned in some cases by careful stoning or by filing with a point file. For either step, the points should be removed and placed in a jig or other holder to ensure that the points are as nearly flat as possible when dressing is completed. The points should be sufficiently flat so that they make contact over at least one-third of the point face. *Breaker points should not be redressed unless the practice is approved by the manufacturer.*

The breaker points should be checked for adequate lubrication. A drop or two of engine oil on the *oiler felt* attached to the point assembly will usually restore the lubrication. Care must be taken to see that no oil is on the breaker points. Excess oil must be removed or it may cause the points to burn black.

Breaker-point spring tension can be checked by hooking a small spring scale to the movable point and applying sufficient force to open the points. The points should not be opened more than $\frac{1}{16}$ in [1.59 mm] because the spring may be weakened. A weak breaker-point spring will cause the points to "float" or fail to close in time to build up the primary field to full strength. This is most likely to occur at high speeds. The breaker-point area is checked for cleanliness. With the exception of the oiler felt, all parts should be dry and clean. An approved solvent may be used to clean the metal parts, but the solvent must not get on the oiler felt.

The primary capacitor (condenser) should be tested with a suitable condenser tester, such as that shown in Fig. 10-75. This instrument includes ranges for 0.1 to 0.4 μF, 0.4 to 1.6 μF, and 1.5 to 4.0 μF. The range is selected to accommodate the capacitor being tested by means of the SELECTOR and the MFD range switch. The MFD (microfarad) switch selects the correct capacitance range for the capacitor being tested. The unit of capacitance is the microfarad (μF).

The manufacturer's instructions should be followed; however, the steps are generally as follows:

(a) See that the selector switch is in the OFF position.

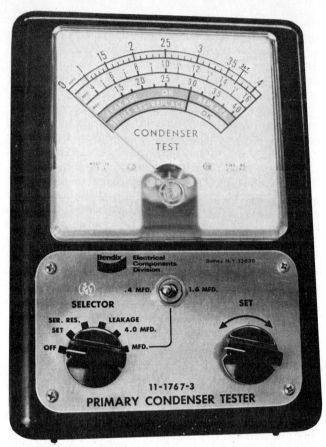

FIG. 10-75 A condenser (capacitor) tester.

(b) Plug the instrument power cord into a power receptacle.

(c) See that the test leads are not shorted or grounded.

(d) Move the selector switch to SET and adjust the instrument for proper setting with the SET knob.

(e) Turn the selector switch off and connect the test leads to the capacitor: one lead to the case and one lead to the insulated lead or terminal of the capacitor.

(f) Set the instrument range for the capacitor being tested.

(g) Check for capacitance, leakage, and series resistance by rotating the selector switch.

Do not handle the tester leads except when the selector switch is in the OFF position. The high voltage can cause a severe shock.

5. Check of the magneto shaft and gears for excessive play and backlash. If these are beyond specified limits, the magneto must be overhauled.

6. Examination of high-tension parts, such as the distributor rotor, distributor block, and terminal block. These parts may be made of Formica, Bakelite, or other heat-resistant insulating material. These parts are sometimes called dielectric parts because they must have high dielectric strength to withstand the ignition voltages. Defects to observe are cracks, dark lines (carbon tracks) indicating flashover or leakage, and burning, also caused by leakage of high-tension current. Over a period of time, a thin coating of dust may collect in the distributor area. This dust can absorb enough moisture from the air to make it conductive, with the result that the high-tension current will use the conductive dust as a bridge to ground. The current will often create a burned path referred to as a **carbon track.** This track becomes more conductive as the current flow continues and eventually acts as a short circuit. The hot spark burns the insulating material and dust and releases carbon, which is conductive. If carbon tracks are found, they should be removed if possible. If they cannot be removed, the part should be replaced. All the high-tension parts should be cleaned with recommended solvents and then dried and waxed with a high-temperature wax.

7. Check of internal timing. This is accomplished as explained in the discussion of magneto timing. By use of a timing light, it can be determined that the breaker points open at the E-gap position. If necessary, the points can be adjusted in position to establish the correct opening time. The timing of the distributor can usually be determined by noting the position of marks on the distributor gear with respect to matching marks on the distributor drive gear.

8. Installation of the magneto. When this is done, the procedure for timing the magneto to the engine must be followed.

Testing Magnetos

Upon completion of a service inspection or an overhaul, a magneto should be tested on a **magneto test stand.** This stand includes a variable drive to permit operation of the magneto at any desired speed, a tachometer, a spark rack, and suitable controls. A typical magneto test stand is shown in Fig. 10-76.

The test stand is equipped with provisions for both base mounting and flange mounting. These mountings are adjustable to permit accurate alignment of the test-stand drive to the magneto shaft. The test stand is reversible, so that the magneto can be rotated either to the right or left, depending on the requirement. The

FIG. 10-76 A magneto test stand.

operator must take particular care to ensure that the magneto is rotated in the correct direction.

After the magneto is mounted on the stand, the high-tension leads are connected from the distributor to the **spark rack.** The spark rack is adjusted for the correct gap specified for the particular magneto being tested. It will be noted that the gap used for the spark rack is much greater than the gap of a spark plug. This is because the current at a particular voltage will jump a much greater distance in unpressurized air than in the compressed air in a cylinder.

The **coming-in speed** of a magneto is a good indication of the magneto's performance. With the spark gap set at the specified distance, the magneto rpm is slowly increased. When the coming-in speed is attained, a steady discharge of sparks will occur at the spark gaps. The magneto speed is increased to maximum specified rpm to test high-speed performance. The tests are conducted for specified periods of time to ensure that performance is reliable under operating conditions.

If the coming-in speed of a magneto is too high, it is known that the magnet is weak, that the internal timing is not correct, that the capacitor is defective, or that there is some other defect in the magneto. The magnet can be tested with a magnetometer (Gauss meter) and, if weak, it can be recharged with a magnet charger of proper design. The capacitor can be tested with a capacitor tester as explained previously.

During the test of a magneto on a test stand, the magneto must not be operated without the high-tension leads connected to the spark rack or without some other means whereby the high-tension current can flow to ground. If the current cannot discharge through normal paths, the voltage will build up to a level which may break down the insulation in the magneto coil and ruin the coil. The gap of the spark rack must not be increased to a distance where the spark cannot jump, because the high-voltage current will seek another path to ground and this will result in damage to the coil or to the distributor.

The same damage may occur during operation of the engine in flight if a spark-plug lead should break or if high resistance should occur in an ignition lead for any other reason.

Spark Plugs

The reconditioning and service of spark plugs was discussed previously; however, certain points in the installation and removal of spark plugs should be reemphasized.

When removing the high-tension lead from a spark plug, care must be taken to avoid damaging the coupling elbow. Usually the elbow can be held firmly with a suitable clamp or wrench while the spark-plug coupling nut is loosened with another wrench. When the nut has been loosened and disengaged from the spark-plug, the terminal sleeve assembly ("cigarette") should be pulled straight out of the spark-plug sleeve to avoid damage. When connecting the coupling to the spark plug, the terminal sleeve should be in alignment with the spark-plug sleeve as it is inserted into the sleeve.

Both the terminal sleeve assembly and the ceramic sleeve in the spark plug must be clean and dry when assembled. The spark-plug sleeve can be cleaned with a swab dipped in alcohol or any other approved solvent, and the terminal sleeve can be cleaned with a cloth dampened with the solvent. After the parts are cleaned and dried, the technician should make sure that there is no lint left in or on the parts. Lint can absorb moisture and form a path for leakage of high-tension current.

To prevent the terminal sleeve from sticking inside the spark-plug sleeve, a high-temperature mold release such as MS-122 fluorocarbon spray can be applied to the terminal sleeve. This is particularly important for hot-running spark plugs.

When installing a spark plug in an engine and when connecting the spark-plug coupling nut, a torque wrench must be used to ensure accurate tightening of the plug and nut. Overtorquing will cause damage and possible seizure of the threads in both the spark plug and the coupling nut. Overtorquing can also cause the coupling elbow to be bent to the point where it will damage the insulation of the high-tension lead. Undertorquing can result in loosening of the parts and subsequent failure.

Harness Testing and Inspection

As previously mentioned, high-tension cable for aircraft ignition systems consists of a few strands of stainless steel wire covered with a thick layer of an insulating material such as silicone rubber. Over this is a layer of glass fiber reinforcement and over the reinforcement is another thick layer of insulating material.

The insulation of the ignition cable is designed to withstand very high voltage without breaking down. Over a period of time, however, leakage of ignition current will occur. Even a new cable will have a small amount of leakage, but this is not important until it increases to the extent that the spark at the spark electrodes is weakened or stopped.

To ensure that the dielectric strength of ignition cable insulation is adequate and that excessive leakage is not occurring, a harness tester called a **megohmmeter** or **megger** is used. Typical testers are the Bendix 11-8950 High-Tension Lead Tester and the Eastern Electronics Cable Tester, Model E5.

A harness tester is an electric unit designed to produce DC voltages up to 15,000 V which can be applied to individual leads in an ignition harness. A typical unit may include gages to measure the applied voltage and leakage current, a voltage control, input leads, output leads, and required control switches. The units include instructions for proper application.

To test ignition leads, all leads are disconnected from the spark plugs and all but the one being tested are grounded to the engine. With the leads being grounded, the tester will show leakage between leads as well as from leads to ground. The high-voltage lead from the tester is connected to the spark-plug terminal of the lead being tested. The ground lead is attached to the engine. The tester may also be grounded to earth through a water pipe or other adequate means.

Manufacturer's instructions are provided for all harness testers and should be followed. Since the units produce very high voltage, it is essential that the operator be most careful when the unit is turned on.

The voltage of the tester is adjusted to the level given in the instructions, which is usually 10,000 V. When the control switch is turned on, this voltage is applied to the lead being tested. Leakage will show on the microammeter and should not exceed 50 μA.

As testing of each lead continues, one or more leads will likely show high leakage because of the position of the distributor rotor in the magneto. If the rotor finger is aligned with the electrode for the lead being tested, the current will jump the gap to the rotor and flow to ground through the magneto coil. When this occurs, the engine crankshaft should be rotated to change the alignment of the distributor rotor so that the lead can be retested.

If the test shows excessive leakage in several cables, it is likely that the distributor block or terminal block is defective. The condition of the block should therefore be thoroughly examined. If the test indicates that the harness is faulty, all the cables should be replaced.

When a distributor block shows a modest amount of leakage, it can sometimes be restored to good condition by cleaning and waxing with an approved high-temperature wax. If leakage persists, the block should be replaced.

A comparatively simple cable tester is illustrated in Fig. 10-77. This unit operates from either 12 V or 24 V DC. The instrument is set for the correct voltage by means of the selector switch. When the tester is properly connected to ignition cables, the indicator light will reveal excessive leakage.

During the inspection of an ignition harness, it is important to note the routing of individual spark-plug cables with respect to engine parts, and particularly to the exhaust manifold. Cables should be routed and supported so that they cannot rub against engine parts or be located near hot parts which could burn the insulation. Sometimes it is necessary to adjust clamps and other supports to move the cable from a position where it can become damaged by abrasion or heat.

Sharp bends should be avoided in ignition leads. If a cable is bent sharply or twisted, the insulation is under stress and can develop weak points through which high-tension current can leak.

FIG. 10-77 An ignition cable (harness) tester.

OVERHAUL OF MAGNETOS

It is not the intent of this section to describe in detail the overhaul of any particular type of magneto, because such instructions are to be found in the appropriate manufacturer's manual. We shall, however, discuss the general requirements for overhaul of a typical magneto.

When a magneto is received for overhaul, all pertinent information such as make, type, and serial number should be recorded on the work order. In addition, the service record of the magneto should be noted, including the time of operation since new or since the last overhaul.

The magneto should be cleaned thoroughly and disassembled according to the instructions given in the appropriate overhaul manual. The magnet should be handled carefully and should have a soft iron **keeper** of the proper shape placed over the poles to prevent loss of magnetism. Care must be taken to ensure that the magnet is not dropped, jarred, or subjected to excessive heat, all of which can cause loss of magnetism.

It is good practice to place all parts of the magneto in a compartmented tray for protection and convenience of handling.

Inspection and Testing

The magnet and magnet shaft are inspected for physical damage and wear. The magnet should then be tested with a magnetometer (Gauss meter) to see that the magnetic strength is adequate for operation. Weak magnets can be returned to the manufacturer for remagnetization or can be remagnetized in the overhaul shop if proper equipment is available.

The **magnetometer** is a device incorporating soft iron shoes to fit the poles of the magnet. When the magnet is correctly positioned on the shoes, the indicator will show the level of magnetism.

The capacitor (condenser) for the primary circuit is often replaced at major overhaul to ensure maximum operational life. However, it is not usually necessary to replace mica capacitors if a capacitance test and leakage test reveal satisfactory condition.

The capacitance test is accomplished by a **capacity** (capacitance) **tester,** as explained previously. This device is used to apply a carefully regulated alternating current to the capacitor and the response of the capacitor is indicated in microfarads on an indicating dial. Care must be observed in using the capacity tester, because the voltage is often at a level which can be injurious or fatal.

Leakage, indicating failure of the dielectric, should be tested in accordance with the manufacturer's recommendations. Usually this involves the application of a direct current of specified voltage to the capacitor with a milliammeter in series. The amount of leakage is indicated by the milliammeter. Any appreciable current leakage is cause for rejection.

It is generally recommended that breaker-point assemblies be replaced at major overhaul. This will ensure best performance and maximum life. Worn points and worn cam followers, even though reconditioned, cannot provide the durability and performance of new assemblies. A cam follower worn beyond certain limits will make it impossible to adjust the breaker points for correct operation.

The breaker cam is inspected for wear and condition. If the wear is beyond specified limits, the cam must be replaced. The cam surface must be smooth and free of any pits, corrosion, or other surface defects.

The distributor rotor is cleaned in an approved solvent and examined for cracks, carbon tracks, or other indications of failure. The solvent used for cleaning must be of a type which will not damage the finish of the rotor. It is usually recommended that after inspection and any other processing specified by the manufacturer the rotor be coated with a high-temperature wax to prevent high-voltage leakage and absorption of moisture.

Shaft bearings and distributor bearings are inspected and serviced in a manner similar to bearings for other engine accessories. Since the bearings are usually of the sealed type, it is recommended that new bearings be installed at major overhaul. Overhaul manuals sometimes provide instructions for the reconditioning of sealed bearings.

The coil of a high-tension magneto includes both a primary and a secondary winding. Some coils include the primary capacitor in the coil. The coil should be tested for current leakage between the primary and secondary windings, for continuity of both windings, and for resistance of each winding. Resistance can be checked with an ohmmeter or multimeter—primary lead to ground for the primary and high-tension contact to ground for the secondary.

Assembly

After all parts are inspected, tested, and processed in accordance with instructions, the magneto is ready for assembly. The sequence of assembly procedures is determined by the make and type of magneto.

The principal factors in assembly are proper handling of parts to avoid damage; use of proper tools; correct torquing of screws, nuts and bolts; and following instructions relating to assembly and timing.

Testing

When the magneto has been completely assembled and timed according to instructions, it should be tested on a magneto test bench. This process has been explained previously in this chapter. It is important to note the specifications for testing is given in the overhaul manual. The direction of rotation, coming-in speed, and width of spark gap must be correct to accomplish the test. If the coming-in speed is at or below the requirement and the magneto continues to produce strong sparks for a specified period, the magneto can be considered satisfactory.

REVIEW QUESTIONS

1. At what point in the four-stroke cycle does the ignition event take place?
2. Name the essential components of an ignition system.
3. Explain the function of ignition shielding.
4. What are the advantages of magneto ignition?
5. Why is it necessary to employ a booster coil,

induction vibrator, or impulse coupling with magneto ignition?

6. Why is a dual magneto ignition system employed with aircraft engines?
7. What is the value of staggered ignition timing?
8. Explain how the electric current is produced in the primary coil of a magneto.
9. What determines the voltage produced by a primary coil?
10. Explain the function of the rotating magnet in a magneto.
11. What are the pole shoes?
12. Explain the static-flux curve for a magneto.
13. Describe the magneto coil assembly.
14. Describe a breaker assembly (contact breaker).
15. What is the advantage of a pivotless breaker assembly?
16. What is the cam follower?
17. Explain the function of the primary condenser (capacitor) in a magneto.
18. Explain the electrical events that occur in a magneto during operation.
19. Explain the E-gap angle in the operation of a magneto.
20. What happens with respect to the secondary voltage of an operating magneto when the secondary circuit is open?
21. What materials are used in the construction of a magneto? Why are such materials used?
22. How do you determine magneto rpm with respect to engine rpm?
23. Explain the polarity or direction of the sparks from a magneto.
24. What is the P lead in a magneto ignition circuit?
25. Describe the construction and function of a distributor.
26. What is the magneto sparking order?
27. Explain coming-in speed.
28. Describe a magneto harness assembly.
29. How would you replace a high-tension lead in a rigid harness manifold?
30. What is unique about the operation of an ignition switch in a magneto system?
31. Tell how you would test an ignition switch after installation to ensure correct operation.
32. If a magneto carries the model designation SF9LN-3, what can you tell about the magneto?
33. Describe three types of magneto ignition boosters.
34. Why is some type of ignition booster necessary?
35. Explain the purpose of the retard breaker on some Bendix magnetos for light-aircraft engines.
36. Explain the operation of a combination starter-ignition switch.
37. If the primary lead from the magneto to the ignition switch should become broken or disconnected, what effect will this have on the operation of the system?
38. What will happen if the ground lead from the ignition switch is broken or disconnected?
39. Describe the timing procedure for the Bendix S-200 magneto.
40. Tell how you would install and time a Bendix S-200 magneto.

41. What are the two key points in the installation and timing of any magneto?
42. How are high-tension leads anchored in the distributor grommet on a Bendix magneto such as the S-200?
43. What precaution must be observed when a magneto is installed on an engine and the switch lead (P lead) is disconnected?
44. Why should the primary lead be disconnected from a magneto when checking the opening time of the breaker points if a direct-current timing light is being used?
45. Give the order for connecting high-tension leads to spark plugs on a six-cylinder opposed engine. Include both top and bottom plugs.
46. Describe an inspection procedure for the S-200 magneto at 50-h intervals.
47. Describe the Bendix D-3000 dual magneto.
48. Describe the breaker-point arrangement for the Bendix D-2000 and D-3000 magnetos.
49. What is the location of the primary capacitors (condensers) in D-2000 and D-3000 magnetos?
50. Describe the Slick series 600 magneto.
51. Describe the timing procedure for the Slick series 600 magneto.
52. How is the breaker cam lubricated?
53. Explain the operation of the impulse coupling.
54. What would happen if the impulse coupling spring should break during operation?
55. What problems are reduced through the use of a low-tension ignition system?
56. Explain the operation of a low-tension ignition system.
57. Why is the primary winding of the transformer in a low-tension ignition system designed with greater resistance than the winding in the magneto coil?
58. Describe the construction of a typical transformer coil for a low-tension ignition system.
59. What would be the effect if the primary winding of a low-tension transformer coil were shorted out?
60. Compare the S-600 low-tension magneto with the S-200 high-tension magneto.
61. Describe the distributor system for the S-600 magneto.
62. Describe the harness assembly for the S-600 low-tension ignition system.
63. What are the resistances of the primary and secondary windings in the transformer coils for the S-600 system?
64. Describe the timing procedure for the S-600 magneto.
65. Give the procedure for timing the S-600 magneto to the engine.
66. Describe the distributor assembly for the Bendix low-tension ignition system used on the R-2800 engine.
67. Explain the construction of one of the detachable lead assemblies.
68. How are the compensated breaker cams mounted and at what speed do they rotate with respect to the crankshaft?
69. By what means is the distributor position established when it is mounted on the engine?

70. Explain the reason for the use of a compensated cam.
71. Describe three types of timing lights and explain the use of each.
72. Describe a gas-turbine ignition system.
73. Why does a gas-turbine ignition system operate only during starting of the engine?
74. Why is it necessary to design gas-turbine ignition systems with a very high energy output compared with an ignition system for a reciprocating engine?
75. What voltage and current frequency are required for the input of the JT9D ignition system?
76. Describe the construction of a shielded aircraft spark plug.
77. What is the purpose of the resistor in a spark plug?
78. What should be done with a spark plug which has been dropped?
79. Explain the difference between a hot and a cold spark plug.
80. What is the advantage of platinum electrodes?
81. Compare massive-electrode and fine-wire spark plugs.
82. Describe the complete procedure for removing, reconditioning, and installing spark plugs.
83. What precautions must be taken in regapping a spark plug?
84. What may result if a long-reach spark plug is installed in an engine designed for a short-reach spark plug?
85. Why are spark plugs tested under pressure?
86. What damage may be done by overtorquing a spark plug?
87. Compare a gas-turbine igniter plug with a spark plug.
88. What kind of a spark is produced by an igniter plug?
89. Where will the technician find specific information for the service and maintenance of an ignition system?
90. List the general items to be checked in the complete inspection of a magneto.
91. Under what condition may a technician redress breaker points?
92. What causes a carbon track?
93. What precaution should be taken in the lubrication of the breaker cam?
94. What precaution should be observed in opening pivotless breaker points?
95. What is the appearance of breaker points in good operating condition?
96. What is the minimum amount of breaker-point surface that should make contact?
97. What should be done if there is too much play in the magneto shaft or too much backlash in the gears?
98. How is the magneto test bench used? Give precautions.
99. Explain the significance of *coming-in* speed.
100. What will happen if the magneto is rotated at high speed without providing a path for high-tension current?
101. What precaution must be taken in removing a high-tension lead from a spark plug?
102. Explain the reason for great care in the insertion of the terminal sleeve assembly in a spark plug.
103. What is the importance of cleanliness and dryness in ignition systems?
104. How is a harness tester used in testing an ignition harness of high-tension lead?
105. What safety precaution must be observed by the operator of a harness tester?
106. What is the maximum leakage permitted in an ignition system being tested?
107. Discuss the importance of proper routing of ignition leads on an engine.
108. Why should sharp bends and twists be avoided in the routing of ignition leads?
109. When a magneto is disassembled for overhaul, what is the purpose of placing a keeper across the poles of the magnet?
110. How is the magnetic strength of the magnet checked?
111. How is a capacitor tested for capacitance?
112. How can you check a magneto coil for continuity?
113. What is usually done with respect to breaker points at magneto overhaul?
114. After an overhauled magneto is assembled, how do you test it for satisfactory operation?

11 ENGINE STARTING SYSTEMS

● TYPES OF STARTING SYSTEMS

Starters for Reciprocating Engines

The types of starting systems which have been installed in aircraft for starting reciprocating engines may be classified as follows: (1) direct hand cranking; (2) direct cranking, either hand or electric; (3) hand inertia; and (4) combination hand and electric inertia.

The first two may be described as direct-cranking systems, and the second two may be referred to as inertia-type cranking systems.

Starters for Gas-Turbine Engines

Starters for gas-turbine engines may be classified as air-turbine (pneumatic) starters, electric starters, fuel-air combustion starters, and cartridge-type starters. The most commonly used starter is the air-turbine type. This type of starter requires a high-volume air supply which may be provided by a ground starter unit, a compressed-air bottle on the airplane, an auxiliary power unit on the aircraft, or compressor bleed from other engines on the aircraft. This starter will be described later in this chapter.

● DIRECT-CRANKING SYSTEMS

Direct Hand-Cranking Starter

The **direct hand-cranking starter** is sometimes described as a **hand-turning gear-type starter.** It consists of a worm-gear assembly that operates an automatic engaging and disengaging mechanism through an adjustable torque overload-release clutch. It has an extension shaft that may be either flexible or rigid, depending on the design. To prevent the transmission of any reverse motion to the crank handle in case the engine "kicks" backward while it is being cranked, a ratchet device is fitted on the hand crankshaft.

This type of starter can be used with a gear ratio of 6:1 for any engine rated at 250 hp [186.43kW] or less. It has a comparatively low weight, and it is simple to operate and maintain. It was extensively used on early airplanes which had low-horsepower engines and no source of electrical power for starting. On seaplanes, where it was very difficult to start the engine by swinging the propeller, it was especially popular. However, it has been entirely supplanted by more efficient designs.

Direct-Cranking Electric Starter

When the direct-cranking method is used, there is no preliminary storing of energy in the flywheel as there is in the case of the inertia-type starters. The starter of the direct type, when electrically energized, provides instant and continuous cranking. The starter fundamentally consists of an electric motor, reduction gears, and an automatic engaging and disengaging mechanism, which is operated through an adjustable torque overload-release clutch. The engine is therefore cranked directly by the starter.

The motor torque is transmitted through the reduction gears to the adjustable torque overload-release clutch, which actuates a helically-splined shaft. This, in turn, moves the starter jaw outward, along its axis, and engages the engine-cranking jaw. Then, when the starter jaw is engaged, cranking starts.

When the engine starts to fire, the starter automatically disengages. If the engine stops, the starter automatically engages again if the current continues to energize the motor.

The automatic engaging and disengaging mechanism operates through the adjustable torque overload-release clutch, which is a multiple-disk clutch under adjustable spring pressure. When the unit is assembled, the clutch is set for a predetermined torque value. The disks in the clutch slip and absorb the shock caused by the engagement of the starter dogs. They also slip if the engine kicks backward. Since the engagement of the starter dog is automatic, the starter disengages when the engine speed exceeds the starter speed.

The most prevalent type of starter used for light and medium engines is a series electric motor with an engaging mechanism, such as a Bendix drive or some other means of engaging and disengaging the starter gear from the engine gear. In all cases, the gear arrangement is such that there is a high gear ratio between the starter motor and the engine. That is, the starter motor turns many times the rpm of the engine. Delco-Remy and Prestolite starters are quite popular for light aircraft, although the Delco-Remy starter is no longer in production.

● INERTIA STARTERS

Although inertia-type starters are not in common use today, the technician may encounter them from time to time on older aircraft. For this reason it is deemed proper to retain a brief description of inertia starters in this text.

Types of Inertia Starters

There are two types of inertia starters: (1) the hand-cranking type, commonly called the **hand inertia starter,** in which the flywheel is accelerated by hand only; and

(2) the electric type, commonly called the **combination inertia starter** and sometimes referred to as the **combination hand and electric inertia starter,** in which the flywheel is accelerated by either a hand crank or an electric motor. All movable parts, including the flywheel, are set in motion during the energizing of an inertia starter of either type.

Principles Governing the Operation of the Inertia Starter

Newton's first law states: **Every body continues in its state of rest or uniform motion in a straight line unless it is compelled to change that state by some external force.** This is also known as a statement of the property of **inertia.**

The cranking ability of an **inertia starter** for an airplane depends on the amount of energy stored in a rapidly rotating flywheel. The energy is stored in the flywheel slowly during the energizing process, and then it is used very quickly to crank the engine rapidly, thus obtaining from the rotating flywheel a large amount of power in a very short time.

Under ordinary conditions, the energy obtained from the flywheel is great enough to rotate the engine crankshaft three or four times at a speed of 80 to 100 rpm. In this manner, the inertia starter is used to obtain the starting torque needed to overcome the resistance imposed upon the cranking mechanism of the engine by reason of its heavy and complicated construction.

The speed at which the engine crankshaft is rotated may be less than the **coming-in speed for the magneto,** which is the minimum crankshaft speed at which the magneto will function satisfactorily. Therefore, if the engine uses magnetos for ignition, as practically all modern engines do, an ignition booster of some type must be provided. It is usually installed on or near the engine and operated while the inertia starter is cranking the engine.

Hand Inertia Starter

In using the hand inertia starter, when a hand crank is placed in the crank socket and the crank rotated, a gear relationship between the crank and the flywheel makes it possible for a single turn of the hand crank to cause the flywheel to turn many times. For example, one revolution of the hand crank may cause the flywheel to revolve 100 or more times, depending on the make, model, etc., of the starter. The speed of all movable parts is gradually increased with each revolution of the hand crank, and most of the energy imparted to the crank is stored in the rapidly rotating wheel in the form of kinetic energy.

Figure 11-1 is a sectional diagram of a hand inertia starter. The crank socket, flywheel, engaging lever, mounting flange, starter driving jaw, torque overload-release clutch, springs, disks, and barrel are labeled on the drawing.

One type of clutch consists of one set of disks fastened to the shaft and another set of disks, made of a different kind of metal, fastened to the barrel. The disks are pressed together by springs. The retaining ring compresses the springs and can be adjusted to set the value of the slipping torque. This feature is important because the normal operation of the clutch is to slip

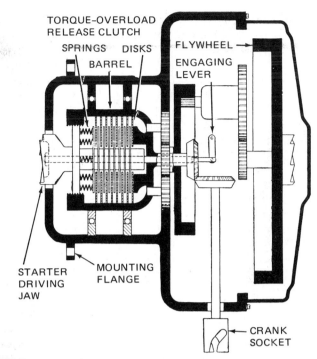

FIG. 11-1 Sectional drawing of a hand inertia starter.

momentarily after the starter and engine jaws are meshed. During the process of slipping, a torque is exerted on the crankshaft until the initial resistance of the engine is overcome and the clutch is again able to hold. The maximum holding torque is called the **breakaway.** This breakaway and the slipping torque depend on the size of the engine being cranked.

Figure 11-2 illustrates an early type of inertia starter with cranking and engaging controls. In this illustration the following parts are labeled: (1) starter crank, (2) extension crank, (3) adapter and universal assembly, (4) lever, (5) starter pull rod, (6) spring link, (7) spring, (8) starter, (9) bolt, (10) nut, (11) bolt, and (12) nut. When the starter crank is inserted through the left side of the airplane, it can be turned to crank the mechanism and start the engine.

The clutch prevents any injury to the starter caused by an engine kickback or a sudden overload. If the engine crankshaft does not rotate for any reason, the clutch slips until the flywheel ceases to turn.

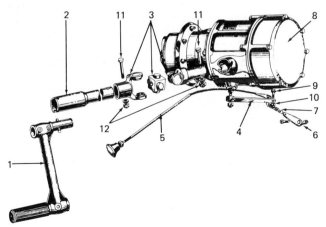

FIG. 11-2 Early type of inertia starter with controls.

Combination Hand and Electric Inertia Starter

A combination hand and electric inertia starter may consist of a hand inertia starter with an electric motor attached, and the gear and clutch arrangement may be like that of the hand inertia starter. The flywheel may be accelerated by either a hand crank or the electric motor. When the starter is energized by hand cranking, the motor is mechanically disconnected and no longer operates.

When the motor is operated, a movable jaw on a helically-splined shaft engages the motor directly to the inertia-starter flywheel in one type of inertia starter called a **jaw-type starter-motor engaging mechanism**, illustrated in Fig. 11-3.

On some starters of this general type, the starter jaw tends to remain at rest when the motor armature starts to rotate, but as the shaft turns, the jaw moves forward along the splined shaft until it engages the flywheel jaw. Ordinarily, there is no trouble with this type of mechanism, but if the jaw binds on the shaft, it will not engage the flywheel, and thus the motor races. When the engine fails to start, the operator must not continue to attempt cranking; otherwise, the teeth may be stripped on either the flywheel or the motor jaw or both. The correct procedure is to wait until the starter flywheel comes to rest before energizing the motor in a second attempt to crank. This avoids both the racing of the motor and the stripping of the teeth.

Electric Motors for Inertia Starters

The typical electric motor for an inertia starter is either a 12- or a 24-V series-wound DC motor with windings of low resistance. The reason for the low resistance of the windings is that when the starter switch is closed, there is a great amount of current drawn in order to deliver a powerful starting torque. When the motor gains speed, the induced electromotive force (emf) in a reverse direction causes a smaller amount of current to flow. For example, an inertia-starter motor, which draws about 350 A at starting, may draw only about 75 A at high speed.

This can be explained in another manner by saying that a counter emf is established when the motor gains speed.

An inertia-starter motor is never operated at full voltage unless there is a load imposed upon it. If there is no load, and if the motor has a small amount of internal friction, it will race and the armature may fly apart ("burst") because of the centrifugal stresses.

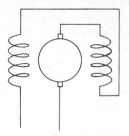

FIG. 11-4 Schematic diagram of a series motor circuit.

Figure 11-4 is a simple schematic diagram of a series motor. The field coils are connected in series with the armature. Since all the current used by the motor must flow through both the field and the armature, it is apparent that the flux of both the armature and the field will be strong. The greatest flow of current through the motor will take place when the motor is being started; hence, the starting torque is high. Series-wound motors are used wherever the load is continually applied to the motor and is heavy when the motor first starts. In addition to starting motors, motors used to operate landing gear, cowl flaps, and similar equipment are of this type.

● DIRECT-CRANKING STARTERS FOR LIGHT-AIRCRAFT ENGINES

Starter Motor

A typical direct-cranking motor is illustrated in Fig. 11-5. The armature winding is of heavy copper wire capable of withstanding very high amperage. The windings are insulated with special heat-resistant enamel, and after being placed in the armature, the entire assembly is doubly impregnated with special insulated varnish. The leads from the armature coils are staked in place in the commutator bar and then soldered with high-melting-point solder. An armature constructed in this manner will withstand severe loads imposed for brief intervals during engine starting.

The field frame assembly is of cast-steel construction with the four field poles held in place by countersunk screws, which are threaded into the pole pieces. Since

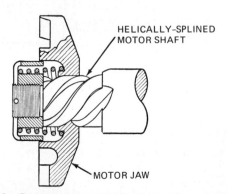

FIG. 11-3 Starter-motor engaging jaw and mechanism.

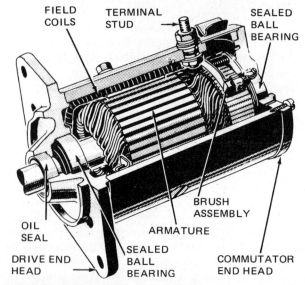

FIG. 11-5 Direct-cranking starter motor.

a motor of this type is series-wound, the field coils must be wound with heavy copper wire to carry the high starting current.

The end frames are of cast aluminum and are attached to the field frames by means of screws threaded into one end. The ball bearings are of the sealed type and are pressed into recesses in the end frames. An oil seal is placed in the drive end frame to prevent engine oil from entering the motor.

The brush assemblies are attached to the commutator end frame. The brushes are not held in brush holders in the particular motor illustrated but are attached with screws to a pivoted brush arm. This arm is provided with a coil spring which holds the brush firmly against the commutator.

The field frame is slotted at the commutator end to provide access to the brushes for service and replacement. A cover band closes the slots to protect the motor from dirt and moisture.

The positive terminal extends through the field frame. It is insulated from the frame by means of a composition sleeve and washers of similar material. The negative side of the power supply comes through the field frame.

Overrunning Clutch

In order to engage and disengage a starter from an engine it is necessary to employ some type of an engaging mechanism. In the discussion of inertia starters, the starter jaw mounted on a helically splined shaft was mentioned as the means for engaging the starter to the engine. Various types of mechanisms have been designed for light-engine starters, one of which is called the **overrunning clutch.**

A typical overrunning clutch arrangement is shown in Fig. 11-6. It will be observed that the clutch consists of an inner collar; a series of rollers, plungers, and springs enclosed in a shell assembly; and the shaft upon which the assembly is mounted. The hardened steel rollers are assembled into notches in the shell. The notches taper inward with the result that the rollers seize the collar when the shell is turned in such a direction that the rollers are moved toward the small end of the notches. Thus, when the shell is turned in one direction, the collar must turn with it; however, when it is turned in the opposite direction, the collar can remain stationary.

Complete Starter Assembly

A starter assembly employing a manually operated switch and a shift lever is shown in Fig. 11-7. This

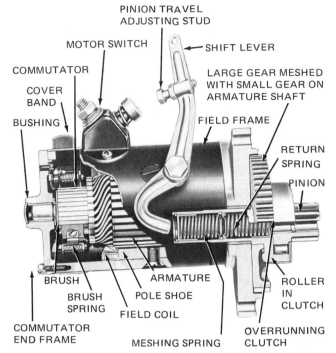

FIG. 11-7 Starter with overrunning clutch, manually operated switch, and clutch lever.

starter assembly makes use of an overrunning clutch similar to that described above, and it also incorporates a pair of gears so that there is a gear reduction between the motor armature and drive pinion. This gear reduction provides an increase in cranking torque at the drive pinion.

The shift lever is operated by means of a cable or wire control from the airplane cockpit. The control has a return spring with sufficient tension to bring the lever to the fully released position when the control is released.

When the control is operated to start the engine, the lower end of the shift lever thrusts against the clutch assembly and causes the overrunning clutch drive pinion to move into mesh with the engine starter gear. If the drive pinion and starter gear teeth butt against each other instead of meshing, the **meshing spring** inside the clutch sleeve compresses to spring-load the drive pinion against the starter gear. Then, as soon as the armature begins to turn, engagement and cranking take place. In the unit illustrated, the overrunning clutch drive pinion is supported on a stub shaft, which is part of the engine. As the drive pinion is moved into mesh, the stub shaft causes the **demeshing spring** inside the sleeve to be compressed. The demeshing spring produces demeshing whenever the shift lever is released. After the engine starts, overrunning clutch action permits the pinion to overrun the clutch and gear during the brief period that the pinion remains in mesh. Thus, high speed is not transmitted back to the cranking motor armature.

Starter Assembly with 90° Adapter Drive

The complete starter assembly, adapter, and clutch assembly for a six-cylinder opposed engine is shown in the exploded view of Fig. 11-8. The starter is mounted on a right-angle drive adapter which is attached to the

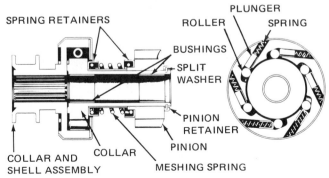

FIG. 11-6 Overrunning clutch.

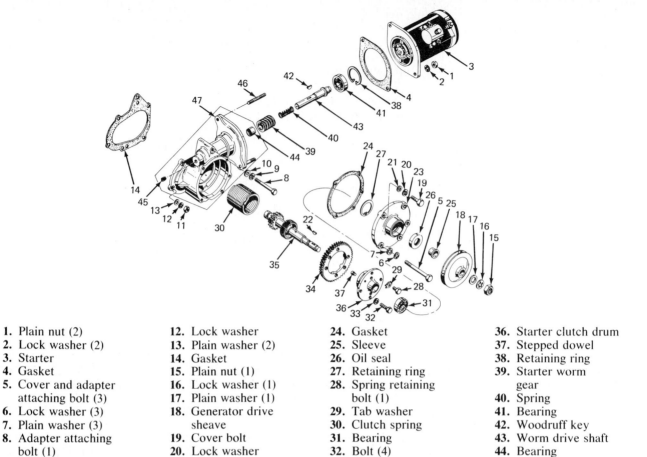

1. Plain nut (2)	12. Lock washer	24. Gasket	36. Starter clutch drum
2. Lock washer (2)	13. Plain washer (2)	25. Sleeve	37. Stepped dowel
3. Starter	14. Gasket	26. Oil seal	38. Retaining ring
4. Gasket	15. Plain nut (1)	27. Retaining ring	39. Starter worm
5. Cover and adapter	16. Lock washer (1)	28. Spring retaining	gear
attaching bolt (3)	17. Plain washer (1)	bolt (1)	40. Spring
6. Lock washer (3)	18. Generator drive	29. Tab washer	41. Bearing
7. Plain washer (3)	sheave	30. Clutch spring	42. Woodruff key
8. Adapter attaching	19. Cover bolt	31. Bearing	43. Worm drive shaft
bolt (1)	20. Lock washer	32. Bolt (4)	44. Bearing
9. Lock washer (1)	21. Plain washer	33. Lock washer	45. Plug (1)
10. Plain washer	22. Woodruff key	34. Starter worm wheel	46. Stud (2)
11. Plain nut (2)	23. Cover	35. Starter shaftgear	47. Adapter

FIG. 11-8   Direct-cranking starter with spring-type clutch and 90° drive.   (Teledyne Continental)

rear end of the crankcase. The tongue end of the starter shaft mates directly with the grooved end of the worm shaft. The worm shaft (43) is supported between a needle bearing (44) at its left end and a ball bearing (41) which is retained in the adapter by a Truarc snap ring. The worm (39) is driven by the shaft through a Woodruff key. The worm wheel (34) is attached by four bolts to a flange on the clutch drum (36) which bears on the shaftgear (35). Two dowels center the wheel on the drum and transmit the driving torque. A heavy helical spring (30) covers both the externally grooved drum and a similarly grooved drum machined on the shaftgear just ahead of the clutch drum. The spring is retained on the clutch drum by an in-turned offset at its rear end which rides in a groove around the drum, just ahead of the flange. The in-turned offset of the clutch spring is notched, and the clutch drum is drilled and tapped for a spring retaining screw. The front end of the spring fits closely in a steel sleeve, pressed into the starter adapter. When the starter is energized, friction between the clutch spring and the adapter sleeve and between the spring and the clutch drum, which is turned by the worm wheel, tends to wind up the spring on the clutch and shaftgear drums, locking them together so that the shaftgear rotates and turns the crankshaft. As soon as the engine starts, the shaftgear is driven faster than the clutch spring and tends to unwind it, thus increasing its inside diameter

so that the shaftgear spins free of the starter drive. The generator drive pulley is mounted on the rear end of the shaftgear and driven through a Woodruff key, so that it always turns at shaftgear speed.

Starter with Bendix Drive

A typical Prestolite starter motor with a Bendix drive is shown in Fig. 11-9. These starters are used on many of the light-aircraft engines now in operation.

The Prestolite starter motor is quite similar to those described previously. An examination of Fig. 11-10 will show that the motor includes the conventional parts for a series motor. These are the end-frames (end heads), field frame, field coils, brushes and brush-plate assembly, armature, bearings, and assembly parts. This particular starter has the electric starter switch (start relay) mounted on the field frame.

FIG. 11-9   Photograph of a typical Prestolite starter.

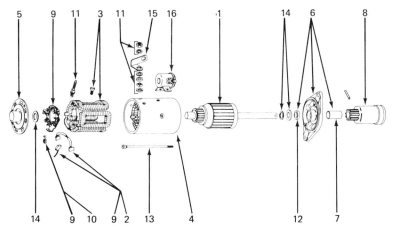

1. Armature
2. Brush set
3. Field coils
4. Field frame
5. Commutator end head
6. Drive end head
7. Drive end bearing
8. Bendix drive
9. Brush plate assembly
10. Brush spring
11. Terminal stud
12. Oil seal
13. Through bolt
14. Thrust washers
15. Connector
16. Starter relay

FIG. 11-10   Exploded view of a Prestolite starter.

The maintenance of the starter motor illustrated is the same as that for any other motor of this type. Inspections include an examination of the brushes and commutator, a check for the presence of oil or grease inside the motor and the presence of lead particles in and near the brush-plate assembly, and a test for the security of all electrical connections. The main terminal studs are examined for tightness in their mountings. Loose terminal studs can result in arcing and ultimate failure of the electrical connection.

During the replacement of any starter, the technician must determine the correct part number for the replacement part. Parts numbers are specified in the manufacturer's overhaul manual and the parts list for the aircraft involved.

STARTERS FOR MEDIUM AND LARGE ENGINES

Early-type radial engines were often equipped with hand inertia starters or electric hand inertia starters. These starters were effective. However, there was some inconvenience associated with their use because of the necessity of accelerating the starter and then engaging it to the engine for the start. If the energy stored in the flywheel was dissipated before the engine started, then it was necessary to accelerate again. The inertia starter was therefore more complex in system, construction, and operation than a direct-cranking electric starter; hence, designers and manufacturers eventually developed direct-cranking starters which were convenient and effective in operation.

The Bendix Type 756 Starter

A typical direct-cranking starter for medium-sized engines with a displacement not to exceed 985 in³ [16.15 L] is shown in Fig. 11-11. This is the Bendix type 756 starter, which is used for starting engines as large as the Pratt & Whitney R-985 nine-cylinder radial engine, as well as a number of the larger opposed engines.

The starter consists basically of a heavy-duty series-wound electric motor, reduction gearing, multiple-disk clutch, automatic engaging and disengaging device, and a driving jaw. Since the motor is series-wound and designed for intermittent duty, it develops a very high starting torque.

The reduction gearing for this starter has a ratio of 69:1; hence, the motor can turn at more than 8000 rpm while the starter is cranking the engine at a little more than 100 rpm. Torque overload protection is provided by means of a multiple-disk, lubricated clutch.

The starter jaws may be designed with either 3 or 12 teeth, depending on the requirements of the engine. The jaw is automatically engaged with the engine jaw when the starter is energized and disengaged when the engine starts. Jaw travel from the retracted position to the extended position is about $\frac{11}{32}$ in [8.73 mm].

The average performance of the type 756 starter is shown in the chart of Fig. 11-12. This chart shows efficiency, voltage, current, jaw rpm, and brake horsepower for various values of torque.

The type 756 starter is designed for use with a standard 24-V DC power supply. Batteries employed with

FIG. 11-11   Bendix type 756 direct-cranking starter.   *(Bendix)*

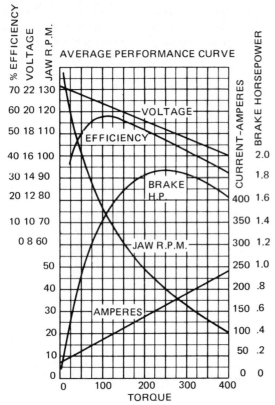

FIG. 11-12 Performance curves for the type 756 starter.

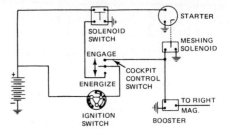

FIG. 11-13 Electric circuit for inertia-type starter.

and, at the same time, energize a booster coil that delivers high voltage to the trailing distributor finger on the right-hand magneto until the pilot releases the switch.

Starter Circuit for a Light Twin Airplane

The battery-control and starter circuit for the Cessna-Model 310 airplane is shown in Fig. 11-14. This is an actual circuit diagram for certain models of the airplane, and it includes the wire identification numbers. It will be observed that the battery solenoid must be closed before power can reach the bus which supplies the starter solenoids. The battery-control switch is in the ground circuit of the battery solenoid.

The starter solenoids are controlled by means of button-type starter switches through which power is supplied to the solenoid windings. Electrical power for the starter switches is taken from the right fuel boost switch or the right auxiliary pump switch.

● TYPICAL STARTING PROCEDURE FOR LARGE ENGINES

For the starting of any particular aircraft engine, it is essential that the technician or pilot carefully follow

the system must have sufficient capacity to supply the starter motor for several starting operations. Since the starter will draw a very high amperage when it is first energized, the power is controlled through a solenoid switch. The solenoid is energized when the starter is closed, thus causing the heavy current-carrying contacts to be closed. The winding of the solenoid is usually of the intermittent type and should not remain energized for more than 1 or 2 min at any one time. The windings heat rapidly and will burn out if the control switch remains on for more than a few minutes.

The clutch adjustment for the starter is of particular importance. If the torque setting is too low, the clutch will slip and will not provide satisfactory rotation of the engine. If the setting is too high, the shock upon engagement will be too great and the engaging jaws may be damaged. The clutch may be tested by means of a prony brake or other torque testing device. It is adjusted by increasing or decreasing the clutch-spring pressure.

● ELECTRIC CIRCUITS FOR STARTING SYSTEMS

System for Electric Inertia Starter

Figure 11-13 is a diagram of a typical inertia-starter electric system. If the ignition switch is placed in the BOTH position, the battery current can flow to the cockpit control switch. If that switch is placed in the ENERGIZE position, the current closes the starter relay and the inertia starter begins to pick up speed. When the inertia starter has attained the desired speed of rotation, the control switch is turned to ENGAGE, thus enabling the current to actuate the meshing solenoid

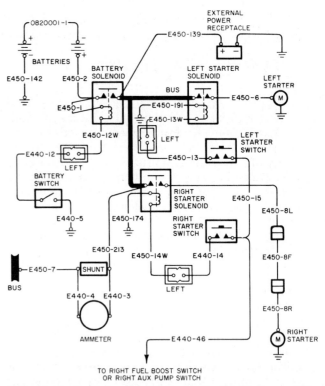

FIG. 11-14 Battery and starter circuits for the Cessna 310 airplane.

the procedures established by the aircraft manufacturer or operator. The procedures described in the following paragraphs are typical of the methods employed by an airline in starting the engines of a Convair Model 340.

Procedure

1. Always obtain a clearance signal from a member of the ground crew before rotating the engine with the starter. Make sure that the starting selector switch is set for the engine being started. Return the switch to OFF after the engines are started.

2. Whenever the left engine is being started or run, the passenger door must be closed and locked. The right engine may be started and run with the passenger steps extended and the nose wheel blocked.

3. The engines must be rotated with the starter while the magneto switch is off until 12 propeller blades have passed a fixed point. If the propeller is seen to stop suddenly during the starting operation, the starter switch must be released immediately. Sudden stoppage indicates a hydraulic lock in the lower cylinders, which must be cleared.

4. During the starting procedure, if the engine has backfired and caused an induction fire, the starting switches must be released immediately, the throttle closed, the mixture returned to IDLE CUTOFF, the fuel boost pump turned off, and the ignition switch turned off. An induction fire is indicated by a rapid rise in carburetor air temperature. This temperature should be watched closely.

Thirty seconds or more after the procedure above has been followed, the starting operation may be repeated.

5. If an engine does not start after a 30- to 45-s rotation with the starter, wait at least 3 min before attempting again to start the engine. Turn the fuel boost pump and ignition switch off during the waiting period. It is necessary to allow the starter to cool because of the rapid heating that takes place while the starter is rotating the engine.

6. If engine oil pressure and primary-compressor oil pressure are not indicated within 30 s after the engine starts, the engine must be stopped immediately. The oil pressure indicators should be closely watched during starting.

7. Steps in starting:
(a) Obtain an "all clear" signal from the ground crew. Start the no. 2 engine first. For pressure carburetors, see that the mixture control is at IDLE CUTOFF.
(b) Turn the fuel boost pump on. Wait 10 to 15 s before proceeding.
(c) See that the ignition switch is off. Set the starting selector switch to the engine being started. Set the throttle to 1½ in [3.81 cm] open.
(d) Press the START and START SAFE switches to ON. Watch the propeller and rotate the engine until 12 blades have passed a fixed point.
(e) Continue holding the switches. Turn the ignition switch to BOTH, and press the BOOST switch to ON. Press the PRIME switch as required. If the engine is hot, allow it to turn one or two revolutions; then apply prime lightly.

(f) As soon as the engine fires, continue to hold the PRIME switch on and adjust the throttle until the engine is running at 500 to 800 rpm on prime alone. Slowly advance the mixture control to AUTO LEAN. If the engine starts to die, return the mixture control to IDLE CUTOFF at once. Do not move the throttle rapidly, since this may cause damage to the carburetor balance diaphragm.
(g) Set the throttle to hold the rpm at 900 to 1000.
(h) Turn the fuel boost pump off.
(i) Immediately check and call aloud the oil pressure for the engine being started.

Any person who is starting an engine of a large airplane must exercise extreme care and attention to avoid damaging the engine. Overheating of the starter, backfiring and induction fires, and lack of oil pressure may all cause damage which will result in extensive delays and costly repairs.

● TROUBLESHOOTING AND MAINTENANCE

In this section we shall discuss the usual troubles found in the operation of typical, conventional starters installed on airplanes with reciprocating engines. Although the instructions are broad in their application, they cover the situations normally encountered. The study of manuals and handbooks issued by manufacturers and those responsible for inspection and maintenance is always recommended in order that the equipment may be serviced in accordance with its own particular design characteristics.

Failure of Starter Motor to Operate

When the starter switch on an airplane is placed in the START position and the starter fails to operate, the trouble can usually be traced to one of the following: (1) electrical power source, (2) starter-control switch, (3) starter solenoid, (4) electric wiring, or (5) the starter motor itself. A check of each of these items will usually reveal the trouble.

The electric power source for a light airplane is the battery. If the starter fails to operate or if a "click" is heard, the battery charge may be low. This can be quickly checked by means of a hydrometer. A fully charged battery will give a reading of 1.275 to 1.300 on the hydrometer scale. When the battery is low, the starter solenoid may click or chatter and the starter motor will fail to turn or it may turn very slowly. The solution to this problem is, of course, to provide a fully charged battery. In many cases the technician will merely connect a fully charged battery externally in parallel with the battery in the airplane. When this is done, it is important to see that the external battery is of the same voltage as the battery of the airplane and that the terminals are connected positive to positive and negative to negative.

If the battery of the airplane is found to be fully charged, the fault may be in the control switch. This can be checked by connecting a jumper across the switch terminals. If the starter operates when this is done, the switch is defective and must be replaced.

If the starter-control switch in the cockpit appears to be functional, the trouble may be in the starter solenoid. The solenoid can be checked in the same manner described for the starter-control switch. In this case, a heavy jumper must be used because the current flow across the main power circuit of the solenoid is much greater than that in the control circuit. If the starter operates when the jumper is connected across the main solenoid terminals, it is an indication that the solenoid is defective. Before the solenoid is replaced, however, the solenoid-control circuit should be thoroughly checked out. The circuit breaker or fuse in the control circuit may be failing, or the wiring may be defective.

If the power source, wiring, control switch, and starter solenoid are all functioning properly, it is apparent that the trouble lies in the starter motor. In this case, the band covering the brushes of the motor should be removed and the condition of the brushes and commutator checked. If the brushes are badly worn or the spring tension is too weak, the brushes will not make satisfactory contact with the commutator. This trouble can be corrected by replacing the brushes and/or the brush springs as required. If the commutator is black, dirty, or badly worn, it should be cleaned with No. 000 sandpaper or the starter should be removed for overhaul. Usually it is not necessary to overhaul a starter between engine overhauls.

Failure of Starter to Engage

If the starter motor turns but does not turn the engine when one attempts to start an engine, it is apparent that the overrunning clutch, the disk clutch, or the engaging mechanism has failed. In this case it is necessary to remove the starter and correct the trouble. If the starter is engaged by means of the Bendix-type mechanism in which the engaging gear is moved into mesh by means of a heavy spiral thread on the screw shaft, cold oil or grease on the thread will cause the gear to stick, thus preventing the gear from meshing. When this occurs, the problem can be solved by cleaning the spiral shaft and applying light oil for lubrication.

● STARTERS FOR GAS TURBINES

The comparatively small gas-turbine engines (under 6000 lb [26 690 N] of thrust) are often equipped with heavy-duty electric starters, or **starter generators.** These are simply electric motors or motor-generator units which produce a very strong starting torque because of the large amount of electrical power which they consume. Even though these starters require expensive auxiliary power units to supply the low-voltage high-amperage direct current necessary for operation, they are satisfactory for the size of the engine.

Because of the high power requirements for starting large gas-turbine engines, gas-turbine starters were developed. Among the turbine starters used for modern gas-turbine engines are low-pressure air turbines, high-pressure air turbines, fuel-air combustion starters, and solid-fuel combustion starters.

Low-Pressure Air-Turbine Starter

The low-pressure air-turbine starter is designed to operate with a high-volume low-pressure air supply, usually obtained from an external turbocompressor unit mounted on a ground service cart or from the airplane's low-pressure air supply. The air supply must produce a pressure of about 35 psig [241.33 kPa] and a flow of more than 100 lb/min [45.36 kg/min].

A drawing of the AiResearch 383042-1, ATS 100, air-turbine starter designed for use on the Boeing 707 airliner is shown in Fig. 11-15. This starter is a lightweight turbine air motor equipped with a rotating assembly, reduction-gear system, splined output shaft, cutout-switch mechanism, and overspeed-switch mechanism. The complete unit is mounted within a scroll assembly and gear housing. The unit includes a mounting flange designed to mate with a standard AND-20002, type XII-S engine drive pad.

A cross-sectional drawing of the starter is shown in Fig. 11-16. A careful study of this drawing will enable the student to understand the construction and operation of the unit. The low-pressure air is introduced into the scroll (5) through a 3-in [7.62-cm] duct which is not shown in the illustration. From the scroll, the air passes through nozzle vanes to the outer rim of the turbine wheel (4). Since this is an inward-flow turbine design, the air expands radially inward toward the center of the wheel and is then expelled through the exducer (3). The exhausted air passes through the screen (1) and out to the atmosphere. The expansion of the air from a pressure of about 35 psig [241.33 kPa] to atmospheric pressure imparts energy to the turbine wheel, causing it to reach a speed of about 55,000 rpm. This low-torque high speed is converted to a high-torque low speed by means of the 23.2 : 1 reduction gearing.

The **rotating assembly** of the starter consists of the turbine wheel, a spacer, a spur gear (9), and a nut. The turbine wheel assembly is an integral wheel and shaft with an exducer pinned on the exhaust end of the shaft against the front face of the wheel. The spacer, gear, and nut are installed on the opposite end of the shaft, which also provides for the installation of the two ball bearings in which the rotating assembly is mounted. The spacer, bearings, and gear are held on the shaft by the nut, which is secured by a rollpin through the shaft.

The **heat barrier and oil seal** (36) are positioned to provide the correct clearance between the turbine wheel assembly and the heat barrier. The heat barrier and the oil seal installed on the heat barrier prevent the passage of compressed air from the scroll into the housing and the passage of lubricating oil from the housing into the scroll.

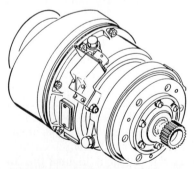

FIG. 11-15 AiResearch air-turbine starter. *(AiResearch)*

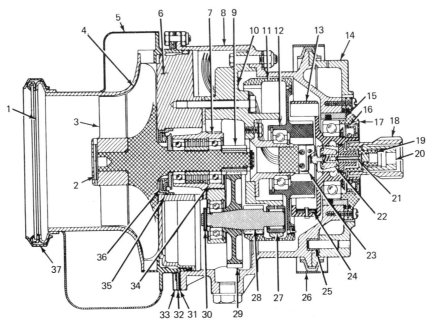

FIG. 11-16 Cross-sectional drawing of an air-turbine starter. *(AiResearch)*

The **oil-seal assembly** (36) consists of a rotor mounted on the turbine wheel shaft and a stator placed in the heat barrier. The rotor serves as an oil slinger. The carbon stator, containing an O-ring packing, is spring-loaded against the rotor, providing sealing against the passage of compressed air or lubricating oil.

The **bearing carrier** (7) supports the two ball bearings in which the rotating assembly is mounted. The bearing carrier and heat barrier with the oil seal are bolted to the gear carrier of the reduction-gear system.

The **reduction-gear system** consists of a **gear carrier** (10), three spur-gear-shaft assemblies (only one assembly shown at 28 and 29), an internal gear (11), and an internal gear hub (12). The **gear carrier assembly** is a matched pair of forgings brazed together and bolted inside the housing. The gear carrier provides for the installation of a ball bearing and a needle bearing in which each of the spur-gear-shaft assemblies rotate and a ball bearing on which the internal gear hub rotates. Each of the three **spur-gear-shaft assemblies** consists of an integral spur gear and tapered shaft (28), planet spur gear (29), bearing, and nut. The planet spur gear has a tapered bore and is friction-mounted on the tapered shaft, being held in position by the bearing and nut. The nut is secured on the shaft by a rollpin through the shaft. The planet spur gears mesh with and are driven by the spur gear of the rotating assembly. The internal gear is mounted over the spur-gear-shaft assemblies in mesh with and driven by the three spur gears of the gear-shaft assemblies. The internal gear hub is installed in and attached to the internal gear by means of a lock ring. The internal gear hub rotates on the ball bearing installed on the gear carrier assembly and is integral with the jaw of the engagement mechanism.

The **engagement mechanism** consists of a drive hub (12) and a drive-shaft assembly (13). The drive hub has a series of ratchet teeth equally spaced about the outside diameter, and these teeth engage the three pawls which are mounted inside the drum of the drive-shaft assembly. The arrangement of the pawl and spring drive

assembly is shown in Fig. 11-17. The drive shaft is internally threaded for installation of the switch actuating governor (22 in Fig. 11-16) and internally splined for installation of the output shaft (18) and provides a sealing surface for the drive-shaft seal (16) which is installed in the housing. Each pawl-spring assembly (24) is a series of leaf-type springs of varying length, as also shown in Fig. 11-17. Each spring assembly is riveted inside the drive-shaft drum.

The operation of the engagement mechanism is such that the drive-shaft pawls are disengaged from the teeth of the ratchet (pawl drive jaw) when the engine is running. Before and during starting, at low rotational speeds, the pawl springs in the drive shaft force the drive-shaft pawls into engagement with the ratchet teeth of the drive hub. This engagement transmits the rotation of the drive hub through the drive-shaft assembly and the output shaft to the engine on which the starter is installed. When engine lightoff occurs and the drive-shaft speed exceeds that of the drive hub, a ratcheting action takes place between the drive-shaft pawls and the ratchet teeth. This ratcheting action serves to disengage the starter from the engine as the engine overspeeds the starter drive hub. As engine speed continues to increase, the starter drive pawls assume the function of flyweights as centrifugal force overcomes the force of

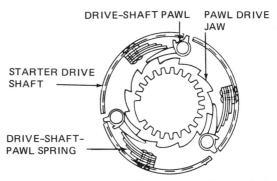

FIG. 11-17 Engaging mechanism for the AiResearch air-turbine starter.

the pawl springs and the pawls are completely withdrawn from engagement with the drive hub.

Before disengagement of the drive-shaft pawls from the drive hub and as the starter output-shaft speed approaches the predetermined cutoff point, centrifugal force causes the flyweights of the switch actuating governor to move outward. This actuates the snap-action switch in the gear carrier of the starter reduction-gear system to open the control circuit. This initiates the sequence of operations to interrupt the flow of compressed air, and the starter then becomes inoperative.

Operation of the Low-Pressure Air-Turbine Starter

Any one of the four engines on a Boeing 707 airliner may be started from the low-pressure external air supply. To furnish the necessary air for low-pressure starting, a turbocompressor outlet duct is connected to the airplane at the pneumatic ground-service connection.

To initiate a start, the **ground-start selector switch** is placed in the LOW PRESS (MANIFOLD) position and the start switch for the engine being started is placed in the GROUND START position. This causes 28-V DC power to flow to the solenoid which controls the **starter low-pressure air shutoff valve.** The valve is operated by a pneumatic actuator which is controlled by the solenoid. As the low-pressure air valve opens, air flows to the starter turbine and rotates the N_2 rotor of the engine through the drive from the engine accessory section where the starter is mounted.

The start switch is held in the GROUND start position until lightoff occurs and the output-shaft speed reaches the calibrated cutoff point. At this point, interruption of the flow of inlet-air to the starter is automatic, and disengagement of the starter from the engine takes place as the engine drives the output shaft to a higher speed. The ground-start switch is then released.

Release of the ground-start switch at any time will shut off the flow of compressed air to the starter, and no attempt must be made for another starting operation until the engine has completely stopped all rotation. To make certain that the engine has stopped turning, the operator should wait for 15 s after the engine tachometer shows zero rpm.

The normal duty cycle for the starter is 30 s *on* followed by 1 min *off.* When the engine is cold and the starter has not been operated for at least 60 min, the time *on* may be extended to 1 min at speeds up to the starter cutoff, which is approximately 2400-rpm output-shaft speed. After a 1-min cooling period another 1-min start attempt may be made, after which a 5-min cooling period must be allowed.

Starter operation is discontinued by releasing the ground-start switch which deenergizes the solenoid in the air-pressure shutoff valve. The operational sequence of the valve and starter is then broken so that the flow of compressed air to the starter is shut off and the operation of the starter is discontinued.

High-Pressure Air-Turbine Starters

A high-pressure air-turbine starter is essentially the same as the low-pressure starter except that it is equipped with an axial-flow turbine in place of the radial, inward-flow turbine previously described. The high-pressure starter is fitted with both low-pressure and high-pressure air connections to provide for operation from either type of air supply.

The usual air supply for the high-pressure starter operation is a high-pressure air bottle mounted in the airplane. This air bottle is charged to a pressure of about 3000 psi [20 685 kPa] and is used for starting one of the engines when an external low-pressure source is not available. On the Boeing 707 airplane, both the no. 2 and no. 3 engines are equipped with high-pressure starters, so either one or both can be started from the high-pressure system. After one of these engines is started, the low-pressure air from the turbocompressors is fed to the airplane manifold, thus making it possible to start the other engines from the low-pressure air supply of the airplane.

Combustion Starters

The two principal types of combustion starters are the fuel-air combustion starter and the cartridge-type starter. The turbine operating section of these starters is similar to or identical with that of air-turbine starters.

A **fuel-air combustion** starter is actually a small turboshaft engine. The engine fuel supply is a small fuel accumulator mounted on the starter, and the air supply is a high-pressure (3000-psi) air bottle near the engine. The fuel accumulator is refilled as needed from the aircraft fuel supply, and the air bottle is recharged by means of a compressor mounted in the airplane.

The air supply to the fuel-air combustion starter is reduced to the proper level for turbine operation by means of a pressure regulator. Typical operating air pressure is about 300 psi [2068.5 kPa].

The fuel-air combustion starter has the advantage of being a completely airplane-contained unit which is used for starting engines when no external service units are available. The operation of the starters is reliable and effective.

The principal disadvantage of these starters is the weight. The fuel-air combustion starter weighs between 50 and 60 lb [22.68 and 27.22 kg], and an air-turbine starter with the same capacity weighs from 25 to 35 lb [11.34 to 15.88 kg].

The **cartridge-type** starter may be considered an air-turbine starter operated by means of hot gases from a solid fuel cartridge instead of compressed air. As a matter of fact, some air-turbine starters can be adapted to cartridge operation merely by installing a cartridge combustion chamber and gas duct to the air-turbine starter.

The advantage of the cartridge-type starter is that it is a self-contained unit and does not require an external power source. The two principal disadvantages are the cost of the fuel cartridges and the erosion of the turbine parts by solid particles in the hot gases.

Inspection and Maintenance of Turbine Engine Starters

Routine inspection of turbine starters are similar to those of other accessories. These include security of mounting, freedom from oil leaks, and security of air and gas ducts, liquid lines, and electric wiring. Special inspections required for particular units are listed in the manufacturer's operation and service instructions.

Maintenance and service include the changing of the

lubricant in the gear housing. Since the starter is attached to the engine, it is likely to be exposed to very high temperatures. For this reason, the lubricant used is of the high-temperature type such as MIL-L-7808C or MIL-L-23699. Even this type of lubricant will lose its lubricating qualities after a time; hence, it must be tested regularly. The presence of metal particles in the lubricant is checked whenever a change is made in order to detect an incipient failure. Very small particles are normal, but particles which produce a sandy feel to the lubricant are a definite indication of internal damage in the starter.

Normally overhaul of turbine starters should not be necessary at less than 2000 h of engine service; however, the life of a starter is largely dependent on the skill and knowledge of the flight engineer or pilot who operates it. Under no circumstances should anyone but a well-trained and -informed individual be permitted to operate the starting system.

● STARTING SYSTEM FOR A LARGE TURBOFAN ENGINE

The starting system for a large airliner, the McDonnell Douglas DC-10 airplane, is described in this section. The system is used with the General Electric CF6 engines which power the DC-10.

From the diagram shown in Fig. 11-18, it will be observed that the energy for the starting system may be supplied by an onboard auxiliary power unit (APU), a ground air supply, or an operating engine on the aircraft. The APU and the engines are connected through

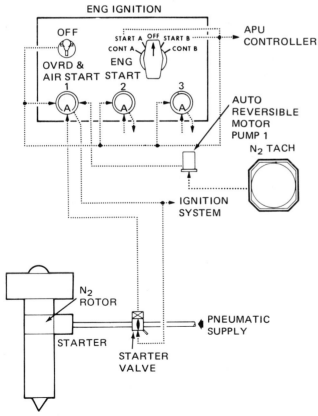

FIG. 11-18 Schematic drawing of the starting system for the Douglas DC-10 airplane. (*McDonnell Douglas Corp.*)

check valves and control valves to the common pneumatic system manifold.

Control and Indicating System

The control and indicating system provides means to actuate the starter shutoff valve, control, and pneumatic supply to the starter, to indicate the position of the starter shutoff valve, and to terminate the starting cycle. The system includes the engine-start switch, the starter shutoff valve, and the pneumatic starter mounted on the engine accessory gearbox.

The **engine-start switch** is located on the forward overhead panel in the flight compartment (cockpit) and controls the operation of the **starter shutoff valve.** The switch is a push-botton type with an integral indicating light in the knob. The switch is actuated when depressed and is held in the ON position by a holding coil until it is released by a speed signal from the N₂ **rpm indicator speed switch.** N_2 rpm is the rotational speed of the high-pressure compressor rotor in the engine. The speed switch is operated in conjunction with the ignition system controls. Power to the switch is provided by the engine ignition switch.

The starter shutoff valve, illustrated in Fig. 11-19, is a diaphragm-actuated butterfly-type pneumatic valve, electrically controlled and pneumatically operated. The valve functions to control the flow of compressed air to the starter. It consists of a valve body housing with an integral, butterfly-type closure element (gate) and appropriate in-line end flanges for direct mounting; a diaphragm-type pneumatic actuator mechanically coupled through a lever arm to the butterfly shaft; a solenoid-operated single-ball selector valve with manual override for control of valve position; a rate-control orifice which provides a controlled opening time; a stainless-steel, sintered wire-mesh filter; and a mechanical pointer for visual indication of valve position. The lower end of the butterfly shaft is equipped with a handle to permit manual opening of the valve in the

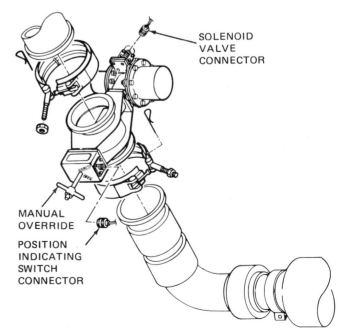

FIG. 11-19 Starter shutoff valve. (*McDonnell Douglas Corp.*)

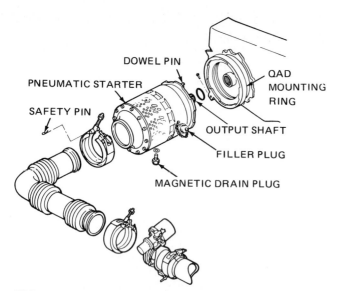

FIG. 11-20 Pneumatic starter and its mounting. *(McDonnell Douglas Corp.)*

event that the actuator supply pressure should be lost.

As shown in Fig. 11-19, a **position indicating switch** is located on the lower end of the butterfly shaft and energizes the light in the engine-start switch when the butterfly valve is not closed.

As previously mentioned, the valve is equipped with a rate-control orifice so that the opening rate is controlled to limit the maximum starter impact torque experienced during running starter engagements. **Running engagements** occur when the starter is engaged while the engine is rotating. These may occur during a restart in flight or a start on the ground when the engine is turning.

Because of the need for dry atmosphere in the solenoid- and servo-valve area of the starter shutoff valve, a pneumatic heater is made integral with the valve. As previously explained, moisture supports corrosion and is detrimental to the operation of electric circuits.

The pneumatic starter is an AiResearch ATS100-350 air turbine. A similar type starter is illustrated in Figs. 11-15 and 11-16. A drawing of the starter and its mounting on the accessory gearbox is shown in Fig. 11-20. The starter is a single-stage turbine consisting of a stator, turbine wheel, reduction gear, ring gear and hub jaw, gear housing, overrunning clutch, splined output shaft, and integral quick-attach-detach (QAD) mounting flange.

The starter gears and bearings are lubricated by a self-contained oil system using the same type of oil as that required by the engine (MIL-L-23699). Fill and drain ports are provided in the housing for servicing

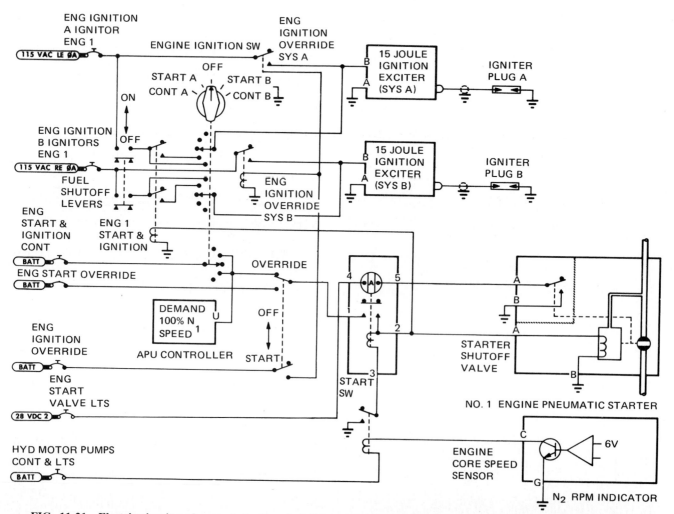

FIG. 11-21 Electric circuit and controls for the air-turbine starter on the DC-10 aircraft. *(McDonnell Douglas Corp.)*

the oil system. A magnetic chip detector is incorporated in the oil drain plug.

The starter output shaft is splined to fit a matching internally splined shaft in the accessory gearbox. The shaft is lubricated by the engine oil system and is provided with a shear section to protect the engine in case of starter malfunction or failure. The shear section is designed to shear at 1400 to 1600 lb·ft [1898.4 to 2169.6 N·m].

Operation

The operation of the DC-10 engine starting system involves energizing the starter with compressed air while the engine is provided with the proper amount of fuel and the ignition system is functioning. The procedure for starting can be understood by studying the diagrams of Fig. 11-21.

The engine-start and ignition-control switch is turned on to provide electric power for the engine-start switch and for ignition. When the start switch is depressed, the holding-coil circuit is completed through the N_2 rpm indicator switch. The energized coil holds the start switch in the ON position until the N_2 compressor rotor speed is 45 percent of maximum, at which time the switch in the N_2 rpm indicator opens. The start switch can be opened manually at any time.

Actuation of the engine-start switch energizes the starter shutoff-valve solenoid, as can be seen in the diagram of Fig. 11-21. This allows inlet-air pressure to be ported to the OPEN chamber of the valve actuator. Since the effective area of the OPEN chamber is greater than the effective area of the CLOSE chamber, the actuator opens the butterfly valve and compressed air can flow to the turbine starter. As the butterfly valve begins to open, the valve position indicating light switch closes and the indicating light in the starter switch comes on.

Air entering the starter inlet flows through the stator and is directed radially inward to propel the turbine wheel to high-speed rotation. Expended air is exhausted into the fan compartment of the engine and exited overboard through the compartment ventilating ports.

The high rotational speed of the turbine is reduced from a high-speed, low-torque characteristic to a low-speed, high-torque characteristic by means of a planetary and spur-gear combination. The high torque delivered by the starter very quickly accelerates the engine to lightoff speed. The starter continues to provide torque to the engine until it reaches self-sustaining speed.

The clutch mechanism for the starter is similar to that described previously and illustrated in Fig. 11-17. It will be noted that as soon as engine speed exceeds starter speed, the clutch mechanism will begin to disengage. Very soon the pawls completely disengage because of centrifugal force.

When the N_2 rotor of the engine has attained approximately 45 percent of full rpm, the switch in the N_2 rpm indicator opens and deenergizes the holding coil of the starter switch. The switch therefore opens and causes the starter shutoff valve to be closed. This, of course, causes the starter to stop and the indicator light in the starter switch to be deenergized.

In the operation of a starter such as that described in this section, the technician, pilot, or flight engineer should follow the procedure set forth by the manufacturer or the airline which operates the aircraft. Airlines have established procedures which have been developed to produce the safest and most effective results.

● REVIEW QUESTIONS

1. Name four types of starters which may be found on reciprocating engines.
2. Describe a direct-cranking electric starter.
3. How is cranking energy stored in an inertia starter?
4. What protective device is installed in a starter to prevent damage when the starter is engaged to the engine?
5. What type of electric motor is used for a starter?
6. If a starter motor is operated without a load, what precaution must be taken?
7. What precaution must be taken with respect to the length of time a starter is operated? Why?
8. Describe an overrunning clutch.
9. Describe a starter system with a 90° drive.
10. Why is reduction gearing required for a direct-cranking starter?
11. How is the torque setting for a direct-cranking starter adjusted?
12. What type of device is used as a switch to handle the heavy current required by a starter?
13. Draw a diagram of a typical starter circuit including the essential components.
14. List the causes for failure of a starter motor to operate.
15. How would you test a starter solenoid for operation?
16. How can the starter-to-engine jaw clearance be increased?
17. Name four types of starters for gas-turbine engines.
18. Describe an air-turbine starter.
19. What pressure and flow of air is required for a low-pressure air-turbine starter?
20. Describe the engagement mechanism for an AiResearch air-turbine starter.
21. If the starting switch is momentarily released while one is starting a gas-turbine engine, what should be done?
22. How long may the air-turbine starter be operated at one time?
23. From what sources may pneumatic energy be obtained to operate the starter on a large fanjet engine?
24. Describe the starter shutoff valve on the starting system for a DC-10 airliner.
25. What is the function of the N_2 rpm indicator switch?
26. What is the purpose of the position indicating switch?
27. What is the purpose of the shear section in the starter output shaft?
28. Give the sequence of events necessary to start an engine in the DC-10 aircraft.
29. At what engine speed does the starter system disengage?
30. What instructions should the operator follow in starting the engines of a large jet airliner?

12 ENGINE OVERHAUL PRACTICES

In the operation of any aviation flight activity, one of the most critical segments of the operation is the maintenance of the airplanes and powerplants, and of these two, the powerplant is certainly a major consideration. Individuals responsible for the maintenance or overhaul of a certificated aircraft engine or any of its parts bears a serious responsibility to the public, to themselves, and to their organization. They must perform their work in such a way that, when they have completed a particular repair or overhaul job, they can say, "This part (or this engine) will not fail."

● THE NEED FOR OVERHAUL

Someone entering the field of aviation maintenance may wonder why engine overhaul is necessary, what it consists of, and how often it must be done and may have a variety of other questions. These questions are not difficult, because experience has provided the answers.

After a certain number of hours of operation, an engine undergoes various changes which make the overhaul necessary. The most important of these are as follows:

1. Critical dimensions in the engine are changed as a result of wear and stresses, thus bringing about a decrease in performance, an increase in fuel and oil consumption, and an increase in engine vibration.
2. Foreign materials including sludge, gums, corrosive substances, and abrasive substances accumulate in the engine.
3. The metal in critical parts of the engine may be crystallized owing to constant application of recurring stresses.
4. One or more parts may actually fail.

Experience with a particular make and model of engine establishes the average period of time this model may be operated without expectation of failure. On the basis of this experience a normal period of overhaul is established. The owner of a private aircraft has two options when the engine on the aircraft has reached its recommended overhaul time, especially if the engine has been operated so that it has not been overheated or overstressed. The owner may disregard the engine overhaul time recommended until it is decided that a major overhaul is required, the decision being based on the performance of the engine, or the owner may decide to have a **top overhaul** performed on the engine. A top overhaul is the complete reconditioning of the cylinders, pistons, and valve-operating mechanism.

Whichever option is selected, the owner must realize that any time the overhaul period of an engine is extended, other factors may become apparent. The value of the aircraft may be depreciated because of the high engine time, or when the engine is eventually overhauled it may be worn beyond repairable limits, thus requiring extensive replacement of parts and greatly increasing the cost of the overhaul. The additional cost may be greater than the amount that was saved by extending the major overhaul period. The advice of a qualified engine specialist should be obtained before deciding to extend the overhaul time.

The actual time that an engine can be operated between overhauls (TBO) is determined largely by the manner in which it has been operated. If the operator is careful and continuously observes the rules for good engine operation, the life of the engine may be extended for several hundred hours. On the other hand, careless operation, such as extended climb at full power, may create a need for engine overhaul in much less than the normal period. If the engine is operated with excessive cylinder-head temperature for a short time, the piston rings may become "feathered" and lose their capacity to seal the cylinder. This results in loss of compression and high oil consumption. A good piston ring has sharp, square edges where contact with the cylinder wall is made. If the ring is overheated, these edges are damaged and the ring is said to be feathered. Cases are known where a new engine has been installed in an airplane and the first flight has resulted in ring feathering because the pilot failed to observe the operating requirements of the engine. In such cases, the cylinders of the engine must be removed and new rings installed.

With respect to engines used in commercial air transport airplanes, the engine overhaul period is determined by experience and is approved by the Federal Aviation Administration. Typical overhaul periods range from 800 to 1500 h. The major overhaul periods for modern jet engines used in commercial service may be more than 15,000 h. Minor repairs and replacements are made at shorter intervals as necessary. With the modular construction of some large turbine engines, overhaul can be accomplished on various modules (sections) as required without replacing the entire engine.

Definition of Overhaul

The dictionary definition of **overhaul** is "the disassembly, inspection, and repair of an integrated mechanism." For an aircraft engine this expands into the complete disassembly of the engine, cleaning and stripping of all parts, thorough inspection of every part, checking of measurements for all bearings and other fitted parts according to the manufacturer's Table of Limits, repair or replacement of parts as required, reassembly of the engine according to the manufacturer's instructions, preparation of a complete record of the overhaul, and a test run of the engine as specified by the manufacturer.

● THE OVERHAUL SHOP

Organization

In order to provide for efficient overhaul operations so engine overhaul can be accomplished with a minimum cost to the engine owner, an overhaul shop must be well organized. The type of organization largely depends on the size of engines to be overhauled and the volume (number of engines to be overhauled in a given period of time). Overhaul shops may be established to handle comparatively few small engines or they may be large companies which overhaul hundreds of small and large engines each year. A sample organization chart for a medium-sized overhaul operation is shown in Fig. 12-1.

Note in the organization chart that the Production Department is serviced by all the other departments because the production of overhauled engines is the business of the organization.

The Accounting Department handles all financial records, including costs of production, payment of invoices, billing of customers, tax reports, payroll preparation, etc. An effective Accounting Department is essential, not only to perform the routine accounting operations but also to keep management informed regarding the financial condition of the organization.

The Purchasing, Receiving, and Shipping Department maintains a smooth flow of engines and materials into the facility and finished engines out to the customer. It will be noted that this department maintains direct communication with some of the subdivisions in Production.

The major functions of the Production Department are indicated in Fig. 12-1. The success of the operation depends on the quality of work done in the subdepartments and the efficiency with which each operates. Note that the organization is responsible for the quality

of the product and must answer to both the Federal Aviation Administration (FAA) and the customer if the overhaul work is not airworthy.

The FAA requires that official inspections for quality of production be accomplished by a separate inspection individual or group who will make an objective appraisal of all work performed. It is the responsibility of the Inspection (or Quality Control) Department to see that all the parts of an engine meet approved specifications before they are installed in the engine. They are also required to check repair operations and assembly procedures. The assembly inspection is particularly important because every nut and bolt must be tightened to the proper torque and then safetied by approved methods. A repair station which is certificated by the FAA must comply with Federal Aviation Regulations, part 145.

The function of the **receiving inspection** is to inventory each engine as it is received, making records of the general condition of the engine, the accessories included with the engine, the serial numbers of the engine and its accessories, and any other information which may be essential to the satisfactory completion of the engine overhaul. These records are necessary for the satisfaction of the customer and to provide information for any subsequent investigation of the history of the engine.

In the **disassembly and cleaning** section the engine is given a preliminary cleaning to remove external dirt and grease. This may be done by careful steam cleaning, by means of a petroleum solvent wash, or by the use of an emulsion cleaner. After this external cleaning, the engine is started through the disassembly procedure in accordance with the manufacturer's instructions in the overhaul manual. In all cases, it is important that the operator in charge of cleaning make sure that the cleaning material being used will not harm the engine or its parts. Manufacturers of cleaning chemicals can furnish information regarding their products and instructions regarding methods for using the products correctly.

For disassembly, the engine should be mounted on a suitable disassembly stand (overhaul stand) so that the engine will be held in a position for convenient handling. As each part is removed from the engine, a metal tag with the work order number should be attached to the part.

After the engine is disassembled, the parts are inspected thoroughly before being given the final cleaning and stripping treatment. Residual material, such as metal particles, on the parts can reveal internal damage which must be examined thoroughly. The purpose of the stripping and cleaning is to remove all grease, paint, carbon, and any other material which coats or covers the bare metal. This is necessary to allow for an adequate inspection.

The **overhaul inspection** division determines what damage and wear have occurred in the engine, thereby providing information needed for repair and replacement of parts. This division utilizes all the inspection methods necessary to ensure that the engine and its parts are brought up to airworthy standards. At this time the quality-control inspectors will provide assistance.

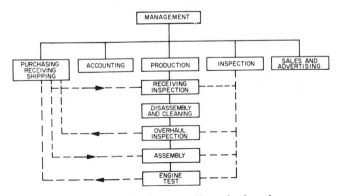

FIG. 12-1 Organization for a medium-sized engine-overhaul shop.

During the overhaul inspection, records are made indicating whether parts must be replaced or what type of repair or rework is required. The various parts may be sent to different sections of the shop for particular types of work. For example, a special section may be devoted exclusively to cylinder repair, another section may handle crankcases, etc. The number of special sections in the shop is determined by the volume of work handled by the shop. The various special shop sections that perform repair and reconditioning of engine parts and accessories may be organized under a Repair Division of the company. Parts to be repaired would be routed from Overhaul Inspection to the Repair Division, and after repair or reconditioning such parts would be routed back to Overhaul Inspection for final inspection before being sent to the Assembly Division.

Replacement parts for an engine are drawn from the stock room and charged to the work order number or are ordered through the Purchasing Department. A shop which overhauls a substantial number of the same type of engine will usually maintain a stock of the most commonly required parts.

The Assembly Division receives the airworthy parts of the engine, including new parts and repaired parts, and proceeds with the assembly of the engine according to the manufacturer's overhaul manual. Airworthiness Directives, Type Certificate Data Sheet, manufacturers' bulletins, and other applicable instructions. During and after assembly the quality-control inspector checks each step of assembly for compliance with pertinent requirements.

Upon completion of assembly and inspection, the engine is installed in an engine test stand and is operated according to the approved run-in schedule. During run-in the operation of the engine is carefully observed for rpm, temperatures, vibration, power output, manifold pressure, etc. If the engine does not perform according to specifications, the reason for the unsatisfactory performance must be determined and corrected.

The larger overhaul organizations will usually have a Sales and Advertising Department. It is the function of this department to bring the overhaul business to the organization by means of advertising, offering bids, and making direct contacts with potential customers. A good Sales Department is often the difference between the success and failure of an overhaul business.

One of the most critical functions of the Sales Department, working in cooperation with the Accounting Department, is to determine the cost of engine overhaul jobs and to offer realistic bids for jobs to be done. Some companies list flat rates for certain engine overhaul jobs, and others bid on the basis of a flat rate for labor only. In the latter case the parts required for the engine are listed and charged for separately.

Another type of overhaul service offers rebuilt engines at a given price provided that a suitable trade-in engine is supplied. In this case, the trade-in engine must meet certain rigidly established specifications before full credit is allowed.

The pricing of an engine overhaul job must be such that all costs and a reasonable amount of profit are provided for. The costs include labor, parts, and overhead. The overhead includes all costs of operating the business, including administration, sales, staff services, cost of buildings and equipment, and various other charges. A certain percentage of the income from each engine overhaul job must be applied to the overhead charge.

Certification of the Overhaul Station

In order to provide overhaul services for certificated aircraft engines, an overhaul agency should be certificated as an FAA certificated repair station with ratings to cover all types of overhaul work performed. A **certificated repair station** is defined by Federal Aviation Regulations as *"a facility for the maintenance, repair, and alteration of airframes, powerplants, propellers, or appliances, holding a valid repair station certificate with appropriate ratings issued by the Administrator."* Certification by the FAA as a certificated repair station permits the repair station to perform major overhaul on all items for which appropriate ratings are held. This certification also permits the repair station to approve the overhauled items for return to service. All agencies currently approved by the FAA are listed in the Consolidated Listings of FAA Certified Repair Stations. Further details regarding the responsibilities and privileges of a certificated repair station are given in part 145 of the Federal Aviation Regulations.

Shop Safety and Preparation for Overhaul

In the preparation of an overhaul shop, it is essential that all required equipment be on hand. A listing of equipment is given later in this chapter. In the operation of the shop, it is important that adequate safety practices be exercised. Safety involves the design and arrangement of the shop, the use of proper tools and equipment, and the observance of safe practices by the overhaul technicians and their helpers. To ensure that all reasonable safety precautions are practiced and that all equipment meets safety requirements, it is advisable that a professional safety engineer be engaged during the establishment and organization of the shop. The safety engineer should be asked to make periodic inspections of the shop and shop operations to check for unsafe practices and conditions and to discover any equipment which has become unsafe.

Among requirements for the safety and equipment in an overhaul shop are the following:

1. The arrangement of the shop should be such that crowding does not exist.

2. Materials must be stored in proper racks and containers.

3. Walkways should be painted on the floor and kept free of obstructions at all times except when necessary to move equipment.

4. Suitable engine overhaul stands must be provided for the types of engines being overhauled. The design of engine stands varies with the type of engines.

5. All machinery must be in good repair and provided with approved safety guards.

6. Electric equipment and power outlets must be correctly identified for voltage and power handling capability.

7. Electric outlets must be designed so that it is

not possible to plug in equipment other than that which is designed for the voltage and type of current supplied at the outlet.

8. Adequate lighting must be provided for all areas.

9. Cleaning areas must be adequately ventilated to remove all dangerous fumes from solvents, degreasing agents, and other volatile liquids.

10. Properly designed engine hoisting equipment must be available for lifting engines.

11. Operators should wear safety shoes to prevent injury to their feet in case heavy equipment or parts are dropped.

12. Operators of lathes, machine tools, drills, and grinders must wear safety goggles.

13. Operators of cleaning equipment must wear protective clothing, rubber gloves, and face shields.

14. The floor of the shop should have a surface which does not become slippery.

15. Grease, oil, and similar materials should be removed from the floor immediately after they are spilled.

16. Proper types of fire extinguishers should be available to shop personnel.

17. Personnel should take care to keep clear of an engine which is being hoisted except as necessary to prevent it from swaying or turning.

18. Cleanliness is of prime importance in an engine overhaul shop. All areas must be kept free of sand, dust, grease, dirty rags, and other types of contaminants.

19. Greasy or oily rags must be stored in closely covered metal cans. The rag cans should be emptied frequently.

20. Flammable liquids must be stored in an approved fireproof area. The approval is normally granted by the fire department which has jurisdiction in the area.

21. The area where engines are run should be paved and kept free of all debris which could be picked up by air currents caused by a propeller or jet intake.

● RECEIVING THE ENGINE

Receiving Inspection

The purpose of the receiving inspection is to determine the general condition of the engine when it is received and to provide an inventory of the engine and all its accessories and associated parts. It is most essential that every part be accounted for because an owner may not be aware that certain parts or accessories have been removed before the engine was delivered to the overhaul facility. The receiving inspection report should be made on a standardized form, and a copy of this form should be provided for the owner.

Included in the receiving inspection report should be the make, model, and serial number of the engine and all its accessories. In addition, all parts such as cylinder baffles, ducting, brackets, etc., should be listed so that there will be no question regarding the responsibility for these parts after the engine is completed.

The general condition of the engine should be carefully noted on the receiving report. If the engine has been in a crash, fire, or other situation where damage may have been caused, it is necessary that such conditions be recorded. The cost of overhauling an engine which has suffered damage, other than normal wear, will be considerably greater than usual. In many cases it is not economically feasible to overhaul such an engine.

Records

Federal Aviation Regulations require that a permanent record of every maintenance (except preventive maintenance), repair, rebuilding, or alteration of any airframe, powerplant, propeller, or appliance shall be maintained by the owner or operator in a logbook or other permanent record satisfactory to the FAA administrator and shall contain at least the following information: (1) an adequate description of the work performed; (2) the date of completion of the work performed; (3) the name of the individual, repair station, manufacturer, or air carrier performing the work; and (4) the signature and the certificate number of the person, if a certificated mechanic or certificated repairman, approving as airworthy the work performed and authorizing the return of the aircraft or engine to service.

All major repairs and major alterations to an airframe, powerplant, propeller, or appliance must be entered on a form acceptable to the FAA administrator. Such form must be executed in duplicate and must be disposed of in such manner as, from time to time, may be prescribed by the administrator.

All major alterations must be entered on form ACA-337, the approved major repair and alteration form. This form must be executed in accordance with pertinent instructions, and the original copy given to the owner of the unit altered or repaired. The repair station should retain a copy for its permanent record, and one copy must be sent to the local FAA office within 48 hours of the time that the powerplant or other unit is returned to service.

The administrator will accept, in lieu of form ACA-337 for major repairs made only in accordance with a manual of specifications approved by the administrator, the customer's work order upon which repairs are recorded by the repair station. The original copy of the work order must be furnished the owner or purchaser, and the duplicate copy must be retained at least 2 years by the repair station.

The owner of a powerplant which has been overhauled by a certificated repair station should be supplied with a copy of a **maintenance release**. This release should accompany the engine until it is installed in the aircraft, and at that time the installing agency will make the release available to the owner for incorporation in the permanent record of the aircraft. The maintenance release may be included as a part of the work order, but it must contain the complete identification of the engine including the make, model, and serial number. The following statement must also be included:

The engine identified above was repaired and inspected in accordance with current Federal Aviation Regulations and was found airworthy for return to service.

Pertinent details of the repair are on file at this agency

under Work Order No. _____

Date _____

Signed _____
                    (Signature of authorizing individual)
for _____
(Agency name)   (Certificate No)
_____
                    (Address)

In addition to the formal records and statements required by FAA regulations, the repair station should maintain a complete record of all repair operations and inspections accomplished on each engine or component overhauled. This record should contain an account of every repair operation, every inspection made, and every replacement of a part. The inspection record should show the dimensions of each part measured and all fits and clearances. These measurements are compared with the manufacturer's **Table of Limits** so that repairs and replacements can be made to restore worn parts to the required specification.

The inspection record should show the types of inspection procedures employed, such as magnetic inspection, fluorescent penetrant inspection, dye penetrant inspection, x-ray inspection, eddy current inspection, and any other approved process employed to determine the air-worthiness of the engine and its parts. If any of these inspections is performed by another agency, the certificate of that agency must be included in the overhaul record.

From time to time, special **service bulletins** are issued by the manufacturer of an engine to require alterations or parts replacements designed to improve the performance and reliability of the engine. During the overhaul of an engine, bulletins must be complied with, and a statement of such compliance must be made in the overhaul record. A typical service bulletin is shown in Fig. 12-2. The FAA issues **Airworthiness Directives** pertaining to aircraft and engines whenever it appears that certain changes should be made in order to correct discrepancies or to improve the reliability of the unit. During the overhaul of an engine, all Airworthiness Directives pertaining to the engine should be checked to make sure that such directives are complied with, and a statement to the effect should be entered in the overhaul record.

In addition to the special bulletins and Airworthiness Directives, the **FAA Engine Specification** or **Type Certificate Data Sheet** covering the particular engine should be checked for **notes** or other information relating to the engine. If the performance of special operations or changes is required, these must be accomplished and a record made to show what has been done. A portion of a Type Certificate Data Sheet is shown in Fig. 12-3.

The preparation of adequate overhaul records cannot be overstressed. These records are particularly important for the protection of the overhaul agency in the event of an engine failure. If the record is complete and properly prepared, the overhaul agency can show that all overhaul work was accomplished in accordance with the manufacturer's overhaul manual and that all required operations were performed. This type of record will usually absolve the overhaul agency of responsibility in case of engine failure.

● CLEANING PROCESSES

During the overhaul of an aircraft engine it is necessary to clean the engine externally before disassembly and also to clean the parts after disassembly. Great care must be taken during the cleaning processes because the engine parts can be seriously damaged by improper cleaning or the application of the wrong types of cleaners to certain parts of the engine. In every case, the person in charge of the cleaning processes should study the engine manufacturer's recommendations regarding the cleaning procedures to be used for various parts of the engine and also comply with the directions provided by the manufacturer of the cleaning agent or process.

In general, we may say that there are three types of cleaning required when an engine is overhauled. These are degreasing, removal of soft types of dirt and sludge, and the removal of hard carbon deposits. The removal of grease and soft types of residues is relatively simple; however, hard carbon deposits require the application of rather severe methods. It is during this type of cleaning that the greatest care must be exercised in order to avoid damaging the engine parts.

External Degreasing

When an engine is first received for overhaul, it usually has oil and dirt on various parts of the exterior. In some cases the dirt and oil mixture is "baked" on to the extent that it is not entirely removable with an ordinary solvent. Usually, however, the oil and loose dirt can be removed by means of a petroleum solvent such as mineral spirits, kerosene, or other approved petroleum solvent. Some operators steam-clean the engine, thus removing oily deposits and caked dirt. When steam cleaning is done, the engine should be washed off with plain hot water afterward to remove all traces of the alkaline steam-cleaning solution. This is necessary in order to prevent the corrosion of aluminum and magnesium parts.

Degreasing after Disassembly

After the engine is disassembled, all oil should be removed from the parts before further cleaning is attempted. The reason for this is that the additional cleaning processes are much more effective after the surface oil is removed. Two of the principal methods for removing the residual lubricating oil and loose sludge are washing in a petroleum solvent and employing a vapor degreaser. Remember that the engine parts should be inspected before the cleaning is done.

Cleaning with a petroleum solvent can be accomplished in a special cleaning booth where the parts are supported on a wooden grill and sprayed with a solvent gun using an air-pressure source of 50 to 100 psi [345 to 690 kPa]. The spray booth should be provided with a ventilating system to carry away the vapors left in the surrounding air, and the operator should wear ad-

TELEDYNE CONTINENTAL® AIRCRAFT ENGINE

service bulletin

M84-10
Supersedes M81-3 R-1

Technical Portions are
FAA-DER Approved

20 August 1984

TO: Aircraft Manufacturers, Distributors, Dealers, Engine Overhaul Facilities, Owners and Operators of Teledyne Continental Motors' Aircraft Engines.

SUBJECT: ENGINE PRESERVATION FOR ACTIVE AND STORED AIRCRAFT

MODELS
AFFECTED: All Models

Gentlemen:

Engines in aircraft that are flown only occasionally tend to exhibit cylinder wall corrosion more than engines in aircraft that are flown frequently.

Of particular concern are new engines or engines with new or freshly honed cylinders after a top or major overhaul. In areas of high humidity, there have been instances where corrosion has been found in such cylinders after an inactive period of only a few days. When cylinders have been operated for approximately 50 hours, the varnish deposited on the cylinder walls offers some protection against corrosion. Hence a two step program for flyable storage category is recommended.

Obviously, even then proper steps must be taken on engines used infrequently to lessen the possibility of corrosion. This is especially true if the aircraft is based near the sea coast or in areas of high humidity and flown less than once a week.

In all geographical areas the best method of preventing corrosion of the cylinders and other internal parts of the engine, is to fly the aircraft at least once a week, long enough to reach normal operating temperatures, which will vaporize moisture and other by-products of combustion. In consideration of the circumstances mentioned, TCM has listed three reasonable minimum preservation procedures, that if implemented, will minimize the detriments of rust and corrosion. It is the owners responsibility to choose a program that is viable to the particular aircrafts' mission.

Aircraft engine storage recommendations are broken down into the following categories:

A. Flyable Storage (Program I or II)

B. Temporary Storage (up to 90 days)

C. Indefinite Storage

TELEDYNE CONTINENTAL MOTORS
Aircraft Products Division

FIG. 12-2 Typical service bulletin.

E-252-28
CONTINENTAL
C90-8F, -8FJ
C90-12F, 12FH, 12FJ, -12FP
C90-14F, -14FH, -14FJ, -16F
0-200-A, 0-200-B, 0-299-C

April 30, 1979

TYPE CERTIFICATE DATA SHEET NO. E-252

Engines of models described herein conforming with this data sheet (which is part of type certificate No. 252) and other approved data on file with the Federal Aviation Agency, meet the minimum standards for use in certificated aircraft in accordance with pertinent aircraft data sheets and applicable portions of the Civil Air Regulations provided they are installed, operated and maintained as prescribed by the approved manufacturer's manuals and other approved instructions.

Type Certificate Holder	Teledyne Continental Motors P. O. Box 90 Mobile, Alabama 36601		
Model	C90-8F	C90-12F, -14F, -16F	0-200-A, -B, -C
Type	4H0A	– –	– –
Rating, standard atmosphere			
Max. continuous hp., r.p.m., at sea level pressure altitude	90-2475	– –	100-2750
Takeoff hp., 5 min., r.p.m., full throttle, at sea level pressure altitude	95-2625	– –	100-2750
Fuel (min. grade aviation gasoline)	80/87	– –	– –
Lubricating oil, ambient air temperature	Oil Grade		
Below 40°F.	SAE 20	– –	– –
Above 40°F.	SAE 40	– –	– –
Bore and stroke, in.	4.062 x 3.875	– –	– –
Displacement, cu. in.	201	– –	– –
Compression ratio	7:1	– –	– –
Weight (dry), lb.	184	188	190
C.G. location (with accessories)			
Fwd. of rear face of mounting lugs, in.	6.2	4.6	– –
Below crankshaft center line, in.	1.5	1.3	1.2
Propeller shaft, SAE No.	1 Flange	– –	– –
Carburetion (see NOTE 4 for injectors)	Marvel-Schebler MA-3SPA (CMC P/N 627367, 629175, 637101 or 637835)	– –	Marvel-Schebler MA-3SPA
	Bendix-Stromberg NA-S3A1 (CMC P/N 530625, 530726, 531126, 530846, 531157)	– –	TCM P/N 627143, 640416 or 633028
Ignition	2 Bendix-Scintilla S4RN-21 or -1227; or Slick-Electro 443 or 4003 magnetos or 1 ea. Bendix-Scintilla S4RN-200 and -204	2 Bendix-Scintilla S4LN-21 or -1227 or 1 ea. S4LN-200 and -204; Slick-Electro 447, 4001 or 4201 magnetos	– –
Timing, °BTC	26 Top, 28 Bottom	– –	24 Top, 24 Bottom
Spark plugs	See NOTE 6	– –	– –
Oil sump capacity, qt.	5 or 6	– –	– –
NOTES	1 thru 6	1,2,3,4,6	1,2,3,4,6

"– –" indicates same as preceding model"

Certification basis   Part 13 of the Civil Air Regulations.
                       Type Certificate No. 252

Production basis   Production Certificate No. 7
                    Production Certificate No. 508 (All models except C90-16F)

Page No.	1	2	3
Rev. No.	28	28	26

FIG. 12-3   Portion of a Type Certificate Data Sheet.

equate protective clothing. A drain should be provided underneath the wooden grill to collect the used cleaning solvent. During the cleaning process particular care should be applied to make sure that all crevices, corners, and oil passages are cleaned.

Vapor degreasing is accomplished with equipment especially designed for this type of cleaning. A vapor degreaser consists of an enclosed booth in which a degreasing solution such as trichloroethylene is heated until it vaporizes. The engine parts are suspended above the hot solution, and the hot vapor dissolves the oil and soft residue to the extent that they flow off the parts and drop into the container below. The vapor degreaser should be operated only by a person who is thoroughly familiar with the equipment and the manufacturer's instructions regarding its use.

Stripping

The stripping process is employed to remove paint and various resinous varnishes which have formed in the engine during its operation. Some stripping solutions are also effective in removing some of the harder carbon deposits. A number of solutions suitable for stripping have been developed by manufacturers of chemical cleaning agents, but each solution must be used according to proper direction, or damage may be inflicted on the engine parts. In all cases the person responsible for cleaning and stripping the parts must be thoroughly familiar with the solution used and must know what engine parts may be cleaned safely with the solution. Alkaline (caustic) solutions will attack aluminum, magnesium, and some of the alloys employed in aircraft engines. Strong alkaline solutions will also remove the Alodine coating that is applied to many of the aluminum parts of engines. The Alodine coating is usually applied to the interior surfaces of crankcases, valve covers, accessory sections, and other surfaces that are not available for cleaning or treatment from the outside. The operator doing the stripping of the engine parts must be careful to follow the proper instructions regarding the cleaning and stripping of all such parts.

In order to prevent electrolytic action in the stripping solution, dissimilar metals should not be placed in the solution at the same time. Magnesium and aluminum parts can be completely destroyed when placed in a cleaning vat with steel parts for more than a few minutes. Cold strippers are available which are neutral in their effect on metals but will produce good results in removing paint and engine varnishes. They also tend to soften hard carbon, thereby aiding the process of soft grit blasting.

A typical cold-stripping process involves the immersion of parts, such as the engine crankcase, in a vat containing the solution for several hours or overnight. The parts are then removed and steam-cleaned to remove all traces of the stripping solution and all material which has been loosened by the solution. After steam cleaning, the parts are washed with clear hot water and all openings, oil passages, crevices, etc., are checked for cleanliness. This is particularly important because the clogging of one oil passage can cause failure of the engine.

It is inadvisable to soak aluminum and magnesium parts in solutions containing soap because some of the soap will become impregnated in the surface of the material even though it is washed thoroughly after soaking. Then, during the engine operation, the heat will cause the soap residue to contaminate the engine oil and cause severe foaming. This will result in loss of oil and possible damage to the engine.

Vapor Blast

Vapor blasting is employed for special cleaning jobs and is accomplished by means of specially designed equipment and materials. The vapor solution is used with a fine abrasive and should be applied in an enclosed booth. The vapor is hot and is applied with a high-pressure air gun.

The use of vapor blasting is limited to parts and areas of parts which will not be damaged by a small amount of material erosion. It may be used on the tops of pistons, on the outside of cylinders, and in almost any area which is not a bearing surface of some type. In all cases, the manufacturer's instructions must be followed.

Blasting with Grit and Other Materials

For the removal of hard carbon from the inside of cylinders and the tops of pistons, soft **grit blasting** offers one of the most satisfactory processes. The blasting material (grit) consists of ground walnut shells, ground fruit seeds, and other organic or plastic materials. The grit is applied by means of a high-pressure air gun in an enclosed booth, as shown in Fig. 12-4. The operator's hands and arms are extended into sleeves and gloves sealed into the machine for full protection while the gun and the parts being cleaned are manipulated. The work can be observed through a window in the front of the booth.

In the application of grit blasting for the removal of hard carbon, paint, and other residues, the operator must use care to avoid damage to highly polished surfaces and to avoid the plugging of oil passages. If crankcases are grit-blasted, it is advisable to plug all the small openings into which the grit may enter. Some operators do not use the grit blast on parts having oil passages.

In general, **sandblasting** is not employed in the cleaning of engine parts except the valve heads. This is because a sandblast will erode the metal so rapidly that serious damage may be done before the operator is aware of it. Furthermore, sand tends to embed itself in soft metals and will later become a source of abrasion in the operating engine. Under some circumstances, such as preparation for metallizing of cylinders, the outside of the cylinder is carefully sandblasted before the application of the metal spray. Special uses

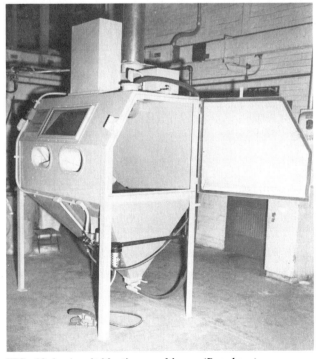

FIG. 12-4 A grit blasting machine. *(Pangborn)*

of sandblasting may be approved for particular purposes; however, this may be done only by a thoroughly trained operator and then only within the limits specified for the process. After engine parts are cleaned and dried, they should be coated with a preservative oil while awaiting inspection and further processing. This is to prevent the rust and corrosion which are likely to occur.

In addition to the grit and other materials employed for cleaning by blasting, such materials as rice hulls, plastic pellets, baked wheat, and glass beads are used. Technicians must asssure that they are using an approved material for the blasting job they are performing.

When using glass beads, the addition of aluminum oxide powder adds to the effectiveness of the process either as a cleaning agent or as a mild form of "shot peening." The size of the glass beads is important and should be checked in the instructions for the particular job being done.

● ENGINE DISASSEMBLY AND CLEANING

Tools and Equipment

Before attempting to overhaul an aircraft engine, the powerplant maintenance technician should see that all the essential tools and equipment required are available. Engine manufacturers design special tools and fixtures to be used for the overhaul of their engines; however, it is not necessary to have every tool listed in the overhaul manual. Many standard tools such as wrenches, gear pullers, arbors, lifting slings, reamers, etc., can be used in place of similar tools which may be available from the manufacturer. If a particular overhaul shop is likely to receive many engines of the same type for overhaul, it is a good plan to equip the shop with most of the tools designed for the overhaul of that type of engine.

Among the tools and equipment needed for engine overhaul are the following:

Engine Shop Equipment

Arbors of several sizes	Drill press
Arbor press	Engine lathe
Chain hoist	Generator test stand
Cleaning equipment	Heating equipment
Crankshaft thread cap	Lifting sling
Crankshaft wrench	Penetrant inspection
Magnetic inspection	equipment
equipment	Test stand and propeller
Magneto test stand	Timing disk and top-
Magneto timing tools	center indicator or
Overhaul stand	Time-Rite equipment
Parts trays	Timing lights
Cylinder fixtures or hold-	Vises
ing blocks	

Engine Overhaul Tools

Connecting-rod align-	Ignition harness tools
ment tools	Piston-ring compressor
Cylinder base wrench	Reamers
Gear pullers	Tap and die tools

Valve-guide tools	Valve-spring compressor
Valve refacing equip-	Valve-spring tester
ment	
Valve-seat grinding	
equipment	

Hand Tools

Box wrenches	Pliers
Bushing reamers	Safety wire tools
Counterbores	Screw extractors
Diagonal cutters	Spanner wrenches
Drifts of several sizes	Spot facers
Drill motor	Stud drivers
End wrenches	Stud pullers
Heli-Coil tools	Torque wrenches

Precision Measuring Tools

Depth gages	Small hole gages
Dial gages	Surface plate
Height gages	Telescope gages
Micrometers	V blocks

Nondestructive Inspection and Testing Equipment

Magnaflux	Ultrasonic
Zyglo	Dye penetrant
Eddy-current	

In addition to the listed tools and equipment, it is likely that the engine will have certain features which require the use of manufacturer's special tools. It is therefore necessary to obtain these special tools for each different type of engine overhauled.

Manufacturer's Manual

Every engine overhaul should be accomplished in accordance with the instructions given in the manufacturer's overhaul manual. Usually the manufacturer's manual will give specific steps for the disassembly of the engine, and by following these steps the technician will avoid many problems. The technician must be sure that the manual and other data referred to for overhaul information apply to the model number, including the dash number, of the engine being overhauled.

Procedures

For a typical small-aircraft engine, the disassembly follows a sequence beginning with the installation of the engine on the overhaul stand. The overhaul manual then specifies that parts be removed or disassembled in the following order:

1. Ignition system
2. Fuel-injection system
3. Induction system
4. Magneto and accessory drives
5. Oil sump
6. Oil cooler
7. Alternator assembly
8. Starter and starter drive adapter
9. Oil pump assembly
10. Cylinders and pistons
11. Crankcase

12. Camshaft assembly

13. Crankcase assembly

The foregoing order of disassembly, established for the Teledyne Continental GTSI0-520 engine, and the details of each operation are given in the overhaul manual. If the technician has carefully noted each step and carried out the operations as given, the engine will be completely disassembled and ready for the first visual inspection.

During disassembly the operator should identify and mark all parts. This is often done by attaching small metal or plastic tags to each part with soft safety wire. The tags are stamped with the work order number. Further identification is needed for certain parts which must be reinstalled in the same location from which they are taken. This applies to cylinders, pistons, connecting rods, valves, etc. The overhaul manual specifies which parts must be replaced in their original locations. These parts may be identified by attaching additional tags to show the location. Nuts and bolts may be strung on safety wire in groups which are identified for location.

Cylinders, pistons, connecting rods, valve lifters, and other parts requiring exact location should be clearly marked to show the location in the engine. For example, the piston from the no. 1 cylinder should be marked with the number 1. Usually this number will be stamped on the head of the piston. Connecting rods have numbers stamped on the big end. Careful marking and identification of the parts will ensure proper assembly and will provide for correct fits and clearances when measurements for wear are made.

It is recommended that all parts be given a preliminary visual inspection as they are removed and before they are cleaned. Parts that are obviously damaged beyond repair should be discarded and marked so that they will not be reused. A record of all parts thus discarded should be made for the overhaul file.

Cleaning

When the engine is disassembled, all small parts should be placed in metal wire baskets in order to prevent the loss of any part during the cleaning process. The first cleaning operation is the removal of oil and sludge by washing in solvent or in a vapor degreaser. This is to prevent the dilution and contamination of the stripping solution during the next phase of cleaning.

As mentioned previously, great care must be taken during the stripping and decarbonizing process. The operator must be familiar with the characteristics of the cleaning solution and must follow the recommendations of the manufacturer. Magnesium and aluminum parts are particularly susceptible to damage in alkaline solutions, and prolonged exposure to such solutions must be avoided.

Care must be exercised in the handling of decarbonizing (stripping) solutions to prevent the material from contacting the skin because these solutions will usually cause severe irritation or burning. The operator should therefore wear goggles, rubber gloves, and protective clothing while working with these solutions.

After the required soaking time in the stripping solution, the engine parts should be steam-cleaned and then washed with clear hot water. Oil passages and other small apertures should be examined to see that all sludge and other residues are removed. The parts are then dried with compressed air, and those requiring further cleaning are routed to the next operation. Steel parts which are clean should be coated with a rust inhibiting oil immediately after the cleaning operation is completed.

Except as permitted by the instructions in the overhaul manual, wire wheels, steel scrapers, putty knives, or abrasives should not be used for cleaning parts or removing carbon. The use of these items may leave scratches on the parts which will lead to stress concentrations and ultimate failure.

Hard carbon is removed from piston heads and from the inside of cylinder heads by means of a soft grit blast. The carbon which has collected inside the piston head and in the ring lands is the most difficult to remove. After the part has been soaked in the decarbonizing solution, the carbon can usually be removed by grit blasting, although it is sometimes necessary to use a soft scraper. The operator must be careful not to damage the piston-ring lands or remove any metal from the small radii between the ring lands and the bottom of the ring grooves. The glazed surfaces on the piston and the piston pin should not be removed.

● INSPECTIONS

Use of the Manufacturer's Overhaul Manual

Although we have explained the importance of using the manufacturer's overhaul manual previously, we must reemphasize this essential policy. The manufacturer of an aircraft engine has all the information relating to the construction and design of engine parts and has made numerous tests of the assembled engine and its parts. The overhaul manual prepared by the manufacturer is therefore the most valid source of information available for a particular make and model of engine.

The overhaul manual provides information on all procedures of a special nature and also on general procedures for the disassembly, cleaning, inspection, repair, modifications, assembly, and testing of engines. These procedures must be followed in order to ensure that the engine is overhauled in a manner which will provide the required reliability.

Table of Limits

The Table of Limits found in the manufacturer's overhaul manual supplies the information necessary to determine the degree of wear for various parts. A portion of the Table of Limits for a Teledyne Continental GTSI0-520 engine is shown in Fig. 12-5. In addition to the Table of Limits, overhaul manuals often include illustrations with reference numbers to identify the location of each measurement.

In the **Table of Limits** it will be noted that three dimensions for each measurement are usually given. These are the serviceable limit, new minimum, and new maximum. If a measurement exceeds the **serviceable limit**, the part is no longer suitable for the engine and must not be used. If the dimension is between the

Ref. No.	Chart No.	Description	Serviceable Limit	New Parts Min.	New Parts Max.
		CYLINDER AND HEAD ASSEMBLY			
1	1	Cylinder bore (lower 4.25 " of barrel) . . . . Diameter:	5.256	5.251	5.253
2	1	Cylinder bore choke (at 5.75" from open end of barrel) . . . . Taper:	− 0.001	0.0010	0.0025
3	1	Cylinder bore out-of-round . . . . :	0.002	0.000	0.001
4	1	① Cylinder bore . . . . Allowable Oversize:	5.261	5.255	5.256
5	1	Cylinder bore roughness . . . . RMS:		15	30
6	1	Intake valve seat insert in cylinder head . . . . Diameter:		0.007T	0.010T
7	1	Intake valve guide in cylinder head . . . . Diameter:		0.001T	0.0025T
8	1	② Exhaust valve guide in cylinder head . . . . Diameter:		0.001T	0.0025T
8	1	③ Exhaust valve guide in cylinder head . . . . Diameter:		0.0012T	0.0027T
9	1	Exhaust valve seat insert in cylinder head . . . . . . Diameter:		0.007T	0.010T
10	1	Intake valve seat . . . . Width:		0.063	0.140
11	1	Exhaust valve seat . . . . Width:		0.063	0.140
		Exhaust valve seat to valve guide axis . . . . Angle:		44°30′	45°00′
		Intake valve seat to valve guide axis . . . . Angle:		59°30′	60°00′
		ROCKER ARMS AND SHAFTS			
12	1	Rocker shaft in rocker arm bushing . . . . Diameter:	0.004L	0.0010L	0.0025L
13	1	Rocker arm bushing in rocker arm . . . . Diameter:		0.002T	0.006T
14	1	Intake valve in guide . . . . Diameter:	0.005L	0.0012L	0.0027L
14	1	Intake valve in guide . . . . Diameter:		0.0010L	0.0036L
15	1	② Exhaust valve in guide . . . . Diameter:	0.007L	0.003L	0.0047L
15	1	③ Exhaust valve in guide . . . . Diameter:		0.0035L	0.0062L
16	1	Intake valve face to stem axis . . . . Angle:		59°45′	60°15′
17	1	Exhaust valve face to stem axis . . . . Angle:		45°00′	45°30′
18	1	Intake valve (Max. tip regrind 0.015″) . . . . Length:	4.789	4.804	4.824
19	1	Exhaust valve (Max. tip regrind 0.015″) . . . . Length:	4.791	4.806	4.826
20	1	Intake and exhaust valve concentricity . . . . :	0.004	0.000	0.002
		Rocker arm foot to end of valve . . . . Clearance:		0.060	0.200
		PISTONS, RINGS AND PINS			
21	1	Piston (bottom of skirt) in cylinder . . . . Diameter:	0.016L	0.008L	0.014L
22	1	First piston ring in groove . . . . Side Clearance:	0.006L	0.0015L	0.0040L
		First piston ring (standard gap) . . . . Tension:	12 lbs.	12.5 lbs.	17.5 lbs.
23	1	Second piston ring in groove . . . . Side Clearance:	0.006L	0.0015L	0.0040L
		Second piston ring (standard gap) . . . . Tension:	8 lbs.	9.5 lbs.	14.5 lbs.
24	1	④ Third piston ring in groove . . . . Side Clearance:		0.0035L	0.0055L
		Third piston ring (standard gap) . . . . Tension:	8 lbs.	9 lbs.	14 lbs.
25	1	⑤ Fourth piston ring in groove . . . . Side Clearance:		0.006L	0.008L
		Fourth piston ring (standard gap) . . . . Tension:	8 lbs.	9 lbs.	13 lbs.
26	1	Fifth piston ring in groove . . . . Side Clearance:		0.006L	0.008L
		Fifth piston ring (standard gap) . . . . Tension:	8 lbs.	9 lbs.	13 lbs.
27	1	First piston ring in cylinder . . . . Gap:		0.033	0.044
28	1	Second piston ring in cylinder . . . . Gap:		0.030	0.046
29	1	④ Third piston ring in cylinder . . . . Gap:		0.030	0.046
30	1	⑤ Fourth piston ring in cylinder . . . . Gap:	0.059	0.033	0.049
31	1	Fifth piston ring in cylinder . . . . Gap:	0.059	0.033	0.049
32	1	Piston pin in piston (Std or 0.0005″ OS) . . . . Diameter:	0.0013L	0.0003L	0.0007L
33	1	Piston pin in connecting rod bushing . . . . Diameter:	0.0040L	0.0022L	0.0026L
34	1	Bushing in connecting rod . . . . Diameter:		0.0025T	0.0050T

FIG. 12-5   Portion of a Table of Limits. *(Teledyne Continental)*

serviceable limit and the **new maximum**, the part can be used but should not ordinarily be reinstalled in an engine being given a major overhaul. The correct dimension for parts installed in a newly overhauled engine should fall between the **new minimum** and new maximum dimensions.

To illustrate the use of the Table of Limits we shall assume that the clearance for the rocker shaft in the rocker-arm bearing is being measured. First we use a telescoping gage in the rocker-arm bearing, as shown in Fig. 12-6, to measure the minimum inside diameter of the bearing. It is well to take the measurement several times using different diameter locations in the bearing. When the telescoping gage has been locked at the correct dimension, we measure the extension of the gage with a micrometer caliper, as shown in Fig. 12-7. The micrometer must not be tightened on the telescoping gage but should touch each end of the gage lightly. The reading of the gage is noted and recorded on the inspection sheet. In this case we shall assume that the reading is 0.7205 in [18.30 mm]. We then measure the rocker shaft in the center at several different diameters with the micrometer and find that the dimension is 0.7195 in [18.275 mm]. We record this dimension in the inspection record and then note the difference between the two dimensions. This is the clearance of the rocker shaft in the bearing, and since it is 0.0010 in [0.0254 mm], we accept it as satisfactory because it is within the limit given in the table.

Some engine parts require that a number of measurements be taken, for example, the cylinders and pistons. All the dimensions shown in the table must be measured and recorded.

It will be observed that some of the dimension figures in the table are followed by the letter L or T. The L stands for *loose* and indicates a clearance between the parts. The letter T stands for *tight* and indicates a "pinch" or "interference" fit. This usually means that one part must be shrunk into the other. In the case of the valve guides, the limits shown are 0.001 to 0.0025 T. That is, the valve guide must have a diameter of 0.001 to 0.0025 in [0.0254 to 0.0635 mm] larger than

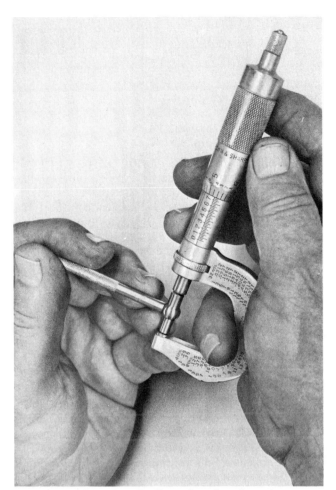

FIG. 12-7 Measuring the dimension of a telescoping gage with a micrometer caliper.

the hole into which it is installed. Installation is usually accomplished by heating the cylinder head and then pressing or driving the cold valve guide into place with the proper type of driving tool.

Manufacturer's Bulletins

As mentioned in an earlier part of this section, the manufacturers of aircraft engines issue bulletins when it becomes apparent that changes or modifications should be made in certain parts of an engine. Such a bulletin may require the installation of a new type of camshaft at the next overhaul, an immediate replacement of all rocker shafts, the drilling of a new oil passage, or any number of operations. The purpose of the change is to increase the reliability and airworthiness of the engine. In many cases the changes described in the bulletins are mandatory, and in other cases the changes are recommended but not required. During the overhaul of an engine, technicians must check all bulletins which have been issued for the make and model of engine with which they are working.

Airworthiness Directives

Airworthiness Directives (ADs) are issued by the FAA to specify changes or modifications required for aircraft, engines, propellers, accessories, or parts. All ADs must be complied with as specified by the directive. These directives are similar in purpose to the manufacturers' bulletins, and in many cases the change

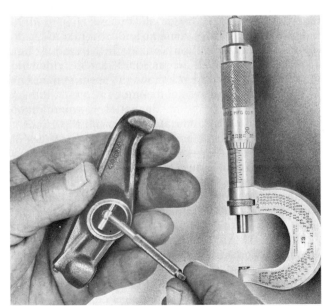

FIG. 12-6 Using a telescoping gage.

required by an AD is also required by a manufacturer's bulletin issued within a short time after the AD is issued. As is the case with the bulletins, all the ADs for an engine must be checked when the engine is overhauled to make sure that required changes and modifications are made.

Engine Specifications and Type Certificate Data Sheets

Each engine installed on an airplane with a standard airworthiness certificate must be type-certificated. Pertinent information regarding certificated engines is listed for each engine on an **Engine Specification Sheet** or a **Type Certificate Data Sheet** issued by the FAA. The information for engines certificated prior to *January 1, 1959*, is given on the Engine Specification Sheet, and information for engines subsequent to that date is shown on the Type Certificate Data Sheet. For engines certificated before World War II, engine specifications are given in a document called the Aircraft Engine Listing.

The overhaul technician should study the specifications given for the engine to be overhauled, paying particular attention to the notes given with the specifications. The notes provide information essential to the proper operation of the engine or its accessories.

Inspection Procedures

The inspection processes and procedures for engine parts may vary somewhat from one overhaul shop to another; however, the primary function of inspection must always be accomplished. This function is to ensure that all parts installed in an engine are airworthy.

Detailed inspection forms should always be used to record the results of inspections, and these forms become a permanent part of the overhaul file. A description of each type of inspection should be recorded together with the results of each inspection.

Inspections fall into three principal categories: (1) visual inspection, which involves a visual examination of parts to note any defects visible to the eye; (2) inspection with special aids, such as magnetic inspection, fluorescent penetrant inspection, dye penetrant inspection, and ultrasonic inspection; and (3) dimensional inspections accomplished with micrometers, gages, and other measuring devices.

Visual inspection is accomplished by direct examination and with the use of a magnifying glass. In each case a strong light should be used to aid in revealing all possible defects. During visual inspection special attention is given those areas of the engine and its parts where experience has shown damage to be most likely. Visual inspection will usually reveal cracks, corrosion, nicks, scratches, galling, scoring, and other disturbances of the metal surfaces.

The various types of damage and defects in engine parts to be noted during visual inspection may be defined as follows:

Abrasion A roughened area where material has been eroded by foreign material being rubbed between moving surfaces.

Brinelling Indentations of a surface caused by high force pressing one material against another. A ball bearing pressed with sufficient force against a softer material will cause a Brinell mark, or indentation.

Burning Damage to a part due to excessive temperature.

Burr A rough or sharp edge of metal, usually the result of machine working, drilling, or cutting.

Chafing The wear caused by two parts rubbing together under light pressure and without lubrication.

Chipping Small pieces of metal broken away from a part as a result of careless handling or excessive stress.

Corrosion Electrolytic and chemical decomposition of a metal, often caused by joining dissimilar metals in a situation where moisture exists. Surface corrosion is caused by moisture in combination with chemical elements in the air.

Crack Separation of metal or other material, usually caused by various types of stress, including fatigue stresses caused by repeated loads.

Dent Similar to a Brinell mark except that a dent is usually found in sheet metal where the metal is deformed on both the side where the force is applied and on the opposite side, usually caused by the impact of a hard object against a surface.

Elongation Stretching or increasing in length.

Flaking The breaking away of surface coating or material in the form of flakes, caused by poor plating, poor coating, or severe loading conditions.

Fretting The surface erosion caused by very slight movement between two surfaces which are tightly pressed together.

Galling The severe erosion of metal between two surfaces which are pressed tightly together and moved one against the other. Galling can be considered a severe form of fretting.

Gouging The tearing away of metal by means of a hard object being moved along a softer surface under heavy force. A piece of hard material caught between two moving surfaces can cause gouging.

Grooving A channel worn into metal as a result of poor alignment of parts.

Growth Elongation caused by excessive heat and centrifugal force on turbine engine compressor and turbine blades.

Indentation Dents or depressions in a surface caused by severe blows.

Nick A sharp-sided depression with a V-shaped bottom, caused by careless handling of tools or parts.

Oxidation Chemical combining of a metal with atmospheric oxygen. Aluminum oxide forms a tough, hard film and protects the surface from further decomposition. However, iron oxides do not form a continuous cover or protect underlying metal; thus, oxidation of steel parts is progressive and destructive.

Peening Depressions in the surface of metal caused by striking the surface with blunt objects or materials.

Pitting, or spalling Small, deep cavities with sharp edges; may be caused in hardened-steel surfaces by high impacts or in any smooth part by oxidation.

Runout Eccentricity or wobble of a rotating part. Eccentricity of two bored holes or two shaft diameters. A hole or bushing out-of-square with a flat surface, usually measured with a dial indicator. Limits stated indicate the full deflection of the indicator needle in one complete revolution of the part or the indicator support.

Scoring Deep scratches or grooves caused by hard particles between moving surfaces. Similar to gouging.

Scuffing, or pickup The transfer of metal from one surface to a mating surface when the two are moved together under heavy force without adequate lubrication.

Magnetic (Magnaflux) Inspection

Magnetic inspection is applied to steel parts which can be magnetized by passing a strong electric current through the parts. As previously explained, an electric current is always accompanied by a magnetic field. When a part is magnetized, the magnetic lines of force travel through the steel. However, if there is a discontinuity such as a crack, the lines of force will tend to leave the metal at the crack, as shown in Fig. 12-8. A small magnetic field is thus created at the surface of the metal, and this field will attract the magnetic particles which are applied to the part during the magnetizing process. The magnetic particles are in the form of a powder, either dry or suspended in a petroleum fluid. When a fluid is used in which the magnetic particles are suspended, the part is magnetized and the fluid is applied to the part by means of a fluid nozzle. As the fluid flows over the part, the magnetic particles will adhere to any area where the magnetic lines of force leave the surface of the metal. The concentration of magnetic particles will reveal the presence of a crack or other possible defect.

Magnetic inspection is accomplished by means of special equipment, such as that manufactured for the purpose by the Magnaflux Corporation. Parts requiring magnetic inspection may be sent to an authorized inspecting agency, or the inspection may be performed by a person who is thoroughly trained in the use of magnetic inspection equipment. Operators must understand the nature of the magnetic fields employed in the magnetic inspection and must use the correct procedures and techniques for each part inspected. They must understand that the magnetic field is perpendicular to the current flowing through the part and that the inspection is not likely to reveal a crack which is parallel to the magnetic field. Because of this, they will realize that it is sometimes necessary to remagnetize a part in order to produce fields in different directions. For example, a crankshaft may be inspected for longitudinal cracks by passing a current through the shaft lengthwise, as shown in Fig. 12-9. In order to detect cracks in other directions it may be necessary to employ a coil or solenoid, as shown in the inspection of the cylinder (Fig. 12-10).

In any case where an operator is using a Magnaflux

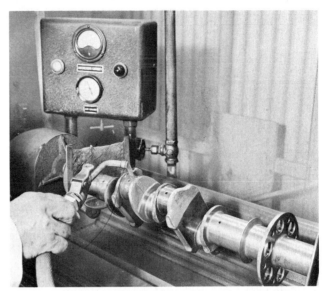

FIG. 12-9 Applying fluid-suspended magnetic particles during the magnetic inspection of a crankshaft.

machine for the inspection of steel parts, a Magnaflux Inspection Data Chart should be consulted to determine the method of magnetization, the current applied, the method of inspection, and the critical areas of the parts being inspected.

Since the current available from a Magnaflux machine may be more than 3000 A, the operator must use care to avoid overheating the parts. If a small part is subjected to 3000 A, it will quickly become heated and the contact areas may be burned. Thus, the part will be rendered useless and must be discarded. If the operator follows the instructions given in the operation manual for the equipment, damage to the parts being inspected will be unlikely.

In addition to the magnetic powder used for detecting defects, Magnaglo powder may be employed. In this case, the inspection after magnetization is made under ultraviolet light. Defects are revealed by the fluorescence of the Magnaglo powder adhering to the crack or other defect.

When performing magnetic inspection on parts that have oil passages or cavities that may collect and be

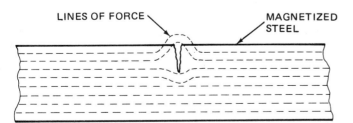

FIG. 12-8 Magnetic lines of force leaving surface of metal.

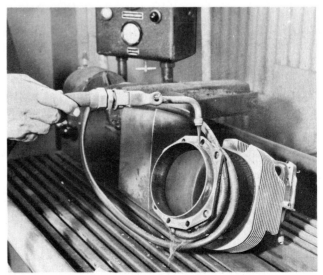

FIG. 12-10 Using a coil for magnetic inspection.

obstructed by the magnetic or fluorescent powder that is used, all the small openings leading to the passages or cavities must be plugged with tight-fitting wood plugs or with hard grease which is soluble in lubricating oil. This is to prevent particles from becoming lodged in places from which they would be difficult to remove and in places that are not subject to visual inspection. After the magnetic particle inspection, the plugs must be removed and the part cleaned thoroughly in solvent and dried with compressed air.

After parts have been subjected to magnetic inspection, they must be demagnetized. If this is not done, the parts will pick up and hold small steel particles which can cause serious damage within the engine during operation. Demagnetization is accomplished by passing the parts slowly through a strong alternating magnetic field and then moving the parts slowly out of the field. The parts will not be demagnetized if current for the demagnetizing coil is turned off while the parts are still in the field. The use of a demagnetizing coil is shown in Fig. 12-11.

Upon completion of the magnetic inspection, the engine parts should be washed in clean petroleum solvent and dried with compressed air. Following this, they should be coated with a thin layer of corrosion inhibiting oil.

Penetrant Inspection

Engine parts made of aluminum alloy, magnesium alloy, bronze, or any other metal which cannot be magnetized are usually inspected by means of a fluoresent penetrant, a dye penetrant, ultrasonic equipment, or eddy-current equipment.

Inspection by means of a penetrant should include the testing of the parts, the tabulation of the nature and extent of the discontinuities found, and the final decision regarding the suitability of the parts for further service. The operator should be a specialist who has been thoroughly trained to evaluate correctly the various indications which may be found.

The process of **fluorescent penetrant inspection** involves the immersion of the thoroughly cleaned parts in the penetrant solution for sufficient time to allow the penetrant to enter all cracks and other discontin-

FIG. 12-11   Demagnetizing a part.

uities. The time may be from 10 to 45 min depending upon the nature of the part. The heating of parts before immersion will increase the sensitivity of the process. After the part is removed from the penetrant, it is placed upon a drain rack for a short time to allow excess penetrant to drain back into the vat, and then the part is washed with warm water and dried.

The development of fluorescent penetrant indications will occur after a relatively short time without any further treatment. However, the development will be hastened and improved by the application of a dry developing powder which is dusted over the surface of the part. All excess powder is brushed off, and then the part is placed under an ultraviolet light in a darkened booth. At all points where the penetrant has seeped out of cracks or other discontinuities, a bright line or spot will appear. It is then up to the inspector to determine if the indication represents a defect which requires that the part be rejected, if the part can be repaired, or if the indication is of no consequence.

Dye penetrant inspection serves the same purpose as fluorescent penetrant inspection; however, it has the advantage of being performed without the special equipment required for the fluorescent penetrant inspection. Dye penetrant inspection is accomplished by application of a highly penetrating dye to the surface of a part. Excess dye is washed off with the specified solvent, and the part is dried. A developer is then brushed or sprayed on the surface of the part and allowed to dry. Since the developer forms a pure white, porous coating on the part, the penetrant will be revealed in a bright red line or spot wherever it has entered a crack or other opening in the surface.

X-ray Inspections

X-ray inspections, though not employed on a routine basis for most engine overhauls, are specified from time to time to detect certain types of metal defects. The x-ray is particularly effective in detecting discontinuities inside castings, forgings, and welds. A powerful x-ray can penetrate metal for several inches and produce an image which will reveal defects within the metal.

Only qualified personnel should attempt to perform x-ray examination of parts and materials. Particular attention must be paid to safety considerations in order to avoid injury from radiation. Qualified x-ray technicians are trained to observe all applicable safety precautions.

Eddy-Current Inspections

Eddy-current inspections are also effective in discovering defects inside metal parts. The eddy-current tester applies high-frequency electromagnetic waves to the metal and these waves generate eddy currents inside the metal. If the metal is uniform in its structure, the eddy currents will flow in uniform pattern and this will be shown by the indicator. If a discontinuity exists, the effect of the eddy currents will be changed and the indicator will produce a reading greater than normal for the particular test. An eddy-current tester is shown in Fig. 12-12.

The use of the eddy-current tester is only of value when operated according to instructions by a techni-

FIG. 12-12 An eddy-current inspection unit.

cian who is well experienced and who knows what types of tests can be made effectively.

Dimensional Inspections

Dimensional inspections are employed to determine the degree of wear for parts of the engine where moving surfaces are in contact with other surfaces. If the wear between surfaces exceeds the amount set forth in the manufacturer's Table of Limits, the parts must be replaced or repaired in accordance with approved methods.

It is well for the inspector to know that parts are usually manufactured with dimensions in sixteenths of an inch in nonmetric engines. This is in accordance with a practice established by the Society of Automotive Engineers many years ago. A correctly dimensioned part will therefore measure $x/16$ in $+ 0.000$ minus an amount up to the tolerance. The decimals for sixteenths of an inch are as follows:

$\frac{1}{16} = 0.0625$	$\frac{7}{16} = 0.4375$	$\frac{12}{16} = 0.7500$
$\frac{2}{16} = 0.1250$	$\frac{8}{16} = 0.5000$	$\frac{13}{16} = 0.8125$
$\frac{3}{16} = 0.1875$	$\frac{9}{16} = 0.5625$	$\frac{14}{16} = 0.8750$
$\frac{4}{16} = 0.2500$	$\frac{10}{16} = 0.6250$	$\frac{15}{16} = 0.9375$
$\frac{5}{16} = 0.3125$	$\frac{11}{16} = 0.6875$	$\frac{16}{16} = 1.0000$
$\frac{6}{16} = 0.3750$		

The dimensions of shafts, crankpins, main bearing journals, piston pins, and similar parts are measured with a micrometer caliper as shown in Fig. 12-13. The micrometer should be equipped with a vernier scale so that measurements can be taken to the nearest ten-thousandth. A micrometer having a ratchet sleeve or stem is recommended, to ensure that the measuring pressure between the anvil and stem is uniform for all measurements.

FIG. 12-13 Measuring a crankpin with a micrometer caliper.

The inside diameters of bushings, bearings, and similar openings are measured with telescoping gages. While the gage is in the opening, it is locked in place to preserve the dimension. The dimension of the telescoping gage is then measured with a micrometer as explained previously. If a hole is too small to receive a telescoping gage, a **small-hole gage** is used. Gages of this type are also called **ball gages**. The ball end of the gage is inserted in the hole and expanded until it fits the hole snugly as shown in Fig. 12-14. It is then removed and measured with a micrometer to obtain the dimension of the hole.

Engine manufacturers often supply **plug gages** or **"go**

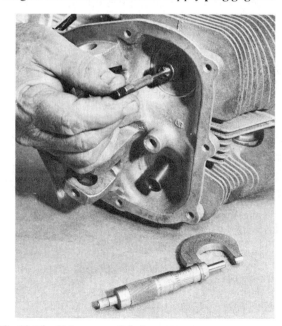

FIG. 12-14 Using a small-hole gage.

and no-go" gages to measure the dimensions of certain holes or openings. These gages are used according to the instructions given by the manufacturer.

Cylinder barrels are measured with a **cylinder bore gage**, as shown in Fig. 12-15. This gage will show the wear, out-of-roundness, and taper. The cylinder bore is measured by sliding the gage from the top to the bottom of the cylinder in the direction of piston thrust and also at 90° to this direction. In this way the out-of-roundness can be checked for the full length of the barrel. Before the gage is placed in the cylinder barrel, the gage needle is set at zero with the basic dimension of the barrel. Deviations from the basic dimension will be shown as the needle moves in a positive or negative direction.

Connecting rods must be measured for bearing and bushing dimensions and also for alignment. The methods used for checking the alignment (twist and convergence of bearing with bushing) will be given later in this section.

The dimensions of small gaps, such as the clearance between piston rings and ring lands, is measured with a **thickness gage**. If a gage of the specified thickness will enter the gap without the use of undue force, it is evidence that the gap is at least as great as the gage dimension. Thickness gages are used to measure side clearances, end clearances, valve clearances, and other similar dimensions.

The **depth gage** is used to provide an accurate indication of the distance between fixed surfaces, such as the distance of the parting surface of an oil pump housing to the end of the gear. The use of a depth gage is shown in Fig. 12-16.

Dial gages are particularly useful for checking the alignment or out-of-roundness of rotating parts. The use of a dial gage in checking the alignment of a crankshaft is described later in this section.

We have mentioned a variety of measuring instruments or gages in this section; however, engine manufacturers often provide special gages to be used with their particular engines. These gages are described in the manufacturer's overhaul manual for the engine concerned and should be used according to the instructions given.

FIG. 12-15   Using a cylinder bore gage.

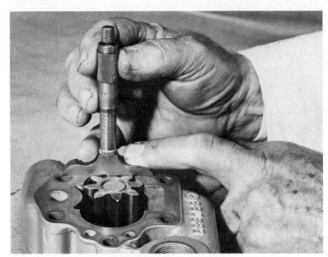
FIG. 12-16   Using a depth gage.

● CYLINDER OVERHAUL

The repair operations required for cylinders depend on the wear and damage which may be found during inspection. Some operations are performed at each overhaul, and others are performed only if required in order to restore the cylinder to an airworthy condition.

If an engine is being overhauled for the first time, it is quite likely that the only operations required on the cylinders will be routine inspections after cleaning, reseating the valves by grinding and lapping, and breaking the glaze on the cylinder walls by means of a cylinder hone.

When cylinders have been overhauled several times, it is probable that the cylinder bore will have to be ground oversize or chrome-plated and ground to standard dimensions. Valve guides will need to be replaced, and if the valve seats have been ground a number of times, it may be necessary to replace the seats. Rocker-shaft bosses which incorporate bushings must be rebushed from time to time to retain the dimension specified in the Table of Limits.

Cylinder Head

After the cylinder has been stripped of all paint and cleaned thoroughly, the cast head should be examined for cracks, pitting, and other damage. Particular attention should be given the area between the valve seats. This area is sometimes subjected to concentrated stresses and cracks are likely to develop. The use of Zyglo or a dye penetrant inspection method is most satisfactory for cast-aluminum cylinder heads.

If cracks are found in the cylinder head, the cylinder must be discarded. A small amount of pitting is not generally serious enough to render the cylinder unsatisfactory for use, but the rough edges of pits should be smoothed by means of a scraper, burnishing tool, or a suitable file.

Cooling Fins

During inspection of the cylinder it may be found that some of the cooling fins have been damaged. The fins on the cylinder head are made as integral parts of the cast-aluminum-alloy head; hence, they are brittle and easily cracked or broken. Broken head fins should be filed smooth at the broken edges to eliminate roughness and sharp edges. If it becomes necessary to cut out a vee notch to stop a head fin crack, a slotted drill bushing to fit over the fin may be used with a $\frac{3}{16}$-in [4.76-mm] twist drill to cut the notch. The apex of the notch and the edges of the cut should be rounded to reduce the possibility of further cracking. If repairs and previous damage have removed as much as 10 percent of the total head fin area, the cylinder assembly is considered beyond repair. In any case, the manufacturer's overhaul manual should be consulted for limitations on fin repair.

Steel barrel fins which have been bent can be straightened with a long-nose plier or a special slotted tool designed for the purpose.

Cylinder Barrel

The condition of the cylinder barrels, pistons, and piston rings is a most vital factor in the performance of a reciprocating engine. If the dimensions and surface conditions of these parts are not satisfactory, combustion gases can escape past the piston rings into the crankcase, and oil from the crankcase can enter the combustion chambers. The engine will lose power and the excess oil in the combustion chambers will foul the spark plugs and cause an accumulation of carbon. For these reasons, it is necessary to overhaul cylinders periodically to ensure that the cylinder and piston assemblies are in condition to perform effectively.

The cylinder barrel, or bore, is inspected for dimensions, corrosion, scoring, and damage. The method for measuring dimensions is illustrated in Fig. 12-15. A small amount of corrosion or scoring can sometimes be removed by means of No. 400 wet-or-dry sandpaper followed by the use of crocus cloth to produce a fine finish. The dimensions of the barrel should be measured after it is sanded; if they are not within limits, the barrel must be honed or ground to a standard oversize. Oversize dimensions are usually 0.010, 0.015, or 0.020 in [0.254, 0.381, 0.508 mm], depending on how much metal must be removed to produce a uniform new surface on all parts of the cylinder wall. The manufacturer's overhaul manual should be consulted to determine the requirements for any particular make and model of engine. It must be noted at this point that the greatest wear of a cylinder barrel takes place near the top because of the high temperatures existent in this area during operation. For this reason, care must be taken to measure the cylinder barrel in this area during inspection.

When regrinding cylinder walls, it is most important to consult the manufacturer's overhaul manual to determine whether the cylinder barrels are **nitrided** (surface-hardened) or chrome-plated. Nitrided barrels usually should not be ground to more than 0.010 in [0.254 mm] oversize because of the danger of grinding through the hardened surface. If worn beyond acceptable limits, chrome-plated barrels should be chemically stripped and replated to standard dimensions.

Grinding and chrome plating should be accomplished by an operator whose process has been approved by the FAA. The advantage of porous chrome or channel chrome plating is that the cylinder barrel will show very little wear between overhauls and will usually remain serviceable with standard dimensions for several thousand hours of operation.

When the cylinders of an engine have been chrome-plated, the piston rings used in these cylinders must not be chrome-plated. The overhaul technician must determine whether cylinders have been chrome-plated and, if so, the piston rings must be unplated cast iron or steel. The plating facility employed to chrome the cylinders can supply the proper piston rings to match its unique plating process.

If it is found in a particular engine that one or more cylinders must be ground to oversize, then all the cylinders in the engine should be given the same treatment. The balance of the crankshaft, piston rod, and piston assembly will be seriously disturbed if oversize pistons are installed in some cylinders and others are standard.

Cylinder grinding is accomplished by means of high-quality precision grinding equipment. The cylinder is firmly mounted on the grinding machine and the grinding wheel is adjusted so that it will take the required cut from the cylinder wall to produce the correct dimension. The operator of the grinding machine must make sure to use the correct type of grinding wheel so that the finish will be as specified. Finish is specified in microinches (μin), usually from 10 to 30. If a surface is ground to 10 μin, the depth of the grinding scratches will not exceed 10 millionths of an inch.

A **cylinder honing machine** is used to produce the final finish of the cylinder walls. The hone usually consists of four high-quality rectangular stones mounted on a fixture. When the cylinder is mounted on the machine and the hone is inserted in a cylinder, the stones make contact with the cylinder walls along their full length. The hone is rotated by means of an electric motor and is moved in and out of the cylinder at a uniform rate while being rotated. The stones remove surface roughness and irregularities, thus producing a smooth crisscross surface which the piston rings can wear in for a good working surface. In some cases, it is recommended that piston rings be lapped to cylinder walls.

Most maintenance technicians are not expected to grind cylinders barrels because this work can be done most satisfactorily in a specially equipped shop set up for this type of work. They must be able to inspect the cylinder barrel and determine what type of treatment is necessary to restore the barrel to satisfactory operating condition. They must check the top, middle, and bottom of the cylinder for out-of-round condition as set forth in the Table of Limits. While doing this, they should also check for the **choke** of the barrel, provided that the barrel is designed with choke. A choked barrel is usually designed with a slightly smaller dimension at the top than at the skirt. Because of higher operating temperatures near the top of the cylinder, a choke bore provides a straight bore during engine operation.

Cylinders in operation for several hundred hours will usually have a step, or ridge, worn near the top of the barrel. This ridge is formed at the point where the top edge of the top piston ring stops when the piston is at top dead center. When a cylinder is reground, the ridge is removed by the grinder; however, if the cylinder barrel is within limits and does not require grinding, the ridge should be removed or smoothed out by hand honing. If the ridge is not removed, the top piston ring will be damaged when the engine is assembled and operated.

When the cylinder barrels of an engine have been ground to a standard oversize or when the barrels have been chrome-plated to standard size, an oversize indicator must be provided on each cylinder. Oversizing aircraft engine cylinders is limited because of the relatively thin cylinder walls. The color and location of the indicator may be specified by the manufacturer. Formerly the coding was as follows: 0.010 in oversize—green; 0.015 in oversize—yellow; 0.020 in oversize—red; chrome-plated standard—orange. These codings were painted just above the base flange of the cylinder. This location is not recommended for opposed engines because the paint at the base of the cylinder interferes with proper cooling.

Studs

The studs by which the exhaust pipes are attached to the cylinder head are usually subject to severe corrosion because of heat and the chemical effects of the exhaust gases. It is therefore often necessary to replace these studs. During inspection the threads of the studs should be carefully examined, and if appreciable erosion is noted, the studs should be replaced. Stud replacement is described in the section covering crankcase repair.

Skirt and Flange

During the handling of cylinders, care must be exercised to avoid damaging the skirt. Approved practice calls for mounting the cylinder on a wooden cylinder block when it is not being worked on. If cylinder blocks are not available, the cylinder may be placed on its side on a wooden rack. If pushrod housings are attached, the cylinder must be placed on its side in a manner to avoid putting a stress on the housings. The cylinder must not be lifted or carried by grasping the pushrod housings.

If the cylinder is handled properly, there is no reason why the skirt should become damaged. Usually such damage is caused by carelessly allowing the skirt to strike a hard metal object. If a small nick or scratch should be found on the skirt, it can be removed by careful stoning and polishing.

The cylinder mounting flange must be examined for cracks, warping, damaged bolt holes, and bending. Warpage of the flange can be checked by mounting the cylinder on a specially designed surface plate and using a thickness gage to determine the amount of warp. Another common method is to place a straightedge across the flange at locations about 45° apart and check the gap under the straightedge with a thickness gage. A very small amount of warp can be removed by lapping the bottom of the flange on the cylinder surface plate, otherwise, any appreciable defects require that the cylinder be discarded.

The cylinder flange should be given an especially careful examination if any of the cylinder hold-down bolts or nuts were found to be loose at the time of disassembly. A loose hold-down nut or bolt will cause exceptional stresses to be imposed on the flange and also on the crankcase.

Valve Guides, Seats, and Valves

If inspection reveals that valve guides are oversize or damaged in any way, it is necessary that they be replaced. Valve guides should be removed according to the manufacturer's instructions, and the method usually requires driving with a special piloted drift or using a valve-guide puller. Since the valve guide has a flange on the outside end, it is necessary to drive it from the inside of the cylinder. Before it is driven or pulled, the inner end of the guide should be cleared of all hard carbon to prevent scoring of the guide hole. Some operators recommend heating the cylinder head to about 450° F [232°C] before removing the valve guides because this tends to expand the valve-guide hole and permit easier removal. In all cases the cylinder should be properly supported on a cylinder-holding fixture while valve-guide removal and replacement are being done.

If equipment is not available to pull or drive the valve guide from inside the cylinder, it is possible to cut away the outer end and flange of the guide by means of a spot facer and then drive the guide to the inside of the cylinder. Care must be taken to avoid cutting into the guide flange seat.

After the valve guides are removed, the guide holes must be inspected for scoring, roughness, and diameter. If the hole size is not within specified limits, it will be necessary to ream or broach it to an oversize dimension. The actual size of the hole should be approximately 0.002 in [0.051 mm] smaller than the outside diameter of the guide to be installed. The Table of Limits for typical opposed engines establishes limits of 0.001 to 0.0025 in [0.025 to 0.064 mm] interference (T) fit. Valve guides are made in oversizes of 0.005 and 0.020 in [0.127 and 0.508 mm]. Valve guides and valve-guide holes are measured as shown in Fig. 12-14.

Guide holes which are within the required dimensional limits and are not scored or damaged in any way do not require repair. It is merely necessary to install new valve guides by driving with a suitable installing tool. The guide must be carefully aligned with the hole before driving. Valve seats must be reground *after* replacing valve guides in every case.

Valve seats can usually be repaired by grinding with specially designed seat grinding equipment. A typical valve-seat grinder is shown in Fig. 12-17. As can be observed in the illustration, the valve-seat grinder consists of an electric motor equipped with an angle drive to permit grinding of the seats at the angle in which they are installed in the cylinder. The grinding wheel is mounted on a collet equipped with a pilot which fits into the valve guide. The pilot holds the grinding wheel in exact alignment with the valve seat. The valve seat is ground by inserting the pilot of the grinder stone on the seat. The stone angle must be correct for the particular seat being ground. Some engines have a 30° seat for the intake valve and a 45° seat for the exhaust valve. The 30° intake-valve-seat angle provides improved gas-flow characteristics. After the grinding stone is in place, the end of the drive unit is placed in the socket and the seat is ground.

COLLET

PILOT

GRINDING STONES

FIG. 12-17 Valve-seat grinder.

When grinding valve seats, the operator must be sure to have the correct stone (angle, hardness, grit size) and to remove only sufficient material to provide a smooth seat of the correct width. The width of the seat is adjusted by using a 15° stone to reduce the outer diameter and a 75° stone to increase the inside diameter. The width and diameter of the seat after it is ground must be within the dimensions specified by the manufacturer.

If a valve seat has been reground so many times that the entire face of the 15° narrowing wheel must be brought into contact with the seat in order to grind to the required dimension, the seat is beyond limits and must be replaced.

A valve seat is removed by cutting it out with a special counterbore suitable for the purpose while the cylinder is mounted in a fixture which holds it at the proper angle under the spindle of a drill press. The drill press is used to rotate the counterbore, thus cutting the valve seat from its recess. This is a precision operation and must be performed by an experienced operator.

After the seat is removed, the inside diameter of the valve-seat recess must be measured to determine which oversize seat is to be installed. The recess must then be cut to the correct oversize dimension as specified in the Table of Limits. This is accomplished by means of a valve recess cutter available from the manufacturer.

The cylinder is then heated to 600 to 650°F [316 to 343°C], and the new valve seat is driven into the recess with the replacement drift. The manufacturer's manual must be consulted to ascertain the correct temperature to be used. After replacement the new valve seats are ground to the proper face dimension as previously explained.

Valves are inspected for burning, erosion, stretch, diameter of stem, cracking, scoring, warping, and thickness of the head edge. Because of the high temperatures to which exhaust valves are exposed, these valves often need to be replaced even though the intake valves are still in serviceable condition. The valves are subjected to many thousands of recurring tension loads, because the valve is opened by cam action and closed by the valve springs. This, of course, tends to stretch the valves, and the effects of stretching are commonly noted with exhaust valves. Figure 12-18 shows how a valve is examined for stretch with a valve stretch gage or contour gage.

The valves are refaced by means of a standard valve

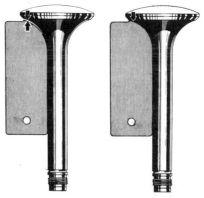

FIG. 12-18 Inspecting valves for stretch with a contour gage.

refacing machine. This operation is illustrated in Fig. 12-19. The refacing machine rotates the valve while the grinding wheel is moved back and forth across the face. The surface of the grinding wheel should be dressed to a smooth finish before the operation is started.

Before the valve is inserted in the chuck of the machine, the stem should be examined for cleanliness and alignment. If the stem is not perfectly smooth and straight, the valve head will wobble and the face will not be true. The operator must be sure to use the proper type of grinding wheel and set the machine for the correct grinding angle. If the manufacturer's instructions call for an **interference fit,** the valve face angle will be as much as 1° less than the corresponding seat in the cylinder. An interference fit creates a thin line contact between the valve face and seat and ensures positive seating of the valve after lapping.

Lapping Valves

After valves and seats are ground, they are **lapped** to provide a gas-tight and liquid-tight seal. Each valve is placed in its particular seat, one at a time, and the valve face is rotated against its seat with an approved lapping compound until there is a perfect fit between the valve and seat. All lapping compound is then carefully removed from both the seat and valve. The valves are then placed in a numbered rack to make certain that they will be installed in the correct cylinder.

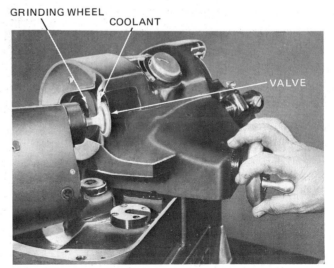

GRINDING WHEEL

COOLANT

VALVE

FIG. 12-19 Refacing a valve.

311

Lapped valves should be inspected to see that the face and seat of each valve makes proper contact according to the limits specified in the overhaul manual. Figure 10-20(A) shows the lapped area of a properly ground valve. The lapped area is midway between the edge of the head and the bottom of the face. The area should have a frosty gray appearance. In Fig. 12-20(B) the lapped area is too near the edge of the face at the top. The valve shown at (C) has been ground too much and the valve edge has been thinned to a **feather edge.** This valve would burn in operation and cause preignition. The burning would soon cause the valve to leak and eventually fail completely.

Testing Valve Springs

Valve springs are visually inspected for evidence of overheating and for pitting caused by corrosion. They are then placed in a valve-spring compression tester, as shown in Fig. 12-21, and the compression load is checked when the spring is compressed according to instructions. The outer spring will normally have a greater compression strength than the inner spring.

Installation of Valves

Valves are installed by inserting the stems through the valve guides from inside the cylinder and then holding them in place while the cylinder is placed over a cylinder block. The block bears against the valve heads and holds the valves on the valve seats. The lower spring seat is installed over the valve stem, and the valve springs are placed on the seat. Assembly instructions should be checked to see if there is a difference between the ends of the springs. The springs are compressed by means of a valve-spring compressor. The

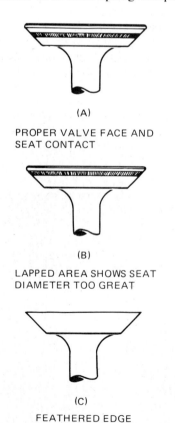

(A)

PROPER VALVE FACE AND
SEAT CONTACT

(B)

LAPPED AREA SHOWS SEAT
DIAMETER TOO GREAT

(C)

FEATHERED EDGE

FIG. 12-20   Correct and incorrect valve face conditions.

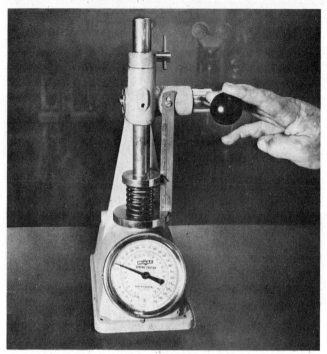

FIG. 12-21   Testing the compression of a valve spring.

valve retaining keys are installed in the groove around the valve stem, and the spring compressor is then released.

Rocker Arms, Shafts, and Bushings

Valve rocker arms are inspected for wear on both ends, cracks, dimension of bushing, corrosion, and any other condition which may render them unsuitable for use. Frequently the rocker arms need no repair, and sometimes the only repairs required are replacement and reaming of the bushing. These are accomplished by means of a suitable arbor and arbor press by which the old bushing is pressed out and a new bushing is pressed in. The bushing hole should be examined for condition before the new bushing is pressed in. Special attention must be given the position of the oil hole in the bushing to make sure that it is aligned with the oil hole in the rocker arm.

Rocker-shaft bushings in the cylinder head must be replaced if they are worn beyond limits. If the bushing is held in place with a dowel pin, the pin must be drilled out before the bushing is removed. Removal of the bushing is accomplished with a drift or arbor. The cylinder must be properly supported while removing the rocker-shaft bushings in order to prevent damage.

After the rocker-shaft bushings have been removed, each bushing hole must be checked for size with either a telescoping gage or a special plug gage. If the bushing hole dimension is above the maximum limit, it is necessary to install an oversize bushing. When it has been decided which oversize bushing is required, the hole is reamed to the correct size for the bushing. The new rocker-shaft bushing is installed with an installation drift or similar tool in accordance with the manufacturer's instructions. If the bushing is held in place with a dowel pin, it will be necesary to drill a new hole in the bushing for the dowel pin and then install a pin of the correct size. Special instructions relating to this operation are provided by the manufacturer for the engines in which dowel pins are employed.

After rocker-shaft bushings are installed in the cylinder head, it is necessary to check them for dimension and ream them to size if required. The final cut should be made with a finish reamer to produce a very smooth surface.

● PISTONS AND RINGS

Pistons

Pistons are inspected for cleanliness, wear, scoring, corrosion, cracks, and any other apparent damage. Oil holes are checked for freedom from carbon. Wear is checked by means of a large micrometer caliper, taking the measurements specified in the Table of Limits. The measurements taken on the pistons are compared with cylinder bore measurements to determine piston clearances.

If a piston is found to have deep scoring on the sides, it must be discarded. Very shallow scoring is not cause for rejection and may be left on the piston. No attempt should be made to remove light scoring with sandpaper or crocus cloth because this may change the contour of the skirt. Light scoring can be reduced by machining only enough to remove the metal that has piled up.

The piston-ring lands are inspected for cracks or for scratches which may lead to cracks. The ring lands must be smooth, and the grooves free from carbon or damage of any kind.

The pin bore is checked with a telescoping gage, and the dimension is checked against the corresponding piston pin to determine the clearance. Piston pins are usually dimensioned to provide a push fit.

Piston Rings

New piston rings are installed when the engine is overhauled. These rings must be approved for the installation and must be inspected for fit. The end clearance (gap) is checked by inserting the ring in the skirt of the cylinder and pushing it in with the head of a piston to ensure that the ring is square with the bore. At a point inside the cylinder even with the flange, the gap of the piston ring is measured as shown in Fig.

12-22. This gap must be within the tolerance given in the Table of Limits. The piston-ring gap is necessary to prevent seizure of the rings in the cylinder as the pistons and rings expand with the high temperatures of operation.

Piston-ring tension is measured at a point 90° from the gap when the ring is compressed to the normal gap. This tension must conform to the approved specifications.

After the piston-ring gap is measured, the rings are installed on the pistons in the proper grooves for inspection of side clearance. The installation of the rings may be accomplished with a ring expander, or it can be done by hand. Some operators prefer hand installation because they feel that it is less trouble. The side clearance of the piston rings is measured with a thickness gage as shown in Fig. 12-23. The specified ring clearance is necessary in order to ensure free movement of the rings in the ring grooves and a free flow of oil behind the rings. The correct fitting of piston rings to pistons and cylinders and the approved type of piston rings installed in an engine are critical because the wear between piston rings and upper cylinder walls is usually greater than the wear in any other part of the engine.

With respect to the installation of piston rings, it is especially important that the technician observe the instructions of the manufacturer. A particular make and model of engine requires a certain combination of piston rings as set forth in the parts manual or list for the engine. It is vital that some piston rings be installed top side up. The word "top" is usually etched on the side of the ring, indicating that this side of the ring must be nearest the top of the piston. Some piston rings which are symmetrical in cross section may be installed with either side up.

Some manufacturers recommend **lapping** for cast-iron piston rings installed in a nonplated cylinder. This is accomplished by installing the piston rings in the piston for a particular cylinder and then inserting the piston with rings into the appropriate cylinder barrel. The proper grade of lapping compound is first applied to the walls of the cylinder barrel. The piston is then

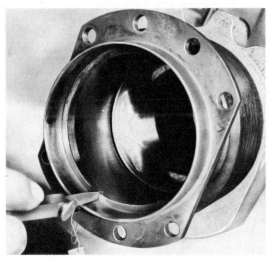

FIG. 12-22 Measuring piston-ring gap.

FIG. 12-23 Measuring piston-ring side clearance.

moved in and out of the cylinder by hand while being rotated. The operation is continued until the surfaces of all rings have the same appearance, indicating that they have made smooth contact with the cylinder walls. After lapping the piston rings, all traces of lapping compound must be removed from both the piston and the cylinder. This is accomplished by means of water or petroleum solvent, depending on the type of compound used. After cleaning, the piston and cylinder should be coated with preservative oil, working it well down into the piston-ring grooves and behind the rings.

● CRANKSHAFTS

The crankshaft of an engine is without question one of the most critical parts. The dimensions of the journals, the balance, and the alignment of the shaft must be within tolerances, or the engine will vibrate and may ultimately fail. It is easily understood that the crankshaft is subjected to extremely rigorous treatment during the operation of an engine because it must bear the constant hammering of the connecting rods as they transfer the force of the piston thrust to the connecting rod journals. The inspection and repair of the crankshaft must therefore be accomplished with great care and precision if it is to perform reliably for the hundreds of hours between overhauls.

Magnetic Inspection

After the crankshaft is disassembled and cleaned, it should be inspected by the magnetic method. It may be magnetized radially by mounting it lengthwise in the Magnaflux machine and magnetized longitudinally by means of a heavy coil. To accomplish a thorough inspection, the shaft should be magnetized by both methods. During the inspection, the radii of crankpins and main bearing journals should be checked carefully for cracks. In many cases these cracks are extremely fine and can be detected only by the magnetic method. The entire shaft should be examined from end to end for cracks or other possible defects. After the magnetic inspection is completed, the shaft must be demagnetized.

Refer back to Figs. 12-9 to 12-11 in the accompanying text for a description of magnetic inspection and the subsequent demagnetization of a part.

Crankpins and Main Journals

Crankpin journals and main bearing journals are checked for dimension with a micrometer caliper. The "mike" should be placed at four different diameters of the journal to make certain that the measurement is correct. If the journal is found to be oval (out-of-round) more than 0.0015 in [0.0381 mm], the crankshaft must be discarded or the journals must be reground to not more than 0.010 in [0.254 mm] undersize and renitrided. During the measurement of the crankpins and main journals, the dimensions must be recorded in the overhaul file. These same dimensions are compared with the bearing measurements at a later time to determine bearing clearances.

During the inspection of crankshaft journals, if a small amount of roughness is noted, the journal surface should be smoothed with crocus cloth. It is best to do this while the shaft is being rotated slowly in a lathe.

Crankshaft journals should be reground undersize by an experienced operator in a shop which is properly equipped for grinding and renitriding. It is recommended that shafts which require regrinding be returned to the manufacturer for this repair.

When a crankshaft is reground, the exact radii of the original journal ends must be preserved in order to avoid the possibility of failure during operation. If a small ledge or step is left in the metal at the radius location, the shaft is likely to develop cracks which may lead to failure of the journal.

Crankshaft Alignment

Crankshaft alignment is checked by mounting the shaft on vee blocks placed on a level surface plate and rotating the shaft while a dial gage is used to measure the **runout.** This operation is shown in Fig. 12-24. The crankshaft runout should be checked at the center main journals while the shaft is supported at the thrust and rear journals. It should also be checked at the propeller flange or at the front propeller-bearing seat. Permissible runout tolerances are given in the Table of Limits.

A crankshaft alignment and runout check should always be performed on an engine that has undergone a sudden stoppage, such as that caused when the propeller of the aircraft strikes the ground or a solid object. To perform the check with the engine still installed in the aircraft, the propeller is removed and a dial indicator is installed firmly on the front of the engine. The "finger" (actuating rod) of the dial indicator is placed so that it rests on the smooth part of the propeller shaft forward of the splines. The shaft is then rotated and any eccentricity of the shaft will be indicated on the dial. If, during the runout check, a crankshaft is found bent, straightening is not recommended.

Counterweights

The crankshafts of many engines are dynamically balanced by means of counterweights and dynamic balances mounted on extensions of the crank cheeks. The size and mounting of these counterweights and balances are such that they dampen out (reduce) the torsional vibration which occurs as the result of the connecting-rod thrust. The proper method for removing and replacing the counterweights differs with various

FIG. 12-24 Checking crankshaft alignment (runout).

types and models of engines; hence, the overhaul technician should make sure that he or she follows the exact procedure outlined in the overhaul manual for the model of engine upon which he or she is working. Refer to Chap. 3.

Sludge Chambers and Oil Passages

Some crankshafts are manufactured with hollow crankpins which serve as sludge removers. Drilled oil passages through the crank cheeks carry the oil from inside the main journals to the chambers in the crankpins. The sludge chambers may be formed by means of spool-shaped tubes pressed into the hollow crankpins, or they may be formed by sludge plugs mounted in each end of the crankpin.

The sludge chambers of a crankshaft must be disassembled and cleaned during overhaul to remove the soft-carbon sludge which has collected. If the sludge chambers are formed by means of tubes pressed into the hollow crankpins, the overhaul technician must make certain that the tubes are reinstalled correctly to avoid covering the ends of the oil passages.

The oil passages in the crankshaft should be cleaned at the time that the shaft was originally cleaned; however, they should be checked again before the crankshaft is declared ready for reassembly in the engine. Soft copper wire passed through the passages will verify that they are clear of dirt or other obstruction.

● CAMS AND VALVE TAPPETS

Camshaft Alignment

The camshaft of an opposed or in-line engine can be checked for alignment in the same manner as that described for a crankshaft. The shaft is mounted in the vee blocks at the end bearings, and the runout is measured at the center bearings.

Camshaft Journals

Camshaft journals are measured with a micrometer caliper, and the measurements are recorded for later comparison with the measurement of the camshaft bearing bores in the crankcase. The clearances are specified in the Table of Limits. The camshaft journals should be examined for scratches, scoring, and other damage. Defects which are not apparent to the unaided eye will usually be revealed during the magnetic inspection.

Cam Surfaces

The surfaces of the cams must be examined with a magnifying glass to detect incipient failure. Any obvious wear, spalling, pitting surface cracks, or other damage is cause for rejection of the camshaft. If examination with a magnifying glass reveals small pores in the surface, this also requires that the shaft be discarded.

Cam Gear

The cam gear is inspected for wear, cracks, and other damage. If any defects are found, the gear must be replaced. At the time of assembly the cam-gear backlash with the crankshaft gear is checked. If the backlash exceeds the value given in the Table of Limits, the gears must be replaced.

Valve Tappets

Valve tappets are inspected for wear, spalling, pitting, scoring, and other damage. Defective tappets must be replaced. The OD of the tappet body is compared with the ID of the bore in the crankcase to determine the clearance. This value must be within approved limits.

Hydraulic lifters or tappets are disassembled and inspected for wear. If any appreciable wear is noted, as evidenced by a feathered edge of worked metal at the shoulder in the tappet body, the entire tappet assembly must be discarded. The hydraulic tappet cylinder and plunger assembly must be checked to see that no burrs or binding exist and that the ball check valve is not leaking. A leaking check valve can be tested as follows: Make sure that the cylinder and plunger assembly are dry, and hold the lifter cylinder between the thumb and middle finger in a vertical position with one hand. Then place the plunger in position so that the plunger just enters the lifter cylinder. When the plunger is depressed quickly with the index finger, it should return to approximately its original position. If the plunger does not return but remains in a collapsed position, the ball check valve is not seating properly. The cylinder should be recleaned and then checked again. If the check valve still does not seat, the unit is defective and the entire cylinder and plunger assembly must be discarded.

Hydraulic cylinder and plunger parts cannot be interchanged from one assembly to another, and if a part of any assembly is defective, the entire assembly is discarded. The cylinder and plunger (hydraulic unit) assemblies can be interchanged in lifter bodies, provided that the clearances are within the approved limits.

● CONNECTING RODS

The inspection and repair of connecting rods involve (1) visual inspection for nicks, cracks, bending, corrosion, and other damage; (2) magnetic inspection; (3) checking of alignment for parallelism and convergence between the bearing end and the piston end; (4) rebushing; and (5) replacement of bearings. Visual inspection is accomplished in the usual manner, directly and with a magnifying glass. A rod which obviously has been bent or twisted must be discarded. Connecting rods should be weighed to see that all weights are the same within $\frac{1}{4}$ oz [7 g].

The requirements for a connected rod to be used in a Continental O-470 engine are shown in Fig. 12-25. This drawing illustrates the dimensions as well as the tolerances for twist and convergence.

Checking Twist and Convergence

The twist of a connecting rod is checked by installing push-fit arbors in both ends and supporting the rod by means of the arbors on parallel steel bars resting on a surface plate. Measurements are then taken with a thickness gage at each supporting point to determine the amount of twist. This check is shown in the pho-

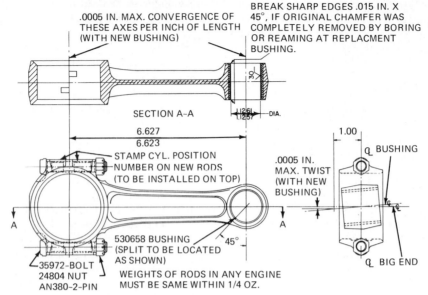

.0005 IN. MAX. CONVERGENCE OF THESE AXES PER INCH OF LENGTH (WITH NEW BUSHING)

BREAK SHARP EDGES .015 IN. X 45°, IF ORIGINAL CHAMFER WAS COMPLETELY REMOVED BY BORING OR REAMING AT REPLACMENT BUSHING.

SECTION A-A

6.627
6.623

STAMP CYL. POSITION NUMBER ON NEW RODS (TO BE INSTALLED ON TOP)

.0005 IN. MAX. TWIST (WITH NEW BUSHING)

530658 BUSHING (SPLIT TO BE LOCATED AS SHOWN)

35972-BOLT
24804 NUT
AN380-2-PIN

WEIGHTS OF RODS IN ANY ENGINE MUST BE SAME WITHIN 1/4 OZ.

1.00

BUSHING

BIG END

FIG. 12-25 Dimensional requirements for a connecting rod.

tograph of Fig. 12-26. The clearance between the support bar and the arbor in the small end of the rod is measured and divided by the number of inches between the center points of the supporting steel bars. The twist must not exceed 0.0005 in [0.0127 mm] per inch of distance between the support point center lines.

The arbor in the large end of the connecting rod may be supported by matched vee blocks rather than on the steel bars. Either method will give the desired results.

To check convergence, the difference in the distance between the arbors is measured at a given distance on each side of the connecting-rod center. This is accomplished by installing a precision measuring arm with a ball end on one end of the arbor into the small end of the connecting rod. The distance of the measuring arm from the center line of the rod is noted, and the measuring arm is adjusted so that the ball just touches the arbor in the big end of the rod. The measuring arm is then moved to the opposite end of the arbor, and the difference in distance is checked with a thickness gage as shown in Fig. 12-27. If the distances are checked at points 3 in [7.62 cm] from the center line of the connecting rod, the total distance between the measuring points is 6 in [15.24 cm]. Therefore, if the difference in the distances between the arbors on each side of the

connecting rod is less than 0.003 in [0.076 mm], the convergence is within limits.

Bushing Replacement

If the piston-pin clearance in the connecting-rod bushing is excessive, it is necessary to replace the bushing. This is accomplished by pressing out the old bushing and pressing in a new bushing with an arbor press. The new bushing should be lubricated before it is installed and must be perfectly parallel with the bore into which it is pressed. After a new piston-pin bushing is installed in the connecting rod, it is usually necessary to ream or bore the bushing to the correct size. This operation is particularly critical because the alignment of the bushing must be held within 0.0005 in [0.0129 mm] per inch as previously explained. The bushing is usually bored with special equipment designed for this purpose. One such device is the **hydrobore** shown in Fig. 12-28. For this method the connecting rod is mounted on a faceplate, as shown, so that it is exactly perpendicular to the axis of cutter bar rotation. The boring tool is then brought into contact with the inner surface of the bushing while turning at

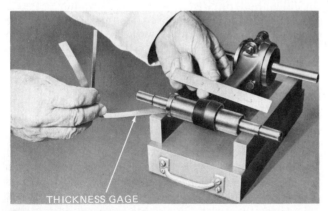

FIG. 12-26 Checking the twist of a connecting rod.

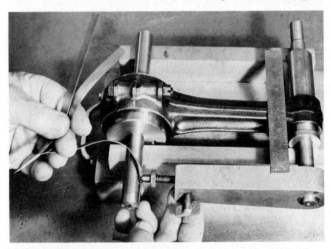

FIG. 12-27 Checking the convergence of connecting-rod ends.

FIG. 12-28 A hydrobore is used to bore connecting-rod bushings.

a slow rate. A small cut is taken with a smooth cutting tool for as many passes as necessary to provide the desired dimension.

Connecting-Rod Bearings

New connecting-rod bearing inserts are installed when an engine is overhauled. Bearing measurements are taken with the bearings installed in the connecting rods before the rods are assembled to the crankshaft, with the connecting-rod bolts reasonably tight. These measurements are then checked against the dimensions of the crankpins previously recorded. The operator must check that the original location of each connecting rod is maintained during the clearance check and at final assembly.

Replacements

At the time of engine overhaul, all connecting-rod bolts, nuts, washers, and cotter pins are replaced. As mentioned above, new connecting-rod bearings are also installed.

● CRANKCASE

Crankcases are inspected for cracks, warping, damage to machined surfaces, worn bushings and bearing bores, loose or bent studs, corrosion damage, and other conditions which may lead to failure in service. Two-piece crankcases are manufactured with matched parts; so if one half must be discarded, the entire crankcase is replaced.

Inspection

Since the majority of crankcases are made of aluminum-alloy castings, a fluorescent penetrant or dye penetrant provides a suitable method for detecting cracks, porosity, and similar defects. These inspections should be made by a person experienced in the use of the method selected.

Visual inspection with a 10-power magnifying glass over the entire surface of the crankcase will reveal most of the defects mentioned previously and will also lead to detection of all but the very fine cracks. If minor nicks and abrasions are discovered, they can usually be removed by filing, stoning, and with crocus cloth.

Replacement of Studs

Studs or stud bolts are metal pins threaded on each end and used for the attachment of parts to one another. One end is provided with coarse (NC) threads and the other with fine (NF) threads. The coarse-threaded end is designed to be permanently screwed into a casting such as the crankcase, and the fine-threaded end has a nut installed for holding an attached part.

Studs which are damaged, bent, or broken must be removed and replaced with new studs. Studs which are not broken can be removed with a **stud remover** or with a small pipe wrench. The stud should be turned slowly to avoid heating the casting. Broken studs which cannot be gripped by a stud remover or pipe wrench are removed by drilling a hole in the center of the stud and inserting a **screw extractor.** The screw extractor may have straight splines or helical flutes. The Easyout extractor has helical flutes, as shown in Fig. 12-29.

After a stud is removed, the coarse-threaded end should be examined to determine whether the stud is standard or oversize. This is indicated by machined or stamped markings on the coarse-threaded end. Identification markings are shown in Fig. 12-30. The replacement stud should be one size larger than the stud removed. If the threads in the case are damaged, or if the old stud was maximum oversize, it will be necessary to retap the hole with a **bottoming tap** and install a Heli-Coil insert for a standard size stud.

Before the new stud is installed, the coarse threads should be coated with a compound specified by the manufacturer. This compound serves to lubricate and protect the threads and may also seal the threads to prevent the leakage of lubricating oil from inside the engine. The stud should be installed with a **stud driver** and screwed into the case a distance specified by the manufacturer.

Heli-Coil Inserts

A **Heli-Coil insert** is a helical coil of wire having a diamond-shaped cross section. When this coil is prop-

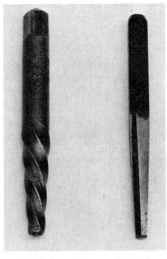

EASYOUT STRAIGHT FLUTED SCREW EXTRACTOR

FIG. 12-29 An Easyout and a straight-fluted screw extractor.

| Typical part no. | Oversize on pitch dia. of coarse thread, in. | Optional identification marks on coarse thread end | | Identification color code |
		Stamped	Machined	
XXXXXX	Standard	None		None
XXXXXXP003	0.003			Red
XXXXXXP006	0.006			Blue
XXXXXXP009	0.009			Green
XXXXXXP007	0.007			Blue
XXXXXXP012	0.012			Green

FIG. 12-30   Identification data for oversize studs.

erly installed in a threaded hole, it provides a durable thread to receive standard studs or screws.

When a threaded hole has been damaged or oversized beyond accepted limits, it can be repaired by retapping and installing a Heli-Coil insert. The tap and installing tools are provided by the manufacturer of the Heli-Coil inserts and must be used according to instructions. The inserts are used in cylinders for spark-plug holes and stud holes and in other parts for stud holes and screw holes.

Heli-Coil inserts may be removed and replaced if they become worn or damaged. The special extracting tool provided by the manufacturer must be used for removal. Heli-Coil inserts and tools are illustrated in Fig. 12-31.

Measuring Bearing Bores

During the overhaul of an engine it is important to determine the clearances of crankshaft main bearings and camshaft bearings. This is accomplished by temporarily assembling the crankcase with the main-bearing inserts installed. The through-bolts which hold the crankcase together are installed and tightened sufficiently to remove all clearance between the parting surfaces. The main bearings and camshaft-bearing bores

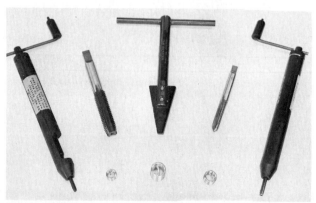

FIG. 12-31   Heli-Coil inserts and tools.

are then measured, and the measurements are compared with the corresponding crankshaft and camshaft measurements to determine the clearances. These clearances are compared with the Table of Limits to ensure that they are satisfactory.

The bores which receive the valve tappet bodies and all other bearing bores in the crankcase section are measured and compared with the parts or shafts to be installed to determine that clearances are within limits. All clearances are recorded in the overhaul file for the engine.

● ACCESSORY SECTION

Since the necessary sections for various engines differ considerably for various models and types of engines, specific instructions for overhaul will not be given in this text. It is stressed, however, that the inspections, repair operations, and assembly operations described in the manufacturer's overhaul manual must be followed carefully.

Gears

Accessory gears are examined for wear, overheating, scoring, pitting, and alteration of the tooth profile. Any appreciable damage is cause for rejection of the part. Steel gears and gear shafts can be magnetically inspected for cracks in accordance with approved methods. After inspection the parts must be demagnetized.

Shafts and Bushings

The fits and clearances of gears on shafts and of shafts in bushings must be checked to see that they are within the tolerances specified in the Table of Limits. If shaft clearances in bushings are excessive, it is necessary to remove, replace, and ream new bushings. In each case these operations must follow the manufacturer's directions. In many cases of bushing replacement it is necessary to drill out dowel pins which hold the bushings in place. This is a critical operation and must be executed with great care to avoid damage to the engine. Manufacturers often supply drill jigs or bushings to guide the drill, thus reducing the possibility of drilling at the wrong angle or in the wrong position. Another precaution which must be emphasized is the alignment of oil holes in bushings. If a bushing is installed where an oil passage emerges, the position of the bushing must be checked as it is pressed into place, and the oil hole in the bushing must be checked after installation to see that it coincides with the hole in the bore.

Treatment of Interior Engine Surfaces

The interior surfaces of engine crankcases, accessory cases and similar parts are provided with a protective coating to eliminate corrosion damage. During the overhaul of an engine it is important that such coatings be examined and restored if necessary. One commonly employed coating is Alodine, and it is easily restored by following the manufacturer's instructions. The restoration process involves cleaning the bare aluminum thoroughly with an approved, nongreasy solvent and then applying the corrective solution as instructed. The solution is allowed to remain for a few minutes until the desired color is obtained. The area is then washed thoroughly and dried. In all cases, the manufacturer's overhaul instructions should be followed for each make and model of engine.

● ENGINE ASSEMBLY

The assembly of an engine must follow a sequence recommended for the particular model of engine being assembled. Since the "core" of an engine is the crankshaft, assembly usually starts with the installation of connecting rods on the shaft.

Before assembly is started, the engine parts are checked for cleanliness because dust and other foreign matter will often collect during the inspection, repair, and storage of parts. Parts may be washed with a petroleum solvent and then dried with compressed air in order to remove accumulated dirt. They should then be coated with a corrosion inhibiting oil.

Use of Safety Wire

During the final assembly of an engine and the installation of accessories, it is often necessary to **safety wire** (lockwire) drilled head bolts, cap screws, fillister head screws, castle nuts, and other fasteners. The wire used for this purpose should be soft stainless steel or any other wire specified by the manufacturer.

The principal requirement for lockwire installation is to see that the tension of the wire tends to tighten the bolt, nut, or other fastener. The person installing safety wire must therefore see that the wire pull is on the correct side of the bolt head or nut to exert a tightening effect.

A length of safety wire is inserted through the hole in the fastener, and the two strands are then twisted together by hand or with a special safety-wire tool, as shown in Fig. 12-32. The length of the twisted portion is adjusted to fit the installation. One end of the wire is then inserted through the next hole, and the two ends are again twisted together. The wires are twisted

FIG. 12-32 Using a safety-wire twisting tool.

tightly with pliers but not so much that the wire is weakened. After the job is completed, the excess wire is cut off to leave a stub end of about $\frac{1}{2}$ in [12.70 mm]. The stub should be bent back toward the nut. Typical examples of lockwiring are shown in Fig. 12-33.

Self-locking Nuts

Self-locking nuts may be used on aircraft engines if all the following conditions are met:

1. Their use is specified by the manufacturer.
2. The nuts will not fall inside the engine should they loosen and come off.
3. There is at least one full thread protruding beyond the nut.
4. Cotter pin or locking-wire holes in the bolt or stud have been rounded off so they will not cut the fiber of the nut.
5. The effectiveness of the self-locking feature has been checked and found to be satisfactory prior to its use.

Engine accessories should be attached to the engine by means of the types of nuts furnished with the engine. On many engines, however, self-locking nuts are furnished for such use by the engine manufacturer for all accessories but the heaviest, such as starters and generators.

On many engines, the cylinder baffles, rocker box

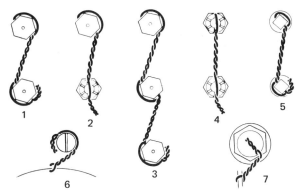

FIG. 12-33 Examples of lockwiring.

covers, drive covers and pads, and accessory and supercharger housings are fastened with fiber insert locknuts which are limited to a maximum temperature of 250° F [121°C] because above this temperature the fiber will char and consequently lose its locking characteristic.

Most engines require some specially designed nuts to provide heat resistance; to provide adequate clearance for installation and removal; to provide for the required degrees of tightening or locking ability which sometimes requires a stronger, specially heat-treated material, a heavier cross section, or a special locking means; to provide ample bearing area under the nut to reduce unit loading on softer metals; and to prevent loosening of studs when nuts are removed.

Torque Values

One of the most important processes a technician must consider in the assembly of an engine or other parts of an aircraft is the **torque** applied when tightening nuts and bolts. Required torque values for various nuts and bolts in an engine are specified in the manufacturer's overhaul and maintenance manuals.

Torque wrenches are designed with a scale so that the technician can read the value of applied torque directly from the scale. The scale is marked in **inch–pounds** (in·lb) or **pound–inches** (lb·in) in the present system in the United States for small to medium bolts and nuts. For bolts and nuts of $\frac{3}{4}$ in [19.05 mm] diameter or larger, it is usually more convenient to employ **foot–pounds** or **pound–feet**. In the SI metric system, the **newton-meter** (N·m) is used as a measure of torque. One pound foot is equal to 1.356 newton-meter.

Torque wrenches are shown in Fig. 12-34. The value shown on the wrench in **pound–inches** is equal to the force applied in pounds at the handle multiplied by the number of inches (length) from the center of the handle to the center of the turning axis over the nut. This is illustrated in Fig. 12-35. If the torque wrench is used with an adapter, as shown in Fig. 12-36, the length of the adapter must be added and the total torque computed. To do this, the torque indicated on the scale T_w

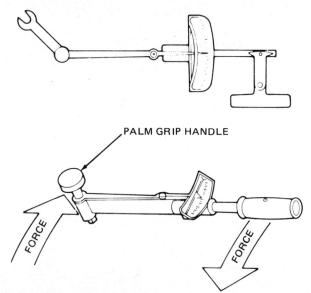

FIG. 12-34   Torque wrenches.

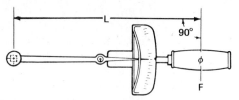

FIG. 12-35   Measurement of torque.

must be divided by the length L to determine the force F being applied to the wrench handle. The force F must then be multiplied by the length of the adapter A plus the length of the wrench L. The equation then becomes

$$T_a = (A + L)\left(\frac{T_w}{L}\right)$$

If the torque wrench is used with an offset adapter, the value A is not the length of the adapter but the distance measured between two lines perpendicular to the axis of the wrench handle, one passing through the rotational axis of the wrench and the other passing through the center of the nut or bolt being turned. This is illustrated in Fig. 12-37.

It must be emphasized that the torque range of a bolt or nut is critical and that an incorrect value of torque applied during assembly will often cause failure. In all assembly operations, the technician must consult the torque value charts supplied by the manufacturer.

Palnuts

Palnuts are made of thin sheet steel and are used as locking devices for plain nuts. After the plain nut is properly torqued, the Palnut is screwed onto the bolt finger-tight. It is then turned about one-fourth turn with a wrench to lock it in place. Excessive turning of the Palnut will damage the threads and render the nut ineffective.

Washers

Flat washers (AN-960) are used under hex nuts to protect the engine part and to provide a smooth bearing surface for the nut. Such washers may be reused provided that they are inspected and found to be in good condition. Washers which are grooved, bent, scratched, or otherwise damaged should be discarded.

Lock washers may be used in some areas but only with the approval of the manufacturer. Where a part must be removed frequently, lock washers should not be used because of the damage which occurs each time the nut or bolt is loosened.

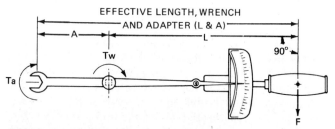

FIG. 12-36   Torque wrench with adapter.

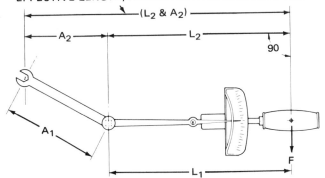

EFFECTIVE LENGTH, WRENCH WITH OFFSET ADAPTER

FIG. 12-37 Torque wrench with offset adapter.

Connecting Rods

New connecting-rod-bearing inserts should be snapped into the rods and rod caps with the tangs of the inserts fitted into the cutouts provided. Each bearing is lubricated with approved oil. Then the rod and cap are installed on the crankpins according to the numbers stamped on the rods and caps. The position numbers should be on the upper side when the engine is in the horizontal position.

Since different manufacturers of engines do not necessarily designate engine cylinder numbers from the same end of the engine, the overhaul technician must make sure to use the correct order when installing parts according to cylinder number. For example, the Lycoming O-480 engine has the cylinders numbered with no. 1 being the right front cylinder, whereas the no. 1 cylinder of the Continental O-470 engine is the right rear cylinder. In both cases the **front** of the engine is the propeller end.

The connecting rods are installed with new bolts, washers, nuts, and corrosion-resistant cotter pins, and each nut is tightened with a suitable torque wrench to the torque specified. A nut must not be "backed up" in an effort to obtain a certain torque value but should always approach the correct value while being tightened. If the cotter-pin hole cannot be aligned with the nut when the nut has been torqued properly, it may be necessary to substitute nuts or bolts until the correct position can be attained. When the cotter pin is installed, one tang is bent back on the side of the nut and the other is bent out over the end of the bolt.

Some manufacturers specify the use of roll pins instead of cotter pins. When these are called for, the pin holes in the connecting-rod bolts must be of the proper diameter for the pins used.

After a connecting rod is installed on the crankpin, the side clearance should be checked with a thickness gage to see that it is within approved limits. The rod should rotate freely on the crankpin, but there should be no noticeable play when tested manually.

Assembling the Crankcase

The crankcase of an opposed engine can be assembled to the crankshaft assembly either while the shaft is mounted in a vertical assembly stand or with the crankcase supported on a workbench. The procedure depends on the type of equipment available in the overhaul shop.

If the crankcase is assembled on a workbench, it must be supported so that the cylinder pads are about 6 in [15.24 cm] above the surface of the table. The right or left section of the case, as designated by the overhaul manual, should be placed on the supports, and all preliminary assembly operations specified should be completed.

The parting flange of the crankcase is coated with a thin layer of an approved sealing compound, care being taken not to apply so much that it will run inside the engine upon assembly. A single strand of No. 50 silk thread is then placed along the parting flange inside the bolt holes as specified by the manufacturer.

Prior to installation of the crankshaft and connecting rod assembly in the crankcase, the front oil seal is installed on the crankshaft. The crankshaft gear can be installed either before or after the connecting rods.

The crankshaft is lifted carefully by two persons so that the correctly numbered connecting rods will be down and the others up. It is placed into the crankcase, care being used to see that the crankshaft seal fits into the seal recess without damage. The upper connecting rods are then laid gently to the side so that they rest on the crankcase flange.

If the engine construction is such that valve tappet bodies cannot be installed after the crankcase is assembled, these must be installed before the camshaft is placed in the assembly position.

The valve tappet bodies are lubricated on the outside and installed in the tappet bores of the opposite half of the crankcase. The camshaft with the cam gear installed is then placed in position and wired in place with brass or soft steel wire.

Before the opposite half of the crankcase is placed in the assembly position, the end clearances of the crankshaft and camshaft are checked with a feeler gage. This is to ensure that the end play is within specified limits.

If required, the valve tappet bodies are installed in the crankcase section in which the crankshaft assembly has been placed. The other section of the crankcase in which the camshaft has been placed is now mated with the first section, care being taken to see that all parts fit together properly and that the cam-gear timing marks are aligned with the timing marks on the crankshaft gear. This automatically times the camshaft to the crankshaft. The two halves are then partially bolted together as specified in the overhaul manual, and the wire holding the camshaft is removed.

It must be emphasized at this time that the foregoing procedures are usually followed for the assembly of the crankcase of an opposed engine. However, there is considerable variation in procedure for different makes and models, and the most desirable method is usually that described by the manufacturer in the overhaul manual.

When an assembly stand is employed which holds the crankshaft in a vertical position, the two halves of the crankcase are assembled simultaneously on the crankshaft and connecting-rod assembly. Regardless of which method is employed, care must be exercised to prevent damage to any part. After the crankcase halves are bolted together, the connecting rods should be supported by means of rubber bands or cords to prevent them from striking the edges of the cylinder pads.

Pistons and Cylinders

Before installation, the piston and piston pin are generously coated with a preservative oil. This oil should be worked into the ring grooves so that all piston rings are thoroughly lubricated. Each piston is numbered and must be installed on the correspondingly numbered connecting rod. The piston is positioned so that the number on the head is in the location specified by the manufacturer, assuming that the engine is in its normal horizontal position. For radial engines, it must be remembered that the cylinder with the master rod is removed *last* during disassembly and installed *first* during assembly. This is to provide adequate support for the master-rod assembly and to hold the link rods and pistons in such a position that the lower piston ring will not be below the skirt of the cylinder.

Before a piston is installed, the crankshaft should be turned so that the connecting rod for the cylinder being installed is in the TDC position. Installation of the piston is accomplished by placing the piston in the proper position over the end of the connecting rod and pushing the piston pin into place through the piston and connecting rod. A piston-ring compressor is then hung over the connecting rod to be in readiness for installation of the cylinder. Prior to installation the cylinders should be checked for cleanliness and the inside of the barrel coated with preservative oil. A base flange packing ring is installed around the skirt at the intersection of the skirt and flange. The correctly numbered cylinder is lifted into position, and the cylinder skirt is placed over the piston head. The ring compressor is then placed around the piston and upper piston rings, and the rings are compressed into the piston grooves. The cylinder can then be moved inward so that the skirt slides over the piston rings as the compressor is pushed back. When all the piston rings are inside the cylinder skirt, the compressor is removed and the cylinder flange stud holes are carefully moved into place over the studs. The base flange packing ring is then checked for position, and the cylinder is pushed into place. Cylinder hold-down nuts are screwed onto the studs and tightened lightly. The upper nuts are installed first in order to provide good support for the cylinder. After all the nuts are in place, they are tightened moderately but not torqued to the full value. A torque handle is installed on the cylinder base wrench, and the nuts are torqued in the sequence specified in the overhaul manual. *It is of the utmost importance that the cylinder hold-down nuts be tightened evenly and to the correct torque value to prevent warping and undue strain on one side of the flange.* Some manufacturers require that all cylinders be installed before the final torquing of cylinder hold-down nuts. Cylinder hold-down nuts are secured by means of safety wire or with Palnuts.

Valve Mechanism

Since the valve mechanisms for engines of different makes and models vary considerably, we shall not describe any particular method of installation, assembly, or valve timing in this chapter. We shall stress, however, that the assembly must be in the sequence described by the manufacturer in order to avoid omitting any required operation. If this is done, the timing of valves will be correct. All parts should be perfectly clean before installation and should be coated with clean lubricant. This is particularly true of the valve lifter cylinder and plunger assembly. It is especially important to see that the pushrod socket is in place in the tappet body before the pushrod is installed.

After the complete valve-operating mechanism has been assembled, the clearance between the rocker arm and valve stem should be checked with the cylinder at TDC on the compression stroke. If the valve clearance is not within the limits specified, it must be adjusted by installing a pushrod of slightly different length or by adjusting the screw in the end of the rocker arm.

Upon completion of the valve-mechanism installation, the rocker cover is installed with a gasket and safetied. If self-locking nuts are specified, no further safetying is required.

● HELICOPTER POWER TRAINS

Because the power train of a helicopter is essentially an extension of the engine, or powerplant, we shall include in this section a brief discussion of the special characteristics of helicopter power systems and the need for meticulous care in the maintenance and overhaul of power trains. Of course, to describe many helicopter types or to cover details of overhaul would require far more space than is available in a textbook. Each make and model of helicopter differs from others and all the special processes and techniques employed for maintenance and overhaul are described in the particular maintenance manual for the model concerned.

Description

An exploded view of a Hughes series 269 helicopter is shown in Fig. 12-38. The various parts of the power train are identified as the **main rotor, main-rotor drive shaft, main transmission, belt-drive transmission, tail-rotor drive shaft, tail-rotor transmission,** and **tail rotor.** The main rotor drive shaft is supported by the **main-rotor mast** and the tail-rotor drive shaft is supported in the **tail boom.**

An exploded view of the power train is shown in the drawing of Fig. 12-39. The engine is mounted under the main transmission with the output shaft pointing aft to mate with the input fitting of the belt transmission. The belt transmission consists of two V-drive pulleys having V grooves for eight V belts. An idler pulley maintains proper tension on the belts. Inside the input pulley is a **sprag clutch** (freewheel drive) which permits the power train to rotate even though the engine is stopped. This feature is necessary to permit autorotation in case of engine failure. It also reduces stress loads which would otherwise occur when engine power is suddenly reduced.

The principal function of the main transmission which is driven from the top pulley of the belt transmission is to change the direction of the drive from horizontal to vertical. This transmission consists of an input shaft with a spiral bevel pinion gear and an output shaft with a ring gear. The shafts are supported in bevel roller

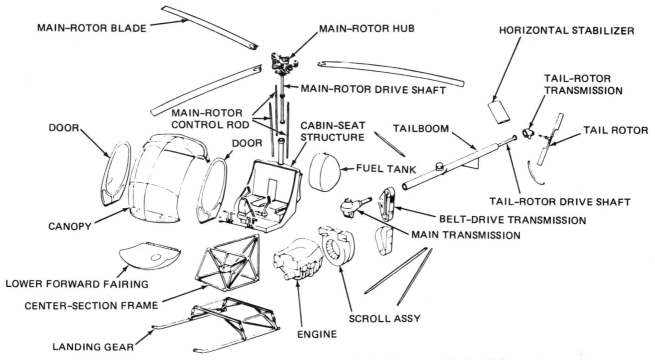

FIG. 12-38 Exploded view of a Hughes series 269 helicopter. *(Hughes Helicopters)*

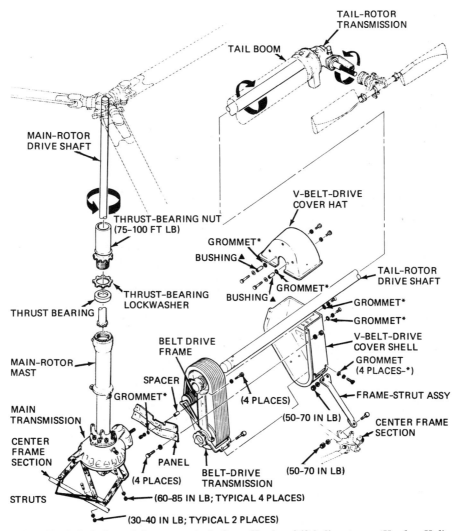

FIG. 12-39 Exploded view of the power train for a Hughes 269 helicopter. *(Hughes Helicopters)*

bearings mounted in the transmission housing. An inspection plug is installed in the housing above the pinion gear to permit examination of the gear condition. The gears are manufactured from forged triple-alloy gear steels and are carburized, case-hardened, and ground to precision tolerances.

A portion of the power train for the Bell Model 204B helicopter is shown in Fig. 12-40. This shows how the main transmission is installed at the front of the Lycoming turboshaft engine. In the drive coupling between the engine and the transmission is a freewheeling (sprag) clutch to allow the rotor systems to continue operating even if the engine stops. This, of course, permits autorotational descent in case of engine failure. The gearing in the transmission includes two planetary systems which provide a total gear ratio of 20.37:1; that is, the engine drive shaft turns more than 20 times as fast as the main rotor. The gear reduction to the tail rotor, including the tail-rotor transmission gearing, provides a gear reduction of 4:1. The tail-rotor drive extends to the rear from the **accessory drive and sump case,** which is below the main transmission.

Maintenance and Overhaul of Helicopter Power Trains

As mentioned previously, the power train of a helicopter includes the coupling to the engine drive shaft, the main transmission, the main-rotor drive shaft, the tail-rotor drive shaft, and the tail-rotor transmission together with necessary attachments and supports. The importance of regular and proper maintenance cannot be overemphasized because the helicopter can remain in flight or make a safe emergency landing only if both the main rotor and the tail rotor are rotating at an

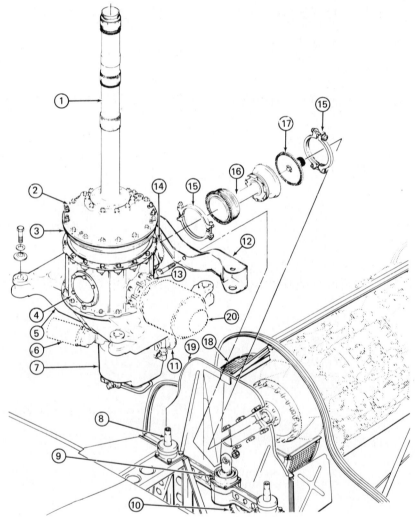

1. Mast	**11.** Tail-rotor drive quill
2. Top case	**12.** Fifth mount support
3. Planetary-ring-gear case	**13.** Generator drive quill
4. Main case	**14.** Input drive quill
5. Support case	**15.** Coupling clamps
6. Hydraulic pump and tachometer drive quill	**16.** Input drive shaft
7. Accessory drive and sump case	**17.** Engine coupling adapter
8. Pylon main mounts	**18.** Screen
9. Fifth pylon mount	**19.** Baffle
10. Tail-rotor drive shaft	**20.** DC generator

FIG. 12-40 **Main transmission and drive shafts for the Bell Model 204B helicopter.** *(Bell Helicopter Co.)*

adequate speed. The tail rotor must be coupled to the main rotor through the transmission so that it will always provide adequate force against the torque of the main rotor and provide directional control.

Helicopter maintenance manuals provide instructions regarding regular inspections and the number of flight hours between these inspections. In addition, a lubrication schedule and instructions are provided. Adherence to both the inspection schedule and the lubrication schedule are essential to the safe operation of the helicopter. It is particularly important that the correct types of lubricants are used when lubricating and servicing a helicopter.

Because of the nature of helicopter flight, all parts of the power train are subjected to intense and repeated stresses. For this reason all such stressed parts are constructed of high-strength alloy steels, machined and finished with a high degree of precision. Heat treating, nitriding, case hardening, and other processes are employed to provide maximum strength, resistance to metal fatigue and wear, and dependability. It is the technician's responsibility to ensure that the quality of all parts is maintained so that the integrity of the power-train system remains intact.

Various parts of the power-train system are subject to replacement after a specified period of operation because experience and tests reveal that the stresses to which parts are subjected cause metal fatigue which may cause failure after a given number of hours of operation. The individual in charge of maintenance must ensure that records are kept to provide information on the number of operating hours for any time-limited part. Such parts must then be discarded when operation time has reached the hours specified in the maintenance manual.

During overhaul of power trains and associated assemblies, procedures similar to those specified for engine overhaul are employed. These include magnetic, x-ray, sonic, and other inspections specified in the overhaul manual. Dimensional limits and tolerances are specified and must be observed carefully.

Particularly important in the inspection of power-train and rotor parts is examination for nicks, scratches, dents, corrosion, cracks, and any other condition which could start a progressive failure of a highly stressed part. For example, the failure and crash of one helicopter was caused by a very small knife cut on the shank of a main rotor blade. The technician had used a pocket knife to remove masking tape and had inadvertently marred the surface enough to cause stress concentration and cracking.

The dimensions and number of defects permitted in certain areas on parts are specified in the overhaul manual. Stress studies by engineers indicate where a part will be subjected to the highest stresses and greatest frequency of recurring stresses. In these areas it is seldom that any dent, scratch, or other defect can be permitted. In some cases, the part can be repaired by stoning and polishing, and in others the part must be discarded.

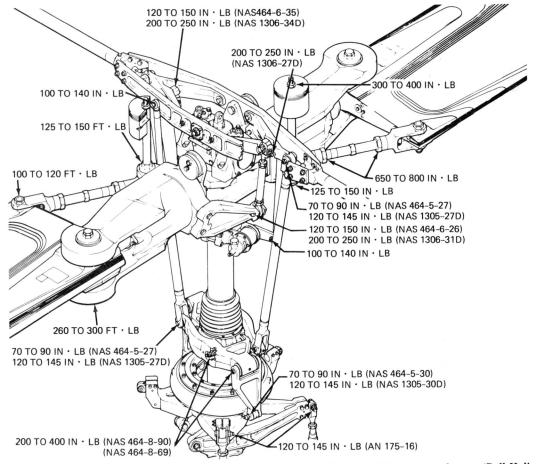

FIG. 12-41 Main-rotor system for the Bell Model 204B helicopter, showing assembly torque values. (Bell Helicopter Co.)

During disassembly, cleaning, inspection, and repair of power trains, the technician must adhere to all specifications and instructions provided by the manufacturer. During assembly, it is particularly important that the torque requirements for nuts and bolts be observed. Figure 12-41 is a drawing with torque values for the Bell Model 204B helicopter main-rotor system. Failure to maintain the torque values in the ranges specified can lead to failure of the part concerned.

● REVIEW QUESTIONS

1. Describe the changes which take place in an aircraft engine during operation and eventually make an overhaul necessary.
2. How and why does the manner of operation of an engine affect the time between overhauls?
3. What is accomplished during the overhaul of an engine?
4. Briefly describe the organization of an overhaul shop.
5. Why is a separate inspection (quality-control) department required in a certificated engine overhaul shop?
6. List the functions of the receiving inspection group.
7. Discuss the need for complete records of an engine overhaul.
8. Make a list of safety rules which should be followed in an engine overhaul shop.
9. Discuss the importance of manufacturer's bulletins, Airworthiness Directives, Engine Specification, and Type Certificate Data Sheet.
10. Describe the cleaning operations required during engine overhaul.
11. What is meant by *degreasing*?
12. Describe vapor degreasing.
13. What precautions must be taken when stripping engine parts in a chemical stripping solution?
14. What is done with engine parts after they are removed from the stripping solution?
15. Describe *vapor blasting*.
16. What is the principal value of *grit blasting*?
17. What precautions must be taken before and during the grit blasting of engine parts?
18. For what purposes can sandblasting be employed in the overhaul of an aircraft engine?
19. Why should engine parts be coated with a preservative oil after they are cleaned?
20. Discuss the need for proper shop equipment and engine overhaul tools in an engine overhaul shop.
21. What is the principal source of information for the overhaul of an engine?
22. Why must engine parts be tagged or marked for identification?
23. What is the danger involved in the use of wire wheels, scrapers, knives, and harsh abrasives in cleaning carbon or paint from engine parts?
24. For what purpose is the Table of Limits used?
25. What is the purpose of a manufacturer's bulletin?
26. List three general categories of inspections for engine parts.
27. List different types of damage and defects to be noted in parts inspections.
28. Briefly describe magnetic inspection.

29. Why is it necessary to demagnetize parts after a magnetic inspection?
30. Describe dye penetrant and fluorescent penetrant inspections.
31. What instrument is used to measure the cylinder bore?
32. Describe the use of a telescoping gage, small-hole gage, and micrometer caliper.
33. Describe the use of a thickness gage.
34. What inspections should be made on a cylinder head?
35. Discuss the repair of cooling fins and the limitations on the amount of damage which can be tolerated.
36. Explain the importance of cylinder barrel condition.
37. For what conditions is a cylinder barrel inspected?
38. What precautions should be observed with respect to grinding a cylinder which has been hardened by nitriding?
39. If a chrome-plated cylinder barrel is found to be worn beyond acceptable limits, how may the barrel be repaired?
40. What is the advantage of a chrome-plated cylinder barrel?
41. What precautions must be observed with respect to the piston rings used in a chrome-plated cylinder barrel?
42. If it is found that one cylinder of an engine needs to be ground to an oversize dimension, what should be done with respect to the other cylinders and pistons?
43. Describe cylinder grinding and honing.
44. What is a choked cylinder barrel, and what is the purpose of choke in a cylinder barrel?
45. Give the color coding for cylinders which have been ground to standard oversize dimensions.
46. Describe a stud.
47. What precautions must be observed with respect to the skirt and flange of a cylinder while handling?
48. For what defects is a cylinder mounting flange inspected and what procedures are followed?
49. Describe the removal of a valve guide.
50. What should be the approximate dimension of a valve guide compared with that of the hole into which it is to be installed?
51. Describe the installation of valve guides.
52. What condition requires the replacement of a valve seat?
53. How is the correct width of a valve seat obtained during the grinding process?
54. Describe the removal and installation of a valve seat.
55. What is the advantage of a 30° valve-face angle over a 45° angle?
56. For what defects are valves inspected?
57. How is a valve stem inspected for stretch?
58. What precaution must be taken before the stem of a valve is inserted into the chuck of a refacing machine?
59. What is meant by *interference fit* of a valve face and seat?

60. What inspections are made on valve springs?
61. Describe the installation of valves.
62. For what defects are rocker arms inspected?
63. Describe the process of removing and installing rocker-arm bushings.
64. What precaution must be taken with respect to the position of the oil hole in a rocker-arm bushing?
65. Describe the installation of rocker-shaft bushings in a cylinder head. What is the function of the dowel pin used in some installations?
66. For what defects are pistons inspected?
67. How may a slightly scored piston be repaired?
68. How would you determine piston-to-cylinder bore clearance?
69. Describe the clearance checks required for piston rings.
70. What may happen if the piston-ring gap is not adequate?
71. Why is it important to check the manufacturer's overhaul manual when installing piston rings?
72. Where does the greatest amount of wear occur in a reciprocating engine?
73. What is meant by *lapping* piston rings in a cylinder?
74. Why is the crankshaft considered a most critical part of a reciprocating engine?
75. When measuring a crankpin or main-bearing journal of a crankshaft, how many measurements should be taken for each crankpin or for each journal?
76. After a crankshaft has been reground to an undersize condition, what should be done with respect to the ground areas?
77. Why is it advisable to return a crankshaft to the manufacturer for regrinding?
78. What is the importance of the radii of journal ends?
79. Describe the procedure for checking crankshaft alignment or runout.
80. After an engine has experienced a sudden stoppage, what inspection should be made with respect to the crankshaft? How is this accomplished?
81. What is the function of counterweights and dynamic balances?
82. How are sludge chambers in a crankshaft cleaned?
83. What precaution must be observed in the installation of sludge-chamber tubes?
84. How are oil passages in the crankshaft cleaned?
85. What conditions require the rejection of a camshaft?
86. For what defects is a cam gear inspected?
87. How would you check the hydraulic unit of a valve tappet assembly for satisfactory operation?
88. List the inspections required for connecting rods.
89. Describe the procedures for checking the twist and convergence of a connecting rod.
90. What condition must be maintained when reaming or boring a connecting-rod bushing?

91. How is the clearance of connecting-rod bearings determined?
92. What parts associated with connecting rods should be replaced at the time of engine overhaul?
93. What should be done if it is found that one section of a two-piece crankcase must be discarded?
94. What inspection methods are used with a crankcase?
95. Describe the replacement of studs.
96. Under what circumstances would it be appropriate to use a bottoming tap?
97. Explain the use of Heli-Coil inserts.
98. Explain how main-bearing and camshaft-bearing bores in the crankcase are measured.
99. Discuss the inspection of accessory bearings.
100. Discuss the replacement of bushings in the accessory section of an engine.
101. What should be done with engine parts immediately before assembly is started?
102. Explain the proper use of safety wire.
103. Under what conditions may self-locking nuts be used on an engine?
104. Explain the use of a torque wrench.
105. How is the torque determined when an extension is used with a torque wrench?
106. What is a Palnut?
107. What limitations are applied to the use of lock washers?
108. What precautions must be taken with respect to the location and position of connecting rods during engine assembly?
109. How is the camshaft held in place during the assembly of the crankcase for an opposed engine?
110. When assembling the crankcase for an opposed engine, what must be done with respect to the position of the crankshaft gear and cam gear? Why?
111. What is done with respect to the parting surfaces of the crankcase during assembly?
112. What is done to prevent the connecting rods from striking and damaging the edges of the cylinder pads during engine assembly?
113. What precautions must be taken with respect to installing the pistons?
114. In assembling a radial engine, what piston and cylinder should be installed first? Why?
115. Describe the procedure for installing a cylinder.
116. Discuss the tightening and torquing of cylinder hold-down nuts.
117. How would you check the valve clearance after the engine is assembled?
118. What are the principal units in a helicopter power train?
119. What are the principal functions of the main transmission?
120. Why is a freewheeling clutch essential in a helicopter power train?
121. What is the likely effect of nicks and scratches on highly stressed parts of the power train?

13 ENGINE INDICATING AND WARNING SYSTEMS

● ENGINE INSTRUMENTS

The safe and efficient operation of aircraft power-plants of all types requires that indicating instruments be installed to measure various values associated with the performance of the engine. Early aircraft engines usually were provided with nothing more than an oil pressure gage, an oil temperature gage, and a tachometer. With these basic instruments the pilot was able to determine with reasonable accuracy whether or not the engine was performing as it should.

As engines became more complex with improved design and higher performance, it became necessary to measure additional factors governing performance. Accordingly, manifold pressure gages, carburetor air temperature gages, cylinder-head temperature gages, fuel pressure gages, fuel flowmeters, exhaust-gas temperature gages, torquemeters, and a number of indicating instruments were developed and employed. Gas-turbine engines required even more gages, including the engine pressure ratio (EPR) indicator, exhaust-gas temperature gage, turbine inlet-temperature gage, and others.

We will examine briefly the operating principles of engine gages, explain how troubles are indicated by the gages, and describe methods of maintenance and repair for indicating systems. We shall not attempt to describe the overhaul of instruments because this type of work must be done by qualified instrument technicians in approved instrument overhaul shops. Additional information regarding the theory, construction, and maintenance of instruments and instrument systems is provided in other texts of this series.

Pressure Instruments

Pressure instruments are usually of the bourdon-tube type, diaphragm type, or bellows type. The particular type of mechanism employed to provide an indication of pressure depends on the requirements of the system and the level of pressure which must be measured.

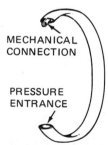

FIG. 13-1   Bourdon tube.

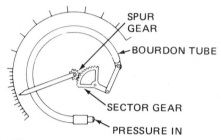

FIG. 13-2   Bourdon-tube indicating mechanism.

Comparatively high pressures usually require the use of the bourdon-tube instrument.

The design of a typical **bourdon tube** is illustrated in Fig. 13-1. The tube is formed in a circular shape with a flattened cross section. The material of the tube is thin metal sheet, such as brass. When air or liquid pressure is applied to the open end of the bourdon tube, the tube tends to straighten out. This action of the tube is utilized to move a sector gear which, in turn, rotates the spur gear attached to an indicating needle shaft. Thus, when the bourdon tube changes its circular form because of changes in the pressure applied to the tube, the sector gear is moved and the indicating needle is rotated. The arrangement of a bourdon-tube indicating mechanism is shown in Fig. 13-2.

The oil pressure gage and some oil temperature gages are typical of bourdon-tube intruments. The face of an oil pressure gage for a light aircraft is shown in drawing Fig. 13-3. This instrument is usually mounted with other engine instruments in a common subpanel. In many

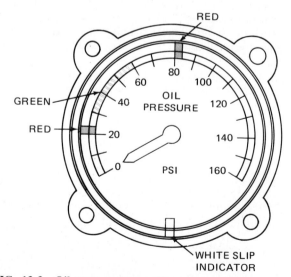

FIG. 13-3   Oil pressure gage with range markings.

cases, oil pressure, oil temperature, and cylinder-head temperature instruments are in one case, called the **engine gage unit.**

The oil pressure gage for small aircraft is directly connected by means of a tube to the engine at a point immediately downstream from the engine oil pump. The gage therefore indicates the pressure being delivered to the engine. The instrument includes a small restriction in the inlet to prevent pressure surges from damaging the instrument. When the aircraft is operated in cold-weather conditions, the tube from the instrument to the engine is filled with light oil. The light oil allows the instrument to react to pressure changes in a normal manner, whereas engine oil would be partially congealed, thus preventing proper operation.

The oil pressure gage is of primary importance when an engine is first started. If there is no indication of oil pressure within the first 30 s, the engine should be stopped and an investigation made to determine the cause. In some cases, when an engine has been idle for an extended period of time, oil has drained from the oil pump and it is necesary to prime the pump by injecting oil into the inlet side. Other causes may be a low oil supply, a clogged oil inlet line, or a broken line.

During operation, the oil pressure gage can provide an indication of engine condition. Below-normal oil pressure is a good sign that the engine has become worn to the extent that an overhaul is required. If the reading of the oil pressure gage fluctuates rapidly, it is an indication of a low oil supply.

All engine instruments are color-coded to direct attention to approaching operating difficulties. The oil-pressure gage face is marked to indicate normal operating ranges. In Fig. 13-3, the gage has a green band from 30 to 60 psi [206.85 to 413.7 kPa]. This represents the normal operating range for the engine. Red lines indicate maximum and minimum allowable operating conditions.

When the limit and range markings are placed on the glass face of an instrument, a white line is placed on the glass and extended to the case of the instrument. This line will be broken and offset if the cover glass should move. When this condition is noted, the cover glass should be rotated until the white line is again in alignment. The glass should then be secured to prevent further rotation. It is obvious that if the cover glass rotates, the limit and range markings will not give a true indication for operation of the engine.

Inspections of the oil gage system include the following:

1. Check for oil leaks at the gage connection and the engine fitting.

2. Check the gage for security of mounting and make sure that there is no looseness of mounting screws.

3. Examine the routing of the oil line from the engine to the gage. See that the line is securely attached with suitable padded clamps and that the line does not rub against any part or structure which could cause wear from vibration.

4. Examine the limit and range markings to ensure that they are adequate and that the face glass has not rotated.

In installations for some large aircraft, the oil pressure indicating system utilizes a **pressure transmitter** mounted near the engine. The transmitter consists of two chambers separated by a flexible diaphragm. One chamber is connected to the engine oil pressure outlet and the other is connected to the oil pressure gage. The chamber on the gage side of the transmitter is filled with a light oil which also fills the line leading to the gage. The advantage of this type of system is that the light oil cannot be contaminated with engine oil and will therefore continue to be effective for extreme cold-weather operation. The transmitter operates by applying engine oil pressure to the diaphragm which, in turn, pressurizes the light oil in the instrument side of the chamber. The pressure is then applied to the oil pressure gage through the pressure gage tube.

Diaphragm-type gages are generally used to measure comparatively low pressures. The construction of an instrument diaphragm is shown in Fig. 13-4. Note that the diaphragm consists of two thin metal discs concentrically corrugated and sealed together at the edges to form a cavity. One side is provided with an opening at the center through which air or gas pressure can be applied to the inside of the diaphragm. As pressure is applied, the two sides spread apart, thus producing a motion which can be used to actuate an indicating needle. How a diaphragm may be linked to the indicating needle is illustrated in Fig. 13-5.

An engine instrument which utilizes a diaphragm to sense pressure is the **manifold pressure (MAP) gage.** This gage was mentioned in the discussions on engine performance and induction systems. The abbreviation, MAP, is used to indicate manifold pressure because manifold pressure is absolute pressure; that is, MAP stands for *manifold absolute pressure.* The manifold pressure gage is, in effect, a barometer because it measures atmospheric pressure when it is not connected to a running engine. When the engine is stopped, the manifold pressure gage should indicate the local barometric pressure. To accomplish this, the MAP gage is provided with two diaphragms. One diaphragm is a sealed *aneroid* cell which responds to ambient atmospheric pressure. The other diaphragm is connected to the intake manifold of the engine and responds to the manifold pressure. The effect is that the gage provides an indication of the *absolute* pressure (pressure above zero pressure) existing in the intake manifold of the engine.

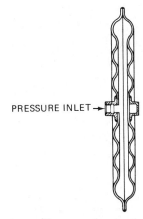

PRESSURE INLET →

FIG. 13-4   Drawing to illustrate the construction of an instrument diaphragm.

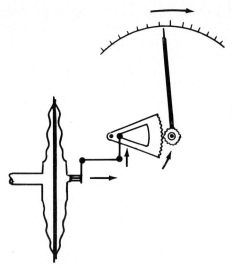

FIG. 13-5 How a diaphragm may be linked to the indicating needle.

Manifold pressure gages may be designed so that the manifold pressure of the engine is applied to the inside of a diaphragm or to the exterior of the diaphragm in a sealed instrument case. In all cases the instrument must be designed so that the chemical constituents in the intake manifold will not get into the operating mechanisms of the instrument. Furthermore, the pressure line from the manifold to the diaphragm or instrument case must be provided with a restriction such as a coiled capillary tube to prevent excessive pressures from reaching the instrument, as in the case of a backfire.

The face of a typical manifold pressure gage is shown in Fig. 13-6 and the interior mechanism is shown in Fig. 13-7. The cover glass on the face or the dial of the instrument has limit and range markings to aid the pilot in operating the engine safely and efficiently. The range marked in blue shows the manifold pressure which may be used while operating in AUTO LEAN and the green band shows the range of pressures which are satisfactory for operation in AUTO RICH. The red line indicates maximum allowable manifold pressure for takeoff power, and the white line shows slippage of the cover glass.

The manifold pressure gage is particularly important for use with supercharged engines and those with which

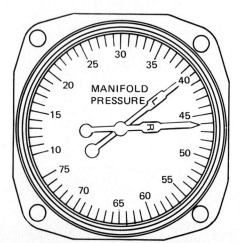

FIG. 13-6 The face of a manifold pressure gage.

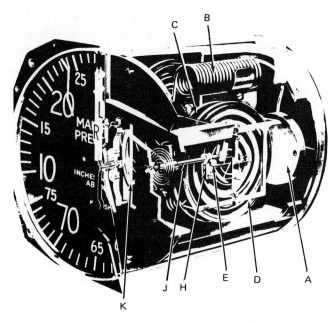

A. pressure connection
B. capillary coil
C. pressure diaphragm
D. aneroid diaphragm
E. bimetallic temperature compensator
H. actuating arm
J. rocking shaft
K. sector gear

FIG. 13-7 Interior mechanism of a manifold pressure gage. (Kollsman)

a constant-speed propeller is employed because the manifold pressure has a direct effect on mean effective pressure (MEP). Excessive MEP can cause severe engine damage. When an engine is being operated under supercharged manifold pressures, the pilot must know that safe operating pressures are not exceeded. As we have explained previously, excessive manifold pressure will cause high cylinder pressures which result in detonation, overheating, and preignition.

When a naturally aspirated (unsupercharged) engine is operated with a constant-speed propeller, the pilot cannot determine the power output of the engine without knowing the manifold pressure, because the engine rpm will be constant even when the power varies greatly.

The manifold pressure gage is calibrated in inches of mercury (inHg). When the instrument is in standard sea-level conditions and is not on an operating engine, the reading of the instrument should be 29.92 inHg [101.34 kPa]. Unsupercharged engines always operate below atmospheric pressure because the engine is drawing air into the cylinders and the friction of the air passages causes a reduction in air pressure. When such an engine is idling, the manifold pressure is likely to be less than one-half atmospheric pressure.

The manifold pressure gage is installed in the instrument panel adjacent to other engine instruments. Ideally, it should be next to the tachometer so that the pilot can quickly read rpm and manifold pressure to determine engine power output. The panel on which the instruments are mounted is usually a shock-mounted subpanel which is attached to the main instrument panel. During inspections all the shock mountings should be examined carefully for damage and signs of deterioration.

The pressure line leading from the engine to the manifold pressure gage may consist of metal tubing, a pressure hose, or a combination of both. On one twin-engine aircraft, the manifold pressure line consists of four sections identified as the *indicator line, fuselage line, nacelle line,* and *nacelle hose.* The various sections are joined by means of standard tubing and hose fittings. In some installations a **purge valve** is connected to the pressure line near the instrument. The purpose of this valve is to remove moisture from the pressure line. To purge the line, the engine is started and operated at idle speed. Since the manifold pressure at idling is less than 15 inHg [50.81 kPa], a strong suction exists in the line. When the purge valve is opened, air flows through the valve and the pressure line to the engine. The valve is held open about 30 s and this effectively removes all moisture in the line. After the purge valve is closed, the manifold pressure gage should show correct manifold pressure for the engine.

During inspections the manifold pressure gage should be checked for proper operation. Before the engine is started the gage should show local barometric pressure. After the engine is started and is idling, the MAP gage reading should drop sharply. When the engine rpm is increased, the gage should be watched to see that it increases evenly and in proportion to power output. In the event that the reading lags or fails to register, the cause can usually be traced to one of the following discrepancies: (1) The restriction in the instrument case fitting is too small, (2) the tube leading from the engine to the gage is clogged or leaks, or (3) the diameter of the tubing is too small for its length. If the reading is jumpy and erratic when the engine speed is increased, the restriction in the case fitting is probably too large. If there is a leak in the gage pressure line, the manifold pressure will read high for an unsupercharged engine. If the engine is supercharged and is operating at a manifold pressure above atmospheric, a leaking pressure line will cause the instrument to read low. If the pressure line is broken or disconnected, the MAP gage will read local barometric pressure. A leaking or broken pressure line on an unsupercharged engine will permit outside air to leak into the manifold and lean the mixture to some extent, depending on the size of the tubing and fitting. On a supercharged engine, a leaking or broken pressure line will allow air or a fuel-air mixture to escape, provided that the engine is operating at a manifold pressure above atmospheric.

Inspections and installation for the manifold pressure gage should be accomplished in a manner similar to any other panel-mounted instrument. Additional information on instruments and their care is given in other texts of this series.

Fuel pressure gage. Several different types of fuel-pressure gages are in use for aircraft engines, each being designed to meet the requirements of the particular engine fuel system with which it is associated. The technician should know the type of gage before making judgments regarding its operation or indications.

Any fuel system utilizing an engine-driven or electric fuel pump must have a fuel pressure gage to ensure that the system is working properly. If a float-type carburetor is used with the engine involved, the fuel pressure gage will be a basic type, probably with a green range marking from 3 to 6 psi [20.69 to 41.37 kPa]. If the engine is equipped with a pressure discharge carburetor, the fuel-pressure-gage range marking will be placed in keeping with the fuel pressure specified for the carburetor. This will probably be between 15 and 20 psi [103.43 and 137.9 kPa]. A red limit line will be placed at each end of the pressure-range band to indicate that the engine must not be operated outside the specified range.

If an engine is equipped with either a direct fuel injection system or a continuous-flow fuel-injection system, the fuel pressure is a direct indication of power output. Since engine power is proportional to fuel consumption and since fuel flow through the nozzles is directly proportional to pressure, fuel pressure can be translated into engine power or fuel-flow rate or both. The fuel pressure gage, therefore, can be calibrated in terms of percent of power. A gage of this type is shown in Fig. 13-8. Note that the instrument indicates a wide range of pressures at which the engine can operate. It is calibrated in pounds per square inch and also indicates percent of power and altitude limitations. The face of the instrument is color-coded blue to show the normal cruise range during which the engine can be operated in AUTO LEAN and green to show the range during which the AUTO RICH mixture setting should be used.

Fuel pressure gages are constructed similar to other pressure gages used for relatively low pressures. The actuating mechanism is either a diaphragm or a pair of **bellows.** A drawing to illustrate a bellows is shown in Fig. 13-9. The advantage of the bellows is that it provides a greater range of movement than does a diaphragm. In a typical fuel gage the mechanism includes two bellows capsules joined end to end. One capsule is connected to the fuel pressure line and the other is vented to ambient pressure in the airplane. The fuel pressure causes the fuel bellows to expand and move toward the air capsule. The movement is transmitted to the indicating needle by means of conventional linkage.

Temperature Gages

The safe and efficient operation of an aircraft engine requires that temperatures in the engine be monitored and controlled within carefully delineated limits. The most critical temperature information needed for the

FIG. 13-8   A fuel pressure gage.

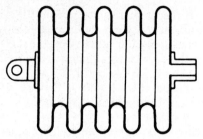

FIG. 13-9 A bellows capsule.

operation of a reciprocating engine relates to oil temperature and cylinder-head temperature. As long as these temperatures are within established limits, it is not likely that any damage to the engine will occur because of heat.

In addition to the oil temperature gage and the cylinder-head temperature gage, carburetor air temperature (CAT) gages and exhaust-gas temperature (EGT) gages are installed in many aircraft. For reciprocating engines, the exhaust-gas temperature gage is the indicator for the **economy mixture indicating system.** Since the temperature of the exhaust gas increases as combustion improves, it reveals to the pilot when the engine is providing most complete combustion, thus getting the most power for the fuel consumed. EGT gages are also employed as exhaust-gas analyzers and may be installed with sensors in all the exhaust stacks of an engine. The purpose of the exhaust-gas analyzer is to indicate the fuel-air ratio of the mixture being burned in the cylinders. The carburetor air temperature gage is important as a means of detecting icing conditions and regulating engine performance, particularly when a float-type carburetor system is turbocharged.

Oil temperature gage. The most common type of oil temperature gage is operated electrically and may utilize either a Wheatstone bridge circuit or a ratiometer circuit. Since the ratiometer circuit provides a more dependable temperature indication under varying input voltage conditions, this type of instrument has become standard for most installations.

The circuit for a **Wheatstone bridge** instrument is shown in Fig. 13-10. The bridge consists of three fixed resistors of 100 Ω each and one variable resistor whose resistance is 100 Ω at a fixed point such as 0° C, or 32° F.

With the bridge circuit connected to a battery power source as shown, it can be seen that if all four resistances are equal, current flow through each side of the bridge will be equal and there will be no differential in voltage between points A and B. When the variable resistor which is in the temperature-sensing bulb is exposed to heat, the resistance increases. When this happens it can be seen that more current will flow through the top portion of the bridge than will flow through the bottom portion. This will cause a voltage differential between points A and B and current will then flow through the galvanometer indicator. The instrument is calibrated to give a correct reading of the temperature sensed by the variable resistor in the sensing bulb. If any one segment of the bridge circuit should be broken, the indicator needle would move to one end of the scale, depending on which circuit was broken.

The circuit for a **ratiometer** instrument is shown in Fig. 13-11. This circuit has two sections, each supplied with current from the same source. As shown in the drawing, a circular iron core is located between the poles of a magnet in such a manner that the gap between the poles and the core varies in width. The magnetic flux density in the narrower parts of the gap is much greater than the flux density in the wider portions of the gap. Two coils are mounted opposite each other on the iron core and both coils are fixed to a common shaft on which the indicating needle is mounted. The coils are wound to produce magnetic forces which oppose each other. When equal currents are flowing through the two coils, their opposing forces will balance each other and they will be in the center position as shown. If the current flowing in coil A is greater than that flowing in coil B, the force produced by coil A will be greater than the force of coil B, and coil A will move toward the wider portion of the gap where a lower flux density exists. This will move coil B into an area of higher flux density a distance sufficient to create a force which will balance the force of coil A. At this point the position of the indicating needle gives the appropriate temperature reading.

Like the sensing bulb for the Wheatstone bridge instrument, the bulb for the ratiometer instrument contains a coil of fine resistance wire which increases in resistance as the temperature rises. The resistance wire is sealed in a metal tube and is connected to pin connectors for the electrical contacts.

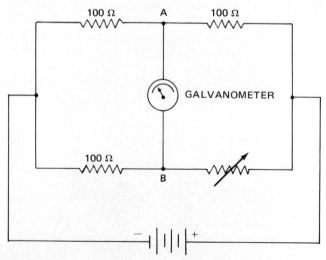

FIG. 13-10 Circuit for a Wheatstone bridge.

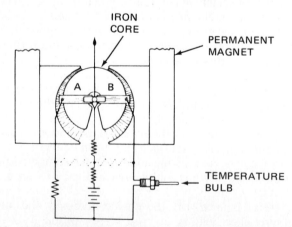

FIG. 13-11 Circuit for a ratiometer-type instrument.

If it is found during inspection that the ratiometer-type instrument is giving an off-scale high reading, then the sensing circuit is broken or there is an open circuit in the balancing coil. The circuits may be checked with an ohmmeter to locate the break. If the instrument is found to be defective, it should be removed and sent to an approved instrument repair facility.

As mentioned previously, the oil temperature gage is probably the most important instrument associated with the operation of a reciprocating engine with the exception of the oil pressure gage. It not only tells the pilot when the engine is warmed up sufficiently for takeoff but also keeps her or him informed regarding the proper operation of the engine in flight.

The temperature-sensing bulb is usually located in the engine oil system at a point immediately *after* the oil has passed through the oil cooler. Some oil coolers have provision for the temperature bulb to be installed near the outlet of the cooler. In any event, the temperature gage measures the temperature of the oil entering the engine. The face of the oil temperature gage is marked with a green band showing the range of normal operating temperatures. A red line is used for both the minimum allowable operating temperature and the maximum safe temperature. The manufacturer usually specifies the temperature limits for engine operation. As a general rule, 40° C (104° F) is considered the minimum safe temperature for operation, 60 to 70° C (140 to 158° F) is the normal operating range and 100° C (212° F) is the maximum allowable temperature for a reciprocating engine.

In the past, many oil temperature gages were of the **vapor pressure** type, and many of these instruments are still in service. A vapor-pressure temperature gage is a simple bourdon-tube instrument connected to a liquid-filled bulb by means of a capillary tube. The bulb is filled with a highly volatile liquid which vaporizes and develops pressure inside the bulb which is transmitted to the bourdon tube through the capillary tube. The pressure of the vapor is in proportion to the temperature; hence, the indicating needle shows the temperature of the medium in which the bulb is placed. Vapor pressure instruments were also used for carburetor air temperature indicators and coolant temperature indicators.

In the installation and removal of a vapor pressure instrument, the technician must take care to ensure that the capillary tube connecting the temperature bulb to the instrument is not kinked or otherwise damaged. For protection against wear and abrasion, the capillary tube is usually encased in a sheath made of fine woven wire.

Since the instrument and temperature bulb for a vapor pressure assembly form a sealed unit, the capillary tube cannot be cut to fit any particular routing. Instead, the tubing is usually longer than required and the excess length is coiled and secured with padded clamps to the aircraft structure. The entire tubing length must be adequately supported at short intervals so that it cannot be worn by rubbing against any part of the structure or other installations.

Thermocouple instruments. The measurement of very high temperatures such as cylinder-head temperature and exhaust-gas temperature requires the use of **ther-**mocouples. A thermocouple is the junction of two dissimilar metals which generates a small electric current that varies according to the temperature of the junction. For this reason, it does not require an external power source. To be operational, the thermocouple must be connected in a circuit, as shown in Fig. 13-12. The dissimilar metals can be **constantan and iron, Alumel and Chromel,** or some other combination of metals or alloys which will produce the required results. Generally, reciprocating engine cylinder-head temperature (CHT) systems use iron and constantan, and turbojet exhaust-gas temperature (EGT) systems use Alumel and Chromel. The complete thermocouple circuit consists of the "cold" junction, the "hot" junction, electric leads made from the same material as the thermocouple, and a galvanometer-type indicating instrument.

In Fig. 13-12 we have indicated that the dissimilar metals constituting the thermocouple are constantan and iron. Constantan is chosen as one of the metals because its resistance is affected very little by changes in temperature. Note that a constantan wire is connected from the hot junction to the cold junction at the instrument, and that an iron wire is connected as the other side of the circuit. When the hot junction is at a higher or lower temperature than the cold junction, a current will flow through the circuit and instrument. The value of the current will depend on the difference in temperature between the two junctions.

The indicating instrument used with a thermocouple system includes a bimetallic thermometer mechanism which provides for inclusion of the ambient temperature at the instrument. If the instrument is not connected to the thermocouple leads or if it is connected and the temperatures of the two junctions are equal, the indicator will show ambient temperature.

The hot junction of the thermocouple system may be constructed in a number of different forms. For reading cylinder-head temperature on reciprocating engines, the thermocouple is metallically bonded to a copper spark-plug washer or is designed to be secured by a screw in a special well in the head of one of the cylinders. The cylinder to which the thermocouple is attached is determined by the manufacturer. It is the usual practice to find what cylinder operates at the highest temperature and to attach the thermocouple to this cylinder.

Thermocouples for reading exhaust-gas temperature (EGT) for either a reciprocating engine or a gas-turbine engine are constructed in the form of probes. The probe containing the thermocouple is inserted into a suitably designed opening in the exhaust pipe or tail pipe so that it is exposed to the stream of exhaust gases. For a reciprocating engine, one thermocouple is usually adequate; but for gas-turbine engines, thermocouple

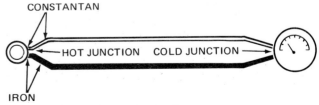

FIG. 13-12 Thermocouple circuit.

probes are installed around the entire periphery of the turbine exhaust case. The thermocouples are connected in series, thus providing an average temperature reading of the points around the exhaust. Some operators install a thermocouple on each cylinder of a reciprocating engine and use a rotary switch to select the cylinder to be checked.

The EGT gage used with reciprocating engines is primarily to furnish temperature readings for the **economy mixture indicating system.** As mentioned before, the temperature of the exhaust gases is highest when the combustion is most complete. The temperature therefore informs the pilot when the engine is operating at best economy.

The installation of the thermocouple probe in an exhaust stack is shown in Fig. 13-13(A). The probe is mounted in a clamp and the clamp is placed around the exhaust stack. This installation does not require a special fitting to be welded in the exhaust stack. An installation showing the welded fitting in the exhaust stack is given in Fig. 13-13(B).

Typical instructions for making a calibration check of the economy mixture indicating system are as follows:

1. Fly the airplane at 7500 ft [2286 m] altitude with an average cruising speed condition at 65 percent power.

2. Lean the mixture slowly and observe the EGT gage. Continue leaning until the EGT reaches a peak and starts to decrease. Enrich the mixture slowly to regain the peak temperature reading.

3. Record the peak reading after the system has stabilized.

4. Lean the mixture to a setting which provides an EGT reading of not less than 25° F below peak EGT [3.89° C].

5. Use the adjusting screw on the face of the instrument and position the pointer at $\frac{4}{5}$ full scale.

6. If the adjustment requires more than ±75° F [23.89° C], adjust the calibration screw on the back of the instrument one turn clockwise for an increase in indicator reading of 25° on the indicator scale or one turn counterclockwise for a decrease of 25°.

The face of a typical EGT gage for a reciprocating engine is shown in Fig. 13-14. This is a double indicator for a two-engine airplane. Note that there are two calibrating screws on the face of the instrument. On the bottom of the back of the instrument are two additional calibrating screws for additional adjustment as required.

The installation, maintenance, and troubleshooting of thermocouple instrument systems are not difficult; however, they do involve careful attention to certain requirements of such systems. Among these are the following:

1. When installing a thermocouple system, make sure that the specified types of thermocouples, leads, and instruments are used.

2. Check the leads for length, resistance value, and condition. Resistance and continuity can be checked with an ohmmeter or multimeter. Set the meter to the correct range, then connect the meter leads, one to each end of the thermocouple lead. The meter will show the resistance of the thermocouple lead. Thermocouple leads are matched to specific installations and must not be altered.

3. Check the indicating instrument to see that it indicates the ambient temperature when not connected to the thermocouple leads. Usually the instrument will include adjusting screws, sometimes called zero adjusters. These are used to adjust for ambient temperature.

4. Install the instrument in the instrument panel according to the manufacturer's instructions. Handle the instrument carefully to prevent damage due to shock. See that the panel is properly shock-mounted.

5. Install and route the thermocouple leads as specified by the aircraft manufacturer. Secure the leads with clamps at specified intervals and see that the leads do not rub on any structure which will cause wear or other damage. At points where there is a junction in the leads, be sure to correctly connect the terminals. *If the leads are reversed, the indicator will give a reverse reading.* See that all connections are tight and secure.

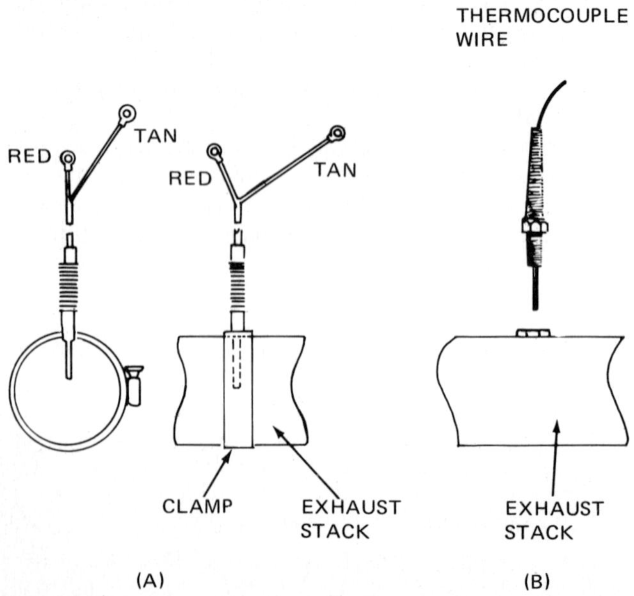

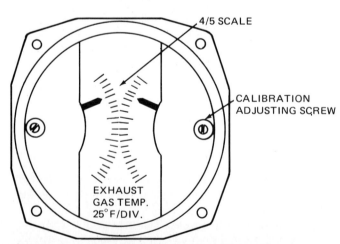

FIG. 13-13 Thermocouple installation in an exhaust stack.

FIG. 13-14 Exhaust-gas temperature (EGT) gage.

6. Install the thermocouple on the specified cylinder. See that the attachment is tightened and secured as required.

The following faults in the operation of a thermocouple system are the result of discrepancies as noted:

1. When the instrument gives a reading of ambient temperature only, there is a break in the thermocouple circuit.

2. When the reading is erratic, there is a loose connection in the circuit or in the instrument.

3. When the reading is reversed (decreases with an increase in temperature), the thermocouple leads are reversed at one of the junctions or at the instrument.

Tachometers

The tachometer is a primary engine instrument designed to provide an accurate indication of engine rpm. For reciprocating engines the reading is in rpm, but for gas-turbine engines the reading is given in percent of maximum allowable rpm.

Early tachometers were of the mechanical type which employed flyweights on a rotating shaft to produce movement of the indicating needle. Figure 13-15 illustrates the use of flyweights in an instrument mechanism. Flyweight tachometers were effective and operated well; however, the instruments were subject to more wear than later types.

Magnetic tachometers. A very common and widely used tachometer mechanism is the magnetic type. This mechanism utilizes a cylindrical magnet rotating in an aluminum **drag cup** to produce movement of an indicating needle. The flexible drive shaft from the engine is connected to the cylindrical magnet. As the magnet rotates, it generates eddy currents in the aluminum drag cup and these currents produce electromagnetic forces which cause the drag cup to rotate in the same direction as the rotating magnet. The drag cup is restrained by a coiled spring so that it can rotate only through the distance allowed by the spring. The amount of drag-cup rotation is proportional to the speed of the rotating magnet; hence, the drag-cup rotation provides a measure of engine rpm. The drag cup is mounted on the same shaft as the indicating needle of the instrument which produces the reading on the dial of the tachometer. The operation of a magnetic tachometer is illustrated in Fig. 13-16.

The flexible shaft for a mechanical drive tachometer is encased in a flexible metal housing. To operate prop-

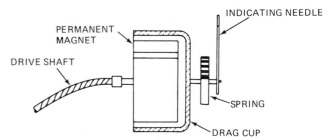

FIG. 13-16   Illustration of the magnetic tachometer principle.

erly, the shaft and housing must be free of kinks, dents, and sharp bends. There should be no bend with a radius of less than 6 in [15.24 cm] and no bend within 3 in [7.62 cm] of either end terminal.

If the tachometer is noisy or the pointer oscillates, check the housing and shaft for defects or improper installation. Disconnect the cable shaft from the tachometer and pull it out of the housing slowly and carefully. Check the shaft for worn spots, kinks, breaks, dirt, and lack of lubrication.

If the condition of the cable is satisfactory, clean it thoroughly and then apply AC type ST-640 speedometer cable grease, Lubriplate 110, or an equivalent grease to the lower two-thirds of the cable to provide a light coat of lubricant. Insert the shaft as far as possible into the housing, making sure the proper end is inserted. Rotate the shaft until the end terminal seats in the engine drive fitting. Insert the upper end of the shaft into the tachometer fitting and see that it seats properly. Screw the housing terminal nut onto the tachometer fitting and torque as specified. A typical instrument usually requires approximately 50 in·lb [5.65 N·m]. Safety the nut as required.

If it is found that the tachometer instrument is not functioning properly, replace the instrument. Repair of the instrument should be done only in an approved instrument overhaul shop.

Electric tachometers. Because of the distances between the engines and the cockpit on large aircraft, the flexible drive shafts used for tachometer operation on small aircraft cannot be used. For this reason, electric tachometers were developed so that the rpm indication from the engine could be transmitted through electric wiring.

An early type of electric tachometer was simply a permanent-magnet DC generator mounted on the engine and a voltmeter, scaled for rpm, mounted on the instrument panel in the cockpit. The voltage produced by the generator is directly proportional to engine rpm; hence, the reading of the voltmeter in the cockpit is an indication of engine rpm.

A commonly employed electric tachometer system utilizes a three-phase generator mounted on the engine to develop three-phase AC current whose frequency is determined by engine rpm. This current is transmitted to a three-phase synchronous motor in the tachometer instrument. The motor speed is determined by the frequency of the current which, in turn, is established by the engine rpm. Thus, instead of using a flexible drive shaft to transmit engine rpm to the instrument in the cockpit, a three-phase electric current performs the same function.

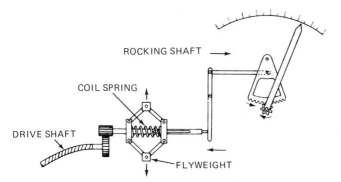

FIG. 13-15   Use of flyweights in a tachometer.

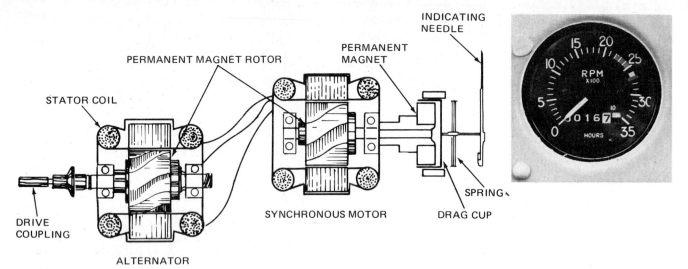

FIG. 13-17 AC electric tachometer system.

Figure 13-17 illustrates the AC electric tachometer system. The alternator, mounted on the engine, produces a three-phase current which is delivered to the synchronous motor. The three-phase current creates a rotating field in the stator of the **synchronous motor**. The rotor of the synchronous motor, being a permanent magnet, aligns itself with the rotating field and turns at the same speed as the alternator. The synchronous motor drives a magnet in the drag cup of the instrument, thus causing the drag cup to turn and move the indicating needle. As explained previously, the drag cup is restrained by a coiled balance spring and will turn to a point where the spring force and the rotating force are balanced. The indicating needle will therefore move a distance proportional to engine speed and provide a reading of rpm.

Many tachometers, particularly those for light aircraft, include an **hour meter** by which the total flight hours of the aircraft are recorded. Such an hour meter is shown in Fig. 13-17.

Hour meters are manufactured in two basic designs. For one type, the meter mechanism is geared directly to the rotating element within the tachometer. This type of mechanism records engine revolutions and provides an accurate record of time only when the engine is operating at the design cruising rpm. For an engine designed to cruise at 2100 rpm, the hour meter will record 1 h of flight time as the engine completes 126,000 revolutions.

The Hobbs hour meter is operated by an electric clock mechanism. This system receives power from the aircraft battery or electric system and may be connected to operate when the master switch is turned on, when the engine is operating, or when the aircraft is in flight. The latter type of system would provide the only true indication of actual flight time; and in this case, the actuating switch is operated by the landing gear. At takeoff, when the weight of the aircraft is taken off the landing gear, the switch closes and turns on the hour meter.

Another system for turning on the Hobbs meter utilizes a pressure-operated switch connected to the oil system. This switch is adjusted so that it will close when the oil pressure reaches a predetermined level.

Tachometers for helicopters are dual instruments. These instruments include one system for indicating main-rotor speed and another system for indicating engine drive output rpm. The face of the instruments includes two indicating hands and two sets of dial markings.

Electronic tachometers. An electronic tachometer involves a system that utilizes electric impulses from the magneto ignition system to produce the engine rpm indication. The tachometer includes a frequency-to-voltage converting circuit that feeds into a standard meter movement. The electric impulses are obtained from a special pair of breaker points in one of the engine magnetos. The impulses, having a frequency proportional to engine rpm, produce a voltage by means of the converting circuit that is also proportional to engine rpm. This voltage fed to the meter movement produces a needle deflection that indicates engine rpm.

The simplest type of electronic tachometer has two connections. One receives the signal pulses from the magneto and the other is connected to ground. In this case the impulses from the magneto are sufficiently strong to operate the instrument with no outside power required.

A more complex electronic tachometer operates on the principle described above; however, it includes an amplifier circuit with one or two transistors to strengthen signal impulses that otherwise might not be sufficient to operate the instrument. This type of tachometer system requires three electrical connections, one being for an input of 12 V from the aircraft electrical system.

Maintenance and inspections. Many of the older light aircraft utilize mechanically driven tachometers. The principal maintenance required for these is service and lubrication of the drive shaft.

The inspection and maintenance of electric tachometer systems is the same as for other electric systems. If either the tachometer generator (alternator) or the instrument is defective, it should be replaced. If the instrument vibrates or shakes, it should be checked for security of mounting and the panel should be checked for integrity of the shock mounts.

During routine inspections, both the tachometer generator at the engine and the instrument in the cockpit should be checked for security of mounting. Examine the electric plugs and wiring for condition. The plugs should be tight in the receptacles. Wiring should be properly laced or clamped to prevent vibration or fluttering which could cause wear. Usually the electric wiring from a number of instruments will be secured together in a harness and the harness will be supported by suitable clamps. Routing of electric wiring or harness should be such that they are not exposed to excessive heat or liquids.

Fuel Flowmeters

Fuel flowmeters are employed on all large aircraft to provide the flight engineer or pilot with important information regarding the efficient operation of the engines. The flow meter is the most accurate way to determine the fuel consumption of all types of engines. The quantity of fuel being burned per hour, when integrated with other factors, makes it possible to adjust power settings for maximum range, maximum speed, or maximum economy of operation. If the fuel consumption of an engine is abnormal for any particular power setting, then the operator is warned that there is something wrong (a fault) in the engine or its associated systems.

Fuel flowmeters may be scaled for pounds per hour, gallons per hour, or kilograms per hour. The most accurate units of measurement for fuel are the pound or the kilogram because these are measures of weight. Since the weight of a gallon of fuel varies with temperature, fuel temperature would have to be considered in making a computation based on gal/h fuel consumption.

In engine test cells, a tubular flowmeter is often used. This is simply a glass or plastic tube with a tapered inside diameter. The tube is mounted vertically with an indicating ball inside. Fuel enters the bottom of the tube and flows out the top and in so doing causes the indicating ball to rise a distance proportional to the rate of fuel flow. Figure 13-18 shows a tubular fuel flowmeter.

Light aircraft with continuous-flow fuel-injection engines usually employ a fuel pressure gage which also serves as a flowmeter. This is possible because the fuel flow in a direct fuel-injection system is proportional to fuel pressure. The instrument can therefore be calibrated either in psi or gallons per hour or both. A gage of this type is shown in Fig. 13-19. Due to its design characteristics, this system is not as accurate as the above-mentioned fuel flowmeters. If a fuel-injection nozzle would become clogged, the gage would indicate high fuel flow rather than the lower indication normally expected.

Fuel-flow indicating systems for large aircraft are of two general types. The first, used in many older aircraft, employs a **synchro system** to transmit a measurement of fuel flow from the sensor and transmitter to the indicator. The principle of the synchro system is illustrated in Fig. 13-20. Fuel flow through the flow sensor causes rotation of the transmitter rotor to a position proportional to the fuel flow rate. The rotor, being energized electrically by 400 Hz AC, generates a three-phase current in the stator which flows through the stator of the indicator. The magnetic field thus produced reacts with the field of the rotor in the indicator and causes it to assume a position corresponding to the position of the rotor in the transmitter. The indicator needle is mounted on the shaft with the rotor; hence, it moves to a position established by the rotor. The indicator needle thereby shows the fuel flow rate on the instrument scale.

Fuel-flow indicating systems for modern jet aircraft are complex electric and electronic circuits and devices. The transmitter is located in the fuel line between the engine-driven fuel pump and the carburetor, or fuel-control unit. The sensing device may be a gate-type unit or it may be a turbine which generates an electric signal. In any case, the signal sent from the transmitter contains the information required regarding fuel flow rate. The electric signal is delivered to the "receiver" of the indicator where it is converted to a form which provides movement of the indicator needle or tape. Additional information regarding electric and electronic indicating systems is given in the associated text covering *Aircraft Electricity and Electronics*.

Fuel Pressure Warning Systems

For aircraft equipped with multiple-tank fuel systems, it is desirable to have a means of warning the

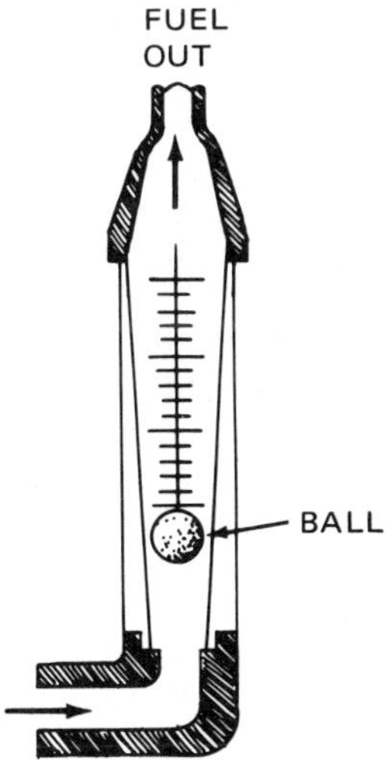

FUEL
OUT

BALL

FUEL IN →

FIG. 13-18   Tubular fuel flowmeter.

FIG. 13-19   Pressure gage also used as a flowmeter.

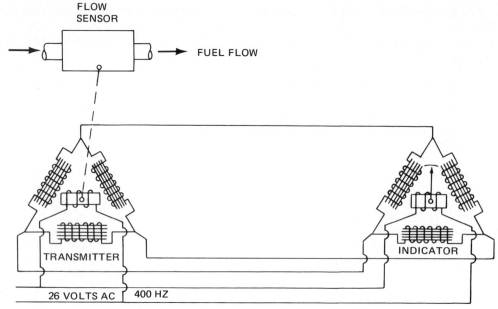

FIG. 13-20 Flowmeter utilizing a synchro transmitting system.

pilot or flight engineer that the fuel in an operating tank is exhausted and that the fuel-selector valve must be set to draw fuel from another tank. The fuel-pressure system consists of a diaphragm-operated pressure switch with one side of the diaphragm connected to the main fuel line downstream from the main fuel pressure pump, electric wiring to a warning light in the cockpit, and possibly a buzzer or other aural alarm. Fuel pressure holds the switch open at normal pressures, but the switch closes when the pressure falls. This turns on the warning light in the cockpit.

Fault Indicating and Isolating Systems

From the simple indicators used on early aircraft, engine indicating systems have developed to include (in addition to basic measurements of temperatures, pressures, and speeds) a multitude of indicators to give the flight crew of an aircraft information about the operation of engines and powerplant systems. Modern jet aircraft incorporate huge panels containing instruments, indicator lights, test switches, control switches, and adjustment knobs by which the flight engineer can continuously monitor the powerplants, powerplant systems, and all other systems in the aircraft and perform tests which will make it possible to isolate the faults which may develop. Many of the lights and other indicators are used for fault indicating and isolating.

An example of a fault isolating system is the FEFI/TAFI system developed by the McDonnel Douglas Corporation for use with the DC-10 airliner. This system incorporates two distinct programs, each set forth in a comprehensive manual. The first program deals with the collection of operational information in flight. This is called the flight engineer's fault isolation (FEFI) program. The second part of the system is called the turn around fault isolation (TAFI) program, wherein technicians employ information developed during operation to isolate faults through the use of the TAFI manual. When a fault has been isolated and identified, the corrective action is immediately indicated. A de-

tailed discussion of the FEFI/TAFI system is provided in Chap. 23.

● FIRE WARNING SYSTEMS

Types of Systems

The need for fire warning and suppression systems in aircraft has led to the development of a number of different types of designs and installations. Extensive research and engineering studies have been done to determine what systems and devices are most effective in providing immediate warning of fire or overheat conditions so that appropriate action can be taken by flight crews. Problems encountered include the location and routing of sensing devices, false warnings due to a variety of causes, difficulties in providing means for testing of systems, and the establishment of effective maintenance and testing procedures.

The types of systems which have been developed generally meet the following requirements:

1. Provide an immediate warning of fire or overheat by means of a red light and an audible signal in the cockpit.
2. Provide an accurate indication that a fire has been extinguished as well as an indication if a fire reignites.
3. Durability and resistance to damage from all the environmental factors which may exist in the location where the system is installed.
4. Incorporation of an accurate and effective testing system to ensure the integrity of the system.
5. A system which is easily removed and installed.
6. A system which operates from the aircraft electric system without inverters or other special equipment and which requires a minimum of power.

Fire and overheat warning systems are installed in many areas of large aircraft; however, only those applicable to aircraft powerplants are discussed in this section. The principles explained are actually applicable to all systems.

High temperatures may be detected by a number of devices. Among these are thermocouples, thermistors (resistors or materials which change resistance substantially with changes in temperature), gases in sealed elements, and bimetallic thermal switches. Fire and overheat detectors are constructed as small units a few inches in length and also as tubular units from 18 in [45.7 cm] to more than 15 ft [4.6 m] for continuous-loop systems. Diameters of the continuous-loop sensing elements may be from less than 0.060 in [1.5 mm] to more than 0.090 in [2.3 mm].

A comparatively simple fire detection system, such as that installed on light aircraft, consists of one or more thermal switches connected in parallel through an alarm circuit which includes a red fire-warning light and an alarm bell. A schematic diagram of a circuit for this type of system is shown in Fig. 13-21(A). The thermal switches in this system, sometimes referred to as spot detectors, are located at the points in the engine nacelle where temperatures are likely to be the highest. They are designed to close when the temperature reaches a level substantially above the normal operating temperature.

The **thermocouple** fire warning system shown in the schematic diagram of Fig. 13-21(B) utilizes a reference thermocouple that provides a comparison of temperatures between the reference zone and the active zone. The reference thermocouple is isolated from the active thermocouples so that the reference signal will not be affected by engine heat. When a fire occurs, the temperature in the area of the active thermocouples will exceed the temperature of the reference thermocouple, causing a current to flow through the sensitive relay. This relay closes and activates the slave relay that closes and directs current to the warning light in the cockpit. The test circuit of the thermocouple system includes a heating element that causes the test thermocouple in the thermal test unit to produce a current that activates the system for test.

Fenwal system. For both reciprocating engines and turbine engines on large aircraft, continuous-loop fire warning systems are employed. The Fenwal continuous fire-detection and overheat system utilizes a **sensing element** which may vary in length from 1.5 ft [0.46 m] to 15 ft [4.57 m], the length being selected to fit the particular engine system design. The sensing element consists of a small (0.089 in [2.26 mm] OD), lightweight, flexible Inconel tube with a pure nickel wire center conductor. The voids and clearances between the tubing and the conductor and the porous ceramic are saturated with a eutectic (low-melting-point) salt mixture. A section of the sensing element is illustrated in Fig. 13-22.

The tube is hermetically sealed at both ends. The nickel conductor terminates in a welded joint at the tip of a hermetically sealed terminal, thus providing an electric contact at each end of the element. End terminals of the element are shown in Fig. 13-23.

The sensing element responds to a specified degree of elevated temperature at any point along its entire length. At the alarm temperature of the element, the resistance of the eutectic salt drops rapidly and permits increased current to flow between the inner and outer conductors. The increased current flow provides the signal which is sensed in the electronic **control unit.** The control unit produces the output signal to actuate

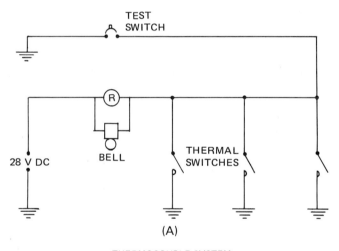

(A)

THERMOCOUPLE SYSTEM

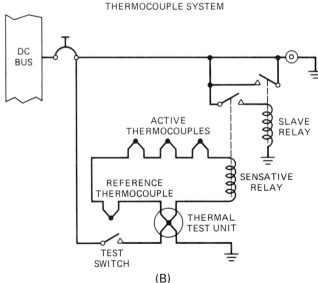

(B)

FIG. 13-21   Fire warning system for small aircraft.

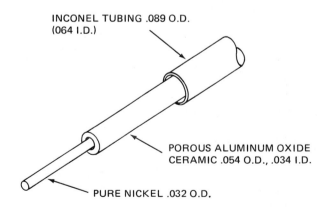

INCONEL TUBING .089 O.D. (064 I.D.)

POROUS ALUMINUM OXIDE CERAMIC .054 O.D., .034 I.D.

PURE NICKEL .032 O.D.

VOIDS BETWEEN TUBING, CERAMIC & WIRE AND POROSITY OF CERAMIC ARE SATURATED WITH A EUTECTIC SALT MIXTURE.

FIG. 13-22   Section of the Fenwal sensing element.   *(Fenwal)*

FIG. 13-23   End terminals of the sensing element.

the alarm circuit. The Fenwal system incorporates a few spot detectors wired in parallel between two separate circuits. This prevents a short circuit in either one from causing a false fire-warning signal.

After corrective action has been taken and an overheat condition no longer exists, the resistance of the eutectic salt in the sensor increases to its normal level and the system returns to a standby condition. Should the fire reignite, the system would again produce an alarm.

Kidde system.   The continuous fire warning system designed by Walter Kidde and Company, Inc., utilizes a **sensing element assembly** comprised of a pair of sensing elements mounted on a preshaped, rigid support tube assembly. The sensing elements are held in place by dual clamps riveted to the tube assembly approximately every 6 in [15 cm]. Asbestos grommets are provided to cushion the sensing elements in the clamps. Four nuts per assembly are used to secure the sensing element end connectors to the bracket assemblies at either end of the support tube assembly. The nuts are secured with safety wire.

The sensing element of the Kidde system consists of an Inconel tube containing a **thermistor** (thermal resistor) material in which are embedded two electric conductors. The sensing elements terminate at both ends in electric connectors with one of the conductors grounded to the shell of the connector.

The design of the Kidde sensing element is shown in Fig. 13-24. The resistance of the thermistor material decreases rapidly as the element is heated. When the resistance decreases to a predetermined point as the result of a fire, the control circuit monitoring the resistance of the element produces an alarm signal in the cockpit.

In the Kidde system described, each sensing loop is connected to its own electronic control circuit mounted on a separate circuit card. This arrangement provides for complete redundancy, thus ensuring operation of the system even though one side may fail. Either the Fenwal or Kidde system will detect a fire when one sensing element is inoperative, even though the press-to-test circuit does not function.

Pneumatic system.   A pneumatic fire and overheat detection system, manufactured by the Systron-Donner Corporation, utilizes a gas-filled, sealed tube to provide the signal to actuate an alarm. The tube is connected to a **responder,** which is essentially an assembly containing one or two diaphragm-type switches.

The sensor consists of two tubes, one inside the other (coaxial), containing two separate gases plus a gas-absorption material inside the inner tube. The outer tube is of stainless steel with an outside diameter of 0.063 in [1.6 mm]. The outside diameter varies with different models. Between the inner and outer tubes an inert gas (helium) is contained under pressure.

The inner tube is made of an inert metallic material and encloses the gas-absorption material. At temperatures below the fire alarm level, the material retains the gas. The cross-sectional drawing in Fig. 13-25 illustrates the construction of the sensor.

The principle of operation is based on the laws of gases. If the volume of a gas is held constant, its pressure will increase as temperature increases; thus, the helium gas between the two tubing walls will exert a pressure proportional to the average absolute temperature along the entire length of the tube. One end of the tube is connected to a small chamber containing a metal diaphragm switch. One side of the diaphragm is therefore exposed to the sensor tube pressure and the other side is exposed to ambient pressure. When the average temperature along the tube reaches the predetermined overheat level, the gas pressure will cause the metal diaphragm to snap over ("oil can") and close the electric contacts. The metal diaphragm forms one of the contacts and the end of a conductor leading from the pin contact in the end of the responder forms the other contact. When the pressure causes the diaphragm to snap over, the diaphragm hits the end of the conductor and completes the electric circuit. This activates the alarm in the cockpit. The operation of the diaphragm is illustrated in Fig. 13-26.

The action of the helium gas in the sensor to provide an overheat alarm is called the **overheat function.** The operation of the sensor to detect a fire is called the **discrete function** and this is performed by the inert gas contained in the gas-absorption material in the inside tube of the sensor. When any portion of the sensor tube is exposed to the fire warning temperature, the gas-absorption material immediately releases a large quantity of gas which quickly increases the pressure in the tube to a level necessary to close the diaphragm switch. This produces the fire alarm signal. When the temperature is reduced to normal levels after action has been taken to suppress the fire, the gas-absorption material reabsorbs the gas and the pressure drops, thus opening the switch and cutting off the alarm. At this

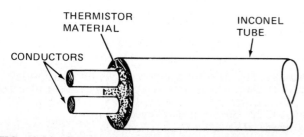

FIG. 13-24   Construction of the Kidde sensing element.

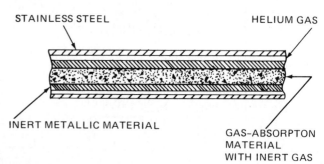

FIG. 13-25   Construction of a pneumatic temperature sensor.

340

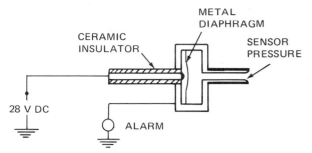

FIG. 13-26 Diaphragm pressure switch.

time the system is ready again to produce overheat or fire warnings.

To provide integrity monitoring for the system, an additional diaphragm switch is incorporated in the responder housing. This switch is normally held closed by the gas pressure in the sensor tube. The test circuit will function as long as the integrity switch remains closed. Figure 13-27 is a schematic diagram to show the operation of the **integrity monitoring** system. Note that the push-to-test switch and the integrity switch are connected in series. When both switches are closed, current can flow from the 28-V DC source through the circuit to provide an alarm signal. If there is no response when the test switch is closed, the flight engineer or pilot will know that the gas pressure in the sensor has been released as a result of some type of injury to the sensor tube.

Routing of Fire Warning Sensors

The routing of overheat and fire warning sensors has been determined by the manufacturer through extensive engineering tests. Areas in the powerplant nacelle are often divided into *zones* and each zone is carefully checked to determine normal maximum operating temperatures and the areas of highest temperatures. Sensors are routed through locations where fires are most likely to occur.

Overheat conditions can occur where no fire exists; however, since overheat is an indication of a dangerous condition, the flight crew must be warned when such conditions are indicated. Overheat conditions may be caused in nacelles of large reciprocating engines by cracks or burn-through in the exhaust manifold, turbosupercharger ducts or housings, or tailpipes. Failure of engine baffling can also produce dangerous hot spots. In turbine engine nacelles, overheat conditions can develop as a result of a cracked compressor housing, leaking-bleed air ducts, burn-through in combustors, and burn-through or other leaks in the exhaust system.

It must be remembered that the routing of overheat and fire warning sensors is designed carefully to provide the most effective system possible. It is therefore essential that the replacement of sensing elements be done strictly in accordance with the original routing.

Installation of Sensors

Various manufacturers provide different types of clamps, clips, grommets, brackets, and other supporting devices for sensing elements. Typical items for installation and attachment of a Fenwal system are shown in Fig. 13-28.

The **bulkhead connector** is used at bulkheads and firewalls. It is a feed-through electric connector which electrically joins two sections of a sensing element where it is necessary for the element to pass through.

The **wire end-fitting assembly** is an electric connector designed according to customer specifications for connection to the electric circuit of the aircraft.

The **support-flange assembly** is used for the attachment of fittings to structure. These units are supplied in a variety of configurations.

Mounting clips are supports for sensing elements and are used with bushings or grommets. The mounting clip is attached to the engine or aircraft structure and provides a quick means of installation or removal of a sensing element. The clip is held closed by a Cam Loc fastener which may be opened or closed quickly with a screwdriver.

Typical instructions for the installation of a Fenwall sensing element are given in the drawings of Fig. 13-29. For any specific type or model, the appropriate manufacturer's instructions should be consulted.

Attachment details for a Kidde sensing-element assembly are shown in the drawing of Fig. 13-30. As explained previously, the assembly consists of a preformed support tube on which two sensing elements are mounted by means of clamps and grommets. Brackets attached to the support tube provide for attachment to the nacelle or aircraft structure. At the ends of the sensing elements are pin and socket receptacles to provide for electric connections.

Inspections and Testing

Instructions for the inspection and testing of fire and overheat warning systems are provided by the manufacturers. Because of the different types and models

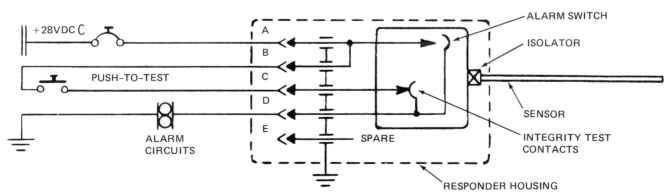

FIG. 13-27 Schematic diagram of alarm and integrity circuits. (*Systron-Donner Corp.*)

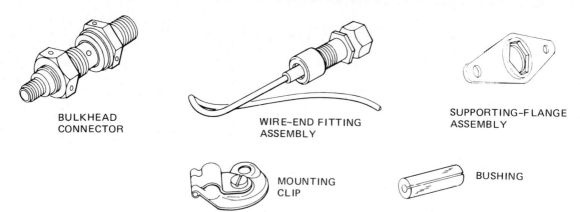

FIG. 13-28 Installation fittings and mounting devices. *(Fenwal)*

of systems, the technician must be sure to consult the appropriate manuals and use the correct specifications in making tests.

Typical inspection instructions are as follows:

1. Remove or open engine cowling as required to gain access to sensing elements.

2. Check end fittings and sensor connectors for tightness and proper safetying.

3. Check bulkhead fittings and supports for security.

4. Examine all clamps and grommets (bushings) and check to see that the sensing element is held securely in place.

5. Examine the entire length of the sensors for wear, sharp bends, kinks, dents, crushed sections, cuts, cracks, proximity to structure, and security of mounting. Discard elements which do not meet the manufacturer's specifications.

6. Check the routing of sensors against the manufacturer's instructions.

7. Check the security of electronic control-circuit units.

The testing of sensing elements will vary according to the design and principle of operation. Elements which produce warning signals through resistance changes (Fenwal and Kidde) are tested by means of a megger and ohmmeter as specified by the manufacturer. The sensing element is disconnected at the ends and the resistance of the conductors is checked. Insulation resistance between the conductors and outer tube is checked with a megger.

Typical testing instructions for a Kidde sensing element are as follows:

1. Allow the sensing element to come to temperature equilibrium with its surroundings in the immediate vicinity of the mercury thermometer.

2. Check the resistance of each element from the center contact at one end to the center contact at the other end. The maximum allowable resistance for each element is specified by the manufacturer. If an element exceeds the specified values, replace it.

3. Using a high-range ohmmeter, connect one lead of the meter to the shell of the sensing element connector and the other lead to the center contact. Observe the resistance reading, using a scale that provides approximate center-scale deflection for accuracy. The

reading must be higher than the megohm-vs.-temperature curves given in the appropriate graph.

4. Using an insulation resistance test set (megger), connect one lead to the shell of the sensing-element connector and the other lead to the center contact. Apply the test voltage of 350 to 500 V DC. The reading must be higher than the megohm-vs.-temperature curves given in the applicable manufacturer's chart.

Operational tests are accomplished by applying heat to an element installed in an operating system. When a controlled flame is applied to any portion of a sensing element, the system should provide an immediate fire warning alarm signal. When the heat is removed from the element, the alarm should stop.

For daily and preflight tests, the push-to-test switches are used. When the switches are pressed, the system being checked should produce an alarm.

Types of Fires

Fires are classified according to the combustible materials involved and the general nature of the fire. Class A fires are those involving ordinary combustible materials such as paper, cloth, and wood. Class B fires involve flammable liquids such as gasoline, oils, kerosene, and jet fuel. Class C fires are those involving electrical wiring and equipment. Methods for extinguishing fires are determined by their nature and class. For example, water and water-based extinguishing agents must not be used for class B and class C fires.

● FIRE SUPPRESSION SYSTEMS

A fire suppression or extinguishing system consists of one or more pressure tanks (containers) containing an effective fire extinguishing agent, an instantaneous release valve, deployment lines to carry the agent to the engine compartment (nacelle) involved with the fire, and associated controls and indicators. Some systems are designed to provide only one release of extinguishing agent per engine and others provide for two releases of agent. The two-shot feature may be attained by utilizing two separate agent containers for each area or by incorporating a cross-feed system so that the extinguishing agent can be drawn from either of two different containers for each engine.

Extinguishing agents used in engine fire extinguishing systems vary in types. Some commonly used agents

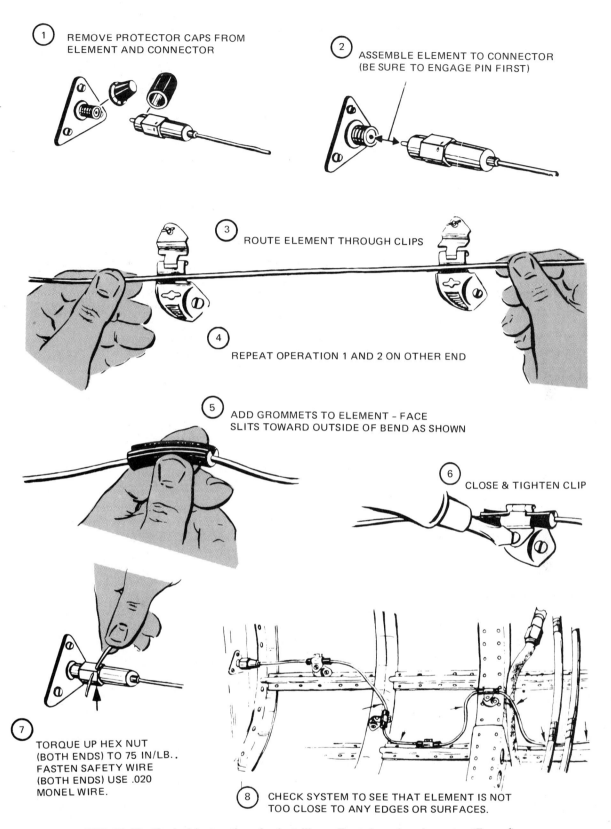

1 REMOVE PROTECTOR CAPS FROM ELEMENT AND CONNECTOR

2 ASSEMBLE ELEMENT TO CONNECTOR (BE SURE TO ENGAGE PIN FIRST)

3 ROUTE ELEMENT THROUGH CLIPS

4 REPEAT OPERATION 1 AND 2 ON OTHER END

5 ADD GROMMETS TO ELEMENT – FACE SLITS TOWARD OUTSIDE OF BEND AS SHOWN

6 CLOSE & TIGHTEN CLIP

7 TORQUE UP HEX NUT (BOTH ENDS) TO 75 IN/LB.. FASTEN SAFETY WIRE (BOTH ENDS) USE .020 MONEL WIRE.

8 CHECK SYSTEM TO SEE THAT ELEMENT IS NOT TOO CLOSE TO ANY EDGES OR SURFACES.

FIG. 13-29 Typical instructions for installing a Fenwal sensing element. *(Fenwal)*

are carbon dioxide (CO_2), Freon, bromotrifluoromethane (CF_3Br), and chlorobromomethane (CH_2ClBr) also known as CB. Some factors in the selection of an agent are cost, ease of handling, weight, toxicity, corrosive effect on metals, and effectiveness in extinguishing fires.

CO_2 is widely used as a fire extinguishing agent around aircraft because it can be used safely for both class B and class C fires. Furthermore, from a safety standpoint, CO_2 has little or no toxic or corrosive effect. It is contained as a liquid in highly pressurized containers and when released through a cone-shaped nozzle im-

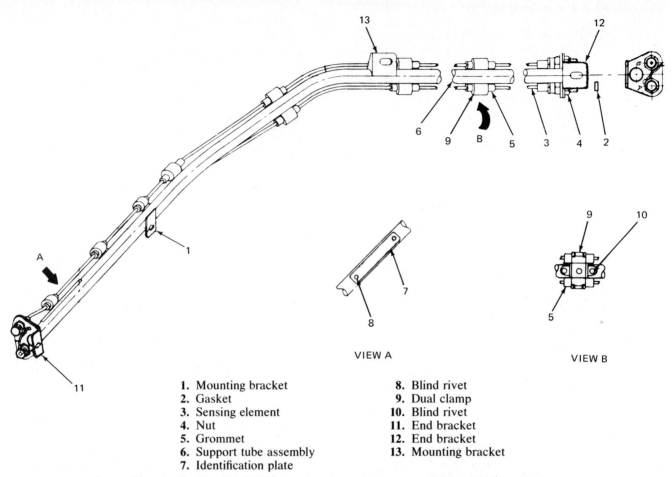

1. Mounting bracket	**8.** Blind rivet
2. Gasket	**9.** Dual clamp
3. Sensing element	**10.** Blind rivet
4. Nut	**11.** End bracket
5. Grommet	**12.** End bracket
6. Support tube assembly	**13.** Mounting bracket
7. Identification plate	

FIG. 13-30   Details of Kidde sensing-element assembly.   *(Walter Kidde and Co.)*

mediately becomes a gas at very low temperature. Some of the gas freezes and produces a "snow" that, together with the gaseous component, is effective in smothering a fire by displacing the oxygen in the air.

System for Light Aircraft

A typical fire extinguishing system for a light twin-engine airplane consists of a pressure bottle (container) mounted in the aft section of each engine nacelle, an explosive cartridge (squib) in the outlet fitting, a discharge hose leading to the forward section of the nacelle, and a smaller hose leading off the main hose to discharge the Freon agent from the center of the nacelle area toward the inboard side of the nacelle. If an overheat condition is detected, a FIRE light will indicate which engine is involved. The extinguisher is activated by opening the guard and pressing the FIRE light. This completes the electric circuit to the squib, causing it to fire and rupture the disk in the outlet. Freon gas is discharged from the container and floods the nacelle to extinguish the fire.

After the bottle discharges, an amber E light comes on to show that the bottle discharged. This light will continue to show the empty condition until the extinguisher bottle is replaced.

Systems for an Airliner

The fire extinguishing equipment for an airliner such as the Douglas DC-10 incorporates three systems with six pressurized agent containers, two for each engine. The containers for the wing-mounted engines are lo-

cated in pairs in the leading edge of the wing between the nacelle and the fuselage on each side of the aircraft. The mounting of the containers is shown in Fig. 13-31. Note that the two containers feed through a Y fitting to the common deployment line. The arrangement of the system for the aft engine and APU compartment is shown in Fig. 13-32.

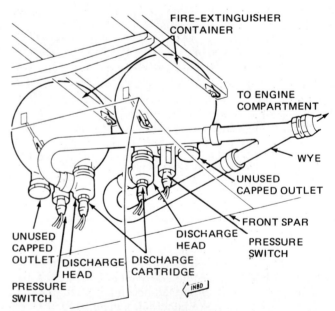

FIG. 13-31   Mounting of extinguishing agent containers. *(McDonnell Douglas Corp.)*

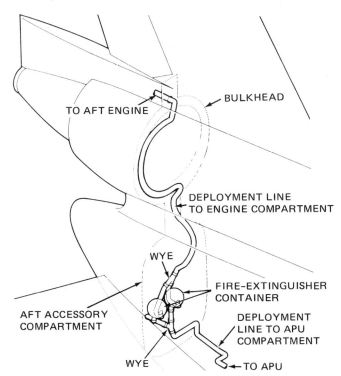

FIG. 13-32 Deployment system for aft engine and APU compartment. *(McDonnell Douglas Corp.)*

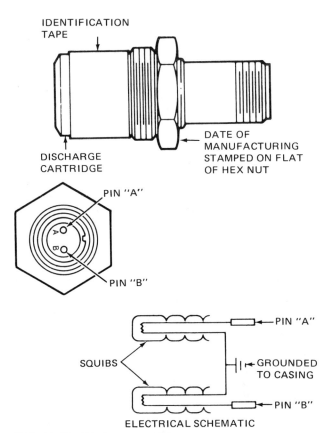

FIG. 13-33 Discharge cartridge.

The six agent containers are hermetically sealed and constructed of stainless steel. They may vary in capacity in accordance with customer requirements. The containers are filled with bromotrifluoromethane (CF_3Br) and pressurized with nitrogen. Each container is provided with two outlet ports with frangible metal disks welded to each outlet neck. The disks serve as pressure-relief valves that will rupture if an overpressure condition develops because of excessive heat. Depending on which disk ruptures on the wing-mounted containers, the discharge of CF_3Br will flow to the engine nacelle or out through two orifices in the cap of the unused outlet neck. For the aft-engine containers, the relief discharge will flow either to the engine nacelle or to the APU compartment.

Each container incorporates a pressure switch that remains open when the container is adequately pressurized. If the pressure drops below the specified level, the switch will close and illuminate the AGT LOW light which is amber in color. This light remains ON until the pressure is restored to normal.

The discharge cartridge is an electrically fired explosive unit that provides force to rupture the disk seal in the outlet port of the container and release the extinguishing agent to the deployment lines. One discharge cartridge is installed on each wing-mounted container and two are installed on the aft-engine container.

The discharge cartridge, shown in Fig. 13-33 is common on most modern large and medium aircraft. It contains two squibs (explosive charges) that are fired by an electric circuit that heats the powder to the ignition point. When the fire extinguisher control handle is pulled down, the electric circuit is closed and the squibs are fired. The force of the explosion ruptures the disk seal, thus releasing the extinguishing agent. A screen in the discharge head prevents particles of the disk from entering the deployment line.

The deployment lines and extinguisher outlets for wing-mounted engines are shown in Fig. 13-34. Note that the deployment lines terminate at five discharge nozzles in the engine nacelle. Two nozzles are located in the fan section and three in the core section. The agent may also be distributed to some engine areas through perforated tubing.

The operation of the systems is controlled through levers and switches on three panels, as shown in Fig. 13-35. When an overheat condition or fire occurs, a red light will appear on the control-lever handle for the

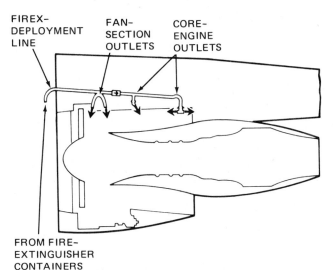

FIG. 13-34 Deployment lines and extinguisher outlets for wing-mounted engines. *(McDonnell Douglas Corp.)*

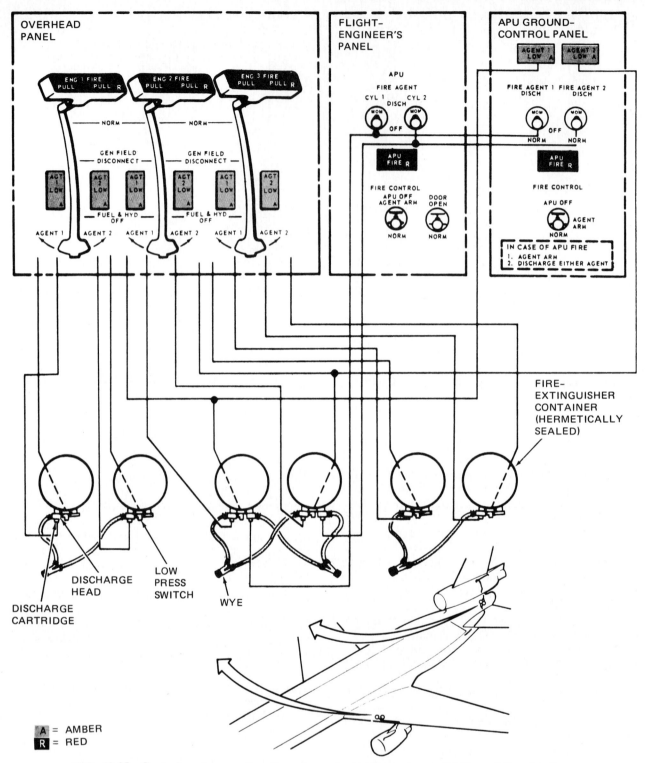

FIG. 13-35 Controls and connections for a fire extinguishing system. *(McDonnell Douglas Corp.)*

engine involved. At the same time, a loud aural alarm will sound. The pilot pulls the illuminated handle down all the way and turns it clockwise to discharge agent 1 or counterclockwise to discharge agent 2.

When the handle is pulled down, the aural alarm is shut off, the generator field for the affected engine is disconnected, and the fuel and hydraulics for the engine are shut off. At the same time, a switch controlling the circuit to the discharge cartridge is closed, thus firing the cartridge and discharging the selected con-

tainer through the deployment lines to the engine nacelle.

The two containers in the rear of the aircraft are interconnected so that the system may be used for the rear engine or the APU. If an APU fire exists, the red APU FIRE indicator will be illuminated and an aural alarm will sound. The flight engineer arms the system with a switch on his or her panel and then closes the switch for the agent container that he or she wishes to use. This fires the discharge cartridge in the container chosen and releases the agent to the APU compartment. The APU system can also be activated from the APU ground control panel in the bottom of the fuselage.

Fire Extinguishing System Indicators

All fire extinguishing systems for aircraft are required to incorporate indicators by which the system condition is shown. The systems described above utilize amber lights to indicate when a container has been discharged or when the pressure is low. Each container includes a pressure switch that is connected to the indicator light. When pressure is low in a container, the switch closes and turns on the amber light. Indicator lights are located on the fire-control panel in the cockpit so that the pilot is warned immediately when any container is low in pressure or is discharged.

For many aircraft fire extinguishing systems, frangible colored disks are employed as condition indicators. These are mounted in the side of the fuselage where they are readily visible to a crew member making a preflight inspection. A *yellow* or *amber* disk is used to indicate when an extinguishing agent has been discharged by use of the cockpit controls and a *red* disk is used to show when the discharge has occurred as a result of overpressure and rupture of the safety disk in the outlet of the agent container. A hose or tubing is connected from the main discharge line to the fitting behind the yellow or amber disk. Discharge of the agent by the cockpit control will direct pressure to the disk and cause it to rupture.

The safety disk fitting in the agent container is connected by a hose or tubing to the red disk fitting. When the safety disk is ruptured because of overpressure, agent pressure ruptures the red disk, thus providing an indication that the container is discharged.

Maintenance of Fire Extinguishing Systems

The maintenance of fire extinguishing systems is described in manufacturers' manuals for specific systems. There are, however, typical practices which apply to many systems. Periodic checks and tests are generally specified to ensure that a system and components are in operable condition.

The condition of lines, hoses, fittings, and components is checked visually and by feel to determine the condition and security of mounting.

Agent containers having pressure gages are checked by observing the pressure gages to see that the pressure is within the prescribed range for pressure vs. temperature. Containers without pressure gages are checked by weighing to determine that the quantity of agent in the container is adequate. Hydrostatic test dates must be stamped on the containers so that the technician will know when to remove them for weight checks.

Discharge cartridges are examined to see that the service life of the unit has not been exceeded. Service life is specified in the number of hours that a cartridge will be effective as long as it has not been exposed to temperatures above a specified limit.

When a discharge cartridge has been removed for any reason it must be reinstalled in the same container from which it was removed. During removal and installation, the terminals of the discharge cartridge should be grounded or shorted to prevent accidental firing. Before connecting the cartridge terminals to the electric system, the system should be checked with a voltmeter to see that no voltage exists at the terminals.

Because of the differences in fire extinguishing systems, instructions for inspection and maintenance will vary. For this reason it is essential that the technician follow the procedures set forth in the manufacturers' manuals.

● REVIEW QUESTIONS

1. Discuss the need for instruments in the operation of aircraft engines.
2. What operational conditions are measured by engine instruments?
3. Describe the mechanisms by which pressure instruments are operated.
4. What is the purpose of the restriction in the pressure port of an oil pressure gage?
5. When starting an engine, what precaution must be taken with respect to oil pressure?
6. What problem is indicated if the oil pressure gage fluctuates rapidly during operation?
7. What colored markings are placed on the face of an oil pressure gage and what are their meanings?
8. What is the purpose of the white line at the edge of the cover glass of an instrument?
9. Describe the operation of a diaphragm-type pressure gage.
10. Describe the purpose and operating principles of a manifold pressure gage system.
11. What device protects the manifold pressure gage from damage due to backfiring?
12. Why is a manifold pressure gage essential in the operation of an engine equipped with a constant-speed propeller?
13. What is the purpose of a purge valve in a manifold pressure gage system?
14. What pressure should a manifold pressure gage indicate when the engine is not running? Why?
15. What may be the trouble when the manifold pressure gage indication lags or fails to register?
16. What causes jumpy or erratic operation of the MAP gage?
17. What is the effect of a broken or leaking MAP gage pressure line for an unsupercharged engine? A supercharged engine?
18. Why is a fuel pressure gage needed with an engine fuel system in which a fuel pump is incorporated?
19. Why is it possible to employ a fuel pressure gage as a fuel-flow indicator for engines with fuel injection?
20. What is the meaning of the blue marking on the face of a fuel pressure (flow) gage?
21. What types of temperature gages are required for the safe operation of aircraft engines?
22. Describe the operating principles of engine oil temperature indicating systems.
23. How does a ratiometer-type instrument differ from a Wheatstone bridge instrument?
24. At what point in an oil system is the temperature-sensing bulb located?

25. What is the reason for an off-scale reading of a ratiometer-type instrument?
26. What precautions must be taken during the removal and installation of a vapor pressure instrument?
27. Explain the operating principles of thermocouple-type temperature-indicating systems.
28. What would cause a thermocouple instrument system to produce erratic indications?
29. What are the important considerations in the installation of thermocouple leads?
30. How can you measure the resistance of thermocouple leads?
31. What is the cause of an inverse reading in a thermocouple temperature indicating system?
32. How would you determine what types of thermocouples to use in a particular gas-turbine engine EGT system?
33. Describe a mechanical tachometer instrument.
34. Describe a magnetic-type tachometer.
35. Describe an electric tachometer system.
36. How would you proceed to troubleshoot an electric tachometer system?
37. How would you install the flexible drive shaft for a tachometer?
38. Explain the use of synchronous motors in tachometer systems.
39. Describe the system by which special breaker points in an engine magneto may be used to provide engine rpm indication.
40. What is the principal value of a fuel flowmeter?
41. Describe the purpose and operating principles of a fuel-pressure warning system.
42. What is meant by *fault isolating* and *fault indicating*?
43. What are the basic requirements of a fire or overheat warning system?
44. Describe the different types of heat detectors.
45. Describe a fire warning system for a light aircraft.
46. Explain the principle of the sensing element employed in the Fenwal continuous fire detection and overheat warning system.
47. Compare the Fenwal sensing element with that of the Kidde system.
48. Describe the pneumatic sensing element.
49. How are fire detection systems checked for operation?
50. Explain the importance of proper routing of fire detection sensors.
51. What conditions should be checked in the inspection of fire warning systems?
52. How do you test the resistance of sensing elements?
53. How do you make a test of fire detection system operation?
54. Describe a basic fire extinguishing system.
55. Explain how the extinguishing agent is released from the container.
56. Describe a discharge cartridge.
57. What methods are used to provide more than one discharge of extinguishing agent for each engine?
58. What methods are used to show that adequate extinguishing agent and pressure are available?
59. Describe typical inspections applicable to fire extinguishing systems.

14 ENGINE CONTROL SYSTEMS

Need for Engine Controls

It is obvious that all aircraft powerplants must be controlled precisely in order to provide the performance required for the many different modes of aircraft operation. The controls provide a means by which the pilot, copilot, or flight engineer can manipulate the engine and engine accessories from the pilot's compartment or the flight engineer's station. Modern aircraft engines, whether gas-turbine types or reciprocating types, have many control systems and mechanisms, some mechanical, some electronic or electrical, some hydraulic and others that may have a combination of actuating forces and devices. It is the purpose of this chapter to discuss the controls that are mechanical in nature.

Desirable Characteristics

Each part or unit of the engine which is subject to control and each engine accessory must be controlled not only as an individual unit but also in consideration of its effect on all the other parts, units, and accessories. Therefore, engine control systems must be accurate in their manipulation, positive, reliable, and effective regardless of the distance between the controlled unit and the control in the pilot's compartment or the flight engineer's station.

● PRINCIPAL TYPES OF CONTROL SYSTEMS

The three general types of mechanical engine control systems are (1) **push-pull rods** with bellcranks and levers, (2) **cable and pulley systems,** and (3) **flexible push-pull wires encased in a coiled-wire sheath.** Frequently, aircraft engine-control systems embody combinations of the above systems. These are basic control systems. They may or may not be included in the hydraulic, electric, or electronic control systems installed in modern airplanes.

It is also possible to classify engine controls as (1) manually operated, (2) semiautomatic, and (3) automatic. The rapid advancement in engine design, the use of a higher power output in many powerplants, the installation of superchargers and turbosuperchargers, the introduction of highly controllable propellers, and the need for accurate fuel schedules in the operation of gas-turbine engines are some of the factors calling for the adoption of automatic or semiautomatic controls, although a method of manual control is almost always retained to provide for the possible failure of automatic or semiautomatic control mechanisms.

Automatic and semiautomatic control may be accomplished by hydraulic, electric, or electronic mechanisms. In this connection, the word "electronic" is distinguished from the words "electric" and "electrical" by the simple fact that any *electronic* circuit must, by definition, contain vacuum tubes or transistors. However, in the broad usage of the words, "electric" and "electrical" are words that include electronic devices as a lesser subdivision of the general field.

Push-Pull Tube Characteristics

The rods or tubing used to transmit control lever movements are generally called **push-pull rods.** The amount of force to be exerted on a push-pull rod determines its diameter and its wall thickness if it is a hollow tube, since it must withstand both compression and tension. Compression tends to increase a bend, and tension tends to decrease it; hence, bends are avoided, for if they were present, they would tend to change the length and therefore the adjustment of the mechanism under control.

Guides for Rods and Tubing. Guides are often provided for the tubing or rods to prevent flexing and to give them mechanical support where needed. The tubing or rod must slide through the guides smoothly and without friction, especially where the rods or tubing are long. Guides may be made of fiber, plastic, hard rubber, a composition such as micarta, or other material of similar characteristics. The plastic called Teflon makes an excellent guide because of its almost complete freedom from friction.

Types of Attachment for Control Rods. There are four types of attachment for the ends of control rods: (1) clevis and pin, (2) ball bearing, (3) ball joint, and (4) threaded. If properly inspected and adjusted when necessary, these various rod ends will not be responsible for the so-called give and slack which causes much of the improper operation of control-system linkage. Clevis-rod ends may be screwed on the tube or flexible joint and locked with a jam nut. A clevis pin, washer, and cotter pin may be used to fasten the rod end to the control arm or bellcrank. When the control-tube movement is not in line with the control arm movement, a ball-joint end is used on the control tube to allow an angle (usually up to 15°) between the two parts. Screw ends permit adjusting the length of the tube by screwing it in or out. Obviously, when a tube is screwed in, it is shorter and less thread is exposed.

Figure 14-1 shows a clevis-rod end, locknuts, a push-pull control rod cut apart in the drawing to reduce the length, the threaded end of a rod, an inspection hole,

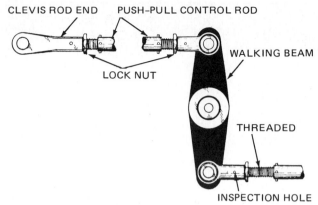

CLEVIS ROD END   PUSH–PULL CONTROL ROD

LOCK NUT

WALKING BEAM

THREADED

INSPECTION HOLE

FIG. 14-1 Beam and clevis-rod end.

and a straight beam which is sometimes called a "walking beam." It is definitely not a *bellcrank*.

Figure 14-2 emphasizes the bellcrank, which has two arms approximately at right angles to each other. A ball-bearing rod end is also illustrated.

Bellcranks. A bellcrank is a double lever or crank arm in which there are two cranks approximately at right angles to each other. The purpose of a bellcrank is to provide a means of changing the direction of motion, that is, to transmit the motion around some obstacle. For example, it may be necessary to change a forward movement to a rearward movement or to change a horizontal movement to a vertical movement. In this manner, it is possible to obtain the desired relative movement between the engine control lever in the pilot's compartment or at the flight engineer's station and the engine unit which it controls.

If the arms of a bellcrank are of equal length, the change of movement is accomplished without any gain or loss in the movement of the linkage; but if the arms are not of equal length, a gain or a loss in movement is obtained. For example, it is possible to produce a relatively great movement of an engine-unit adjustment by means of a comparatively small movement of the cockpit control. The reverse is also true. A comparatively great movement of the cockpit control may produce a relatively small movement of the engine-unit adjustment. The results obtained depend on the original design of the bellcrank, its installation, and how it is rigged to the engine units. Regardless of the transmission of movement, it must be supported by bearings that will reduce the friction to a minimum. Ball bearings are generally used for this purpose.

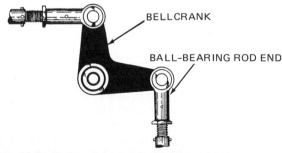

BELLCRANK

BALL-BEARING ROD END

FIG. 14-2 Bellcrank and ball-bearing rod end.

Cable-and-Pulley Control Systems

A **pulley** is essentially a wheel, usually grooved, mounted in a frame or block so that it can readily turn upon a fixed axis. Two or more wheels may be mounted in the same frame, either on the same axis or on different axes. When two or more wheels are mounted in the same frame, the pulley is described as having two or more **sheaves.**

A single **fixed pulley** is a lever of the first class and has a mechanical advantage of only 1. A single pulley gains neither force nor speed. Its only purpose is to change the direction in which the force is applied.

A single **movable pulley** is a lever of the second class and has a mechanical advantage of 2, because its diameter is twice its radius; hence, it can be used to gain force, to gain speed, or to change direction. Pulleys can be combined in different ways to obtain various results in force, speed, and change of direction. A detailed discussion of the laws governing pulleys is beyond the scope of this text.

The installation of cable and pulley systems in a multiengine airplane is often a complicated procedure. In a single-engine airplane, where the distance between the control levers and the engine units is short, a push-pull control rod may be used; but the great distances and the need for changes of direction in a multiengine airplane make it essential to use cables and pulleys.

Figure 14-3 illustrates various types of pulley clusters. All the types illustrated and even more may be found on a typical multiengine airplane.

Pulleys for aircraft control systems are usually made of a phenolic or micarta composition, plastic, or aluminum alloy. They are often supported on antifriction bearings. Typical aircraft pulleys are shown in Fig. 14-4.

The cables are usually of flexible steel, attached to the outer edges of certain pulleys at some points in the

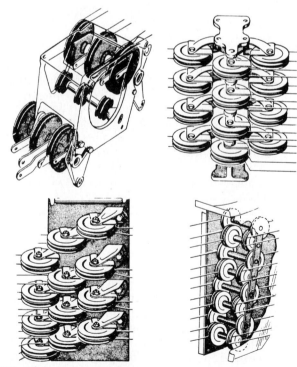

FIG. 14-3 Pulley clusters.

FIG. 14-4 Aircraft pulleys. *(Tansey Aircraft Pulley Co.)*

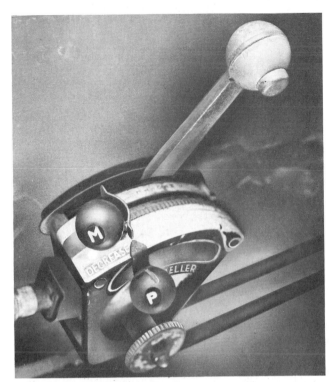

FIG. 14-5 A control quadrant.

system, guided and supported by **fairleads** (phenolic or micarta blocks with holes in them to admit the cables) at other points in the system, and also guided and supported by other pulleys which may also change the direction of the cable movement.

In some powerplant installations the cables extend from the control levers to the engine units; but in other installations the cables extend only from the control levers to the fire wall of each engine nacelle and are there connected to push-pull control rods that transmit the movement to the engine units.

Various devices are used for adjusting the tension and the length of the cables, such as turnbuckles, adjustable pulley clusters, and adjusting links. The manufacturer's instructions for each installation prescribe how the cables will be adjusted so that the control handles in the pilot's compartment reach their fore-and-aft stops just after the stops in the engine sections are reached. Thus, **springback** is provided to assure the complete range of control operation.

The Control Quadrant. The word "quadrant" means a fourth part, usually the fourth part of a circle, or 90°. The engine **control quadrant** in an airplane is usually a control lever which pivots back and forth over a base through a 90° arc, but it may also refer to the base on which the lever is mounted. For example, the **throttle quadrant** is the base upon which the throttles are mounted in the pilot's compartment or at the flight engineer's station of a multiengine airplane. A control quadrant is shown in Fig. 14-5.

The Control Pedestal. The **control pedestal** is a frame or mount on the floor of the cockpit or pilot's compartment in a multiengine airplane where some of the engine controls are connected. The control pedestal supports one or more control quadrants. Figure 14-6 illustrates a control pedestal for a twin-engine airplane.

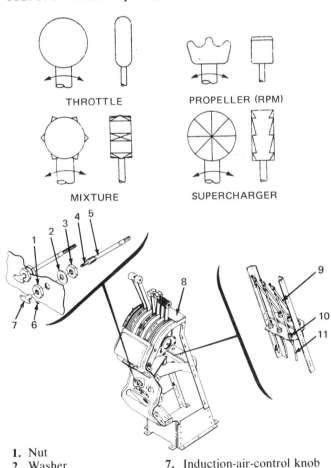

1. Nut
2. Washer
3. Nut
4. Control wire
5. Control-wire housing
6. Roll pin
7. Induction-air-control knob
8. Control pedestal
9. Teleflex push-pull unit
10. Swivel
11. Teleflex control conduit

FIG. 14-6 Control pedestal with approved shapes for control knobs. *(Cessna Aircraft Co.)*

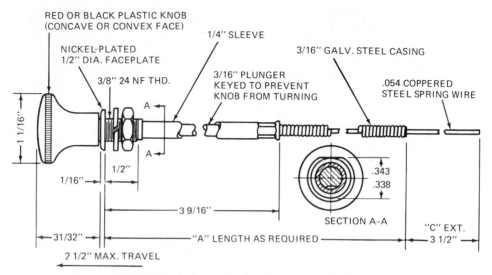

FIG. 14-7   A push-pull control.   *(Arens Controls)*

The control quadrant on this pedestal has controls for the throttle, engine rpm (propeller), and carburetor mixture.

It will be observed that the quadrant levers are attached to Teleflex push-pull cable units which are mounted so that they can swivel with the movement of the lever. Note that the operating knobs on the control levers are required to have shapes easily recognized by feel. The shapes of the knobs are illustrated in Fig. 14-6.

Control Assemblies. The control assemblies in the pilot's compartment or cockpit of a single-engine airplane are much simpler than those installed in a multiengine airplane. A single-engine airplane may have only one control quadrant, which is used to control the throttle, the fuel-air mixture, the propeller, and possibly a supercharger. Individual control levers may be provided for such powerplant accessories as the carburetor air heater, the cowling shutters, the oil cooler shutters, and possibly the radiator in the case of a water-cooled engine.

In a multiengine airplane, the complexity of the controls installed in the pilot's compartment may make it necessary to have two or more control quadrants mounted on a control pedestal where most of the engine controls are located to afford a central point of operation for the convenience of the operator.

A problem encountered in the operation of a control quadrant is the tendency of the controls to creep out of the position in which they are placed. This is partly overcome by providing a reasonable amount of friction in the quadrant assemblies, but it is also met by mechanical locking devices, which temporarily hold the controls in the positions where they are placed by the operator until he or she is again ready to move them.

Flexible Push-Pull Controls

Flexible push-pull controls are used for a variety of remote-control situations in aircraft. This is particularly true for light aircraft, where the distance through which the control must operate is not so great that friction becomes a problem.

The construction of a simple push-pull control is shown in Fig. 14-7. This control is mounted on the instrument panel and may be used for carburetor air heat, cabin heater, or other similar function. The knob is attached to a **plunger** into which a steel spring wire is secured. The plunger is inserted into the **sleeve** which serves as a guide. At the forward end of the sleeve is a threaded section and a **faceplate** to provide for mounting on the panel. The assembly is inserted through a hole in the panel and is mounted firmly by means of a locknut and washer on the rear of the panel.

The steel wire of the control is enclosed in a $\frac{3}{16}$-in [4.76-mm] galvanized-wire casing which is in the form of a coil or spiral. The casing guides the control wire to the operating mechanism.

The method by which the control should be mounted in the aircraft is shown in Fig. 14-8. Note that each end of the flexible casing is securely mounted with terminal mounting clamps so that there will be no slack or play in the operation. Intermediate clamps must be used on both sides of bends to prevent flexing and a resultant loss of movement. The control should be adjusted so that there is a small amount of springback at the full in position of the knob to ensure that the operated lever has moved through its full range.

Vernier Push-Pull Control. A vernier push-pull control is constructed to provide for a coarse adjustment

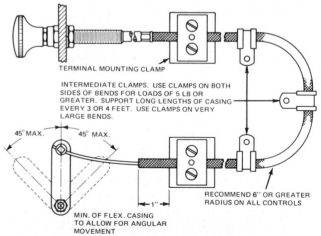

FIG. 14-8   Mounting of a push-pull control.   *(Arens Controls)*

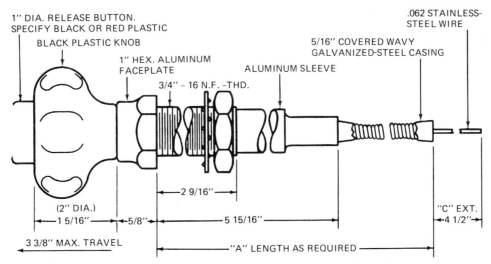

FIG. 14-9   **Vernier push-pull control.**   *(Arens Controls)*

and a fine (vernier) adjustment, especially for the throttle and propeller controls. A unit of this type is shown in Fig. 14-9. When it is desired to make a fine adjustment, the plastic knob is rotated to the right for increase and to the left for decrease. To make a coarse adjustment, the lock button is depressed and the knob is pushed in or pulled out. The fine adjustment is used when precise power settings are made. The approximate engine rpm is set by moving the propeller control in or out as required. Then the final adjustment is made by rotating the knob. After the rpm is set, the manifold pressure is adjusted with the throttle control, first with the coarse adjustment and then with the vernier.

Control for Use with Quadrant. Push-pull wire-type controls are often used with control quadrants on the pedestal. A simplified diagram to show how this is accomplished is shown in Fig. 14-10. The end of the control unit is connected to the arm of a quadrant lever instead of a knob. The sleeve is secured to a bracket with a swivel fitting to allow a few degrees of side movement as the quadrant lever position is changed. The engine end of the control assembly is connected

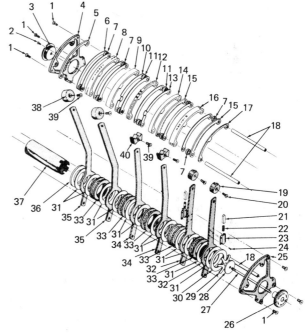

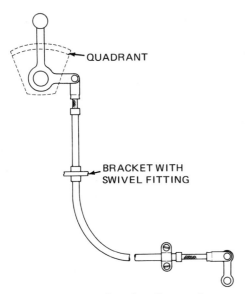

FIG. 14-10   **Lever-operated push-pull control.**

1. Screw	21. Ratchet stop
2. Lock screw	22. Spring
3. Retainer	23. Housing
4. Left mounting plate	24. Screw
5. Spacer	25. Right mounting plate
6. Guide	26. Friction knob
7. Spacer	27. Center stud
8. Spacer	28. Washer
9. Spacer	29. Spring
10. Left pitch-lever stop	30. Spacer
11. Spacer	31. Friction disk
12. Spacer	32. Mixture lever
13. Right pitch-lever stop	33. Spacer
14. Spacer	34. Propeller pitch lever
15. Rack	35. Throttle lever
16. Spacer	36. Spacer
17. Spacer	37. Hub
18. Guide rod	38. Throttle-lever knob
19. Mixture-lever knob	39. Screw
20. Screw	40. Pitch-lever knob

FIG. 14-11   **Exploded view of a control quadrant.**   *(Cessna Aircraft Co.)*

to the controlled unit with a clevis or other suitable device.

Construction of a Control Quadrant. To illustrate the detailed construction of a control quadrant for a modern light twin airplane, an exploded view is shown in Fig. 14-11. This is the control quadrant assembly for a Cessna Model 310 airplane. The quadrant includes two levers for the throttles, two for rpm (propeller pitch) control, and two for the mixture control. Observe that spacers, washers, and friction disks are installed between the levers to provide for friction to prevent creeping of the levers. This friction is adjusted by means of the friction knob at the right side of the quadrant.

● ENGINE-CONTROL FUNCTIONS

A variety of engine mechanisms require control from the cockpit of the airplane, and many of these can be controlled manually by means of the devices previously explained. We shall not attempt to explain the electric or electronic controls often used with modern powerplants but shall confine our discussion to the units and devices controlled manually.

Throttle

The throttle-control knob or lever is probably the most conspicuous of all controls, and by some persons it is considered to be the most important because it controls engine power. The throttle lever is marked with the letter T or the word "throttle" and its direction of operation is indicated by the words "open" and "closed". The throttle knob is a thick disk, flat on each side, to comply with the standard requirements of the FAA. Throttle movement is always such that forward movement will open the throttle valve and rearward movement will close the valve.

Propeller Control

The propeller control is also of great importance in engine power adjustment for airplanes equipped with constant-speed or controllable propellers. The propeller-control lever and linkage must be precise in its operation to provide for accurate control of rpm. The control lever is linked to the propeller governor, thus establishing the rpm at which the engine is to operate. Engine power is determined by the combined settings of rpm and manifold pressure in accordance with the throttle and propeller-control-lever positions.

The propeller-control-lever knob is quadrant-shaped and scalloped on the top. This is to provide a distinctive shape which the pilot will quickly recognize. Forward movement of the lever will increase rpm, and rearward movement will decrease rpm.

Mixture Control

The mixture-control lever on the quadrant is used to adjust the fuel-air mixture through the mixture-control lever on the carburetor or fuel-control unit. The lever positions are usually marked for FULL RICH, LEAN, and IDLE CUTOFF. Some airplanes have markings for AUTO RICH and FULL RICH to provide two rich settings.

The FULL RICH setting provides a maximum rich condition which is not affected by the automatic mixture control.

The mixture-control lever is arranged so that a forward movement will provide a richer adjustment. The knob on the quadrant lever is distinctively shaped for recognition. It is a thick disk like the throttle knob; however, it has a raised diamond pattern on the periphery.

Carburetor Air Heat

The control lever or knob for the carburetor air heat may or may not be mounted on the pedestal. It is usually not a lever of the type used for the previously described controls but may be a push-pull knob on the front of the pedestal or on the instrument panel. The function of this control is to operate a gate valve in the air-induction system to provide either cold air or heated air for the carburetor. Heated air is required for the engine when there is danger of carburetor icing.

Miscellaneous Engine Controls

In addition to the controls described in the foregoing paragraphs, engines may require several other controls depending on the design of the engine and aircraft. Among others are controls for the cowl flaps, oil coolers, superchargers, and intercoolers. All such controls must be marked to show their functions and must be easily accessible to the pilot.

● ENGINE-CONTROL SYSTEMS FOR LARGE AIRCRAFT

The engine-control systems for large aircraft are rather complex; however, they utilize cable and pulley mechanisms in much the same manner as those for smaller aircraft. To provide an example of control systems for large aircraft, we shall describe the throttle (thrust lever) system of the Boeing 720 jet airliner.

General Description

The Boeing 720 airliner is equipped with four separate engine throttle systems to provide for individual control of each engine. The starting of each engine is accomplished by the use of a single lever to energize the ignition system and to initiate the flow of fuel to the engine. Another lever assembly controls both forward and reverse thrust by regulating fuel flow and actuating the thrust reverser. An interlocking mechanism prevents simultaneous initiation of forward and reverse thrust for each engine.

The throttle system consists of an engine-start lever and a thrust lever assembly for each engine, connected by a series of throttle-control cables and mechanical linkages to the fuel-control units on the engine. This system is illustrated in Fig. 14-12. A thrust-lever friction brake applies a braking force to all thrust-lever assemblies during forward thrust operation.

The engine-start lever is connected by cables to an engine-control drum-and-shaft assembly in a nacelle strut. The arrangement of this assembly and the associated linkages is shown in Fig. 14-13. The drum-and-shaft assembly is connected by a rod and bellcrank installation on the right side of the engine to the fuel-

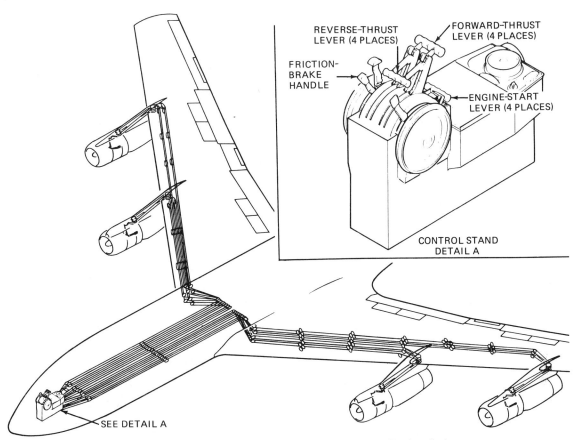

FIG. 14-12 Throttle system for a jet airliner. *(Boeing Co.)*

control unit. Advancing the engine-start lever actuates an ignition switch to energize the ignition system. Further movement of the start lever opens a pilot shutoff valve in the fuel-control unit.

The thrust-lever assembly is connected by cables to the drum-and-shaft assembly as shown in the illustration. Movement of the thrust lever is therefore transmitted through the bellcrank and rod linkages to the fuel-control unit.

Actuation of the thrust-lever assembly regulates fuel flow in the fuel-control unit, subject to the automatic limiting features incorporated in the unit. For reverse thrust, the lever assembly movement actuates the thrust reverser before increasing fuel flow. It should be noted that the direction of travel of the thrust-control cables and drums is the same for decreasing forward thrust as it is for increasing reverse thrust.

Thrust-Lever Assembly

Four thrust-lever assemblies on the control-stand (pedestal) quadrant control the forward and reverse thrust of the engines. Each thrust-lever assembly consists of a forward thrust lever, a reverse thrust lever, a reverse thrust-control link, a pawl, a brake drum, and a thrust-control drum. The forward thrust lever, with the reverse thrust lever attached to it, is mounted on the brake drum. One end of the control link is riveted to the reverse thrust lever, and the opposite end is attached to the thrust-control drum. Various positions of the assembly are shown in Fig. 14-14.

As either thrust lever is advanced from the idle position, the control link rotates the thrust-control drum to actuate the fuel-control unit to increase thrust. The

forward thrust idle position is against an idle stop on the quadrant, and full forward thrust is obtained before the lever is all the way forward. The reverse thrust lever, when in the idle position, is against an idle stop on the forward thrust lever.

An interlock mechanism prevents simultaneous actuation of the forward and reverse thrust levers to ensure positive forward or reverse thrust control. The ability of each lever to move depends on the position of the other lever. If the forward thrust lever is more than 2° from the idle position, the reverse thrust lever cannot be moved more than 12° from idle. However, if the reverse thrust lever is advanced more than 12° from idle, the forward thrust lever cannot be moved. The interlock between the levers is a pawl riveted to the forward thrust lever with the pawl between the thrust lever and the control link. When the forward thrust lever is 2° or less from the idle position, the pawl is aligned with a lockout hole in the web of the thrust lever cover. As the reverse thrust lever is returned to the idle position, the control link pushes the pawl from the hole to unlock the forward thrust lever. When the forward thrust lever is more than 2° from the idle position, the pawl is not aligned with the lockout hole. The web then opposes the force of the control link on the pawl so that the reverse thrust lever cannot be moved more than 12° from idle.

Engine-Start Levers

Four **engine-start levers** on the control-stand quadrant are used to start the engines. Each lever controls ignition and fuel as explained previously. The start lever is provided with a spring-loaded detent catch

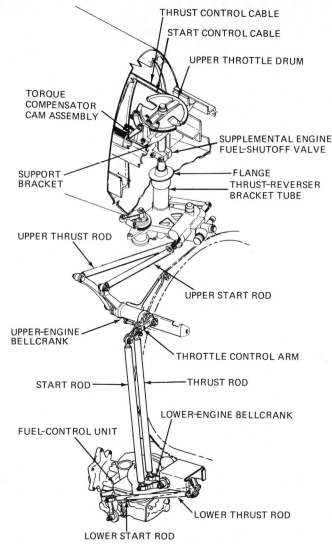

THRUST CONTROL CABLE

START CONTROL CABLE

UPPER THROTTLE DRUM

TORQUE COMPENSATOR CAM ASSEMBLY

SUPPLEMENTAL ENGINE FUEL-SHUTOFF VALVE

SUPPORT BRACKET

FLANGE THRUST-REVERSER BRACKET TUBE

UPPER THRUST ROD

UPPER START ROD

UPPER-ENGINE BELLCRANK

THROTTLE CONTROL ARM

START ROD

THRUST ROD

LOWER-ENGINE BELLCRANK

FUEL-CONTROL UNIT

LOWER THRUST ROD

LOWER START ROD

FIG. 14-13 Drum-and-shaft assembly with linkages. *(Boeing Co.)*

which may be released by lifting the knob. The detent secures the lever in the CUTOFF and IDLE positions. An additional detent is provided between these two positions. This catch is provided to ensure that a throttle left insecurely in the IDLE position will not creep to the CUTOFF position and cause an unintentional engine shutdown. A stop gate and detent are provided at the START position.

Throttle-Control Cables

Throttle-control cables are 7 by 7, $\frac{3}{32}$-in [2.38-mm] flexible steel with standard swaged attachment fittings and Boeing Aircraft fittings. They serve as thrust and start cables to connect the fuel-control unit on each engine to the respective engine-start-lever and thrust-lever assembly. The cables are routed under the floor from the control stand through the lower nose compartment and above the forward cargo compartment ceiling. From the cargo compartment the cables are routed along the wing's leading edge to the nacelle struts. In each strut the cables are routed to the drum-and-shaft assembly, which is linked to the fuel-control unit.

Thrust-Lever Friction Brake

A thrust-lever friction brake on the control-stand quadrant applies a variable braking force to all thrust levers during forward thrust operation. The friction brake is used to select manually the proper amount of braking force in order to prevent throttle creep during flight. The friction brake consists of a brake handle mechanically linked to two leaf springs and four brake shoes.

The brake handle, mounted to the right of the thrust levers, is connected by a brake link and an eyebolt to a brake crank. Bolted to the crank are the leaf springs and the brake shoes. As the brake handle is advanced, friction between the brake shoes and the brake drums is increased. A ratchet locks the brake handle in any position.

Engine Control Drum-and-Shaft Assembly

The engine-control drum-and-shaft assembly in the nacelle strut of each engine is a mechanical link in the throttle system which provides for independent control of the fuel-control unit by the thrust and engine-start levers. The assembly consists of concentric engine-start and thrust-control shafts, a thrust-control drum, engine-start drum, engine thrust-control crank, and an engine-start crank. These are illustrated in Fig. 14-13. The start drum and the start crank are mounted on each end of the engine-start shaft. The thrust drum and thrust crank are mounted on each end of the thrust-control shaft, which is mounted inside the start shaft. The assembly is supported by a bracket on the strut aft of the forward engine mount. The start lever contacts the strut bracket to provide a mechanical stop for the engine-start-control system at cutoff. A lug on the strut bracket provides a stop for the thrust system at both 100 percent forward and reverse thrust.

● POWER CONTROL SYSTEMS FOR HELICOPTERS

Because of the special nature of helicopter flight operation, the control of the powerplant in a helicopter must be closely coordinated with the flight controls. This is accomplished by coordinating the power control with the collective pitch control and placing a rotating grip for throttle or power control on the collective pitch-control stick.

The collective pitch control is used to increase or decrease the pitch of the main-rotor blades collectively; that is, all the blades increase or decrease pitch at the same time as the collective pitch control is raised or lowered. When collective pitch is increased, engine power must be increased at the same time if engine rpm is to be maintained. When the helicopter is on the ground with the engine idling, the main-rotor pitch is essentially flat, or has a zero angle of attack on the rotor blades. When it is desired to take off, the pilot increases collective pitch and engine power simultaneously.

Power Control System

A typical helicopter power-control system is shown in Fig. 14-15. This particular system is for the turbine-

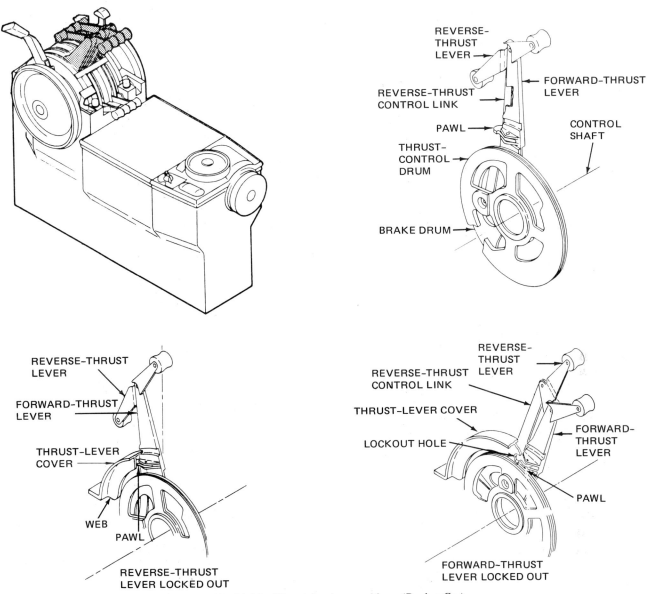

FIG. 14-14   Thrust-lever assembly. *(Boeing Co.)*

driven Bell Model 204B helicopter. The illustration clearly reveals the complex design required and the coordination with the collective pitch control. This is required whether the system is designed for operation with a turbine engine or a piston engine.

Maintenance and inspection for a helicopter power-control system is accomplished in accordance with the appropriate manufacturer's manual and is similar to those for other control systems.

● INSPECTION AND MAINTENANCE

Rods and Tubing

1. Operate the system slowly, and watch for evidence of any strain on the rods and tubing that will cause bending or twisting.

2. Examine each rod end that is threaded, and observe whether or not the rod is screwed into the socket body far enough to be seen through the inspection hole.

3. Eliminate any play by making certain that all connections are tight.

4. Examine the guides to see if the rods bind too much on the guides, but do not mistake any binding for springback. Replace any guides that cause binding.

5. Adjust the length of screw-end rods by screwing into the control end or backing out. Retighten the locknuts.

6. If any rod is removed, mark it to show its location on reassembly.

7. Replace any ball-bearing rod ends that cause lost motion.

Bellcranks

1. Examine for security of mounting, for wear, and for proper lubrication.

2. Replace any bellcrank bearings that are causing lost motion.

3. Mark each part that is removed, and reassemble in the correct order.

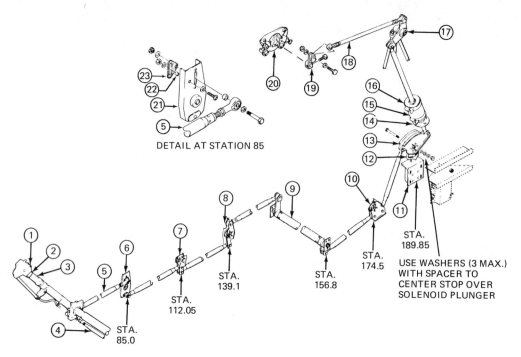

DETAIL AT STATION 85

STA. 189.85

USE WASHERS (3 MAX.)
WITH SPACER TO
CENTER STOP OVER
SOLENOID PLUNGER

STA. 174.5

STA. 156.8

STA. 139.1

STA. 112.05

STA. 85.0

1. Idle stop release switch
2. Friction adjustment
3. Twist grip control
4. Adjustable tube to copilot control
5. Adjustable tube
6. Bellcrank
7. Bellcrank
8. Bellcrank
9. Torque tube
10. Bellcrank
11. Solenoid assembly
12. Flight idle stop
13. Bellcrank assembly
14. Boot support
15. Boot
16. Boot retainer assembly
17. Bellcrank
18. Adjustable rod
19. Control arm
20. Fuel-control power-lever shaft and stops
21. Adjustable bellcrank
22. Spacer
23. Serrated plate

FIG. 14-15 Helicopter power-control system. *(Bell Helicopter Co.)*

Cables

1. Inspect for frayed or broken cables.

2. Clean the cables where they pass through fairleads or over pulleys, and then cover them with the prescribed rust preventive.

3. Maintain the correct tension constantly, not merely at the prescribed inspection intervals. Use a **tensiometer** to obtain an accurate reading of the cable tension, and check it against the approved tension given in the manufacturer's instructions.

4. Inspect all turnbuckles, adjusting links, and other devices used to maintain tension in the cables. Safety such devices after each adjustment.

5. If any cables are removed, mark them to show their location in the system. When new cables are installed, apply any required code markings in the same manner as they were on the original cables.

Pulleys

1. Inspect each pulley to be sure that it is properly mounted and securely fastened.

2. Operate the system to observe how the cables pass through the pulleys. Where necessary, adjust or replace to obtain the proper clearance and tension. Correct any misalignment.

3. Check pulley bearings or bushings for excessive play, and replace pulleys in which the bearings are found to be defective. Pulleys which do not have prelubricated bearings or impregnated bushings should be oil-lubricated.

4. Check pulley grooves for wear, either on the sides or on the bottom of the groove. Replace pulleys when wear is sufficient to interfere with satisfactory operation.

Flexible Push-Pull Controls

1. Inspect push-pull controls for wear and smoothness of operation.

2. If controls stick or are hard to operate, disconnect the control wire or cable from the controlled unit and pull the wire out of the casing or conduit. Clean all grease, rust, or dirt from the wire, and lubricate it with low-temperature grease. If it appears that the casing contains dirt or rust, it can be cleaned by pushing a cleaning wire through it two or three times.

3. Inspect the casing for damage at bends or where abrasion may have occurred. Replace casings which are broken or badly worn.

4. Lubricate the plungers and sleeves with oil or other lubricant specified in the maintenance manual.

5. Adjust the travel of the control to allow the amount of springback required for the controlled unit.

Miscellaneous Inspections

1. Examine the control unit in the cockpit or pilot's compartment or at the flight engineer's station for free operation, for the correct travel of the control levers through their extreme range of movement, for security of mounting, and for lost motion.

2. With a helper at the control unit in the pilot's

compartment, go to the point of application of each of the engine controls and check there for any lost motion while the helper operates the control levers. Then, while the helper continues to operate the control levers slowly, trace back each control system to the control unit in the pilot's compartment, observing each subordinate part or unit en route.

3. Throughout all control systems, look for proper safetying, broken or misaligned pulleys, missing or loose nuts or bolts, dirty connections, and lack of lubrication. Either remedy each defective condition as it is found, or record it on a check sheet and then remedy the unsatisfactory conditions in a systematic manner; that is, you may accomplish all the lubrication jobs in one session.

4. All through the engine-control system, examine for any play. Be sure that all mounting bolts, shaft bolts, rods, check nuts, etc., are tight. Having accomplished an inspection with the engines stopped, operate the engines and check all the controls for each engine, one engine at a time. When this is done, check the operation with all engines operating.

5. Pay particular attention to the adjustment of the throttle levers, especially in a multiengine airplane. They should be together at equal positions of the throttle valves in the carburetors. This may require an individual adjustment of each throttle valve.

6. Check the fuel-air mixture-control system to be certain that it operates in accordance with the prescribed limits given in the manufacturer's instruction book.

7. Operate the carburetor air-heating system, and check the controls carefully to be certain that all play is eliminated. See that the carburetor air heater valve in the carburetor intake is tightly closed when the control lever is in the OFF, CLOSED, or COLD position. Be sure that hot air does not leak past the valve and mix with the cold air in the induction system, leading to loss of engine power and increased fuel consumption.

8. Review the "squawk sheets" filled out by the flight crew for the past several flights, and be certain that all defects have been remedied.

● REVIEW QUESTIONS

1. Define the purpose of engine-control systems.
2. Name three basic types of engine-control systems.
3. Describe the components of a push-pull rod system.
4. What means is used to prevent the flexing of long rods?
5. Name the types of attachments for push-pull rod systems.
6. What condition requires the use of pulley and cable systems?
7. Describe a typical pulley used in a control system.
8. What method is used to support long cables between pulleys?
9. Describe a control quadrant.
10. Why is a control pedestal employed in multiengine aircraft?
11. Why are controls rigged to allow springback in the control lever?
12. How are the control levers prevented from creeping during operation?
13. Describe a flexible push-pull control.
14. What are the requirements for mounting a flexible push-pull control in an airplane?
15. What is meant by *vernier* as applied to a push-pull control?
16. What engine systems or devices are usually controlled from the cockpit?
17. Briefly describe the throttle-control system for a typical jet airliner.
18. Explain the need for coordinated power control in a helicopter.
19. With what flight control in a helicopter is the engine power control coordinated?
20. Name the principal points of inspection for a push-pull rod system.
21. What inspections should be given cables?
22. Describe service and maintenance for flexible push-pull controls.
23. How can the technician be certain that threaded rod end fittings have sufficient thread engagement?

15 ENGINE INSTALLATION, TEST, INSPECTION, AND TROUBLESHOOTING

● TYPES OF INSTALLATION

Since the engine installation for an airplane depends on the type of airplane and the type of engine installed, specific instructions that will cover all installations cannot be given. Types of engine installations include (1) radial engines, (2) horizontal engines, (3) in-line air-cooled engines, (4) in-line water-cooled engines, (5) turboprop engines, and (6) turbojet engines. Within each of these classifications are a number of different configurations, so it is obvious that the airplane manufacturer's instructions must be followed in each case.

For conventional reciprocating engines there are many similarities in engine installation, and for the purposes of this chapter we shall confine the discussion to the general practices to be observed for radial and opposed-type engines. Examples of procedures to be employed with specific engines will be given to emphasize the details involved.

● ENGINE INSTALLATION

Preparation for Installation

If an engine has been stored in an engine case, special instructions relating to unpacking the engine will usually be included in the case. The case should be placed in the correct position (top side up) so that the engine case cover can be lifted off. After the attaching bolts are removed, the cover is carefully lifted to avoid damage to the engine.

If the engine has been properly preserved and packed to prevent corrosion damage while moving from the factory overhaul shop to the purchaser, it will be sealed in a plastic envelope. The magnetos will probably be mounted on the engine together with the ignition harness. The carburetor may be in a separate package within the case.

The engine will be bolted to the supports built into the case, and it will be necessary to remove bolts and other attachments before the engine can be removed. The technician in charge of unpacking the engine must exercise great care to prevent damage and the loss of small parts. He or she should first locate all paperwork, such as overhaul records and unpacking instructions, and then proceed according to instructions.

Lifting eyes are usually provided for the attachment of a hoist to the engine. In the case of radial engines, the lifting eye is screwed on to the end of the propeller shaft after the removal of the protective cap. In-line and opposed engines have lifting eyes attached to the crankcase. Turbine engines in aircraft are usually handled by means of a wing- or pylon-mounted hoist used in conjunction with a transportation dolly into which the engine is lowered during engine removal.

When the engine cover is removed, a hoist should be attached to the lifting eye and sufficient tension placed on the hoisting cable to remove most of the weight from the mounting brackets. The mounting bolts should then be removed, and the engine hoisted and placed on a suitable stand. This is necessary because there are usually a number of fittings and parts which must be assembled to the engine before it is ready to be installed in the airplane.

When the engine is firmly mounted on the stand, all shipping and preservative plugs are removed. These plugs, with the exception of drain plugs, should be replaced immediately with the proper fitting for engine operation. Fittings include oil pressure fitting, manifold pressure lifting, crankcase vent fitting, oil temperature bulb fitting, and others. The desiccant plugs should be removed from the spark-plug holes and from the oil sump. At this time the engine should be rotated a few times to permit drainage of the preservative oil.

New spark plugs and washers of the correct type should be installed, and the plugs tightened to the following torques: 14-mm plugs, 200 to 245 in·lb [22.6 to 27.69 N·m]; 18-mm plugs, 260 to 300 in·lb [29.38 to 33.9 N·m]. Before the spark plugs are installed, a small amount of approved antiseize thread lubricant should be applied to the upper threads of the plugs. Care must be taken to prevent thread lubricant from getting on the electrodes of the spark plugs.

When fittings having pipe threads are installed, the threads of the fittings should be lightly coated with an approved thread lubricant and the fittings should be installed with proper torque to prevent damage to the threads.

After installation of the spark plugs, the ignition harness elbows are attached to the plugs. Care must be taken to insert the sleeves ("cigarettes") into the spark plugs in good alignment to prevent damage. The elbow nuts are then screwed into place by hand until finger-tight. After this they are torqued only enough to prevent them from turning during operation. Excessive torque will bend and collapse the spark-plug elbows.

If the carburetor is of the pressure type and has been packed and preserved separately, it is necessary to *condition* it before placing it in operation. This is accomplished by following these instructions: Remove the inlet strainer and all plugs leading to fuel chambers, and drain all the preserving oil from the carburetor. Wash the inlet strainer in clean fuel, and replace it. Also, replace the plugs.

With the throttle lever in the WIDE OPEN position and

the manual mixture control in the FULL RICH position, inject clean fuel through the fuel-inlet connection at the proper pressure for the carburetor until clean fuel flows from the discharge nozzle. Close the throttle, and place the manual mixture control in the IDLE CUTOFF position. Disconnect the fuel supply line, and install a plug in the fuel inlet to retain the fuel in the carburetor. Permit the carburetor to soak with the fuel for at least 8 h before placing it into service.

During the preceding operation, care must be taken to see that fuel does not enter the air sections of the diaphragm chambers.

Before the engine is installed in the airplane, it should be inspected thoroughly. Both the engine manual and the airplane manual should be consulted to make sure that all fittings, baffles, and accessories are securely fastened and safetied as necessary. A careful check at this time may save much time and trouble later.

Installation in the Airplane

The installation of the engine in the airplane should follow the directions given by the manufacturer. An example of the instructions given for the installation of an engine in a light twin airplane is given here to show the details of a typical operation. These instructions are given in the manufacturer's manual for the Cessna 310 airplane. Reference numbers are omitted.

a. Hoist the engine to a point just above the nacelle.

b. Install the IO-470-D engine on engine mounts as follows:

1. Temporarily install mount pads, spacers, and pins to all four engine mounts using AN7-24 bolts and suitable washers. See that the pins are in a position which will align them with the slots at the side of the holes in the engine mount brackets.

2. Be certain that the propeller-control conduit and bracket are outside the engine mounts.

3. Lower the engine slowly into place on the engine mount brackets.

4. Remove the temporary bolts and washers. Install the spacer, mount pad, washer, and bolt. Torque the bolts to 40 ± 3 ft·lb [54.24 ± 4.07 N·m].

c. Fasten the engine ground strap to the left aft engine mount bracket.

d. Connect wires and cables as follows:

1. Attach the wire bundle to the right manifold intake pipe with three clamps and the generator cables to the left intake manifold pipe with two clamps.

2. Route the oil temperature wires forward along the right side of the engine, below the cylinders, and behind the intake manifold pipe. Use a sump attachment bolt to attach a clip to the engine crankcase. Fasten the oil temperature wires to this clip with a wire clamp. Attach the connector to the oil temperature bulb, located directly below the oil cooler, and safety. Connect the ground wire to the nacelle just below the right forward engine mount bracket.

3. Route the magneto ground wires forward through the right aft engine baffle.

WARNING: These magnetos DO NOT have internal grounding springs. Ground the magnetos to the crankcase to prevent accidental firing. DO NOT connect the aircraft ground wires to the magnetos at this time.

4. Connect the starter cable to the starter.

5. Attach the tachometer generator electric connector to the tachometer generator, and safety.

6. Route the cylinder-head temperature bulb wires forward along the left side of the engine, below the cylinders, and behind the intake manifold pipe. Install the cylinder-head temperature bulb in the no. 4 cylinder. Attach the cylinder-head temperature-bulb ground wire to the nacelle just below the left aft engine mount bracket.

7. Attach the generator cables to the generator. Ground the generator cable shielding to the lower stud of the air/oil separator bracket.

8. Attach the wiring at the radio noise filter.

9. Attach the wiring at the fuel pressure switch.

e. Connect lines and hoses as follows:

1. Connect the crankcase breather line.

2. Connect the vacuum line to the vacuum pump.

3. Connect the air/oil separator exhaust line at the air/oil separator.

4. Connect the fuel vapor return line and fuel-pressure-gage line to the fuel-air control unit.

5. Connect the manifold pressure hose and the two manifold drain lines.

6. Route the oil pressure hose aft through a lightening hole in the canted bulkhead, and attach to the oil pressure fitting at the inboard nacelle rib. Fasten the oil pressure hose to the canted bulkhead with a clamp.

7. Connect the fuel-pump supply line at the fuel pump.

8. Connect the ram air tube at the fuel strainer shroud, and clamp it.

Note: If anti-ice and/or oil-dilution systems are installed, connect the anti-ice hose at the fitting below the no. 1 cylinder and connect the oil-dilution hose at the crankcase fitting just below the fuel pump.

f. Attach the propeller-control bracket with two clamps. Connect the propeller control at the governor. Rig the propeller control in accordance with appropriate instructions.

g. Connect the induction air control, the mixture control, and the throttle control. Rig these controls as directed in the pertinent section of this manual.

h. Install the exhaust stack centering springs.

i. Install the propeller in accordance with propeller instructions.

j. Connect the battery cable.

k. Inspect the installation for safeties, loose connections, and missing bolts, clamps, screws, or nuts; for proper routing of cables, hoses, and lines; and for correct connection.

l. Make a magneto-switch ground-out and continuity check. Connect the magneto ground wires to the magnetos.

m. Service the engine in accordance with lubrication instructions.

n. Install the engine cowling.

Installation Inspection

After an engine installation is completed, it is required that a complete installation inspection be made. The following checklist names the inspections to be accomplished after the installation of the IO-470-D engine in the Cessna 310 airplane:

1. Propeller mounting bolts safetied
2. Engine mounts secure
3. Oil-temperature-bulb electric connector secure and safetied; ground wire connection tight
4. Oil pressure-relief valve plug safetied
5. Tachometer generator electric connector secure and safetied
6. Starter cable connection secure and insulating boot in place
7. Cylinder-head temperature bulb installed and ground wire connection tight
8. Generator cable connections secure and cable shielding grounded
9. All wiring securely clamped in place
10. Fuel pump connections tight
11. Manifold pressure hose connections tight
12. Oil pressure connections clamped and tight
13. Fuel-injection nozzles tight
14. Fuel-injection lines clamped and tight
15. Fuel manifold secure
16. All flexible tubing in place and clamped
17. Crankcase breather-line connections secure
18. Air/oil separator exhaust line and return oil hose connections secure
19. Vacuum line and vacuum-pump outlet hose connections secure
20. Oil-dilution hose connections tight
21. Propeller anti-ice hose connections tight
22. Engine controls properly rigged
23. Oil drain plugs tight and safetied
24. Oil quantity check, 12 qt in each engine
25. Hoses and lines secure at fire wall
26. Fuel-air control unit and air-intake box secure
27. Shrouds installed on engine-driven fuel pump, fuel filter, and fuel-control unit; ram air tubes installed and clamped
28. Induction system clamps tight
29. Exhaust system secure
30. Spark plugs tight, ignition harness connections tight, and harness properly clamped
31. Magneto ground wires connected and safetied
32. Engine nacelle for loose objects (tools, rags, etc.)
33. Cowling and access doors for security

The foregoing listings of instructions for a specific engine and airplane are given to emphasize the many important details involved in engine installation. The installer and the inspector must follow the checklists to make sure that no operation is left incomplete.

Installation in the Test Stand

If an engine has just been overhauled but not tested and run in as required, it should be installed in a suitable test stand and run in according to the test schedule to make sure that it is operating to specifications. Small engines are sometimes run in on an airplane, but the standard run-in procedure must be modified to some extent if this is done.

An engine test stand should be mounted in a test cell equipped with necessary controls, instruments, and special measuring devices required for fuel consumption, power output, oil consumption, heat rejection to the oil, and standard engine performance data. The following instruments and devices will usually be required:

1. Fuel tank with adequate capacity (at least 50 gal [189.27 L])
2. Oil tank with capacity of 10 gal [37.85 L] or more
3. Fuel flowmeter
4. Scales for weighing oil
5. Cylinder-head temperature gages
6. Manifold pressure gage
7. Tachometers (one which counts revolutions)
8. At least two oil temperature gages (inlet and outlet)
9. Fuel pressure gage
10. Oil pressure gage
11. Manometer to test crankcase pressure
12. 12-V battery or other power source
13. Fuel pressure pump, either manual or electric
14. Engine test propeller ("test club")
15. Magneto switch
16. Suitable starter controls
17. Control panel for mounting instruments and controls
18. An accurate clock for checking run-in time
19. Throttle control
20. Mixture control

The control room of the test cell should be provided with a safety-glass window located so that the operator has a good view of the engine during the run-in procedure. The safety-glass window should be adequate to prevent any flying object from entering the control room. The area in which the engine is installed should be protected with gates or doors to prevent personnel from entering the propeller area while the engine is running; however, provision must be made for the operator to gain access to the rear part of the engine in order to make necessary adjustments.

The actual installation of the engine in the test stand depends partly on the design of the test stand. The following installation steps may be considered typical:

1. See that the test stand is equipped with all items necessary for testing the make and model of engine to be installed.
2. Hoist the engine into place, and align mounting brackets.
3. Install mounting bolts, washers, lock washers, and nuts. Tighten the nuts to proper torque.
4. Install short exhaust stacks with gaskets to cylinder exhaust ports.
5. Connect the fuel-supply line to the engine-driven pump inlet or to the carburetor as required. Make sure the supply pressure is correct for the carburetor or fuel

unit. The pressure will vary for gravity-fed float-type carburetors, pump-fed float-type carburetors, injection carburetors, or direct fuel-injection units.

6. Connect the oil-supply and return lines if the engine is the dry-sump type.

7. Attach cylinder-head temperature sensing units (thermocouples or temperature bulbs) as required.

8. Connect the magneto switch wires to the magnetos.

9. Connect the pressure line for the oil pressure gage.

10. Connect the pressure line for the fuel pressure gage.

11. Connect the oil-temperature-gage line (electric or capillary).

12. Connect the manifold pressure line.

13. Connect the throttle control.

14. Connect the mixture control.

15. Connect the electric cable to the starter.

16. Connect the tachometer cable or electric lines as required.

17. Install a suitable cooling shroud for the engine.

18. Install the test propeller (test club). Make sure that the test propeller is of the correct rating for the engine being tested.

19. Service the engine with the proper grade of lubricating oil. If the engine is to be stored for a time after the test, use a preservative-type lubricating oil.

20. Perform a complete inspection of the installation to make sure that all required installation procedures have been completed.

● ENGINE TEST AND RUN-IN

Preoiling

Before the engine is actually started for the first time, it should be preoiled to remove trapped air in oil passages and lines and to make sure that all bearing surfaces are lubricated.

Preoiling can be accomplished in several ways, one of these being the use of a pressure oil container. With this method, the oil pressure unit is connected to the oil inlet line for the engine. Oil under pressure (about 20 psi [137.9 kPa]) is then applied to the inlet while the engine is rotated through several turns with the starter. This is continued until a generous flow of oil comes from the oil *out*, or *return line*. Prior to preoiling, one spark plug should be removed from each cylinder to reduce the load on the starter. The engine should not be rotated with the starter for periods longer than 30 s.

If an oil pressure tank is not available, the engine may be preoiled by connecting the normal test-cell oil supply to the engine inlet after making sure that the line from the tank to the engine is free of air. It will be necessary to rotate the engine more than is required with the pressure tank; however, preoiling can be accomplished satisfactorily in this manner.

Starting Procedure

Before the engine is actually started, the engine area should be checked for loose objects which could be picked up by the propeller. The engine itself should be checked for tools, nuts, washers, and other small items which may be lying loose.

The following steps are typical of an engine starting in the test cell:

1. Turn on the master power switch.

2. Open the throttle about one-tenth of the total distance.

3. Turn on the fuel pump, and check the fuel pressure.

4. Press the starter switch, and allow the engine to turn through three or four revolutions to check for liquidlock.

5. Turn the magneto switch ON for the magneto having the impulse coupling or induction vibrator.

6. Place the mixture control into the FULL RICH or IDLE CUTOFF position, depending on the type of fuel control.

7. As soon as the engine starts running smoothly, adjust the throttle for the desired rpm, usually 1000 rpm or less for a newly overhauled engine. If the engine is equipped with Bendix or Continental fuel injection, the mixture control must be moved to FULL RICH as soon as the engine starts.

8. Immediately check for oil pressure. If oil pressure does not register within 30 s, shut down the engine and check for the trouble.

9. If the engine operates properly, shut off the fuel boost pump.

10. As soon as the engine is operating smoothly, turn the magneto switch off *momentarily* in order to determine whether the engine can be shut off with the switch in case of emergency.

11. If the engine is operating satisfactorily, continue with the test run as specified by the manufacturer.

A log should be kept and the instrument readings recorded every 15 min. The log sheet should also include the date of the test, the engine number, and the type and nature of the test, along with the total number of hours of engine operation. All periods during the test run when the engine was not in operation should be recorded, along with the reason. If for any reason it should be necessary to replace any part, the complete reason for the rejection of the part should also be recorded.

Run-in Test Schedule

The manufacturer's overhaul manual provides instructions and a run-in schedule for newly overhauled engines. The purpose of the run-in is to permit newly installed parts to burnish or "wear in," piston rings to seat against cylinder walls, and valves to become seated. The run-in also makes it possible to observe the engine operation under controlled conditions and to ensure proper operation from idle to 100 percent power. The time during which an engine is operated at full power is referred to as a **power check.** The purpose of this check is to assure satisfactory performance.

The engine run-in should be accomplished with the engine installed in a test cell equipped as specified in the manufacturer's overhaul manual. The engine should be equipped with a correctly designed and rated club propeller or a dynamometer which will apply the specified load to the engine. Calibrated instruments must be available in the test cell to measure such parameters as cylinder-head temperature, oil temperature, manifold pressure, intake air temperature, turbocharger intake air pressure, turbocharger air-outlet pressure, turbocharger exhaust outlet pressure, and any other parameters specified by the manufacturer.

Slave (external) oil filters should be installed for both the engine and the turbocharger oil systems. These are to trap metal particles which are often present in a newly overhauled engine. It is particularly important that metal particles be prevented from entering the turbocharger and turbocharger-control units. For this purpose, the oil filter should have the capability of removing all particles having a dimension of 100 μm (0.01 mm) or greater and should have an area such that there is no restriction to oil flow.

A typical run-in schedule for a direct-drive (no propeller reduction gear) engine with the prescribed propeller load requires 10-min periods of operation at 1200, 1500, 1800, 2000, 2200, and 2400 rpm. Following this, the engine is operated at normal rated horsepower for 15 min. An oil-consumption run is made after the standard run-in schedule has been completed. The oil-consumption run requires 1 h of operation at normal rated power. The quantity of oil consumed is determined by weighing the oil in the system or in an external tank both before and after the oil-consumption run.

The run-in schedule for the supercharged Continental GTSIO-520-H engine is given below. This schedule is provided here as an example only; it is not to be used in place of the schedule published in the overhaul manual.

In the schedule shown, the 5-min period during which the engine is operated at 3400 rpm is the **power check**. If the engine will not come up to normal operating rpm when operating with a test club or fixed-pitch propeller, the engine is considered a "weak engine" and corrective action must be taken.

Period	Time, min	rpm	Turbocharger outlet pressure, inHg
1	5	1200	
2	10	1500	
3	10	2100	
4	10	2600	
5	10	2800	
6	10	3000	
7	10	3200	
8	5	3400 ± 25	42.0 − 43.0 (100% power)
9	5	3000	34.8 − 35.8 (68.5% power)
10	5	2600	33.5 − 34.5 (44.8% power)
11	10	600 ± 25	Cooling period (idle)

It must be noted at this point that some engine manufacturers recommend that the engine run-in be accomplished in slow flight rather than on the ground. This is because better cooling of the engine can be obtained in flight.

Oil-Consumption Run

If desired, an oil-consumption check run may be made at the end of the test in the following manner. However, the instructions in the overhaul manual should be followed for any particular make and model of engine. Record the oil temperature. Stop the engine in the usual manner. Place a previously weighed container under the external oil tank, and remove the drain plug. Allow the tank to drain for 15 min. Replace the drain plug. Weigh the oil and container. Record the weight of oil (i.e., total weight less weight of container). Replace the oil in the tank. Start the engine, warm up to the specified rpm ± 20, and operate at this speed for 1 h. At the conclusion of 1 h of operation and with the oil temperature the same as recorded at the time of previous draining (it is important to keep this oil temperature as constant as possible), again drain the oil in the same manner as before. The difference in oil weights at the start and at the end of the run will give the amount of oil used during 1 h of operation. The oil consumption for 1 h on an overhauled engine with new rings installed should not exceed 1 lb [0.45 kg] for GO-480-D engines, 1.5 lb [0.68 kg] for GO-480-C1B6 and -G1B6 engines, 1.8 lb [0.82 kg] for GSO-480 engines, and 3 lb [1.36 kg] for GTSIO-520-C engines.

● ENGINE PRESERVATION

If an engine is to be stored for a time after having been run in, it should be preserved against corrosion. This is particularly important with respect to the interior of cylinders, where the products of combustion will start corrosion of the bare cylinder walls within a very short time.

Preservation Run-in

As previously mentioned, an engine which is to be stored should be run in with a preservation oil as the lubricant. In addition, if possible, the last 15 min of operation should be done with clear (unleaded) gasoline at about two-thirds of full rpm. This will tend to remove the accumulation of corrosive residues which are in the cylinders and combustion chambers.

Interior Treatment

Upon completion of the run-in, all necessary adjustments of valves, magneto timing, and other operations which will require turning of the crankshaft should be accomplished before preservation is begun. The engine can then be started and idled, using clear gasoline, while preservative oil is sprayed into the carburetor inlet. The spray should not be heavy enough to choke the engine and should continue only until heavy white smoke issues from the exhaust. The engine should then be shut down immediately.

When the engine is stopped, the preservative oil should be drained from the crankcase or the sump. Spark plugs are then removed from the cylinders, and preservative oil is sprayed into the cylinders as the engine is rotated, several times for each cylinder. The rotation can be accomplished with the starter.

After all cylinders have been sprayed, dehydrator plugs containing **silica gel** are installed in the spark-plug holes and in the sump drain. The dehydrator plugs will absorb the moisture within the engine, thus reducing the tendency for the interior to corrode.

The short exhaust stacks should be removed and preservative oil sprayed into the exhaust ports. The ports should then be covered with metal plates and gaskets to seal them against moisture.

Exterior Treatment

All openings into the engine should be sealed with airtight plugs or with waterproof tape. If the carburetor is removed for separate preservation, a dehydrator bag can be placed in the carburetor opening before it is sealed. The bag should be tied to an exterior fitting so that it can be easily removed when the engine is prepared for operation.

After the engine is completely sealed, it may be sprayed lightly with preservative oil or other approved coating. If the engine is to be stored for as long as 6 months, it should be sealed in a waterproof plastic bag. The bag is first placed over the mounting bolts in the engine case, and the engine is installed in the case with the mounting bolts sticking through the bag. The bag is sealed at the engine mounting bolts when the bolts are tightened.

After a number of dehydrator bags are placed in the bag with the engine, the bag should be sealed according to the directions furnished. An indicator is also placed in the bag with the desiccant exposed through a window to show when the humidity level in the bag has reached a point where it is necessary to represerve the engine. When the desiccant in the bags or dehydrator plugs loses its color and begins to turn pink or white, the preservation is no longer effective and must be redone.

Inspection after Storage

When an engine is removed from storage after having been preserved, certain inspections should be made to ascertain that it has not been damaged by corrosion. The exterior inspection consists of a careful examination of all parts to see if corrosion has taken place on any bare metal part or under the enamel. Corrosion under enamel will cause the enamel to rise in small mounds or blisters above the smooth surface.

Interior inspections should be accomplished in all areas where it is possible to insert an inspection light. The most vulnerable area is inside the cylinders where the bare steel of the cylinders has been exposed to the combustion of fuel. Inspection of the cylinders is accomplished by removing the spark plugs from the cylinders and inserting an inspection light in one of the spark-plug holes. The inside of the cylinder can then be seen by looking through the other spark-plug hole. If rust is observed on the cylinder walls, it is necessary to remove the cylinder and dispose of the rust. If the cylinder walls are badly pitted, it will be necessary to regrind the cylinders and install oversize pistons and rings.

The rocker box covers should be removed to inspect for corrosion of the valve springs. Pitted springs will be likely to fail in operation owing to stress concentrations caused by the pitting.

When the exhaust port covers are removed to permit installation of the exhaust stacks, the ports and the valve stems can be examined for corrosion. A small amount of corrosion on the cast aluminum inside the exhaust port is not considered serious; however, if the valve stem is rusted, the rust must be removed or the valve replaced.

● PREFLIGHT INSPECTION

In order to provide an example of an actual preflight engine-check procedure recommended by a manufacturer, the following instructions are taken from the owner's handbook for a light twin airplane. The technician performing an engine maintenance run should follow the preflight checklist just as a flight crew would. On most modern aircraft, many of the aircraft systems tie directly into the engine in one way or another. For this reason, maintenance-run personnel must know the entire aircraft as well as the engine.

Preflight

Before each flight, visually inspect the airplane, and/or determine that

1. The tires are satisfactorily inflated and not excessively worn.
2. The landing-gear oleos and shock struts operate within limits.
3. The propellers are free of detrimental nicks.
4. The ground area under the propeller is free of loose stones, cinders, etc.
5. The cowling and inspection opening covers are secure.
6. There is no external damage or operational interference to the control surfaces, wings, or fuselage.
7. The windshield is clean and free of defects.
8. There is no snow, ice, or frost on the wings or control surfaces.
9. The tow bar and control locks are detached and properly stowed.
10. The fuel tanks are full and are at a safe level of proper fuel.
11. The fuel tank caps are tight.
12. The fuel system vents are open.
13. The fuel strainers and fuel lines are free of water and sediment by draining all fuel strainers once a day.
14. The fuel tanks and carburetor bowls are free of water and sediment by draining sumps once a week.
15. There are no obvious fuel or oil leaks.
16. The engine oil is at the proper level.
17. The brakes are working properly.
18. The radio equipment is in order.
19. The weather is satisfactory for the type of flying you expect to do.
20. All required papers are in order and in the airplane.
21. Upon entering the airplane, ascertain that all controls operate normally, that the landing gear and other controls are in proper position, and that the door is locked.

Starting

The following procedure is for one particular engine installation. In all cases, starting should be accomplished as instructed in the operator's manual.

Before starting the engine, the pilot should set the parking brake and turn on the master switch and the electric fuel pumps. Each should be individually checked for operation. When the engine is cold (under 40°F [4.44°C]), prime three to five strokes, making sure that the fuel valves are on, the cross-feed is off, fuel pressures are normal, and the fuel quantity has been checked. Push the mixture controls to FULL RICH, with carburetor heat off, and open the throttles about ¼ in [6.35 mm]. If the engines are extremely cold, they should be pulled through by hand four to six times.

Next turn all ignition switches on and engage the starter on the left engine first. After the engine starts, idle at 800 to 1200 rpm and start the right engine. If the battery is low, before starting the right engine, run the left engine over 1200 rpm to cut in the generator. This will produce extra power for starting the right engine. If the engine does not start in the first few revolutions, open the throttle on that engine while the engine is turning over with the ignition on. When the engine starts, reduce the throttle.

If this procedure does not start the engine, reprime and repeat the process. Continue to load the cylinders by priming or to unload by turning the engine over with the throttle open. If the engine still does not start, check for malfunctioning of the ignition or fuel systems.

Priming can be accomplished by pumping the throttle controls. However, care must be used because excessive pumping may overprime the engine, making starting difficult. Priming of engines with fuel injection is accomplished as explained elsewhere in this text.

When the engines are warm (over 40°F [4.44°C]), do not prime but turn the ignition switches both on before engaging the starter. The engines should start after rotating through about four compression strokes.

Warm-up and Ground Check

As soon as the engines start, the oil pressure should be checked. If no pressure is indicated within 30 s, stop the engine and determine the trouble.

Warm up the engines at 800 to 1200 rpm for not more than 2 min in warm weather or 4 min in cold weather. If electric power is needed from the generator, the engines can be warmed at 1200 rpm, at which point the generator cuts in. As soon as the engine is warmed up and operating smoothly, a **magneto grounding check** should be made. This is accomplished by moving the ignition switch from the BOTH position to the OFF position momentarily while the engine is running at a low rpm (idle speed or slightly above). The engine should stop firing immediately when the ignition switch is placed in the OFF position. This test assures that the engine can be shut down in case of emergency. The ignition switch should be returned to the BOTH position as soon as it is determined that the magnetos can be turned off.

The magneto operation should be checked with the engine running at approximately 2000 rpm, or at any other speed recommended by the manufacturer. MAP should be approximately 15 inHg (50.81 kPa). The magneto operational check is made by operating the engine briefly on each magneto separately and noting the drop in engine rpm. With the engine operating on both magnetos at the desired rpm, the ignition switch is moved from the BOTH position to the R (right) position. The engine rpm will drop slightly because it is operating on only one set of spark plugs. The rpm drop should not exceed 125 rpm unless otherwise specified in the operator's manual. The ignition switch is turned back to the BOTH position as soon as the rpm drop has been noted. The engine is then allowed to run with the switch in the BOTH position for enough time to allow the engine rpm to return to the original value for operation with both magnetos. This clears (defouls) the inactive spark plugs. The switch is then turned to the L (left) position and the rpm drop noted. The engines are warm enough for takeoff when the throttles can be opened without the engine faltering.

Carburetor heat should be checked during the warm-up to make sure that the heat-control operation is satisfactory and to clear out the carburetor if any ice has formed. It should also be checked in flight occasionally when outside-air temperatures are between 20 and 70°F [−6.67 and 21.11°C] to see if icing is forming in the carburetor. In most cases when an engine loses manifold pressure without apparent cause, the use of carburetor heat will correct the condition.

The propeller controls should be moved through their normal ranges during the warm-up to check for proper operation, then left in the full HIGH RPM (low-pitch) positions. Full feathering checks on the ground are not recommended because of the excessive vibration caused in the powerplant installations. However, feathering action can be checked by momentarily pulling the propeller controls into the feathering position and allowing the rpm to drop not lower than 1500, then returning the controls to a normal operating position.

The electric fuel pumps should be turned off after starting or during warm-up to make sure that the engine-driven pumps are operating. Prior to takeoff the electric pumps should be turned on again to prevent loss of power during takeoff as a result of fuel-pump failure.

Stopping the Engine

Aircraft engines are today usually equipped with carburetors or fuel-control units having an IDLE CUTOFF control by which the fuel flow into the engine may be stopped. After a flight and a few minutes of taxiing, the engine can usually be stopped almost immediately by placing the mixture control in the IDLE CUTOFF position. If the engine is exceptionally hot as indicated by the cylinder-head temperature gage and the oil temperature gage, it is good practice to allow the engine to idle for a short time before shutdown.

After the engine is stopped, the magneto switch should be placed in the OFF position and all other switches should then be turned off.

● ONE-HUNDRED-HOUR AND ANNUAL INSPECTIONS

Federal Aviation Regulation 43 requires that every aircraft used for hire have a thorough inspection after every 100 h of flight and must have a similar inspection within the twelve months preceding the time of a flight for hire. The powerplant section of these inspections includes the following:

1. Cowling: for cracks, security of fastening devices, and other defects.

2. All systems: for proper operation, improper installation, poor general condition, defects, and insecure attachment.

3. Accessories: for defects and mounting.

4. Exhaust stacks: for cracks, evidence of exhaust leakage, other defects and attachment. An aircraft using a jacket around the engine exhaust as a source of cabin heat should be visually inspected and checked for carbon monoxide leakage.

5. Lines, hoses, clamps, and fittings: for leaks, poor condition, and looseness.

6. Engine controls: for proper operation, defects, improper travel, and improper safetying or lack of safetying.

7. Flexible vibration dampeners: for poor condition and deterioration.

8. Engine mount: for cracks, looseness of mounting, and looseness of engine to mount.

9. Engine: for visual evidence of excessive oil, fuel, or hydraulic leaks; tightness, security, and safetying of studs and nuts; cylinder compression; evidence of metal particles or other foreign matter on oil screens and drain plugs. If the engine has weak cylinder compression in any cylinder, the condition must be corrected.

10. Spark plugs: reconditioned or replaced.

11. Magneto: for timing and condition of breaker mechanisms.

12. Under the provisions of FAR 43 the engine must be run in accordance with its applicable run checklist and a compression check be performed while the engine is still warm. A record of the compression check must be placed in the engine log book.

For any particular aircraft, the manufacturer's maintenance manual provides instructions for the specific inspections that should be made for the make and model of aircraft and engine.

● ENGINE TROUBLESHOOTING

A good understanding of engine theory and the function of all parts and systems is essential to skillful engine troubleshooting. Engine operational malfunctions can usually be traced to one or more of three basic causes. These are (1) ignition malfunctions, (2) fuel system malfunctions, and (3) engine parts malfunctions. We shall consider each of these types of malfunction separately.

Ignition Malfunctions

Ignition troubles may be traced to defective magnetos, defective transformers in a low-tension system, improper timing, spark plugs which are burned or otherwise damaged, poor insulation on the high-voltage leads, short-circuited or partially grounded primary or switch (P) leads, burned breaker points, or loose connections.

Missing at high speeds. Misfiring of the engine at high speeds can be caused by almost all the foregoing defects in varying degrees. If the engine is operating at high speeds and high loads, the manifold pressure and the cylinder pressure will be high. As previously explained, it requires more voltage at the spark-plug gap

to fire the plug when the pressure at the gap is increased. This means that at high engine loads the ignition voltage will build up higher than at low engine loads. This higher voltage will seek to reach ground through the easiest path, and if there is a path easier to follow than through the spark-plug gap, the spark plug will not fire and the spark will jump through a break in the insulation or it may follow a path where dampness has reduced the resistance.

If the airplane is operating at high altitudes, the high-voltage spark will be still more likely to leak off the high-tension leads instead of going through the spark plug. The lower air pressure at high altitudes permits the spark to jump a gap more readily than it will at the higher pressure near sea level.

A weak breaker-point spring will also cause misfiring at high speeds. This is because the breaker points do not close completely after they are opened by the cam. This condition is called **floating points** because the cam follower actually does not maintain contact with the cam but floats at some points between the cam lobes.

Engine fails to start. If the engine will not start, the trouble can be in a defective ignition switch. Since aircraft engines have dual ignition systems, it is rare that both systems would become inoperative at the same time. It is possible, however, that, in some magneto switches, both magnetos could be grounded through a short circuit inside the switch.

If it is the recommended practice to start the engine on one magneto only and the engine will not start, an attempt should be made to start the engine on the other magneto. If the cause of the trouble is in the first magneto system, the engine will fire on the other magneto.

The checking of magnetos during the engine test will usually reveal malfunctions in one magneto or the other before a complete failure occurs. The defective magneto can then be removed and repaired before serious trouble occurs.

Defective spark plugs. Defective spark plugs are usually detected during the magneto check. If one spark plug fails, the engine rpm will show an excessive rpm drop when it is checked on the magneto supplying the defective plug. The bad plug may be located by the **cold-cylinder** check. This is accomplished as follows: Start and run the engine for a few minutes on both magnetos. Perform a magneto check, and determine which magneto indicates a high rpm drop. Stop the engine, and let it cool until the cylinders can be touched without burning the hand. Start the engine again, and operate on the one magneto for which the high rpm drop was indicated. Run the engine for about 1 min at 800 to 1000 rpm, and then shut it down. Immediately feel all cylinders with the hand, or use a magic wand to determine which is the cold cylinder. This cylinder will have the defective spark plug. The **magic wand** is a temperature gage operated by means of a thermocouple on the end of the wand. When the thermocouple is placed against the cylinder, a temperature indication is produced on the dial of the instrument.

The foregoing procedure for locating a defective spark plug is particularly important for a large engine having 14 to 18 cylinders. The time and expense involved in removing and replacing all the spark plugs to correct

Indication	Cause	Remedy
Engine will not start.	No fuel in tank.	Fill fuel tank.
	Fuel valves turned off.	Turn on fuel valves.
	Fuel line plugged.	Starting at carburetor, check fuel line back to tank. Clear obstruction.
	Defective or stuck mixture control.	Check carburetor for operation of mixture control.
	Pressure discharge-nozzle-valve diaphragm ruptured.	Replace discharge-nozzle valve.
	Primer system inoperative.	Repair primer system.
Engine starts, runs briefly, then stops.	Fuel tank vent clogged.	Clear the vent line.
	Fuel strainer clogged.	Clean fuel strainer.
	Water in the fuel system.	Drain sump and carburetor float chamber.
	Engine fuel pump inoperative or defective.	Replace engine-driven fuel pump.
Black smoke issues from exhaust. Red or orange flame at night.	Engine mixture setting too rich.	Correct the fuel-air mixture adjustment.
	Primer system leaking.	Replace or repair primer valve.
	At idling speed, idle mixture too rich.	Adjust idle mixture.
	Float level too high.	Reset carburetor float level.
	Defective diaphragm in pressure carburetor.	Replace pressure carburetor.

one defective plug can be avoided by pinpointing the defective plug through a cold-cylinder check.

Fuel System Troubles

Fuel systems, carburetors, fuel pumps, and fuel-control units can cause a wide variety of engine malfunctions, some of which may be difficult to analyze. A thorough understanding of the system and its components is essential if the technician hopes to resolve the problems of a particular system effectively. The charts shown herein list some of the most common problems encountered with fuel systems and suggest remedies.

The charts do not cover all symptoms which may develop with fuel systems and carburetors because of the many different designs involved. The technician, in each case, should analyze the type of system upon which he or she is working and become familiar with the operation of the carburetor or fuel-control unit used in the system. Fuel-injection systems involve some problems that are unique to such systems. The accompanying chart lists problems that may be encountered with one particular fuel-injection system.

Oil System Troubles

Oil system troubles are usually revealed as leaks, no oil pressure, low oil pressure, fluctuating oil pressure, high oil pressure, and high oil consumption. The correction of oil leaks is comparatively simple in that it involves tracing the leak to its source and then applying the indicated repair procedure. If oil has spread over a large area of the engine, it is sometimes necessary to wash the engine with solvent and then operate it for a short period to find the leak.

A check for oil pressure when an engine is first started is always a standard part of the starting procedure. If the oil pressure does not show within about 30 s, the engine is shut down. Lack of oil pressure can be caused by any one of the following conditions: no oil in the tank; no oil in the engine oil pump (hence, no prime); an air pocket in the oil pump; an oil plug left out of a main oil passage; an inoperative oil pump; an open pressure-relief valve; a plugged oil-supply line; or a broken oil line. If a condition of no oil pressure exists, the technician should start for the most likely cause first and then check each possibility until the trouble is located.

Low oil pressure can be caused by a variety of discrepancies including the following: oil pressure-relief valve improperly adjusted, broken oil relief-valve spring, sticking pressure-relief valve, plug left out of an oil passage, defective gasket inside the engine, worn oil pressure pump, worn bearings and/or bushings, dirty oil strainer, excessive temperature, wrong grade of oil, and leaking oil-dilution valve. The cause of low oil pressure is often more difficult to ascertain than some of the other oil problems; however, a systematic analysis of the problem by the technician will usually lead to a solution. One of the first questions technicians must ask is, "Did the condition develop gradually, or did it show up suddenly?" They should also check to see how many hours of operation the engine has had. Another most important consideration is the actual level of the oil pressure. If it is extremely low, the technician will look for an "acute" condition, and if it is only slightly low, the likely cause will be different.

High oil pressure can result from only a few causes. These are an improper setting of the relief valve, a sticking relief valve, an improper grade of oil, the low temperature of oil and engine, and a plugged oil passage. The cause can usually be located easily except in the case of a newly overhauled engine where a relief-valve passage may be blocked. It should be noted that oil pressure will be abnormally high when a cold engine is first started and is not yet warmed up.

High oil consumption is usually the result of wear or leaks. If blue oil smoke is emitted from the engine breather and exhaust, it is most likely that the piston

Indication	Cause	Remedy
Engine will not start. No fuel flow indication.	Fuel-selector-valve in wrong position.	Position fuel-selector-valve handle to main tank.
	Dirty metering unit screen.	Clean screen.
	Improperly rigged mixture control.	Correct rigging of mixture control.
Engine acceleration is poor.	Idle mixture incorrect.	Adjust fuel-air control unit.
Engine will not start. Fuel flow gage shows fuel flow.	Engine flooded.	Clear engine of excessive fuel.
	No fuel to engine.	Loosen one line at fuel manifold nozzle; if no fuel shows, replace fuel manifold.
Engine idles rough.	Restricted fuel nozzle.	Clean nozzle.
	Improper idle mixture.	Adjust fuel-air control unit.
Very high idle and full-throttle fuel pressure present.	Relief valve stuck closed.	Repair or replace injector pump.
Engine runs rough.	Restricted fuel nozzle.	Clean nozzle.
	Improper pressure.	Replace pump.
Low fuel pressure at high power.	Leaking turbocharger discharge pressure.	Repair leaking lines and fittings.
	Check valve stuck open.	Repair or replace injector pump.
Low fuel flow gage indication.	Restricted flow to metering valve.	Clean fuel filters and/or adjust mixture control for full travel.
	Inadequate flow from fuel pump.	Adjust fuel pump.
Fluctuating or erroneous fuel flow gage indication.	Vapor in system.	Clear with auxiliary fuel pump.
	Clogged ejector jet in vapor-separator cover.	Clean jet.
	Air in fuel flow gage line.	Repair leak and purge line.
High fuel flow gage indication.	Altitude compensator stuck.	Replace fuel pump.
	Restricted nozzle or fuel manifold valve.	Clean or replace as required.
	Recirculation passage in pump restricted.	Replace fuel pump.
Fuel discharging into engine compartment. Relief valve probably not operating.	Leaking diaphragm.	Repair or replace injector pump.
No fuel pressure.	Check valve stuck open.	Repair or replace injector pump.
Unmetered fuel pressure.	If high, internal orifices are plugged.	Clean internal orifices in injector pump.
	If low, relief valve stuck open.	Repair or replace injector pump.

rings are worn, with the result that **blowby** occurs: The pressure built up in the crankcase causes the oil spray inside the crankcase to be blown out the breather. High breather pressures occur on some engines because of a buildup of sludge in the breather tube; this may be detected by excessive propeller-shaft-seal leakage. Cleanliness of the breather tube is usually a 100-h item on engines where this may be a problem. The worn rings will also allow oil to pass the piston and enter the combustion chamber, where it is burned. This, of course, produces blue smoke at the exhaust.

Another cause of high oil consumption is a worn master rod bearing in a radial engine; this permits an excessive amount of oil to be sprayed from the bearing and into the cylinder bores. If the scavenger pump is defective, the oil will not be removed from the sump as rapidly as required, and this will also lead to excessive oil consumption.

Operating an engine at high-power settings and high temperatures will increase the oil consumption. If an apparently normal engine is using more oil than it should, the pilot should be questioned regarding operation of the engine in flight. The pilot of an aircraft should observe the reading of the oil pressure gage frequently during operation. If the gage should begin to fluctuate, the flight should be terminated as soon as possible because the indication is that there is a low oil supply.

Induction Systems

The designs of induction systems for reciprocating engines vary considerably for different aircraft and engine combinations. The simplest types of induction systems include an air filter in the forward-facing air-scoop, a carburetor air-heating system, ducting to the carburetor, the carburetor, and the intake manifold or intake pipes that carry the fuel-air mixture to the valves. Other systems include turbochargers, superchargers, alternate-air systems and carburetor deicing systems. For a particular aircraft and engine combination, the technician should consult the operator's and maintenance manuals for the aircraft. Induction systems are described in Chap. 7 of this text.

Problems that may arise in a typical induction system include the following:

1. Dirty and/or damaged air filter
2. Worn, loose or damaged air ducting
3. Loose or defective air temperature bulb
4. Defective air heater valve
5. Defective alternate-air valve
6. Loose carburetor mounting
7. Defective carburetor mounting gasket
8. Leaking packings or gaskets at intake pipes
9. Leaking intake manifold

It must be remembered that any crack or other opening that allows air to enter an intake manifold, in the case of naturally-aspirated engines, will cause the fuel-air mixture to be excessively lean and may bring about engine damage and adversely affect engine performance. Leaks in the intake manifold of a supercharged or turbocharged engine will allow the fuel-air mixture to escape, thus reducing manifold pressure and engine power output.

Inspection of a typical induction system includes the following:

1. Check the air filter for condition, cleanliness, and security. Service the air filter according to instructions.
2. Check the air ducting to the carburetor for wear damage, cracks, and security of mounting.
3. Check the air-heater valve and ducting for wear, cracks and security of mounting. Check the valve door bearings for wear and lubricate according to instructions.
4. Check the carburetor air temperature bulb for security.
5. Check the carburetor mounting for security. Tighten any loose bolts or capscrews.
6. Check the carburetor mounting gasket for possible air leakage. Replace the gasket if damaged.
7. Check the intake pipes and/or manifold for condition and security.
8. Check the intake-pipe packing nuts for tightness. Check the packings or gaskets for shrinkage or damage.
9. Check the alternate-air system for condition.

Backfiring

The condition known as **backfiring** occurs when the flame from the combustion chamber burns back into the intake manifold and ignites the fuel-air mixture before it enters the engine. It often occurs when starting a cold engine because of poor (slow) combustion. The fuel-air mixture in the cylinder is still burning at the time the intake valve opens and the flame burns back through the intake valve. This sometimes causes a fire in the induction system.

Any defect in the carburetor or fuel-control system which causes an excessively lean mixture can lead to backfiring. If the condition persists after an engine is warmed up, it will be necessary to follow a systematic procedure to locate the cause.

Another cause of backfiring is sticking intake valves. This does not usually occur with a new or newly overhauled engine, but it is likely to be encountered with an older engine operated at high temperatures. If a sticking intake valve remains open, it can cause the engine to stop and may cause considerable damage to the induction system.

Ignition troubles often cause backfiring. If high-tension current leaks at the distributor block, it can cause firing of plugs out of time, with the result that the mixture may fire in a cylinder when the intake valve is open. If a newly overhauled engine is being started for the first time and backfiring persists, the technician should check for ignition timing and for proper connection of the spark-plug leads. Ignition out of time can also cause afterfiring through the exhaust.

Afterfiring

Afterfiring is the burning of fuel-air mixture in the exhaust manifold after the mixture has passed through the exhaust valve. It is characterized by explosive sounds as well as by large flames trailing outward from the exhaust stacks (torching). Afterfiring is usually caused by excessive fuel (rich mixture) in the exhaust to the extent that the fuel continues to burn after the mixture leaves the cylinder. The condition may be caused by overpriming, excessively rich mixture, poor ignition, and improper timing. Since there are comparatively few causes for afterfiring, the condition is usually easy to correct.

Compression Testing

Since engine compression is a primary factor in the proper operation of an engine and its ability to develop full power, the technician will be called upon from time to time to check the compression of engine cylinders. The procedure is not difficult, but it does require that the technician have a good understanding of the procedure to be followed. As stated earlier, the compression check is a 100-h inspection requirement.

The compression test reveals whether the valves, piston rings, and pistons are adequately sealing the combustion chamber. If leakage of pressure is excessive, it is obvious that the cylinder cannot develop its full force.

A simple method for checking the compression of the cylinders for small engines is "pulling through" the propeller. First the engine should be operated until it is thoroughly warmed up. Immediately after stopping the engine (ignition switch off), the propeller is pulled through and the back pressure of each cylinder is noted. The person pulling the propeller stands clear of the blades to avoid injury if the engine should happen to fire. An experienced technician can feel any appreciable difference in the compression of the cylinders. If the compression for any cylinder seems to be weak, the exhaust stacks are checked for a hissing sound as the propeller is rotated. If such a sound is heard, it is known that an exhaust valve is leaking.

If only one cylinder shows poor compression, it will probably be advisable to correct the trouble by making the repair required. To determine which cylinder has weak compression, one spark plug is removed from the no. 1 cylinder. The thumb is then placed over the spark-plug hole, and the engine is rotated until pressure on the thumb indicates that the piston is approaching TDC on the compression stroke. This fixes the position of the crankshaft with respect to firing order. The engine is then rotated, and the number of each cylinder in the order of firing is noted as back pressure on the propeller indicates that it is passing the compression stroke. The number noted when the weak cylinder reaches the compression stroke indicates the cylinder to be repaired.

In some cases the weak compression is caused by a valve which has an accumulation of carbon on the stem and is therefore not closing completely. This condition can sometimes be corrected by removing the rocker box cover and tapping the valve-stem end of the rocker arm sharply with a soft hammer to remove the carbon from the stem or the valve seat. The success of the operation is checked by pulling the engine through. If the valve still leaks, it will be necessary to remove the cylinder and recondition the valves and seats.

The compression of large engines is usually checked by means of a **compression tester**. This is a device which applies a controlled air pressure to the cylinder through a spark-plug hole so that the leakage can be determined by comparing the readings of two gages. The procedure for checking the compression of a cylinder is as follows:

1. Warm up the engine to operating temperature.

2. Remove the front spark plug from the cylinder to be checked, and place the piston of the cylinder at TDC on the compression stroke.

3. Connect the compression tester to an air pressure source of 100 to 120 psi [689.5 to 827.4 kPa].

4. Install the cylinder pressure-line fitting into the spark-plug hole. Attach the air pressure line to the fitting in the cylinder.

5. Have two persons hold the propeller blade so that the piston will remain at the TDC position.

6. Adjust the pressure regulator of the compression tester to show the correct pressure (about 80 psi [551.6 kPa]) on the fixed pressure gage.

7. Open the air valve slowly to allow time for the persons holding the propeller to make sure that the piston is at TDC. If the piston is not at TDC, the propeller will tend to turn one way or the other because of the air pressure on the piston. If the propeller turns in the direction of normal rotation, the crankshaft is past TDC.

8. Turn the air valve full on, and note the reading of the cylinder pressure gage. The difference between this pressure and the pressure of the fixed pressure gage will show the amount of leakage from the cylinder. If the difference in pressure exceeds that shown in the table for the engine, the cylinder should be repaired. Generally, if a cylinder shows a pressure loss of 25 percent or more, it is considered unsatisfactory.

The compression tester employed in the above procedure operates on the principle of pressure differential across a metering orifice, as shown in Fig. 15-1. If a regulated pressure of 80 psi is applied on one side of the orifice and there is no leakage on the other side of the orifice, both pressure gages will read 80 psi. However, as leakage increases on the cylinder side of the orifice, the cylinder pressure gage reading will decrease. When the difference between the two readings exceeds a certain value, the cylinder compression is considered unsatisfactory.

It must be noted that the procedure for checking cylinders with compression testers may vary to some extent, depending on the design of each tester. The procedure described in this section is to provide an understanding of how the operation is accomplished and is not to be used with a particular tester unless it happens to agree with the instructions for the tester being employed.

Troubleshooting Charts

Aircraft manufacturers include troubleshooting procedures and troubleshooting charts of various types in the aircraft maintenance manuals. The information provided refers to a particular type and make of aircraft and the type of engine which powers the aircraft. Differences exist in procedures because of the type of engine fuel system, whether the engine is supercharged or not, and other factors. Some instructions apply to the majority of engines while other instructions may apply to only one type of engine and aircraft combination. In all cases, technicians should utilize the charts and instructions prepared for the aircraft upon which they are working.

The troubleshooting chart for general engines is typical of those prepared for a light aircraft. Some of the instructions involve fuel-injection systems and the turbocharger. Other instructions in the chart are applicable to any piston-type aircraft engine.

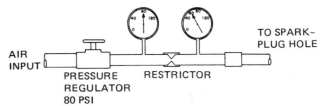

FIG. 15-1 Drawing to illustrate the principle of a compression tester.

TROUBLESHOOTING CHART—GENERAL ENGINE

Indication	Cause	Remedy
Engine will not start. Engine cranking. All circuit breakers and switches in correct position.	Lack of fuel.	Check fuel valves. Service fuel tanks.
	Engine overprimed.	Clear engine. Follow correct starting procedure.
	Induction system leaks.	Correct leaks.
	Starter slippage.	Replace starter.
Engine will not run at idling.	Propeller lever set for DECREASE RPM.	Place propeller lever in HIGH RPM position for all ground operations.
	Improperly adjusted carburetor or fuel-injection system.	Readjust system as required.
	Fouled spark plugs.	Change spark plugs.
	Air leak in intake manifold.	Tighten loose connection or replace damaged part.
Engine misses at high speed.	Broken valve spring.	Replace valve spring.
	Plugged fuel nozzle.	Clean or replace.
	Warped valve.	Replace valve.
	Hydraulic tappet worn or sticking.	Replace tappet.
	Weak breaker spring in magneto.	Repair magneto.
Engine runs too lean at cruising power.	Improper manual leaning procedure.	Manual lean in accordance with operator's manual.
	Low fuel flow.	Check and clean fuel strainer.
	Carburetor or fuel-injection system malfunction.	Correct malfunction.
Engine runs rough at high speed.	Loose mounting bolts or damaged mount pads.	Tighten or replace mountings.
	Plugged fuel nozzle.	Clean or repair.
	Propeller out of balance.	Remove and repair propeller.
	Ignition system malfunction.	Troubleshoot ignition system and repair.
Engine idles rough.	Improperly adjusted carburetor or fuel-injection system.	Adjust system as required.
	Fouled spark plugs.	Clean or replace spark plugs.
	Improperly adjusted fuel controls.	Adjust fuel controls.
	Discharge-nozzle air vent manifold restricted or defective.	Clean or replace.
	Dirty or worn hydraulic lifters.	Clean or replace.
	Burned or warped exhaust valves, seats. Scored valve guides.	Repair or replace.
Engine runs rich at cruising power.	Restriction in air-intake passage.	Remove restriction.
Spark plugs continuously foul.	Piston rings worn or broken.	Overhaul engine.
	Spark plugs have wrong heat range.	Install proper range spark plugs.
Sluggish engine operation and low power.	Improper rigging of controls.	Rerig controls.
	Leaking exhaust system to turbo.	Correct exhaust system leaks.
	Restricted air intake.	Correct restriction.
	Turbo wheel rubbing.	Replace turbocharger.
	Ignition system malfunction.	Troubleshooting ignition system and correct malfunction.
	Carburetor or fuel-injection system malfunction.	Troubleshoot and correct malfunction.
	Engine valves leaking. Piston rings worn or sticking.	Overhaul engine.
High cylinder-head temperature.	Octane rating of fuel too low.	Drain fuel and fill with correct grade.
	Improper manual leaning procedure.	Use leaning procedure set forth in the operator's manual.
	Bent or loose cylinder baffles.	Inspect for condition and correct.
	Dirt between cooling fins.	Remove dirt.

Indication	Cause	Remedy
	Exhaust system leakage.	Correct leakage.
	Excessive carbon deposits in combustion chambers.	Overhaul engine.
Oil pressure gage fluctuates.	Low oil supply.	Determine cause of low oil supply and replenish.
Engine oil leaks.	Damaged seals, gaskets, O rings, and packings.	Repair or replace as necessary to correct leaks.
Low compression.	Excessively worn piston rings and valves.	Overhaul engine.
Engine will not accelerate properly.	Unmetered fuel pressure too high.	Adjust engine fuel pressure according to specifications.
	Turbocharger waste gate not closing properly.	Refer to turbocharger and controls manual.
	Leak in turbocharger discharge pressure.	Repair or replace as necessary.
Slow engine acceleration on a hot day.	Mixture too rich.	Lean mixture until acceleration picks up. Then return control to FULL RICH.
Engine will not stop at IDLE CUTOFF.	Manifold valve not seating tightly.	Repair or replace manifold valve.
Manifold pressure overshoot on engine acceleration.	Throttle moved forward too rapidly.	Open throttle about half way. Let manifold pressure peak, then advance throttle to full open.
Slow engine acceleration at airfields with ground elevation above 3500 ft [1066.80 m].	Mixture too rich.	Lean mixture with manual mixture control until operation is satisfactory.
When climbing to 12,000 ft [3657.60 m], engine quits when power reduced.	Fuel vaporization.	Operate boost pump when climbing to high altitudes. Keep boost pump on until danger of vapor is eliminated.

● REVIEW QUESTIONS

1. List the essential steps in unpacking and removing an engine from the shipping case.
2. Describe a general procedure for removing an engine from preservation and placing it in service.
3. What torque should be used for the installation of 14-mm spark plugs? 18-mm plugs?
4. Describe the installation of a fitting having pipe threads.
5. After a pressure-type carburetor has been in preservation, what treatment should be applied to it before service?
6. What inspections should be made following an engine installation?
7. What information should be obtained during the engine test run in a test cell?
8. Describe the preoiling procedure for a newly overhauled engine.
9. How may it be determined that a reciprocating engine with a dry sump is preoiled sufficiently?
10. List the steps in a typical engine starting procedure.
11. Why does one allow the engine to turn through three or four revolutions before turning on the ignition switch?
12. Describe a *power check*.
13. What is a "weak engine"?
14. Describe the procedure for an engine run-in after overhaul.
15. Why is it recommended in some cases that the engine run-in be done in flight?
16. Describe an *oil-consumption run*.
17. Describe a method for preservation of an engine after it has been run in on a test stand.
18. Why is unleaded gasoline useful in engine preservation?
19. What material is used to absorb moisture inside an engine during storage?
20. When the desiccant in an engine preservation bag turns pink or white, what should be done?
21. What sections of the engine should be inspected for corrosion or rust after a long period of storage?
22. What must be done if rust is found on the cylinder walls?
23. What should be done if an engine is overprimed during an attempt to start?
24. What important check should be made immediately after an engine has started?
25. If the battery is found to be low in charge after starting one engine on a twin-engine airplane, what can be done to start the other engine?

26. Describe the checking of magnetos after engine warm-up.
27. How can you check the feathering system for the propeller?
28. When should carburetor heat be checked?
29. Why should electric fuel pumps be turned off after starting the engine and then turned on again before takeoff?
30. When an engine is equipped with a carburetor which has an idle cutoff, what is the procedure for shutting down the engine?
31. What engine inspections are made during the 100-h or annual inspections of an aircraft?
32. List six ignition malfunctions.
33. What are the likely causes when an engine misfires at high rpm?
34. Why is an engine more likely to misfire at high altitudes than at sea level?
35. Describe a *cold-cylinder check*.
36. List remedies for fuel system discrepancies which prevent the engine from starting.
37. What may be the trouble if an engine starts, runs briefly, and then stops?
38. What condition causes black smoke to be emitted from the engine exhaust?
39. What condition causes blue or whitish smoke to be emitted from the engine exhaust and the engine breather?
40. List five possible causes for low engine oil pressure.
41. List five causes for high oil pressure.
42. Compare the oil pressure of a newly started cold engine with that of the same engine after it has been warmed up.
43. If the oil pressure gage fluctuates rapidly, what problem is indicated?
44. Explain reasons why an engine will *backfire*.
45. What damage can be caused by backfiring?
46. If an engine which has been running normally suddenly begins to backfire violently, what is a likely cause?
47. Explain the reasons for *afterfiring*.
48. Describe a simple method for checking the compression of a small-aircraft engine.
49. Describe the use of a *compression tester*.

16 PROPELLER FUNDAMENTALS

DESCRIPTION AND NOMENCLATURE

The **aircraft propeller** normally consists of two or more blades and a central hub by means of which the blades are attached to a shaft driven by the engine. The purpose of the propeller is to pull the airplane through the air. It does this by means of the thrust obtained by the action of the rotating blades on the air.

Nomenclature

In order to explain the theory and construction of propellers, it is necessary first to define the parts of various types of propellers and give the nomenclature associated with propellers. Figure 16-1 shows a fixed-pitch one-piece wood propeller designed for light aircraft. Note carefully the *hub, hub bore, bolt holes, neck, blade, tip,* and *metal tipping.* These are the common terms applied to a wood propeller.

The cross section of a propeller blade is shown in Fig. 16-2. This drawing is shown to illustrate the **leading edge** of the blade, the **trailing edge,** the cambered side, or **back,** and the flat side, or **face.** From this illustration it is apparent that the propeller blade has an airfoil shape similar to that of an airplane wing.

Since the propeller blade and the wing of an airplane are similar in shape, each blade of an aircraft propeller may be considered as a rotating wing. It is true that it is a small wing, which has been reduced in length, width, and thickness, but it is still a wing in shape. At one end this midget wing is shaped into a shank, thus forming a propeller blade. When the blade starts rotating, air flows around the blade just as it flows around the wing of an airplane, except that the wing, which is approximately horizontal, is lifted upward while the blade is "lifted" forward.

The nomenclature for an **adjustable,** or **ground-adjustable,** propeller is illustrated in Fig. 16-3. This is a metal propeller with two blades clamped into a steel **hub assembly.** The hub assembly is the supporting unit for the blades, and it provides the mounting structure by which the propeller is attached to the engine propeller shaft.

The propeller hub is split on a plane parallel to the plane of rotation of the propeller to allow for the installation of the blades. The blade **root** consists of machined ridges which fit into grooves inside the hub. When the propeller is assembled, the sections of the hub are held in place by means of **clamping rings** secured by means of bolts. When the clamping-ring bolts are properly tightened, the blade roots are held rigidly so that the blades cannot turn and change the blade angle.

Figure 16-4 shows two views and various cross sections of a propeller blade. The **blade shank** is that portion of the blade which is near the butt of the blade. It is usually thick to give it strength, and it is cylindrical where it fits the hub barrel, but the cylindrical portion of the shank contributes little or nothing to thrust. In an attempt to obtain more thrust, some propeller blades are designed so that the airfoil section (shape) is carried all the way along the blade from the tip to the hub. In other designs, the airfoil shape is carried to the hub by means of blade **cuffs,** which are thin sheets of metal or other material, which function like cowling, as shown in Fig. 16-5.

In Fig. 16-4 the tip section, the center of the hub, and the blade butt are shown. The **blade butt,** or **base,** is merely the end of the blade which fits into the hub.

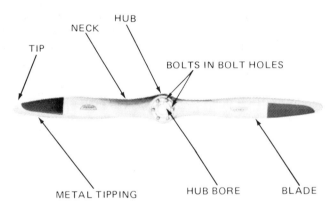

FIG. 16-1   Fixed-pitch one-piece wood propeller.

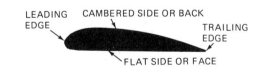

FIG. 16-2   Cross section of a propeller blade.

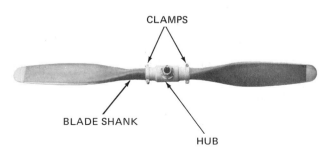

FIG. 16-3   Nomenclature for a ground-adjustable propeller.

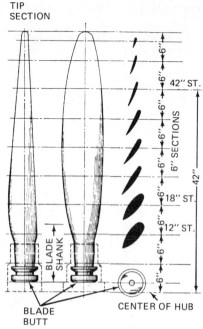

FIG. 16-4   Propeller blade, showing blade construction and blade sections.

PROPELLER THEORY

The Blade-Element Theory

The first satisfactory theory for the design of aircraft propellers was known as the **blade-element theory**. This theory was evolved in 1909 by a Polish scientist named Dryewiecki; hence, it is sometimes referred to as the **Dryewiecki theory.**

This theory assumes that the propeller blade from the end of the hub barrel to the tip of the propeller blade is divided into various small, rudimentary airfoil sections. For example, if a propeller 10 ft [3 m] in diameter has a hub 12 in [30.48 cm] in diameter, then each blade is 54 in [137.16 cm] long and can be divided into fifty-four 1-in [2.54-cm] airfoil sections. Figure 16-6 shows one of these airfoil sections located at a radius r from the axis of rotation of the propeller. This

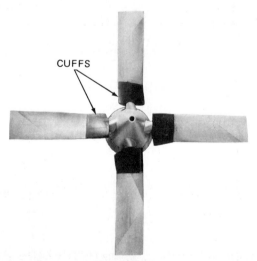

FIG. 16-5   Propeller with blade cuffs.

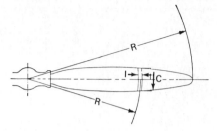

FIG. 16-6   The blade element of a propeller.

airfoil section has a span of 1 in and a chord C. At any given radius r, the chord C will depend on the plan form or general shape of the blade.

According to the blade-element theory, the many airfoil sections, or **elements,** being joined together side by side, unite to form an airfoil (the blade) that can create thrust when revolving in a plane about a central axis. Each element must be designed as part of the blade to operate at its own best angle of attack to create thrust when revolving at its best design speed.

The thrust developed by a propeller is in accordance with Newton's third law of motion: **For every action, there is an equal and opposite reaction, and the two are directed along the same straight line.** In the case of a propeller, the first action is the acceleration of a mass of air to the rear of the airplane. This means that, if a propeller is exerting a force of 200 lb [889.6 N] to accelerate a given mass of air, it is, at the same time, exerting a force of 200 lb tending to "pull" the airplane in the direction opposite that in which the air is accelerated. That is, when the air is accelerated rearward, the airplane is pulled forward. The quantitative relationships among mass, acceleration, and force can be determined by the use of the formula for Newton's second law: $F = ma$, or **force is equal to the product of mass and acceleration.** This principle is discussed further in the chapter on the theory of jet propulsion.

A **true pitch propeller** is one that makes use of the blade-element theory. Each element (section) of the blade travels at a different rate of speed, that is, the tip sections travel faster than the sections close to the hub. When the elements (sections) are arranged so that each is set at the proper angle to the relative airstream, they all advance the same distance during any single revolution of the propeller.

Blade Stations

Blade stations are designated distances along the blade as measured from the center of the hub or from some reference line marked near the tip. In Fig. 16-4 the stations are measured from the center of the hub. If a blade is divided into 6-in [15.2-cm] intervals, as it is in this case, the 6- to 12-in [15.2- to 30.5-cm] blade section is between the 6- and 12-in stations; the 12- to 18- [30.5- to 45.7-cm] blade section is between the 12- and 18-in stations; etc.

This division of a blade into sections separated by stations provides a convenient means of discussing the performance of the propeller blade, locating blade markings, finding the proper point for measuring the blade angle, and locating antiglare areas.

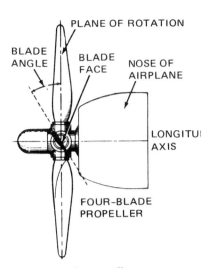

FIG. 16-7   Four-blade propeller.

Blade Angle

Technically, the **blade angle** is defined as the angle between the face or chord of a particular blade section and the plane in which the propeller blades rotate. Figure 16-7 is a drawing of a four-blade propeller, but only two blades are fully shown in order to simplify the presentation. The blade angle, the plane of rotation, the blade face, the longitudinal axis, and the nose of the airplane are all designated in this illustration. The plane of rotation is perpendicular to the crankshaft.

In order to obtain thrust, the propeller blade must be set at a certain angle to its plane of rotation, in the same manner that the wing of an airplane is set at an angle to its forward path. While the propeller is rotating in flight, each section of the blade has a motion that combines the forward movement of the airplane with the circular or rotary movement of the propeller. Therefore, any section of a blade has a path through the air that is shaped like a spiral or a corkscrew, as illustrated in Fig. 16-8.

An imaginary point on a section near the tip of the blade traces the largest spiral, a point on a section midway along the blade traces a smaller spiral, and a point on a section near the shank of the blade traces

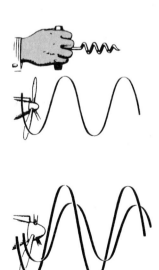

FIG. 16-8   Path of the propeller through the air.

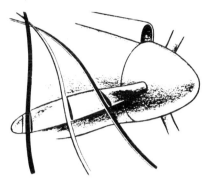

FIG. 16-9   Pitch distribution.

the smallest spiral of all. In one turn of the blade, all sections move *forward* the same distance, but the sections near the tip of the blade move a greater *circular* distance than the sections near the hub.

If the spiral paths made by various points on sections of the blade are traced, with the sections at their most effective angles, then each individual section must be designed and constructed so that the angles become gradually less toward the tip of the blade and larger toward the shank. This gradual change of blade section angles is called **pitch distribution** and accounts for the pronounced *twist* of the propeller blade, as illustrated in Fig. 16-9. Since the blade is actually a twisted airfoil, the blade angle of any particular section of a particular blade is different from the blade angle of any other section of the same blade.

The blade angle is measured at one selected station when the blade is set in its hub, depending on the propeller diameter. If the blade angle at this station is correct, then all blade angles should be correct if the blade has been carefully designed and accurately manufactured.

The blade angle is so important that a change in blade angle of only 1° will affect the rpm of a direct-drive engine between 60 and 90 rpm. The effect that it would have on an engine with propeller-gear reduction would depend on the gear ratio. In some installations, a deviation of 1° or less of blade-angle setting is occasionally permitted by specific instructions, but this is by no means a common practice.

The reason for a variation of the propeller blade angle from the hub to the tip of a blade can be understood by examining Fig. 16-10. The three triangles show

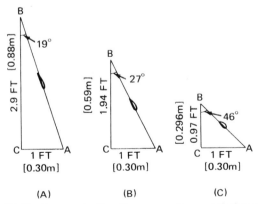

FIG. 16-10   Demonstrating the reason for a variation of propeller blade angle from root to tip.

the relative movement of the airplane and a particular section of the propeller blade during flight at 150 mph [241.40 km/h] with the propeller turning at 2000 rpm. The triangle at *A* represents the movement of the blade section 36 in [91.44 cm] from the propeller hub.

The diameter of the circle traversed by the blade section at 36 in from the hub is $3 \times 2\pi$ ft, or 18.85 ft [5.75 m]. With the propeller turning at 2000 rpm, the section of the blade travels 37,700 ft/min [11 490.96 m/min] or about 628 ft/s [191.41 m/s].

If the airplane is traveling at a true airspeed of 150 mph [241.40 km/h], it is moving through the air at 220 ft/s [67.06 m/s]. This means, then, that the airplane travels 220 ft while the blade section is moving 628 ft. From these data we can determine that, while the airplane is moving 1 ft [0.30 m], the propeller blade section at 36 in from the hub is moving slightly less than 2.9 ft [0.88 m]. This is illustrated in triangle *A*, which shows blade section distance in its plane of rotation as *BC* and the airplane distance as *CA*. The actual track of the propeller blade section is *BA*, and the relative wind direction is along the line *AB*. The angle of attack of the propeller blade is the difference between the angle of *AB* with respect to the plane of rotation and the propeller blade angle.

From a table of tangents or by measuring, we can find that the angle *ABC* is a little more than 19°. If the blade angle is set at 22°, the angle of attack of the blade will be somewhat more than 3°.

The triangle *B* in Fig. 16-10 represents the travel of a blade section at 24 in [60.96 cm] from the hub when the airplane is moving at 150 mph TAS (true airspeed) and the propeller is turning at 2000 rpm. By using the same methods of computation, we find that the angle at *B* is about 27°, and that to provide an angle of attack of 3° the blade angle would have to be set at 30°. Under the same conditions, we will find that a blade section of 12 in [30.48 m] from the hub will move at an angle of 46° from the plane of rotation as shown in triangle *C*.

It is apparent that a fixed-pitch propeller will be efficient only through a narrow range of operating conditions. Fixed-pitch propellers are therefore designed to operate most efficiently at the cruising speed of the airplane on which they are installed.

Blade Angle and Angle of Attack in Flight and on the Ground

The **angle of attack of a propeller blade section** is the angle between the face of the blade section and the direction of the relative airstream, as illustrated in Fig. 16-11. The direction of the relative airstream depends on the direction that the airfoil moves through undisturbed air and the velocity of forward movement.

The relative air motion is along the pitch angle; hence, the angle of attack is added to the pitch angle to obtain the blade angle as the sum of the two angles.

In Fig. 16-11 the blade airfoil section *M* of the rotating propeller travels from *A* to *B* when the airplane is parked on the ground. The trailing edge of the propeller determines the plane of rotation represented by the line *AB*.

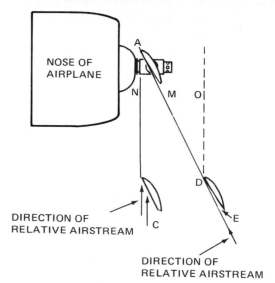

FIG. 16-11   Blade angle with aircraft at rest and in flight.

The **relative wind** is the direction of the air with respect to the movement of the airfoil, as shown in Fig. 16-12. When the airplane is in the air, the relative wind results from the forward motion of the airplane and the circular motion of the propeller blade sections. When the engine is run on the ground, there is no forward movement of the airplane (on the assumption that the airplane is standing) but there is a relative wind caused by the air flowing through the propeller. There is also a certain amount of pitch angle of the air motion with regard to the blade sections. The angle of attack is therefore represented in Fig. 16-11 by the angle *C*. This is the angle at which the propeller section meets the relative airstream.

Also with respect to Fig. 16-11, in flight the airplane moves forward, and as it moves forward from *N* to *O*, the airfoil section *M* will travel from *A* to *D*. The trailing edge follows the path represented by the line *AD*, which represents the relative airstream. The angle of attack then becomes smaller and is represented by *E* in the illustration.

The normal angle of attack in flight of the propeller blades for many airplanes varies from 0 to 15°. Refer-

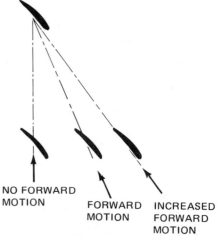

FIG. 16-12   Relative wind with respect to propeller blade.

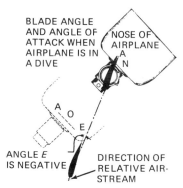

FIG. 16-13 High speed results in a negative angle of attack.

ring to Fig. 16-13, we see that in a power dive the acceleration due to the force of gravity may give the airplane a speed greater than the speed which the propeller tends to reach. The angle of attack, represented by the letter *E*, is then negative and tends to hold back the airplane.

In a steep climb with the forward speed reduced, the angle of attack is increased, as shown in Fig. 16-14. Whether the airplane is in a power dive or in a steep climb, the aerodynamic efficiency of the propeller is low.

Note that the propeller blade angles in the foregoing illustrations appear to be high, but for the purpose of illustration the angles have been exaggerated in these views.

● PROPELLER OPERATION

The actual operation of propellers in flight depends on a number of factors. Among these are aircraft design, engine design, the nature of aircraft operation desired, conditions of flight, performance required, and various others. If an airplane is required to fly great distances with a maximum of economy, the propeller should be of an automatic type which will adjust blade angle to meet the power and speed values selected by the pilot. For airplanes used in short, local flights, the extra cost of an automatic propeller system is not usually justified.

Pitch

The **effective pitch** is the actual distance the airplane moves forward during one revolution (360°) of the propeller in flight. *Pitch* does not mean the same as *blade angle,* but the two terms are commonly used inter-

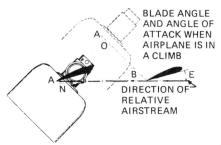

FIG. 16-14 Increased angle of attack as airplane climbs.

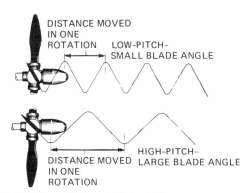

FIG. 16-15 Low pitch and high pitch.

changeably because they are so closely related. Figure 16-15 shows two different pitch positions. The heavy black airfoil drawn across the hub of each represents the cross section of the propeller to illustrate the blade setting. When there is a small blade angle, there is a low pitch and the airplane does not move very far forward in one revolution of the propeller. When there is a large blade angle, there is a high pitch and the airplane moves forward farther during a single revolution of the propeller.

A **fixed-pitch propeller** is a rigidly constructed propeller on which the blade angles may not be altered without bending or reworking the blades. A **controllable-pitch** propeller is one provided with a means of control for adjusting the angle of the blades during flight.

The propeller action somewhat resembles the operation of a gearshift in an automobile when the pitch is controllable. When only fixed blade-angle propellers were used on airplanes, the angle of the blade was chosen to fit the principal purpose for which an airplane was designed. If it was designed for fast climbing and quick takeoff, it had a comparatively low blade-angle propeller, although it was no more suitable for high-speed flying or diving than the low gear on an automobile is adapted to high-speed driving on the highway.

With a fixed blade-angle propeller, an increase in engine power causes increased rotational speed, and this causes more thrust, but it also creates more drag from the airfoil and forces the propeller to absorb the additional engine power. In a similar manner, a decrease in engine power causes a decrease in rotational speed and consequently a decrease in both thrust and drag from the propeller.

When an airplane with a fixed blade-angle propeller dives, the forward speed of the airplane increases. Since there is a change in the direction of the relative wind, there is lower angle of attack, thus reducing both lift and drag and increasing the rotational speed of the propeller.

On the other hand, if the airplane goes into a climb, the rotational speed of the propeller decreases, the change in the direction of the relative wind increases the angle of attack, and there is more lift and drag and less forward speed for the airplane.

The propeller can absorb only a limited amount of excess power by increasing or decreasing its rotational speed. Beyond that, the engine will be damaged. For this reason, as aircraft engine power and airplane speeds both increased, the engineers found it necessary to design propellers with blades that could rotate in their sockets into different positions to permit changing the blade-angle setting to compensate for changes in the relative wind brought on by the varying forward speeds. This made it possible for the propeller to absorb more or less engine power without damaging the engine. The first step in this development produced a propeller with two different blade-angle settings. One gave a low angle for takeoff and climb. The other gave a high blade angle for cruising and diving.

Geometrical Pitch and Zero-Thrust Pitch

A distinction is made between effective pitch and other kinds of pitch. The **geometrical pitch** is the distance an element of the propeller would advance in one revolution if it were moving along a helix (spiral) having an angle equal to its blade angle. Geometrical pitch can be calculated by multiplying the tangent of the blade angle by $2\pi r$, r being the radius of the blade station at which it is computed. For example, if the blade angle of a propeller is 20° at the 30-in [76.2-cm] station, we can apply the formula

$$2\pi \times 30 \times 0.364 = 68.61 \text{ in } [174.27 \text{ cm}]$$

The tangent of 20° is 0.364. The geometrical pitch of the propeller is therefore 68.58 in. This is the distance the propeller would move if it were going forward through a solid medium.

The **zero-thrust pitch,** also called the **experimental mean pitch,** is the distance a propeller would have to advance in one revolution to give no thrust.

The **pitch ratio** of a propeller is the ratio of the pitch to the diameter.

Slip

Slip is defined as the difference between the geometrical pitch and the effective pitch of a propeller. It may be expressed as a percentage of the mean geometrical pitch or as a linear dimension.

The **slip function** is the ratio of the speed of advance through undisturbed air to the product of the propeller diameter and the number of revolutions per unit time. This may be expressed as a formula: V/nD, where V is the speed through undisturbed air, D is the propeller diameter, and n is the number of revolutions per unit time.

The word "slip" is used rather loosely by many people in aviation to refer to the difference between the velocity of the air behind the propeller (caused by the propeller) and that of the aircraft with respect to the undisturbed air well ahead of the propeller. It is then expressed as a percentage of this difference in terms of aircraft velocity.

If there were no slippage of any type, and if the propeller were moving through an imaginary solid substance, then the geometrical pitch would be the calculated distance that the blade element at two-thirds the blade radius would move forward in one complete revolution of the propeller (360°).

Types of Propellers

In designing propellers, the engineers try to obtain the maximum performance of the airplane from the horsepower delivered by the engine under all conditions of operation, such as takeoff, climb, cruising, and high speed. Practically all propellers may be classified under four general types, as follows:

1. **Fixed pitch.** The propeller is made in one piece. Only one pitch setting is possible because of its design. It is usually a two-blade propeller and is often made of wood, although other materials, such as aluminum alloy, steel, and even phenolic compounds, are used.

2. **Adjustable pitch.** The pitch setting can be adjusted only with tools on the ground when the engine is not operating. This type usually has a split hub. The propeller may be removed from the engine when the pitch is being adjusted on some airplanes, but on others this is not necessary. It has at least two blades and often more. Wood may be used in its construction, but it is generally made of steel or aluminum alloy.

3. **Controllable pitch.** The pilot can change the pitch of the propeller in flight or while operating the engine on the ground by means of a pitch-changing mechanism that may be mechanically (manually), hydraulically, or electrically operated. The blades may be made of aluminum alloy, steel, wood, or one of the resin-bonded materials.

4. **Constant speed.** The constant-speed propeller utilizes a hydraulically or electrically operated pitch-changing mechanism controlled by a governor. The setting of the governor is adjusted by the pilot with the rpm lever in the cockpit. During operation, the constant-speed propeller automatically changes its blade angle to maintain a constant engine speed. In straight and level flight, if engine power is increased, the blade angle is increased to make the propeller absorb the additional power while the rpm remains constant. The pilot selects the engine speed required for any particular type of operation.

Terms Used in Describing Pitch Change

The principal terms used in describing propeller-pitch change are: (1) **two-position,** which makes available only two pitch settings; (2) **multiposition,** which makes any pitch setting within reasonable limits possible; (3) **automatic,** which provides a pitch-setting control by some automatic device; (4) **automatic, selective,** which enables pilots to select and control, during flight, the exact conditions at which they want the automatic features to operate; and (5) **automatic, nonselective,** which provides for pitch control entirely independent of the pilot.

Forces Acting on a Propeller in Flight

The forces acting on a propeller in flight are (1) **thrust,** which is the component of the total air force on the propeller and is parallel to the direction of advance and induces bending stresses in the propeller; (2) **centrifugal force,** which is caused by the rotation of the propeller and tends to throw the blade out from the central hub and produces tensile stresses; and (3) **torsion,** or **twisting forces,** in the blade itself, caused by the fact

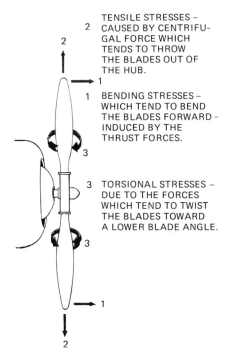

2 TENSILE STRESSES – CAUSED BY CENTRIFUGAL FORCE WHICH TENDS TO THROW THE BLADES OUT OF THE HUB.

1 BENDING STRESSES – WHICH TEND TO BEND THE BLADES FORWARD – INDUCED BY THE THRUST FORCES.

3 TORSIONAL STRESSES – DUE TO THE FORCES WHICH TEND TO TWIST THE BLADES TOWARD A LOWER BLADE ANGLE.

FIG. 16-16   Forces and stresses on propeller blades during flight.

that the resultant air forces do not go through the neutral axis of the propeller, producing torsional stresses.

Stresses to Which Propellers Are Subjected at High Speeds

Figure 16-16 illustrates the three general types of stresses to which propellers rotating at high speeds are subjected. These stresses are bending stresses, tensile stresses, and torsional stresses, explained in detail as follows:

1. **Bending stresses** are induced by the thrust forces. These stresses tend to bend the blade forward as the airplane is moved through the air by the propeller. Bending stresses are also caused by other factors, such as the drag caused by the resistance of the air, but these are of small importance in comparison with the bending stresses caused by the thrust forces.

2. **Tensile stresses** are caused by centrifugal force. This always tends to throw the blade out from the central hub. The hub resists this tendency, and hence the blades "stretch" slightly.

3. **Torsional stresses** are produced in rotating propeller blades by two twisting moments. One of these stresses is caused by the air reaction on the blades and is called the **aerodynamic twisting moment.** The other stress is caused by centrifugal force and is called the **centrifugal twisting moment.** During the ordinary propeller operation, the centrifugal force tends to turn the blades to a lower angle. In some propeller-control mechanisms, this centrifugal twisting moment is employed to aid in turning the blades to a lower angle when necessary to obtain greater propeller efficiency in flight, thus putting a natural force to work.

Torsional stresses increase with the square of the rpm. For example, if the rpm is doubled, the stresses will be four times as great.

Tip Speed

Flutter or vibration may be caused by the tip of the propeller blade traveling at a rate of speed approaching the speed of sound, thus causing excessive stresses to develop. This condition can be overcome by operating at a lower speed or by telescoping the propeller blades, that is, reducing the propeller diameter without changing the blade profile.

Tip speed is actually the principal factor determining the efficiency of high-performance airplane propellers of conventional two- or three-blade design. It has been found by experience that it is essential to keep the tip velocity below the velocity of sound, which is about 1116.4 ft/s [340.28 m/s] at standard sea-level pressure and temperature and varies with temperature and altitude. At sea level the velocity of sound is generally taken as about 1120 ft/s, but it decreases about 5 ft/s [152.4 cm/s] for each increase in altitude of 1000 ft [304.80 m].

The efficiency of high-performance airplane propellers of conventional two- or three-blade design may be expressed in terms of the ratio of the tip speed to the velocity of sound. For example, at sea level, when the tip speed is 900 ft/s [274.32 m/s], the maximum efficiency is about 86 percent, but when the tip speed reaches 1200 ft/s [365.76 m/s], the maximum efficiency is only about 72 percent.

It is often necessary to gear the engine so that the propeller will turn at a lower rate of speed in order to obtain tip ratios below the velocity of sound. For example, if the engine is geared in a 3:2 ratio, the propeller will turn at two-thirds the speed of the engine.

When the propeller turns at a lower rate of speed, the airfoil sections of the blades strike the air at a lower speed also, and they therefore do not do so much work in a geared propeller as they would do in one with a direct drive. It is necessary in this case to increase the blade area by using a larger diameter or more blades.

Ratio of Forward Velocity to Rotational Velocity

The efficiency of a propeller is also influenced by the ratio of the forward velocity of the airplane in feet per second to the rotational velocity of the propeller. This ratio can be expressed by a quantity called *the V over nD ratio*, which is sometimes expressed as a formula: *V/nD,* where *V* is the forward velocity of the airplane in feet per second, *n* is the number of revolutions per second of the propeller, and *D* is the diameter in feet of the propeller. Any fixed-pitch propeller is designed to give its maximum efficiency at a particular value of forward speed of the airplane, which is usually the cruising speed in level flight, and at a particular engine speed, which is usually the speed employed for cruising. At any other condition of flight where a different value of the *V/nD* ratio exists, the propeller efficiency will be less.

Propeller Efficiency

Some of the work performed by the engine is lost in the slipstream of the propeller, and some is lost in the production of noise, neither of which can be converted to horsepower for turning the propeller. We have already examined the effect of tip speed on propeller efficiency. In addition, it is well known that the maximum propeller efficiency that can be obtained in practice under the most ideal conditions, using conventional engines and propellers, has been only about 92 percent, and in order to obtain this efficiency it has been necessary to use thin airfoil sections near the tip of the propeller and very sharp leading and trailing edges. Such airfoil sections are not practical where there is the slightest danger of the propeller picking up rocks, gravel, water spray, or similar substances that might damage the blades.

The **thrust horsepower** is the actual amount of horsepower that an engine-propeller unit transforms into thrust. This is less than the **brake horsepower** developed by the engine, since propellers are never 100 percent efficient.

In the study of propellers, two forces must be considered: **thrust** and **torque**. The thrust force acts parallel to the axis of rotation of the propeller, and the torque force acts parallel to the plane of rotation of the propeller. The thrust horsepower is less than the torque horsepower. The **efficiency** of the propeller is the ratio of the thrust horsepower to the torque horsepower:

$$\text{Propeller efficiency} = \frac{\text{thrust horsepower}}{\text{torque horsepower}}$$

Feathering

The term **feathering** refers to the operation of rotating the blades of a propeller to an edge-to-the-wind position for the purpose of stopping the rotation of the propeller whose blades are thus feathered and of reducing drag. Therefore, a **feathered blade** is in an approximate in-line-of-flight position, streamlined with the line of flight. Some, but not all, propellers can be feathered.

Feathering is necessary when an engine fails or when it is desirable to shut off an engine in flight. The pressure of the air on the face and back of the feathered blade is equal, and the propeller will stop rotating. If it is not feathered when its engine stops driving it, the propeller will "windmill" and cause drag.

Another advantage of being able to feather a propeller is that a feathered propeller creates less resistance (drag) and disturbance to the flow of air over the wings and tail of the airplane. Furthermore, a feathered propeller prevents additional damage to the engine if the failure was caused by some internal breakage, and it also eliminates the vibration which might damage the structure of the airplane.

The importance of feathering the propeller for an engine which has failed on a multiengine airplane cannot be overemphasized. If the propeller cannot be feathered in such a case, it is likely that the engine will "run away," that is, overspeed to the point where great damage may be caused. The lubrication system of the engine may fail because of the excessive speed, and this will cause the engine to "burn up." The heat generated may set the engine on fire, in which case the airplane itself may be destroyed. The excessive speed of the engine may cause the propeller to lose a blade,

thus bringing about an unbalanced condition which will cause the engine to be wrenched from its mounting. Numerous cases of runaway engines which caused airplane crashes are on record.

Summing up some of the advantages of feathering a propeller when an engine failure occurs, not only is there less drag but there is a better performance on the part of the remaining engines, a better speed, a higher ceiling, a better powerplant control, and the airplane can be flown safely to a point where an emergency landing can be made.

Reverse Thrust

When propellers are reversed, their blades are rotated below their positive angle (that is, rotated through "flat" pitch) until a negative blade angle is obtained which will produce a thrust acting in the opposite direction to the forward thrust normally given by the propeller.

This feature is helpful for handling multiengine flying boats in restricted areas and for landing operations of large airplanes because it reduces the length of landing runs, which in turn reduces the amount of braking needed and materially increases the life of the brakes and tires. Reverse propeller thrust is used in almost every case where a four-engine transport-type aircraft is landed. This applies to turboprop-equipped aircraft as well as to conventional types. Jet-propelled transport aircraft utilize jet engines which are equipped with jet reversing devices for the purpose of reducing the landing roll.

● PROPELLER CLEARANCES

Certain minimum clearances have been established with respect to the distances between an aircraft's propeller and the ground, the water, and the aircraft structure. These clearances are necessary to prevent damage during extreme conditions of operation and to reduce aerodynamic interference with the operation and effectiveness of the propeller. The minimum clearances are set forth in Federal Aviation Regulations.

Ground Clearances

Aircraft equipped with tricycle landing gear must have a minimum of 7 in [17.78 cm] clearance between the tip of the propeller and the ground when the aircraft is in the taxiing or takeoff position with the landing gear deflected. Aircraft with tail-wheel landing gear must have a minimum of 9 in [22.86 cm] clearance between the tip of the propeller and the ground under the same conditions, that is, with the aircraft in the position where the tip of the propeller would come nearest the ground during operation. This would normally be during takeoff for an aircraft equipped with tail-wheel landing gear.

Water Clearance

Sea planes or amphibious aircraft must have a clearance of at least 18 in [45.72 cm] between the tip of the propeller and the water unless it can be shown that the aircraft complies with the regulations regarding water-spray characteristics set forth in FAR 25.239.

Structural Clearances

The tip of the propeller blades must have at least 1 in [2.54 cm] radial clearance from the fuselage or any other part of the aircraft structure. If this is not sufficient to avoid harmful vibration, then additional clearance must provided.

Longitudinal clearance (fore and aft) of the propeller blades or cuffs must be at least ½ in [12.70 mm] between propeller parts and stationary parts of the airplane. This clearance is with the propeller blades feathered or in the most critical pitch configuration.

Between the spinner or rotating parts of the propeller, other than the blades or cuffs and stationary parts of the aircraft, there must be positive clearance. The stationary part of the aircraft in this case would probably be the engine cowling or a part between the cowling and the spinner.

● HELICOPTER ROTORS

Propeller fundamentals and theory apply to helicopter rotors to a large extent. Both the *main rotor* (rotary wing) and the *tail rotor* are designed with airfoil shapes similar to those incorporated in propellers.

A helicopter rotor has characteristics of a controllable-pitch propeller. With the main rotor of a helicopter, however, the pitch of rotor blades on the same rotor usually varies considerably from blade to blade, depending on the flight operation being performed. One reason for this is that there is a substantial difference in the airspeed between the advancing blade and the retreating blade when the helicopter is moving horizontally.

Tail rotors are more nearly like propellers than are main rotors. The tail rotor provides side thrust to overcome the torque of the main rotor and give directional control to the helicopter.

● REVIEW QUESTIONS

1. Name the principal parts of a single-piece wood propeller.
2. Compare the cross section of a propeller blade with the cross section of a wing.
3. What is meant by a *ground-adjustable* propeller?
4. Define *blade root*.
5. Explain the purpose of *blade cuffs*.
6. What is the *blade-element theory*?
7. What law of motion explains propeller thrust?
8. Why does the blade angle change from the hub to the tip?
9. Why is a fixed-pitch propeller limited in its range of operation?
10. Discuss the angle of attack of a propeller blade with respect to airplane speed.
11. Explain the difference between a ground-adjustable propeller and a controllable propeller.
12. What is the *geometrical pitch* of a propeller?
13. Define propeller *slip*.
14. Explain the operation of a *constant-speed* propeller.
15. Explain the effect of each of the three principal forces acting on a propeller in flight.
16. What is the limiting factor with respect to the rpm at which a propeller may be operated?
17. Explain *propeller efficiency*.
18. Give the reasons why it is essential to use feathering propellers on multiengine aircraft.
19. Describe the process of propeller reversing.
20. What is the value of reverse thrust?
21. What radial tip clearance is required between a propeller and the fuselage of an aircraft?
22. Why should a propeller be placed in its highest pitch configuration when checking for longitudinal clearance?
23. What are the ground and water clearances required for the tip of a propeller installed on an aircraft?
24. Compare helicopter rotors with propellers.

17 PROPELLERS FOR LIGHT AIRCRAFT

● WOOD PROPELLERS

In the early days of aviation, all propellers were made of wood, but the development of larger and higher horsepower aircraft engines made it necessary to adopt a stronger and more durable material; hence, metal is now extensively used in the construction of propellers for all types of aircraft. Some propeller blades have been made of plastic materials, specially treated wood laminations, and plastic-coated wood laminations. For most purposes, however, the metal propellers have been most satisfactory where cost is not a primary consideration.

The average aviation technician today will seldom be required to repair wood propellers. For this reason we do not cover the details of wood propeller repair. It is our purpose to give the technician basic information about wood propellers including construction, inspection, balancing, removal, and installation.

Construction

The first consideration in the construction of a wood propeller is the selection of the right quality and type of wood. It is especially important that all lumber from which the propeller laminae (layers) are to be cut should be kiln-dried. A wood propeller is not cut from a solid block but is built up of a number of separate layers of carefully selected and well-seasoned hardwoods, as illustrated in Fig. 17-1.

Many types of wood have been used in making propellers, but the most satisfactory are sweet or yellow birch, sugar maple, black cherry, and black walnut. In some cases, alternate layers of two different woods have been used to reduce the tendency toward warpage. This is not considered necessary, however, because the use of laminations of the same type of wood will effectively reduce the tendency for a propeller to warp under ordinary conditions of use.

The spiral or diagonal grain of propeller wood should have a slope of less than 1 in 10 when measured from the longitudinal axis of the laminae.

Propeller lumber should be free from checks, shakes, excessive pinworm holes, unsound and loose knots, and decay. Sap stain is considered a defect. The importance of selecting a high grade of lumber cannot be too strongly emphasized in order to reduce the effect of the internal variations present in all wood.

As shown in Fig. 17-1, the laminations of wood are given a preliminary shaping and finishing and then are stacked together and glued with high-quality glue. Pressure and temperature are carefully controlled for the prescribed time. After the glue has set according

to specifications, the propeller is shaped to its final form using templates and protractors to ensure that it meets design specifications.

After the propeller is shaped, the tip of each blade is covered with fabric to protect the tip from moisture and reduce the tendency to crack or split. The fabric is thoroughly waterproofed. Finally, the leading edge and tip of each blade is provided with a sheet-brass shield to reduce damage due to small rocks, sand, and other materials encountered during takeoff and taxiing. The metal tipping and leading-edge shield are shown in Fig. 17-2.

The center bore of the hub and the mounting-bolt holes are very carefully bored to exact dimensions. This is essential to good balance upon installation.

During the final production stages, the propeller is balanced, both horizontally and vertically. A propeller on a balance stand is shown in Fig. 17-3. **Horizontal balance** can be adjusted by adding or removing solder from the tips. When finished, the solder must blend in with the contour of the tip. The metal at the tip is vented by drilling a few 0.040-in [1.016-mm] holes at the extreme tip. These holes help to eliminate any moisture which might condense under the metal tipping.

Vertical imbalance is corrected by attaching a metal weight to the light side of the hub. The size of the weight is determined by first applying putty at a point 90° from the horizontal center line. When the propeller balances, the putty is removed and weighed. A metal plate is then cut to a size which will approximate the weight of the putty. The weight of the metal plate must be adjusted for the weight of the attaching screws and solder which are used for the attachment. The plate is attached to the hub at the 90° location with countersunk screws. The heads of the screws are soldered and then smoothed with a file. Finally, varnish is applied to match the finish of the rest of the propeller.

Propeller Track

For the **track** of a propeller to be correct, corresponding points on the two blades must lie in the same plane perpendicular to the axis of rotation. Checking the track of either a wood propeller or a metal propeller can be done on a propeller surface table, as shown in Fig. 17-4. Each corresponding point on each blade should have the same height from the table surface as indicated by the height gage.

When a propeller is installed on an airplane engine in the aircraft, the track can be checked by rotating the tip of the propeller past a fixed reference point attached to the aircraft. This method is shown for a

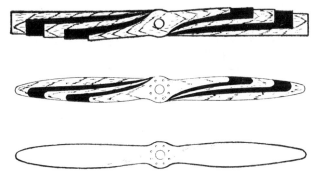

FIG. 17-1 Construction of a typical wood propeller.

metal propeller in Fig. 17-5. The track of one blade should normally be within $\frac{1}{16}$ in [1.59 mm] of the other blade. Constant-speed and controllable propellers should be placed in low pitch before checking for track unless otherwise specified by the manufacturer.

● INSTALLATION AND REMOVAL

Types of Hubs

The propeller is mounted on its shaft by means of several attaching parts. The types of hubs generally used to mount a wood propeller on the engine crankshaft are (1) a forged steel hub fitting a splined crankshaft, (2) a tapered forged steel hub for connecting to a tapered crankshaft, and (3) a hub bolted to a steel flange forged on the crankshaft.

Hubs Fitting a Tapered Shaft. On some models having a hub fitting a tapered shaft, the hub is held in place by a retaining nut that screws on the end of the shaft. A locknut safeties the retaining nut, and a puller is required for removing the propeller from the shaft. The locknut screws into the hub and bears against the retaining nut. The locknut and the retaining nut are then safetied together with either a cotter pin or a lockwire.

A newer design has a snap ring instead of a locknut. When the propeller is to be removed, the retaining nut is backed off and bears against the snap ring, and the propeller is thus started from the shaft. Holes in the retaining nut and the shaft are provided for safetying.

Hubs Fitting a Splined Shaft. A retaining nut that screws on the end of the shaft is used to hold a hub fitting a splined shaft, as shown in Fig. 17-6. Front and rear cones are provided to seat the propeller properly on the shaft. The **rear cone** is made of bronze and is of one-piece construction. It seats in the rear-cone seat of the hub. The **front cone** is a two-piece split-type steel cone. A groove around its inner circumference makes it possible to fit the cone over a flange of the propeller retaining nut.

The front cone seats in the **front-cone seat** of the hub when the retaining nut is threaded into place. In

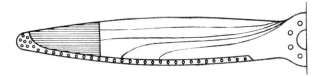

FIG. 17-2 Metal-tipped propeller blade.

FIG. 17-3 Propeller on a balance stand.

order that the front cone will act against the snap ring and pull the propeller from the shaft when the retaining nut is unscrewed from the propeller shaft, a snap ring is fitted into a groove in the hub forward of the front cone. This snap ring must not be removed when the splined-shaft propeller is removed from its shaft because the snap ring provides a puller for the propeller.

When a hub which has a bronze bushing instead of a front cone is used, a puller may be required to start the propeller from the shaft.

A **rear-cone spacer** is provided with some designs to prevent the front cone from bottoming on the forward ends of the splines. If the rear cone is too far back, the front cone will come in contact with the splines before the propeller is secure.

The principal purpose of a retaining nut is to hold the propeller firmly on its shaft. A secondary purpose, in some designs, is to function as a puller with the snap ring to aid in removing the propeller.

Integral-Hub Flange-Type Crankshaft. The integral-hub flange-type crankshaft is manufactured with the propeller mounting hub forged on the front end of the crankshaft, as shown in Fig. 17-7. The flange includes integral bushings that fit into counterbored recesses in the rear face of the propeller hub. The recesses are concentric with the bolt holes. A stub shaft on the front end of the crankshaft forward of the flange fits the propeller bore and ensures that the propeller will be correctly centered.

Installing a Propeller on a Tapered Shaft

Before installing any propeller, wipe the shaft and the inside of the hub with a clean, dry rag until they

FIG. 17-4 Checking propeller track on a surface table.

CHECK DISTANCE HERE FOR BOTH BLADES

REFERENCE BAR ATTACHED TO WING STRUTS

FIG. 17-5 Checking the track of a metal propeller on an airplane.

are free of dirt, grease, and other foreign substances. Using a fine file or a handstone with skill and discretion, remove any burrs or rough spots which might prevent the hub from slipping all the way on the shaft. Apply a thin coating of light engine oil to the shaft before the propeller is installed.

Before a propeller is installed on a tapered shaft, the fit of the propeller to the shaft should be checked. This may be done with Prussian blue. First, both the hub and the shaft are cleaned and all roughness is removed. Then a thin coating of Prussian blue is applied to the shaft. The propeller hub is installed, and the retaining nut is tightened to the proper torque for normal installation. The hub is then removed, and the degree of surface contact inside the hub is shown by the transfer of Prussian blue. If the contact area is 70 percent or more, the fit is satisfactory. Some authorities recommend a fit of 85 or 90 percent. If the area of contact is less than 70 percent, the hub may be fitted to the shaft by lapping with a fine lapping compound. When the correct fit is attained, the lapping compound must be completely removed from both the hub and shaft, and then both surfaces should be coated with light engine oil.

The installation of a propeller on a tapered shaft depends on the type of hub. If a locknut is used, lift the propeller into position. Be sure that the key on the shaft lines up with the keyway on the hub. Slide the propeller well back on the shaft. Unless there is some-

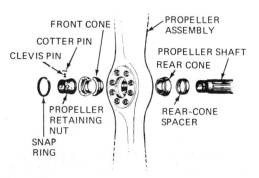

FRONT CONE
COTTER PIN
CLEVIS PIN

PROPELLER ASSEMBLY

PROPELLER SHAFT
REAR CONE

PROPELLER RETAINING NUT
SNAP RING

REAR-CONE SPACER

FIG. 17-6 Installation parts for a splined-shaft propeller.

FIG. 17-7 Integral-hub flange-type crankshaft.

thing wrong, the hub will not bind as it slides on the tapered shaft. Screw the retaining nut on the end of the shaft. Note that a shoulder on the retaining nut bears against a shoulder in the hub and forces the hub on the shaft. Use the wrenches designated by the manufacturer for the final tightening, but do not apply any extra leverage.

Next, screw the locknut into the hub. Be careful in starting the nut to ensure that there is no "cross threading," because the thread on this nut is comparatively fine. Pull the locknut tight, but do not tighten it as much as the retaining nut. One of the locking-wire holes must be in line with a hole in the retaining nut. Finally, use either a lockwire or a cotter pin to secure the retaining nut and the locknut.

If the propeller is designed with a snap-ring puller, it is merely necessary to place the propeller in the proper position on the shaft, install and tighten the retaining nut, install the snap ring, and install the safety clevis pin or bolt.

Removal of a Propeller from a Tapered Shaft

Although there are not many tapered-shaft propellers still in use, they will be encountered by the technician, particularly if he or she is employed by a fixed-base operator that maintains older, privately owned aircraft. It is therefore appropriate to discuss the removal of such propellers briefly.

The most prevalent problem in removing a tapered-shaft propeller is that the hub often sticks to the shaft, perhaps due to overtorquing during installation or because the shaft was not properly lubricated. If, after the retaining nut has been removed, the propeller cannot be removed from the shaft with reasonable force, it will be necessary to employ a propeller puller. Propellers that are designed with a snap ring and puller nut are easier to remove because the nut applies pulling force against the snap ring as the nut is backed off.

The snap ring and retaining nut should be lubricated well to reduce friction between the snap ring and the shoulder of the nut. The nut is rotated by placing a bar through the holes in the nut and applying sufficient pressure on the bar to turn the nut.

Installation of a Propeller on a Splined Shaft

To install a propeller on a splined shaft first install the rear-cone spacer if there is one on the assembly (see Fig. 17-6). Install the rear cone on the propeller shaft. Match the wide spline of the shaft with the wide groove of the hub and slide the propeller well back against the rear cone. Next, assemble the front cone to the retaining nut, and screw the nut to the propeller shaft. A bar about 3 ft [0.9 m] long is placed through the holes in the nut for the final tightening as specified in the maintenance manual. It is not necessary to pound the tightening bar. The snap ring is then installed in its groove in the hub. The retaining nut is safetied with a clevis pin and a cotter pin. If the propeller retaining nuts have elongated locking holes, a washer is placed under the cotter pin. The clevis-pin head should be to the inside, and the washer and cotter pin should be outside.

Removing a Propeller from a Splined Shaft

In order to remove a propeller from a splined shaft, remove the cotter pins and clevis pin that secure the propeller retaining nut and then unscrew the propeller retaining nut. The front cone over the flange of the retaining nut presses against the snap ring in the hub and pulls the propeller away from the shaft for a short distance. When the propeller is loose, it is usually slipped off easily by hand; but if this is not possible with a reasonable amount of force, remove the snap ring, nut, and front cone. Then clean the threaded portion of the shaft and nut; lubricate the cone, nut, and shaft with clean engine oil; reassemble; and finally apply force to unscrew the nut. The rear cone and spacer are left with the engine if a new propeller is to be installed. A propeller puller is used to start the propeller from the shaft if there is a bronze bushing instead of a front cone.

Installation of a Propeller on an Integral-Hub Flange Type of Shaft

To install a propeller on an integral-hub flange type of shaft, first place the propeller on the stub shaft and rotate the propeller sufficiently to mate the recesses on the back of the propeller with the bushings on the flange. Then make certain that there are no metal chips or other particles of foreign substances on the threads of the bolts, and coat the threads with a light engine oil. Next, insert the bolts in the bolt holes. If necessary, use a soft-headed hammer to drive the bolts through the hub, but strike lightly. Apply the hub bolt nuts: draw them up evenly, tightening each only a small amount each time, tightening back and forth from one nut to another, thus avoiding any tendency to throw the propeller out of track. Finally, use a torque wrench to tighten the nuts to the values prescribed by the manufacturer, being careful not to injure the surface of the wood propeller hub. A propeller on an integral-hub flange shaft is removed simply by unscrewing and removing the retaining bolts and lifting the propeller from the shaft.

Inspection of a New Propeller Installation

When a new fixed-pitch propeller has been installed and operated, the hub bolts should always be inspected for tightness after the first flight and after 25 h of flying. Thereafter, the bolts should be inspected and checked for tightness at least every 50 h of operation, unless otherwise specified in the appropriate maintenance manual. No definite time interval can be specified for wood propellers, since bolt tightness is affected by changes in the wood caused by the moisture content in the air where the airplane is flown and stored. During wet weather, some moisture is apt to enter the propeller wood through the drilled holes in the hub. The wood swells, but since expansion is limited by the bolts extending between the two flanges, some of the wood fibers are crushed. Later, when the propeller dries out during dry weather, a certain amount of propeller hub shrinkage takes place and the wood no longer completely fills the space between the two hub flanges. Accordingly, the shrinkage of the wood also results in loose hub bolts.

Safety Precautions for Propeller Maintenance

The following rules are brief statements of important safety practices to be followed by all personnel engaged in the inspection, maintenance, overhaul, repair, or manufacture of all aircraft propellers:

1. Wear safety shoes.

2. When lifting a propeller manually, bend the knees, keep the shoulders back, and lift with the leg muscles and not with the back and abdominal muscles more than absolutely necessary. Take a deep breath and hold it until the propeller is held in a comfortable position. Stand with the feet in line and reasonably separated. If the propeller is too heavy for the number of personnel assigned to lift it, they should get more personnel or use a hoist. This section of the rules is intended to prevent hernias (ruptures), back injuries, and similar disabling effects.

3. Do not stand under a propeller in position for hoisting or allow others to walk under it. Do not sling or hoist the propeller over the heads of others. Move it slowly. Be sure it is securely fastened to the hoisted device.

4. Stand clear of the rotating blades of a propeller when it is in a horizontal position on the assembly plate, when it is being balanced on a stand, or when it is rotating on an airplane.

5. If a pit is used for balancing propellers, keep it covered when not in use.

6. See that all personnel handling propellers are careful not to create hazardous conditions that might hurt themselves or others. Do not allow anyone who is sleepy, inattentive, or careless to work around propellers.

7. Use only approved cleaning solvents for cleaning propellers parts. If such solvents must be used in accordance with special safety practices, instruct the necessary personnel in their care and use and then enforce the rules.

MAINTENANCE AND REPAIR OF PROPELLERS

Importance of Propeller Inspection, Maintenance, and Repair

Since the purpose of the propeller is to pull the airplane through the air, it is beyond question a vital part of the airplane that requires the highest degree of accuracy, quality, and efficiency in its inspection, maintenance, and repair.

General Nature of Propeller Repairs

When objects, such as stones, dirt, birds, etc., strike against the propeller blades and hub during flight or during takeoff and landing, they may cause a bend, cut, scar, nick, scratch, or some other defect in the blade or hub. If the defect is not repaired, local stresses are established which may cause a crack to develop, resulting eventually in the failure of the propeller or hub. For this reason propellers are carefully examined at frequent intervals, and any defects that are discovered are repaired immediately according to methods and procedures that will not further damage the propeller.

The terminology of propeller inspection, maintenance, and repair is very precise. Repairs and alterations are rigidly classified and assigned to certain types of repair agencies for accomplishment. After the work is assigned to the correct individuals or organizations, the propeller must be carefully cleaned before work is performed on it. Then the necessary inspections, repairs, alterations, and maintenance procedures may be accomplished. They are carefully regulated to prevent a technician from doing more harm than good.

Other inspection and maintenance operations include the checking of blade angles, field checking of a propeller for track, marking and coating of blades for identification purposes, servicing of the front cones, preparation of propeller shafts for installation, examining of hubs and blades for looseness, proper disposition of attaching parts, lubrication of the propeller assembly, and correct procedures after an accident involving a propeller.

Authorized Repairs and Alterations

The technician contemplating repair, overhaul, or alteration should be thoroughly familiar with the approved practices and regulations governing the operation which he expects to perform. All repairs or alterations of propellers must be accomplished in accordance with the regulations set forth in Federal Aviation Regulations and the pertinent manufacturer's manual.

Repairs and alterations on propellers are divided into four main categories: (1) major alterations, (2) minor alterations, (3) major repairs, and (4) minor repairs.

A **major alteration** is an alteration which may cause an appreciable change in weight, balance, strength, performance, or other qualities affecting the airworthiness of a propeller. Any alteration which is not accomplished in accordance with accepted practices or cannot be performed by means of elementary operations is also a major alteration.

A **minor alteration** is any alteration not classified as a major alteration.

A **major repair** is any repair which may adversely affect any of the qualities noted in the definition of a major alteration.

A **minor repair** is any repair other than a major repair.

Classification of Repair Operations

Changes such as those in the following list are classified as major alterations unless they have been authorized in the propeller specifications issued by the Federal Aviation Administration.

1. Changes in blade design
2. Changes in hub design
3. Changes in governor or control design
4. Installation of a governor or feathering system
5. Installation of a propeller deicing system
6. Installation of parts not approved for the propeller
7. Any change in the design of a propeller or its controls

Changes classified as minor alterations are those similar to the types listed below.

1. Initial installation of a propeller spinner
2. Relocation or changes in the basic design of brackets or braces of the propeller controls
3. Changes in the basic design of propeller control rods or cables

Repairs of the types listed below are classified as propeller major repairs, since they may adversely affect the airworthiness of the propeller if they are improperly performed.

1. Any repairs to or straightening of steel blades
2. Repairing or machining of steel hubs
3. Shortening of blades
4. Retipping of wood propellers
5. Replacement of outer laminations on fixed-pitch wood propellers
6. Inlay work on wood propellers
7. All repairs to composition blades
8. Replacement of tip fabric
9. Repairing of elongated bolt holes in the hub of fixed-pitch wood propellers
10. Replacement of plastic covering
11. Repair of propeller governors
12. Repair of balance propellers of rotorcraft
13. Overhaul of controllable-pitch propellers
14. Repairs involving deep dents, cuts, scars, nicks, etc., and straightening of aluminum blades
15. Repair or replacement of internal elements of blades

Propeller repairs such as those listed below are classified as propeller minor repairs.

1. Repairing of dents, cuts, scars, scratches, nicks, and leading-edge pitting of aluminum blades if the repair does not materially affect the strength, weight, balance, or performance of the propeller

2. Repairing of dents, cuts, scratches, nicks, and small cracks parallel to the grain of wood blades

3. Removal and installation of propellers

4. The assembly and disassembly of propellers to the extent necessary to permit (*a*) assembly of propellers partially disassembled for shipment and not requiring the use of balancing equipment, (*b*) the accomplishment of routine servicing and inspection, and (*c*) the replacement of parts other than those which normally require the use of skilled techniques, special tools, and test equipment

5. Balancing of fixed-pitch and ground-adjustable propellers

6. Refinishing of wood propellers

Persons and Organizations Authorized to Perform Repairs and Alterations on Propellers

The regulations governing the persons and organizations authorized to perform propeller repairs and alterations are subject to change, but in general, maintenance, minor repairs, or minor alterations must be done by a certificated repair station holding the appropriately rated technician (A&P) or a person working under the direct supervision of such a technician, or an appropriately certificated air carrier. Major repairs or alterations on propellers may be performed only by an appropriately rated repair station, manufacturer, or air carrier in accordance with the regulations governing their respective operations.

Requirements governing persons or organizations authorized to perform maintenance and repairs on propellers are set forth in Federal Aviation Regulations.

It should be remembered that minor repairs and alterations are those which are not likely to change the operating characteristics of the propeller or affect the airworthiness of the propeller. All other repairs and alterations are major in nature and must be accomplished by the properly authorized agencies.

General Inspection for Defects

Wood propellers are inspected for such defects as cracks, bruises, scars, warp, oversize holes in the hub, evidence of glue failure, evidences of separated laminations, sections broken off, and defects in the finish. The tipping should be inspected for such defects as looseness or slippage, separation of soldered joints, loose screws, loose rivets, breaks, cracks, eroded sections, and corrosion. Frequently, cracks do appear across the leading edge of the metal tipping between the front and rear slits where metal has been removed to permit easier forming of the tip curvature. These cracks are considered normal and are not cause for rejection.

The steel hub of a wood or composite propeller should be inspected for cracks and wear. When the hub is removed from the propeller, it should be magnetically inspected. Any crack in the hub is cause for rejection. The hub should also be inspected for wear of the bolt holes.

All propellers, regardless of the material of which they are made, should undergo regular and careful inspection for any possible defect. Any doubtful condition such as looseness of parts, nicks, cracks, scratches, bruises, or loss of finish should be carefully investigated and the condition checked against repair and maintenance specifications for that particular type of propeller.

Causes for Rejection

Propellers worn or damaged to such an extent that it is either impossible or uneconomical to repair them and make them airworthy should be rejected and scrapped. The following conditions are deemed to render a wood propeller unairworthy and are therefore cause for rejection.

1. A crack or deep cut across the grain of the wood

2. Split blades

3. Separated laminations, except the outside laminations of fixed-pitch propellers

4. More screw or rivet holes, including holes filled with dowels, than are used to attach the metal leading-edge strip and tip

5. An appreciable warp

6. An appreciable portion of wood missing

7. A crack, cut, or damage to the metal shank or sleeve of blades

8. Broken lag screws which attach the metal sleeve to the blade

9. An oversize shaft hole in fixed-pitch propellers

10. Cracks between the shaft hole and the bolt holes

11. Cracked internal laminations

12. Excessively elongated bolt holes

General Repair Requirements

Propellers should be repaired in accordance with the best accepted practices and the latest techniques. Manufacturers' recommendations should always be followed if such recommendations are available. It is recognized that the manuals may not be available for some of the older propellers; in such cases, the propellers should be repaired in accordance with standard practices and FAA regulations.

When a propeller is repaired or overhauled by a certificated agency, the Air Agency Certificate number or the name of the agency should be marked indelibly on the repaired propeller. It is recommended that a decalcomania giving both the repair agency's name and Air Agency Certificate number be used for this purpose. If the original identification marks on a propeller are removed during overhaul or repair, it is necessary that they be replaced. These include the name of the manufacturer and the model designation.

Repairs for Minor Damage

Small cracks parallel to the grain in a wood propeller should be filled with an approved glue thoroughly worked into all portions of the cracks, dried, and then sanded smooth and flush with the surface of the propeller. This treatment is also used with small cuts. Dents or scars which have rough surfaces or shapes that will hold a filler and will not induce failure may be filled with a mixture of approved glue and clean, fine sawdust, thoroughly worked and packed into the defect, dried, and then sanded smooth and flush with the surface of the propeller. It is very important that all loose or foreign matter be removed from the place to be filled so that a good bond of the glue to the wood is obtained.

Repairs for Major Damage of Wood Propellers

As explained previously, the aviation technician rarely is involved with the major repair of a wood propeller. For this reason, such repairs are not described in this section. If an A&P technician should be confronted with the need to have a wood propeller reconditioned or repaired, a properly certificated propeller repair station would be able to perform the necessary work. Help and advice in such a matter can be obtained from the FAA General Aviation District Safety Office.

● FIXED-PITCH METAL PROPELLERS

Description

A fixed-pitch metal propeller is usually manufactured by forging a single bar of aluminum alloy to the required shape. Typical of such propellers is the McCauley Met-L-Prop shown in Fig. 17-8. The propeller shown carries the basic model numbers 1A90, 1B90, or 1C90.

The propeller shown in the illustration is provided with a center bore for the installation of a steel hub or adapter to provide for different types of installations. The six hub bolt holes are dimensioned to fit a standard engine crankshaft flange. The propeller is anodized to prevent corrosion.

Advantages

The advantages of a single-piece fixed-pitch metal propeller are (1) simplicity of maintenance, (2) durability, (3) resistance to weathering, (4) light weight, (5) low drag, and (6) minimum service requirements. The propellers are efficient for a particular set of operating conditions.

● COMPOSITE PROPELLER BLADES

Although this chapter deals primarily with propellers for comparatively small aircraft, some of the propellers described are employed on midsize aircraft that are used for business and commuter airline service. This is true of propellers manufactured with composite blades inasmuch as a substantial number of midsize aircraft are equipped with propellers having composite blades.

Composite blade construction is generally described as involving the use of various plastic resins reinforced with fibers or filaments composed of glass, carbon, Kevlar, or boron. The resin matrix may be of epoxy, polyester, or polymide. Glass fiber with epoxy had been used extensively for many years to manufacture a wide variety of light-weight high-strength structures. Graphite or carbon filament with epoxy was developed more recently and has proved even stronger and more

FIG. 17-8 McCauley Met-L-Prop.

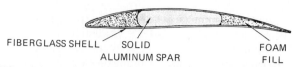

FIG. 17-9 Cross section of spar/shell propeller blade. *(Hamilton Standard)*

durable than glass fiber composites. Other combinations of fibers and resins have followed, with the result that there has been a continual improvement in propeller design, particularly with respect to weight.

Among propeller companies that manufacture propellers having composite blades are Hamilton Standard, Hartzell, and Dowty Rotol. Blade construction and design varies among the companies; however, similar results are achieved.

A composite propeller blade designed and manufactured by the Hamilton Standard Company consists of a solid aluminum-alloy spar around which a fiberglass shell with the correct airfoil shape is placed. The space between the spar and the shell is filled with a plastic foam that provides a firm support for the shell. The outer surface of the shell is given a coating of polyurethane. The term **spar/shell** is used to describe this type of construction. A cross section of the spar/shell propeller blade is shown in Fig. 17-9. The spar/shell design is one of several modified-monocoque types of propeller blades.

Another type of modified-monocoque composite propeller blade is illustrated in Fig. 17-10. This blade consists of a laminated Kevlar shell into which is placed a foam core. A cross section of the blade is shown in Fig. 17-11. The **Kevlar shell** consists of both unidirectional and multidirectional layers of material bonded with epoxy to form the shell laminate. The leading and trailing edges of the blade are reinforced with solid unidirectional Kevlar as shown in the drawing. Two unidirectional Kevlar shear webs are placed between the camber and face surfaces of the shell to provide resistance to flexing and buckling. The polyurethane foam that fills the spaces inside the shell supplies additional resistance to any distortion that could be caused by the operating stresses that the propeller encounters.

The drawing of Fig. 17-10 illustrates the construction and configuration of the blade. This illustration shows the metal cap that is bonded to the leading edge of the blade. This cap serves to reduce the erosive and damaging effects of sand, gravel, rain, and other materials that may be encountered during operation. A Kevlar

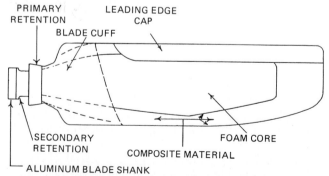

FIG. 17-10 Blade construction. *(Hartzell Propeller)*

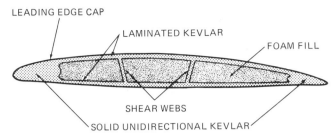

FIG. 17-11 **Cross section of a propeller blade.** *(Hartzell Propeller)*

cuff is attached at the base of the blade to improve aerodynamic efficiency.

Figure 17-12 illustrates how the composite blade is attached to (retained by) the aluminum-alloy blade shank. The Kevlar blade shell is flared at the butt end to conform to the shape of the blade shank. The blade clamp then holds the flared end of the shell firmly against the shank. This arrangement provides the secondary retention. The primary retention is accomplished by a winding of Kevlon roving impregnated with epoxy resin. The primary retention winding holds the shell tightly against the blade shank. This assures that the composite blade cannot separate from the shank even under the most extreme operating stresses.

The drawing of Fig. 17-12 shows how the ball bearing between the blade clamp and the shoulder of the hub spider supports the centrifugal load of the rotating propeller and permits the blade to rotate axially to change pitch. The needle bearings that support the blade shank on the pilot tube are also illustrated.

● GROUND-ADJUSTABLE PROPELLERS

As previously mentioned, a ground-adjustable propeller is designed to permit a change of blade angle when the airplane is on the ground. This permits the adjustment of the propeller for the most effective operation under different conditions of flight. If it is desired that the airplane have a maximum rate of climb, the propeller blades are set at a comparatively low angle so that the engine can rotate at maximum speed to produce the greatest power. The propeller blade, in any case, must not be set at an angle which will permit the engine to overspeed. When it is desired that the engine operate efficiently at cruising speed and at high altitudes, the blade angle is increased.

A ground-adjustable propeller may have blades made of wood or of metal. The hub is usually of two-piece steel construction with clamps or a large nut to hold the blades securely in place. When it is desired to change the blade angle of a ground-adjustable propeller, the clamps or blade nuts are loosened and the blade is rotated to the desired angle as indicated by a propeller protractor. The angle markings on the hub are not considered accurate enough to provide a good reference for blade adjustment; hence, they are used chiefly for checking purposes.

Installation

The installation for ground-adjustable propellers follows the practices previously described for fixed-pitch propellers. The following steps may be considered typical for such an installation:

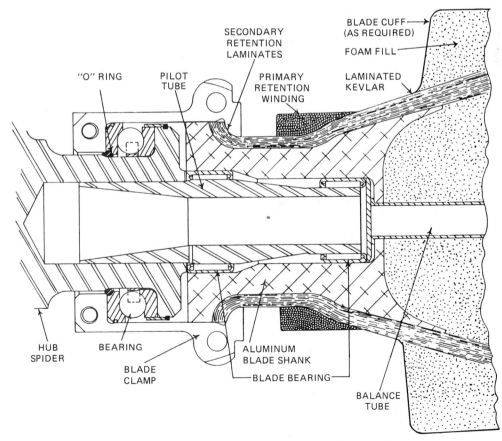

FIG. 17-12 **Blade retention system.** *(Hartzell Propeller)*

1. Make sure that the propeller being installed has been approved for the engine and aircraft on which it is being installed.

2. See that the propeller has been inspected for proper blade angle and airworthiness.

3. See that the propeller shaft and the inside of the propeller are clean and covered with a light coat of engine oil.

4. Install the rear-cone spacer (if used) and the rear cone.

5. Lift the propeller into place carefully, and slide it onto the shaft, making sure that the wide splines are aligned and that the splines are not damaged by rough handling of the propeller.

6. See that the split front cone and the retainer nut are coated with engine oil, assemble them, and install as a unit. (This step in the procedure will vary according to the design of the retaining devices. With some propellers the front-cone halves are installed, then the retainer nut, and finally a snap ring.)

7. Tighten the retainer nut to the proper torque as specified by the manufacturer or according to other pertinent directions. Usually a 3-ft [0.9-m] bar will enable the technician to apply adequate torque for small propeller installations.

8. Install the safety pin or other safetying device.

Adjustment

Adjustment of the blade angle for a ground-adjustable propeller may be done on a propeller surface table, as shown in Fig. 17-13. The propeller is mounted on a mandrel of the correct size, and the blade angle is checked with a large propeller protractor as shown in the illustration. The blade clamps or retaining nuts are loosened so that the blades can be turned; and after the correct angle is established, the blades are secured in the hub by the clamps or blade nuts. The blade angle must be checked at a specified blade station as given in the pertinent instructions.

The method for checking the blade angle when the propeller is installed on the engine is the same as that

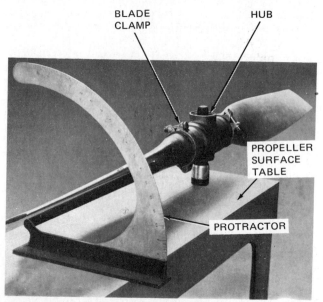

FIG. 17-13 Adjustment of blade angle on a ground-adjustable propeller.

used for other propellers and will be described later in this chapter.

● CONTROLLABLE-PITCH PROPELLERS

As the name implies, a **controllable-pitch propeller** is one on which the blade angle can be changed while the aircraft is in flight. Propellers of this type have been used for many years on aircraft where the extra cost of such a propeller was justified by the improved performance obtained.

Advantages

The controllable-pitch makes it possible for the pilot to change the blade angle of the propeller at will in order to obtain the best performance from the aircraft engine. At takeoff the propeller is set at a low blade angle so that the engine can attain the maximum allowable rpm and power. Shortly after takeoff the angle is increased slightly to prevent overspeeding of the engine and to obtain the best climb conditions of engine rpm and airplane speed. When the airplane has reached the cruising altitude, the propeller can be adjusted to a comparatively high pitch for a low cruising rpm or to a lower pitch for a higher cruising rpm and greater speed.

Two-Position Propeller

A **two-position propeller** does not have all the advantages mentioned in the foregoing paragraph; however, it does permit a setting of blade angle for best takeoff and climb (low-pitch, high rpm) and for best cruise (high-pitch, low rpm).

One of the best-known two-position propellers was manufactured by the Hamilton Standard Propeller Division of the United Aircraft Company. This propeller was used extensively for training and utility aircraft during World War II and is still used on some agricultural aircraft.

A schematic diagram of a two-position propeller pitch-changing mechanism is shown in Fig. 17-14. The principal parts of this assembly are the hub assembly, the counterweight and bracket assembly, and the cylinder and piston assembly. The blade angle is decreased by the action of the cylinder and piston assembly when engine oil enters the cylinder and forces it forward. The cylinder is linked to the blades by means of a

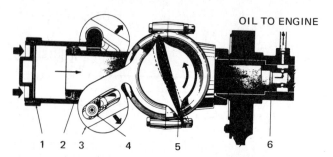

1. Propeller cylinder
2. Propeller piston
3. Propeller counterweight and bracket
4. Propeller counterweight shaft and bearing
5. Propeller bland
6. Engine propeller shaft

FIG. 17-14 Drawing of a two-position propeller pitch-changing mechanism.

bushing mounted on the cylinder base and riding in a slot in the counterweight bracket. As the cylinder moves outward, the bracket is rotated inward, and since the bracket is attached to the base of the blade, the blade is turned to a lower angle.

When the oil is released from the cylinder by means of a three-way valve, the centrifugal force acting on the counterweights moves the counterweights outward and rotates the blades to a higher angle. At the same time, the cylinder is pulled back toward the hub of the propeller.

The basic high-pitch angle of the propeller is set by means of four blade-bushing index pins which are installed in aligned semicircular notches between the counterweight bracket and the blade bushing when the two are assembled. The pitch range is set by adjusting the counterweight adjusting screw nuts in the counterweight bracket.

Initial horizontal balance of the two-position counterweight-type propeller is adjusted by balancing washers installed in the base of the blades. Initial vertical balance is accomplished by means of balancing washers installed in the space provided in the barrel supporting block inside the hub barrel. Final balance is adjusted by installing lead wool in or removing it from the hollow assembly bolts.

A counterweight-type propeller may also be designed as a constant-speed propeller to be controlled by a propeller governor. In this case, the governor controls the flow of oil to and from the propeller cylinder in accordance with engine rpm. The governor is adjusted for the desired engine rpm by means of a control in the cockpit.

The foregoing brief discussion of counterweight-type propellers has been given to provide a basic understanding of their operation. Since these propellers will not often be encountered on many aircraft today, it is not deemed necessary to describe them in complete detail.

Beechcraft Series 215 Propeller

The Beechcraft series 215 is a controllable-pitch propeller on which the blade angle is changed by means of an electric motor mounted on the **fixed sleeve** (1), Fig. 17-15. A pinion gear actuated by the motor through a gearbox drives the **ring gear** (3) in the illustration. The ring gear controls the position of the **pitch-control bearing** (2) by means of internal threads in the **ring-gear hub** (4). These threads engage lugs on the pitch-control bearing, thus causing it to move forward and rearward as the ring gear rotates. The lugs on the pitch-control bearing also engage slots in the fixed sleeve and prevent the external race of the bearing assembly from rotating with the propeller. The **control bolts** (7) attached to the inner race of the bearing assembly extend forward into the propeller hub and are attached to the **yoke** (11). The **bushing** (10) attached to the yoke engages a slot in the propeller blade butt. The fore-and-aft movement of the pitch-control bearing, as the ring gear is turned, is transmitted through the control bolts to the yoke. The yoke moves fore and aft and rotates the blades by means of the bushing (10). An exploded view of the 215 propeller hub assembly is shown in Fig. 17-16.

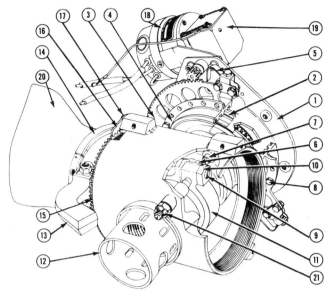

1. Fixed sleeve
2. Pitch-control bearing
3. Ring gear
4. Ring-gear hub
5. Spring stop
6. Yoke locknut
7. Control bolt
8. Ring-gear attaching bolt
9. Actuator bolt
10. Bronze bushing
11. Yoke
12. Propeller-retention nut
13. Counterweight arm
14. Balance ring
15. Blade-retention nut
16. Blade-retaining-nut lock
17. Blade-retaining nut-lock bolt
18. Propeller-motor mounting bracket
19. Propeller-motor assembly
20. Propeller blade
21. Safety low-pitch stop nut

FIG. 17-15   Pitch-changing mechanism for a Beechcraft series 215 propeller. *(Beech Aircraft Co.)*

Each **blade assembly** for the series 215 propeller consists of an aluminum-alloy blade, a steel sleeve, blade bearings, and a blade-retention nut. Thirty-one $\frac{7}{16}$-in [11.11-mm] diameter steel ball bearings ride in the bearing races of each blade. The blade-retention nut, which remains as a permanent assembly on each blade, holds the blade ball bearings in their races and threads into the propeller hub to secure the entire blade assembly. A single slot on each blade butt is provided for the bronze bushings.

The **hub assembly** consists of a pitch-control mechanism and a hub body. The hub body is constructed to make one piece. It is threaded to receive the blade-retention nuts, and the center bore is splined to fit a 20-spline crankshaft. The hub body may be considered the foundation of the entire propeller. The propeller-retention nut is designed so that, when it is loosened, the outer surface of the ridge on the aft end of the nut comes into contact with the snap ring in the hub body and acts as a puller to aid in removing the propeller. The hub body receives the centrifugal load of the rotating propeller blades through the threads for the blade-retention nuts.

The **pinion-and-pitch-control-gear assembly** consists of the internally threaded ring-gear hub, ring gear, stationary sleeve assembly, and a split lock ring. The pinion meshes with the ring gear and is an integral part of the motor which is mounted on the fixed sleeve. The ring gear is attached to the ring-gear hub which

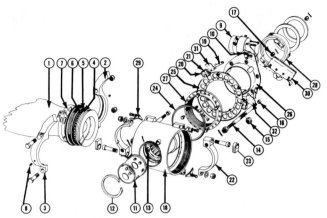

1. Blade assembly
2. Counterweight
3. Counterweight
4. Blade-bearing race
5. Ball bearing
6. Balde-bearing race
7. Balance-ring assembly
8. Hex-socket set screw
9. Motor assembly
10. Internal hex-head screw
11. Propeller-retention nut
12. Snap ring
13. Safety low-pitch stop nut
14. Pitch-control bolt
15. Yoke locknut
16. Pitch-control-bolt sleeve
17. Actuator-bearing attaching nut
18. Hub
19. Ring-gear clip tooth
20. Spring stop
21. Spring stop bracket
22. Yoke
23. Blade bushing
24. Gear retainer ring
25. Ring gear
26. Stationary sleeve
27. Gear hub
28. Actuator-bearing assembly
29. Blade-retention-nut lock
30. Constant-speed switch assembly
31. Constant-speed switch actuator cam
32. Switch-actuator cam

FIG. 17-16 Exploded view of the hub assembly for a series 215 propeller. *(Beech Aircraft Co.)*

fits over the stationary sleeve assembly and is held in place by means of the gear retainer ring. The sleeve contains slots so that the lugs on the outer race of the actuator bearing project through the sleeve and engage the internal threads of the ring-gear hub. The entire assembly is secured to the engine by bolting the stationary sleeve assembly to a plate on the nose case of the engine.

The **drive mechanism** consists of an electric motor, necessary gearing, and pinion gear. The electric motor is mounted on the fixed sleeve and provides the power for operating the blade-pitch actuating mechanism.

During operation of the propeller, the adjustment is controlled from the airplane cockpit by a three-position toggle switch. This switch is held in the INCREASE or DECREASE RPM position until the desired rpm is obtained, then it is returned to the center OFF position. When the engine is started, the propeller is adjusted to the HI RPM (low-pitch) position by means of the toggle switch. This same position is used for takeoff. The desired cruising rpm is attained by moving the control switch to the LO RPM (high-pitch) position and releasing it when the proper rpm is indicated. For approach and landing, the propeller is placed in the maximum high rpm position.

The removal, installation, inspection, and maintenance of the Beechcraft 215 propeller follows standard procedures established for other similar propellers.

Special instructions are provided in the manufacturer's manual.

● **CONSTANT-SPEED PROPELLERS**

As previously explained, a **constant-speed** propeller is controlled by a speed governor which automatically adjusts propeller pitch to maintain a selected engine speed. If the rpm of the propeller tends to increase, the governor senses the increase and responds by causing the propeller blade angle to increase. Also, when the propeller rpm tends to decrease, the governor causes a decrease in propeller blade angle. An increased blade angle will cause a decrease of rpm, and a decreased blade angle will cause an increase of engine rpm.

The pitch-changing devices for constant-speed propellers include electric motors, hydraulic cylinders, centrifugal force acting on flyweights, or a combination of these methods. The means by which these methods are applied will become clear as we examine some of the typical constant-speed propellers designed for light airplanes.

● **BEECHCRAFT SERIES 278 PROPELLER**

The Beechcraft series 278 propeller, shown in Fig. 17-17, is a hydraulically operated all-metal constant-speed propeller designed for use with Continental series O-470 engines. This engine has a flange-type crankshaft, and the propeller installation is a simple, bolted, propeller-to-engine attachment. The propeller hub, blades, and spinner constitute the complete propeller assembly.

Propeller Hub

The hub assembly for the Model 278 propeller, shown in the exploded view of Fig. 17-18, includes the hub body, machined from welded and copper-brazed steel forgings; the blade actuating mechanism; and retaining parts for the propeller blades. A piston and shaft assembly rides in the center of the hub, which is honed smooth and acts as a stationary cylinder. The piston shaft extends forward from the piston and is attached to a yoke, forward of the hub body, which links it with two actuator bolts. These bolts protrude from guides on opposite sides of the hub. Within the hub the actuator bolts are linked together by a split yoke which fits around the outside of the stationary cylinder. The halves of the yoke are tied together by two pitch-control bolts with bronze actuator bushings fitted on their heads. Since the actuator bushings also fit into slots in the propeller blade butts, the axial piston motion moves the piston shaft and actuator bolt assembly, which in turn transmits a rotary motion to the propeller blades to change their pitch.

FIG. 17-17 Beechcraft Model 278 propeller.

394

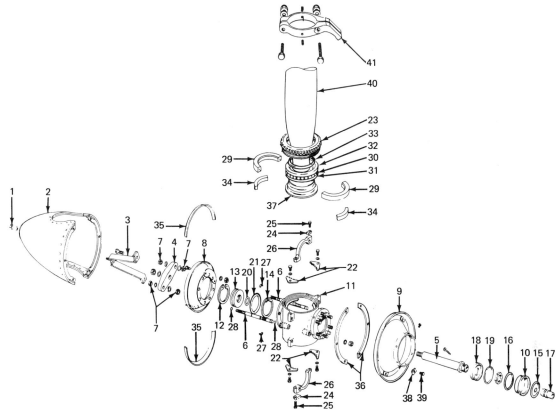

1. Nut
2. Spinner shell assembly
3. Spinner-retainer bracket assembly
4. Propeller-front yoke
5. Propeller-piston shaft assembly
6. Propeller-control bolt
7. Nut
8. Propeller-spinner front bulkhead
9. Propeller-spinner aft bulkhead
10. Oil-retainer aft partition
11. Propeller-hub brazed assembly
12. Snap ring
13. Propeller-piston guide
14. Snap ring
15. O ring
16. O ring
17. Oil-transfer-line connector tube
18. Propeller piston
19. O ring
20. O ring
21. O ring
22. Blade-retention-nut lock
23. Retainer-propeller-blade nut
24. Sliding-actuator bushing
25. Actuator bolt
26. Propeller yoke
27. Control-bolt lock
28. O ring
29. Balance ring assembly
30. Ball bearing
31. Thrust-bearing-inboard-race assembly
32. Thrust-bearing-outboard-race assembly
33. Thrust-bearing gasket
34. Balance weight
35. Propeller-shell-bulkhead forward pad
36. Propeller-shell bulkhead aft pad
37. Propeller-blade retainer
38. T nut
39. Bolt
40. Propeller blade
41. Counterbalance weight set

FIG. 17-18   Exploded view of the Model 278 propeller.

Blades

Each propeller blade assembly consists of an aluminum-alloy blade, steel sleeve, blade bearings, and a blade-retention nut. Thirty-one $\frac{1}{16}$-in [1.59-mm] diameter steel ball bearings ride in the bearing races of each blade. The blade-retention nut, which remains as a permanent assembly on each blade, holds the blade ball bearings in their races and threads into the propeller hub to secure the entire blade assembly. A single slot on each blade butt is provided for the bronze actuator bushings described before.

The Spinner

A polished aluminum-alloy spinner streamlines the propeller installation and contributes to engine cooling. Except for the blade openings, the spinner shell completely encloses the propeller hub assembly. A retaining bolt at the front of the attaching bracket on the forward portion of the propeller hub flange is used to secure the spinner to the propeller. The spinner is supported by two aluminum-alloy bulkheads attached to the propeller hub. The forward bulkhead assists with spinner alignment, while the aft bulkhead bottoms the spinner and secures it from rotational slippage. Balancing bolts installed in nut plates on the aft bulkhead are used to balance the propeller assembly, and drain holes in the flange at the aft portion of the spinner shell prevent accumulation of moisture which could cause an out-of-balance condition.

Principles of Operation

The blade-angle changes of the propeller are dependent on the balance between governor-boosted oil pressure and the inherent centrifugal tendency of the propeller blades to maintain a low-pitch angle. The balance differential is maintained by the governor, which either meters oil pressure to or allows oil to drain from the propeller cylinder in the quantity necessary to maintain the proper blade angle for constant-speed operation. A drawing of the governor is shown in Fig. 17-19.

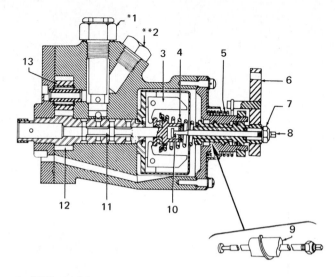

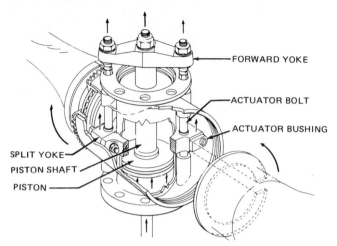

FIG. 17-20 Operation of a Model 278 propeller. *(Beech Aircraft Co.)*

1. Differential-pressure-relief valve
2. High-pressure-relief valve
3. Flyweights
4. Speeder spring
5. Control-lever spring
6. Speed-adjusting control lever
7. Locknut
8. Lift-rod adjustment
9. Speed-adjusting worm
10. Pilot-valve lift rod
11. Pilot valve
12. Governor-pump drive gear
13. Governor-pump idler gear

FIG. 17-19 Woodward propeller governor.

Within the governor, the L-shaped **flyweights** are pivoted on a disk-type **flyweight head** coupled to the engine gear train through a hollow drive-gear shaft. The **pilot-valve plunger** extends into the hollow shaft and is so mounted that the pivoting motion of the rotating flyweights will raise the plunger against the pressure of the **speeder spring** or allow the spring pressure to force the plunger down in the hollow shaft. The positions assumed by the plunger determine the flow of oil from the governor to the propeller. Governor oil is directed to a transfer ring on the engine crankshaft and thence into the crankshaft tube which carries it into the rear side of the piston cylinder arrangement in the propeller hub. The linear motion of the piston is changed to the rotary motion of the blades, as shown in Fig. 17-20. Since the centrifugal twisting force of the propeller blades is transmitted to the propeller piston, the governor-boosted oil pressure must overcome this force to change the engine rpm. Forward motion of the piston increases pitch and decreases engine rpm, while rearward motion of the piston decreases pitch and increases engine rpm.

The action of the pitch-changing mechanism is clearly shown in Fig. 17-20. As governor oil pressure enters the cylinder to the rear of the piston, the piston moves forward. This motion is transmitted through the **piston shaft** to the **forward yoke**. **Actuator bolts** are attached to each end of the yoke, and these bolts, being attached to the **split yoke,** carry the motion of the forward yoke to the split yoke. Each **actuator bushing** mounted on the split yoke fits into a groove on the butt of each blade, and when the bushings are moved forward by the split yoke, the blades are forced to rotate.

During operation of the propeller in flight, the governor flyweights react to engine rpm. If the engine is turning faster than the selected rpm, the flyweights will move outward and cause the pilot valve in the governor to move upward or toward the governor head. With this valve position, the oil pressure from the governor pump is directed to the propeller and the propeller piston moves forward to increase the blade angle and decrease the rpm.

When the engine is "on speed," the governor flyweights are in a neutral position and the pilot valve seals the oil pressure in the propeller system so that there is no movement in either direction. The oil pressure prevents the piston from moving backward; hence, the blade angle cannot decrease.

If engine rpm falls below the selected speed ("underspeed" condition), the flyweights of the governor move inward and allow the pilot valve to move toward the base of the governor. This position of the pilot valve opens a passage which permits the oil to flow from the propeller to the engine, thus allowing the blade angle to decrease and the rpm to increase. The blade angle tends to decrease because of the centrifugal twisting force as explained previously.

Propeller Installation

The installation of the Model 278 propeller on the crankshaft flange of the engine is comparatively simple. The following procedure is recommended by the manufacturer:

1. Check the condition of the propeller and piston oil transfer tube at the engine crankshaft flange.

2. Align the two guide pins in the propeller hub with the corresponding holes in the engine crankshaft flange, placing the no. 1 blade on the side of the TC (top center) mark on the engine flange, and install the propeller.

3. Install the retaining nuts and washers or retaining bolts and washers. Torque the propeller-retention nuts to 600 to 800 in·lb [67.8 to 90.4 N·m] or the 278-395 retention bolts to 600 to 700 in·lb [67.8 to 79.1 N·m].

4. Safety the nuts or bolts to the guide pins with safety wire.

5. Position the spinner on the propeller according to the marks made on removal, and install the spinner retaining nut. Tighten the nut until the flange at the rear of the spinner shell touches the rubber strip on

FIG. 17-21 McCauley Model 2A36C18 propeller.

the rear bulkhead completely around. Tighten the nut an additional two or three turns, and safety it with a cotter pin.

Maintenance and Repair

The maintenance and repair of the Beechcraft Model 278 propeller should be performed according to the established methods employed for metal propellers. These methods comply with FAA regulations and with the instructions given in the manufacturer's overhaul manual. Specific instructions for the overhaul of the propeller are provided by the manufacturer, and overhaul should be accomplished in a properly certificated propeller repair station or by the manufacturer.

● McCAULEY CONSTANT-SPEED PROPELLER

A McCauley Model 2A36C18 propeller (Fig. 17-21) is an all-metal constant-speed propeller controlled by a single-acting governor. The blades are made of forged aluminum alloy, and the hub parts are made of steel.

A schematic diagram of the propeller hub mechanism is shown in Fig. 17-22. A careful study of this drawing will reveal the *cylinder* at the front of the propeller hub, the *piston* inside the cylinder, the hollow *piston rod* through which oil flows to and from the cylinder, the *blade actuating pin*, the *low-pitch return boost spring*, the *hub assembly*, and the *blade assembly*. During operation, when the piston is fully forward, the blades are in the low-pitch position. If the engine overspeeds, the governor will direct governor oil pressure through the crankshaft into the hollow piston rod of the propeller. The oil flows through the piston rod and into the cylinder, forcing the piston to move back. The piston rod is linked to the blade butts through the *link assemblies* and the blade actuating pins, and as the piston rod moves backward, the blades are forced to rotate in the hub. This increases the pitch and reduces the engine speed. If the engine rpm falls below the value selected by the governor control, the governor pilot valve will move downward and open the passages which allow the oil in the propeller piston to

return through the piston rod to the engine. The piston is pushed forward by the low-pitch return boost spring and by the centrifugal twisting of the rotating blades. When the propeller is "on speed," the oil pressure in the cylinder is balanced against the two forces tending to turn the blades to low pitch.

The detailed construction of the hub assembly, pitch-changing mechanism, and the blade assemblies is shown in the exploded view of Fig. 17-23.

● BEECHCRAFT SERIES 279 FULL-FEATHERING PROPELLER

The Beechcraft Model 279 propeller is a hydraulic constant-speed full-feathering propeller with a flanged hub. It employs an engine-driven hydraulic governor for control. The governor is double-acting in that it directs oil under boost pressure to the propeller for both increase and decrease rpm adjustments. The governor is similar to that described for the Model 278 propeller and is adjusted by means of a control in the cockpit.

Operation

The operation of the Model 279 propeller can be understood by examining Fig. 17-24. It will be observed that the hub of the propeller contains a sleeve inside of which is a movable cylinder. Inside the movable cylinder is a stationary piston mounted at the end of the **piston rod**, and inside the piston rod is the **oil distributor tube** which directs oil to either side of the piston, depending on action of the governor. When governor-boosted oil pressure is directed to the forward side of the piston, the cylinder moves forward to increase the blade angle. If the oil pressure is directed to the rear side of the piston, the cylinder moves back and decreases the blade angle.

The yoke, mounted on the forward end of the movable cylinder, is linked to two rod assemblies extending around opposite sides of the hub. These rod assemblies are attached to arms mounted on the shank of each blade. As the cylinder is moved forward or backward, the linkage causes the blades to rotate. The blade-angle range is from full low-pitch to a full-feathered position. The two rod assemblies, called the **propeller pitch-control link assemblies**, are shown in Fig. 17-25. As the yoke is moved forward by the movable cylinder, the blades will rotate toward a higher blade angle until they finally reach the full-feathered position.

Construction

The construction and parts arrangement of the Beechcraft Model 279 propeller are illustrated in Fig. 17-26. When the propeller hub parts are assembled, the **piston** (19) is attached to the **piston rod** (13). The **cylinder** (26) is inside the **hub** (1). The **piston and rod assembly** is stationary inside the hub, and the cylinder is free to move forward and rearward. In the forward end of the cylinder is the **oil retainer partition** (21) which seals the cylinder. The **yoke** (27) is mounted on the forward end of the cylinder with the **propeller pitch-control link assemblies** (31) attached to each side. The ends of these link assemblies are connected to the **blade-actuator-arm assemblies,** as shown in Fig. 17-27, which

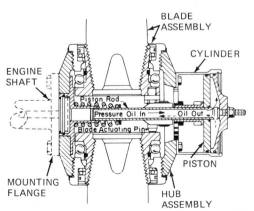

FIG. 17-22 Drawing of the hub mechanism for the Model 2A36C18 propeller. *(McCauley Industrial Corp.)*

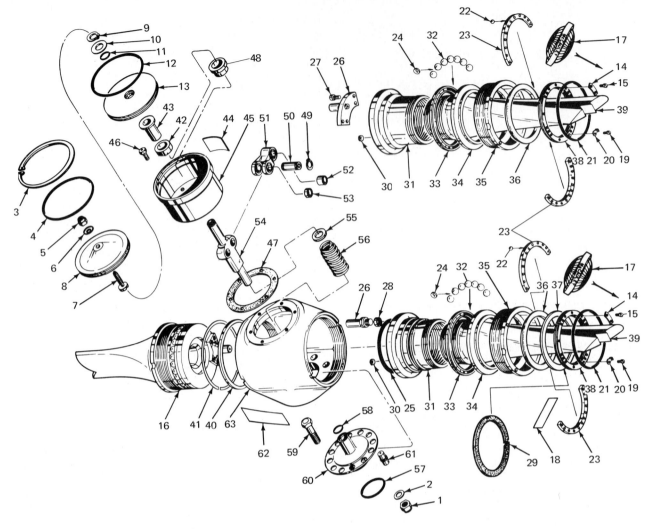

1. Nut	20. Preload nut lock	37. Outer preload bearing	52. Blade actuating pin
2. Plain washer	21. O-ring packing	race	bearing
3. Internal retaining ring	22. Bearing ball	38. Preload nut	53. Piston-rod-bearing
4. O-ring packing	23. Preload bearing retainer	39. Blade	54. Piston rod
5. Self-locking nut	24. Ball separator	40. Retention-nut lock ring	55. Plain washer
6. Dyna seal	25. O-ring packing	41. Balancing shim	56. Low-pitch return boost
7. Low-pitch stop screw	26. Blade actuating pin	42. High-pitch stop-spacer	spring
8. Cylinder head	27. Knurled-socket-head	stock	57. O-ring packing
9. Bowed retaining ring	cap screw	43. Piston-rol sleeve	58. O-ring packing
10. Piston washer	28. Actuating-pin washer	44. Spring kit-	59. Hub mounting bolt
11. O-ring packing	29. Gasket	installation decal	60. Hub and piston guide
12. O-ring packing	30. Ferrule-staking plug	45. Cylinder assembly	flange
13. Piston	31. Blade-retention ferrule	46. Bolt	61. Hub-alignment dowel
14. Balance weight	32. Bearing ball	47. Cylinder gasket	62. Propeller-
15. Screw	33. Inner race	48. Cylinder bushing	installation-instructions
16. Blade assembly	34. Outer race	49. External retaining ring	decal
17. Decal	35. Blade-retention nut	50. Piston-rod pin	63. Propeller hub
18. Decal	36. Inner preload bearing	51. Link assembly	
19. Screw	race		

FIG. 17-23 Exploded view of the McCauley constant-speed propeller.

illustrates the construction of the blade assemblies at the butt.

The propeller blades are made of aluminum alloy and are shrunk into steel sleeves called **blade retainers.** The blade retainers cannot be removed without danger of damaging the blades. The arrangement and construction of the blade bearings are clearly shown in the illustration. When the blade is assembled, the bearing

assembly is located between the flange of the blade retainer and the propeller blade-retention nut.

Feathering System

The feathering system for the Model 279 propeller consists of a feathering pump, a reservoir, a feathering time-delay switch, and a propeller feathering light located approximately in the center of the instrument

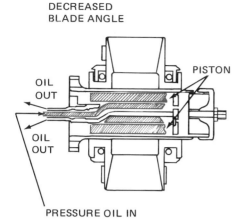

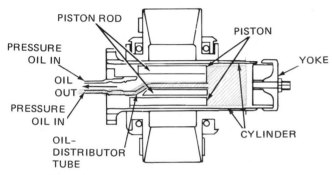

FIG. 17-24 Operation of the Beechcraft Model 279 propeller.

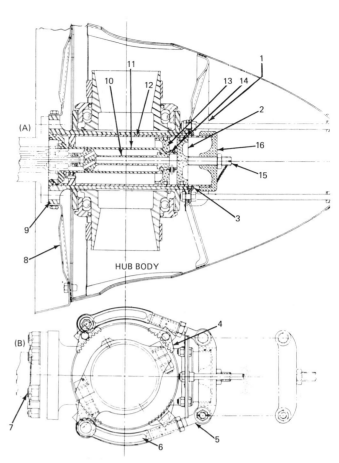

1. Front bulkhead	6. Propeller pitch-control link assembly	11. Piston rod
2. Partition	7. Stud	12. Cylinder
3. Flange	8. Bulkhead	13. Piston
4. Lock	9. Hub	14. Nut
5. Rod end	10. Oil tube	15. Screw
		16. Yoke

FIG. 17-25 Operating mechanism of the Model 279 propeller. (A) Cross section. (B) Plan view with 90° turn. *(Beech Aircraft Co.)*

panel. The propeller is feathered by moving the control in the cockpit against the low-speed stop. This causes the **pilot-valve lift rod** in the governor to hold the pilot valve in a **decrease rpm** position regardless of the action of the governor flyweights. This causes the propeller blades to rotate through high pitch to the feathered position.

Maintenance and Repair

The inspection, maintenance, and repair of the Beechcraft Model 279 propeller should be accomplished according to standard practices described elsewhere in this chapter and in accordance with the manufacturer's instructions. With the exception of minor repairs, routine inspections, and normal service, all work should be accomplished in a properly certificated repair station.

● HARTZELL FULL-FEATHERING PROPELLER

Hartzell constant-speed, full-feathering propellers utilize hydraulic pressure to reduce the pitch of the blades and a combination of spring and counterweight force to increase the pitch. If the pitch is increased to the limit, the blades are in the feathered position.

Operation

Figure 17-28 is a schematic drawing of the Hartzell Model HC-82XF-2 propeller hub assembly to illustrate the pitch-changing mechanism. When the engine speed is below that selected by the pilot, the governor pilot valve directs governor oil pressure to the propeller. This pressure forces the cylinder forward and reduces the propeller pitch. When the cylinder moves forward, it also compresses the feathering spring.

If engine speed increases above the rpm selected, the governor opens the oil passage to allow the oil in the propeller cylinder to return to the engine. The feathering spring and the counterweight force cause the blades to rotate to a higher pitch position.

Feathering is accomplished by releasing the governor oil pressure, allowing the counterweights and feathering spring to feather the blades. This is done by pulling the governor pitch control back to the limit of its travel, thus opening up a port in the governor to allow the oil from the propeller to drain back to the engine. The time necessary to feather depends on the size of the oil passage from the propeller to the engine and on the force exerted by the spring and counterweights. The larger the passages through the governor and the heavier the springs, the quicker is the feathering action. The elapsed time for feathering is usually between 3 and 10 s.

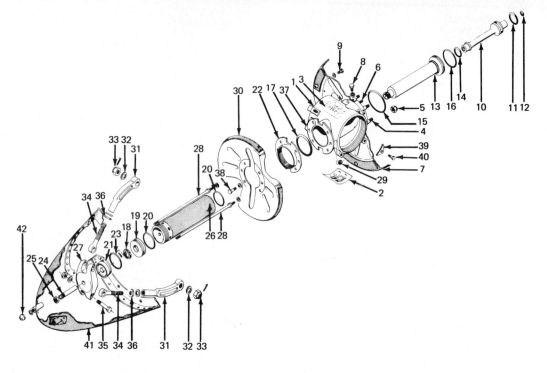

1. Hub body	10. Oil-transfer tube	22. Oil-retainer flange	32. Washer
2. Placard for low-pitch adjustment	11. O ring	23. O-ring	33. Nut and cotter pin
3. Placard for assembly information	12. Seal	24. Pitch-adjustment screw	34. Link-rod end assembly
4. Propeller-hub stud	13. Propeller piston rod	25. Check nut	35. Bolt, washer, and nut
5. Hub retaining nut	14. Seal	26. Oil-cylinder assembly	36. Check nut and washer
6. Propeller-hub dowel pin	15. Seal	27. Pitch-control yoke	37. Hub placard
7. Spinner-aft-bulkhead assembly	16. Seal	28. Spinner-retainer rod	38. Bolt and washer
8. Bolt and washer	17. Packing	29. Nut and washer	39. Blade-retention-nut lock
9. Balance bolts and washer	18. Propeller piston rod	30. Spinner-front-bulkhead assembly	40. Bolt and washer
	19. Piston	31. Propeller pitch-control link assembly	41. Spinner assembly
	20. Packing		42. Nut
	21. Oil retainer partition		

FIG. 17-26 Exploded view of the Model 279 propeller hub assembly.

Unfeathering the propeller is accomplished by re-positioning the governor control to the normal flight range and restarting the engine. As soon as the engine cranks over a few turns, the governor starts to un-feather the blades and soon windmilling takes place, thus speeding up the process of unfeathering. In order to facilitate cranking of the engine, the feathering blade angle is set at 80 to 85° at the three-fourths station on the blades. In general, restarting and unfeathering can be accomplished within a few seconds.

Special unfeathering systems may be installed with the Hartzell propeller when it is desired to increase the speed of unfeathering. Such a system is shown in Fig. 17-29. During normal operation the accumulator stores governor oil pressure; when the propeller is feathered, this pressure is trapped in the accumulator because the accumulator valve is closed at this time. When the propeller control is placed in the normal position, the pressure stored in the accumulator is ap-plied to the propeller to rotate the blades to a low-pitch angle. It must be remembered that when the propeller is feathered, there is no pressure available from the governor because the engine is stopped. The pressure stored in the accumulator is used in place of the pres-sure which would normally be supplied by the gov-ernor.

Propeller Governor

We have discussed the propeller governor previ-ously and explained its operation to some degree in explaining the operation of constant-speed propellers; however, it will be beneficial to examine the illustration of Fig. 17-30 in order to gain a more complete under-standing of governor operation.

The governor is geared to the engine in order to sense the rpm of the engine at all times. The speed sensing is accomplished by means of rotating **flyweights** in the upper part of the governor body. As shown in the drawing, the flyweights are L-shaped and hinged at the outside where they attach to the **flyweight head.** The **toe** of each flyweight presses against the race of a bearing at the upper end of the **pilot valve.** Above the bearing are the **speeder-spring seat** and the **speeder spring** which normally holds the **pilot-valve plunger** in the down position. Above the speeder spring is the **adjusting worm,** which is rotated by means of the **speed-adjusting lever.** The speed-adjusting lever is connected to the propeller control in the cockpit. As the speed-

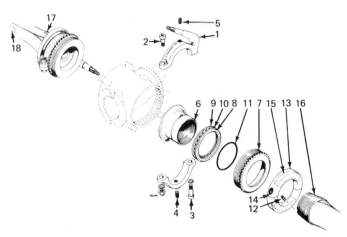

1. Blade-actuator-arm assembly
2. Bolt
3. Bolt, washers, and nut
4. Headless oval-point set screw
5. Headless cup-point set screw
6. Propeller-blade retainer
7. Propeller-blade-retention nut
8. Blade-thrust bearing race (outer)
9. Blade-thrust bearing race (inner)
10. Ball (bearing)
11. Gasket
12. Headless cup-point set screw
13. Balance ring assembly
14. Hex-socket-head cap screw
15. Balance weight
16. Propeller blade
17. Bearing-information placard
18. Beechcraft-propeller placard

FIG. 17-27 Exploded view of the Model 279 blade assembly.

adjusting lever is moved, it rotates the adjusting worm and increases or decreases the compression of the speeder spring. This, of course, affects the amount of flyweight force necessary to move the pilot-valve plunger. If it is desired to increase the rpm of the engine, the speed-adjusting control lever is rotated in the direction to increase speeder-spring compression. It is therefore necessary that the engine rpm increase in order to apply the additional flyweight force to raise the pilot-valve plunger to an on-speed position.

In the top drawing of Fig. 17-30 the governor is in the overspeed condition. The engine rpm is greater than that selected by the control, and the flyweights are pressing outward. The toes of the flyweights have

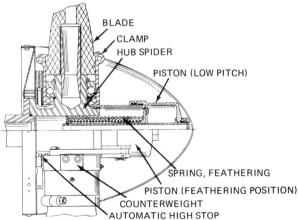

FIG. 17-28 Drawing of the Hartzell Model HC-82XF-2 feathering propeller hub assembly. *(Hartzell Propeller)*

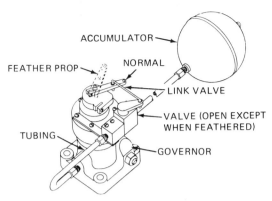

FIG. 17-29 Unfeathering system for the Hartzell propeller.

raised the pilot-valve plunger to a position which permits oil pressure from the propeller to return to the engine. The propeller counterweights and feathering spring can then rotate the propeller blades to a higher angle, thus causing the engine rpm to decrease.

When the governor is in an underspeed condition, that is, with engine rpm below the selected value, the governor flyweights are held inward by the speeder spring and the pilot-valve plunger is in the down position. This position of the valve directs governor oil pressure from the governor gear pump to the propeller cylinder and causes the propeller blades to rotate to a lower pitch angle. The lower pitch angle allows the engine rpm to increase.

The governor shown in the drawing is equipped with a lift rod to permit feathering of the propeller. When

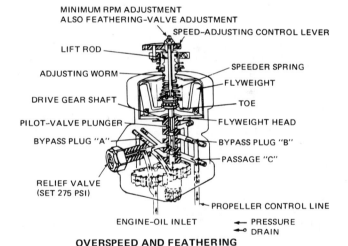

OVERSPEED AND FEATHERING

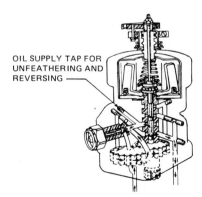

FIG. 17-30 Operation of the Woodward propeller governor.

the cockpit control is pulled back to the limit of its travel, the lift rod in the governor holds the pilot-valve plunger in an overspeed position. This causes the blade angle of the propeller to increase to the feathered position regardless of flyweight or speeder-spring force.

It is important to observe the effect of the speeder spring on governor operation. If the speeder spring should break, the pilot-valve plunger would be raised to the overspeed position, which would call for an increase of propeller pitch. This, of course, would allow the propeller to feather. If the speeder spring should break in a governor for a nonfeathering, constant-speed propeller, the propeller blades would rotate to maximum high-pitch angle.

Propeller governors similar to the one described are also arranged for double-acting operation where governor pressure is directed to the propeller through different passages for both increase rpm and decrease rpm. This is accomplished merely by utilizing the oil passages in a different manner. A study of the drawing of Fig. 17-30 will show that some of the passages are plugged, and if the use of passages is changed, the governor may be adapted to different types of systems. The arrangement for any particular propeller system is shown in the manufacturer's manual for the propeller under consideration.

Propeller Synchronizer System

Some twin-engine aircraft with constant-speed propellers are equipped with an automatic synchronizer system to match the rpm of the two engines. This system may be turned ON or OFF, thus providing for manual or automatic control.

A typical synchronizer system includes a master governor, a slave governor, magnetic pulse pickups on both governors, an electronic control, box assembly, an actuator motor, a trimmer assembly, a flexible drive shaft, a control switch, and an indicator light. A schematic diagram of such a system is shown in Fig. 17-31.

During operation of the system, the two governors send pulse signals from the magnetic pickups to the control-box circuit. If the pulse signals are not exactly at the same frequency, the control circuit will rotate the actuator motor in a direction to equalize the governor signals. As the actuator motor rotates, it turns the flexible shaft leading to the trimmer assembly which is attached to the control lever on the slave governor. The trimmer moves the governor-control lever to make the appropriate rpm adjustment for the slave engine.

The automatic synchronizer system is placed in operation by first adjusting the engine rpm manually for synchronization at the desired cruising value. The synchronizer switch is then turned ON. The system has a limited range of synchronization to prevent the possibility of the slave engine's losing more than a limited amount of rpm in case the master engine is feathered with the synchronizer ON.

Inspection, maintenance, and repair of a synchronizer system are described in the *Aircraft Maintenance Manual* and the *Governor Maintenance Manual*.

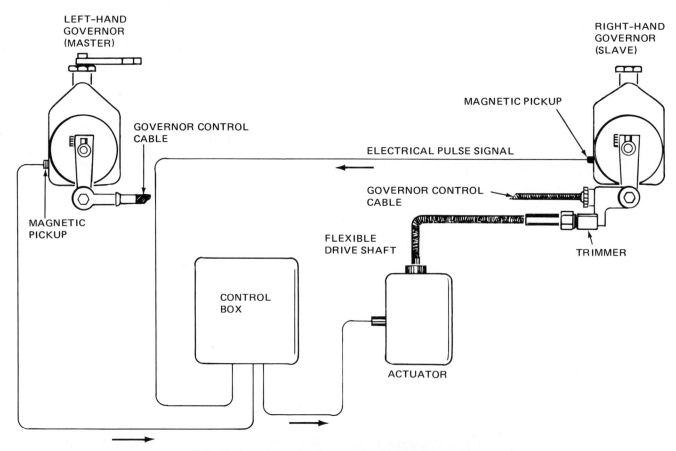

FIG. 17-31 Schematic diagram of a synchronizer system.

MAINTENANCE AND REPAIR OF METAL PROPELLERS

Hollow and Solid-Steel Propellers

Damaged steel propeller blades should not be repaired except by the manufacturer. Welding or straightening is not permissible on such blades, even for very minor repairs, except by the manufacturer because of the special process employed and the heat treatment required. A blade developing a crack of any nature in service should be returned to the manufacturer for inspection. When a blade is considered non-repairable, a notice of rejection should be made out by the manufacturer and sent to the nearest inspector of the Federal Aviation Administration.

Inspection of Steel Blades

The inspection of steel blades may be either visual or magnetic. The visual inspection becomes easier to accomplish if the steel blades are covered with engine oil or rust-preventive compound. The full length of the leading edge, especially near the tip; the full length of the trailing edge; the grooves and shoulders on the shank; and all dents and scars should be examined with a magnifying glass to decide whether defects are scratches or cracks.

In the magnetic inspection of steel blades and propeller parts, the blade or part to be inspected is mounted in a machine, and then the blade is magnetized by passing a current through the blade or part, using a power supply of 2000 to 3000 A at 6 V. Either a black or a red mixture of an iron-base powder and kerosene is poured over the blade or part at the time that it is magnetized. North and south magnetic poles are established on either side of any crack in the metal. The iron filings arrange themselves in lines within the magnetic field thus created. A black or a red line, depending on the color of the mixture, will appear wherever a crack exists in the blade or part.

Repair of Minor Damage to Steel Blades

Minor injuries to the leading and trailing edges only of steel blades may be smoothed by handstoning, provided that the injury is not deep.

Aluminum-Alloy Propellers

A seriously damaged aluminum-alloy propeller blade should be repaired only by the manufacturer or by repair agencies certificated for this type of work. Such repair agencies should follow manufacturer's instructions.

Definition of Damaged Propellers

A damaged metal propeller is one that has been bent, cracked, or seriously dented. Minor surface dents, scars, nicks, etc., which are removable by field maintenance technicians are not considered sufficient to constitute a damaged propeller.

If the model number of a damaged blade appears on the manufacturer's list of blades which cannot be repaired, the blade should be rejected.

Blades Bent in Face Alignment. The extent of a bend in the face alignment of blades should be carefully

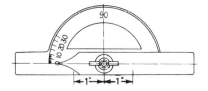

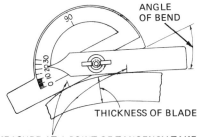

FIG. 17-32 Measuring the angle of bend.

checked by means of a protractor similar to the one illustrated in Fig. 17-32. Only bends not exceeding 20° at 0.15-in [3.81-mm] blade thickness to 0° at 1.1-in [2.79-cm] blade thickness may be cold-straightened. After straightening, the affected portion of the blade must be etched and thoroughly inspected for cracks and other flaws. Blades with bends in excess of this amount require heat treatment and must be returned to the manufacturer or an authorized agent for repair.

Manufacturers often specify the maximum bends which can be repaired by cold-straightening on specific models of propellers. Figure 17-33 is a chart which shows the maximum allowable bend for cold repair of the McCauley Models 1A90, 1B90, and 1C90 fixed-pitch metal propellers. From the chart, for example, it can be determined that if the propeller is bent at the 16-in [40.64-cm] radius, the maximum degree of bend which can be straightened cold is 9°. At the 32-in [81.28-cm] radius the blade can be repaired by cold-straightening if the bend is as great as 18.5°.

Blades Bent in Edge Alignment. Blades which are bent in edge alignment should not be repaired by anyone except the manufacturer or a certificated repair station holding the appropriate rating.

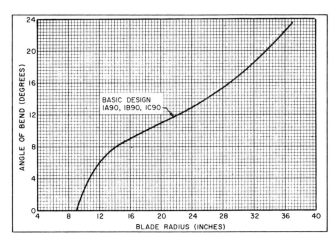

FIG. 17-33 Chart to show the maximum allowable bend for a cold repair.

Inspection and Treatment of Defects

Scratches and suspected cracks should be given a local etch, as explained below, and then examined with a magnifying glass. *The shank fillets of adjustable-pitch blades and the front half of the undersurface of all blades from 6 to 10 in [15.24 to 25.40 cm] from the tip are the most critical portions.*

Adjustable-pitch blades should be etched locally on the clamping portion of the shank at points $\frac{1}{4}$ in [6.35 mm] from the hub edge in line with the leading and trailing edges and should be examined with a magnifying glass for circumferential cracks. The shank must be within drawing tolerance. Any crack is cause for rejection. The micarta shank bearing on controllable and hydromatic propeller blades should not be disturbed except by the manufacturer. Blades requiring removal of more material than that specified as permissible in this chapter under the heading "Repair of Pitted Leading Edges" should be scrapped.

Local Etching

To avoid dressing off an excess amount of metal, checking by local etching should be accomplished at intervals during the progress of removing cracks and double-back edges of metal. Suitable sandpaper or fine-cut files may be used for removing the necessary amount of metal, after which, in each case, the surfaces involved should be smoothly finished with No. 00 sandpaper. Each blade from which any appreciable amount of metal has been removed should be properly balanced before being used.

When aluminum-alloy blades are inspected for cracks or other failures and for bends, nicks, scratches, and corrosion, the application of engine oil to the blades helps the inspector to see the defects, especially with a magnifying glass. If there is any doubt about the extent of the defects, local etching is then performed.

Purposes. Local etching has four principal purposes: (1) it shows whether visible lines and other marks within small areas of the blade surfaces are actually cracks instead of scratches; (2) it determines, with a minimum removal of metal, whether or not shallow cracks have been removed; (3) it exposes small cracks that might not be visible otherwise; and (4) it provides a simple means for inspecting the blades without removing or disassembling the propeller.

The caustic soda solution is a 20 percent solution prepared locally by adding to the required amount of water as much commercial caustic soda as the water will dissolve and then adding some soda pellets after the water has ceased to dissolve the caustic to be sure that the solution is saturated. The quantity required depends on the amount of etching to be done. This caustic soda solution should reveal the presence of any cracks.

The acid solution is a 20 percent nitric acid solution prepared locally by adding 1 part commercial nitric acid to each 5 parts of water. This acid solution is used to remove the dark corrosion caused by the application of the caustic soda solution to the metal.

Keep the solutions in glass or earthenware containers. Do not keep them in metal containers, since they attack metal. If any quantity of either the caustic soda or the acid solution is spilled, flush the surface it hits with fresh water, especially if it is a metal surface.

Procedures. Clean and dry the area of the aluminum-alloy blade to be locally etched. Place masking tape around the area under suspicion to protect the adjoining surfaces. Smooth the area containing the suspected defect with No. 00 sandpaper. Apply a small quantity of the caustic soda solution with a small swab to the suspected area. After the suspected area becomes dark, wipe it off with a clean cloth dampened with clean water, but do not slop too much water around the suspected area or the water will remove the solution from the defect and spoil the test. The dark stain that appears on an aluminum-alloy blade when the caustic solution is applied is caused by the chemical reaction between the copper in the alloy and the caustic soda (sodium hydroxide). If there is any defect in the metal, it will appear as a dark line or other mark. Examination under a microscope will show small bubbles forming in the dark line or mark.

It may require several applications of the caustic soda to reveal whether or not a shallow defect has been removed since a previous local etching was performed and a defect discovered. Immediately after the completion of the final test, all traces of caustic soda must be removed with the nitric acid solution. The blade is rinsed thoroughly with clean water, and then it is dried and coated with clean engine oil.

The inspection of aluminum-alloy propeller blades for cracks and flaws may be accomplished by means of a chromic acid anodizing process. This is superior to the caustic etching process and should therefore be used if facilities are available.

The blades should be immersed in the anodizing bath as far as possible, but all parts not made of aluminum alloy must either be kept out of the chromic acid bath or be separated from the blade by nonconductive wedges or hooks. The anodizing treatment should be followed by a rinse in clear, cold, running water for 3 to 5 min, and the blades should be dried as soon as possible after the rinse, preferably with an air blast. After the blades are dried, they should stand for at least 15 min before examination. Flaws, such as cold shuts and inclusions, will appear as fine black lines. Cracks will appear as brown stains caused by chromic acid bleeding out onto the surface.

The blades may be sealed for improved corrosion resistance by immersing them in hot water (180 to 212°F [82 to 100°C]) for $\frac{1}{2}$ h. In no case should the blades be treated with hot water before the examination for cracks, since heating expands any cracks and allows the chromic acid to be washed away.

Inspection of aluminum-alloy propeller blades for cracks and other defects may also be accomplished by means of fluorescent penetrant process or the dye penetrant process. These methods for the inspection of nonferrous metals are explained in Chap. 12.

Treatment of Minor Surface Defects

Dents, cuts, scars, scratches, nicks, etc., should be removed or otherwise treated by means of fine sandpaper and fine-cut files provided that their removal or

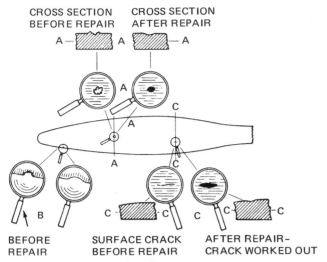

FIG. 17-34 Repair of surface defects.

CROSS SECTION BEFORE REPAIR   CROSS SECTION AFTER REPAIR

BEFORE REPAIR

SURFACE CRACK BEFORE REPAIR

AFTER REPAIR— CRACK WORKED OUT

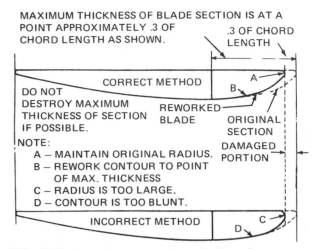

MAXIMUM THICKNESS OF BLADE SECTION IS AT A POINT APPROXIMATELY .3 OF CHORD LENGTH AS SHOWN.

.3 OF CHORD LENGTH

CORRECT METHOD

DO NOT DESTROY MAXIMUM THICKNESS OF SECTION IF POSSIBLE.

REWORKED BLADE

ORIGINAL SECTION

DAMAGED PORTION

NOTE:
A – MAINTAIN ORIGINAL RADIUS.
B – REWORK CONTOUR TO POINT OF MAX. THICKNESS
C – RADIUS IS TOO LARGE.
D – CONTOUR IS TOO BLUNT.

INCORRECT METHOD

FIG. 17-35 Rework of a propeller's leading edge.

treatment does not materially weaken the blade, materially reduce its weight, or materially impair its performance or reduce the blade dimensions below the minimum established by the manufacturer. Minimums will usually be given in the manufacturer's service and overhaul manual.

The metal around the longitudinal surface cracks, narrow cuts, and shallow scratches should be removed to form shallow saucer-shaped depressions, as illustrated in Fig. 17-34.

The metal at the *edges* of defects requires careful treatment. Metal at the edges of wide scars, cuts, nicks, etc., should be rounded off, and the surfaces within the edges should be smoothed out, as shown in Fig. 17-34 by that portion of the drawing marked with the letter *B*. Blades that require the removal of metal to a depth of more than $\frac{1}{8}$ in [3.18 mm] and a length of more than $1\frac{1}{2}$ in [3.81 cm] overall should be rendered unserviceable. The dimensions given here are not applicable to all propellers. *The manufacturer's manual should be consulted to ensure that correct information is used in the repair of specific models of propellers.*

The **raised edges of scars** require a slightly different treatment. The raised edges at wide scars, cuts, nicks, etc., should be carefully smoothed to reduce the area of the defect and the amount of metal to be removed, as illustrated in Fig. 17-34 in that part of the drawing which is marked with the letter *A*. It is not permissible to peen down the edges of any defect. With the exception of cracks, it is not necessary to remove completely or "saucer out" all of a comparatively deep defect. Properly rounding off the edges and smoothing out the surface within the edges are sufficient, since it is essential that no unnecessary amount of metal be removed.

Number of Defects Allowable in Blades

More than one defect falling within the above limitations is not sufficient cause alone for the rejection of a blade. A reasonable number of such defects per blade is not necessarily dangerous, if within the limits specified, unless their location with respect to each other is such as to form a continuous line of defects that would materially weaken the blade.

Repair of Pitted Leading Edges

Blades that have the leading edges pitted from normal wear in service may be reworked by removing sufficient material to eliminate the defects. In this case, the metal should be removed by starting at approximately the thickest section as shown in Fig. 17-35, and working well forward over the nose camber so that the contour of the reworked portion will remain substantially the same, avoiding abrupt changes in section or blunt edges. Blades requiring the removal of more material than the permissible reduction in width and thickness from the minimum drawing dimensions should be rejected.

For repairing blades, the permissible reductions in width and thickness from the minimum original dimensions allowed by the blade drawing and blade manufacturing specifications are shown in Fig. 17-36 for locations on the blade from the shank to 90 percent of the blade radius. The outer 10 percent of blade length may be modified as required.

Tolerances Listed in Blade Manufacturing Specifications

Tolerances listed in the blade manufacturing specifications govern the width and thickness of new blades. These tolerances are to be used with the pertinent blade drawing to determine the minimum original blade di-

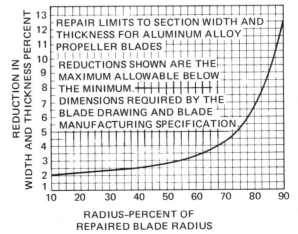

REPAIR LIMITS TO SECTION WIDTH AND THICKNESS FOR ALUMINUM ALLOY PROPELLER BLADES

REDUCTIONS SHOWN ARE THE MAXIMUM ALLOWABLE BELOW THE MINIMUM.

DIMENSIONS REQUIRED BY THE BLADE DRAWING AND BLADE MANUFACTURING SPECIFICATION.

REDUCTION IN WIDTH AND THICKNESS PERCENT

RADIUS-PERCENT OF REPAIRED BLADE RADIUS

FIG. 17-36 Propeller blade repair limits.

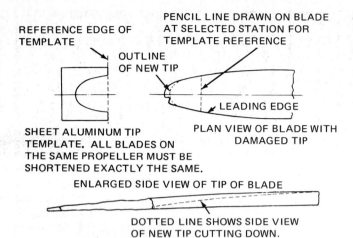

FIG. 17-37 Repair of a damaged propeller tip.

mensions to which the reductions of Fig. 17-36 may be applied.

Shortening of Blades to Remove Defects

When the removal or treatment of defects on the tip necessitates shortening a blade, each blade used with it must likewise be shortened. Such sets of blades should be kept together. Figures 17-37 and 17-38 illustrate acceptable methods.

With some propeller blades, the length may be reduced substantially and the propeller can then be given a new model number in accordance with the manufacturer's specifications. The reduction in length may require an increase in the blade angle, and the length must agree with the specification for the new model number.

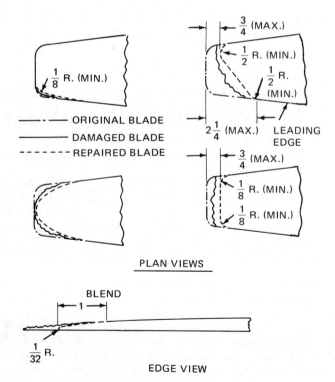

FIG. 17-38 Repair of a damaged square tip.

Causes for Rejection

Unless otherwise specified in this text, a blade having any of the following defects must be rendered unserviceable: (1) irreparable defects, such as a longitudinal crack, cut, scratch, scar, etc., that cannot be dressed off or rounded out without materially weakening or unbalancing the blade or materially impairing its performance; (2) general unserviceability due to removal of too much stock by etching, dressing off defects, etc.; (3) slag inclusions in an excessive number or cold shuts in an excessive number or both; and (4) transverse cracks of any size.

● REPAIR OF COMPOSITE PROPELLER BLADES

Certain repairs of composite propeller blades can be made in the field in accordance with instructions set forth in the appropriate repair manual. Repairs that can be made in the field are those described as minor; that is, these repairs correct minor damage. Minor damage is that which does not affect the airworthiness or operation of the propeller. This type of damage consists of nicks, dents, scratches, gouges, depressions, chordwise cracks in the leading edge cap, and debonding of the leading edge cap.

The manufacturer's repair manual describes the characteristics and dimensions that determine whether a particular blade injury can be classified as minor. An injury that exceeds the specifications for size and depth is major damage and requires that the propeller be removed and sent to the factory or a factory-approved repair facility.

Repairs for minor damage to a composite propeller blade are usually accomplished by cleaning the damaged area, removing the paint, and sanding. The damaged area is then filled with a mixture of chopped glass fiber and epoxy. After the epoxy has hardened, the area is sanded to conform to the contour of the blade. Final finish is accomplished with approved primer and polyurethane paint.

● CHECKING BLADE ANGLES

The blade angles of a propeller may be checked by using any precision protractor which is adjustable and is equipped with a spirit level. Such a protractor is often called a **bubble protractor.**

Universal Propeller Protractor

The blade angles of a propeller may be accurately checked by the use of a **universal propeller protractor,** which is the same instrument used to measure the throw of control surfaces. An accurate check of blade angles cannot be made by referring to the graduations on the ends of the hub barrels or on the shanks of the blades of propellers; such references are suitable only for rough routine field inspections and emergency blade settings.

A **protractor** is merely a device for measuring angles. The propeller protractor consists of an aluminum frame in which a steel ring and a disk are mounted, as shown in Fig. 17-39. The principal, or "whole-degree," scale is on the disk. The vernier, or "fractional-degree,"

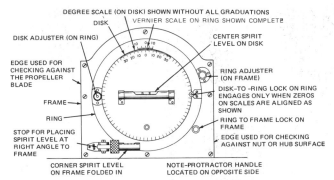

FIG. 17-39　Propeller protractor.

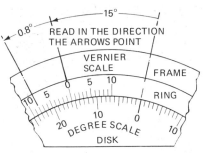

FIG. 17-40　Reading the protractor scale.

scale is on the ring. The zeros on these two scales provide reference marks which can be set at the two sides of an angle, thereby enabling the operator to read from zero to zero to obtain the number of degrees in the angle.

Two adjusting knobs provide for the adjustment of the ring and disk. The ring adjuster is in the upper right-hand corner of the frame; when it is turned, the ring rotates. The disk adjuster is on the ring; when this knob is turned, the disk rotates.

There are two locks on the protractor. One is the disk-to-ring lock, located on the ring. It is a pin that is held by a spring when engaged, but it engages only when the pin is pulled out and placed in the deep slot and when the zeros on the two scales are aligned. Under these conditions, the ring and disk rotate together when the ring adjuster is turned and when the ring-to-frame lock is disengaged. To hold the spring-loaded pin of the disk-to-ring lock in the released position, it is first pulled outward and then turned 90°.

The other lock, the ring-to-frame lock, is on the frame. It is a right-hand screw with a thumb nut. The disk can be turned independently of the ring by means of the disk adjuster when the ring is locked to the frame and the disk-to-ring lock is released.

There are two spirit levels on the protractor. One is the center, or disk, level. It is at right angles to the zero graduation mark on the whole-degree scale of the disk; hence, the zero graduation mark will lie in a vertical plane through the center of the disk whenever the disk is "leveled off" in a horizontal position by means of the disk level.

The other level is the corner spirit level, located at the lower left-hand corner of the frame and mounted on a hinge. This level is swung out at right angles to the frame whenever the protractor is to be used. It is used to keep the protractor in a vertical position for the accurate checking of the blade angle.

As shown in Fig. 17-39, the degree scale for the protractor is on the center disk and the vernier scale is on the ring just outside the disk. The vernier graduations have a ratio of 10 to 9 with the degree graduations; that is, 10 graduations on the vernier scale will match with 9 graduations on the degree scale. As shown in Fig. 17-40, the reading between the 0° point on the degree scale and the 0° point on the vernier scale is somewhat more than 15°. To find the amount in tenths of a degree, the vernier scale is read in the same direction from the 0° point on the vernier scale to the point where a vernier-scale graduation coincides approximately with a degree-scale graduation. In the

drawing of Fig. 17-40, this is a point 8 graduations to the left of the 0° point on the vernier scale and is read 0.8°. The total angle shown by the protractor is therefore 15.8°.

As pointed out above, the number of tenths of a degree in the blade angle is found by observing the number of vernier-scale spaces between the zero of the vernier scale and the vernier-scale graduation that comes closest to being in perfect alignment with a degree-scale graduation line. Always read tenths of degrees on the vernier scale in the same direction as the degrees are read on the degree scale.

How to Measure the Propeller Blade Angle

To measure the propeller blade angle, determine how much the flat side of the blade slants from the plane of rotation. If a propeller shaft is in the horizontal position when the airplane rests on the ground, the plane of propeller rotation, which is perpendicular to the axis of rotation or the propeller shaft, is vertical. Under these conditions, the blade angle is simply the number of degrees that the flat side of the blade slants from the vertical, as illustrated in Fig. 17-41.

However, an airplane may rest on the ground with its propeller shaft at an angle to the horizontal. The plane of propeller rotation, being perpendicular to the propeller axis of rotation, is then at the same angle to the vertical as the propeller shaft is to the horizontal, as represented by angle A in Fig. 17-42.

Under these conditions, the number of degrees that the flat side of the propeller blade slants from the vertical is the blade angle minus the ground angle of the airplane, or angle B in the same illustration. To obtain the actual blade angle, the ground angle of the airplane (which is also the angle at which the plane of rotation slants from the vertical) must be added to the angle at which the flat side of the blade slants from the vertical in the opposite direction, as represented by angle C in the same illustration.

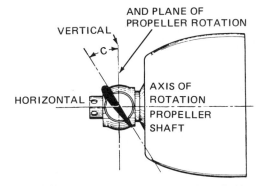

FIG. 17-41　Measuring a propeller's blade angle.

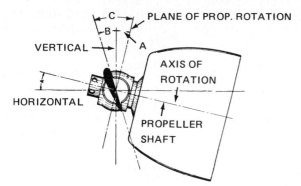

FIG. 17-42 Effect of ground angle on the measurement of blade angle.

Angles *A* and *B* are measured with a universal protractor and added in two related operations. It is then possible to read the total angle, or blade angle *C*, from the degree and vernier scales of the protractor.

Checking and Setting Blade Angles when the Propeller is on the Shaft

The following steps are recommended when a universal propeller protractor is available for checking propeller blade angles while the propeller is installed on the shaft of the engine. If it is necessary to use another type of protractor, the procedure to be followed will be modified.

1. Mark the face of each blade with a lead pencil at the blade station prescribed for that particular blade.

2. Turn the propeller until the first blade to be examined is in a horizontal position with its leading edge up.

3. Using a universal propeller protractor, swing the corner spirit level out as far as it will go from the face of the protractor.

4. Turning the disk adjuster, align the zeros of both scales and lock the disk to the ring by placing the spring-loaded pin of the disk-to-ring lock in the deep slot.

5. See that the ring-to-frame lock is released. By turning the ring adjuster, turn both zeros to the top. Refer to Fig. 17-39, which is a picture of the universal propeller protractor.

6. Hold the protractor in the left hand, by the handle, with the curved edge up. Place one vertical edge of the protractor across the outer end of the propeller retaining nut or any hub flat surface which is parallel to the plane of propeller rotation, which means that it is placed at right angles to the propeller-shaft center line. Using the corner spirit level to keep the protractor vertical, turn the ring adjuster until the center spirit level is horizontal. The zeros of both scales are now set at a point that represents the plane of the propeller rotation. This step can be understood better by referring to the illustration of the measurement of the blade angle in Fig. 17-43.

7. Lock the ring to the frame to fix the vernier zero so that it continues to represent the plane of propeller rotation.

8. Release the disk-to-ring lock by pulling the spring-loaded pin outward and turning it to 90°. This completes what is often called the **first operation,** shown in the left part of Fig. 17-43.

9. Change the protractor to the right hand, holding it in the same manner as before, and place the other vertical edge of the protractor, which is the edge opposite the first edge used, against the blade at the mark which was made by a pencil on the face of the blade. This is the beginning of the **second operation** and is illustrated by the picture of the protractor to the right in Fig. 17-43. Keep the protractor vertical by means of the corner spirit level. Turn the disk adjuster until the center spirit level is horizontal, as shown in the illustration. In this manner, the angle at which the flat side of the blade slants from the vertical is added to the angle at which the plane of rotation slants from the vertical in the opposite direction.

10. Read the number of whole degrees on the degree scale between the zero of the degree scale and the zero of the vernier scale. Read the tenths of a degree on the vernier scale from the vernier zero to the vernier-

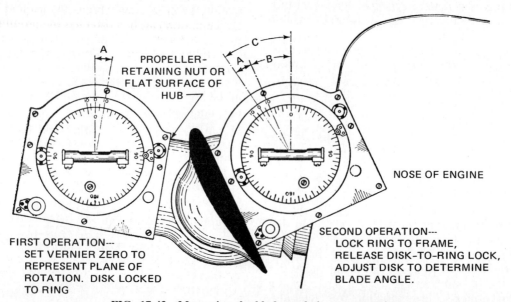

FIG. 17-43 Measuring the blade angle in two operations.

scale graduation that comes the closest to lining up with a degree-scale graduation. In this manner the blade angle is determined.

11. Obtain the required blade angle by making any necessary adjustments of the blade or the propeller pitch-changing mechanism.

12. Repeat this procedure for each of the remaining blades to be checked.

● MISCELLANEOUS INSPECTION AND MAINTENANCE

Special Inspections after Accidents

If the propeller strikes or is struck by any object, examine it for damage. Disassemble any propeller that has been involved in an accident, and carefully inspect the parts for damage and misalignment before using the propeller again. Examine all steel parts and otherwise serviceable steel propeller blades for minor injuries by means of a magnetic inspection supervised by competent personnel. Have aluminum-alloy blades which are otherwise serviceable given a general etching by competent operators.

Any accident which severely damages the propeller may also damage the engine. It is good practice, therefore, to check the alignment of the crankshaft after an accident in which the propeller has been damaged. Crankshaft alignment **runout** may be checked as follows:

1. Remove the propeller.

2. Install a dial gage on a mounting attached to the nose of the engine with the finger of the dial gage touching the smooth area on the outer rim of the flange on either the aft side or the forward face of the flange.

For a spline or taper shaft, install a dial gage on the nose of the engine and place the finger of the gage on the smooth surface forward of the spline or taper.

3. Rotate the propeller shaft through a complete revolution, and observe the movement of the gage indicating needle. If the shaft runout is out of limits according to the manufacturer's specifications, the engine must be removed and overhauled.

Front Cones

Since hub front-cone halves are machined in pairs, the original mated halves are always used together in the same installation. If one half becomes unserviceable, both halves are rejected. Before installation and use, the two halves of a front cone are held together by a thin section of metal left over from the manufacturing process. This metal must be sawed through with a hacksaw and the two separated halves gone over carefully with a handstone to remove all rough and fine edges and to round off the sharp edges where the cones were cut apart. After this process is completed, the two halves are always taped together when not installed.

Front-Cone Bottoming

A front cone sometimes **bottoms** against the outer ends of the propeller-shaft splines; that is, the apex of the front cone hits the ends of the splines before the cone properly seats in its cone seat in the hub. The hub is loose because it is not seated properly and held tight by the cones, even though the retaining nut may be tight.

Whenever a splined hub is found to be loose, even though the retaining nut is tight, an inspection is made for front-cone bottoming unless there is a more probable cause of the trouble. Also, this condition may be manifested by excessive propeller vibration during preflight operations.

Inspection for Front-Cone Bottoming

To check for front-cone bottoming, first apply a thin coating of Prussian blue to the apex of the front cone. Then install the propeller on the shaft and tighten the propeller retaining nut. Next, remove the retaining nut and front cone. See if the Prussian blue has been transferred to the ends of the splines of the propeller shaft. If it has not been transferred, the front cone is not bottoming and the Prussian blue can be cleaned off.

If the Prussian blue has been transferred to the ends of the shaft splines, install a steel spacer behind the rear cone to correct the condition of front-cone bottoming. Spacers for this purpose are generally made in any shop adjacent to the place where the work is being performed and are $\frac{1}{8}$ in [3.18 mm] thick.

The presence of the spacer moves the entire propeller assembly forward, causing the front cone to seat in the hub before its apex hits the end of the shaft spline. After the installation of the spacer, the Prussian-blue test should be made again. If bottoming is still indicated, inspect the hub-shaft end and all attaching parts for excessive wear or any other condition that might cause a failure to fit. Worn or defective parts should be replaced.

Rear-Cone Bottoming

Occasionally a situation will exist where the front edge of a rear cone may bottom against the ends of the splines in the propeller hub. This condition is caused by wear of both the cone and the cone seat in the hub due to prolonged service and will prevent the cone and cone seat from being firmly engaged. If inspection shows that the front of the rear cone is touching the splines, the condition can be corrected by carefully removing not more than $\frac{1}{16}$ in [1.59 mm] of material from the front edge (apex) of the cone or replacing the cone with a new one.

Shaft and Hub Splines

The splines on the propeller shaft and inside the propeller hub should be carefully inspected for damage and wear. Wear of the splines should be checked with a single key no-go gage made to plus 0.002 in [0.05 mm] of the base drawing dimensions for spline land width. If the gage enters more than 20 percent of the spline area, the part should be rejected.

Balancing Controllable Propellers

Upon completion of repairs, the horizontal and vertical balance of a propeller must be checked. If any unbalanced condition is found, correction must be made according to the manufacturer's instructions. Balancing methods include the installation of weights in the shanks of the blades, lead wool packed into holes drilled in the ends of the blades, lead packed into hollow bolts, and various others. In any event, the manufacturer's recommendations must be followed for any specific type of propeller. For some propellers, only the manufacturer is permitted to perform the balancing operations.

Anti-icing and Deicing Systems

Propeller anti-icing may be accomplished by spraying isopropyl alcohol along the leading edges of the blades. The anti-icing fluid is carried in a reservoir in the airplane in sufficient quantities for any possible demand. The fluid is pumped from the reservoir to the propeller by an electrically driven **anti-icing pump** which is controlled from the cockpit. The propeller is equipped with a **slinger ring** having nozzles aligned with the leading edge of each blade. When the pump is turned on, the fluid is forced out the nozzles of the slinger ring by centrifugal force and carried along the leading edges of the propeller blades.

Another anti-icing and deicing system which is extensively used is the heating of the blades by electric heating elements. These elements are centered in boots to the leading edge of the blades or are mounted within the blade. Power for the heating elements is transferred through slip rings at the rear of the propeller hub.

Servicing, inspection, maintenance, and repair of anti-icing and deicing systems for propellers will vary with different propellers and aircraft. It is therefore essential that the technician performing these operations follow the instructions furnished by the manufacturer of the aircraft and the manufacturer of the propeller.

● REVIEW QUESTIONS

1. Why is a wood propeller made of laminations?
2. Describe briefly the construction of a wood propeller.
3. Why is sheet metal installed on the tip and leading edges of a wood propeller?
4. What is the purpose of the small holes drilled in the metal tipping at the tip of a wood propeller?
5. Why is fabric sometimes applied to the tips of wood propellers?
6. Describe the procedure for checking and correcting horizontal and vertical balance of a wood propeller.
7. How would you check the track of a propeller?
8. What is the purpose of the cones used with the installation of a propeller on a splined shaft?
9. Why is a rear-cone spacer used with some installations?
10. How does the retaining nut for a propeller serve as a puller when the propeller is removed?
11. Explain how the fit of a tapered hub and shaft is checked.
12. What precaution must be taken when the hub bolts of a wood propeller are tightened?

13. What authorization is required with respect to the major repair or alteration of a propeller?
14. What type of repair is that which does not materially affect the strength, weight, balance, or performance of a propeller?
15. What defects should be noted during the inspection of a wood propeller?
16. What materials are commonly employed in the construction of composite propeller blades?
17. Describe the construction of a typical composite propeller blade.
18. List the causes for rejection of a wood propeller.
19. Describe the repair of small cracks parallel to the grain in a wood propeller.
20. What is the principal advantage of a ground-adjustable propeller?
21. How is the blade-angle adjustment made on a ground-adjustable propeller?
22. Describe the pitch-changing mechanism for the Beechcraft series 215 propeller.
23. Describe the feathering system for the Beechcraft series 279 propeller.
24. Explain the operation of the governor used with the Hartzell full-feathering propeller.
25. Who is authorized to repair damaged steel propeller blades?
26. How may minor injuries to the leading and trailing edges of a steel blade be repaired?
27. Define a damaged propeller blade.
28. What inspection should be made on an aluminum-alloy propeller blade which has been straightened?
29. Describe the method for inspecting cracks in an aluminum-alloy blade.
30. What is the purpose of local etching?
31. What material is used for local etching?
32. What minor damage in a composite propeller blade may be repaired in the field?
33. Describe a typical repair of minor damage in a composition propeller blade.
34. Describe the treatment of minor surface defects.
35. Discuss the shortening of blades to remove defects.
36. What device is used for checking propeller blade angle?
37. Describe the procedure for checking blade angle when the propeller is mounted on the aircraft.
38. After a propeller is damaged in an accident, what inspections should be made?
39. What inspection should be made on the engine after the propeller has been damaged in an accident?
40. How would you check a propeller for *front-cone bottoming*?
41. How may the condition known as *rear-cone bottoming* be corrected?
42. Describe methods for propeller anti-icing and deicing.
43. What material is commonly used as a propeller anti-icing fluid?

18 PROPELLERS AND TURBOPROPELLERS FOR LARGER AIRCRAFT

The number of large aircraft utilizing propellers for thrust has been decreasing steadily since the advent of turbojet and turbofan engines; however, there are still many propeller aircraft operating throughout the world, and it is necessary for the technicians working on such aircraft to be familiar with the operation, service, and maintenance of the propellers. In this chapter we shall examine typical large-aircraft propellers which have been used extensively in the past and are still in use to a substantial degree.

Among the propellers which we shall examine are the Hamilton Standard constant-speed counterweight propeller, the Hamilton Standard hydromatic reversing propeller, the Allison Turbopropeller used with the Allison Model 501-D13 turboprop engine, the Hartzell reversible propeller used with the PT6A turboprop engine, and the Dowty Rotol turbopropeller used on the Swearingen Metroliner with a Garrett turbine engine. These propellers are still being utilized throughout the world on a variety of aircraft.

Although there are other makes and models of propellers in use, the principles explained in this section should provide the technician with sufficient understanding to interpret and use operation and service instructions for any propeller.

● HAMILTON STANDARD COUNTERWEIGHT PROPELLERS

Two-Position Propeller

In Chap. 17 we described the operation of two-position counterweight propellers for small aircraft. These propellers are also designed for larger aircraft, so we will review the basic principles.

Although the Hamilton Standard counterweight propellers are no longer in production, there are still some aircraft which utilize these propellers. For this reason, it is useful for the aviation technician to have some knowledge regarding their operation.

As explained in Chap. 17, the two-position propeller is controlled from the cockpit by means of a three-way valve operated by the pilot. The valve directs engine oil pressure to the propeller cylinder to cause the pitch to decrease. When the valve position is changed to allow the oil to flow back to the engine, the counterweights rotate the blades to high pitch.

The range of pitch change for the two-position propeller is about 10°. The low-pitch position is used for takeoff so that the engine can develop its maximum rpm. After takeoff and climb, the propeller is placed in the high-pitch position for cruising.

The operation of the two-position propeller is illustrated in Fig. 17-14. An examination of this illustration will help the student to understand the principles of operation.

Constant-Speed Counterweight Propeller

The constant-speed counterweight propeller is essentially the same as the two-position propeller with the exception of the blade-angle range and the controlling mechanism. The constant-speed propeller may have a range of either 15 or 20° depending on the type of installation. This range is determined by the adjustment of the stops in the counterweight assembly. If the propeller is set for a 20° range, a return-spring assembly is installed in the piston to assist the counterweights in returning the cylinder to the rearward position when the governor calls for increased pitch.

The control of the constant-speed propeller is accomplished by means of a propeller governor similar to those described previously. The governor operates by means of flyweights which control the position of a pilot valve. When the propeller is in an underspeed condition, with rpm below that for which the governor is set, the governor flyweights move inward and the pilot valve then directs engine oil pressure to the propeller cylinder through the engine propeller shaft. This moves the cylinder forward and reduces the propeller pitch. When the engine is in the overspeed condition, the governor action is opposite and the pilot valve allows oil to drain from the cylinder back to the engine.

The propeller control is in the cockpit of the airplane and is marked for INCREASE RPM and DECREASE RPM. **Increase rpm** means lower pitch, and **decrease rpm** means higher pitch. When the propeller is operating in an onspeed condition, the blade angle is usually somewhere between the extreme ranges. Control of the governor is accomplished by rotating a shaft through a cable linkage. Rotation of the shaft changes the compression of the governor speeder spring which controls the flyweight position.

● THE HAMILTON STANDARD HYDROMATIC PROPELLER

In addition to the counterweight-type propellers, the Hamilton Standard Division of the United Aircraft Corporation, now United Technologies Corporation, manufactured the hydromatic constant-speed propeller. The original constant-speed propeller was further developed into the constant-speed, full-feathering Model 23E50 propeller and the reversing Model 43E60 propeller.

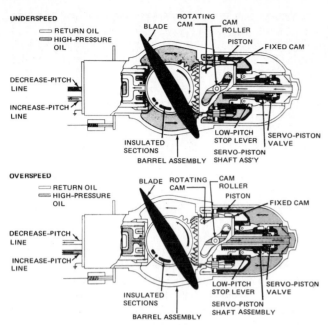

FIG. 18-1 Operation of the hydromatic pitch-changing mechanism. *(Hamilton Standard)*

It will be noted that the propeller is composed of three major assemblies: the **hub assembly,** the **dome assembly,** and the **low-pitch stop-lever assembly.** In the non-reversing type of constant-speed hydromatic propeller which has the full-feathering capability, the **distributor valve** is in the place of the low-pitch stop-lever assembly.

The **hub assembly** includes the blade assemblies, the spider, and the barrel assembly. At the butt of the blade assemblies are sector gears which mesh with the gear in the dome assembly. The spider is a forged and machined steel unit incorporating arms which extend into and provide bearing support for the blade shanks. The barrel assembly provides the structure to contain the spider and blade butts and holds the hub assembly together.

The **dome assembly** incorporates the piston, fixed cam, rotating or moveable cam, and support for other required units. The bevel pitch-changing gear which meshes with the sector gears on the blade butts is on the rotating cam.

The **low-pitch stop-lever assembly** provides the means for maintaining a minimum low-pitch for the propeller blades until the reversing process is called for. Figure 18-1 illustrates the operation of the hydromatic reversing propeller.

Principles of Operation

The pitch-changing mechanism for both underspeed and overspeed conditions is shown.

The forces acting to control the blade angle of the propeller are **centrifugal twisting moment** and **high-pressure oil.** The centrifugal twisting moment tends to turn the blades to a lower angle. The high-pressure oil is directed to the propeller to change the blade angle in either direction.

The pitch-changing mechanism consists of a piston which moves forward and backward in the dome cylinder. Rollers on the piston engage cam slots in the cams. The cams are so arranged, one within the other,

that the rotation of one cam is added to the rotation of the other, thus doubling the movement that would be obtained from one cam alone. The outer cam is stationary, and the inner cam is rotated. This inner cam carries the bevel gear which meshes with the blade gears.

The blade angle of the propeller is controlled by means of a **double-acting governor** during constant-speed operation. The governor directs high-pressure oil to either side of the propeller piston as necessary to maintain the constant engine speed. If an underspeed condition exists, the oil is directed to the inboard side of the piston. This causes the piston to move outward and rotate the cam in a direction which will decrease the pitch of the blades. The opposite action takes place if an overspeed condition exists. Only the center portion of the cam slots is utilized during constant-speed operation.

To **reverse** the blade angle of the propeller, the blades are rotated through the low-pitch position. The reversing operation makes use of the high-pressure oil from the feathering pump to force the piston to the full-forward position.

When it is desired to reverse the propellers after landing, the pilot pulls the throttles backward to the reverse position. The throttles actuate switches which set the system into operation.

The **pilot valve** of the governor is provided with two oil chambers which permit an artificial overspeed or underspeed condition to be imposed upon the governor. A solenoid valve on the governor is used to control the oil flow which actuates the pilot valve for the artificial conditions. When the solenoid is energized, as in reversing, it directs high-pressure oil to the positioning chamber and causes the pilot valve to move downward into the underspeed condition. The oil pressure overrides the normal control of the governor flyweights. When the pilot valve is in the DOWN position, it directs oil to the rear side of the propeller piston, causing it to move forward and reduce the propeller blade angle.

Normally the low-pitch angle of the propeller is limited by the **low-pitch stop levers;** however, when the propeller is reversed, these stop levers are released through the action of a servo piston in the low-pitch stop-lever assembly. The servo piston is actuated when the high-pressure oil from the feathering pump and governor pump reaches a pressure of 250 psi [1723.75 kPa] and opens the servo valve, directing oil to the servo piston. The servo piston then travels forward, removing the lever wedge and allowing the release of the stops, thus permitting the propeller piston to move into the full-forward, or reverse-pitch, position. During reversing, the forward high-angle section of the cam slots is utilized.

Unreversing the propeller is accomplished by moving the throttle forward from the reverse position. This movement actuates switches which deenergize the governor solenoid and start the auxiliary or feathering pump. The unreversing operation is essentially the same as the feathering operation, since both call for an increase of positive pitch.

Feathering of the propeller is accomplished by directing high-pressure oil from the feathering pump to the forward side of the propeller piston. This is accomplished through the governor pilot valve. The high-pressure oil positions the pilot valve in an artificial overspeed condition, thus causing the oil flow to go to the forward side of the propeller piston.

The pilot starts the feathering operation by pushing in on the feathering button. This button is part of a cockpit-mounted feathering solenoid-switch assembly which, when closed, will remain closed until the propeller is feathered. At this time a pressure of approximately 650 psi [4481.75 kPa] is built up at the forward side of the propeller piston and thus at the governor cutout switch. The cutout switch is in the ground circuit of the feathering-switch holding coil and is opened by the oil pressure at approximately 650 psi. When the cutout switch opens, the feathering switch is released and the feathering operation is complete. During feathering, the rearward high-angle section of the cam slots is utilized.

Unfeathering is accomplished by holding the feathering button in the OUT position for about 2 s. This creates an artificial underspeed condition at the governor and causes high-pressure oil from the feathering pump to be directed to the rear of the propeller piston. As soon as the piston has moved outward a short distance, the blades will have sufficient angle to start rotation of the engine. When this occurs, the unfeathering switch can be released and the governor will assume control of the propeller.

The Propeller Governor

The double-acting propeller governor used with the reversing propeller, illustrated in Fig. 18-2, consists of a set of spring-loaded flyweights driven by the engine, a pilot valve actuated by the flyweights and high-pressure oil, a gear pump, and an electrically driven governor head which regulates the governor for constant-speed operation.

The centrifugal force acting on the flyweights is the primary regulating force of the governor. When the propeller overspeeds, the flyweights move outward and raise the pilot valve. This opens passages which direct governor oil pressure to the outer or forward side of the propeller piston. The piston moves rearward and rotates the blades to a higher pitch, thus reducing the engine speed.

When an underspeed condition exists, the speeder spring holds the governor flyweights in an inward po-

sition. This places the pilot valve in a DOWN position and reverses the conditions of oil flow described for the overspeed condition. The propeller piston is moved forward and the blade angle is decreased, thus allowing for engine speed to increase.

In connection with the governor is a solenoid valve which controls high-pressure oil to the positioning chambers of the pilot valve. When the solenoid valve is open, the pilot valve is caused to move to the DOWN position. When the solenoid valve is closed, high-pressure oil from the feathering pump will raise the valve. As stated previously, the UP position of the pilot valve is an overspeed condition and the DOWN position of the valve is an underspeed condition.

Typical Hydromatic Propeller Assembly

An exploded view of the principal components and installation requirements for a Hamilton Standard Model 23E50 constant-speed full-feathering propeller is shown in Fig. 18-3. This drawing illustrates the additional parts required when the propeller is to be installed on an engine shaft.

It will be noted that this model includes a **distributor valve,** which is utilized only for the nonreversing, constant-speed propeller with full-feathering capability. The function of the distributor valve is to direct high-pressure oil to the rear of the piston in the dome when it is desired to unfeather the propeller.

The feathering process on the 23E50 propeller is nearly the same as on the 43E60 propeller except that the piston moves *rearward* in the dome of the 23E50 propeller to decrease pitch whereas it moves *forward* in the 43E60 propeller to decrease the pitch. When the feather button in the cockpit is depressed, a holding circuit to hold the button in is completed through the pressure cutout switch. Depressing the button also completes the circuit to the feather pump motor, which picks up oil from the main oil system and pumps it to the propeller governor through an external oil line. The feather pressure pump repositions the pressure transfer valve in the governor, allowing the high-pressure oil to bypass the governor and pass through the propeller shaft to the distributor valve and into the rear side of the piston. Auxiliary pressure moves the piston forward until it reaches the feather stop. The feathering process is then completed by the system pressure building above the pressure-cutout-switch setting. When the cutout switch opens, the holding circuit to the switch is broken and the button automatically pops out, stopping the feather motor. Normally, when the system pressure increases above 300 psi [2068.4 kPa] the switch will open.

To unfeather the propeller, the button is also pressed, but the button must be manually held in to overcome the pressure cutout switch. With the button held in, the feather pump pressure is approximately 600 psi [4136.8 kPa]. The high pressure that is exerted on the rear side of the distributor valve moves the distributor valve forward, rerouting the pressure to the front of the piston, moving the piston to the rear, and rotating the blades out of the feather position. As soon as the blades move from the full-feather position the propeller will begin to rotate (windmill) and the governor will resume normal governing operation.

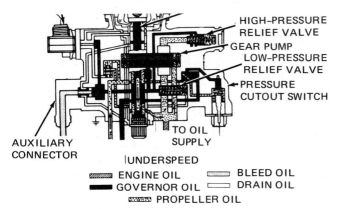

FIG. 18-2 Drawing of the propeller governor.

HIGH-PRESSURE
RELIEF VALVE
GEAR PUMP
LOW-PRESSURE
RELIEF VALVE
PRESSURE
CUTOUT SWITCH

AUXILIARY
CONNECTOR

TO OIL
SUPPLY

UNDERSPEED

ENGINE OIL — BLEED OIL
GOVERNOR OIL — DRAIN OIL
PROPELLER OIL

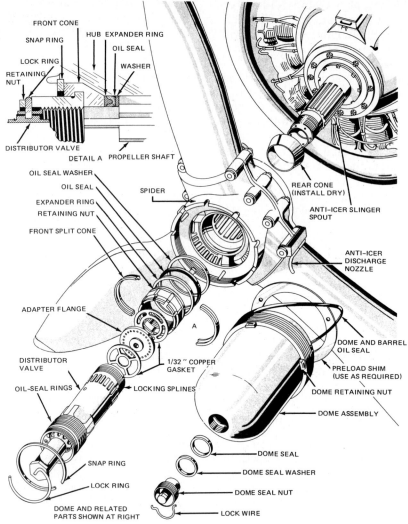

FRONT CONE
SNAP RING
HUB EXPANDER RING
OIL SEAL
LOCK RING
RETAINING NUT
WASHER
DISTRIBUTOR VALVE
DETAIL A   PROPELLER SHAFT
OIL SEAL WASHER
OIL SEAL
EXPANDER RING
RETAINING NUT
FRONT SPLIT CONE
SPIDER
ADAPTER FLANGE
DISTRIBUTOR VALVE
OIL-SEAL RINGS
1/32" COPPER GASKET
LOCKING SPLINES
A
SNAP RING
LOCK RING
DOME AND RELATED PARTS SHOWN AT RIGHT
REAR CONE (INSTALL DRY)
ANTI-ICER SLINGER SPOUT
ANTI-ICER DISCHARGE NOZZLE
DOME AND BARREL OIL SEAL
PRELOAD SHIM (USE AS REQUIRED)
DOME RETAINING NUT
DOME ASSEMBLY
DOME SEAL
DOME SEAL WASHER
DOME SEAL NUT
LOCK WIRE

FIG. 18-3 Typical hydromatic propeller assembly.

As can be seen from Fig. 18-3, the sequence of events in installation would be as follows:

1. Install the rear cone, dry, on the splined propeller shaft.

2. Install the propeller (hub and blades) on the shaft.

3. Install the split front cone on the retaining nut and install both with required seals on the shaft. Torque according to instructions.

4. Install the distributor valve in the dome and secure with a snap ring and lock ring.

5. Place the adapter flange with copper gasket on each side of the inside gear at base of dome. Install the dome on the propeller with necessary preload shims and tighten the dome retaining nut to the correct torque.

6. With the propeller in low pitch, fill the dome with engine oil and install the dome-seal nut with dome-seal washer and dome seal. Secure with lockwire.

Installation instructions and some maintenance instructions will vary with different types of propellers. The technician must always follow the instructions provided for the particular installation.

Automatic Synchronization

Any system used for synchronizing rotating units requires that a reference speed be established. In the synchronizing system used with the Hamilton Standard hydromatic reversing propeller, one of the aircraft engines is used as a "master" to set the pace for the other engines. Figure 18-4 is a schematic diagram of the synchronization circuit for the Hamilton Standard reversing propeller.

On the four-engine installation, the system is arranged so that either one of two engines may be used as the master engine. This provides a safety factor in case the engine being used should fail. On the Douglas DC-6 airplane, the system is arranged so that either the no. 2 or 3 engine may be used as a master engine. The other engines are called "slave" engines. The master engine is selected by means of a toggle switch.

The "speed signal" for each engine is taken from the three-phase tachometer generators or alternators. This signal is in the form of a three-phase alternating current having a frequency proportional to engine speed. When this current is applied to the stator winding of a three-phase motor, it will establish a rotating field.

The output of the tachometer generator of the master engine is fed to the stators of the **differential motors** for the slave engines. The output of the tachometer generator of each slave engine is fed to the rotor of the differential motor for that engine. If the master engine and a slave engine are operating at the same

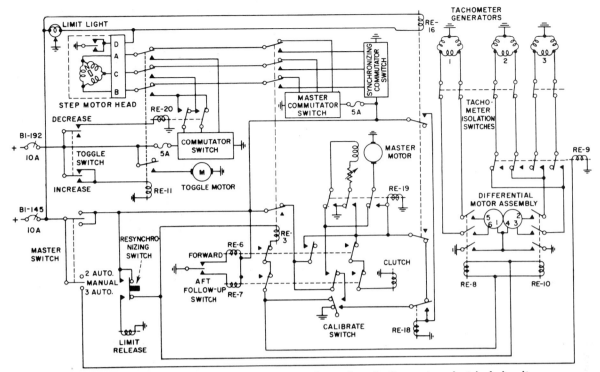

FIG. 18-4 Schematic diagram of the automatic synchronization system electrical circuit.

speed, the fields produced in the differential motor rotor and stator will be rotating at the same speed. Under these circumstances the rotor of the differential motor will remain stationary. If the master engine and the slave engine are turning at different speeds, the rotor and stator fields in the differential motor will be rotating at different speeds. This will cause the rotor to turn at a speed necessary to lock the two fields together. Thus, when any slave engine is not synchronized with the master engine, the differential motor for the slave engine will rotate. Each differential motor for a slave engine drives a commutator switch which is connected to the stator of the governor step motor. The governor step motor is the actuating unit which adjusts the force applied to the governor speeder spring for higher or lower engine rpm.

The **commutator switch** consists of three sets of contacts actuated by a cam with lobes spaced 120° apart. The contacts are arranged so that a single center contact is alternately pressed against the contact at each side. Thus, the center contact will be alternately positive and negative. The output of the commutator switch is connected to the three terminals of the delta-wound stator in the governor step motor. The effect is to produce a field in the stator which rotates by steps and turns the rotor in step with the field.

In some synchronizing systems, the differential motor, which rotates as a result of a difference in engine speeds, is replaced by a mechanical differential-gear system. The output of the tachometer generators is directed to synchronous motors. For each slave engine, one motor, connected to the differential gear, is driven by the master signal. The other motor is driven by the slave-engine signal. If the engines are synchronized, there is no differential in speed and no rotation at the differential-gear output shaft. The output shaft is connected to the commutator-switch drive and drives

the switch in the same manner as the differential motor previously described.

The operation of the automatic synchronization system may be summarized as follows.

The master engine is selected by the pilot to provide a reference speed for the slave engines. The tachometer alternator current from the master engine induces a rotating field in the stators of differential motors for each slave engine. This field reacts with the rotor field of each differential motor, and when there is a difference in speed, the differential motor will rotate the commutator switch for the "off-speed" slave engine. The output of the commutator switch causes the governor step motor to rotate and adjust the governor for a change of propeller blade angle to correct the engine speed.

The propeller-control system has a number of safeguards to provide for operation in case some part of the synchronization system fails. Control can be taken away from the synchronization system at any time either by individual toggle switches or by a master switch. The system is designed to prevent the slave engines from following the master engine in case the master engine changes speed more than 3 percent. This is accomplished through devices called **limited-band mechanisms,** which permit an automatic change of only 60 to 70 rpm at the engine.

● THE ALLISON TURBOPROPELLER

The Allison (formerly Aeroproducts) A6441FN-606 and -606A turbopropeller shown in Fig. 18-5 is a hydraulically controlled four-blade unit incorporating an integral hydraulic system. The hydraulic governing system operates independently of any of the other systems and maintains precise control during all operating conditions. Electrical power is supplied to the propel-

415

FIG. 18-5   The Allison A6441FN-606 turbopropeller.
(Allison Gas-Turbine Operations, General Motors Corp.)

ler for synchronization and ice control. The propeller is designed for use with the Allison Model 501-D13 turboprop (prop jet) engine as shown in Fig. 18-6.

Some of the important design features of the turbopropeller are as follows:

1. Four hydraulic pumps, driven by propeller rotation, provide the required flow and pressure to maintain propeller control.

2. An electrically driven feather pump provides hydraulic pressure for static operation as well as for feathering or unfeathering.

3. Autofeathering, manual, and emergency feathering systems are provided.

4. Safety devices are incorporated for protection against excessive drag in the event of engine or propeller malfunctions.

5. The propeller provides uniform variation of thrust throughout the **beta range** with power lever movement. The Greek letter **beta** is used to denote blade angle. Beta range denotes that range of operation in which beta (blade angle) is scheduled.

6. The propeller is guaranteed up to at least 4500 shaft horsepower (shp) [3355.65 kW] at takeoff and 2500 shp [1864.24 kW] at maximum reverse.

7. The propeller system is designed for primary reliability and fail-safe operation.

8. The propeller can be feathered at any rotational speed up to the propeller design limitation of 142 percent of rated rpm.

FIG. 18-6   Turbopropeller mounted on an Allison 501-D13 engine. *(Allison Gas-Turbine Operations, General Motors Corp.)*

9. There is no limit to the number of times a propeller can be feathered or unfeathered in flight.

10. The propeller is designed to have an increase blade-angle change rate of 15°/s and a decrease blade-angle change rate of 10°/s at 1020 propeller rpm.

11. The feather pump will provide a static blade-angle change of 3°/s.

12. When signaled to feather during operation, the initial blade-angle change rate is well in excess of 15°/s but decreases as the rpm decreases.

Leading Particulars

Number of blades	4
Propeller diameter	13.5 ft [4.11 m]
Blade chord	18.5 in [46.99 cm]
Governing speed	1020 prop rpm
Total weight installed including hydraulic fluid and grease	1038 lb [470.83 kg] max

Aerodynamic Blade Angles at 42-in R Station

Maximum reverse	−4°
Ground idle	1.5°
Start	7°
Flight idle	20°
Feather	94.7° + 0.2 − 0.1
Beta follow-up at takeoff	31.5° + 0.5 − 1.0
Mechanical low-pitch stop	18.25 to 18.5°
Hydraulic low-pitch stop	20°

Major Assemblies and Components

The propeller consists of five major assemblies mounted on the engine propeller shaft plus the control components and assemblies mounted in the aircraft. These major assemblies are (1) *hub*, (2) *blade and retention*, (3) *regulator*, (4) *feather reservoir*, and (5) *spinner*. The engine-driven propeller alternator, electronic controls for synchronization and phase synchronization, ice-control accessories, and necessary switches, relays, etc., complete the installation. The major assemblies are illustrated in Fig. 18-7.

Hub Assembly. The hub assembly is the principal structural member of the propeller and consists of the **hub, torque units,** and **master gear assembly.** A cross-sectional drawing of the hub assembly is shown in Fig. 18-8. Note that each blade has a separate pitch-changing device called the **torque unit** and that the four blades are all meshed with the **master gear.** The master gear therefore ensures that all blades are at the same angle.

The hub, machined from a steel forging, provides mounting for the pitch-control mechanisms and the sockets for the retention of the blades. Splines and cone seats are machined on the inner diameter of the hub. The splines and cone seats provide the means of properly positioning the hub on the propeller shaft. The hub contains hydraulic fluid passages which deliver oil, under pressure, from the regulator to the torque units, master gear assembly, and feather reservoir. The hub is shot-peened to increase the fatigue life, and it is cadmium-plated to prevent surface corrosion.

Each hub socket contains a **torque unit** which consists principally of a fixed spline, a piston, and a cylinder. A retaining bolt, which secures the fixed spline to the hub, incorporates a tube to provide a passage

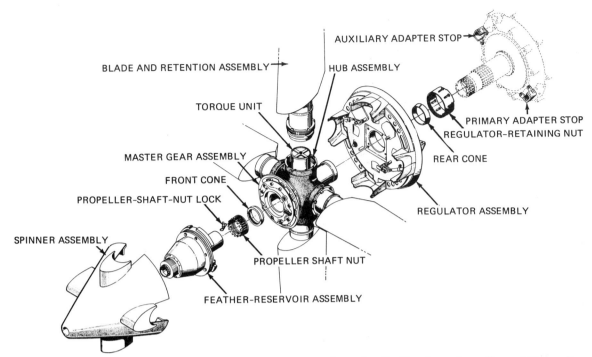

FIG. 18-7 Major assemblies of the Allison turbopropeller. *(Allison Gas-Turbine Operations, General Motors Corp.)*

for fluid to the outboard or **decrease-pitch** side of the piston. A port in the base of the fixed spline provides an oil passage to the inboard or increase-pitch side of the piston. Helical splines, machined on the components of the torque unit, convert linear motion of the piston to rotation of the cylinder. Cylinder rotation is transmitted to the blade through an indexing ring and matching splines on the cylinder and in the blade root. Figure 18-9 illustrates the construction of the torque-unit assemblies.

The **master gear assembly** consists of a housing, master gear, mechanical low-pitch stop (MLPS), mechanical pitch lock, air shutoff, and feedback mechanism parts. The housing is splined and bolted to the front face of the hub and provides the mounting surface for the feather reservoir. The master gear coordinates the movement of the torque units so that all blades are maintained at the same aerodynamic blade angle. Splined to the housing are the parts which constitute the MLPS and the mechanical pitch lock. Feedback mechanism parts, actuated by blade-angle changes, operate the cooling-air shutoff mechanisms, beta light switch, and feedback shaft which mechanically positions a part of the control linkage of the hydraulic governor.

As the blade angle approaches the feathered position, the master-gear-assembly feedback mechanism

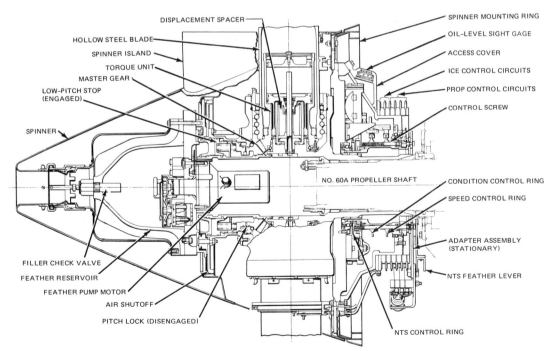

FIG. 18-8 Cross section of the hub assembly of the Allison A6441FN-606 turbopropeller. *(Allison)*

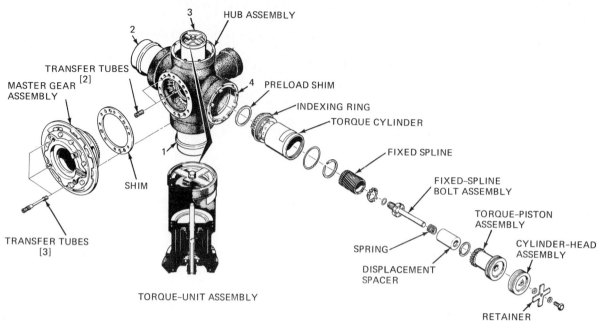

FIG. 18-9 Hub assembly showing the torque unit. *(Allison)*

drives the air shutoff shutter to its closed position. The shutter begins to move toward the closed position at 60° blade angle, and when the blade angle reaches approximately 75°, all airflow through the spinner is shut off. The air shutter is closed to prevent rapid cooling of the propeller operating mechanisms.

The beta light switch controls the cockpit beta light circuit. When the blade angle is below 18.5°, the beta light switch is closed by the master-gear-assembly feedback mechanism. When the switch is closed, the beta light is on in the cockpit to indicate that the propeller is operating in the beta range. At blade angles above 18.5°, the beta light switch is open and the beta light is out. This indicates that the propeller is not operating in the beta range; that is, the governor is in control of rpm.

Blade and Retention Assemblies. Each blade and retention assembly consists of a blade cuff, deicing element, cuff deicing slip ring, and integral ball-bearing set and is retained in its hub socket by a blade retaining nut.

The hollow steel blade consists of a **camber sheet** and **thrust member** which are brazed together. The camber sheet, formed from sheet steel, completes the camber surface of the blade. The thrust member, machined from a steel forging, constitutes the blade shank thrust face, longitudinal strengthening ribs, and leading- and trailing-edge reinforcements. Prior to brazing, the interior surfaces of the thrust member and camber sheet are ground and polished for maximum fatigue strength. The camber sheet and thrust member are brazed together along the ribs and leading and trailing edges. This basic design provides maximum aerodynamic contours coupled with excellent structural qualities. The external surface of the blade is zinc-plated and passivated electrochemically to obtain maximum corrosion protection. In addition to this, a portion of the blade tip and leading edge is "dull" chrome-plated for abrasion resistance. Integral ballbearing races are machined, locally hardened, and ground on the root

portion of the blade. The outer races are ground and precision-fitted with balls and separators to obtain a blade-bearing retention with maximum service life and optimum load distribution.

The foamed plastic structure of the cuff is covered by a fiberglass shell, neoprene fabric, and ice-control element which is flush-mounted.

Regulator Assembly. The regulator assembly which is mounted on the rear of the hub consists of the cover and housing assembly, pumps, valves, and hydraulic fluid required to provide controlled flow to the hub pitch-change and -control mechanisms. These parts, except for portions of the adapter assembly, are contained in the regulator cover and housing. The regulator assembly, except for the adapter assembly, rotates with the propeller. A drawing of the regulator assembly is shown in Fig. 18-10. Observe in particular the **shoes** (speed shoe and condition shoes) which straddle the control rings mounted on the front of the engine. As the rings are moved forward and backward in response to control from the airplane through the control screws and gears, the shoes are also moved backward and forward as they rotate around the edges of the rings. The movement of the shoes transmits signals to the components of the regulator.

When the propeller assembly is placed on the propeller shaft, two engine-mounted adapter stops engage with tangs on the adapter assembly. This prevents rotation of the adapter assembly when the propeller rotates. The adapter assembly is supported by two ball-bearing assemblies in the regulator. Seals, located in the regulator cover and housing, contact the adapter assembly to retain the hydraulic fluid.

The **adapter assembly** primarily consists of the mechanical control components which are coupled to the valves and pump power gear in the regulator and the accessory plate outside the regulator. Mechanical control movements from the engine or cockpit are transferred into the regulator to control the hydraulic operation of the propeller. This mechanical control is

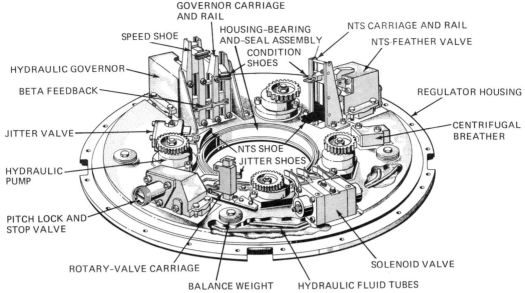

FIG. 18-10 **Propeller regulating assembly.** *(Allison)*

accomplished by movement of the propeller condition lever or negative torque signal (NTS) feather lever.

The accessory plate, which is attached to the adapter assembly, supports the brush block assemblies, the rotary actuator, a solenoid stop, and adapter stop tangs. Two brush block assemblies are provided, one for ice control and the other for control of the solenoid valve, beta light, and feather pump. Electric power is supplied to and conducted through the brush block assemblies to the slip rings mounted on the regulator cover. Electric power is also supplied to the rotary actuator.

Feather Reservoir Assembly. The feather reservoir assembly, installed on the forward face of the hub, consists of a feather motor, pump, pressure-control valve, check valve, and filler check valve mounted in a housing and cover assembly. A sufficient quantity of hydraulic fluid is available in the feather reservoir to accomplish feathering in the event that the regulator hydraulic supply is depleted. This assembly also supplies hydraulic flow for stationary pitch-change operation and for completion of the feathering cycle. Air is admitted through the spinner nose and directed over the cover for cooling the hydraulic fluid.

Spinner Assembly. The **spinner** is a one-piece aluminum-alloy assembly. It provides for the streamlined flow of air past the blade cuffs and over the feather reservoir, hub, and regulator. The spinner is positioned and supported by dowels on the regulator spinner mounting ring and retained to the feather reservoir by a nut in the nose of the spinner. Spray-mat-type ice-control circuits are on the external surface of the spinner. A schematic diagram of the anti-icing and deicing circuits for the spinner and cuffs is shown in Fig. 18-11.

Mechanicial Control System

The propeller must receive the following intelligence in order for it to function as desired: (1) power-lever position, (2) emergency handle position, (3) feather solenoid energized or deenergized, and (4) NTS system actuated or not actuated.

The foregoing intelligence is provided the propeller by means of two linkage systems. One system is attached to the propeller condition-control lever, and the other to the propeller NTS feather lever.

The power lever and emergency handle are each connected by aircraft linkage to the engine **coordinator.** Linkage from the coordinator is attached to the propeller condition-control lever. The purpose of the coordinator is to receive signals from the emergency handle and the power lever and transmit these signals electrically and mechanically to the propeller-control system and to the fuel-control unit. Thus the fuel scheduling and propeller are controlled together to provide correct conditions for all power settings. The **power lever** operates through a range of 0 to 90° and has five basic positions. These are (1) MAXIMUM REVERSE, 0°; (2) GROUND IDLE, 9°; (3) START, 15 to 20°; (4) FLIGHT IDLE, 34°; and (5) TAKEOFF, 90°.

The coordinator contains a cam assembly called a **discriminator** which permits the power lever to control the linkage to the propeller if the emergency handle is in its normal (push-in) position. The discriminator permits the emergency handle to override the action of the power lever in positioning the linkage to the pro-

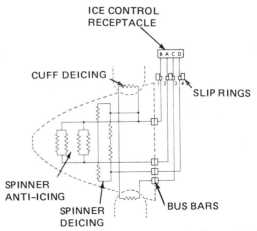

FIG. 18-11 **Schematic diagram of the anti-icing and deicing circuits.**

peller any time the emergency handle is pulled. The propeller condition-control lever thus "informs" the propeller regulator of the power-lever and emergency handle positions.

The feather solenoid and the engine NTS system are connected to the NTS linkage, which is attached to the propeller NTS feather lever. The NTS feather lever informs the propeller regulator of feather solenoid and NTS system requirements on the hydraulic system.

The **emergency handle** provides the means for shutting down an engine by a single action. Pulling the emergency handle will actuate mechanical linkage and electric circuits to shut off the fuel and feather the propeller.

The **feather solenoid,** secured to the reduction-gear assembly, is mechanically attached to the engine NTS linkage. When the solenoid is energized, it actuates the NTS linkage, which results in the propeller NTS feather lever being moved to the feather position. Thus, the propeller is signaled to feather if the propeller NTS system is not blocked out owing to power-lever setting. The NTS system is blocked out below the 21° position of the coordinator.

The reduction-gear NTS system is also mechanically attached to the engine NTS linkage. If negative torque is from -300 to -420 shp [-223.71 to -313.19 kW], the NTS system actuates the engine NTS linkage, which results in the propeller NTS feather lever being moved to the feather position. Thus the propeller is signaled to feather. As the propeller blade angle increases, the negative torque produced by the propeller decreases to a point at which the reduction-gear NTS system no longer is actuated. The NTS system returns to the NORMAL position, which removes the feather signal to the propeller. The propeller hydraulic governor then regains control of the propeller and decreases the blade angle. If the negative torque again actuates the NTS system, the foregoing cycle repeats itself. The cycling of the NTS system from the actuated to the nonactuated position is called **NTSing.** If NTSing occurs during an approach with low-power settings, a slight advance of the power lever will stop it. If NTSing is the result of engine failure, it can be stopped by feathering the propeller.

Hydraulic System

The propeller will function satisfactorily when the proper type of fluid is used in the system. Approximately 3 gal [11.36 L] of hydraulic fluid is required per propeller. This fluid is contained in the regulator reservoir and the feather reservoir. A visual oil-level sight gage is provided for quick, easy hydraulic-fluid level check. The propeller does not consume fluid. However, if replenishment is required, fluid may be added through a filler tube and check valve at the front of the feather reservoir without removing the spinner. The use of a pressure filling device is recommended because this will provide rapid filling and ensure the use of only clean, filtered hydraulic fluid.

Any hydraulic fluid which drains into the regulator is called **uncontrolled drain.** Fluid which drains into the feather reservoir is called **controlled drain.** Except for a small amount of internal leakage, all oil pumped by the hydraulic pumps is directed to controlled drain.

The airflow over the exterior of the feather reservoir provides for the cooling of the hydraulic fluid. The feather reservoir overflows the cooled fluid back into the regulator reservoir where the pumps again pick it up for recirculation. Thus, the controlled drain provides for cooling of the hydraulic fluid and keeps the feather reservoir full. In the event of a regulator hydraulic component malfunction which requires feathering of the propeller, the feather reservoir and pump can supply the flow of fluid required for feathering.

Four gear-driven hydraulic pumps, located in the propeller regulator, are driven only when the propeller is rotating. Their output volume is a function of propeller rpm. Thus, if a propeller is signaled to feather during flight, the output of these pumps decreases as rpm decreases. An auxiliary source of hydraulic pressure is required to feather the propeller completely. If an auxiliary source of hydraulic pressure were not available, the propeller would not complete feathering but would continue to rotate at a high blade angle with low rpm and low drag.

The auxiliary source of hydraulic pressure is provided by an electrically driven pump. Operation of this pump is required for feathering, unfeathering, and static checks of the hydraulic system.

The components of the system mounted in the regulator are (1) four hydraulic pumps, (2) NTS feather valve, (3) hydraulic pump, (4) jitter valve, (5) solenoid valve, (6) pitch lock and stop valve, and (7) the centrifugal breather.

The four hydraulic pumps supply the flow of hydraulic fluid for normal operation of the hydraulic system.

The NTS feather valve responds to either an NTS signal or a feather signal by porting all fluid pumped by the hydraulic pumps to **increase-beta** (blade-angle) passages. Thus a rapid rate of blade-angle change required for feathering or NTS response can be effected.

The **hydraulic governor** is a hydromechanically controlled device used for both beta-range and governing-range operation. In the governing range the hydraulic governor effects blade-angle changes required to maintain a constant engine rpm of 13,820. In this range of operation the propeller is classified as "nonselective and automatic." In the beta range the hydraulic governor provides the means by which blade angles may be selected by the power lever and the propeller is classified as "controllable and multiposition." The unit in the governor which acts automatically to regulate engine speed is the **speed-sensitive element.** This unit is affected by centrifugal force and spring pressure. When engine speed is excessive, the centrifugal force on the speed-sensitive element overcomes spring pressure and the element moves the distributor valve to route hydraulic fluid under pressure to the pitch-change (torque) units at the base of the propeller blades.

The **jitter valve** is used to "pulse" certain elements within the hydraulic governor to increase their sensitivity and response rate.

The **solenoid valve** is controlled by a phase synchronizer module to provide a means of trimming propeller speed by biasing the fluid flow from the hydraulic governor to the torque units.

The **pitch-lock and stop-valve assembly** contains the

necessary valves required for the control of the mechanical low-pitch stop and pitch-lock mechanisms in the master gear assembly. It also incorporates a valve element which provides for the transition of the propeller operation from the beta range to the governing range, or vice versa. The assembly contains a speed-sensitive unit which causes the blade angle to be locked if sufficient overspeed occurs. This prevents a further decrease in propeller pitch and helps to prevent a greater increase in engine speed.

The **feather pump** is mounted in the feather reservoir and provides an auxiliary source of hydraulic pressure for static ground operation of the propeller and for any feathering or unfeathering operation. During normal operation, the feather pump is not used. The pump is driven by an electric motor which is actuated for approximately 60 s by the action of the feather pump motor timer. When the propeller is unfeathered, the motor is energized as long as the feather button is held in the unfeather position.

At the outlet of the feather pump is a check valve which allows the feather pump to deliver fluid into the hydraulic system but prevents the four regulator pumps from pumping fluid back through the feather pump.

During normal operation of the propeller, centrifugal twisting moment (CTM) tends to decrease blade angle. Thus, greater pressures are required to increase the blade angle during propeller rotation than when the propeller is static. The **feather pressure-control valve** controls the discharge pressure of the feather pump as a function of rpm; that is, the higher the rpm, the greater the discharge pressure from the feather pump.

Safety Features

Because of the design characteristics of turboprop (propjet) powerplants wherein high drag forces are associated with either engine or propeller failure, certain safety devices must be incorporated in the turbopropeller to ensure safe operation. These safety devices reduce the asymmetric thrust to values compatible with airframe structural limitations and controllability.

Pitch Lock

The **pitch-lock** mechanism is a speed-sensitive device designed to protect against blade-angle pitch-down in the event of a propeller-control-system malfunction which results in an overspeed condition. Pitch-lock protection is needed only during governing-range operation; hence, it is blocked out below 28° coordinator power settings, below 21.5° blade angle, or above 73.5° blade angle.

Mechanical Low-Pitch Stop (MLPS)

The **mechanical low-pitch stop** is designed to prevent the inadvertent entry of the blades into the beta range during flight in the event of propeller-control-system malfunction. The MLPS with a setting of 18.25 to 18.5° "backs up" the normal hydraulic low-pitch stops.

Hydraulic Low-Pitch Stop (HLPS)

The **hydraulic low-pitch stop** is designed to prevent the propeller hydraulic governor from decreasing the blade angle into the beta range during governing operation. The HLPS has a setting of 20° blade angle from 30 to 68° coordinator. From 68 to 90° coordinator HLPS is scheduled from 20 to approximately 31.5° blade angle. This scheduling of the HLPS is called **beta follow-up** and means that the propeller pitch cannot decrease below 20 or 30° during constant-speed operation controlled by the governor.

Negative Torque Signal (NTS) System

The **NTS system** operates by signals from the reduction-gear assembly. The stationary ring gear is mounted in the gear case with helical splines, and if torque on the ring is opposite to normal operation, the ring moves and actuates switches to produce the negative torque signal. The device functions at a specified torque, as mentioned previously. If sufficient negative torque is produced when the propeller tends to drive the engine—owing to "lean" fuel schedules at flight-idle power settings, air gusts on the propeller, loss of engine power, temporary fuel interruptions, or similar conditions—a signal will be produced. The propeller responds to this signal by increasing blade angle as necessary to limit the negative torque. The NTS system within the propeller is blocked out below 21 to 24° coordinator power setting.

Manual Feather

Manual feather is accomplished by depressing the FEATHER button in the cockpit. When this is done, the propeller is signaled to feather by the feather solenoid, the electric feather pump motor is energized for approximately 60 s by a timer, and fuel flow is stopped via an electric fuel-cutoff actuator on the engine. Thus, the propeller will feather and the engine will stop running when the FEATHER button is depressed. If this button is depressed when operating below a 21 to 24° coordinator power setting, fuel flow will be stopped but the propeller will not feather because the NTS system is blocked out of operation.

Auto Feather

The **auto-feather** control system is incorporated in the engine reduction-gear assembly. Whenever the propeller thrust is greater than positive 500 lb, the auto-feather system cannot function. When propeller thrust is less than positive 500 lb, the auto-feather-control system will produce auto feather if the auto-feather system is armed. The arming circuit of any given engine auto-feather system includes several switches in series with each other. These switches are a cockpit manually positioned switch, a 75° coordinator power lever switch, and necessary feather-button switches. On a four-engine installation, an engine auto-feather circuit is armed by a switch in each of the other engine feather buttons. On a two-engine installation, the feather circuit is armed by a switch in the opposite engine FEATHER button.

The auto-feather system operates in conjunction with signals from the **thrust-sensitive signal** (TSS) system which is contained in the engine reduction-gear assembly. The TSS system is designed to provide an auto-feather signal in case of power loss during takeoff. The TSS system is independent of the NTS system, since the TSS has a minimum positive **thrust** setting while the NTS system has a negative **torque** setting. If the thrust developed by the propeller is in excess of a positive 500 lb, two switches (parallel) in the thrust-sensitive-switch assembly on the reduction-gear assembly are open. In the event of an engine failure on takeoff, the propeller thrust decreases. When it drops below a positive 500 lb, the thrust-sensitive-switch assembly switches are closed. Thus, if this switch assembly is properly armed and engine failure occurs, the auto-feather circuitry is energized to effect a feather shutdown.

Emergency Feather

Emergency feathering is accomplished by manually pulling the emergency handle. When this handle is pulled, switches are tripped and mechanical linkage is moved which signals the propeller to feather and fuel flow to stop. The switches control electric circuits which pull the FEATHER button into its feather position via a "pull-in" coil. Thus, pulling the emergency handle effects engine shutdown electrically and mechanically. Emergency feathering can be accomplished under any condition of engine operation regardless of the power setting.

Solenoid Stop

The principal function of the solenoid stop is to prevent a loose or disconnected condition-control lever from entering the beta range inadvertently and scheduling low blade angles which would cause excessive drag during flight.

Synchronizing and Phase Synchronizing

There are five major components which make up the synchronizing and the phase synchronizing systems: (1) propeller alternator, (2) rotary actuator, (3) solenoid valve, (4) electronic synchronizer module, and (5) electronic phase synchronizer module. The effect of these components on the hydraulic system is considered with reference to the following factors of operation:

$K1$ Speed error, hydraulic
$K2$ Acceleration sensitivity
$K3$ Rate of change of trim on hydraulic governor with respect to speed error
$K4$ Rate of change of trim on hydraulic governor with respect to phase error

The normal propeller governing is obtained by means of a hydromechanical speed-sensitive governor in the regulator. This governor, utilizing the $K1$ factor of operation, provides satisfactory stability for all normal operation. No external accessories are required for $K1$ operation.

To obtain optimum stability, it is necessary to use a control factor which is proportional to the acceleration of the engine. Acceleration intelligence, or the $K2$ factor, is obtained from the engine-driven propeller alternator and the electronic circuits in the phase synchronizer module. The resulting voltage is used to power the solenoid valve in the regulator. Since governors are speed-sensitive devices, some rpm variation (± 25 engine rpm) is to be expected when only the $K1$ factor is utilized. The solenoid valve biases or trims the flow of fluid from the hydraulic governor as required to quicken the recovery from offspeed by damp-

ening out any overcorrection made by the hydraulic governor.

The synchronizing function, the $K3$ factor, requires that the governing rpm of the slave propeller be reset to the governing rpm of the master propeller. This is accomplished by the slave and master propeller alternators sending speed intelligence to the slave synchronizer module. This module compares the speed of the slave relative to the master and pulses the slave rotary actuator accordingly. The rotary actuator, when pulsed, actuates the necessary mechanical controls to effect a governing rpm change on the slave propeller. When the rpm of the slave propeller is the same as the rpm of the master propeller, the propellers are synchronized.

The phase synchronizing function, the $K4$ factor, establishes the proper phase relationship by resetting the slave governor at a constant rate whenever the propellers are not at the optimum phase angle. This is accomplished by the slave and master propeller alternators sending phase-angle intelligence to the slave synchronizer module. The synchronizer module sends the phase-error information to the slave phase synchronizer module, which converts this phase-error information into the proper $K4$ signal voltage. This control voltage is mixed with the $K2$ voltage already present. The result is that the solenoid valve is pulsed so that it biases or trims the flow from the hydraulic governor for effective reset action to provide the optimum phase relationship between the slave and master propellers.

Anti-icing and Deicing System

The propeller has components on which ice would form during icing conditions in flight. It is necessary to either anti-ice or deice these components if icing conditions are encountered. **Anti-icing** is a term used to identify components which are heated continuously, while **deicing** is a term used to identify components which are heated in sequence by a timer. "Runback" from an anti-iced component could result in undesirable ice formations. Components immediately aft of the anti-iced areas are deiced. Thus, deicing is used to prevent accumulation of these undesirable ice formations by heating the components long enough to remove the ice but not long enough to produce runback. Approximately the front two-thirds of the spinner is anti-iced. The deiced components consist of blade cuffs, spinner islands, and that part of the spinner which is not anti-iced. Refer to Fig. 18-11 for anti-icing and deicing circuits.

Propeller-Control Schedule

The basic requirement for any aircraft powerplant is to provide controllable thrust throughout the range of operating conditions. In a turboprop installation, the major problem becomes one of matching the extreme shp variations of a gas-turbine engine in such a manner as to provide reliably controlled thrust at all times.

For flight operation, a gas-turbine engine is demanded to deliver power within a relatively narrow band of operating rotational speeds. This requires the control system to extract from the propeller controllable thrust within close rpm limits. During flight, the speed-sensitive governor of the propeller automatically controls the blade angle as required to maintain a constant rpm of 13,820 on the engine.

Three factors tend to vary the rpm of the engine during operation. These factors are **shp, airspeed,** and **air density.** If the rpm is to remain constant, the blade angle must vary directly with shp, directly with airspeed, and inversely with air density. The speed-sensitive governor provides the means by which the propeller can adjust itself automatically to varying power and flight conditions while converting the shp to thrust.

To maintain a reasonable thrust schedule for ground operation, adequate control cannot be obtained in the region of 0 to 20 percent thrust with a pure shp scheduling system as used in governing. For ground operation, controllable thrust can be obtained by scheduling and coordinating fuel flow and blade angle according to the dictates of the power lever while allowing the engine to remain within rpm limitations.

The ground range of operation of the propeller is called the **beta range** and extends from 0 to 30 percent coordinator power setting. In this range a definite blade angle is scheduled by the power lever. This angle is from -4 to $+20°$ at the 42-in [106.68-cm] station. The governing range of the propeller extends from 30 to 90° coordinator power setting during which the propeller blade angle may increase to more than 40°. In this range the hydraulic governor of the propeller automatically controls the blade angle as required in order to maintain a constant engine rpm of 13,820 and a propeller rpm of 1020.

The power-lever taxi range (0 to 34° coordinator) is used only for ground operation. The power-lever flight range (34 to 90° coordinator) is used for all flight operation, ground power checks, and takeoff.

As previously explained, the propeller system incorporates safety features to prevent any possibility of the propeller entering into the beta range of operation while the aircraft is airborne. This is necessary because beta-range blade angles in flight would result in extremely high drag, making directional control of the aircraft difficult if not impossible.

If the power lever is retarded rapidly to flight idle, a small amount of coordinator pointer overtravel may occur but the propeller remains in the governing range down to 30° coordinator. The 4° of coordinator pointer difference between the propeller governing range and the power-lever flight range is a "safety cushion" against the propeller's transitioning into the beta range of operation as a result of rapid power-lever movement to flight idle.

The function of the **solenoid stop** is to prevent the propeller condition control from moving into the beta range if the engine linkage to this lever becomes disconnected in flight. The electric circuit which controls the solenoid stop includes a 34° power-lever-actuated switch that is in series with a landing-gear scissor switch. The solenoid stop is deenergized any time the power lever is above 34° or when there is no weight on the landing gear. Thus, the solenoid stop is never energized in flight. If the linkage should become disconnected, the condition-control lever tends to "free fall" into the beta range, but it cannot enter the beta range because the solenoid stop is contacted at 31.5° coordinator. At this lever position the propeller remains in governing

range, the propeller operation remains normal, and the flight crew would be unaware of the condition.

If there is weight on the landing gear, the solenoid stop will be energized when the power lever is retarded below 34° coordinator. Thus, normally the solenoid stop is positioned so that the flight crew can easily move the power lever and consequently the condition-control lever below 31.5° coordinator. In the event that a malfunction occurs which results in the solenoid stop not being energized, the solenoid stop is contacted at 31.5° coordinator. If this occurs, the flight crew can exert a little extra force on the power lever to override the stop and move the condition-control lever into the beta range.

Summary of Turbopropeller Features

Because of the many intricate details involved in the construction and operation of the Allison Model A6441FN-606 propeller, it is well to summarize the principal features of this propeller.

1. The propeller is a full-feathering constant-speed four-blade unit capable of converting more than 4500 shp [3363.75 kW] into thrust. It is designed for an SAE 60 propeller shaft.

2. A complete hydraulic system with four pressure pumps is included in the hub section of the propeller to provide the power for the various ranges of operation from maximum reverse thrust to takeoff power. Hydraulic fluid is carried in the rotating regulator section and in the feathering reservoir.

3. The constant-speed feature is provided by means of a speed-sensitive governor. In the governing range the unit is set to provide an engine rpm of 13,820 and a propeller rpm of 1020. The speed-sensitive element of the governor is balanced between centrifugal force and spring pressure. When overspeed occurs, the governor provides for hydraulic fluid under pressure to be directed to the blade torque units.

4. The blade torque units consist of pistons mounted in helical splines. When fluid pressure is applied to one side of a piston, it moves outward or inward in the helical splines, thus producing a rotation of the blade. The four torque units are synchronized by means of a master gear.

5. The principal cockpit controls for the propeller are the power lever, FEATHER button, emergency handle, and synchronizing controls. The power lever is used to set the range of operation through the coordinator and the condition-control lever. The coordinator signals the fuel-control unit as required for all ranges of operation in accordance with the movement of the power lever.

The FEATHER button is used for manual feathering of the propeller through the NTS system. It provides for complete feathering and for stopping the engine when the NTS system is not blocked out.

The emergency handle is used to stop the engine at any time it is deemed necessary. It provides for feathering of the propeller and shutting off of fuel in any of the operating ranges.

6. The propeller is provided with a negative torque signal system to reduce propeller drag in case of engine power loss. This system begins to feather the propeller when negative torque on the propeller shaft reaches −300 to −420 shp [−223.71 to −313.19 kW].

7. A thrust-sensitive signal activates an automatic feathering system if positive thrust decreases below positive 500 lb [226.80 kg]. This is to reduce drag in case of engine failure on takeoff.

8. The propeller is controlled through the power lever to operate in two principal ranges. These are the **beta range,** from 0 to 30° power-lever setting, and the **governing range,** from 30 to 90° power-lever setting. In the beta range the power lever controls the propeller blade angle and fuel flow through the coordinator. In the propeller governing range, the governor in the propeller controls engine speed at 13,820 rpm and the power lever provides for fuel scheduling through the coordinator and the fuel-control unit.

9. The propeller has a **maximum-reverse** setting to provide for reverse thrust after landing. Maximum reverse is produced by moving the power lever to the 0° position. This causes rotation of the propeller blades to a negative pitch setting and applied engine power through the fuel control.

The complete propeller system is shown in Fig. 18-12. This illustration shows how the power lever and the emergency handle both operate through the coordinator to move the condition-control lever at the propeller. It can also be observed that the NTS signal comes from the reduction-gear section of the engine and is delivered through the NTS feather lever to the propeller.

● HARTZELL TURBOPROPELLERS

In Chap. 17 we described the Hartzell Model HC-82XF-2 constant-speed full-feathering propeller. The principle of operation was illustrated in Fig. 17-23. In this section we shall describe the HC-B3TN propellers, which operate in a manner similar to the propeller described in Chap. 17. However, the HC-B3TN propellers, shown in Fig. 18-13, have special features which adapt them to turboprop engines.

The HC-B3TN-2 propeller is a constant-speed, feathering, three-blade model designed primarily for

FIG. 18-13 Hartzell HC-B3TN propeller. *(Santa Monica Propeller Service)*

use with the United Aircraft of Canada PT-6 turboprop engine. The HC-B3TN-3 propeller is also designed for use with the PT-6 engine; however, this model is reversible. The HC-B3TN-5 model is a constant-speed, feathering, reversible propeller designed for use with the Garrett-AiResearch TPE-331 turboshaft (turboprop) engine.

The propellers mentioned above are all similar in design to the model described in Chap. 17. They operate on the same principles, and the same general maintenance practices should be employed. The blade shanks of the HC-B3TN propellers each have two sets of needle bearings in order to reduce friction. These can be seen in the drawing of Fig. 18-14. The blade shanks, designated as T, can be used only with turboprop engines. The propeller flange, designated as N, has eight $\frac{9}{16}$-in [14.29-mm] diameter bolt holes on a $4\frac{1}{2}$-in [11.43-cm] diameter bolt circle.

The reversible models of the propeller have lengthened piston cylinders in order to provide for the additional movement beyond the low-pitch position to attain the required reverse-pitch angle. Low pitch and reverse pitch are accomplished by means of governor oil pressure directed through a passage in the engine shaft to the propeller piston. This causes the piston to move outward (forward). The piston is connected to

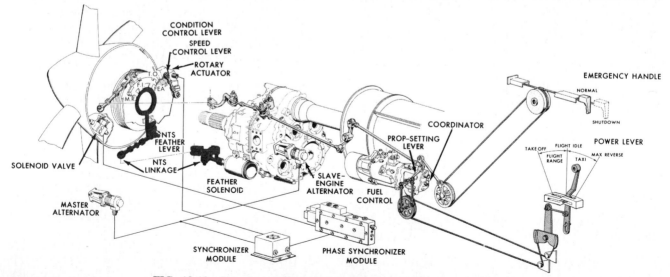

FIG. 18-12 Drawing of the complete Allison propeller-control system.

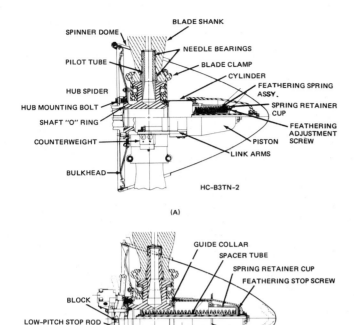

FIG. 18-14 Cutaway drawing of the propeller operating system. *(Hartzell Propeller Co.)*

the blade shanks by means of **link arms,** thus producing rotation of the blades as the piston moves.

The nomenclature of the piston and cylinder in the propellers under discussion here is unusual in that the **piston** is outside the **cylinder.** This is the reverse of common terminology; hence, students must be careful to make sure they understand which unit is referred to. Note in Fig. 18-14 that the piston forms the outer case of the operating assembly and the cylinder is inside.

In the drawing it will be seen that there are three co-axial springs inside the piston-cylinder assembly. These constitute the **feathering-spring assembly.** When the piston moves forward and the pitch is reduced, the springs are compressed. When the governor allows engine oil to flow from the propeller back to the engine, the springs and the force produced by the **blade counterweights** pull the piston back toward the hub and the blade pitch increases. If the propeller control is set for **feathering,** the feathering-spring assembly will pull the piston back to the position where the blades are feathered. The total pitch-change range is 110°.

As mentioned earlier, the propellers discussed in this section are controlled by a speed-sensing governor which maintains the engine rpm selected by the pilot. The governor is installed on the engine and provides oil pressure varying from 0 to 385 psi [2654.58 kPa]. Governor oil pressure is used to decrease propeller pitch and to place the blades in the reverse-pitch position.

Feathering the propeller blades is accomplished by opening a valve in the governor to permit oil to return from the propeller piston to the engine. This allows the feathering spring to force the oil out of the propeller

and move the piston rearward to the feathered position. The HC-B3TN-5 propeller has spring-loaded, centrifugal responsive latches which prevent feathering when the propeller is stationary but permit feathering when the propeller is rotating. The -2 and -3 models have no latches, so they can be feathered when the engine is not running. Unfeathering, in the cases of the -2 and -3 propellers, is accomplished by starting the engine to allow the governor to pump oil into the propeller piston cylinder. In the case of the -5 propeller, unfeathering is best accomplished by means of an electric pump.

The reversible propellers are provided with hydraulic low-pitch valves, commonly called "beta (β)" valves. The **beta valve** prevents the governor from moving the piston beyond a prescribed low-pitch position. This prevents the propeller from reversing unless the reverse position is selected by the pilot. Reversing is accomplished by manually moving the control which adjusts the low-pitch stop to the reverse-pitch position. During this operation, the governor is adjusted automatically to produce oil pressure for the reversing operation. Return from reverse is accomplished by manually moving the control to place the low-pitch stop in the normal low-pitch position. This allows the oil to drain from the piston cylinder as the forces of the springs and counterweights move the piston rearward.

The operation of the complete propeller-control system may be understood by examining the schematic diagram of Fig. 18-15. This diagram shows the control system as installed for the PT-6 turboprop engine.

THE DOWTY-ROTOL TURBOPROPELLER

The Dowty-Rotol R321 four-bladed turbopropeller is similar in operating principle to the Hartzell turbopropeller described previously in this section. The propeller blades are rotated to low pitch and reverse pitch

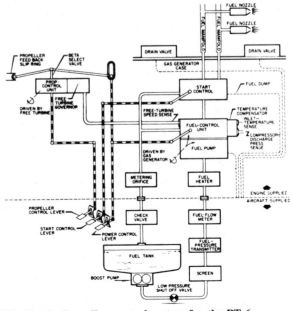

FIG. 18-15 Propeller-control system for the PT-6 turboprop engine. *(Pratt & Whitney)*

by means of oil pressure. When the oil pressure is released, the blades rotate to high pitch and feathered position by means of a combination of spring and counterweight forces. The propeller-control system balances the opposing forces to provide the correct pitch for constant-speed operation in flight. The operating principle of the pitch-changing mechanism is an important safety factor because it assures that the propeller will be feathered quickly in case of oil pressure loss during takeoff. This eliminates propeller drag and makes the aircraft easier to control.

A cutaway drawing to illustrate the Dowty-Rotol propeller operating mechanism and construction is provided in Fig. 18-16. The propeller hub consists of two identical steel forgings that bolt together to retain the propeller blades. A 21-ball thrust bearing between the hub shoulder and blade root shoulder provides for low-friction rotation of each blade and absorbs the centrifugal force.

An **operating pin** is mounted on the face of each blade root by means of a socket screw. The operating pin incorporates a needle bearing to reduce friction between the pin and the **crosshead.** The operating pins fit into the crosshead slot to provide blade rotation as the crosshead moves forward and backward.

An oil-transfer tube, called the **beta tube,** screws into the center of the crosshead shaft as shown in the simplified drawing of Fig. 18-17 and terminates in the **beta sleeve** in the engine. This tube carries oil pressure to the rear side of the piston that is attached to the forward end of the crosshead. The oil pressure forces the piston forward, causing the crosshead to rotate the propeller blades to low pitch and reverse pitch. When the piston is in the full-forward position it bears against the full-reverse stop and the propeller blades are in the full-reverse position, which is $-12.5°$ to $-14.5°$. When oil pressure is released the oil flows out through the beta tube, the piston moves to the rear, and the propeller blades rotate toward high pitch because of feather-spring pressure and counterweight force. Complete release of the pressure results in the piston moving to

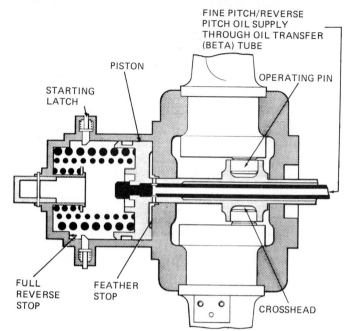

FIG. 18-17 Simplified drawing of a pitch-changing mechanism. *(Dowty-Rotol)*

the feather stop at the rear of the cylinder. With this piston position the propeller is fully feathered and the blade pitch is approximately $+85°$ as measured from the blade setting line.

Starting latches are incorporated in the cylinder to assure that the propeller blades will remain in low pitch (approximately 0°) during engine starting. The starting latches engage a groove on the piston, thus holding it in the forward (low-pitch) position until propeller speed reaches 300 rpm. At this time the latches are disengaged by centrifugal force. The arrangement of the starting latches is shown in Fig. 18-17. To assure that the latches are engaged for starting, the propeller is placed in the low-pitch (0° or slightly reversed) position at engine shutdown.

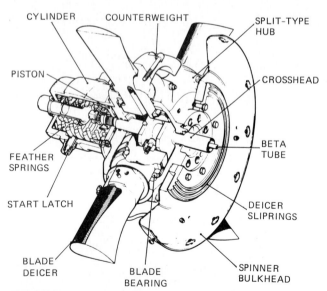

FIG. 18-16 Cutaway illustration of the operating mechanism for a Dowty-Rotol turbopropeller. *(Dowty-Rotol)*

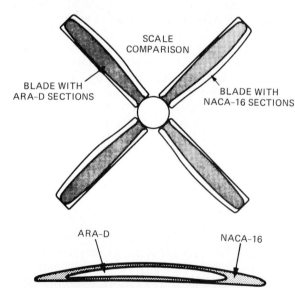

FIG. 18-18 Comparison of propeller blade designs. *(Dowty-Rotol)*

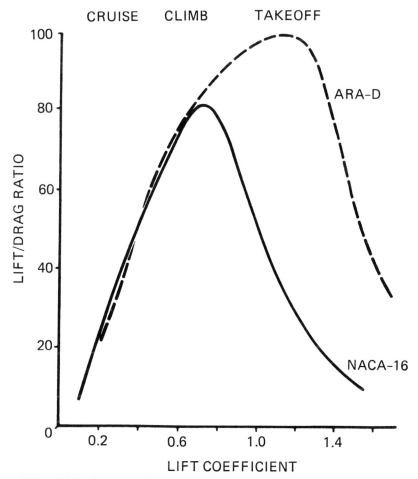

FIG. 18-19   Curves to show propeller blade performance.   *(Dowty-Rotol)*

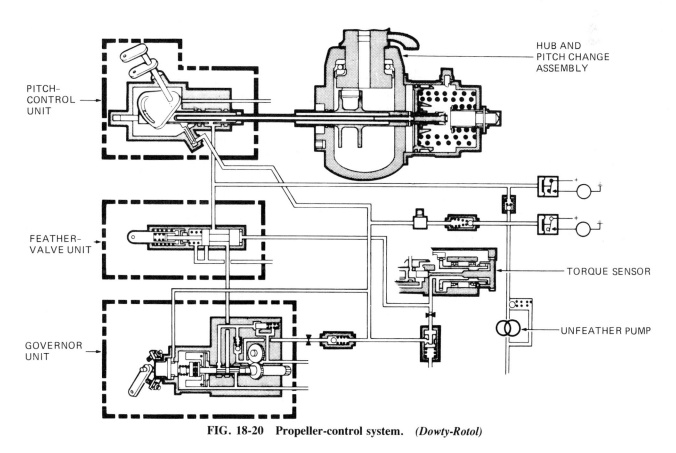

FIG. 18-20   Propeller-control system.   *(Dowty-Rotol)*

Propeller Blade Design

The blades of the Dowty-Rotol R321 propeller are of a recent advanced-technology aerodynamic design that substantially improves performance. The formerly used blades were of the NACA-16 design shown in Fig. 18-18. The new configuration is a Dowty-Rotol development designated ARA-D. The improved performance of a ARA-D design is shown in the chart of Fig. 18-19.

Propeller-Control System

The control system for the Dowty-Rotol R321 propeller as installed on the Fairchild Swearingen Metroliner is illustrated in Fig. 18-20. The **pitch-control** mechanism is adjusted by means of two levers to select the operating mode for the propeller. The constant-speed governing mode is selected for in-flight operations and the beta mode is selected for ground operations. A low-blade-angle stop for approach and landing is provided by the **beta valve.**

The **feather valve** can be operated either manually or hydraulically. Thus, the pilot can feather the propeller when it is deemed necessary. In case of engine failure a negative torque signal (NTS) will be received by the **torque sensor,** which then moves to operate the feather valve hydraulically and feather the propeller. As explained in this chapter, negative torque is developed when the propeller is windmilling and driving the engine.

The **propeller governor** is a flyweight-controlled mechanism similar to the governors described previously. It is manually adjusted by the pilot to select the desired engine rpm. The selected rpm is maintained constantly during in-flight operations.

● REVIEW QUESTIONS

1. Give a brief description of the counterweight-type constant-speed propeller.
2. When a propeller control in the cockpit is placed in the DEC RPM position, what takes place with respect to pitch?
3. Describe the pitch-changing mechanism of the hydromatic reversing propeller.
4. How is reverse pitch obtained?
5. Describe the feathering operation of the hydromatic propeller.
6. Explain the operation of the propeller governor.
7. What is the function of a master engine with respect to engine synchronizing?
8. What is the purpose of the governor step motor in the synchronizing system?
9. Briefly describe the synchronization system for the Hamilton Standard reversing propeller.
10. What is meant by the *beta range* in the operation of the Allison turbopropeller?
11. Describe the physical features of the Allison turbopropeller?
12. Give the number of hydraulic pumps in the system, and tell how they are driven.
13. What is the propeller rpm at governing speed?
14. Describe the construction of the propeller blades.
15. At what position of the power lever does the governing range begin?
16. At what position of the power lever is the engine started?
17. Describe the regulator assembly.
18. Where is the hydraulic fluid for the propeller hydraulic system stored?
19. Describe a torque unit.
20. What forces act on the governor to produce speed control?
21. What is the function of the master gear?
22. Describe the method by which control is transmitted from the nonrotating engine to the rotating propeller.
23. Describe the feather reservoir assembly.
24. Explain the purpose of the negative torque signal system.
25. Through what means is electrical power transferred to the rotating propeller assembly?
26. How would you stop the engine and propeller quickly if the NTS system were blocked out?
27. What is the purpose of the thrust-sensitive signal?
28. How is hydraulic fluid added to the propeller supply?
29. Explain the purpose of the solenoid stop.
30. What is the function of the coordinator?
31. Briefly describe the principle of operation of the Hartzell propeller for turboprop engines.
32. Explain the operation of the feathering spring.
33. Discuss the function of the governor with the Hartzell turbopropeller operation.
34. Explain the importance of the beta valve in a reversible propeller.
35. Describe the principle of operation for the Dowty-Rotol turbopropeller.
36. Explain the need for and the operation of the starting latches.
37. Discuss the propeller blade design for the Dowty-Rotol R321 turbopropeller.
38. What is the function of the torque sensor?

19 PRINCIPLES OF JET PROPULSION

BACKGROUND OF JET PROPULSION

Discovery of the Jet-Propulsion Principle

No one knows who first discovered the jet-propulsion principle, but the honor is sometimes given to a man named Hero, who lived in Alexandria, Egypt, about 150 B.C. He invented a toy whirligig turned by steam, as illustrated in Fig. 19-1, and called his invention an **aeolipile,** but apparently he did not discover any very useful purpose for his discovery.

The historical records are not very definite in describing the aeolipile. If it resembled the picture in Fig. 19-1, it was a primitive form of a jet or reaction engine. On the other hand, some authorities describe it as being operated by hot air instead of steam. The heating of air in a vertical tube induced a flow of air in several tubes arranged radially around a horizontal wheel, and rotation resulted from the creation of an impulse effect if this version of the story is accepted. In that case, Hero's invention was a gas turbine.

About 1500, Leonardo da Vinci, who was a successful jack-of-all-trades, sketched a device that could be placed in a chimney where the upward movement of hot gases would turn a spit for roasting meat. In 1629, Giovanni Branca, another Italian, perfected a steam turbine that applied the jet principle and could be used to operate primitive machinery.

Figure 19-2 is a drawing of an invention called **Newton's carriage,** a jet-propelled steam carriage. Although Newton himself may have supplied only the idea, there are authorities who attribute the design of the carriage to a Dutchman, Willem Jako Gravesande.

Turbine Development

The first patent covering a gas turbine was granted to John Barber of England in 1791. It included all the essential elements of the modern gas turbine except that it had a reciprocating-type compressor. In 1808, a patent was granted in England to John Bumbell for a gas turbine which had rotating blades but no stationary, guiding elements. Thus the advantages gained today by the multistage type of turbine were missed.

In 1837, a Frenchman named Bresson was granted a patent for a machine in which a fan delivered air under pressure to a combustion chamber where the air was mixed with a gaseous fuel and burned. The hot products of combustion were then cooled by excess air and directed in the form of a jet against a turbine wheel. This was essentially a gas turbine, but there is apparently no record of its practical application.

In 1850, W. F. Fernihough was granted a patent in England for a turbine operated by both steam and gas, but as long as steam was used, the development of a true gas turbine was held back. However, a man named Stolze designed what was probably the first true gas turbine in 1872 and tested working models in the years between 1900 and 1904. Stolze used both a multistage reaction gas turbine and a multistage axial compressor.

Sir Charles Parsons, the great English inventor, obtained a patent in 1884 for a steam turbine, in which he advanced the theory that a turbine could be converted into a compressor by driving it in an opposite direction with an external source of power. Parsons believed that compressed air could be discharged into a furnace or combustion chamber, fuel injected, and the products of combustion expanded through a turbine. This idea of a compressor was essentially the same as that which we have today except for the shape of the blades.

Charles G. Curtis is generally credited with the filing of the first patent application in the United States for a complete gas turbine. His application was filed in 1905, although previously, in 1902, he filed an application for a rotary compressor, blower, and pump combination and actually obtained the patent in 1914. There is some argument about how much Curtis did to develop the gas turbine, but he is credited without dispute, with the invention of the Curtis steam engine, and he was one of the pioneers in the development of steam turbines.

We come to one of the great names in jet-propulsion development. Sanford A. Moss, who eventually became one of the leading engineers of the General Electric Company, completed his thesis on the gas turbine in 1900 and submitted it to the University of California in application for his master's degree. The contributions of Moss to the development of engines of all types are so extensive that to describe them completely would fill several volumes. However, a few of his outstanding contributions will be mentioned. Figure 19-3 is a sketch of Moss drawn shortly before his death.

In 1902, experiments were conducted at Cornell University with what was probably the first gas turbine developed in the United States. A combustion chamber designed by Moss was used with a steam-turbine bucket wheel, which functioned as the gas-turbine rotor. A steam-driven compressor supplied compressed air to the combustion chamber. The engine was not a success from the practical standpoint because the power required to drive the compressor was greater than the power delivered by the gas turbine, but from the experiments Moss learned enough to enable him to start

FIG. 19-1 Hero's aeolipile.

the General Electric Company's gas-turbine project the next year.

In the following years there were various turbine inventions and developments in the United States and in Europe, but the next outstanding one was the construction of the first General Electric turbosupercharger by Moss during World War I. The products of combustion of the engine exhaust drove a turbine wheel at constant pressure, and the turbine wheel, in turn, drove a centrifugal compressor that supplied the supercharging.

Strictly speaking, the first General Electric turbosupercharger was based on French patents by Rateau; hence, Moss and the General Electric Company are entitled to credit for developing the running model, although the credit for the idea behind it belongs to Rateau of France.

It is interesting to consider that the turbosupercharger was developed as an offshoot of a gas turbine, the turbosupercharger then went through a long stage of development, and finally the engineers took the knowledge that they acquired from working with the turbosupercharger and applied it to jet propulsion.

Frank Whittle began work on gas turbines while he was still an RAF air cadet. He applied for a patent in England in 1930 for a machine having a blower compressor mounted at the forward end and a gas turbine at the rear end of the same shaft, supplied by energy from the combustion chamber. Discharge jets were located between the annular housings of the rotary elements and in line with several combustion chambers distributed around the circumference.

On May 14, 1941, flight trials began with a Gloster E28/39 experimental airplane equipped with Whittle's engine, which was known as the W1. The flight tests were successful, thus greatly increasing the interest of both government and manufacturers and setting the stage for the tremendous progress which was to come.

FIG. 19-2 Newton's steam carriage.

FIG. 19-3 Sanford A. Moss

While Whittle and his associates were working on the development of the W1 engine in England, the Heinkel Aircraft Company in Germany was also busy with a similar task. The German company was successful in making the first known jet-propelled flight on August 27, 1939, with a Heinkel He 178 airplane powered by a Heinkel HeS 3B turbojet engine having a thrust of 880 to 1100 lb [3914.24 to 4892.8 N].

The pioneer jet-propelled fighter planes built in England by the Gloucester (Gloster) Aircraft Company, Ltd., and in the United States by the Bell Aircraft Corporation were powered by a combustion, gas-turbine, jet-propulsion powerplant system developed from Frank Whittle's designs and built by the General Electric Company. Today, Frank Whittle, whose picture appears in Fig. 19-4, is internationally recognized as one of the great leaders in his field.

FIG. 19-4 Frank Whittle. (*American Hall of Aviation History, Northrop University*)

Only a few of the many important inventors and engineers who contributed to the modern jet-engine program have been mentioned, but the work of everyone of them has been based fundamentally on Newton's laws of motion.

● REACTION PRINCIPLES

Newton's Laws of Motion

The thrust produced by a turbojet engine may be explained to a large extent by two of the three laws of motion promulgated by Newton. These may be stated as follows:

First Law. A body at rest tends to remain at rest and a body in motion tends to continue in motion in a straight line unless caused to change its state by an external force.

Second Law. The acceleration of a body is directly proportional to the force causing it and inversely proportional to the mass of the body.

Third Law. For every action, there is an equal and opposite reaction.

The second and third laws of motion are the most applicable to the development of thrust and the equations for jet thrust may be derived directly from the second law. This law may be stated in words differing from those given above; however, the meaning is the same: **A force is created by a change of momentum, and this force is equal to the time rate of change of momentum.** The **momentum** of a body is defined as *the product of its mass and velocity.*

From Newton's second law, it is apparent that if we change the velocity of a body (or a mass), a force is required. In equation form, this principle is expressed $F = dM/dt$, where F is the force, M is the momentum, and t is the time. This equation means that the force created by a change in the velocity of a body is equal to the amount of change in velocity divided by the change in time when appropriate units of measurement are used in the equation.

To understand the operation of equations dealing with **acceleration,** the nature of acceleration must be understood. Acceleration may be defined as a change in velocity. Generally, the word "acceleration" is used to indicate an increase in velocity. This is *positive* acceleration. A decrease in velocity is *negative* acceleration, or **deceleration.**

It is known through studies in the laws of physics that a free-falling body will accelerate at 32.2 ft/(s)(s)—or s² [9.8 m/(s)(s)]. This means that the velocity of the body will increase 32.2 ft/ each second that it is in a free-falling state, discounting the effects of air friction. Thus, 32.2 ft/s² is called the **acceleration of gravity,** and the letter g is used to indicate this value.

If an automobile weighs 3000 lb [1360.78 kg] and we wish to know how much force is necessary to cause it to accelerate from 0 to 60 mph [96.56 km/h] in 5 s, we can proceed as follows:

Since 60 mph is equal to 88 ft/s [26.82 m/s], to reach this velocity, the automobile must accelerate at 17.6 ft/s each second. Since the automobile would accelerate at 32.2 ft/s² if the force applied were equal to its weight, we know that the needed force is less than 3000 lb. The force needed is then determined by multiplying 3000 by 17.6/32.2. The force needed for the required acceleration is then 1639.75 lb [7293.61 N].

The formula used for the foregoing problem may be stated

$$F = \frac{Wa}{g}$$

where F = force, lb
W = weight, lb
a = acceleration, ft/s²
g = 32.2

The problem can then be stated in equation form:

$$F = \frac{3000 \times 17.6}{32.2}$$

Then,　　　　$F = 1639.75$ lb

The basic equation which we may derive from Newton's second law is $F = Ma$, where F is force, M is mass (weight), and a is acceleration. **Mass** is a basic property of matter, whereas **weight** is the effect of gravity on a mass. When considering matter on the surface of the earth, mass and weight are approximately the same.

The equation for thrust may be written

$$F = \frac{w}{g}(V_2 - V_1)$$

where F = force, lb
w = flow rate, lb/s of air and fuel
g = acceleration of gravity (32.2 ft/s²)
V_2 = final velocity of gases
V_1 = initial velocity of gases

The approximate thrust of an engine may be determined by considering the weight of the air only because the fuel weight is a very small percentage of the air weight. However, to obtain an accurate indication of thrust produced by the acceleration of the fuel-air mixture, it is necessary to include both fuel and air in the equation. The equation then becomes

$$F = \frac{w_a}{g}(V_2 - V_1) + \frac{w_f}{g}(V_j)$$

where w_a = airflow through the engine, lb/s
w_f = fuel flow, lb/s
V_j = velocity of gases at the jet nozzle, ft/s

In this equation, V_j represents the acceleration of the fuel because the initial velocity of the fuel is the same as that of the engine.

In actual practice, all the pressure of the gases flowing from the nozzle of a jet engine cannot be converted to velocity. This is particularly true of a jet nozzle in which the velocity of the gases reaches the speed of sound. In these cases, the static pressure of the gases at the jet nozzle is above the ambient air pressure. This difference in pressure creates additional thrust proportional to the area of the jet nozzle. The thrust generated at the jet nozzle is indicated by the equation

$$F_j = A_j (P_j - P_{am})$$

where F_j = force (thrust)
  A_j = area of jet nozzle, ft^2
  P_j = static pressure at jet nozzle, lb/ft^2
  P_{am} = static pressure of the ambient air.

When the jet nozzle thrust is added to the reaction thrust created by acceleration of gases in the engine, the equation for the net thrust becomes

$$F_n = \frac{w_a}{g} (V_2 - V_1) = \frac{w_f}{g} (V_j) + A_j (P_j - P_{am})$$

To better understand what takes place in a typical jet engine, it is well to examine the pressures and temperatures which exist within the engine during operation. Figure 19-5 is a drawing showing the values of pressure and temperature within a twin-spool turbofan engine. It will be noted that there is a substantial rise in both pressure and temperature through the two compressor sections. Also, the pressure has increased more than 13 to 1 and the temperature has increased from 59 to 715°F [15 to 379°C] in the diffuser section. The air from the compressor enters the burners where fuel is added and burned. Pressure drops slightly while velocity and temperature increase markedly. As the hot gases pass through the turbine section, pressure, temperature, and velocity all decrease. This is because the

turbines extract energy from the high-temperature, high-velocity gases.

The effects of heat in a jet engine, or any engine, are explained by the **laws of thermodynamics.** The first law states that *energy is indestructible*. Heat energy is imparted to the air in a gas-turbine (jet) engine by the compressor, and additional heat is added when the fuel is burned. The heat energy is changed to thrust and the gases are cooled as they pass through the turbine section and out the jet nozzle. In any energy cycle, the total quantity of energy involved is exactly equal to the energy which can be accounted for in any of the forms in which it may occur throughout the cycle, i.e., mechanical energy, heat energy, pressure energy, etc.

The second law of thermodynamics in one form states that *heat cannot be transferred from a colder body to a hotter body*. The cooling of an engine involves this principle, in that heat is transferred from hotter bodies or substances to cooler bodies or substances.

The pressures and temperatures of gases follow the principles set forth in Boyle's law and Charles' law. Boyle's law states that *the volume of a confined body of gas varies inversely as its absolute pressure, the temperature remaining constant.*

Charles' law states that *the volume of a gas varies in direct proportion to the absolute temperature.* This law explains the expansion of gases when heat is added by the burning of fuel in the engine.

From the foregoing discussion, it is seen that many physical laws are involved in the operation of a gas-turbine engine. The operation of some of these laws is apparent when we examine the **Brayton cycle,** also known as the **constant-pressure cycle,** which defines the events that take place in the turbine engine. Figure 19-6(A) is a diagram to represent the events of the Brayton cycle. The actual values represented by the curves may vary considerably with different engines; however, the principle is the same for all. Point 1 in

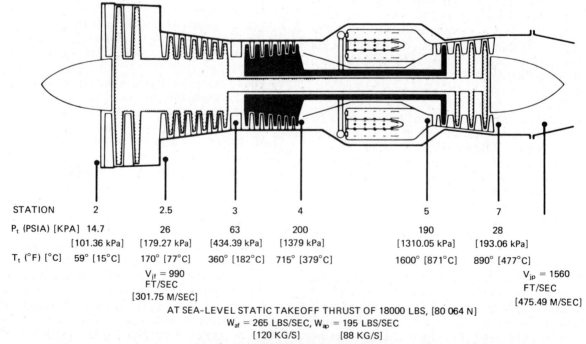

STATION	2	2.5	3	4	5	7
P$_t$ (PSIA) [KPA]	14.7 [101.36 kPa]	26 [179.27 kPa]	63 [434.39 kPa]	200 [1379 kPa]	190 [1310.05 kPa]	28 [193.06 kPa]
T$_t$ (°F) [°C]	59° [15°C]	170° [77°C]	360° [182°C]	715° [379°C]	1600° [871°C]	890° [477°C]

V$_{jf}$ = 990 FT/SEC [301.75 M/SEC]

V$_{jp}$ = 1560 FT/SEC [475.49 M/SEC]

AT SEA-LEVEL STATIC TAKEOFF THRUST OF 18000 LBS, [80 064 N]

W$_{af}$ = 265 LBS/SEC, W$_{ap}$ = 195 LBS/SEC
[120 KG/S]   [88 KG/S]

FIG. 19-5 Pressures and temperatures within a twin-spool turbofan engine. *(Pratt & Whitney)*

432

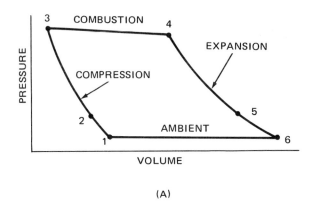

(A)

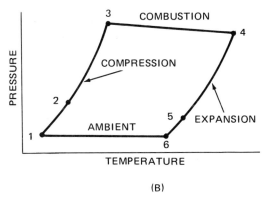

(B)

FIG. 19-6 The Brayton cycle.

the drawing indicates the condition of the air in front of the engine before it is affected by the inlet duct of the engine. After the air enters the inlet duct, it is diffused and the static pressure increases. This is indicated by point 2, which represents the air condition at the entrance to the compressor. Through the compressor, the air volume is decreased and the pressure is increased substantially as shown by the curve from point 2 to point 3. At point 3, fuel is injected and burned, causing a rapid increase in volume and temperature. Because of the design of the combustion chamber, the pressure drops slightly as the velocity of the hot gas mixture increases to the rear. At point 4, the heated gases enter the turbine where energy is extracted, causing a decrease in both pressure and temperature. The curve from point 5 to point 6 represents the condition in the exhaust nozzle as the gases flow out to ambient pressure. The difference between the positions of point 1 and point 6 indicates the expansion of the air caused by the addition of heat from the burning fuel.

The temperature diagram for the Brayton cycle is shown in Fig. 19-6(B). It will be noted that the temperature increases because of compression. During combustion, the pressure drops slightly and the temperature increases at a rapid rate as a result of the burning of fuel. During expansion of the gases through the turbine, the temperature and pressure of the gases are reduced to the point where they enter the atmosphere. At this time, the temperature is still considerably above the temperature of the ambient atmo-

sphere, but it decreases rapidly as the gases leave the jet nozzle and go into the atmosphere behind the engine. Note in Fig. 19-6(A) and (B) that, although the pressure drop is slight, combustion occurs at a constant pressure.

Converting Thrust to Horsepower

Because power is determined by using the product of a force and a distance, it is not possible to make a direct comparison of thrust and horsepower in a jet engine. When the engine is driving an airplane through the air, however, we can compute the equivalent horsepower being developed. When we convert the foot pounds per minute of 1 hp to mile pounds per hour, we obtain the figure of 375. That is, 1 hp is equal to 33,000 ft·lb min or 375 mile·lb/h. Thrust horsepower (thp) is then obtained by using the following formula:

$$thp = \frac{thrust\ (lb) \times airspeed\ (mph)}{375}$$

If a jet engine is developing 10,000 lb [44 480 N] thrust and is driving an airplane at 600 mph [965.61 km/h], the thp is found thus:

$$thp = \frac{10,000 \times 600}{375} = 16,000\ [11\ 931.2\ kW]$$

It should be noted that **thrust** in the SI system is measured in newtons. One newton [N] is equal to 0.224 82 lb force.

Effects of Reaction

In its simplest form, the jet-propulsion engine may be considered as a device for taking in air, adding energy to that air (by burning fuel), and producing thrust from the accelerated gases. Whether this is considered from the momentum, the pressure, or the energy standpoint is not too important for the beginner.

An example of reaction is a child in a boat. If the child rows the boat toward the shore, steps on the bow, and then attempts to leap ashore, the force with which the child pushes forward acts to push the boat backward. The child usually misses the shore and lands in the water, simply because a person who wants to leap successfully, will stand on something fixed, which resists this acting force.

In a similar manner, the end of a garden hose lying on the lawn is driven backward by the reaction of the water when the nozzle is turned on and it shoots forth from the nozzle.

Another example is the reaction of water against the arms of a swimmer or the oars of a boat. The reaction of the water against the propellers of a steam vessel makes the ship move. The reaction of the air against the wings of a bird enables it to fly. The reaction of the air against the propellers of an airplane as they drive the airplane forward and the reaction against the airplane itself follow Newton's second law of motion.

When a pistol or a rifle is fired, it recoils in accordance with Newton's third law of motion; and although it might seem at first thought that this recoil results from the push of the expelled gases on the surrounding air, this is an erroneous assumption. The kick of the

433

gun, whether the bullet breaks through some dense substance or flies through ordinary atmosphere or shoots through a vacuum which has no density at all, is exactly the same in each case.

If the jet of a jet-propelled airplane is regarded as having the same effect as a stream of millions of bullets in multiple force, then the continuous kick which propels the airplane does not result from the jet pushing against the air behind it but follows Newton's second law of motion.

The application of jet powerplants to aircraft has made it commonplace to fly faster than the speed of sound, once considered impossible. The principal limiting factor to the speed of aircraft at present is the heating effect caused by the friction of air against the surface of the aircraft.

Thrust Developed by a Conventional Propeller

The fixed blades of a conventional propeller are continually scooping air from the front and throwing it to the rear. In effect, the propeller is boring its way through the atmosphere like a giant disk. It is the reaction to the mass of air being thrown violently to the rear that drives the airplane forward. However, the conventional propeller reaches a limit in the speed at which it can revolve. In order to explain how jet propulsion breaks through this limitation, we shall redesign, in theory, a conventional piston engine and propeller combination.

Theoretical Conversion of the Conventional Powerplant into a Jet Engine

If the conventional airplane engine and propeller are enclosed in a nacelle, thus narrowing down the stream of air which reacts to the propeller, we have created one form of jet propulsion, as illustrated in Fig. 19-7. Next, the large propellers of conventional design are replaced with a small compressor, the purpose of which is to produce a large mass of compressed air, as shown in Fig. 19-8. A combustion chamber is installed, and fuel is injected to expand the compressed air and increase its velocity, as indicated in the diagram of Fig. 19-9. Finally, the external source of power, which is the airplane engine of conventional form, is removed, and in the path of the exhaust gases a turbine is installed to drive the compressor, as shown in Fig. 19-10. In principle, at least, the conventional engine and propeller have been converted in this manner into a jet engine. The principal difference in operation is that the conventional engine and propeller accelerate a large volume of air a small amount while the jet engine accelerates a smaller volume of air a large amount.

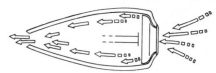

FIG. 19-8   Propeller replaced by a compressor.

● FACTORS AFFECTING THRUST

Air Density

The **density** of air is expressed as the weight per unit volume. For example, 1 ft³ [28.31 L] of dry air at standard sea-level pressure (29.92 in Hg [101.34 kPa]) and temperature (59°F [15°C]) weighs 0.076475 lb [0.035 kg]. From this we can determine that 13 ft³ [37.8 L] of air weighs approximately 1 lb [0.45 kg]. Since the weight (mass) of air consumed by a jet engine is a primary factor in determining thrust, any condition which affects the weight (density) of the air consumed by the jet engine will also affect the thrust.

The conditions affecting the weight of a given volume of air are **pressure, temperature,** and **humidity.** At a constant temperature, the density of air will vary in proportion to absolute pressure; that is, as pressure increases, density also increases. On the other hand, density decreases as temperature increases.

The effect of humidity is not so pronounced as the effects of pressure and temperature; however, humidity must be taken into account when an accurate evaluation of thrust is required. This is particularly true during operation at or near sea level in areas where humidity is likely to be high. The effect of increased humidity is to decrease air density.

Engine RPM

For any particular engine, the thrust increases rapidly as the rpm approaches the maximum design speed of the engine. Among the reasons for this are the design of the compressor and turbine, the increase in mass airflow, and the increase in the difference between intake velocity V_a and jet nozzle velocity V_j.

Airspeed

As intake airspeed increases, thrust decreases because there is a decrease in the difference between V_a and V_j. However, when an aircraft is in flight, the increase in airspeed is accompanied by ram effect. For this reason the net change in thrust is small because the ram effect offsets the effect of change in airspeed.

Altitude

The decrease in air density due to lower pressures at high altitudes results in a decrease in thrust. Because of the decrease in drag at altitude, the performance of

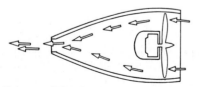

FIG. 19-7   Jet propulsion from a propeller.

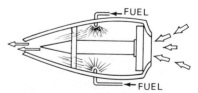

FIG. 19-9   Burning fuel used to expand gases.

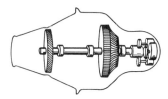

FIG. 19-10   Turbine driving the compressor.

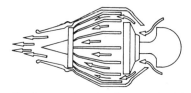

FIG. 19-12   Continuous burning in a gas-turbine engine.

the aircraft is actually improved even though net thrust decreases.

When all factors are combined, it is found that the jet aircraft performs most efficiently at high speed and high altitudes.

● TYPES OF TURBOJET ENGINES

The Centrifugal-Flow Jet Engine

A cross-sectional drawing of a centrifugal turbojet engine is shown in Fig. 19-11. The term **centrifugal** means that the compressor is of the centrifugal type and that the air is compressed by centrifugal force. The air enters near the center of the impeller and is thrown outward to the **diffuser** as the impeller rotates. The diffuser converts the kinetic energy of the air leaving the compressor into potential energy (pressure) by exchanging velocity for pressure. An advantage of the centrifugal-flow compressor is its high pressure rise per stage. A centrifugal compressor is either a double-entry type, shown in Fig. 19-11, or a single-entry type, illustrated in the schematic drawing of Fig. 19-12.

The early jet engine was equipped with a starter, fuel pumps, an overspeed governor, a fuel control, and other accessories. The ducting at the front of the engine allows air to flow into the compressor and thence through the diffuser section to the combustion chambers into which fuel is introduced through nozzles. This arrangement is shown in Fig. 19-11. The high-velocity gases resulting from the fuel combustion enter a nozzle diaphragm which directs the flow against the blades (buckets) of the turbine wheel. Since the turbine is mounted on a common shaft with the compressor, the compressor is driven by the turbine through a splined coupling.

In operation, first the starter motor rotates the compressor and turbine and brings the compressor to a speed which drives a large volume of air through the combustion chambers. When the airflow is sufficient, fuel is injected into the chambers through the spray nozzles and is ignited by means of igniter plugs. It should be understood that the jet engine is not an alternate firing engine. The spark igniters are used only for the initial firing, and the fuel in all the combustion chambers burns continuously like a blowtorch, as indicated in the drawing of Fig. 19-12.

The turbojet engine actually burns only a small percentage of the air taken into the compressor. The larger volume of air flows around the outside of the burning flame and serves as a cooling blanket to prevent the high temperature from burning the combustion chamber. The design of a combustion chamber is shown in Fig. 19-13.

The Axial-Flow Turbojet Engine

In an **axial-flow** jet engine the air flows axially, that is, in a relatively straight path in line with the axis of the engine, as shown in Fig. 19-14. The axial-flow compressor consists of two elements: a rotating member called the rotor, and the stator that consists of rows of stationary blades. The principle of operation of the axial-flow turbojet engine is the same as that of the centrifugal-flow engine; however, the axial-flow engine has a number of advantages. Among these are: (1) The air flows in an almost straight path through the engine, and hence less energy is lost as a result of changing direction; (2) the pressure ratio (ratio of compressor inlet pressure to compressor discharge pressure) is greater because the air can be compressed through as many stages as the designer wishes; (3) the engine frontal area can be smaller for the same volume of air consumed and (4) there is high peak efficiency.

A cutaway illustration of a typical axial-flow turbojet engine is shown in Fig. 19-15. This is the Westinghouse J-34 engine, which was used for many years in naval aircraft. The principal parts of the engine are shown with proper nomenclature.

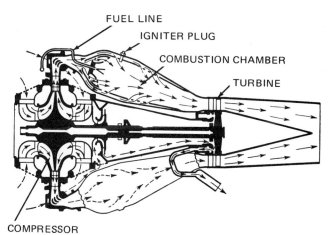

FIG. 19-11   Cross-sectional drawing of a centrifugal turbojet engine.

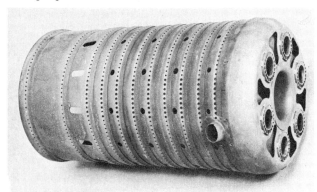

FIG. 19-13   A combustion chamber.

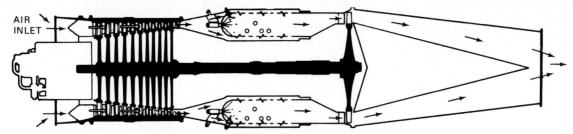

FIG. 19-14 Drawing of an axial-flow turbojet engine.

The J-34 engine has an 11-stage axial compressor which discharges into a double annular combustion chamber. An **annular** chamber consists of one burning compartment which surrounds the turbine shaft. The inner liner of the combustion chamber surrounds the turbine shaft and protects it from the heat of the combustion. Cooling air flows between the inner liner of the combustion chamber and the turbine shaft. The fuel nozzles are arranged in a ring at the forward end to provide for a uniform quantity of fuel spray around the chamber. In the double annular chamber, there are two concentric rings of fuel spray nozzles separated by the combustion chamber liner.

The Dual-Compressor Axial-Flow Engine

A dual-compressor jet engine utilizes two separate compressors, each with its own driving turbine. This type of engine is also called a "twin-spool" or "split-compressor" engine.

A drawing to illustrate the construction of the dual-compressor engine is shown in Fig. 19-16. The forward compressor section is called the low-pressure compressor (N_1) and the rear section the high-pressure compressor (N_2). It will be observed from the drawing that the low-pressure compressor is driven by a two-stage turbine mounted on the rear end of the inner shaft and the high-pressure compressor is driven by a single-stage turbine mounted on the outer coaxial shaft. The high-pressure rotor turns at a higher speed than the low-pressure rotor.

One of the principal advantages of the split compressor is greater flexibility of operation without danger of stall. The low-pressure compressor can operate at the best speed for the accommodation of the low-pressure, low-temperature air at the forward part of the engine. During high-altitude operation where air density is low, the N_2 compressor rpm will increase as compressor load decreases. This makes N_1 in effect a supercharger for N_2. The high-pressure compressor is speed-governed to operate at the proper speeds for the most efficient performance in compressing the high-temperature, high-pressure air toward the rear of the compressor section. The use of the dual compressor makes it possible to attain pressure ratios of more than 20 : 1, whereas the single axial compressor produces pressure ratios of 6 or 7 : 1 unless variable stator vanes are employed.

Turbofan Engines

A **turbofan** engine may be considered a cross between a turbojet engine and a turboprop engine. The turboprop engine drives a conventional propeller through reduction gears to provide a speed suitable for the propeller. The propeller accelerates a large volume of air in addition to that which is being accelerated by the engine itself. The turbofan engine accelerates a smaller volume of air than the turboprop engine but a larger volume than the turbojet engine.

A drawing to illustrate the arrangement of a forward turbofan engine with a dual compressor is shown in Fig. 19-17. The fan's rotational speed on this engine is the same as the low-pressure (N_1) compressor speed.

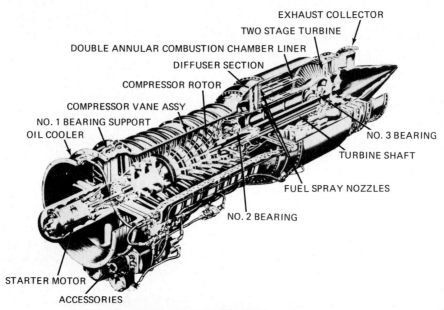

FIG. 19-15 Westinghouse J-34 turbojet engine.

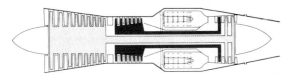

FIG. 19-16 Arrangement of a dual-compressor (twin-spool) turbojet engine. *(Pratt & Whitney)*

During operation the air from the fan section of the forward blades is carried outside the engine through ducting. It may either be dumped overboard or be carried to the rear of the engine and mixed with the exhaust. The effect of the turbofan design is to greatly increase the power-weight ratio of the engine and to improve the thrust specific fuel consumption (tsfc).

Turbofan engines may be constructed with a forward fan as in the Pratt & Whitney JT3D or with an aft fan as in the General Electric CJ-805-23 engine. The GE aft fan is a one-stage counterrotating free turbine mounted in a housing attached to the main turbine casing. The turbine is driven by the engine exhaust and has fan extensions on each blade. These fan blades extend into a coaxial duct which surrounds the main engine exhaust cone. Airflow enters the forward end of the duct and is expelled coaxially with the engine exhaust to produce additional thrust.

The High-Bypass Turbofan Engine

During recent years, the high-bypass turbofan engine has become one of the principal sources of power for large transport aircraft. Among such engines are the Pratt & Whitney JT9D, the General Electric CF6, and the Rolls-Royce RB. 211. These engines are used respectively in the Boeing 747, the Douglas DC-10, and the Lockheed L-1011 aircraft.

A **high-bypass** engine utilizes the fan section of the compressor to bypass a large volume of air compared with the amount which passes through the engine. The bypass ratio for the Pratt & Whitney JT9D and the Rolls-Royce RB.211 is approximately 5 : 1. This means that the weight of the bypassed air is 5 times the weight of the air passed through the core of the engine. The bypass ratio for the General Electric CF6 engine is approximately 6.2 : 1; however, some models have a variable bypass ratio and the amount of bypassed air may be more or less than stated above.

The principal advantages of the high-bypass engines are greater efficiency and reduced noise. The high-bypass engine has the advantages of the turboprop engine but does not have the problems of propeller control. The design is such that the fan can rotate at its most efficient speed, depending on the speed of the aircraft and the power demanded from the engine. Photographs of high-bypass engines are shown in Fig. 19-18.

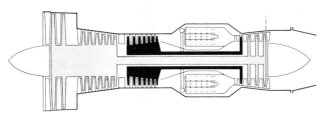

FIG. 19-17 Arrangement of a forward turbofan engine. *(Pratt & Whitney)*

PRINCIPAL PARTS OF A JET ENGINE

The main sections of a turbine engine are the **compressor section**, **combustion section**, and **turbine section**. In this discussion the diffuser and nozzle diaphragm are discussed as separate units. Generally, the diffuser is a component part of the compressor section and the nozzle diaphragm is a part of the turbine section.

The Compressor

As previously mentioned, there are two types of compressors with respect to airflow, the **centrifugal** type and the **axial** type. Many of the early engines utilized centrifugal compressors; however, present designs are mostly of the axial-flow type.

The configuration of a centrifugal-flow compressor is shown in Fig. 19-19. This is a double-sided turbine with air inlets at both the front and the rear. Air reaches the rear inlet of the compressor by flowing between the compressor outlet adapters, as shown in the drawing of Fig. 19-11.

Although the centrifugal compressor is not so expensive to manufacture as the axial-flow compressor, its lower efficiency eliminates the advantages of lower cost. Among the successful centrifugal engines manufactured in the United States was the Allison J-33 engine shown in Fig. 19-20.

The compressor rotor and one-half the stator case for an axial-flow turbojet engine are shown in Fig. 19-21. The parts shown are designed for the Westinghouse J-34 turbojet engine. It will be observed that the compressor blades are shaped like small airfoils and that they become smaller in each stage moving from the front of the compressor to the rear. The stator blades are also shaped like small airfoils, and they too become smaller toward the high-pressure end of the compressor. The purpose of the stator blades is to change the direction of the airflow as it leaves each stage of the compressor rotor and give it proper direction for entrance to the next stage. They also eliminate the turbulence that would otherwise occur between the compressor blades. The ends of the stator blades are fitted with shrouds to prevent the loss of air from stage to stage and to the interior of the compressor rotor.

During the operation of the compressor the air pressure increases as it passes each stage, and at the outlet into the diffuser it reaches a value several times that of the atmosphere, the actual pressure being over 70 psi [482.65 kPa].

The arrangement of a dual-axial compressor is shown in Fig. 19-22. This compressor design, as explained previously, makes it possible to obtain extremely high pressure ratios with reduced danger of compressor stall because the low-pressure compressor is free to operate at its best speed and the high-pressure compressor rotor is speed-regulated by means of the fuel-control unit. Compressor stall occurs when the air velocity in the first compressor stages is reduced to a level where the angle of attack of the compressor blades reaches a stall value. This condition is aggravated during acceleration owing to air "choke" at the outlet.

ROLLS ROYCE
RB·211

GENERAL ELECTRIC CF6

PRATT & WHITNEY JT9D

FIG. 19-18   High-bypass turbofan engines.

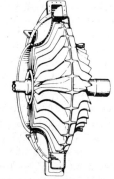

FIG. 19-19   Drawing of a centrifugal-flow compressor.

FIG. 19-20   Allison J-33 turbojet engine.

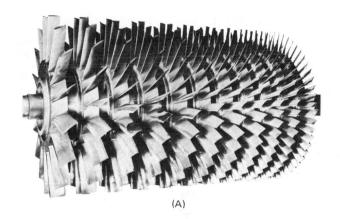

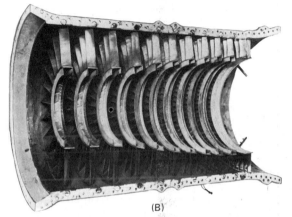

FIG. 19-21   (A) Rotor and (B) stator of an axial-flow compressor.

In the dual-compressor engine, the compressor drive shafts are coaxial; that is, one shaft is mounted on bearings inside the other. The high-pressure compressor rotor is driven by a single turbine mounted on the aft end of the outer shaft, and the low-pressure compressor rotor is driven by a two-stage turbine which is mounted on the end of the inner drive shaft aft of the first turbine.

Important features of compressor design developed by the General Electric Company for the J-79 (CJ-805) and the J-85 (CJ-610) engines are the **variable inlet guide vanes and variable stator vanes.** Variable stator vanes are also used in the GE CF6 high-bypass engine and the Pratt & Whitney JT9D. The variable vanes are automatically regulated in pitch angle by means of the fuel-control unit. The regulating factors are compressor inlet temperature and engine speed. The effect of the variable vanes is to provide a means for controlling the direction of compressor interstage airflow, thus ensuring a correct angle of attack for the compressor blades and reducing the possibility of compressor stall.

The Diffuser

The **diffuser** for a typical gas-turbine engine is that portion of the air passage between the compressor and the combustion chamber or chambers. The purpose of the diffuser is to reduce the velocity of the air and prepare it for entry to the combustion area. As the air velocity decreases, its static pressure increases in accordance with Bernoulli's law. As the static pressure increases, the ram pressure decreases. The diffuser is the point of highest pressure within the engine.

Combustion Chambers

The combustion section of a turbojet engine may consist of individual combustion chambers ("cans"), an annular chamber which surrounds the turbine shaft, or a combination consisting of individual cans within an annular chamber. The latter type of combustor is called the **can-annular** type or simply the **cannular** type. The combustion section and fuel nozzles provide proper mixing of the fuel and air.

A typical can-type combustion chamber consists of an outer shell and a removable liner with openings to permit compressor discharge air to enter from the outer chamber. Approximately 25% of the air that passes through the combustion section is actually used for combustion, the remaining air being used for cooling. A fuel nozzle is located at the front end of the combustion chamber through which fuel is sprayed into the inner liner. The flame burns in the center of the inner liner and is prevented from burning the liner by a blanket of excess air which enters through holes in the liner and surrounds the flame. All burning is completed before the gases leave the chamber. Figure 19-23 is an illustration of a single can-type combustor.

The high-bypass turbofan engines mentioned pre-

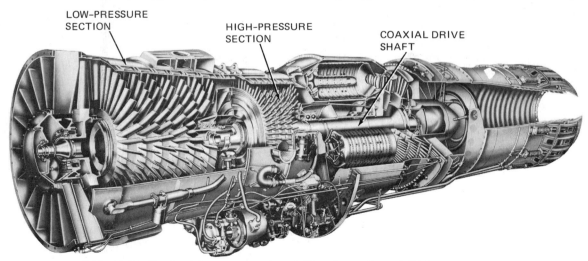

FIG. 19-22   Dual-axial compressor in a J-57 engine.   *(Pratt & Whitney)*

439

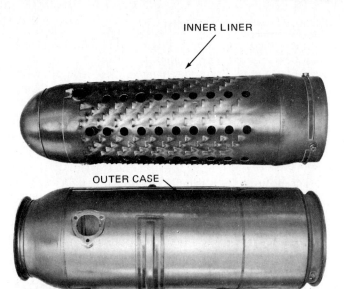

FIG. 19-23 Single can-type combustor.

viously employ annular combustion chambers. These have proved to be efficient and effective in producing smoke-free exhaust. The general configuration of the combustion chamber for the Pratt & Whitney JT9D engine is shown in Fig. 19-24. This is a two-piece assembly consisting of an inner and an outer liner. At the front are 20 fuel-nozzle openings with **swirl vanes** to help vaporize the fuel. Two of the openings, on opposite sides of the combustion chamber, are designed to hold the igniter plugs.

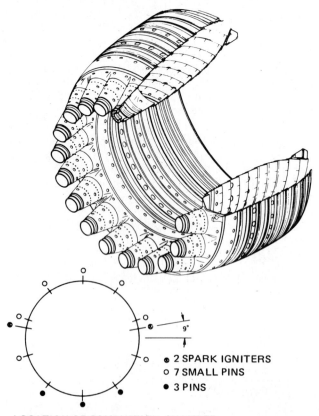

⊗ 2 SPARK IGNITERS
○ 7 SMALL PINS
● 3 PINS

LOCATION OF COMBUSTION-CHAMBER
RETAINING PINS

FIG. 19-24 Combustion chamber for the JT9D turbofan engine. *(Pratt & Whitney)*

Additional description of the JT9D combustion chamber will be given in Chap. 21.

Turbine-Nozzle Diaphragm

The **turbine-nozzle diaphragm** is a series of airfoil-shaped vanes arranged in a ring at the rear of the combustion section of a gas-turbine engine. Its function is to control the speed, direction, and pressure of the hot gases as they enter the turbine. A nozzle diaphragm is shown in Fig. 19-25. The vanes of the nozzle diaphragm must be designed to provide the most effective gas flow for the particular turbine used in the engine.

The vanes in a turbine-nozzle diaphragm are of airfoil shape to control the high-velocity gases in the most effective manner. When mounted in the nozzle ring, the vanes form convergent passages which change the direction of the gas flow, increase the gas velocity, reduce the pressure, and reduce the temperature of the gases. The heat and pressure energy of the gases are reduced as velocity energy is increased.

The total outlet area of the turbine nozzle is the sum of the areas of the cross sections of the passages between the vanes. The outlet area is less than the inlet area of the nozzle; the gas velocity is thus greater at the outlet than at the inlet. Note dimensions A, B, C, and D in Fig. 19-26, which shows the arrangement of the nozzle vanes and their effect on the gases. It will be noted that the direction of the gas flow is changed to impinge upon the turbine blades at the most effective angle.

It can readily be seen that the turbine vanes are exposed to the highest temperatures in the engine. Even though the gases are at a higher temperature during the fuel-burning process, the combustion chamber is protected from these high temperatures by a surrounding blanket of air. In order to withstand the extreme temperatures (1700 to 2000°F [927 to 1093°C]), the turbine-nozzle vanes and support rings must be made of high-temperature alloys and must be provided with cooling. Furthermore, they must be mounted and assembled in a manner to permit expansion and contraction without causing warpage or cracking.

Cooling is accomplished by making the vanes hollow and flowing compressor bleed air through them. A cross section of some typical air-cooled vanes is shown in Fig. 19-27. The air flows into the vanes and then out through holes in the leading and trailing edges where

FIG. 19-25 Turbine-nozzle diaphragm.

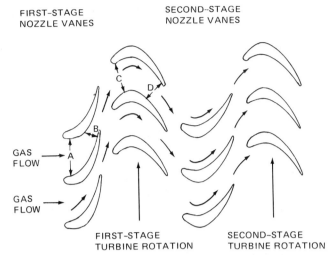

FIRST-STAGE
NOZZLE VANES

SECOND-STAGE
NOZZLE VANES

GAS FLOW →

GAS FLOW →

FIRST-STAGE
TURBINE ROTATION

SECOND-STAGE
TURBINE ROTATION

FIG. 19-26   Arrangement of nozzle vanes and turbine blades.

it mixes with the exhaust gases. Air cooling of this type is called **convection** or **film cooling.**

In some engines, the nozzle vanes are constructed of sintered, high-temperature alloys to provide walls with a certain degree of porosity. Cooling air is directed to the inside of the vanes, after which it flows out through the porous walls. This is called **transpiration cooling.**

The high temperatures in the nozzle and turbine area increase the corrosion rate of even the most corrosion-resistant materials. For this reason, first-stage turbine vanes and turbine blades are often provided with a corrosion-resistant coating. One such treatment is called *Jo-Coating.*

Because of the expansion and contraction of the materials caused by the high temperature, the nozzle vanes must be mounted in the inner and outer rings in a manner to prevent warping. This is accomplished by making the mounting holes in the support rings slightly larger than the end of the vanes. In other cases, the

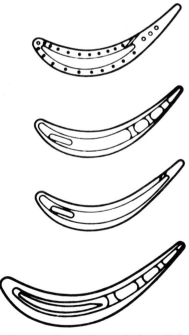

FIG. 19-27   Cross sections of typical air-cooled vanes.

vanes are welded into the rings but the rings are cut in sections to allow for expansion. Some of the modern engines have the turbine-nozzle vanes attached only to the outer ring; they can thus expand and contract without warpage.

Turbines

A turbojet engine may have a single-stage turbine or a multistage arrangement. The function of the turbine is to extract kinetic energy from the high-velocity gases leaving the combustion section of the engine. The energy is converted to shaft horsepower for the purpose of driving the compressor. Approximately three-fourths of the energy available from the burning fuel is required for the compressor. If the engine is used for driving a propeller or a power shaft, up to 90 percent of the energy of the gases will be extracted by the turbine section.

Turbines are of three types: the **impulse** turbine, the **reaction** turbine, and a combination of the two called a **reaction-impulse** turbine. Turbojet engines normally employ the reaction-impulse type.

The difference between an impulse turbine and a reaction turbine is illustrated in Fig. 19-28. The pressure and speed of the gases passing through the impulse turbine remain essentially the same, the only change being in the direction of flow. The turbine absorbs the energy required to change the direction of the high-speed gases. A reaction turbine changes the speed and pressure of the gases. As the gases pass between the turbine blades, the cross-sectional area of the passage decreases and causes an increase in the gas velocity. This increase in velocity is accompanied by a decrease in pressure according to Bernoulli's law. In this case the turbine absorbs the energy required to change the velocity of the gases. Typical turbines are illustrated in Fig. 19-29.

Since the nozzle vanes and turbine blades in gas-turbine engines are subjected to extremely high temperatures, it is essential that they be constructed of high-temperature alloys and that some type of special

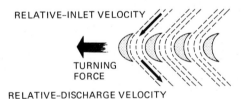

RELATIVE-INLET VELOCITY

TURNING FORCE

RELATIVE-DISCHARGE VELOCITY

RELATIVE-INLET VELOCITY = RELATIVE-DISCHARGE VELOCITY

IMPULSE

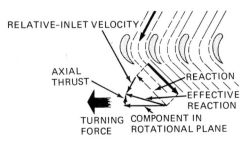

RELATIVE-INLET VELOCITY

AXIAL THRUST

REACTION

EFFECTIVE REACTION

TURNING FORCE

COMPONENT IN ROTATIONAL PLANE

REACTION

FIG. 19-28   Comparison of impulse and reaction turbines.

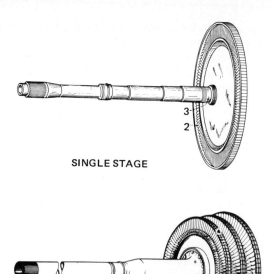

SINGLE STAGE

THREE STAGE

FIG. 19-29 Typical turbines.

cooling be provided. If the vanes and blades in the turbine area cannot withstand the temperatures to which they are subjected, burning and stress-rupture cracks will develop. It must be remembered that the efficiency of an engine becomes greater as the temperature of the gases at the burner outlet becomes greater. The development of high-temperature alloys containing cobalt, columbium (niobium), nickel, and other elements has made it possible to increase the operating temperature of gas-turbine engines, thus increasing the power available from the engines. A further development is the application of coatings to the vanes and blades to

withstand heat and prevent high-temperature corrosion.

Since the temperature of the burning gases in a gas-turbine engine decreases substantially as the gases pass through each turbine stage, it is usually necessary to provide special cooling for the first stage only. In some engines, the second-stage nozzle vanes and turbine blades are also air-cooled. As mentioned previously, the first-stage nozzle vanes are made with interior passages through which air is passed for cooling. First-stage turbine blades are also made with air passages, as shown in Fig. 19-30. This illustration shows cross sections of typical turbine blades. The method by which cooling air is directed through the first-stage turbine blades of the Pratt & Whitney JT9D engine is shown in Fig. 19-31. The cooling arrangement for the Rolls-Royce RB·211 turbine blades is shown in Fig. 19-32. Air for cooling is bled from the high-pressure compressor in each case.

Turbine blades are made in *shrouded* and *unshrouded* configurations. The shrouded blade has an extension cast on the tip to mate with the extension on the adjacent blade and form a continuous ring around the blade tips. This ring aids in preventing the escape of exhaust gases around the tips of the blades. The blades shown in Fig. 19-32 are of the shrouded type. Because of the added weight of the shrouds, shrouded turbine blades are particularly susceptible to **growth** caused by centrifugal force and overtemperature conditions.

The attachment of turbine blades to the turbine disk is usually accomplished by means of "fir-tree" slots broached in the rim of the disk and matching fir-tree bases cast or machined on the blades. After the blades are inserted in the slots on the rim of the turbine disk, they are held in place by means of rivets or metal tabs

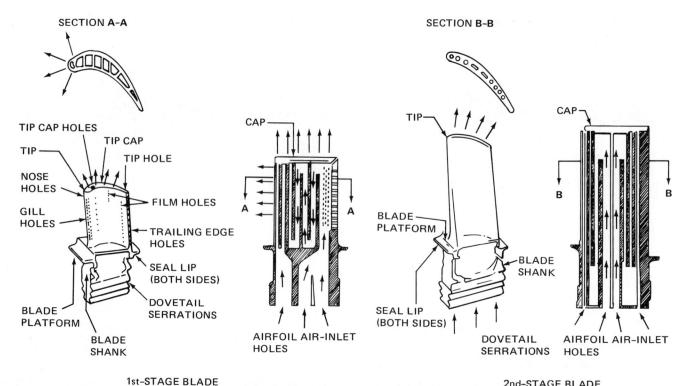

FIG. 19-30 Air-cooled turbine blades.

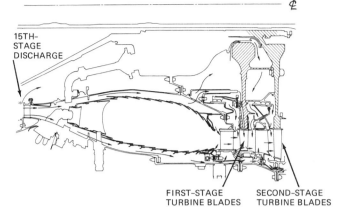

FIG. 19-31 Cooling of first-stage turbine blades on a JT9D turbofan engine. *(Pratt & Whitney)*

which are bent over the base of the blade. In the General Electric CF6 engine, the turbine blades are held in the disk rim by means of a single-piece blade retainer bolted to the face of the disk. These retainers not only hold the blades in place but also serve as seals for the faces of the disks to prevent the loss of cooling air.

The balance of a turbine wheel assembly is critical because of the high rotational speed during operation. For this reason all turbine blades are **moment-weighed** and marked to ensure that the correct blade is placed in each slot on the rim of the turbine disk. During the assembly of the blades to the disk, the technician must take great care to ensure that the proper blades are installed.

The Exhaust Nozzle

The normal function of the exhaust nozzle, or cone, is to control the velocity and temperature of the exhaust gases. Although a certain amount of thrust would be produced even if there were no exhaust nozzle, the thrust would be comparatively low and the direction of flow would not be properly controlled. When a convergent nozzle is used, the velocity of the gases is increased and the flow is directed so the thrust is in line with the engine.

The cross-sectional area of the nozzle outlet is most critical. If the area is too large, the engine will not develop maximum thrust, and if the area is too small, the exhaust temperature will be excessive at full-power conditions and will damage the engine. On some early gas-turbine engines, the exhaust-nozzle area was adjusted by means of small fittings called "mice." These

were installed at the nozzle exit as required to give the correct nozzle area.

If the exhaust gases of a turbojet engine reach supersonic speed it is necessary to employ a convergent-divergent (C-D) exhaust duct in order to obtain maximum thrust. To ensure that a constant weight or volume of a gas will flow past any given point after sonic velocity is reached, the rear part of a supersonic duct must be enlarged to accommodate the additional weight or volume that will flow at supersonic speeds. The rate of increase in area in a divergent duct is just sufficient to allow for the increase in the rate of change in volume of the gases after they become sonic.

Airliners include thrust-reversing devices to assist the brake system in slowing the aircraft after landing. The two most commonly used types of thrust reversers are the **aerodynamic blockage system** illustrated in Fig. 19-33 and the **mechanical blockage system** illustrated in Fig. 19-34. The thrust reverser shown in Fig. 19-33 is equipped with vanes and deflectors by which the exhaust gases are deflected outward and forward and are controlled through the thrust lever in the cockpit. After the aircraft has landed, the pilot moves the levers to the rear of the IDLE position. This causes the deflecting vanes to move into the main stream of the gasflow through the engine and reverse the gasflow direction. At the same time as the thrust levers are moved further rearward, the fuel flow to the engine is increased. When the thrust levers are in the FULL REVERSE position, the engine power output is approximately 75 percent of full-forward thrust capability. The system illustrated in Fig. 19-33 consists of a number of cascade vanes and solenoids with a pneumatic motor operating through gears and shafts to move the cascade vanes to the deployed position. Some aircraft utilize a separate thrust-reverser lever from the throttle; however, the two are connected to the same cable and clutch systems to prevent the thrust reverser from being deployed above the IDLE position. On other aircraft, the throttle and thrust lever are the same lever.

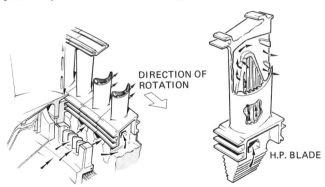

FIG. 19-32 Cooling of turbine blades for an RB-211 turbofan engine. *(Rolls-Royce)*

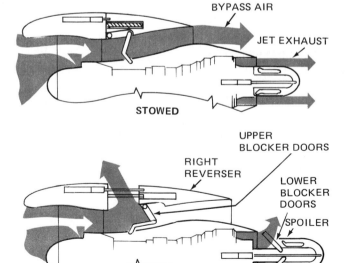

FIG. 19-33 Thrust-reverser action on a DC-10 airplane. *(McDonnell Douglas Corp.)*

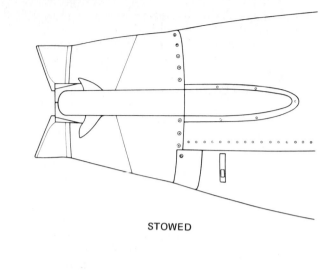

STOWED

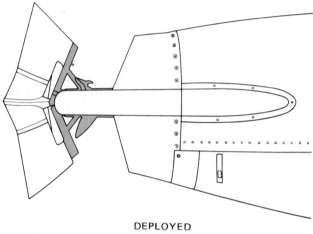

DEPLOYED

FIG. 19-34 Thrust-reverser action on a business aircraft. (Dee Howard Co.)

The mechanical blockage reverser shown in Fig. 19-34 consists of two blocker doors, or "clamshells," which, when stowed, form the rear part of the engine nacelle. When they are deployed as shown in the lower part of the illustration, they form a barrier to the exhaust gases and deflect them to produce a reverse thrust.

This thrust reverser is hydraulically operated and electrically controlled. The reverser cannot be deployed unless engine rpm is less than 65 percent.

Engine Noise

Because of the many complaints of excessive noise caused by gas-turbine engines and because of the danger of physical injury from such noise, engine and aircraft manufacturers together with government agencies have been actively engaged in experimenting with and modifying the engines in an effort to reduce noise to an acceptable level. Some of the methods for reducing noise are described later in this section.

Sound may be defined as *that which can be heard*. The reason sound can be heard is because it consists of a series of pressure waves in the air. A sound can consist of a combination of many waves in a wide range of frequencies or it can consist of a *pure* tone which is a single-frequency wave that follows the sine-wave pattern.

Noise may be defined as *unwanted and usually irritating sound*. The noise produced by a turbine engine consists of all frequencies audible to the human ear with intensities reaching levels which can be physically destructive. The intensity of sound is measured in **decibels** (dB). One decibel is one-tenth of a **bel**, the basic unit. A barely audible sound has an intensity of 1 bel, whereas the intensity of the sound produced by a jet engine may attain a value of 155 dB (15.5 bels) near the engine at takeoff power.

On the decibel scale, the intensity of sound increases on what is described mathematically as a logarithmic progression. This means that if the sound level in decibels is doubled, the intensity of the sound will be equal to the square of the original sound. If the sound level in decibels is tripled, the intensity of the sound will be equal to the cube of the original sound.

A scale indicating the decibel value of certain sounds is given in Fig. 19-35. It can be seen from this scale that any sound over 100 dB is very intense. The maximum level of sound that can be evaluated by the human ear is approximately 120 dB. Above this level, the ear can feel increasing intensity but cannot hear the

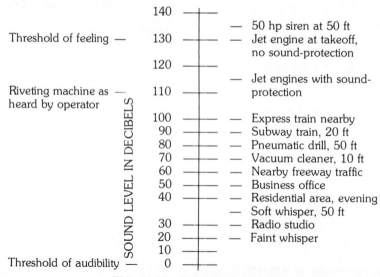

FIG. 19-35 Scale of sound values.

difference. Also, above this level, ear damage can occur.

Noise from a turbojet engine is caused by a number of forces, but it basically stems from the "torturing" of the air passing through the engine. Initially, the air is violently broken up and chopped into segments as it enters the inlet duct, passes through the inlet guide vanes, and encounters the compressor blades. Much of the sound thus created is in a wide range of frequencies, but a single frequency is also heard. This is the familiar "whine," caused by the compressor blades chopping the incoming air. A sound of this type is called "discrete," because it has an identifiable frequency and can be recognized in relation to other sounds. The most intense sound at high-power-engine settings comes from the exhaust nozzle. This sound is caused by the shear turbulence between the relatively calm air outside the engine and the high-velocity jet of hot gases emanating from the nozzle. The noise caused by the jet exhaust is termed **broad-band** noise because it includes many frequencies.

In a turbofan engine, considerable noise is caused by the secondary airflow from the fan section of the engine. This air has a lower velocity than the primary jet exhaust; hence, the noise is not as intense. An additional factor affecting the jet exhaust of the turbofan engine is that energy is extracted from the primary exhaust stream to drive the fan, and this results in decreased velocity for the exhaust jet. Thus, the noise produced is lower in intensity.

Reduction of Noise. As previously mentioned, experimentation has been going on for many years in an effort to reduce jet-engine noise. One of the first devices was a **multiple-tube** jet nozzle for commercial jet aircraft instead of the single exhaust nozzles used by military aircraft and early commercial aircraft. The effect of this type of nozzle is to reduce the size of the individual jet streams and increase the frequency of the sound. The higher frequency sound attenuates (reduces) more rapidly as distance from the source is increased; hence, the sound at 500 or 1000 ft [152.4 or 304.8 m] from the aircraft is not as intense as it would be with the single-nozzle engine.

Another exhaust nozzle is the **corrugated-perimeter** type. This nozzle has an effect similar to the multiple-tube, type nozzle. In all modified nozzles, it is essential that the total outlet area be maintained to provide maximum jet efficiency. Multiple-tube and corrugated-perimeter nozzles are shown in Fig. 19-36.

The development of turbofan engines has made possible additional reduction of engine noise. In the turbofan engine, both the primary airflow and the secondary airflow are reduced in velocity compared with the turbojet. As explained previously, the reduced air velocity results in a decrease in noise intensity.

The high-bypass engines such as the General Electric CF6, the Pratt & Whitney JT9D, and the Rolls-Royce RB·211 produce noise of lower intensity than the turbojet or low-bypass turbofan engines for several reasons. The air discharge velocities are lower compared with those of other engines, there are no inlet guide vanes in the front of the fan section, and the engines are provided with sound-absorbing liners in-

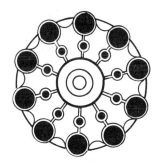

MULTIPLE-TUBE TYPE
END VIEW

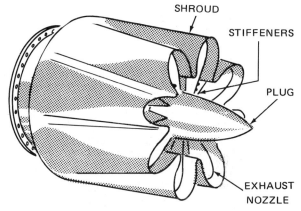

FIG. 19-36 Noise supressors.

side the fan ducts and exhaust nozzle. The arrangement of the noise-absorbent linings in the Rolls-Royce RB·211 engine is shown in Fig. 19-37. Advanced-design engines such as the Pratt & Whitney JT9D-7R4 and 2037 and the Garret ATF3-6 have reduced noise further by the use of new fan blade design, improved exhaust systems, and more effective soundproofing materials.

Protection Against Noise. It is essential that crew members working around turbine engines be provided with approved ear protectors. The most common types of protectors are over-the-ear devices in which earphones are installed for communication purposes. These protectors are muffs which completely enclose the ears, thus protecting them from noise while permitting the use of earphones for voice communication. All persons working in the areas where turbine engines are operated should ensure that they have approved ear protectors for use when engines are operating.

● GAS-TURBINE ENGINE SYSTEMS

We shall briefly mention here the principal systems necessary for the operation of gas-turbine engines and shall describe these systems more fully when we deal with specific types of engines. The systems normally associated with gas-turbine engines are the lubrication system, fuel and fuel-control system, ignition system, starter system, and the air-bleed and -supply system.

Lubrication Systems

The lubrication systems for gas-turbine engines are similar in many respects to lubrication systems for piston engines. An oil-supply tank is mounted on or near the engine, and the oil is supplied to the bearings and

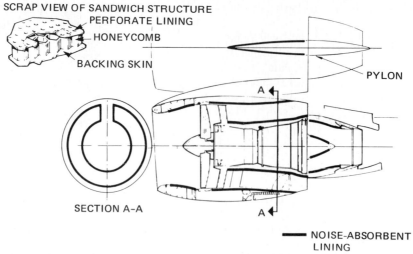

SCRAP VIEW OF SANDWICH STRUCTURE

PERFORATE LINING

HONEYCOMB

BACKING SKIN

PYLON

SECTION A-A

NOISE-ABSORBENT LINING

FIG. 19-37   Noise-absorbent linings in a turbofan engine. *(Rolls-Royce)*

gears in the engine under pressure from the main pressure pump. The system includes pressure-relief valves to prevent excessive pressures and a filter to remove foreign particles from the lubricant. Inside the engine the oil is drained to one or more sumps where it is picked up by scavenge pumps and returned to the oil tank. On many engines the oil is routed through fuel-oil coolers which serve to heat the fuel and cool the oil. The hot oil is passed through one set of passages, and the cold fuel flows through adjacent passages so that the heat of the oil can pass through the walls to the fuel.

The lubricant used with the majority of modern high-performance engines is of the synthetic type which will withstand much higher temperatures than will the petroleum-type lubricants previously used. Specifications for two of the synthetic lubricants are MIL-L-7808 and MIL-L-23699.

Additional information on lubrication systems is given in Chap. 8. Systems for particular engines are described in the sections concerning the engines.

Fuel and Fuel-Control Systems

The fuel system for a turbojet engine incorporates boost pumps, engine-driven pumps, filters, flow indicators, shutoff valves, fuel ice indicators, drain valves, fuel heater, main fuel-control unit, fuel nozzles, and an afterburner fuel control for engines equipped with afterburners.

The most critical part of the fuel system is the **fuel-control unit.** Among the parameters (measurements or quantities) fed to the fuel-control unit are power-lever (throttle) position, compressor inlet temperature (CIT), engine speed, compressor discharge pressure (CDP) which may also be burner pressure, exhaust-gas temperature (EGT), and possibly some others. The fuel-control unit integrates the values of these parameters and adjusts the fuel metering orifice accordingly. The fuel flow to the nozzles is therefore controlled according to all the conditions of operation so that excess fuel cannot be supplied to the engine regardless of the throttle position. If the throttle is moved from the idle position to the full-power position, the fuel-control unit will schedule fuel only as it can be properly used by

the engine. As engine speed and airflow increase, the fuel flow increases to provide a proper rate of acceleration without danger of overheating the engine or causing a "rich blowout."

When the throttle is moved to the idle position from a cruise or full-power position, the fuel flow will decrease gradually to prevent a "lean flameout."

For many years the majority of fuel controls have been of the hydromechanical type. This means that their operation was primarily accomplished by means of hydraulic and mechanical devices such as pistons, valves, speed sensors, gears, levers, cams, and pressure chambers. In recent years, with the advent of dependable solid-state electronic technology, fuel-control systems have been developed that include computers and associated electronic input devices that are much more accurate and trouble-free than the hydromechanical types. At the time of this writing only a few electronic fuel-control systems are in use and these are incorporated in the most recent aircraft, such as the Boeing 757 and 767 airliners.

Descriptions of both hydromechanical and electronic fuel-control systems are provided in Chap. 20.

Ignition Systems

Ignition systems for gas-turbine engines are not designed for continuous operation and therefore differ considerably from those manufactured for piston engines. The spark produced by the gas-turbine ignition system has many times the energy provided by a piston-engine system. One of the reasons this is necessary is that the mixture of fuel and air in a gas-turbine combustor is moving at a high velocity and is not uniformly distributed. The ignition system is therefore designed to produce a large, flaming spark.

The spark igniters (spark plugs) for a jet engine are usually located at two positions. If the combustion chamber is of the can type or the cannular type, the spark igniters are located in a position to ignite the fuel in two of the cans. The flame is then carried to the other chambers through **flame tubes.**

Information regarding the detailed arrangement of ignition systems is given in Chap. 10 and in sections describing specific engines.

446

Starting Systems

Gas-turbine engine starting systems have been discussed in the chapter on "Engine Starting Systems." However, we shall briefly review some of the principal points to be remembered.

Gas-turbine engines utilize a variety of starting systems. However, the pneumatic-type starter is most commonly used on large engines in commercial use. Pneumatic starters may be supplied with air from a ground cart, from a high-pressure bottle mounted in the aircraft, or from an auxiliary power unit (APU) mounted in the aircraft.

Air-Bleed and -Supply Systems

Compressed air from the compressor section of the gas-turbine engine is used for a number of purposes. It must be understood that the compression of the air as it moves through the compressor causes a substantial rise in temperature. For example, the air at the last stage of the compressor may reach a temperature of over 650°F [343.33°C] as a result of compression. This heated air is routed through the compressor inlet struts to prevent icing, and it is also used for various other heating tasks, such as operation of the fuel heater, aircraft heating, thermal anti-icing, etc.

Some engines are provided with automatic air bleeds which operate during the starting of the engine to prevent air from piling up at the high-pressure end of the compressor and "choking" the engine. This permits easier starting and accelerating without the danger of compressor stall.

Compressor air is also utilized within the engine to provide cooling for the turbine wheel and the turbine-nozzle vanes. These vanes are hollow to provide passages for the cooling air which is carried through the engine from the compressor to the area surrounding the nozzle diaphragm. Even though the compressed air is heated by compression well above its initial temperature, it is still much cooler than the burning exhaust gases and can therefore provide cooling.

The airflow from the compressor is also used within the engine to pressurize internal areas and control oil flow through the labyrinth seals and around bearings. Fig. 19-38 illustrates the type of bearing sump used on most modern turbine engines. Oil seals are used to help retain the oil in the bearing cavity and they may be of a carbon type or of the labyrinth type shown in the figure. One great advantage of a labyrinth-type seal is that there is virtually no wear because the rotating and stationary members of the seal do not touch. As noted in the drawing, two seal chambers exist: the oil seal next to the bearing and the air seal forming a pressurized chamber between the air and oil seals. The oil seal must retain the oil in the sump with the aid of the air seals. Oil for lubrication of the bearing flows through the oil jet, over the bearing, and out through the oil drain. The pressurization chamber is pressurized by air supplied from the engine bleed-air system. The volume of air supplied must maintain an adequate flow of air inward across the oil seal to blow the oil inward to the sump while some of the air is leaking outward between the rotating and stationary members of the air seal. The theory is that if air is flowing inward, oil cannot flow outward. A vent is provided at the top of the sump to ensure inward airflow to the sump. An

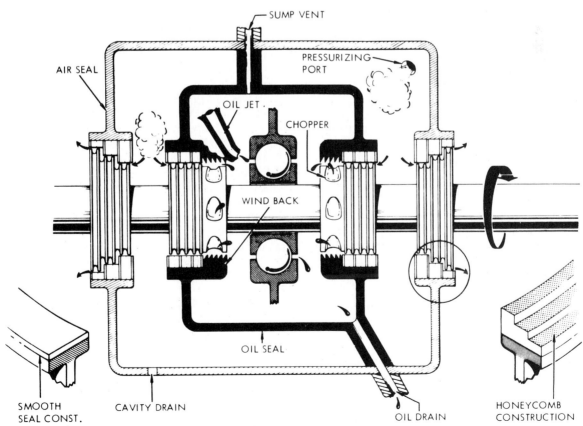

FIG. 19-38 Typical bearing sump.

overboard drain is provided in the pressurized chamber to evacuate any oil that may leak past the oil seal. This prevents oil from getting past the air seal to the compressor, contaminating the bleed air used for cabin pressurization. Oil that flows from the oil drain is picked up by a scavenge pump and returned to the main tank.

The engine compressors often supply pressure for the operation of accessories within the airplane. This air is called **customer air supply.** It can be used for operating a variety of pneumatic devices.

● TURBOPROP AND TURBOSHAFT ENGINES

Turboprop and **turboshaft** engines are similar to turbojet engines; however, they are designed to extract power from the exhaust-gas stream to drive propeller shafts and mechanical power shafts. This is accomplished by installing additional turbines in the exhaust-gas stream. These engines are described in other sections of this text.

● ADVANCED MANUFACTURING PROCESSES

Manufacturers of gas-turbine engines are continuously developing improved methods of manufacture to reduce costs, increase reliability, reduce weight, simplify maintenance, and increase the performance efficiency of the engines. The methods and processes employed today involve the application of technologies that have been developed as a result of years of research and experimentation and have resulted in the production of engines that are exceptionally reliable and can be operated for thousands of hours without the need for overhaul or repair.

Laser-Beam Welding

A **laser beam** is concentrated energy in the form of pure, coherent light. The diameter of the beam can be adjusted to the size necessary for a particular operation, even down to the size of a pin point. When used with automated controls, the laser beam can make welds much more rapidly and precisely than can be done with other methods.

Inertia Bonding

Inertia bonding is described as a solid-state welding technique which forms very strong weld joints with metallurgical properties equivalent to the base metal. The inertia bonds are more reliable than mechanical joints and have superior fatigue strength.

In inertia bonding, a rotating part is forced against its stationary mating part and the resulting heat generated by friction and pressure causes the metal to bond without actually melting.

Powder Metallurgy

In **powder metallurgy,** powdered metal is formed into engine parts with appropriate dies under heat and pressure. The resulting parts are described as near-net-shape because they require very little machining and finishing. The process therefore greatly decreases the cost of production and the wastage of material. **Hot**

isostatic pressing (HIP) is one of the most advanced processes for forming powdered alloys into finished parts by the application of extreme pressure and heat. Turbine disks of many modern engines are manufactured with powdered alloys and the HIP process.

Automated Investment Casting

Investment casting has been in use for many years in a variety of applications; however, it is now being accomplished automatically, thus greatly reducing cost and assuring the quality and uniformity of the cast products. The process is employed extensively for the production of turbine blades and vanes.

Investment casting is a process consisting of a number of distinct steps that result in a finished part that is an accurate copy of the original pattern in dimensions (Fig. 19-39). The pattern for a part to be produced by investment casting is constructed of a hard wax formed in the exact shape of the part to be cast. The pattern and mold are often designed so that many identical parts can be produced in one operation.

The wax pattern is dipped automatically by robot arms into a vat containing a suitable slurry. The pattern is then lifted out of the slurry and exposed to a spray of dry refractory material called "stucco" that adheres to the slurry. The dipping in the slurry and the application of the stucco is alternately performed by the robot devices until a mold of the desired thickness has been formed around the pattern. After the mold has solidified over the pattern, the mold and pattern are subjected to heat. The wax pattern melts and leaves the required cavity within the mold. The mold is "fired" by exposure to high temperature to give it the strength and resistance to heat that is required for casting.

If the part to be cast requires internal passages, such as the cooling-air passages for turbine blades and vanes, these are created by placing correctly shaped cores inside the mold in precisely located positions. The cores are made of a heat-resistant ceramic or similar material that can be removed from the part by leaching, that is, chemical dissolving.

Before casting, the finished mold is heated to a temperature at or near the temperature of the molten alloy to be cast. This is to prevent the rapid cooling and solidification of the alloy that would result in unsatisfactory grain structure in the part being cast. The alloy is poured into the mold, after which the mold and alloy are cooled under precisely controlled conditions. Upon cooling, the mold breaks away and leaves the cast part or parts ready for further finishing.

In recent years, directionally solidified and single-crystal castings have been developed by Pratt & Whitney and the Howmet Turbine Components Corporation. These are castings that are slowly cooled from the bottom of the mold. A **directionally solidified** casting develops crystals during cooling that extend from one end of the part to the other, thus eliminating the crosswise grain structure found in ordinary castings. Since the crystals in a turbine blade cast with the directionally solidified method are parallel to the centrifugal force loads applied during operation, the blade can withstand considerably more force than a blade with a granular structure.

Directionally solidified investment casting is accom-

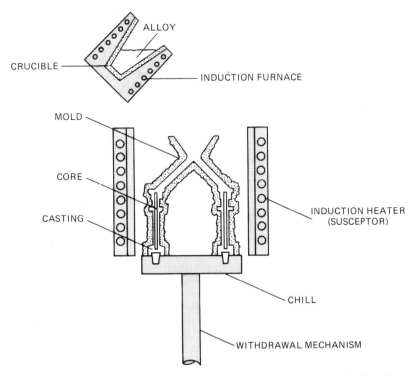

FIG. 19-39 Drawing to illustrate the investment casting process. *(Howmet Turbine Components Corp.)*

plished by means of a furnace, shown schematically in Fig. 19-39. The mold is placed on a heat sink (chill plate) within an induction furnace. The molten alloy is poured into the heated mold, and as heat is reduced, the molten alloy begins to solidify as the heat is drawn off by the water-cooled chill plate. The mold is then slowly withdrawn from the furnace and solidification of the alloy continues upward. The effect of this procedure is to cause the metal crystallization to take place in a linear manner through the length of the casting.

Single-crystal castings are produced in a manner similar to that of the directionally stabilized casting. To produce a single-crystal casting, a "seed" crystal is placed at the bottom of each mold cavity. This causes a single crystal to be formed as the mold is withdrawn from the furnace. Single-crystal turbine blades provide maximum strength and are used in such engines as the Pratt & Whitney JT9D-7R4 and 2037 engines. The turbine blades are coated with alloys that resist heat, oxidation, sulfidation, and high-temperature corrosion.

● REVIEW QUESTIONS

1. Describe Hero's aeolipile.
2. In what way was Newton's carriage associated with jet propulsion?
3. Discuss the contributions of Sanford Moss to the development of the turbine engine.
4. How did Frank Whittle's developments contribute to the growth of the gas-turbine industry in the United States?
5. When and in what country was the first known jet-aircraft flight made?
6. What laws explain the development of thrust in a jet engine?
7. What is required to change the velocity of a mass?
8. Explain *acceleration*.
9. What is meant by the *acceleration of gravity?*
10. Give the formula for thrust when the change in gas velocity is considered.
11. Explain the thrust produced at the jet nozzle.
12. Describe the changes in the pressures and temperatures of gases as they go through a jet engine.
13. Give two laws of thermodynamics.
14. State Boyle's law and Charles' law.
15. Describe the Brayton cycle.
16. Give some examples of reaction forces.
17. How can the thrust of a jet engine be converted to horsepower?
18. Compare jet-engine thrust with propeller thrust.
19. Explain how air density affects jet-engine thrust.
20. What conditions affect air density?
21. Explain the effects of engine rpm, airspeed, and altitude on the thrust produced by a jet engine.
22. Describe the construction of a centrifugal jet engine.
23. Explain the operation of an axial-flow compressor.
24. Describe the arrangement of a dual compressor.
25. Compare the speed of the high-pressure compressor and the low-pressure compressor rotor in a dual-rotor engine.
26. Describe a turbofan engine.
27. Compare a high-bypass engine with a low-bypass engine.
28. What are the advantages of a high-bypass engine?
29. Why are some engines equipped with variable stator vanes in the compressor?
30. What is the function of the diffuser?
31. Describe three different types of combustion chambers.
32. What prevents overheating and burning of the combustion-chamber liners?
33. What type of combustion chambers are used in the high-bypass engines?

34. Explain the function of the turbine-nozzle diaphragm.
35. How are turbine-nozzle vanes designed for cooling?
36. Explain the difference between impulse and reaction turbines.
37. What provisions are made to prevent expansion from damaging turbine-nozzle rings?
38. What is one method of preventing high-temperature corrosion on turbine-nozzle vanes and turbine blades?
39. What is gained by operating turbine engines at very high temperatures?
40. What features of a turbine engine make it possible to operate at very high temperatures?
41. Why is it necessary to provide special cooling for only the first stages of a turbine?
42. What is the advantage of shrouded turbine blades?
43. How are turbine blades attached to turbine wheels?
44. What is done with turbine blades to ensure accurate balance when the blades are changed?
45. Explain the importance of the turbine-exhaust nozzle-area.
46. What type of exhaust nozzle is employed in an engine where exhaust gases reach supersonic speeds?
47. Describe the operation of a thrust reverser.
48. Where are thrust-reverser controls located in an airplane?
49. Compare the thrust reverser for a high-bypass engine with that of a reverser for a plain jet engine.
50. Discuss the type of noise produced by a turbojet engine.
51. What is the principal reason that noise is produced by a jet engine?
52. Describe methods by which noise intensity has been reduced in gas-turbine engines.
53. What precautions should be taken by technicians working in the vicinity of operating turbine engines?
54. What units are usually included in the lubrication system for a gas-turbine engine?
55. What type of lubricant is most commonly used in modern gas-turbine engines?
56. Name the units usually incorporated in the fuel system for a modern gas-turbine engine.
57. How does the ignition system for a gas-turbine engine differ from the ignition system for a piston engine?
58. What types of starters are usually employed with gas-turbine engines?
59. Discuss the functions of air-bleed and -supply systems for a turbine engine.
60. Discuss some of the advanced technology processes used in the manufacture of gas-turbine engines and engine components.

20 GAS-TURBINE CONTROL SYSTEMS

The control of fuel flow for reciprocating engines was described in Chaps. 4–6. It was shown that the fuel flow to the engine is controlled by the total mass of the air entering the induction system of the engine. The carburetor meters the fuel to provide the correct mixture to be burned in the cylinders. The quantity of the mixture is controlled by the opening of the throttle in accordance with the output power desired by the pilot.

The mass of the air which flows through a gas-turbine engine is so great that the direct air-measuring concept for fuel metering cannot be employed. Instead, several parameters (measurements) which affect the operation of the engine are fed into the fuel-control unit and their net effect, as computed by the control unit, determines the quantity of fuel flow to the fuel nozzles of the engine.

● PRINCIPLES OF FUEL CONTROL

Many different types and models of fuel-control units have been designed and used for gas-turbine engines. The simplest type of control can be a manually controlled valve; however, this is not practical, because the pilot would have to watch several gages and make frequent adjustments in order to keep the engine operating and to prevent damage. If the quantity of fuel is not correct for the velocity and pressure of the air, the engine cannot function. If too much fuel enters the combustion chamber or chambers, the turbine section may be damaged by excessive heat, the compressor may stall or surge because of back pressure from the combustion chambers, or a **rich blowout** may occur. A rich blowout takes place when the mixture is too rich to burn. If too little fuel enters the combustion chambers, a **lean dieout** occurs.

Among the various factors which must be considered in the regulation of fuel flow for a gas-turbine engine are ambient air pressure (P_{amb}), compressor inlet-air temperature (CIT), engine rpm, velocity of the air through the compressor, compressor inlet-air pressure (P_{t2}), compressor discharge pressure (P_{s4}), (also essentially the same as burner pressure), turbine inlet temperature (T_{t5} or T_{t6}), tail-pipe temperature, and throttle- or power-lever setting. Since some of the foregoing factors are interrelated, the parameters applied to the fuel-control unit can be reduced to ambient pressure, compressor inlet temperature (CIT or T_{t2}), burner pressure (P_{s4}), high-pressure compressor rotor speed (N_2), and power-lever (throttle) position. In an engine with a dual-compressor rotor, the first (low-pressure) rotor speed is N_1 and the second (high-pressure) rotor rpm is N_2.

A number of different methods of operation are used in fuel-control units; these are described in part by classification. For many years the majority of fuel-control units have been *hydromechanical* in operation and are called **hydromechanical** fuel-control units. This means that the operation is *hydraulic* and *mechanical*. A number of late-type engines are now controlled by means of electronic fuel control systems. These systems include computers that precisely measure all parameters and provide signals that result in maximum efficiency, reliability, and fuel economy. In this chapter we shall examine a variety of hydromechanical controls before considering the electronic systems.

The principal sections of a fuel-control unit are the fuel metering section and the computing section. The **fuel metering section** consists of a fuel metering valve across which a constant fuel pressure differential is established. Fuel flow through the valve depends on its degree of opening and this is controlled by the computing section and the power-lever position.

The **computing section** of the control unit accepts signals from the engine and the pilot to determine how much fuel should be delivered to the fuel nozzles in order to prevent damage from excessive heat, stalling, surge, or blowout (flameout). If the power lever is moved all the way forward, the computing section will begin to increase the opening of the metering valve but only enough to allow a gradual acceleration of engine speed as airflow through the engine increases sufficiently to keep the engine relatively cool and to provide a suitable mixture for combustion. This permits maximum acceleration without engine damage. If the burner pressure approaches a level which could cause surge or stall, the computer will limit the fuel flow until airflow is increased sufficiently. In some fuel-control units an air-bleed control system is included. This system causes bleed valves to open and allow excessive back pressure to be reduced in the compressor, thus avoiding stall or surge.

Figure 20-1 shows how one type of fuel-control unit responds to engine conditions, air pressure and temperature, and power-lever (throttle) position. In general, the following sequence takes place for a typical engine:

1. The engine is accelerated by the starter until the correct starting rpm is attained.
2. The power lever is advanced to IDLE position to provide fuel (positions 1 to 2 on the curve).
3. As engine rpm increases, fuel flow decreases from the acceleration level to the point where it is correct to sustain idle speed (position 3).

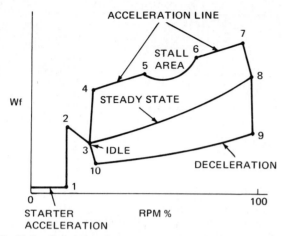

FIG. 20-1 Curves to show operation of a fuel-control unit.

4. The power lever is advanced to the full-power position and fuel flow is increased as indicated by position 4.

5. As engine rpm increases, fuel flow increases to position 5 and then it is decreased to avoid the stall and surge area. In the range of speeds between position 5 and position 6, the fuel must be reduced to prevent back pressure from the combustion chamber which could cause compressor stall or surge.

6. At position 6, the fuel flow is restored to the normal maximum acceleration level until the engine approaches maximum speed (position 7).

7. Shortly before maximum speed is reached, the governor reduces fuel flow to prevent overspeeding. The fuel flow is stabilized, as shown at position 8, at a quantity which will maintain 100 percent engine rpm.

8. When it is desired to reduce engine rpm to idle speed, the power lever is moved to the IDLE position. Fuel flow is immediately reduced to the value indicated by position 9 on the curve. This is a value which will permit maximum deceleration but will not let the fuel-air mixture drop to a ratio where a lean dieout may occur.

9. The engine rpm decreases, as indicated by the curve between positions 9 and 10, and the fuel flow decreases slowly.

10. When the engine rpm approaches idle speed, the fuel flow increases to the point where idle speed can be maintained. This is indicated by the curve between positions 10 and 3 in the diagram.

The curves of Fig. 20-1 will vary considerably as air pressure and temperature change. These values are applied to the computing section of the fuel control to modify fuel flow as conditions dictate.

● THE HAMILTON STANDARD JFC25 FUEL-CONTROL UNIT

General Description

The Hamilton Standard JFC25 fuel-control unit in its various configurations is designed for use with the Pratt & Whitney J57 and JT3C turbojet engines and for the JT3D turbofan engine. The control unit is of the hydromechanical type, being operated hydraulically and mechanically.

A diagram of the JFC25 fuel-control unit is given in

Fig. 20-2. The fuel metering section of the control is shown in the upper right portion of the drawing and it includes the **pressure regulating valve**, the **throttle valve**, and the **minimum pressure and shutoff valve**. The computing section of the unit includes the **power lever** (throttle lever), the **burner-pressure limiting valve**, the **speed-sensing governor**, and various cams, gears, valves, and levers which respond to the input signals furnished by the burner-pressure sensor and the power-lever position.

In the JFC25 fuel-control unit, the burner pressure (P_b), the engine speed (N_2), and the throttle-valve opening provide for a range of fuel-air ratios suitable for engine operation under all conditions. The fuel-control unit ensures that the ratio of fuel flow in lbh (W_f) to the burner pressure (P_b) remains within a range to prevent rich blowout, lean dieout, surge, or stall. It must be pointed out that some fuel-control units employ compressor discharge pressure (CDP or P_{t4}) as a principal parameter for fuel control. Actually, the compressor discharge pressure variation is essentially the same as burner pressure variation.

Operation

Fuel from the main fuel pump enters the fuel control through the fuel inlet shown in the upper right portion of the drawing of Fig. 20-2. It passes through the coarse filter and out to the pressure regulating valve and the throttle valve. Fuel to operate the servo systems in the unit passes through a fine filter to remove small particles which could interfere with operation. The fine filter is self-cleaning through the washing action of the fuel flow. The coarse filter incorporates a relief valve which will open if the filter should become clogged, thus preventing reduction of fuel flow.

The **pressure regulating system** includes the pressure regulating servo sensor and the pressure regulating valve. The **pressure-regulating-valve** assembly consists of a cylinder with a reset piston whose function is to adjust the spring pressure against the valve sleeve. The **pressure regulating servo sensor** compares the pressures on either side of the throttle valve and moves the reset piston to increase or decrease spring pressure on the regulating valve to maintain the correct pressure differential across the throttle valve. If the fuel pressure is too high, the regulating valve opens and bypasses fuel back to the inlet side of the main pump. Actually, during operation, the pressure regulating valve will almost always be partially open and bypassing fuel will not be required for operation.

Since a constant-pressure differential is maintained across the throttle valve, fuel flow is proportional to the amount of throttle opening. The throttle opening is controlled by the computer section, as mentioned previously.

After fuel passes through the throttle valve, it is subject to control by the **minimum pressure and shutoff valve**. This valve is shown in the extreme upper right section of the drawing of Fig. 20-2. It serves to ensure that the fuel pressure is sufficient to operate the servo units in the fuel control and adequate for proper functioning of the fuel nozzles. The valve also serves as the fuel shutoff valve, being actuated by servo pressure from the **sequence valve**.

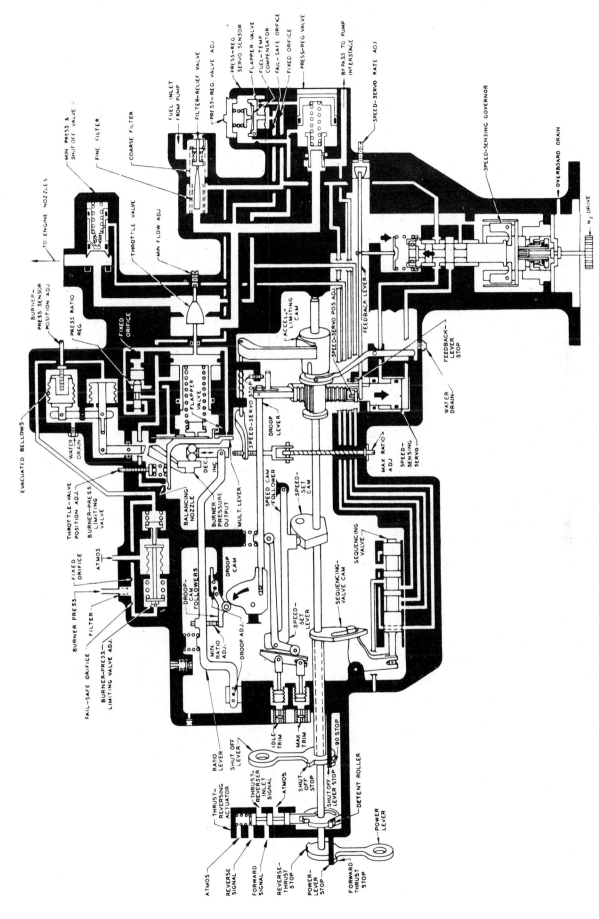

FIG. 20-2 Schematic drawing of the Hamilton Standard JFC25 fuel-control unit. (Hamilton Standard)

453

As can be seen in the drawing, the minimum pressure and shutoff valve consists of a plunger or sleeve-type valve spring-loaded to the *closed* position. It is opened by throttle downstream pressure, which overcomes the spring pressure and the pump interstage pressure. The valve is closed by spring pressure and high-pressure fuel directed from the sequence valve. The sequence valve is controlled by the fuel shutoff lever.

Because of the spring force acting on the minimum pressure valve, fuel pressure must build up to a predetermined level before fuel can flow to the engine nozzles. This pressure is needed to operate the servos in the control unit.

The **throttle valve** is a contoured cone operating in a sharp-edged orifice. It is made to close by a spring and high-pressure fuel and to open as the result of throttle servo pressure exerted against a large-area servo piston. In a steady-state operating condition, the forces acting in each direction on the throttle valve are in equilibrium. If a reduction in fuel flow is needed, the servo pressure is reduced and the servo piston moves to close the valve. When increased fuel is required, the servo pressure is increased to overcome spring force and fuel pressure, thus moving the valve to a more open position.

As explained previously, the fuel pressure differential across the throttle valve is controlled to remain constant. Thus, the fuel flow through the throttle valve is directly proportional to the area of the throttle-valve opening.

To prevent a complete shutdown of fuel flow while the engine is operating, the throttle valve is provided with an adjustable **minimum flow stop.** This stop is set to provide fuel sufficient to maintain engine operation at minimum rpm.

The **burner-pressure sensor assembly,** shown at the top in Fig. 20-2, consists of a pair of matched bellows, one evacuated and the other open to burner pressure as regulated by the burner-pressure limiting valve. In addition, the assembly includes a water drain, a linking lever, an adjustment, and the housing. Because of the evacuated bellows which provides a standard reference, the force exerted by the assembly is proportional to burner pressure.

Note in the drawing that the **burner-pressure limiting valve** will act to vent burner pressure to the atmosphere should the burner pressure become excessive. This is accomplished by means of a valve which balances the force of a spring against the force produced by the bellows. High pressure causes the bellows to overcome the force of the spring and open the valve.

In the linkage between the burner-pressure sensor and the throttle-valve piston, the rollers are placed between the **burner-pressure output lever** and the **multiplying lever.** The position of these rollers is controlled by the **ratio lever,** thus increasing or decreasing the effect of the burner-pressure output on the multiplying lever. As the rollers are moved downward in the drawing, the **flapper valve** tends to close and cause a buildup of fuel pressure on the piston which controls the throttle valve, thus increasing the opening of the valve. The total effect of the burner-pressure signal and the speed signal produces the correct W_f/P_b ratio for the proper operation of the engine.

The **speed-sensing governor** will probably be recognized as a typical flyweight speed-sensing unit similar to those used in many other mechanisms where speed is a factor. The governor is shown at the bottom right of Fig. 20-2. It includes a sleeve assembly with two filter screens, a valve, a flyweight assembly, and a head support plate. The drive shaft of the governor is splined to mate with an accessory drive on the engine. The drive is powered from the N_2 compressor; hence, the speed sensing provides a signal proportional to the N_2 speed of the engine. The governor flyweights provide a force which positions a spring-loaded pilot valve. This valve controls fuel servo pressure to and from the **speed-sensing servo** seen at the bottom center of the drawing.

When engine speed changes, the flyweight force varies and the pilot valve is positioned to port the speed servo piston chamber to high-pressure or low-pressure fuel and this results in a movement of the piston. The motion of the piston rotates the three-dimensional **acceleration-limiting cam** and also repositions the governor pilot valve through a **feedback lever.** This causes the speed-sensing system to attain a state of equilibrium when the speed selected by the pilot is attained.

The action of the power lever may be understood by carefully studying Fig. 20-2. When the power-lever position is changed, the **speed-set cam** is rotated. This applies force to the **speed cam follower** which transmits the movement to the **ratio lever.** This lever moves the rollers between the burner-pressure output lever and the multiplying lever which actually control the position of the throttle valve. This action was explained previously.

The action of the acceleration limiting cam is to ensure that excessive fuel not be provided for the engine during acceleration. Minimum fuel is controlled by means of the minimum flow adjustment of the throttle valve as explained previously.

The **shutoff lever** operates through the **sequencing valve** which controls pressure to the minimum pressure and shutoff valve, as mentioned earlier in this section. When the shutoff lever is placed in the shutoff position, the **sequencing-valve cam** is rotated, thus moving the sequencing valve to a position which ports high-pressure fuel to the minimum pressure and shutoff valve, thus causing it to close and shut off the fuel.

Summary of Operating Principles

The operating system of the fuel control may be considered to consist of a **metering section** and a **computing section.** The metering section selects the rate of fuel flow to be supplied to the engine nozzles to provide the selected engine output in keeping with the conditions sensed by the computing section of the control.

High-pressure fuel is supplied from the engine-driven fuel pump. This fuel passes through the filter system and provides for coarse filtering of the engine fuel and fine filtering of the servo fuel used to operate the control.

The fuel next encounters the pressure regulating valve and pressure regulating sensor, which are designed to provide a constant pressure across the throttle valve. The pressure regulating sensor continuously compares the fuel pressure entering the throttle valve with the pressure downstream from the valve and moves to adjust the pressure regulating valve to maintain the correct pressure differential. Fuel in excess of that required to maintain the correct pressure differential is bypassed back to the inlet side of the fuel pump.

The degree of opening of the throttle valve is controlled by the computing system. This system responds to burner pressure (P_b), engine speed (N_2), and power-lever position. Burner pressure is sensed by the burner-pressure sensor and engine speed is sensed by the speed-sensing governor. Signals from the power lever and the speed-sensing governor are modified by the three-dimensional cam and the droop system.

The **three-dimensional cam**, also called the **acceleration-limiting cam**, ensures that acceleration occurs at a level which will not permit overheating, surge, stall, or blowout. The **droop system** ensures that fuel flow will decrease as engine speed approaches the selected value, thus preventing overspeed and providing for an accurately controlled steady-state condition.

In connection with the power lever on the JFC25 fuel control is a cam which controls the **thrust-reversing actuator valve.** This valve directs pressure to the thrust-reverser system when the power lever is placed in the REVERSE THRUST position.

Certain models of the JFC25 fuel-control unit are equipped with a special cam and switch to initiate water injection for takeoff. Fuel flow is automatically increased while water injection is taking place.

● THE HAMILTON STANDARD JFC60 FUEL-CONTROL UNIT

Description

A description of the JFC60 fuel-control unit is included in this section because this unit is in extensive use. The JFC60 fuel-control unit is employed with the Pratt & Whitney JT8D turbofan engine which powers such aircraft as the Boeing 727, the Boeing 737, and the Douglas DC-9. It is also used on some foreign aircraft.

The JFC60 fuel-control unit operates in a manner similar to the JFC25 described previously; however, this unit incorporates a system for sensing compressor inlet temperature (CIT or T_{t2}) as an additional parameter for engine control. The unit has two control levers operating through coaxial shafts, one being the power lever which controls thrust and thrust reversal, and the other being the fuel shutoff lever.

The unit accurately governs engine steady-state speed and also provides for control of acceleration and deceleration to avoid overtemperature, stalling, surge, and flameout. The unit functions with a metering system and a computing system which operate in a manner similar to that described for the JFC25 fuel-control unit. A diagram of the JFC60 fuel-control unit is shown in Fig. 20-3.

Operation

Figure 20-3 shows the fuel inlet at the lower right of the drawing. The fuel passes through a coarse filter before entering the throttle valve. Note that fuel can be bypassed by the **pressure regulating valve** which senses fuel pressure upstream and downstream of the **throttle valve** in order to make a continuous comparison of the pressure differential across the valve. Thus a constant-pressure differential of 40 psi [275.8 kPa] is maintained across the throttle valve.

Entering fuel from the main fuel pump passes through a fine filter to provide clean fuel for servo operation in the control unit. Both the coarse filter and the fine filter have relief valves to permit fuel flow if they should become clogged.

The **computing section** of the control unit consists of the **speed-sensing governor** shown near the center of the drawing, the **CIT sensing system** shown in the lower left of the drawing, the **CDP (P_{s4}) sensing system** at the upper left of the drawing, and the **power lever** shown also at the upper left of the drawing. It will be noted that the governor flyweights move the N_c or N_2 **pilot valve** which controls servo pressure to the three-dimensional **acceleration-limiting cam.** Thus, the cam is moved axially in accordance with engine speed. The CIT sensing system controls a pilot valve which directs servo pressure to the **CIT servo** which moves the CIT servo lever, and this, in turn, rotates the three-dimensional acceleration-limiting cam by means of a rack and pinion gear. The effect of changes in air density is thus supplied to the control unit.

The compressor discharge-pressure sensor feeds signals to the unit through the **CDP output lever.** It can be seen that this lever acts to apply pressure to the **throttle-valve rollers** which act against the **multiplying lever.** It is now apparent that all the parameters affecting engine operation are linked together to apply force to the multiplying lever which moves the throttle-valve pilot valve and thus directs pressure to or from the **throttle-valve actuating piston** to change its position and adjust the fuel flowing to the engine. This operation is very similar to that described for the JFC25 control explained previously. Both controls act to maintain the proper W_f/P_{s4} ratio for correct engine operation.

The JFC60 fuel-control unit is equipped with a **windmill bypass and shutoff valve,** which is shown in the upper left portion of the drawing. The shutoff feature is similar to that described previously for the JFC25 unit; however, the windmill bypass system is an addition to take care of fuel flow when the engine is shut down in flight and is windmilling. The windmill bypass and shutoff valve is ported to the main pressure regulating valve. Fuel pressure on the spring side of the valve piston is now ported to pump interstage pressure, which permits the pressure regulator to act as a relief valve to dispose of all windmilling fuel pressure.

The feature of the JFC60 fuel control which permits acceleration without allowing excessive fuel flow is the three-dimensional (3D) cam, contoured to provide a schedule of W_f/P_{s4} in relation to compressor inlet temperature as a limiting factor throughout the acceleration range. The 3D cam is adjusted by engine speed and CIT, as can be seen in the drawing. The output of the cam is used to move a pair of rollers between the compressor discharge-pressure lever and the multiplying lever to effect changes in throttle opening. The result of the foregoing actions is to permit engine ac-

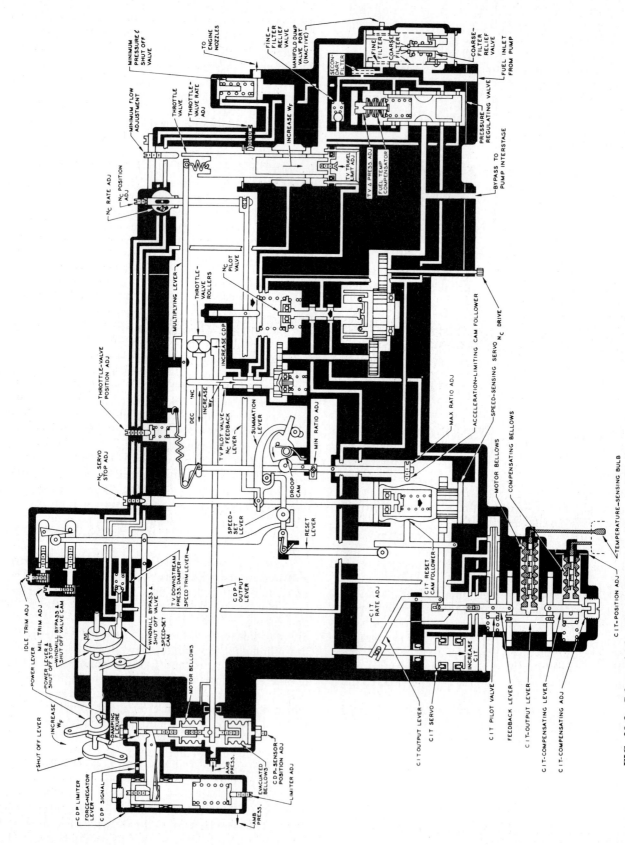

FIG. 20-3 Schematic drawing to illustrate the operation of the Hamilton Standard JFC60 fuel-control unit. (*Hamilton Standard*)

celeration within the temperature and surge limits of the engine.

Deceleration control is provided by the constant radius portion of the droop cam and by adjustment of the roller position linkage to limit the travel of the rollers toward decreasing fuel flow, thereby effecting a minimum W_f/P_{s4} ratio. This provides a linear relationship between fuel flow and compressor discharge pressure, which results in blowout-free deceleration.

Engine-speed control is accomplished by comparing the actual speed, as indicated by the position of the speed servo, to the desired speed value required for the power selected by the pilot through the power-lever positioning of the **speed-set cam.** The speed-set cam is shown in the upper left portion of Fig. 20-3 on the power-lever shaft. The power lever actuates the speed-set cam to select a **governor droop line.** The position of the droop line is biased by compressor inlet temperature. Deviation of actual speed from the desired speed causes movement of the **speed servo.** This movement of the speed servo is transmitted through a lever and results in the repositioning of the droop cam. The rollers in the multiplication system are positioned through the action of the droop cam to be a function of the speed error. The repositioning of the rollers then provides the required steady-state W_f/P_{s4} ratio setting.

The **temperature-sensor bellows assembly,** shown in the lower left portion of Fig. 20-3, consists of the motor and compensating bellows unit, a feedback lever, a pilot valve, an output lever and pushrod, a compensating lever and pushrod, and the temperature-sensor housing. Compressor inlet temperature is sensed by a liquid-filled bulb mounted in the compressor inlet and connected to a liquid-filled bellows in the control. The liquid expands with increased temperature and the extra volume travels through a capillary tube to the liquid-filled motor bellows in the control. The bellows length changes and through levers displaces a four-way pilot valve, which results in movement of the temperature servo piston. The servo piston is connected through a linkage to a rack which meshes with the spline on the 3D cam and the motion of the piston rotates the cam. The feedback lever is attached to the rack, and as the rack moves to rotate the cam, it also repositions the pilot valve in order to return the valve to the steady-state position. The rotation of the 3D cam, acting through a linkage, resets the governor droop and accelerating line. Any ambient air- or fuel-temperature variation which acts on the motor bellows and capillary tube also acts on a **compensating bellows** and dead-ended capillary tube, causing the fixed pivot of the motor-bellows lever system to move to a new position so that the net result of the variation is not sensed at the pilot valve.

In the foregoing description, note that the operation of the JFC60 fuel-control unit serves the same functions as those described for the JFC25 control.

● THE HAMILTON STANDARD JFC68 FUEL-CONTROL UNIT

The JFC68 fuel-control unit serves the functions previously described for other hydromatic controls as well as an additional function of supplying reference pressures and hydraulic pressure for the **engine vane control.** The JFC68 control is designed for use with the Pratt & Whitney JT9D turbofan engine and is considerably more complex than the controls previously described. Note, however, that the same engine operating parameters are used to provide input signals to the control and that the same principles of operation are employed.

The JFC68 fuel control is mounted on the main-engine fuel pump and operates in connection with the engine vane control (EVC3) to regulate the thrust of the engine. The control utilizes the W_f/P_{s4} ratio as a control parameter, as do the other fuel controls described in this section.

Metering System

The JFC68 fuel-control metering system consists of regulated fuel pressure applied across a window-type throttle valve. This is essentially the same system used in the JFC60 control. The engine fuel is filtered through a coarse filter and the servo fuel is filtered through a fine filter. The fuel metering system can be seen in the right center section of the diagram of Fig. 20-4. The main units of the system are the **pressure-regulating-valve sensor,** the **pressure regulating valve,** and the **throttle valve.** Note that the throttle valve (near the center of the drawing) consists of a hollow sleeve with openings to allow fuel to flow out to the engine. The area of the openings depends on the axial position of the sleeve as determined by the computing section.

Fuel not necessary to maintain the pressure differential across the throttle valve is bypassed by the pressure regulating valve back to the pump interstage. Pump interstage pressure is maintained inside the case of the control.

Computing System

The computing system of the control unit utilizes engine N_2 speed, burner pressure (P_{s4}), compressor inlet temperature (T_{t2} or CIT), ambient pressure (P_{amb}), and power-lever position to schedule fuel to the engine. P_{s4} is sensed by the engine **burner-pressure sensor.** This unit consists of two bellows, one of which is evacuated and the other exposed to burner pressure on the outside and ambient pressure on the inside. This unit is shown in the lower right section of the drawing of Fig. 20-4. It will be noted that the bellows movement is applied to a flapper valve which directs servo pressure to and from each side of a servo. The servo acts on a lever which controls the position of rollers between the *ratio lever* and the *multiplying lever.* The multiplying lever moves the **throttle-valve pilot valve** to direct servo pressure which moves the throttle valve. The force on the multiplying lever is balanced by the **throttle-valve feedback spring** as a function of actual fuel flow. It will be noted that the **throttle-valve pilot valve** and other spool-type valves in the fuel control are continuously rotated. This is to keep them free from dirt particles which might become caught between operating surfaces and cause the valves to stick. Thus the unit is said to be able to "eat dirt."

Acceleration control is provided through the three-

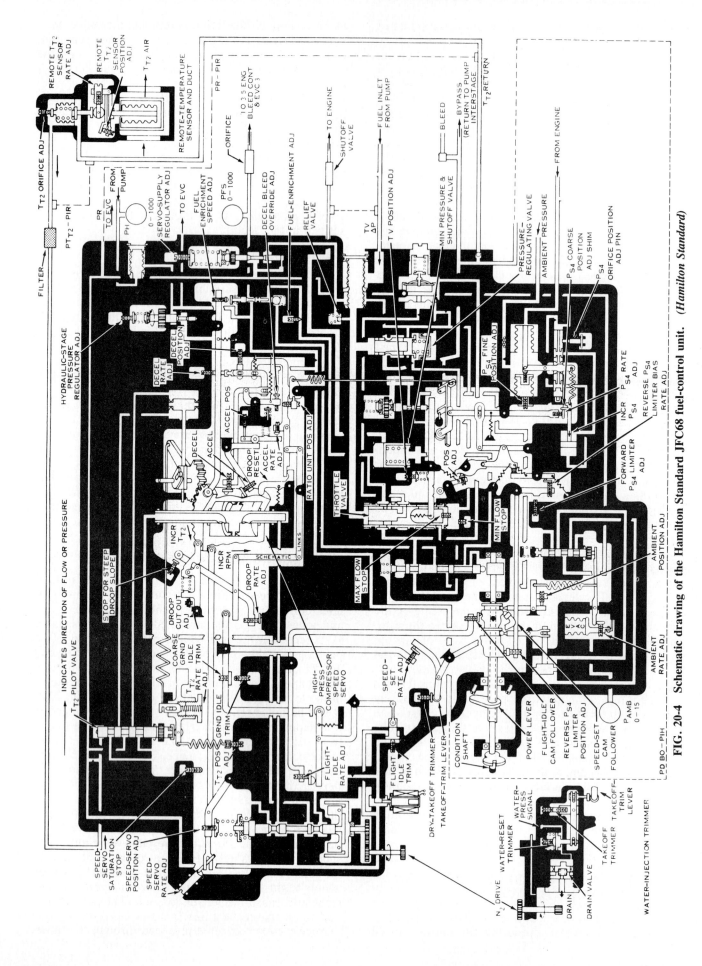

FIG. 20-4 Schematic drawing of the Hamilton Standard JFC68 fuel-control unit. (*Hamilton Standard*)

dimensional cam which is positioned axially by servo pressure from the N_2 speed governor through the operation of a servo piston inside the cam body. The cam is positioned radially by servo pressure from the T_{t2} (CIT) pilot valve acting on a servo that rotates a sector gear meshed with gear teeth on the cam. The three-dimensional cam is shown in the upper center portion of the drawing of Fig. 20-4.

The 3D cam is contoured to define a schedule of W_f/P_{s4} versus engine speed for each engine-inlet-temperature value. The operation of the cam is such that it permits engine accelerations which avoid the over-temperature and surge limits of the engine without adversely affecting engine accelerating time. It will be noted that the cam produces its effects through three cam followers: one for acceleration control, one for deceleration control, and one for droop reset. The outputs of these followers are fed through a series of linkages to the ratio-unit spring which is connected to the throttle-valve actuating system. It can also be seen that the **deceleration cam follower** acts through a linkage and valve to control the **compressor bleed actuator.**

The **power lever** and the **condition lever** are shown in the lower left portion of the drawing above the **ambient-pressure sensor.** The power lever rotates the **speed-set cam,** which determines the point at which the governor droop linkage overrides the **acceleration-limiting cam** (3D cam) to decrease the W_f/P_{s4} ratio with increasing engine speed. This will continue until steady-state W_f/P_{s4} is attained. The governor droop-reset contour is provided on the 3D cam to maintain a constant droop slope at all operating conditions.

During acceleration, the 3D cam provides a biased temperature schedule down to the new power-lever setting droop line or the minimum flow line, whichever occurs first. Deceleration continues until steady-state operation is attained. Both the acceleration and deceleration schedules are provided by surfaces on the 3D cam; hence, both schedule as a function of engine speed.

As explained previously, engine speed is controlled by the speed governor. The governor continually compares actual engine speed (N_2) with desired speed as selected by the pilot through the power lever and speed-set cam. The power lever rotates the speed-set cam to set the desired governor droop line while the cam is moved axially as a function of P_{amb} to provide the biasing of the selected speed. The **ambient-pressure sensor** is shown in the lower left of the drawing of Fig. 20-4. It can be seen that the sensor bellows moves to adjust a lever which positions a pilot valve. This valve directs servo pressure to and from a servo which moves the speed-set cam. Feedback to the ambient-pressure sensor is provided by means of a lever riding in a groove on the **landing-idle cam.**

The condition lever rotates the **idle selection cam** and the cam which actuates the **windmill bypass and shutoff valve.** The windmill function of the windmill bypass and shutoff valve directs pump interstage pressure to the back of the **pressure regulating valve.** The pressure regulating valve is thus permitted to open and bypass fuel to pump interstage should the engine windmill during an in-flight engine shutdown.

Functions of the Fuel Control in Operation

The actual functions of the JFC68 fuel control during operation of the engine may be understood more clearly by following the diagram of Fig. 20-5. This is similar in many respects to the fuel-control curves shown previously; however, this diagram of curves is based on the ratio W_f/P_{s4} and engine speed N_2, whereas the other diagram was based on fuel flow W_f and engine speed.

In Fig. 20-5, position 1 represents the engine condition at the time acceleration by the starter has proceeded to the point where fuel can be injected into the combustion chamber. At this time, the W_f/P_{s4} ratio immediately moves to position 2. The acceleration cam is in control of fuel flow and the engine accelerates along the line 2 to 3 to 4. If the power lever is in the idle position, the fuel ratio will decrease to points 5 and 6 due to governor droop, and the engine will now be in a steady-state condition. When the power lever is advanced, the fuel ratio increases from point 6 to point 7 where the acceleration cam is again in control. The fuel-air ratio increases to the maximum permitted and then remains constant as the engine accelerates to position 8 where the governor droop again takes effect until a steady-state condition is attained at position 9.

The line from position 9 to position 10 represents the fuel decrease when the power lever is returned to the idle position. The throttle valve closes to the minimum deceleration condition and the engine decelerates to position 11. From this point to position 12, governor droop takes effect and increases fuel flow to allow the engine speed to reach the steady-state condition at position 12. If the thrust reverser is actuated at position 12, the fuel flow increases to position 13 and the engine accelerates to position 14. Governor droop again takes effect and reduces fuel flow to position 15 where the **compressor discharge-pressure (CDP) limiter** reduces fuel still further until the steady-state condition is reestablished at position 16 (9). When power is again reduced, the function line follows the same path as before. With the power lever in the idle position, the deceleration continues to position 18 and then fuel flow is increased to establish the idle speed at position 19. When the engine is shut off by means of the condition lever, fuel flow is stopped as indicated by position 20. The engine then decelerates to the zero rpm condition.

The various conditions described in the foregoing are summarized in the table shown with Fig. 20-5.

EVC3 Engine Vane Control

An important unit which operates in connection with the JFC68 fuel control on the JT9D engine is the EVC3 **engine vane control** shown in Fig. 20-6. This control is designed to regulate the variable high-pressure compressor stator vanes of the engine by scheduling the position of the vanes in accordance with requirements dictated by the Mach number of compressor inlet airflow. The Mach number in this case may be considered as the velocity of the airflow adjusted for temperature.

The stator vanes are small airfoils and as such they and the rotor blades are subject to stall when the angle of attack of the airstream becomes too great. Varying

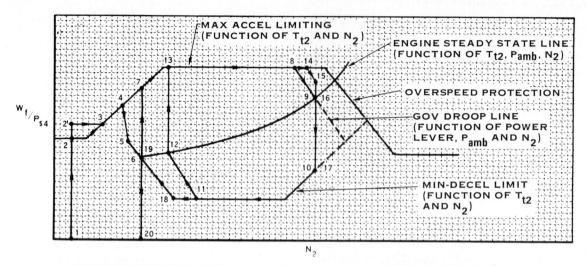

PATH	CONTROLLING FUNCTION	ENGINE OPERATING PARAMETER
1—2—3—4	ACCEL CAM	STARTING
1—2—3—4	ACCEL CAM AND SELECTOR SHAFT	COLD-START ENRICHMENT (MANUALLY SELECTED WHEN REQ'D.)
4—5—6	GOVERNOR DROOP	START TO MIN IDLE
6—7	POWER LEVER	STEP INPUT TO ACCEL
7—8	ACCEL CAM	ACCELERATION
8—9	GOVERNOR DROOP	ACCEL TO STEADY STATE
9—10	POWER LEVER	STEP INPUT TO DECEL
10—11	DECEL CAM	DECELERATION
11—12	GOVERNOR DROOP	DECEL TO STEADY STATE (LANDING IDLE)
12—13	POWER LEVER	STEP INPUT TO ACCEL (ACTUATE THRUST REVERSER)
13—14	ACCEL CAM	ACCEL (THRUST REVERSER ON)
14—15	GOVERNOR DROOP	ACCEL TO CDP LIMITING
15—16	CDP LIMITER	CDP LIMITING TO STEADY STATE
16—17	POWER LEVER	STEP INPUT TO DECEL
17—18	DECEL CAM	DECELERATION
18—19	GOVERNOR DROOP	DECEL TO MIN IDLE
19—20	SELECTOR LEVER	SHUT OFF

FIG. 20-5 Curves to illustrate the operation of the Hamilton Standard JFC68 fuel-control unit. *(Hamilton Standard)*

the angle of the stator blades in accordance with the airflow velocity and temperature prevents stalling of the rotor blades and stator vanes.

The engine vane control (EVC) consists of a **pressure ratio sensor, flapper valve, servo-operated three-dimensional cam, actuator pilot valve,** and necessary linkage. In addition, the control contains two signal valves which supply hydraulic pressure, as a function of pressure ratio, to two remotely mounted engine actuators. The control signal valves are also designed to supply a prescribed cooling flow through the supply lines to the engine bleed-control valves.

The EVC schedules a hydraulic signal for operation of the engine stator vanes as a function of compressor inlet pressure ratio, $P_{t3} - P_{s3}/P_{t3}$. This is accomplished by a two-bellows null-type **pressure ratio sensor** of a force-vector design. This sensor is a torque-balance system arranged so that a ratio of two signal pressures may be determined, as shown in the drawing of Fig. 20-7. In this application, the total pressure P_{t3} and the static pressure P_{s3} are measured by two bellows assemblies to provide force outputs that are proportional to the total pressure and the differential pressure $(P_{t3} - P_{s3})$ from which the pressure ratio $P_{t3} - P_{s3}/P_{t3}$ is derived.

The bellows are arranged so that when they are in the null position, forces from the bellows act at right angles to one another through **tension links** that are attached to a **common pivot.** The resultant forces from the bellows are counteracted by a connecting link between the common pivot and the **feedback-lever pivot.** At the null position, the vector sum of the bellows' forces and the counteracting force in the connecting link are balanced.

At the null position of the system, the **common pivot's** and the **feedback lever's** rotational axes are on the same center line. When the pressure ratio changes, the force balance is upset and the common pivot rotates from its null position. This rotation is constrained about the feedback-lever pivot through the connecting link. The lateral part of the common pivot's rotation moves the flapper valve from the null position, thereby causing a high servo pressure on one side of the servo piston. The piston moves in proportion to flapper-valve displacement. This movement of the servo piston, in turn, displaces the three-dimensional cam which rotates the feedback lever, realigning the rotational axes of the common pivot and the feedback link. Realignment restores the null position of the flapper valve, restoring the force balance in the vector system.

Scheduling of the engine stator vanes is achieved by the three-dimensional cam that is contoured to a prescribed schedule. The cam displaces the actuator pilot valve, sending a high-pressure signal (P_{a1} or P_{a2}) to the remotely mounted **stator-vane actuator** (SACS1), which hydraulically positions the engine stator vanes. The

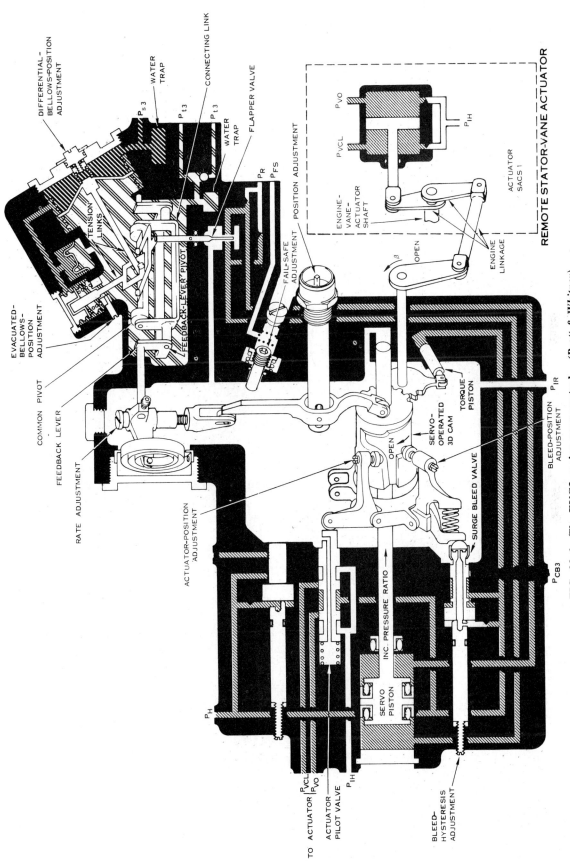

FIG. 20-6 The EVC3 engine vane control. *(Pratt & Whitney)*

461

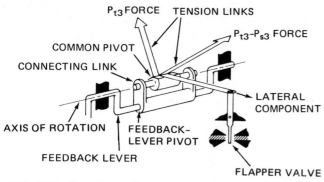

FIG. 20-7 Drawing to illustrate operation of the pressure ratio sensor.

SACS1 also provides a feedback signal, through a mechanical linkage, to rotate the three-dimensional cam and reestablish the null position of the actuator pilot valve.

The three-dimensional cam also incorporates two additional contours that actuate the start and surge bleed-control valves. These valves, in turn, provide high-pressure signals (P_{cb3} and $P_{cb3.5}$) to remotely mounted engine bleed-valve actuators. The control valves are also designed to supply a prescribed cooling bleed flow through the supply lines to the engine bleed-valve actuators.

Interrelation of Control Units

Figure 20-8 illustrates the interrelation between the main fuel control and the engine vane control and also the position of the stator-vane actuator in the system. A careful study of this diagram and the various controlling and actuating forces will help the student understand the many factors governing the operation of a large engine such as the JT9D.

● THE WOODWARD FUEL-CONTROL UNIT

The Woodward fuel-control unit, manufactured in a number of configurations by the Woodward Governor

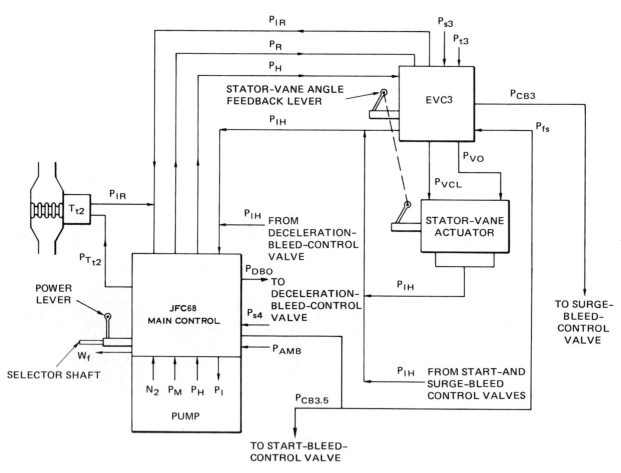

N_2—High-pressure compressor speed
P_{amb}—Ambient pressure
P_{DBO}—Deceleration-bleed override
P_F—Fine-filtered main-stage pressure
P_H—Controlled hydraulic-stage pressure

P_I—Controlled body pressure (pump-interstage pressure)
P_{IH}—Pump interstage from hydraulic supply
P_{IR}—Pump interstage from regulated supply
P_M—Main-stage inlet pressure

P_R—Regulated fine-filtered servo supply
P_S—Servo-control pressure
P_{s4}—Engine-burner pressure
T_{t2}—Engine-inlet total temperature
W_f—Metered fuel flow
P_{Tt2}—Signal pressure from T_{t2}-sensor

P_{A1}—Metered vane-actuator pressure
P_{A2}—Metered vane-actuator pressure
P_{cb3}—Bleed-control pressure
$P_{cb3.5}$—Start-bleed pressure
P_{FS}—Speed-signal pressure

FIG. 20-8 Drawing to illustrate the interrelation between the fuel-control unit and the EVC. *(Hamilton Standard)*

Co., Rockford, Illinois, is designed to control fuel flow and the position of variable compressor stator vanes for large turbofan engines. Figure 20-9 shows two views of such a control unit.

In this section we will describe the functions and the principal components of the unit. Detailed information should be obtained from the manufacturer's manual.

Basic Description

The Woodward fuel control discussed in this section is designed for use with the General Electric CF6 high-bypass turbofan engine. The unit is hydromechanical and is controlled by four parameters: (1) engine speed, (2) compressor inlet temperature (CIT), (3) compressor discharge pressure (CDP), and (4) power-lever position. These same parameters are utilized in the control units described previously.

The function of variable-stator-vane control is the same as that described for the JFC68 fuel control and the EVC3 engine vane control. The arrangement is different but the purpose is the same. In the Woodward control, the vane-control function is within the fuel-control unit.

Fuel Metering

Fuel metering in the Woodward control is accomplished by means of a variable **fuel metering port** across which is maintained a constant pressure differential.

FIG. 20-9 Photographs of the Woodward 3034 fuel-control unit. *(Woodward Governor Co.)*

This is the same principle described for the other controls discussed in this chapter. In each case, the opening of the fuel metering port (throttle valve) is determined by the computing section of the control.

Fuel is directed to the control fuel inlet at pump pressure and in a volume greater than is needed for any condition of engine operation. Excess fuel is returned to the pump inlet by means of the **fuel bypass valve**. This valve is controlled to provide the correct differential pressure across the fuel metering port.

Figure 20-10 shows the interrelations among the various components and systems of the fuel-control unit.

Note in the diagram that a pressure-sensing line upstream of the fuel metering port carries engine pump pressure to the **differential pilot valve** at the lower side of the **sensing land**. Another sensing line carries pressure downstream from the fuel metering port to the upper side of the same sensing land. The differential pressure on the sides of the sensing land produces a force balanced by the differential pilot-valve spring shown at the top of the valve. When the force of the spring and that caused by the differential pressure are equal, the pilot valve is in the neutral position and the differential of pressure across the fuel metering port is correct. Should the differential pressure become too great, the pilot-valve plunger will move upward and permit bleeding of **differential servo pressure** from the bypass valve, thus permitting the bypass valve to open further to bypass more fuel. This reduces the pump pressure and hence the differential pressure. When the differential pressure has been corrected, the pilot-valve plunger returns to the neutral position. If the differential pressure becomes too low, the pilot-valve plunger moves downward and permits pump pressure to build up the differential servo pressure applied to the bypass valve. This closes the bypass valve sufficiently to decrease the bypassing of fuel and increase pump pressure, thus increasing the differential pressure across the fuel metering port.

Speed Governing and Sensing

Speed governing is attained through the **speed governor** which is flyweight-operated as in other control units. The power lever (throttle lever) applies force to the governor **speeder spring** through the **throttle cam** and **speed-setting shaft**. When the flyweight centrifugal force and the speeder-spring force are balanced, the engine is in an on-speed condition. If the engine is not on speed, the governor will direct pressure to or from the **governor servo piston** to change the opening of the fuel metering port.

Note in Fig. 20-10 that the adjustment of the speeder spring by means of the power lever is modified by inputs from the **idle servo system** and the CIT parameter. The total effect of the speed governor is to continually compare engine speed with desired speed and make adjustments when the speed is not correct.

Speed sensing is accomplished by means of the **tachometer**, which is also a flyweight-operated unit driven by the engine. The tachometer senses engine speed (N_2) and rotates the 3D cam by means of a servo geared to the 3D cam shaft. The 3D cam may also be referred to as the **fuel-scheduling and variable-stator-vane control cam.**

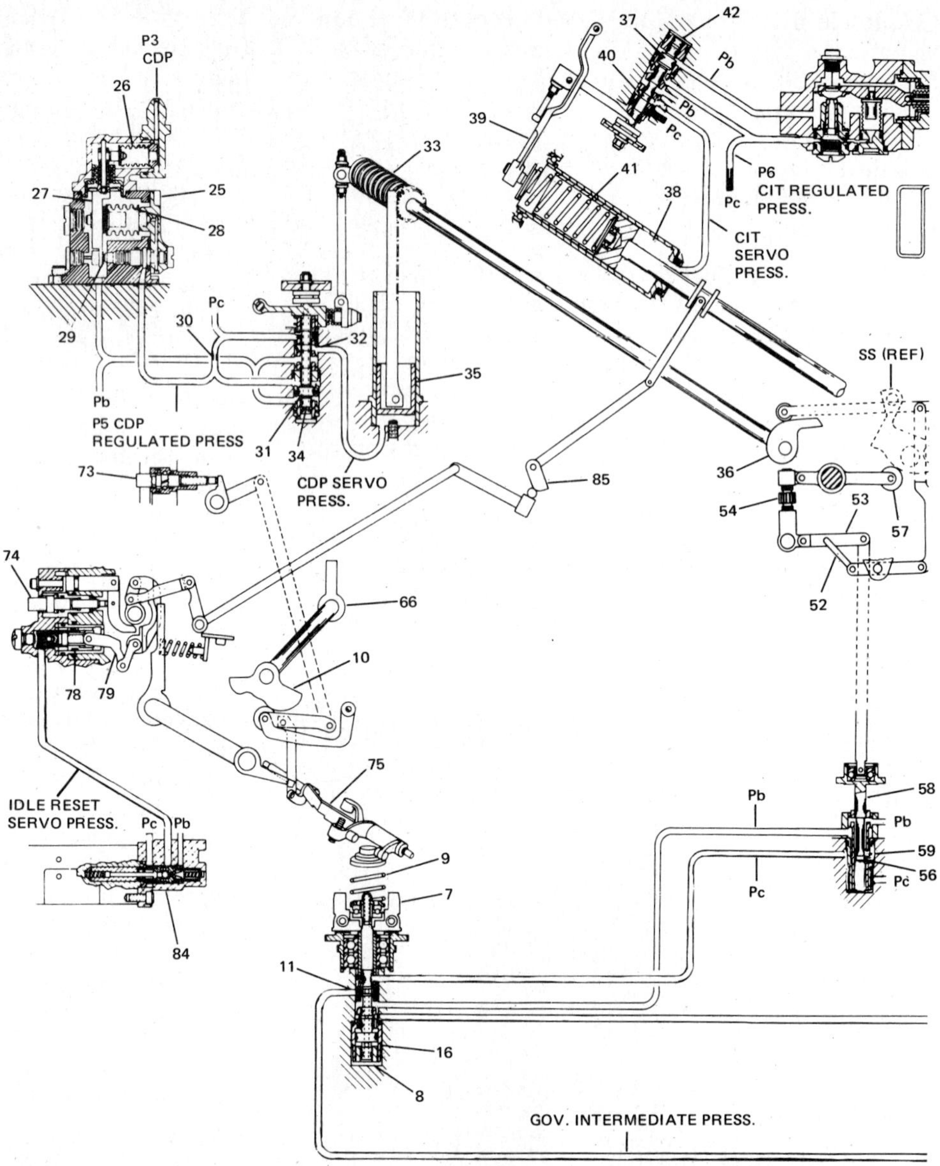

1. P_c pressure-regulating port
2. P_c pilot-valve plunger
3. P_c spring
4. P_{cr} Regulating port
5. P_{cr} Spring
6. P_{cr} Pilot-valve plunger
7. Governor flyweight
8. Governor pilot-valve plunger
9. Governor speeder spring
10. Throttle cam
11. Governor pilot-valve control port
12. Buffer piston
13. Governor servo piston
14. Fuel valve rotor
15. Buffer spring
16. Governor compensating land
17. Fuel metering port
18. Differential pilot-valve sensing land
19. Differential pilot-valve plunger
20. Differential pilot-valve spring
21. Bypass valve plunger
22. Differential pilot-valve control land
23. Bypass valve spring
24. Specific gravity adjusting screw
25. CDP sensor
26. CDP sensing bellows
27. CDP sensor lever
28. P_5 sensing bellows
29. P_5 regulating valve
30. P_5 orifice
31. CDP plunger sensing land
32. CDP pilot-valve control port
33. CDP restoring spring
34. CDP plunger
35. CDP servo piston
36. CDP scan
37. CIT pilot-valve sensing land
38. CIT servo piston

39. CIT feedback linkage
40. CIT pilot-valve control land
41. CIT reference spring
42. CIT pilot-valve plunger
43. 3D cam

44. Tachometer flyweight
45. Tachometer reference spring
46. Tachometer pilot-valve plunger
47. Tachometer feedback cam

48. Tachometer feedback lever
49. Tachometer servo piston
50. 3D camshaft
51. Computer summing lever

FIG. 20-10 Schematic drawing of a Woodward fuel-control unit. *(Woodward Governor Co.)*

Compressor Inlet-Temperature Sensing

Compressor inlet temperature (CIT) is sensed by means of a separate unit mounted on the engine fan frame. This unit is called the **CIT sensor** and its function is to direct pressure to and from the **CIT pilot valve** located in the fuel-control unit. The pilot valve, in turn, controls the CIT servo which moves the fuel-scheduling 3D cam to adjust for inlet air temperature. A schematic drawing of the CIT sensor is shown in Fig. 20-10.

Compressor Discharge-Pressure Sensing

The compressor discharge pressure is sensed by a bellows-type unit. The effect of CDP is to control hy-draulic pressure which, in turn, controls the **CDP pilot valve**. This valve regulates hydraulic pressure to the **CDP servo** geared to the **CDP camshaft**. As the shaft rotates, it turns the CDP cam, thus applying the CDP input to the **computer summing lever**. The CDP input is therefore a part of the total effect of all parameters applied to control the position of the 3D cam and the followers which provide a response.

Computer Action

It can now be seen that the operation of three cams—the **3D cam**, the **fuel cam**, and the **CDP cam**—provides an integrated effect which is applied to the summing lever to feed a signal force to the **limit pilot valve**. This valve then controls the hydraulic pressure supply *to*

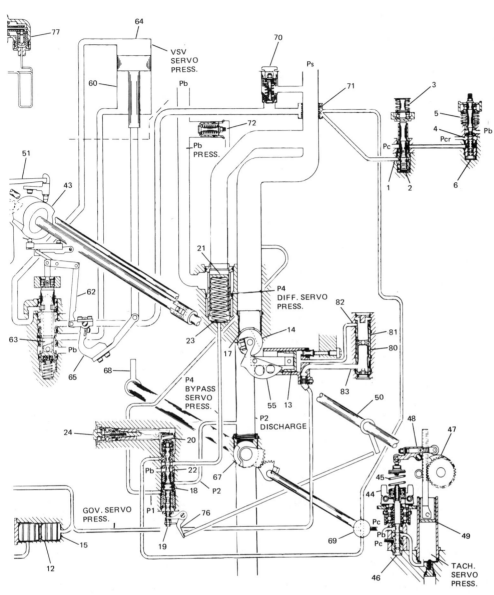

60. VSV servo
61. VSV follower
62. VSV restoring link
63. VSV plunger
64. VSV servo piston
65. VSV restoring arm
66. Throttle lever
67. Shutdown valve rotor
68. Shutoff lever
69. Pump unloading rotor
70. P_f relief valve
71. P_f filter
72. P_{cr} relief valve
73. Maximum-speed adjustment screw
74. Ground idle adjustment screw
75. Speed-setting shaft
76. Overspeed trip lever
77. CIT sensor
78. Speed-setting servo piston
79. Idle speed stop (flight)
80. Jump and rate limiter
81. Piston
82. Port
83. Port
84. Solenoid
85. Cam

P_s, supply pressure
P_2, outlet pressure
P_c, control pressure
P_2, compressor discharge pressure
P_b, bypass pressure
P_4, differential servo pressure
P_5, P_3, regulated pressure
P_3, servo pressure
P_6, CIT regulated pressure
P_{cr}, regulated case pressure
P_f, filtered P_s

52. Computer summing point
53. Limit pilot-valve floating lever
54. Cam summing link
55. Fuel cam
56. Limit pilot-valve acceleration port
57. Fuel cam follower
58. Limit pilot-valve plunger
59. Limit pilot-valve deceleration port

and the drain *from* the **governor pilot valve.** The governor action is thus limited as a function of the positions of the cams.

Variable-Stator-Vane (VSV) Control

Note in Fig. 20-10 that the variable-stator-vane pilot valve is affected by a follower on the 3D cam. Thus, the hydraulic pressure to the **VSV servo** is controlled by the position of the 3D cam. The servo is an actuator which moves the variable stator vanes as necessary to provide the correct angle of attack for the air flowing through the compressor. Two VSV actuators are mounted on the compressor.

Idle System

The **ground-flight idle system** is designed to provide the correct idle speed for ground operation of the en-

gine or for flight. The system is solenoid-activated by an electric signal from the pilot or flight engineer. When the solenoid is energized, the solenoid valve ports bypass fuel pressure to the **idle-reset servo piston** which moves to reset the **idle stop** to the *low idle* or *ground idle* position. In the deenergized condition, the solenoid-valve ports control pressure to the idle-reset servo piston, which then moves the idle stop to the high idle position for flight conditions.

Fuel Shutoff Lever

The fuel shutoff-lever shaft is concentric with the power-lever shaft and functions to rotate the fuel shutoff valve to the open or closed position. At the same time, it rotates the **fuel-pump unloading rotor valve,** thus permitting fuel pressure and spring force to move the differential pilot-valve plunger down. This applies control fuel pressure to the bypass valve. The bypass-valve spring plus the control pressure acts to permit bypass of excess fuel pressure and at the same time maintains operating pressure within the fuel-control unit during coastdown or windmilling of the engine.

● FUEL-CONTROL SYSTEM FOR A TURBOSHAFT ENGINE

A typical turboshaft engine incorporates a **gas producer (gas generator)** and a **power-turbine system** within the engine. The gas producer consists of a compressor, the combustion chamber, fuel nozzle or nozzles, and the gas-producer turbine. This section produces the high-velocity, high-temperature gases which furnish the energy to drive the **power turbine.** The power turbine usually incorporates two or more stages (turbine wheels) which extract the energy from the gases and deliver power to the output shaft.

A fuel-control system for a turboshaft engine is often comprised of two sections, one which senses and regulates the gas-producer part of the engine and the other which senses the operation and requirements of the power-turbine section. The complete system controls engine power output by controlling the gas-producer speed which, in turn, governs the power output. The gas-producer speed levels are established by the action of the **power-turbine fuel governor** which senses power-turbine speed. The power-turbine speed is selected by the operator as the load requires, and the power needed to maintain this speed is automatically maintained by power-turbine governor action on metered fuel flow.

The **power-turbine governor** incorporates rotating flyweights which continually sense power-turbine speed. Through the speed sensing, the governor produces actions which direct the **gas-producer fuel control** to schedule the correct amount of fuel for the required operation.

Fuel flow for engine control is established as a function of *compressor discharge pressure* (P_c), engine speed (N_1 for the gas producer and N_2 for the power turbine), and gas-producer throttle-lever position. It will be noted that these same parameters are employed in the control of other turbine engines and some controls utilize additional parameters for fuel control. Turbojet engines utilize turbine inlet temperature (TIT) as an important factor in fuel control; however, this is not required for the engine control under discussion.

Gas-Producer Fuel Control

The fuel-control system described here is the Bendix system employed on the Allison series 250 turboshaft engine and is illustrated in Fig. 20-11. Note that the gas-producer fuel control and the power-turbine governor are interconnected so that each may affect the operation of the other as required.

The **gas-producer fuel control** is similar in many respects to other fuel controls described previously. Its primary function is the same. Fuel entering the control encounters a bypass valve which maintains a constant differential between fuel pump pressure (P_1) and metered fuel pressure (P_2). Excess fuel is bypassed back to the entering side of the pump. The constant pressure differential is applied across the metering valve; hence,

fuel flow will be in proportion to the opening of the metering valve. The degree to which the metering valve is open is controlled by the action of the governor bellows and the acceleration bellows and is modified by the action of the derichment valve during starting. The maximum range of movement of the metering valve is controlled by the minimum flow stop and the maximum flow stop. The unit also incorporates a maximum-pressure-relief valve, a manually operated shut-off valve, and a bellows-operated start-derichment valve.

The operation of the gas-producer fuel control is based on the control of various air pressures by the speed governors and the use of these pressures to move the metering valve as required. The control may therefore be classed as pneumatic or pneumomechanical.

The simplified drawing of Fig. 20-12 illustrates how air pressure may be controlled to operate a metering valve. In the drawing, air pressure (P_c), which may be compared to compressor discharge pressure, is applied to the controller and flows through an air bleed. The rate of flow through the air bleed will determine the difference between P_c and modified pressure P_x. The rate of flow is determined by the position of the governor valve. If the governor valve is completely closed, there will be no flow through the air bleed and P_c will equal P_x. Since P_x would be at a comparatively high level, the pressure in the bellows chamber would cause the bellows to collapse and the metering valve to open through the linkage to the metering valve.

When the governor valve is closed, the P_x pressure will be much less than P_c. This will allow the bellows to expand and close the metering valve.

The metering valve in the gas-producer fuel control is operated by lever action in accordance with the movement of the **governor bellows** and the **acceleration bellows.** Note that the governor bellows and acceleration bellows are affected by variations in the P_x and P_y pressures. P_x and P_y pressures are derived by passing P_c pressure through two air bleeds. The rate of airflow through these bleeds is controlled by action of the governor as modified by throttle position and the influence of the **power-turbine governor.**

Before lightoff and acceleration, the metering valve is set at a predetermined open position by the acceleration bellows under influence of ambient pressure. At this point, ambient pressure and P_c pressure are the same because the compressor is not operating.

The **start-derichment valve** is open during lightoff and acceleration until a preestablished P_c is reached. The open derichment valve vents P_y pressure to atmosphere, thus allowing the governor bellows to move the metering valve toward the **minimum flow stop.** This keeps fuel flow at the lean fuel schedule required for starting and acceleration. As compressor rpm increases, the derichment valve is closed by P_c acting on the derichment bellows. Observe the drawing of Fig. 20-11 and note that P_c pressure is applied to the inside of the derichment bellows. When the derichment valve is closed, control of the metering valve is returned to the normal operating schedule in which the effects of P_x and P_y as regulated by the governor are operating through the governor bellows and acceleration bellows.

During acceleration the P_x and P_y pressures are equal

to the modified compressor discharge pressure (P_c) up to the point where the **speed-enrichment orifice** is opened by the governor flyweight action. This action bleeds P_x pressure while P_y pressure remains at a value equal to P_c. Under the influence of the P_y-minus-P_x pressure drop across the governor bellows, the metering valve moves to a more open position, thus increasing fuel flow as required for acceleration.

Gas-producer rpm (N_1) is controlled by the gas-producer control governor. The governor flyweights operate the **governor lever** which controls the governor bellows (P_y) bleed at the **governing orifice**. The flyweight operation of the governor lever is opposed by a variable spring load which is changed in accordance with the position of the throttle acting through the **spring-scheduling cam**. Opening the governor orifice bleeds P_y pressure and allows P_x pressure to control the governor action on the bellows. The P_x action on the bellows moves the metering valve to a more closed position until metered flow is at steady-state requirements.

The **governor-reset** section of the gas-producer fuel control is utilized by the power-turbine governor to override the speed-governing elements of the fuel control to change the fuel schedule in response to load conditions applied to the power turbine. The diaphragm and spring in the governor-reset assembly apply force to the governor lever to modify the effect of the governor springs.

Power-Turbine Governor

The power-turbine section of the engine calls upon the gas-generator section for more or less energy, depending on the load requirements. It is the function of the **power-turbine fuel governor** to provide the actuating force directed to the gas-producer fuel control which responds by increasing or decreasing fuel as required to produce the needed gas energy.

As shown in Fig. 20-11, the power-turbine speed is scheduled by the **power-turbine governor lever** and the **power-turbine speed-scheduling cam**. The cam, operated by the throttle, sets a governor-spring load which opposes the force of the **speed flyweights.** As the desired speed is approached, the speed weights, operating against the governor spring, move a link to open the power-turbine governor orifice. The speed flyweights also open the overspeed bleed (P_y) orifice but at a higher speed than where the regular governor orifice (P_g) is opened.

The power-turbine governor, like the gas-producer fuel control, utilizes controlled air pressure to accomplish its purposes. Compressor discharge pressure (P_c) enters the **air valve**, which is a pressure regulator. The output of the air valve is regulated pressure (P_r). P_r pressure is applied to one side of the diaphragm in the governor-reset section of the gas-producer fuel control. Governor pressure (P_g), developed when P_r pressure passes through the P_g bleed, is applied to the opposite side of the diaphragm. When the governor orifice is closed, P_r and P_g are equal and produce no effect on the governor-reset diaphragm. When the governor orifice is opened by action of the flyweights, P_g pressure is reduced. The effect of $P_r - P_g$ on the diaphragm is to produce force through the **governor-reset rod** to the gas-producer governor lever (power-output link) to supplement the force of the flyweights in the gas-producer governor. This opens the P_y orifice and bleeds P_y pressure, thus causing the gas-producer governor bellows to move the fuel metering valve to a more closed position. This, in turn, reduces gas-producer speed. Gas-producer speed cannot exceed the gas-producer fuel-governor setting.

The governor-reset diaphragm is preloaded to establish the active P_r-minus-P_g range. This is accomplished by means of a spring, as shown in Fig. 20-11.

The overspeed orifice in the power-turbine governor bleeds P_y pressure from the governing system of the gas-producer fuel control. This gives the system a rapid response to N_2 overspeed conditions.

● ELECTRONIC ENGINE CONTROLS

Introduction

Because of the need to control precisely the many factors involved in the operation of modern high-bypass turbofan engines, airlines and manufacturers have worked together to develop electronic engine-control (EEC) systems that prolong engine life, save fuel, improve reliability, reduce flight-crew workload, and reduce maintenance costs. The cooperative efforts have resulted in two types of electronic engine controls, one being the *supervisory engine-control system* and the other the *full-authority engine-control system*. The supervisory control system was developed and put into service first and has proved effective in military aircraft and modern airliners. A supervisory electronic engine-control system is employed with the JT9D-7R4 engines installed in the Boeing 767 airliner.

Essentially, the **supervisory engine-control system** includes a computer that receives information regarding various engine operating parameters and adjusts a standard hydromechanical fuel-control unit to obtain the most effective engine operation. The hydromechanical unit responds to the EEC commands and actually performs the functions necessary for engine operation and protection.

The **full-authority EEC** is a system that receives all the necessary data for engine operation and develops the commands to various actuators to control the engine parameters within the limits required for the most efficient and safe engine operation. This type of system is employed on advanced-technology engines such as the Pratt & Whitney series 2000 and 4000.

Supervisory Electronic Engine-Control System

The digital supervisory electronic engine-control system employed with the JT9D-7R4 turbofan engine includes a hydromechanical fuel-control unit such as the Hamilton Standard JFC68 described earlier, a Hamilton Standard EEC-103 electronic engine-control unit, a hydromechanical air-bleed and vane control, a permanent-magnet alternator to provide electrical power for the EEC separate from the aircraft electrical system, and an engine inlet pressure/temperature probe to sense P_{t2} and T_{t2}. The hydromechanical units of the system control such basic engine functions as automatic starting, acceleration, deceleration, high-pressure rotor speed (N_2) governing, variable compressor

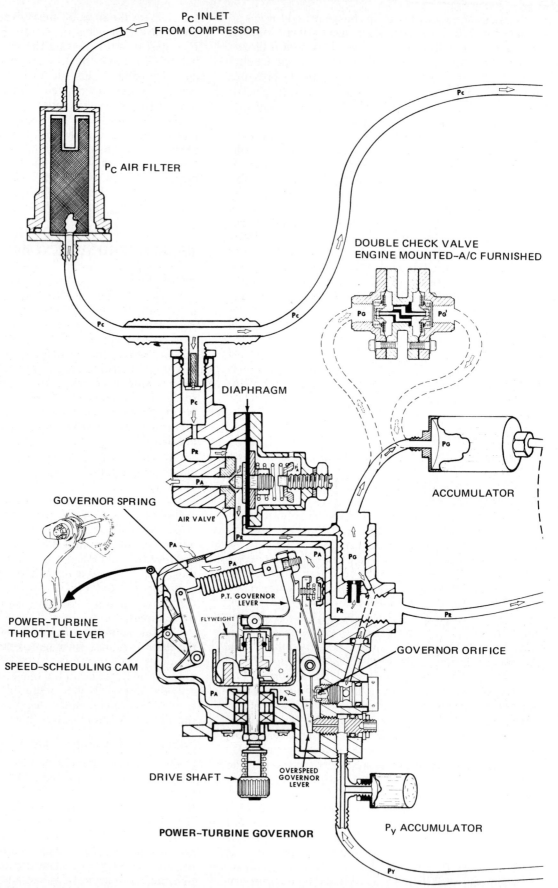

FIG. 20-11 Schematic drawing to illustrate a Bendix fuel-control system for a turboshaft engine. *(Allison Gas-Turbine Operations, General Motors Corp.)*

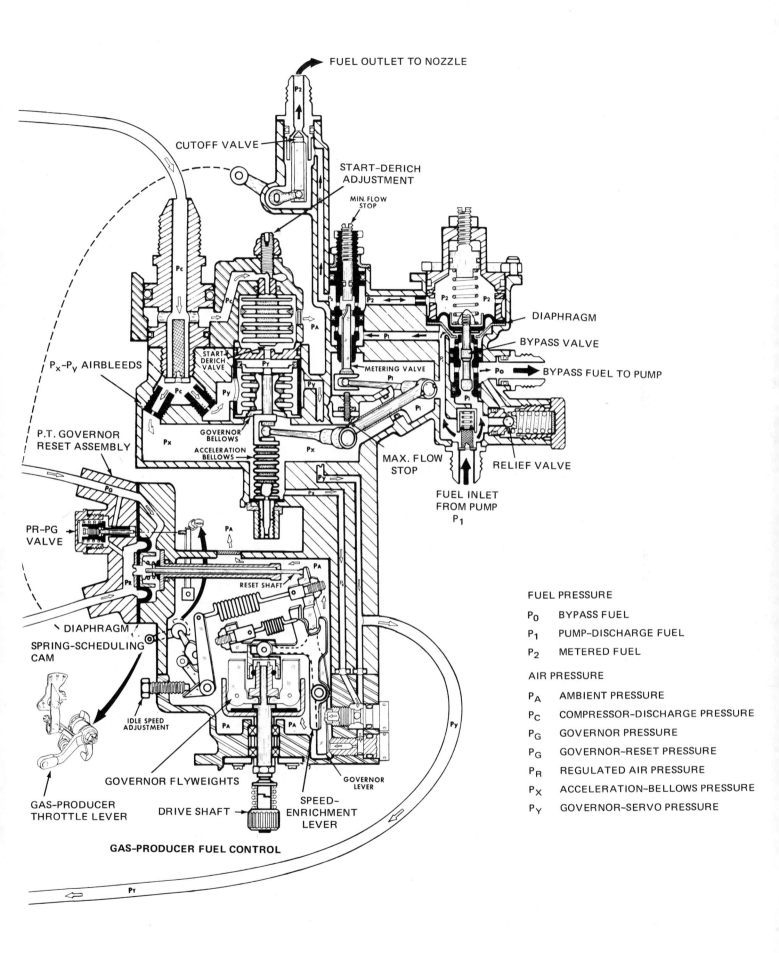

FUEL OUTLET TO NOZZLE

CUTOFF VALVE

START–DERICH
ADJUSTMENT

MIN. FLOW
STOP

DIAPHRAGM

BYPASS VALVE

P_x–P_y AIRBLEEDS

START–
DERICH
VALVE

METERING VALVE

BYPASS FUEL TO PUMP

P.T. GOVERNOR
RESET ASSEMBLY

GOVERNOR
BELLOWS

ACCELERATION
BELLOWS

MAX. FLOW
STOP

RELIEF VALVE

PR–PG
VALVE

FUEL INLET
FROM PUMP
P_1

DIAPHRAGM

SPRING-SCHEDULING
CAM

RESET SHAFT

IDLE SPEED
ADJUSTMENT

GAS-PRODUCER
THROTTLE LEVER

GOVERNOR FLYWEIGHTS

DRIVE SHAFT

SPEED-
ENRICHMENT
LEVER

GOVERNOR
LEVER

GAS-PRODUCER FUEL CONTROL

FUEL PRESSURE

P_0 BYPASS FUEL

P_1 PUMP-DISCHARGE FUEL

P_2 METERED FUEL

AIR PRESSURE

P_A AMBIENT PRESSURE

P_C COMPRESSOR-DISCHARGE PRESSURE

P_G GOVERNOR PRESSURE

P_G GOVERNOR-RESET PRESSURE

P_R REGULATED AIR PRESSURE

P_X ACCELERATION-BELLOWS PRESSURE

P_Y GOVERNOR-SERVO PRESSURE

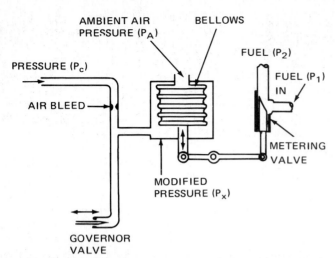

FIG. 20-12 Simplified drawing to show the use of pneumatic control for a fuel metering valve.

stator-vane position, modulated and starting air-bleed control, and burner pressure (P_b) limiting. The EEC provides precision thrust management, N_2 and exhaust-gas temperature (EGT) limiting, and cockpit display information of engine pressure ratio (EPR) limit, EPR command, and actual EPR. It also provides control of modulated turbine-case cooling and turbine-cooling air valves, and transmits information regarding parametric and control system condition for possible recording. Such recorded data is utilized by maintenance technicians to aid in eliminating faults in the system.

The supervisory electronic engine control, by measuring EPR and integrating thrust-lever (throttle) angle, altitude data, Mach number, inlet air pressure (P_{t2}), inlet air temperature (T_{t2}), and total air temperature in the computation, is able to maintain constant thrust from the engine regardless of changes in air pressure, air temperature, and flight environment. Thrust changes

occur only when the thrust-lever angle is changed and the thrust remains consistent for any particular position of the thrust lever. Takeoff thrust is produced in the full-forward position of the thrust lever. Thrust settings for climb and cruise are made by the pilot as the thrust lever is moved to a position to obtain the correct EPR for the thrust desired. The EEC is designed so that the engine will quickly and precisely adjust to a new thrust setting without the danger of overshoot in N_2 or temperature. It adjusts the hydromechanical fuel-control unit through a torque-motor electrohydraulic servo system.

In a supervisory EEC system, any fault in the EEC that adversely affects engine operation causes an immediate reversion to control by the hydromechanical fuel-control unit. At the same time, the system provides an annunciator light signal to the cockpit to inform the crew of the change in operating mode. A switch in the cockpit enables the crew to change from EEC control to hydromechanical control if it is deemed advisable.

The supervisory EEC is integrated with the aircraft systems as indicated in the diagram of Fig. 20-13. The input and output signals are shown by the directional arrows. Although the EEC utilizes aircraft electrical power for some of its functions, the electrical power for the basic operation of the EEC is supplied by the separate, engine-driven permanent-magnet alternator mentioned earlier.

The output signals of the supervisory EEC that affect engine operation are the adjustment of the hydromechanical fuel-control unit and commands to solenoid-actuated valves for control of modulated turbine-case cooling and turbine-cooling air.

Full-Authority Electronic Engine Control

A full-authority electronic engine control performs all functions necessary to operate a turbofan engine efficiently and safely in all modes, such as starting,

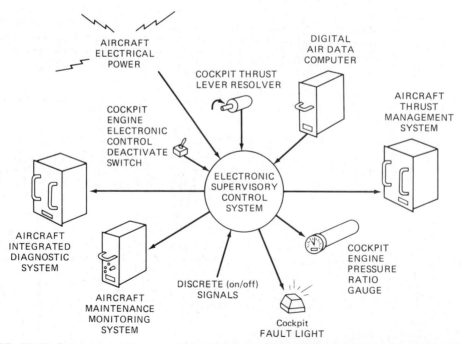

FIG. 20-13 Integration of a supervisory electronic engine control with aircraft systems. (ASME)

accelerating, decelerating, takeoff, climb, cruise, and idle. It receives data from the aircraft and engine systems, provides data for the aircraft systems, and issues commands to engine-control actuators.

The information provided in this section is based on the Hamilton Standard EEC-104, an electronic engine control designed for use with the Pratt & Whitney 2037 advanced-technology engine. The unit is shown in the photograph of Fig. 20-14. This is a dual-channel unit having a "crosstalk" capability so that either channel can utilize data from the other channel. This provision greatly increases reliability to the extent that the system will continue to operate effectively even though a number of faults may exist. Channel A is the primary channel and channel B is the secondary, or backup, channel.

The following abbreviations and symbols are used in this section to identify functions, systems, and components:

ACC	Active clearance control
BCE	Breather compartment ejector
EEC	Electronic engine control
EGT	Exit- (exhaust-) gas temperature
EPR	Engine pressure ratio
FCU	Fuel-control unit
LVDT	Linear variable differential transformer
N_1	Low-pressure spool rpm
N_2	High-pressure spool rpm
P_{amb}	Ambient air pressure
P_b	Burner pressure
PMA	Permanent-magnet alternator
P_{s3}	Static compressor air pressure
P_{t2}	Engine inlet total pressure
$P_{4.9}$	Exhaust-gas pressure
TCA	Turbine-cooling air
TLA	Throttle-lever angle
TRA	Throttle-resolver angle
T_{t2}	Engine inlet total air pressure
$T_{4.9}$	Exhaust-gas temperature (EGT)
W_f	Fuel flow

Fig. 20-15 is a block diagram to indicate the relationships among the various components of the EEC system. Input signals from the aircraft to the EEC-104

FIG. 20-14 Hamilton Standard EEC-104 electronic engine control. *(Hamilton Standard)*

include throttle-resolver angle (which tells the EEC the position of the throttle), service air-bleed status, aircraft altitude, total air pressure, and total air temperature. Information regarding altitude, pressure, and temperature is obtained from the air-data computer as well as the P_{t2}/T_{t2} probe in the engine inlet.

Outputs from the engine to the EEC include overspeed warning, fuel flow rate, electrical power for the EEC, high-pressure rotor speed (N_2), stator-vane-angle feedback, position of the 2.5 air-bleed proximity switch, air/oil cooler feedback, fuel temperature, oil temperature, automatic clearance-control (ACC) feedback, turbine-cooling air (TCA) position, engine tailpipe pressure ($P_{4.9}$), burner pressure (P_b), engine inlet total pressure (P_{t2}), low-pressure rotor speed (N_1), engine inlet total temperature (T_{t2}), and exhaust-gas temperature (EGT or $T_{4.9}$). Sensors installed on the engine provide the EEC with measurements of temperatures, pressures, and speeds. These data are used to provide automatic thrust rating control, engine-limit protection (overspeed, overheat, and over-pressure), transient control, and engine starting.

Outputs from the EEC to the engine include fuel-flow torque-motor command, stator-vane-angle torque-motor command, air/oil cooler-valve command, 2.5 air-bleed torque-motor command, ACC torque-motor command, oil-bypass solenoid command, breather-compartment-ejector solenoid command, and turbine-cooling-air solenoid command. The actuators that must provide feedback to the EEC are equipped with linear variable differential transformers (LVDT) to produce the required signals.

During operation of the engine-control system, fuel flows from the aircraft fuel tank to the centrifugal stage of the dual-stage fuel pump. The fuel is then directed from the pump through a dual-core oil/fuel heat exchanger which provides deicing for the fuel filter as the fuel is warmed and the oil is cooled. The filter protects the pump main-gear stage and the fuel system from fuel-borne contaminants. High-pressure fuel from the main-gear stage of the fuel pump is supplied to the fuel-control unit which, through electrohydraulic servo valves, responds to commands from the EEC to position the fuel metering valve, stator-vane actuator, and air/oil cooler actuator. Compressor air-bleed and active-clearance-control (ACC) actuators are positioned by electrohydraulic servo valves that are controlled directly by the EEC using redundant torque-motor drivers and feedback elements. The word "redundant" means that units or mechanisms are designed with backup features so that a failure in one part will not disable the unit and operation will continue normally. Actuator-position feedback is provided to the EEC by redundant linear variable differential transformers (LVDT) for the actuators and redundant resolvers for the fuel metering valve. Fuel-pump discharge pressure is used to power the stator-vane, 2.5 air-bleed, air/oil cooler, and ACC actuators. The EEC activates turbine-cooling air, engine breather ejector, and the aircraft-provided thrust reverser through electrically controlled dual-solenoid valves.

The EEC and its interconnected components are shown in the drawing of Fig. 20-16. It will be noted that the EEC is mounted on the top left side of the

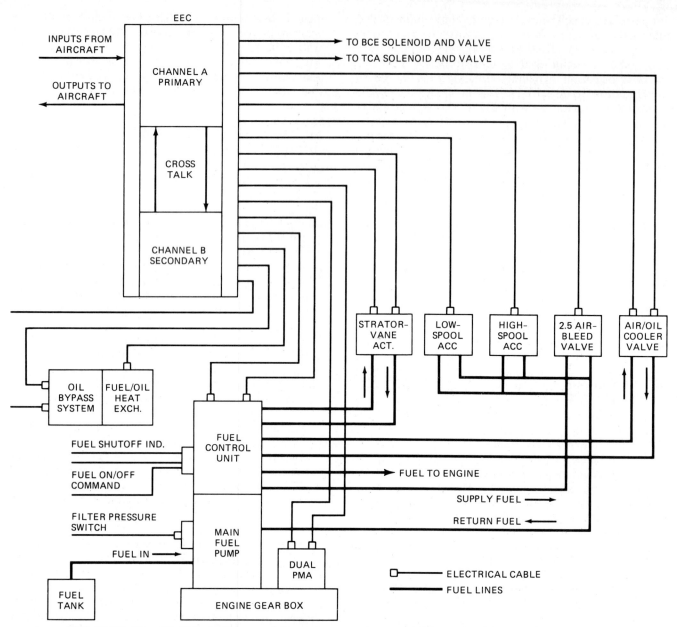

FIG. 20-15 Simplified block diagram of the engine-control system with the Hamilton Standard EEC-104.

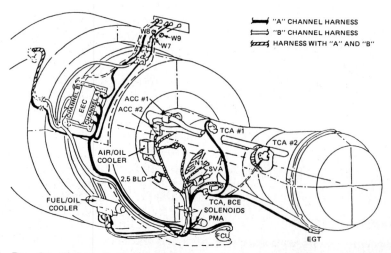

FIG. 20-16 Drawing to show electronic engine-control units on an engine. *(Hamilton Standard)*

engine fan case. The mounting is accomplished with vibration isolators (shock mountings) to protect the unit.

The benefits of employing full-authority electronic engine control result in substantial savings for the aircraft operator. Among these benefits are reduced crew workload, reduced fuel consumption, increased reliability, and improved maintainability.

Flight-crew workload is decreased because the pilot utilizes the EPR gage to set engine thrust correctly. An EPR gage is shown in the drawing of Fig. 20-17. To set thrust, the pilot only has to set the throttle-lever angle to a position that results in lining up the engine-pressure-ratio command from the EEC with the reference indicator that is positioned by the thrust-management computer. The EEC will automatically accelerate or decelerate the engine to that EPR level without the pilot having to monitor the EPR gage.

Reduced fuel consumption is attained because the EEC controls the engine operating parameters so that maximum thrust is obtained for the amount of fuel consumed. In addition, the active-clearance-control (ACC) system assures that compressor and turbine blade clearances are kept to a minimum, thus reducing pressure losses due to leakage at the blade tips. This is accomplished by the ACC system as it directs cooling air through passages in the engine case to control engine case temperature. The EEC controls the cooling airflow by sending commands to the ACC system actuator.

Engine trimming is eliminated by the use of the full-authority EEC. When an engine is operated with a hydromechanical fuel-control system it is necessary periodically to make adjustments on the fuel-control unit to maintain optimum engine performance. To trim the engine it is necessary to operate the engine on the ground for extensive periods of time at controlled speeds and temperatures. This results in the consumption of substantial amounts of fuel plus work time for maintenance personnel and downtime for the aircraft. With the full-authority EEC, none of these costs is experienced.

The fault-sensing, self-testing, and correcting features designed into the EEC greatly increase the reliability and maintainability of the system. These features enable the system to continue functioning in flight and provide fault information that is used by maintenance technicians when the aircraft is on the ground. The modular design of the electronic circuitry saves maintenance time because circuit boards having defective components are quickly and easily removed and replaced.

Garrett Digital Fuel Controller

The electronic engine control designed for operation with the Garrett TFE-731-5 turbofan engine is called a Digital Fuel Controller (DFC) and is a full-authority system. The DFC is shown in the photograph of Fig. 20-18.

The DFC for the TFE 731-5 engine performs the following functions:

1. Maintains required thrust with varying altitude, airspeed, and inlet air temperature (T_{t2})
2. Maintains adequate surge margin throughout the operating range and during acceleration and deceleration of the engine
3. Provides automatic fuel enrichment during starts
4. Provides schedules for minimum and maximum fuel flow
5. Provides temperature limiting at all times
6. Automatically detects overspeed and actuates the fuel cutoff valve
7. Provides for synchronizing engine speeds in multiengine applications
8. Provides for automatic transfer to a backup mode if electrical power is reduced below minimum or if critical failures are detected
9. Provides for use of alternate values for noncritical faults

The DFC utilizes a single power-lever (throttle) position with dual (ground-flight) idle thrust. The power-lever-angle (PLA) input to the controller establishes the engine thrust. The simplified block diagram of Fig. 20-19 shows the inputs to the DFC and the outputs to the engine. Inputs are power-lever angle, engine inlet pressure, engine inlet temperature, engine spool speeds, and interturbine temperature (ITT). The discrete (on-off) inputs, such as mode select switch, are derived

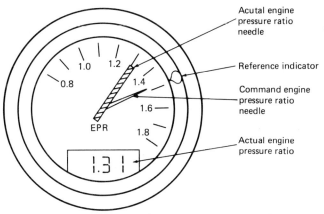

FIG. 20-17 Drawing of an engine pressure ratio (EPR) gage.

FIG. 20-18 Garrett digital fuel controller. *(Garrett Corp.)*

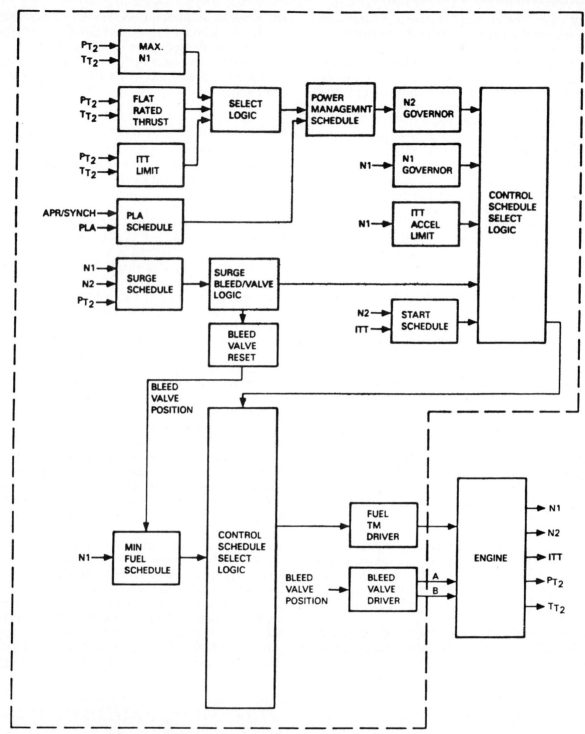

FIG. 20-19 Simplified block diagram of a digital fuel control. *(Garrett Corp.)*

from the cockpit and from the engine itself. The DFC outputs include a proportional drive for regulating fuel flow through a hydromechanical fuel-control unit, command to the air-bleed-valve solenoids, and command to the overspeed solenoid.

The DFC includes an extensive, built-in test feature capable of isolating faults to a line replacement unit (LRU) for interfacing components and for self-diagnosing the controller itself. It is capable of retaining fault history and annunciating faults on the cockpit panel display.

● **REVIEW QUESTIONS**

1. Why is a fuel-control unit for a gas-turbine engine more complex than a carburetor for a piston engine?
2. Why is a manual fuel control not suitable for a gas-turbine engine?
3. What is meant by a *rich blowout*? A *lean dieout*?
4. Name the engine operating conditions (parameters) which must be controlled to ensure efficient and safe performance of a gas-turbine engine.

5. Which of the parameters requested in the previous question are generally employed in the operation of a fuel-control unit?

6. What terms are employed to classify fuel-control units as to type and function?

7. What is meant by a *hydromechanical* fuel-control unit?

8. Describe the function of a *fuel metering section* in a hydromechanical fuel-control unit.

9. What is the function of the *computing section*?

10. Describe the functions of a hydromechanical fuel-control unit from the time the engine is started, operated at 100 percent power, and stopped.

11. What engine operating parameters are sensed by the Hamilton Standard JFC25 fuel-control unit?

12. Describe the means by which the JFC25 fuel-control unit senses and utilizes burner pressure.

13. Describe the functions of the minimum pressure and shutoff valve.

14. What device controls the throttle valve in the JFC25 fuel-control unit?

15. What is the purpose of the minimum flow stop?

16. Explain the function of the rollers between the burner-pressure output lever and the multiplying lever.

17. How is the engine speed sensed and how is the signal used for the JFC25 fuel-control unit?

18. When the power-lever position in the cockpit is changed, what happens in the fuel-control unit?

19. What engine operating parameters are sensed by the Hamilton Standard JFC60 fuel-control unit?

20. Explain the function of the two control levers in the JFC60 fuel-control unit.

21. What pressure differential is maintained across the throttle valve in the JFC60 fuel-control unit?

22. Name the functional units and systems included in the computing section of the JFC60 fuel-control unit.

23. What is the function of the windmill bypass and shutoff valve?

24. Describe the metering system for the JFC68 fuel-control unit.

25. What engine operating parameters are sensed by the computing section of the JFC68 fuel-control unit?

26. What is the purpose of continually rotating the spool-type valves in the JFC68 fuel-control unit and how is the rotation accomplished?

27. Describe the control of fuel which takes place in the JFC68 fuel control during the starting, acceleration, deceleration, and stopping of the JT9D engine.

28. Describe the EVC3 engine vane-control unit.

29. How is the EVC3 unit controlled by the JFC68 fuel-control unit?

30. Name the principal parts of the EVC3 unit engine vane control.

31. Explain the functional relation between the engine vane control and the stator-vane actuator.

32. What engine operating parameters are utilized by the Woodward fuel control for the General Electric CF6 gas-turbine engine?

33. Describe the means by which the differential fuel pressure is established across the fuel metering port in the Woodward fuel-control unit.

34. Describe the function of the speed governor in the Woodward fuel-control unit.

35. How is the variable-stator-vane pilot valve controlled in the Woodward fuel-control unit?

36. Explain the ground/flight idle system.

37. Describe the operation of the fuel shutoff system in the Woodward fuel-control unit.

38. Describe the function of the *gas-producer* fuel control for a turboshaft engine.

39. Explain the use of air pressure to control the fuel metering valve in a pneumatic control system.

40. Describe the function of a power-turbine governor.

41. What is the difference between a *supervisory* and *full-authority* electronic engine control?

42. Describe the operation of a supervisory electronic engine control.

43. What are the principal input parameters to a full-authority electronic engine control?

44. What output commands are produced by a full-authority electronic engine control?

45. What are the benefits derived from the use of a full-authority electronic engine control?

21 TURBOFAN ENGINES

INTRODUCTION

During the past 45 years since the gas turbine first became a reality, many changes and improvements have been made in the design and performance of these engines. Among the first practical engines were the German Jumo, the British W1, and the US GE I-A, which was an American version of the British engine. These were all pure jet engines in that they did not drive propellers; neither did they employ the bypass principle used almost universally with modern engines.

In this section we shall examine the construction and general configuration of some of the most commonly used engines for large and medium-sized aircraft. Airliners in the United States are usually equipped with Pratt & Whitney JT3D, JT8D, or JT9D engines; General Electric CF6 engines; or Rolls Royce RB·211 engines. Small- and medium-sized aircraft usually are equipped with jet or turbofan engines in the 500- to 5000-lb [2224- to 22 240-N] thrust range.

The engines for large aircraft are all axial-flow, multiple-compressor turbofan engines. The turbofan engines of this type have proved to be most efficient and economical for the operation of large airliners.

In recent years, turbofan engines have been improved by changing the aerodynamic design of turbine and compressor blades and vanes, utilizing improved alloys for blades and vanes, modifying fan blade design, and changing the internal engine-cooling systems, plus a number of other modifications. As a result, the engines are more fuel-efficient, develop more thrust for their weight, and operate more quietly. Some of these improvements will be discussed later in this chapter.

THE PRATT & WHITNEY JT3D ENGINE

The Pratt & Whitney JT3D engine is a low-bypass axial-flow turbofan, commonly used for such aircraft as the Boeing 707 and 720 and the Douglas DC-8. It was developed from the J57 jet engine, which was originally used primarily for military aircraft.

The original commercial model of the J57 was the JT3C, a pure jet engine used for a number of years on the Boeing 707 and Douglas DC-8 aircraft. These engines were rated in the 10,000-lb [4536-kg] thrust class, whereas the modern fan versions can produce more than twice the thrust of the first engines. This has been accomplished by design refinements, the use of alloys in the hot sections which can withstand higher temperatures, operating at higher temperatures and higher

rpm, and the addition of the low-bypass fan section. A photograph of a JT3D engine is shown in Fig. 21-1.

Specifications

Among the specifications listed for gas-turbine engines are **thrust, power-weight ratio, fuel consumption, oil consumption,** and **air mass flow.** Thrust is listed in pounds thrust (lbt) or newtons (N). The power-weight ratio, which is actually a thrust-weight ratio, is listed in pounds thrust per pound of engine weight (lbt/lb) or in kilograms of thrust (kgp) per kilogram of engine weight (kgp/kg); this value could also be expressed in newtons per kilogram (N/kg). Fuel consumption is listed in pound per thrust per hour (lb/lbt/h), or lb/(lbt)(h); grams per kilogram of thrust per hour (g/kgp/h), or g/(kgp)(h). Oil consumption is listed in pounds per hour (lb/h) or grams per hour (g/h). Air mass flow is listed in lb/h or kg/h.

The specifications given here are to provide general information and are not to be construed as applying to every JT3D engine. It must be remembered that modifications are made frequently and that advanced models of the engines will have different specifications in some parameters.

Length	136.6 in	346.96 cm
Diameter (fan)	53 in	134.62 cm
Frontal area	15.3 ft^2	1.42 m^2
Weight	4260 lb	1932 kg
Takeoff thrust	19,000 lb	84 512 N
Power-weight ratio	4.46 lbt/lb	4.46 kgp/kg
Fuel consumption (max. cont.)	0.52 lb/lbt/h	525 g/kgp/h
Oil consumption	1.5 lb/h	680 g/h
Compressor ratio	13.5:1	
Bypass ratio	1.43:1	
Fan pressure ratio	1.82:1	
Air mass flow, fan	280 lb/s	127 kg/s
Air mass flow, core	195 lb/s	88 kg/s

General Configuration

The arrangement of the structural components of the JT3D engine may be understood by examining the drawing of Fig. 21-2. The forward fan consists of two stages attached to the forward end of the low-pressure compressor, so that the fan turns at the same rate and is driven by the same turbine section as the low-pressure compressor. The **fan section** includes fixed inlet guide vanes and fixed outlet guide vanes. Air from the fan section divides to provide the bypass thrust (secondary air) and the main-engine core supply (primary

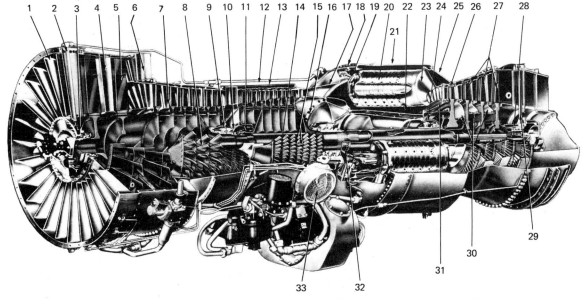

1. Inlet guide vanes
2. Breather tube
3. No. 1 bearing
4. No. 1 stator vane
5. Fan second-stage rotor blade
6. Fan-outlet guide vane
7. Compressor-stator-vane shroud
8. Forward compressor rotor
9. No. 2 bearing
10. No. 2½ bearing
11. No. 3 bearing
12. Compressor-intermediate case
13. Rear compressor case
14. Rear compressor rotor
15. No. 4 bearing
16. Diffuser section
17. Fuel manifold
18. Front-combustion-chamber outer case
19. Combustion chamber
20. Combustion chamber inner liner
21. Rear-combustion-chamber outer case
22. Combustion-chamber inner case
23. Combustion-chamber outlet duct
24. No. 5 bearing
25. Turbine nozzle case
26. Rear-compressor-drive turbine
27. Forward-compressor-drive turbine
28. No. 6 bearing
29. Support rod
30. Knife-edge seals
31. Turbine inner case
32. Main accessory drive gears
33. Air-bleed port

FIG. 21-1   Pratt & Whitney JT3D turbofan engine.

air) in the ratio of 1.43:1, as given in the specification chart.

The **low-pressure compressor** (N_1) consists of eight stages, including the two fan stages, and is driven through the inner coaxial shaft by the last three stages of the turbine. The low-pressure compressor rotor is constructed of titanium-alloy disks and blades. The rotor is supported by the no. 1 roller bearing at the front and the no. 2 ball bearing at the rear.

The **high-pressure compressor** (N_2) is mounted on the forward end of the outer drive shaft and is supported by the no. 3 roller bearing and the no. 4 duplex ball

bearing. The no. 2½ bearing supports the inner drive shaft inside the outer shaft. The N_2 compressor consists of seven stages of steel blades mounted in steel disks and is driven by the first stage of the turbine. Steel is used in this compressor section because the temperature of the highly compressed air attains a level of more than 700°F [371°C] at the compressor outlet. During operation, the N_2 compressor rotates at roughly 1½ times the rpm of the low-pressure compressor.

The compressor rotors are housed in the compressor casings. The casings consist of a total of five sections. These are the **front-compressor case and vane assembly,**

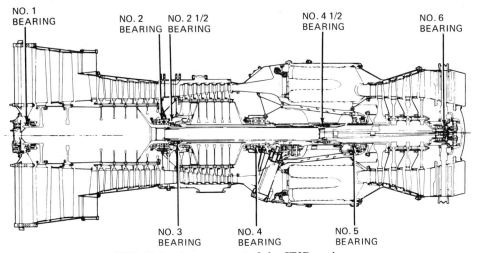

FIG. 21-2   Arrangement of the JT3D engine.

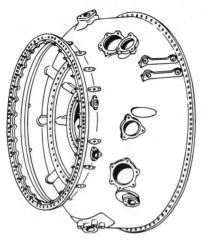

FIG. 21-3 Diffuser case.

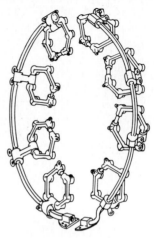

FIG. 21-5 Fuel manifold for the JT3D engine.

the **fan discharge case assembly,** the **front-compressor rear case assembly,** the **compressor-intermediate case,** and the **rear-compressor case.** The stator blades for the compressor sections are mounted inside the various sections of the compressor casings. The compressor-intermediate case includes automatically operated **bleed valves** which open at appropriate times to prevent compressor surge and stall.

The **diffuser case,** shown in Fig. 21-3, is mounted aft of the rear compressor and forward of the combustion case. Its principal function is to change the direction of the high-velocity, high-pressure air and direct it to the combustion chambers. It forms one of the main supporting structures of the engine and includes the mounting for the no. 4 double ball bearing.

The **combustion section** of the JT3D engine is of the **cannular** type. An annular case houses eight combustion chambers, or cans, as shown in Fig. 21-4, and each of these chambers is equipped with six fuel nozzles. Fuel is supplied through a circular fuel manifold inside the forward end of the main combustion chamber. The fuel manifold is shown in Fig. 21-5. The combustion section includes the **front-combustion-**

chamber outer case, the **rear-combustion-chamber outer case,** and the **combustion-chamber inner case.**

Immediately to the rear of the combustion section is the **combustion-chamber outlet duct,** illustrated in Fig. 21-6. This assembly is designed to direct the hot gases from the individual combustion chambers to the first-stage turbine nozzle. It is constructed of heat-resistant sheet metal and is secured in place by the same bolts which attach the combustion-chamber outer case to the turbine-nozzle outer case.

The first-stage **nozzle vanes** are cast from a heat- and corrosion-resistant alloy and are provided with a protective coating to prevent high-temperature corrosion 'and erosion. The second-, third-, and fourth-stage vanes are also fabricated from high-temperature alloys.

The first-stage **turbine rotor** drives the rear, or high-pressure (N_2), compressor. A drawing of this turbine is shown in Fig. 21-7. The 130 turbine blades are attached to the rim of the turbine disk by means of a fir-tree arrangement. When the blades are installed, they are secured in place by means of rivets. The blades are made of forged nickel alloy and are of the reaction-impulse design. At the outer end of each blade is a shroud section which mates with a similar shroud section on the adjacent blade. When all the blades are installed in the disk, a solid ring is formed by the shroud sections. This prevents the loss of gases at the tips of the blades and adds strength to the total assembly.

The blade and tip shroud design makes it impossible to remove one blade only. If a blade must be removed, it is necessary to remove the rivets from the blades 90°

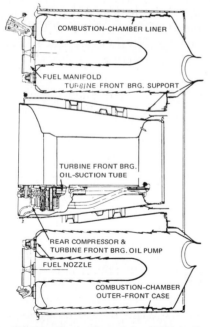

FIG. 21-4 Combustion section of the JT3D engine.

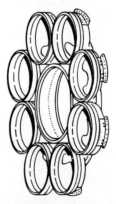

FIG. 21-6 Combustion-chamber outlet duct.

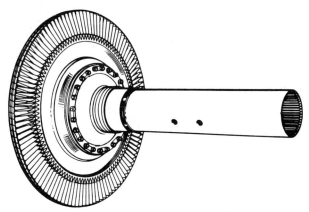

FIG. 21-7 First-stage turbine rotor.

on each side of the blade to be removed and to slide the blades together until the center blade will clear the disk.

The **turbine disk** is machined from corrosion-resistant, high-temperature nickel alloy and is attached to the outer drive shaft by means of a drilled flange on the front face. Balance holes are also provided in the flange. The fir-tree slots which retain the turbine blades are cut in the rim of the disk.

The **rear-compressor drive-turbine rotor** (first-stage turbine) **shaft** is provided with internal splines on the forward end to mate with external splines on the **rear-compressor rear hub**. The shaft and hub are secured in the mated position by means of the **rear-compressor drive-turbine shaft coupling**. The position of the turbine in the turbine case is determined by a spacer ring which bears against a shoulder inside the forward end of the turbine shaft. The spacer is placed between the rear-compressor rear hub and the shoulder inside the turbine shaft.

The front, or low-pressure, compressor of the JT3D engine is driven by the second, third, and fourth stages of the turbine. These three turbine stages are held together by 10 tie rods and the drive shaft is secured to the front disk by means of bolts. The drive shaft extends forward through the center of the **rear-compressor drive shaft** and is supported inside the shaft by means of the **no. 4½ bearing** called the **turbine intershaft bearing**. The shaft is joined to the **front-compressor rear hub** by means of splines and a coupling.

The turbines are supported by means of the **no. 5 roller bearing** and the **no. 6 roller bearing**. The no. 5 bearing is called the **turbine front bearing** and the no. 6 bearing is called the **turbine rear bearing**. The no. 5 bearing supports the outer or rear-compressor drive shaft just forward of the first-stage turbine. The no. 6 bearing is to the rear of the fourth turbine stage and supports the front-compressor-drive turbine rotor by means of the rear-turbine hub. The bearing is mounted in the **no. 6 bearing-support rods.**

The **turbine exhaust case** is a principal structural section of the engine attached to the **turbine-nozzle outer case** by a bolted flange. Its functions are to carry the exhaust gases to the tail pipe and to support the no. 6 bearing. As mentioned previously, the no. 6 bearing is mounted to the rear of the turbine assembly and is supported by the no. 6 bearing support which, in turn, is supported by the no. 6 bearing-support rods. These

rods extend through streamlined struts from the bearing support to the exhaust case.

The JT3D engine is equipped with two thrust reversers, one for the primary jet exhaust and one for the fan exhaust. The effect of the reversers is to reverse the direction of the exhaust and fan discharges and thus to cause the engine to deliver reverse thrust.

Lubrication System

The lubrication system for the JT3D engine, illustrated in Fig. 21-8, is of the pressure type with an externally mounted oil tank which holds 6 gal [22.8 L] of synthetic lubricating oil, MIL-L-23699 or MIL-L-7808. These are high-temperature oils designed to withstand the temperatures of modern gas-turbine engines. When the oils are hot, they give off toxic fumes, so the technician must be careful not to breathe the fumes.

The gear-type oil pump is a duplex unit containing a pressure section and a scavenge section. The pump is shown at the bottom center of the schematic diagram in Fig. 21-8.

Note that oil flows from the tank to the pump, from whence it is directed through a main oil filter and individual oil filters to all the bearing cavities. Scavenge pumps move the oil from the bearing sumps to the main gearbox sump, from whence the main scavenge pump delivers the oil through a cooler back to the main tank. Oil from the no. 4 bearing cavity flows through the accessory driveshaft housing and lubricates the accessory drive bearings.

The entire lubrication system is interconnected by a vent line to remove trapped air pressure. The accessory box is then vented overboard through a centrifugal air/oil separator which prevents atomized oil particles from being dumped overboard.

Other Engine Systems

The fuel system and ignition system for the JT3D engine have been partially described in the preceding sections of this text. As mentioned previously, the fuel delivered to the engine is controlled by a hydromechanical unit which computes the amount of fuel required for each type of operation.

The ignition system is of the high-energy type explained previously. Because of the high voltages employed in the ignition system, it is essential that the technician avoid coming in contact with high-voltage parts of the system to avoid what could be a lethal shock.

● THE PRATT & WHITNEY JT8D GAS-TURBINE ENGINE

Introduction

The Pratt & Whitney JT8D is a twin-spool axial-flow gas-turbine engine similar in general construction to the JT3D. A principal difference is that the JT8D is equipped with a secondary air duct which encases the full length of the engine. A simple diagram to illustrate the general configuration of the engine is shown in Fig. 21-9. A photograph of the engine is provided in Fig. 21-10.

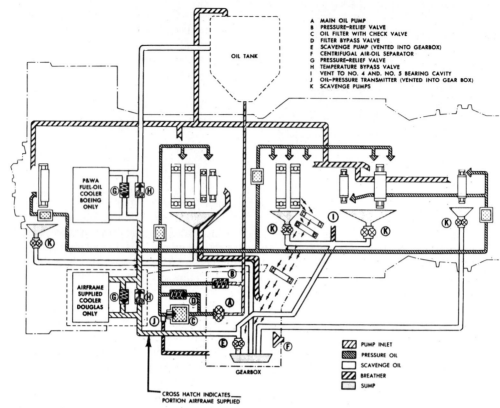

FIG. 21-8   Lubrication system for the JT3D engine.

Specifications

The specifications for a typical JT8D engine given here are for information only. As explained previously, the various models of a particular type of engine may have differing specifications in some areas.

Length	120.0 in	304.80 cm
Diameter	43.0 in	109.22 cm
Frontal area	10.1 ft^2	0.94 m^2
Weight	3300 lb (approx.)	1497 kg
Takeoff thrust	14,500 lbt	64 496 N
Power-weight ratio	4.50 lbt/lb	4.52 kgp/kg
Fuel consumption	0.57 lb/lbt/h	570 g/kgp/h
Oil consumption	3.0 lb/h	1360 g/h
Compressor ratio	16.9:1	
Bypass ratio	1.03:1	
Fan pressure ratio	1.91:1	
Air mass flow, fan	159 lb/s	70.5 kg/s
Air mass flow, core	163 lb/s	73.9 kg/s

General Configuration

The general configuration of the JT8D engine may be understood by examining Figs. 21-9 and 21-10. The bearing arrangement in the JT8D is similar to that of

the JT3D. The no. 1 and no. 2 bearings support the low-pressure compressor (N_1) rotor, the no. 3 and no. 4 bearings support the high-pressure compressor (N_2) rotor, the no. $4\frac{1}{2}$ bearing is the intershaft bearing supporting the inner drive shaft within the hollow outer drive shaft, and the no. 5 and no. 6 bearings support the turbine.

A more complete understanding of the bearings and their function can be obtained from Table 21-1.

The JT8D engine has six general sections. These are the **air-inlet section**, the **compressor section**, the **combustion section**, the **turbine and exhaust section**, the **accessory drives**, and the **fan discharge section**. These will be discussed in more detail as we proceed. Engine sections are shown in Fig. 21-11.

In modern gas-turbine engines it is common practice to identify the flanges by which the engine sections are

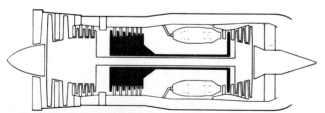

FIG. 21-9   Arrangement of the Pratt & Whitney JT8D engine.

FIG. 21-10   Photograph of the JT8D engine.   (*Pratt & Whitney*)

TABLE 21-1

Bearing Name	Location	Number	Type
Front compressor, front	Inlet case; front-compressor front hub	1	Roller
Front compressor, rear	Compressor case, front; front-compressor rotor rear hub	2	Duplex ball
Rear compressor, front	Compressor case, rear; main accessory drive gear	3	Ball
Rear compressor, rear	Diffuser case; rear-compressor rotor rear hub	4	Duplex ball
Turbine intershaft	In line with midpoint of combustion chamber case; outer race within rear-compressor drive-turbine shaft; inner race and rollers on front-compressor drive-turbine shaft	$4\frac{1}{2}$	Roller
Turbine front	In line with combustion-chamber-case rear flange; inner race on rear-compressor drive-turbine shaft	5	Roller
Turbine rear	Turbine exhaust case; front-compressor drive-turbine rear hub	6	Roller

bolted together. Figure 21-12 identifies the flanges for the JT8D engine.

Air-Inlet Section

As shown in the drawing of Fig. 21-11, the **air-inlet section** is the most forward section of the engine. Basically, it forms the air inlet for the engine and houses the inlet guide vanes. The purpose of the vanes is to direct the incoming air at the proper angle to the first fan stage. The vane at the bottom of the air inlet is thicker than the others in order to accommodate engine tubing.

As indicated previously, the no. 1 bearing front support assembly is mounted in the center of the compressor inlet case. Behind the front support is the

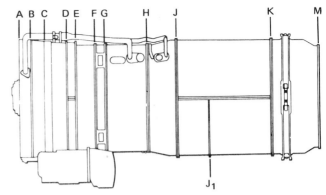

FIG. 21-12 Flanges by which the JT8D is assembled.

no. 1 bearing rear support. Mounted on the front of the inlet case, in the center, is the **front accessory drive support.** The air-inlet, or fan-inlet, case is shown in Fig. 21-13.

As can be seen in the drawing, the compressor inlet case contains 19 inlet guide vanes. Eighteen of these are identical, but the bottom vane is thicker to accommodate engine tubing as mentioned previously. The vanes are brazed between the inner and outer shroud cases, which are made of titanium. The outer shroud case has a double wall; that is, a second ring is brazed outside the inner ring of the outer case to create an annular passage into which anti-icing air flows from the compressor. This air passes through the hollow vanes to the center of the assembly where it discharges forward through the front of the inner shroud case.

The front accessory drive support is constructed of cast magnesium and includes a four-stud N_1 tachometer pad on the upper front face. A pressure oil passage in the support carries oil from the rear of the outer flange into the center, then rearward through the no. 1 bearing oil nozzle. An oil scavenge passage carries oil from a pump boss on the lower rear face of the support cavity back toward the outside of the support, then to another opening in the rear of the outer flange.

The N_1 tachometer-drive gear shaft and the scavenge-pump gear shaft are driven by the **front accessory drive gear shaft** located in the front hub of the front-compressor rotor. The no. 1 bearing oil scavenge pump

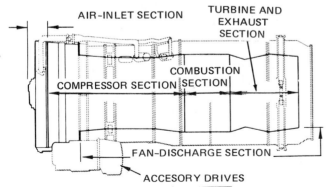

FIG. 21-11 Sections of the JT8D engine.

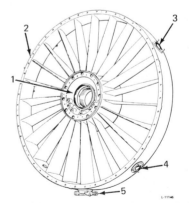

1. NO. 1 BEARING FRONT SUPPORT
2. FAN-INLET CASE
3. BOSS, INLET CASE
4. BOSS, INLET CASE
5. CONNECTOR ASSEMBLY

FIG. 21-13 Air-inlet case.

is mounted on the pump boss inside the front accessory drives support.

Fan and Compressor Sections

The **front-compressor section** and the **fan section** of the JT8D engine are both part of the same rotating assembly. The fan is actually the outer ends of the first two compressor blades. These two blades are enclosed by the front- and rear-fan cases. Figure 21-14 illustrates the arrangement of the JT8D engine with respect to position of units and airflow.

The N_1, or low-pressure, compressor partially compresses the air entering the engine before the air is delivered to the N_2, or high-pressure, compressor. While primary air flows through the core of the engine, secondary air from the fan is passing through the fan duct surrounding the engine. In this engine, the primary air and secondary air are almost equally divided by weight.

As shown in the drawing and explained earlier in this section, the front compressor is driven by the second, third, and fourth turbine stages through the inner or **front-compressor drive shaft.** Thus, the front compressor rotates independently of the rear compressor.

The rear (N_2) compressor is driven by the first-stage turbine through the **rear-compressor drive shaft.** This shaft is splined onto the rear-compressor rear hub and is retained by the turbine-shaft coupling.

The N_2 compressor is driven by only one turbine stage; however, it actually does more work on the primary air than does the N_1 compressor. This fact may raise questions in the reader's mind about why the first stage of the turbine can produce so much more power than the other stages. The answer is that the first-stage turbine is exposed to the hot gases from the engine at the time that the gases have their maximum energy in the forms of velocity, pressure, and heat. Moreover, the last three stages of the turbine not only must start the compression of the primary air but must also drive the fan section to accelerate and partially compress the secondary airflow to greatly increase engine thrust. In the JT3C turbojet engine, the N_1 compressor is driven by two turbine stages. In the JT3D and JT8D turbofan engines, three turbine stages are used to drive the N_1 compressor and fan. Thus, it appears that the power developed by one extra turbine stage is necessary to drive the fan section.

The front-compressor rotor-and-stator assembly consists of the six-stage front-compressor rotor, the front-compressor front and rear cases, and the vanes and shrouds for stages one through five. The sixth-stage vanes are in the compressor-intermediate case. The front-compressor rotor has a front hub which serves as the first-stage disk, a rear hub which serves as the fourth-stage disk, four rotor disks, six stages of blades

secured in the hubs and disks, five rotor-disk spacers, and two sets of tie rods.

The rotor-disk spacers each have two knife-edge airseals on their outer diameters. These knife edges rotate just inside matching seal rings on the inside diameter of the vane and shroud assemblies. The knife-edge airseal of the second-to-third-stage spacer is incorporated in the rear flange of the spacer and matches the ring inside the second-stage vanes. As explained previously, the purpose of the knife-edge seals is to prevent loss of air between stages. It must be remembered that there is a distinct increase in air pressure between each set of stages and that the higher pressure air will flow around the ends of the stator vanes and compressor blades through any available space to the preceding stage. The purpose of the airseals is to reduce the leak space to a minimum.

The first-stage compressor blades include the fan section and are dovetailed into matching grooves in the front-hub rim. They are retained at the leading edge of the blade root by a tab that prevents rearward movement and a positioning ring that prevents forward movement. The second-stage blades, and also the fan, are held in the disk by a pin-joint attachment with a flared rivet. A pin-joint root attachment is illustrated in Fig. 21-15.

The third through the sixth stages of compressor blades are fastened to the disks by dovetail-root sections which fit into broached slots in the disk rim. Tablocks fit in the bottom of the blade root and disk slots and are bent inward at the tab to effect blade retention.

The front hub and the second-stage disk are made of titanium. The third-stage disk is steel, and the rear hub and fifth- and sixth-stage disks are titanium. The first- through the sixth-stage compressor blades are titanium.

The **compressor-intermediate case** is located to the rear of the front compressor. It forms the outer wall of the inner engine from the fan discharge to the diffuser-case front flange, defined as flange H. The sixth-stage stator vanes, constructed of steel, and the no. 3

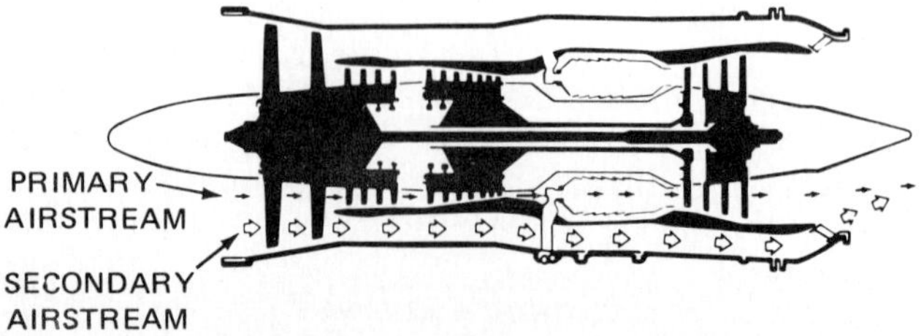

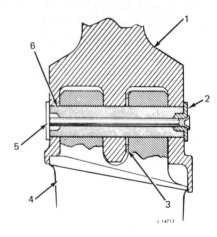

1. DISK
2. WASHER
3. WASHER
4. BLADE
5. RIVET
6. PIN

L 14713

FIG. 21-15   A pin-joint root attachment.

PRIMARY AIRSTREAM

SECONDARY AIRSTREAM

FIG. 21-14   JT8D engine arrangement, showing airflow.

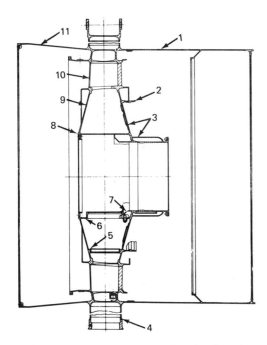

1. Fan-discharge-rear-compressor inner duct
2. Rear-compressor-rotor front-airseal ring
3. No. 3 bearing housing support
4. Compressor-intermediate case
5. Gearbox-drive-bearing-housing lower support
6. Gearbox-drive-bearing-housing upper support
7. Seal-bleed manifold segment
8. Mounting flange for no. 2 bearing housing
9. No. 2 bearing-housing support
10. Sixth-stage vane
11. Fan-discharge-front-compressor inner duct

FIG. 21-16 Cross section of the compressor-intermediate case and fan case.

bearing housing are welded inside the case. The no. 2 bearing housing is bolted to the front face of the case. A steel support and a support bushing below it are positioned at the bottom center of the case to accommodate the **main accessory drive gear-shaft-bearing housing**. Inside the bearing housing is the **main accessory drive bevel-gear shaft** with a roller bearing at the top and a ball bearing at the bottom.

The **compressor-intermediate fan case** and the compressor-intermediate case are shown in Fig. 21-16. It will be noted that the sections labeled **fan-discharge rear-compressor inner duct** and **fan-discharge front-compressor inner duct** form the inner lining of the fan duct. The compressor-intermediate fan case incorporates streamlined struts between the intermediate case and the outer diameter of the engine. The larger six o'clock strut accommodates the accessory gearbox main drive shaft.

The rear compressor utilizes a rotor having seven stages of disks and blades. These are separated by disk spacers and six steel stator-vane stages. The blade stages are numbered 7 through 13 from front to rear. The vane stages, behind their blade stages, are numbered correspondingly, 7 through 12 in the rotor and stator assembly. The thirteenth stator stage and the compressor exit vanes are located in the front end of the diffuser case.

The rear compressor (N_2) is held together by twelve tie-rods. These hold the disks, spacers, and front and rear hubs together axially. A triple knife-edge airseal is secured to the front of the seventh-stage disk and a four-edge airseal is integral with the rear of the thirteenth-stage disk. Blade attachment to the disks is accomplished by a dovetailed lock at the blade root, with the exception of the seventh stage, which uses a pin-joint attachment.

The seventh-, eighth-, and ninth-stage blades of the rear-compressor rotor are titanium. The tenth- through the thirteenth-stage blades are steel. Steel is used for these parts because of the high temperature of the compressed air at this point. The rear-compressor rotor disks are steel except for the thirteenth-stage disk, which is made of nickel alloy. The thirteenth-stage disk incorporates an air-sealing configuration on the rear with four knife edges. The knife edges match the steel thirteenth-stage air-sealing ring positioned inside the diffuser case.

Diffuser Section

The function of the diffuser section has been explained previously. In the JT8D engine, the air passes through the last row of rear-compressor blades at a high velocity. The motion is both rearward and circular in pattern around the engine. Two rows of radial, straightening exit-guide vanes at the entrance to the **diffuser case** slow the circular whirl pattern and convert the whirl-velocity energy to pressure energy. After passing through these straightening vanes, the air still has a strong rearward velocity. This velocity is so high that it would be nearly impossible to maintain a flame in its airstream. A gradually increasing cross section of the air passage decreases the velocity of the airflow and at the same time converts the velocity energy to pressure energy.

The forward part of the diffuser case houses the rearmost section of the rear compressor. The exit stator assembly of the compressor is bolted to flanges in the front openings of the case. This unit contains an inner shroud, outer shroud, and small vanes brazed in place.

In the divergent section of the diffuser case are nine hollow struts having small circular openings on either side which supply compressor discharge air to a manifold around the case. The manifold provides the discharge air for anti-icing and airframe use through two ports on its outer perimeter. The construction of the diffuser case is illustrated in Fig. 21-17.

Nine fuel-nozzle-support mounting pads are located radially near the rear of the case between the hollow struts. Behind the mounting pads are mounting lugs for the front of the individual combustion chambers. The pressure-sensing boss for P_{t4} pressure is located near the two o'clock position on the right-side outer surface of the diffuser case.

The no. 4 bearing compartment is located in the center of the diffuser case near the rear, as indicated in Fig. 21-17. Heat shields are bolted and lockwired in front of the bearing compartment to minimize the temperature inside the compartment. In addition, a tubing system brings eighth-stage discharge air to the annulus between the second and third labyrinth-seal units and bleeds air from the annulus between the first and second labyrinth units to the fan discharge path. The tubes,

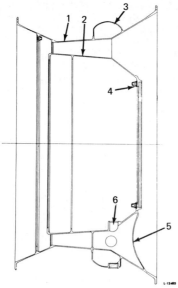

1. OUTER DIFFUSER CASE
2. INNER DIFFUSER CASE
3. DIFFUSER-CASE AIR MANIFOLD
4. NO. 4 BEARING-SUPPORT BOLT CIRCLE
5. DIFFUSER-CASE STRUT
6. NO. 4 BEARING AIR-BLEED-TUBE OPENING

FIG. 21-17 Diffuser case.

secured to the openings in the no. 4 bearing airseal-ring assembly, hold down the bearing compartment temperature by bleeding hot air before it can reach the compartment.

The scavenge pump for the no. 4 and no. 5 bearings is located inside the diffuser case in the rear portion of the no. 4 bearing housing. It has two stages and is driven by a gear mounted on the rear-compressor rear hub.

Combustion Section

The combustion section of the JT8D engine is of the cannular type as explained previously. This means that

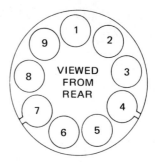

FIG. 21-18 Numbering of combustion chambers.

individual combustion chambers are arranged around the engine inside a single annular chamber. This is similar to the arrangement used in the JT3D engine.

The combustion chambers in the JT8D engine are numbered clockwise as viewed from the rear of the engine. The no. 1 chamber is at the top of the engine, as shown in Fig. 21-18. In Fig. 21-19, the individual combustion chambers are located between the **fuel nozzles** and the **combustion-chamber rear support**. They are held in place by means of a mounting pin at the front. All the chambers have one male and one female flame tube which interconnect to provide rapid flame propagation when the engine is first started. Chambers 4 and 7 each have a spark-igniter opening. Nine interconnector tubes connect the male and female flame tubes of the chambers.

The annular combustion chamber in which the individual chambers are located is formed by the concentric combustion-chamber outer case and combustion-chamber inner case. The **combustion-chamber inner case** is secured to the diffuser-case inner rear flange and to the outer flange of the no. 5 bearing housing. It forms the inner wall of the combustion chamber and serves to position the no. 5 bearing through the bearing housing. The inner and outer cases are major structural members of the engine.

Inside the rear section of the combustion area is the

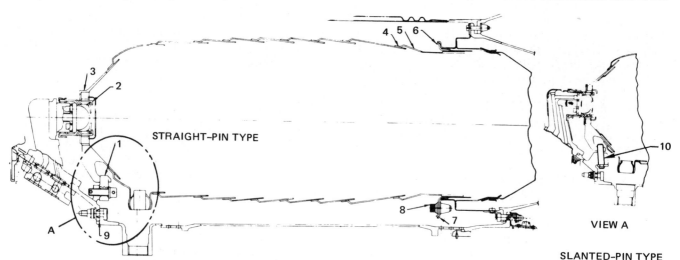

1. Mounting pin
2. Fuel-nozzle heatshield
3. Fuel-nozzle stator
4. Combustion-chamber detail rear liner
5. Diverging portion of rear liner
6. Combustion-chamber rear positioning guide
7. Combustion-chamber rear support
8. Combustion-chamber support bolt (two required at each location)
9. Diffuser-case rear flange
10. Mounting pin

FIG. 21-19 Cross section of a combustion chamber.

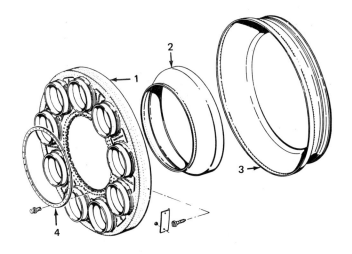

1. COMBUSTION-CHAMBER REAR SUPPORT
2. COMBUSTION-CHAMBER INNER-OUTLET DUCT
3. COMBUSTION-CHAMBER OUTER-OUTLET DUCT
4. REAR-SUPPORT REINFORCING PLATE

FIG. 21-20 Combustion-chamber rear support.

welded **combustion-chamber rear support.** This is illustrated in Fig. 21-20. This unit is a large circular plate with nine openings around a simple central opening. The nine smaller openings support the rear ends of the individual combustion chambers. The rear outer flange is equipped with bolts attached to the support by stops and rivets.

Behind the combustion-chamber rear support are the **combustion-chamber inner and outer outlet ducts** which form the passage through which the hot gases are carried to the first-stage turbine nozzle. Included are air-deflector ducts which divide the cooling air in both the inner and outer duct into two streams. The hot gases pass through nine support openings as they are guided to the turbine nozzle.

Inside the combustion-chamber inner case are the **turbine-shafts heat shields.** These are bolted to the rear of the no. 4 bearing support. Also inside the inner case are the oil-scavenge-pump heatshield assembly, the no. $4\frac{1}{2}$ bearing heatshield assembly, the no. 5 bearing oil-nozzle assembly, the oil scavenge pump shield, and the no. 5 bearing oil scavenge tube. These are illustrated in Fig. 21-21.

The **combustion-chamber outer case** is attached by bolts to the rear flange of the diffuser case and the front flange of the turbine-nozzle case and encloses the combustion chamber. It forms the outer wall of the main engine and the inner wall of the fan annular duct.

Fuel drain valves are located on the bottom center-line of the outer combustion case, one at the front and one at the rear. A fuel drain manifold carries drain fuel outside the outer fan duct. Both flanges of the combustion-chamber inner duct turn inward and the front flange is scalloped. The case is constructed of corrosion- and heat-resistant steel with nickel-cadmium and baked-on aluminum enamel at the flanges to resist corrosion.

Turbine and Exhaust Section

The turbine section includes the turbine front case, the turbine rear case, an inner case and seal, four stages of turbine-nozzle vanes, the two turbine rotors with four stages, and the coaxial drive shafts.

The **turbine front case** is a comparatively short section attached to the rear of the combustion-chamber outer case. It is of decreasing diameter and extends to the first-stage turbine. The first-stage turbine vanes are mounted in the rear of the turbine front case and immediately to the rear of the nozzle vanes is the first-stage turbine. The case is constructed of corrosion- and heat-resistant steel.

Attached to the rear flange of the turbine front case is the **turbine rear case.** This section is also constructed

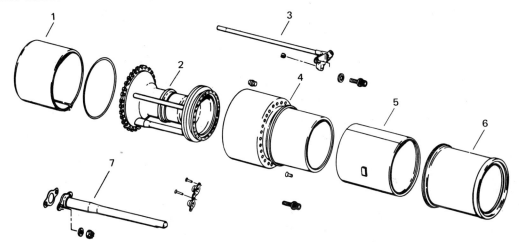

1. Oil-scavenge-pump heatshield assembly
2. No. 4½ bearing heatshield assembly
3. No. 5 bearing oil-nozzle assembly
4. Oil-scavenge-pump shield
5. Turbine-shafts-bearing heatshield asssembly
6. Turbine-shafts-bearing heatshield-shield assembly
7. No. 5 bearing oil-scavenge tube

FIG. 21-21 Heat shields for turbine shafts and bearings.

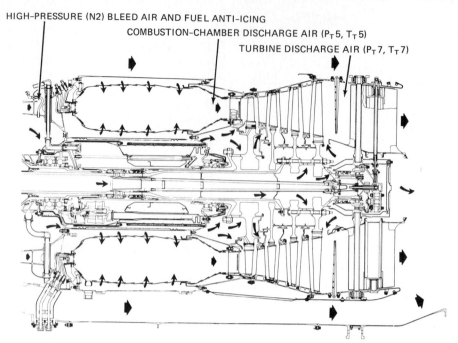

FIG. 21-22   Combustion and turbine sections, showing airflow and gasflow.

of steel and increases in diameter from front to rear in order to accommodate the second, third, and fourth turbine vanes. The rear flange of this case is bolted with the **fourth-stage turbine outer seal ring** to the front flange of the **turbine exhaust case.**

The arrangement of the turbine and exhaust section can be understood by a study of Fig. 21-22. The turbine-nozzle vanes are mounted in the inside diameter of the turbine front case and turbine rear case. The blade-shroud-seal rings can be observed between the nozzle vanes. It will be remembered that the turbine blade tips form a complete ring or shroud to reduce gas losses. The outer rim of the shroud forms a knife-edge seal which mates with the blade-shroud-seal rings.

Attached to the outer flange of the no. 5 bearing housing and extending rearward is the **turbine-nozzle inner case and seal assembly.** Riveted to the rear inner flange of this assembly is the **turbine-rotor inner first-stage airseal** which matches the integral shoulders on the front of the first-stage turbine disk. A groove in the rear outer flange of the turbine-nozzle inner case and seal assembly accommodates the inner rear shroud of the first-stage turbine-nozzle vanes. Segmented multiple turbine vane shroud nuts are riveted to the forward outer flange.

The first-stage **turbine-nozzle vanes** are coated and air-cooled. In later models of the JT8D engine, the cooling air is passed inside the vanes and out through air-exit holes in the airfoil trailing edge on the concave side.

Figure 21-22 shows the installation of the second-, third-, and fourth-stage turbine-nozzle vanes. All the nozzle vanes are attached to the inside of the turbine rear case.

As explained previously, the rear, or high-pressure, compressor (N_2) is driven by the first-stage turbine. In early engines, the turbine disk and the drive shaft were one unit, whereas in later engines the first-stage turbine is a separate unit bolted to the shaft flange as shown

in Fig. 21-23. As seen in the drawing, the turbine blades are attached to the turbine disk by means of fir-tree slots in the rim of the disk. They are held in the slots by means of rivets.

The front-compressor drive-turbine rotor includes the front-compressor drive-turbine shaft, three turbine stages, and the spacers and airseals between the disks. Twelve tie-rods secure the disks and spacers to each other and to the rear flange of the rotor shaft. The

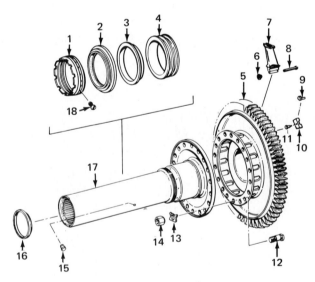

1. No. 5 bearing inner-race retaining nut	10. Counterweight
2. Seal seat	11. Rivet
3. Bearing spacer	12. Tiebolt
4. Labyrinth seal	13. Keywasher
5. First-stage turbine disk	14. Tie-rod nut
6. Washer	15. Positioning plug
7. First-stage turbine blade	16. Turbine-shaft spacer
8. Rivet	17. Rear-compressor-drive turbine shaft
9. Counterweight	18. Retaining screw

FIG. 21-23   Rear-compressor drive-turbine rotor assembly.

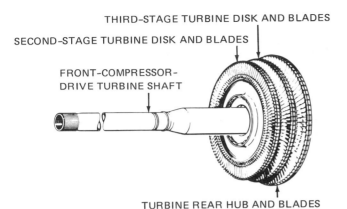

FIG. 21-24 Front-compressor drive-turbine rotor assembly.

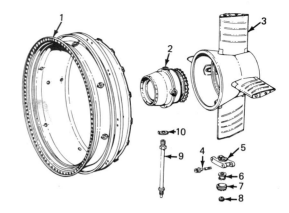

1. Turbine-exhaust case
2. No. 6 bearing housing
3. Turbine-exhaust duct
   and fairing assembly
4. No. 6 bearing-support
   rod boss
5. Bolt
6. Locking nut
7. Extremally threaded ring
8. Plug
9. No. 6 bearing strut
10. Keywasher

FIG. 21-25 Turbine exhaust case and fairing assembly.

turbine bearings pressure and scavenge-oil tubes assembly is located inside the shaft.

The blades for the second, third, and fourth turbines are secured in the disks by fir-tree slots and rivets. The number of blades in each of the three disks is 88, 92, and 74, respectively, from the second stage to the fourth stage. Lugs on the second- and third-stage turbine rotor inner airseals mate with slots on the second- and third-stage disks to prevent rotation between the disks and seals.

The turbine disks for the front-compressor drive turbine can be assembled in only one position because one of the twelve tie-rod holes in each disk is offset. This ensures a proper balance of the complete assembly. A drawing of the turbine assembly is shown in Fig. 21-24.

In later engines of the JT8D type, the front-compressor drive-turbine rotor assembly is designed as a unit or module with the stator assembly. The complete module, which includes the turbine case, nozzle vanes, and turbines, is called the **front-compressor drive-turbine rotor and stator assembly.** This makes it possible to remove and replace the entire assembly without disturbing the other components of the engine. The change can be made with the engine installed in the aircraft.

The **turbine exhaust case,** constructed of welded steel, is the most rearward section of the basic inner engine. It is bolted to the rear flange of the turbine case and decreases from front to rear in order to increase the velocity of the exhaust gases. The turbine exhaust case is the principal support for the no. 6 bearing housing, the thermocouples for temperature sensing, and the pressure-sensing manifold. The **turbine exhaust case and fairing assembly** with the no. 6 bearing-support rods acts as a structural member to support the no. 6 bearing and associated units as shown in Fig. 21-25.

Accessory and Components Drives

The JT8D incorporates two units on the engine to accommodate accessory and components drives. The first is the **front accessory drive housing** which is mounted on the front of the inlet case. This unit incorporates an external pad for mounting the N_1 tachometer generator. The other unit is the **accessory and components drive gearbox** mounted under the engine at the fan discharge intermediate case. It has one drive pad for both the fuel pump and fuel control plus drive pads for the alternator, constant-speed drive, starter, and hydraulic pump. Other units driven from this section are the N_2 tachometer, oil pressure pump, and oil scavenge pumps. An oil pressure-relief valve, oil strainer, and bypass valve are included in this section.

Engine Systems

The engine systems for the JT8D engine are similar to those described previously in other sections of this text. The technician is advised to consult the overhaul and maintenance manuals for specific instructions regarding a particular system.

● THE PRATT & WHITNEY JT8D SERIES 200 ENGINE

The manufacturers and operators of all modern gas-turbine engines continuously seek to improve the performance, reliability, and cost-effectiveness of the engines. This is particularly true of the Pratt & Whitney JT8D turbofan, which is the most widely used engine in existence. The most recent version of the engine is the JT8D-200, a model of which is shown in Fig. 21-26. The original JT8D engine produced 14,000 lb [62,272 N] of thrust and this performance was gradually increased to 17,400 lb [77 395 N]. The JT8D-200 engines produce thrust of from 19,250 lb [85 624 N] to 20,850 lb [92 741 N].

The primary considerations in the improvement of gas-turbine engines has been to reduce fuel consumption, increase reliability, and reduce noise. In the JT8D-200 engines fuel consumption has been reduced through the use of improved materials; more effective aerodynamic design of fan, compressor, and turbine blades and vanes; better cooling techniques; more effective gas seals; and other refinements. Noise reduction has been accomplished through improved fan design, better sound-absorbing materials, and the use of a multiple-tube exhaust discharge. The design of the exhaust discharge system can be seen in the cutaway view of Fig. 21-26.

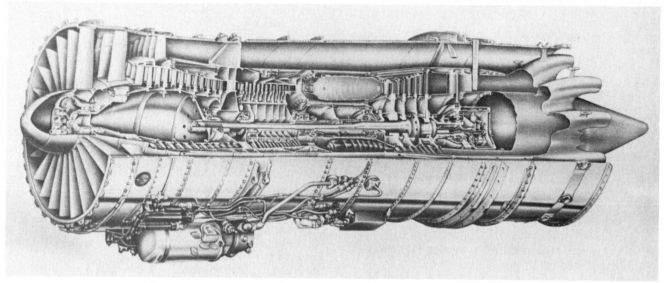

FIG. 21-26   The JT8D series 200 engine.   *(Pratt & Whitney)*

● THE PRATT & WHITNEY JT9D TURBOFAN ENGINE

The Pratt & Whitney JT9D turbofan engine shown in Fig. 21-27 is one of the new generation of gas-turbine engines designed chiefly to power the wide-body aircraft such as the Boeing 747 and Douglas DC-10. It is designated a high-bypass engine because the fan section bypasses more than 5 times as much air as passes through the core, or main part, of the engine.

The JT9D is a twin-spool engine, driving the fan and low-pressure compressor by means of the four most rearward turbine stages through the center coaxial shaft. The high-pressure compressor is driven by the two most forward turbine stages.

The combustion chamber is of the full annular type with 20 fuel nozzles at the front of the assembly.

Performance

The specifications given here are for the JT9D-3A engine and are not to be construed as true for all models. Many different models of the JT9D engine have been developed and the technician is cautioned to make certain of the "dash-number" of the engine. Some models of the JT9D engine are rated at more than 50,000 lbt:

| Length | 154.2 in | 391.67 cm |
| Diameter | 95.6 in | 242.82 cm |

Frontal area	50 ft²	4.6 m²
Weight	8470 lb	3842 kg
Takeoff thrust (wet)	45,000 lb	200 160 N
Power-weight ratio	5.12 lbt/lb	5.13 kgp/kg
Fuel consumption	0.36 lb/lbt/h	640 g/kgp/h
Oil consumption	3.0 lb/h	1360 g/h
Compressor ratio	22:1	
Bypass ratio	5.17:1	
Fan pressure ratio	1.51:1	
Air mass flow, fan	1271 lb/s	576 kg/s
Air mass flow, core	246 lb/s	111 kg/s

General Configuration

The basic design of the JT9D turbofan engine may be understood by an examination of Fig. 21-28. This is a simplified drawing to show the principal sections of the engine. The engine comprises five rotating assemblies, or modules, that are independently balanced so that they may be used interchangeably with other engines of the same model. The five rotating modules are the **fan assembly, low-pressure compressor (N_1) assembly, high-pressure compressor (N_2) assembly, high-compressor drive-turbine assembly,** and **low-compres-**

FIG. 21-27   Pratt & Whitney JT9D turbofan engine.

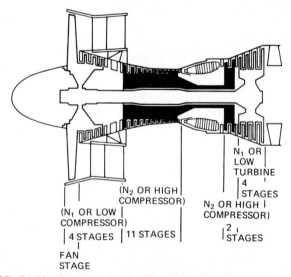

FIG. 21-28   Basic arrangement of the JT9D engine.

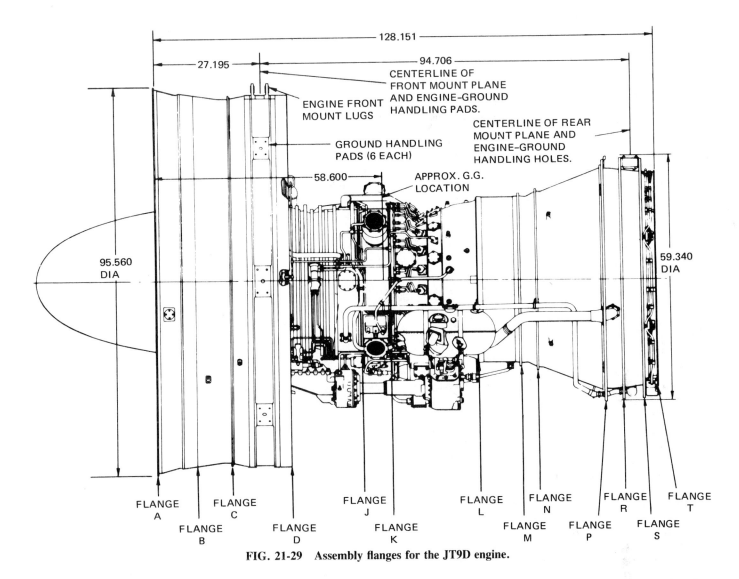

FIG. 21-29 Assembly flanges for the JT9D engine.

sor drive-turbine assembly. When the engine is assembled, there are only two principal rotating assemblies. These may be termed the **fan and low-pressure compressor and drive turbine** and **the high-pressure compressor and drive turbine.**

The flanges by which the JT9D engine is assembled are shown in Fig. 21-29. This drawing provides views of mounting lugs, handling pads, accessory drives, oil tank, air bleeds, fuel lines, and other features.

A cross-sectional drawing of the JT9D engine is shown in Fig. 21-30. This drawing shows the internal construction of the engine. From this drawing it can be seen that the low-pressure compressor consists of the fan rotor plus three additional stages mounted on a single disk which is attached to the drive shaft just forward of the no. 1 bearing.

The high-pressure inlet guide vanes and the nos. 5, 6, and 7 stator vanes are variable to provide for adjustment of airflow direction in accordance with air velocity and pressure. The no. 2 bearing supports the forward end of the high-compressor shaft. As can be seen in the drawing, the section of the high-compressor shaft assembly is comprised of front and rear conical hubs between which are mounted the disks and spacers that support the compressor blades. The assembly thus has the appearance of a drum with conical ends.

As shown in the drawing, the high compressor is driven by the no. 1 and no. 2 turbine stages. The no. 3 bearing supports the assembly at the rear. The four rearmost stages of the turbine drive the low compressor and the fan. This assembly is supported at the rear by the no. 4 bearing.

Fan and Low-Compressor Assembly

As mentioned previously, the fan stage forms the first stage of the low compressor and also provides the high bypass of fan air. The fan has 46 blades balanced in pairs; they are thus replaceable in balanced sets. The blades have dovetail roots and are retained axially in the disk by means of shear locks.

The fan case and outlet duct are lined with sound-absorbing material to reduce noise. Additional noise-reduction factors are the relatively low tip speed of the fan blades and the fact that there are no inlet guide vanes ahead of the fan.

The fan-disk hub is splined to a coupling which is, in turn, splined to the low-turbine shaft. The fan disk is positioned axially by means of a spacer between the fan and the coupling. Axial positioning of the fan is an important factor in adjusting fan blade-tip clearance. The fan case is tapered so that the axial movement of the fan will vary the tip clearance.

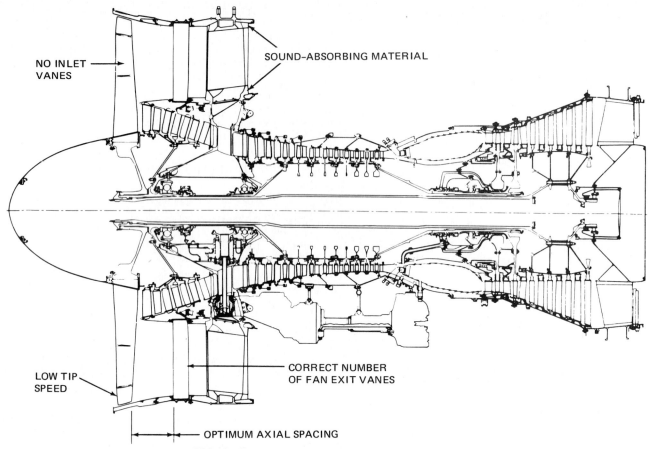

NO INLET VANES

SOUND–ABSORBING MATERIAL

LOW TIP SPEED

CORRECT NUMBER OF FAN EXIT VANES

OPTIMUM AXIAL SPACING

FIG. 21-30   Cutaway drawing of the JT9D engine.

The construction of the fan and fan hub, together with the low-pressure compressor hub and coupling, provide for easy removal of the fan, low turbine, and low compressor. The complete assembly is held in place by means of a large coupling nut at the forward end of the turbine shaft.

Each fan blade has a rubber air-sealing strip cemented to the butt end to reduce leakage of air from the airfoil section of the blade around the root end.

An enlarged drawing of a cross section of the low-pressure compressor is shown in Fig. 21-31, which shows the nos. 2, 3, and 4 compressor blades between the stationary vanes. It also shows how the compressor blades are attached to the compressor hub. It must be remembered that the no. 1 compressor stage is the fan, which is not shown in the drawing.

A drawing of the anti-icing system for the first-stage stator vanes is provided in Fig. 21-32, which shows the hot-air duct that brings air from the ninth stage of the compressor forward to the first-stage vane. It can be seen that the air flows into the hollow vane and out through holes near the trailing edge of each vane.

Intermediate Case

The **intermediate case** is the section of the engine between the low compressor and the high compressor. It serves a major structural function and supports the no. 1 and no. 2 bearings. The **tower shaft** which provides for accessory drives engages the bevel ring gear, which is mounted on the forward end of the high-compressor front hub. The shaft passes through the lower strut of the case. This can be seen in Fig. 21-30.

A cross section of the bearing compartments in the intermediate case is shown in Fig. 21-33. This drawing reveals some of the details including various types of oil and airseals.

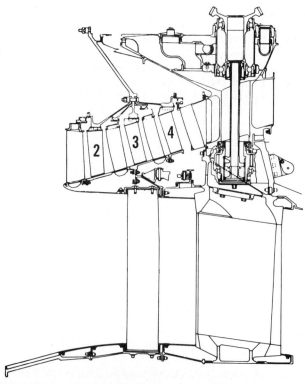

FIG. 21-31   Cross section of the low-pressure compressor.

490

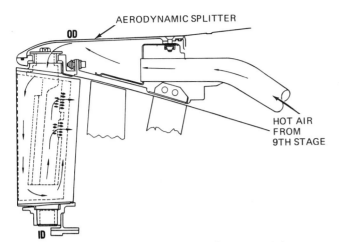

FIG. 21-32 Anti-icing system for the first-stage stator.

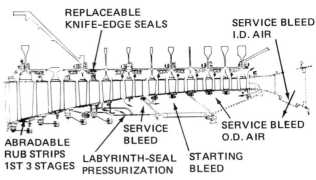

FIG. 21-34 Section of the high-pressure compressor.

High Compressor

A section of the high-pressure compressor (N_2) is shown in Fig. 21-34. In the drawing it can be seen that the inlet guide vanes and the nos. 5, 6, and 7 stators are variable. As explained in another section of this text, the **variable vanes** are designed to change pitch as necessary to avoid surge and stall as the air pressures and velocities change.

Since the high compressor increases air pressure to more than 300 psi [2068.5 kPa], the airseals in the section must be most effective. Abradable rub strips are placed between rotors and stators to ensure tip seal.

Stages 8 through 11 are enclosed by two outer cases which serve as manifolds for bleed air from stages 7 through 9. Stages 12 through 15 are enclosed by the forward section of the **diffuser case.** This forms a manifold around the rear stages of the compressor for fifteenth-stage air bleed.

The arrangement for adjustment of the variable vanes is shown in Fig. 21-35. Each stage of vanes is operated through a unison ring which makes all vanes in the stage rotate the same amount. All four unison rings are connected to a bellcrank rotated by means of a hydraulic cylinder. The hydraulic cylinder receives operating energy in the form of hydraulic pressure from the engine vane control (EVC), which was described in Chap. 20.

Combustion Section

The combustion chamber for the JT9D engine is a two-piece annular type having an inner and outer liner. The two liners which form the combustion chamber are shown separated in Fig. 21-36. A cutaway view of the assembled chamber was shown in Fig. 19-24.

The cutaway drawing of Fig. 21-37 shows a cross section of one side of the combustion chamber together with a section of the diffuser case and the **first-stage turbine-nozzle assembly.** In this drawing can be seen one of the 20 fuel nozzles bolted externally to the diffuser case and extending into the forward end of the combustion chamber. Surrounding the fuel nozzle in the nozzle opening are swirl vanes to assist in vaporizing the fuel as it is discharged into the combustion chamber.

Each fuel nozzle is equipped with three fittings. One of these is for primary fuel, one for secondary fuel, and one for water. The fuel and water are carried in three steel manifolds which surround the diffuser case. Individual lines and fittings carry the fuel and water to the nozzles.

At the rear of the combustion chamber in the drawing of Fig. 21-37 can be seen the first-stage nozzle vane. There are 66 vanes in the nozzle assembly classified for nozzle-area sizing and individually replaceable. They are hollow and internally cooled with fifteenth-stage

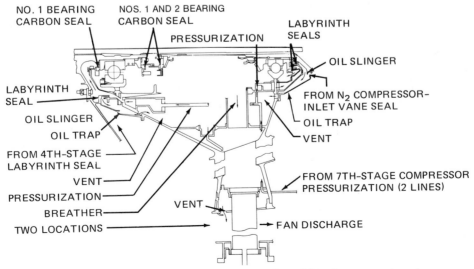

FIG. 21-33 Cross section of the no. 1 and no. 2 bearing compartments.

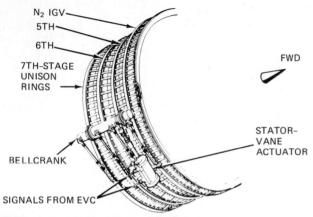

FIG. 21-35   Variable vanes arrangement.

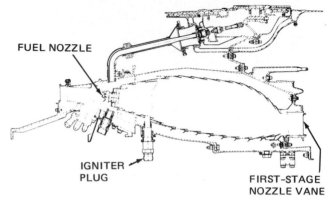

FIG. 21-37   Cross section of the combustion chamber.

air. The vanes in their support can be removed from the engine as a complete assembly. The rear of the combustion section is designed with an outer combustion chamber which can be telescoped forward to permit individual nozzle-vane replacement.

As shown in the drawings, the diffuser case comprises the forward end of the combustion section. Figure 21-38 shows the diffuser case and Fig. 21-39 some of the functions of the case.

The no. 3 bearing housing is supported inside the case by means of ten struts. Four of the struts have openings inside the engine to bleed fifteenth-stage clean air to four openings on the case. Two of the struts carry seventh-stage cooling air into the cooling jacket around the no. 3 bearing and carry the bearing breather pressure out. Two other struts carry seventh-stage cooling air out from the bearing area. It will also be noted that oil pressure for the no. 3 bearing is carried through one of the struts and that scavenge oil is removed through the six o'clock strut.

The diffuser case holds the mountings for the two spark igniters. These igniter plugs are mounted on each side of the case and extend into the inside of the combustion chamber on each side, 9° above the centerline. An igniter plug can be seen in Fig. 21-37.

Cooling of the first-stage nozzle and first-stage turbine is accomplished by fifteenth-stage air which is bypassed around the combustion chamber into a cooling duct. A series of jets at the aft end of the duct direct the air into 116 holes drilled tangentially through the first-stage disk at the broached end. The air then passes

into the 116 hollow first-stage blades and exits from the blade tips and from small holes near the leading edges.

The entire turbine area is cooled by flowing air through numerous holes and passages, which eventually lead the air to openings where it can join the exhaust stream.

Turbine Section

As explained previously, the turbine section of the JT9D engine is comprised of two turbine assemblies. The first is the high-compressor drive turbine and the second is the low-compressor drive turbine. The **high-compressor drive turbine** consists of the first two stages of the turbine which are located immediately to the rear of the combustion section. A cross section of the high-compressor drive-turbine assembly is shown in Fig. 21-40.

In the drawing it can be seen that the first-stage turbine disk is attached to the second-stage disk by means of a bolted flange. Knife-edge airseals are provided to prevent air leakage between the turbine blades and vanes. The airseals and blades operate against abradable shrouds at the outside diameter of the blades and the inside diameter of the vanes. Abradable shrouds permit rubbing without damage to the parts. The second-stage disk is splined to the high-compressor drive shaft (hub) as shown in the drawing.

The nozzle vanes and turbine blades are cooled by means of air which flows through interior passages, as

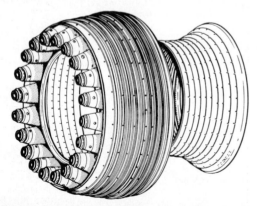

FIG. 21-36   Combustion chamber with inner and outer liners separated.

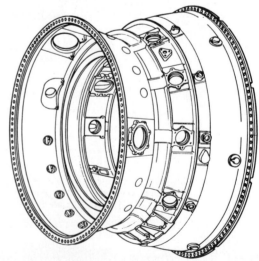

FIG. 21-38   Diffuser case.

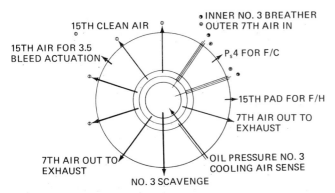

FIG. 21-39 Some functions of the diffuser case.

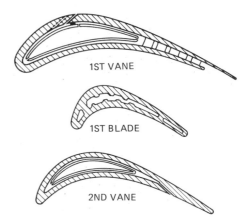

FIG. 21-41 Cooling features of turbine-nozzle vanes and blades.

shown in Fig. 21-41. The turbine blades are Jo-coated and the vanes are chromallized to reduce gas erosion and high-temperature corrosion.

The first-stage turbine has 116 blades mounted in fir-tree slots broached in the rim of the disk. The blades are held in place by 29 blade retaining plates riveted to the disk and to each other by 116 long rivet pins. The plates contain the cooling airflow from the disk so that it flows into the blades and not out the fir-tree slots. The turbine blades are moment-weighed and are installed individually in the disk. The retaining plates are also classified by weight.

In the first-stage turbine, the blades are not shrouded, which makes it possible to exit the blade-cooling air through holes in the tips of the blades.

In the second-stage turbine there are 138 blades mounted in fir-tree slots. The blades are retained by means of flare-type rivets. As seen in Fig. 21-40, the second-stage blades are shrouded with integral knife-edge seals to reduce gas leakage at the tips. The blades are moment-weighed, as is the case with all the turbine and compressor blades.

As explained previously, the **low-pressure compressor drive turbine,** often termed the **low-compressor drive turbine,** drives the fan and three stages of the N_1 compressor. This turbine section includes the third, fourth, fifth, and sixth turbine stages and is illustrated in Fig. 21-42. The drive-shaft arrangement can be seen in the drawing of Fig. 21-30.

The low-compressor drive turbine incorporates four disks to which the turbine blades are attached in fir-tree slots. The blades are held in place by rivets. The fifth-stage disk is integral with the rear-turbine hub. The other three disks are assembled with the fifth-stage disk by means of through-bolts, as shown in Fig. 21-42.

Note in the drawing that the nozzle vanes for the four turbine stages described here are attached to the inside of the **turbine rear case.** Shrouds at the ends of the vanes mate with knife-edge seals mounted between the pairs of disks. The tips of the turbine blades for all four stages are shrouded and equipped with knife-edge seals. The seals are abradable so that they can make contact with the rubbing strips without causing damage.

The complete low-turbine section can be removed as a unit and is therefore one of the rotating modules of the engine. The module is balanced and can be interchanged with similar modules on other engines of the same type.

The **turbine exhaust case** is illustrated in Fig. 21-43. This is a complete welded structure with 15 struts between the outer case and the no. 4 bearing housing. Within the bearing housing is the no. 4 bearing compartment surrounded by seventh-stage air for cooling and heat protection.

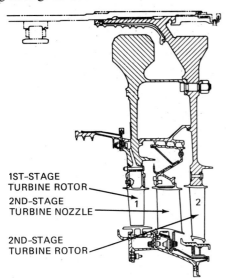

FIG. 21-40 Cross section of a high-pressure compressor drive turbine.

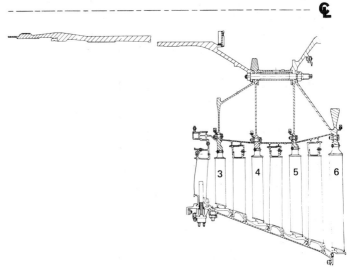

FIG. 21-42 Section of the low-compressor drive turbine.

493

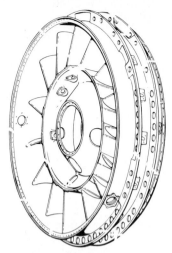

FIG. 21-43 Turbine exhaust case.

Six pressure-sensing probes for measuring P_{t7} are manifolded together between the last two flanges of the case. Pressure oil for the no 4 bearing flows through one of the struts and scavenge oil flows out through another.

The rear mountings for the engine are located on the top of the turbine exhaust case. This requires that the case be designed and built for high structural integrity.

To reduce engine distortion resulting from heat and stress, a thrust frame is installed between the intermediate case and the turbine exhaust case. Through this frame, engine thrust is transmitted directly from the intermediate case to the aircraft structure. This is accomplished through a fitting called a "puck" which forms the attachment from the engine to the aircraft.

Accessory Section

The accessory section of the JT9D engine consists of two gearboxes mounted under the engine and driven through a tower shaft. The tower shaft incorporates a spur-gear pinion which meshes with the **accessory drive** mounted on the front hub of the high-compressor hub.

Power from the tower shaft is carried to the **angle gearbox** and from thence through a horizontal shaft to the main gearbox located at the bottom of the engine near the forward end of the combustion section. The location and arrangement of the accessory gearboxes can be seen in the drawing of Fig. 21-30.

Conclusion

The foregoing description does not cover many details of the JT9D engine and its construction, but it does provide sufficient detail so that the technician can understand the general design and operation. To obtain complete information about any engine, the technician should consult the maintenance and overhaul manuals for the engine concerned.

Systems for the JT9D Turbofan Engine

Some of the systems for the JT9D engine have been partly described earlier. The fuel system includes the Hamilton Standard JFC68 fuel control, a TRW fuel pump with booster, an engine vane control, filters, a fuel-oil cooler, and other units common to the majority of gas-turbine fuel systems.

The lubrication system for the JT9D engine is similar to those for other gas-turbine engines. A pressure pump delivers oil to the various bearings where the oil is sprayed on the bearings from jets. The oil and oil spray is contained in the bearing compartments by means of air pressure acting through the seals. The oil is returned to the oil tank by scavenge pumps in each area. Air is separated from the oil and is vented overboard. Relief valves provide protection against extreme pressure. The system is equipped with a main filter plus a number of "last-chance filters" to ensure that the oil delivered to the bearings is clean.

The ignition system has been described earlier. As mentioned previously, there are actually two systems of the high-energy type. The electrical energy is stored in capacitors and then released to provide an extremely high-energy spark for the ignition of fuel. The energy is required only for starting; however, it may be turned on at times when there is a possibility of flameout.

● THE PRATT & WHITNEY JT9D-7R4 TURBOFAN ENGINE

The Pratt & Whitney JT9D-7R4 turbofan engine that powers the Boeing 767 and Airbus A310 aircraft is the product of years of research and development plus the experience in the operation of earlier engines for millions of flight hours. Advanced technology has been applied in many parts of the engine with the result that substantial improvements in fuel consumption, reliability, engine life, maintainability, and noise emission have been accomplished. A cutaway view of the JT9D-7R4 engine is shown in Fig. 21-44. Some of the improved features are pointed out in the illustration.

The basic design of the -7R4 engine is essentially the same as that of the JT9D described in the foregoing section of this text; however, the changes and additions made on the -7R4 have provided substantial gains for the operators of the engine. Fuel savings up to 8 percent and thrusts up to 56,000 lb [249 088 N] are possible with some models of the engine.

Fan Section

The fan of the -7R4 engine has been modified by the use of wide-chord single-shroud fan blades and the use of a reduced number of fan blades and exit guide vanes. These features, together with the convergent-divergent fan exit nozzle, contribute to significant fuel savings during cruise operation. The aft-positioned blade shroud reduces losses associated with operation of the fan at high rotational speeds. The illustration of Fig. 21-45 shows the difference between the fan blade used on earlier models of the JT9D and the blade used on the -7R4 model. The modification of the fan section has produced fuel savings as well as reduced noise emission.

Low-Pressure Compressor

The JT9D-7R4 engine is equipped with the four-stage compressor utilized in the JT9D-59A, -70A, and -7Q models of the engine. Previous models employed a three-stage low-pressure compressor as illustrated in Fig. 21-46. The four-stage design increases the overall engine pressure ratio and airflow capacity, providing improved performance, lower turbine operating tem-

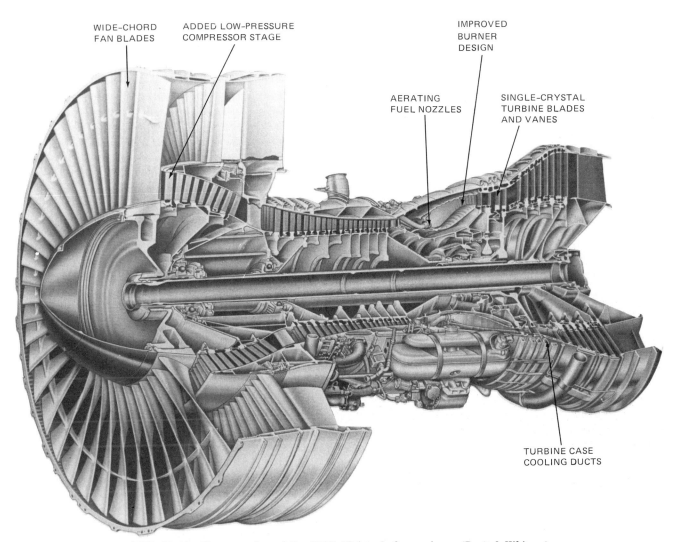

FIG. 21-44 Cutaway view of the JT9D-7R4 turbofan engine. *(Pratt & Whitney)*

peratures, growth capability, and improved surge margin.

High-Pressure Compressor Stators

The high-pressure compressor stators in the -7R4 engine are coated with an alloy having the trade name ''SermeTel'' that has proved very beneficial. This is a long-lasting, thin-film coating that provides improved protection against corrosion and erosion in the compressor section, thereby extending parts life and reducing maintenance and overhaul costs. The coating also contributes to lower fuel consumption.

Burner Design

The bulkhead configuration burner design used in later series JT9D engines is also employed in the -7R4. This burner improves performance and durability as compared to earlier designs. Because it provides lower temperatures at the outer diameter of the annulus, it reduces airseal expansion, thus retaining clearances and minimizing turbine rubbing. The fuel nozzles are of the aerating type; that is, air under pressure is mixed with the fuel before it sprays into the burner. This type of fuel nozzle reduces emission levels and eliminates coking.

Turbine Blades

The blades of the high-pressure turbine are single-crystal castings. The special Pratt & Whitney alloy used and the single-crystal feature substantially increase the mechanical properties of the blades—such as creep strength and resistance to low-cycle fatigue, oxidation, and high-temperature corrosion—and provide increased heat resistance. The use of these blades in the high-pressure turbine lowers maintenance costs and saves fuel by reducing cooling-air requirements.

Turbine Cooling

Cooling air is supplied to the turbine blades and vanes (airfoils) at high pressures through injection nozzles near the disk rim. As part of the -7R4 series turbine tailoring, the injection nozzles are sized to meet the cooling requirements of the particular engine model. By using only the amount of air from the engine airflow that is required for adequate cooling, maximum efficiency is attained while maintaining durability.

The -7R4 series engines employ a modulated turbine-case cooling system to improve performance and save fuel during cruise operation. The system starts operating after takeoff and climb because during these operations the turbine blade clearance must be greater than at cruise. During cruise operation, air flows through

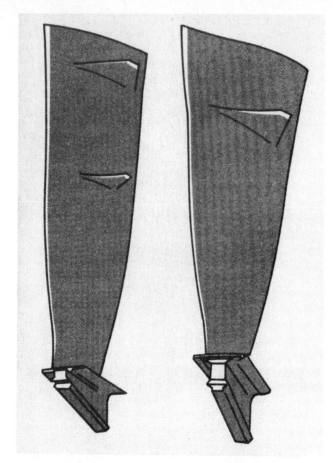

DOUBLE-SHROUDED
FAN BLADE,
46 PER ASSEMBLY

JT9D-7R4 SINGLE-SHROUD,
WIDE-CHORD BLADE,
40 PER ASSEMBLY

FIG. 21-45   Comparison of fan blades. *(Pratt & Whitney)*

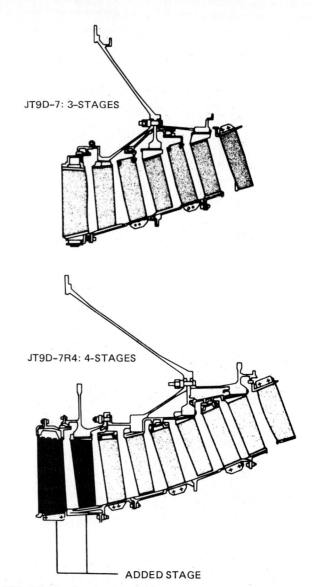

JT9D-7: 3-STAGES

JT9D-7R4: 4-STAGES

ADDED STAGE

FIG. 21-46   Comparison of low-pressure compressors.

cooling ducts around the turbine case and the resultant cooling shrinks the case to minimize blade tip clearances, thus increasing the turbine's effectiveness and saving fuel. The case cooling system utilized to control blade tip clearance is termed **active clearance control (ACC)**.

Air Seals

Lightweight carbon airseals for the no. 2 and no. 3 bearing compartments in the -7R4 engine have replaced the labyrinth seals used in earlier engines. This change has resulted in a reduction of weight by eliminating the external plumbing required for seal-pressurization air ducting. The use of carbon seals improves engine performance and reduces oil consumption because there is less air leakage from the bearing compartments.

Electronic Engine Control

Although it is not a part of the engine, the electronic engine control (EEC) utilized with the -7R4 engine contributes substantially to the engine's performance. The engine control used with the -7R4 engine is the Hamilton Standard EEC-103. This is the **supervisory engine control** described in Chap. 20. The precise control of engine functions by the EEC-103 in conjunction with the hydromechanical fuel-control unit increases engine efficiency significantly, thereby reducing fuel con-

sumption. In addition, the EEC reduces flight-crew workload, maintenance costs, and aircraft downtime.

● THE PRATT & WHITNEY 2037 TURBOFAN ENGINE

The Pratt & Whitney 2037 turbofan engine shown in Fig. 21-47 is termed an advanced-technology engine because it incorporates all the advanced design features and materials that contribute to engine efficiency, reliability, low fuel consumption, and maintainability. The engine powers the Boeing 757 airliner and is tailored to meet all the power requirements of that aircraft.

Specifications

The general specifications for the Pratt & Whitney 2037 turbofan engine are as follows:

Length	133.7 in	339.60 cm
Diameter, fan tip	78.5 in	199.39 cm
Diameter, fan case	85.0 in	215.90 cm
Diameter, turbine exhaust case	49.0 in	124.46 cm

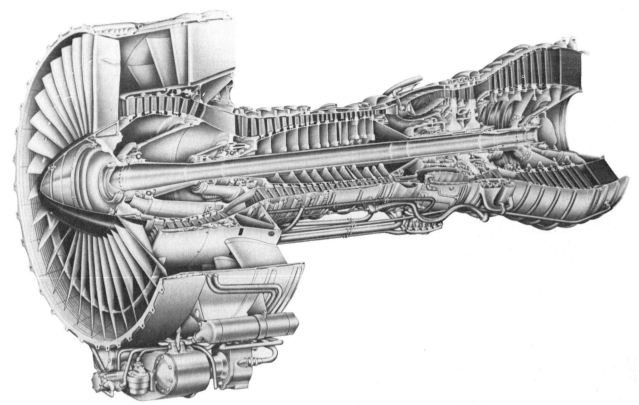

FIG. 21-47 Pratt & Whitney 2037 turbofan engine. *(Pratt & Whitney)*

Weight	6,675 lb	3 027 kg
Takeoff thrust, sea-level static	37,000 lb	164 576N
Fuel consumption, thrust specific, 35,000 ft [10 668 m], 0.8 Mach	0. 563 lb/lbt/h	0.574 g/kgp/h
Overall pressure ratio	30:1	
Bypass ratio (cruise)	5.8:1	
Fan pressure ratio	1.70:1	
Total airflow	1,193 lb/s	541 kg/s
Low-pressure compressor stages	Fan plus 4	
High-pressure compressor stages	12	
Low-pressure turbine stages	5	
High-pressure turbine stages	2	

Description

The Pratt & Whitney 2037 is a twin-rotor, high-bypass-ratio, axial-flow turbofan engine designed for use with commercial and military transport aircraft. It will be seen from Fig. 21-47 that its configuration is similar to that of the JT9D, although the 2037 is somewhat smaller.

The single-stage fan and four-stage low-pressure compressor are driven by a five-stage turbine at the rear of the engine. The twelve-stage high-pressure compressor is driven by a two-stage turbine through a hollow shaft that is outside and coaxial with the low-pressure compressor drive shaft. The two rotors are supported by five main bearings.

The fan has 36 wide-chord, single-shroud blades similar to those on the JT9D-7R4 engine. The aft location of the shroud reduces airflow interference. The design of the fan section contributes to maximum efficiency and minimum noise.

The airfoils (blades and vanes) are described as "controlled-diffusion" airfoils. These permit higher Mach numbers without loss in efficiency and are used in both the compressors and turbines. The thicker leading and trailing edges of the airfoils, shown in Fig. 21-48, also enhance performance retention (continued serviceability) by increasing resistance to airborne particulate erosion.

Variable stators are employed in the first five stages of the high-pressure compressor to increase compressor efficiency, control surge, and eliminate compressor stall. These stators provide stability over the entire operating envelope by controlling the aerodynamic loading on the compressor stages. The disks and spacers in the high-pressure rotor are joined by electron-beam

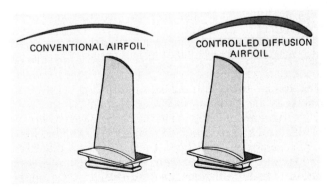

FIG. 21-48 Comparison of airfoils.

FIG. 21-49 Single-crystal turbine blade.

welding. This technique, used instead of bolted joints, eliminates leakage and improves clearance control by stiffening the rotor.

The engine has an annular combustor (burner) with 24 single-pipe airblast fuel nozzles. The combustor design provides improved heat transfer from the lower lip which reduces thermal stress. In addition, the design provides uniform discharge of the cooling air to the inner surface of the downstream burner wall.

The high-pressure turbine blades are single-crystal castings made from PW1480 alloy. A blade of this type is shown in Fig. 21-49. These blades provide high creep strength, thermal and low-cycle fatigue life, and resistance to oxidation and high-temperature corrosion. The outer airseals for the high-pressure turbine are of an abradable, nonmetallic material. These seals improve performance by reducing blade tip clearances during operation. They also improve performance retention and reduce the risk of turbine blade damage.

The high-pressure turbine disks and the last high-pressure compressor disk are produced from fine-mesh PW1100 powder metal. Compared to conventional forged disks, the powder-metal disks offer higher tensile strength, greater stress-rupture resistance, and longer low-cycle fatigue life.

It will be noted in the illustration of Fig. 21-47 that low pressure turbines increase in diameter substantially from the first stage to the last stage. This flowpath configuration is designed to provide the best possible aerodynamic performance. The large mean diameter and smooth gasflow transition from the high-pressure turbine is more efficient than the smaller mean-diameter turbines used on earlier engines.

Engine Control

The Pratt & Whitney 2037 engine is controlled by the Hamilton Standard EEC-104 **full-authority electronic engine control** described in Chap. 20. All of the engine variables—including starting, fuel flow, variable stator positioning, acceleration, and deceleration—are regulated by the EEC. The automated thrust thus provided eliminates the need to continually readjust throttle settings during takeoff and climb. The EEC reduces engine maintenance time by eliminating the requirement for engine trimming.

● THE GENERAL ELECTRIC CF6 TURBOFAN ENGINE

The General Electric CF6 turbofan engine is also one of the new generation of high-bypass engines designed for the large, wide-bodied airliners. The principal models at present are the CF6-6 and the CF6-50 which are designed for use in the Douglas DC-10 airliner. The CF6-50 is also installed in the Airbus Industrie A-300B airplane. A photograph of the CF6-50 is shown in Fig. 21-50.

The differences between the CF6-6 and the CF6-50 are in the numbers of compressor and turbine stages. The CF6-6 engine has a single-stage low-pressure booster immediately to the rear of the fan. This is called a booster stage. The compressor for the CF6-6 has 6 rows of variable stator blades and 10 rows of fixed stator blades. The compressor rotor has 16 stages and is driven through the outer drive shaft from the high-pressure turbine. The low-pressure turbine has five stages and drives the fan and the one-stage booster.

The CF6-50 engine has a three-stage low-pressure compressor immediately to the rear of the fan. This increases the air pressure considerably before it enters the high-pressure compressor. The high-pressure compressor has 5 rows of variable stator vanes and 10 rows of fixed vanes. The rotor consists of 15 stages of blades and is driven by the outer drive shaft from the high-pressure turbine. The low-pressure turbine has four stages instead of the five stages in the CF6-6 engine.

Specifications

Some of the differences in performance of the two types of engines can be understood by examining the specifications given below. It must be noted that there have probably been recent changes in some of the specifications, but those given will provide a good understanding of the performance of each type.

	CF6-6	CF6-50
Length	193 in	190 in
Diameter	94.0 in	94.0 in
Frontal area	48.2 ft²	48.2 ft²
Weight	7450 lb	8225 lb
Takeoff thrust	40,000 lbt	50,000 lbt
Power-weight ratio	5.45 lbt/lb	5.85 lbt/lb
Fuel consumption	0.35 lb/lb/h	0.39 lb/lbt/h
Oil consumption	3.0 lb/h	3.0 lb/h
Compressor ratio	26.6:1	29.9:1
Bypass ratio	6.2:1	4.4:1

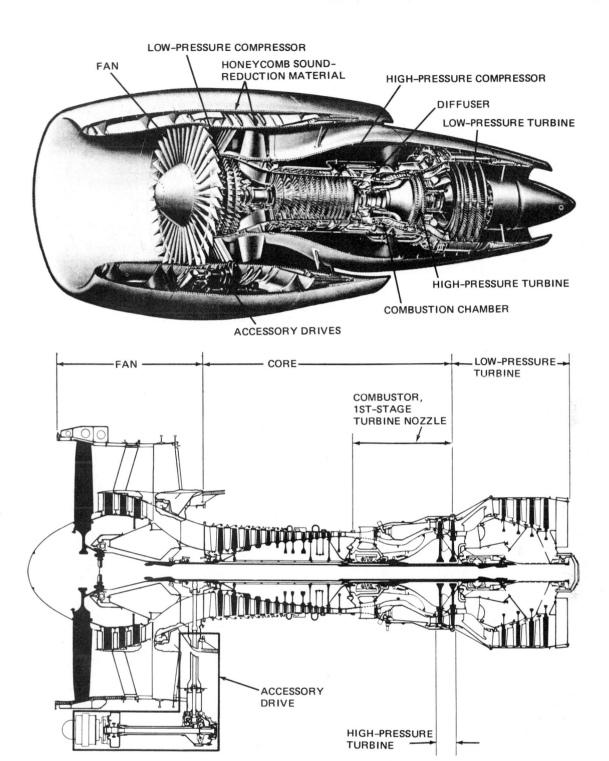

FIG. 21-50 General Electric CF6-50 turbofan engine.

Fan pressure ratio	1.64:1	1.69:1
Air mass flow, fan	1160 lb/s	1178 lb/s
Air mass flow, core	183 lb/s	270 lb/s

It will be noted from the specifications above that both the CF6-6 and the CF6-50 engines have very high pressure ratios. It will further be noted that the CF6-50 utilizes a greater percentage of the air through the core of the engine and achieves a 25 percent increase in takeoff thrust. This is accomplished in part because of the greater pressure ratio. The effect of the greater total volume of air passing through the core and fan is also to increase the total thrust substantially.

General Configuration

As can be seen from the photograph of Fig. 21-50 and the drawings of 21-50 and Fig. 21-51, the CF6-50 engine can be considered as a dual-compressor engine similar in many respects to the Pratt & Whitney JT9D previously discussed. It will be noted, however, that the performance of the CF6-50 is a little higher than that of the JT9D and is similar to that of the JT9D-7R4. This is largely because of the compressor design which produces the high pressure ratio of 29.9:1.

The General Electric CF6-6 and CF6-50 engines are classed as dual-rotor high-bypass-ratio turbofan en-

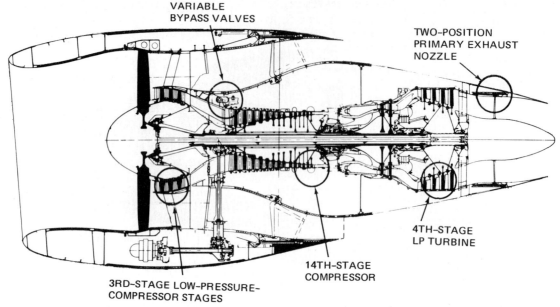

VARIABLE
BYPASS VALVES

TWO-POSITION
PRIMARY EXHAUST
NOZZLE

4TH–STAGE
LP TURBINE

14TH–STAGE
COMPRESSOR

3RD–STAGE LOW-PRESSURE-
COMPRESSOR STAGES

FIG. 21-51   Cutaway drawing of the CF6 engine.

gines. They incorporate variable stators in the compressors, annular combustors, air-cooled core-engine turbine, and a coaxial front fan with a low-pressure compressor driven by a low-pressure turbine. The engine includes a fan thrust reverser and a core-engine thrust reverser or spoiler to produce reverse thrust during the landing roll.

Engine Description

The CF6 engine is basically divided into five principal sections, as shown in Fig. 21-51. These are similar to the main sections of other gas-turbine engines and are classified according to function.

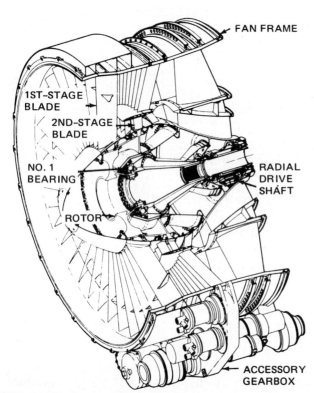

FAN FRAME

1ST-STAGE
BLADE

2ND-STAGE
BLADE

NO. 1
BEARING

RADIAL
DRIVE
SHAFT

ROTOR

ACCESSORY
GEARBOX

FIG. 21-52   Drawing of the CF6 fan section.

The **fan section** of the CF6 engine is shown in Fig. 21-52. The CF6-50 engine fan section includes the three compressor stages which may be defined as the **low-pressure compressor.** This compressor supercharges the incoming air so that the engine pumps 55 percent more air than the basic engine. Variable air-bypass valves are provided aft of the low-pressure compressor to discharge air into the fan stream to establish proper flow matching between the low- and high-pressure spools during transient operation.

The fan rotor assembly is made of forged titanium disks and blades with aluminum platforms and spacers. Details of the fan blade installation are shown in Fig. 21-53. Note that the blade roots are inserted into dovetail slots in the rims of the disks. Platforms are installed between the fan blades to provide a smooth airflow at the blade roots.

Construction of the fan rotor is shown in Fig. 21-54. The location of the no. 1 and no. 2 bearings is shown at the front and rear of the **forward shaft.**

Compressor

The compressor rotor for the CF6-6 engine is shown in Fig. 21-55. The design and construction of the rotor are similar to those described previously.

The stator for the CF6 engine comprises two sections, the **compressor front-stator assembly** and the **compressor rear-stator assembly.** In the CF6-50 engine, the front section mounts stator blades for the first 11 stages of compression and the rear section mounts stator blades for the twelfth through fourteenth stages of compression. The inlet guide vane for the high-pressure compressor and the stator vanes for the first six stages are variable. As explained previously, the variable vanes are employed to provide for correct direction of airflow in the compressor under all operating conditions. This reduces the possibility of compressor surge and stall. Signals from the fuel-control unit schedule high-pressure fuel to the **variable-stator-vane actuators** to adjust the vane angles as required.

The high-pressure compressor rotor is a combined

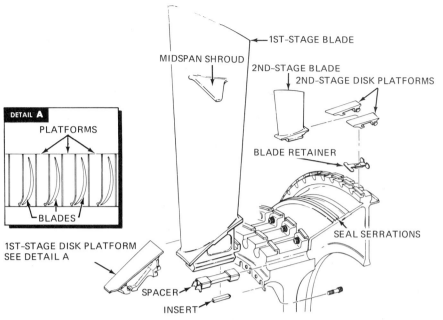

FIG. 21-53 Fan blade installation.

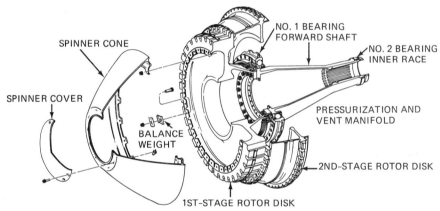

FIG. 21-54 Construction of the fan rotor.

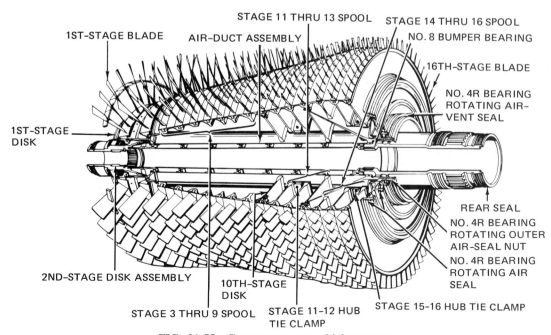

FIG. 21-55 Compressor rotor, high-pressure.

spool-and-disk structure utilizing axial dovetails to retain the compressor blades in stages 1 and 2 and circumferential dovetails for the remaining stages. The blades for the first two stages are inserted axially in the slots cut into the disk rims and are held in place by tab-type **blade retainers.** The blades in the circumferential slots are inserted into the grooves through entry slots and then moved along the groove so that the dovetails' shoulders engage the locking lugs and hold the blades in place. The blade dovetails are stronger than the blades; hence, a failure will not cause as much damage as would be the case if the blade and dovetail should come out of the slot.

On the CF6-50 engines, the compressed air for aircraft use (customer air) is bled from the inside diameter of the air passage through hollow stage-8 stator vanes. It passes through the vane bases and then through round holes in the casing skins into a pair of manifolds. Engine air is extracted at stage 7 for turbine midframe cooling. This air passes through semicircular slots in adjacent stage-7 vane bases and then through round holes in the casing skins to the manifold.

Engine air is extracted at stage-10 stator for the second-stage high-pressure turbine cooling and nose cowl anti-icing. Similarly, this air passes through semicircular slots in the stage-10 vane base and then through round holes in the casing skins to a pair of manifolds.

In the compressor casing, the variable inlet guide vanes and stages 1 through 6 are mounted in the conical part of the forward section. The variable vane bearing seats are formed by radial holes and counterbores through circumferential supporting ribs. The variable vane trunnions are supported in glass cloth bushings impregnated with Teflon. These bushings form bearings which endure through thousands of hours of operation. Furthermore, Teflon has a self-lubricating characteristic; hence, the bushings never need lubrication.

The fixed stages of the stator vanes are mounted in the cylindrical portion of the forward casing. These vanes are forged in one piece and are mounted in circumferential tracks machined in the inner surface of the casing. The forward casing is machined from a one-piece titanium forging.

Compressor Rear Frame

The compressor rear frame of the CF6 engine is a principal support structure for the engine. In the for-

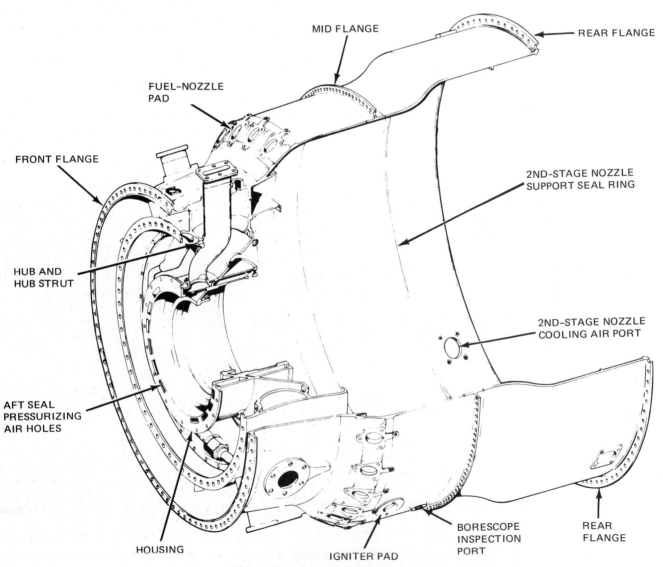

FIG. 21-56 Compressor rear frame.

502

ward-center portion is a bearing housing containing two bearings for the rear support of the compressor rotor. The front flange of the compressor rear frame is bolted to the rear flange of the compressor stator. The bearing housing is supported by 10 streamlined struts extending outward through the outer shell of the frame. In addition to providing bearing support, the struts are utilized as passages for air and oil.

The compressor rear frame, which provides support for the rear of the compressor, houses the combustor, 30 fuel nozzles, and 2 igniters. The configuration of the compressor rear frame is shown in Fig. 21-56. Note that the unit is attached to the compressor by two forward flanges and to the high-pressure turbine section by one rear flange.

Combustor

The **annular combustor** for the CF6 engine is illustrated in Fig. 21-57. It consists of four sections which are riveted together into one unit and spot-welded to prevent rivet loss. The four sections are the **cowl assembly** which serves as a diffuser, the **dome**, the **inner skirt**, and the **outer skirt**. The complete assembly fits around the compressor rear-frame struts where it is mounted at the cowl assembly by 10 equally spaced radial mounting pins.

The cowl assembly is designed to provide the diffuser action required to establish uniform and predictable flow profiles to the combustion liner in spite of irregular flow profiles which may exist in the compressor discharge air. Forty box sections welded to the cowl walls form the aerodynamic diffuser elements as well as a truss structure to provide the strength and stability of the cowl's ring section. The combustor mounting pins are completely enclosed in the compressor rear-frame struts and do not impose any drag losses in the diffuser passage. Mounting the combustor at the cowl assembly provides accurate control of diffuser dimensions and eliminates changes in the diffuser flow pattern which could be caused by expansion of the assembly because of heat.

A cross section of the combustor is shown in Fig. 21-58. The **dome** is the forward section inside the cowl assembly; it is the part which forms the actual forward unit of the combustion area. In the dome are 30 vortex-inducing axial swirler cups, one for each fuel nozzle. In the drawing of Fig. 21-56 the fuel nozzle and swirl cups can be seen. The swirl cups are designed to pro-

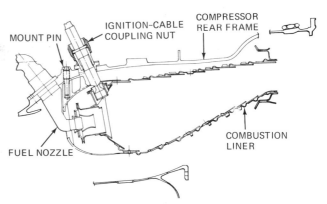

FIG. 21-58 Cross section of the combustor.

vide the airflow patterns required for flame stabilization and proper fuel-air mixing. The dome design and swirl cup geometry, coupled with the fuel-nozzle design, are the principal factors which contribute to smokeless combustion. The axial swirlers serve to lean out the fuel-air mixture in the primary zone of the combustor, and this helps to eliminate the formation of the high-carbon visible smoke which normally results from overrich burning in this zone. The dome is continuously film-cooled, as shown by the airflow in Fig. 21-58.

The combustor skirts may be compared with combustion-chamber liners, as described for some other engines. Each skirt consists of circumferentially stacked rings which are joined by resistance weld and brazed joints. The liners are continuously film-cooled by primary combustion air which enters each ring through closely spaced circumferential holes. The primary-zone hole pattern is designed to admit the balance of the primary combustion air and to augment the recirculation for flame stabilization. Three axial planes of dilution holes on the outer skirt and five planes on the inner skirt are employed to promote additional mixing and to lower the gas temperature at the turbine inlet. Combustion-liner and turbine-nozzle airseals provided on the trailing edge of the skirt allow for expansion due to heat (thermal growth) and accommodate manufacturing tolerances. The seals are coated with wear-resistant material.

High-Pressure Turbine

The high-pressure turbine for the CF6 engine, shown in Fig. 21-59, consists of only two stages located immediately to the rear of the combustor. This turbine drives the high-pressure compressor through the outer coaxial shaft.

The nozzle-vane arrangement for the first- and second-stage turbine stages is shown in Fig. 21-60. The first-stage vanes are cooled by convection, impingement, and film cooling. The details of vane construction and cooling features are provided in Fig. 21-61. The first-stage nozzle assembly is bolted at its inner diameter to the first-stage support and receives axial support at its outer diameter from the second-stage nozzle support.

The first-stage nozzle support is a sheet-metal and machined ring weldment. In addition to supporting the first-stage nozzle, it forms the inner flow path wall from the compressor rear frame to the nozzle. The first-stage

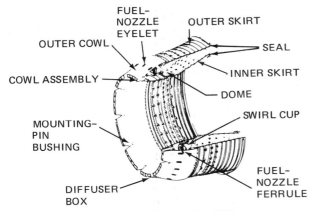

FIG. 21-57 Annular combustor for the CF6 engine.

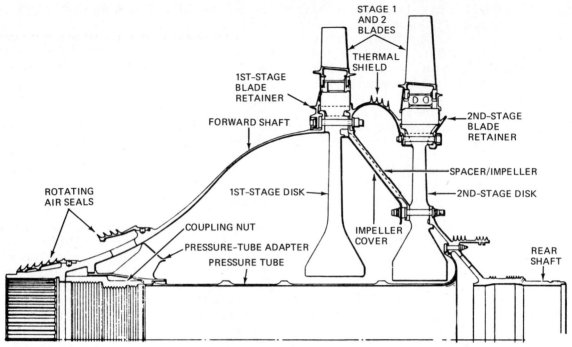

FIG. 21-59   The CF6 high-pressure turbine.

nozzle support also retains the baffle and the pressure balance seal support.

The first-stage vanes are coated to improve erosion and oxidation resistance. The vanes are cast individually and welded into pairs to decrease the number of gas leakage paths and to reduce the time required for field replacement. These welds are partial-penetration welds to allow easy separation of the two vanes for repair and replacement of individual halves. The vanes are cooled by compressor discharge air which flows through a series of leading-edge holes and fill holes located close to the leading edge on each side. Air

flowing from these holes forms a thin film of cool air over the length of the vane. Internally, the vane is divided into two cavities, and air flowing into the aft cavity is discharged through trailing-edge slots.

The second-stage nozzle is cooled by convection. Air from the thirteenth stage of the compressor flows through the vanes, exiting through holes in the trailing edges of the vanes and also out the inner end of the vanes to help cool the turbine rotor. The second-stage **nozzle support** is a conical ring, as shown in Fig. 21-60. The support is bolted rigidly between the flanges of the compressor rear frame and the turbine midframe.

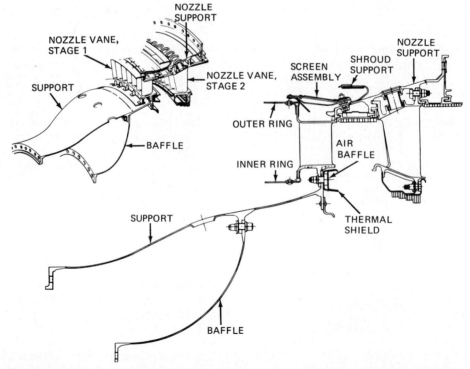

FIG. 21-60   Nozzle-vane arrangement and supports for the first two stages of the turbine.

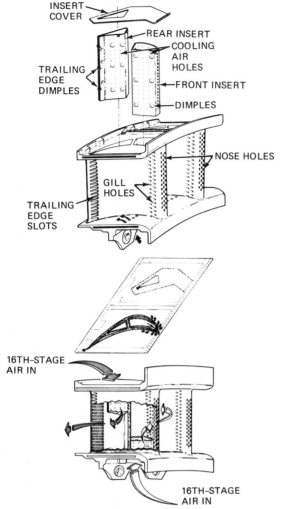

FIG. 21-61 Nozzle-vane construction and cooling features.

The nozzle support holds the nozzle-vane segments, cooling-air feeder tubes, and the first- and second-stage turbine shrouds. The nozzle segments are cast and then coated for protection against erosion and oxidation. The inner ends of the vane segments form a mounting circle for the interstage-seal attachment. The **interstage seal** is composed of six segments of about 60° each which bolt to the vane segments. The function of the seal is to minimize the leakage of gases between the inside diameter of the second-stage nozzle and the turbine rotor. The sealing diameter has four consecutive steps for maximum effectiveness of each sealing tooth. These sealing teeth are shown in Fig. 21-62, which is a cross section of the high-pressure turbine. The teeth

are between the turbine disks on the **catenary thermal shield.**

The **turbine shrouds** form a portion of the outer aerodynamic flow path through the turbine. The shrouds are located axially in line with the turbine blades and form a pressure seal to prevent high pressure gas leakage or bypass at the blade-tip ends. The sealing surfaces are of conventional honeycomb filled with nickel aluminide compound. This construction of the sealing surfaces makes it possible for the blade tips to rub the surface without causing damage. The first-stage turbine shrouds consist of 24 segments and the second stage has 11 segments.

The high-pressure turbine rotor consists of a conical forward turbine shaft, two turbine disks, two stages of turbine blades, a conical rotor spacer, a catenary-shaped thermal shield, the aft stub shaft, and precision-fit rotor bolts. Drawings of the high-pressure turbine are shown in Figs. 21-62 and 21-63. The rotor of the turbine is cooled by a continuous flow of compressor discharge air drawn from holes in the nozzle support. This flow cools both the inside of the rotor and both disks before passing between the paired dovetails and out to the blades.

The conical **forward turbine shaft** transmits energy for rotation of the compressor. Torque is transmitted through the female spline at the forward end of the shaft. Two **rotating airseals** attached to the forward end of the shaft maintain compressor discharge pressure in the **rotor-combustion-chamber plenum** to furnish part of the corrective force necessary to minimize the unbalanced thrust load on the **high-pressure rotor thrust bearing.** The thrust bearing is a ball-type bearing designed to take the thrust loads of the high-pressure rotor system. This bearing does not take any of the axial bearing loads because these loads are handled by roller bearings.

The **turbine rotor spacer** can be seen between the

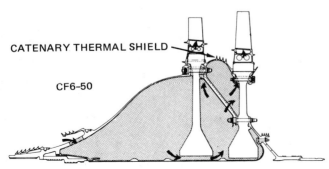

FIG. 21-62 Cross section of the high-pressure turbine.

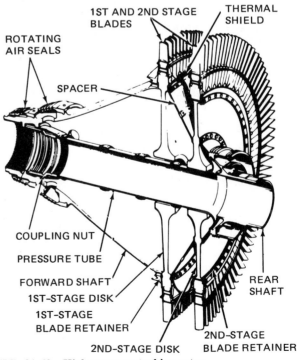

FIG. 21-63 High-pressure turbine rotor.

turbine disks in the drawings. This spacer is a cone which serves as the structural support member between the turbine disks and also transmits the torque from the second-stage turbine to the forward shaft.

The turbine blades are brazed together in pairs with side-rail doublers added for structural integrity. Channel-shaped **squealer tip caps** are inserted into the blade tips and are held in place by crimping the blade tip and brazing. Both stages of blades are cooled by compressor discharge air which flows through the dovetail and between the blade shanks into the blades. The construction of the first-stage blades to allow for cooling is shown in Fig. 21-64. First-stage blade cooling is a combination of internal convection and external film cooling. The convection cooling of the midchord region is accomplished through a labyrinth. The leading-edge circuit provides internal convection cooling by impingement of air against the inside surface and by flow through the leading-edge and gill holes. Convection cooling of the trailing edge is provided by air flowing through the trailing-edge exit holes. Second-stage blades are entirely cooled by convection. All the cooling air is discharged from the tips of these blades.

The rear shaft of the high-pressure turbine bolts to the second-stage disk and supports the aft end of the turbine rotor. The shaft end is supported in a roller bearing.

The **blade retainers** serve two primary functions. They prevent the blades from moving axially under gas loads and they seal the forward face of the first-stage rim dovetail and the aft face of the second-stage rim dovetail from the leakage of cooling air. An additional function is to cover the rotor-bolt ends at the rotor rim and thereby prevent a substantial drag loss. These retainers are single-piece construction held on by the same bolts that attach the forward shaft and thermal shield to the turbine disks.

The **pressure tube** serves to separate the high-pressure rotor internal-cooling air supply from the region of the fan midshaft which is concentric to the rotor. It is threaded into the front shaft and bolted to the rear shaft.

Low-Pressure Turbine Section

The low-pressure turbine for the CF6 engine has two different configurations. The CF6-6 model has five stages and the CF6-50 has four stages. In either case, the low-pressure turbine utilizes a rotor supported between roller bearings mounted in the turbine midframe and the turbine rear frame. A horizontally split low-pressure turbine casing containing stator vanes is bolted to these frames to complete the structural assembly. This provides a rigid, self-contained module which can be precisely and rapidly interchanged on the engine without requiring a subsequent engine test run. The **low-pressure turbine shaft** engages the long **fan drive shaft** through a spline drive and is secured by a lock bolt. The forward flange of the turbine midframe is bolted to the aft flange of the compressor rear frame, after installation of the high-pressure turbine, to complete the engine assembly. A cross section of the CF6 low-pressure turbine is shown in Fig. 21-65.

Since the temperature of the exhaust gases is reduced considerably through the high-pressure turbine, the low-pressure turbine blades do not require cooling. Compressor-seal leakage air is used to cool the first two low-pressure turbine disks to reduce thermal gradients.

The increased flow and higher wheel speed of the CF6-50 engine would result in very low stage loadings if a five-stage low-pressure turbine were used on the engine. According to engineering studies, the four-stage turbine performs more efficiently.

The **first-stage low-pressure turbine nozzle** consists of 14 segments, each containing 6 vanes, as shown in Fig. 21-66. The segments are supported at their inner and outer ends by the turbine midframe and low-pressure (LP) turbine casing, respectively. This provides low vane-bending stress and freedom to expand or contract thermally without thermal loads and stresses. The other stages of LP turbine vanes are cantilevered from the low-pressure turbine (LPT) stator casing. These can be seen in the drawing of Fig. 21-67, which illustrates one-half the LPT stator casing. The major parts

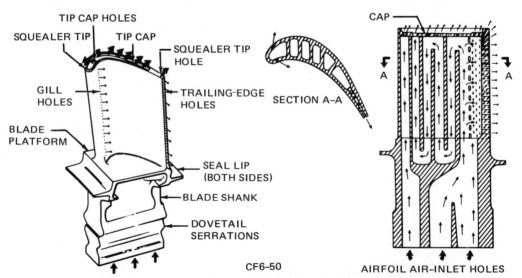

FIG. 21-64 Construction of the first-stage turbine blades.

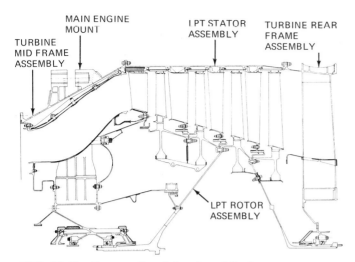

FIG. 21-65 Cross-sectional drawing of the low-pressure turbine.

of the **stator** are the nozzle stages, a split casing, shrouds, and interstage airseals. Each of the nozzle stages is composed of cast segments of six vanes each, any segment of which can be replaced with simple tools. The shrouds are in segments held in place by projections mating with slots formed by the casing and nozzles. The interstage seals are bolted to the inside-diameter flange of the nozzles. Both shrouds and seals have abradable honeycomb sealing surfaces to allow close clearance without the risk of rotor damage caused by unusual rubs. The interstage seals partially restrain the inner vane ends to provide damping results which result in low vane stresses.

The **low-pressure turbine rotor assembly** is shown in Fig. 21-68. Note that the turbine blades have interlocking blade shrouds to produce a continuous shrouded surface around the diameter of each turbine stage. This reduces blade stress and helps to seal gases within the desired flow path. The shroud interlocks are in contact at all rotor speeds due to an interference fit caused by pretwisting the blades. This provides adequate damping at low speeds and damping proportional to rotor speed at higher speeds. Shroud interlocks are hard-coated for wear resistance.

The disks for the low-pressure turbine are made of Inconel 718 alloy and each has integral torque-ring ex-

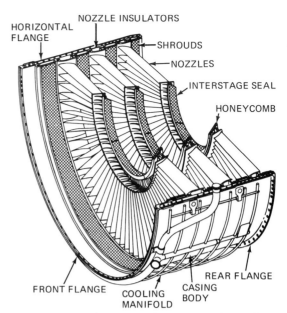

FIG. 21-67 Low-pressure turbine stator casing and vanes.

tensions. These are attached to the adjacent disks by close-fitting bolts. Bolt holes through the disk webs have been eliminated by locating them in the flanges where stresses are low.

Figure 21-69 shows how the front and rear shafts are attached to the disks. This arrangement forms a stiff rotor structure between bearings.

The turbine blades are attached to the disks by means of multitang dovetails. Replaceable rotating airseals mounted between the disk flanges mate with stationary seals to provide interstage air sealing.

Because of its low length-to-diameter ratio and the accurate fitting of bolts to fasten the disks together, the low-pressure turbine rotor is highly stable. In the event of a midshaft failure, the LPT rotor would move aft until blade rows interfered with the stators. This would prevent any dangerous overspeed.

In the CF6-50 engine, an external impingement system is employed to cool the low-pressure rotor casing. This system employs fan discharge air bled from the inner fan flowpath. Utilizing fan air instead of stage-9 air, as used for the CF6-6 engine, results in a lower chargeable airflow and provides a further improvement of specific fuel control.

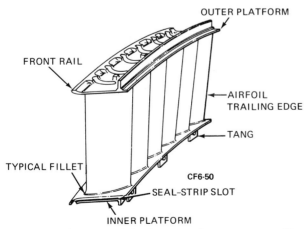

FIG. 21-66 Segment of the first-stage low-pressure turbine nozzle.

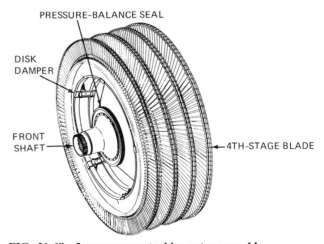

FIG. 21-68 Low-pressure turbine rotor assembly.

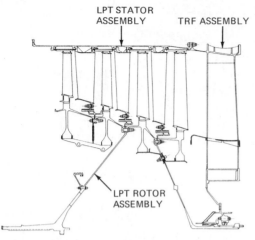

FIG. 21-69 Cross section of the low-pressure turbine.

The LPT module, consisting of turbine and case, is designed for easy removal and reinstallation, either with the engine installed or removed from the aircraft. The only quick-engine-change (QEC) items requiring removal are the thrust spoiler and drives, the EGT electrical harness, and condition monitoring leads, if installed. The LPT module may be removed and replaced in 4 elapsed hours with a total of 28 worker-hours.

Support Structures

The supporting structures of the CF6 engine are the fan frame, compressor rear frame, turbine midframe, and turbine rear frame. These structures, bolted together, provide the rigid casing necessary to support the rotating and combustion elements of the engine.

The **fan frame** is illustrated in Fig. 21-70. It supports the forward end of the compressor, the fan rotor, fan stator, forward engine mount, radial drive shaft, transfer gearbox, horizontal drive shaft, and accessory gearbox. The fan frame has 12 stainless-steel struts equally spaced at the leading edge. The struts bolt to the aft outer casing with eight bolts per strut. The six and twelve o'clock struts are the leading edges of the upper and lower pylons, shaped to fair into the pylon side-walls. The radial shaft is enclosed within the lower

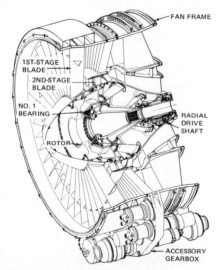

FIG. 21-70 Drawing of the fan section.

pylon and is bolted to the six o'clock strut. The fan frame houses the A sump which includes the nos. 1, 2, and 3 bearings. Sump pressurizing air enters through the leading edge of struts 4, 5, 9, and 10 in the fan stream section.

The forward engine mount attaches to the aft flange of the fan frame and a thrust mount linkage connects to the no. 1 strut casing.

The **compressor rear frame,** shown in Fig. 21-71, is made up of the main-frame structural weldment, the inner combustion casing and support, the compressor discharge airseal, and the B sump housing. Axial and radial loads are taken in the rigid inner-ring structure and transmitted in shear into the outer casing. The frame is constructed of Inconel 718 because of its high strength at elevated temperatures, its excellent corrosion resistance, and good repairability compared with other high-temperature nickel alloys.

It will be remembered that the compressor front and rear casing components are located between the fan frame and the compressor rear frame (see Fig. 21-56). The compressor casing provides considerable structural support, but it is not considered a main-support structure.

An examination of Fig. 21-71 will show that the compressor rear frame includes a section at the front which serves partially as a diffuser. It also houses the combustion chamber described previously.

Bearing axial and radial loads and a portion of the high-pressure turbine first-stage nozzle loads are taken in the inner ring, or hub, and transmitted through the 10 radial struts to the outer shell. The inner ring of the frame is a casting which contains approximately half the radial strut length. The cross-sectional shape of the hub is a box to provide structural rigidity.

The outer strut ends are casting which complete the formation of the struts when welded to the hub. Combining the hub and outer strut casting in this manner forms a smooth strut-to-ring transition with the minimum concentration and no weld joints in the transition area. The 10 radial struts are airfoil-shaped to reduce aerodynamic losses and are sized to provide adequate internal area for sump service lines and bleed airflow. The hub and outer strut end assembly is then welded into the outer shell, which is a sheet-metal and machined ring weldment defining the outer-flow annulus boundary as well as providing the structural load path between the high-pressure compressor casing and turbine midframe.

To provide for the differential thermal growth between sump service tubing and the surrounding structure, the tubes are attached only at the sump, and slip joints are used where tubes pass through the outer strut ends.

The **turbine midframe** is illustrated in Fig. 21-72. It consists of the outer casing reinforced with hat section stiffeners, the link mount castings, strut end castings, cast hub, eight semitangential bolted struts, the C sump housing, and a one-piece flowpath liner. The frame casing and cast hub operate cool enough at all conditions to permit the use of Inconel 718. The operating temperature does not exceed 1100°F [593.33°C].

The struts are made of René 41, which is a high-temperature alloy developed by the General Electric

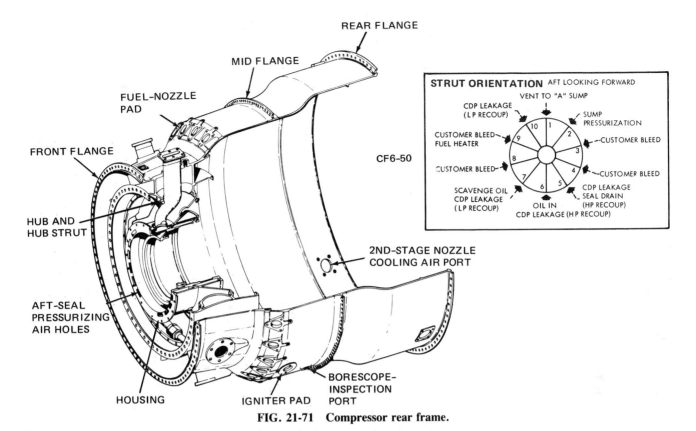

FIG. 21-71 Compressor rear frame.

Company. The selection of bolted structural joints at the inner and outer ends of the struts was made to satisfy the unique requirements arising from the location of the midframe between the high-pressure and low-pressure turbines. The liner and fairing assembly is a one-piece unit.

The turbine midframe supports the no. 5 and no. 6 bearings. The bearing-support cone and sump housing is bolted to the forward flange of the frame's inner hub for ease of replacement, maintenance, and manufacture. Tubing through the frame's structural struts for sump service is secured to the sump by bolted flanges or B nuts to make the pump completely separable from the frame structure. The sump housing is of double-

walled construction so that the wall of the sump bathed in oil is cooled by fan discharge air, keeping its temperature below 350°F [176.67°C].

The **turbine rear frame,** illustrated in Fig. 21-73, is the rearmost structural member of the engine. This unit can be divided into the main-frame structure, the inner flowpath liner, the D sump, and the sump service piping. It supports the no. 7 bearing in the D sump.

The assembly of the turbine rear frame is similar to that of the midframe but without the use of bolted struts. It is a welded Inconel 718 structure with eight equally spaced, partially tangential struts supported on two axially spaced rings at both the hub and outer ring. With the struts and 500°F [260°C] lower gas temperature than the turbine midframe, the rear frame can be designed without flowpath-liner or strut fairings. The main advantage of this type construction is the acces-

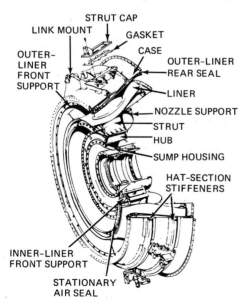

FIG. 21-72 Turbine midframe.

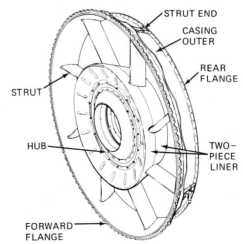

FIG. 21-73 Turbine rear frame.

sibility of structural welds for visual inspection without disassembly.

Referring back to Fig. 21-51, we can see the turbine rear frame at the extreme rear of the engine. It is separated from the turbine midframe by the low-pressure turbine stator assembly, which is illustrated in Fig. 21-67.

Engine Systems

The systems for the CF6 engine are similar to those described previously. All systems utilize components of proven technology and reliable performance.

The fuel system is shown in Fig. 21-74. The principal units are the fuel pump, fuel-oil heat exchanger, main engine control, CIT sensor, variable stator vane actuator, pressurizing valve, and fuel nozzle.

The lubrication system is of the pressure type, with oil controlled in the various sumps by means of airseals and pressurization from the engine compressor. Scavenge oil is drained to the scavenge pump, from whence it is returned to the oil tank.

The electric system is shown in the diagram of Fig. 21-75. It will be noted that the ignition system is energized by means of 400-Hz 115-V alternating current. Other circuits are supplied from the 28-V DC source.

The General Electric CF6-80 Engine

The General Electric CF6-80 engine is essentially an upgraded version of the CF6-50 engine. Both engines are in approximately the same thrust class; however, design improvements have been made in length, weight, and maintainability at low cost to produce the CF6-80 model. Improvements in the CF6-80 design have re-

sulted in greater dependability, durability, and efficiency.

The CF6-80A model engine is utilized in the Boeing 767 aircraft and the CF6-80A1 engine is used to power the Airbus Industrie A310. Differences in the CF6-80A and -80A1 models are in the location of the accessory gearbox. The CF6-80A has a core-mounted gearbox which provides a smaller frontal area, thus improving the aerodynamic efficiency of the engine. The -80A1 model has the gearbox mounted on the fan stator case which provides better access for maintenance. Each model of engine was designed specifically for the aircraft for which it was to be used. General Electric manufactures both the engine and nacelle for the Airbus A310 but supplies the engine only for the Boeing 767.

The CF6-80 engine produces 48,000 lb [213 504 N] thrust and has a bypass ratio of 4.66:1. The compressor pressure ratio is 28:1.

● THE ROLLS-ROYCE RB·211 TURBOFAN

The Rolls-Royce RB·211 turbofan engine is another example of the modern high-bypass high-performance engines designed for use on large, wide-bodied transport aircraft. The RB·211 engine, illustrated in Fig. 21-76, is a three-shaft engine, designed primarily to power the Lockheed L-1011 airliner.

The three-shaft configuration is employed to permit the fan to rotate independently so that it will not limit the optimum rotational speed of the intermediate compressor. The fan is driven by the rearmost three stages

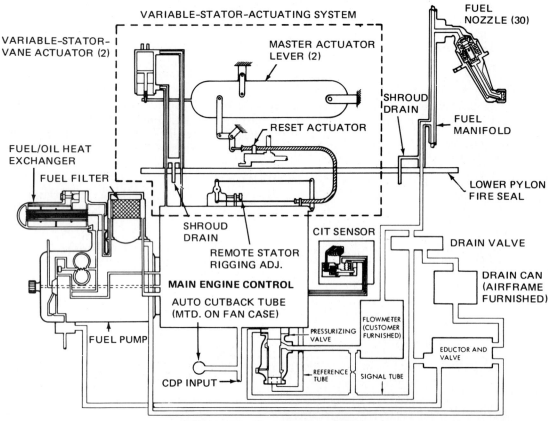

FIG. 21-74 The CF6 fuel system.

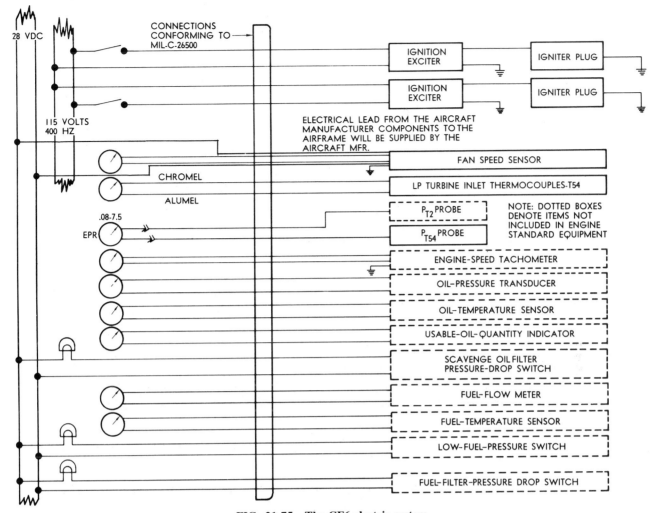

FIG. 21-75   The CF6 electric system.

of the turbine through the inside coaxial shaft, the intermediate compressor is driven by one stage of the turbine, and the high-pressure compressor is driven by the first stage of the turbine through the outer of the three coaxial shafts. The drive arrangement can be seen in Fig. 21-77.

The combustion chamber is a short, annular type fitted with 18 fuel vaporizers (nozzles).

Performance

As is true of other engines, the specifications given for one model are not necessarily the same for others. The RB·211 engine was designed for growth; hence, it is likely that later models will have higher thrust ratings as well as variations in fuel consumption and pressure ratios.

Length	119.4 in	303.28 cm
Diameter	85.5 in	217.17 cm
Frontal area	30.0 ft^2	2.79 m^2
Weight	6353 lb	2882 kg
Takeoff thrust	42,000 lb	186 816 N
Power-weight ratio	6.4 lbt/lb	6.4 kgp/kg

FIG. 21-76   Cutaway photograph of the Rolls-Royce RB·211 turbofan engine.   (Rolls-Royce)

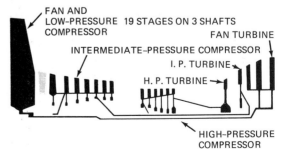

FIG. 21-77   Drive arrangement in the RB·211 engine.

511

Fuel consumption	0.36 lb/lbt/h	360 g/kgp/h
Oil consumption	2.0 lb/h	910 g/h
Compressor ratio	25:1	
Bypass ratio	5:1	
Fan pressure ratio	1.42:1	
Air mass flow, fan	1095 lb/s	497 kg/s
Air mass flow, core	230 lb/s	104 kg/s

General Configuration

The general arrangement of the Rolls-Royce RB·211 engine can be understood by studying the illustrations of Figs. 21-76 to 21-78. The use of the three-shaft design makes it possible to produce an engine which is more rigid and shorter and has fewer compressor and turbine stages.

A major consideration in the design of the RB·211 engine is stiffness or rigidity. Figure 21-78 shows the major load-carrying structures.

Bearing Arrangement

The RB·211 engine utilizes eight bearings to support the rotating assemblies (Fig. 21-79). The inner shaft through which the fan is driven is supported by a bearing at the front and a bearing at the rear of the engine. Note that there are two bearings near the front. The second of these bearings supports the forward end of the intermediate compressor shaft. The rear end of this same shaft is supported by the third bearing.

The three thrust ball bearings are grouped in the **intermediate casing.** The **low-pressure thrust bearing** is an intershaft bearing that forms a cluster with the **intermediate-pressure bearing,** as shown in the drawing. The fifth and sixth bearings support the high-pressure compressor system and are located at a distance from the hot combustion zone to keep the operating temperature at a satisfactory level.

Compressor Sections

The **fan section** of the RB·211 engine is the low-pressure compressor. The inner ends of the fan blades give the air flowing into the core of the engine its first boost. The air then flows through a single stage of variable stator vanes into the intermediate-pressure section.

The fan section consists of 33 titanium blades held in the fan disk by means of fir-tree roots. The disk is faced with an anti-iced nose cone.

The **intermediate-pressure compressor** is a seven-stage unit built up of disks welded together by means of the electron-beam process. The blades are held in the disks by means of dovetail roots.

The **high-pressure compressor** has six stages. It consists of two electron-beam-welded drums bolted through the third-stage disk. The blades are dovetailed in the disks.

Compressor-to-shaft joints are made by means of **curvic couplings** rather than spline couplings to facilitate modular assembly and disassembly. A curvic coupling resembles a radial ring gear.

Combustion Section

The **combustion chamber** for the RB·211 engine is of the annular design, similar to those described previously for other engines. A cross section of the combustion chamber is shown in Fig. 21-80. This particular chamber is shorter than others, which makes a two-bearing support system possible, with no bearings located near the combustion area.

The annular combustion chamber provides a clean aerodynamic extension of the compressor outlet casing and permits consistent cooling over a small surface area. The combustion cooling rings are of an advanced design which provides for high cooling efficiency.

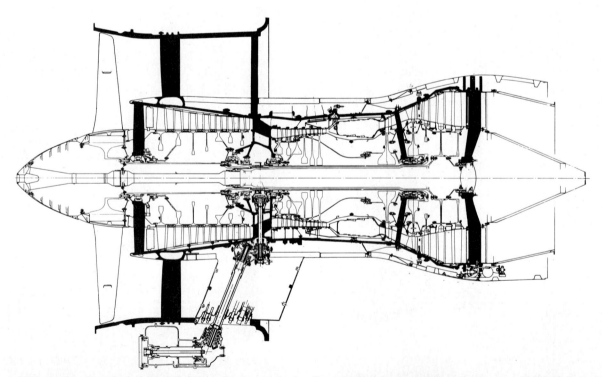

FIG. 21-78   Drawing of the RB·211 engine to emphasize the principal load-carrying structures.   (Rolls-Royce)

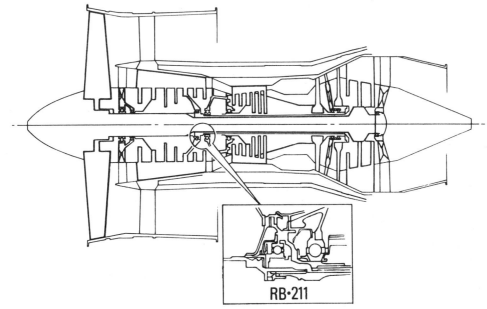

FIG. 21-79 Bearing arrangement in the RB·211 engine. (Rolls-Royce)

The fuel is atomized in a high-velocity airstream by the fuel-injection nozzles before it enters the combustion zone. The design gives good atomization of the fuel over the entire operating range and eliminates any tendency towards overrichness in the center of the combustion chamber, thus reducing heat radiation and smoke generation.

Turbine Section

The high-pressure (HP) turbine blades of the RB·211 engine are impingement-cooled by air bled from the HP compressor. Air is fed to the turbine blades through pre-swirl nozzles which accelerate the air in the direction of disk rotation. The cooling air enters the finned interior of the blades, after which it flows at high velocity through a slot to impinge against the leading-edge inner surface. This results in a high degree of heat transfer from the metal to the air. The air is then exhausted through the blade training edge. A discussion of this cooling system is given in Chap. 19 and is illustrated in Fig. 19-32.

As mentioned previously, the turbine section of the engine consists of three parts. These are (1) a one-stage HP turbine, a one-stage intermediate-pressure (IP) turbine, and a three-stage LP and fan turbine. These can be seen in Figs. 21-77 and 21-78.

Engine-Control System

The control system for the RB·211 engine is shown in Fig. 21-81. The system provides the following characteristics of performance:

1. Flat-rated takeoff thrust. When the system is once set for takeoff, no further power-lever adjustments are necessary.

2. An altitude schedule which maintains correct thrust during climb without power-lever adjustments, provided that the required scheduled climb airspeed is held to a reasonable tolerance.

3. Accurate measurement and indication of thrust independent of air temperature. These indications reveal any engine damage or deterioration.

4. Surge-free engine handling.

5. Freedom to advance the power levers fully open for go-around or other emergency. Automatic limiters protect the engine from overboost, overtemperature, and overspeed.

Modular Construction

The modular design of the RB·211 engine is illustrated in Fig. 21-82. This construction permits sectional repair and maintenance of the engine, thus making it

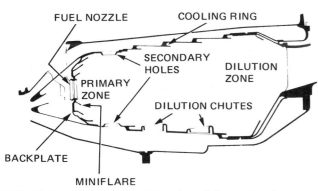

FIG. 21-80 Cross-sectional drawing of the combustion chamber.

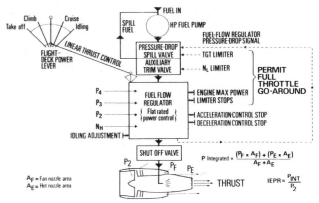

FIG. 21-81 Control system for the RB·211 engine.

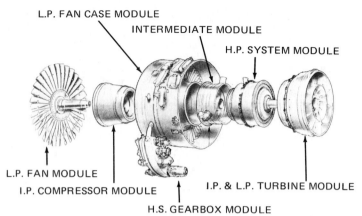

L.P. FAN CASE MODULE

INTERMEDIATE MODULE

H.P. SYSTEM MODULE

L.P. FAN MODULE

I.P. COMPRESSOR MODULE

H.S. GEARBOX MODULE

I.P. & L.P. TURBINE MODULE

FIG. 21·82   The modules that facilitate maintenance for the RB·211 engine.

possible to set up service-life criteria for each module rather than for the engine as a whole. Each module can be removed and replaced in the engine while the engine is still installed in the aircraft.

● THE GARRETT-AIRESEARCH TFE731 TURBOFAN ENGINE

A comparatively recent medium-power turbofan engine for business and general aviation aircraft is the TFE731 engine manufactured by the Garrett Turbine Engine Company, a division of the Garrett Corporation. A photograph of the engine is shown in Fig. 21-83.

The TFE731 engine introduces some rather innovative design features when compared with larger engines in general. The fan is driven through a reduction gear to avoid overspeeding and yet allow the low-pressure compressor to operate at the most efficient speed. This feature is illustrated in Fig. 21-84. The low-pressure compressor is of the axial type, whereas the high-pressure compressor is a centrifugal unit. The total pressure ratio developed is 15:1. The combustion chamber is a reverse-flow annular type which accounts for the reduced length of the engine.

Specifications

The general specifications of the TFE731 engine are listed below; it must be remembered, however, that

FIG. 21-83   Photograph of the Garrett TFE731 engine.   (Garrett Turbine Engine Co.)

these may vary somewhat, depending on the particular model of the engine.

Length	49.7 in	126.24 cm
Diameter	28.2 in	71.63 cm
Frontal area	5.2 ft^2	0.48 m^2
Weight	625 lb	283 kg
Takeoff thrust	3500 lb	15 568 N
Power-weight ratio	5.60 lbt/lb	560 kgp/kg
Fuel consumption	0.49 lb/lbt/h	490 g/kgp/h
Oil consumption	0.90 lb/h	400 g/h
Compressor ratio	15:1	
Bypass ratio	2.67:1	
Fan pressure ratio	1.54:1	

General Description

We shall not attempt to describe all the details of the TFE731 engine here; the student is urged to study the features of the engine as shown in Fig. 21-84 and Fig. 21-85. Note that a part of the fan output is directed inward to the low-pressure compressor; hence, the fan provides a first stage of compression. The fan speed is reduced by a ratio of 0.555:1 with respect to the low-pressure compressor. This permits the fan to rotate at a maximum speed without having the fan tips exceed the speed of sound. In a particular operating situation, the fan can turn at more than 10,000 rpm, the low-pressure compressor at more than 18,000 rpm, and the high-pressure compressor at almost 29,000 rpm.

The low-pressure compressor consists of the fan plus four axial stages and is driven by the three most rearward turbine stages. The high-pressure compressor is driven by the first turbine stage.

Figure 21-85 shows all the principal features of the engine construction including the fan, compressors, reduction gearing, bearings, combustion chamber, and turbine. The modular design of the engine for easy maintenance is illustrated in Fig. 21-86.

The Garrett TFE731-5 Turbofan Engine

The Garrett TFE731-5 engine shown in the cutaway view of Fig. 21-87 is an upgraded version of the TFE-731 described in the foregoing section of this text. The engine has undergone changes in four major areas: fan section, fan gearbox, high-pressure turbine, and low-pressure turbine. The general configuration remains the same as that of previous models; however, performance has been increased substantially.

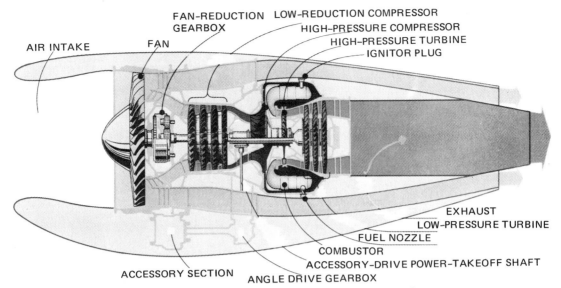

FIG. 21-84 Operational drawing of the TFE731 engine.

The fan section performance has been improved by increasing its capacity with a larger fan. This has resulted in increasing the bypass ratio from 2.67:1 to 3.48:1. Engine takeoff thrust is increased from 3500 lb [15 568 N] to 4500 lb [20 016 N]. The fan gearbox has been modified to accommodate the new fan requirements.

The high-pressure turbine has been improved by the use of a new alloy (MAR-M247) in both the blades and vanes. This makes possible a small increase in turbine operating temperature and a resultant increase in thrust. Modifications have been made in the three-stage low-pressure turbine to increase its load capacity as necessary to drive the larger fan and low-pressure compressor.

The TFE731-5 engine is controlled by a full-authority electronic fuel control that is described as a *digital fuel controller* (DFC). This control is described in Chap. 20. The use of the DFC system results in higher engine efficiency, reduced flight-crew workload, and lower maintenance costs.

THE GARRETT ATF3-6 TURBOFAN ENGINE

The Garrett ATF3-6 engine is a new and unique design based on features developed by the Garrett Turbine Engine Company over a period of many years. The engine weighs approximately 1100 lb [499 kg] and develops 5050 lb [22 462 N] at sea level. The bypass ratio is 2.67:1 and the thrust specific fuel consumption is 0.506 lb/h/lb.

The ATF36 is a three-spool engine with separate turbines to drive the fan, low-pressure compressor, and high-pressure compressor. This configuration can be seen in the cutaway photograph of Fig. 21-88. The

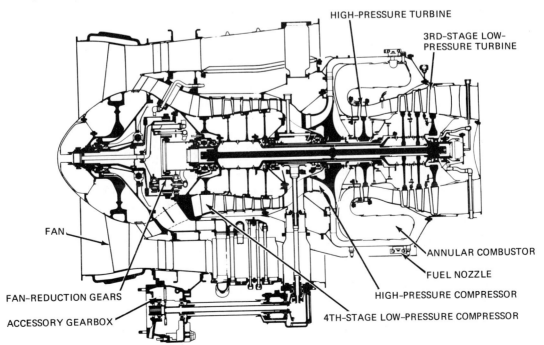

FIG. 21-85 Cutaway drawing of the TFE731 engine.

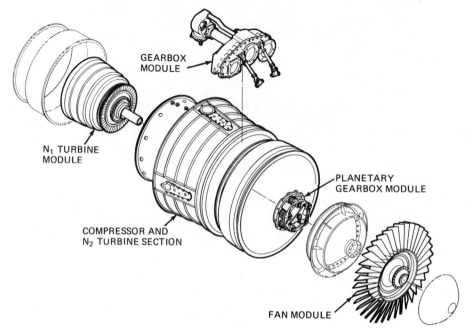

FIG. 21-86 Modular construction for easy maintenance.

fan drive shaft is surrounded by and is coaxial with the low-pressure compressor drive shaft. The three-stage fan turbine is mounted on the rear end of the drive shaft and is located immediately to the rear of the two-stage low-pressure turbine. The five-stage axial-flow low-pressure compressor is mounted on the hollow drive shaft with the low-pressure turbine. The entire low-pressure spool is supported by bearings on the outside of the fan-spool shaft.

The high-pressure centrifugal compressor is mounted to the rear of the high-pressure turbine on the same

shaft to form the high-pressure spool. This spool is located immediately to the rear of and is separate from the fan turbine. The high-pressure spool drives the engine accessories through the accessory gearbox, which is mounted on the rear of the engine. Reduction gears lower the high turbine rpm to a level suitable for the accessories.

The airflow and gasflow during operation of the engine is shown in Fig. 21-89. Airflow starts at the inlet cowl and enters the single-stage fan. Air leaving the fan is split, the major portion flowing through the by-

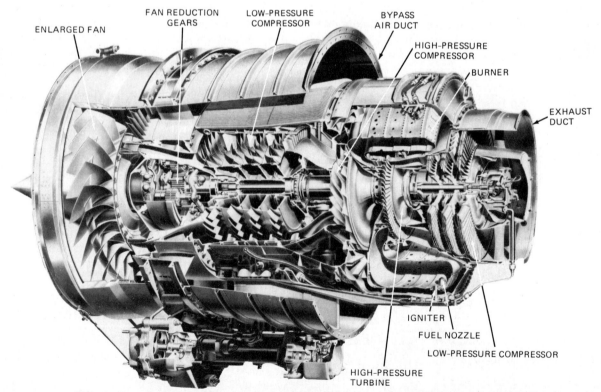

FIG. 21-87 Cutaway photograph of the TFE731-5 engine. (*Garrett Turbine Engine Co.*)

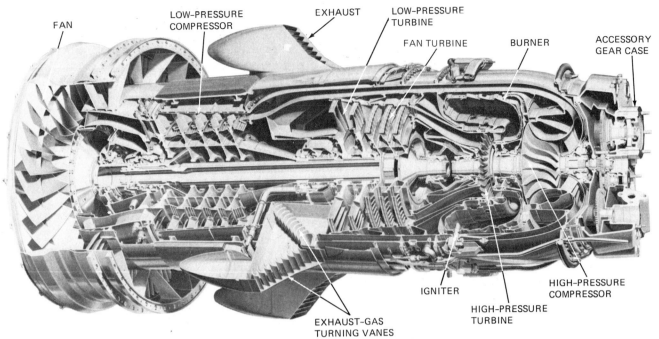

FIG. 21-88   The Garrett ATF3-6 turbofan engine.   *(Garrett Turbine Engine Co.)*

pass duct and the balance of the air entering the low-pressure compressor through variable inlet guide vanes. Air leaving the low-pressure compressor is split into eight ducts and is carried to the rear of the engine where it is turned 180° and directed to the inlet of the high-pressure compressor. High-pressure discharge air from the high-pressure compressor flows into the diffuser and then into the annular burner section. Fuel is injected into the burner and the fuel-air mixture is ignited to provide the high-velocity gases necessary to drive the turbines. The burner is a reverse-flow type in which the gasflow changes direction approximately 180°. The gases are directed inward from the burner to enter the high-pressure turbine. Leaving this turbine, the gases flow forward and outward to deliver energy to the fan turbine and low-pressure turbine.

Gas from the turbines flows forward to enter eight turning-vane modules. These modules turn the gasflow to the rear and discharge it into the bypass airstream. The exhaust gases mix with the bypass air and the combined air and gas is discharged at a high velocity through a common nozzle. This design assures maximum nozzle efficiency and low noise emission.

The ATF3-6 is of modular construction, as shown in Fig. 21-90. This type of design is to simplify maintenance by making it possible to remove sections for repair or overhaul without having to disassemble the complete engine. While the engine is installed in the aircraft, those sections that require the most frequent service and repair, such as the high-pressure section and accessory drives, can be removed and overhauled without disturbing the other sections of the engine. With modular engine design it is not necessary to perform complete engine overhaul on a periodic schedule. This permits module replacement or repair on an "as required" basis.

The ATF3-6 engine is controlled by a full-authority electronic fuel control similar to that employed with the TFE731-5 engine. The control system and interconnected units are shown in Fig. 21-91.

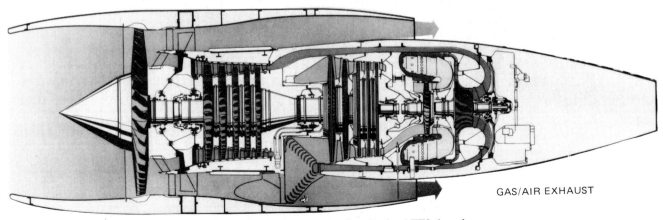

FIG. 21-89   Airflow and gasflow in the ATF3-6 engine.

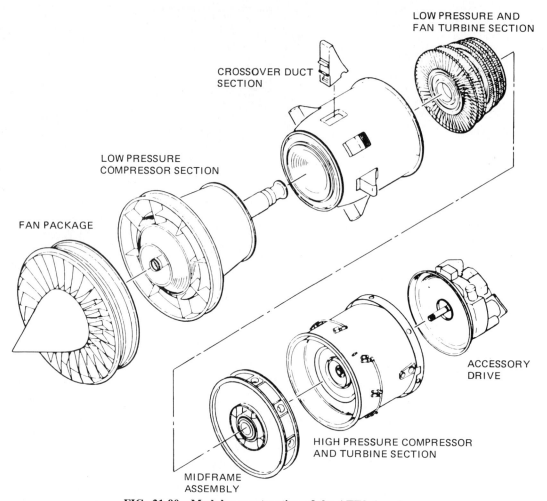

LOW PRESSURE AND
FAN TURBINE SECTION

CROSSOVER DUCT
SECTION

LOW PRESSURE
COMPRESSOR SECTION

FAN PACKAGE

ACCESSORY
DRIVE

HIGH PRESSURE COMPRESSOR
AND TURBINE SECTION

MIDFRAME
ASSEMBLY

FIG. 21-90   Modular construction of the ATF3-6 engine.

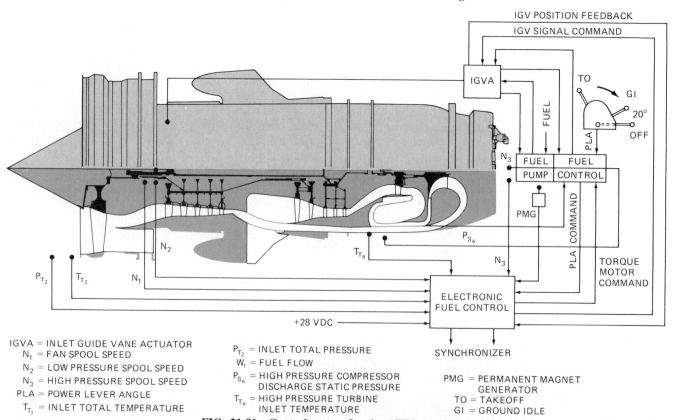

IGVA = INLET GUIDE VANE ACTUATOR
N_1 = FAN SPOOL SPEED
N_2 = LOW PRESSURE SPOOL SPEED
N_3 = HIGH PRESSURE SPOOL SPEED
PLA = POWER LEVER ANGLE
T_{T_2} = INLET TOTAL TEMPERATURE

P_{T_2} = INLET TOTAL PRESSURE
W_f = FUEL FLOW
P_{S_6} = HIGH PRESSURE COMPRESSOR
DISCHARGE STATIC PRESSURE
T_{T_8} = HIGH PRESSURE TURBINE
INLET TEMPERATURE

PMG = PERMANENT MAGNET
GENERATOR
TO = TAKEOFF
GI = GROUND IDLE

FIG. 21-91   Control system for the ATF3-6 engine.

● REVIEW QUESTIONS

1. Explain the difference between a pure jet engine and a turbofan engine.
2. What is the difference between a low-bypass engine and a high-bypass engine?
3. Briefly describe the Pratt & Whitney JT3D engine.
4. What is the bypass ratio of the JT3D engine?
5. Describe the combustion section of the JT3D.
6. What is the purpose of the coating applied to the turbine-nozzle vanes?
7. What is the N_2 compressor and by which turbine section is it driven?
8. What is the effect of the shroud at the ends of the turbine blades?
9. What must be done to remove a turbine blade from a disk in the JT3D engine?
10. How are turbine blades retained in the disk rims?
11. Where is the no. $4\frac{1}{2}$ bearing located and what is its function?
12. How is the rear-compressor drive shaft joined to the front-compressor rear hub?
13. Give a brief description of the lubrication system for the JT3D engine.
14. Compare the Pratt & Whitney JT3D engine with the JT8D engine.
15. List the main sections of the JT8D engine.
16. Describe the compressor inlet case for the JT8D.
17. Which turbine stages drive the front, or N_1, compressor?
18. Why can the first-stage turbine develop more power than other turbine stages?
19. What is the function of the knife-edge seals in the compressor?
20. Why are the last stages of the high-pressure compressor made of steel?
21. What is the location of the scavenge pump for the no. 4 and no. 5 bearings?
22. Compare the combustion section of the JT8D engine with the combustion section of the JT3D.
23. Describe the turbine section for the JT8D.
24. Discuss the improvements made on the JT8D-200.
25. Describe the exhaust system for the JT8D-200 engine and give the advantages of such a design.
26. How does the Pratt & Whitney JT9D engine compare with the JT3D and the JT8D engines?
27. What is the difference in bypass ratios for the high-bypass and low-bypass engines?
28. List the principal modules of the JT9D engine.
29. Describe the low-pressure compressor for the JT9D.
30. Which compressor stator vanes are variable?
31. What features of the JT9D engine help to reduce engine noise?
32. By what means are the first-stage compressor vanes protected from icing?
33. Describe the combustion section for the JT9D.
34. By what means is cooling provided for the first-stage turbine nozzle and turbine blades?
35. Describe the turbine section for the JT9D.
36. Describe the accessory drive section of the JT9D.
37. How has the fan section of the Pratt & Whitney JT9D-7R4 engine been modified to improve performance?
38. What improvements have been made in the low-pressure compressor of the JT9D-7R4 engine?
39. Give the advantage of single-crystal turbine blades and vanes.
40. What is the purpose of the turbine-case cooling system employed on the JT9D-7R4 engine?
41. What is the advantage of carbon airseals over the labyrinth type?
42. Describe the basic design of the Pratt & Whitney 2037 engine.
43. What benefits are derived from controlled-diffusion airfoils?
44. Compare powder-metal turbine disks with conventional forged disks.
45. Describe the differences between the General Electric CF6-6 and CF6-50 engines.
46. What means are used to provide durable and trouble-free trunnions for the variable vanes in the CF6 engine?
47. Describe the combustor for the CF6.
48. How are the nozzle vanes for the first-stage nozzle assembly cooled?
49. What are the two functions of the blade retainers on the disks of the high-pressure turbine?
50. Briefly describe the low-pressure turbine.
51. Explain the advantage of the modular concept as demonstrated by the low-pressure turbine module for the CF6 engine.
52. Name the supporting structures of the CF6 engine.
53. What parts of the CF6 engine are supported by the fan frame?
54. Describe the compressor rear frame for the CF6.
55. What is the function of the turbine midframe?
56. What are the units of the turbine rear frame?
57. List the principal units of the CF6 fuel system.
58. Compared with other gas-turbine engines described previously, what is unique about the Rolls-Royce RB·211 engine?
59. What is the particular advantage of the RB·211 engine design?
60. How many bearings are utilized in the RB·211 engine to support rotating units?
61. Describe the fan section of the RB·211 engine.
62. How does the combustion chamber for the RB·211 engine compare with other combustion chambers?
63. Describe cooling for the high-pressure turbine blades of the RB·211 engine.
64. Explain the advantage of the modular construction of the RB·211 engine.
65. What is unusual about the compressor arrangement in the Garrett-AiResearch TFE731 engine?
66. What is the advantage of the reverse-flow combustion chamber in the TFE731 engine?
67. Compare the performance of the TFE731 engine with other engines in respect to power-weight ratio and fuel consumption.
68. What improvements have been made in the TFE731-5 engine?
69. Describe the Garrett ATF3-6 engine.
70. Describe the airflow and gasflow in the ATF3-6.
71. Describe the exhaust system for the ATF3-6.
72. What is the advantage of modular construction?

22 TURBOPROP AND TURBOSHAFT ENGINES

● INTRODUCTION

The gas-turbine engine in combination with a reduction-gear assembly and a propeller has been in use for many years and has proved to be a most efficient power source for aircraft operating in speed ranges of 300 to 450 mph [482.70 to 724.05 km/h]. These engines provide the best specific fuel consumption of any gas-turbine engine, and they perform well from sea level to comparatively high altitudes (over 20,000 ft [6096 m]).

Although a variety of names has been applied to gas-turbine and propeller combinations, the most widely used name is **turboprop,** which will generally be used in this section. Another popular name is "propjet."

The power section of a turboprop engine is similar to that of a turbojet engine; however, there are some important differences, the most important of which is the turbine section. In the turbojet engine the turbine section is designed to extract only enough energy from the hot gases to drive the compressor and accessories. The turboprop engine, on the other hand, has a turbine section which extracts as much as 75 to 85 percent of the total power output to drive the propeller. For example, the Allison Model 501 engine extracts 3460 hp [2580.12 kW] for the propeller and produces 726 lb [3229.25 N] thrust. The total **equivalent shp** (shp plus thrust or eshp) is given as 3750 eshp [2796.38 kW]. This means, of course, that the turbine section will usually have more stages than that of the turbojet engine and that the turbine blade design is such that the turbines will extract more energy from the hot gas stream of the exhaust. In a turboprop engine, the compressor, the combustion section, and the compressor turbine comprise what is often called the **gas generator** or **gas producer.** The gas generator produces the high-velocity gases which drive the **power turbine.**

Another important feature of the turboprop engine is the reduction-gear assembly through which the propeller is driven. The gear reduction from the engine to the propeller is of a much higher ratio than that used for reciprocating engines because of the high rpm of the gas-turbine engine. For example, the gear reduction for the Rolls-Royce Dart engine is 10.75:1 and the gear reduction for the Allison Model 501 engine is 13.54:1.

Because the propeller must be driven by the turboprop engine, a rather complex propeller-control system is necessary to adjust the propeller pitch for the power requirements of the engine. At normal operating conditions, both the propeller speed and engine speed are constant. The propeller pitch and the fuel flow must then be coordinated in order to maintain the constant-speed condition; that is, when fuel flow is decreased, propeller pitch must also decrease.

● THE ALLISON MODEL 501-D13 ENGINE

Although the Allison 501-D13 turboprop engine has largely been superseded by later designs, its description is retained in this text because some are still in use on older transport aircraft. Furthermore, the 501 is a good engine to provide instruction on turboprop theory, and some technical schools use it as a training aid.

The Allison 501-D13 turboprop (propjet) engine is the commercial version of the Allison T-56 engine used for military aircraft. The 501-D13 engine is rated at 3750 eshp [5027 kW] at standard day conditions with an rpm of 13,820 and a turbine inlet temperature of 971°C (1780°F). The engine consists of a **power section, a reduction gear,** and a **torquemeter assembly.** The power section and reduction gear are connected and aligned by the torquemeter assembly, and added rigidity is provided by two struts. The propeller is mounted on a single-rotation SAE 60A propeller shaft. Figure 22-1 illustrates the configuration and general construction of the engine.

The following definitions apply to the engine:

Front The propeller end.

Top Determined by the breather located at the forward end of the power section.

Right and left Determined by standing at the rear of the engine and facing forward.

Rotation The direction of rotation is determined when standing at the rear of the engine and facing forward. The power-section rotor turns in a counterclockwise direction, and the propeller rotation is clockwise.

Accessories rotation Determined by facing the mounting pad of each accessory.

Combustion liner numbering The combustion liners are numbered from 1 through 6 in a clockwise direction when the engine is viewed from the rear. The no. 1 liner is at the top vertical center line.

Compressor and turbine stages numbering Numbered beginning from the forward end of the power section and progressively moving rearward. The compressor has 1 through 14 stages, and the turbine 1 through 4 stages.

Main rotor bearings numbering Numbered beginning at the forward end of the power section, moving

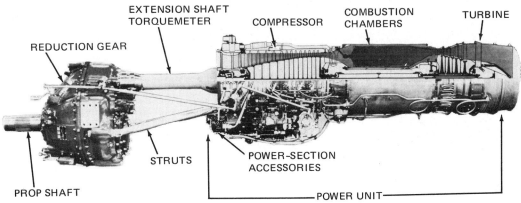

FIG. 22-1 Allison Model 501-D13 turboprop engine. *(Allison Gas-Turbine Operations, General Motors Corp.)*

rearward, with the no. 1 being at the forward end of the compressor rotor, no. 2 at the rear end of the compressor rotor, no. 3 at the forward end of the turbine rotor, and no. 4 at the rear of the turbine rotor.

Igniter plug location There are two igniter plugs installed in the engine, one in combustion liner no. 2 and one in combustion liner no. 5.

Engine Performance

The performance of the Allison 501-D13 engine is far superior to a conventional reciprocating engine of the same weight, and its fuel consumption per horsepower is comparable to that of a reciprocating engine. The following performance data emphasize the value of this engine as a powerplant for commercial aircraft:

Power output   3750 eshp, or 3460 shp + 726 lbt
   [2796 kW, or 2580 kW + 3230 N] at 13,820 rpm
Maximum continuous power   2170 eshp, or 1820 shp
   + 145 lbt [1618 kW, or 1357 kW + 645 N]
                     at 13,820 rpm
Weight without propeller   1756 lb [797 kg]
Power-weight ratio   2.14 eshp/lb [3.51 kW/kg]
Fuel consumption (sfc, cruise)   0.55 lb/hp/h
                     [334.5 g/kW/h]
Oil consumption   3.0 lb/h [1.36 kg/h]

Engine Description

Compressor Assembly. The compressor assembly consists of a compressor air-inlet housing assembly, compressor rotor assembly, compressor housing, diffuser, and diffuser oil-scavenge pump assembly.

The **compressor air-inlet housing** is a magnesium-alloy casting designed to direct and distribute air into the compressor rotor. It also provides the mounting location for the front-compressor bearing, the engine breather, the accessories drive housing assembly, the anti-icing air valves, the torquemeter housing, and the inlet anti-icing vane assembly.

The inlet anti-icing vane assembly is mounted on the aft side of the air-inlet housing and is utilized to impart the proper direction and velocity to the airflow as it enters into the first stage of the compressor rotor. These vanes may ice up under ideal icing conditions; hence, provisions are made to direct heat to each of the vanes. Air which has been heated by compression may be extracted from the outlet of the compressor (diffuser) and directed through two tubes to the anti-icing valves

mounted on the compressor air-inlet housing. The inlet anti-icing vanes are hollow and mate with inner and outer annuli (passages) into which the hot air is directed. From the annuli the air flows through the vanes and exits into the first stage of the compressor through slots provided in the trailing edge of the vanes. The construction of the compressor air-inlet housing is shown in Fig. 22-2.

The **compressor rotor** is an axial-flow type consisting of 14 stages. It is supported at the forward end by a roller bearing (bearing no. 1) and at the rear by a ball bearing (bearing no. 2). Figure 22-3 shows the arrangement of the compressor rotor, and Fig. 22-4 shows an inspector checking the seal diameter with a dial gage.

The entire **compressor housing assembly** is fabricated from steel and consists of four quarters permanently bolted together in halves. The split lines of the housing are located 45° from a vertical centerline. The compressor vane assemblies are installed in channels in the compressor housing and are securely located and held in position by bolts. The inner ring of the vane assemblies supports the interstage airseals which form a labyrinth seal, thus preventing air from one stage bleeding back to the previous stage. Between each row of vane assemblies (stator blades) the housing is coated with a special type of sprayed aluminum to provide a minimum compressor rotor-blade clearance, thus increasing compressor efficiency. The outlet vane assembly consists of an inner and an outer ring supporting two complete circles of vanes. These are used to straighten airflow prior to its entering the combustion section. The compressor housing is ported around the circumference to provide for air bleed. Four bleed-air valves are mounted on the outside of the compressor housing at the fifth stage and four at the tenth stage. Those valves at the fifth stage are manifolded together as are those at the tenth stage. These valves are used to unload the compressor during the start and acceleration or when operating at low-speed taxi idle.

The **compressor diffuser,** which is of welded-steel construction, is bolted to the flange at the aft end of the compressor housing assembly and forms the mid-structural member of the engine. One of three engine-to-aircraft mountings is located at this point. Six airfoil struts form passages which conduct compressed air from the outlet of the fourteenth stage of the compressor to the forward end of the combustion liners. These struts also support the inner cone which pro-

521

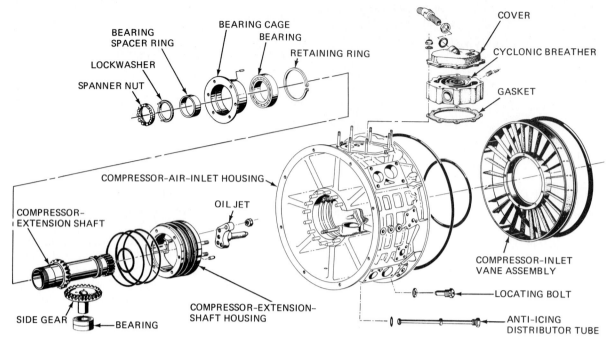

FIG. 22-2 Compressor air-inlet housing.

vides the mounting for the rear-compressor bearing, the seals, the rear compressor-bearing oil nozzle, the diffuser oil scavenge pump, and the forward end of the inner combustion chamber. Air is extracted from ports on the diffuser for anti-icing and operation of the fifth- and tenth-stage bleed-air valves. Bleed air is also extracted from this point by the airframe manufacturer for aircraft anti-icing, for crossfeeding from one engine to another for engine starter operation, and for operation of the oil cooler augmentor during ground operation. The six fuel nozzles are mounted on and extend into the diffuser. A fire shield is provided at the rear split line to protect the cooler portions of the engine from the high-temperature sections.

Combustion Assembly. The combustion section of the 501-D13 engine, illustrated in Fig. 22-5, consists of an outer and an inner combustion casing that form an annular chamber in which six **combustion liners** (cans) are located. These liners are spaced evenly around the center axis of the engine and serve to contain the flame of the burning fuel. Fuel is sprayed continuously from a nozzle during operation into the forward end of each combustion liner. During the starting cycle, two igniter

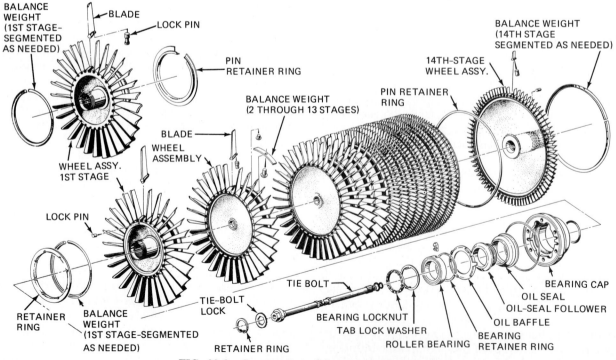

FIG. 22-3 Construction of the compressor rotor.

FIG. 22-4 Inspecting the seal diameter of the rotor.

is designed to locate and house the forward-turbine bearing, the seal assembly, the front-turbine oil jet, and the turbine front oil scavenge pump. The casing is divided into six equal passages by six airfoil struts. Each of these passages provides the means of locating and supporting the aft end of a combustion liner. Located around the outer casing are 18 holes with one thermocouple assembly mounted in each. Thus, three thermocouple assemblies are available at the outlet of each combustion liner. These 18 thermocouple assemblies are dual, and thus two complete and individual circuits are available. One is used to provide a temperature indication, referred to as turbine inlet temperature (TIT), to the cockpit, and the other is used to provide a signal to the electronic fuel-trimming system. The circuits measure the average temperature of all 18 thermocouples, thus providing an accurate indication of the gas temperature entering the turbine at all times.

The **turbine rotor assembly,** shown in Fig. 22-6, consists of four turbine wheels splined on a turbine shaft. The entire assembly is supported by a roller bearing at the forward end and a roller bearing at the aft end. A turbine coupling-shaft assembly connects the turbine rotor to the compressor rotor, thus transmitting the power extracted by the four-stage turbine assembly to the compressor, driven accessories, reduction-gear assembly, and propeller. All four stages of blades are attached to the wheel rims in broached serrations of five-tooth fir-tree design. The first-stage turbine wheel has the smallest blade area, with each succeeding stage becoming larger.

The **turbine vane casing** encases the turbine rotor assembly and retains the four stages of turbine vane assemblies. It is the structural member for supporting the turbine rear-bearing support. The vanes are airfoil design and serve two basic functions. They increase the gas velocity prior to each turbine wheel stage and also direct the flow of gases so that they will impinge

plugs, located in combustion liners 2 and 5, ignite the fuel-air mixture. All six liners are interconnected near their forward ends by **crossover tubes.** Thus, during the starting cycle after ignition has taken place in the no. 2 and no. 5 combustion liners, the flame propagates to the remaining liners. Liners which do not utilize an igniter plug have at the same location a **liner support assembly** which positions the combustion liner and retains it axially. The outer combustion-chamber casing provides the supporting structure between the diffuser and the turbine section. Mounted on the bottom of the outer combustion-chamber casing are two combustion-chamber drain valves which drain fuel after a false start and at engine shutdown.

Turbine Unit Assembly. The turbine unit assembly includes six major components. These are the (1) turbine inlet casing, (2) turbine rotor, (3) turbine oil scavenge pumps, (4) turbine vane casing, (5) turbine vane assemblies, and (6) turbine rear-bearing support.

The **turbine inlet casing** is attached at its forward end to the outer and inner combustion-chamber casings. It

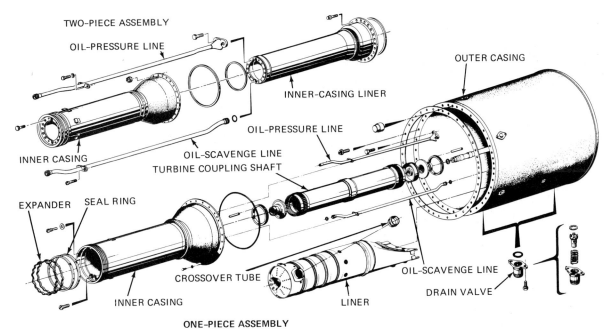

FIG. 22-5 Combustion section.

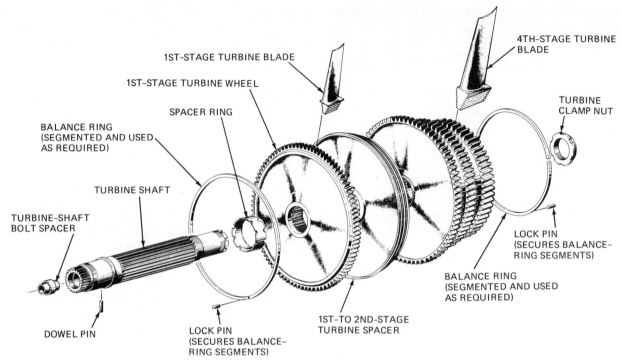

FIG. 22-6 Turbine rotor assembly.

Labels in figure:
1ST-STAGE TURBINE BLADE
1ST-STAGE TURBINE WHEEL
SPACER RING
BALANCE RING (SEGMENTED AND USED AS REQUIRED)
TURBINE SHAFT
TURBINE-SHAFT BOLT SPACER
DOWEL PIN
LOCK PIN (SECURES BALANCE-RING SEGMENTS)
1ST-TO 2ND-STAGE TURBINE SPACER
BALANCE RING (SEGMENTED AND USED AS REQUIRED)
LOCK PIN (SECURES BALANCE-RING SEGMENTS)
TURBINE CLAMP NUT
4TH-STAGE TURBINE BLADE

upon the turbine blades at the most effective angle. Figure 22-7 shows the installation of vane assembly sections.

The **turbine rear-bearing support** attaches to the aft end of the turbine vane casing. It houses the turbine rear bearing, the turbine rear oil scavenge pump and support, and the inner exhaust cone and insulation. It also forms the exhaust nozzle for the engine.

Accessories Drive Housing Assembly. The **accessories drive housing assembly** is a magnesium-alloy casting mounted on the bottom of the compressor air-inlet housing. It includes the necessary gear trains for driving all power-section-driven accessories at their proper rpm in relation to engine rpm. Power for driving the gear trains is taken from the compressor extension shaft by a vertical shaftgear. The accessories driven through this assembly are the speed-sensitive control, speed-sensitive valve, oil pump, fuel control, and fuel pump.

Reduction-Gear Assembly

The principal function of the **reduction-gear assembly** is to provide a means of reducing power-section rpm to the range of efficient propeller rpm. It also provides pads on the rear of the case for mounting the starter, cabin supercharger, alternator, tachometer generator, propeller alternator, hydraulic pump, DC generator, and an oil pump.

The reduction gear has an independent lubrication system which includes a pressure pump and two scavenge pumps. The oil supply is furnished from the oil tank, which also supplies the main engine.

The reduction-gear assembly is remotely located from the power section, as illustrated in Fig. 22-1, and is attached to the power section by the torquemeter shaft and two tie struts. The remote location of the reduction-gear assembly offers the following advantages: (1) better air-inlet ducting which increases engine efficiency and performance, (2) an opportunity for readily mounting the gearbox offset up or down for high- or low-wing aircraft, (3) additional space for mounting driven accessories without affecting frontal area, (4) an engine in the minimum frontal area, and (5) an ability to utilize an electronic torquemeter.

The reduction-gear assembly, illustrated in Fig. 22-8, has an overall reduction-gear ratio of 13.54:1, accomplished through a two-stage step-down gear train. The primary step-down is accomplished by a spur-gear train having a ratio of 3.125:1, and the secondary step-

FIG. 22-7   Installing vanes in the turbine section.

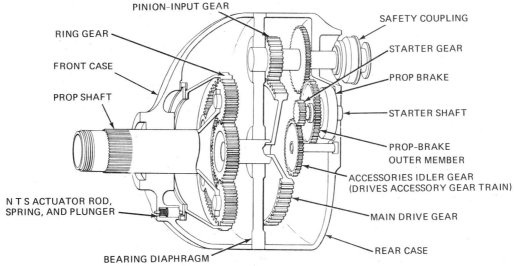

FIG. 22-8 Reduction-gear assembly.

down is provided by a planetary-gear train with a ratio of 4.333:1. The propeller shaft rotates in a clockwise direction when viewed from the rear of the engine. In addition to the reduction gears and accessories drives, the reduction-gear assembly includes four other major components: (1) the propeller brake, (2) the negative torque signal (NTS) system, (3) the thrust-sensitive signal (TSS), and (4) the safety coupling.

Propeller Brake

The **propeller brake** is designed to prevent the propeller from windmilling when it is feathered in flight and to decrease the time for the propeller to come to a complete stop after ground shutdown. It is a friction-type brake consisting of a stationary inner member and a rotating outer member which, when locked, acts upon the primary-stage reduction gearing. During normal engine operation, reduction-gear oil pressure holds the brake in the released position. This is accomplished by oil pressure which holds the outer member away from the inner member. When the propeller is feathered or at engine shutdown, reduction-gear oil pressure drops off and the effective hydraulic pressure decreases, thus allowing a spring force to move the outer member into contact with the inner member.

Negative Torque System. The **negative torque system** (NTS) was explained in the section describing the turbopropeller; however, a brief summary of the operation will be given here. As mentioned before, the NTS is designed to prevent the aircraft from encountering excessive propeller drag when the air is driving the propeller. The system is a part of the reduction-gear assembly, is completely mechanical in design, and is automatic in operation. A negative torque value in the range of 250 to 370 hp [186.43 to 275.91 kW], transmitted from the propeller to the reduction gear, causes the planetary ring gear to move forward through openings in the reduction-gear front case. Only one rod is used to actuate the propeller NTS linkage. When the NTS is actuated, the propeller increases blade angle until the abnormal propeller drag and resultant excessive negative torque is relieved. The propeller will never go to the feather position when actuated by the NTS

system but will modulate through a small blade-angle range such that it will not absorb more than approximately 250 to 370 hp. As the negative torque is relieved, the propeller returns to normal governing.

Thrust-Sensitive Signal. The **thrust-sensitive signal** (TSS) provides for initiating automatic feathering at takeoff in case of engine failure. The system must be armed prior to takeoff if it is to function, and a blocking circuit is provided to prevent automatic feathering of more than one propeller. The system is armed by the **auto-feather arming switch** and a power-lever-actuated switch. The setting of the power-lever switch is such that, if operation is normal, the propeller will be delivering considerably more than 500 lb [2224 N] positive thrust. This prevents auto feather except when engine failure occurs.

The system is designed to operate (if armed) when the propeller is delivering less than 500 lb positive thrust. The propeller shaft tends to move in a forward axial direction as the propeller produces positive thrust. Axial travel is limited by a mechanical stop applied to the ball-thrust-bearing outer race. When the propeller is producing normal thrust of more than 500 lb, a spring in front of the thrust bearing is compressed. When thrust decreases to below 500 lb, the spring can expand and move the thrust bearing forward. This movement is multiplied through mechanical linkage and transmitted mechanically to a pad on the left side of the reduction-gear front case. An electrical switch assembly mounted on the case energizes the feathering circuit.

Safety Coupling. The **safety coupling** could readily be classified as a "backup" device for the NTS system. The coupling has a negative torque setting of approximately 1700 hp [1268 kW], and if the NTS system or propeller fails to function properly, the coupling disengages the reduction gear from the power section, resulting in a considerable reduction in drag. The safety coupling is located and attached to the forward end of the torquemeter shaft, which transmits power from the engine to the reduction gear. During normal operation the safety coupling connects the torquemeter shaft to the reduction gear by helical splines, straight splines, and an intermediate member. When the negative torque is sufficient, the force developed by the helical splines moves the intermediate member against the force of five washerlike circular springs and permits the intermediate member to disengage its helical splines from the inner member. The splines remain disengaged with a ratcheting action until the negative torque is reduced and the splines can reengage.

Torquemeter Assembly and Tie Struts

The torquemeter housing, illustrated in Fig. 22-9, provides alignment, and two tie struts provide the necessary rigidity between the power section and the reduction-gear assembly. The tie struts are adjustable through two eccentric pins located at the reduction-gear end. These pins are splined to provide a positive locking method after proper alignment is established by the torquemeter housing. The torquemeter provides the means of accurately measuring shaft horsepower input into the reduction-gear assembly. It has an indicated accuracy of ± 35 hp [26.11kW] from 0 to maximum allowable power which represents ± 1 percent actual horsepower at standard-day static takeoff power. The torquemeter consists of the following major parts: (1) the **torquemeter inner shaft,** which is the torque shaft; (2) the **torquemeter outer shaft,** which is the reference shaft; (3) the **torquemeter pickup assembly** consisting of a magnetic pickup; (4) the **torquemeter housing;** (5) the **phase detector;** and (6) the **indicator.**

The principal operation of the torquemeter is that of measuring electronically the angular deflection (twist) which occurs in the torque shaft relative to the zero reference shaft. The actual degree of angular deflection is sensed by the pickup assembly and transmitted to the phase detector. The phase detector converts the pickup signal into an electric signal and directs it to the indicator located on the instrument panel.

The torquemeter functions because of the arrangement of the torque shaft, which twists in proportion to torque inside the reference shaft. The reference shaft is a tube surrounding and rigidly attached to the torque shaft at the power-section end. At the reduction-gear end the reference tube is free to rotate outside the torque shaft, and a displacement of the magnetic pickup occurs to produce the torque signal. It can be seen from the drawing of Fig. 22-9 that a torque force applied at the power end of the shaft assembly will cause the torque shaft to twist. Since there is no torque on the reference shaft, the twist of the inner shaft will evidence itself as an angular displacement between the flanges at the drive end.

Lubrication Systems

The Model 501-D13 engine power section and reduction-gear assembly have separate and independent lubrication systems which utilize a common oil supply from the engine oil tank. This tank is designed and manufactured by the airplane manufacturer.

The engine manufacturer supplies the airframe manufacturer with information regarding the amount of oil flow required by the reduction-gear assembly and the power section and also the heat rejection from the reduction gear and the power section. The airframe manufacturer then designs an aircraft oil-supply system which will provide the required volume of oil flow and the necessary oil cooling. The airframe manufacturer also supplies the following indications in the cockpit: (1) power-section oil pressure, (2) reduction-

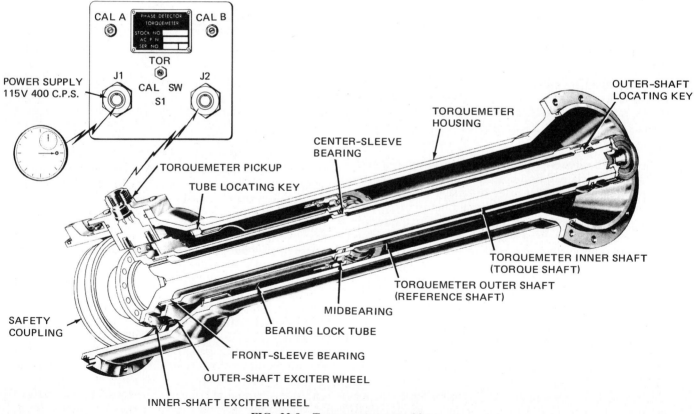

FIG. 22-9 Torquemeter assembly.

gear oil pressure, (3) oil inlet temperature, and (4) oil quantity.

Power-Section Lubrication System. Figure 22-10 is a schematic diagram of the engine oil system in the power section. This system includes the following units with their locations as indicated:

1. **Main oil pump.** This includes the pressure pump, a scavenge pump, and the pressure-regulating valve, which is located on the forward side of the accessories-drive housing assembly.

2. **Oil filter.** Located on the forward side of the accessories-drive housing.

3. **Check valve.** Located in the oil filter assembly.

4. **Bypass valve** (filter). Located in the accessories-drive housing assembly.

5. **Three scavenge pumps.** Located in the diffuser, turbine inlet casing, and the turbine rear-bearing support.

6. **Scavenge-relief valve.** Located in the accessories drive housing assembly.

7. **Breather.** Located on top of the air-inlet housing.

Oil is supplied from the aircraft tank to the inlet of the pressure pump whence it passes through the oil filter before being delivered to other parts of the engine. System pressure (filter outlet pressure) is regulated to 50 to 75 psi [344.75 to 517.13 kPa] by the pressure regulating valve. A bypass valve is incorporated in the system in the event that the filter becomes clogged sufficiently to restrict oil flow. A check valve prevents oil from seeping into the power section whenever the engine is not running.

The scavenge pump, which is incorporated in the main oil pump, and the three independent scavenge pumps are so located that they will scavenge oil from the power section in any normal attitude of flight. The scavenge pump located in the main oil pump scavenges oil from the accessories-drive housing. The other three scavenge pumps scavenge oil from the diffuser and from the front and rear sides of the turbine. The outputs of the diffuser scavenge pump and the front turbine scavenge pumps join that of the main scavenge pump. The output of the rear-turbine scavenge pump is delivered to the interior of the turbine-to-compressor tie bolt and the compressor rotor tie bolt. This oil is directed to the splines of the turbine coupling-shaft assembly and to the splines of the compressor extension shaft. Thus, the output of the rear-turbine scavenge pump must be rescavenged by the other three scavenge pumps. A scavenge-relief valve is located so that it will prevent excessive pressure buildup in the power-section scavenge system. The combined flows of scavenged oil from the power section and reduction-gear scavenge systems must be cooled and returned to the supply tank. A magnetic plug is located on the bottom of the accessories-drive housing, and another at the oil scavenge outlet on the forward side of the main oil pump assembly. The purpose of these plugs is to attract and hold steel particles which may be in the oil. If particles of appreciable size are found on the plugs at inspection, it is an indication of internal failure and requires that the engine be checked thoroughly to locate the source of the steel.

Reduction-Gear Lubrication System. The reduction-gear oil system is a dry-sump system which feeds from the same oil-supply tank used for the power-section system. The system includes the following items, with each of their respective locations as indicated:

1. **Pressure pump.** Located on the left rear side of the reduction gear.

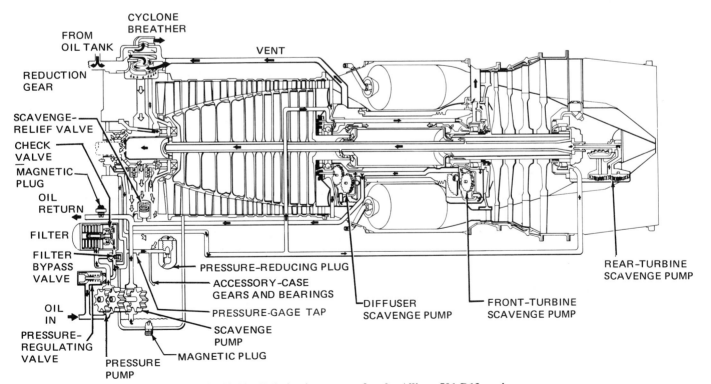

FIG. 22-10   Lubrication system for the Allison 501-D13 engine.

2. **Filter.** Located in the pump body assembly.

3. **Filter bypass valve.** Located in the pump body assembly.

4. **Check valve.** Located in the pump body assembly.

5. Two **scavenge pumps.** One located in the bottom of the rear case and the other in the front case below the prop shaft.

6. Two **pressure-relief valves.** One for the pressure system and the other for the scavenge system. The scavenge-relief valve is located in the common outlet of the scavenge pumps, and the other is in the rear-case housing near the oil filter outlet.

The foregoing components of the reduction-gear oil system can be located in the diagram of Fig. 22-11. Oil flows from the pressure pump through a filter and to all parts within the reduction gear which require lubrication. In addition, oil pressure is used as hydraulic pressure for the operation of the propeller brake assembly. A filter bypass valve ensures continued oil flow in the event that the filter becomes clogged. A check valve prevents oil flow by gravity into the reduction gear after engine shutdown. A relief valve set at 180 psi [1241.1 kPa] to begin opening and to be fully open at 250 psi [1723.75 kPa] prevents excessive system pressure. This valve is not to be construed as being a regulating valve, since its only function is that of limiting pressure.

The location of the scavenge pumps provides for scavenging in any normal attitude of flight. The output of the two scavenge pumps returns the oil by a common outlet to the aircraft system. A relief valve, which is set at the same values as the one in the reduction-gear pressure system, limits the maximum scavenge pressure. A magnetic plug is installed on the bottom rear of the reduction-gear assembly to provide a means for draining the case and to detect possible failure in the gear system.

Air-Bleed Systems

The Allison Model 501-D13 engine is provided with two air-bleed systems, one for anti-icing and the other for reducing the compressor load during reduced speed operations. These have been mentioned previously; however, a more detailed description is given here.

Anti-icing System. The **anti-icing system** consists of necessary lines and valves to permit heated compressor air to flow through the air-inlet units requiring heat for anti-icing. The system is manually controlled and is turned on by means of a switch in the cockpit. When the switch is turned on, the **anti-icing solenoid** is deenergized to cause anti-icing airflow to the inlet anti-icing vane assembly, the compressor air-inlet housing struts, the fuel-control temperature-probe deicer, and the upper half of the torquemeter housing shroud. Note that the solenoid is *deenergized* when anti-icing air is on. The purpose of this arrangement is to provide a fail-safe system. In case 24-V DC power fails, the anti-icing system will operate to furnish protection even though it may not be needed.

Acceleration Bleed System. The **acceleration bleed system** is also called the fifth- and tenth-stage bleed-air system because the system bleeds air from the fifth and tenth stages of the compressor during engine start and acceleration and at low-speed taxi. The system unloads the compressor from 0 to 13,000 rpm in order to prevent engine stall and surge. It includes four pneumatically operated bleed-air valves located at the fifth stage and four located at the tenth stage, a speed-sensitive valve mounted on the forward side of the accessories housing assembly, and the necessary manifolding and plumbing. The bleed-air valves at the fifth stage are manifolded together, with the outlet being provided through the nacelle forward of the engine spray baffle assembly. The tenth-stage bleed-air valves empty into another manifold, which is ducted to the aft side of the engine spray baffle assembly. The speed-sensitive valve is a flyweight type which responds to engine rpm. When the engine is running at less than 13,000 rpm, the valve is so positioned that all bleed-air-valve piston heads are vented to the atmosphere. This allows the compressor fifth- and tenth-stage pressures to move the pistons to their open position, thus

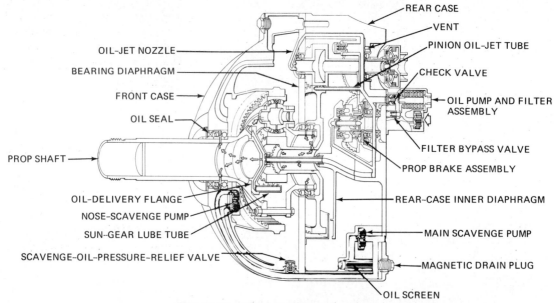

FIG. 22-11   Reduction-gear oil system.

bleeding air overboard. When the engine is running at 13,000 rpm or more, the speed-sensitive valve directs fourteenth-stage air to the bleed-valve piston heads. Since fourteenth-stage pressure is always greater than fifth- or tenth-stage pressures, the bleed-air-valve pistons move to the closed position, thus preventing air bleed from the fifth and tenth stages. During low-speed taxi operation the fifth- and tenth-stage bleed-air valves will be in the open position because engine rpm at this time is below 13,000.

Speed-Sensitive Control

The **speed-sensitive control,** shown in Fig. 22-12, is a flyweight-type unit used to actuate three microswitches which control certain engine functions at predetermined speeds. The control is mounted on the forward side of the accessories drive housing assembly and is driven from the accessory gear train. The control actuates the microswitches to provide for the following actions to take place at 2200, 9000, and 13,000 rpm.

At 2200 rpm:

1. The fuel-control cutoff-valve actuator opens the cutoff valve at the outlet of the fuel control to provide fuel for starting the engine.

2. Ignition is turned on.

3. The drip valve is energized to the closed position.

4. The fuel-pump paralleling valve is closed to place the fuel pumps in parallel. The engine fuel-pump light should go on to show operation of the secondary pump.

5. The primer valve opens if the primer button is held in the ON position.

At 9000 rpm:

1. The ignition turns off.

2. The drip valve is deenergized but remains closed because of fuel pressure.

3. The paralleling valve opens and places the fuel pumps in series.

At 13,000 rpm:

1. The electronic fuel-trimming system is changed from temperature limiting with a maximum temperature of 830°C (1526°F) to temperature limiting with a maximum temperature of 978°C (1792°F).

2. The maximum possible ''take'' of fuel by the temperature datum valve is reset to 20 percent rather than the previous 50 percent.

Ignition System

The Model 501-D13 engine utilizes a Bendix TCN-24 high-energy ignition system of the capacitor discharge type and operates on 14 to 30 V DC input. The operation of these systems has been described previously for other gas-turbine engines. The system includes an exciter and an ignition relay mounted on top of the compressor housing, the lead assemblies, and two igniter plugs. During the starting cycle as rpm reaches 2200, the speed-sensitive control automatically completes an electric circuit to the ignition relay. This closes the circuit to the exciter, thus providing electric energy to the igniter plugs. When engine rpm reaches 9000, the ignition circuits are deenergized through the action of the speed-sensitive control. Operation of the ignition system requires that the fuel and ignition switch in the cockpit be in the ON position.

The Fuel System

The fuel system for the Model 501-D13 engine includes all the units illlustrated in Fig. 22-13 together with associated components.

The engine operates on gas-turbine kerosene conforming to Allison specification EMS-64A or JP4 (MIL-T-5624), which has similar characteristics, or on other fuels designed for gas-turbine engine operation.

The fuel system must deliver metered fuel to the six fuel nozzles of the engine as required to meet all possible conditions of engine operation either on the ground or in flight. This imposes a number of requirements on the fuel system and its controlling units. Some of these requirements are as follows:

1. The capability of starting under all ambient conditions.

2. Adjustment for rapid changes in power.

3. A means of limiting the maximum allowable turbine inlet temperature (TIT).

4. A system which will enable the operator to select a desired power setting (determined by TIT) and have it automatically maintained regardless of altitude, free-air temperature, forward speed, and fuel Btu content.

5. A system which incorporates an rpm-limiting device in the event of propeller-governor malfunction.

6. A system which must control fuel flow during the rpm range in which the engine compressor is susceptible to stall and surge.

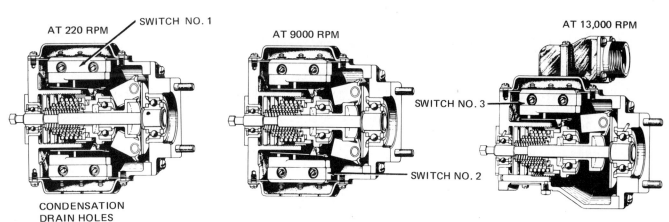

FIG. 22-12 Speed-sensitive control.

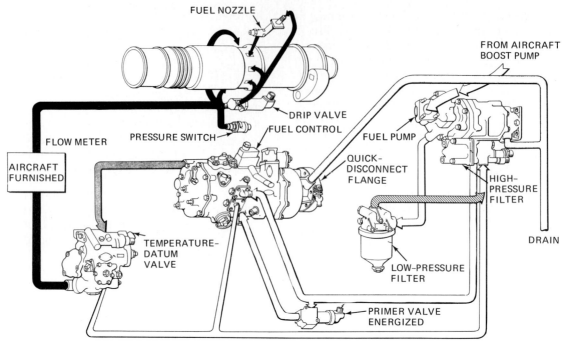

FIG. 22-13 Fuel system for the Allison 501-D13 engine.

7. A system which coordinates propeller-blade angle with fuel flow during ground operation.

8. A system which is capable of operating, if necessary, on the hydromechanical fuel control. However, if this is necessary, closer monitoring of power-lever and engine instruments will be necessary.

Fuel Pump and Low Pressure Fuel Filter. Fuel from the aircraft fuel system is supplied to the engine fuel pump where it enters into a boost element and is then directed to the **low-pressure filter.** The filter is a paper-cartridge type incorporating two bypass valves (relief valves) which open in the event that the filter becomes clogged from fuel contamination. The paper cartridge is of the type that must be replaced at certain specified inspection intervals.

The fuel-pump assembly includes, in addition to the boost element, two spur-gear-type high-pressure pumps, commonly referred to as the **primary** and **secondary** elements. During normal operation these pumps are in series. However, during engine starting (2200 to 9000 rpm) the pumps are placed in parallel by the action of a paralleling valve in the high-pressure filter. The paralleling of the pumps is used to increase the fuel flow during low rpm. Failure of either the primary or secondary pump will not affect normal operation because either pump has sufficient capacity of fuel flow for takeoff power.

High-Pressure Fuel Filter. The **high-pressure fuel filter** assembly consists of two check valves, a paralleling valve, a fuel filter, a pressure switch, and a bypass valve. The high-pressure fuel filter accomplishes six principal functions: It (1) filters the output of the primary and secondary pumps, (2) connects the two pumps in parallel during the starting cycle, (3) connects the two pumps in series during normal operation with the primary pump supplying high-pressure fuel flow to the power section, (4) automatically enables the secondary pump to take over upon failure of the primary pump,

(5) provides a means of checking primary and secondary pump operation during the starting procedure, and (6) provides a means of indicating primary pump failure with the engine fuel-pump light.

Fuel-Control Unit. The fuel control for the Allison Model 501-D13 engine is a Bendix AP-B3 hydromechanical unit which senses air density, acceleration, rpm, TIT, and power-lever position and schedules fuel accordingly to provide proper operation without surge, stall, excessive temperatures, etc. It supplies a controlled fuel flow for starting and maintains a proper rate of flow for acceleration. The control actually schedules 20 percent more fuel than is normally required for engine operation in order to provide the temperature datum control with fuel for trim purposes.

● THE ROLLS-ROYCE DART TURBOPROP ENGINE

General Description

The Rolls-Royce Dart engine has been in use for many years on a variety of aircraft including the Vickers Viscount and the Fairchild F-27 Friendship. The engine has proved to be rugged, dependable, and economical with overhaul periods extending to more than 2000 h.

The Dart engine utilizes a single-entry two-stage centrifugal compressor, a can-type through-flow combustion section, and a three-stage turbine. The general design of the engine can be seen in the drawing of Fig. 22-14. This drawing shows the arrangement of the propeller, reduction gear, air inlet, compressor impellers, combustion chambers, turbine, and exhaust.

An external view of the Dart engine is shown in Fig. 22-15. This view shows the propeller shaft, air-intake casing, oil cooler, compressor casings, combustion chambers, turbine casing, and exhaust unit. The engine

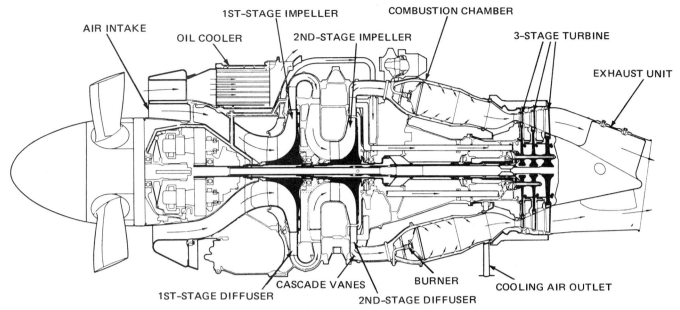

FIG. 22-14   Arrangement of the Rolls-Royce Dart engine.   *(Rolls-Royce)*

is approximately 45 in [112.3 cm] in diameter and 98 in [248.92 cm] in length.

Engine Data

The general data and performance for the Dart Model 528 engine are as follows:

Power output	1825 shp plus 485 lbt [1368 kW plus 2157 N]
Compression ratio	5.62:1
Engine rpm	15,000
Weight (without propeller)	1415 lb [642 kg]
SFC (specific fuel consumption)	0.57 lb/eshp/h [346.72 g/kW/h]
Power-weight ratio	1.51 eshp/lb [2.13 kW/kg]

Internal Features

The cutaway photograph of the Dart engine shown in Fig. 22-16 reveals the internal construction of the engine. At the forward end is the **reduction-gear as-sembly** which reduces the propeller-shaft speed to 0.093 of the speed of the engine. The reduction-gear housing is integral with the **air-intake casing.**

Immediately to the rear of the reduction-gear assembly is the compressor section, which includes two centrifugal impellers. Both impellers are clearly visible in the illustration. Accessory drives are taken from the reduction-gear assembly and through a train of gears aft of the second-stage compressor impeller.

Seven interconnected **combustion chambers** are located between the compressor section and the turbine. These combustion chambers are skewed or arranged in a spiral configuration to shorten the engine and take advantage of the direction of airflow as it leaves the compressor.

A **three-stage turbine** is located to the rear of the combustion chambers. As in other turboprop engines, this turbine is designed to extract as much energy as possible from the high-velocity exhaust gases.

Reduction-Gear Assembly

The reduction-gear assembly shown in Fig. 22-17 is of the compound type having high-speed and low-speed

FIG. 22-15   External view of the Dart engine.   *(Rolls-Royce)*

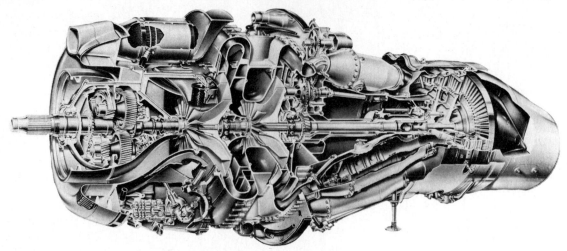

FIG. 22-16   Cutaway view of the Dart engine.   *(Rolls-Royce)*

trains. The high-speed train consists of a high-speed pinion connected to the main shaft and driving three layshafts through helical gear teeth. To isolate the main shaft couplings from propeller vibrations, a torsionally flexible shaft is used to couple the high-speed pinion to the main shaft. The three layshafts are mounted in roller bearings supported by panels in the gear casing.

The **low-speed gear train** consists of helical gears formed on the front end of the layshafts which drive the internal, helically toothed **annulus gear.** This annulus gear is bolted to the propeller-shaft driving disk. As a result of driving through the helical gears, the layshafts tend to move axially. This movement is limited by limit shafts mounted coaxially within the layshafts. Each limit shaft is prevented from moving by a ball thrust race at the rear.

The **propeller shaft** is supported by roller bearings housed in the front panel and the domed front casing. Axial thrust is taken on a ball bearing mounted behind the front roller bearing. A labyrinth-type seal assembly, pressurized by compressed air, surrounds the propeller shaft where it passes through the front cover and prevents loss of lubrication oil to the atmosphere.

To permit propeller oil to be transferred from the stationary casing to the rotating propeller shaft, a transfer seal assembly is used. It consists of babbitt lined with bronze bushings fitting closely around an adapter

located inside the rear end of the propeller shaft. Tubes screw into the adapter and convey the oil to the pitch-control mechanism.

Torquemeter

Under normal operating conditions, the helical teeth of the gear train produce a forward thrust in each layshaft which is proportional to the propeller-shaft torque. This load is hydraulically balanced by oil pressure acting on a piston assembly incorporated in the forward end of each layshaft. The necessary oil pressure is obtained by boosting engine oil pressure with a gear pump mounted on the layshaft front-bearing housing and driven from a gear attached to the propeller shaft.

The forward thrust of the layshafts resulting from the greater torque of the low-speed gear train is partially balanced by the rearward thrust produced by the lesser torque of the high-speed gear train. The residual forward thrust is balanced by the torquemeter oil pressure. A gage in the cockpit indicates this pressure, which is a measure of the torque transmitted by the gear. The engine power is calculated from the reading of the gage.

Auxiliary Drives

The auxiliary drives receive power from a bevel gear splined to the rear of the lower limit shaft which meshes with another bevel gear supported in plain bearings in the rear panel of the reduction-gear case. Through the auxiliary drives, the oil pumps, fuel pumps, and the propeller-control unit are driven.

Compressor

The compressor for the Dart engine comprises two stages, one immediately to the rear of the other, as shown in Fig. 22-16. The first-stage impeller is 20 in [50.8 cm] in diameter and has 19 blades, while the second-stage impeller is 17.6 in [44.7 cm] in diameter with 19 blades.

The compressor casings include the **front-compressor casing,** the **intermediate casing,** and the **second-stage outlet casing.** The front-compressor casing and the second-stage casing carry the diffuser-vane rings, and the intermediate casing carries the interstage guide vanes internally and the engine mounting points externally.

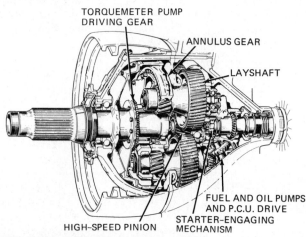

TORQUEMETER PUMP DRIVING GEAR

ANNULUS GEAR

LAYSHAFT

FUEL AND OIL PUMPS AND P.C.U. DRIVE

STARTER-ENGAGING MECHANISM

HIGH-SPEED PINION

FIG. 22-17   **Reduction-gear assembly.**

Each rotating assembly consists of an impeller and rotating guide vanes (RGVs). The assemblies are splined onto separate shafts and individually balanced. The split shaft facilitates bearing alignment and makes it unnecessary to disturb the balance during engine buildup. The guide vanes and impellers are locked to the shafts by nuts and cup washers.

Passages are machined through the first-stage rotating guide vanes and between the impeller vanes to permit water-methanol injection. The first-stage shaft is supported at the front by a roller bearing and at the rear by a ball bearing. The second-stage shaft is supported at the front by helical splines inside the rear of the first-stage shaft and at the rear by a ball bearing.

Surrounding each rotating assembly is a **diffuser-vane ring.** Each ring consists of a number of fixed vanes forming divergent channels.

Between the compressor stages is a set of guide vanes. Air leaving the first-stage compressor passes between these vanes before entering the second-stage RGVs. The vanes are so angled that they impart a whirling velocity to the airstream.

Combustion Section

The combustion section consists of seven individual combustion chambers such as that shown in Fig. 22-18, arranged in an inward spiral (skewed) with respect to the engine main shaft to shorten the engine and promote a smooth gasflow. The chambers are numbered counterclockwise, viewed from the rear, with no.1 being at the top. Each combustion chamber consists of an expansion chamber, an air casing, the flame tube, and interconnectors.

The **expansion chambers,** forming the forward ends of the combustion chambers, are fitted to the compressor outlet elbows by two link bolts, the seating between the chamber and elbow being formed by a spherical joint ring. At the rear they are attached to the air casing on a bolted flange. Each expansion chamber provides the location for a fuel burner (nozzle), and provision is made for fuel drain connections where necessary. High-energy igniter plugs are carried in the no. 3 and no. 7 chambers.

The **air casings** are bolted to the expansion chambers at the front; however, they are inserted in the discharge nozzles at the rear with a slip fit sealed by piston rings. This permits expansion and contraction of the casings. Each air casing carries two interconnectors, three flame-tube locating pins, and fuel drain connections where necessary. Because of the various positions of interconnectors and fuel drain connections, the casings are not interchangeable.

The **flame tubes** are fabricated in sections from a high-temperature metal-alloy sheet, the joints being welded and riveted. The tube is located at the front in the air casing by three pins and is supported at the rear by a spherical seating inside the discharge nozzle. The head of each tube carries a set of fixed swirl vanes to assist in efficient mixing of fuel and air.

The **interconnectors** are necessary to equalize the gas pressure and provide a means of passing the flame during light-up from the no. 3 and no. 7 chambers to the other chambers. Each interconnector consists of two concentric tubes to connect the air casings and the flame tubes by independent passages. To provide an expansion joint, the outer tubes carry sealing rings seated in bores in the air casings. A three-bolt flange forms the joint between each interconnector connecting adjacent combustion chambers.

Turbine Section

The turbine section of the Dart engine consists of three turbine wheels fitted with blades and of the nozzle box assembly, which contains three sets of nozzle guide vanes (NGVs). The compressor drive shaft and the inner reduction-gear drive shaft are coaxial and are attached with bolted flanges to the three turbine wheels.

The **nozzle box** is a welded two-piece casing into which are fitted the seven combustion-chamber discharge nozzles. It is surrounded by a heat shield. On the front flange of the nozzle box is fitted the nozzle box mounting drum, which, together with the inner cone and turbine bearing housing, is bolted to the intermediate casing. Flanges on the inside of the nozzle box and inner cone and interstage labyrinth-seal platforms provide the location of the nozzle guide vanes.

The **nozzle guide vanes** form a series of nozzles in which the gases are accelerated. They are of airfoil section and cast hollow to maintain as nearly as possible a constant sectional thickness to reduce thermal stress.

There are 70 high-pressure (HP) vanes hooked into flanges machined on the inner cone and nozzle box outer casing, and 14 of these are used as locators. The inner location is provided by slots in the flange of the inner cone, and the outer location is via locating pegs fitted through the nozzle box casing and engaging in the guide-vane outer platforms.

Fifty-six intermediate-pressure (IP) vanes are supported in grooves in the nozzle box outer casing by the tongues on the outer platforms hooking into the grooves in the nozzle box. They are positioned axially by two rings, and the turbine interstage seal is carried on their inner platforms. At the leading edge of 12 of the vanes, provision is made for fitting the thermocouples. Twenty-eight of the vanes are used for locators.

The three **turbine wheels** are secured to the turbine and inner drive shaft by taper bolts. Each wheel consists of a steel disk to which is fitted Nimonic-alloy turbine blades, and each blade carries its own shroud. To reduce losses at the blade tips, seals are formed on

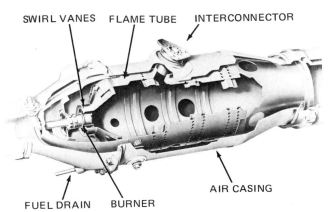

FIG. 22-18 Combustion chamber for the Dart engine.

SWIRL VANES   FLAME TUBE   INTERCONNECTOR

FUEL DRAIN   BURNER   AIR CASING

the shrouds of the HP and IP blades. The root of each blade is of fir-tree shape and fits into a corresponding slot broached in the rim of the disk. The blades are locked to the disk by locking tabs. Labyrinth-type seals are fitted between the stages of the turbine to control the disk-cooling airflows.

Exhaust Unit

The **exhaust unit** which is bolted to the nozzle box consists of two concentric cones joined by three support fairings. Each fairing is secured by setscrews to a sole plate on the outer cone. The interior of the inner cone is vented to the exhaust-gas stream by three circumferentially positioned holes called **pressure balance holes.** Fuel-drain holes are incorporated in the assembly to prevent the accumulation of fuel.

When the engine is installed, the exhaust unit is arranged within a conical shroud with its discharge end centrally located in the jet pipe inlet. An annular gap formed between the discharge end of the unit and the jet pipe inlet creates an ejector effect which draws air into the stream. This air is drawn from the combustion compartments between the exhaust-unit outer cone and its surrounding shroud. A flow of cooling air is thus provided over the whole combustion compartment.

Oil System

The oil system for the Dart engine is shown in Fig. 22-19. The oil tank is an integral part of the engine, consisting of the annular chamber surrounding the first-stage air inlet. The oil cooler is mounted at the top of the tank as shown. During operation, oil is drawn from the standpipe at the bottom of the tank to flow past an oil temperature bulb and thence to the pressure pump. The pump applies pressure to the oil and forces it to all parts of the engine requiring lubrication.

There are four scavenge pumps in the engine oil system. These pumps scavenge oil from the reduction-gear section, the interstage bearing, second-stage compressor rear bearing, and accessory gearbox drive gears and from the turbine bearings. Oil from the scavenge pumps is delivered by a common external pipe on the left side of the air-intake casing to the oil cooler. The oil cooler discharges into the oil tank where the oil is directed over a **deaerator tray** which spreads it out thinly to permit release of included air.

Air released from the oil in the tank passes through a hollow intake web into the reduction-gear section. From there it passes through the hollow high-speed pinion shaft and compressor shafts to the compressor-turbine coupling and out to the auxiliary gearbox drive housing. The first gear of the auxiliary gearbox drive carries a centrifugal breather. The air released by the breather passes to atmosphere through a cast pocket in the top of the rear compressor casing. Any air in the compressor interstage bearing housing is passed to the breather through the holes in the compressor shaft.

High-pressure oil with a maximum of 70 psi [482.65 kPa] is taken to the propeller-control unit (PCU), where the pressure is increased to 670 psi [4619.65 kPa] maximum by the PCU pump. The increased pressure supply is directed by the control-valve assembly of the PCU to the **pitch-change and stop-withdrawal** mechanism of the propeller. The pitch-change and stop-

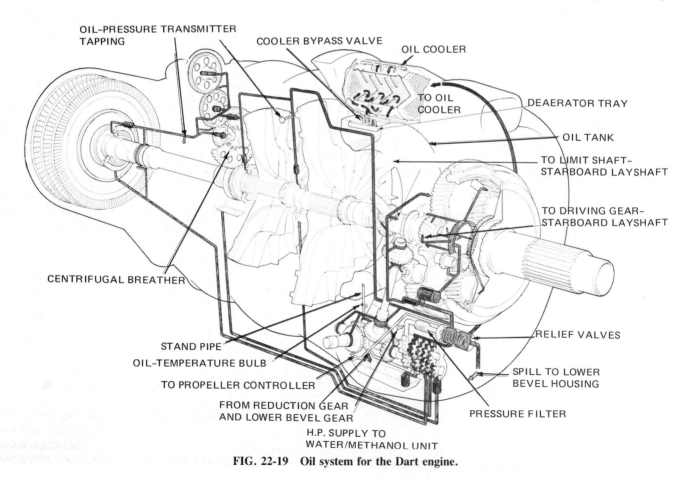

FIG. 22-19 Oil system for the Dart engine.

withdrawal oil supplies are transferred by drilled passages in the air-intake casing and reduction gear to the propeller shaft. In the propeller shaft are spring-loaded sealing bushings to maintain the flow separation on transfer to the concentric oil tubes in the shaft.

The oil system includes features considered standard for engine oil systems, such as filters, pressure-relief valves, oil quantity indicator (dipstick), scavenge oil filters, oil cooler, oil pressure transmitter, oil temperature bulb, and oil-pressure warning light.

Fuel System

The fuel system for the Dart engine is designed to satisfy the basic requirements of the engine for all types of operation. The system must provide full atomization of the fuel over the complete range of fuel flow, control fuel flow according to engine demand, provide engine over-speeding control, ensure a specific flow for a given throttle position, compensate fuel flow for altitude conditions, limit flow to suit the engine power rating, provide a correct idling fuel flow, and provide for complete fuel shutoff when it is desired to stop the engine.

The operation of the fuel pump and fuel-control unit can be understood by examining Fig. 22-20. The fuel pump consists of an engine-driven rotor carrying seven plungers spring-loaded against a circular cam plate. The output of the pump is varied by changing the angle of the cam plate relative to the rotor through the action of a servo piston. The piston assembly is carried in an alloy body which incorporates the inlet and outlet ports. These communicate with the revolving rotor through a fixed valve plate containing two kidney-shaped ports.

As the rotor of the pump revolves around the cam plate, each plunger in turn is extended and receives low-pressure fuel. It then delivers the fuel at high pressure as the plunger is pushed in during its rotation around the inclined face of the cam plate. Since the pump is driven at a fixed ratio to engine speed, the pump output at maximum stroke is proportional to engine speed. Since, for any given rpm, the engine fuel requirement does not coincide with the maximum pump output, the pump stroke must be varied independently of rpm. This variation in fuel flow to suit engine demand is attained by altering the cam-plate angle.

The pump servo, consisting of a spring-loaded piston in a cylinder connected to the cam plate, is integral with the pump. Movement of the servo piston alters the cam-plate angle and the plunger stroke, thus changing fuel flow. The servo piston receives high fuel pressure on both sides, that on the spring side first passing

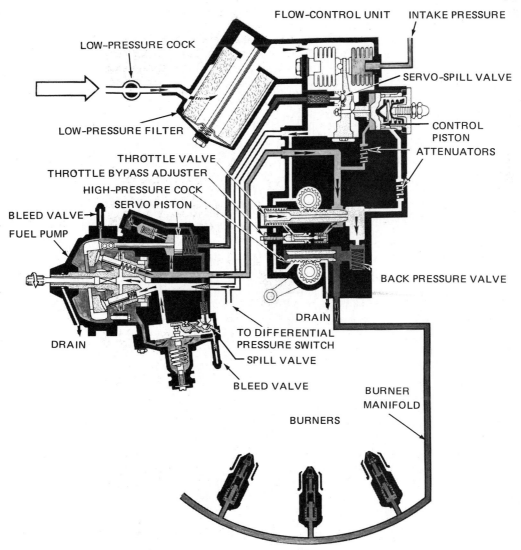

FIG. 22-20 Fuel-control unit and variable pump.

through an orifice. Fuel flow from the spring side of the piston is controlled by a **spill valve.** When the spill valve is open, the fuel pressure is relieved and the pressure on the opposite side of the piston moves the piston in a direction to reduce the angle of the cam plate. This decreases the pump output.

Engine overspeed is controlled by the diaphragm-type governor in the fuel pump. As shown in the diagram, the pump rotor contains passages through which fuel flows into the pump body by centrifugal force. This being true, it is apparent that the pressure within the pump body will vary according to engine rpm. Since the governor diaphragm is exposed on one side to the pump centrifugal pressure, the diaphragm will move when pressure becomes excessive. This is the case when the engine reaches an overspeed condition. As the diaphragm moves, it pushes a lever which releases a spill valve controlling fuel pressure on the spring side of the servo piston and thus reduces the pump cam-plate angle which, in turn, reduces fuel flow. The reduction in fuel flow continues until the engine speed stabilizes at the predetermined overspeed rpm set by the tension spring.

A secondary function of the overspeed governor spill valve is to prevent excessive fuel pressures in the system. Thus, it acts as a relief valve. The overspeed governor spill-valve rocker arm is loaded by a spring which, through its leverage, will maintain the spill valve in a closed position unless there is an excessive rise in pump delivery pressure. If this occurs, the spill valve opens and reduces pump delivery pressure.

In the fuel-flow-control unit a spill valve controls the pump servo according to throttle position. This valve is kept informed of the throttle position by a spring-loaded control piston which senses the fuel flow via pressure signals from upstream and downstream of the throttle valve. Attenuators in the pressure-sensing lines dampen out any pressure fluctuation from the fuel pump. The control piston movement is transmitted by a push-rod to the flexibly mounted lever housing the spill valve.

Under stabilized conditions the fuel-pressure differential across the control piston balances the control-piston spring force. The spill-valve position is thus automatically adjusted so that the pump servo piston selects the correct pump stroke for fuel flow. When the throttle is opened, the pressure differential across the throttle valve decreases and the control piston senses this decrease. The piston moves to close the spill valve, thus causing the pump output to increase until fuel flow is correct for the new throttle setting. The system then stabilizes in the new position.

Fuel-flow adjustment for variations in altitude is accomplished through the action of the intake pressure aneroid bellows shown in the diagram. As altitude increases, the bellows exerts pressure on the spill valve which reduces the pump output. The bellows is so designed that no further action of the bellows to increase fuel flow can take place when ambient pressure reaches 14.7 psi [101.36 kPa.]. This is to prevent the engine from being provided with excessive fuel.

Fuel flow from the control unit passes through the high-pressure cock and thence to the burners in the combustion chambers. These burners, or nozzles, are designed to provide a hollow conical spray of fuel at the forward end of each combustion chamber. The burners include thread-type filters.

Water-Methanol System

In operation under high-ambient-temperature conditions, there is a reduction in engine mass airflow and the fuel flow is reduced by trimming in order to maintain the turbine working temperatures within acceptable limits. This results in a reduction of engine shp which can be restored to takeoff level by injecting a water-methanol mixture into the first-stage compressor through drilled passages in the rotating guide vanes and impeller. Water and methanol from the aircraft tank are fed by a tank pump and electrically operated feed cock to the metering valve of the water-methanol unit. The cockpit selector switch operates both the feed cock and tank pump, and a cockpit light indicates when water and methanol are being supplied. The feed cock is interconnected with the propeller feathering system so that the water-methanol supply is automatically shut off when the propeller is feathered.

The water-methanol mixture used in the Dart engine consists of water containing between 36 and 38 percent of methanol (methyl alcohol) by weight. This is approximately equivalent to 43.8 volumes of methanol and 56.2 volumes of water. The water and methanol must meet rigid specifications of quality and purity.

Starting and Ignition Systems

The starter system for the Dart engine is typical of electric-motor starter systems. The starter motor is energized through relays controlled by a starter switch in the cockpit. The system is interconnected with a vibrator-type, high-energy ignition system to provide for ignition when the engine is started. Overspeed and safety relays are placed in the system to provide for cutoff of the system when the starter reaches the maximum allowable speed.

● THE UNITED AIRCRAFT OF CANADA PT6A TURBOPROP ENGINE

The PT6A turboprop engine, manufactured by United Aircraft of Canada, Ltd., is used in various configurations for a number of aircraft. Typical of these aircraft is the Beech King Air and the De Havilland Twin Otter. The engine provides 550 shp [410 kW] and upward, depending on the particular model and its application. A photograph of the engine is shown in Fig. 22-21.

An exploded view showing the principal components of the PT6A engine is given in Fig. 22-22. The air intake

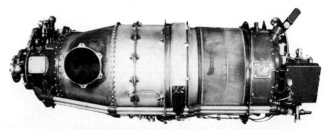

FIG. 22-21 The PT6A turboprop engine. *(Pratt & Whitney Aircraft of Canada)*

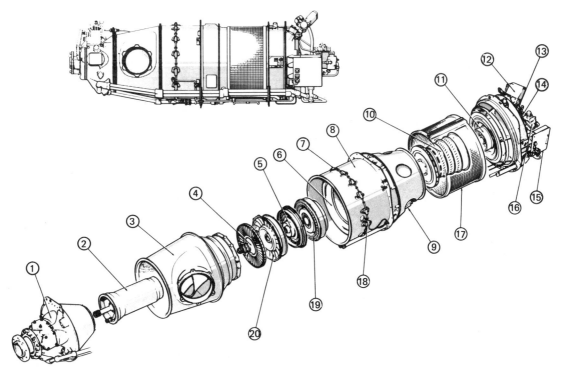

1. Propeller reduction gearbox
2. Power-turbine support housing
3. Exhaust duct
4. Power turbine
5. Compressor turbine
6. Combustion-chamber liner
7. Fuel manifold
8. Gas generator case
9. Compressor bleed valve
10. Compressor assembly
11. Compressor-inlet case
12. Oil to fuel heater
13. Dipstick and filler cap
14. Accessory gearbox
15. Ignition-current regulator
16. Fuel-control unit and pump
17. Air-inlet screen
18. Ignition glow plug
19. Compressor-turbine guide vanes
20. Power-turbine guide vanes

FIG. 22-22 Exploded view of the PT6A engine.

for the engine is at the rear and the exhaust outlet is near the front. The engine includes the principal sections described for other gas-turbine engines: air inlet, compressor, diffuser, combustion chamber, turbine section, and exhaust. A cross section of the PT6A engine is shown in Fig. 22-23.

The PT6A engine is described as a lightweight, **free-turbine** engine, designed for use in fixed-wing or rotary-wing aircraft. The term *free turbine* means that the turbine which drives the output shaft is not mechanically connected to the turbine which drives the compressor.

Inlet air enters the engine through an annular plenum chamber formed by the compressor inlet case. From the inlet the air is directed inward to the three-stage axial compressor and from there to the single-stage centrifugal compressor. The two compressor sections are constructed as one unit. Air from the centrifugal compressor is thrown outward through diffuser pipes and turned 90° to a forward direction before being led through straightening vanes to the combustion chamber.

The **combustion chamber** is formed by the gas-generator case and the rear end of the exhaust duct assembly. The **combustion-chamber liner,** located in the combustion chamber, is an annular, reverse-flow unit with various-sized perforations which provide for the entry of compressed air. The flow of air changes direction to enter the combustion-chamber liner where it reverses direction and mixes with the fuel.

Fuel is injected into the combustion-chamber liner

by way of 14 simplex nozzles supplied by a common manifold. The fuel-air mixture is initially ignited by two ignition glow plugs when starting the engine. The glow plugs protrude into the combustion-chamber liner where they are exposed to the fuel-air mixture. The burned gases expand from the combustion-chamber liner and are reversed in direction before passing through guide vanes to the compressor turbine. The vanes guide the expanding gases to ensure that they impinge upon the turbine blades at the correct angle with minimum loss of energy. The gases continue to expand and pass forward through a second set of vanes to the power turbine.

The compressor and power turbines are located in the approximate center of the engine with their shafts extending in opposite directions. This provides for simplified installation and inspection procedures. Engine exhaust gases are discharged through an exhaust plenum to the atmosphere through exhaust ports.

The **accessory gearcase** is located at the rear of the engine immediately behind the air inlet. The gearcase incorporates appropriate accessory drives and mounting pads. The accessories are driven from the compressor by means of a coupling shaft which extends the drive through a conical tube in the oil tank center section. The oil tank is integral with the compressor inlet case and has a capacity of 2.3 gal (U.S.) [8.71 L].

The propeller shaft is driven from the power turbine through a two-stage, planetary reduction gear having a ratio of 15:1. The power turbine turns at 33,000 rpm; hence, the propeller turns at 2200 rpm.

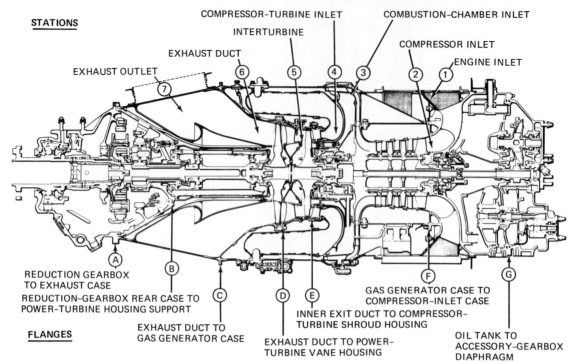

STATIONS

EXHAUST OUTLET
EXHAUST DUCT
INTERTURBINE
COMPRESSOR-TURBINE INLET
COMBUSTION-CHAMBER INLET
COMPRESSOR INLET
ENGINE INLET

REDUCTION GEARBOX
TO EXHAUST CASE
REDUCTION-GEARBOX REAR CASE TO
POWER-TURBINE HOUSING SUPPORT

FLANGES

EXHAUST DUCT TO
GAS GENERATOR CASE

EXHAUST DUCT TO POWER-
TURBINE VANE HOUSING

INNER EXIT DUCT TO COMPRESSOR-
TURBINE SHROUD HOUSING

GAS GENERATOR CASE TO
COMPRESSOR-INLET CASE

OIL TANK TO
ACCESSORY-GEARBOX
DIAPHRAGM

FIG. 22-23   Cross section of the PT6A engine, showing flanges.

Compressor Inlet Case

The **compressor inlet case,** shown in Fig. 22-24, is a circular aluminum-alloy casting, the front of which forms a plenum chamber for the passage of compressor inlet air. The rear portion consists of a hollow compartment which forms the integral oil tank. The intake is screened to preclude the ingestion of foreign objects.

The no. 1 bearing, bearing support, and airseal are contained within the compressor inlet-case centerbore. The bearing support is secured to the inlet-case centerbore flange by four bolts. A special nut and shroud washer retains the no. 1 bearing outer race in its support housing. A puller groove is provided on the rear face of the no. 1 bearing split inner race to facilitate its removal. The compressor assembly and the no. 1 bearing area are shown in Fig. 22-25. An oil nozzle,

fitted at the end of a cored passage, provides lubrication to the rear of no. 1 bearing at approximately the one o'clock position. Other cored passages are provided for pressure and scavenge oil.

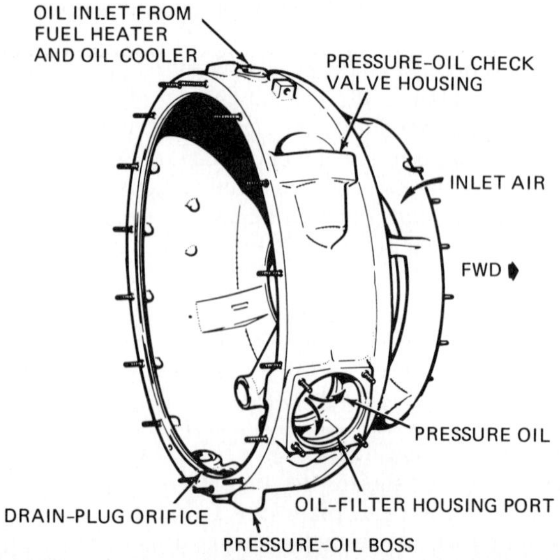

The oil pressure-relief valve and the engine main oil filter, with check-valve and bypass-valve assemblies, are located on the right side of the inlet case at the one and three o'clock positions, respectively. A fabricated conical tube complete with preformed packings is fitted in the center of the oil tank compartment to provide a passage for the coupling shaft which extends the compressor drive to the rear accessories. The pressure oil pump, driven by an accessory drive, is located in the bottom portion of the integral oil tank and is secured by four bolts to the accessory diaphragm.

Compressor Rotor and Stator Assembly

The **compressor rotor and stator assembly** shown in the drawing of Fig. 22-25 consists of a three-stage axial rotor, three interstage spacers, three stators, and a single-stage centrifugal impeller and housing. The first-stage rotor blades are made of titanium to improve impact resistance while the second and third stages are made of stainless steel. The rotor blades are dovetailed into their respective disks with a clearance between the blade and disk which causes a clicking sound during compressor rundown. The clearance is allowed to accommodate metal expansion due to the high temperatures.

Axial movement of the rotor disks is limited by the interstage spacers placed between the disks. The airfoil cross section of the first-stage blades differs from those of the second and third stages, the latter two being identical. The length of the blades differs in each stage, decreasing from the first to the third stage.

The first- and second-stage stator assemblies each contain 44 vanes and the third stage contains 40 vanes. Each set of stator vanes is held in position by a circular

OIL INLET FROM
FUEL HEATER
AND OIL COOLER

PRESSURE-OIL CHECK
VALVE HOUSING

INLET AIR

FWD

PRESSURE OIL

DRAIN-PLUG ORIFICE

OIL-FILTER HOUSING PORT

PRESSURE-OIL BOSS

FIG. 22-24   Compressor inlet case.

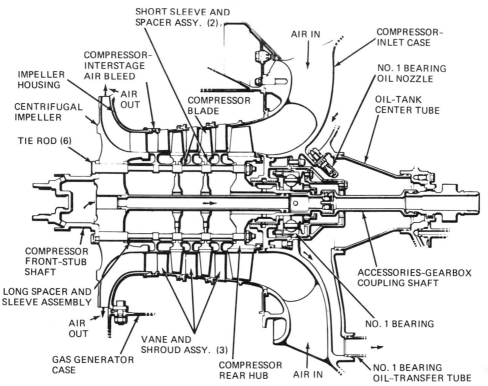

FIG. 22-25 Compressor assembly and no. 1 bearing area.

ring with the vane outer ends protruding through and brazed to the ring. Part of each ring also provides the shrouds for the adjacent set of compressor blades.

The **compressor front stubshaft,** the **centrifugal impeller,** and the **impeller housing** are positioned in that order, followed alternately by an interstage spacer, a stator assembly, and a compressor rotor disk. These are stacked and securely held together by six numbered tie-rods. A series of slots and lugs, provided on the impeller housing and compressor shroud assemblies, interlock and prevent rotation of the assembly. The impeller housing is in turn secured in the gas generator case by eight eccentric bolts. The compressor front stubshaft consists of a hollow steel forging machined to accommodate the no. 2 bearing air-seal and the no. 2 bearing assembly. The no. 2 bearing is a roller type which supports the front of the compressor and the attached turbine in the gas-generator case.

The **compressor rear hub** is an integral part of the first-stage compressor rotor disk. It consists of a steel forging machined with an extended hollow shaft to accommodate the no. 1 bearing airseal and the no. 1 bearing. The no. 1 bearing which supports the rear of the compressor assembly in the inlet case is a ball-type bearing. A short, hollow steel coupling, with ball lock and internal splines at each end, extends the compressor drive to the **accessory input-gear shaft.** The ball lock, incorporated at the front end of the coupling, prevents end thrust on the two accessory input-gear-shaft roller bearings.

The complete compressor and stator assembly is fitted into the center rear portion of the **gas-generator case** which forms the compressor housing. The assembly is secured in this position by the impeller housing at the front and the compressor inlet at the rear.

Gas-Generator Case

The **gas-generator case,** shown in Fig. 22-22, is attached to the front flange of the compressor inlet case and encloses both the compressor and the combustion section. The case consists of two stainless-steel sections fabricated into a single structure. The rear inlet section provides housing support for the compressor assembly. The no. 2 bearing with two airseals is positioned in the center bore of the gas-generator case. The bearing has a flanged outer race secured in the support housing by four bolts. The front and rear airseal stators with their spiral-wound gaskets are each secured in the center bore of the case by eight bolts. An oil nozzle with two jets, one in the front and the other in the rear of the bearing, provides lubrication. The 14 radial vanes brazed inside the double-skin center section of the gas-generator case provide a pressure increase to the compressed air as it leaves the centrifugal impeller. The compressed air is then directed through 70 straightening vanes welded inside the gas-generator-case diffuser and out to the combustion-chamber area through a slotted diffuser outlet baffle.

The front section of the gas-generator case forms an outer housing for the combustion-chamber liner. It consists of a circular stainless-steel structure provided with mounting bosses for the 14 fuel-nozzle assemblies and common manifold. Mounting bosses are also provided at the six o'clock position for the fuel dump valve and combustion-chamber front and rear drain valves. Two ignition glow plugs are located at the four and eight o'clock positions. The plugs protrude into the combustion-chamber liner to ignite the fuel-air mixture. Three equally spaced pads, located on the outer circumference of the gas-generator case, provide accommodation for flexible-type engine mounts. The compressor bleed-valve outlet port is located at the eight o'clock position.

Combustion-Chamber Liner

The **combustion-chamber liner,** as shown in Fig. 22-26, is of the reverse-flow type and consists primarily of an annular, heat-resistant steel liner open at one end. A series of straight, plunged, and shielded perforations allows air to enter the liner in a manner designed to provide the best fuel-ratios for starting and sustained combustion. Direction of airflow is controlled by cooling rings especially located opposite the perforations. The perforations ensure an even temperature distribution at the compressor turbine inlet. The domed front end of the combustion-chamber liner is supported inside the gas-generator case by 7 of the 14 fuel-nozzle sheaths. The rear of the liner is supported by sliding joints which fit into the inner and outer exit duct assemblies. The duct assemblies form an envelope which effectively changes the direction of the gasflow by providing an outlet in close proximity to the compressor turbine vanes. The outlet duct and heat-shield assembly is attached to the gas-generator case by seven equally spaced support brackets located in the case. The heat shield forms a passage through which compressor discharge air is routed for cooling purposes. A scalloped flange on the rear of the outer duct locks in the support brackets and secures the assembly. The center section of the assembly is bolted with the compressor-turbine guide-vane support to the center bore of the gas-generator case.

Turbine Section

The turbine rotor section, shown in Fig. 22-27, consists of two separate single-stage turbines located in the center of the gas-generator case and completely enveloped by the annular combustion-chamber liner. It must be noted that the two turbines are mounted on shafts which extend in opposite directions. The rear shaft drives the compressor and the forward shaft drives the propeller through the reduction-gear assembly.

Compressor-Turbine Vanes

The compressor-turbine vane assembly (guide vanes) consists of 29 cast-steel vanes located between the combustion-chamber exit ducts and the compressor turbine. The vanes are cast with individual dowel pins on the inner platforms which fit in the outer circumference of the vane support. Sealing is ensured by a ceramic fiber cord packing located on the outer diameter of the support which is secured to the center bore of the gas-generator case by eight bolts. The outer platforms of the vanes are sealed with a chevron packing and fit into the shroud housing and the exit duct. They are secured to the center bore by 12 bolts. The shroud housing extends forward and forms a runner for two **interstage sealing rings.** The interstage sealing rings provide a power seal and an internal mechanical separation point for the engine. Fourteen compressor-turbine shroud segments located in the shroud housing act as a seal and provide a running clearance for the compressor turbine.

Compressor Turbine

The compressor turbine consists of a two-plane, balanced turbine disk with blades and classified weights. This turbine drives the compressor in a counterclockwise direction. The two-plane, balanced assembly is secured to the front stubshaft by a simplified center lock bolt. A master spline is provided to ensure that the disk assembly is always installed in a predetermined position to retain proper balance. The disk embodies a reference circumferential groove to provide for checking disk growth when required. The 58 blades in the compressor-turbine disk are secured in fir-tree serrations machined in the outer circumference of the disk and held in position by individual tubular rivets. The blades are made of cast-steel alloy and embody **squealer tips.** A squealer tip is designed to cause a minimum amount of pickup if the blade should come into contact with the shroud segments during operation.

The required number of classified weights is determined during balancing procedures. These weights are riveted to the appropriate flanges machined on the turbine disk. A small rotor machined on the rear face of the turbine disk provides a sealing surface to control the flow of turbine-disk cooling air.

Interstage Baffle

The compressor turbine is separated from the power turbine by an **interstage baffle** which prevents dissipation of turbine gases and consequent transmission of heat to the turbine disk faces. The baffle, shown in Fig. 22-27, is secured to and supported by the power-turbine ring. The center section of the baffle includes small circular-lipped flanges on the front and rear faces. The flanges fit over mating rotor seals machined on the respective turbine disk faces to provide control of cooling airflow through the perforated center of the baffle.

Power-Turbine Guide Vanes

Between the compressor turbine and the power turbine is a ring which holds the **power-turbine guide vanes.**

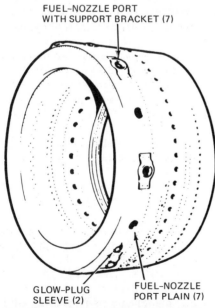

FUEL–NOZZLE PORT
WITH SUPPORT BRACKET (7)

GLOW–PLUG SLEEVE (2)   FUEL–NOZZLE PORT PLAIN (7)

FIG. 22-26   Combustion-chamber liner.

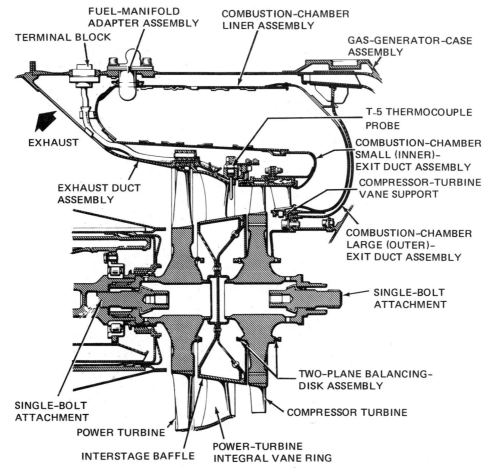

FIG. 22-27 Turbine rotor section.

Depending on the model of the engine, the vanes are separately cast and fitted into the interstage baffle rim by means of integral dowel-pin platforms or they are cast integrally with the turbine-vane ring. The position of the vane ring can be seen in Fig. 22-27. The **stator housing** with the enclosed vane assembly is bolted to the exhaust duct and supports two sealing rings at flange D (see Fig. 22-23). The rings are self-centering and held in position by retaining plates bolted to the rear face of the stator housing.

Power Turbine

The **power-turbine disk** assembly consists of a turbine disk, blades, and classified weights and drives the reduction gearing through the power-turbine shaft in a clockwise direction. The disk is manufactured with close tolerances and incorporates a reference circumferential groove to permit disk-growth check measurements. The turbine disk is splined to the turbine shaft and secured by a single center lock bolt and keywasher. This arrangement is shown in Fig. 22-27.

A master spline ensures that the power-turbine disk is always installed in a predetermined position to retain the original balance. The required number of classified weights is determined during balancing procedures and they are riveted to a special flange located on the rear face of the turbine disk. The power-turbine blades differ from those of the compressor turbine in that they are cast complete with notched and shrouded tips. The 41 blades are secured by fir-tree serrations machined in the rim of the turbine disk and held in place by means of individual tubular rivets. The blade tips rotate inside a double knife-edge shroud and form a continuous seal when the engine is running. This reduces tip leakage and increases turbine efficiency.

Support Bearings

The power-turbine disk and shaft assembly is supported and secured in the **power-turbine shaft housing** by two bearings. The no. 3 bearing is a roller type and can be seen at the rear of the power-turbine shaft in Fig. 22-28. The no. 4 bearing is a ball type and is shown in the drawing at the forward end of the power-turbine shaft. The no. 3 bearing includes an inner race and a flanged outer race. The inner race is secured on the shaft together with the power-turbine airseal and power-turbine disk by the center lock bolt. The outer race is secured inside the power-turbine shaft housing by four bolts and tab-washers.

The no. 4 bearing includes a split inner race and a flanged outer race. The split inner race is secured on the front of the power-turbine shaft, together with the first-stage reduction-gear coupling and a coupling positioning ring, by a keywasher and spanner nut. A puller groove is incorporated on the front half of the split inner race to facilitate removal. The flanged outer race is secured inside the power-turbine shaft housing by four bolts and tab-washers.

Reduction Gearbox

The PT6A engine is equipped with a gearbox at the front for the reduction of engine speed to a level suit-

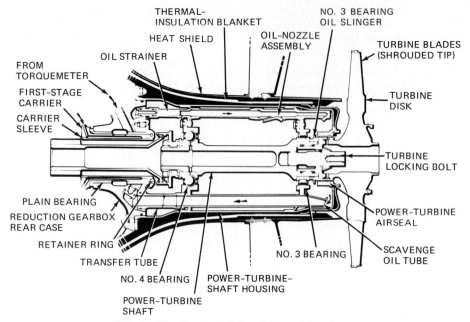

FIG. 22-28 **Power-turbine shaft and bearings.**

able for driving a propeller or a power shaft. In this section we are concerned only with the engine which drives a propeller and in which the reduction gear ratio is 15:1. This means that the engine speed of 33,000 rpm is reduced to 2200 rpm at the propeller drive shaft.

The reduction gearbox shown in Fig. 22-29 consists of two magnesium-alloy castings bolted to the front flange of the exhaust duct. The first stage of reduction is contained in the rear case.

The first-stage reduction sun gear consists of a short, hollow steel shaft which has an integral spur gear at the front end and is externally splined at the rear end. The external splines engage the retainer coupling by which the first-stage sun-gear shaft and the power-turbine shaft are joined.

The first-stage ring gear is located in helical splines provided in the first-stage reduction-gearbox rear case. The torque developed by the power turbine is transmitted through the sun gear and planet gears to the ring gear, which is opposed by the helical splines. Thus, the ring gear cannot rotate but the planet-gear carrier is caused to rotate. The ring gear moves axially a short distance because of the helical splines, and this movement is used to operate the torquemeter. The torquemeter will be described later.

The second stage of reduction gearing is contained

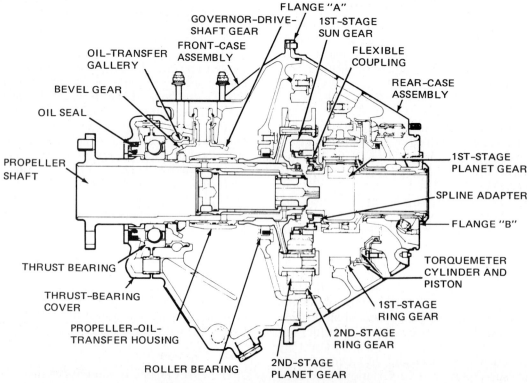

FIG. 22-29 **Two-stage reduction gearbox.**

in the **reduction-gearbox front case**. The first-stage planet carrier is attached to the second-stage sun gear by a flexible coupling which also serves to dampen any vibrations between the two rotating masses. The second-stage sun gear drives five planet gears in the second-stage carrier. A second-stage ring gear is fixed by splines to the reduction-gearbox front case and is secured by three bolted retaining plates. The second-stage carrier is in turn splined to the propeller shaft and secured by a retaining nut and shroud washer. A flanged bearing assembly, secured by bolts to the case, provides support for the second-stage carrier and the propeller shaft. An oil transfer tube and nozzle assembly fitted with preformed packings is secured within the propeller shaft to provide lubrication for the no. 4 bearing.

The accessories located on the reduction-gearbox front case are driven by a bevel gear mounted on the propeller shaft behind the thrust-bearing assembly. The propeller thrust loads are absorbed by a flanged ball bearing located in the front face of the reduction-gearbox center bore. The bevel drive gear, adjusting spacer, thrust bearing, and seal runner are stacked and secured to the propeller shaft by a single spanner nut and key-washer. The thrust-bearing cover is secured to the front of the reduction gearbox and incorporates a removable oil-seal retaining ring to facilitate removal of the oil seal.

Torquemeter

As mentioned previously, a **torquemeter** is used to determine the torque force being exerted by the engine. The quantity indicated by the torquemeter is used to determine power output.

The torquemeter for the PT6A engine operates in a manner similar to those previously described. The mechanism consists of a torquemeter cylinder, torquemeter piston, valve plunger, and spring (see Fig. 22-30).

Rotation of the reduction-gear first-stage ring gear is resisted by the helical splines which impart an axial movement to the ring gear and this movement is transmitted to the torquemeter piston. This in turn moves a valve plunger against a spring, opening a metering orifice and allowing an increased flow of oil to enter the torquemeter chamber. This movement will continue until the oil pressure in the torque chamber balances the force on the ring gear caused by the torque being absorbed by the gear. Any change in power-control-lever setting will recycle the sequence until a state of equilibrium is again achieved.

Hydraulic lock in the torquemeter is prevented by allowing the oil to bleed continuously from the pressure chamber into the reduction-gear casing through a small bleed hole provided in the top of the torquemeter cylinder.

Because the external oil pressure within the reduction gearbox may vary and affect the total oil pressure on the torquemeter piston, the internal pressure is measured. The difference between the torquemeter oil pressure and the reduction-gearbox internal oil pressure accurately indicates the torque being produced. The two pressures are internally routed to bosses located on the top of the reduction-gearbox front case where connections can be made to suit individual cockpit instrumentation requirements.

Power-Turbine Support Housing

The **power-turbine support housing** consists of a fabricated steel cylindrical unit attached to the reduction-gearbox rear case by 12 studs. The housing provides support for the power-turbine shaft assembly and two bearings, as shown in Fig. 22-28. A labyrinth-type seal, secured at the rear of the housing, prevents oil leakage into the power-turbine section. An internal oil transfer tube has four oil nozzles and provides front and rear lubrication to the no. 3 and no. 4 bearings. A scavenge tube, secured inside the housing at the six o'clock position, transfers bearing scavenge oil to the front of the engine.

Exhaust Duct

The **exhaust duct**, shown in Fig. 22-31, consists of a divergent, heat-resistant steel unit provided with two

1. Gearbox pressure
2. Torquemeter pressure
3. Oil-control piston
4. Control spring
5. Metering orifice
6. Piston
7. Torquemeter chamber
8. First-stage planet gear
9. First-stage ring gear
10. Helical splines
11. Casting
12. Cylinder
13. Bleed hole

FIG. 22-30 Torquemeter.

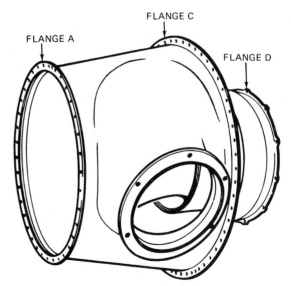

FIG. 22-31 Exhaust duct assembly.

outlet ports, one on each side of the case. The duct is attached to the front flange of the gas-generator case and consists of inner and outer sections. A reinforcing ring is provided at the rear of the exhaust duct at flange D. This consists of a scalloped stainless-steel ring machined in two halves and coupled to form a circle by two equally spaced clevis-pin joints. The power-turbine stator housing is secured to flange D by 12 bolts which screw into and also secure the reinforcing ring. The outer conical section which has two flanged exhaust outlet ports forms the outer gas path and also functions as a structural member to support the reduction gearbox. The inner section forms the inner gas path and provides a compartment for the reduction-gearbox rear case and the power-turbine support housing. A removable sandwich-type heat shield insulates the power-turbine support housing from the hot exhaust gases. A short no. 3 bearing cover and spacer are secured at the rear of the power-turbine support housing by a retaining ring. A drain passage, located at the six o'clock position in the exhaust duct, enables residual fuel accumulation in the exhaust duct during engine shutdown to drain into the gas-generator case where it is discharged overboard through the front drain plug.

Accessory Gearbox

Accessories for the PT6A engine are driven from the **accessory gearbox** located at the rear of the engine. The gearbox shown in Fig. 22-32 consists of two magnesium-alloy castings, both of which are attached to the rear flange of the compressor-inlet case by 16 studs. The front casting, which incorporates front and rear preformed packings, forms an oil-tight diaphragm between the oil tank compartment of the inlet case and the accessory drives. The diaphragm also provides support for the accessory drive-gear bearings, seals, and the main pressure oil pump secured to the diaphragm by four bolts. The diaphragm is attached to the accessory gearbox housing by four countersunk screws and nuts located at the fourth, eighth, four-

teenth, and eighteenth positions in clockwise rotation, assuming that the first is at the twelve o'clock position.

The rear casting of the gearbox forms a cover and provides support bosses for the accessory drive bearings and seals. The internal oil scavenge pump is secured inside the housing and a second scavenge pump is externally mounted. Mounting pads and studs are provided on the rear face for the combined starter-generator, the fuel-control unit with the sandwich-mounted fuel pump, and the N_g tachometer. (N_g is the speed of the compressor turbine.) A large access plug located below the starter-generator mounting pad provides passage for a puller tool which must be used to disengage and hold the ball-locked coupling shaft and input-gear shaft during disassembly. Three additional pads are available for optional requirements. Accessory drives are supported on similar roller bearings fitted with garter-type oil seals. An oil-tank filler cap with an integral dipstick is located at the eleven o'clock position on the rear housing to facilitate servicing of the oil system. A centrifugal oil separator mounted on the starter-generator drive-gear shaft separates the oil from the engine breather air in the accessory gearbox housing. A cored passge in the accessory diaphragm connects the oil separator to an external mounting pad located at the twelve o'clock position on the rear housing. A carbon face seal located on the front of the gear shaft in the accessory diaphragm prevents pressure leakage through the bearing assembly.

Air Systems

The PT6A engine has three separate air bleed systems: a compressor bleed control, a bearing compartment airseal and bleed system, and a turbine-disk cooling system. A fourth system is available as an optional source of high-pressure air for use in operating auxiliary air frame equipment. A blanked mounting flange located on the gas-generator case is provided for external connections.

Compressor Bleed Valve. The compressor bleed valve automatically opens a port in the gas-generator case to spill interstage compressed air ($P_{2.5}$) thereby providing antistall characteristics at low engine speeds (less than 80 percent N_g). The port closes gradually as higher engine speeds are attained.

The compressor bleed valve, located on the gas-generator case at the seven o'clock position and secured by two bolts, consists of a piston-type valve in a ported housing. This valve is illustrated in Fig. 22-33. The piston assembly is supported in the bore of the housing and is guided by a seal support plate, guide pin, and guide-pin bolt, the latter holding the piston assembly together. A rolling diaphragm permits the piston full travel in either direction, to open or close the port, while at the same time effectively sealing the compartment at the top of the piston. A port in the gas-generator case provides a direct passage for the flow of compressor interstage air ($P_{2.5}$) to the bottom of the bleed-valve piston.

Compressor discharge air (P_3) is tapped off and applied to the bleed valve through a nozzle (fixed orifice) in the bleed-valve cover, then passed through an intermediate passage and out to the atmosphere through

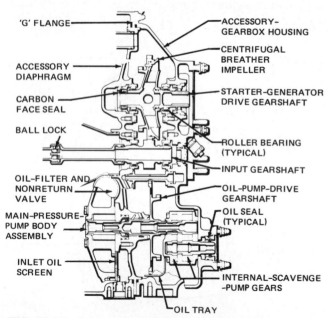

FIG. 22-32   Cross section of the accessory gearbox.

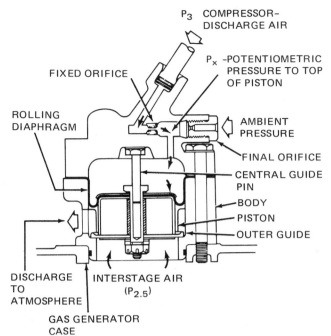

FIG. 22-33 Schematic drawing of the compressor bleed valve.

a metering plug (convergent-divergent orifice). The control pressure (P_x) between the two orifices acts upon the upper side of the bleed-valve piston, so that when P_x is greater than $P_{2.5}$, the bleed valve closes. In the closed position, the interstage air port is sealed by the seal support plate which is forced against its seat by the effect of P_x. Conversely, when P_x is less than $P_{2.5}$, the bleed valve opens and allows interstage pressure ($P_{2.5}$) to be discharged to the atmosphere. The piston is prevented from closing off the P_x feed line to the

upper section of the valve chamber and from damaging the piston assembly by the stop formed by the hexagon of the guide pin, which is screwed into the valve cover, and the hexagon of the piston guide-pin bolt.

For calibration purposes, P_x is measured by installing a suitable fitting in the valve cover and adjustment is made by varying the diameter of the metering plug and/or nozzle orifice. The calibration should be such that P_x exceeds $P_{2.5}$ when the overall compressor ratio (P_3/P_2) is 3.70. P_2 is the compressor inlet pressure.

Bearing Compartment Seals, Turbine Cooling, and Air-Bleed Systems. Pressure air is utilized to seal the first, second, and third bearing compartments and also to cool both the compressor and free turbines. The airflow for these purposes is shown in Fig. 22-34. The pressure air is used in conjunction with airseals which establish and control the required pressure gradients. It should be remembered that air pressure is used in turbine engines to prevent oil from leaking into areas where it is not required or where it would be detrimental to the operation of the engine. The airseals used on the engine consist of two separate parts. One part takes the form of a plain rotating surface. The corresponding part consists of a series of stationary expansion chambers (the **labyrinth**) formed by deep annular grooves machined in the bore or a circular seal. The clearance between the rotating and stationary parts is kept as small as possible consistent with mechanical safety.

Compressor interstage air is utilized to provide a pressure drop across the airseal located in the front of the no. 1 bearing. The air is led through perforations in the rim of the compressor long spacer and sleeve assembly into the center of the rotor. It then flows rearward through passages in the three compressor

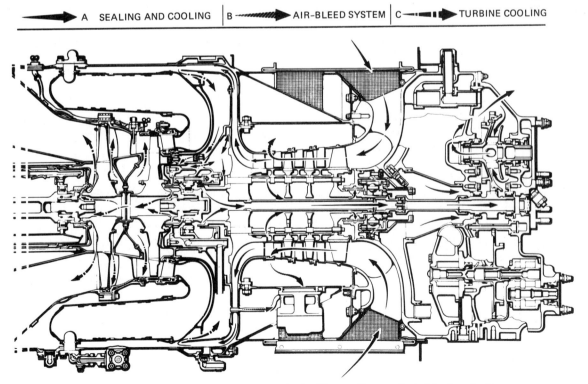

FIG. 22-34 Bearing compartment seals, turbine cooling, and air-bleed systems.

disks and out to an annulus machined in the center of the airseal via passages in the compressor rear hub. The pressure air is allowed to leak through the labyrinth and thereby provides the required pressure seal.

The no. 2 bearing is protected by an airseal at the front and at the rear of the bearing. Pressure air for this system is bled either from the centrifugal impeller tip or, depending on engine speed, from the labyrinth seal connecting it to the turbine-cooling air system. The air flows through passages in the no. 2 bearing support, equalizing the air pressure at the front and rear of the bearing compartment and thereby ensuring a pressure seal in the front and rear labyrinths.

The compressor- and power-turbine disks are both cooled by compressor discharge air bled from the slotted diffuser baffle area down the rear face of the outer exit duct assembly. It is then metered through holes in the compressor-turbine vane support into the turbine baffle hub where it divides into three paths. Some of the air is metered to cool the rear face of the compressor-turbine disk and some to pressurize the bearing seals. The balance is led forward through passages in the compressor-turbine hub to cool the front face of the compressor turbine. A portion of this cooling air is also led through a passage in the center of the interstage baffle where the flow divides. One path flows up the rear face of the power-turbine disk while the other is led through the center of the disk hub and out through drilled passages in the hub to the no. 3 bearing airseals and front face of the power-turbine disk.

The cooling air from both turbine disks is dissipated into the main gas stream flow to the atmosphere. The bearing cavity leakage air is scavenged with the scavenge oil into the accessory gearbox and vented to the atmosphere through the centrifugal breather.

Lubrication System

The lubrication system for the PT6A engine is shown in Fig. 22-35. As explained previously in this section, the oil tank is integral with the accessory gearbox. The lubricant used for the engine is type I, MIL-L-7808, or type II, MIL-L-23699. The Pratt & Whitney specification for the oil is PWA 521. The oil-tank capacity is 2.3 gal (U.S.) or 1.92 gal (Imp.) [8.71 L] with an expansion space of 0.7 gal (U.S.) or 0.58 gal (Imp.) [2.64 L].

During operation, oil from the oil tank is picked up by the **main oil pressure pump** and forced through the **main oil filter.** The filter is a disposable cartridge type or a cleanable metal type. If the filter should be clogged, the oil flows through a bypass valve. A check valve is incorporated in the end of the filter housing, thus making it possible to change the filter without draining the oil tank.

The **oil pressure-relief valve** is located on the gearcase above the filter cover. The relief valve and filter are shown in Fig. 22-36. The pressure-relief valve is adjusted at the factory and no further adjustment is usually required. Oil released by the relief valve is returned to the tank. Figure 22-36 shows the location of the oil-tank filler cap, oil-tank drain plug, and other units on the accessory gearcase.

As can be seen in Fig. 22-35, lubricating oil is carried to all parts of the engine through a system of tubes and passages. Thus, pressure oil reaches all bearings, gears, and other moving parts requiring lubrication. The oil reduces friction, cools the engine parts, and cleans the engine by removing foreign particles either deposited in the bottom of the oil tank or picked up by the oil filter. Scavenge oil is drained into sumps and is pumped back to the oil tank. Oil from the reduction-case scavenge pump and the accessory-case scavenge pump is routed through an oil cooler if the engine installation in the aircraft requires such a cooler. The oil cooler is supplied by the airframe manufacturer.

Some installations include a fuel heater through which the fuel is heated by engine oil. The **fuel heater** is a simple heat exchanger with an automatic temperature control. A ball valve is closed by a vernatherm plunger when the fuel is below the desired temperature and this causes the hot oil from the engine to flow through the heater. Heat from the oil passes through the metal walls of the heater passages to the fuel.

Instruments for monitoring the oil system include a

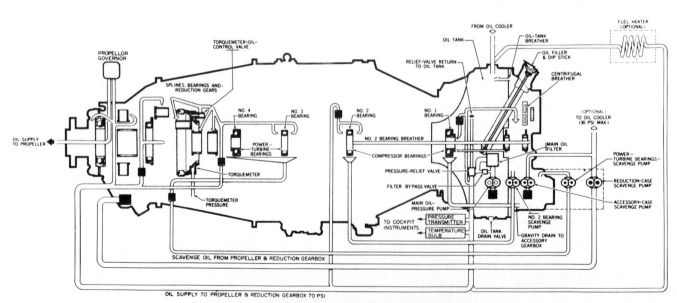

FIG. 22-35 Lubrication system for the PT6A engine.

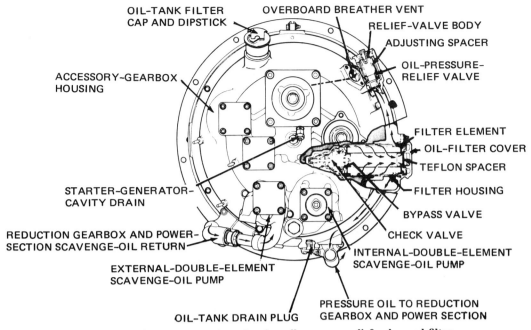

FIG. 22-36 Accessory gearbox showing oil pressure-relief valve and filter.

temperature gage and a pressure gage. The normal operating pressure is 65 to 85 psig [448.18 to 586.08 kPa] with a minimum of 40 psig [275.8 kPa]. Normal operating temperature of the oil is 74 to 80°C (166 to 176°F).

Ignition System

The PT6A engine employs **glow plugs** for ignition rather than the high-energy spark ignitors employed on the majority of gas-turbine engines. The system provides the engine with ignition capable of quick light-ups at extremely low ambient temperatures. The basic system consists of an ignition-current regulator with a selectable circuit to the two sets of ballast tubes, two shielded ignition cable and clamp assemblies, and two ignition glow plugs.

The **ignition-current regulator** is usually secured by three bolts to the accessory gearbox; however, it may be airframe-mounted if necessary. The regulator box has a removable cover and contains four ballast tubes. The circuit for the regulator and system is shown in Fig. 22-37. Each ballast tube consists of a pure iron filament surrounded by helium and hydrogen gases enclosed in a glass envelope sealed to an octal base. The iron filament has a positive coefficient of resistance; that is, the resistance increases as temperature increases. Thus, when current flows through the filament, the temperature rises and the resistance increases to reduce the current flow. At low temperatures the ballast-tube resistance is low, which compensates for power losses due to low temperature. As the temperature rises, resistance increases, thus stabilizing the glow-plug current. Each glow plug is wired in series with two parallel-connected ballast tubes and either one or both of the glow plugs can be selected for light-up. Ballast tubes provide an initial surge of current when the system is turned on and a lower stabilized current after approximately 30 s: This characteristic provides rapidly heated glow plugs for fast lightups.

The ignition glow plug consists of a helical heating element fitted into a short plug body. The plugs are secured to the gas-generator case in threaded bosses provided at the four and eight o'clock positions. The heating element lies slightly below the end of the plug body. Four holes, equally spaced on the periphery of the plug body, lead into an annulus provided below the coil. During starting, the fuel sprayed by the fuel nozzles runs down along the lower wall of the combustion-chamber liner into the annulus. The fuel is vaporized and ignited by the hot coil element which heats up to approximately 1316°C [2400°F]. This is the yellow-heat range, which is sufficient to ignite fuel vapors instantly. The four air holes bleed compressor discharge air from the gas-generator case into the plug body and then past the hot coil into the combustion chamber to produce a "hot streak" or a torching effect which ignites the remainder of the fuel-air mixture in the chamber. The

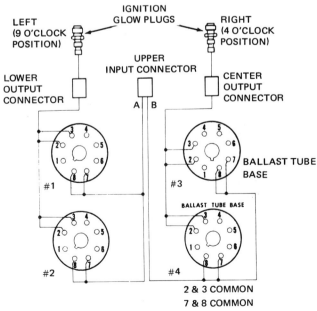

FIG. 22-37 Ignition-current regulator circuit.

air also serves to cool the coils when the engine is running with the ignition system turned off.

Fuel System

The fuel system for the PT6A turboprop engine includes a gear-type or a vane-type fuel pump with a 74-μm inlet filter, a **fuel-control unit** (FCU), a **temperature compensator,** a **power-turbine governor,** an **automatic fuel dump valve,** and a **fuel manifold adapter assembly** which includes the fuel nozzles. The function of the system is to provide the correct amount of fuel for all operating conditions and to limit fuel as necessary to avoid damage to the engine, stall, and flameout.

The fuel pump is mounted on the accessory gearbox and is driven through a splined coupling. Another splined coupling shaft extends the drive to the FCU, which is bolted to the rear face of the fuel pump. Fuel from the aircraft boost pump enters the fuel pump through the 74-μm filter and flows into the pump chamber. Fuel from the pump is delivered at a high pressure to the FCU through a 10-μm filter. Both the inlet filter and the outlet filter are provided with bypass valves to permit fuel flow if the filters should become clogged. The fuel-control unit is classed as a pneumatic type since the primary actuating element is a pair of bellows which respond to varying air pressures. The controlling parameters for the fuel-control unit are *power-lever position, compressor discharge pressure* (P_3), *compressor inlet temperature* (T_2), *and compressor rpm* (N_g). An additional governor supplies an input signal to prevent overspeeding of the power turbine when the propeller governor is not in control of power-turbine speed (N_f).

The **temperature compensator** receives inputs from the inlet temperature and compressor discharge pressure. The effect of the temperature is to modify the P_3 pressure delivered to the fuel-control **differential bellows** through the N_g governor. The N_g governor varies the pressures delivered to the differential bellows, depending on whether the N_g speed is correct as required by the setting of the power-control lever.

The differential bellows moves the metering valve to deliver fuel to the fuel nozzles. The pressure differential across the metering valve is maintained at a constant value; hence, fuel flow is proportional to the opening of the metering valve.

An **automatic fuel-dump valve** is provided to drain fuel from the fuel manifold in the combustion chamber when the engine is shut down. Drains are also provided to release fuel from the bottom of the combustion chamber. The drains eliminate the possibility of fuel collecting in the combustion chamber at shutdown.

● THE LYCOMING T53 TURBOSHAFT ENGINE

The Lycoming T53 turboshaft engine was initially manufactured in 1958 and since that time the family of T53 engines and others has grown in number as well as performance capabilities without compromising reliability. T53 engines are manufactured for use as turboshaft engines for helicopters and as turboprop engines for fixed-wing aircraft. The engine is also used in marine and industrial fields.

For this section, we shall discuss the T53-L-13B turboshaft engine illustrated in Fig. 22-38. The performance and specifications of the engine are as follows:

Rated shaft power (30 min)	1400 hp [1044 kW] at s. level, stand. day (59°F) [15°C]
Maximum shaft power (continuous)	1250 hp [932 kW]
Specific fuel consumption at rated power	0.58 lb/shp/h[352.8 g/kW/h]
Maximum allowable oil consumption	0.14 gal/h [0.38 kg/h]
Maximum gas-producer rpm	25,600
Maximum output rpm	6640
Reduction-gear ratio (power turbine—output shaft)	3.2105:1
Compression ratio	7.2:1
Diameter	23.00 in [58.42 cm]
Length	47.60 in [120.90 cm]
Frontal area	2.88 ft² [0.27 cm²]
Weight	540 lb [245 kg]
Power-weight ratio	2.59 shp/lb [4 kW/kg]
Oil specification	MIL-L-23699 and -7808, or equivalent
Fuel specification	MIL-F-5264 and MIL-F-46005A

Engine Description

The T53-L-13B engine, shown in Fig. 22-39, is a free-power turbine designed primarily for helicopter applications. However, because of its durability and reliability, other uses have developed. *Free turbine* means that the power turbines are not mechanically coupled to, or limited by the speed of, the compressor drive turbines. The engine consists of the *air-inlet housing,* a *carrier and gear assembly,* a five-stage axial one-stage centrifugal *compressor,* an *air diffuser,* a reverse-air-flow combustion chamber, a two-stage *gas-producer turbine,* a two-stage *power turbine,* an *exhaust diffuser,* and a *power shaft.* The complete gas-producer assembly, or N_1, system consists of five axial stages (first and second stages transonic) and one centrifugal impeller and the two gas-producer turbine rotors. The gas-producer turbine rotors drive the com-

FIG. 22-38   Lycoming T53-L-13B turboshaft engine. *(Avco Lycoming Div., Avco Corp.)*

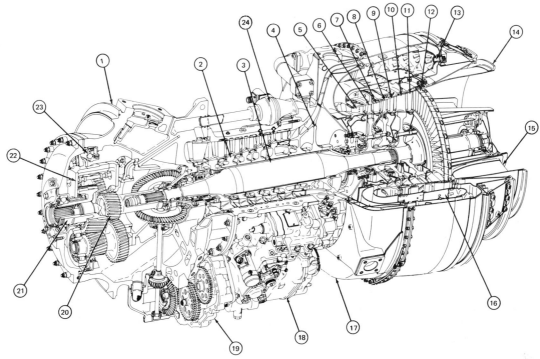

1. Inlet housing
2. Axial compressor
3. Power shaft
4. Centrifugal compressor
5. First-stage gas-producer-turbine nozzle
6. First-stage gas-producer-turbine rotor
7. Second-stage-turbine nozzle
8. Second-stage gas-producer-turbine rotor
9. First-stage power-turbine nozzle
10. First-stage power-turbine rotor
11. Second-stage power-turbine nozzle
12. Second-stage power-turbine rotor
13. Fuel-injector nozzle
14. Exhaust diffuser
15. Exhaust-diffuser strut
16. Combustor liner
17. Air diffuser
18. Fuel reg. and power-turbine governor
19. Accessory gearbox
20. Sun gear
21. Output shaft gear
22. Output-reduction carrier and gear assembly
23. Torquemeter system
24. Anti-icing valve

FIG. 22-39 Cutaway drawing of the Lycoming T53 engine.

pressor rotor, and, coaxial with it, the power-turbine rotors drive the power shaft, sun gear, output reduction carrier, and gear assembly which constitute the N_2 system.

Directional references for the engine are as follows:

Front	End of engine from which power is extracted.
Rear	End of engine from which exhaust gas is expelled.
Right and Left	Determined by viewing the engine from the rear.
Bottom	Determined by the location of the accessory drive gearbox (six o'clock).
Top	Directly opposite, or 180° from, the accessory drive gearbox (twelve o'clock). (The engine lifting eyes are located at the top of the engine.)
Direction of rotation	Determined as viewed from the rear of the engine. The direction of rotation of the gas-producer turbines is counterclockwise. The power turbines and the output gearshaft rotate in a clockwise direction.
O'clock	Position expressed as viewed from the rear of the engine.

Operational Description

The engine is started by an electric starter, geared to the compressor rotor shaft, which mechanically starts rotation of the compressor rotor. At the same time, the starting-fuel solenoid valve and the ignition system are also energized. This causes the fuel pump to drive fuel to the four starting nozzles at the two, four, eight, and ten o'clock positions in the combustor, and the adjacent ignitor plugs ignite and sustain combustion of the fuel jets until the starting system is deenergized.

At 8 to 13 percent N_1 (compressor) speed, the distributor valve opens, allowing fuel to flow from the fuel regulator through 22 atomizers into the already inflamed combustor where it accelerates both the compressor and power rotors. At approximately 40 percent N_1 speed, the flow of air and combustion are self-sustaining and the starter, starting-fuel valve, and ignitors are deenergized. Control of power and speed is then taken over by the fuel regulator and power-turbine governor throughout the operating regime.

Airflow and Gasflow Description

Atmospheric air is drawn into the annular passageway of the inlet housing, passing rearward across the variable inlet guide vanes, which direct the air to the engine compressor section. The combination of one rotating member (compressor rotor) and one stationary member (stator) constitutes one stage of compression.

The fifth-stage stator includes a second set of vanes which serve as exit guide vanes for the axial compressor and direct the air at the proper angle to the leading edge of the centrifugal compressor vanes.

The centrifugal compressor further accelerates the air as it passes radially to the diffuser-housing passageway. Three sets of vanes in the air diffuser convert the air velocity into pressure and redirect the airflow rearward to the combustor.

At this point the air enters the combustor section, surrounds the combustor liner, and passes into the annular combustion area through slots, louvers, and holes in the liner. Upon entering the combustion area, flow direction is reversed. At the same time, the air performs the dual function of cooling the combustor liner and supporting combustion. Combustion is made possible by introducing fuel into the combustion area through 22 atomizers (nozzles). The atomized fuel mixes with the air and burns.

The hot gas flows forward in the combustion area to the deflector, which reverses its flow. Flowing again rearward, the gas is directed across the two-stage gas-producer turbine system. The first gas-producer nozzle directs (impinges) the high-energy gas onto the first-stage turbine, through the second gas-producer nozzle, and onto the second gas-producer turbine. The power system also utilizes the two-stage turbine concept. On leaving the second gas-producer turbine the gas, still possessing a high work potential (approximately 40 percent), flows across the first power-turbine nozzle onto the first power turbine, through the second power-turbine nozzle, and onto the second power turbine. On passing from the second power turbine, the gas is exhausted into the atmosphere through the exhaust-diffuser passageway.

Cooling and Pressurization

As with other gas-turbine engines, the T53 engine requires controlled airflow for cooling internal components subject to high temperatures and for pressurizing the main-bearing seals and the intershaft oil seal at the forward end of the power shaft. The arrows in Fig. 22-40 indicate the flow of cooling air inside the engine. Compressor air is bled through ports at the periphery of the centrifugal compressor impeller, cooling the diffuser-housing front face and pressurizing the no. 2 main-bearing forward seal and aft oil seal, and is then ported through the rear-compressor shaft between the rotor assembly and the power shaft. This air flows both forward and aft: some between the carbon elements of the no. 1 main-bearing seal, then forward to the aft face of the intershaft seal; some emerging at the aft end of the rear-compressor shaft to cool the rear face of the second gas-producer rotor, the forward face of the first power-turbine rotor, and the first-stage power-turbine nozzle, then into the airstream; the remainder through holes into the power shaft and aft into the interior of the second power-turbine rotor assembly. It is then ported through the turbine hub and spacer to cool the rear surface of the first power-turbine rotor, the forward surface of the second power-turbine rotor, and both faces of the second power-turbine nozzle. It is then discharged into the exhaust stream.

Compressed cooling air is directed through various ports and channels to all the moving components and the stationary nozzles subjected to the highest temperatures within the turbine section of the engine in a complex and fairly comprehensive system. The exhaust diffuser and the rear face of the second-stage power-turbine rotor are cooled only by circulation of ambient air through the hollow diffuser struts, impelled by the venturi action where it joins the exhaust.

Compressor and Impeller Housing Assemblies

The axial and centrifugal magnesium compressor housings each consist of two matched halves. Mounted within the axial housings are five stages of stator vanes

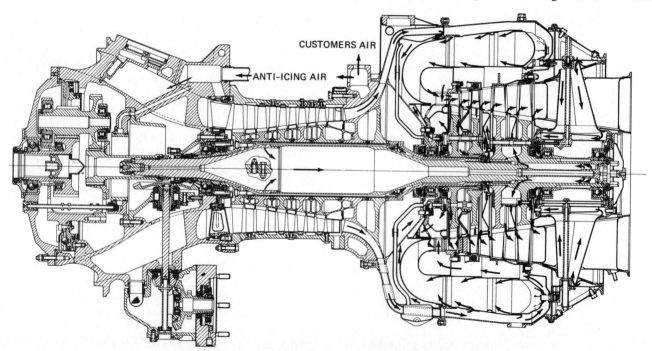

FIG. 22-40 Cooling and pressurization airflow.

and five rows of steel inserts to maintain radial blade clearances against erosion. Mounted externally on the axial housing are the starting-fuel solenoid valve and a two-piece steel band which is part of the interstage bleed system.

The centrifugal impeller housing provides mounting facilities for the anti-icing valve, ignition exciter, and interstage bleed-actuator assembly. In addition, the hollow housing is utilized as a manifold by which compressor-discharge-pressure bleed air may flow to the anti-icing valve and a customer air-bleed port located externally at the twelve o'clock position.

Anti-icing

Pressurized hot air from the annular manifold in the centrifugal compressor housing flows forward through the manually controlled airflow shutoff regulator into the hollow annulus on top of the inlet housing (see Fig. 22-40). This hot air is then directed through five of the six hollow inlet housing support struts to anti-ice the air-inlet area. Hot scavenge oil, draining through the lower strut into the accessory drive gearbox, anti-ices the bottom of the air-inlet area. Hot air also flows through the inlet guide vanes for anti-icing and then into the compressor area. In the event of electrical power failure, anti-icing becomes continuous.

Air-Diffuser Housing

The air-diffuser housing is comprised of an inner and outer shell separated by three rows of vanes. The outer shell provides three additional engine mount pads and an aft lifting eye. The airflow, upon exiting the impeller, is turned, slowed, and smoothed out by the three vane stages prior to entering the combustor housing. A portion of the air exiting the third row of vanes is bled into an air manifold in the outer shell for anti-icing, customer bleed, and operation of the interstage air-bleed actuator. The inner shell supports the compressor rotor's rear no. 2 main bearing and seals, the deflector assembly, the first-stage gas-producer nozzle, and the second-stage gas-producer cylinder. Two oil tubes provide paths for pressure and scavenge oil into and out of the no. 2 bearing housing from pressure-scavenge ports on the outer shell.

Compressor Rotor Assembly

The compressor rotor assembly consists of a steel first-stage rotor disk, a welded titanium sleeve constituting the second- through fifth-stage rotor disks, a one-piece centrifugal titanium impeller, and a steel rear shaft for mounting the two driving turbines. The four sections are bolted together and balanced as an assembly. The compressor rotor assembly is supported by two main bearings.

Inlet Housing

The magnesium inlet housing assembly is composed of two principal areas. The outer housing, supported by six hollow struts, forms the outer wall of the air-inlet area and houses the anti-icing annulus. The outer housing also provides facilities for mounting the engine and accessories. Located within the inlet housing assembly are: compressor rotor forward no. 1 main bearing and seal, variable inlet guide vane assembly, forward power-shaft support bearing, accessory carrier and support assembly, sun gear and output reduction carrier and gear assembly, and torquemeter system.

Combustor Turbine Assembly

The combustor turbine assembly consists of the exhaust-diffuser support cone assembly, fuel manifold assembly, fire-shield assembly, exhaust-diffuser assembly, second power-turbine rotor and bearing housing assembly, V-band coupling, atomizing combustion-chamber assembly, second-stage power-turbine nozzle, first-stage power-turbine rotor, and first-stage power-turbine nozzle. The second power-turbine rotor and bearing housing assembly consists of the turbine disks and blades, the no. 3 and no. 4 main bearings, the no. 3 main-bearing seal, and bearing housing. The **exhaust diffuser** contains hollow struts through which cooling air is supplied to the no. 3 and no. 4 main-bearing housings and the rear face of the second-stage power-turbine disk. The **combustion-chamber assembly** consists of the **combustion-chamber housing** and the **combustion-chamber liner.**

Description of Interstage Bleed System

The interstage bleed system is supplied with the engine to improve compressor-acceleration characteristics. The system automatically relieves the compressor of a small amount of air to prevent compressor stall in the low-speed range and during compressor acceleration or deceleration. A schematic drawing of the system is shown in Fig. 22-41.

The air-bleed actuator operates by means of compressor discharge air extracted from a port on the right-hand side of the air diffuser. Compressor discharge air entering the actuator assembly passes through a filter to the underside of the relay-valve diaphragm. A small portion of this air, which is under the diaphragm, is bled through an orifice in the base of the relay-valve assembly to an external line which directs it to a slide valve located on the fuel-regulator housing. When the slide valve is in the open position, this air is vented overboard, reducing pressure at the top surface of the diaphragm. Simultaneously, air is being bled overboard through the open relay valve. This reduces pressure at the bottom surface of the diaphragm. This equalization of pressure on both surfaces of the diaphragm causes it to remain in a neutral position, holding the relay valve in its open position. With the relay valve held in the open position, the major portion of the compressor discharge air that enters the actuator assembly is vented to the atmosphere. When the compressor discharge actuating pressure is vented, the actuator spring, located on top of the actuator piston, expands and pushes the piston downward, causing the bleed band to open and remain open as long as the slide valve on the fuel regulator is in the open position.

It follows, therefore, that when the slide valve is closed, the bleed band will be closed. This is accomplished by a buildup of pressure on the top side of the relay-valve diaphragm which forces the relay valve down, closing the overboard vent. With the overboard vent closed, the compressor discharge pressure is now routed into the actuator piston assembly, overcoming the spring load and forcing the piston assembly to move

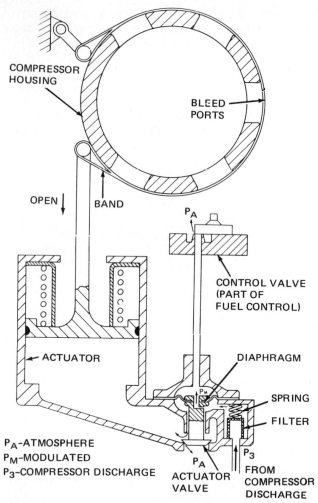

COMPRESSOR HOUSING

BLEED PORTS

OPEN

BAND

P_A

CONTROL VALVE (PART OF FUEL CONTROL)

ACTUATOR

DIAPHRAGM

P_M

SPRING

FILTER

P_A - ATMOSPHERE
P_M - MODULATED
P_3 - COMPRESSOR DISCHARGE

P_A
ACTUATOR VALVE

P_3
FROM COMPRESSOR DISCHARGE

FIG. 22-41 Schematic drawing of the interstage bleed system.

upward. This, in turn, causes the bleed band to close around the compressor bleed ports.

The entire sequence of operation is controlled by the fuel regulator which senses gas-producer speed, fuel flow, and pilot demand, thereby ensuring proper opening and closing of the interstage air bleed.

Power-Turbine Governor and Tachometer Drive Assembly

The power-turbine governor and tachometer drive assembly is mounted at the ten o'clock position on the exterior of the inlet housing and is driven through shafts and gearing from the power shaft. The drive assembly provides mounts and drives for the power-turbine tachometer generator and the power-driven rotary torquemeter boost scavenge pump. The drive assembly also drives the power-turbine governor and incorporates a strainer and metering cartridge for lubrication of the drive-gear train. A torquemeter-relief valve, located on the upper portion of the housing, allows for adjustment of the torquemeter boost oil pressure.

Accessory Drive Gearbox Assembly

The accessory drive gearbox assembly is mounted at the six o'clock position on the exterior of the inlet housing. It is driven through a shaft and gearing from the compressor forward shaft. The power-driven ro-

tary oil pump, oil filter, fuel regulator, compressor rotor tachometer generator, and starter generator are mounted on the gearbox. A magnetic chip-detector drain plug is installed in the bottom of the gearbox.

Variable Inlet Guide-Vane System

To provide the desired compressor surge margin, the angle of attack of the inlet air to the first compressor rotor must be within the stall-free operating range of the transonic airfoil (first two stages of the compressor). Since this stall-free operating range varies with compressor speed, it becomes necessary to vary the angle of attack as a function of compressor speed. This is accomplished by varying the angular position of the inlet guide vane. The variable inlet guide-vane assembly is located in front of the first compressor rotor and consists of a series of hollow blades positioned by a synchronizing ring. Operation of the guide vanes is controlled by an actuator assembly mounted externally at the inlet to axial compressor housing split line. The actuator is connected to the synchronizing ring by a control rod which attaches to a piston within the actuator. Depending on inlet air temperature and compressor speed, high-pressure fuel from the regulator assembly is ported through external hoses to one side of the piston which positions the control rod. The inlet guide-vane position is transmitted back to the regulator assembly by an external feedback rod to nullify the high-pressure-fuel signal.

Lubrication System

The lubrication system for the T53 engines is designed for operation up to 25,000 ft [7620 m] above sea level with oil conforming to MIL-L-23699 or -7808 specification. It will provide satisfactory lubrication at oil inlet temperatures ranging from −65 to 200°F [−53.89 to 93.33°C].

The lubrication system consists of a vane-type pump incorporating a pressure and scavenge element, a main 40-μm oil filter with bypass capabilities, an oil manifold, and external oil hoses. A schematic diagram of the system is shown in Fig. 22-42.

Main Fuel-Regulator Assembly

Control of fuel flow is a critical function for any gas-turbine engine and must be achieved with precision in order to get satisfactory and efficient turbine operation throughout the operating regime. The critical objective is to maintain balance between combustion air and fuel uniting in the combustion chamber, but there are too many variables for even the most skilled pilot to achieve this with only a throttle lever. Too much fuel, relative to air, causes hot starts and surging. Too little fuel results in no start or, if running, a flameout. The fuel-flow and -control system is shown in Fig. 22-43.

If the fuel regulator is to add the correct amount of fuel to the air in the combustor, it must continuously sense how much air there is in the combustor. This depends on compressor speed (N_1), air temperature (T_1), and air pressure (P_1), which vary with altitude.

The engine speed cannot change as rapidly as the throttle is moved, so the fuel regulator is governed by the gas-producer governor (N_1) and the power-turbine governor (N_2) to damp the changes of fuel flow de-

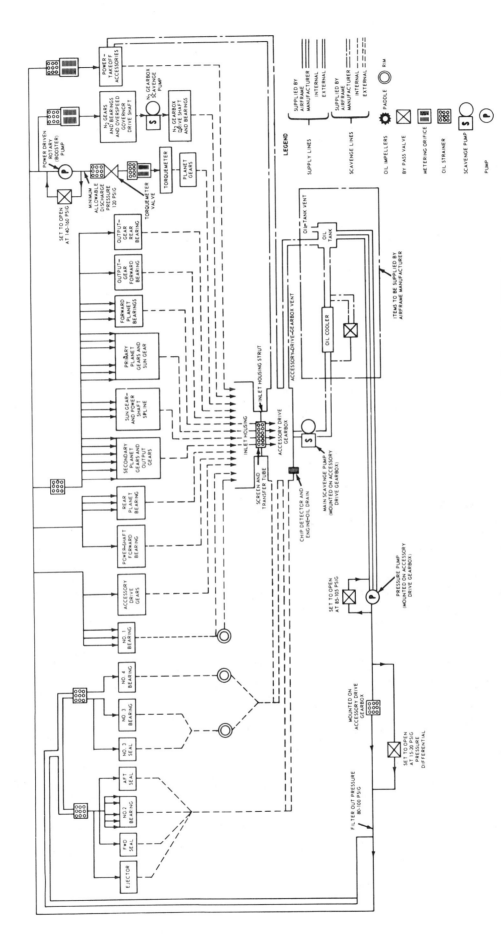

FIG. 22-42 Schematic diagram of the lubrication system for the Lycoming T53 engine.

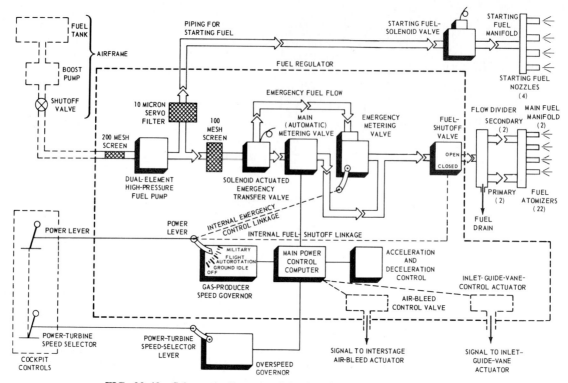

FIG. 22-43 Schematic diagram of the fuel-flow and -control system.

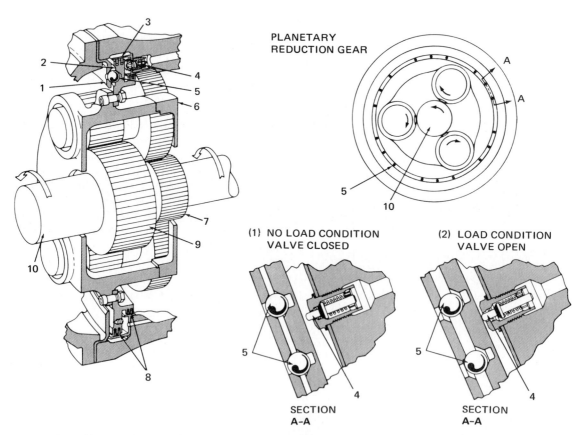

1. Fixed plate
2. Movable plate
3. Torquemeter cylinder
4. Torquemeter valve
5. Torquemeter balls
6. Carrier and gear assembly
7. Sun gear
8. Torquemeter sealing rings
9. Output shaft gear
10. Output shaft

FIG. 22-44 Hydromechanical torquemeter.

554

manded by throttle movement. This delays the changes in fuel flow just long enough for them to be acceptable at the changing engine speeds without causing the improper engine operation of an unbalanced fuel-air mixture.

The hydromechanical fuel-regulator assembly is made up of a Model TA-2S fuel regulator and a Model PTG-3 power-turbine governor.

It consists essentially of the following components:

1. Dual-element fuel pump
2. Speed input servo and gas-producer governor (N_1)
3. Power-turbine speed governor (N_2)
4. Acceleration and deceleration fuel-flow control
5. Manual (emergency) control system
6. P_1 multiplier assembly (altitude compensator)
7. T_1 motor assembly (temperature compensator)
8. Transient air-bleed control
9. Variable inlet guide-vane actuating system

Because of the complexity of the fuel regulator, as indicated in the list, provision is made for last-resort manual control by the pilot in case of serious malfunction of the automatic control.

Torquemeter

The torquemeter is a hydromechanical torque measuring device located in the reduction-gear section of the engine inlet housing (see Fig. 22-44). It uses engine oil as the means for determining and measuring engine torque which is read in the cockpit as psi oil pressure. Although this system uses engine oil, it is not a part of the lubrication system.

The mechanical portion of the torquemeter consists of two circular plates. One is attached to the inlet housing and is identified as the stationary plate. The second, or movable plate, is attached to the reduction-gear assembly. The movable plate contains front and rear torquemeter sealing rings, which enable it to function as a piston in the rigidly mounted cylinder-chamber assembly of the torquemeter. The cylinder assembly houses the variable-opening torquemeter (poppet) valve and the movable plate maintains the fixed-orifice metered bleed, which functions in relation with the poppet valve. The movable plate is separated from the stationary plate by steel balls positioned in matched conical sockets machined in the surfaces of both plates.

When the engine is not operating [section A-A(1) in Fig. 22-44], the torquemeter assembly's movable plate is in a position forward and clear of the torquemeter's valve plunger, allowing the spring-loaded valve to remain in the closed position. With the engine operating and a load applied to the output shaft, the torque developed in the engine to drive the shaft is transmitted from the sun gear through the reduction-gear assembly. The attached movable plate therefore tends to rotate with the assembly. However, this mechanically limited rotary movement positions the steel balls against the conical sockets of both plates, resulting in the movable plate being axially directed rearward in the assembly [section A-A(2), Fig. 22-44].

The plate, moving rearward, contacts the torque-meter-valve plunger, opening the valve and allowing oil to flow into the cylinder. This contact is maintained during all engine operation and the size of the valve opening varies as the plate moves rearward or forward. As torque continues to increase and the torquemeter valve opens further, the oil pressure developed in the cylinder exerts pressure against the piston (movable plate), restraining its rearward movement. With the engine operating in a steady-state condition, the cylinder oil pressure and movement of the plate hold in an equalized position, maintaining a constant pressure in the cylinder. This pressure is proportional to engine torque, measured and read in the cockpit.

Main Electric Cable Assembly

The main electric cable assembly, shown in Fig. 22-45, furnishes all necessary interconnecting wiring between the main disconnect plug and the eight branched electric connectors. The eight electric accessories served by this cable are the gas-producer tachometer generator, oil temperature bulb, manual fuel-regulator solenoid valve, ignition exciter, anti-icing solenoid valve, starting-fuel solenoid valve, chip-detector and power-turbine tachometer generator. The main disconnect plug mates with an electric receptacle of the airframe wiring, establishing electric continuity to the various airframe components.

Ignition Exciter

Ignition of the four igniter plugs is powered by a coil which builds up the originally introduced 28-V primary voltage to approximately 2500 V, on the same basis as automobile ignition.

Radio-frequency energy is generated within the exciter during normal operation. An inductive capacitive filter has been incorporated at the input to prevent this energy from being fed back onto the 28-V input line. Radio-frequency interference on this line could be detrimental to the operation of other electric accessories. This filter is tuned to radio frequencies and does not offer any appreciable opposition to the flow of 28-V direct current.

● THE ALLISON SERIES 250 GAS-TURBINE ENGINE

The Allison Series 250 gas-turbine engine, developed and manufactured by Allison Gas-Turbine Operations, General Motors Corporation, is manufactured in a number of different configurations, including turboshaft and turboprop models. Four of the turboshaft engines are shown in Fig. 22-46 and two of the turboprop engines are shown in Fig. 22-47. Another arrangement in which a Model 250 engine drives a propeller is shown in Fig. 22-48. In this installation, a turboshaft engine drives a separate gearbox upon which the propeller is installed. The arrangement is called a **turbine pac.**

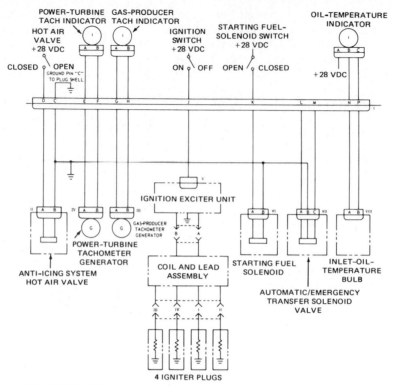

FIG. 22-45 Schematic diagram of the main electric cable assembly.

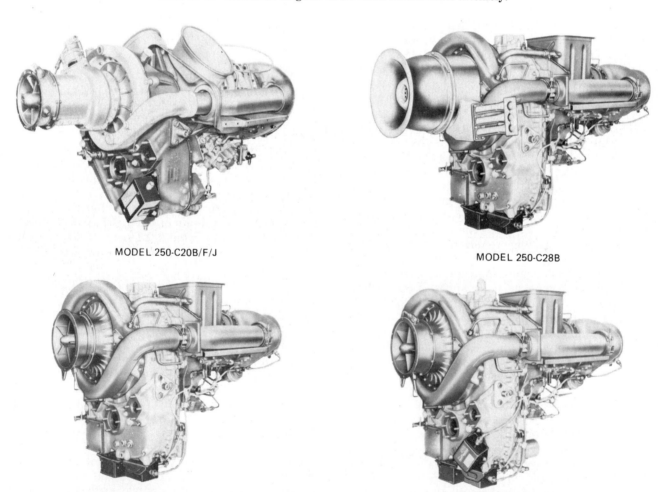

MODEL 250-C20B/F/J

MODEL 250-C28B

MODEL 250-C28C

MODEL 250-C30/P/R

FIG. 22-46 Various models of the Allison 250 turboshaft engine. (*Allison Gas-Turbine Operations, General Motors Corp.*)

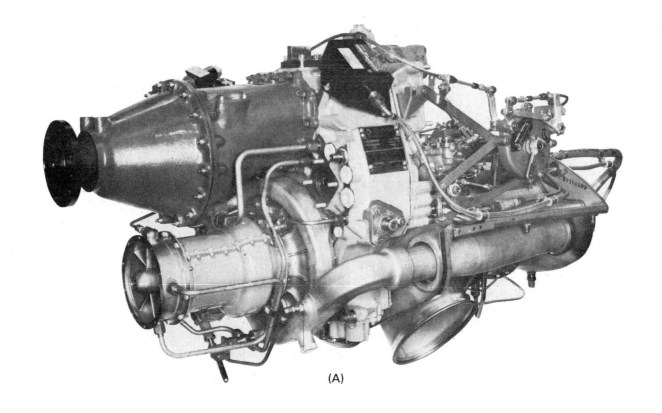

(A)

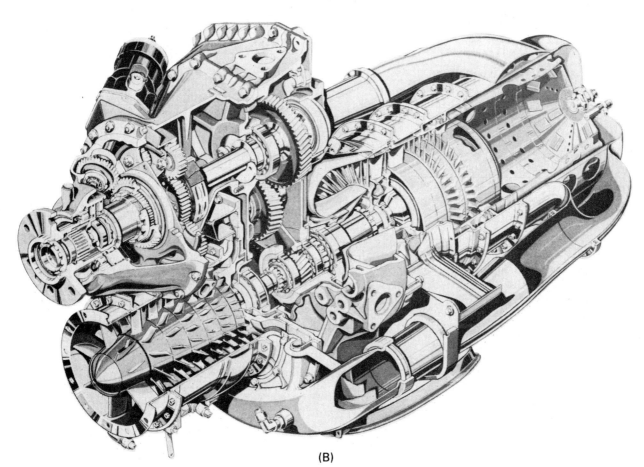

(B)

FIG. 22-47 Two turboprop models of the Allison 250 engine. *(Allison Gas-Turbine Operations, General Motors Corp.)*

FIG. 22-48 The Allison "turbine pac" arrangement.

The Allison 250-C20 engine described in this section is composed of four major sections, or modules. These are the **compressor section,** the **turbine section,** the **combustion section,** and the **accessory gearbox section.** The compressor is at the front of the engine, the combustion chamber is at the rear, and the turbine section is near the center. Figure 22-49 is a cutaway drawing to illustrate the construction.

Performance and Specifications

	250—C20	250—C20B
Takeoff power	400 hp [298 kW]	420 hp [313 kW]
Jet thrust	40 lb [178 N]	42 lb [187 N]
Gas-producer rpm	52,000	53,000
Output-shaft rpm	6016 rpm	
Weight	155 lb [70.3 kg]	Same
SFC	0.630 lb/shp/h 0.650 lb/shp/h [383.21 mg/W/h 395.38 mg/W/h]	
Length	40.7 in [103.4 cm]	Same
Height	23.2 in [58.9 cm]	Same
Width	19.0 in [48.3 cm]	Same

It must be explained that the foregoing specifications apply only to the engine models shown. Some models of the Series 250 engine can produce more than 700 hp [522 kW].

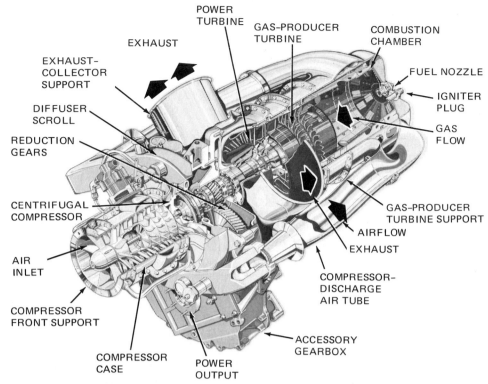

FIG. 22-49   Cutaway view of the Allison 250-C20 turboshaft engine.

Compressor

The compressor section for the engine under discussion consists of a compressor front support, case assembly, rotor wheels with blades (for axial compressor section), centrifugal impeller, front diffuser assembly, rear diffuser assembly, diffuser-vane assembly, and diffuser scroll. Some models of the Allison 250 engine do not include the six-stage axial compressor section and all compression is accomplished by means of the centrifugal compressor.

Air enters the engine through the compressor inlet and is compressed by the axial and centrifugal compressor sections. It then passes through the scroll-type diffuser into two external ducts which convey it to the combustion section at the rear of the engine. The compressor is driven directly by the gas-producer turbine at speeds of more than 50,000 rpm. A drawing to illustrate airflow and gasflow through the Allison 250-C20B engine is shown in Fig. 22-50.

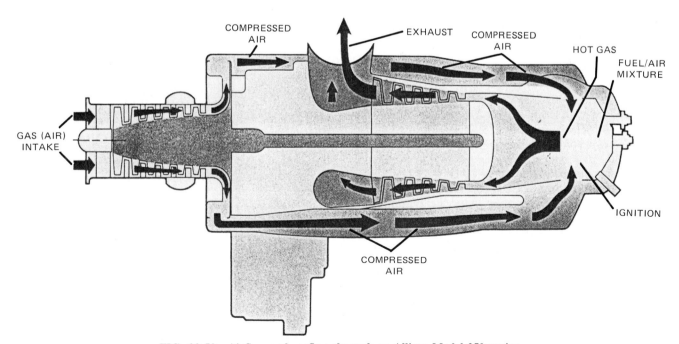

FIG. 22-50   Airflow and gasflow through an Allison Model 250 engine.

Combustion Section

The combustion section, located at the rear of the engine, consists of the **outer combustion case** and the **combustion liner**. A fuel nozzle and a spark igniter are mounted in the aft end of the outer combustion case. Compressed air from the two external ducts enters the combustion liner at the aft end through holes in the liner dome and skin. The air is mixed with the fuel sprayed from the fuel nozzle and combustion takes place. The hot gases move forward out of the combustion liner to the first-stage gas-producer turbine nozzle.

Turbine Section

The **turbine** consists of a two-stage **gas-producer rotor**, a two-stage **power-turbine rotor**, a **gas-producer turbine support**, a **power-turbine support**, and a **turbine and exhaust collector support**. The turbine is mounted between the combustion section and the power and accessory gearbox. The two-stage gas-producer turbine drives the compressor and the accessory gear train. The power turbine drives the output shaft through the reduction-gear train. The expanded gas, having passed through the turbine stages, discharges in an upward direction through the twin ducts of the turbine and exhaust collector.

Power and Accessories Gearbox

The main power and accessories drive gear trains are enclosed in a single gear case. This gear case serves as the structural support of the engine and all engine components including the engine-mounted accessories are attached to it. A two-stage helical and spur-gear set is used to reduce the rotational speed from 33,290 rpm at the power turbine to 6016 rpm at the output drive spline. Accessories driven by the power-turbine gear train are the power-turbine governor and an airframe-furnished power-turbine tachometer generator. The gas-producer gear train drives the compressor, fuel pump, gas-producer fuel control, and an airframe-furnished gas-producer tachometer generator. The starter drive and a spare drive are also in the gas-producer drive train.

Fuel System

The fuel system with Bendix controls is illustrated in Fig. 22-51. The components of the system are a **fuel pump**, a **gas-producer fuel control**, a **power-turbine governor**, and a **fuel nozzle**. The fuel control and the governor are located schematically in the system between the fuel pump and fuel nozzle. The actual flow of fuel in the Bendix system involves only the pump, gas-producer fuel control, and fuel nozzle.

An alternate fuel system for the Allison 250 engine is furnished by the Chandler Evans Company (CECO). This system uses hydromechanical controls and fuel enters both the gas-producer fuel control and the power-turbine fuel governor.

Lubrication System

The lubrication system for the Allison 250-C20 turbine engine is shown in Fig. 22-52. This is a dry-sump system utilizing a pressure pump and scavenge pumps.

Two metal chip detectors are included in the system to aid in discovering wear problems within the engine. The oil tank and the oil cooler are both airframe-furnished.

● THE GARRETT TPE331 TURBOPROP ENGINE

The Garrett TPE331 turboprop engine, designated the T76 for military purposes, is illustrated in Fig. 22-53. The engine has been in use for many years; however, improvements are constantly being made with resultant increases in power, efficiency, and reliability. These improvements include changes in compressor design, new airfoil configurations for turbine blades and vanes, and new superalloys for hot section parts. The most recent model, the TPE331-14, can produce more than 1200 shaft horsepower [894.84 kW] for take-off, whereas original models were rated at 655 shp [488.43 kW]. Specific fuel consumption (sfc) has improved from 0.556 to 0.49 lb/hp/h [338.2 to 298.12 g/kW/h].

Description

The TPE331 engine is a single-spool machine; that is, it has one main rotating assembly that includes both the compressor and the turbine. At the forward end of the shaft is the gear that drives the propeller reduction and accessory gears. A two-stage centrifugal (radial) compressor is located on the main shaft in the forward section of the engine. To the rear of the compressor section is the three-stage turbine that extracts power from the hot, high-velocity gases and delivers the power through the main shaft to both the compressor and the propeller reduction gears. The turbine is surrounded by the reverse-flow annular combustor. The engine is of modular construction to simplify repair and maintenance procedures.

Operation

During operation of the TPE331 engine, air flows from the inlet to the first stage of the compressor. From the first stage the air flows outward and then is routed through ducting back toward the center of the second-stage compressor. From the second stage the pressurized air flows outward and back around the outside of and into the annular combustor. Atomized fuel injected through nozzles in the rear of the combustor is ignited and the resulting hot gases flow forward, then turn inward and flow to the rear through the three-stage turbine where they are ejected out the rear of the engine.

Engine Systems

The propeller-control system for the TPE331 engine is similar to those for other turboprop engines. During flight the system automatically adjusts propeller pitch to maintain a constant propeller speed. As engine power is increased through movement of the power lever, propeller pitch increases as the propeller governor moves to release oil from the pitch-changing mechanism in the propeller hub. For ground operations, the propeller-control system provides for reverse thrust and a beta mode of operation for taxiing. As explained earlier

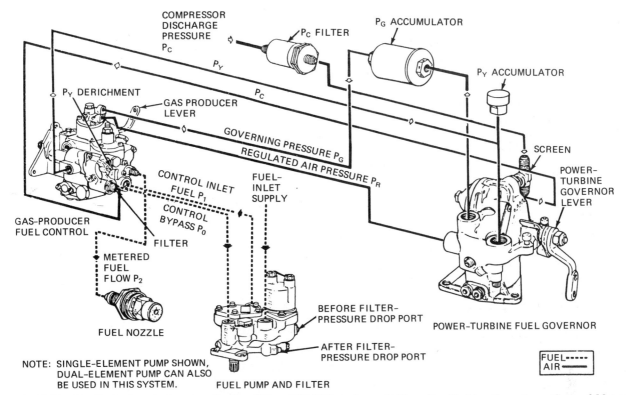

FIG. 22-51 Bendix fuel-control system for the Allison 250-C20 engine. *(Allison Gas-Turbine Operations, General Motors Corp.)*

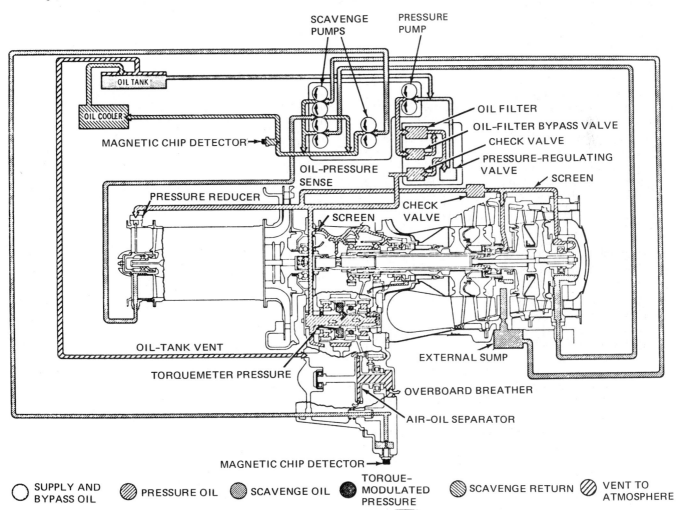

FIG. 22-52 Lubrication system for the Allison 250-C20 engine.

FIG. 22-53   Cutaway view of the Garrett TPE331-14 turboprop engine.   *(Garrett Turbine Engine Co.)*

in this text, the **beta** mode is employed for low power and reduced rpm operations on the ground.

As mentioned before, the propeller pitch-changing mechanism is actuated hydraulically and is designed for both feathering and reverse thrust. Hydraulic pressure moves the propeller toward low pitch; however, high pitch and feathering are accomplished by means of coil springs and counterweight force.

Pilot-operated controls include the power lever, engine rpm lever, manual feather control, and unfeather switch. The system includes a torque sensor to detect negative torque, that is, a situation where the engine is being rotated by the windmilling propeller. When a negative torque signal (NTS) is produced, the system automatically feathers the propeller to stop engine rotation and reduce drag.

Anti-icing for the TPE331 engine involves an electrically actuated system controllable by the pilot. The inlet anti-icing system employs warm bleed air from the second-stage compressor. This air is directed through a forward manifold to an anti-icing shield surrounding the outer side of the air inlet. The air then flows rearward and is discharged into the engine nacelle. The inner wall of the air inlet is a part of the gearbox case and is warmed by engine oil.

The fuel system for the TPE331 engine is designed to pressurize and regulate the fuel as it is directed to the nozzles that atomize it as it is injected into the rear of the combustor. The fuel-control unit is coordinated

with the propeller control to provide the correct amount of fuel to meet the speed and power requirements of the engine. The system includes a fuel-control unit, solenoid valve. flow divider, fuel nozzle and manifold assembly, and an oil/fuel heat exchanger. The system automatically controls fuel flow for variations in power-lever position, compressor discharge pressure, and inlet-air temperature and pressure. The solenoid valve is actuated automatically to open during starting and to close on engine shutdown. It is manually closed when the propeller is feathered.

● REVIEW QUESTIONS

1. Compare the power section of a turboprop engine with that of a turbojet engine.
2. List three advantages of a turboprop engine.
3. Describe the construction and explain the function of the compressor air-inlet housing for the 501-D13 turboprop engine.
4. What provision is made to prevent ice from forming in the compressor air-inlet housing?
5. Describe the construction of the compressor rotor.
6. Why are air-bleed valves required in the compressor housing?
7. What is the function of the combustion liners?
8. What are the two principal functions of the thermocouples in the turbine section of the 501-D13 engine?

9. Describe the turbine rotor assembly.
10. Name the accessories mounted on the reduction-gear assembly.
11. What advantages are gained by locating the reduction-gear assembly remotely from the power section of the engine?
12. What is the purpose of the propeller brake?
13. Explain the operation of the *negative torque system*.
14. Under what condition does the *thrust-sensitive signal* take effect?
15. Describe the operation of the torquemeter.
16. Name the components of the power-section lubrication system.
17. Explain the functions of the *speed-sensitive control*.
18. Describe the ignition system for the 501-D13 engine.
19. Give a brief description of the fuel system for the 501-D13 engine.
20. Describe the compressor of the Rolls-Royce Dart engine.
21. What type of combustion section is employed in the Dart engine?
22. Name the principal parts of the Dart combustion chamber.
23. Describe the construction of the reduction-gear assembly for the Dart engine.
24. What type of torquemeter is employed in the Dart engine?
25. What is the function of the interconnectors between the combustion chambers?
26. What provision is made for the expansion and contraction of the combustion chambers?
27. What material is used for the turbine blades?
28. Describe the arrangement of the oil tank for the Dart engine.
29. How are the turbine blades attached to the turbine disks?
30. Describe the exhaust unit.
31. What is the purpose of the *deaerator tray* in the oil system?
32. Describe the operation of the fuel pump.
33. Through what unit is the water-methanol mixture injected into the engine?
34. What type of starting system is employed for the Dart engine?
35. What is meant by the term *free turbine*?
36. Describe the airflow through the PT6A engine.
37. Describe the compressor for the PT6A engine.
38. What part of the engine is considered to be a gas generator?
39. Where is the oil tank for the PT6A engine located?
40. What provision is made for the removal of the no. 1 bearing?
41. Of what materials are the compressor rotor blades made?
42. Describe the gas-generator case for the PT6A engine.
43. How is the combustion-chamber liner supported in the gas-generator case?
44. Describe the turbine section and the operation of the two separate turbine systems.
45. What is the purpose of the interstage baffle?
46. What provision is made for checking turbine-disk growth?
47. How are the turbine blades attached to the turbine disks in the PT6A engine?
48. What is the propeller reduction-gear ratio on the PT6A turboprop engine?
49. Describe the reduction-gear assembly.
50. Describe the operation of the torquemeter.
51. What provision is made to avoid the accumulation of fuel in the exhaust duct during shutdown?
52. Describe the accessory gearbox.
53. What are the three principal air-bleed systems for the PT6A engine?
54. Explain the purpose of the compressor bleed valve.
55. Explain the principle of airseals.
56. Describe the lubrication system for the PT6A engine.
57. How does the ignition system for the PT6A engine differ from those used on the majority of other turbine engines?
58. Describe the fuel system for the PT6A engine.
59. What are the controlling parameters for the pneumatic fuel-control unit?
60. Give a brief general description of the Lycoming T53 gas-turbine engine.
61. Describe the compressor for the T53 engine.
62. Describe the airflow through the combustion chamber.
63. Explain how the turbine rotors are cooled.
64. What materials are used in the construction of the compressor rotor assembly?
65. What is the function of the interstage bleed system for the T53 engine?
66. Describe the operation of the air-bleed actuator.
67. What accessories for the T53 engine are driven from the accessory drive gearbox?
68. Describe the operation of the variable inlet guide-vane system.
69. Compare the lubrication system for the T53 engine with those of other gas-turbine engines.
70. What type of fuel-control unit is used with the T53 engine?
71. Compare the operation of the torquemeter for the T53 engine with that of the PT6A engine.
72. Give a brief description of the Allison 250-C20 turboshaft engine and principal components.
73. What is the principal structural component of the engine?
74. What units control the power output of the engine?
75. Describe briefly the lubrication system for the Allison 250 turboshaft engine.
76. Describe the Garrett TPE331 turboprop engine.
77. Describe the engine systems for the TPE331 engine.

23 GAS-TURBINE OPERATION, INSPECTION, AND MAINTENANCE

Because of the great variety of gas-turbine engines, it is not possible to set forth standard procedures which will apply to all such engines. There are, however, certain common characteristics among gas-turbine engines and a number of features which lend themselves to accepted standards of procedure and workmanship. In this section, we shall describe some general practices and provide specific examples of operations, inspections, and maintenance practices.

The most important considerations for the technician responsible for the operation and maintenance of gas-turbine engines are an adherence to principles of good workmanship and attention to detail, plus a consistent practice of following the instructions provided in the operator's and manufacturer's manuals.

● STARTING AND OPERATION

Starting Gas-Turbine Engines

A gas-turbine engine should be started only when all conditions required for the safety of the engine and nearby property and personnel are met. The engine pod or nacelle should be checked for loose material, tools, or other items which could be ingested by the engine and cause **foreign-object damage** (FOD) to vanes, blades, and other interior parts. The best type of surface on which to operate gas-turbine engines mounted in aircraft is smooth concrete, which is free of all items or material which could be drawn into the engine intake. The aircraft must be positioned so that the exhaust heat and high-velocity gases will not cause damage to other aircraft, ground service equipment, or vehicles and will not cause injury to personnel. Figure 23-1 shows the temperatures and velocities of the exhaust stream to the rear of a JT9D engine at both idle and takeoff speeds. It can be seen that even at idle speed, the temperatures and gas velocities are hazardous. At takeoff power, the engine produces dangerous temperatures and high gas velocities at more than 200 ft [60.96 m] to the rear of the jet nozzle. It will also be observed in the drawings that a 25-ft [7.62-m] radius at the front of the engine is considered a danger zone because of the velocity of the air flowing to the engine inlet. This area should be meticulously clean to prevent the air from carrying dirt and foreign objects into the engine.

Small gas-turbine engines are often equipped with starter-generators. These units are electric motors which apply starting torque for the engine during starting; then, when the engine rpm attains a self-sustaining level, the starter motor becomes a generator to supply the aircraft electric system. A high-capacity series winding in the motor is employed while the starter is cranking the engine and a parallel or shunt field winding of many turns is employed when the unit is serving as a generator. Starter-generators should not be operated longer at any one time than specified in the operator's manual for the engine. The starters draw a high level of current when starting and therefore heat up very rapidly; they are thus subject to damage by overheating if their operation is not limited to the time intervals specified.

Large gas-turbine engines are generally equipped with air-turbine starters which receive their air supply from a ground-service unit or from an auxiliary power unit (APU) installed in the aircraft. Other types of starters that may be used for large gas-turbine engines are small gas-turbine engines (fuel-air starters) geared to the main engine and cartridge-type starters which receive their energy from gas generated by an explosive cartridge. Some of these starters were discussed in Chap. 11.

The principles involved in the starting of a gas-turbine engine are relatively simple. It is merely necessary to rotate the engine at a speed sufficient to provide adequate air volume and velocity for starting, provide high-intensity ignition means in the combustion chamber, and introduce fuel through the fuel nozzles in an amount that will not produce excessive heat but will provide sustained combustion and further acceleration of the engine.

In Chap. 11 we have described the principles of gas-turbine engine starting; however, it is well to review starting practices here in connection with the operation and maintenance of gas-turbine engines. Starting gas-turbine engines can be manual or automatic, depending on the particular installation. The majority of starts for today's engines are accomplished automatically.

The manual start of a gas-turbine engine usually includes the following steps:

1. Connect the starting power unit to the aircraft.
2. Turn on the master power switch.
3. Turn on the fuel boost pump switch.
4. Turn on the ignition switch.
5. Press the starter switch and accelerate the engine until at least 10 percent of maximum rpm is attained.
6. Slowly advance the throttle (power lever) and watch the exhaust-gas temperature indicator until a rapid temperature rise is indicated. This signifies a lightup. Do not advance the throttle further until the temperature stabilizes. *The temperature must not be permitted to exceed the maximum allowable specified for the engine.*

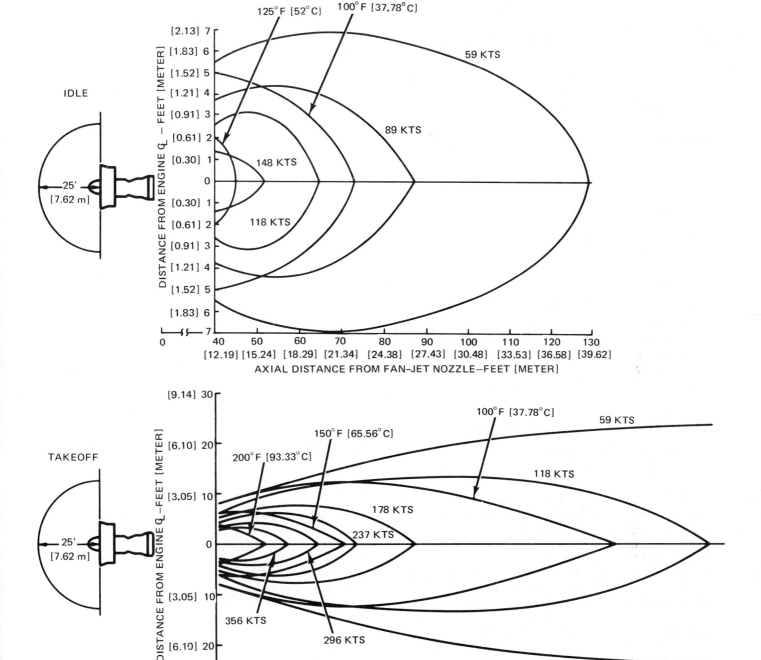

FIG. 23-1 Temperatures and velocities of gases to the rear of an operating JT9D engine.

7. When the temperature has stabilized, it will begin to decrease as engine rpm increases. Advance the throttle slowly again until the engine rpm reaches idle speed but do not permit the temperature to exceed the level specified.

8. Release the starter switch when it is certain that the engine has reached a self-sustaining speed.

To start a gas-turbine engine with an automatic system, operators must follow the procedure established for the particular aircraft and engine with which they

are working. An automatic start may be accomplished as follows with some variations depending on the aircraft and engine combination:

1. See that an air supply is available either by means of a ground service unit or an onboard APU or from an operating engine on the aircraft.

2. Turn on the aircraft electrical power.

3. Turn on the fuel pump switch.

4. Place the power lever in the IDLE position.

5. Rotate the engine selector switch to the position for starting the engine selected.

6. Press the starter switch. It is normally equipped with a holding coil so that it will remain closed until the engine is rotating at self-sustaining speed. At this point, current to the holding coil is cut off and the switch opens.

The engine accelerates; when it reaches a level of rpm somewhat above 10 percent of maximum, fuel will be supplied under pressure to the fuel nozzles. The ignition is turned on automatically at the time that the starter switch is depressed or shortly thereafter, but it is always on before fuel flow is supplied. Soon after fuel flow begins (within 20 s), fuel is ignited and the engine accelerates at an increased rate. Fuel flow is controlled automatically by the fuel-control unit and is not permitted to exceed the amount required for the correct rate of acceleration. The engine cannot have a hot start or a hung start unless there is a defect in the starting system or fuel control.

When the engine has accelerated to 35 percent or more of maximum rated rpm, a centrifugal switch cuts off electrical power to the holding coil of the start switch and the switch opens. The engine accelerates to idle speed and remains in that condition until the power lever is moved by the operator.

It is important that the operator of a gas-turbine engine observe the instruments pertaining to engine operation while performing the starting procedure. A fuel-pressure gage should indicate correct fuel pressure before the engine is started. The exhaust-gas temperature (EGT) gage should be watched closely at lightup to ensure that the temperature has not exceeded the maximum value allowed. Oil pressure should show on the oil-pressure gage shortly after the engine starts to rotate.

Typical Airline Start Procedure

Technicians and flight-crew members employed by an airline are given specific instructions and checklists which are to be followed before and during the start of an engine. The procedures to be followed in the starting of a JT8 engine on a 727 aircraft for a major airline are as follows:

1. Perform a *walk-around* inspection according to the appropriate checklist.

2. Perform a control cabin prestart check as set forth in the checklist.

3. Start the auxiliary power unit (APU).

    (a) Turn essential bus-selector switch to APU.
    (b) Switch AC meter to APU.
    (c) Arm automatic fire shutdown switch.
    (d) Reset APU fire detection system.
    (e) Turn master switch ON.
    (f) After 10 s, turn master switch to START.
    (g) Observe CRANK light; turn START switch ON.
    (h) Check APU frequency and voltage (408 Hz/115 VAC).

4. Obtain ALL CLEAR signal.

5. Check pneumatic pressure (35 psi [241.33 kPa] minimum).

6. Set engine START switch to GROUND START.

7. When N_2 rpm is 20 percent, place START lever in IDLE position.

8. Observe starting exhaust-gas temperature (EGT). This should not be more than 350°C (662°F) when outside air temperature (OAT) is less than 59°F [15°C] or not more than 420°C (788°F) when OAT is more than 59°F [15°C].

9. START switch release at 40 percent N_2 speed.

10. Check engine parameters after idle speed stabilizes.

EGT	300 to 420°C when air bleed or power extraction is used.
N_2 rpm	54 to 59.4 percent
Oil pressure	40 to 55 psi [275.8 to 379.23 kPa] (44 to 46 psi desired [303.38 to 317.17 kPa])
Oil temperature	40 to 60°C (120°C maximum)
Fuel flow	Approximately 1000 lb/h [453.59 kg/h]

Engine Trimming and Adjustment

Trimming a gas-turbine engine is the process of adjusting the fuel-control unit so that the engine will produce its rated thrust at the designated rpm. The thrust is determined by measuring the **engine pressure ratio (EPR)**, which is the ratio of turbine discharge pressure to engine inlet pressure (P_{t7}/P_{t2}). On engines equipped with variable compressor vanes, it is necessary to check the vane angles and the operation of the **engine vane control** (EVC) during the trim process. The trimming of a gas-turbine engine may be compared with the tuning of a piston engine for optimum performance. Gas-turbine engines that utilize computer-controlled fuel systems do not require trimming because trimming adjustments are made automatically by the fuel-control computer.

Gas-turbine engines manufactured by Pratt & Whitney Aircraft are tested at the factory and adjusted to produce rated thrust. The engine speed (N_2) which is required for the engine to deliver rated thrust is stamped on the engine data plate or recorded on the engine data sheet of the engine log book. The information is supplied in both rpm and percent of maximum rpm. Because of manufacturing tolerances and slight variations which occur during the manufacture of engines, no two engines are exactly alike, and very rarely will two engines of the same model produce rated thrust at exactly the same rpm. The rpm for rated thrust stamped on the data plate will therefore vary from engine to engine.

Engine trimming is required from time to time because of changes that take place during the life of an engine. Dust and other particulate matter will adhere to the surfaces of the compressor rotor blades and stator vanes and lead to a slight resistance to airflow. Erosion of the leading edges of blades and vanes caused by dust, sand, and other material changes the characteristics and performance of the compressor. The turbine blades and vanes, exposed to very high temperatures, are subject to corrosion, erosion, and distortion. All the foregoing factors tend to cause the engine thrust to decrease over a period of time; hence, trimming is necessary to restore the rated performance of the engine. Generally speaking, when an engine indicates high exhaust-gas temperature (EGT) for a particular engine pressure ratio (EPR), it means that the engine is out of trim.

The procedure for trimming a particular engine is specified by the agency which operates the aircraft in which the engine is installed. In collaboration with engine manufacturers, airlines develop and publish correct procedures that should be followed carefully by the technicians who perform the job of trimming.

The following general principles for trimming are for information only.

1. Head the airplane as nearly as possible into the wind. Wind velocity should not be more than 20 mph [32.19 km/h] for best results. See that the area around the aircraft is clean and free from items which could enter the engine or cause other problems during the engine run.

2. Install the calibrated instruments required for trimming. One of these is a presssure gage to read turbine discharge pressure (P_{t7}) or **engine pressure ratio** (EPR). Another important instrument is the calibrated tachometer to read N_2 rpm.

3. Install a part-throttle stop or fuel-control trim stop as specified in the trim instructions.

4. Record ambient temperature and barometric pressure. These are necessary to correct performance readings to standard sea-level condition. The pressure and temperature information is used to determine the desired turbine discharge pressure or EPR by means of the **trim curve** published for the engine.

5. Start the engine and operate at idle speed for the time specified to ensure that all engine parameters have stabilized. Operate the engine at trim speed as established by the trim stop on the fuel control for about 5 min to stabilize all conditions. The overboard air-bleed valves should be fully closed and all accessory air bleed must be turned off.

6. Observe and record the P_{t7} or EPR to determine how much trimming is required or whether any is required. If trimming is required, adjust the fuel-control unit to give the desired P_{t7} or EPR. When this is attained, record the engine rpm, the exhaust-gas temperature, and the fuel flow.

7. The observed rpm is corrected for speed bias by means of a temperature-rpm curve to provide a new engine trim speed in percent corrected to standard conditions.

It must be pointed out that these procedures will vary considerably, depending on the type and model of engine being trimmed. The purpose of trimming for all engines, however, remains the same: to provide optimum engine performance without exceeding the limits of rpm and temperature established for the engine. As explained previously, engines equipped with computer-controlled fuel controls do not require periodic trimming because the adjustments are made automatically by the computer.

Engine Testing

The testing of a new or overhauled gas-turbine engine to ensure correct performance is accomplished on an instrumented test stand and procedures for testing are developed and published by the engine manufacturer. These procedures must be followed precisely to ensure that correct information is obtained regarding the performance of the engine.

The operation of an engine on a test stand is usually accomplished with a **bellmouth air inlet.** The purpose of this type of inlet is to eliminate any loss of air pressure at the compressor inlet. The reason for loss of pressure with a straight inlet and the effect of the bellmouth inlet is illustrated in Fig. 23-2. Since a large volume of air is drawn into the engine, a rapid increase in air velocity must take place as the air nears the inlet. Moreover, to supply the demand, air must flow from areas outside the direct frontal position of the engine. Much of the airflow will have to change direction almost 90° as it comes from the sides of the inlet and enters the compressor. This results in a pressure drop with a straight inlet duct. The bellmouth duct guides the air in such a way that there is essentially no pressure drop to the compressor inlet. If the bellmouth duct is protected by a screen, a certain amount of pressure drop will occur which must be taken into consideration when measuring the performance of the engine.

When testing a gas-turbine engine, it is common practice to measure certain essential parameters in order to evaluate the engine performance correctly. Among these parameters are the following:

1. Ambient air temperature T_{am}
2. Ambient air pressure P_{am}
3. Exhaust total pressure P_{t7}
4. N_1 rpm N_1
5. N_2 rpm N_2
6. Exhaust-gas temperature   EGT
7. Fuel flow pounds per hour (pph) W_f
8. Thrust F_a
9. Low-pressure compressor outlet pressure P_{s3}
10. High-pressure compressor outlet pressure P_{s4}

These parameters are usually adequate to determine engine performance, but others may be recorded if desired or necessary.

When an engine is assembled as a complete powerplant for a quick engine change (QEC), it is necessary to consider the equipment installed on the engine because it may affect some of the performance measure-

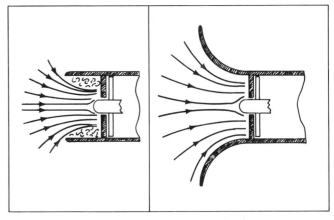

FIG. 23-2   Effect of a bellmouth air-intake duct.

ments. Oil flow and temperature will be changed as a result of the engine oil cooler and the engine pump. Likewise, fuel flow and pressures will be affected by the engine-driven fuel pump.

Because standard performance of an engine occurs only under standard conditions, air pressure and temperature must be corrected to standard conditions. This is accomplished by means of correction factors designated by the Greek letters **delta** (δ) and **theta** (θ). *Delta* is used to correct for pressure and *theta* provides the correction for temperature.

The values for *delta* and *theta* may be found on an appropriate chart or they may be computed as follows:

$$\delta = \frac{P}{P_0} = \frac{P}{29.92}$$

$$\theta = \frac{T}{T_0} = \frac{t(°F) + 460}{519}$$

where P = observed barometric pressure (in HG abs)
P_0 = standard-day barometric pressure
T = temperature, °R (°F + 460)
T_0 = standard-day temperature, 519°R

If Kelvin degrees are used to indicate absolute temperature, then the Celsius or centigrade scale is used. °C + 273 converts centigrade to Kelvin. Standard-day temperature in degrees Kelvin is 288.

To apply δ and θ to correct measurements, the following methods are employed:

$$N_2 \text{ (corrected)} = \frac{N_2 \text{ (observed)}}{\sqrt{\theta_{t2}}}$$

$$\text{EGT °R (corrected)} = \frac{\text{EGT (observed)} + 460}{\sqrt{\theta_{t2}}}$$

$$W_f \text{ (corrected)} = \frac{W_f \text{ (observed)}}{d_{t2}\sqrt{\theta_{t2}}}$$

$$F_n \text{ (corrected)} =$$

$$\frac{F_n \text{ (observed)} + \text{Inst corrected} + \text{cell corrected}}{\delta_{t2}}$$

The values observed and computed as shown for N_2 rpm, exhaust-gas temperature, and fuel flow are recorded and plotted on a chart, as shown in Fig. 23-3. Note that the pressure ratios for the low-pressure compressor and high-pressure compressor are recorded. These pressure ratios are indicated as P_{s3}/P_{am} for the low-pressure compressor and P_{s4}/P_{am} for the high-pressure compressor. The net thrust (F_n) of an engine can be determined directly from the thrust meter in the cell and can also be found from the EPR and an EPR conversion table for the engine.

The foregoing discussion is presented as an example of how the performance of an engine can be determined. In actual practice, it will be found that other tests may be performed and other parameters measured. For any particular type or model of engine, specific instructions are made available by the manu-

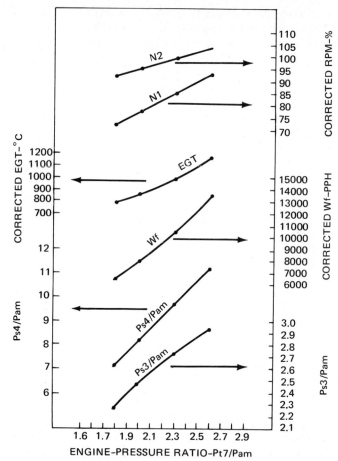

FIG. 23-3 Typical operational curves for a gas generator.

facturer for the testing of the engine in a test cell or on the aircraft.

Operational Checks

To ensure that a gas-turbine engine is in satisfactory operating condition, engine and aircraft manufacturers specify certain **operational checks** to be routinely performed by maintenance personnel. The particular types of checks and the procedures to be followed vary, depending on the type of engine and aircraft involved. In this section, to provide an example of typical checks, the checks recommended for the General Electric CF6-50 engine are described briefly.

Dry motoring check: The dry motoring check may be required during or after inspection or maintenance to ensure that the engine rotates freely, that instrumentation functions properly, and that starter operation meets speed requirements for successful starts. This check is also used to prime and leak-check the lubrication system when maintenance has required replacement of system components.

To perform a dry motoring check, the technician engages in the following procedure:

1. Ascertains that all conditions required prior to a normal start are met. These conditions can be established by conducting a normal prestart inspection.
2. Positions engine controls and switches as follows:

   (a) Ignition, OFF.
   (b) Fuel shutoff lever, OFF.

(c) Throttle, IDLE.

(d) Fuel boost, ON.

3. Energizes the starter and motors the engine as long as necessary to check instruments for positive indications of engine rotation and oil pressure.

4. Deenergizes the starter and makes the following checks during coastdown:

(a) Listens for unusual noises. Checks for roughness. Normal noise consists of clicking of compressor and turbine blades, gear noise, and seal rubs.

(b) Inspects the lubricating system lines, fittings, and accessories for leakage.

(c) Checks the oil level in the oil tank.

Wet motoring check: When it is necessary to check the operation of fuel-system components after removal and replacement or to perform a depreservation of the fuel system, the wet motoring check is employed. This is accomplished as follows:

1. Position the engine controls and switches as for a dry motoring check.

2. Energize the starter.

3. When core engine speed (N_2) reaches 10 percent, place the fuel shutoff lever to ON and check for oil pressure indication.

4. Continue motoring the engine until the fuel flow is 500 to 600 lb/h [226.80 to 272.16 kg/h] or for a maximum of 60 s. Observe the starter operating limits.

5. Place the fuel shutoff lever to OFF and continue motoring the engine for at least 30 s to clear the fuel from the combustion chamber. Check to see that fuel flow drops to zero.

6. Deenergize the starter and during coastdown check for unusual noises as mentioned for a dry motoring check.

7. Inspect the fuel system lines, fittings, and accessories for leakage.

8. Check the concentric fuel shroud for leakage. No leakage is permitted.

9. Inspect the lubrication system for leakage.

10. Check the oil level in the oil tank.

Idle check: The idle check consists of checking for proper engine operation as evidenced by leak-free connections, normal operating noise, and correct indications on engine-related instruments. Engine drain lines must be disconnected from drain cans to check for leakage.

After the engine is started according to approved procedure, the following steps are taken:

1. Stabilize engine at *ground idle*.

2. Check fan speed (N_1), core-engine speed (N_2), oil pressure, and exhaust-gas temperature (EGT) to see that they are within the proper ranges according to the *ground-idle speed chart* and engine specifications. Engine speeds will vary according to compressor inlet temperature (T_{t2}).

3. Visually inspect fuel, lubrication, and pneumatic lines, fittings, and accessories for leakage.

4. Deenergize flight-idle solenoid. During operations above ground idle, do not exceed the open-cowling limitations imposed by the airframe manufacturer.

5. Stabilize at *flight idle* and check the same parameters checked for ground idle. See that they are within the limitations set forth on the *flight-idle speeds charts*.

Power assurance check: The power assurance check is performed to make sure that the engine will achieve takeoff power on a hot day without exceeding rpm and temperature limitations. During the tests, the engine being tested is not used to supply power for any aircraft systems—electric, hydraulic, or other. The engine is tested at 50 percent, 75 percent, and maximum power.

During engine operation for the power assurance checks, exhaust-gas temperature (EGT) must be observed constantly to avoid the possibility of overtemperature. Should the temperature approach maximum allowable, the throttle must be retarded sufficiently to hold the EGT within limits. In the operation of the engine, the throttle should always be moved slowly.

To perform the power assurance test, the following steps are taken:

1. Set the engine power at nominal N_2 speed as indicated on the appropriate chart for the total air temperature (TAT). For example, the nominal N_2 curve on the chart may coincide with the 91.8 percent line at 10°C for the 50 percent power setting. The throttle will therefore be adjusted to produce 91.8 percent rpm when the TAT is 10°C for a 50 percent power setting.

2. Four minutes after the throttle lever is set, record the average readings of TAT, N_1 speed, N_2 speed, EGT, EPR (engine pressure ratio), and fuel flow (W_f). Correct W_f for local barometric pressure in accordance with instructions.

$$\text{Corrected } W_f = \frac{\text{observed } W_f \times 29.92}{\text{actual barometric pressure}}$$

3. Using N_1 (where N_1 = target N_1 − observed N_1) as a correction factor, adjust readings according to the parameter adjustments set forth in the operations manual.

In the operation of gas-turbine engines of any type, it must be emphasized that temperatures and rpm for both N_1 and N_2 must be watched carefully. If it is expected that a beyond-limits condition is developing, the operator should take immediate action by retarding the throttle or shutting the engine down.

Before shutting down a hot engine, it should be operated at ground-idle speed for about 3 min to permit temperature reduction and stabilization. As soon as the engine is shut down, the EGT gage should be observed to see that EGT starts to decrease. If EGT does not decrease, it indicates an internal fire and the engine should be dry-motored at once to blow out the fire.

During coastdown after the engine is shut down, a technician should listen for unusual noises in the engine such as scraping, grinding, bumping, squealing, etc.

TYPICAL GAS-TURBINE ENGINE INSPECTIONS

Because of the great variety of gas-turbine engines in existence, no attempt will be made in this text to give instructions about inspection and maintenance of specific gas-turbine engines. We shall, however, examine some of the conditions common to the majority of gas-turbine engines and provide information supplied by manufacturers and operators as examples of typical methods and processes. It must be remembered that a particular method or process approved for one type of engine may not be satisfactory for another type of engine. *It is essential, therefore, that all inspection and maintenance practices be done in accordance with manufacturer's and operator's maintenance manuals.*

The inspections established for gas-turbine engines fall into a number of classifications and are dependent on the type of operation to which the engines are subjected. For example, an airline whose routes are long will need an inspection schedule different from a local-service airline where flights are short and takeoffs and landings are frequent. Inspections and maintenance procedures are scheduled, therefore, with consideration given for the number of **flight cycles** an engine has experienced as well as for the total hours of engine operation. A flight cycle is one takeoff and landing.

To illustrate the difference in the number of flight cycles which may be imposed on an engine in different types of operation, we may consider a commuter operation from Los Angeles to San Francisco as compared with an overseas flight to the orient. A commuter airplane may accumulate 12 h of operation time while completing 15 flight cycles. On the other hand, a transpacific flight to Hong Kong may put 18 h of operation time on the engines while completing only three flight cycles. It is clear that the wear, erosion, and heat damage will be much greater on the commuter airplane engines than it will for the transpacific engines. Accordingly, inspection and maintenance operations will have to be scheduled more often for the commuter airplane engines.

Periodic inspections are required after a given number of operation hours or flight cycles or a combination of both. These inspections may be classed as *routine,* *minor,* or *major.* When these inspections are to be performed is established by the operator of the aircraft in accordance with information supplied by the engine manufacturer and with the results of operational experience.

In addition to the periodic inspections performed on a regular basis, airlines often specify turnaround inspections which may be designated as "light checks," "A" checks, or some other nomenclature. Each airline has its own classifications for inspections; the only way for a person to know what is to be inspected and how it is to be inspected is to get the information from the check sheet furnished by the company.

Routine Operational Inspections

A typical airline may designate standard service operations and inspections by such names as no. 1 service, no. 2 service, "A" check, and "B" check. These various operations will include a number of standard operations plus special operations as needed.

A no. 1 service may be performed by station personnel and by the flight engineer each time the airplane lands or after several landings, depending on the time of flight. Usually the service will include correction of critical log items as well as regular service (fuel and resupply) plus a **walk-around** inspection. The walk-around inspection includes visible inspection of all items which can be observed from the ground. The engine inspection at this time includes a look at the engine inlet and fan, observation of any fuel or oil leakage from engine pods, and an examination of the tail pipe and turbine section with a flashlight.

The no. 2 service may include the following for the engines:

1. Review of the flight log and cabin log
2. Check of engine oil quantity
3. Visual inspection of the engines with cowls open

The "A" check discussed here is performed after approximately 100 h of operation. Inspections and service relating to the engines are as follows:

1. Service oil tanks to FULL. Enter in the inspection records the number of quarts added for each engine.
2. Service the constant-speed drive (CSD) as required.
3. Check engine inlet, cowling, and pylon for damage. Check for irregularities and exteriodrain leakage.
4. Inspect the engine exhaust section for damage using a strong inspection light. Note condition of rear turbine.
5. Check the thrust-reverser ejectors and reverser buckets for security and damage.
6. Check the reverser system with ejectors extended for cracks, buckling, and damage.

The "B" check is more comprehensive than the "A" check; it includes the following:

1. Check engine nose cowl, inlet chamber, guide vanes, and first-stage compressor blades using a strong inspection light.
2. Check engine, installations, midsection, and cowling. Spray cowling latches with approved lubricant.
3. Check the fire extinguisher indicator disks.
4. Perform oil filtering in accordance with maintenance manual.
5. Remove oil screen and check for carbon and metal.
6. Install oil screen and torque-screen cover nuts to proper value (approximately 25 to 30 lb·in [2.83 to 3.39 N·m]).
7. Check oil quantity within 2 h after engine shutdown and add approved oil as required. Enter oil added on work-control record.
8. Check constant-speed drive (CSD) oil. Add approved oil as required but do not overfill.
9. Check starter oil. Add approved oil to level of filler port. Make a record of oil added.

10. Check the ignition system as follows:

  (a) Position four air-bleed switches on the air-conditioning panel to OFF.
  (b) Move start lever to IDLE.
  (c) Position start-control switch on overhead panel to FLIGHT position.
  (d) See that the igniter at the no. 7 combustion chamber is firing.
  (e) Return controls to OFF position.
  (f) Position start-control switch to GROUND position.
  (g) Move start lever to the START position.
  (h) See that the igniters are firing by the use of an approved tester.
  (i) Return controls to the OFF position.
  (j) Close engine cowling and check security of latches and inspection plates.

In checking the ignition systems for a modern gas-turbine engine, it is important that body contact not be made with the high-energy output. The voltage is such that the current flow through the body could be fatal.

11. Check reversers and deflector doors as follows:

  (a) Place reverser in reverse-thrust position. Install lock clamps and warning tags on reverse levers.
  (b) Using a strong inspection light, check the tail pipe and fairing, the reverser clamshells, the turbine exit area, outlet guide vanes, and rear turbine blades.
  (c) Check deflector doors and fittings for cracks.
  (d) Check for delamination of the door inner and outer skin using an inspection light and testing by hand.
  (e) Check the inner and outer skin for dents, cracks, and punctures.
  (f) Check the deflector door forward link to the support pivot bolt for tightness. The bolt should not turn by hand.
  (g) Check the bolts which secure the forward link support assembly to the reverser structure with a wrench.
  (h) Check the deflector drive pivot bolt. It should not turn by hand.
  (i) Check the deflector door stops for excessive looseness and lubricate door-link and rod end bearings.

These notes are given as examples of possible inspection procedures and are not necessarily appropriate for any particular aircraft. It must be remembered that all aircraft and engine combinations have specific procedures established and approved. The approved procedures should be followed in all cases.

Nonroutine Inspections

During the operation of a gas-turbine engine, various events may occur which require an immediate special inspection to determine whether the engine has been damaged and what corrective actions must be taken.

Among some of the events which require special inspections are foreign-object ingestion, bird ingestion, ice ingestion, overlimit operation (temperature and rpm), excessive G loads, and any other event that could cause internal or external engine damage.

Nonroutine inspections require the same inspection techniques utilized for daily and periodic inspections. These include unaided visual inspections, inspection with lights, use of magnifiers, application of fluorescent or dye penetrants, use of a borescope or chamberscope, and use of radiography techniques. Usually the maintenance manual for the engine will specify which technique is most effective for any particular inspection.

Borescope and Chamberscope

The borescope was used for many years as a device for examining the inside of cylinder bores on reciprocating engines. When not in use for extended periods of time these engines were subject to corrosion on the cylinder walls and, if not corrected before the engine was placed in operation, the piston rings would soon wear out. This, of course, caused the engines to lose power and consume excessive oil. The borescope was effective in combating the problem.

The borescope was derived from similar instruments developed by the medical profession. Cystoscopes and similar devices have been employed by doctors for many years to examine interior body cavities. The same technique is now effective in examining the interior parts and surfaces of gas-turbine engines.

The borescope may be compared with a small periscope (see Fig. 23-4). At one end is an eyepiece with one or more lenses attached to the light-carrying tube. At the end of the tube is a mirror, lens, and strong light. The tube is inserted through engine **borescope ports** located in the engine case at points necessary to examine all critical areas inside the engine. The ports are normally closed with removable plugs. A drawing to illustrate one type of borescope system is shown in Fig. 23-5.

When borescope inspections are to be performed, the technician should identify the plugs as they are removed to be sure that they are reinstalled in the same ports from which they were removed. Upon reinstallation, the threads and pressure faces of the plugs should be lightly coated with an antiseize compound such as MIL-T-5544 or equivalent.

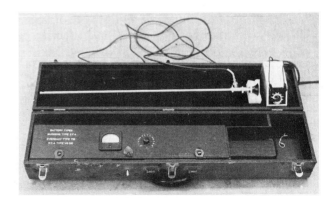

FIG. 23-4 One type of borescope.

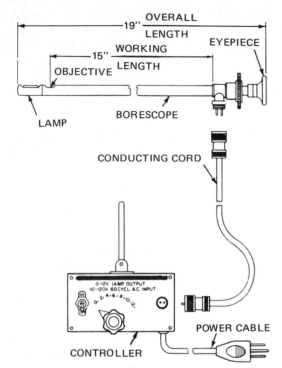

FIG. 23-5   A borescope system.

The **chamberscope** is a specially designed borescope for use in examining the interior of combustion chambers and combustion-chamber liners. Provision is usually made in the design of chamberscopes and borescopes to photograph the interior parts of engines. This is accomplished by means of a camera adapter together with a strobe light for interior illumination.

Foreign-Object Damage (FOD)

Foreign-object damage to a gas-turbine engine may consist of anything from small nicks and scratches to complete disablement or destruction of the engine. The flight crew of an aircraft may or may not be aware that FOD has occurred during a flight. If damage is substantial, however, it will be indicated by vibration and changes in the engine's normal operating parameters. Damage to the compressors or turbines usually results in an increase in EGT, a decrease in EPR, and a change in the rpm ratio between the core engine and the fan section (N_2/N_1 ratio).

When FOD has occurred, the inspections required depend on the nature of the foreign object or objects. If an external inspection indicates substantial damage to the fan section or to inlet guide vanes, the engine will have to be removed and overhauled. If the damage to the forward sections of the engine is slight, a borescope inspection of the interior of the engine may make it unnecessary to remove the engine. Damage to vanes, fan blades, and compressor blades can be repaired if it does not exceed certain limits specified by the manufacturer. If the engine operates normally after repairs are made, it can be placed back in service.

Inspections for Overlimit Operation

Even though technicians and flight crews take every precaution possible to prevent overlimit operation of an engine, such operations sometimes occur. Often the cause is a malfunction of the engine fuel control or a malfunction in the engine. In any case, when overlimit operations do occur, it is necessary to perform certain inspections to determine what damage may have been caused.

At starting, the most critical parameter for the engine is EGT. The technician or crew member starting the engine must watch the EGT gage carefully and continuously when the throttle is moved to the IDLE position. As soon as lightoff occurs, there is a rapid rise in EGT; but if all systems are working properly, the EGT should not overshoot. If it does, the person starting the engine should immediately retard the throttle to reduce fuel flow to the combustion chamber.

The technician who starts a gas-turbine engine should be familiar with the operating limitations. Figure 23-6 is a temperature-limit chart for starting a large, high-bypass engine. Note that any temperature above 675°C (1247°F) is cause for special attention. Temperatures that fall in area A require special inspections, and temperature-time values that fall in area C are cause for engine overhaul.

After an engine has been started and the operation is stabilized at ground idle, higher temperatures can be permitted during taxiing and preparation for takeoff. The chart of Fig. 23-7 shows temperature-time limitations for operations other than starting. The charts of Figs. 23-6 and 23-7 are applicable to one engine only and are not typical of all engine limitations.

If a gas turbine has been operated above the limits set for EGT but at a level not high enough to call for

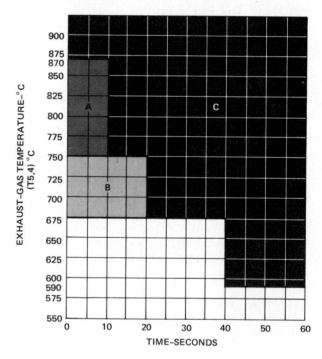

NOTE: ANY OPERATION IN AREAS "A", "B", OR "C" MUST BE VERIFIED AND INDICATION SYSTEM CALIBRATED'

Starts in area A must be recorded and immediate corrective action must be taken prior to further start attempts. A borescope inspection must be performed, prior to further start attempts.

Starts in area B must be recorded. Persistent starts in this area are cause for corrective action

Any operation in area C requires removal of engine to overhaul.

FIG. 23-6   Temperature-limit chart for starting.

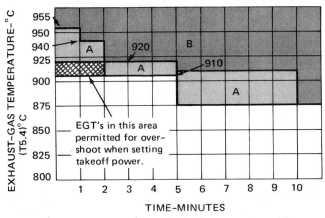

FIG. 23-7 Temperature-time limits for operations other than starting.

removal and overhaul, a borescope inspection is usually called for. An external visual inspection of the hot section of the engine should also be made. In this inspection, the hot section of the engine is checked for indications of burn-through or metal distortion due to excessive heat. If such indications are found, the section must be disassembled for further inspection and repair.

When borescope inspections are called for after operation at excessive temperature, the following are inspected:

1. Combustion chamber and liner assembly to determine if cracks and burned areas exceed those permissible as specified by the manufacturer.

2. Fuel nozzles for excessive carbon buildup or plugged orifice.

3. First-stage high-pressure turbine (HPT) nozzle for cracks, burned areas, warping, and plugged cooling-air passages. Serviceable limits for defects specified in the maintenance manual.

4. Second-stage HPT nozzle for defects as listed.

5. HPT rotor for cracks, tears, nicks, dents, and metal loss. Cracks in the turbine blades are cause for removal and replacement. Dents and nicks within certain limits may be permitted in the second-stage blades as specified by the manufacturer.

6. Turbine midframe liner for cracks, nicks, dents, burns, bulges, and gouges. Limitations for these defects are specified by the manufacturer. Bulges associated with heat discoloration are cause for rejection.

7. First-stage low-pressure turbine (LPT) nozzle for cracks, nicks, dents, burns, etc., as for other turbine sections.

8. LPT stator assembly as above.

9. LPT rotor assembly as above. No cracks are permitted in any turbine blades. Limited dents and nicks are allowed.

Overspeed inspections: Overspeed inspections for a typical high-bypass fan engine is primarily concerned with rotating assemblies. One manufacturer specifies the following inspections if the fan section has been operated at speeds from 116 to 120 percent rpm:

1. Check the fan rotor for freedom of rotation.

2. Check the first-stage fan shroud for excessive rub.

3. Inspect the low-pressure compressor with a borescope.

4. Inspect the inlet and the exhaust nozzles for particles.

5. Inspect all four stages of the LPT with a borescope for blade and vane damage. Inspect the fourth-stage blades through the exhaust nozzle.

If the fan speed has exceeded 120 percent, the fan rotor, fan midshaft, and LPT rotor must be removed, disassembled, and inspected in accordance with instructions.

If the core-engine rotor (high-pressure compressor and high-pressure turbine) has been operated at speeds from 107 to 108.5 percent, the following inspections are specified:

1. Inspect the exhaust nozzle for particles.

2. Inspect the core compressor with a borescope for blade and vane damage.

3. With a borescope, inspect the HPT for blade damage.

If the core-engine rotor has been operated above 108.5 percent, the engine must be removed, disassembled, and inspected according to instructions.

Other Special Inspections

In addition to inspections discussed so far, there are certain events which happen occasionally that require special attention. Among these are fire damage, operation with no oil pressure, accident damage, and engine stall. Each inspection required depends on the nature and severity of the event, but they often require removal of the engine for disassembly and possible major overhaul.

When an engine has been operated with no oil pressure for more than 2 min, the engine must be removed for overhaul. If an engine is involved in an accident, the nature and severity of the accident is taken into account. The matter is frequently referred to the engine manufacturer's technical representative.

Fan Blade Shingling

Fan blade shingling is the overlapping of the midspan shrouds of the fan blades. When the blades of a rotating fan encounter resistance which forces them sideways an appreciable distance, shingling will take place. Figure 23-8 provides a simplified illustration of the situation which causes shingling.

Shingling can be caused by engine stall, bird strike, FOD, or engine overspeed, in which case the fan must be inspected at both the upper and lower surfaces of the midspan shrouds for chafing, scoring, and other damage adjacent to the interlock surfaces. All blades that are overlapped or show indication of overlapping must be removed and inspected according to the appropriate manual. The blades and shrouds should be checked with a dye or fluorescent penetrant to detect cracks. No cracks are permitted in the fan blades. Blade tips are examined for curl and the lightening holes are checked for cracks and deformation. All blades that have overlapped and are found to be serviceable may be reinstalled, but they must be removed within an additional 50 h of operation for inspection.

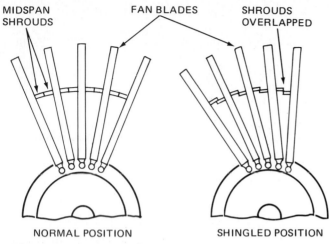

MIDSPAN SHROUDS   FAN BLADES   SHROUDS OVERLAPPED

NORMAL POSITION   SHINGLED POSITION

FIG. 23-8 Shingling of fan blades.

Additional inspections in the fan area include the abradable material for damage due to rubbing of fan blade tips and damage to the fan-speed sensor head for damage due to blade contact.

● TURBINE ENGINE TROUBLESHOOTING

The troubleshooting of turbine engines follows, in general, the procedures traditionally employed for reciprocating engines; however, new and improved techniques have been developed which aid considerably in identifying and solving technical problems. **Troubleshooting** may be defined as the detection of fault indications and the isolation of the fault or faults causing the indication. When the fault is isolated or identified, the correction of the fault is simply a matter of applying the correct procedures.

Manufacturers and operators of gas-turbine engines work together to develop information and techniques regarding the operation of the engines to establish techniques for troubleshooting. Numerous systems have been developed by which faults are detected and analyzed to isolate and correct. We shall not attempt to describe all such systems; however, a discussion of some typical systems and techniques will give the technician an understanding of the procedures involved.

Fault Indicators

Fault indicators include any instruments or devices on an aircraft which will give a member of the crew information about a problem developing in the operation of the engine. These indicators may be divided into two groups: the standard engine instruments used to monitor the operation of the engines; and special devices designed to detect indications of trouble which may not be revealed by the engine instruments. Typical engine instruments for a gas-turbine engine are exhaust-gas temperature (EGT) gages, percent rpm gages (N_1 and N_2), engine pressure ratio (EPR) gage, oil temperature gages, oil pressure gages, and fuel gages. Where turboshaft or turboprop engines are installed, torque indicating gages are often included. These instruments are all effective in detecting faults.

In addition to the standard instruments, built-in troubleshooting equipment (BITE) systems are often installed. These systems include special sensors and transducers which produce signals of vibration and other indications that are indicative of problems developing.

The FEFI/TAFI System

The FEFI/TAFI system has been developed by McDonnell Douglas Corporation to be used with the DC-10 airliner as a means of fault detection and isolation for not only the engines but also the entire aircraft and its operating systems. FEFI stands for **flight environment fault isolation** and TAFI stands for **turn-around fault isolation.** As the names indicate, the FEFI is employed by the flight crew to develop information by which the ground crew can isolate faults and apply corrective measures. The TAFI utilizes the FEFI information and follows procedures designed to isolate faults most effectively.

The FEFI/TAFI approach to fault isolation is founded on the premise that the majority of system failures cause an observable pattern of fault symptoms. The indications may be any type recognizable by the flight crew or ground maintenance personnel, such as abnormal instrument readings, warning lights, warning flag, sound, vibration, or a combination of one or more of these.

Very often a unique pattern of indications can be directly attributed to one specific cause. When this occurs, it is only necessary to take the appropriate corrective action, such as replacement of a component or another specific task. When a failure pattern can be established from any one of several different sources, it is necessary to make further checks to identify the actual cause. This is the function of TAFI as utilized by the technician.

The two manuals involved in the FEFI/TAFI system are shown in Fig. 23-9. The FEFI manual contains fault symptom pages for each airplane system including powerplant systems. The manual is prepared specifically for the flight crew as an on-aircraft document for use in the flight environment. When a particular fault is detected, the manual is used to obtain a **fault code.** The fault code is entered in the flight log as an integral part of the maintenance-discrepancy record.

As mentioned previously, the TAFI manual is designed for use by ground maintenance personnel. It contains provisions to key the fault code included in the flight-log report to a specific corrective action. The manual also provides the maintenance crew with definitive fault-isolation sequences called "fault trees," component locations, supplemental data, and simplified schematics.

The FEFI manual can be considered the "eyes" of the FEFI/TAFI system. It is used to equate a particular fault symptom to a fault code and report that code as a part of the flight-log maintenance-discrepancy entry. There is a secondary benefit in that it permits ground maintenance personnel, through an equivalent fault symptom page in the TAFI manual, to view the fault as the flight crew did in the actual flight environment.

The chapters in the FEFI manual are grouped and identified identical to the sequence in the Douglas Aircraft *Flight-Crew Operating Manual.* In addition, the manual is indexed to provide for easy location of the page involved for any fault.

FIG. 23-9 FEFI/TAFI manuals. *(McDonnell Douglas Corp.)*

The page of the FEFI manual dealing with engine flameout and compressor stall is shown in Fig. 23-10. Note that each engine-fault description is followed by the number 71-00 and a letter code and a number which indicates which engine is affected. The number 71-00 is the Air Transport Association specification number identifying powerplant, general, in accordance with specification ATA-100. The letter code identifies the particular fault indicated and the number following shows which engine is affected.

As an example of the use of the FEFI manual, assume that engine 3 of an aircraft has suddenly lost power in flight and that the fuel flow gage shows adequate fuel pressure. The flight-crew member will refer to the page shown in Fig. 23-10 and locate the description of the fault. One will then code the fault 71-00 EA 3 and record the information in the flight log. In addition, one will give any other information one may have regarding the suspected cause of the fault, if any. The ground crew, after noting the coding of the fault,

will turn to the 71-00-00 section of the TAFI manual and locate the page corresponding to the fault code. From this the fault description is noted and the *Maintenance Action* page of the TAFI manual is consulted to determine the fault-isolation sequence. The sequence for code 71-00 EA is shown on the page illustrated in Fig. 23-11.

The TAFI manual includes **locator pages** in each section which show where various items of engine equipment and accessories are located. One such page is shown in Fig. 23-12.

Schematic drawings of fluid and electric systems are often needed to aid in fault isolation. The TAFI manual includes simplified schematic diagrams wherever they are needed for a particular system. One of the pages related to powerplant troubleshooting is shown in Fig. 23-13.

Fault Isolation

As previously explained, manufacturers and operators of engines have worked together to develop information to aid in identifying and isolating engine faults. From this information, troubleshooting charts are developed to establish the most logical sequence of actions to isolate faults. The first item to be checked in case of a fault is the item most likely to be the cause. If this is found satisfactory, the next most likely item is checked, and so on until the fault is isolated. The sequence of items to be checked is listed in the troubleshooting chart for each model and type of engine. The Maintenance Action page illustrated in Fig. 23-11 shows how a troubleshooting chart is used for one particular fault. The chart shown has been referred to as a "fault tree."

The technician working on the isolation of engine faults should refer to the specific troubleshooting chart or procedure for the engine being checked. When an engine accessory is found to be the cause of the fault, the accessory troubleshooting charts should be used to determine the trouble and make corrections.

Some typical engine indications and faults are given as follows:

Indication: No engine rotation when attempting a start

Probable faults:

1. Defective air valve
2. Defective control circuit to air valve
3. Seized compressor rotor, starter, or gearbox
4. Seized oil or scavenge pump
5. Foreign-object damage or blockage

Indication: Low engine rotational speed during starting

Probable faults:

1. Low starter air pressure
2. Pneumatic leaks
3. Malfunction of air valve
4. Compressor or fan blades rubbing on airseals
5. Internal engine damage

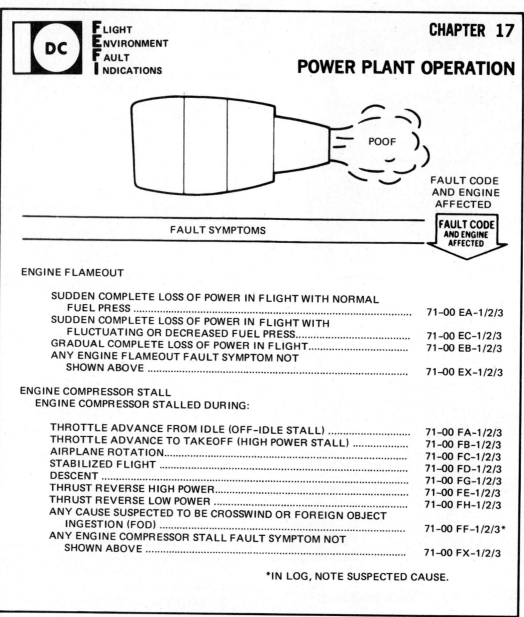

FIG. 23-10 FEFI fault code page. *(McDonnell Douglas Corp.)*

Indication: Slow acceleration after start (hung start)

Probable faults:

Same as for low engine rotational speed

Indication: Engine rotates but will not light off

Probable faults:

1. No ignition—defective igniter unit or circuitry
2. No fuel flow to nozzles—defective fuel-control unit or main-engine control (MEC)

Indication: Slow acceleration after lightoff

Probable faults:

1. Defective fuel-control unit
2. Air bleeds not operating on schedule
3. Same causes as for low rotational speed

Indication: No starting-fuel flow

Probable faults:

1. Air in fuel lines
2. Fuel shutoff valve not open
3. Fuel shutoff system defective
4. Defective fuel pump

Indication: Low starting-fuel flow

Probable faults:

1. Air in fuel lines
2. Aircraft fuel shutoff valve in wrong position
2. Aircraft boost pressure low
4. Fuel shutoff system not properly rigged
5. Fuel flow indicating system defective
6. Compressor discharge pressure (CDP) system not functioning correctly

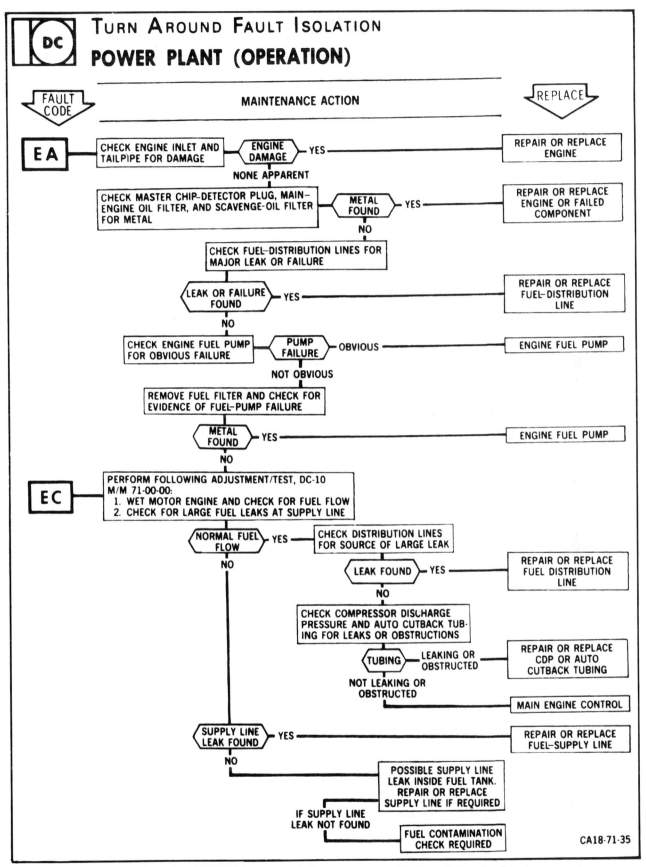

FIG. 23-11 TAFI Maintenance Action sequence. *(McDonnell Douglas Corp.)*

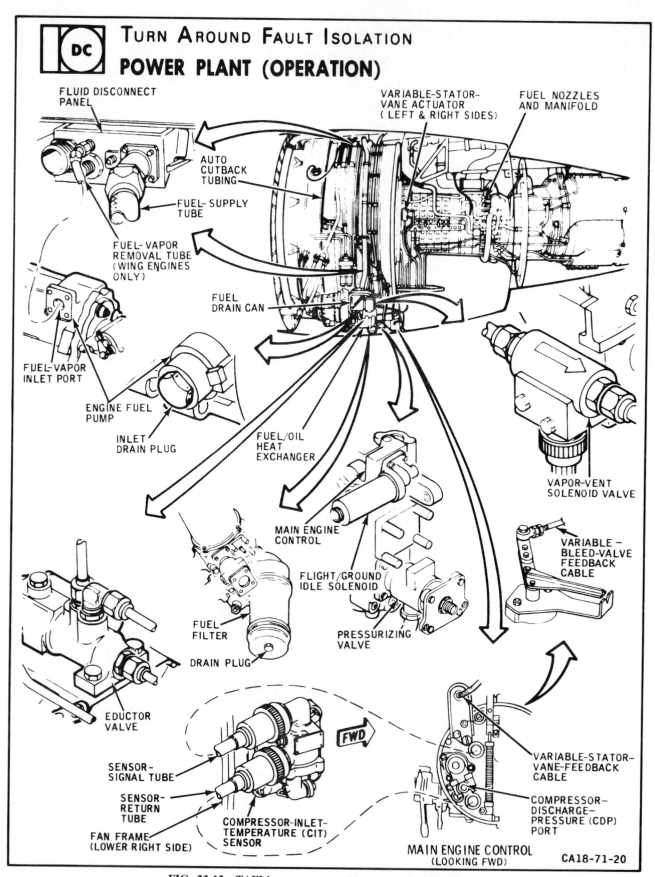

FIG. 23-12 TAFI locater page. (*McDonnell Douglas Corp.*)

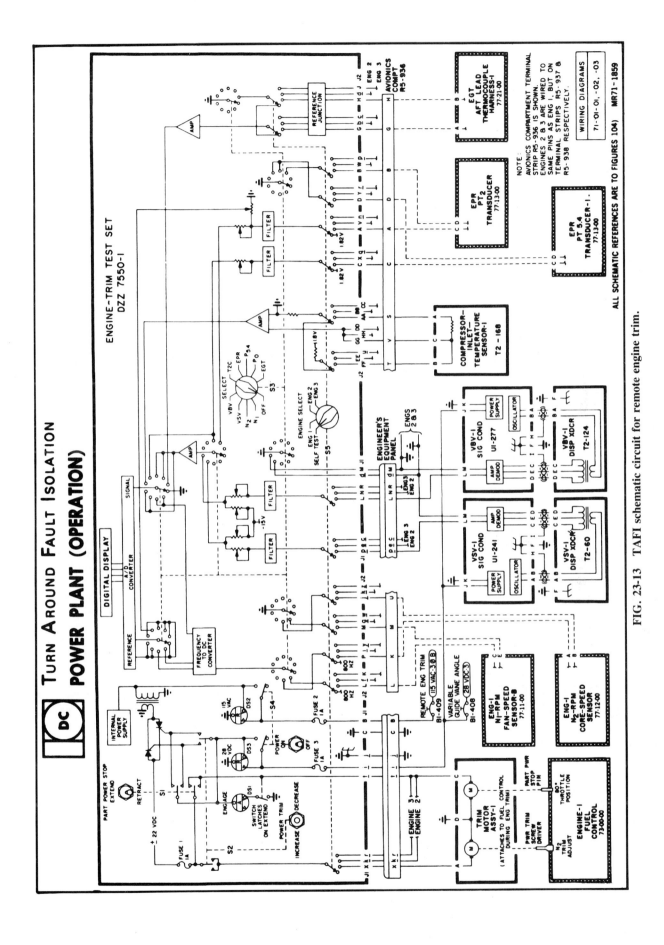

FIG. 23-13 TAFI schematic circuit for remote engine trim.

7. Compressor inlet temperature (CIT) system not functioning correctly

8. Main-engine control (MEC) not functioning correctly

9. Fuel pump defective

Indication: Excessive starting-fuel flow

Probable faults:

1. Flow indicating system defective
2. Fuel-control unit defective
3. CIT sensor malfunctioning

Indication: No lightoff; fuel flow and ignition OK

Probable faults:

1. Manifold drain valve open
2. Fuel nozzle near igniter plug has improper spray pattern

Indication: EGT too high at starting; hot start

Probable faults:

1. EGT-indication system defective
2. Low maximum motoring speed
3. Variable vane system not rigged correctly
4. Variable bypass valve not scheduling correctly
5. Damaged blades or vanes
6. Fuel-control unit malfunction—allows excessive fuel flow
7. CIT sensor malfunctioning

Indication: EGT too high at stabilized rpm

Probable faults:

1. EGT-indication system defective
2. Compressor bleed system malfunctioning
3. Dirty compressor blades
4. Damaged compressor blades or vanes
5. Damaged or eroded turbine blades

Indication: Fan section does not rotate

Probable faults:

1. Fan rotor or low-pressure turbine seized
2. Indication system defective

Indication: Core-engine idle rpm low

Probable faults:

1. Indicating system defective
2. Fuel shutoff-valve system improperly rigged
3. CDP system malfunction
4. Fuel-control unit malfunction
5. CIT sensor malfunction

Indication: Core-engine idle rpm too high

Probable faults:

1. Throttle rigging incorrect
2. Idle solenoid malfunction
3. Fuel-control unit malfunction

Indication: Fluctuating rpm

Probable faults:

1. Fuel boost pressure unstable
2. Variable vane rigging incorrect
3. Variable vane scheduling incorrect
4. Variable bypass-valve scheduling incorrect
5. Fuel-control unit malfunction

Indication: Sluggish response to power lever

Probable faults:

1. Variable vane rigging incorrect
2. Variable vane scheduling incorrect
3. Variable bypass-valve rigging incorrect
4. Variable bypass-valve scheduling out of adjustment
5. CDP tube to fuel-control unit leaking
6. Damaged vanes or blades
7. Fuel-control unit out of adjustment or malfunctioning

Indication: Compressor stall on acceleration

Probable faults: Same as for previous indication

Indication: Fan speed low for power-lever position

Probable faults:

1. Power-lever rigging incorrect
2. Fuel shutoff lever rigging incorrect
3. Fuel leak
4. CDP tube to fuel-control unit leaking
5. Variable vane rigging incorrect
6. Variable vane scheduling out of adjustment
7. Variable bypass-valve rigging incorrect
8. Variable bypass-valve scheduling out of adjustment

Indication: Fan vibration excessive

Probable faults:

1. Instrumentation pickups malfunctioning
2. Damaged fan or turbine blades
3. Fan rotor system out of balance

Indication: Engine pressure ratio (EPR) incorrect

Probable faults:

1. Indicator or transmitter malfunction
2. EPR probe damaged
3. Fan-speed instrumentation defective
4. Internal engine damage

These listings cover many of the problems which may occur in the operation of gas-turbine engines; however, there are more than listed here, so the technician should consult the charts for specific engines and accessories in order to diagnose faults and their sources correctly. Note that there are relatively few causes for fault indications and these may be listed as follows:

Engine damage: blades, vanes, combustion chamber, etc.
Fuel-control unit or main-engine control out of adjustment or malfunctioning
Sensing systems malfunction: CDP, CIT, EGT, EPR, vibration pickups, rpm pickups, thermocouples, etc.
Engine-control rigging: power lever, fuel shutoff valve, variable vane control, reverser control, etc.
Aircraft fuel system defects: pumps, screens, valves, fuel contamination, plumbing, ice.

This list is not all-inclusive; however, it does identify the most common sources of faults.

● JETCAL ANALYZER/TRIMMER SYSTEM

As mentioned previously, a variety of detectors, sensors, instruments, and systems have been developed to detect incipient and existing faults and to provide information regarding the adjustment and calibration of operating units. Some systems are installed aboard the aircraft so that fault indications can be obtained during flight, and others are designed as ground-support equipment. One such system is the Jetcal Analyzer/Trimmer manufactured by Howell Instruments, Incorporated, Fort Worth, Texas. The complete unit is illustrated in Fig. 23-14.

Description

The Jetcal analyzer consists of an instrument case and an accessory case mounted on wheels for mobility, as shown in Fig. 23-14. By means of the case handle, the unit can be easily moved about.

Housed in the instrument case is a **probe controller assembly** and a portable **trimmer assembly.** The trimmer assembly may be removed from the instrument case to be used as a separate instrument in the aircraft cockpit during trimming procedures. When the trimmer is used separately from the analyzer, the interconnect cable is disconnected and the trimmer is powered by the analyzer power cable and a power-cable adapter. Cables, cable adapters, and heater probes are stored in the accessory case.

The Jetcal analyzer is used to check continuity of the EGT system, insulation resistance of the EGT system, resistance of thermocouple harnesses, accuracy of engine thermocouples, engine thermocouple temperature spread during engine trim, accuracy of the EPR system, accuracy of aircraft RPM systems that use standard tachometer generators, and calibration of the aircraft EGT indicator; to bench-test individual thermocouples; to test overheat detection systems; to monitor EGT and rpm during engine trim and pressure and EPR during engine trim; and to correct temperature and rpm readings to standard-day conditions. The standard-day corrections by the analyzer make it unnecessary to convert the engine parameters to standard-day conditions mathematically or by means of charts.

The instrument and control panels for the Analyzer/Trimmer are shown in Fig. 23-15. Note that the temperature, EPR, and rpm information is presented in digital displays.

Except for engine trimming operations, the Jetcal analyzer can perform its functions without the necessity of running the engines. This, of course, saves both time and fuel in the routine maintenance and troubleshooting of the engine systems.

Complete instructions for the operation of the analyzer are provided inside the case cover. In addition, the manufacturer furnishes a service manual which gives step-by-step instructions for use of the analyzer and instructions for calibration and maintenance of the unit.

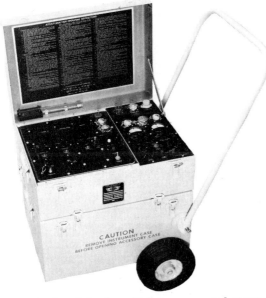

FIG. 23-14 Jetcal Analyzer/Trimmer case and accessory case. *(Howell Instruments)*

FIG. 23-15 Instrument and control panels for the Jetcal Analyzer/Trimmer. *(Howell Instruments)*

TURBINE ENGINE REPAIR AND MAINTENANCE

It is not possible to describe all the practices and procedures for the repair, maintenance, and overhaul of gas-turbine engines in a textbook because these are covered in the various manuals used by the technician for particular engines and accessories. We shall, however, describe a few of the typical practices which may be employed by the technician for certain conditions.

Fan Blades

Fan blades receive damage from time to time because of foreign objects being drawn into the inlet of the engine. Small rocks cause nicks which are usually repairable as specified in the maintenance manual. Typically, a small nick may be repaired if it is within the dimensions specified. Figure 23-16 is an example of repair limits for the first-stage fan blade of a JT8D engine. The cuts made in the process of repairing the blade are termed "flyback cuts."

If fan blade damage is such that all damaged sections can be removed within the limitations shown in the drawing, the blade can be continued in service for a maximum of 20 h. The repair must adhere to any com-

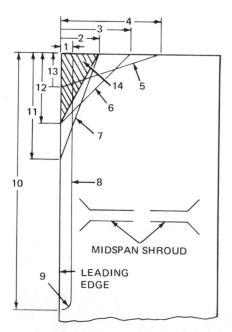

1. 0.250 in [6.35 mm] maximum
2. 0.750 in [19.05 mm] maximum
3. 1.250 in [31.75 mm] maximum
4. 2.00 in [50.8 mm] maximum
5. Cut no. 1
6. Cut no. 2
7. Cut no. 3
8. Cut no. 4
9. 0.250 in [6.35 mm] radius minimum
10. 11.00 in [27.94 cm]
11. 3.8 or 3.9 in [96.52 or 99.06 mm] maximum depending on model of engine.
12. 2.150 in [54.61 mm] maximum
13. 1.150 in [29.21 mm] maximum
14. Nonrepaired tip maximum bend area

FIG. 23-16 Repair limits for fan blades. *(Pratt & Whitney)*

bination of limits shown for cuts 1, 2, 3, and 4; and the blade may be repaired up to the maximum dimension defined by the envelope created by all four cuts.

In the drawing it will be noted that the leading edge of the blade can be cut back a distance of 0.250 in [6.35 mm] for a distance of 11 in [27.94 cm] along the blade. Toward the tip, greater depth cuts can be made as shown.

For blades with FOD confined to the blade tip only, repair may be made and the blade continued in service provided that the repair adheres to the limits shown for cut 1, 2, or 3 and that the blade is repaired up to the maximum dimension of only one of the permissible cuts.

In the repair of fan blades, certain conditions are specified. For example, all repair cuts must have a length-to-depth ratio greater than 4:1. Contours must be smooth and continuous with a minimum radius of 0.250 in [6.35 mm]. The leading-edge contour after repair should conform as nearly as possible to the original. Repaired areas must be checked with a dye or fluorescent penetrant to ensure that there are no cracks.

The repair of fan blades while the fan rotor is installed in the engine requires that the area to be reworked be completely masked off to ensure that no metal splatter can strike any other blade or disk surface. Cutting is accomplished with a 2-in [5.08-cm] cutting wheel mounted in an air chuck operating at 18,000 rpm maximum. A minimum of 0.060 in [1.52 mm] material must be left for hand filing and polishing to ensure removal of any heat-affected areas.

Shingled blades may be unshingled and continued in service for a maximum of 20 h provided that they can be unshingled without further damage and that inspection shows that the midspan shroud of a shingled blade has not hit the airfoil section of an adjacent blade or the radius between the airfoil and the midspan shroud of the adjacent blade. Blades showing evidence of having been hit in this manner must be removed from service before further flight. After 20 h of service, shingled blades should be removed and subjected to overhaul-type inspection.

Compressor Blades

Compressor blades are subject to the same type of damage encountered by fan blades and the repair procedures are similar. Figure 23-17 is adapted from the maintenance manual for the Pratt & Whitney JT8D engine and shows some of the permissible repairs for compressor blades. Note that there are definite limits to the depth of a cut that is allowed in removing a nick, scratch, or other damage caused by the ingestion of a foreign object. The limits vary in accordance with the part of the blade where the damage is located. The portions of the blade which have higher stresses may not be cut as deeply as the portions subjected to lower stresses during operation. When accomplishing blade repairs, care must be taken to maintain the original profile of the blade within reasonable limits.

The foregoing examples of blade repair are provided for information only, to illustrate typical practices. For a specific engine, the appropriate specifications given in the maintenance manual must be used.

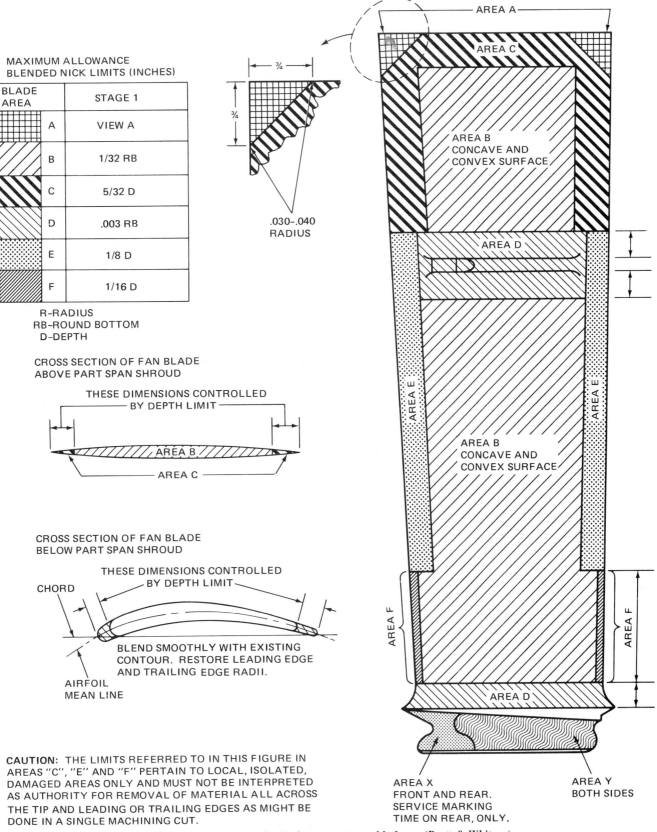

**MAXIMUM ALLOWANCE
BLENDED NICK LIMITS (INCHES)**

BLADE AREA		STAGE 1
▦	A	VIEW A
▨	B	1/32 RB
◤	C	5/32 D
▨	D	.003 RB
▦	E	1/8 D
▨	F	1/16 D

R–RADIUS
RB–ROUND BOTTOM
D–DEPTH

CROSS SECTION OF FAN BLADE
ABOVE PART SPAN SHROUD

THESE DIMENSIONS CONTROLLED
BY DEPTH LIMIT

AREA B

AREA C

CROSS SECTION OF FAN BLADE
BELOW PART SPAN SHROUD

THESE DIMENSIONS CONTROLLED
BY DEPTH LIMIT

CHORD

BLEND SMOOTHLY WITH EXISTING
CONTOUR. RESTORE LEADING EDGE
AND TRAILING EDGE RADII.

AIRFOIL
MEAN LINE

¾

¾

.030–.040
RADIUS

AREA A

AREA C

AREA B
CONCAVE AND
CONVEX SURFACE

AREA D

AREA E

AREA E

AREA B
CONCAVE AND
CONVEX SURFACE

AREA F

AREA F

AREA D

AREA X
FRONT AND REAR.
SERVICE MARKING
TIME ON REAR, ONLY.

AREA Y
BOTH SIDES

CAUTION: THE LIMITS REFERRED TO IN THIS FIGURE IN AREAS "C", "E" AND "F" PERTAIN TO LOCAL, ISOLATED, DAMAGED AREAS ONLY AND MUST NOT BE INTERPRETED AS AUTHORITY FOR REMOVAL OF MATERIAL ALL ACROSS THE TIP AND LEADING OR TRAILING EDGES AS MIGHT BE DONE IN A SINGLE MACHINING CUT.

FIG. 23-17   Repair limits for compressor blades.   *(Pratt & Whitney)*

Turbine Nozzles and Vanes

First-stage high-pressure turbine nozzles and vanes receive the highest temperatures during operation since they are exposed to the gases as they exit from the combustion chamber. The high temperatures lead to expansion cracks, stress-rupture cracks, some burning, and other damage. Stress-rupture cracks usually appear along the leading edge of the blade. Manufacturers and operators have determined what amounts of damage are acceptable and will not affect the safe operation of the engine. The following gives serviceable limits for a high-bypass engine in the first-stage nozzle vanes.

Inspection (Borescope)	Serviceability Limits
1. Axial cracks in the trailing edge (concave side only) or in slots adjacent to trailing edge	Any number 0.30 in [7.62 mm] in length allowed provided they are 0.06 in [1.52 mm] apart; or two per vane 0.80 in [20.32 mm] long, provided they are 0.30 in [7.62 mm] apart; or one 1.50 in [3.81 cm] long with two 0.50 in [12.70 mm] long provided they are 0.30 in apart and do not extend forward of the leading-edge gill holes
2. Axial cracks in the leading edge	Any number 0.50 in [12.70 mm] long if separated by at least 0.25 in [6.35 mm] or any number of cracks interconnecting the cooling holes provided the total length of interconnecting cracks does not exceed 0.60 in [15.24 mm]
3. Radial cracks in the concave surface between the inner and outer platforms	Any number 0.50 in [12.7 mm] long; or two per vane 0.80 in [20.32 mm] long provided they are at least 0.30 in [7.62 mm] apart
4. Radial cracks in the convex surfaces between the inner and outer platforms	One crack allowed 0.80 in [20.32 mm] long
5. Blocked cooling air passages	Five nose holes and four gill holes in each row. A minimum separation of one open hole shall exist between blocked holes. Three trailing-edge slots may be blocked provided that blocked slots are not adjacent
6. Nicks, scores, and scratches	Any number, any length allowed if not over 0.03 in [0.76 mm] deep. Nicks up to 0.10 in [2.54 mm] deep and 0.25 in [6.35 mm] long allowed on airfoil trailing edge
7. Buckling or bowing of the trailing edge	Any number up to 0.30 in [7.62 mm] from original contour
8. Axial cracks in concave surface	Two per vane extending aft from the first row of gill holes to (not through) the slot in the trailing edge
9. Axial cracks in convex surface	Two per vane between the midchord strut and the trailing edge, 0.25 in [6.35 mm] radially apart, total not to exceed 1.0 in [2.54 cm]. One per vane, length not to exceed 1.0 in; width not to exceed 0.50 in [12.70 mm] aft of the midchord strut. A maximum of three vanes per assembly, not adjacent
10. Burns in the trailing edge (loss of metal)	The total area removed from the trailing edge not to exceed 3.00 in² [19.35 cm²] per assembly. Accumulative area is determined by summing individual vane area radial height by axial length
11. Burns and cracks on the convex and concave sides	Not to exceed an area of 1.50 in [3.81 cm] long and 1.0 in [2.54 cm] wide per vane. Maximum of four vanes per 90° arc
12. Burns or spalling on vane leading edge (charred only, no holes through airfoil)	$\frac{1}{2}$ in [12.70 mm] diameter per vane, maximum of four vanes affected per 90° arc
13. Craze cracking	Any amount. (Craze cracking is defined as superficial surface cracks which have no visual width or depth)

Turbine Blades

Serviceability limits for turbine blades are much more stringent than are those for nozzle vanes. This is particularly true for first-stage blades because of the high temperatures involved. The centrifugal stresses to which turbine blades are subjected require that the blades be free of cracks in any area and that no nicks or dents exist in the root area. A limited number of small nicks and dents can be permitted in the areas of the blade away from the root area. No burning or distortion is permitted.

Repairs for Turbine Nozzles, Vanes, and Blades

When a borescope inspection reveals that there is damage or deterioration in the hot sections of the engine, the areas involved must be disassembled sufficiently to remove the defective parts. Parts requiring repair are replaced with new or reworked parts from the factory or an overhaul facility.

Replacement of turbine blades must be done with blades having the correct moment-weight designation. This is to ensure that the turbine rotor will be in balance when assembled. The maintenance manual for the engine specifies the correct arrangement of blades according to their moment-weight markings.

Turbine Engine Overhaul

The overhaul of gas-turbine engines is accomplished by the manufacturer or at approved overhaul stations. The process is similar in many ways to the overhaul of piston engines; however, there are processes required which are not necessary for piston engines. In addition, the overhaul facility for gas-turbine engines requires many special tools and some equipment specially designed for work on particular types and models of engines.

The overhaul of a gas-turbine engine includes a complete disassembly and inspection. Nonrepairable parts are discarded, and those salvageable through rework or recycling are sent to an appropriate facility. Repairable parts are processed as necessary and then given a rigid inspection and/or test to ensure that they are serviceable.

Overhaul instructions for a gas-turbine engine are extensive and detailed. This is illustrated by the fact that the instructions for one large high-bypass engine requires 7 ft of shelf space. Manufacturers and operators continually update the instructions to keep current with operating experience and new developments in design and construction.

The average certificated aviation maintenance technician is not required to perform turbine engine overhaul but can perform various field repairs as specified by the maintenance manual. The principal consideration for the technician is to be sure to have the correct manuals, bulletins, and other instructions when servicing and repairing a gas-turbine engine.

● REVIEW QUESTIONS

1. What precautions should be taken with respect to the surrounding area before starting a gas-turbine engine?
2. Describe the hazards that exist in the area around an operating gas-turbine engine.
3. What limitations should be observed when using an electric starter?
4. What three primary conditions are necessary in order to start a gas-turbine engine?
5. Describe a typical procedure for starting a gas-turbine engine with an automatic system.
6. What engine instrument must be observed at light-up and why?
7. What is meant by *trimming* an engine?
8. Define *engine pressure ratio* (EPR).
9. Briefly describe a typical procedure for trimming a gas-turbine engine.
10. Why is a bellmouth inlet duct used for testing gas-turbine engines?
11. What parameters are usually measured in the testing of a gas-turbine engine?
12. What corrections must be applied to test results to ensure accurate evaluation of engine performance?
13. Describe a *dry motoring check*.
14. What is the purpose of a *wet motoring check*?
15. Describe an *idle check*.
16. What is the purpose of a *power assurance check*?
17. What problem is indicated if the EGT of an engine does not drop immediately after shutdown?
18. What action should be taken by the operator if the EGT does not drop at shutdown?
19. What investigation should be made by the technician during coastdown after an engine is shut down?
20. What is meant by a *flight cycle*?
21. What are periodic inspections?
22. Describe a typical procedure for checking the ignition system of a gas-turbine engine.
23. What precaution should be taken in checking an ignition system?
24. What events may occur to require special or nonroutine inspections?
25. Describe a *borescope* and explain its purpose.
26. What is a *chamberscope*?
27. During engine operation, what are the symptoms of foreign object damage (FOD)?
28. What inspections are likely to be required if a gas-turbine engine has been operated over temperature limits?
29. What inspections may be called for after overspeed operation?
30. What is the usual procedure if a gas-turbine engine has been operated more than 2 min with no oil pressure?
31. Describe the condition known as *fan blade shingling*.
32. What engine instruments are useful as fault indicators?
33. Describe the McDonnell Douglas FEFI/TAFI fault-indication and -isolation system.
34. What is a "fault tree"?
35. What is the general procedure for isolating a fault after an indication has been reported?
36. What would be possible faults leading to low engine rotation speed during starting?
37. What are likely faults causing failure of an engine to light off even though rotation is satisfactory?
38. What are the most common causes of faults in the operation of gas-turbine engines?
39. List the functions of the Jetcal Analyzer/Trimmer.
40. Discuss the possible repairs of fan blades after FOD.
41. Briefly describe how repairs are made on fan and compressor blades.
42. What limitations of serviceability are applicable to cracks in first-stage nozzle vanes?
43. What condition must be observed in the replacement of turbine blades?

INDEX